José Mira Alberto Prieto (Eds.)

Connectionist Models of Neurons, Learning Processes, and Artificial Intelligence

6th International Work-Conference on Artificial and Natural Neural Networks, IWANN 2001
Granada, Spain, June 13-15, 2001
Proceedings, Part I

Springer

Volume Editors

José Mira
Universidad Nacional de Educación a Distancia
Departamento de Inteligencia Artificial
Senda del Rey, s/n., 28040 Madrid, Spain
E-mail: jmira@dia.uned.es

Alberto Prieto
Universidad de Granada
Departamento de Arquitectura y Tecnología de Computadores
Campus Fuentenueva, 18071 Granada, Spain
E-mail: aprieto@atc.ugr.es

Cataloging-in-Publication Data applied for

Die Deutsche Bibliothek - CIP-Einheitsaufnahme

International Work Conference on Artificial and Natural Neural Networks <6, 2001, Granada>:
6th International Work Conference on Artificial and Natural Neural Networks ; Granada, Spain, June 13 - 15, 2001 ; proceedings / IWANN 2001. José Mira ; Alberto Prieto (ed.). - Berlin ; Heidelberg ; New York ; Barcelona ; Hong Kong ; London ; Milan ; Paris ; Singapore ; Tokyo : Springer
Pt. 1. Connectionist models of neurons, learning processes, and artificial intelligence. - 2001
(Lecture notes in computer science ; Vol. 2084)
ISBN 3-540-42235-8

CR Subject Classification (1998): F.1, F.2, I.2, G.2, I.4, I.5, J.3, J.4, J.1

ISSN 0302-9743
ISBN 3-540-42235-8 Springer-Verlag Berlin Heidelberg New York

Springer-Verlag Berlin Heidelberg New York
a member of BertelsmannSpringer Science+Business Media GmbH

http://www.springer.de

Printed in Germany

Typesetting: Camera-ready by author, data conversion by Olgun Computergrafik
Printed on acid-free paper SPIN 10839312 06/3142 5 4 3 2 1 0

Lecture Notes in Computer Science 2084

Edited by G. Goos, J. Hartmanis and [illegible]

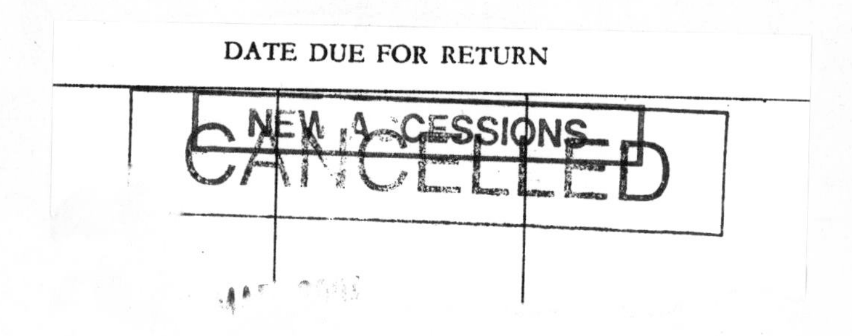

Springer
Berlin
Heidelberg
New York
Barcelona
Hong Kong
London
Milan
Paris
Singapore
Tokyo

Preface

Underlying most of the IWANN calls for papers is the aim to reassume some of the motivations of the groundwork stages of biocybernetics and the later bionics formulations and to try to reconsider the present value of two basic questions. The first one is: "What does neuroscience bring into computation (the new bionics)?" That is to say, how can we seek inspiration in biology? Titles such as "computational intelligence", "artificial neural nets", "genetic algorithms", "evolutionary hardware", "evolutive architectures", "embryonics", "sensory neuromorphic systems", and "emotional robotics" are representatives of the present interest in "biological electronics" (bionics).

The second question is: "What can return computation to neuroscience (the new neurocybernetics)?" That is to say, how can mathematics, electronics, computer science, and artificial intelligence help the neurobiologists to improve their experimental data modeling and to move a step forward towards the understanding of the nervous system?

Relevant here are the general philosophy of the IWANN conferences, the sustained interdisciplinary approach, and the global strategy, again and again to bring together physiologists and computer experts to consider the common and pertinent questions and the shared methods to answer these questions.

Unfortunately, we have not always been successful in the six biennial meetings from 1991. Frequently the well-known computational models of the past have been repeated and our understanding about the neural functioning of real brains is still scarce. Also the biological influence on computation has not always been used with the necessary methodological care. However IWANN 2001 constituted a new attempt to formulate new models of bio-inspired neural computation with the deeply-held conviction that the interdisciplinary way is, possibly, the most useful one.

IWANN 2001, the 6th International Work-Conference in Artificial and Natural Neural Networks, took place in Granada (Spain) June 13-15, 2001, and addressed the following topics:

1. ***Foundations of connectionism.*** Brain organization principles. Connectionist versus symbolic representations.
2. ***Biophysical models of neurons.*** Ionic channels, synaptic level, neurons, and circuits.
3. ***Structural and functional models of neurons.*** Analogue, digital, probabilistic, Bayesian, fuzzy, object oriented, and energy related formulations.
4. ***Learning and other plasticity phenomena.*** Supervised, non-supervised, and reinforcement algorithms. Biological mechanisms of adaptation and plasticity.
5. ***Complex systems dynamics.*** Optimization, self-organization, and cooperative processes. Evolutionary and genetic algorithms. Large scale neural models.

6. ***Artificial intelligence and cognitive processes.*** Knowledge modeling. Natural language understanding. Intelligent multi-agent systems. Distributed AI.
7. ***Methodology for nets design.*** Data analysis, task identification, and recursive hierarchical design.
8. ***Nets simulation and implementation.*** Development environments and editing tools. Implementation. Evolving hardware.
9. ***Bio-inspired systems and engineering.*** Signal processing, neural prostheses, retinomorphic systems, and other neural adaptive prosthetic devices. Molecular computing.
10. ***Other applications.*** Artificial vision, speech recognition, spatio-temporal planning, and scheduling. Data mining. Sources separation. Applications of ANNs in robotics, economy, internet, medicine, education, and industry.

IWANN 2001 was organized by the Universidad Nacional de Educación a Distancia, UNED (Madrid), and the Universidad de Granada, UGR (Granada), also in cooperation with IFIP (Working Group in Neural Computer Systems, WG10.6), and the Spanish RIG IEEE Neural Networks Council.

Sponsorship was obtained from the Spanish CICYT and the organizing universities (UNED and UGR).

The papers presented here correspond to talks delivered at the conference. After the evaluation process, 200 papers were accepted for oral or poster presentation, according to the recommendations of reviewers and the authors' preferences. We have organized these papers in two volumes arranged basically following the topics list included in the call for papers. The first volume, entitled "Connectionist Models of Neurons, Learning Processes, and Artificial Intelligence" is divided into four main parts and includes the contributions on:

I. Foundations of connectionism and biophysical models of neurons.
II. Structural and functional models of neurons.
III. Learning and other plasticity phenomena, and complex systems dynamics.
IV. Artificial intelligence and cognitive processes.

In the second volume, with the title, "Bio-inspired Applications of Connectionism", we have included the contributions dealing with applications. These contributions are grouped into three parts:

I. Bio-inspired systems and engineering.
II. Methodology for nets design, and nets simulation and implementation.
III. Other applications (including image processing, medical applications, robotics, data analysis, etc.).

We would like to express our sincere gratitude to the members of the organizing and program committees, in particular to F. de la Paz and J. R. Álvarez-Sánchez, to the reviewers, and to the organizers of preorganized sessions for their invaluable effort in helping with the preparation of this conference. Thanks also to the invited speakers for their effort in preparing the plenary lectures.

Last, but not least, the editors would like to thank Springer-Verlag, in particular Alfred Hofmann, for the continuous and excellent cooperative collaboration from the first IWANN in Granada (1991, LNCS 540), the successive meetings in Sitges (1993, LNCS 686), Torremolinos (1995, LNCS 930), Lanzarote (1997, LNCS 1240), Alicante (1999, LNCS 1606 and 1607), and now again in Granada.

June 2001

José Mira
Alberto Prieto

Invited Speakers

Oscar Herreras, Dept. of Research. Hospital Ramón y Cajal (Spain)
Daniel Mange, Logic Systems Laboratory, IN-Ecublens (Switzerland)
Leonardo Reyneri, Dip. Elettronica, Politecnico di Torino (Italy)
John Rinzel, Center for Neural Science, New York University (USA)

Field Editors

Igor Aizenberg, Neural Networks Technologies Ltd. (Israel)
Amparo Alonso Betanzos, University of A Coruña (Spain)
Jose Manuel Benitez Sanchez, Universidad de Granada (Spain)
Enrique Castillo Ron, Universidad de Cantabria (Spain)
Andreu Català Mallofré, Univ. Politècnica de Catalunya (Spain)
Carolina Chang, Universidad Simón Bolívar (Venezuela)
Carlos Cotta, University of Málaga (Spain)
Richard Duro, Universidade da Coruña (Spain)
Marcos Faundez-Zanuy, Univ. Politècnica de Catalunya (Spain)
Carlos Garcia Puntonet, Universidad de Granada (Spain)
Gonzalo Joya, Universidad de Málaga (Spain)
Christian Jutten, Inst. National Polytechnique de Grenoble (France)
Dario Maravall, Universidad Politécnica de Madrid (Spain)
Eduardo Sánchez, Universidad de Santiago de Compostela (Spain)
José Santos Reyes, Universidade da Coruña (Spain)
Kate Smith, Monash University (Australia)

Reviewers

Igor Aleksander, Imperial College of Sci. Tech. and Medicine (UK)
José Ramón Álvarez-Sánchez, UNED (Spain)
Shun-ichi Amari, RIKEN (Japan)
A. Bahamonde, Universidad de Oviedo en Gijón (Spain)
Senén Barro Ameneiro, Univ. Santiago de Compostela (Spain)
J. Cabestany, Universidad Politécnica de Cataluña (Spain)
Marie Cottrell, Université Paris 1 (France)
Félix de la Paz López, UNED (Spain)
Ana E. Delgado García, UNED (Spain)
Ángel P. Del Pobil, Universidad Jaime I de Castellón (Spain)
José Dorronsoro, Universidad Autónoma de Madrid (Spain)
José Manuel Ferrández, Universidad Miguel Hernandez (Spain)
Kunihiko Fukushima, Osaka Univ (Japan)
Tamas Gedeon, Murdoch University (Australia)
Karl Goser, Univ. Dortmund (Germany)
Manuel Graña Romay, Universidad Pais Vasco (Spain)
J. Hérault, Inst. N. P. Grenoble (France)
Óscar Herreras, Hospital Ramón y Cajal (Spain)
Gonzalo Joya, Universidad de Málaga (Spain)
Christian Jutten, Inst. National Polytechnique de Grenoble (France)
Shahla Keyvan, University of Missouri-Rolla (USA)
Daniel Mange, IN-Ecublens (Switzerland)
Darío Maravall, Universidad Politécnica de Madrid (Spain)
Eve Marder, Brandeis University (USA)
José Mira, UNED (Spain)
J.M. Moreno Aróstegui, Univ. Politécnica de Cataluña (Spain)
Christian W. Omlin, University of Western Cape South (Africa)
Julio Ortega Lopera, Universidad de Granada (Spain)
F.J. Pelayo, Universidad de Granada (Spain)
Franz Pichler, Johannes Kepler University (Austria)
Alberto Prieto Espinosa, Universidad de Granada (Spain)
Leonardo Maria Reyneri, Politecnico di Torino (Italy)
John Rinzel, New York University (USA)
J.V. Sánchez-Andrés, Universidad de Alicante (Spain)
Francisco Sandoval, Universidad de Málaga (Spain)
J.A. Sigüenza, Universidad Autónoma de Madrid (Spain)
M. Verleysen, Universite Catholique de Louvin (Belgium)

Table of Contents, Part I

Foundations of Connectionism and Biophysical Models of Neurons

Structural and Functional Models of Neurons

Learning and Other Plasticity Phenomena, and Complex Systems Dynamics

Artificial Intelligence and Cognitive Processes

Table of Contents, Part II

Bio-inspired Systems and Engineering

Methodology for Nets Design, Nets Simulation and Implementation

Image Processing

Medical Applications

Robotics

General Applications

Dendrites: The Last-Generation Computers

O. Herreras[1], J. M. Ibarz[1], L. López-Aguado[1], and P. Varona[2]

[1]Dept. Investigación, Hospital Ramón y Cajal, 28034 Madrid, Spain
[2]Dept. Ingeniería Informática., U. Autónoma, Madrid, 28049, Madrid, Spain
Oscar.herreras@hrc.es

Abstract. Neuronal dendrites perform complex computational work with incoming signals. An amazing variety of functions can be achieved by using relatively few geometrical and electrogenic elements. The subtleness of their performance is such that experimental research, necessarily invasive, often yields contradictory results. Biophysical models help to understand better what dendrites can actually do and what they mean for the global processing and, perhaps, teach us how to *observe* such delicate performing devices.

Ins and Outs: Why the Wisests Neurons Speak so Little?

The neural code is primarily written in a binary language made up of temporal sequences of all-or-nothing action potentials (APs), although significant graded analog-like messages co-exist. Individual neurons receive a large number of electrical signals that are processed using a complex collection of non-linear mechanisms. In the light of the simplicity of the main digital output one may wonder what the advantages can be of individual neurons being such powerful and subtle computing machines. The simplest answer would be that the temporal pattern is not all. Certainly, frequency modulation of AP sequences may perform with equivalent accuracy as the fine tuning capabilities of graded communication that prevail in non neuronal cells. Though the AP firing rate is highly variable, different neuron types fire within certain ranges. A most interesting kind is that "distinguished" as of high order or associative neuron, which fire at extremely low rate. These, like sage people, are silent most of the time, just "listening" to a number of different inputs massively delivering synaptic "messages" while only occasionally fire a lonely AP. Do receiving neurons consider irrelevant all this transit? Is the fate of most presynaptic APs to become useless noise? Certainly not. Redundancy is a great success on information transfer, but some cells appear to stubbornly neglect almost every input. No doubt, the informative content of their rare APs is high and we are only beginning to understand how the history of long silences is also included on them. If the input is large and varied, and the output so rare and critical for the animal behavior, no doubt the intermediate processing must be extraordinarily careful, elaborated and subtle.

So far as we know, a good deal of this processing is performed on the elongated cell processes that constitute the dendritic arborization of the neurons. Their geometrical peculiarities, the variety of active channels on their membranes, and the

J. Mira and A. Prieto (Eds.): IWANN 2001, LNCS 2084, pp. 1-13, 2001.

structural and functional plasticity they can undergo constitute the most powerful and versatile biological computing machine ever known.

Some Implications of Structural Factors

Anatomical studies have described a large variety of dendritic trees in different cell types, from unique processes to extensively branched trees [4]. Based on modeling studies, it has been suggested that the overall architecture of the dendritic tree that constitutes the morphoelectronic skeleton somehow determines the overall firing pattern [14]. The large number of exceptions clearly argue against this hypothesis. No doubt, the unquestionable great success of dendrites is to provide a larger surface usable for synaptic contacts at the expense of little volume consumption. The specific requirements of connectivity are the clue. Thousands of small appendages known as synaptic spines serve the same purpose, easily doubling the total membrane surface, but most importantly they constitute specialized regions where synaptic contacts are made, yet they are not universal. In fact, dendritic spines are the preferred membrane loci for excitatory synapses, while inhibitory connections are typically made on the parent branches and somata. This natural sorting reflects an evolutionary trend for electrical isolation of individual excitatory inputs (the "positive messages"), in contrast to the open loci for the (modulatory?) inhibitory inputs. Besides, the spines constitute a strongly isolated biochemical compartment where local enzymatic cascades may trigger modifications resulting in lasting changes of synaptic efficiency, the so-called synaptic plasticity. This may result from ion changes caused by synaptic activity or in conjunction with molecular messages incoming from the control center at the nucleus and hooked "en passage" by selective biochemical tags displayed at the entrance of the spine [2].

For a long time it was presumed that the synaptic weight diminished with the distance to the trigger zone (the axon initial segment, IS), the distal inputs acting as mere modulators of other located more proximally. On this view, the informative load of different inputs would depend on the synaptic locus. While this may be true in some cases, specially when dendritic membranes behave as passive cables, some wiring patterns of the brain pointed toward different solutions. It is not uncommon the stratification or zonation of inputs on different dendritic subregions. On the above premise, this would imply that different types of afferent information to a specific neuron might have a decisive or a modulatory role already based on the synaptic loci. While this might be true, and even a solution to a specific requirement, at least for some synapses it is demonstrated that its location within the dendritic tree is structurally compensated so that their relative functional weight at the soma is rather constant. This can be achieved by varying the size of the postsynaptic site [1] (and presumably the number of receptors), or the dimensions of the spine neck (allowing the passage of more or less current to the parent dendrite). The control of the synaptic weight along different dendritic loci may also be externally settled by the properties of the afferent presynaptic cells. For instance, inhibitory synapses at perisomatic, axonic and dendritic loci of hippocampal CA1 pyramidal cells arise from different populations of interneurons and have marked functional impact in terms of subthreshold integration and subcellular electrogenesis.

Passive but Not Simple

The classical concepts of neuron processing and dendritic integration were obtained from studies on motoneurons [3]. The relative functional simplicity of these cells lies on the fact that their dendrites have limited active properties. Even if virtually all neuron membranes have some active channels on their structure, the study of electrical signal processing in passive models [16,17] provided the basics upon which more complex dynamics can be analyzed.

Neurons are current generators with their membranes acting like lumped units composed of batteries in series with a resistor and in parallel to a capacitor. This simple equivalent circuit already endows neurons with a variety of computational resources. But they would not be effective if geometry is not suitable. Perhaps, the quintessential feature allowing any processing of electrical signals is that narrow elongated dendrites cause neurons to be electrically non-uniform. They can be functionally compartmentalized to process different electrical signals. Due to their reduced dimensions the input impedance is much larger than somata, and hence small currents cause much larger voltages, resulting in the economy of transmitter. Also, because the large resistance of the membranes, a significant fraction of this current will reach the soma/axon.

The membrane capacitance is a primary feature for integration. Because of it, the dendrites are time-delay lines for synaptic potentials, i.e., the peak time becomes delayed and the time course increases away from the synaptic site. Consequently, near and far inputs will be processed differently. Locally, synaptic inputs are much briefer than the resultant voltage at the soma, and their temporal summation is limited. At the soma, however, this will depend on the distance of the inputs. Summation of two synapses at the soma is more effective when they are located far from it, and still further if they are on different dendrites. In the first case, summation is expected to be sublinear because the mutual interaction, decreasing each other's driving force. Sublinear summation decreases as synapses are located further apart [17].

Membrane conductance is heterogeneous along their morphology in all neuron types, either because uneven channel density or irregular synaptic bombardment. Nonuniform membrane conductances cause subtle modulations on the time-course of synaptic inputs, some are even counterintuitive. Model studies with neurons made leakier in distal dendrites (as for pyramidal cells) showed interesting results [17]. Distal synapses are more efficient as their local voltage saturates less because the lower input resistance. The relaxation time is slower, hence the tail of synaptic inputs is longer. There is always a voltage gradient in the dendritic tree (isopotentiality is never reached). And strikingly, the current required to initiate the AP at the soma is lower, i.e., less inputs are required.

Dendritic geometry is also involved in spike initiation. Apparently, the voltage threshold (Vt) should be exclusively defined by the properties of excitable channels. However, this is only true in isopotential structures. In fact, the loss of current from the activated region toward the adjacent compartments makes the Vt higher in non-isopotential structures (to compensate for the current loss). Hence, the morphology of the dendrite plays an important role on the setting of the Vt. Again, modeling studies [17] showed that Vt is smaller at the center of a sealed cable, while the opposite occurs for the threshold of current (It), which is smaller at the end. In the case of branched structures electrically equivalent to the same cable cylinder, the lowest Vt is not at the center. Thick branches have a similar Vt except at its end, where it becomes unable to initiate AP. This is even more dramatic for thin branches, that can only generate AP near the branch point. However, when both branches are activated simultaneously the Vt decreases and both branches can initiate the AP at any location. The reason is that during a distributed input the structure is more isopotential and there is less longitudinal loss of current.

Active Dendritic Properties: What Does It Change?

To say it fast and clear: everything. Certainly, active properties constitute a giant reservoir of functional possibilities of which we are only beginning to grasp a few. The classic view of voltage-dependent sodium and potassium channels serving the only purpose of building a solid message capable of being safely delivered to the target cells can be left to rest. A large unfinished list of voltage dependent channels [12] makes neuronal membranes to behave like complex non-linear devices. Even some traditional parameters that have been traditionally considered passive cell properties as the membrane resistivity are in fact voltage-dependent. The possibilities unclosed by active properties should not be however used to wipe out the knowledge accumulated from the study of passive cable structures, but instead be contemplated as the basis upon which more complex functions are built. Some clear examples of the effect of geometrical and passive properties on APs have been outlined above.

Amongst the functions for active channels in dendrites the subthreshold amplification of excitatory postsynaptic potentials (EPSPs) is the simplest and most intuitive. This mechanism may serve to compensate for cable filtering along dendrites that would impair far located synapses from activating the soma. Sodium and calcium channels are known to participate on this process. It is also known that potassium channels are simultaneously counteracting this amplification, so that a delicate balance is in fact defining the shape of EPSPs at the soma. Under different physiological situations is then possible that the same synaptic input causes a totally different voltage envelope at the soma. The neglect of these channels that necessarily contribute to the shape of EPSPs have devaluated a number of studies aimed to finely "quantify" synaptic transmission. In our view, a new image emerges of EPSPs as being information primers that are continuously re-shaped as they spread within the dendrites on its way to a trigger zone. The obligatory recruitment of active currents allows the up or down regulation by specific control of dendritic Na^+/Ca^{++} or K^+ channels. The quantitative importance of these channels is such that during activation of far located synapses most of the current that finally reaches the soma may actually be generated by them, the synaptic current being lost in the way [6]. On this line of thinking, some authors have hypothesized that active dendrites may operate as expansive nonlinear devices [13]. This attractive idea suggests that active channels are used to boost strong inputs while small inputs are further weakened, which in the end can be understood as an operational filter selecting "important" incoming messages. This type of processing could be simultaneously carried out in different dendritic regions of the same tree, so that a single neuron would actually be made up of multiple subdendritic processors of equivalent relative weight on the final response of the cell. In hippocampal pyramidal cells, the proximal region of apical dendrites displays slow active subthreshold responses (labeled boosting currents in Fig 1.), clearly discriminated from synaptic currents, that may serve as an obligatory link between synaptic currents and dendro-somatic APs [6] (Fig 1). Using a protocol of mild dendritic damage we found that the selective switch off of these slow active currents uncouples far synaptic currents from spike electrogenesis at the soma/IS, avoiding cell output [7]. This connecting role for slow active currents may constitute the expansive nonlinear device mentioned above [13].

Other reported functions of active dendritic channels may go from the synaptic release of transmitter at specialized dendro-dendritic synapses, the shunting of EPSPs by backpropagating APs, and the increase of internal Ca ions that may initiate enzymatic cascades leading to lasting synaptic plasticity. Special mention deserves their involvement on the generation of bursting behavior in many neuron types. Bursting in hippocampal pyramidal cells and thalamocortical cells may be a direct consequence of Ca^{++} channel activation by synaptic input [9].

Spike Up or Spike Down: Re-edition of a Historical Mistake?

No doubt, the most outstanding possibility unclosed by the presence of active channels in dendrites is that they can generate APs. There is an evident reluctance to accept such a possibility in its full extent. Strikingly, AP dendritic firing is known from the beginning of the experimental neurophysiology in the first half of the past century, apparently without bothering prejudice from the salient researchers at the moment. Its reborn is being arduous, however. Dislikers argue that dendritic AP firing swipe off synaptic integration. Being this an extreme possibility, it is by no means a common circumstance, and dendritic firing may be used by different neurons in many ways. It can be a signaling mechanism between spines, or the result of cooperative activity on a group of them. If synaptically initiated in dendrites, and of decremental nature [11], dendritic spikes may constitute an internal language between subdendritic regions and the soma. They constitute a powerful amplifying mechanism for distal synapses. When backpropagated from the axon/soma they might be used as signaling retrograde signals to reset the activation state of V-dependent channels, as an unspecific message to notify spines of successful activation, to modulate synaptic currents by shunting, or to selectively reinforce some synapses by Hebbian coincidence.

The most controversial role for dendritic APs is to play a true decisive role for cell output, as an alternative (or complement) to the classic trigger zone in the IS. In our opinion, a major historical mistake has completed again around dendrites. First was the unjustified elevation of motoneuron cell properties (of passive dendrites) as

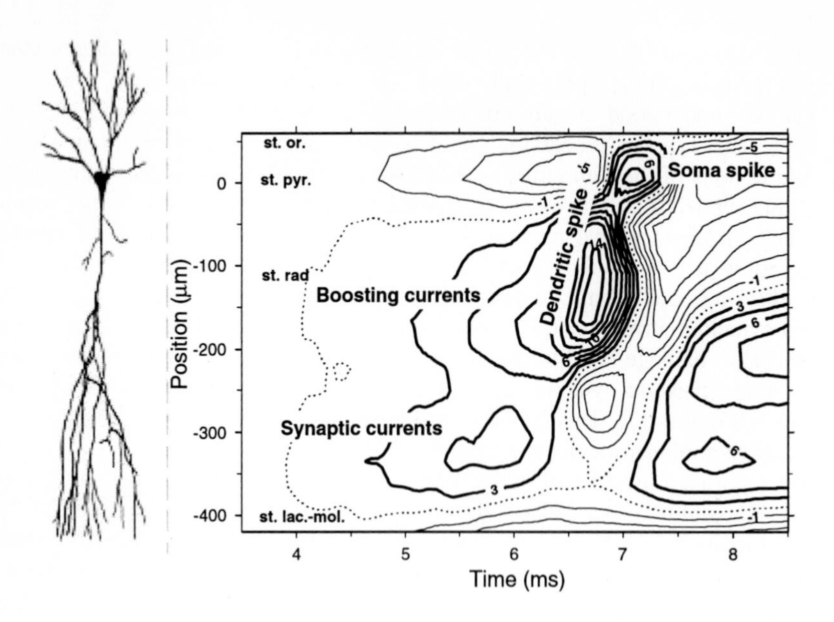

Fig. 1. Contour plot of inward (thick) and outward (thin lines) currents in CA1 pyramidal cells obtained by current-source density analysis *in vivo* showing the spatiotemporal distribution of current flow during synaptic activation of Schaffer collaterals (~ -350 µm). The AP is initiated in the apical shaft (dendritic spike) once the proximal boosting currents grow over the threshold, and then propagate into the soma and axon.

representative of a non existent "generic" neuron type. Massive research on this cell type, and the reputation of the leading researchers caused the unfortunate bias. The mistake nowadays is caused by technical reasons and fashion. Dendritic properties were first studied in central neurons in adult animals *in vivo*. Later on, they were "re-discovered" in the 90's, but the studies were performed on juvenile animals, on tissue slices maintained *in vitro*, and using the award-winning technique of whole-cell recordings. Unfortunately, all three factors have important drawbacks for the study of active dendritic properties, and so the controversy began.

The hot issue is whether pyramidal cell dendrites can initiate an AP that spread regeneratively toward the cell soma/axon upon synaptic activation or else, merely conduct APs backpropagated from an initial locus at the IS. The practical relevance of this issue is that in the first case, the dendritic tree may be envisaged as a multicompartmental parallel processing machine, each unit endowed with all potentialities of the entire single cell. In other words, dendrites would not be merely collecting devices with partial processing capabilities, they would also be able to decide cell output, either as individual branches, subdendritic regions or entire trees. Historically, the results obtained *in vivo* since the 60's agreed that these dendrites can initiate an AP that will finally get out through the axon (Fig.1). Evoked field potentials, current-source density (CSD) analysis and intracellular recordings were rather clear in this respect (see ref 5; also 6 and refs. therein). Subsequent work made *in vitro* during the 80's failed to show this initial locus in dendrites. For some

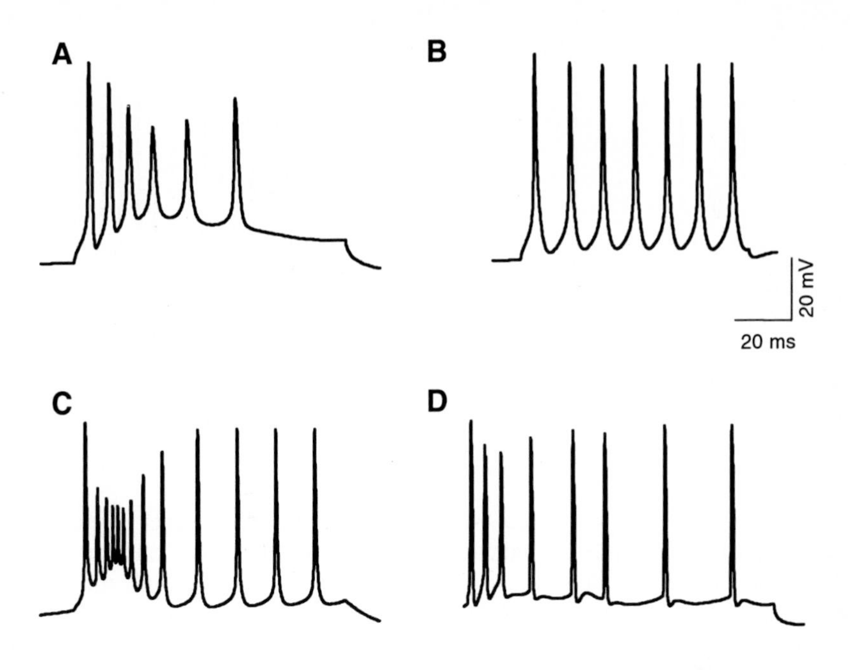

Fig. 2. Response of different CA1 pyramidal model cells to depolarizing pulses injected at the soma. A and B are equivalent cylinder models (refs 24 and 20), while C and D are branched dendritic models (refs 23 and 21). A,B: 0.9 nA; C,D: 0.5 nA. Note large differences in the firing pattern, rates and accommodation.

researchers, this was considered disproving, for others, naturally this only demonstrated the functional deficits of the posttraumatic state of sliced tissue. Subsequent careful *in vitro* reports, however, partially or entirely confirmed the findings *in vivo* [23]. With the arrival and rapid explosion of the whole-cell technique for recording from single cells, the supporters of backpropagating APs become majority and in the 90's the ability of dendrites to initiate/conduct spikes decreased further. Dendritic AP initiation got renewed air when some researchers realized that the juvenile animals used with this technique have immature dendrites and their poor Na^+ channel allotment strongly decreased the possibility of dendrites to support initiation and only barely active conduction. Next, the efforts were directed to demonstrate that the triggering zone was the IS [19], but again the results were inconclusive for the same reasons. In parallel, as the *in vitro* techniques were progressively refined and researchers gradually took consciousness of the unphysiological and invasive character of this approach, more chances were admitted for dendrites to generate propagating dendritic spikes. A weird situation occurred in which some dendrites apparently "learnt" how to fire spikes during 90's. At present, the predominant view of *in vitro* workers is that the AP can be initiated in dendrites during large (abnormal?) inputs, and that most fail to invade the soma [18].

The blatant error this time has been the (unconscious?) neglect of previous sound data obtained with time-resistant optimized non-invasive techniques due to a

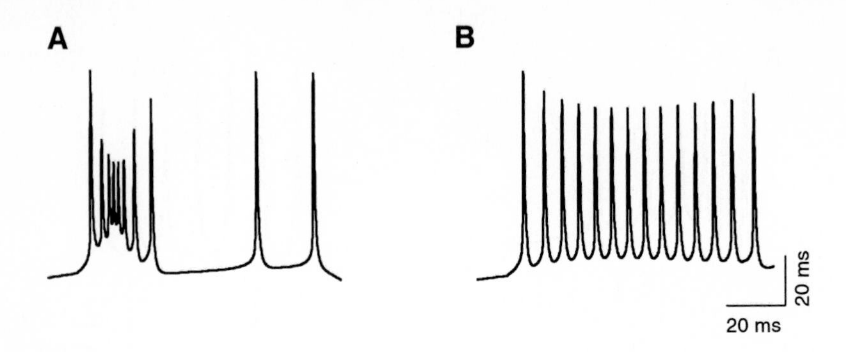

Fig. 3. The presence of active dendrites strongly modifies the physiology of the soma. Response of a model neuron (ref. 23) to a 0.5 nA depolarizing pulse at the soma with active (A) or passive dendrites (B).

dangerous tendency to get direct visualization of known events stretching the capabilities of fashionable techniques beyond reasonable positions. In spite of sophistication these techniques cannot avoid strong interference with the observed object under study. Finally, we should not fall in a third historical mistake by emphasizing the polemic initiation of spikes in pyramidal cells to the point of translating the scientific community the erroneous impression that the debate is on a generic neuronal property. As mentioned above, there is not such a thing as a generic neuron prototype. The variety of neuronal types and active properties is such that all possible ranges of dendritic spike activity can be found.

Modeling Dendrites: What Can It Say to Physiologists?

The role of computer models has always gone beyond a mere help to understand neuronal properties. No one can deny the great step forward achieved by theoretical studies on passive dendrite neurons [16] for the understanding of cable properties and synaptic integration, and the initiation and propagation of APs [15]. Most predictions advanced by these studies have been accomplished, and because of that the experimental results are assimilated better and faster. However, the computation of active dendrites has been troublesome because the paucity and unreliability of quantitative data. We even consider that dendritic modeling is often confusing and misleading. The arrival of easy-to-handle modeling tools allows experimentalists to undertake a job formerly restricted to theoreticians. In parallel, a clear trend can be observed in the neurophysiological literature from analytical studies signed by the later group toward models replicating the most recently obtained data, usually those in the same report. These studies are carried out mainly by physiologists. The obvious risk is that simple replication of highly partial experimental data adds nothing new as it can be achieved in many different ways, while it may (and in fact does) cause the erroneous impression that the model treatment supports or reinforces the

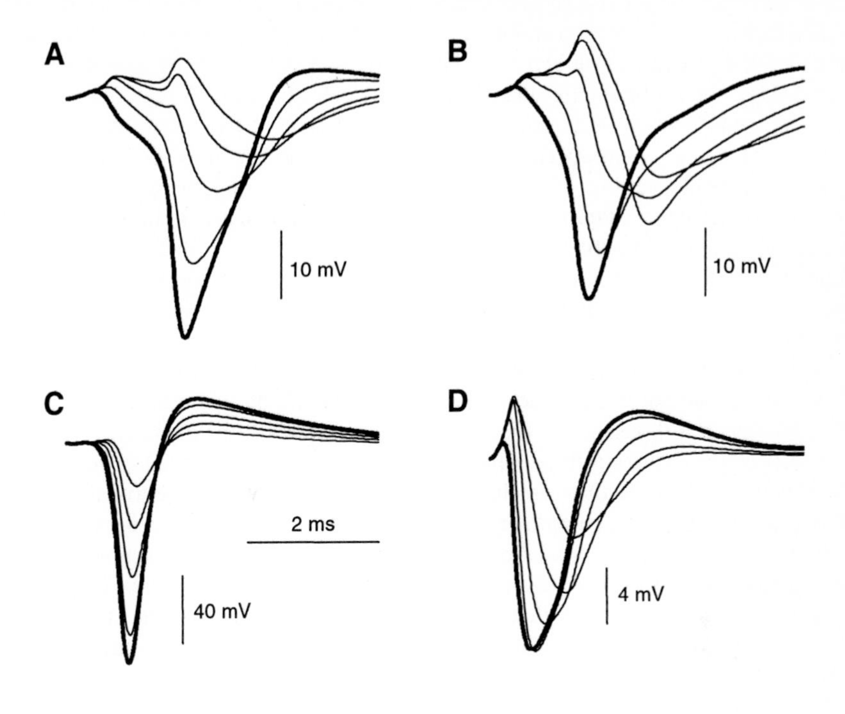

Fig. 4. Macroscopic modeling. The waveform of the model antidromic population spike (PS) is plotted from the soma layer (thick tracing) down to 200 μm below within the apical dendritic tree in 50 μm steps for a CA1 model made up of units as in Fig. 2.

interpretations chosen by the authors. Every mixed experimental/computational paper suffers to some extent of this bias (including ours).

In order to highlight this problem and the unwanted consequences we have made a comparative study of the most relevant models available so far on the neuronal type most frequently used in both, experimental and modeling studies, the CA1 pyramidal cell. We have checked up to 5 different models of this neuron (taken from refs. 8,20,21,22,24, some with slight modifications to allow parallel testing). All performed reasonably well when used to replicate the specific physiological details of interest for the authors. However, when put in a common benchmark their properties were amazingly different. Figure 2 shows a clear example during responses to standard depolarizing pulses. Neither, the threshold current, the firing rate, or the properties of adaptation coincided. Besides, the amplitude and waveform of the AP, the extent and properties of dendritic backpropagation, the synaptic generation of APs and other more subtle responses also differed.

The causes for this lack of uniformity are easy to track down because we can re-examine every element of the models. In short, some authors deemed unnecessary to use a branched dendritic model and employed an equivalent cylinder instead. Others considered irrelevant the use of active dendrites or used active properties *ad hoc* so as to reproduce precisely the behavior of a very specific physiological result, leaving unattended the global neuron behavior. No doubt, this slavish use of neuronal models in physiology should be totally unadvised: they cannot be used for any other purpose and their predictive value is simply none¡. Even worse, the interpretation of the experimental results reproduced by these arbitrary models may be based on the erroneous or faulty setting of the model parameters, leading to false conclusions. Figure 3 shows a clear example of these pitfalls. The same neuron (shown in Fig 2C) has been modeled with active (3A) or passive (3B) dendrites, a difference that some authors apparently do not consider important. The distinct somatic responses to identical somatic depolarizing pulses is patent. The example underscores the importance of the interaction between subregions: the presence of active dendrites causes a gross change of the somatic physiology. Using our own neuron model (Fig. 2D) we found that major differences can also be obtained by tiny adjustments of even one parameter in a thousand, as the presence or absence of one channel, or simply a slight shift of its activation time constant or the equilibrium potential of just one ion [22]. We firmly believe that whichever the goal, when using a neuronal model in physiology, the amazing complexity of interacting nonlinear mechanisms demands the use of the most detailed model possible. Extreme respect for the accuracy of every model parameter is the first lesson for physiologists doing modeling. The positive (and still somber) message is that modeling helps to account for the discrepancies between experimental results. And the unwanted question: If minute parameter changes drives model neurons crazy, how could we expect experimental uniformity using techniques with variable degree of interference with the studied object?

Our Magic Solution: Modeling a Macroscopic Invariant Index

At the moment, any attempts to model a single neuron that can reproduce with a reasonable accuracy the large variety of electrical responses that a neuron is capable of *in vivo*, is heading to failure. The reason is simple, there is a giant number of parameters whose value, range, and dynamic properties are taken, in the best of cases, using a rough approximation. Many physiologists may not agree to this assertion. We are habituated to take for granted the accuracy of experimental results obtained with a dispersion below 1-5 percent. However, it is easy to check, for instance, how after a tremendous amount of work the dynamic voltage range of activation for a specific channel is given with a statistical dispersion of 2-3 mV (great achievement). The surprise is that larger average differences can be found in two different reports. And the final shock is that just 1 mV would be enough to drive a pretended realistic neuron model out of the physiological range for some major electrical feature. The gathering of experimental data of the required precision is very slow, and for many parameters the level of accuracy clearly escapes the possibilities of the finest experimental technique. As an experimentally-based team, we believe that the sooner we recognize these limitations the faster we will start to solve them out.

Such despairing scenario we met in our initial attempts to build up a CA1 pyramidal model that could perform reasonably well, meaning that it could reproduce not only

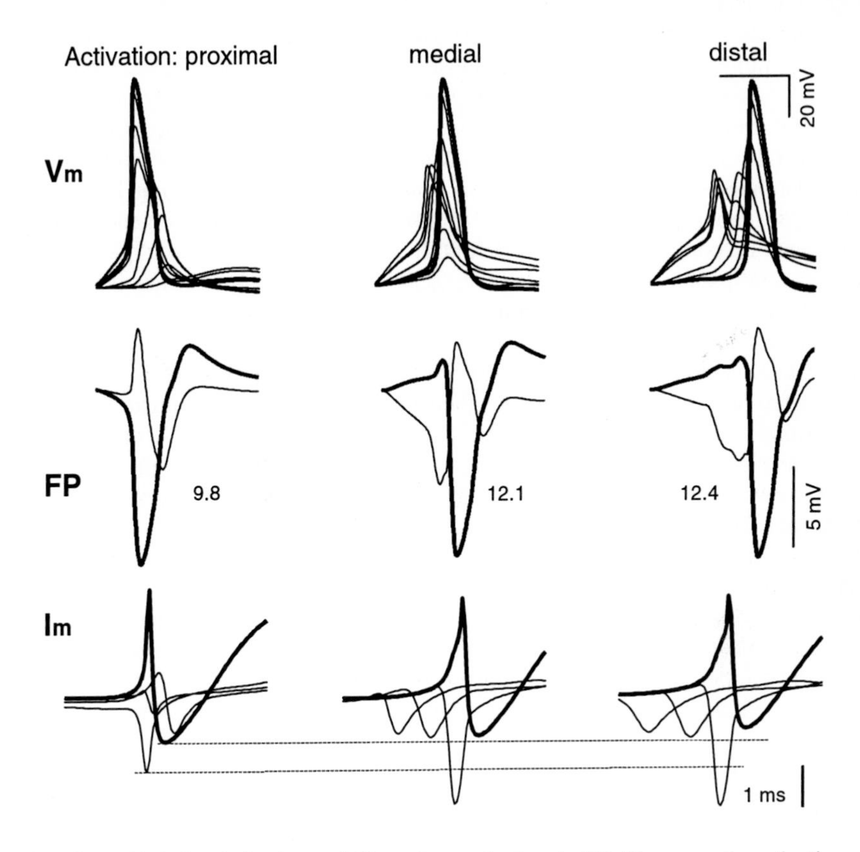

Fig. 5. Multiple levels in the modeling of an orthodromic PS. The synaptic activation was at different loci of the dendritic axis. Upper: Single cell intracellular voltage. Middle: field potential of the compound PS. Lower: transmembrane current.

our favorite electrical behavior but most of the know physiology of that cell type. Instead of abandoning, we devised a possible solution in using a global experimental steady index that comprise in itself the entire parametric space of individual neurons and whose recording was independent of them, i.e., the technique should not interfere with single cell parameters. And such a magic index does exist and it is already famous and widely used, the population spike (PS), an extracellular compound action potential built by the simultaneous currents of all activated neurons along their entire morphology. The spatiotemporal map of the PS includes all ionic and capacitive currents and any possible interaction between distant subregions. The collection of this map *in vivo* is technically simple. The rationale was that minor errors caused by faulty parametric definition in unitary models may go undetected, but would however be easily noticed if potentiated in a macroscopic signal made up of thousand times identical electrical behavior. The lack of precision of experimental data at the unitary and subcellular level would thus be replaced by the steadiness and reproducibility of an antidromic PS. We then began to model the entire CA1 region (several hundred thousand neurons) with detailed pyramidal units, and studied the macroscopic factors that affect the addition of extracellular currents (tissue resistivity, synchrony of activation, number of neurons, tridimensional arrangement of dendritic processes, effects of volume propagation). We continued by analyzing the unitary parameters in a trial/error routine so as to obtain spatiotemporal model PS maps that approached the experimentally recorded potentials [22]. Apart from producing a new more accurate and comprehensive interpretation of this widely employed population index, some predictions about the contribution of dendritic spikes were generated and experimentally confirmed [10]. A definitive tuning of this macroscopic model is however as utopian as for a single neuron model. However, the electrical performance of the prototype pyramidal unit optimized this way is far more realistic than any other in the market (see ref. 22).

A further advantage of this macroscopic model is that it can be used as a benchmark for all different unitary models, so that their performance can be compared. We have initiated this comparison with the already available models of Fig. 2, and the initial results are shown in Fig. 4. Amazing, even bizarre differences can be observed between the model PS obtained with the different model units (note scaling and waveforms). That of Fig. 4D corresponds to our unitary pyramidal cell in a relatively well optimized stage. The model fields are already in close agreement to the experimental ones [22].

Further advantages of this macroscopic model is that it enables the restriction of unitary parameters at different levels (from channel properties, their density and distribution to the gross features of single cell electrogenesis) by direct comparison of the model and experimental PS. Every single parameter will have a noticeable effect on the macroscopic PS. We found that minor changes on the channel assortment of dendrites may cause large deviations of the model PS. In a way, we believe that this model is endowed with a true prospective character that cannot be claimed by any single cell model. Ultimately, in the final stages of refinement, it could be used even to detect when experimental recordings were successful or failed. Figure 5 shows an example of the multiple simultaneous parameters that can be obtained in a single run, and how the different levels can be used to re-adjust the neuron units. It is shown that with relatively low Na^+ channel density a dendritic spike can be initiated in different places of the somatodendritic axis depending on the location and the strength of the input, a result that has been experimentally checked by us.

Let's Them Compute

It would be wise to restrain the excitement arised around the field of dendrites and modeling in physiology. We have gone from the view of passive integrate-and-fire neurons to amazing biological computing machines made up of a number of dendrites each one capable of intricate processing, as if the capabilities of a whole brain could be packed into a single cell, all in just one decade. Evolution would be jealous, no doubt! However, neurons do what they do, nothing else. Modeling has taught some of us a very hard lesson. Experimental research is made by using imperfect techniques controlled by imperfect brains. There is only one thing to do: let's *them* compute.

References

1 Alvarez FJ, Dewey DE, Harrington DA and Fyffe RE (1999) *Cell-type specific organization of glycine receptor clusters in the mammalian spinal cord.* J. Comp. Neurol. 379:150-70.

2 Eberwine JH (1999) *Dendritic localization of mRNAs and local protein synthesis.* In "Dendrites", Stuart, Spruston and Häusse, eds, Oxford, NY, pp 68-84.

3 Eccles JC. *The physiology of nerve cells.* The Johns Hopkins Press, Baltimore, 1957.

4 Fiala JC and Harris KM (1999). Dendrite structure. In "Dendrites", ibid. pp.1-34.

5 Fujita Y and Sakata H (1962) *Electrophysiological properties of CA1 and CA2 apical dendrites of rabbit hippocampus.* J. Neurophysiol. 25: 209-22, 1962.

6 Herreras O (1990) *Propagating dendritic action potential mediates synaptic transmission in CA1 pyramidal cells in situ.* J. Neurophysiol. 64:1429-41.

7 Herreras O and Somjen GG (1993) *Effects of prolonged elevation of potassium on hippocampus of anesthetized rats.* Brain Res. 617:194-204.

8 Hoffman DA, Magee JC, Colbert CM and Johnston D (1997) *K^+ channel regulation of signal propagation in dendrites of hippocampal pyramidal neurons.* Nature 387:869-75.

9 Llinás R (1988) *The intrinsic electrophysiological properties of mammalian neurons. Insights into central nervous system function.* Science 242:1654-64.

10 López-Aguado L, Ibarz JM and Herreras O. (2000) *Modulation of dendritic action currents decreases the reliability of population spikes.* J. Neurophysiol. 83:1108-14.

11 Lorente de Nó R and Condouris GA (1959) *Decremental conduction in peripheral nerve. Integration of stimuli in the neuron.* P.N.A.S. 45:592-617.

12 Magee JC (1999) *Voltage –gated ion channels in dendrites.* In "Dendrites", ibid. pp 139-60.

13 Mel BW (1999) *Why have dendrites? A computational perspective.* In "Dendrites", ibid. pp 271-89.

14 Mainen ZF and Sejnowski TJ (1996) *Influence of dendritic structure on firing pattern in model neocortical neurons.* Nature 382, 363-66.

15 Moore JW, Stockbridge and Westerfield M (1983) *On the site of impulse initiation in a neurone.* J. Physiol. 336:301-11

16 Rall W. *The theoretical foundation of dendritic function.* Segev, Rinzell and Shepherd, eds. MIT Press, 1995.

17 Segev I and London M *A theoretical view of passive and active dendrites.* In "Dendrites", ibid., pp 205-30.

18 Spruston N, Stuart G and Häusser M Dendritic integration. In "Dendrites", ibid. pp231-70.

19 Stuart G, Schiller J and Sakmann B (1997) *Action potential initiation and propagation in rat neocortical pyramidal neurons.* J Physiol. 505:617-32.

20 Traub RD WongRKS, Miles R and Michelson H (1991) *A model of a CA3 hippocampal pyramidal neuron incorporating voltage-clamp data on intrinsic conductances.* J. Neurophysiol. 635-50.

21 Traub RD, Jefferys JGR, Miles R, Whittington MA and Tóth K (1994) *A branching dendritic model of a rodent CA3 pyramidal neurone.* J Physiol 481:79-95.

22 Varona P, Ibarz JM, López-Aguado L and Herreras O. (2000) *Macroscopic and subcellular factors shaping CA1 population spikes.* J. Neurophysiol. 83:2192-208.

23 Vida I, Czopf J, Czéh G (1995) *A current-source density analysis of the long-term potentiation in the hippocampus.* Brain Res. 671:1-11.

24 Warman EN, Durand DM and Yuen GLF (1994) *Reconstruction hippocampal CA1 pyramidal cell electrophysiology by computer simulation.* J Neurophysiol. 71:2033-45.

Homogeneity in the Electrical Activity Pattern as a Function of Intercellular Coupling in Cell Networks

E. Andreu[1], R. Pomares[1], B. Soria[1], and J. V. Sanchez-Andres[2]

[1] Instituto de Bioingeniería. Universidad Miguel Hernández. Campus de San Juan. Aptdo. Correos 18, 03550, San Juan, Alicante, Spain
{etel.andreu, rpomares, bernat.soria}@umh.es
[2] Departamento de Fisiología.Facultad de Medicina. Universidad de La Laguna. Campus de Ofra. 38320, La Laguna, Tenerife, Spain
juanvi@umh.es

Abstract. The aim of this paper is to study changes in the electrical activity of cellular networks when one of the most important electrical parameters, the coupling conductance, varies. We use the pancreatic islet of Langerhans as a cellular network model for the study of oscillatory electrical patterns. The isolated elements of this network, beta cells, are unable to oscillate, while they show a bursting pattern when connected through gap-junctions. Increasing coupling conductance between the elements of the networks leads to the homogeneity of the fast electrical events. We use both experimental data obtained from normal and transgenic animal cells and computational cells and networks to study the implications of coupling strength in the homogeneity of the electrical response.

1 Introduction

In previous papers we have shown that the pancreatic Islet of Langerhans is a good experimental model for the study of oscillatory electrical activity [1,2,3]. When estimulated with intermediate and high glucose concentrations, pancreatic beta cells show a square-waved electrical pattern both in vivo [4] and in vitro [5]. All the cells in an islet burst synchronously [6]. It is necessary to note that this bursting behaviour is only present when cells are electrically coupled and the oscillatory capability is not conserved when they are isolated [7,8]. Besides, the geometrical simplicity of the cells (round shaped) is a great advantage for the computational simulation.

In this highly periodical electrical behaviour two electrical phases can be distinguished, silent phase: cells remain in a hyperpolarized state, and active phase: cells are depolarized and spiking. Synchronicity between cells is restricted however to the square-waved electrical pattern, being the fast superimposed spiking of the active phase not synchronous.

Several papers and models have focused in the generation and electrical pattern of the bursting phenomena, and also great interest has been devoted to study the implications of coupling in the emergence of the oscillatory pattern[1,2,3,6,7,8].

However, the study of the fast spiking properties and its relation with the

J. Mira and A. Prieto (Eds.): IWANN 2001, LNCS 2084, pp. 14-20, 2001.

network properties (coupling between the elements) has not been analyzed in detail yet.

In this paper we try to study the implications of coupling strength in homogeneity of the fast spiking phenomena. We use a valuable experimental model that exhibits an increased coupling , a transgenic mice that overexpress connexin32 [9], and compare its electrical pattern with that obtained from normal mice. Besides, we simulate both conditions in our computational model of the islet, changing the global coupling conductance of our cell network.

2 Methods

2.1. Experimental Methods

Intracellular activity of β-cells from mouse islets of Langerhans was recorded with a bridge amplifier (Axoclamp 2A) as previously described [10]. Recordings were made with thick wall borosilicate microelectrodes pulled with a PE2 Narishige microelectrode puller.

Microelectrodes were filled with 3 M potassium citrate, resistance around 100 MΩ. The modified Krebs solution used had the following composition (mM): 120 NaCl, 25 NaHCO3, 5 KCl and 1 MgCl2, and was equilibrated with a gas mixture containing 95% O2 and 5% CO2 at 37°C.

Islets of Langerhans were microdissected by hand. Pancreatic Islets were obtained from normal albino mice (control) and transgenic mice overexpressing connexin32, these islets show a very high coupling conductance.

Data recorded were directly monitored on an oscilloscope and stored both in hard disk and tape for further analysis . Analysis of the data were performed using Origin 4.0, and Axoscope1.1. We recorded simultaneously neighbor cells of an islet, and distance between cells was estimated directly in the recording chamber.

2.2. Computational Methods

Mathematical simulations were carried out on a SUN Sparc Station 5 platform running Solaris 2.3 (UNIX) operating system.

XPPauto2.5. (B. Ermentrout, Pittsburgh Univ.; http://www.math.pitt.edu) was used to obtain the simplest ordinary differential equation (O.D.E.) and to model the single cell behavior. Gear integration method has been employed (tolerance: 1·10-5; variable step; minimum : step 0.001; maximum step: 25).

XSim (P. Varona and J.A. Siguenza, IIC, UAM; http://www.iic.uam.es/xsim) was used to simulate clustered β–cells. This package allows the modeling of complex neural networks. A modified Heun's method was used for the integration of differential equation systems, using variable and different width adaptive time-steps ($1 \cdot 10^{-5}$ to 10).

2.2.1 Cell model

The element of the network was described by a set of ordinary differential equations, being de membrane potential defined by:

$$C_m \, dV/dt = -(I_{KATP} + I_{KV} + I_{Ca} + I_{KCa}) \quad (1)$$

Where C_m is the membrane capacitance, V is the membrane potential, I_{KATP} is the ATP dependent potassium current, I_{KV} is the voltage-dependent potassium current, I_{Ca} is the calcium dependent calcium current and I_{KCa} is the calcium-dependent potassium current. All this currents show a non-linear behavior [3]

The significant difference of our model with previous beta cells models is that we exclude a cellular slow variable that leads the oscillation, being the single cell model unable to show an oscillatory behaviour when operating in the range of parameters that can be considered physiological.

2.2.2 Network model

Up to 100 single elements were coupled in a network that we will consider as our model islet. Each element receives as input the link with the output of neighboring cells, the coupling function being computed as a non variable pre-fixed parameter (global coupling conductance) multiplied by the electrotonic strength between the cells in contact.

This coupling current, I_{i-c}, of the *i* cell is driven by the voltage difference between this cell and each one of its neighbors at a precise time step and reevaluated in each computing step for every site and unit. Then, changing the coupling conductance total value we can simulate both the control network (normal mice, moderate coupling) and the transgenic network (high coupling)

$$I_{i-c} = \sum_{j=1}^{n} g_{ij} \Delta V_{ij} \quad (2)$$

3 Results

The oscillatory electrical activity in pancreatic beta cells of control and transgenic mice was recorded at stimulatory glucose concentrations (11 mM) and also that condition was reproduced in the model (through imposing a value for the ATP-dependent potassium conductance) for normal coupling conductance (g_{ij}=250 pS) and increased coupling conductance (g_{ij}=900 pS), representative recordings of the electrical activity of normal and transgenic cells are shown in figure 1.

While in the normal coupling condition both experimental and model results show a high heterogeneity in the spikes, the high coupling condition shows only a small variability in the spikes amplitude and duration

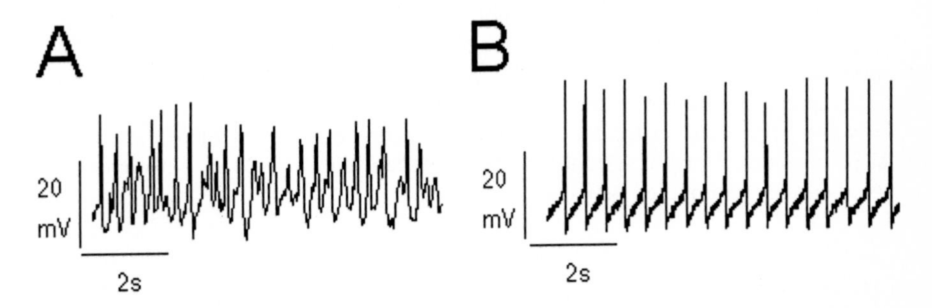

Fig. 1. Electrical activity during the active phase for normal cells (A) and transgenic cells (B)

To quantify changes in the heterogeneity of the spikes, we studied the histograms of spikes amplitudes both in the experimental and model cases (see figure 2) While several populations of spikes can be detected in the normal coupling condition (figure 2, A top and bottom), only three populations can be distinguished in the high coupling condition (figure 2, B top and bottom).

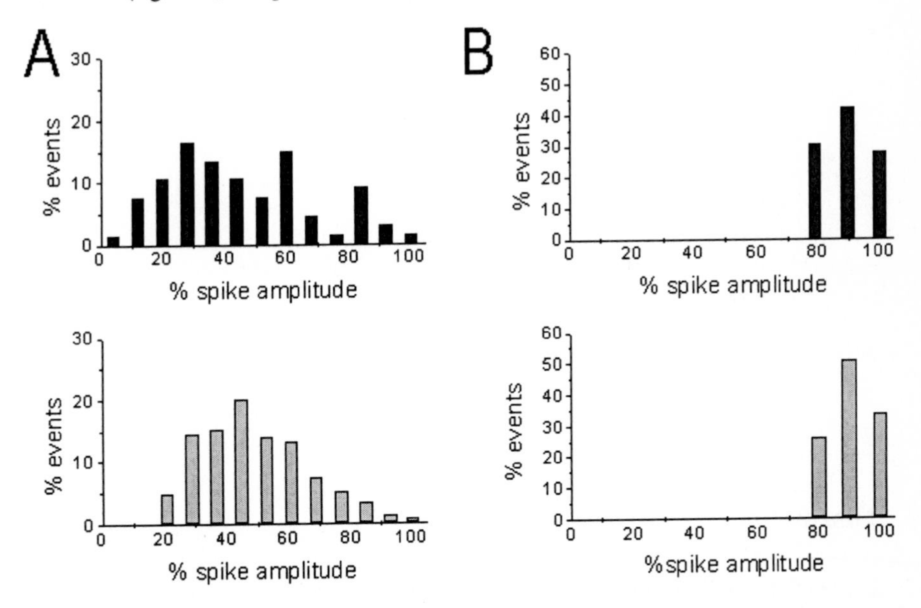

Fig. 2. Spike amplitude histograme of normal (A, dark bars top) and transgenic (B, dark bars top) cells from real islets. The same plots for normal coupled (A, light bars bottom) and highly coupled (B, light bars bottom) cells of the computational model.

In order to identify the different populations we have applied a single criterium, only spikes amplitude has been taken into account for calculating them. The maximum spike amplitude is measured in each situation, and then all the spikes amplitudes are normalized to that maximum.

Afterwards, intervals of 10% of the normalized amplitude have been considered to plot the spikes histogram. Those intervals, however are rather arbitrary, as far as presumablely more than a 25% in the spikes amplitude should be considered to separate spikes population in the physiological range.

However, the same criterium is applied to experimental and model results, and must be considered just an election to make evident the difference between normal coupled and highly coupled networks.

Finally, we have plotted the interval scatterograms of normal and transgenic cells. To obtain the scatterograms we plot the pairs ISI_n vs ISI_{n+1}, the interspike intervals (ISI) have been extensively used to characterize electrophysiological data [11,12]. Thys type of diagram shows also clearly the different spiking characteristcs of cells from normal (figure 3 A) and highly coupled networks (3B)

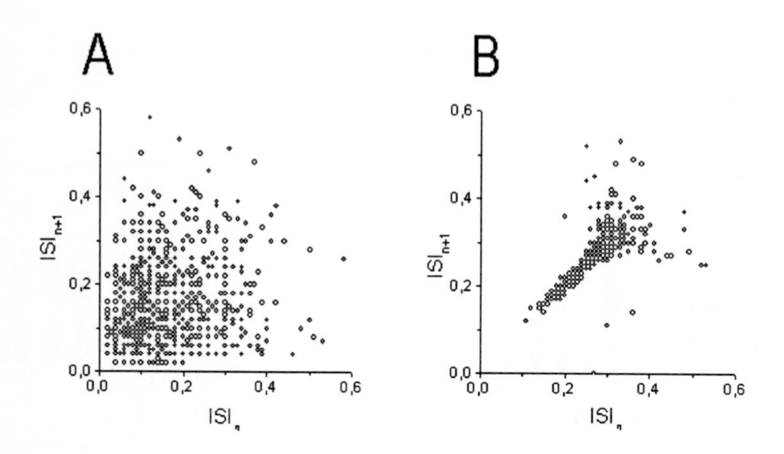

Fig. 3. Interval Scattergrams for normally coupled cells (A) and highly coupled cells (B)

4 Discussion

Intercellular coupling plays a key role in emergence of oscillatory behaviour and also in the synchronization of electrical events in in gap-junction connected cell networks.

In this study we have shown another role of gap-junction mediated communication between cells, the modification of the firing properties of each element in the network.

In our experimental tissue, the pancreatic islet, spikes are calcium action potentials, originated by the entry of calcium through the voltage-activated calcium channels. This calcium entry makes possible the subsequent exocytosis of insulin. Thus, modifying the firing properties, the final function (insulin secretion) is also altered.

We have shown that an increased communication between cells in the networks gives place to an increased homogeneity in the spikes amplitude. Only the bigger spikes are present in those cells and model cells that belong to a highly connected network.

There are several systems where intercellular electrotonic coupling plays a key role [13,14], in fact it has been already proposed that electrical coupling might provide a flexible mechanism for modifying the behavior of an oscillatory neural network[15].

However it is very difficult to test those hypotheses and models given the lack of highly specific modulators of the gap-junction channels. Nevertheless, our experimental approach using transgenic gap-junction-modified networks and realistic computational models has allowed us to test the implications of gap-junction coupling variability on the firing characteristics of an oscillatory system.

Acknowledgments: We are indebted to Pablo Varona for introducing us into XSim. We also thank Dr. P. Meda for giving us the opportunity to work with his transgenic Cx32 mice. This study was supported by grant GV99-146-1-04 of the Generalitat Valenciana.

References

1. Andreu E., Bolea, S., Soria, B., Sánchez-Andres, J.V. Optimal Range of Input Resistance in the Oscillatory Behaviour of the pancreatic beta cell. Lect. Notes Comp. Sci, 930 (1995) 85-89
2. Andreu E., Pomares, R., Soria, B., Sanchez-Andres, J.V. Balance between intercellular coupling and input resistance as a necessary requierement for oscillatory electrical activity in pancreatic beta cells. Lect. Notes Comp. Sci.,1240 (1997) 146-453
3. Andreu, E., Soria, B., Sanchez-Andres, J.V. Intercellular cooperation and the emergence of oscillatory synchronous electrical activity in the pancreatic islet of Langerhans. Submitted to Progress in Biophysics and Molecular Biology.
4. Sanchez-Andres, J.V., Gomis, A., Valdeolmillos, M. The electrical activity of mouse pancreatic beta cells recorded in vivo shows glucose-dependent oscillations. J. Physiol. 486.1 (1995)223-228
5. Dean, P. M., Mathews, E.K. Electrical activity in pancreatic islet cells. Nature, 219 (1968) 389-390
6. Valdeolmillos, M., Gomis, A., Sanchez-Andres, J.V. In vivo synchronous membrane potential oscillations in mouse pancreatic beta cells: lack of co-ordination between islets. J. Physiol., 493 (1996) 9-18.
7. Perez-Armendariz, M., Roy,C., Spray, D.C., Bennett, M.V.L. Biophysical properties of gap junctions between freshly dispersed pairs of mouse pancreatic beta cells. Biophysical Journal, 59 (1991) 76-92.
8. Smolen, P., Rinzel, J., Sherman, A. Why pancreatic islets burst but single beta cells do not. The heterogeneity hypothesis. Biophysical Journal, 64 (1993)1668-1680.

9. Charollais, A., Gjinovci, A., Huarte, J., Bauquis, J., Nadal, A., Martin, F., Andreu, E., Sanchez-Andres, J.V., Calabrese, A., Bosco, D., Soria, B., Wollheim, C.B., Herrera, P.L., Meda, P. Junctional communication of pancreatic beta cells contributes to the control of insulin secretion and glucose tolerance. J.Clin.Invest, 106 (2000) 235-243.
10. Sanchez-Andres, J.V., Ripoll, C., Soria, B. Evidence that muscarinic potentiation of insulin release is initiated by an early transient calcium entry. FEBS Lett., 231 (1988) 143-147
11. Bair, W., Koch, C., Newsome, W., Britten, K., Power spectrum analysis of bursting cells in área MT in the behaving monkey. J. Neurosci., 14 (1994) 2870-2892
12. Rapp, P.E., Goldberg, G., Albano, A.M., Janicki, M.B., Murphy, D., Niemeyer, E., Jimenez-Montaño, M.A., Using coarse-grained measures to characterize electromyographic signals. Int. J.Bifur.Chaos, 3 (1993) 525-541.
13. Yarom, Y. Oscillatory Behavior of Olivary Neurons. In The Olivocerebellar System in Motor Control. Ed. Strata, P.,209-220. Berlin-Heidelberg. Springer-Verlag (1989)
14. Bleasel, A.F., Pettigrew, A.G. Development and properties of spontaneous oscillations of the membrane potential in inferior olivary neurons in the rat. Brain.Res.Dev.Brain.Res., 65 (1992) 43-50.
15. Kepler, T.B., Marder,E., Abbot, L.F. The effect of electrical coupling on the frequency of model neural oscillators. Science, 248 (1990) 83-85.

A Realistic Computational Model of the Local Circuitry of the Cuneate Nucleus

Eduardo Sánchez[1], Senén Barro[1], Jorge Mariño[2], and Antonio Canedo[2]

[1] Grupo de Sistemas Intelixentes (GSI)
Departamento de Electrónica e Computación, Facultade de Físicas,
Universidade de Santiago de Compostela,
15706 Santiago de Compostela, Spain
{elteddy, elsenen}@usc.es
http://elgsi.usc.es/index.html
[2] Departamento de Fisioloxía, Facultade de Medicina,
Universidade de Santiago de Compostela,
15706 Santiago de Compostela, Spain
{fsancala, xmarinho}@usc.es

Abstract. Intracellular recordings obtained under cutaneous and lemniscal stimulation show that the afferent fibers can establish excitatory and inhibitory synaptic connections with cuneothalamic neurons [5]. In addition, distinct types of recurrent collaterals with the capability of either exciting or inhibiting both cuneothalamic neurons and interneurons were also discovered [6]. With these data we have generated hypothesis about which circuits are implicated and also developed realistic computational models to test the hypothesis and study the cuneate properties [17, 18]. The results show that the cuneate could perform spatial and temporal filtering and therefore detect dynamic edges.

1 Introduction

The cuneate nucleus is a structure of the somato-sensory system that is located within the brain stem, in the most rostral area of the spinal chord. With regard to its spatial organization, three areas can be distinguished: caudal, middle and rostral [2]. The middle zone, where the experimental work reported in this study has been carried out, can in turn be divided into two areas [7]: a core (central region), and a shell (peripheral region). The core is basically made up of cuneothalamic neurons, which are also referred to as projection or relay neurons. The shell is basically made up of local neurons, also known as interneurons or non-thalamic projection neurons.

The main cuneate inputs originate from primary afferent (PAF) and corticocuneate fibers. The PAFs establish synaptic contact with projection cells and interneurons [7]. The great majority of the projection neurons in the middle cuneate send their axons to the contralateral ventroposterolateral thalamic (VPL) nucleus via the medial lemniscus [11, 2, 7].

To study the neuronal properties of the middle cuneate, intracellular and whole-cell techniques have recently been used *in vivo* [13, 12, 3, 4]. It was found

J. Mira and A. Prieto (Eds.): IWANN 2001, LNCS 2084, pp. 21–29, 2001.

that cuneothalamic neurons possess two modes of activity: oscillatory and tonic [4]. During deep anaesthesia, the cuneothalamic neurons show oscillatory bursting activity. This behavior changes to the tonic mode when injecting depolarizing currents or stimulating the peripheral receptive fields. A low-threshold calcium current (I_T) and a hyperpolarization-activated cation current (I_h) have been postulated to explain the bursting mode of activity [4]. All these data were considered to develop realistic computational models of the cuneate neurons [15, 16].

In this paper we shifted our focus to provide a model of the local circuitry of the cuneate. The presentation is organized as follows: (1) a brief description of the experimental methods, (2) a summary of the results, (3) an explanation of the hypothesis derived by these data, (4) a description of the computational model, (5) the simulation results, and (6) a final discussion.

2 Experimental Methods

All experiments conformed to Spanish guidelines (BOE 67/1998) and European Communities council directive (86/609/EEC), and all efforts were made to minimize the animals used. Data were obtained from cats (2.3-4.5 kg), which were anesthetized (α-cloralosa, 60 mg/Kg, i.p.), paralyzed (Pavulon, 1 mg/ kg/h, i.v) and placed under artificial respiration [4].

Cuneothalamic projection cells were identified as antidromically activated by medial lemniscus stimulation according to standard criteria, including the collision test as well as confirmation that the critical interval in the collision was not due to soma refractoriness [4]. For cutaneous stimulation, a plate of 20mmx5mm was specifically designed to place 6 electrode pairs. Electrodes were 3 mm separated from each other. With this organization a total length of 15 mm could be stimulated, which its beyond the average size of the cutaneous receptor fields.

3 Experimental Results

Experiments were performed to analyze neural responses under cutaneous and lemniscal stimulation. In the former case, we are looking for primary afferent effects. In the later case, we are interested in recurrent collateral influences.

1. Cutaneous stimulation
   Stimulation over the center of a cuneothalamic receptive field originates excitatory postsynaptic potentials (Fig. 1). After stimulation of points located far away from the center, inhibitory postsynaptic potentials were found.
2. Lemniscal stimulation
   This kind of protocol has induced cuneothalamic responses that consist on a pair of spikes. The first one appears to be an antidromic spike. The second one is clearly synaptic. The explanation can be found on the existence of

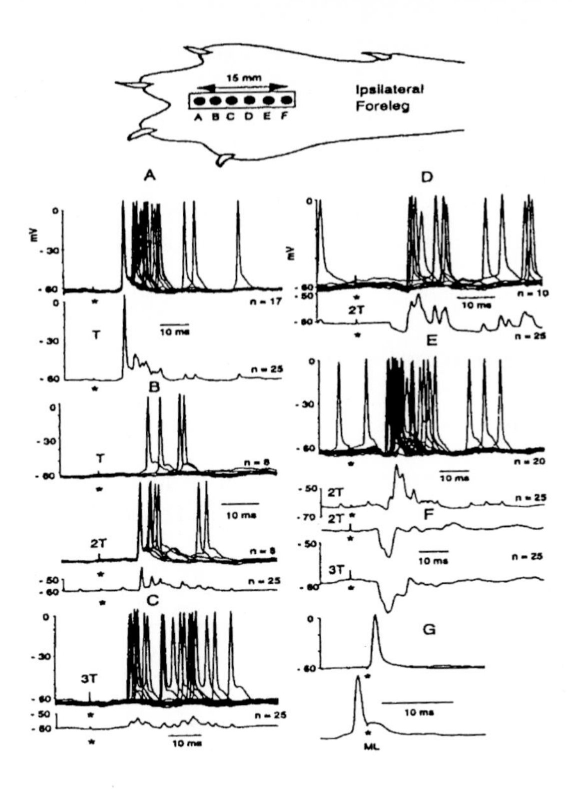

Fig. 1. Recordings under cutaneous stimulation. In the upper part, a drawing pictures the stimulation points in the cat's pow. In A, excitatory responses with lowest latency and stimulation intensity (T) are shown. In points B and C, higher stimulation (2T in B; 3T in C) is required to obtain a similar response, although with higher latency. D and E show inhibitory-excitatory sequences. Finally, point F, in the periphery, shows only inhibitory responses under stimulation. In G, we show the antidromic identification (adapted from Canedo and Aguilar, 2000a).

recurrent collaterals from neighbour neurons that were previously activated by the lemniscal stimulation.

Lemniscal stimulation can also induce inhibition over cuneothalamic neurons (Fig. 2). The inhibition response appears after the antidromic spike and can be explained based on the existence of inhibitory recurrent collaterals between cuneothalamic neurons.

4 Some Hypothesis

The experimental data shown above is the basis of the following hypothesis:

- Afferent excitation and inhibition. If cuneothalamic neurons have center-surround receptive fields, the cuneate can detect edges. This functional feature can be found also in other sensory systems.

- Recurrent lateral inhibition. This mechanism allows a neuron to inhibit the response of less activated neighbours. This effect is only observed over neighbour cells with non-overlapped receptive fields. In this way, only neurons with higher afferent input in a certain region are finally activated.
- Recurrent lateral excitation. Its function could be to synchronize the activity of neighbouring cells with overlapped receptive fields. In this way, a focus of activity would be generated to improve both signal localization and signal quality.

5 The Computational Model

5.1 Neuron Level

The multi-compartmental approach was used to model the neurons [14]. The membrane potential for each compartment is computed by using the following expression:

$$C\frac{\partial V}{\partial t} = -I_m - I_{syn} - I_{inject} - \frac{(V' - V)}{R'_a} - \frac{(V'' - V)}{R''_a} \tag{1}$$

where C is the membrane capacitance, I_m the sum of the ionic currents, I_{syn} the synaptic current I_{inject} the electrical stimulation, R_a the axial resistance, and

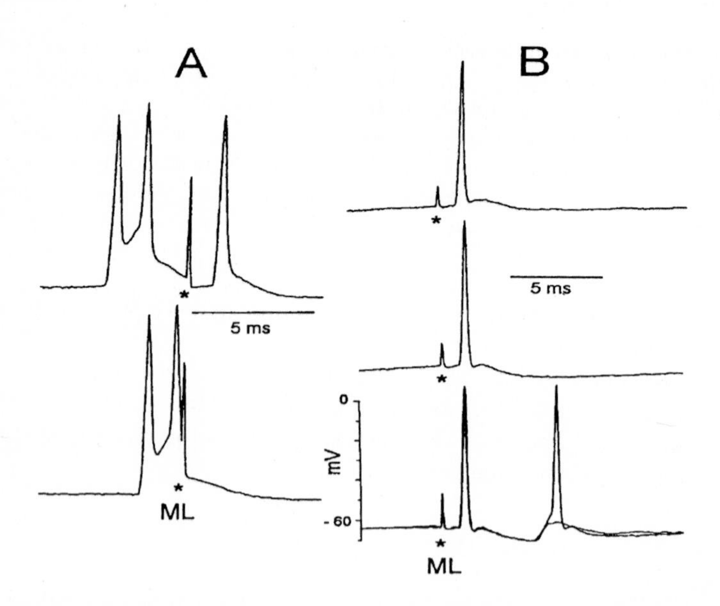

Fig. 2. Recordings under lemniscal stimulation and possible effects of inhibitory recurrent connections. In A, the antidromic identification is shown. In B, an appropriate level of stimulation at the medial lemniscus can induce an inhibition with a rebound that generates spike (lower picture in B) (Adapted from Canedo et al., 2000b).

$\frac{(V'-V)}{R'_a}$ and $\frac{(V''-V)}{R''_a}$ denotes the axial current between each compartment and its adjacent ones.

We have used the following compartments and ionic currents:

- Primary afferents. Two compartments for soma and axon. The membrane current is the sum of a fast sodium current and a delayed rectifier potassium current. Therefore, $I_m = I_{Na} + I_K$.
- Cuneothalamic cells. Three compartments for soma and two dendritic branches. The most remarkable ionic channels are a high-threshold calcium current I_L, a calcium-activated potassium current I_{ahp}, a hyperpolarization-activated cationic current I_h, and a low-threshold calcium current I_T.
- Interneurons. Three compartments for soma, axon and dentritic tree. The ionic currents are very similar to those dicussed for cuneothalamic cells with two major differences: (1) a new slow potassium current I_A, and (2) the substitution of I_{ahp} with other calcium-dependent potassium current I_C.

5.2 Membrane Level

All ionic currents were described by following the Hodgkin-Huxley model [9]. The mathematical description of all this currents can be found in [17, 19].

5.3 Synaptic Level

To describe the conductance associated with the synaptic current we have used the *alpha* function. This conductance is the reasonable of generating postsynaptic potentials after the activation of dendritic receptors by appropriate neurotransmitters [10]. The mathematical expression is:

$$g_{syn}(t) = \alpha \frac{t-t'}{\tau} e^{-\frac{t-t'-\tau}{\tau}} \tag{2}$$

where α is a constant, t' the initial time, and τ the time constant. The synaptic conductance value will be zero if the presynaptic potential is less than a certain threshold and it will be equal the *alpha* function above such threshold. The threshold represents the amount of neurotransmitter needed to activate the postsynaptic receptors.

The conductance value g_{syn} determines the dynamics and the value of the synaptic current I_{syn}:

$$I_{syn}(t) = g_{syn}(t)(V - V_{syn}) \tag{3}$$

where V_{syn} denotes the resting potential of the postsynaptic membrane. In our model we introduce both excitatory and inhibitory connections. We assume the later being glutamatergic and the former being gabaergic. The excitatory connections are modeled by setting V_{syn} to a positive value. The inhibitory connections are setting to a negative value.

5.4 Simulation Tools

For the simulations we have used *Neuron.* The simulation uses an integration method developed by Hines [8], which it is a fast procedure that optimizes performance by combining the advantages of Crank-Nicholson and Euler methods. The machine used for the simulations was a PC with a 600 Mhz Pentium III processor.

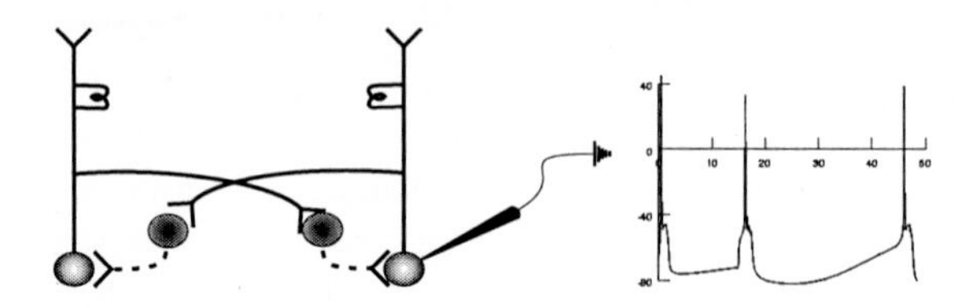

Fig. 3. Lemniscal stimulation and the effect of the inhibitory recurrent connections.

6 Simulation Results

We have developed two different circuits. The first one was used to validate th circuit with some experimental data. This circuit was implemented with 2 cuneothalamic projection neurons, 4 interneurons and 7 sensory receptors. With this configuration we can describe the complete receptive fields of both cuneothalamic cells. We only present here the simulation that shows the dynamics of the inhibitory recurrent collaterals under lemniscal stimulation. We have applied a 2 ms pulse and observe the response of two projection cells that show inhibitory recurrent connections between them. These connections were configured with a constant time τ=3, maximum synaptic conductance g_{sin}^{max}=0.1 and synaptic potential V_{syn}=-90. These values were adjusted to obtain the most similar response between the simulations and the experimental data. Figure 3 shows the neuronal activity after stimulation in t=15ms. Current stimulation initially generates an antidromic spike and, afterwards a synaptic induced inhibition. This inhibition induces an hyperpolarization that activates the cationic current I_h, which in turns depolarizes the membrane and, finally, triggers the activation of a low-threshold calcium current I_T. The calcium potential generates a sodium spike. We believe that this mechanism can explain the experimental data.

In the second circuit we have developed a simplified cuneate with 7 cuneothalamic neurons and 14 sensory receptors. To simplify the computational cost we have replaced the interneurons with appropriate inhibitory gabaergic connections over the cuneothalamic neurons. The connections were set up in such a way that the contribution of the inhibitory periphery is more than the contribution of the excitatory center. We have analyzed in other simulations not shown here, that this is the only configuration that allows edge detection. In Fig. 4 we show the

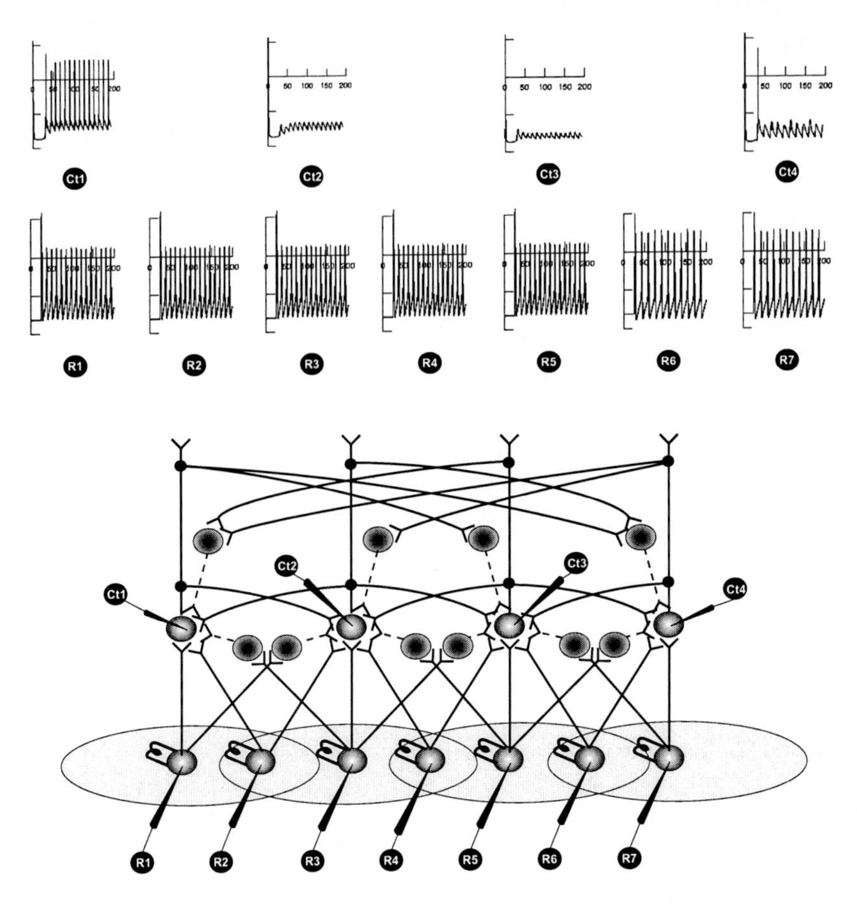

Fig. 4. Maximum edge detection. The concurrent activity of both afferent and recurrent connections permits the detection of the maximun edge in the left side of the stimuli.

circuit and the stimulus chosen to test it. Receptors R6 and R7 were activated less than the others. What we get is that only Ct1 was activated. Although Ct4 shows a strongest postsynaptic activity than its neighbours, its activity is not higher enough to elicit response. This behaviour can be explained by the effect of the recurrent lateral inhibition exerted from Ct1 over its neighbours. With such a configuration, the cuneate would be able to detect the most significant events that appear in the environment.

And finally we want to introduce a new temporal filtering mechanism as a hypothesis. By considering some experimental evidence we believe that autoinhibitory connections can exist in the cuneate and it can perform a type of temporal filtering process. What would happen if we integrate such a mechanism within the circuitry developed so far (Fig. 4)? The results are shown in Fig. 5, in which we have used a uniform stimulus. We observe that edge detection is performed first and, after a certain time interval, the cuneothalamic neurons stop firing. If the stimulus persists, the neurons enter in an oscillatory cycle that could serve as a pattern to signal that no relevant change is present in the environment.

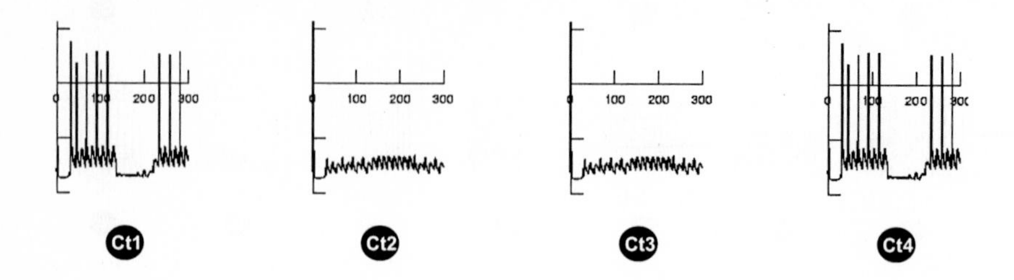

Fig. 5. Temporal filtering and response of cuneothalamic neurons under a persistent stimulus with large surface. The new circuit can detect edges and the temmporal filtering mechanism is the responsible of generating oscillatory activity to signal the presence of a persitent stimuli.

7 Discusion

The main idea behind this study is that the cuneate performs a basic preprocessing of afferent input signals that consists on detecting changes from the outside. Detecting the edges of an stimulus is an example of detecting stimuli with different intensity patterns applied over the skin. The maximum edge detection is an example of detecting differences between selected edges. Detection of dynamic activity is an example of detecting changes in time. All these functional features reinforce the idea that the nervous system is designed to detect changes in the environment and to discard those signals that do not contribute with relevant information [1].

Acknowledgments

We would like to thank to **Laboratorios de Neurociencia y Computación neuronal (LANCON)**, the environment in which this work has been developed.

References

1. Barlow H. B.: The coding of sensory messsages. In "Current problems in animal behaviour". Cambridge University Press. Thorpe W. H. and Zangwill O. L. (Eds.) (1961) 330–360
2. Berkley Karen J., Badell Richard J., Blomqvist A., Bull M.: Output Systems of the Dorsal Column Nuclei in the cat. Brain Research Review. Vol. 11 (1986) 199–225

3. Canedo, A.: Primary motor cortex influences on the descending and ascending systems. Progress in Neurobiology. Vol. 51 (1997) 287–335
4. Canedo A., Martinez L., Mariño J.: Tonic and bursting activity in the cuneate nucleus of the chloralose anesthetized cat. Neuroscience. Vol. 84 **2** (1998) 603–617
5. Canedo A., Aguilar J.: Spatial and cortical influences exerted on cuneothalamic and thalamocortical neurons of the cat. European Journal of Neuroscience. Vol. 12 **2** (2000) 2513–2533
6. Canedo A., Aguilar J., Mariño J.: Lemniscal recurrent and transcortical influences on cuneate neurons. Neuroscience. Vol. 97 **2** (2000) 317–334
7. Fyffe Robert E., Cheema Surindar S., Rustioni A.: Intracelular Staining Study of the Feline Cuneate Nucleus. I. Terminal Patterns of Primary Afferent Fibers. Journal of Neurophysiology. Vol. 56. **5** (1986) 1268–1283
8. Hines M.: A program for simulation of nerve equations with branching geometries. International Journal of Biomedical Computation. Vol. 24 (1989) 55–68
9. Hodgkin A., Huxley A., Katz B.: Measurements of current-voltage relations in the membrane of the giant axon of Loligo. Journal of Physiology (London). Vol. 116 (1952) 424–448
10. Jack J. J. B., Redman S. J.: The propagation of transient potentials in some linear cable structures. J. of Physiology (London). Vol. 215 (1971) 283–320
11. Kuypers H. G. J. M., Tuerk J. D.: The distribution of the cortical fibers within the nuclei cuneatus and gracilis in the cat. J. Anat. Lond. Vol. 98 (1964) 143–162
12. Mariño J., Martínez L., Canedo A.: Coupled slow and delta oscillations between cuneothalamic and thalamocortical neurons in the chloralose anesthetized cat. Neuroscience Letters. Vol. 219 (1996) 107–110
13. Mariño J., Martínez L., Canedo A.: Sensorimotor integration at the dorsal column nuclei. News In Physiological Sciences. Vol. 14 (1999) 231–237
14. Rall W.: Theoretical significance of dendritic tree for input-output relation. In Neural Theory and Modeling. Stanford University Press, Stanford. Reiss, R. F. (Ed). (1964) 73–97
15. Sánchez E., Barro, S., Canedo, A., Martìnez, L., Mariño, J.: Computational simulation of the principal cuneate projection neuron. Workshop Principles of Neural Integration. Instituto Juan March de Estudios e Investigaciones (1997). Madrid.
16. Sánchez E., Barro, S., Canedo, A., Martìnez, L., Mariño, J.: A computational model of cuneate nucleus interneurons. Eur. J. Neurosci. Vol. 10. **10** (1998) 402
17. Sánchez E., Barro S., Mariño J, Canedo A., Vázquez P.: Modelling the circuitry of the cuneate nucleus. In Lecture Notes in Computer Science. Volume I. Springer Verlag. Mira J. and Sánchez Andrés J. V. (Eds). (1999) 73-85.
18. Sánchez E., Barro S., Mariño J, Canedo, A.: Selección de información y preprocesamiento de señales somatosensoriales en los núcleos de las columnas dorsales. Revista de Neurología. Viguera C. (Ed). (1999) 279.
19. Sánchez E.: Modelos del cuneatus: de la biología a la computación. Ph. D. Thesis. Santiago de Compostela. (2000).

Algorithmic Extraction of Morphological Statistics from Electronic Archives of Neuroanatomy

Ruggero Scorcioni and Giorgio A. Ascoli

Krasnow Institute for Advanced Study and Department of Psychology, George Mason University, Mail Stop 2A1, 4400 University Dr., Fairfax, VA 22030 (USA)
{rscorcio, ascoli}@gmu.edu

Abstract. A large amount of digital data describing the 3D structure of neuronal cells has been collected by many laboratories worldwide in the past decade. Part of these data is made available to the scientific community through internet-accessible archives. The potential of such data sharing is great, in that the experimental acquisition of high-resolution and complete neuronal tracing reconstructions is extremely time consuming. Through electronic databases, scientists can reanalyze and mine archived data in a fast and inexpensive way. However, the lack of software tools available for this purpose has so far limited the use of shared neuroanatomical data. Here we introduce L-Measure (LM), a free software package for the extraction of morphological data from digitized neuronal reconstructions. LM consists of a user-friendly graphical interface and a flexible core engine. LM allows both single-neuron study and the statistical analysis of sets of neurons. Studies can be limited to specific regions of the dendrites; morphological statistics can be returned as raw data, as frequency histograms, or as cross-parameter dependencies. The current version of LM (v1.0) runs under Windows and its output is compatible with MS-Excel.

Introduction

Dendritic trees are the principal loci of neuronal input, signal integration, and synaptic plasticity. The study of dendritic morphology is an essential step in any attempt to correlate structure, activity, and function in the nervous system at the cellular level [1,2]. In particular, dendritic morphology can be quantified with geometrical and topological parameters such as lengths, angles, branching order, and asymmetry [3], as well as with parameter interdependence (e.g. Sholl analysis of number of branches versus distance from the soma). In classical neuroanatomical studies, groups of neurons belonging to the same morphological class can be characterized statistically (mean, standard deviation, etc.).

In the past few years, computer-interfaced motorized microscopes have allowed the collection and digital storage of large amounts of 3D morphological data from intracellularly stained neurons [4]. Complete neuronal reconstructions are still extremely time consuming (up to one month/cell), and large statistical studies are remarkably resource demanding. More and more frequently, laboratories and investigators share digital data with the neuroscience community after publishing the analysis relevant to their interest and research. Other groups with a different research

J. Mira and A. Prieto (Eds.): IWANN 2001, LNCS 2084, pp. 30-37, 2001.

focus can then reanalyze the data and extract additional information. The process of neuroanatomical data sharing is thus important because it allows (a) full replicability; (b) time and money saving; and (c) the accumulation of large amount of data beyond the capability of single research groups, and thus higher data-mining potential [5,6]. Digital files of neuronal morphology contain all the information necessary to extract any geometrical and/or topological measurement accurately and quantitatively. Yet, to date, no freeware tool is available to the neuroscience community to analyze digital morphological data on a routine basis. Typically, analysis of morphology is useful in classical neuroanatomical studies [7-9], in computational modeling [2,10-12], and for correlating neuronal structure and electrophysiological activity [13-17]. Recently we created a dedicated software package, called L-Neuron, for the simulation of dendritic morphology. L-Neuron also employs statistical data extracted from digitized neuronal reconstructions to generate virtual neurons [18].

Here we present L-Measure (LM), a computer program that extracts morphological statistics from single-cell digital neuroanatomical files. LM is freely distributed as a stand-alone tool, as well as part of the L-Neuron software http://www.krasnow.gmu.edu/L-Neuron/ (case sensitive)

System Requirements, Software Description, File Formats, and Source of Experimental Data

LM v1.0 runs on Windows platforms and requires the Microsoft Java Virtual Machine, which can be freely downloaded from http://www.microsoft.com/java/. LM has two main components. The engine, written in C++, is dedicated to the numerical extraction of the statistical data from the input files. The graphical user interface, written in java, reads commands and options selected by the user, and translates them to the engine. The engine can also be used directly via command line (e.g. for large automated batch processes). LM outputs are visualized on the screen and/or saved in a file (e.g. in Microsoft Excel format). The engine and the interface are implemented using Object Oriented methodology to ensure fast upgrades and simplify future addition of new options. The LM graphical interface is shown in Figure 1.

The morphology of a neuron is typically described as a set of substructures: trees, bifurcation, branches, and compartments. Each neuron is formed by a soma (the cell body), one axon and one or more dendritic trees. A tree is a part of the neuron structure exiting the soma from a single point. A bifurcation is a splitting point in a tree. A branch is a part of a tree between two consecutive nodes (bifurcations, terminations, tree stems), and it is formed by a series of consecutive compartments. A compartment is a cylindrical representation of the smallest neuronal element. Figure 2 (panels A, B and C) shows examples of schematized neuron and (panels D, E, F) 2D projection of real digitized neuron.

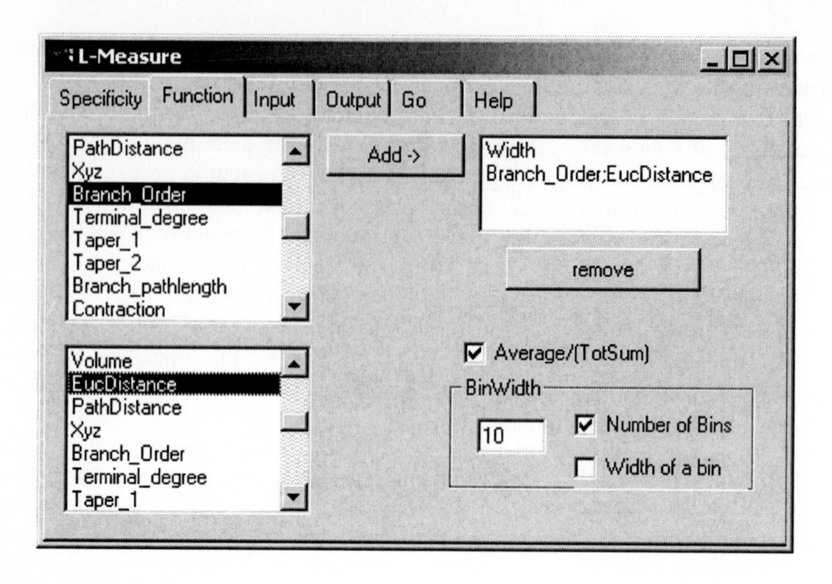

Fig. 1. L-Measure graphical user interface. An online tutorial describing the interface functions is available online at http://www.krasnow.gmu.edu/L-Neuron/.

The LM engine is based on an algorithm that processes neurons sequentially. Each neuron is parsed, and converted in a common internal format. All trees are parsed starting from the soma towards the terminals. Generally, morphometric functions are evaluated at every compartment. Analysis can involve either a single parameter (e.g., the compartment diameter), or a distribution of one parameter versus a second one (e.g., diameter versus Euclidean distance from the soma). For example, to evaluate the statistical distribution of diameter values, LM starts from the soma, and compartment by compartment, collects diameter values until it reaches the end of all trees. At the end of the parsing process the list of values is processed and printed or saved to a file.

Input files of LM are digital morphometric data files, typically in SWC format [6,19]. The SWC format describes neuronal morphology with one ASCII line per cylindrical compartment. Compartments are described by: identifier (typically a progressive integer number starting from 1), type (1: for soma, 2: axon, 3: apical dendrites, 4: basal dendrite, etc.), coordinates of the cylinder's end point (x,y,z in μm), radius (μm) and the identifier of the next compartment in the path to the soma (the "parent").

Typically, neurons are described by digital files of 500 to over 9,000 compartments. LM also supports other common input formats, including Neurolucida [4], Eutectics and its variations (e.g., http://cascade.utsa.edu/bjclab/). LM can also convert any of these formats into SWC (for a detailed format description see [19]).

Among the databases of neuronal morphology available on the web, it is important to cite: SWC Southampton Archive containing 124 rat hippocampal neurons http://www.cns.soton.ac.uk/~jchad/cellArchive/cellArchive.html, Claiborne's archive containing 54 rat hippocampal neurons http://cascade.utsa.edu/bjclab/, and the Virtual Neuromorphology Electronic Database, containing 390 virtual neurons and 9 real neurons distributed between motoneuron and Purkinje cells:

http://www.krasnow.gmu.edu/L-Neuron/. Additional digital data can be obtained from several laboratories upon request (e.g., [20])

Function List and Examples of Applications

LM provides a predefined set of functions to study neuronal morphology. Different functions extract data at different levels of granularity. Entire neurons or single trees can be characterized by their total dendritic length, surface, or volume, number of bifurcations, terminal tips, etc. At the branch level, examples of available functions include branch length, volume, surface, branching order (number of nodes in the path to the soma), number of compartments forming the branch, taper rate (i.e., diameter 'shrinkage'), etc. Bifurcations can be characterized by the ratio between the two daughter diameters, the angles between the two daughters and between the parent and the daughter, etc. In single compartments, LM can measure: diameter, type, length, surface, volume, Euclidean distance from the soma, path distance from the soma, etc. The total number of functions available is over 40. A complete list of functions, including definitions and illustrations is available in the 'help' of LM as well as online at http://www.krasnow.gmu.edu/L-Neuron/.

An important option in LM, 'Specificity', allows the user to specify which part of the neuron should be analyzed. For example, it is possible to selectively include in the analysis only branches that have a given branching order (Figure 3A & 3D: branching order equals 2), only terminal compartments (Figure 3B and 3E), or only a given type of tree (Figure 3C & 3F: basal dendrites). Additionally, the Boolean operators AND and OR are available to enhance flexibility and selectivity.

With this option and the available LM functions, it is possible to perform virtually any type of morphometric analysis, including classical Sholl measurements (i.e., examining compartments within a shell of Euclidean distances from the soma), as well as to extract all parameters needed to generate virtual neuron with the L-Neuron software [12,18].

Different types of analysis can be carried out with LM. Morphological data can be extracted as: (a) raw data, (b) summary by neuron, or (c) summary by a set of neurons. Raw data consists of a list of values (or pairs of values in correlation studies of one parameter versus a second one) to be further processed or visualized by external software. In a summary by neuron, LM averages data in a neuron-by-neuron basis. In a summary by a set of neurons, all data from a group are statistically analyzed together.

In summary studies (b and c), LM can implement three kinds of statistical elaborations: (A) simple numerical extraction, (B) distribution of values in a frequency histogram, and (C) distribution of values of one parameter versus a second parameter.

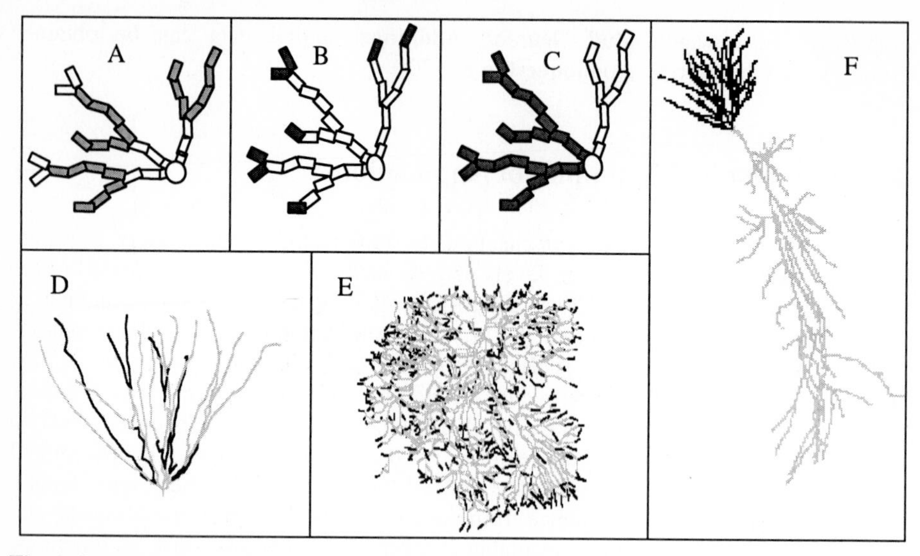

Fig. 2. Panels A, B, and C are simplified compartmental representations of real neurons. Panel D shows a real granule cell (n222 in Southampton archive [6]), E represents a Purkinje cell (cell #1 from [21]) and F represents a CA1 pyramidal cell (n183 in Southampton archive [6]). Compartments highlighted in panels A and D have branching order of 2; compartments highlighted in panels B and E are terminal tips; compartments highlighted in panels C and F are basal dendrites.

The simple extraction of summary data (A) returns the standard statistical parameters of average, standard deviation, etc., for a given function. An example of this output is shown in Table 1, where compartmental length is analyzed.

Function Name	Total Sum	Compartment included	Comp. rejected	Min	Avg	Max	Std Dev
Length	5238.9	611	5	0.039	8.57	51.62	9.28

Table 1. Statistical extraction of function 'length' applied to granule cell n222 of Southampton archive [6]. All values but the number of compartments are in μm.

In this example, Total Sum reports the sum of the values of the selected function over the entire structure, in this case corresponding to the total length of the neuron. The second and third values represent the compartments included (611) and rejected under "specificity" criteria (the 5 compartments rejected corresponds to the somatic compartments). Minimum, maximum and standard deviation characterize the range and variability of data.

When producing histogram distributions of data frequencies, the user selects the number of histogram bins. It is possible to fix either the total number of bins, or the bin width. The output is a tab-separated ASCII text (MS-Excel compatible) that can be visualized with external software. An example of this output, in which the amplitude angle of bifurcation is considered, is shown in Figure 3.

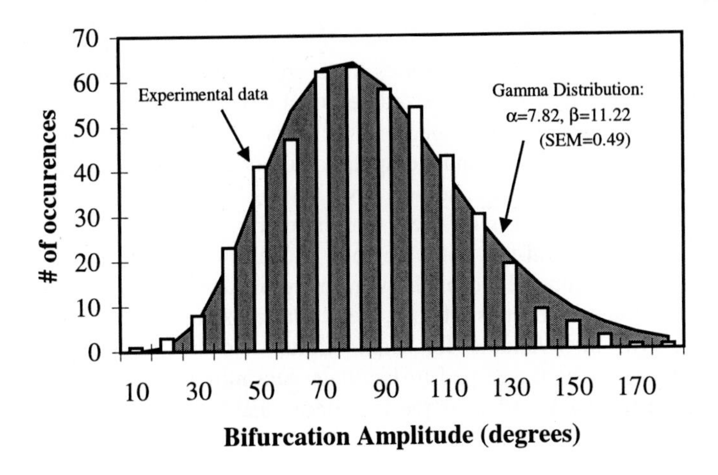

Fig. 3. Histogram plot of Bifurcation Amplitude angle on Purkinje cell # 1 from [21]. Shaded curve shows the best fitting Gamma distribution.

More complex relationships can be studied through dual parameter distributions. In this case, a first parameter is "plotted" against a second one. For example, one can study the distribution of average diameter versus branching order or versus path distance, etc. Another example is shown in Figure 4 displaying the distribution of number of bifurcations over Euclidean distance (Sholl analysis) for 3 motoneurons from [22], by single neuron. The inset shows the same study with the 3 neurons treated as a set.

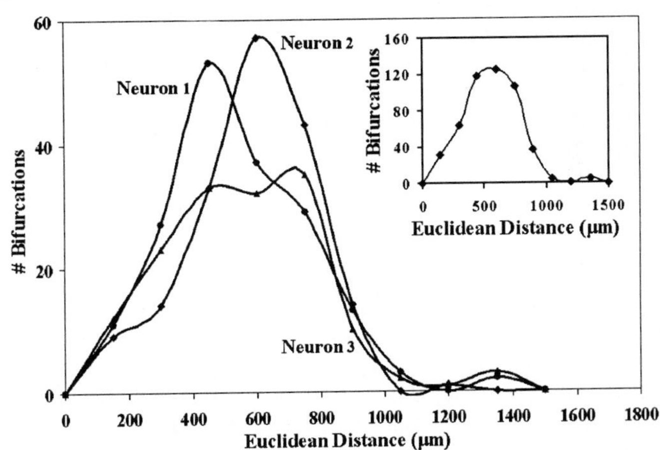

Fig. 4. Sholl analysis of 3 different motoneurons (group analysis in inset). The number of bifurcations within a distance range from the soma (spherical "shell") is plotted against the distance itself.

Conclusion

In the past decade, many laboratories have created their own ad hoc software to extract morphometric data from digitized neuronal structures. The increase in publicly available data from different neuronal classes opened new possibilities of neuroanatomical data mining and computational modeling [5,10,12,14]. However, digitized data are usually dispersed in a variety of formats, and different research groups are usually interested in different morphometric aspects.

L-Measure is a general-use software tool that allows fast, accurate and flexible statistical studies of neuronal morphology from digitized files.

To enhance the capabilities of LM, new additions are planned for the second version. LM will be available for Unix (Linux and Irix) operating systems. A graphical output will be developed to facilitate data mining. Particular attention will be given to user feedback such as requests of additional functions or features.

Acknowledgment

Authors are grateful to Dr. Jeffrey L. Krichmar for continuous discussion and assistance with software implementation. This work was supported in part by Human Brain Project grant 5-R01-NS39600-02, jointly funded by the National Institute of Neurological Disorders and Stroke and the National Institute of Mental Health (National Institutes of Health).

References

[1] Stuart, G., Spruston, N., Hausser, M.: Dendrites. Oxford Univ. Press, Oxford (1999)
[2] Verwer, W.H., van Pelt, J.: Descriptive and Comparative Analysis of Geometrical Properties of Neuronal Tree Structures, J. Neurosci. Methods **18** (1986) 179-206
[3] Uylings, H.B.M., Ruiz-Marcos, A., van Pelt, J.: The Metric Analysis of Three-Dimensional Dendritic Tree Patterns: a Methodological Review, J. Neurosci. Methods **18** (1986) 127-151
[4] Glaser, J.R., Glaser, E.M.: Neuron Imaging with Neurolucida -- a PC-Based System for Image Combining Microscopy, Comput. Med. Imaging Graph. **14**(5) (1990) 307-317
[5] Gardner, D., Knuth, K.H., Abato, M., Erde, S.M., White, T., De Bellis, R., Gardner, E.P.: Common Data Model for Neuroscience Data and Data Model Exchange. J. Am. Med. Inform. Assoc. **8**(1) (2001) 17-33
[6] Cannon, R.C., Turner, D.A., Pyapali, G.K., Wheal, H.V.: An On-Line Archive of Reconstructed Hippocampal Neurons, J. Neurosci. Methods **84** (1998) 49-54
[7] Claiborne, B.J., Amaral, D.G., Cowan, W.M.: Quantitative, Three-Dimensional Analysis of Granule Cell Dendrites in the Rat Dentate Gyrus, J. Comp. Neurol. **302**(2) (1990) 206-219
[8] Cullheim, S., Fleshman, J.W., Glenn, L.L., Burke, R.E.: Membrane Area and Dendritic Structure in Type-Identified Triceps Surae Alpha Motoneurons., J. Comp. Neurol. **255**(1) (1987) 68-81
[9] Bannister, N.J., Larkman, A.U.: Dentritic Morphology of CA1 Pyramidal Neurons from Rat Hippocampus: I. Branching Patterns. J. Comp. Neurol. **360**(1) (1995) 150-160
[10] Ascoli, G.A.: Progress and Perspective in Computational Neuroanatomy, Anat. Rec. **257** (1999) 195-207

[11] van Pelt, J.: Effect of Pruning on Dendritic Tree Topology. J. Theor. Biol. **186**(1) (1997) 17-32
[12] Ascoli, G.A., Krichmar, J.L., Scorcioni, R., Nasuto, S.J., Senft, S.L.: Computer Generation and Quantitative Morphometric Analysis of Virtual Neurons. Anat. Embryol. (Berl) (2001) (in press)
[13] Mainen, Z.F., Sejnowski, T.: Influence of Dendritic Structure on Firing Pattern in Model Neocortical Neurons. Nature **382** (1996) 363-366
[14] Symanzik, J., Ascoli, G.A, Washington, S.D., Krichmar, J.L.: Visual Data Mining of Brain Cells. Comp. Sci. Stat. **31** (1999) 445-449
[15] Washington, S.D., Ascoli, G.A., Krichmar, J.L.: A Statistical analysis of Dendritic Morphology's Effect on Neuron Electrophysiology of CA3 Pyramidal Cells, Neurocomputing **32-33** (2000) 261-269
[16] Vetter, P., Roth, A., Hausser, M.: Propagation of Action Potentials in Dendrites Depends on Dendritic Morphology. J. Neurophysiol. **85**(2) (2001) 926-937
[17] Nasuto, S.J., Krichmar, J.L., Ascoli, G.A.: A Computational Study of the Relationship between Neuronal Morphology and Electrophysiology in an Alzheimer's Disease Model. Neurocomputing (2001) (in press)
[18] Ascoli, G.A., Krichmar, J.L.: L-Neuron: A Modeling Tool for the Efficient Generation and Parsimonious Description of Dendritic Morphology, Neurocomputing **32-33** (2000) 1003-1011.
[19] Ascoli, G.A., Krichmar, J.L., Nasuto, S.J., Senft, S.L.: Generation, Description and Storage of Dendritic Morphology Data, Philos. Trans. R. Soc. Lond. B (2001) (in press)
[20] Ishisuka, N., Cowan, W.M., Amaral, D.G.: A Quantitative Analysis of the Dendritic Organization of Pyramidal Cells in the Rat Hippocampus. J. Comp. Neurol. **362** (1995) 17-45
[21] Rapp, M., Segev, I., Yarom, Y.: Physiology, Morphology and Detailed Passive Models of Guinea-Pig Cerebellar Purkinje Cells., J. Physiol. **474** (1994) 101-18

What Can We Compute with Lateral Inhibition Circuits?

José Mira and Ana E. Delgado

Departamento de Inteligencia Artificial.
Facultad de Ciencias y EUI. UNED. Madrid
{jmira, adelgado}@dia.uned.es

Abstract. In this paper we review the lateral Inhibition (LI) as a connectionist conceptual model invariant in front of level translation, moving up from physical to symbol and then to the knowledge level where LI is transformed into a "pattern of reasoning". Afterwards, we explore the possibilities of this pattern as a component of modelling fault-tolerant cooperative processes. LI can be considered as an adaptive mechanism that integrates excitation and inhibition in a dynamic balance between complementary information sources.

1. Introduction

There are some neuronal structures appearing at the neurogenesis process all through the different levels of integration (molecular, dendro-dendritic contacts, neurons, groups of neurons –columns, layers– and functional systems), structures which are also present at the global behaviour level (perception, reasoning, motor control, …). This lead us to think that, as a consequence of its adaptive value, these structures have been selected as basic functional modules to process information within the nervous system (NS).

Lateral inhibition (LI) appears to us as one of the most significant among above mentioned structures. LI is present in all sensorial modalities (vision, auditory pathway, proprioception, …) and through all levels of integration (receptors, primary areas, associative plurisensorial areas and global perception). LI is also present in neurogenesis where nascent neurons express a gene and deliver LI to their neighbours.

In the visual pathway of the higher mammals we find LI in the retina, lateral geniculate body and cortex. Ernst Mach found it in perception [1] where the perceived brightness is different from the actual physical intensity because there is an accentuation of the differences in brightness actually present at the edges (Mach bands). Hartline and Ratliff set down a mathematical formulation starting from electrophysiological data of the marine arthropod limulus [2]. Hubel and Wiesel, using the concept of receptive field (RF) made an extensive use of the LI concept in their interpretation of the response of visual neurons [3]. Moreno-Díaz proposed a general formulation at analytical level including the non-linear case and later Mira et al [4] move an step forward in the generalization of the LI from the analytic to the algorithmic level, as a cooperative decision mechanism [5], [6] and as a mechanism of cumulative computation based on the "principle of opposites" [7].

J. Mira and A. Prieto (Eds.): IWANN 2001, LNCS 2084, pp. 38-46, 2001.

At present LI is still found as underlying mechanisms of new experimental finding in neurophysiology [8] and, also from the engineering perspective, LI has proved to be very useful as information processing tool for artificial vision and robotics [9].

In this paper we revisit the concept of recurrent and non-recurrent LI considered now as a structural model at knowledge level; all other previous analytic and algorithmic formulations are embedded as particular cases. We start reviewing the LI circuits at the physical level from the original work of Hartline and Ratliff.

2. LI at the Physical Level

We use the methodological scheme of levels and domains of description to revisit the concept of LI starting with the physical level; that is, in terms of the anatomical schemes of local connectivity necessary to embody recurrent and non-recurrent LI.

The original work of Hartline and Ratliff [1] on LI in the compound eye of the limulus summarizes the most relevant characteristics of the stationary case as follows:

1. The frequency of the response of an element when no inhibition is present, depends in a linear way on the logarithm of the luminous intensity to which is subject, once a given threshold is surpassed.
2. The magnitude of inhibition exerted upon an element, depends in the first approach, in a linear way, on the response of the element same, once a given threshold exceeded.
3. The inhibitory action is present only when an element is subject to excitation.
4. The inhibitory action of an element upon other decreases when the distance between both elements increases. In the first approach we may assume that this decrease is proportional to the reverse of their distance.
5. The inhibitory action depends linearly on the number of lighted elements.
6. Response is only present in elements when excitation is higher than inhibition.
7. In stationary state, self-inhibition is functionally equivalent to a decrease of excitation.

The anatomical structure used as basis is shown in figure 1. Each LI "elementary functional unit" is described in terms of two elements: a receiver and an inhibitor. The receiver is a transducer which transforms the luminous intensity into frequency of a train of pulses. The second element, the inhibitor modifies this frequency when inhibition is present. This requires plexus of horizontal interconnections between the own inputs to the inhibitors and that of the neighbor elements that are connected. In this way, LI is achieved by means of an electrochemical coupling ("resistive") between the units that are functionally connected.

The same structural model can be used in the visual pathway of the higher mammals where we find LI in the retina, lateral geniculate body and cortex. The following assumptions are recurrent.

- LI takes place through coupling and interaction factors, $K(x,y;\alpha,\beta)$, which represent the inhibitory influence of the element (α,β) upon element (x,y).
- The global effect upon an element subject to LI influences is cumulative in space and time. The results of the concurrent and successive interactions are compiled by means of addition, integration or counting (measure functions). In some cases we can accept that the number of elements in each functional group (layer, cluster,

column, barrel, ...) is higher enough to enable us a continuous formulation. In some other cases we need specific algorithms of accumulation.

The weighting surface, *K*, can be interpreted as kernel of an integral equation or weighting function of a convolution integral. In the discrete cases *K* correspond to parametric or numerical matrix obtained sampling these *K* surfaces.

- We interpreted the signals at the different functional groups as analogical variables whose physiological counterparts are slow potentials or frequencies of a train of spikes.

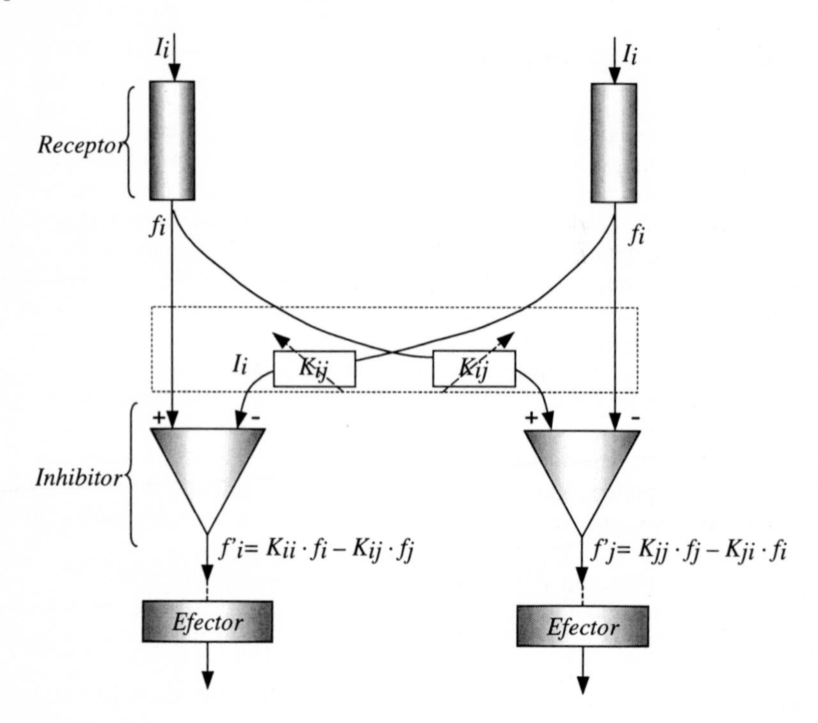

Fig. 1. Computational structure of non-recurrent LI.

If we name $I(\alpha,\beta)$ the input signal on the element located at the coordinates (α,β) and $\Phi(x,y)$ the signal at the output of the element located in position (x,y) we can formulate LI as:

Non Recurrent

$\Phi(x,y)$ = Accumulation of direct excitation $I(x,y)$ with that coming from the interaction with the neighbouring elements, $I(\alpha,\beta)$ through the interaction factors $K(x,y;\alpha,\beta)$.

Recurrente

$\Phi(x,y)$ = Accumulation of direct response $\Phi(x,y)$ with that coming from the interaction with the neighbouring elements, $\Phi(\alpha,\beta)$ through the interaction factors $K^*(x,y;\alpha,\beta)$.

For non recurrent linear case we would have:

$$\Phi(x,y) = \iint_R K(x,y;\alpha,\beta) \cdot I(\alpha,\beta)\, d\alpha\, d\beta \tag{1}$$

and for recurrent case

$$\Phi(x,y) = \iint_{R^*} K^*(x,y;\alpha,\beta) \cdot \Phi(\alpha,\beta)\, d\alpha\, d\beta \tag{2}$$

as shown in figure 2.

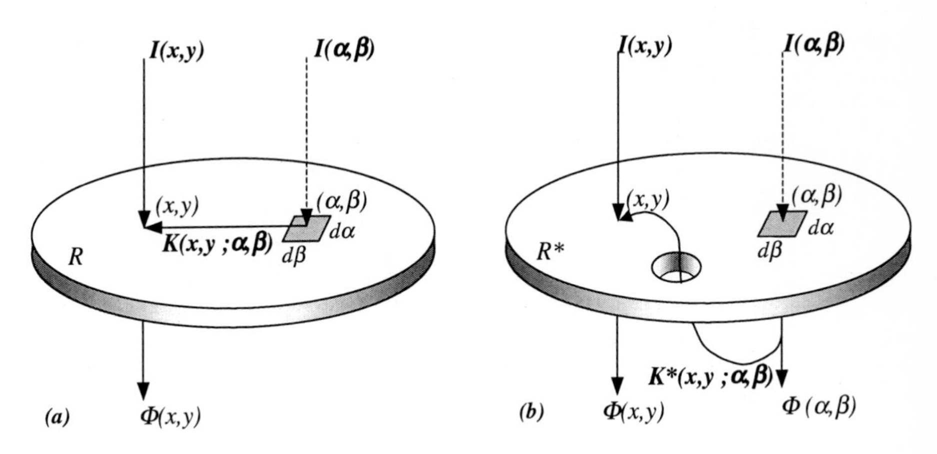

Fig. 2. Formulation of LI. ***(a)*** Non recurrent. ***(b)*** Recurrent.

Once the LI model for the stationary case has been reported as above, we must specify the form of kernels K and K^*. Anatomically all kernels of LI correspond to three-dimensional receptive fields organized in a center-peripheral way as shown in figure 3 (a). Here the neuron is assumed to be centered in volume ON and time is implicit. If we incorporate time in a explicit way we would required a fourth dimension difficult to draw. For our own facility we will use in this case the spatial coordinates (x,y,z) and time as the fourth coordinate, what will consequently give us a cylindrical scheme as in figure 3 (b). Now the input and output areas are cylinders organized with a structure of FIFO memories and with a high capacity to overlap. The specific shape, size, structure and adaptive changes of these spatio-temporal receptive fields convey a relevant part of the LI computation. The overlapping of RF's takes care of intercrossing of inputs, connectivity and feedback. So, as we detail the shape and position of the ON and OFF volumes, we are designing the filter. The other part of the LI computation is directly dependent on the specific algorithm performed with the data coming from these volumes.

The LI analogical formulation correspond to a band-pass spatio-temporal recursive filter of order n. That is to say, LI is a detector of spatio-temporal contrast complemented with the possibilities of (1) non-linear expansions of the input and output spaces (multiplicative & divisive inhibition) (2) non-linear conductances (3) thresholding and (4) "tissue recruitment" covered by dynamic reconfiguration of the ON and OFF volumes [6], [10].

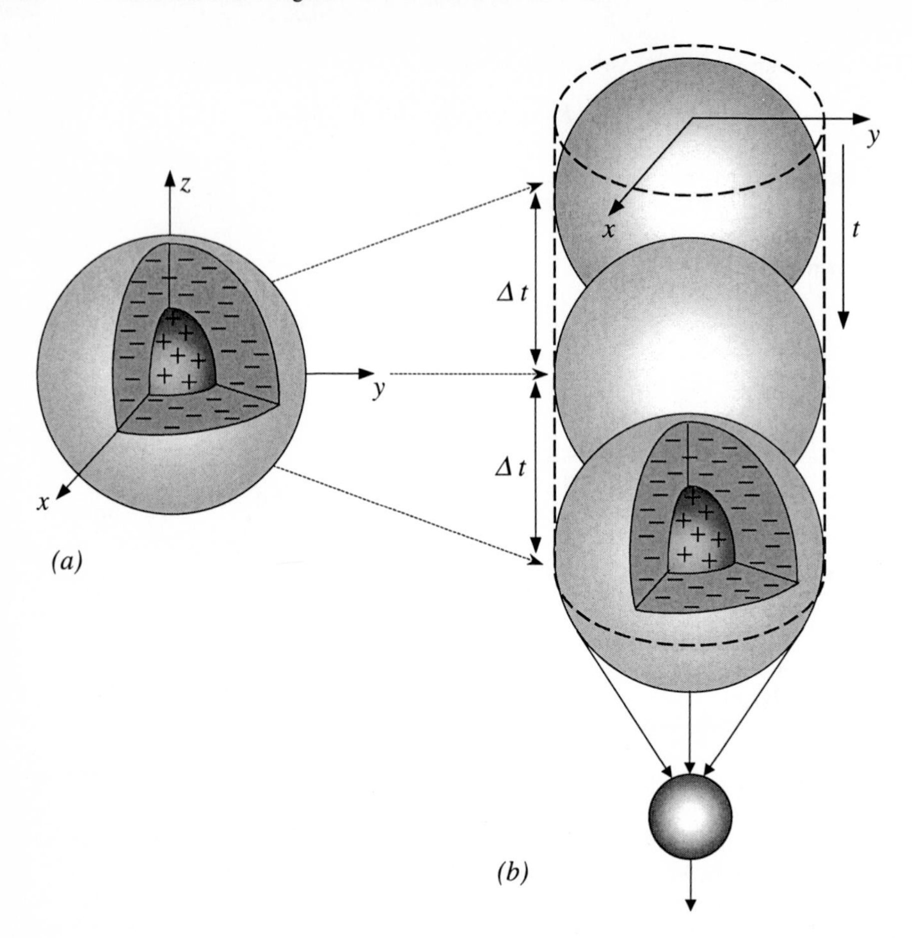

Fig. 3. (a) Three-dimensional LI kernel. A central excitatory volume is surrounded by an inhibitory one. Both are elastically transformable. (b) Shape of spatio-temporal RF's. Each time interval (synaptic delay) correspond to an instantaneous pattern of excitation/inhibition values in the real three-dimensional kernel shown in part (a) of this figure.

At the physical level each unit samples information both from a volume V_i in the input space and another volume V_i^* in the output space. Both spaces are of physical signals and store the temporal sequences of data and subsequent responses during n time intervals. The interaction factors $K(x,y;\alpha,\beta)$ are not only distance dependent but also depends on the intensity level in the element that exerts the inhibitory action. This dependence can be monotonic ("the more, the more", "the more, the less"), and linear in particular, or can be non linear. For example, increasing up to a certain maximum value and decreasing beyond this point giving rise to patterns of maximum efficiency of the sensors, as shown in figure 4 [11]. Also, when the influence of $I(\alpha,\beta)$ arrives to destination, $K(x,y;\alpha,\beta) \cdot I(\alpha,\beta)$, the inhibitory action on the element $I(x,y)$ can be, additive, multiplicative, or divisive.

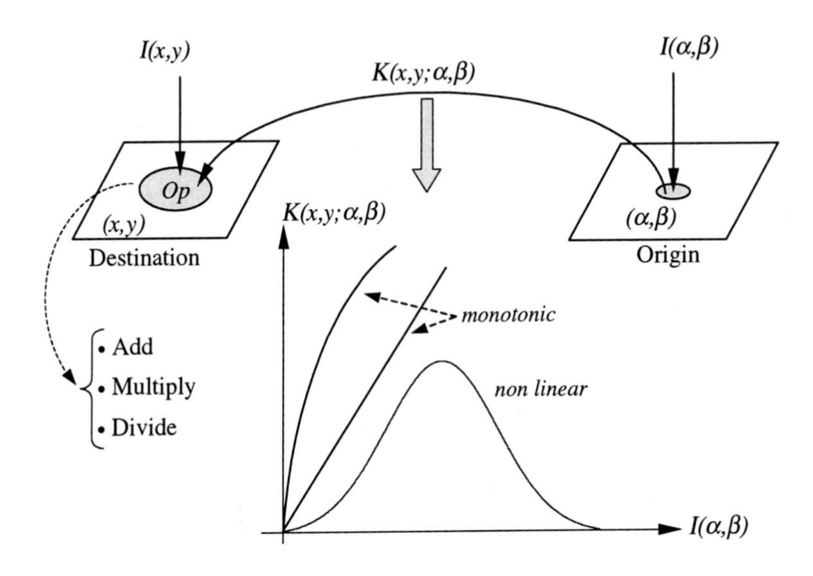

Fig. 4. Different laws of dependency of the interaction factors $K(x,y;\alpha,\beta)$ with the intensity of the signal on origin, $I(\alpha,\beta)$, and different modes of analytical inhibition on destination (add, multiply and divide).

3. LI at the Symbol Level

The analysis of LI at physical level cannot go beyond the language of time-dependent signals with the specification of their meaning. For example, signals coming from sensors (light, temperature, position, ...) and results acting as effectors (orientation, velocities, control signals, ...).

We now abstract the LI formulation to reach the symbol level and then explore the potential repertories of emergent behaviors in programs built-up using LI models as components. If we move from anatomical processors to physiological processes maintaining the formal model underlying LI circuits, we are able to transform these circuits detecting spatio-temporal contrasts into a structural program of local calculus as shown in figure 5. The computation is still connectionist as it works with labeled lines and the processed data being sampled on the same structure of FIFO memories with data fields organized in terms of center/periphery volumes but now the input and output spaces are, in general, spaces of representation.

Each one of the elements of the LI processing network performs a given calculation on the central area of the receptive field in the input space (be it A) and again another on the periphery (be it B). The result of these two accumulation processes obtained from the input data is compared in the condition field of an IF-rule with the corresponding ones from the output space (D and E). If the matching condition is satisfied, then the corresponding action is taken. This action is also written in the output space to be considered in the next time interval as feedback data included in the central and periphery parts of the feedback kernel.

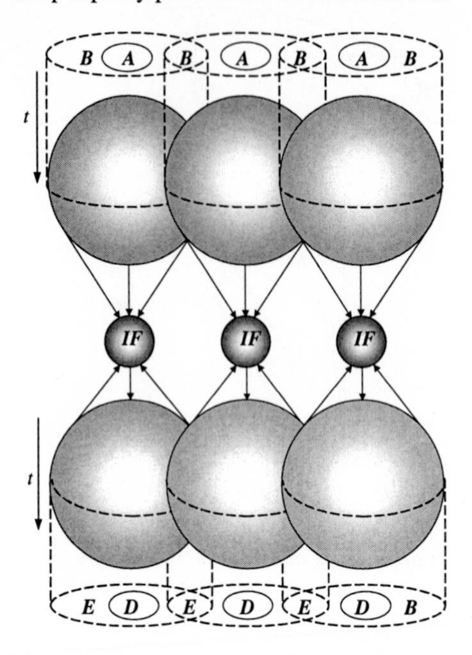

Fig. 5. Interpretation of LI at the symbol level, where "neurons" are inferential elements computing different accumulation algorithms on the central and periphery part of the receptive fields.

Here LI is a structural model that, using the principle of contraries, links different parametric operators. To complete the model we have to specify these operators without being limited by analytics, but using any set of logic and/or relational operators of a high level programming language. For example if we select for A and B multipliers and for C adders, we have the previous analog model of LI interpreted as a band-pass filter followed by a threshold. Alternatively, we can use for A and B selection of maximum value in the center and the periphery of the RF, and for C the rule that contrast these maxims, we have a simple case of cooperative computation ("the winner takes all") where each module compares its assessment with the assessment of the neighboring modules concerning the same information and a consensus is reached through a process of maximum selection and diffusion. This interpretation of LI produce results like that of figure 6.

We can also use probabilistic or fuzzy operators within LI scheme. Connectionism does not impose limitations upon the nature of the local operators, except for that related to the complexity of the adaptive part.

The natural bridge between connectionist and symbolic computation lies in the rise of computational capacity of neural models and the decrease of the "size" of the inferential primitives of the symbolic models. In this way, there is an intermediate sub-symbolic world where it is not necessary to differentiate between the neural inference (a rule) and the primitive inferences used in AI, such as "select", "match", "abstract", "refine", ... Also it is not necessary to distinguish between the underlying formal model (graphs or modular automata that corresponds to lateral inhibition circuits). So now it is possible to pass through the three levels (physiology, symbols and knowledge) observing which libraries of modelling components (LI, reflex arcs,) can neurobiologists offer to AI specialists so that they can improve their modelling methodology.

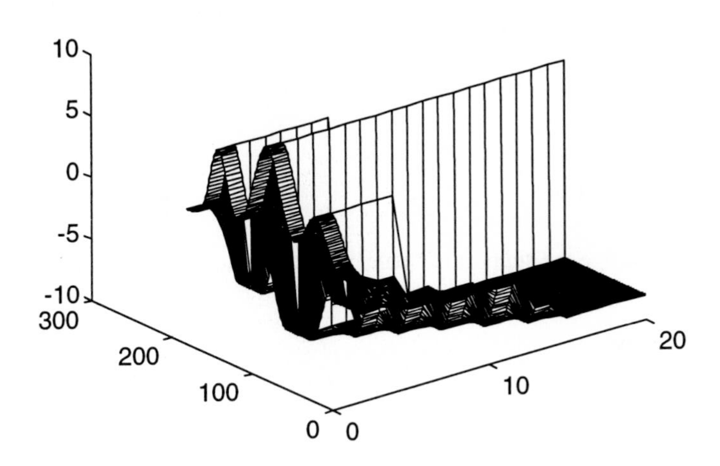

Fig. 6. Interpretation of LI as a cooperative process of first, local selection of maximum and then diffusion. The results correspond to a simulation with 200 modules and receptive fields that overlaps 30 neighbors. Convergence is attained after 9 iterations

4. LI at knowledge Level: A General Principle of Self-Organization

If we abstract again LI at the symbol level and move to the knowledge level, LI is transformed into a cooperative-competitive scheme of social nature. The elements are now "agents" sharing information (overlapping their decisions with those of their neighbors) through dialogue and consensus algorithms and proposing new collective decisions. Thus, LI can be considered as a general principle of self-organization.

Computational modeling at this level is far from being easy and accessible, but its interest is notorious, both in neuroscience and in knowledge engineering (KE). In neuroscience the formulation of these models of social interaction has been associated to the problem of physical damage and the high residual function always preserved (although at lower scale factors) by means of residual tissue reorganization. In KE the difficulty is related to the libraries of problem solving methods with scarce cooperative components. And there we are now, in an attempt to formulate a nontrivial center-periphery organization of pools of cooperative-competitive "agents" that contribute to the cooperative solution of a task. This contribution is factorial and structurally dependent through LI schemes. So, to eliminate one or more "agents" means to eliminate one or more factors thus causing a deficit in the performance of the task, dependent on the structural role and status of that element in the inferential scheme of the task. Nevertheless, this drive out of some agents does not imply to lose the global organization of the task, which depends on the always remaining structure.

We consider LI as an adequate source of inspiration in line with other structural and organizational principles underlying real life: self-organization, plasticity and structural coupling, co-operativity, fault tolerance, genetic and evolutive aspects, etc. Non-trivial formulations for some of these principles, will contribute in a significative way to neuroscience and also to the design of new methods to solve in a computational way the sort of problems which are still genuinely humans.

Acknowledgment

We acknowledge the support of Spanish (CICYT) under project TIC-97-0604.

References

1. Ratliff, F.: Mach Bands: Quantitative Studies on Neural Networks in the Retinal. Holden-Day, Inc. (1965).
2. Hartline, K.K., Ratliff, F.: Spatial Summation of Inhibitory Influences in the Eye of Limulus. Science, 120 (1954) 781
3. Hubel, D.H., Wielse, T.N.: Receptive Fields, Binocular Interaction and Functional Architecture in the Cat's Visual Cortex. J. Physiol. 160 (1962) 106-154.
4. Moreno, R., Rubio, E.: A Model of Non-Linear Processing in Cat's Retina. Biol. Cybernetics 37 (1980) 25-31.
5. Delgado, A.E. et al: A Neurocybernetic Model of Modal Co-operative Decisión in the Kilmer-McCulloch Space. Kybernetes. Vol. 3 (1989) 48-57.
6. Mira, J. et al: Cooperative Organization Connectivity Patterns and Receptive Fields in the Visual Pathway: Application to Adaptive Thresholding. In: Mira, J., Sandoval, F. (eds): From natural to Artificial Neural Computation. LNCS, Vol. 930. Springer-Verlag, Berlin Heidelberg New York (1995) 15-23.
7. Fernández, M.A. et al.: Local Accumulation of Persistent Activity at Synaptic Level: Application to motion analysis. In: Mira, J., Sandoval, F (eds.): From Natural to Artificial Neural Computation. LNCS, Vol. 930. Springer-Verlag, Berlin Heidelberg New York (1995) 137-143.
8. Sánchez, E.: Modelos del Cuneatus. PhD Thesis. Univ. de Santiago (2000).
9. Franceschini, N., Pichou, J.M., Blanes, C.: From Insect Vision to Robot Vision. Phil. Trans. R. Soc. London. B 337 (1992) 283-294.
10. Mira, J. et al.: Cooperative Processes at the Symbolic Level in Cerebral Dynamics: Reliability and Fault Tolerance. In: Moreno, R., Mira, J. (eds.): Brain Processes, Theories and Models. MIT Press (1996) 244-255.
11. Braitenberg, V.: Vehicles. Experiments in Synthetic Psych. The MIT Press. Mass (1998).

Neuronal Models with Current Inputs

Jianfeng Feng

COGS, Sussex University, Brighton BN1 9QH, UK
http://www.cogs.susx.ac.uk/users/jianfeng

Abstract. We propose a novel approach based upon a superposition of 'colored' and 'white' noise to approximate current inputs in neural models. Numerical results show that the novel approach substantially improves the approximation within widely, physiologically reasonable regions of the rising time of α-wave inputs.

1 Introduction

Although single neurone model with random inputs has been widely studied *in computo*, most of such studies are done under the assumption that inputs are instantaneous [1–5, 13]. This assumption is certainly an approximation to the physiological data: any physical process takes time to accomplish. The rising time of excitatory postsynaptic potentials, for example, ranges from a few milliseconds to a few hundred milliseconds (see [10] at page 184). Theoretically in [14] the authors have found the interesting impact of EPSP rising time courses in synchronising neuronal activities.

In this paper we address the following two important issues: how the time courses of EPSPs and IPSPs affect the output of a single neuron model–the integrate-and-fire model and the Hodgkin-Huxley model–and how to employ the Ornstein-Uhlenbeck type process to approximate the models with current inputs. The second issue is important since we all know that theoretically it is very difficult to deal with a system with Poissonian form inputs. For example it has been investigated for decades to use the Ornstein-Uhlenbeck process to approximate neuronal models with Poisson inputs.

We take into account two most commonly encountered noninstantaneous inputs: α-wave and square-wave inputs. We find that the mean and variance of α-wave and square-wave inputs are both weaker than that of Poisson inputs, as one might expect, and therefore the mean of interspike intervals (ISIs) of efferent spike trains of the integrate-and-fire model and the Hodgkin-Huxley model is greater than that of Poisson inputs. The classical approach, the usual approximation, is first applied to approximate the models with current inputs. By the usual approximation, we mean to approximate a stochastic process by a diffusion process, i.e.

$$dx_t = b(x_t, t)dt + \sigma(x_t, t)dB_t$$

where B_t is the standard Brownian motion ('white' noise) and b, σ are appropriately defined functions. For the integrate-and-fire model, the approximation to

J. Mira and A. Prieto (Eds.): IWANN 2001, LNCS 2084, pp. 47–54, 2001.

the mean firing time of efferent ISIs is always valid when the ratio between inhibitory and excitatory inputs is low, but not for the CV. In other words, in the regions of small ratio between inhibitory and excitatory inputs and in order to tell the difference between inputs from different sources, we have to consider the high order statistics of efferent spike trains. Combining our previous results [6] and the results in this paper, we further conclude that in the regions of high ratio between inhibitory and excitatory inputs, the mean firing time is very sensitive to small perturbations. For the Hodgkin-Huxley model we find that the usual approximation is not satisfying even at a very low ratio between inhibitory and excitatory inputs. This also reveals an essential difference between some simple, linear model such as the integrate-and-fire model and the biophysical, nonlinear model such as the Hodgkin-Huxley model. We then propose a novel scheme to approximate the models with current inputs: to replace the 'white' noise in the usual approximation by a superposition of 'color' and 'white' noise. Numerical results show that the novel scheme considerably improves the approximation, for both the integrate-and-fire and the Hodgkin-Huxley model, within widely, physiologically reasonable regions of model parameters.

This is the third of our series of papers aiming to elucidate how more realistic inputs, in contrast to conventional i.i.d. Poisson inputs which has been intensively studied in the literature, affect the outputs of simple neuronal models and thus possibly to provide a full spectrum of the behaviour inherent in these models, thus documenting more thoroughly the restrictions and potential of the models. In [5] we have considered the behaviour of the integrate-and-fire model subjected to independent inputs with different distribution tails; in [6] we have taken into account the behaviour of the integrate-and-fire model with correlated inputs. On the other hand we have generalised these considerations to biophysical models and some intriguing properties have been found[2, 7].

Due to space limit, we only summarize results here and refere the reader to [8] for details.

2 Integrate-and-Fire Model with Synaptic Inputs

Suppose that when the membrane potential V_t is between the resting potential V_{rest} and the threshold V_{thre}, it is given by

$$dV_t = -\frac{1}{\gamma}(V_t - V_{rest})dt + I_{syn}(t)dt \tag{1}$$

where $1/\gamma$ is the decay rate. When V_t is greater than V_{thre} a spike is emitted and V_t is reset to V_{rest}. The model is termed as the integrate-and-fire model. Synaptic input $I_{syn}(t)$ is modelled by

$$I_{syn}(t) = a\sum_{k=1}^{\infty} f(t - T_k^E)I_{\{t \geq T_k^E\}} - b\sum_{k=1}^{\infty} f(t - T_k^I)I_{\{t \geq T_k^I\}}$$

where $T_k^E = \sum_{i=1}^k t_i^E$ ($T_k^I = \sum_{i=1}^k t_i^I$) for i.i.d. random sequences $t_i^E, t_i^I, i = 1, 2, \cdots$, $a > 0, b > 0$ the magnitude of single EPSP and IPSP and I is the indicator function. In the remaining part of the paper we always assume that the distribution of t_i^E is identical to t_i^I and so when we speak of a property of EPSP inputs we simply imply the same property is true for IPSP inputs. t_1^E is assumed to be exponentially distributed with intensity λ.

Example 1 When

$$f(t - T_k^E) = \delta_{(t - T_k^E)}(0)$$

then $N(t) = \int_0^t I_{syn}(s)ds$ (instantaneous inputs) is the Poisson process with intensity λ.

The Poisson process input is an approximation to cell's synaptic inputs. It assumes that the responds to the input instantaneously. There are some other optimal properties such as optimising the mutual information etc. for Poisson inputs[12].

Example 2 When

$$f(t - T_k^E) = \alpha^2(t - T_k^E)\exp(-\alpha(t - T_k^E)) \quad t > T_k^E$$

we have α-wave inputs or an α-synapse. α-wave input is, of course, again an approximation to actual current inputs. We refer the reader to [13] for a discussion on the choice of this function. In contrast to Poisson inputs now the inputs take time to rise and then decay.

The rising time is $1/\alpha$, which is the characteristic time of α-wave synapse. Here we emphasise that for the same neurone the time course of input EPSPs might be very different: for example, the rising time for an increased-conductance EPSP due to the opening of a channel could be a few milliseconds; but the rising time for a decreased-conductance EPSP is a few hundreds milliseconds (see [10] at page 184). When α is small, α-wave inputs can be thought of as an approximation to continuous current inputs; when α is large they approximate Poisson inputs.

Example 3 When

$$f(t - T_k^E) = \frac{1}{\delta} I_{\{T_k^E < t < T_K^E + \delta\}}$$

we have square wave inputs and its duration time is δ.

A slightly more general model than the integrate-and-fire model defined above is the integrate-and-fire model with reversal potentials [11, 15] defined by

$$dZ_t = -\frac{1}{\gamma}(Z_t - V_{rest})dt + \bar{I}_{syn}(t)dt \tag{2}$$

where V_{rest} is the resting potential. Synaptic inputs are given by

$$\bar{I}_{syn}(t) = \bar{a}(V_E - Z_t)\sum_{k=1}^{\infty} f(t - T_k^E) I_{\{t \geq T_k^E\}} + \bar{b}(V_I - Z_t)\sum_{k=1}^{\infty} f(t - T_k^I) I_{\{t \geq T_k^I\}}$$

$\bar{a}, \bar{b}$ are the magnitude of a single EPSP and IPSP respectively, V_E and V_I are the reversal potentials. Z_t (membrane potential) is now a birth-and-death process with boundaries V_E and V_I. Once Z_t is below V_{rest} the decay term $Z_t - V_{rest}$ will push the membrane potential Z_t up; whereas when Z_t is above V_{rest} the decay term will hyperpolarise it. By choosing different reversal potentials and characteristic times of f, $\bar{I}_{syn}(t)$ corresponds to different kind of synapses such as NMDA, AMPA GABA$_A$ and GABA$_B$.

3 A Novel Approach

The method presented in the previous subsection, i.e. the usual approximation, are well known in the literature. But as we have shown before, it fails to approximate the true behaviour of the model with current inputs when either the ratio between inhibitory and excitatory inputs approaches one or the rising time is slow. It is then a natural question to ask that what we have missed in the usual approximation.

Look at the variance $\sigma(t, \lambda)$ given by

$$\sigma(t,\lambda)^2 = \lambda[t - \frac{11}{4\alpha} + \frac{4}{\alpha}e^{-\alpha t} + 2te^{-\alpha t} - \frac{5}{4\alpha}e^{-2\alpha t} - \frac{3t}{2}e^{-2\alpha t} - \frac{\alpha t^2}{2}e^{-2\alpha t}] \quad (3)$$

we see that the leading term we omit in the usual approximation is $11/4\alpha$. Since in the usual approximation only the derivative of $\sigma(t, \lambda)$ is used, the constant term disappears. We therefore want to find a process $\eta^\alpha(t)$ satisfying the property that

$$\langle (B_t - \eta^\alpha(t))^2 \rangle = t - \frac{11}{4\alpha} + O(t\exp(-\alpha t))$$

We choose an Ornstein-Uhlenbeck process given by

$$\begin{cases} d\xi^\alpha(t) = -\frac{\alpha}{2}\xi^\alpha(t)dt + dB_t \\ \xi^\alpha(0) = 0 \end{cases} \quad (4)$$

Let $\eta^\alpha(t) = c\xi^\alpha(t)$, where c is a constant satisfying

$$\langle (B_t - \eta^\alpha(t))^2 \rangle = t + c^2 \int_0^t \exp(-\alpha(t-s))ds - c\int_0^t \exp(-\frac{\alpha}{2}(t-s))ds \quad (5)$$

We find a new scheme $\tilde{i}_{syn}(t)$ to approximate $I_{syn}(t)$ defined by

$$\begin{aligned} d\tilde{i}_{syn}(t) &= (ap\lambda_E - bq\lambda_I)dt + \sqrt{a^2p\lambda_E + b^2q\lambda_I}dB_t \\ &\quad -\sqrt{a^2p\lambda_E + b^2q\lambda_I}\frac{4-\sqrt{5}}{2}d\xi^\alpha(t) \end{aligned} \quad (6)$$

For specification we write down the full integrate-and-fire model again here

$$\begin{cases} dv_t = -\frac{1}{\gamma}(v_t - V_{rest})dt + (ap\lambda_E - bq\lambda_I)dt \\ \qquad +\sqrt{a^2p\lambda_E + b^2q\lambda_I}[dB_t - \frac{4-\sqrt{5}}{2}d\xi^\alpha(t)] \\ d\xi^\alpha(t) = -\frac{\alpha}{2}\xi^\alpha(t)dt + dB_t \\ \xi^\alpha(0) = 0 \end{cases} \tag{7}$$

In [8] numerical simulations are shown with $\alpha = 1$ and $\alpha = 0.01$. We see that a substantial improvement is achieved with the new scheme defined by Eq. (7).

We mention a few words on the novel approach presented here. Instead of the widely used Brownian motion approximation to the α-wave, we have to used a superposition of 'white' and 'color' noise approximation, i.e. the term $B_t - \eta^\alpha(t)$. The fact that this calibration improves the usual approximation is, however, not surprising at all. Due to the current input, we naturally expect that there are temporal correlations in inputs. We have tried different ways to approximate the auto-correlation of $I_{syn}(t)$ which is

$$\begin{aligned} &\langle (I_{syn}(t) - \langle I_{syn}(t)\rangle)(I_{syn}(s) - \langle I_{syn}(s)\rangle)\rangle \\ &= -\frac{2\lambda g^2}{\alpha} + \lambda g^2 s + \frac{2\lambda g^2}{\alpha}\exp(-\alpha s) + \lambda g^2 s\exp(-\alpha s) + \lambda g^2\exp(-\alpha(t-s))[\\ &\quad -\frac{3}{4\alpha} - \frac{1}{4}(t-s) + \frac{2}{\alpha}\exp(-\alpha s) + s\exp(-\alpha s) + (t-s)\exp(-\alpha s) \\ &\quad -\frac{5}{4\alpha}\exp(-2\alpha s) - \frac{3s}{2}\exp(-2\alpha s) - \frac{3}{4}(t-s)\exp(-2\alpha s) \\ &\quad -\frac{\alpha s}{2}(t-s)\exp(-2\alpha s) - \frac{\alpha s^2}{2}\exp(-2\alpha s)] \end{aligned} \tag{8}$$

for $t \geq s$. When $t = s$ Eq. (8) is Eq. (3). Nevertheless, it is a hard problem due to terms taking the form of $t\exp(-t)$ in the auto-correlation of $I_{syn}(t)$. We then simply approximate the first order term $t - 11/(4\alpha)$ and omit all terms containing $\exp(-t)$ (see Eq. (3)). Numerical simulations show that the approximation scheme presented here improves considerably the usual approximation. Furthermore it is valid for $0 < \alpha \leq \infty$, i.e. when $\alpha = \infty$ Eq. (7) gives exactly the usual approximation.

The first exit time of a linear dynamic system with a 'color' noise perturbation has been widely discussed in the literature and different analytical approaches to estimate it have been put forward. We will discuss it in a subsequent publication.

Finally we point out that much as we confine ourselves to the α-wave inputs, the approach presented here is readily generalized to any form of current inputs by calculating the constant c in Eq. (5).

4 Biophysical Models

We apply results in the previous section to biophysical models. In fact the generalisation is almost straightforward since essentially we have approximated synaptic inputs in the previous section.

We consider the following Hodgkin-Huxley model with parameters given in the literature[2].

$$CdV = -g_{Na}m^3h(V - V_{Na})dt - g_k n^4(V - V_k)dt - g_L(V - V_L)dt + I_{syn}(t)dt \quad (9)$$

where

$$\frac{dn}{dt} = \frac{n_\infty - n}{\tau_n}, \qquad \frac{dm}{dt} = \frac{m_\infty - m}{\tau_m}, \qquad \frac{dh}{dt} = \frac{h_\infty - h}{\tau_h}$$

and

$$n_\infty = \frac{\alpha_n}{\alpha_n + \beta_n}, \qquad m_\infty = \frac{\alpha_m}{\alpha_m + \beta_m}, \qquad h_\infty = \frac{\alpha_h}{\alpha_h + \beta_h}$$

$$\tau_n = \frac{1}{\alpha_n + \beta_n}, \qquad \tau_m = \frac{1}{\alpha_m + \beta_m}, \qquad \tau_h = \frac{1}{\alpha_h + \beta_h}$$

with

$$\alpha_n = \frac{0.01(V+55)}{\exp(-\frac{V+55}{10}) - 1} \qquad \beta_n = 0.125\exp(-\frac{V+65}{80})$$

$$\alpha_m = \frac{0.1(V+40)}{1 - \exp(-\frac{V+40}{10})} \qquad \beta_m = 4\exp(-\frac{V+65}{18})$$

$$\alpha_h = 0.07\exp(-\frac{V+65}{20}) \qquad \beta_h = \frac{1}{-\exp(\frac{V+35}{10}) + 1}$$

The parameters used in Eq. (9) are $C = 1, g_{Na} = 120, g_K = 36, g_L = 0.3, V_k = -77, V_{Na} = 50, V_L = -54.4$. All parameters in synaptic inputs are the same as in the previous sections except that $a = b = 1.$, since when $a = b = 0.5$ the firing time is too long (cf. Fig. 1 in [2]). The initial values for m, n, h and the membrane potential are $0.0529, 0.317, 0.5961$ and -65 respectively.

Figures in [8] plot a comparison with different inputs with $\alpha = 1$. Again it is evident to see that the novel approach of the previous section gives a much better approximation than the usual approximation. Comparing to the results obtained from the integrate-and-fire model, we see that both the usual approximation and the novel approach gives worse results. In other words, the noninstantaneous input has more impact on the biophysical model than that on the integrate-and-fire model, which is basically a linear model.

From figures in [8] we might conclude that with current inputs, the efferent spike trains of the Hodgkin-Huxley model is quite regular with a CV less than .5. However, when standard deviation, s, of output interspike interval is plotted against mean firing time, m, we obtain approximately straight lines

$$s = km - r \quad (10)$$

This suggests an effective refractory period of about $m = r/k$. We note that for inputs of the usual approximation, it is about 10.46msec and for α-wave inputs is 11.25 msec. This implies that, once the effective refractory period is subtracted from each interspike interval, CV is about 0.65 for Poisson inputs, and 0.8 for α-wave inputs. The CV and refractory period of the usual approximation are both smaller than that of α-wave. The conclusions above agrees with our basic belief that: all neurons fire irregularly when subjected to sufficient low intensity random input, and almost all neurons fire regularly if driven very hard.

The Hodgkin-Huxley model is numerically solved using an algorithm for stiff equations from NAG library (D02NBF) with step size of 0.01. Further small step sizes are used and we conclude no significant improvements are observed. The spike detecting threshould used in the simulations for the Hodgkin-Huxley model is 0 mV, as we employed before[2].

5 Discussion

We have presented a theoretical and numerical approach for studying the impact of noninstantaneous inputs on the output of neuronal models. For α-wave and square-wave, and any noninstantaneous inputs, analytical and numerical results are obtained for the usual approximation which essentially reveals the difference between the instantaneous and noninstantaneous inputs. When the ratio between inhibitory and excitatory inputs is low and the rising time is short, the usual approximation produces satisfying results for the integrate-and-fire model; but not for the Hodgkin-Huxley model. We then proposed a new approximate scheme based upon a superposition of 'white' and 'color' noise to approximate neuronal models with current inputs. Numerical simulations show that the new scheme considerably improves the approximation. Since α-wave inputs are much more close to actually biological reality than instantaneous inputs and are widely applied in modeling neural activities, we conclude that in studying neuronal activities subjected to synaptic inputs it is reasonable to replace the classical Ornstein-Uhlenbeck process by the following process

$$\begin{cases} \tilde{i}_{syn}(t) = [ap\lambda_E - bq\lambda_I]t \\ \qquad\qquad + \sqrt{a^2p\lambda_E + b^2q\lambda_I}[B_t - \frac{4-\sqrt{5}}{2}\xi^\alpha(t)] \\ d\xi^\alpha(t) = -\frac{\alpha}{2}\xi^\alpha(t)dt + dB_t \\ \xi^\alpha(0) = 0 \end{cases} \tag{11}$$

This also opens up new theoretical problems such as to estimate the first exit time of neuronal models subjected to a superposition of 'white' and 'color' noise inputs as defined by Eq. (11). A few issues we are going to further explore are

- It worths further studying the effect of more biologically realistic inputs such as AMPA, NMDA, GABA_A and GABA_B on the output of neuronal models. AMPA and GABA_A are fast and equivalent to the case of a large α, while NMDA and BABA_B are slow and a small α.
- We have observed in the present paper that α plays a more important role in the nonlinear model (the Hodgkin-Huxley model) than the 'linear' model (the integrate-and-fire model). The usual diffusion approximation gives poor results in estimating the mean first exit time and CV for some parameter values in the case of the intergrate-and-fire neuron, but for most parameter

values in the case of the Hodgkin-Huxley neuron. In fact, how the correlation in inputs (the color noise term) affects the output of a nonlinear system is extensively studied in the past few years, see for example [9]. We expect that α could play a role of a 'time switcher' in neuronal models: the input could be subthreshold or superthreshold, by controlling α alone.

Acknowledgement. We are grateful to D. Brown, S. Feerick, and anonymous referees for their comments on the manuscript. The work was partially supported by BBSRC and a grant of the Royal Society.

References

1. Brown D., and Feng J. (1999) Is there a problem matching model and real CV(ISI)? *Neurocomputing* **26-27** 117-122.
2. Brown D., Feng J., and Feerick, S. (1999) Variability of firing of Hodgkin-Huxley and FitzHugh-nagumo neurones with stochatic synaptic input. *Phys. Rev. Lett.* **82** 4731-4734.
3. Feng, J.(1997), Behaviours of spike output jitter in the integrate-and-fire model. *Phys. Rev. Lett.* **79** 4505-4508.
4. Feng J., and Brown D.(1998). Spike output jitter, mean firing time and coefficient of variation, *J. of Phys. A: Math. Gend.*, **31** 1239-1252.
5. Feng J, and Brown D. (1998). Impact of temporal variation and the balance between excitation and inhibition on the output of the perfect integrate-and-fire model *Biol. Cybern.* **78** 369-376.
6. Feng J., and Brown D.(2000). Impact of correlated inputs onthe output of the integrate-and-fire mdoels *Neural Computation* **12** 711-732.
7. Feng J., and Brown D.(1999). Integrate-and-fire model and Hodgkin-Huxley model with correlated inputs *Phys. Rev. Lett.* (submitted).
8. Feng J., Li, G.B. (2001) Integrate-and-fire and Hodgkin-Huxley models with current inputs *J. Phys. A.* **34** 1649-1664.
9. Garcia-Ojalvao J., and, Sancho J.M. (1999) *Noise in Spatially Extended Systems* Springer-Verlag: New York.
10. Kandel E.R., Schwartz J.H., and Jessell T.M.(1991) *Principles Of Neural Science*, 3ed edition, Prentice-Hall International Inc.
11. Musila M., and Lánský P.(1994). On the interspike intervals calculated from diffusion approximations for Stein's neuronal model with reversal potentials, *J. Theor. Biol.* **171**, 225-232.
12. Rieke, F., Warland, D., de Ruyter van Steveninck, R., and Bialek W. (1997) Spikes. The MIT Press.
13. Tuckwell H. C. (1988), *Stochastic Processes in the Neurosciences.* Society for industrial and applied mathematics: Philadelphia, Pennsylvania.
14. Van Vereeswijk C., Abbott L.F., and Ermentrout G.B. (1994), *Jour. Computat. Neurosci.* **1** 313-321.
15. Wilbur W.J., and Rinzel J. (1983), A theoretical basis for large coefficient of variation and bimodality in neuronal interspike interval distributions. *J. theor. Biol.* **105** 345-368.

Decoding the Population Responses of Retinal Ganglions Cells Using Information Theory

J.M. Ferrández[1,2], M. Bongard[1,3], F. García de Quirós[1], J.A. Bolea[1], J. Ammermüller[3], R.A. Normann[4], and E. Fernández[1]

[1] Instituto de Bioingeniería, U. Miguel Hernández, Alicante,
[2] Dept. Electrónica y Tecnología de Computadores, Univ. Politécnica de Cartagena,
[3] Dept. Neurobiologie, Univ. Oldenburg, Germany
[4] Dept. Bioengineering, University of Utah, Salt Lake City, USA
Corresponding Author: jm.ferrandez@upct.es

Abstract. In this paper information theory is used to analyze the coding capabilities of populations of retinal ganglions cells for transmitting information. The study of a single code versus a distributed code has been contrasted using this methodology. The redundancy inherent in the code has been quantified by computing the bits of information transmitted by increasing number of neurons. Although initially there is a linear growth, when certain numbers of neurons are included, information saturates. Our results support the view that redundancy is a crucial feature in visual information processing.

1. Introduction

Understanding how information is coded in different sensory systems is one of the most interesting challenges in neuroscience today. Much work is focused on how various sensors encode different stimuli in order to transmit information to higher centers in the neural hierarchy in a robust and efficient manner. Technology is now available that allows one to acquire data with more accuracy both in the temporal and the spatial doomains. However this process produces a huge neural database which requires new tools to extract the relevant information embedded in neural recordings. However, it allows one to postulate some rules about the way different neural systems do this coding.

A first step in studying network coding consists in quantifying how much information single cells encode. One can then compare these data with the information encoded by population of cells. It is also interesting to determine to what extent the information conveyed is redundant, or how many neurons are required to transmit an external stimulus. Distributed representations will generally provide additional advantages over single cell coding, as fault tolerance is enhanced if some units or some connections are damaged.

While the code is partially understood in the auditory [1] and olfactory systems [2], the visual system still presents a challenge to neuroscientists due to its intrinsic

J. Mira and A. Prieto (Eds.): IWANN 2001, LNCS 2084, pp. 55–62, 2001.

complexity. Different studies have used different analysis tools to approach the decoding objective. Fitzhugh [3] applied a statistical analyzer to the neural data in order to estimate the characteristics of the stimulus, whereas Warland used linear filters [4] for the decoding task. Other approaches used to get insights into the coding process are discriminant analysis (Fernández et al [5]), principal component analysis (Tovee et al. [6]) and supervised and non-supervised neural networks (Ferrández et al. [7]).

Information theory provides an answer to the following question "How much does a given population of cells tell about a variety of stimuli". It uses entropy to estimate of the disorder of the system, assuming that a system with more variance is able to carry more information than a system with zero variance. Initially it was focused on symbols for determining the best code to transmit data. Recent neural studies [8] replace symbols with a list of action potential firing times.

In this paper, information theory is used to measure the information conveyed by single cells and by population of cells. The information conveyed by different numbers of cells is also computed in order to determine the dependence of information on the number of units involved on the coding, or to determine if a minimum number of neurons is required for detecting a given stimulus. Our results show that a population of cells code visual stimuli, and that the information is redundant, providing in this way robustness to visual code.

2. Methods

Recordings were obtained on isolated turtle (*Trachemy scripta elegans*) retinas. The turtles were dark adapted for a few hours, before they were sacrificed and decapitated following ECC rules [5]. The eyes, stored at 4°C, were enucleated and hemisected under dim illumination, and the retinas were removed under bubbled Ringer solution. Care was taken to keep the outer segments intact when removing the pigment epithelium. The retina was flatted on an agar plate, and fixed using a Millipore filter with a square window where the electrodes were placed. The agar plated with the retina was placed on a beam splitter with the ganglion cells facing up. Oxygenated Ringer solution nourished the preparation.

Light stimulation was applied using a halogen light lamp, with wavelength selection provided by narrow band-pass interference filters. Intensity was fixed by using neutral density filters. For each stimulus, either the wavelength, the intensity, or the spot diameter was varied. 250 ms duration stimuli were applied, using a lens to focus the stimulus on the photoreceptor layer of the whole retina.

The Utah microelectrode array [9] was used for obtaining the extracellular recordings. The electrode arrays contain 100 electrodes that were built from silicon on a square grid with 400 microns pitch. The distal 50 microns of the needles, metallized with platinum, form the active site of each electrode. The array was mounted on a micromanipulator, and each electrode was connected to a 5000 gain bandpass (filtered from 250 to 7500 Hz) differential amplifier. The analog signals were digitized and stored in a computer.

The pre-processing consisted of fixing a threshold (based on an estimate of the noise of the electrode) in order to extract the kinetics of electrical activity above this threshold. Spike waveform prototypes were separated by template matching. For each stimulus and for each electrode, the time of the first and second spike, the number of spikes, and the interspike interval during the flash interval was also computed.

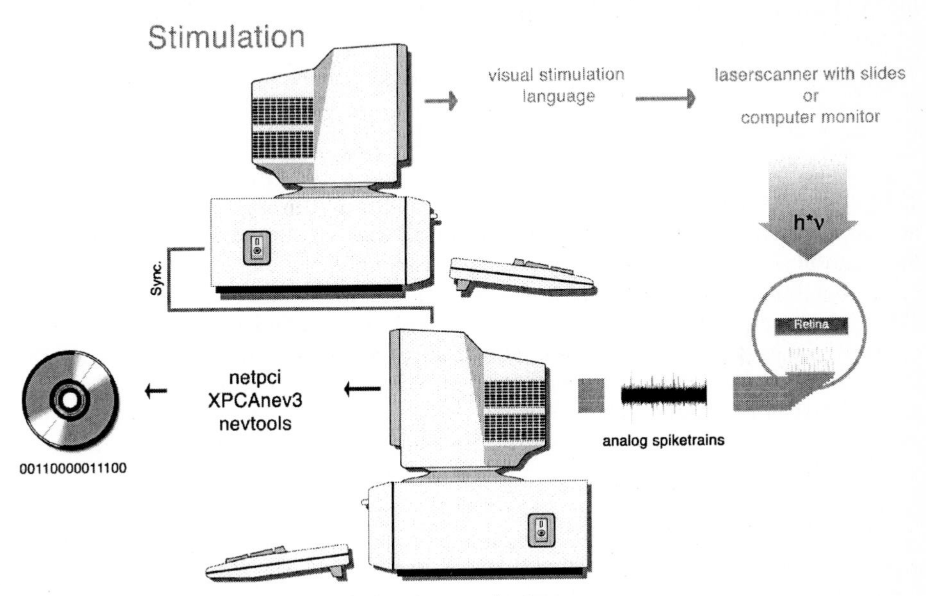

Fig 1. Recording method

3. Information Theory

Information theory had its origins in 1929 where the entropy of a system was computed for making binary measurements. Later, Shannon published "The Mathematical Theory of Communication" [10] where thermodynamic entropy was used for computing different aspects about information transmission. Basically Shannon information theory studies the capacity of channels for encoding, transmitting and decoding different messages, regardless of the associated meaning.

In neural coding, information theory may be used as a tool for determining in a quantitative way the reliability of the code by analyzing the relationship between stimuli and responses. This approach allows one to answer questions about the relevant parameters that transmit information as well as addressing related issues such as the redundancy, the minimum number of neurons need for coding certain group of stimuli, the efficiency of the code, and the maximum information that a given code is able to transmit. An increasing number of neuroscientists are using information theory as a basic tool to give insights about their neural recordings in different systems. [11][12][13][14]

In the present work, the population responses of the retina under several repetitions of flashes were discretized into bins where the firing rates from the cells of the population implement a vector **n** of spikes counts, with an observed probability P(**n**). The probability of the occurrence of different stimuli has a known probability P(s). Finally the joint probability distribution is the probability of a global response **n** and a stimulus s, P(s,**n**).

The information provided by the population of neurons about the stimulus is given by:

$$I(t) = \sum_{s \in S} \sum_{n} P(s,n) \log_2 \frac{P(s,n)}{P(s)P(n)} \qquad (1)$$

This information is a function of the length of the bin, *t,* used for digitizing the neuronal ensembles. The complete P(s, **n**) for all kind of stimuli is not available on the recordings, so given a N table with the pairs stimulus-response it can be calculated the probability of certain response **n**, and a stimulus s in the observed data, P_N(s,**n**) for a limited dataset. In practice P_N(s,**n**) is used instead of P(s,n) in equation (1) producing an overestimation due to the limited number of samples used for the computation of this probability. Sometimes a subtracting term is applied to correct this deviation, which consists of a large value for small population and a decreasing value as the population of registered neurons grows. With more than 10 cells in the population this value tends to be insignificant.

In this paper, equation (1) was computed to determine the information contained in the firing trains of isolated retinal ganglion cells for 7 different stimuli and this data was compared to the information carried out by the whole population. The information contained in an increasing set of ganglion cells was also computed in order to determine if the information grows linearly with the number of cells or if it saturates at a given value.

4. Results

In each experiment we recorded neural activity from about 70 from the 100 electrodes. Often, single unit separation was difficult so that we selected 25-30 prototypes which were unequivocal in terms of both amplitude and shape. The firing rate was extracted and digitized into bins, and the stimulus-response table was constructed. Furthermore the table of the join probability was computed. The information of single cells versus the information of the whole population was then determined using equation (1) for discriminating between 7 different intensities. In Figure 2, the information, in bits, is plotted for single cells in the population versus the whole population information. It can be seen that some cells are more discriminative than others. These cells are usually the cells with more variability in their firing rate. The average information for single cells is 2.8 bits (dotted line), versus the whole population coding which arrives 4.8 bits. This quantity is not the highest limit because it depends on the experiment (in different experiments several cells are recorded with different recognizing capabilities). In another experiment, Figure 3, the cells are more discriminative, averaging 3.4 bits. There are cells which are very good for discriminating while others carry little information. Again the population information is significantly higher than that for individual cells.

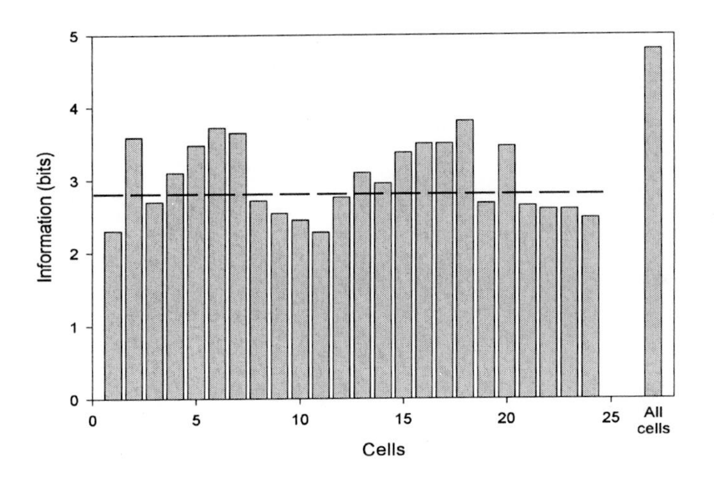

Fig. 2. Single cell coding versus population coding

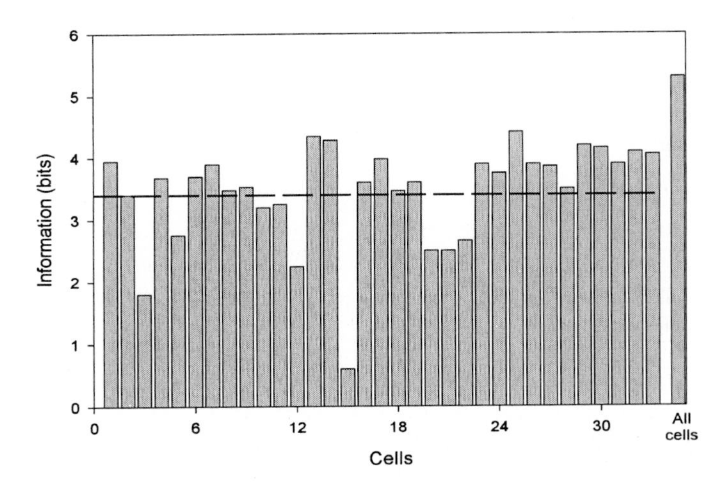

Fig. 3. Single cell coding versus population coding

In Figure 4, the information for an increasing cell population is plotted, with their recognition rates. It can be seen that initially, information rises linearly with the number of cells in the population and then saturates. This means that adding more new cells to the analysis does not provide more information, because with 10 cells the intensities are recognized with practically 100% success. The degree to which the rise of information with the number of cells deviates from linearity is related to the redundancy of the code. So this graphic may represent the redundancy inherent in the firing rate of the population, a characteristic that achieves robustness, or fault tolerance to the system. In another experiment (Figure 5), the behavior is exactly the same, although it takes more units to reach a higher recognition rate, percentage of succesfull stimations, and the information is slightly higher than in the first experiment.

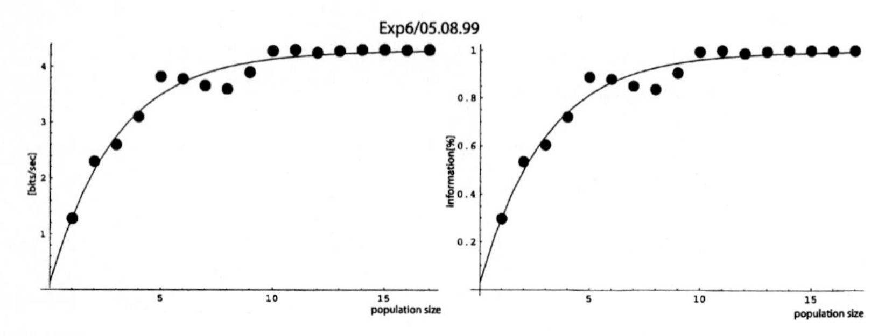

Fig. 4. Information and recognition rates for an increasing population of retinal ganglion cells in discriminating intensities.

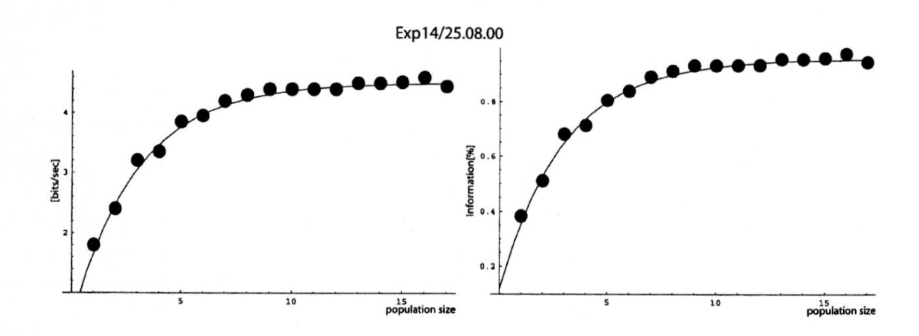

Fig. 5. Information and recognition rates for an increasing population of retinal ganglion cells in discriminating intensities.

5. Conclusions

Information theory has been used for computing the information transmitted by populations of retinal ganglion cells. This method can be used to obtain insights about the parameters and the nature of the coding used by different systems for the transmission of information. Our results show that the information contained in the population firing rates is significantly higher than the information contained in the firing rates of single cells. The overall information registered depends on the experiment, number of cells recorded, discriminative character of this cells, location of the multielectrode array, etc. However the overall behavior does not change significantly.

By analyzing the information contained in the firing rates of increasing population sets, it can be concluded that if initially the number of bits transmitted grows linearly with the size of the population, this curve is clearly non-linear because when certain numbers of cells is reached the information saturates. Under our stimulation protocols, there is always a certain number of cells which are able to transmit almost 100% of the information. Adding more cells to the set does not provide more information. This redundancy could be useful to achieve the robustness required for the retina to maintain its performance under adverse environments.

These results demonstrate the utility and richness of knowledge obtainable from "many-neuron" ensemble recording technologies. We propose that information theory, in addition to other approaches currently used (neuronal networks, principal component analysis, discriminant analysis) could be very useful to get insight into the mechanisms underlying retinal coding. Other aspect which can be analyzed using this approach are color coding as well as coding of natural scenarios. Once the visual code is better understood, spiking retinal models can be constructed that could be used to drive retinally or cortically based visual prostheses.

Acknowledgements

This research is being funded by CICYT SAF98-0098-C02, a grant from the National Organization of the Spanish Blind (ONCE) and MAT2000-1049 to E.F., JM.F., F.J.G. and J.A.B., NSF grant # IBN 9424509 and SFB 517 to J.A., Fellowship from the Spanish Ministerio de Educacion to M.B. and by a State of Utah Center of Excellence Grant and a NSF grant IBN94-24509 to R.A.N.

References

1. Secker H. and Searle C.: "Time Domain Analysis of Auditory-Nerve Fibers Firing Rates". J. Acoust. Soc. Am. 88 (3) pp. 1427-1436, 1990.
2. 2 Buck L.B..: "Information Coding in the vertebrate olfactory system". Ann. Rev. Neurosci. 19, pp. 517-544, 1996.
3. Fitzhugh, R. A.: "A Statistical Analyzer for Optic Nerve Messages". J. Gen. Physiology 41, pp. 675-692, 1958.
4. Warland D., Reinagel P., Meister M.: " Decoding Visual Information from a Population of Retinal Ganglion Cells". J. Neurophysiology 78, pp. 2336-2350, 1997.

5. Fernández E., Ferrández J.M., Ammermüller J., Normann R.A.: "Population Coding in spike trains of sinultaneosly recorded retinal ganglion cells Information". Brain Research, 887, pp. 222-229, 2000.
6. Tovee M.J., Rolls E.T., Treves A. and Bellis R.P..: "Encoding and the Responses of Single Neurons in the Primate Temporal Visual Cortex". Journal of Neurophysiology, 70(2), pp. 640-654, 1993.
7. Ferrández J.M., Bolea J.A., Ammermüller J. , Normann R.A. , Fernández E.: "A Neural Network Approach for the Analysis of Multineural Recordings in Retinal Ganglion Cells: Towards Population Encoding". IWANN 99, Lecture Notes on Computer Science 1607, pp. 289-298, 1999.
8. Rieke F., Warland D., van Steveninck R., Bialek W.: "Spikes: Exploring the Neural Code". MIT Press. Cambridge, MA, 1997.
9. Jones, K.E., Campbell, P.K. and Normann, R.A.:"A glass/silicon composite intracortical electrode array". Annals of Biomedical Engineering 20, 423-437 (1992).
10. Shannon C.E..: "A Mathematical Theory of Communication". Bell Sys Tech J. 27. pp. 379-423 , 1948.
11. Panzeri S., Treves A., Schultz S., Rolls E.T.: "On Decoding the Responses of a Population of Neurons from Short Time Windows". Neural Computation 11. pp. 1153-1577 , 1999.
12. Borst A., Theunissen E.: "Information Theory and Neural Coding". Nature Neuroscience 2(11). pp. 947-957 , 1999.
13. Rolls E.T. Treves A., Tovee M.J.: "The representational capacity of the distributed encoding of information provided by populations of neurons in primate temporal visual cortex". Exp. Brain Research. 114 pp. 149-162 , 1998.
14. Reinagel P., Reid R.C..: "Temporal Coding of Visual Information in the Thalamus". Journal of Neuroscience 20(14). pp. 5392-5400 , 2000.

Numerical Study of Effects of Co-transmission by Substance P and Acetylcholine on Synaptic Plasticity in Myenteric Neurons

Roustem Miftakov[1] and James Christensen[2]

[1] Dept. Radiology, University of Iowa, Iowa City, IA 52242, USA
[2] Dept. Internal Medicine, University of Iowa, Iowa City, IA 52242, USA

Abstract. The biological effect of synaptic co-transmission by substance P (SP) and acetylcholine (ACh) in neurons of the myenteric nervous plexus has been studied numerically. The model, based on known neuroanatomical, biochemical, and pharmacological experimental data about the dynamics of co-transmission of SP and ACh, includes the activation of voltage-dependent ion channels and the second messenger system through the selective binding of neurotransmitters to NK_1 and muscarinic receptor subtypes located on the postsynaptic membrane. The simulations reproduce the electrophysiological and pharmacodynamical behaviors that observed in natural experiments.

1 Introduction

The precise control of the actions of multiple neurotransmitters at specific molecular targets leads to efficiency in intercellular communication in a biological neural network, such as the myenteric nervous plexus (MNP). In chemical synapses it involves the co-release of co-localized neurotransmitters into the synaptic cleft and the subsequent activation of postsynaptic receptors. This fact provides the basic working concept for plasticity in the MNP, which depends mainly on synaptic efficiency and is related to one or another form of memory and learning. The above theory rests on the similarity between experimentally observed long-term potentiation/depression phenomena and the abstract Hebbian hypothesis (1949), which postulates that the synapse strengthens when the pre- and postsynaptic elements are active simultaneously. The rule has been extended to include synaptic weakening due to uncorrelated synaptic linking, localization, "slow "and "fast" neural activity dynamics, and its time evolution.

Nerve-signal transduction within the MNP during coordinated reflexes involves several substances, including acetylcholine, tachykinins, 5- hydroxytryptamine, nitric oxide, and vasoactive intestinal peptide. The tachykinins, are a family of structurally related neuropeptides, include three major mammalian endogenous ligands, namely substance P (SP), neurokinin A (NKA) and neurokinin B (NKB), all widely distributed within subpopulations of nerve cells in the MNP. Neurohistochemical staining and radio-ligand binding technique demonstrate that 59 – 66 % of cells in the MNP of the guinea-pig ileum show pre-synaptic co-localization of SP and ACh (Costa et al, 1996; Grider et al, 1994; Furness et al, 1995; Holzer, et al, 1997).

J. Mira and A. Prieto (Eds.): IWANN 2001, LNCS 2084, pp. 63–71, 2001.

Substance P exhibits a wide range of biological effects by its selective binding to two pharmacologically distinct ionotropic and metabotropic receptors - NK_1 and NK_3. Molecular cloning has revealed the tachykinin receptors to be members of the G-protein coupled receptor superfamily which employs ionositol 1,4,5-triphosphate IP_3 signaling cascades (Kimball and Mulholland, 1995). SP at least partially mediates slow excitatory synaptic potentials (EPSPs) via NK_1 receptors and produces neurogenic contractions/relaxations of intestinal smooth muscle (Johnson et al, 1996; Sarosi, et al, 1998). Acetylcholine, on the other hand, initiates fast EPSP through the activation of ligand gated Ca^{2+} - channels.

The aim of this study is to investigate numerically the effects of high and low frequency stimulation on co-transmission by SP and ACh and the dynamics of the synaptic response of neurons of the MNP.

2 Mathematical Model

The sequence of chemical transformations preceding the electrophysiological events recorded on the postsynaptic membrane, which are required for the mathematical formulation of a model are summarized. Consider two type II neurons of the MNP connected by a synapse containing SP and ACh as neurotransmitters. The main trigger in the initiation of their release is the rapid increase in the concentration of intracellular Ca^{2+} ions during the short period of depolarization of the presynaptic membrane. The activation of reactive Ca^{2+} centers on the vesicular stores of ACh and SP causes quantal exocytosis and subsequent diffusion of the neurotransmitters into the synaptic cleft. It is important to note that: i) the dynamics are such that cholinergic transmission precedes SP transmission, and, ii) ACh and SP interact in a synergistic, rather than an additive manner. In the synaptic cleft part of the ACh is utilized by the acetylcholinesterase enzyme and a part of the SP undergoes enzymatic inactivation by neutral endopeptidase, aminopeptidase, and angiotensin converting enzymes through the formation of substance-enzyme complexes. Fractions of both ACh and SP diffuse to the postsynaptic membrane where they bind to muscarinic and neurokinin type 1 receptors, respectively. Subsequent events follow two different pathways.

The formation of (ACh-R) complex consequently activates ion channels with the generation of the fast EPSP. Part of the released ACh is enzymatically degraded to choline. Tachykinin NK_1 receptors bind to SP and form the SP-R_{NK} complexes which activate the guanine-nucleotide G protein system cascade with the activation of phospholipase C, the cleavage of phosphatidyl inositol 1,4,5-triphosphate and, subsequently, the release of the two fragments, inositol 1,4,5-triphosphate (IP_3) and diacylglycerol (Dg). The latter molecules have synergistic effects on the second messenger system. The former binds to the IP_3 receptor (R_{ER}) on the endoplasmic reticulum causing the rapid release of Ca^{2+} into the cytoplasm. The resulting increase in the concentration of calcium permits the formation of a Ca^{2+} - calmodulin complex that serves as a co-factor in the activation of protein kinase C (PkC). This also requires the presence of a diacylglycerol molecule. The final step in the cascade is the phosphorylation/dephosphorylation of intracellular proteins (Pr) by protein-phosphatase (PPhos) with the consequent alteration of permeabilities of K^+, Ca^{2+}-K^+, and Ca^{2+} channels on the membrane.

A mathematical description of the model is based on the following principal assumptions: i) the dynamics of depolarization of the presynaptic nerve terminal satisfies the modified Hodgkin-Huxley equations; ii) the synapse (the presynaptic terminal, synaptic cleft, and postsynaptic membrane) is a null-dimensional three-compartment open pharmacokinetic model; iii) all the chemical reactions of neurotransmitter transformation are adequately described by first order Michaelis-Menten kinetics; iv) Ca^{2+} ions entering the cell during excitation have equal access to the reactive centres on ACh and SP vesicles. (For full details see Miftakhov and Christensen, 2000).

The dynamics of the wave of depolarization (φ^*) at the presynaptic terminal is defined by:

$$C^{a,m}\ \partial \varphi^*/\partial t = \frac{1}{2R_a}\ \partial/\partial s\ (a^2(s)\ \partial\varphi^*/\partial s) - I_{ionic} \qquad (1)$$

where $C^{a,m}$ is the specific capacitance of the nerve fiber, R_a is the membrane resistance, a(s) is the diameter of the axon, and s is the Lagrangian coordinate.

The total ionic current, I_{ionic} is:

$$I_{ionic} = g_{Na} m^3 h(\varphi^* - \varphi_{Na}) + g_K n^4(\varphi^* - \varphi_K) + g_{Cl}(\varphi^* - \varphi_{Cl}) \qquad (2)$$

Here: g_{Na}, g_K and g_{Cl} represent the maximal conductances of Na^+, K^+ and Cl^- channels, respectively, and *m, n, h* are the probabilities of opening of these channels; φ_{Na}, φ_K, and φ_{Cl} are the reversal potentials of sodium, potassium and chloride (leak) currents. The activation and deactivation of the Na^+, K^+ and Cl^- channels are described as:

$$dy/dt = \alpha_y\ (1-y) + \beta_y\, y, \quad (y = m, n, h) \qquad (3)$$

Here α is the rate at which the channels switch from a closed to an open state and β is the rate for the reverse. Both values depend only on the membrane potential and are given by the expressions:

$$\alpha_m = \frac{0.11\ \theta(\varphi^* - 25)}{1-\exp(-(\varphi^* + 255)/10)},\quad \beta_m = 4.1\ \theta \exp(-\varphi^*/18)$$
$$\alpha_n = \frac{0.005\ \theta(\varphi^* - 10)}{1-\exp((10 - \varphi^*)10)},\quad \beta_n = 0.075\ \theta \exp(-\varphi^*/80) \qquad (4)$$
$$\alpha_h = 0.02\ \theta \exp(-\varphi^*/20),\quad \beta_h = \frac{0.5\ \theta}{1 + \exp((-\varphi^* + 30)/10)}$$

θ - temperature (C°).

The system of kinetic equations of the cycle of chemical transformations of ACh and SP at the synapse is as follows:

$$
\begin{aligned}
&d[Ca^{2+}]/dt = [Ca^{2+}]_{out}\,\varphi^* - k_{+5}[Ca^{2+}]\\
&d[SP_v]/dt = -1/2\,k_v\,[Ca^{2+}][SP_v]\\
&d[ACh_v]/dt = -1/2\,k_c[Ca^{2+}]\,[ACh_v]\\
&d[ACh_f]/dt = k_c[Ca^{2+}][ACh_v] - k_d[ACh_f] + k_{+6}\,[S]\\
&d[SP_f]/dt = k_v\,[Ca^{2+}][SP_v] - k^*_{+2}\,[SP_f]\\
&d[ACh_c]/dt = k_d\,[ACh_f] - k_{+p}[ACh_c] \qquad (5)\\
&d[SP_c]/dt = k_{+2}\,[SP_f] + [SP_c\text{-}Enz](k^*_{-3} + k^*_{-3}\,[SP_c]) -\\
&- [SP_c](k^*_{+3}[E^0] + k^*_{+5}[R_{NK}]) + k^*_{-5})[SP_c\text{-}R_{NK}]\\
&d[ACh_p]/dt = k_{+p}\,[ACh_c] - [ACh_p]\,(k_{+1}\,[R^0] + k_{+2}\,[E^0]) +\\
&+ [ACh_p]\,(k_{+1}[ACh\text{-}R] + +k_{+2}[AChE]) + k_{-1}[ACh\text{-}R] + k_{-2}[AChE]\\
&d[ACh\text{-}R]/dt = k_{+1}\,[ACh_p][R^0] - [ACh\text{-}R]\,(k_{-1} + k_{+4} + k_{+1}[ACh_p])\\
&d[SP_c\text{-}R_{NK}]/dt = k^*_{+5}[SP_c][R_{NK}] - [SP_c\text{-}R_{NK}](k^*_{-5} + k^*_{+6} +\\
&+ k^*_{+7}([G^0] - [SP_c\text{-}R_{NK}\text{-}G])\\
&d[AChE]/dt = k_{+2}\,[E^0][ACh_p] - [AChE]\,(k_{-2} + k_{+3} + k_{+2}\,[ACh_p])\\
&d[S]/dt = k_{+4}[ACh\text{-}R] - k_{+6}\,[S] + k_{+3}\,[AChE]\\
&d[SP_c - Enz]/dt = k^*_{+3}[SP_c][Enz^0] - [SP_c\text{-}Enz](k^*_{-3} + k^*_{+4} + k_{+3}[SP_c])\\
&d[R_{NK}]/dt = (k^*_{-5} + k^*_{+6})[SP_c - R_{NK}] - k^*_{+5}[SP_c][R_{NK}]\\
&d[SP_c\text{-}R_{NK}\text{-}G]/dt = k_{+7}[SP_c\text{-}R_{NK}]([G^0] - [SP_c\text{-}R_{NK}\text{-}G]) - k^*_{-7}[SP_c - R_{NK}\text{-}G]\\
&d\,[PhC]/dt = k^*_{+10}[SP_c - R_{NK}\text{-}G]([PhC]_{inact} - [PhC]) - k^*_{+15}[PhC]\\
&d[PIP_3]/dt = -(k^*_{+11} + k^*_{+12})[PIP_3][PhC]\\
&d[Dg]/dt = k^*_{+12}\,[PhC][PIP_3] - k^*_{+18}[Dg][Ca^{2+}]^4_{ier}\,([PhC]_{inact} - [PhC])\\
&d[IP_3]/dt = k^*_{+11}[PIP_3][PhC] - k^*_{+17}[IP_3][R_{ER}] + (k^*_{-17} - k^*_{-19})[IP_3\text{-}R_{ER}]\\
&d\,[IP_3\text{-}R_{ER}]/dt = b_5\,[IP_3\text{-}R_{ER}\text{-}\,Ca^{2+}] - (a_2 + a_5)[Ca^{2+}]_{ier}[IP_3\text{-}R_{ER}]\\
&d\,[Ca^{2+}]_{ier}/dt = \tau^{-1}(\,V_0 - c_1V_1\,[IP_3\text{-}R_{ER}\text{-}\,Ca^{2+}]^3([Ca^{2+}]_{ier}\\
&- [Ca^{2+}]_{ER}) - V_3[Ca^{2+}]^2_{ier}/([Ca^{2+}]_{ier}^2 - k^{*2}_{+13}))\\
&d[PkC]/dt = k^*_{+8}\,[Dg][Ca^{2+}]^4_{ier}([PhC]_{inact} - [PhC]) - k^*_{-8}\,[PkC]\\
&d\,[Pr]_p/dt = k^*_{+9}\,[PkC]([Pr] - [Pr]_p) - k^*_{-9}\,[PPhos][Pr]_p\\
&[R] = [R^0] - [ACh\text{-}R];\ [E] = [E^0] - [AChE];\ [Enz] = [Enz^0] - [SP_c\text{-}Enz]\\
&[G] = [G^0] - [SP_c\text{-}R_{NK}\text{-}G];\ c_0 = [Ca^{2+}]_{ier} + c_1\,[Ca^{2+}]_{ER};\ [R_{ER}] = c_2[IP_3]
\end{aligned}
$$

Here: $k^{(*)}_{-,+}$ are the constants of backward and forward chemical reactions; k_c is the affinity constant; $k_{d,v}$ are the constant of diffusion; k_{+p} is the velocity of diffusion of ACh on the subsynaptic membrane; $c_{(i)}$, $b_{(i)}$, V_0, V_1 are parameters related to the activation of the release of endoplasmic Ca^{2+}; ACh_v is vesicular acetylcholine and its free (ACh_f), and postsynaptic (ACh_p) fractions; R is receptors; E is acetylcholinesterase enzyme (superscript 0 indicates the initial concentration); ACh-R is acetylcholine-receptor complex; AChE is acetylcholinesterase enzyme complex; S is products of chemical reactions; $(Ca^{2+})_{out}$ is the external calcium ion content; SP_v is vesicular substance P; SP_f is the free fraction of the SP; SP_c is the SP content in the cleft; Enz represents, collectively, neutral endopeptidase, aminopeptidase and angiotensin converting enzymes; SP_c-Enz is substance P - enzyme complex; R_{NK} is neurokinin receptors; SP_c- R_{NK} is substance P - neurokinin receptor complex; G is guanine-nucleotide G protein; PhC is phospholipase C; PIP_3 is phosphatidyl inositol 1,4,5-triphosphate; IP_3 is inositol 1,4,5-triphosphate; Dg is diacylglycerol; R_{ER} is IP_3 receptor on the endoplasmic reticulum; $Ca^{2+}{}_{ier}$ is the calcium of the endoplasm; $Ca^{2+}{}_{ER}$ is the total amount of calcium stored in the endoplasmic reticulum; PkC is protein kinase C; Pr is intracellular protein; PPhos is protein-phosphatase.

The initial conditions assume that the neurons are in the resting state. The boundary conditions assume that the excitation of the system is due to the application of an excitatory stimulus of an amplitude 100 mV and duration of 0.2 s. The description provided above constitutes the mathematical formulation of the model of the co-localization and co-transmission of ACh and SP in a pre-synaptic nerve terminal and the co-existence of NK_1 and muscarinic receptors on the post-synaptic membrane of a neuron. The system of equations (1) – (5) was solved numerically.

3 Results

The results of the numerical calculations are presented below. The dynamics of ACh conversion under normal physiological conditions in a synapse has been extensively studied by Miftakhov and Wingate (1993). For the sake of space, we refer the reader to that paper and here we shall concentrate on the analysis of the effects of low and high frequency stimulation on substance P transformations. Concentrations of co-transmitted fractions of ACh will be given in brackets for comparison.

3.1 Effect of Randomly Applied High Frequency Stimuli

Consider the random application of stimuli at frequency 1 Hz to the soma of a neuron. The action potential elicited by the external stimulation has an amplitude of 72mV in the vicinity of the terminal ending. The depolarization of the presynaptic membrane activates the short-term influx of calcium ions into the terminal through voltage-dependent Ca^{2+} channels. The concentration of cytosolic Ca^{2+} rises to 1.9 – 2.41μM (fig. 1). Some of the ions are immediately absorbed by the buffer system while others diffuse towards the vesicular stores of neurotransmitters. The binding of Ca^{2+} ions to the active centers on the vesicles initiates SP_v release. The velocity of SP_v release is not constant but depends on the concentration of cytosolic Ca^{2+}. As a response to a single excitation, about 0.2 μM of vesicular SP is released ([ACh_v] = 78.2 mM - 1% of vesicular ACh). The maximum concentration of 2.9 μM of SP is being released in response to a train of twelve action potentials in 24 s ([SP_v] = 77.1 μM).

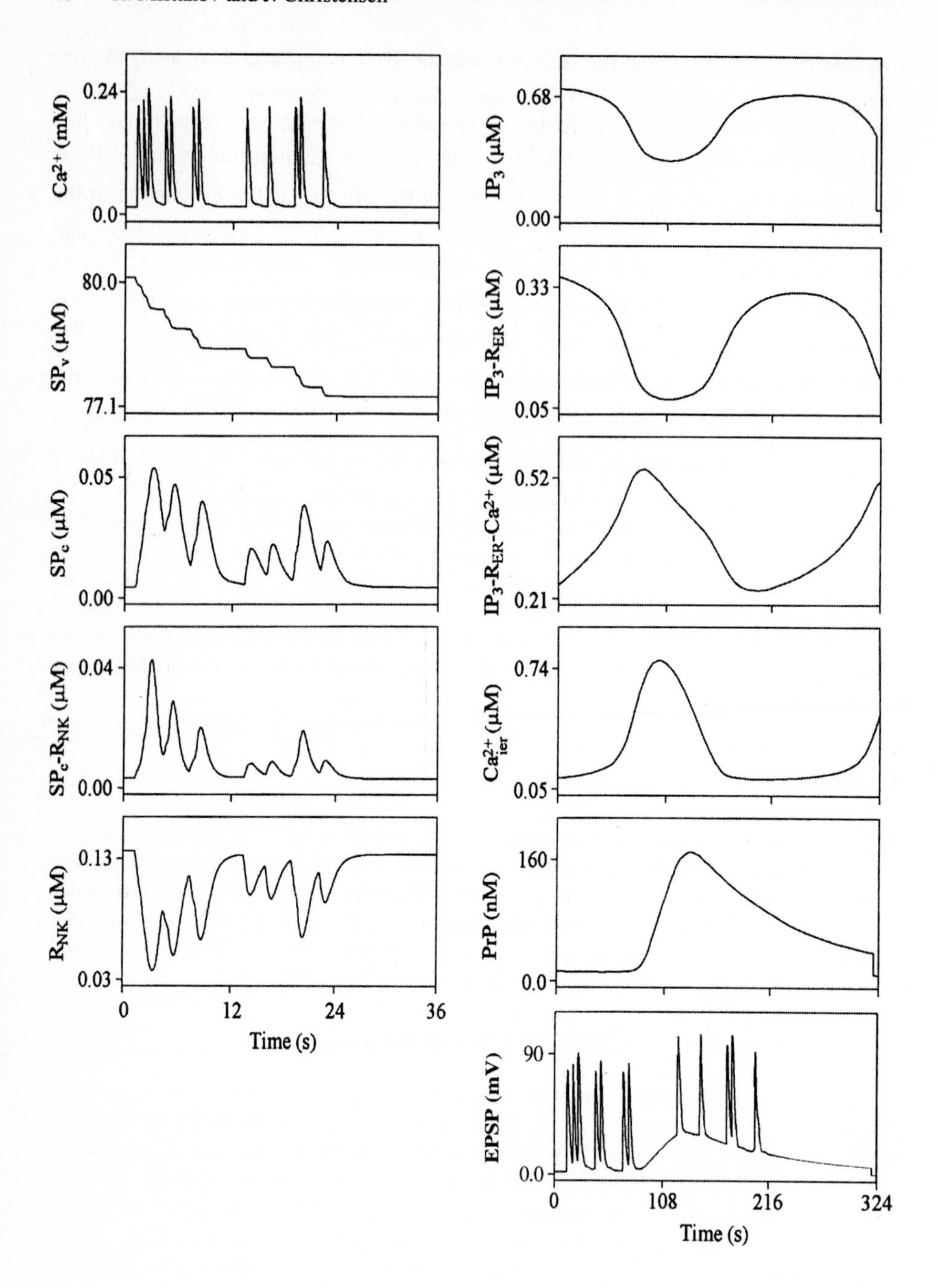

Fig. 1. Response of a post-synaptic membrane to a coherent release of SP and ACh.

The free fraction of neurotransmitter diffuses into the cleft. The concentration of SP_c in the cleft varies in time and does not remain constant but changes depending on the dynamics of binding to the receptors and activity of enzymes. Thus the max[SP_c] = 0.056 μM (max[ACh_c] = 5.38 mM) and min[SP_c] = 0.019 μM are registered at different times in the process. The main part of the SP_c reacts with NK_1 receptors on the postsynaptic membrane where a max[SP_c – R_{NK}] = 0.042 μM and a min[SP_c – R_{NK}] = 0.012 μM are formed. The maximum number of activated receptors equals 0.027μM (6.99 μM).

Another part of the SP_c undergoes fission by enzymes with the formation of SP_c – Enz complex. The complex quickly dissociates into free enzyme and reaction products that are reabsorbed by the nerve terminal to be drawn into a new cycle of substance P synthesis.

With the formation of (SP_c-R_{NK}-G), the activation of the second messenger system in the underlying neuron begins with the transformation of the inactive form of phospholipase C to its active form and the production of inositol 1,4,5-triphosphate and diacylglycerol (Dg). The concentration of Dg constantly rises to reach the maximum steady state value, 50 nM. These changes are concomitant with a decrease in the [IP_3]. Dg serves as a co-factor in the phosphorylation of the intracellular proteins. IP_3 is important in the initiation of the release of calcium ions from the endoplasmic reticulum. It binds to the receptors on the endoplasmic reticulum and forms the (IP_3-R_{ER})-complex (max[IP_3-R_{ER}] = 0.3 μM). The subsequent occupation of Ca^{2+} - activating sites by the complex induces a flux of Ca^{2+}_{ier}, which reaches the maximum of 0.745 μM at a certain moment in the dynamic process. An increase in free cytosolic calcium and the binding of four molecules of C^{2+} to calmodulin, together with Dg, triggers the activation of protein kinase C. As a result, the maximum 166 nM of phosphorylated protein (PrP) is formed. It declines exponentially to reach picomolar concentrations at t = 324 s. The duration of the presence of phosphorylated protein is 244s. The latter changes the permeability of K^+, Ca^{2+}-K^+, and Ca^{2+} ion channels on the membrane, producing slow EPSP's of amplitude 20-25 mV and duration 244 s, and thus altering the firing pattern of the adjacent neuron. The final effects of ACh, in contrast, is the generation of fast EPSP's of very short span, 50-100 ms and high amplitude 90 mV. Acting coherently they superimpose on the long-term potentiation curve of the slow EPSP.

3.2 Effect of Low Frequency Stimuli

Low frequency stimulation of a neuron with a train of 20 impulses duration of 10 s at frequency 0.025 Hz revealed a quantitative difference in the dynamics of the production of PrP. The qualitative dynamics of transformations of other fractions of SP involved in the process of release, binding to receptors, and activation of the second messenger system is similar to that described previously. However, the neuron responds with a periodic pattern of generation of IP_3 with the subsequent production of (IP_3-R_{ER}) and the activation and release of free calcium ions from the intracellular stores (fig. 2). Thus, the maximum 0.7 μM of inositol 1,4,5-triphosphate is produced and it slowly fluctuates at a frequency 0.01 Hz and an amplitude 0.35 μM. The max(IP_3-R_{ER}) = 0.33 μM is formed, which results in a periodic release of Ca^{2+}_{ier}. Its concentration quickly reaches the maximum value 0.745 μM. Free calcium is quickly removed from the cytoplasm and a min[Ca^{2+}_{ier}] = 0.01μM is achieved. As a result , in response to excitation by the first nine stimuli a max[PrP] = 78 nM is being produced. The concentration of PrP declines exponentially to its resting level and to the following four impulses only a small amount (0.12 μM) of phosporylated protein is being created. There is a slight increase in the concentration of PrP as a response to the succeeding stimuli when a maximum 0.27 μM and 0.18 μM of PrP are produced. As anticipated, slow excitatory signals of variable amplitude 5-25 mV and duration 200-244 s are generated on the postsynaptic membrane of the soma of the adjacent neuron. The co-release of ACh and the activation of muscarinic receptors causes the production of fast EPSP's of a constant amplitude 90 mV and duration 50 ms. The result of the cooperative effect of SP and ACh is an irregular summatory long-term EPSP of variable amplitude.

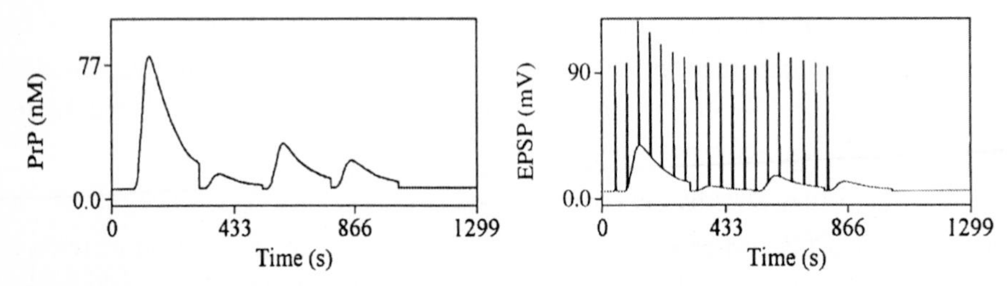

Fig. 2. Production of phosphorylated protein and summated EPSPs as a result of low frequency stimulation.

4 Conclusion

Synaptic strengthening, which is induced by high and low frequency repetitive stimulations, adds a considerable amount of computational power to neuronal structures in the biological networks. The strengthening is related to the phenomenon of long-term potentiation, which is a result of complex biochemical processes that causes a stable increase of the efficiency and plasticity of a synapse. Until recently, models of the dynamics of synaptic plasticity were based on the analysis of molecular mechanisms of conversion of a single neurotransmitter, mainly GABA or NMDA, assuming that only one of the above could initiate either fast or slow excitatory/inhibitory post-synaptic potentials. The model described simulates the co-localization of SP and ACh in a pre-synaptic membrane of a neuron of the MNP and proposes an explanation of the intrinsic mechanisms of synaptic plasticity. Thus, the release of SP causes slow EPSPs and strengthens the synapse, shifting the system to an excitable state. The neuron responds to a coherent supplementary release of ACh with the generation of the high amplitude fast EPSPs which may sustain its firing activity for the period of action of SP. The results of the simulations demonstrate that not only the frequency of pertubations *per se* but the dynamics of the intrinsic biochemical pathways, e.g., the second messenger system and the receptor types, is important in the effective "cognitive" function of the myenteric nervous plexus. A comparison of the patterns of electrical activity (fig.1, 2) reproduced during the numerical simulations shows good qualitative and quantitative agreement with those in experimental recordings made on isolated neural cells of the MNP (Moore et al, 1997).

References

1. Costa, M., Brookes, S.J.H., Steele, P.A., Gibbins, I., Burcher, E., and Kandiah, C.J. Neurochemical classification of myenteric neurons in the guinea-pig ileum. Neuroscience, **75** (1996) 949-967
2. Furness, J.B., Young, H.M., Pompolo, S., Bornstein, J.C., Kunze, W.A.A., and McConalogue, K. Plurichemical transmission and chemical coding of neurons in the digestive tract. Gastroenterol., **108** (1995) 554-563
3. Grider, J.R., and Bonilla, O.M. Differential expression of substance P, somatostatin, and VIP in neurones from cultured myenteric ganglia. Am. J. Physiol., **267** (1994) G322-327
4. Holzer, P., and Holzer-Petsche, U. Tachykinins in the gut. Part. I. Expression, release and motor function. Pharmacol-Ther., **73** (1997) 173-217
5. Johnson, S.M., Bornstein, J.C., Yuan, S.Y., and Furness, J.B. Analysis of contributions of acetylcholine and tachykinins to neuro-neuronal transmission in motility reflexes in the guinea-pig ileum. Brit. J. Pharm., **118** (1996) 973-983
6. Kimball, B.C., Mulholland, M.W. Neuroligands evoke calcium signalling in cultured myenteric neurones. Surgery, **118** (1995) 162-169
7. Miftakhov, R. N., and Wingate, D.L. Mathematical modeling of the enteric nervous network. 1:Cholinergic neuron. J. Med. Eng. & Phys., **16** (1994) 67-738
8. Miftakhov, R.N., and Christensen, J. Cotransmission by substance P and acetylcholine in the myenteric plexus. In: Proc. Int. ICSC Congress on Intel. Syst. & Appl. (ISA'2000), Dec. 12-15, 2000, Wollongong, Australia
9. Moore, B.A., Vanner, S., Bunnett, N.W., and Sharkey, K.A. Characterization of neurokinin – 1 receptors in the submucosal plexus of guinea pig ileum. Am.J. Physiol. **273** (1997) G670 - G678.
10. Sarosi, G.A., Kimball, B.C., Barnhart, D.C., Weizhen, Z., and Mulholland, M.W. Tachykinin neurpeptide-evoked intracellular calcium transients in cultured guinea-pig myenteric neurons. Peptides, **19** (1998) 75-84

Neurobiological Modeling of Bursting Response During Visual Attention

Reza Rajimehr, Leila Montaser Kouhsari

Institute for Studies in Theoretical Physics and Mathematics, Tehran

Abstract: Thalamic neurons have an exclusive property named bursting response. Bursting response seems to have a critical role in producing saliency map and encoding conspicuity of locations during visual attention. Attention window is developed in thalamus due to its retinotopic organization. The global competitive network in thalamic reticular nucleus (NRT) determines which thalamic cells should produce bursting response. These cells are corresponded with attention window or attended location in the visual field. The computational procedure of bursting response is studied by the means of neurobiological modeling of thalamic neurons and their bimodal behavior (periodic bursting and periodic spiking patterns). The effect of NRT on thalamic relay neurons is considered in each mode of thalamic response. We also achieved the results of modeling through computer simulation of bimodal behavior in thalamic neurons.

Keywords: Thalamic relay neuron, Bursting response, Visual attention, Model

1 Introduction

As Jahnsen and Llinás (1,2,3) showed, we understand that all thalamic relay neurons (TRN) have two distinct modes of behavior. First when the cell is near its normal resting potential (- 60 mv), second when it is hyperpolarized and its potential is near - 70 mv. The former responds to an injected current by firing at a rate between 25 and 100 spikes per second, which increases with greater values of the injected current, and later responds to an injected current, after a short delay, with fast burst of spikes and firing rates nearer 300 spikes per second. Because of aftereffect, the cell will not produce another bursting response although the injected current is constant. Ca^{2+} - dependent K^{+} channels are responsible for bursting behavior. Other channels related to this phenomenon are low threshold and high threshold Ca^{2+} currents, early K^{+} current and Na^{+} current. The rhythmic behavior of these ion currents is described as follows.

When the cell membrane is hyperpolarized for at least 150 msec, a low threshold Ca^{2+} current is de-inactivated, which normally inactivated at the membrane resting potential (or at potentials more positive than the membrane resting potential). This current is probably localized at the level of the soma. The low threshold spike leads to the activation of a high threshold Ca^{2+} current (probably localized at the dendritic level). Influx of Ca^{2+} increases the excitatory state of the relay neuron. In this state,

J. Mira and A. Prieto (Eds.): IWANN 2001, LNCS 2084, pp. 72-80, 2001.

the neuron produces action potentials due to Na^+ current. Each action potential is followed by a phase of after-hyperpolarization in result of Ca^{2+} - dependent K^+ currents. This after-hyperpolarization is weak so the neuron can fire bursts of action potentials at higher frequencies. The early K^+ current prolongs the duration of the hyperpolarization. The responses of thalamic neurons to excitatory afferents depend on the value of the resting membrane potential, i.e. on the activity of the inhibitory afferent pathways. This periodic pattern will continue provided that the input current to the neuron is available. The periodic bursting pattern based on ion currents is studied and simulated in this article.

Taylor and Alavi suggested NRT as a *Winner Take All (WTA)* network during visual attention. NRT has a global competitive network in which long-range inhibition occurs so corresponding neurons with attention window can inhibit other neurons (4). These neurons send inhibitory projections to the thalamus.

Crick believes that inhibition of NRT on corresponding thalamic neurons in attention window enables these neurons to produce *bursting response*. In Crick's theory (5), the retinotopic area for attention window is thalamus in which a number of thalamocortical pathways are activated more than other pathways during spot light attention. As a result, transmission of information is facilitated through these activated thalamocortical projections. Thalamus and NRT coupled nets have a role in producing saliency map and implementing the WTA network.

2 Model

RT GABAergic neurons hyperpolarize TR neurons, which inhibitory post-synaptic potential (IPSP) is made by K^+ current (K^+ dependant IPSP is mediated by GABA B receptors). Hyperpolarization increases the threshold of TR neuron.

Let $?_i$ denote the threshold of TR neuron, which is altered by two independent factors:

$$\boldsymbol{\theta}_{i_{total}}(t) = \boldsymbol{\theta}_i^1(t) + \boldsymbol{\theta}_i^2(t) \tag{1}$$

Where $\boldsymbol{\theta}_i^1(t)$ shows the alternation of the threshold by NRT inhibition and $\boldsymbol{\theta}_i^2(t)$ shows the effect of neuronal habituation on the threshold. The initial total threshold of TR neuron is m ($\boldsymbol{\theta}_i^1(0) = 0$ and $\boldsymbol{\theta}_i^2(0) = \boldsymbol{m}$).

$\boldsymbol{\theta}_i^1(t)$ is calculated by stationary solution to below differential equation:

$$\boldsymbol{\varepsilon}\frac{d\boldsymbol{\theta}_i^1}{dt} = -\boldsymbol{\theta}_i^1 + (\mathbf{G_k}(t) - \mathbf{G_k}(0)) \tag{2}$$

Where $G_k(t)$ is the amount of K^+ outward current or cellular channel permeability to K^+, which decreases the amount of intra-cellular K^+ concentration. Increasing of $G_k(t)$ enhances the neuronal threshold. $G_k(t)$ is itself altered by below differential equation:

$$\varepsilon \frac{dG_k}{dt} = -G_k + \psi(t).\phi.Z_i \tag{3}$$

Where $F.Z_i$ is the inhibition of NRT on TR neuron. $?(t)$ is a periodic square function, which alters between 0 and 1. At the beginning ($t=0$), NRT sends signals to thalmus and $?(0)$ equals 1 so $G_k(t)$ increases. As we will discuss later, transient hyperpolarization of the membrane (by high amo unt of $G_k(t)$) for at least 150 msec increases the cellular channel permeability to Ca^{2+}(G_{ca}) and de-inactivates a low threshold transient Ca^{2+} current. The low threshold spike leads to the activation of a high threshold Ca^{2+} current in turn. Increase in the intra-cellular Ca^{2+} concentration enhances the saturation point of TR neuron's activity (A) as an index of maximal firing rate (the amount of A directly depends on $G_k(t)$ with considering phase locking between them). When A reaches a value named A_{max}, then $?(t)$ becomes zero so $G_k(t)$ gradually decreases until A attracts to the point of A_{min}. In this phase, hyperpolarization is removed and K^+ ions enter to the neuron by the help of electrogenic pumps. At the point of A_{min} , $?(t)$ equals 1, which causes $G_k(t)$ to increase. $?(t)$ is not changed until A becomes A_{max}. Ca^{2+}-dependent K^+ currents are activated at the time of A_{max} because of phase locking between A and $G_k(t)$. The early K^+ current also prolongs the duration of hyperpolarization. The procedure of described oscillation continues if NRT sends inhibitory signals continuously. Oscillatory behavior of the bursting phenomenon is due to after-hyperpolarization, which follows action potentials at the point of A_{max} .
Hyperpolarization of the TR neuron due to K^+ current can lead to Ca^{2+} current activation in this neuron. The development of cellular channel permeability to Ca^{2+} can be shown as:

$$\varepsilon \frac{dG_{ca}}{dt} = -G_{ca} + (\theta_i^1(t) - \theta_i^1(0)) \tag{4}$$

Influx of Ca^{2+} into TR neuron increases its excitatory state. In other words, the saturation point of neuronal activity is enhanced by below differential equation:

$$\varepsilon \frac{dA}{dt} = -A + \eta.(G_{ca}(t) - G_{ca}(0)) \tag{5}$$

Where $?$ is a positive constant value indicating the amount of increase in A. If $?$ has a small value, the alternation of A will be less than $\theta_i^1(t)$ so the bursting response is produced only in the presence of remarkable NRT inhibition (see Eq.(12)).

Oscillatory diagrams of $G_k(t)$, $\theta_i^1(t)$, $G_{ca}(t)$ *and* $A(t)$ in TR neuron during NRT inhibition are shown in Fig. (1.a).

Let $g(Y_i)$ denote the output of TR neuron as a sigmoidal function, which is increased by A (for $Y_i = \theta_{i\ total}(t)$). A is the saturation point of the sigmoidal function of TR neuron's output and $g(Y_i)$ is the amount of firing rate. At the A_{max}, the firing rate has a great value near bursting frequency, which becomes maximal at the A_{up} (A_{up} is the upper limit of the oscillation, which is made after A_{max} due to the dynamical property of the system). Then the amount of A and firing rate decreases. This is the same as *aftereffect of bursting response* or *gradually reset* of TR neurons and their specific conductances because of the after-hyperpolarization effect (1). The apparition time of after-hyperpolarization (in result of early K^+ current) depends on A. For $A > A_{\max}$, this aftereffect begins after a delay in increasing phase of $G_k(t)$ *and* decreasing phase of A, which is calculated as:

$$\Delta t = t_{(A_{up})} - t_{(A_{\max})} \tag{6}$$

In this period, TR neuron produces some action potentials (in result of Na^+ current) with variable firing rate. $g(Y_i)$ will equal zero from $t_{A_{\max}}$ in decreasing phase of A in its oscillation to $t_{A_{\max}}$ in increasing phase of A in later period, hence TR neuron will not produce the response in this time interval and the firing occurs after a delay when A gets close to the A_{max}. This response pattern of TR neuron is named *Hopf/Hopf bursting* (6) because A increases the firing rate from A_{max} to A_{up} . The periodic bursting pattern is shown in Fig. (2.a).

If the inhibition of NRT (Z_i) is low, bursting response pattern is not produced because Z_i can not shift A from its lower limit to upper limit. In this case, $G_k(t)$, $\theta_i^1(t)$, $G_{ca}(t)$ *and* $A(t)$ have a limit point attractor (see Fig. (1.b)) so $g(Y_i)$ will be constant and TR neuron will produce periodic spiking pattern with low firing rate provided that $Y_i = \theta_{i\ total}(t)$ (see Fig. (2.b)). For $A < A_{\max}$, ?t will be zero and refractory period exists after each action potential. In the TR neuron as a dynamical system, Z_i acts as a bifurcation parameter by which the neuronal response pattern is fluctuated between two stable states (cycle and point basins of attraction).

The amount of critical Z_i , which the state of neuronal response is fluctuated around it, could be achieved by a dynamical analysis. Consider that G_k directly influences A (provided that the time constant values of $\theta_i^1(t)$ and $G_{ca}(t)$ functions are small) so Eq.(7) is obtained by substituting Eq.(3) in Eq.(5):

$$A = A_0 + \eta.(G_k(t) - G_k(0)).(1 - e^{-t/\varepsilon}) \tag{7}$$

A reaches A_{max} when:

$$G_k(t) = G_k^{A_{mx}} = \frac{A_{\max} - A_0}{\eta.(1 - e^{-t/\varepsilon})} + G_k(0) \tag{8}$$

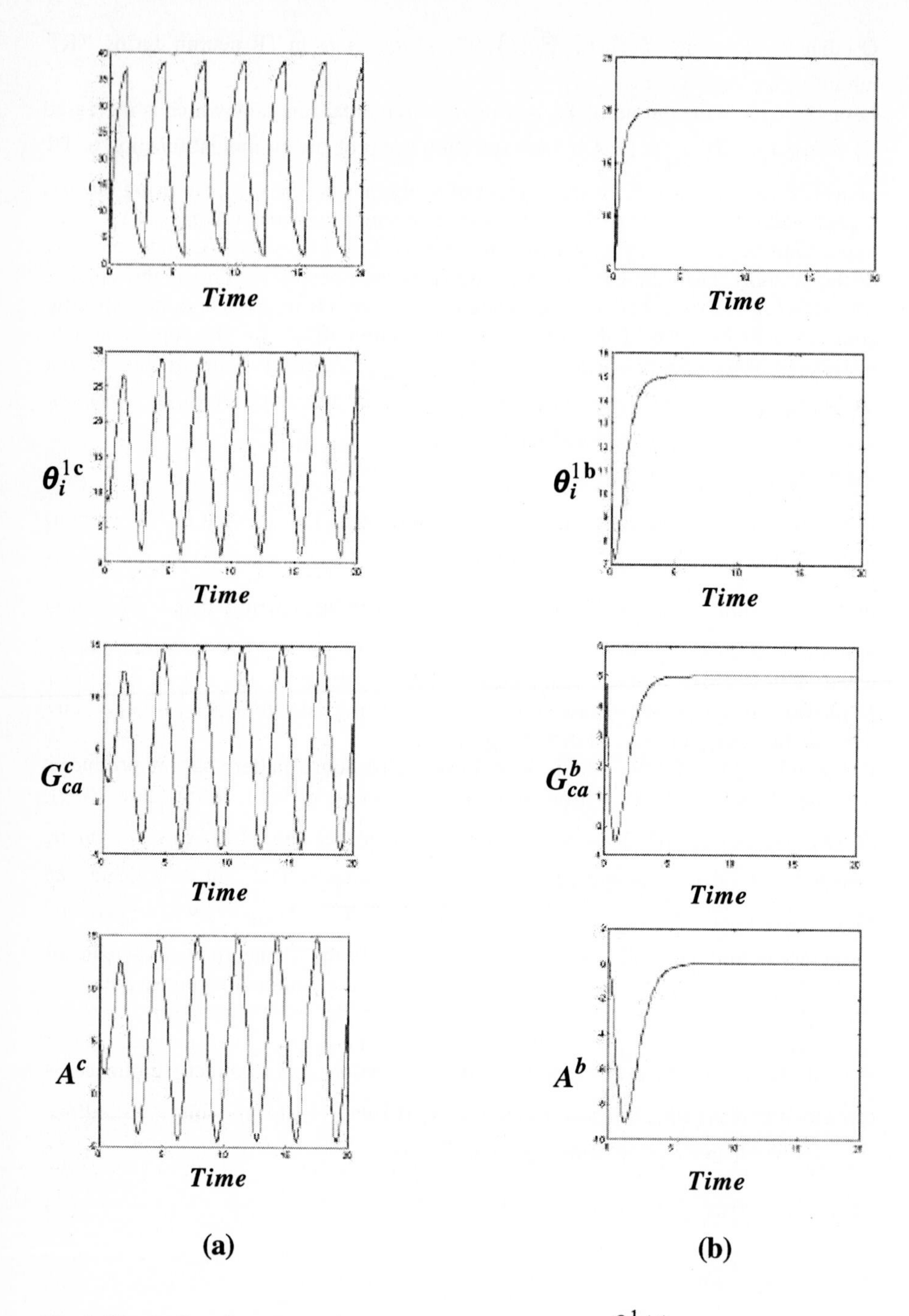

Fig. 1. The results of computer simulation for values of $G_k(t)$, $\theta_i^1(t)$, $G_{ca}(t)$ *and* $A(t)$ in different basins of attraction (cycle (a) and point (b)).

Then $G_k(t)$ decreases exponentially $(?(t) = 0)$ until $G_k(t) = G_k(0)$. In this point, the sign of derivative of A is changed so that upper limit of A in its oscillation is produced. Upper limit point of A is computed as follows:

$$A_{up} = A_{\max} + \eta . Max\left[G_k^{A_{mx}} . e^{-t/\varepsilon} - G_k(0) \right] . (1 - e^{-t/\varepsilon}) \quad (9)$$

Then A begins to decrease until reaches A_{min} . At this point, $G_k(t)$ will be:

$$G_k(t) = G_k^{A_{mn}} = \frac{A_{\min} - A_{up}}{\eta .(1 - e^{-t/\varepsilon})} + G_k(0) \quad (10)$$

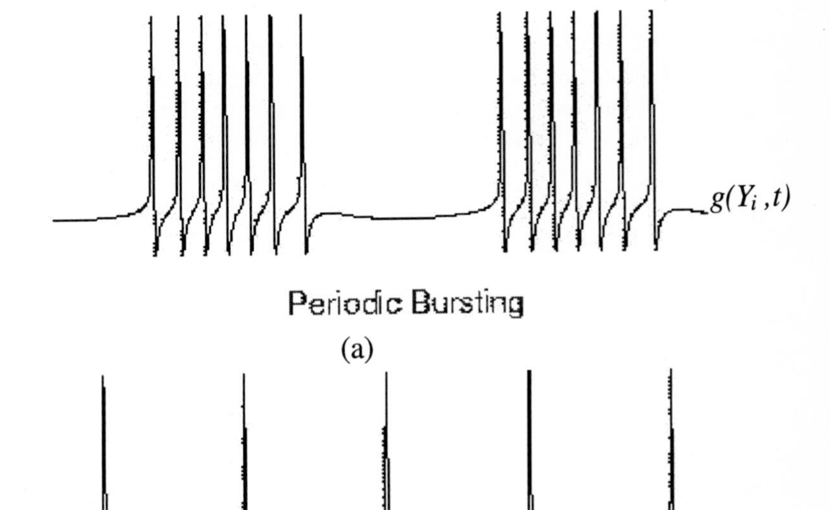

Fig. 2. The neuronal response of TR neuron. (a) When a periodic bursting pattern (fast burst of spikes at the A_{max} and A_{up} with an aftereffect) is produced by the NRT inhibition. (b) When a periodic spiking pattern is produced. (Source: Adapted from Izhikevich, 2000.)

As mentioned before, $?(t)$ becomes 1 and $G_k(t)$ is increased by $F.Z_i$ at the point of A_{min}. Lower limit point of A in its oscillation, in which the sign of derivative of A is changed, is calculated by:

$$A_{low} = A_{\min} - \eta . \left| Min\left[\left(G_k^{A_{mn}} + \phi . Z_i .(1 - e^{-t/\varepsilon}) \right) - G_k(0) \right] . (1 - e^{-t/\varepsilon}) \right. \quad (11)$$

A is moved from A_{low} to A_{max} and then to A_{up} provided that:

$$Z_i^c = \frac{\dfrac{A_{\max} - A_{\min}}{\eta.(1 - e^{-t/\varepsilon})} + G_k(0) - G_k^{A_{mn}}}{\phi.(1 - e^{-t/\varepsilon})} \tag{12}$$

Z_i^c is the same as *critical value* for the bifurcation of the neuronal response. This system has a limit point attractor for $Z_i < Z_i^c$ and a limit cycle attractor for $Z_i > Z_i^c$.

If the firing of TR neuron continues, the *habituation effect* will be appeared (the sensitivity of the neuron for responding to its input will decreased). This effect could be described by minimal increase in the threshold of the TR neuron. The alternations of the threshold due to habituation effect will be given by:

$$\alpha \frac{d\theta_i^2}{dt} = -\theta_i^2 + g(Y_i, t) \tag{13}$$

Where a is a time constant, which is larger than e so θ_i^2 will be alternated slowly. $\theta_i^2(t)$ has also oscillatory behavior, which causes to change the total threshold in Eq.(1). $\theta_i^2(t)$ decreases in the inter-spike interval. A possible mechanism for neuronal habituation is decrease in post-synaptic receptors (down regulation of receptors). The number of post-synaptic receptors after habituation could be computed by the following function:

$$N = N_0 . e^{-\alpha . g(Y_i) . t} \tag{14}$$

Where N_0 is the initial number of receptors. Down regulation of GABA receptors in LGN (7) decreases the inhibitory effect of NRT on TR neurons. This phenomenon dominantly occurs in the magnocellular stream during transient visual attention (8). The results of computer simulation for our bursting model are shown in Fig. (1).

3 Discussion

This neurobiological modeling of bursting response has been described in this article, which its main part is studying the events related to Ca^{2+}-dependant K^+ channels. The basic concepts of our bursting model are extracted from Macgregor's model for single neurons (9).

NRT and thalamus have a basic role in producing bottom-up saliency map during pre-cueing visual attention task. When an abrupt onset cue is presented in the visual field, it makes an intensity change, which is detected by the retinal and thalamic cells. The corresponding spatial location in the NRT becomes salient by the competitive network. The magnification of the neuron is an index of saliency in NRT. The winner neuron in the competition sends a signal to corresponding thalamic neuron, which causes it to produce bursting response (very rhythmic and synchronous action potential discharges are generated during bursting phase (10)). As a result, cued location becomes salient in the thalamus like NRT. This saliency encodes to the cortex after NRT and thalamic computations. On the basis of Niebur et al's model (11), the synchronous thalamocortical oscillations of the neurons whose receptive fields overlap with the focus of attention suppress the response of other cortical neurons associated with non-attended visual stimuli.

Oscillatory behavior of bursting response has a biphasic response time pattern. Each period of this oscillation consists of two parts, which the saturation point of the response function (A) increases in one of them and decreases in another. The first part could be considered as *stimulus onset asynchrony (SOA)* and the second part as *inhibition of return (IOR)*. After cue presentation, A increases until reaches A_{up} . If target is presented at this time, the most facilitation for transformation of information (the best priming of attended location) is made because the firing rate of thalamic neurons is maximal. As a result, the performance will be enhanced for this SOA. A suitable SOA is needed for producing a good saliency map. Physiological studies have shown that the activation of Ca^{2+}-dependent K^{+} conductance elongates 200 msec (12). It means that the hyperpolarization must be elongated in order to have the bursting response. On the other hand, the best SOA for target detection is 200 msec confirmed by the psychophysical experiments (13). The time constant (e) in the model could be determined so that this SOA is satisfied. This is a neurobiological implication for SOA in our model.

SOA is different for cued targets presented at various distances from the fixation point. The further the target from fixation, the longer the SOA at which the reaction time function reaches asymptote (14). This phenomenon can not be absolutely explained by bottom up saliency map representation in thalamus because there is no difference among bursting patterns of thalamic neurons in response to signals of the cue presented at different spatial locations (15). It seems that another mechanism should exist besides the described mechanism for the SOA. The longer SOA for further cues from the fixation point could be due to the delay needed for the shifting of the attentional spot light. Cue causes corresponding thalamic neurons to produce bursting response by the bottom-up mechanism and leads attentional spot light to move to salient location by the top-down mechanism. The shifting of the spot light is analog (16,17). In other words, the spot light of attention, which is about 1 degree of visual angle in size (18) moves across the visual field in an analog fashion and travels through space at a constant velocity of about 1 degree per 8 msec (14) however there is also evidence that it can vary in size and velocity of movement. The size of attentional focus depends on the attentional load. Perhaps, when attending to a single item, the extra attentional resources would allow a further tightening of the attentional field (i.e., a finer resolution of attention) (19). We suggested the possible neuronal mechanism for the shifting of the spot light in the discussion part, which pulvinar has a critical role in this procedure.

The amount of the saturation point (A) decreases during the second part of the oscillation period so the thalamic neuron will not produce any response in the inter-spike interval. This is the inhibitory aftereffect of bursting response, which could be considered as the IOR. The performance of target detection will be poor when the target is presented at the time of IOR. As Jahnsen and Llinás showed before (2,3), the aftereffect of bursting response elongates 80-150 msec.

References

1) Llinás R., Jahnsen H. (1982). Electrophysiology of mammalian thalamic neurons. Nature (London), 297, 406-8.
2) Jahnsen H., Llinás R. (1984). Electrophysiological properties of guinea-pig thalamic neurons: an in vitro study. J Physiol, 349, 205-26.
3) Jahnsen H., Llinás R. (1984). Ionic basis for the electro-responsiveness and oscillatory properties of guinea-pig thalamic neurons in vitro. J Physiol, 349, 227-47.
4) Taylor J.G., Alavi F.N. (1995). A global competitive neural network. Biol Cybern, 72, 233-48.
5) Crick F. (1984). Function of the thalamic reticular complex: The searchlight hypothesis. Proc Natl Acad Sci, 81, 4586-90.
6) Izhikevich E.M. (2000). Neural excitability, spiking and bursting. Int J Bifur Chaos, 10, 1171-266.
7) Hendry S.H., Miller K.L. (1996). Selective expression and rapid regulation of GABAA receptor subunits in geniculocortical neurons of macaque dorsal lateral geniculate nucleus. Vis Neurosci, 13, 223-35.
8) Steinman B.A., Steinman S.B., Lehmkuhle S. (1997). Transient visual attention is dominated by the magnocellular stream. Vis Res, 37, 17-23.
9) MacGregor R.J. (1987). Neural and brain modeling. Academic Press, San Diego, California.
10) Bal T., McCormick D.A. (1993). Mechanisms of oscillatory activity in guinea-pig nucleus reticularis thalami in vitro: a mammalian pacemaker. J Physiol (London), 468, 669-91.
11) Niebur E., Koch C.,Rosin C. (1993). An oscillation-based model for the neuronal basis of attention. Vis Res, 33, 2789-802.
12) Sherman S.M., Koch C. (1986). The control of retinogeniculate transmission in the mammalian lateral geniculate nucleus. Exp Brain Res, 63, 1-20.
13) Nakayama K., Mackeben M. (1989). Sustained and transient components of focal visual attention. Vis Res, 29, 1631-47.
14) Tsal Y. (1983). Movements of attention across the visual field. J Exp Psychol Hum Percept Perform, 9, 523-30.
15) Hammond C. (1996). Cellular and molecular neurobiology. Academic Press, San Diego, California.
16) Chastain G. (1992). Time-course of sensitivity changes as attention shifts to an unpredictable location. J Gen Psychol, 119, 105-11.
17) Shepherd M., Muller HJ. (1989). Movement versus focusing of visual attention. Percept Psychophys, 46, 146-54.
18) Eriksen B.A., Eriksen C.W. (1974). Effects of noise letters upon the identification of a target letter in a nonsearch task. Percept Psychophys, 16, 143-9.
19) Intriligator J., Cavanagh P. The spatial resolution of visual attention. Cog Psychol, In Press.

Sensitivity of Simulated Striate Neurons to Cross-Like Stimuli Based on Disinhibitory Mechanism

Konstantin A. Saltykov, Igor A. Shevelev

Department of Sensory Physiology, Institute of Higher Nervous Activity and Neurophysiology, 117485 Moscow, Russia

Abstract. Sensitivity of simulated striate neurons to the crosses of different shape and orientation was studied in a two-layered neural network which consisted of formal neurons with end-stopping inhibitory and disinhibitory zones in the receptive field. We evaluated the output responses of neurons under variation of a cross size, shape, orientation. It was shown that disinhibitory mechanism could explain neuronal sensitivity to cross-like figures. The tuning of the simulated neurons calculated from their output responses reproduced the characteristic features of the natural units in the cat striate cortex. It was shown which combinations of the relative shape, localization and weight of the excitatory, end-stopping and disinhibitory zones of the simulated receptive field allow to imitate a selective or an invariant tuning to a cross-like figure of the real neurons in the cat primary visual cortex.

1 Introduction

Until recently, it was considered that neurons of the cat primary visual cortex (area 17) are tuned exceptionally for detection of orientation of contour elements. In our recent works we have found that more than 50% of these neurons show, on average, a 3-fold increase in response to an optimal cross-like, a corner or an Y-like figure of specific shape and orientation as compared with the preferred single bar flashing in the receptive field (RF) [7], [8], [9], [10]. Similar data have been obtained on stimulation of the central and peripheral parts of the RF with two orthogonal grids [11] .

Up to now it has been unknown what properties of neuronal net are responsible for sensitivity to the cross-like figures? These can be an excitatory convergence from orientation detectors of the previous functional level, inhibitory and disinhibitory intracortical interaction, and recurring excitation from the RF periphery. Some of these mechanisms were checked experimentally under conditions of local blockade of inhibition [2], [9] , but others are difficult to be controlled experimentally.

In the present study the sensitivity of neurons to a cross of different shape and orientation was simulated in a two-layered neural network, which consists of formal neurons with end-stopping inhibitory and disinhibitory zones in their RF. The aim of the study was to analyze the necessity and sufficiency of a disinhibitory mechanism for displaying the sensitivity to cross. Selective excitatory convergence and recurring excitation were simulated earlier [6].

J. Mira and A. Prieto (Eds.): IWANN 2001, LNCS 2084, pp. 81-86, 2001.

2 Methods

2.1 Description of the Model

The neural net was simulated using the "Neuroimitator" software ("Neuroma-Rd Ltd.", Russia). "Stimuli" were presented at the square input matrix (Fig. 1a, 1) consisting of 289 (17 x 17) formal neurons (from this point on shortening we shall denote them simply as neurons). This model level may be compared with the layer of ganglion cells of the retina or of the principal cells of the Lateral Geniculate Body. The elements of the input matrix did not possess orientation selectivity and were not directly connected. The scatter of neuronal characteristic parameters and their spontaneous activity was not introduced for the model simplification. The excitatory and inhibitory connections converged from the neurons of the first level to the cross-detector of the second level (Fig. 1a, 2).

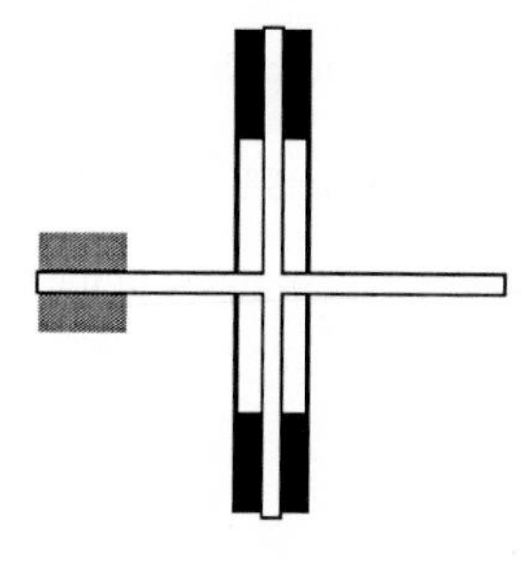

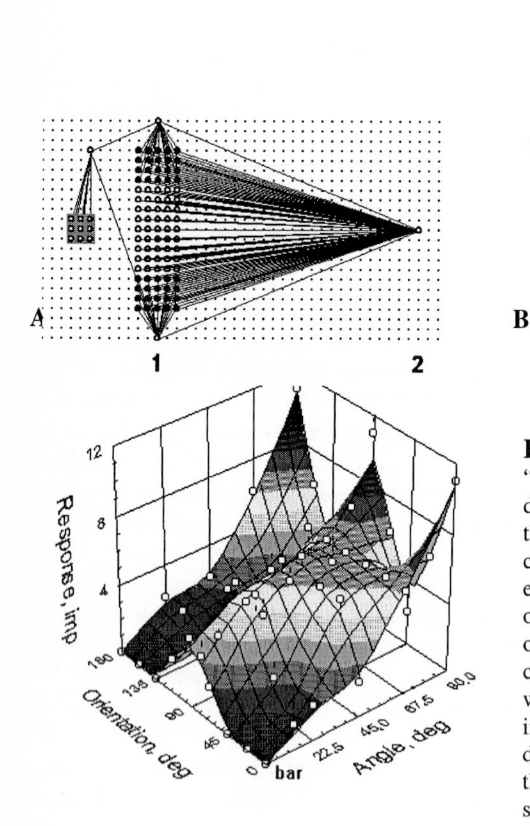

Fig. 1. (**A**) Neuronal net of a simulated "neuron" selective to orthogonal cross due to the disinhibitory mechanism. 1 – the first layer of a model (input matrix consisting from 289 neurons); 2 – an element of the second layer – a detector of the cross-like figure. (**B**) The scheme of the RF of this cross-detector. The RF consists of an excitatory zone shown in white, two zones of end-stopping inhibition (shown in black) and lateral disinhibitory zone (gray). The location in the RF of cross-like figure with optimal shape and and orientation is shown. (**C**) 3-D plot of the unit response to all possible combination of the cross shape (angle between its lines, deg.) and orientation (deg.). At zero angle between lines of the cross the conventional orientation tuning to a single bar can be seen.

2.2 Stimulation

The presentation of the light stimuli was imitated by activation of those cells of the input matrix (Fig. 1a) that form in total a centered bar or a cross. RF was tested by bars or crosses with eight standard orientations from 0° to 157.5° with the step of 22.5°. The angle between the lines of the cross was changed from 45° to 90° with step of 22.5°. Cross-like figure was composed by two identical single bars crossing at the center of the RF.

2.3 Measuring of the Responses

The number of impulses in the on-response was used for estimation of tuning of a "neuron" to the parameters of a bar or a cross-like figure. Three-dimensional plot of such tuning was analyzed, i.e., the dependence of the response value on the combination of all used orientations and shapes of the figure (Fig. 1C). The sensitivity index (SI) to a cross-like figure was estimated as the cross/bar response ratio for the optimal stimuli.

3 Experimental Results

3.1 Sensitivity to Cross Based on the Disinhibitory Mechanism

The RF of a neuron sensitive to cross includes a central excitatory zone, two end-stopping inhibitory zones and a lateral disinhibitory zone (Fig. 1B). The end-stopping inhibitory zones decrease the response to a bar and a cross via interneurons, and the disinhibitory zone blocks up these interneurons activating the detector of a cross. First, let us consider tuning of a detector with relatively small-sized end-stopping inhibitory zones and a small-sized disinhibitory zone. We call them as small if they inhibit or disinhibit the response to a stimulus of only one orientation. The response to a vertical bar must be partly suppressed by the end-stopping inhibition. Now, if the bar changed its orientation, rotating around its center by 22.5°-67.5° its ends will come out from the end-inhibitory zones, but the response will decrease still further as the activated part of the excitatory zone decreases. The response will be the least when the bar orientation is orthogonal to the length of the RF excitatory zone. As a result, the curve of the orientation tuning of the neuron to a bar will be uni-modal and sharp (Fig. 1C, "bar").

Stimulation of this RF by a direct cross of the optimal orientation (Fig. 1B) abruptly changes the response (Fig. 1C). At the vertical orientation of one of the cross lines the response would be suppressed by the end-stopping inhibition. In the meantime the second bar of the cross activates the lateral disinhibitory zone of the RF taking off the end-stopping inhibition. Owing to this the response increases as compared with that to a bar, and the sensitivity index (cross/bar response ratio) exceeds 2.0. That is higher than in the earlier considered detectors of cross with a simple excitatory convergence of signals from two bar-detecting neurons with different orientation preference [6].

In this neuron the sensitivity to cross is selective: it depends on a figure shape and orientation (Fig. 1C). Even a moderate (by 22.5°) turn around the center of the excitatory RF zone leads to a shift of the cross horizontal bar from the disinhibitory zone that sharply decreases the response. On the other hand, a change in the cross shape (angle between the bars) also leads to the response decrease in this RF, since at the vertical position of one of its bars the second bar do not get into the disinhibitory zone.

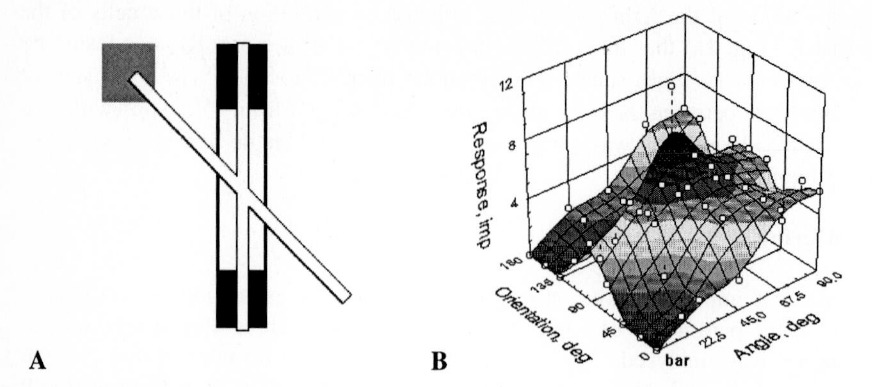

Fig. 2. (**A**) The scheme of the RF selective to the cross with an angle between its bars 45° (disinhibitory zone is shifted as compared with the scheme of Fig. 1B). (**B**) Tuning of the simulated neuron to a cross. The arrow indicates a position of the relief maximum.

The simulation showed that the position of a relatively small-sized disinhibitory zone in the RF defines the shape of the preferred cross. If the disinhibitory zone is shifted part way up or down from its middle position (Fig. 2A), the neuron becomes sensitive to the oblique cross with the angle between its bars from 22.5 to 67.5°(Fig. 2B). That means that the sharp tuning to a cross shape is defined by the position of the small disinhibitory zone in a certain place along the long axis of the excitatory RF zone, whereas the sharp tuning to cross orientation is established by the width of the end-stopping inhibitory zone.

3.2 Invariance to a Cross Shape and Orientation

Variation of the geometrical and weight parameters of the simulated RF zones showed that the wide excitatory and wide end-stopping inhibitory zones of the RF in combination with a high disinhibitory zone (Fig. 3A) provide for the invariance both to orientation and shape of a cross (Fig. 3B). The invariance appears here by the following reasons: i) the activated area of the RF excitatory zone is approximately identical under the action of crosses of different shape and orientation and ii) the end-inhibition and disinhibition are nearly equally expressed for practically all cross shapes and orientations.

4 Discussion and Conclusion

4.1 Functional Significance

Sensitivity of neurons of the cat primary visual cortex (area 17) to the second-order image features – line crossings and branches [7], [8], [9], [10] may be of a significant functional meaning. These features are the key attributes practically of all images and are essential for their recognition. Experimental data show that the detectionof these

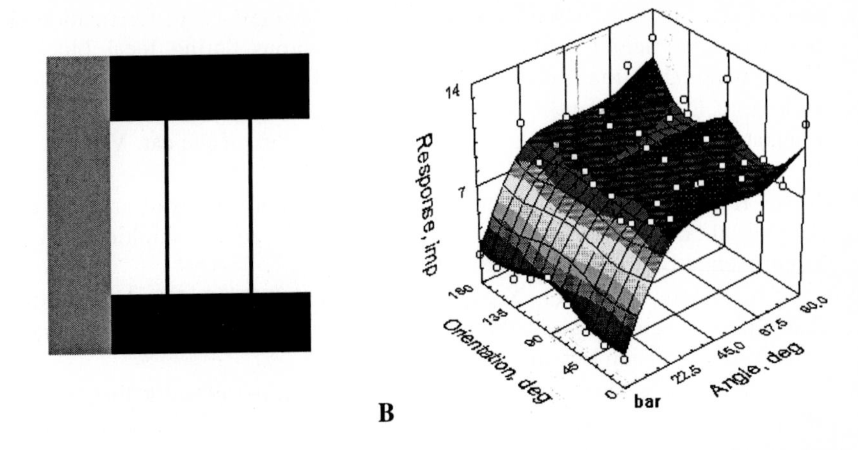

Fig. 3. (**A**) The scheme of the RF of the simulated neuron with relatively wide end-stopping zones of the RF (black) and long disinhibitory zone (gray). (**B**) Invariant tuning of this neuron to a cross shape and orientation.

features takes place at the early stage of cortical processing, i.e., in the primary visual cortex.

4.2 Network Mechanisms

We concluded earlier [6], [7], [9], [10] that the mechanism based on the convergence to the studied neuron of two cells of the previous functional level with different orientation preferences can explain the tuning to a cross only in 1/5 part of neurons with low sensitivity to a cross (cross/bar response ratio is lesser than 2.0).

The present study made it possible to characterize some network mechanisms responsible for sensitivity to a cross-like figure. The RF schemes with end-stopping inhibition and disinhibition seem to be natural. So, it is known that the presence of the end-stopping inhibition in the RFs of neurons of the visual cortex decreases the response to an optimally oriented bar of a sufficient length [1], [2], [4]. On the other hand, there are evidences of the presence of a disinhibitory mechanism outside the classical RF [3], [5] taking off the end-stopping and/or side inhibition [3].

It seems essential that in this highly simplified RF model, variation of the space characteristics of its zones reproduced all types of tuning to a cross-like figure: selective sensitivity to its shape and orientation [7], [8] and the invariance to one or both of these features [7], [10]. It should be stressed that the simulation of the sensitivity to crosses carried out in the present study may be considered only as one of the first approaches to the problem. These mechanisms need future experimental and simulation study.

References

1. DeAngelis, G.C., Freeman, R.D., & Ohzawa, I.: Length and width tuning of neurons in the cat's primary visual cortex, Vol. 71. J.Neurophysiology, (1994) 347-374

2. Eysel, U.T., Shevelev, I.A., Lazareva, N.A., & Sharaev, G.A.: Orientation tuning and receptive field structure in cat striate neurons during local blockade of intracortical inhibition, Vol. 84. Neuroscience, (1998) 25-36
3. Gilbert, C.D., & Wiesel, T.N.: The influence of contextual stimuli on the orientation selectivity of cells in primary visual cortex of the cat, Vol. 30. Vision Research, (1990) 1689-1701
4. Grieve, K.L., & Sillito, A.M.: The length summation properties of layer VI cells in the visual cortex and hypercomplex cell end zone inhibition, Vol. 84. Experimental Brain Research, (1991) 319-325
5. Li, C.Y., Zhou, Y.X., Pei, X., Qiu, F.T., Tang, C.Q., & Xu, X.Z.: Extensive disinhibitory region beyond the classical receptive field of cat retinal ganglion cells, Vol. 32. Vision Research, (1992) 219-228
6. Saltykov, K.A., & Shevelev, I.A.: A model of neuronal network for detection of single bars and cross-like figures, Vol. 29. Neuroscience and Behavioral Physiology, (1999) 385-392
7. Shevelev, I.A.: Second-order features extraction in the cat visual cortex: selective and invariant sensitivity of neurons to the shape and orientation of crosses and corners, Vol. 48. BioSystems, (1998) 195-204
8. Shevelev, I.A., Novikova, R.V., Lazareva, N.A., Tikhomirov, A.S., & Sharaev, G.A.: Sensitivity to cross-like figures in the cat striate neurons, Vol. 69. Neuroscience, (1995) 51-57
9. Shevelev, I.A., Jirmann, K.-U., Sharaev, G.A., & Eysel, U.T.: Contribution of GABAergic inhibition to sensitivity to cross-like figures in striate cortex, Vol. 9. NeuroReport, (1998a) 3153-3157
10. Shevelev, I.A., Lazareva, N.A., Sharaev, G.A., Novikova, R.V., & Tikhomirov, A.S.: Selective and invariant sensitivity to crosses and corners in cat striate neurons, Vol. 84. Neuroscience, (1998b) 713-721
11. Sillito, A.M., Grieve, K.L., Jones, H.E., Cudeiro, J., & Davis, J.: Visual cortical mechanisms detecting focal orientation discontinuities, Vol. 378 Nature, (1995) 492-496

Synchronisation Mechanisms in Neuronal Networks

Santi Chillemi[1], Michele Barbi[1], and Angelo Di Garbo[1, 2]

[1]Istituto di Biofisica CNR, Area della Ricerca di Pisa, via Alfieri 1, Ghezzano, Pisa (Italy)
chillemi@ib.pi.cnr.it, barbi@ib.pi.cnr.it, digarbo@ib.pi.cnr.it
[2]Istituto Nazionale di Ottica Applicata, Largo E. Fermi 6, Firenze (Italy)

Abstract. The synchronisation behaviour of a network of FitzHugh-Nagumo (FHN) oscillators are investigated. As the FHN units are meant to represent interneurons, only inhibitory chemical coupling is considered. Each unit receives synaptic-like input from a given fraction of the other units, randomly chosen. The synchronisation properties of the neuronal population are investigated against the coupling features (number of connections per neuron, synaptic current duration and coupling strength) and the network heterogeneity. Critical parameter values leading to the network synchronisation are found to exist.

1 Introduction

Many processes lead to oscillations of the membrane potential of neurons in the brain. Experimental evidence for these phenomena was found in the visual cortex [1], the olfactory system [2], the hippocampus [3] and the auditory system [4]. Moreover, there is increasing experimental evidence of the functional role, in normal or pathological condition, of the synchronisation phenomena occurring in a neuronal population.

The information coding during a cognitive function, like the perception of an object, requires that its component features (colour, motion, depth, etc.) are processed by separate parts of the cortex. An important question is how the brain rebinds the information on the individual features of the object into a well defined perception. Experimental results suggest that this binding process can be explained by assuming that the neurons, encoding for the different features of the visual scene, synchronise their firing activities [5]. How this synchronisation is achieved is not yet clear, however recent studies on the neocortex have shown that both chemical and electrical synapses (or gap junctions) play an important role in determining the coherence properties of neurons [6, 7]. Moreover, there is increasing evidence that interneurons contribute to brain rhythms by synchronising the firing activities of many pyramidal cells [8].

Epileptic seizures show that the synchronisation of a large neuronal population can also have pathological character. How this phenomenon occurs is still not clearly understood but experimental and theoretical results reveal that both chemical and electrical synapses contribute to the seizure generation and maintenance [9].

Some theoretical investigations have shown that chemical synaptic excitation usually desynchronises rather than synchronise the firing of mutually coupled neurons

J. Mira and A. Prieto (Eds.): IWANN 2001, LNCS 2084, pp. 87-94, 2001.

[10, 11, 12, 13]. On the other hand, in vitro studies suggest that synchronous oscillations can be generated by networks of inhibitory GABAergic interneurons [14, 15]. Therefore, it seems reasonable that chemical inhibitory synapses are critical to produce synchrony and to stabilise rhythmic activities in cortex as well as in hippocampus and in other brain areas.

In this paper we examine the synchronisation properties of a network of coupled FitzHugh-Nagumo (FHN) oscillators [16]. Each single unit is meant to represent a one-compartment model of an interneuron, and it receives inhibitory synaptic inputs from a subset of the other units, randomly assigned. The time course of the postsynaptic current is described by the so-called *alpha-function* [17, 20].

The studies of synchronisation phenomena in networks of inhibitory neurons, appeared in the literature, can be classified in two groups according to how precisely the single neuron and synaptic coupling are modelled. Examples of the first group, where complex biophysical models were used, are found in [10, 11, 19]. The reduced models, like the Integrate and Fire one, were used in [13]. The FHN model can be classified in between the two groups and, to our knowledge, this is the first work on a network of FHN units interacting through inhibitory synapses has been done.

We will investigate in particular how the coupling features (number of synapses formed by each neuron, synaptic current duration, coupling intensity) and the heterogeneity of the network, influence its synchronisation behaviour. Theoretical studies on networks of neural models with inhibitory coupling, suggest that a crucial role for the synchronisation is played by the synaptic time course [10, 11, 13]. Our results with FHN model confirm these findings and also demonstrate that network synchronisation requires a quite small average number of synaptic contacts.

2 Model Description

The FHN model is defined by the two first order differential equations:

$$\varepsilon \frac{dv}{dt} = v(v-a)(1-v) - w \tag{1a}$$

$$\frac{dw}{dt} = v - d\,w - b\,, \tag{1b}$$

where $v(t)$ is the fast voltage variable, $w(t)$ the slower recovery one, ε, a, d, b are the parameters. The periodic firing regime is reached, for the parameter values $\varepsilon = 0.005$, $a = 0.5$, $d = 1$ used here, as parameter b increases through $b_H = 0.264$ (supercritical Hopf bifurcation). In our simulations, the value $b = 0.265$ corresponding to suprathreshold oscillations will be used.

The network consists of N FHN units each of them receiving synaptic input from M ($\leq N$-1) other ones. The network equations are thus:

$$\varepsilon \frac{dv_i}{dt} = F(v_i, w_i) + \sigma C_i(t) + I_i, \qquad \frac{dw_i}{dt} = G(v_i, w_i) \tag{2a}$$

$$C_i(t) = \frac{\alpha^2}{p(N-1)} \sum_{j \neq i} \sum_{k=1}^{n_j} (t - t_{j,k}) e^{\alpha(t_{j,k} - t)} \tag{2b}$$

where the functions $F(v_i, w_i)$ and $G(v_i, w_i)$ are the expressions in the right hands of equations (1a) and (1b) respectively, $t_{j,k}$ are the times when the j-th oscillator fires and n_j is the index of its last firing, σ represents the strength of the chemical coupling, the factor $p = \alpha/e$ in equation (2b) makes the single postsynaptic potential of unity amplitude and α characterises its duration (more precisely, the time interval between the two passages at level $1/e$, is $\Gamma \cong 3\alpha^{-1}$). For any run, the M indices in the sum over j in equation (2b) are chosen uniformly from those of the others N-1 units and the bias currents I_i , introduced to differentiate the units from each other, are extracted from a gaussian distribution with mean $\langle I_i \rangle$ and standard deviation σ_I. Lastly, the firing condition is that v_j crosses from below a threshold set at 0.6.

To investigate the synchronisation behaviour, in amplitude, phase and frequency, of the network, different indicators can be used [17]. Here, we employ a synchronisation measure closely related to the experimental protocols, namely a measure based on the normalised crosscorrelation of the spike trains of all FHN pairs [18]. For any realisation of the connections these crosscorrelations are computed at zero time lag with a time bin $T_B = 0.05\langle T \rangle_S$, where $\langle T \rangle_S$ is the mean firing period of the network. Specifically, for any spike train we define a variable $x(j)$ (j = 1, 2,..., n_{bin}) as 1 or 0 depending on whether almost a spike is fired or not in the j-th bin. Then the coherence measure for two spike trains, i and j, is defined as

$$r_{ij} = \left[\sum_{j=1}^{n_{bin}} x(j) y(j)\right] \Big/ \sqrt{\sum_{j=1}^{n_{bin}} x(j) \sum_{j=1}^{n_{bin}} y(j)} \, . \tag{3}$$

For each run the synchronisation measure of the network, R_S, is obtained by averaging r_{ij} over all FHN pairs. The R values (and also those of the mean firing period $\langle T \rangle$) reported here were obtained by further averaging R_S over five random realisations of the network connections. R = 0 implies a complete incoherence between the firing times of the FHN units and R = 1 a complete synchronisation. For each unit the initial conditions were assigned randomly and belonging to a neighbourhood of the fixed point of the unperturbed FHN equations.

For the numerical integration of equations (2.2) a 4th-order Runge-Kutta method with integration step Δt = 0.003 was used. Simulations performed with Δt = 0.0015 did not exhibit qualitative changes. For all numerical simulations the total integration time was 180 time units. All quantities were computed after a transient time of 150 time units.

3 Results

Let us begin by investigating how the synchronisation measure R and $\Gamma / \langle T \rangle$ depend on the synaptic time course. To this purpose we plot both parameters against α in figure 1. Precisely, their values obtained using three different population sizes (N = 25, 50 and 100) are reported in each panel, while different panels refer to different values of the number M of synaptic connections per each neuron; these are, in percentage of the network size N: 28 % for the left panels, 60% for the middle panels, 80% for right panels, respectively.

Starting from the top panels, they show that there is a well defined window of

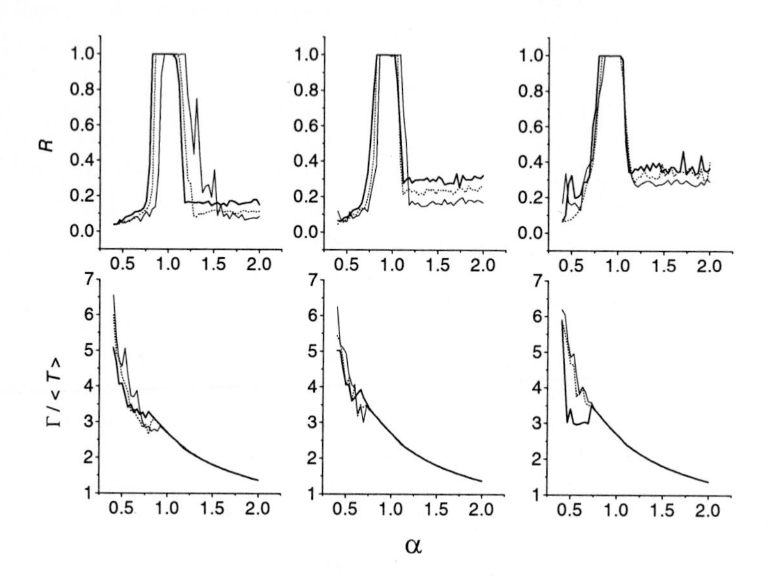

Fig.1 Plots of coherence index R (top panels) and of $\Gamma / \langle T \rangle$ (bottom panels) against α. In each panel the solid line refers to N = 25, the dotted line to N = 50, the bold solid line to N = 100. The M/N value (in percentage) is: left panels (28 %), middle panels (60%), right panels (80%). Strength of synaptic coupling σ = -0.0004. $I_i = 0$ for each FHN unit.

α values where the network gets a complete synchronisation state. Moreover, as the number of FHN units increases, each synchronisation window shifts to the left up to an apparent saturation. Also, with increasing the number of synaptic connections M, the shift of the synchronisation window decreases.

Looking now at the bottom panels of figure 1, we see that the synchronisation of the network occurs for those α values for which the ratio $\Gamma / \langle T \rangle$ falls approximately into the interval {2.5, 3}. To explain this result we must consider the dynamical behaviour of each FHN unit. Indeed, the firing activity of a unit exhibits two qualitatively different behaviours for $\Gamma / \langle T \rangle$ outside the synchronisation interval (data not shown). Typically, for $\Gamma / \langle T \rangle > 3$ the interspike interval of a single FHN exhibit small fluctuations and even suppression phenomena occur, whereas for $\Gamma / \langle T \rangle < 2.5$ it is constant. In the former case, and for a given realisation of the connections, the set of phase differences among the active units is distributed over a small fraction of $\langle T \rangle_S$. Instead for $\Gamma / \langle T \rangle < 2.5$ some pairs of units are almost in phase, but other pairs are in antiphase. These behaviours suggest that there is a critical region of α values where a matching condition between the involved time scales holds, what leads to the complete synchronisation of the network. Our results are in keeping with that found in [19, 21] using more realistic neuron models.

An important question to ask ourselves is the following: how does the synchronisation behaviour of the network depend on M and σ? Experimental results suggest that the synchronisation is turned off as soon as the inhibitory GABA synapses among the neurons are abolished [3]. Thus we should expect our network to synchronise only if the number of inhibitory synapses for neuron is above some threshold. Also, there should be a window of σ values for which the network gets the synchronous firing of its units. In fact, a weak coupling does not provide the FHN units with a synaptic input able to change their phases. On the other hand, a strong coupling may lead to the suppression of the firing activity of some units. In both cases we expect very low values of the synchronisation measure. With this predictions in mind, let us come now to the numerical results.

In the top panel of figure 2 the R values obtained from our simulations for three values of N are plotted against the percentage of synaptic contacts to each

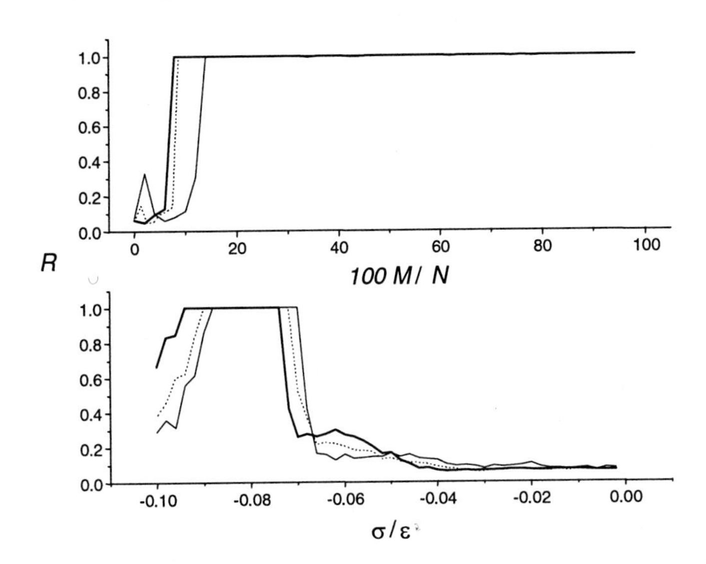

Fig.2 The coherence index is plotted against the percentage of synaptic contacts in the top panel and against the coupling strength measured in ε units in the bottom panel. For each panel the solid line refers to $N = 50$, the dotted line to $N = 80$ and the bold solid line to $N = 100$. For the plots in the top panel $\sigma = -0.0004$ and $\alpha = 0.94$, while for those in the bottom $\alpha = 0.94$ and $100(M/N) = 60$. $I_i = 0$ for each FHN unit.

FHN unit. Inspection of these plots shows that minimum percentages of about 10% are needed to get complete synchronisation. Moreover, as the size N of the network increases, the corresponding threshold values decrease and seem to saturate for higher values of N. Similarly, the bottom panel of figure 2 shows that, for a given percentage of contacts, there is a window of σ values where the network reaches a complete synchronisation. These findings, although qualitative, imply that the synchronisation phenomena occurring in real neuronal populations are an emergent phenomenon coming from the balance between network connectivity and coupling intensity and time scale of the synaptic process.

Up to now all currents I_i in equation (2a) were assumed to vanish. To get a more realistic description of a neuronal network we introduce heterogeneity in it. In general, one should expect that the network synchronisation cannot be maintained at all if the different FHN units have different firing frequencies. But let us inspect the panels of figure 3 where the coherence parameter is plotted for the three cases: $\langle I_i \rangle = 0$ and $\sigma_I = 0.000025$; $\langle I_i \rangle = 0.0002$ and $\sigma_I = 0.000025$; $\langle I_i \rangle = 0$ and $\sigma_I = 0.00001$.

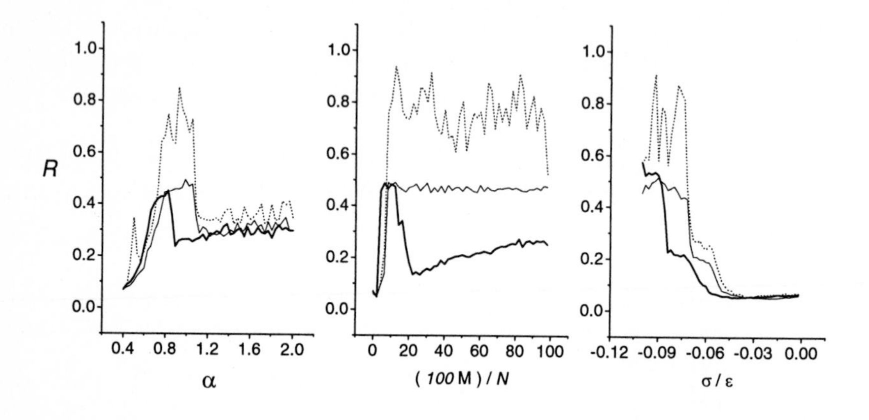

Fig. 3 Synchronisation measure after introducing the inhomogeneity in the network. Left panel: $\sigma = -0.0004$ and 60% of synaptic connections; middle panel $\sigma = -0.0004$ and $\alpha = 0.94$; right panel: $\alpha = 0.94$ and 60% of synaptic connections. For each panel the solid line corresponds to $\langle I_i \rangle=0$ and $\sigma_I = 0.000025$, the bold solid line to $\langle I_i \rangle=0.0002$ and $\sigma_I = 0.000025$ and the dotted line to $\langle I_i \rangle=0$ and $\sigma_I = 0.00001$. N=100 in all three panels.

Comparison of the synchronisation measures plotted in the left panel of figure 3 with the corresponding plots of figure 1, relative to an homogeneous network, shows that, as expected, the network coherence is reduced by the presence of heterogeneity. However, a window of α values where the coherence is higher exists. The width of this window is reduced by increasing the value of $\langle I_i \rangle$.

An interesting effect can be seen by examining, in the middle panel of figure 3, the synchronisation behaviour against the percentage of synaptic connections. Namely, while for $\langle I_i \rangle = 0$ and $\sigma_I = 0.000025$ a minimum number of synaptic contacts per neuron is still needed to get high values of R, just raising $\langle I_i \rangle$ to the value 0.0002 reduces dramatically the window of connectivity values with higher values of the network coherence.

The effects of changes of the coupling strength in the inhomogeneous network are reported in the right panel of figure 3. By comparing them with the corresponding ones in the bottom panel of figure 2 we see that for $\langle I_i \rangle = 0$ and $\sigma_I = 0.000025$ a window of ε values, yielding high coherence, still exists. However, as soon as $\langle I_i \rangle$ is increased to 0.0002, the window cannot anymore be individuated in the parameter range used. Finally, the results reported in figure 3 clearly show that by keeping $\langle I_i \rangle$ constant the synchronisation of the network increases as σ_I is decreased; however, the corresponding plots exhibit stronger fluctuations than those for $\langle I_i \rangle = 0$ and $\sigma_I =$ 0.000025. This behaviour is probably due to the fact that there is a lower balance between the effects of synaptic inputs and the heterogeneity. This circumstance may lead to dynamical states that exhibit high sensitivity to parameter changes.

4 Conclusions

As shown by recent experimental and theoretical studies, networks of inhibitory interneurons can synchronise their firing activities. The duration of inhibitory synaptic currents seems to be a critical parameter for the synchronisation to occur. Theoretically, the synchronisation phenomena have been investigated by using two kinds of neural units: detailed biophysical models and simpler operational ones. In this paper we investigated the synchronisation properties of a network of FHN units interacting by inhibitory synapses modelled by the *alpha-function*. The connections among neurons were assigned randomly.

In an homogeneous network ($I_i = 0$) synchronisation was found to occur for the α values for which the ratio $\Gamma / \langle T \rangle$ falls approximately within the interval {2.5, 3}. These results agree qualitatively with the ones obtained with more realistic neuron models [19, 21]. Moreover, we found that synchronisation requires that the percentage of synaptic connections among neurons is above a threshold value (of ~7% for large size neuronal populations). Also, as expected, the network inhomogeneity reduces the synchronisation level. However, for moderate heterogeneity, regimes of partial synchronisation are still possible. These findings, although qualitative, imply that coherence phenomena occurring in a real neuronal population are an emergent property coming from the balance between several parameters: degree of connectivity, coupling intensities and time scales involved in the synaptic process.

References

1. G. M. Gray and W. Singer, "Stimulus-specific neuronal oscillations in orientation columns of cat visual cortex", *Proc. Natl. Acad. Sci. USA,* vol. 86, pp. 1698-1702, 1989.
2. W. J. Freeman, "Measurement of oscillatory responses to electrical stimulation in olfactory bulb of cat", *J. Neurophys.*, vol. 35, pp. 762-779, 1972.
3. M. A. Whittington, R. D. Traub, J. G. R. Jefferys, "Synchronised oscillations in interneuron networks driven by metabotropic gluttamate receptor activation", *Nature*, vol. 373, pp. 612-615, 1995.
4. R. Galambos, S. Makeig, P. J. Talmachoff, "A 40 Hz auditory potential recorded from the human scalp" *Proc. Natl. Acad. Sci. USA,* vol. 78, pp. 2643-2647, 1981.

5. C. M. Gray, P. Koning, A.K. Engel and W. Singer, "Oscillatory responses in cat visual cortex exhibit inter-columnar synchronisation which reflects global stimulus properties", *Nature,* vol. 338, pp. 334-337, 1989.
6. M. Galarreta and S. Hestrin, "A network of fast-spiking cells in the cortex connected by electrical synapses", *Nature,* vol. 402, pp. 72-75, 1999.
7. J. R. Gibson, M. Belerlein, B. W. Connors, "Two networks of electrically coupled inhibitory neurons in neocortex", *Nature,* vol. 402, pp. 75-79, 1999.
8. M. Thomson, "Neurotrasmission: chemical and electrical interneuron coupling", *Current Biology,* vol. 10, pp. 110-112, 2000.
9. J. L. P. Velazquez and P. L. Carlen, "Gap junctions, synchrony and seizures", *Trend in Neuroscience,* vol. 23, pp. 68-74, 2000.
10. X.J. Wang, J. Rinzel, "Alternating and synchronous rhythms in reciprocally inhibitory model neurons", *Neural Comput.,* vol. 4, pp. 84-97, 1992.
11. X.J. Wang, J. Rinzel, "Spindle rhythmicity in the reticularis thalami nucleus: synchronisation among mutually inhibitory neurons", *Neuroscience,* vol. 53, pp. 899-904, 1993.
12. D. Hansel, G. Mato, C. Meunier, "Synchrony in excitatory neural networks", *Neural Comput.,* vol. 7, pp. 307-335, 1995.
13. C. van Vreeswijk, L.F. Abbott, G.B. Ermentrout, "When inhibition, not excitation synchronises neural firing", *J. Comput. Neurosci.,* vol. 1, pp. 313-322, 1995.
14. Fisahn, F.G. Pike, E.H. Buhl, O. Paulsen, "Cholinergic induction of network oscillations at 40 Hz in the hippocampus in vitro", *Nature,* vol. 394, pp. 186-189, 1998.
15. L.S. Benardo, "Recruitment of GABAergic inhibition and synchronisation of inhibitory interneurons in rat neocoterx", *J. Neurophysiol.,* vol. 77, pp. 3134-3144, 1997.
16. R. A. FitzHugh, "Impulses and physiological states in theoretical models of nerve membrane", *Biophys. J.*, vol. 1, pp. 445-466, 1961.
17. S. Chillemi, M. Barbi, A. Di Garbo, "Synchronisation in a network of FHN units with synaptic-like coupling", *Lecture Notes in Computer Science* **1606**:230-239 (1999).
18. W. P. Welsh, E. J. Lang, I. Sugihara, R. R. Llinàs, "Dynamic organisation of motor control within the olivocerebellar system", *Nature*, vol. **374**, 453-457 (1995)
19. X. J. Wang, G. Buzsaki, "Gamma oscillation by synaptic inhibition in a hippocampal interneuronal network model", *J. of Neuroscience*, vol. **16**, 6402-6413 (1996).
20. *Methods in neuronal modeling*, edited by C. Koch and I. Segev, MIT press (1998)
21. J. A. White, C. C. Chow, J. Ritt, J. Soto-Trevino, N. Kopell, "Synchronisation and oscillatory dynamics in heterogeneous, mutually inhibited neurons", *J. Comput. Neurosci.*, vol. **5**, 5-16 (1998).

Detection of Oriented Repetitive Alternating Patterns in Color Images
A Computational Model of Monkey Grating Cells

Tino Lourens, Hiroshi G. Okuno, and Hiroaki Kitano

Japan Science and Technology Corporation
ERATO, Kitano Symbiotic Systems Project
M. 31 Suite 6A, 6-31-15 Jingumae, Shibuya-ku, Tokyo 150-0001, Japan
{tino, okuno, kitano}@symbio.jst.go.jp

Abstract. In 1992 neurophysiologists [20] found a new type of cells in areas V1 and V2 of the monkey primary visual cortex, which they called grating cells. These cells respond vigorously to a grating pattern of appropriate orientation and periodicity. Three years later a computational model inspired by these findings was published [9]. The study of this paper is to create a grating cell operator that has similar response profiles as monkey grating cells have. Three different databases containing a total of 338 real world images of textures are applied to the new operator to get better a insight to which natural patterns grating cells respond. Based on these images, our findings are that grating cells respond best to repetitive alternating patterns of a specific orientation. These patterns are in common human made structures, like buildings, fabrics, and tiles.

1 Introduction

The accidental discovery of orientation selective cells, which they called *simple* and *complex cells*, in the primary visual cortex by Hubel and Wiesel [5,6] triggered a wave of research activities in neurophysiology. Activities where aimed at a quantitative description of the functional behavior of such cells (e.g., works form De Valois et al. [17,16,1], and von der Heydt [18]). Computational models of simple cells that contain linear filters followed by half-wave rectification are widely accepted, e.g., Movshon [13], Andrews and Pollen [2]. One way of modeling the responses of simple cells is to use Gabor filters [3,7]. Complex cells behave similarly as simple cells, but modeling these cells requires an additional step: spatial summation. Morrone and Burr [12] modeled this summation by taking the amplitude of the complex values of the simple cell operator.

Almost a decade ago, von der Heydt et al. [19,20] reported on the discovery of a new type of neuron in areas V1 and V2 of the monkey visual cortex, they called them *grating cells*, because of their strong responses to grating patterns, but weakly to bars and edges. They estimated that these cells makeup around 4 and 1.6 percent of the population of cells in V1 and V2, respectively, and that around 4 million grating cells of V1 subserve the center 4° of vision. The cells

J. Mira and A. Prieto (Eds.): IWANN 2001, LNCS 2084, pp. 95-107, 2001.

preferred spatial frequencies between 2.6 and 19 cycles per degree with tuning widths at half-amplitude between 0.4 and 1.4 octaves. They found that the cells are highly orientation selective and that the response is dependent on the number of cycles. A minimum of 2-6 cycles is required to evoke a response and leveled off at 4-14 cycles (median 7.5). In this study we propose a computational model for these grating cells that meet the response profiles of their measurements.

The paper is organized as follows: in Sect. 2 computational models of simple and complex cells are briefly introduced. These models are known from the literature, but since they form part of the grating cells they are included for completeness and clarity. A computational model of grating cells is given in Sect. 3. In the same section we will propose a chromatic sensitive class of grating cells, which is not found (yet) by neurophysiologists, but that is biologically plausible. The chromatic type of cells is based upon the color opponent theory and constructed in analogy with color sensitive orientation selective cells. Section 4 elaborates on the results of the model compared to the measured responses by applying the model to the same test patterns used by von der Heydt et al. In the same section the results of this model are compared with an existing model for grating cells. In Section 5 we apply the model to three databases to get better insights in the response of grating cells to real world images. The last section gives the conclusions.

2 Simple and Complex Cell Operators

The receptive field profiles of simple cells can be modeled by complex-valued Gabor functions:

$$\hat{G}_{\sigma,\lambda,\gamma,\theta}(x,y) = \exp\left(i\frac{\pi x_1}{\sqrt{2}\sigma\lambda}\right)\exp\left(-\frac{x_1^2+\gamma^2 y_1^2}{2\sigma^2}\right) , \quad (1)$$

where $x_1 = x\cos\theta + y\sin\theta$ and $y_1 = y\cos\theta - x\sin\theta$. Parameters σ, λ, γ, and θ represent scale, wavelength ($\frac{2}{\sqrt{2}\sigma\lambda}$ is the spatial frequency), spatial aspect ratio, and orientation, respectively. These Gabor functions have been modified such that their integral vanishes and their one-norm (the integral over the absolute value) becomes independent of σ, resulting in $G_{\sigma,\lambda,\gamma,\theta}(x,y)$, for details we refer to Lourens [11]. They provide a transform of the image $I(x,y)$ via spatial convolution. Afterwards, only the amplitudes of the complex values are retained for further processing:

$$\mathcal{C}_{\sigma,\lambda,\gamma,\theta} = ||I * G_{\sigma,\lambda,\gamma,\theta}|| \ . \quad (2)$$

This representation, which models the responses of complex cells, is the basis of all subsequent processing. A high value at a certain combination of (x,y) and θ represents evidence for a contour element in the direction orthogonal to θ. Orientations are sampled linearly $\theta_i = \frac{i\cdot 180}{N}, i = 0\dots N-1$, and the scales are sampled $\sigma_j = \sigma_{j-2} + \sigma_{j-1}$, for $j = 2\dots S-1$, where σ_0 and σ_1 are known constants.

3 Grating Cells

Von der Heydt et al. [19] proposed a model of grating cells in which the activities of displaced semi-linear units of the simple cell type are combined by a minimum operator to produce grating cell responses. This model responds properly to gratings of appropriate orientation and to single bars and edges. However, the model does not account for correct spatial frequency tuning (the model also responds to all multiples of the preferred frequency).

Kruizinga and Petkov [9,14] proposed a model based on simple cell (with symmetrical receptive fields) input. Responses of simple cells are evaluated along a line segment by a maximum operator M. A quantity q, which is 0 or 1, is used to compensate for contrast differences and a point spread function is used to meet the spatial summing properties with respect to the number of bars and length. Except for the spatial summing properties this model does not meet the other criteria of the grating cells. Also it does not always account for correct spatial frequency tuning; it is unable to discriminate between alternating light and dark bars of preferred frequency and alternating bars of about three times this preferred frequency.

In this paper we propose a grating cell operator that meets both the response profiles of the grating cells and appropriate frequency tuning. This operator uses complex cell input responses, modeled by the operator given in (2). Likewise as in the other models, we evaluate the responses along a line segment perpendicular to the length of the bars of a grating. Unlike the other models, we let the variable length depend on the similarity of the responses of the complex cells, but the length is at least $2B_{\mathrm{min}}$ and at most $2B_{\mathrm{max}}$ bars. Instead of a maximum operator we use an averaging operator. No contrast normalization mechanism is incorporated in the grating operator, since we believe that this is compensated for at the stage of the center-surround cells already, see, e.g., Kaplan and Shapley [8]. This implies that the input data is already normalized for contrast.

3.1 Grating Cell Operator

The initial grating response is calculated as follows:

$$\mathbf{G}^{\mathrm{avg}}_{\sigma,\lambda,\gamma,\theta,l}(x,y) = \frac{\rho}{2l+1} \sum_{i=-l}^{l} \mathcal{C}_{\sigma,\lambda,\gamma,\theta}(x+x_i, y+y_i) \ , \tag{3}$$

where ρ is a response decrease factor. This factor is a measure for the deviation from the optimal frequency and uniformity of the complex cell responses. This factor will be discussed below. Parameter l denotes the length over which summation of the complex cell responses will take place.

The variable length $2l$ over which the responses of the complex cells will be evaluated is between the minimum number of bars $2B_{\mathrm{min}}$ and maximum number of bars $2B_{\mathrm{max}}$. Since the operations are performed on a discrete grid we decompose the maximum length from the center of the line segment, in x- and y-direction:

$$l_x = \frac{B_{\max}\sqrt{2}\sigma\cos\theta}{\lambda} \quad \text{and} \quad l_y = \frac{B_{\max}\sqrt{2}\sigma\sin\theta}{\lambda} \; . \tag{4}$$

Similarly, we decompose the minimum length ($B_{\min}$) into m_x and m_y. The preferred bar width (in pixels) equals $\sigma\sqrt{2}\lambda$.

Depending on the preferred orientation θ, the evaluation will take place in x- or y-direction. Hence, parameters x_i, y_i, $l_{\min}$ and $l_{\max}$ are orientation dependent:

$$\begin{array}{l} \text{if } \left|\frac{l_y}{l_x}\right| \leq 1 \text{ and } l_x \neq 0 \\ \quad x_i = i; \;\; y_i = \lfloor i\frac{l_y}{l_x} + 0.5\rfloor; \;\; l_{\max} = |\lfloor l_x + 0.5\rfloor|; \;\; l_{\min} = |\lfloor m_x + 0.5\rfloor| \\ \text{else} \\ \quad x_i = \lfloor i\frac{l_x}{l_y} + 0.5\rfloor; \;\; y_i = i; \;\; l_{\max} = |\lfloor l_y + 0.5\rfloor|; \;\; l_{\min} = |\lfloor m_y + 0.5\rfloor| \end{array} \tag{5}$$

where $\lfloor x\rfloor$ denotes the nearest integer value smaller than or equal to x. Length l of (3) is determined by the maximum, minimum, and average response of the complex cells along the line:

$$\begin{array}{l} l = \min_i(l_i), \;\; l_{\min} < i \leq l_{\max}, \;\; i \in \mathbb{Z} \text{ and} \\ \text{if } \frac{\mathbf{G}^{\max}_{\sigma,\lambda,\gamma,\theta,i}(x,y) - \mathbf{G}^{\text{avg}}_{\sigma,\lambda,\gamma,\theta,i}(x,y)}{\mathbf{G}^{\text{avg}}_{\sigma,\lambda,\gamma,\theta,i}(x,y)} \geq \Delta \text{ or } \frac{\mathbf{G}^{\text{avg}}_{\sigma,\lambda,\gamma,\theta,i}(x,y) - \mathbf{G}^{\min}_{\sigma,\lambda,\gamma,\theta,i}(x,y)}{\mathbf{G}^{\text{avg}}_{\sigma,\lambda,\gamma,\theta,i}(x,y)} \geq \Delta \\ \quad l_i = i - 1 \\ \text{else} \\ \quad l_i = l_{\max} \end{array} \tag{6}$$

Constant Δ, which is a uniformity measure, is a value larger than but near 0. We used $\Delta = 0.25$ in all our experiments. The maximum and minimum $\mathbf{G}$ responses are obtained as follows

$$\mathbf{G}^{\Omega}_{\sigma,\lambda,\gamma,\theta,l}(x,y) = \Omega^{l}_{i=-l}\left(\mathcal{C}_{\sigma,\lambda,\gamma,\theta}(x + x_i, y + y_i)\right) \; , \tag{7}$$

where Ω denotes the min or max operator.

The determination of length l depends on the uniformity of responses of the complex cells along a line perpendicular to orientation θ. If the responses of these cells are not uniform enough the summation will be shorter than $l_{\max}$ and consequently the responses will be less strong. We model a linearly increasing response between $B_{\min}$ and $B_{\max}$:

$$\rho_l = \frac{\frac{l}{l_{\max}}B_{\max} - B_{\min}}{B_{\max} - B_{\min}} = \frac{l - l_{\min}}{l_{\max} - l_{\min}} \; . \tag{8}$$

The modeled response also depends on the uniformity of the complex cell responses. Since ρ_l gives a strong decrease for short lengths (it equals 0 for $l_{\min}$), we do not decrease the response for a length between the minimum number of bars and the minimum number plus one:

$$\text{if } (l_s \leq l_{\min}) \;\; \rho_u = 1 \;\; \text{else} \;\; \rho_u = 1 - \frac{\mathbf{G}^{\max}_{\sigma,\lambda,\gamma,\theta,l_s}(x,y) - \mathbf{G}^{\min}_{\sigma,\lambda,\gamma,\theta,l_s}(x,y)}{2\Delta\mathbf{G}^{\text{avg}}_{\sigma,\lambda,\gamma,\theta,l_s}(x,y)} \tag{9}$$

where $l_s = l - \frac{l_{\max}}{1+\lfloor B_{\max}\rfloor}$ is the length that is one bar shorter in length than l. The evaluation on a slightly shorter length ensures that both criteria in (6) are less than Δ, which implies that $\rho_u \geq 0$. Multiplying factors ρ_l and ρ_u results in the response decrease factor $\rho = \rho_l \rho_u$ from (3).

A weighted summation is made to model the spatial summation properties of grating cells with respect to the number of bars and their length and yields the grating cell response:

$$\mathcal{G}_{\sigma,\lambda,\gamma,\theta,\beta} = \mathbf{G}^{\mathrm{avg}}_{\sigma,\lambda,\gamma,\theta,l} * G\text{ß}_{\frac{\sigma\beta}{\lambda}} \ , \tag{10}$$

where Gß is a two-dimensional Gaussian function. In combination with $B_{\min}$ and $B_{\max}$, parameter β determines the size of the area over which summation takes place. Parameters $B_{\min} = 0.5$ and $B_{\max} = 2.5$ together with a β between 2 and 4 yield good approximations of the spatial summation properties of grating cells. In the experiments we used $\beta = 3$.

3.2 Color Opponent Type of Grating Cells

No evidence is given by von der Heydt et al. [20] that there exist color-opponent grating cells. However, it can be made plausible that such type of cells exist. *Opponent* color-sensitive cells are found at the first levels of processing after the photo receptors. Some of them are orientation selective and have an elongated area which give *excitatory* responses to one color and *inhibitory* responses at one or two flanks to another (opposite) color. These opponent color pairs are red-green and blue-yellow or vice versa [10].

Already in the 1960s, Hubel and Wiesel found cells that responded to edges of a specific orientation, but only recently there is evidence that there are cells that are color and orientation selective. Cells of this type are found in area V4 [22] and in area V2 where 21-43% of the neurons have these properties [4,15].

Color-opponent complex cell operators (Würtz and Lourens [21]) are modeled as follows

$$\mathcal{C}^{e,i}_{\sigma,\lambda,\gamma,\theta} = \left|\left|I^e * G_{\sigma,\lambda,\gamma,\theta} - I^i * G_{\sigma,\lambda,\gamma,\theta}\right|\right| \ , \tag{11}$$

where e, i denotes a "red-green" or "blue-yellow" color opponent pair[1] red-green or blue-yellow (yellow=(red+green)/2) and I^a denotes the color channel a from an (r, g, b) image I. In analogy to achromatic grating cell operator of (10), the color-opponent grating cell operator is

$$\mathcal{G}^{\tau}_{\sigma,\lambda,\gamma,\theta,\beta} = \mathbf{G}^{\mathrm{avg},\tau}_{\sigma,\lambda,\gamma,\theta,l} * G\text{ß}_{\frac{\sigma\beta}{\lambda}} \ , \tag{12}$$

where τ is color opponent pair "red-green" or "blue-yellow". The initial grating response from (3) should be modified to

$$\mathbf{G}^{\mathrm{avg},\tau}_{\sigma,\lambda,\gamma,\theta,l}(x,y) = \frac{\rho^{\tau}}{2l^{\tau}+1} \sum_{i=-l^{\tau}}^{l^{\tau}} \mathcal{C}^{\tau}_{\sigma,\lambda,\gamma,\theta}(x+x_i, y+y_i) \ , \tag{13}$$

[1] The order of a color pair is arbitrary since $\mathcal{C}^{e,i}_{\sigma,\lambda,\gamma,\theta} = \mathcal{C}^{i,e}_{\sigma,\lambda,\gamma,\theta}$.

where l^τ and ρ^τ are evaluated as before, but $\mathbf{G}^{\mathrm{avg},\tau}$ should be used instead of $\mathbf{G}^{\mathrm{avg}}$. Similarly, the maximum and minimum $\mathbf{G}$ operators are evaluated by using antagonistic color channels.

The orientations of the grating cells are combined with an amplitude operator (for both $\mathcal{G}$ and $\mathcal{G}^\tau$)

$$\mathcal{G}\mathrm{all}_{\sigma,\lambda,\gamma,\beta} = \sqrt{\sum_{i=0}^{N-1} \left(\mathcal{G}_{\sigma,\lambda,\gamma,\theta_i,\beta}\right)^2} \ , \tag{14}$$

where N denotes the number of orientations and $\theta_i = i\pi/N$. The achromatic and two color opponent (red-green and blue-yellow) channels are combined by taking the amplitudes to yield the final grating operator

$$\mathcal{G}\mathrm{all}_{\sigma,\lambda,\gamma,\beta} = \sqrt{\left(\mathcal{G}_{\sigma,\lambda,\gamma,\beta}\right)^2 + \left(\frac{1}{2}\mathcal{G}^{r,g}_{\sigma,\lambda,\gamma,\beta}\right)^2 + \left(\frac{1}{2}\mathcal{G}^{b,y}_{\sigma,\lambda,\gamma,\beta}\right)^2} \tag{15}$$

at a single scale. Since $\mathcal{C}^{e,i}_\sigma = \mathcal{C}^{i,e}_\sigma$, and hence also $\mathcal{G}^{e,i}_{\sigma,\lambda,\gamma,\theta,\beta} = \mathcal{G}^{i,e}_{\sigma,\lambda,\gamma,\theta,\beta}$, one channel for every opponent pair is sufficient.

4 Properties of Grating Cells

Von der Heydt et al. [20] describe responses to different synthetic grating patterns. In this section the properties of our grating cell operator are evaluated for different settings of parameters λ and γ. The results of this operator for different settings are compared with the measured data and the response properties of the model of Kruizinga and Petkov [9,14].

Von der Heydt et al. performed different tests to obtain the properties of periodic pattern selective cells in the monkey visual cortex. In the first test they revealed the *spatial frequency* and *orientation* tuning of the grating cells. From the second test they obtained the response properties to an increasing number of *cycles* of square-wave gratings. Their third test described the response properties for *checkerboard* patterns by varying the check sizes. The fourth experiment tested the responses to so-called "Stresemann" patterns. These patterns are gratings where every other bar is *displaced* by a fraction of a cycle.

They also tested the responses to contrast. The contrast profiles of the magno and parvo cells [8] show similarities with the profiles given by von der Heydt et al. [20]. In our experiments we will assume that contrast normalization on the input data took place by means of these magno and parvo cells, i.e. contrast normalization is applied to the input image, already. Hence, the test for contrast responses will be omitted in this study.

Von der Heydt et al. also found the so-called *end-stopped grating* cells which respond only at the end of a grating pattern. In this study this type of grating will not be considered.

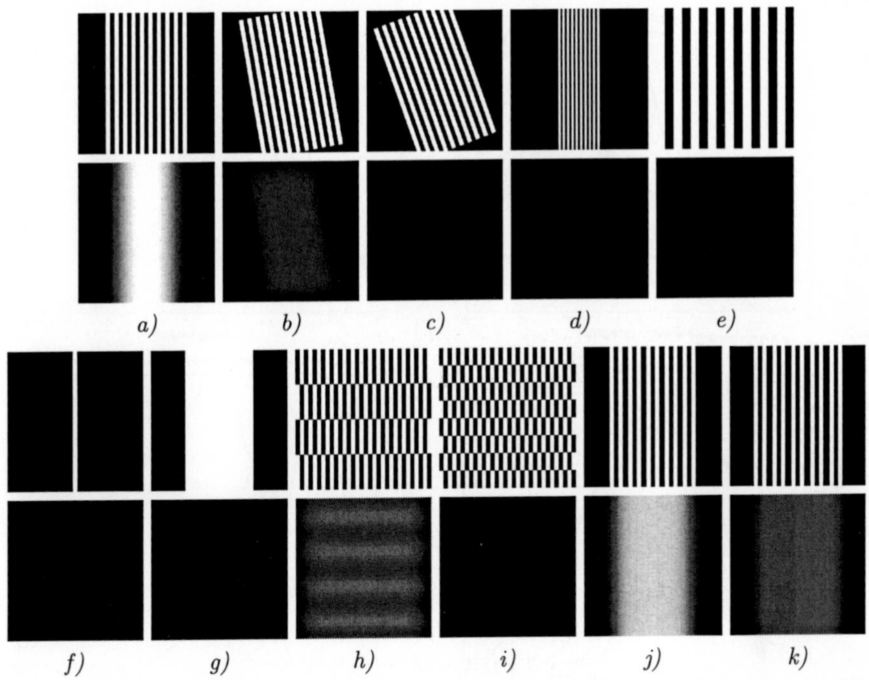

Fig. 1. Responses to square gratings with different orientations and frequencies. Top row gives the stimulus and bottom row the responses of the modeled grating operator. *a)* Grating cells respond vigorously to grating patterns of preferred orientation and frequency. Responses decrease when the pattern differs from this pattern. *b)* and *c)* Responses strongly decrease if the gratings are rotated slightly (10 degrees) and completely vanish at 20 degrees. *d)* and *e)* Doubling or halving the frequency results in zero responses. *f)* and *g)* Grating cells hardly respond to single bars or edges. *h)* and *i)* Increasing the checks in a grating pattern results in a response decrease. *j)* and *k)* Stresemann patterns show similar behavior as in h) and i): the stronger the deviation from a) the weaker its response. The used parameters are $\lambda = 1.00$, $\gamma = 0.25$, $\beta = 3.00$, $B_{\min} = 0.5$, and $B_{\max} = 2.5$.

4.1 Responses to Test Patterns

Figure 1 illustrates that the modeled grating cell operator of (10) shows similar behavior compared to the measurements carried out by von der Heydt et al. [20]. Grating cells respond vigorously to grating patterns of preferred orientation and frequency, but any deviation from this pattern results in a decrease in response.

4.2 Orientation and Frequency Profiles

In this section the properties of grating cells will be modeled as accurate as possible by tuning parameters λ and γ to yield similar responses as measured by von der Heydt et al. [20]. In this paper these profiles are denoted with "vdH ..." in the legends of the figures.

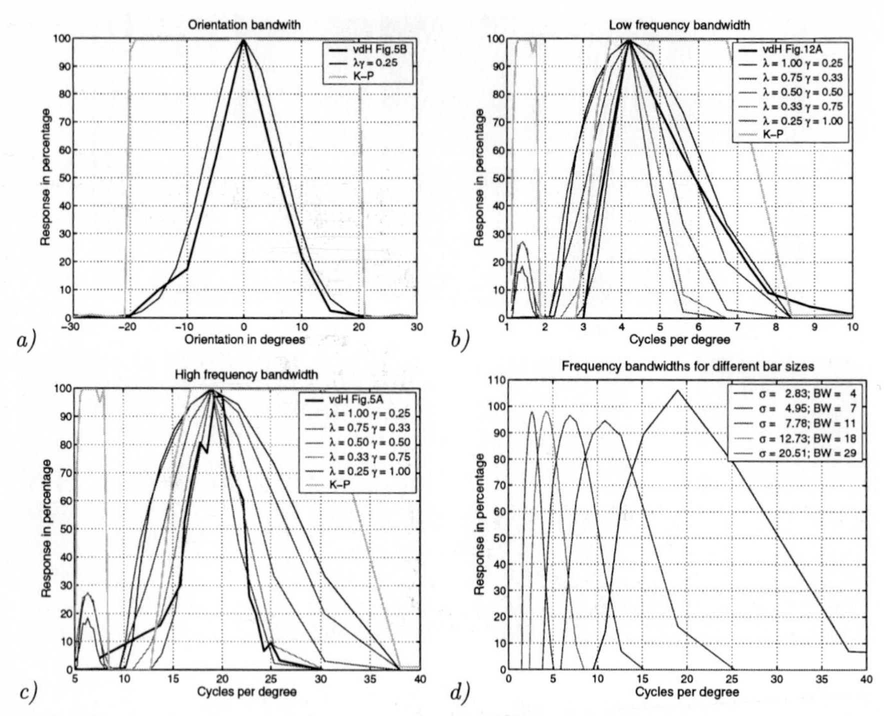

Fig. 2. Measured and modeled response profiles of grating cells. *a)* Orientation, *b)* low frequency, and *c)* high frequency profiles. *d)* Frequency profiles for different preferred bar width sizes (BW = 4, 7, 11, 18, and 29 pixels). Cycles per degree have arbitrary units. Parameters used for d are $\lambda = 1$ and $\gamma = 0.25$.

The orientation bandwidth for grating cells is small, von der Heydt et al. [20] found that the half-maximum responses are at $\pm 6°$. In our model this orientation tuning corresponds approximately to $\lambda\gamma = 0.25$ (Fig. 2a).

Von der Heydt et al. found grating cells with both low and high frequency bandwidth. The response curves of these cells (Fig. 2b and c) are different. Hence, it might be appropriate to model multiple grating cell operators to cover the main bulk of grating cell responses. The modeled grating operators have a smaller bandwidth than the measured grating responses for the low frequency sensitive grating cell, while the same operators (with preferences for higher frequencies) show bandwidths that are similar or slightly larger than the measured responses of grating cells with high frequency preferences. Most appropriate for the low frequency is the model with parameters $\lambda = 1.00$ and $\gamma = 0.25$ while for the high frequency sensitive grating cells $\lambda = 0.33$ and $\gamma = 0.75$ seems a good choice.

However, a problem that occurs for models with γ larger than approximately 0.4 is the response to frequencies that are about a factor 3 larger than the preferred frequency (Fig. 2b and c). If we assume that the preferred bar width is 8 pixels than all bars with a width between 5 and 14 pixels have a response that is stronger than 10 percent of the maximum response ($\lambda = 1$). For $\lambda = 0.33$, bars

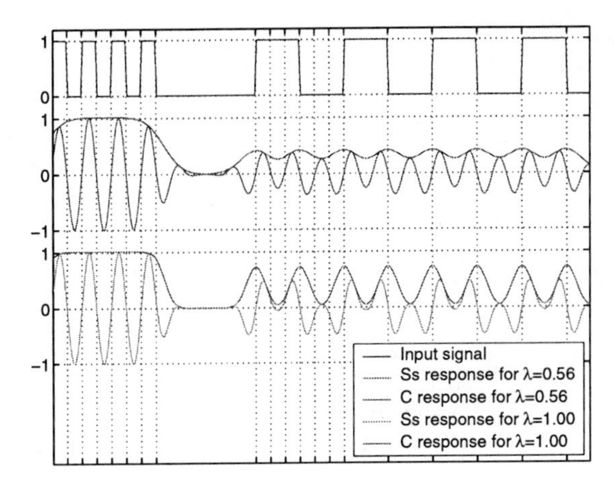

Fig. 3. Input signal (top). Response profile for symmetrical simple and complex cell operators for $\lambda = 0.56$ (middle) and $\lambda = 1.00$ (bottom).

with widths from 6 to 12 and from 20 to 29 pixels have a response stronger than 10 percent of the maximum response. Hence, we suggest to use $\lambda = 1.00$ and $\gamma = 0.25$ for the high frequency sensitive cells also, even though the responses in the second interval never exceed 30 percent of the maximum response.

The frequencies (in cycles per degree) are arbitrary units for the model, since the frequency is determined by the size of the image and the distance of the observer to the image. Hence, modeled grating cells with a preferred bar width of 8 pixels can correspond to both 4.2 and 19 cycles per degree. However, when the preferred bar width is small differences occur in the profile. This is due to the discretization properties of the Gabor filters. Figure 2d illustrates that when the preferred with is 4 pixels the maximum response is over 100 percent. Also, in such case, only 7 measurements can be performed, since it only responds to bars with a width between 1 and 7 pixels.

Figure 2d illustrates the bandwidths for different preferred bar widths, respectively 4, 7, 11, 18, and 29 pixels. The left figure illustrates that these 5 "scales" cover the full range of preferred frequencies (2.6 to 19 cycles per degree) found by von der Heydt et al. If a preferred bar width of 4 pixels is equivalent to 19 cycles per degree than 2.6 cycles per degree corresponds to a bar width of $4 \times 19/2.6 = 29.2$ pixels. The use of these five scales covers the full range well, since the lowest response, between two preferred frequencies, drops at most 25 percent from the maximum response.

The grating cell operator of Kruizinga and Petkov, is available online (`http://www.cs.rug.nl/users/imaging/grcop.html`) and was used with a *bandwidth* of 1.0 and a *periodicity* that equals two times the preferred bar width. The response profile (in our figures denoted by "K-P") of this grating operator shows globally two states: inactive or vigorously firing, which is caused by their normalization quantity q. The choice of $\lambda = 0.56$ gives strong responses to two intervals, which is already caused by the simple cell operators. The interval with

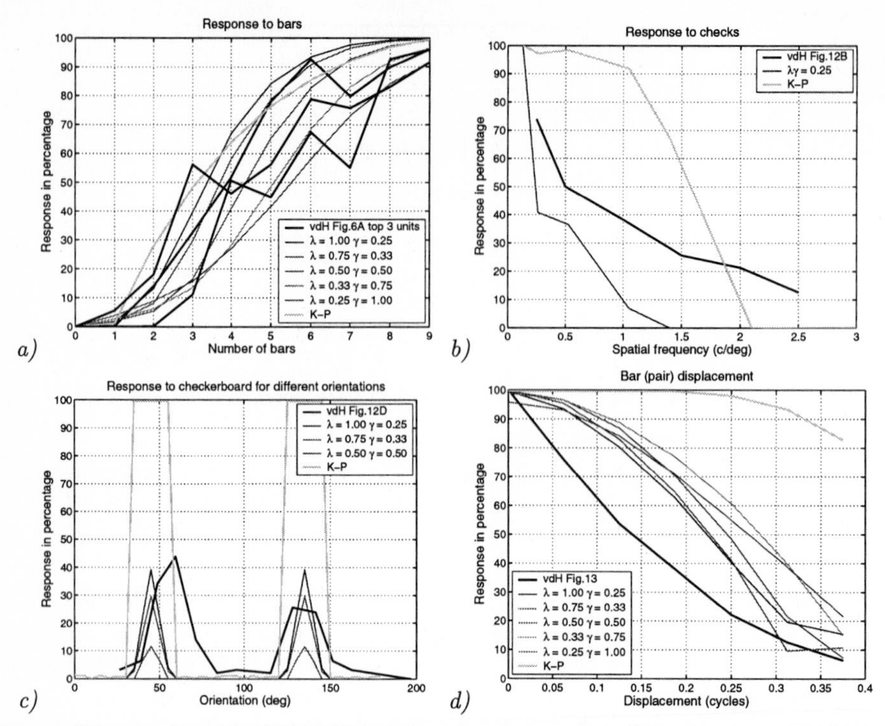

Fig. 4. Measured and modeled properties of grating cells. *a)* Response profile for increasing number of bars. *b)* Response to checks. *c)* Sensitivity to different orientations of a checkerboard pattern. *d)* Responses to increasing shift of a pair of bars, the so-called Stresemann pattern.

the highest frequency (5-11 pixels if the preferred bar width is 8) responds to bars indeed, while the simple cell operator responds to an up and down going edge of a bar in the low frequency interval (19-29 pixels). In the middle of such a low frequency bar there is still some response from both edges, causing a response profile that is similar to that of gratings with preferred frequency. An illustration of this behavior is given for a one-dimensional signal in Fig. 3. This behavior does not occur for $\lambda = 1.00$, as is illustrated in the same figure.

4.3 Profiles for Different Textures

Figure 4a illustrates that the measured results of the grating cells show increasing response with increasing number of bars. In the same figure only one modeled curve is shown, and although modeled grating cells with different parameters show different response curves they all are similar to the responses of the measured cells.

The modeled cells are not as robust to checks as the measured cells as illustrated in Fig. 4b. On the contrary the modeled cells are slightly less sensitive to shifts of bars (Fig. 4d). The responses to different orientations depends on the orientation bandwidth. If $\lambda\gamma = 0.25$ the orientation bandwidth is similar to that of a measured grating cell, but its response to a checkerboard pattern is low (about three times less) compared to that of the measured grating cells. On the other hand when $\lambda\gamma = 0.35$ the responses are comparable, but in this case the orientation bandwidth is wider than that of the measured cell.

5 Oriented Repetitive Alternating Patterns

It is clear that the model for grating cells responds to grating patterns, but the question that rises is to what kind of real world patterns these cells respond. The latter is important will the operator be successfully applied in an artificial vision system. We used three (Brodatz, ColumbiaUtrecht, and the VisTex) freely available databases containing different textures.

The Brodatz database contains 111 images D1 to D112, where D14 is missing. We cut the central part (512×512 pixels) from the grey scale images that are sized 640×640. The ColumbiaUtrecht database contains 61 images sample01 to sample61. Here we used the central part (256×256 pixels) from the color images that are sized 640×480. In the VisTex database we used 166, 512 square sized color images.

The grating cell operator was applied for $N = 16$ orientations, this number is necessary since the half maximum response of the grating cells is at $\pm 6°$, see also Fig. 2a. We combined all scales with a maximum operator:

$$\mathcal{G}\mathrm{all}_{\lambda,\gamma,\beta} = \max_{i=0}^{S-1} \mathcal{G}\mathrm{all}_{\sigma_i,\lambda,\gamma,\beta}, \tag{16}$$

where $S = 5$ is the number of scales and $\sigma_0 = 4\lambda/\sqrt{2}$, $\sigma_1 = 7\lambda/\sqrt{2}$, and $\sigma_j = \sigma_{j-2}+\sigma_{j-1}$ for $j \geq 2$ (as given earlier). The minimum bar width at σ_0 is 4, since smaller widths lead to strong inaccuracy due to the discreteness of the Gabor filters.

The grating cell operator is very selective and responded in only five (samples 38, 46, 49, 51, and 57) images in the ColumbiaUtrecht database. The operator responded in 32 images of the Brodatz database. In the VisTex database the operator responded to three (buildings, fabric, and tile) out of 18 categories and within the categories it responded to about half of the images.

Based on the results from the three databases we conclude that grating cells respond well to man-made objects that have oriented alternating repetitive patterns. A few of these examples are illustrated in Fig. 5.

6 Conclusions

We presented a new model for grating cells that has similar response profiles as monkey grating cells measured by von der Heydt et al. [20]. Unlike the previous models of grating cells (von der Heydt et al. [19] and Kruizinga-Petkov [9,14]) the new model accounts for proper spatial frequency tuning. The Kruizinga-Petkov model is an oriented texture operator, since it responds well to oriented

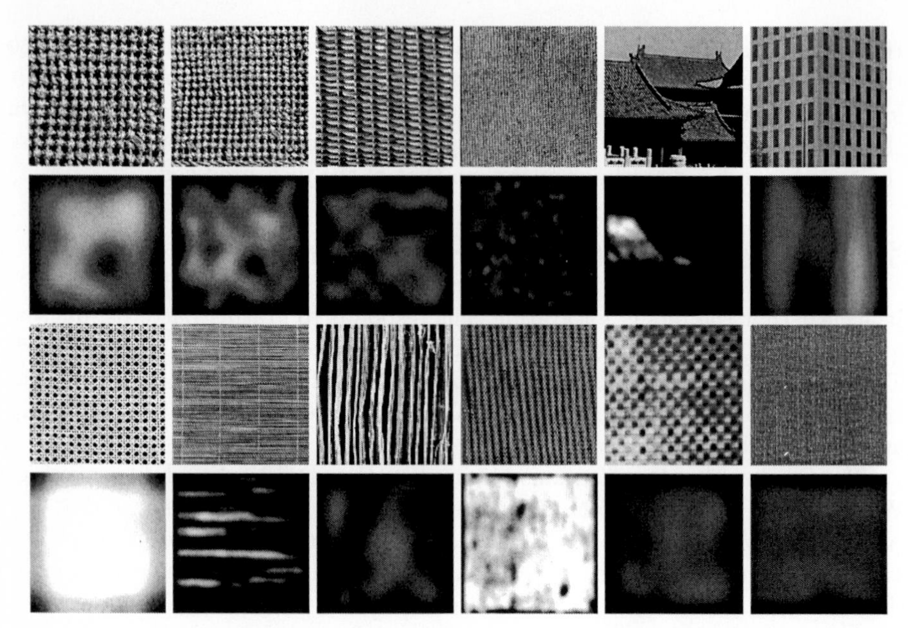

Fig. 5. Images from the databases (odd row) and their response profile (even row) of the grating cells from (16).

texture. Although it is inspired by the work of von der Heydt et al. [20], it is not an accurate model of grating cells because the response profiles differ rather strongly from that of the measured grating cells.

We applied the new model to 338 real world images of textures from three databases. Based upon these results we conclude that grating cells respond to oriented texture, to oriented repetitive alternating patterns to be precise, but are insensitive to many other textures. In general, grating cells are not suitable for texture detection. The grating cell operator responds well if the complex cell responses perpendicular to the preferred orientation show similar strong responses. In such case it is impossible to detect or extract relevant edges in these areas by using complex cells. It therefore seems that grating cells could play a key role in separating form from texture by giving inhibitive feedback responses to the complex cells. A field which we want to explore in the near future.

References

1. D.G. Albrecht, R.L. de Valois, and L.G. Thorell. Visual cortical neurons: are bars or gratings the optimal stimuli? *Science*, 207:88–90, 1980.
2. B. W. Andrews and D. A. Pollen. Relationship between spatial frequency selectivity and receptive field profile of simple cells. *J. Physiol.*, 287:163–176, 1979.
3. John G. Daugman. Uncertainty relation for resolution in space, spatial frequency, and orientation optimized by two-dimensional visual cortical filters. *J. Opt. Soc. Amer.*, 2(7):1160–1169, 1985.

4. K. R. Gegenfurther, D. C. Kiper, and S. B. Fenstemaker. Processing of color, form, and motion in macaque area v2. *Visual Neuroscience*, 13:161–172, 1996.
5. D. Hubel and T. Wiesel. Receptive fields, binocular interaction, and functional architecture in the cat's visual cortex. *Journal of Physiology, London*, 160:106–154, 1962.
6. D.H. Hubel and T.N. Wiesel. Sequence regularity and geometry of orientation columns in the monkey striate cortex. *J. Comp. Neurol.*, 154:106–154, 1974.
7. J. Jones and L. Palmer. An evaluation of the two-dimensional gabor filter model of simple receptive fields in cat striate cortex. *Journal of Neurophysiology*, 58:1233–1258, 1987.
8. E. Kaplan and R. M. Shapley. The primate retina contains two types of ganglion cells, with high and low contrast sensitivity. *Proc. Natl. Acad. Sci. U.S.A.*, 83:2755–2757, April 1986.
9. P. Kruizinga and N. Petkov. A computational model of periodic-pattern-selective cells. In J. Mira and F. Sandoval, editors, *Proceedings of the International Workshop on Artificial Neural Networks, IWANN '95*, volume 930 of *Lecture Notes in Computer Science*, pages 90–99. Springer-Verlag, June 7-9 1995.
10. M. S. Livingstone and D. H. Hubel. Anatomy and physiology of a color system in the primate visual cortex. *J. Neurosci.*, 4:309–356, 1984.
11. T. Lourens. *A Biologically Plausible Model for Corner-based Object Recognition from Color Images.* Shaker Publishing B.V., Maastricht, The Netherlands, March 1998.
12. M. C. Morrone and D. C. Burr. Feature detection in human vision: A phase-dependent energy model. *Proc. of the Royal Society of London*, 235:335–354, 1988.
13. J. A. Movshon, I. D. Thompson, and D. J. Tolhurst. Receptive field organization of complex cells in the cat's striate cortex. *Journal of Physiology*, 283:53–77, 1978.
14. N. Petkov and P. Kruizinga. Computational models of visual neurons specialised in the detection of periodic and aperiodic oriented visual stimuli: bar and grating cells. *Biological Cybernetics*, 76(2):83–96, 1997.
15. H. Tamura, H. Sato, N. Katsuyama, Y. Hata, and T. Tsumoto. Less segregated processing of visual information in v2 than in v1 of the monkey visual cortex. *Eur. J. Neuroscience*, 8:300–309, 1996.
16. K.K. De Valois, R.L. De Valois, and E.W. Yund. Responses of striate cortical cells ito grating and checkerboard patterns. *J. Physiol. (Lond.)*, 291:483–505, 1979.
17. R.L. De Valois, D.G. Albrecht, and L.G. Thorell. Cortical cells: bar and edge detectors, or spatial frequency filters. In S.J. Cool and E.L. Smith III, editors, *Frontiers of Visual Science*, Berlin, Heidelberg, New York, 1978. Springer.
18. R. von der Heydt. Approaches to visual cortical function. In *Reviews of Physiology Biochemistry and Pharmacology*, volume 108, pages 69–150. Springer-Verlag, 1987.
19. Rüdiger von der Heydt, Esther Peterhans, and Max R. Dürsteler. Grating cells in monkey visual cortex: Coding texture? Visual Cortex. In B. Blum, editor, *Channels in the visual nervous system: neurophysiology, psychophysics and models*, pages 53–73, London, 1991. Freund Publishing House Ltd.
20. Rüdiger von der Heydt, Esther Peterhans, and Max R. Dürsteler. Periodic-Pattern-selective Cells in Monkey Visual Cortex. *The Journal of Neuroscience*, 12(4):1416–1434, April 1992.
21. R. P. Würtz and T. Lourens. Corner detection in color images through a multiscale combination of end-stopped cortical cells. *Image and Vision Computing*, 18(6-7):531–541, April 2000.
22. S. Zeki. *A Vision of the Brain.* Blackwell science Ltd., London, 1993.

Synchronization in Brain – Assessment by Electroencephalographic Signals

Ernesto Pereda[1,2,*] and Joydeep Bhattacharya[3]

[1] Laboratory of Biophysics, Department of Physiology, Faculty of Medicine, University of La Laguna, Ctra. La Cuesta-Taco S/N, 38320 and
[2] Department of Systems Engineering, Institute of Technology and Renewable Energies (ITER), Granadilla, 38611 S/C de Tenerife, Spain
[3] Commission for Scientific Visualization, Austrian Academy of Sciences Sonnenfelsgasse 19/2, A-1010 Vienna, Austria

Abstract. The interdependencies between the electroencephalograms of several cortical areas are measured in human adult subjects during different experimental situations by means of methods based on the theory of dynamical systems. Multiple cortical areas are found to be synchronized together while the brain was engaged in higher information processing task. An additional experiment showed that the interhemispheric synchronization between two central derivations significantly increases from awake state to slow wave sleep.
The results show that these new approaches succeed in disclosing the distinct connectivity patterns among neuronal assemblies conforming the cortical brain, thus stressing their ability in analyzing the functioning of the brain.

1 Introduction

Synchronization is one of the most basic phenomena in the brain, where neuronal populations of distant cortical areas communicate with each other in order to find a meaning in a complex and noisy environment [1]. The field of neuroscience can be regarded as an intrinsically multidisciplinary one in which two different approaches for the study of the brain come together: the cell and molecular biology of nerve cells on the one hand, and the biology of cognitive processes on the other [2]. Although studies of single cells are of great help for understanding the functioning of natural neural networks, to monitor the activity of large populations of neurons as well as their mutual connectivity it is necessary to make use of multielectrode arrays set-up that allows the recording and analysis of the simultaneous activities of inter-related brain areas. There is increasing evidence that cortical areas do perform unique elementary functions, but any complex function requires the integrated action of many areas distributed throughout both cerebral hemispheres; thus, multiple cortical areas may not only become coactive during higher-information processing tasks, but may also become interdependent

* Corresponding author. E-mail for correspondence: pereda@iter.rcanaria.es

J. Mira and A. Prieto (Eds.): IWANN 2001, LNCS 2084, pp. 108–116, 2001.

and/or functionally connected [3]. So, the importance of the assessment of the interdependency (in weaker sense of synchronization) between multiple cortical areas cannot be overemphasized. Although modern imaging studies are found to be extremely popular and useful in the localization of brain functions [4], they are not ideally suitable to detect the co-operation among distant cortical areas. The multivariate records of simultaneous electroencephalographic signals (EEG) have the potential to assess higher brain functioning especially when the measurement of the degree of synchronization is of primary concern. There remains a vast opportunities for application of physical methods to develop greater use of EEG signals, arguably the most information-rich and revealing signal generated by the human body, the least understood, the most difficult, yet one of the least expensive and noninvasive.

Conventional mathematical tools used to detect synchronization are the cross correlation and the coherence in time and frequency domain, respectively. They measure the degree of linear interdependence between two signals. But, it has been shown that human EEG might possess non-linear structure [5], so it is advisable to make use of more advanced methods which can detect any non-linear (possibly asymmetric) relationship between two signals. The theory of dynamical systems has recently paved the way to accomplish the aforementioned goal [6–8]. In this report we have aimed at determining the applicability of these newly derived methods while studying the human brain activity in different experimental situations, which might offer the possibility of gathering deeper insight into the co-operation between large-scale cortical areas.

2 Methods

2.1 Phase Synchronization

Here, we briefly summarize the essential facts related with phase synchrony and refer the interested reader to Ref. [9] for detailed discussions. For two mutually coupled periodic oscillators, the phase dynamics can be formulated as

$$\frac{d\phi_1}{dt} = \omega_1 + \eta_1 f_1(\phi_1, \phi_2), \; \frac{d\phi_2}{dt} = \omega_2 + \eta_2 f_2(\phi_2, \phi_1) \tag{1}$$

where $\phi_{1,2}$ are phases of two oscillators, the coupling terms $f_{1,2}$ are 2π-periodic, and $\eta_{1,2}$ are the coupling coefficients. The generalized phase difference or relative phase is defined as

$$\Phi_{m,n} = m\phi_1 - n\phi_2 \equiv m\omega_1 - n\omega_2 + \eta F(\phi_1, \phi_2) \tag{2}$$

where $F(.)$ is also 2π periodic. Generally, phase locking requires $|m\phi_1(t) - n\phi_2(t) - \epsilon| < const$, where ϵ corresponds to some average phase shift, and m, n are some positive integers. It is to be noted that even for synchronized state, the relative phase Φ is not constant but oscillates around ϵ, and this oscillation ceases if the coupling function F depends on Φ only. Since in the present study, signals came from the same physiological system, i.e. the brain, only 1:1 synchronization was considered here; thus $m = n = 1$ and they are dropped for clarity.

For noise-free coupled oscillators, phase synchrony is synonymous with frequency locking, but this is not the case for coupled noisy and/ or chaotic systems. Recently, it has been shown [10] that for coupled chaotic systems, the irregular amplitudes affect the phase dynamics in a similar fashion as noise; thus, both types of systems can be treated within a common framework. For weak noise, Φ is not constant but slightly perturbed and phase slips of $\pm 2\pi$ occur and Φ is stable only between two phase slips. Thus, the relative phase ($\Phi(t)$) is unbounded in strict sense, and the distribution of $\Phi(t)$ mod 2π becomes smeared, but nevertheless unimodal. For strong and unbound noise (i.e. Gaussian noise), phase slips occur in an irregular manner, so only short segments of nearly stable phase are possible and the relative phase difference series perform a random walk; here, the synchronization region is completely smeared and phase synchrony can only be detected in statistical sense.

To quantify the degree of phase synchrony between two signals $\{x_{1,2}(t)\}$, the procedure [11] consists of the following steps: (i) compute instantaneous phases ϕ_i $(i = 1, 2)$ of each signal by analytic signal approach using Hilbert Transform: $\phi_i(t) = tan^{-1}[(\int_{-\infty}^{\infty} \frac{x_i(\tau)}{\pi(t-\tau)} d\tau)/x_i(t)]$, (ii) find relative phase: $|\Phi = |\phi_1 - \phi_2|$, (iii) obtain the distribution function of the relative phase mod 2π by partitioning the the series $\Phi(t)$, and finally, (iv) compute the index, $\rho = (H_{max} - H)/H_{max}$, where H is the Shannon's entropy of the distribution function, and $H_{max} = ln(P)$ where P is the number of partitions. The degree of phase synchrony increases monotonically with ρ.

2.2 Similarity Index

In the case of two weakly coupled non-identical chaotic systems $\mathbf{X}$ and $\mathbf{Y}$, their interdependence can give rise to a functional relationship between the state space vectors of both systems, a phenomenon usually termed as 'generalized synchronization' [12]. Due to this relationship, two close state vectors of $\mathbf{X}$ are simultaneous to two vectors of $\mathbf{Y}$ which are also close and bear the same time indices. In practical terms, let $\{x(k)\}$ and $\{y(k)\}$ $(k = 1, \ldots, n)$ be two simultaneously recorded time series from two systems, $\mathbf{X}$ and $\mathbf{Y}$ respectively. The corresponding state spaces are reconstructed using time delay embedding [13], so that the state vectors are obtained from consecutive scalar values of the series, i.e. $\mathbf{x}(k) = (x(i), x(i - \tau), \ldots, x(i - (m - 1)\tau))$, where m is the embedding dimension and τ is the time delay between successive elements of the state vector. The values of m and τ can be estimated from the time series by different methods [14]. For each state space vector pair $\mathbf{x}(k)$ and $\mathbf{y}(k)$, R nearest neighbors are selected from their individual state spaces. Let, their time indices be denoted as $r_k(i)$ and $s_k(i)$, $i = 1, \ldots, R$. By nearest neighbors we mean that the Euclidean distances between $\mathbf{x}(k)$ and $\mathbf{x}(r_k(i))$ are smaller than the distance between $\mathbf{x}(k)$ and any other state space vector. The squared mean distance from these neighbors is given by

$$D_k^R(\mathbf{X}) = \frac{1}{R}\sum_{i=1}^{R} \| \mathbf{x}(k) - \mathbf{x}(r_k(i)) \|^2 \tag{3}$$

The set of mutual neighbors of $\mathbf{x}(k)$ are those vectors $\mathbf{x}(s_k(i))$, which bear the time indices of R closest neighbors of $\mathbf{y}(k)$. The conditional mean squared distance is defined as

$$D_k^R(\mathbf{X}|\mathbf{Y}) = \frac{1}{R}\sum_{i=1}^{R} \| \mathbf{x}(k) - \mathbf{x}(s_k(i)) \|^2 \tag{4}$$

If the two systems are independent, $D_k^R(\mathbf{X}|\mathbf{Y}) \gg D_k^R(\mathbf{X})$, while for strongly coupled systems $D_k^R(\mathbf{X}|\mathbf{Y}) \approx D_k^R(\mathbf{X})$. To assess the strength of interdependency, the following measure, called similarity index ($S.I.$) is computed

$$S(\mathbf{X}|\mathbf{Y}) = \frac{1}{N'}\sum_{k=1}^{N'} \frac{D_k^R(\mathbf{X})}{D_k^R(\mathbf{X}|\mathbf{Y})} \tag{5}$$

where N' is the total number of state vectors. $S(\mathbf{Y}|\mathbf{X})$ is obtained in an analogous way. This index equals to 1 for identical signals, and it is close to 0 for independent signals. If $S(\mathbf{X}|\mathbf{Y}) \leq S(\mathbf{Y}|\mathbf{X})$, the system X has more influence on the system Y than vice versa. Further, if both $S(\mathbf{Y}|\mathbf{X})$ and $S(\mathbf{X}|\mathbf{Y})$ are high, the coupling is strongly bi-directional. Since both $S(\mathbf{X}|\mathbf{Y})$ and $S(\mathbf{Y}|\mathbf{X})$ are in principle different, this procedure allows the detection of non-symmetrical dependencies, overcoming the limitations of linear tools. It may be noted that no causal relationship is implied here, one can only infer that one systemis more *active* and the other is more *passive* considering the length scale set by the average radii of the spatial clouds. The active system is assumed to possess higher degrees of freedom or larger effective dimension than the passive system.

Remark: The primary substrate of synchrony between neuronal populations is the existence of groups of neurons interconnected by mutual excitatory and inhibitory synaptic connections. Thus, if ρ or S between two signals coming from two electrodes is high, the functional integration manifested by the synaptic connections between large populations of neurons in the associated cortical regions will also be high.

3 Results

Phase synchronization Spontaneous EEG signals were recorded for 90s from 20 male subjects (10 musicians, each with at least 5 years of musical training and 10 non-musicians with no musical training) in different conditions: at resting states with open and closed eyes, listening to various pieces of music, and listening to a text of neutral content. EEG signals were band-pass filtered to

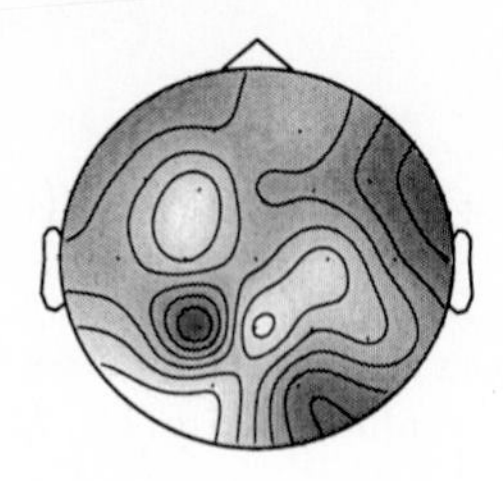

Fig. 1. Averaged raw EEG data (one epoch of 256 samples) for the 19 channels analyzed (one of the musicians listening to Bach). Dark zones represent high amplitude positive EEG voltages whereas light zones represent negative voltages.

extract the components associated with standard frequency bands: δ (0.025-4 Hz), θ (4-7 Hz), α (7-13 Hz), β (13-30 Hz), and γ (30-50 Hz).

Fig. 1 shows a topographic representation of the EEG channels for a musician subject while listening to Bach. From such plots, it is only possible to get information regarding local synchronization among the neurons of a certain cortical area. In order to extract further information on long range synchrony, for each filtered series, a non-overlapping window of 6s duration was used and for each window, values of ρ were measured considering all possible combinations between 19 electrodes. To compare the degree of phase synchrony between two groups, a normalization procedure was carried out to obtain synchrony values comparable between near and distant electrode pairs. Given ρ_{ij} (ρ for electrode pair i and j), let μ_{ij} and ν_{ij} be the mean and variance computed from the set of non-musicians; the relative phase synchrony values are computed as: $\sigma_{ij} = (\rho_{ij} - \mu_{ij})/\sqrt{\nu_{ij}}$. If $\sigma_{ij} \geq 2.33 (\leq -2.33)$, it can be inferred that the degree of phase synchrony between electrode pair i and j, was significantly higher for musicians than non-musicians (or non-musicians vs. musicians, respectively).

Fig. 2 shows the comparison expressed in terms of σ for different conditions. At resting states (Fig. 1(a,b), no significant differences exist in the degree of phase synchrony between two groups in any frequency band, whereas during listening to music (Fig. 1 (c,d)), only the γ-band synchrony was emerged to produce distinctively significant increase for musicians as compared to non-musicians. Further, this enhanced synchrony was found to exist over many electrodes, covering multiple cortical regions. Interestingly, listening to text did not produce any difference between the two groups.

No significant differences at resting conditions are not too surprising since musical education as such cannot yet be so important for the brain to produce a different synchrony when no task was apparently involved. This is true for both eyes closed and eyes opened conditions. But the special role of γ-band for musicians while listening to music demands further attention. Musical training on any instrument not only improves manual skill of both hands but also increases the ability to discriminate different sounds and to mentally follow and reconstruct acoustic architechnotonic structures. Any complex music (like those considered in this study) is interwoven with a variety of acoustic elements such as pitch,

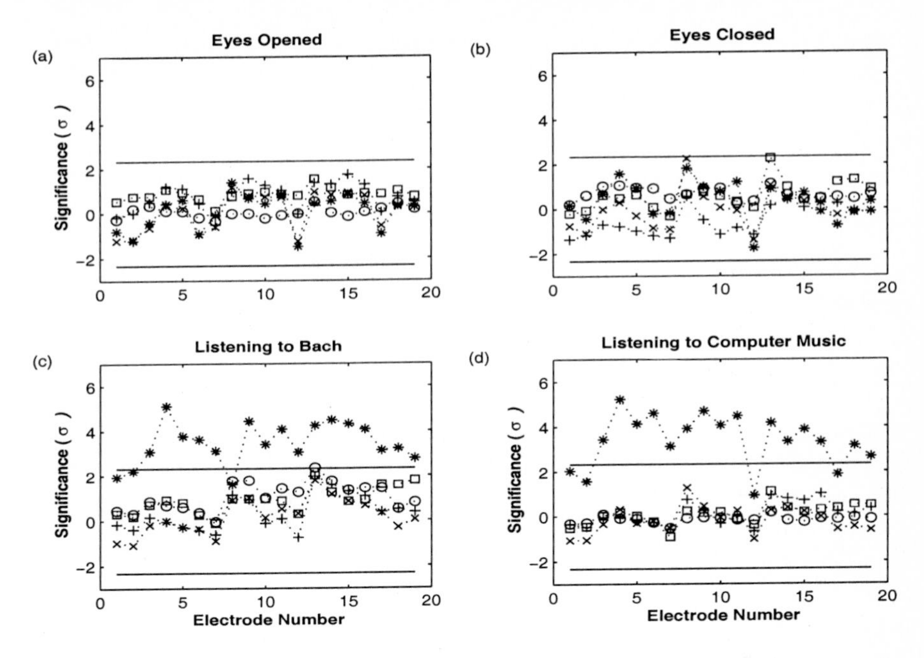

Fig. 2. Phase synchrony for 10 musicians relative to 10 non-musicians in terms of σ at different conditions: with eyes closed (a), with eyes opened (b), listening to Bach (c) and listening to Computer Music (d). Results (averaged over windows, subjects within each group, and for all possible combinations for each electrode (F1, F2, F7, F3, Fz, F4, F8, T3, C3, Cz, C4, T4, T5, P3, Pz, P4, T6, O1 and O2 numbered as 1, ..., 19 according to the standard 10-20 electrode placement system [15]) are shown in five frequency bands: δ (marked by 'o'), θ (marked by square), α (marked by '+'), β (marked by 'x'), and γ (marked by '*'). Horizontal line ($\sigma \pm 2.33$) denotes the 99% significance level. In resting conditions, two groups do not distinguish from each other in terms of synchrony, whereas γ-band synchrony was significantly higher for musicians during listening to music,

loudness, timbre, melodical and harmonical structure, rhythm in an intricate way. Musicians, definitely, have higher abilities to characterize these different elements than non-musicians. Considering the crucial role of oscillations in γ-band in binding several features into a single perceptual entity [1], these higher abilities of discriminating different acoustical elements can be surmised as one reason for the high synchrony for musicians in γ-band which is supposed to integrate the information associated with these various acoustical elements in a dynamical fashion. Further, attentively listening to any music is always led by anticipation. These anticipations are derived from long term musical memories which contain a large collections of musical patterns. During listening to music, high ability of anticipation of musicians calls for an extensive cultivation of the repertoire of stored musical patterns, and we think that this is another possible reason for the high synchrony over distributed cortical regions. Further, we found that the left hemispheres indicated a stronger phase synchrony than right for musicians during listening to music, whereas its opposite for non-musicians; this fact is in line with the evidence that the left side of the brain is better equipped for hosting long-term musical memory [16].

Similarity Index The EEGs of two central monopolar derivations (C3-A2 and C4-A1, 30 Hz high-frequency filter, 50 Hz notch filter, chin EMG and EOG) from ten healthy human subjects (male and female, mean age: 25.3 years) were registered at 256 samples/s, digitized using a 12 bits analogue-to-digital converter and stored for further analysis.

Recordings were carried out in a sound-proof, air-conditioned sleep room during an after lunch rest of about 2-3 hours. Sleep stages were visually scored from paper recordings according to the criteria of Rechtschaffen and Kales [17].

From each subject, different pairs of simultaneous artifact-free segments of 4096 samples (16 seconds) were selected in rest with open and close eyes (OE and CE) and each of the sleep stages (I, II, III, IV and rapid eye movement, REM). The stationarity of each pair of segments was quantified by dividing them in 8 nonoverlapping sequences of 256 samples. The mean and the variance of each sequence were calculated, and the standard deviations (SD) of the distribution of the means and the variances were obtained. The two pairs of segments with the lowest SD for these two distributions were used for the analysis. Prior to the calculations, all the signals were transformed to zero mean and unit SD to take into account that the range of variation of the data was in general different for each record. The possible linear trend in each segment was removed by least squared fit.

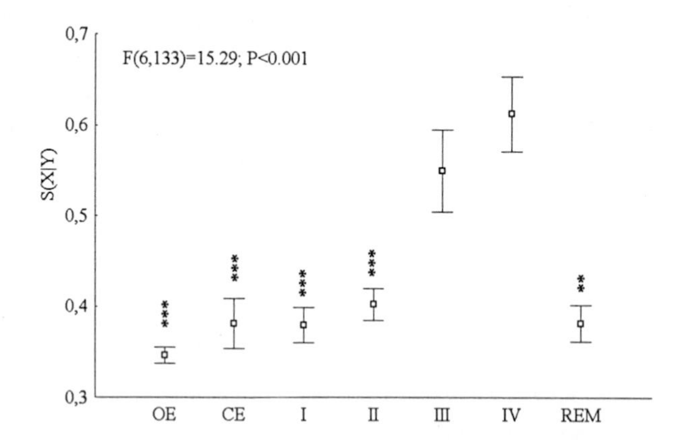

Fig. 3. Mean values ± standard error (whiskers) of $S(\mathbf{X}|\mathbf{Y})$ for the different situations analyzed .It is shown the value of the F distribution (test ANOVA) for the global difference among the stages. Asterisks indicate comparisons against stage IV (test post-hoc) with *, P<0.05; **, P<0.01; ***, P<0.001.

Fig. 3 shows the value of $S(\mathbf{X}|\mathbf{Y})$ for the different stages analyzed. Because there existed no asymmetry in the values of the index for both hemispheres (i.e.,

$S(\mathbf{X}|\mathbf{Y})$ did not differ from $S(\mathbf{Y}|\mathbf{X})$), this figure contains all the relevant information. It is clear that the degree of interdependence between the analyzed derivations increased significantly from awake to deep sleep, with the stage IV showing the greatest values of the index and therefore the greatest interdependence as compared to the other stages.

4 Discussion

Widespread phase synchronization of γ-band was found to be significantly higher in musicians than non-musicians. This enhancement most likely was elicited by music because no differences were found between the two groups at rest or during listening to a neutral text. This higher degree of synchrony in response to musical stimuli has never been previously demonstrated in data from humans at the scale of the whole brain, and this long-range nature of interaction indicates that γ-band phase synchrony may be viewed as a mechanism, which subserves large-scale cognitive integration and not confined only to local visual-feature binding.

As for the S.I., it showed that there exists a clear bilateral interdependence between the central derivations studied here, which increased from awake to deep sleep. This result confirms that of a previous report in which the interdependence between symmetric EEG channels was assessed by using the coherence function [18]. Further, it seems to confirm the bilateral origin of the basic mechanisms of the sleep. It must be stressed that, although this index showed no asymmetry between both derivations, it overcome the traditional spectral and statistical methods in answering a question (i.e., whether there exists any asymmetry in the interdependence) that would remain undisclosed otherwise.

To summarize, the results presented here have clearly shown that the use of recent methods of multivariate signal analysis can provide important information about the cooperative work of the brain in different situations.

Acknowledgements

The authors are truly indebted to Prof. Hellmuth Petsche and Prof. Julián González for helpful comments and suggestions. E. Pereda received the financial support of the Fondo de Investigaciones Sanitarias (PI 00/0022-02) and the Cabildo Insular de Tenerife.

References

1. Singer, W., Engel, A.K., Kreiter, A.K., Munk, M.H.J., Neuenschwander, S., Roelfsama, P.R., *Neuronal assemblies: necessity, signature and detectibility*, Trends. Cognit. Sci. 20, 252-261, 1997.
2. Kandel, E.R., Squire, L.R., *Neuroscience: Breaking Down Scientific Barriers to the Study of Brain and Mind*, Science 290 (5494), 1113-1120, 2000.

3. Bressler, S.L., *Large-scale cortical networks and cognition*, Brain. Res. Rev. 20, 288-304, 1995.
4. Posner, M.I., Raichle, M.E., *Images of Mind*, Scientific American Library, New York, 1997.
5. Pereda, E., Gamundi, A., Nicolau, M.C., Rial, R., González, J., *Interhemispheric differences in awake and sleep human EEG: a comparison between non-linear and spectral measures*, Neurosci. Lett. 263, 37-40, 1999.
6. Arnhold, J., Grassberger, P., Lehnertz, K., Elger, C.E., *A robust method for detecting interdependencies: application to intracranially recorded EEG*, Physica D 134, 419-430, 1999.
7. Pereda, E., Rial, R., Gamundi, A., González, J., *Assessment of changing interdependencies between human electroencephalograms using nonlinear methods*, Physica D 148, 147-158, 2001.
8. Bhattacharya, J., Petsche, H., *Musicians and the gamma band: a secret affair?*, Neuroreport 12, 371-374, 2001.
9. Pikovsky, A., Rosenblum, M., Kurths, J., *Synchronization: A Universal Concept in Nonlinear Sciences*, Cambridge University Press, 2001.
10. Rosenblum, M., Pikovsky, A., Kurths, J., *Phase synchronization of chaotic oscillators*, Phys. Rev. Lett. 76, 1804-1807, 1996.
11. Tass, P., Rosenblum, M., Weule, J. et al., *Detection of n:m phase locking from noisy data: Application to magnetoencephalography"*, Phys. Rev. Lett. 81, 3291-3294, 1998.
12. Rulkov, N.F., Sushchik, M.M., Tsimiring, L.S., Abarbanel, H.D.I., *Generalized synchronization of chaos in directionally coupled chaotic systems*, Phys. Rev. E 51, 980-994, 1995.
13. Takens, F., *Detecting strange attractors in turbulence*, Dynamical Systems and Turbulence, D.A. Rand and L.S. Young (eds.), Lecture Notes in Mathematics, vol. 898, Springuer-Verlag, Berlin, pp. 366-381, 1981.
14. Kantz, H., Schreiber, T. *Nonlinear Time Series Analysis*, Cambridge University Press, Cambridge, 1997.
15. Jasper, H.H., *Report of the committee on methods of clinical examination in electroencephalography*, Electroencephal. Clin. Neurophysiol. 10, 371-375, 1958.
16. Ayotte, J., Peretz, I., Rousseau, I., Bard, C., Bojanowski, M., *Patterns of music agnosia associated with middle cerebral artery infarcts*, Brain 123, 1926-1938, 2000.
17. Rechtschaffen, A., Kales, A. *A Manual of Standarized Terminology, Techniques and Scoring System for Sleep Stages of Human Subjects*, UCLA Brain Information Service, Brain Research Institute, Los Angeles, 1968.
18. Achermann, P., Borbély, A. A., *Coherence analysis of the human sleep electroencephalogram*, Neuroscience 85 1195-1207, 1998.

Strategies for the Optimization of Large Scale Networks of Integrate and Fire Neurons

Manuel A. Sánchez-Montañés

E.T.S. de Ingeniería Informática,
Universidad Autónoma de Madrid, 28049 Madrid, Spain

Abstract. In this paper we summarize several techniques that allow a dramatic speed-up of the simulation of networks of integrate and fire (I & F) neurons. The speed-up these methods allow is investigated in simulations where the computation time is measured as a function of network activity. Several topics are discussed such as the current limits of real time (largest network that can be simulated in real time) and the computational convenience of mean-field approximations. We conclude that the simulation of large models of I & F neurons in real time is feasible within the current technology.

1 Introduction

Many models require the simulation of lots of neurons while conserving temporal resolution of individual spikes, such as models of binding by synchronization. Their complexity requires a lot of computation time that makes difficult and slow the study of the parameter space. These models are difficult to be simplified by mean-field approximations since these methods loose temporal and spatial resolution. On the other hand, a satisfactory validation of a neural model requires the use of realistic input using real data from cameras, microphones, etc. [6, 10]. Moreover, biological systems work in real time dealing with the stimuli as "they come" (there is no preprocessing of input statistics). Therefore, a neuronal model should be able to work in real time with real world stimuli in order to be satisfactorily validated [10]. For all these reasons the optimization of networks of spiking neurons is critic for complex neural models.

2 Network Model

A good compromise between biological realism and computational cost consists in modeling each neuron as a conductance-based I & F unit [4]:

$$C\frac{dV_i(t)}{dt} = g_L\left(E_L - V_i(t)\right) + \sum_{j=1}^{M}\left(E_j - V_i(t)\right)\sum_{k=1}^{N} g^j_{k\to i}(t) \tag{1}$$

where V_i is the membrane potential of the neuron, C is the membrane capacitance, g_L is the leak conductance, E_L is the resting potential, E_j is the equilibrium potential for the synapses of type j, and $g^j_{k\to i}(t)$ is the synaptic conductance associated to the connection of type j from neuron k to neuron i. When V_i

J. Mira and A. Prieto (Eds.): IWANN 2001, LNCS 2084, pp. 117–125, 2001.

reaches the voltage threshold, V_{th}, the neuron fires a spike and its potential is reset to V_{reset} and clamped during τ_{ref} ms (absolute refractory period). Note that with this general definition of conductance-based I & F neuron we can model spike-triggered currents as self-connections.

Usually the dynamics of the conductance $g^j_{k\to i}(t)$ is defined as the summation of the contributions of all the past presynaptic spikes:

$$g^j_{k\to i}(t) = w^j_{k\to i} \sum_{t^k_{sp}(l)<t-d^j_{k\to i}} K^j(t - t^k_{sp}(l) - d^j_{k\to i}) \tag{2}$$

where $w^j_{k\to i}$ is the weight of the connection, $t^k_{sp}(l)$ is the time when the $l\,th$ spike in neuron k occurred, $d^j_{k\to i}$ is the delay in the synapse (time necessary for the presynaptic spike signal to reach the end of the synapse) and $K^j(t)$ is the kernel which defines the synaptic current dynamics.

3 Techniques to Optimize the Simulation

3.1 Efficient Integration of the Membrane Potential

There are different methods to integrate numerically non-linear equations [7]. However, many of these techniques are difficult to be applied to systems that combine continuous dynamics with discrete events, which is the case of our network. For this reason, in addition to simplicity and speed benefits, I & F models are usually integrated using a forward Euler method [7].

In fig. 1 we can see the numerical integration of the membrane potential when the neuron receives high-frequency input from an excitatory synapse. The exact solution (not shown) is indistinguishable from the numerical integration using the forward Euler method with a time step Δt of $0.01\,ms$ (fig. 1B). However, due to the nonlinearities in the dynamics, the use of large time steps makes the system unstable (fig. 1 C). In addition this method has the stiffness problem [7]. These problems could be easily avoided if we use a backward Euler method [7] which for our system (eq. 1) leads to the following algorithm:

$$V_i(t+\Delta t) = \frac{V_i(t) + \frac{\Delta t}{C} E_L\, g_L + \frac{\Delta t}{C} \sum_{j=1}^{M} E^j \sum_{k=1}^{N} g^j_{k\to i}(t+\Delta t)}{1 + \frac{\Delta t}{C} g_L + \frac{\Delta t}{C} \sum_{j=1}^{M} \sum_{k=1}^{N} g^j_{k\to i}(t+\Delta t)} \tag{3}$$

The computation of $\sum_{k=1}^{N} g^j_{k\to i}(t+\Delta t)$ can be performed in a straightforward way as it will be shown in section 3.2. Thus this method allows the use of large time steps while keeping the system stable, with no additional computing cost (fig. 1 D).

3.2 Exact and Efficient Computation of the Synaptic Dynamics

If we compute eq. 2 using directly its definition we should record the times of every spike in the presynaptic neuron, and perform a summation over all of them,

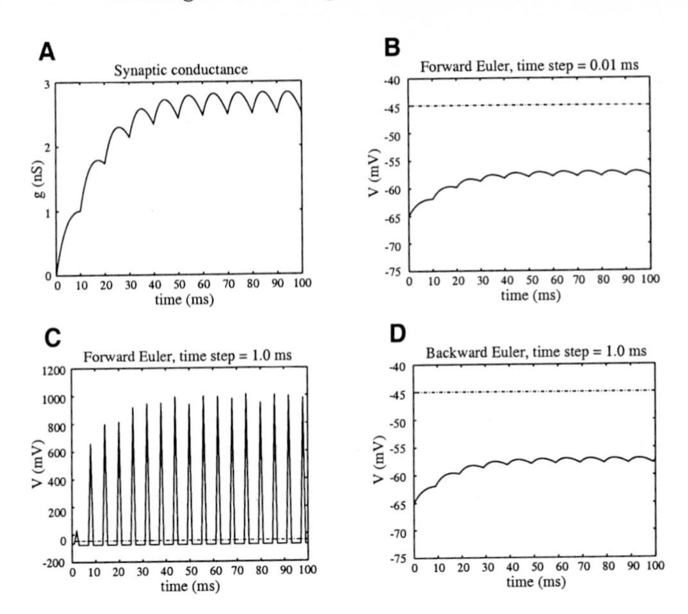

Fig. 1. Forward Euler method versus backward Euler method. The neuron receives input from an excitatory synapse with alpha-function dynamics, firing at 100 Hz. **A**: dynamics of $g_{syn}(t)$. **B**: V integrated using the forward Euler method, with a time step Δt of $0.01\,ms$. **C**: V integrated using the forward Euler method, $\Delta t = 1.0\,ms$. **D**: V integrated using the backward Euler method, $\Delta t = 1.0\,ms$. The parameters of the model are $C{=}0.4\,nF$, $g_L{=}20\,nS$, $E_L{=}-65\,mV$, $V_{th}{=}-45\,mV$, $V_{reset}{=}-75\,mV$, $t_{ref}{=}5\,ms$, $E_{exct}{=}0\,mV$, $\tau_{exct}{=}10\,ms$ and $w_{exct}{=}1\,nS$.

this for each synapse and every time step, leading to a large computational cost. An alternative is to integrate a differential equation whose dynamics is equivalent to eq. 2 [3]. However, the most efficient way to compute this summation is through the Z-transform [9], computing $\sum_{k=1}^{N} g^j_{k\to i}(t)$ as a recurrent discrete-time equation [8, 1]. For clarity reasons we define $G^j_i(t) \equiv \sum_{k=1}^{N} g^j_{k\to i}(t)$. Note that the values of the variables $G^j_i(t)$ completely determine the dynamics of the neuron (eq. 1). In addition we define $\Gamma^j_i(n\Delta t) \equiv \sum_{k=1}^{N} w^j_{k\to i}\, s_k((n-1)\Delta t - d^j_{k\to i})$ where $s_k(n\Delta t)$ is 1 if the neuron k has fired in the nth time step, being 0 otherwise (note that we are discretizing both the spike times $t^k_{sp}(l)$ and the synaptic delays $d^j_{k\to i}$). Then:

- If $K(t) = e^{-t/\tau}$ (exponential function),

$$G^j_i(n\Delta t) = c_1\, G^j_i((n-1)\Delta t) + \Gamma^j_i((n-1)\Delta t) \tag{4}$$

  with $c_1 = e^{-\Delta t/\tau}$.
- If $K(t) = \frac{e}{\tau}\, t\, e^{-t/\tau}$ (alpha function),

$$G^j_i(n\Delta t) = c_1\, G^j_i((n-1)\Delta t) + c_2\, G^j_i((n-2)\Delta t) + c_3\, \Gamma^j_i((n-1)\Delta t) \tag{5}$$

  with $c_1 = 2e^{-\Delta t/\tau}$, $c_2 = -e^{-2\Delta t/\tau}$, and $c_3 = \frac{e\Delta t}{\tau} e^{-\Delta t/\tau}$.

– If $K(t) = \frac{\tau_1 \tau_2}{\tau_1 - \tau_2}(e^{-t/\tau_1} - e^{-t/\tau_2})$ (difference of exponentials),

$$\left\{ \begin{array}{l} G1_i^j(n\Delta t) = c_1\, G1_i^j((n-1)\Delta t) + \frac{\tau_1 \tau_2}{\tau_1 - \tau_2}\, \Gamma_i^j((n-1)\Delta t) \\ \\ G2_i^j(n\Delta t) = c_2\, G2_i^j((n-1)\Delta t) + \frac{\tau_1 \tau_2}{\tau_1 - \tau_2}\, \Gamma_i^j((n-1)\Delta t) \end{array} \right\} \quad (6)$$

$$G_k^j(n\Delta t) = G1_k^j(n\Delta t) - G2_k^j(n\Delta t) \quad (7)$$

with $c_1 = e^{-\Delta t/\tau_1}$ and $c_2 = e^{-\Delta t/\tau_2}$.

In fig. 2 we can see how this method works for the alpha-function case. The Z-transform method can also be applied to more complicated dynamics such as the NMDA synapse [5]. Note that the variables G_i^j, $G1_i^j$ and $G2_i^j$ are local to the postsynaptic neuron i, so there is no need of storing and updating the $g_{k\to i}^j(t)$ of each synapse every time step. These variables are affected by the presynaptic activity through $\Gamma_i^j(t)$, which is just the summation of the weights of the active synapses, that is, the synapses where a spike is just arriving to the postsynaptic neuron. Thus only when a neuron spikes, its contribution to the $\Gamma_i^j(t)$ of the postsynaptic neurons should be calculated. Therefore this algorithm computes the contributions of only active synapses, making the total simulation time dependent on their mean activity. Synaptic delays can also be included with no computational cost using similar techniques.

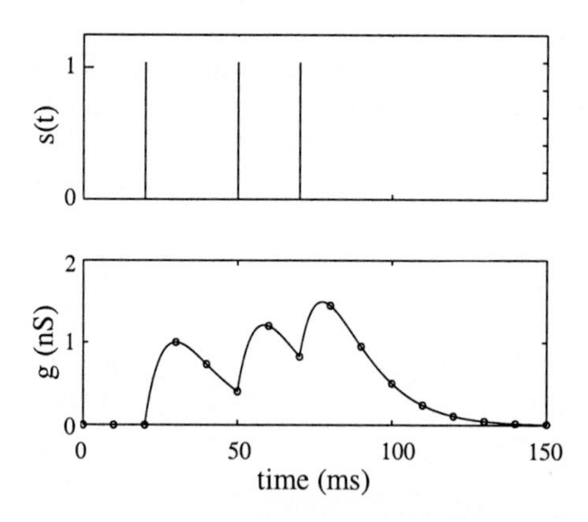

Fig. 2. Exact calculation of the $g(t)$ time-course using the Z-transform for a synapse with alpha-function dynamics. **Top**: spike train in the presynaptic neuron. **Bottom**: exact $g(t)$ calculated from eq. 2 (solid line) and using the algorithm based on the Z-transform, eq. 5 (circles). Note that this technique is exact even when large time steps are used (Δt=10 ms in this case). τ_α=10 ms, w=1 nS.

4 Implementation

We have developed a *C* code which simulates networks of I & F neurons incorporating the described optimization techniques. We did benchmarks under several conditions in a Pentium III 600 MHz running under Linux. All the benchmarks were performed 5 times under the same CPU conditions, observing no significant changes over repeated simulations. The code was intended to be a general-use tool for simulating arbitrary networks, so we did not include additional optimizations apart from those described in the paper. Hence the results shown here are generalizeable to any network model of I & F neurons.

5 Results

5.1 Computation Time as a Function of Network Activity

When the described optimizations are implemented, the time needed to compute one time step (ΔT_{comput}) basically depends on the number of neurons N_{neur}, the number of synapses N_{syn}, and the synaptic activity rate r_{syn} (average number of spikes transmitted per synapse in one time step) as:

$$\Delta T_{comput} = k_1 N_{syn} r_{syn} + k_2 N_{neur} + k_3 \tag{8}$$

r_{syn} depends on the integration time step Δt (given in ms) and the mean activity transmitted per synapse F_{syn} (expressed in Hz) as $r_{syn} = 0.001 \Delta t \, F_{syn}$. Thus r_{syn} ranges from 0 (no activity at any synapse) to 1 (every synapse transmits a spike at every time step).

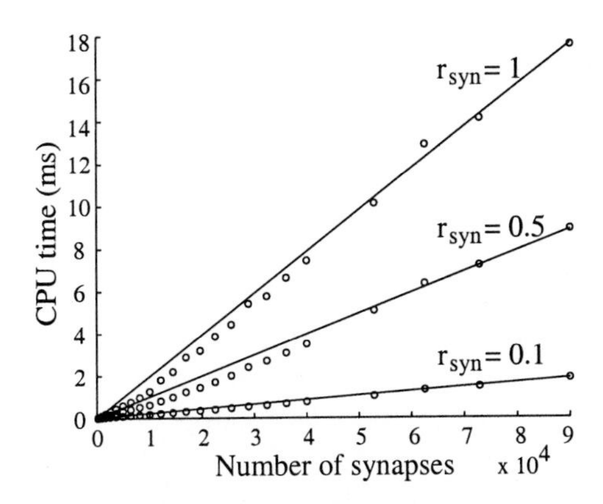

Fig. 3. Time needed by an optimized code to compute one time step of the simulation as a function of the number of synapses and the synaptic activity rate r_{syn} (average number of spikes transmitted by a synapse in one time step). The data can be satisfactorily fitted to $k_1 N_{syn} r_{syn} + k_2 N_{neur} + k_3$ (solid line). $k_1 = 1.9\,10^{-4}$, $k_2 = 4.5\,10^{-4}$ and $k_3 = -0.0037$.

We performed simulations of a model with all-to-all connectivity (see fig. 3). The observed computation time can be accurately described by eq. 8 (see fig. 3). The coefficients k_1, k_2 and k_3 depend on the details of the implementation and processor characteristics, but the qualitative behavior remains. When the level of activity in the network is low, it is possible to compute large networks very fast. For example, with a time step of $1\,ms$ and synapses firing at 10 Hz (the synaptic activity rate r_{syn} is 0.01) a network of 600 neurons and 360000 synapses can be computed in real time (each time step takes $1\,ms$ of computer time). In the other extreme, when the level of activity per time step in the network is very high, the speed of the simulation is equivalent to a simulation where the synaptic dynamics is not optimized (each synapse is updated every time step). Thus the optimization of the synaptic dynamics allows a speed-up of up to several orders of magnitude when the level of activity in the network is low.

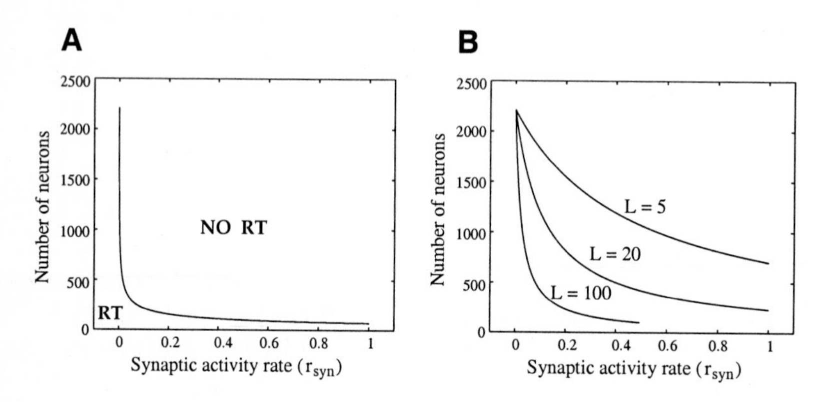

Fig. 4. A: size of the largest all-to-all network that can be simulated in real time (RT). **B**: size of the largest network with local topology that can be simulated in real time. Each neuron is connected to L neurons.

5.2 Where Is the Current Limit to Real Time?

The size of the largest network that can be simulated in real time basically depends on the number of synapses and the average activity that passes through them per time step. In figure 4 A the limit for an all-to-all network is shown as a function of its activity. For low activity levels large networks can be simulated in real time. For example, if the time step is $1\,ms$ and $F_{syn} = 10\,Hz$ ($r_{syn} = 0.01$) networks of up to 600 neurons and 360000 synapses can be simulated in real time. For networks with local or sparse topology (a neuron connects to L neurons in average), the size of the largest network that can be simulated in real time as a function of the activity level goes down much slower than for an all-to-all topology (fig. 4 B). For example, if the time step is $1\,ms$, $F_{syn} = 10\,Hz$, and $L = 10$, networks of up to 2000 neurons can be simulated in real time.

5.3 Is It Computationally Convenient to Use Units with Continuous Output?

In many theoretical works the dynamics of the neurons does not include spiking mechanisms, in order to make the model simpler. The main advantage of this is that sometimes it makes analytical studies possible. On the other hand, a technique to reduce the size of a model of spiking neurons consists in doing some type of mean-field approach to the dynamics of the network. The resultant equations from both methods can be written as:

$$\frac{do_i(t)}{dt} = -\tau^{-1} o_i(t) + \sum_{j=1}^{M} \sum_{k=1}^{N} w^j_{k\to i} f^j(o_k(t - d^j_{k\to i}) \qquad (9)$$

where $o_i(t)$ is a variable that defines the state of the model's unit (neuron / ensemble of neurons), τ is the time constant of the unit, $w^j_{k\to i}$ is the weight of the connection of type j between unit k and unit i, $f^j(x)$ is a function that describes how the activity in the presynaptic unit affects the postsynaptic dynamics, and $d^j_{k\to i}$ is the delay in the corresponding connection.

We would like to explore whether, on the other hand, this has computational advantages. The time necessary to calculate a time step in these kind of models can be written analogously to eq. 8, but now the synaptic activity rate r_{syn} is 1 (due to the continuous nature of the unit output). Thus the ratio between the computation time necessary to simulate the same period of time for the I & F network and for the simplified continuous network can be written as

$$\frac{T^{IF}_{comput}}{T^{S}_{comput}} = \frac{\Delta t^S}{\Delta t^{IF}} \frac{k_1^{IF} N^{IF}_{syn} r^{IF}_{syn} + k_2^{IF} N^{IF}_{neur} + k_3^{IF}}{k_1^S N^S_{conn} + k_2^S N^S_{units} + k_3^S} \qquad (10)$$

which is the speed-up we obtain by simplifying the model. Δt^{IF} and Δt^S are the time steps used for simulating the I & F model and the simplified model respectively; N^S_{units} and N^S_{conn} are the number of units and connections in the simplified model respectively. We would like to analyze this speed-up with networks with much more synapses than neurons. Then, the terms multiplied by N^{IF}_{syn} and N^S_{conn} predominate in this equation. We also define the reduction factor as $R \equiv \frac{N^{IF}_{syn}}{N^S_{conn}}$. Because eqs. 1 and 9 are integrated using an equivalent number of operations we can assume that $k_1^{IF} \simeq k_1^S$. Then,

$$\frac{T^{IF}_{comput}}{T^{S}_{comput}} = 0.001\, R\, \Delta t^S\, F^{IF}_{syn} \qquad (11)$$

where F^{IF}_{syn} is the average frequency of activation of the synapses in the integrate and fire model (given in Hz) and Δt^S is expressed in ms.

In figure 5 we can see the reduction factor beyond which the simplified model is computationally more efficient than the I & F one, assuming that $\Delta t^S = 0.1\, ms$. For low activities, the I & F model is faster than the simplified model

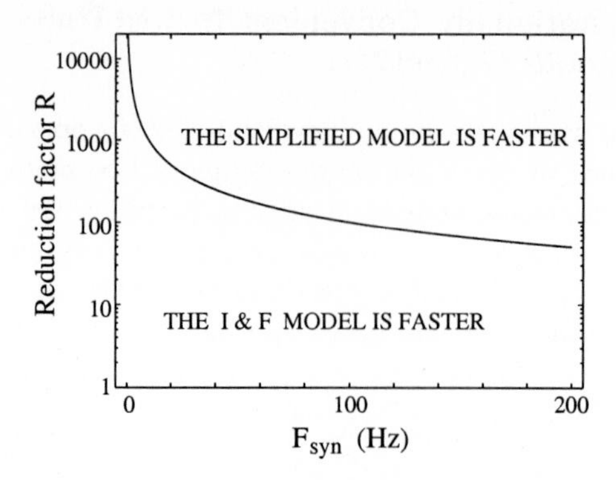

Fig. 5. Reduction factor beyond which a mean field approximation is computationally more efficient than the I & F model, as a function of synaptic activity (upper region: the simplified model is faster than the I & F model; lower region: vice versa). The time step used to integrate the approximation is $0.1\,ms$. The reduction factor is defined as the ratio between the number of connections in the I & F model and the number of connections in the simplified model.

even using large reduction factors. For $F_{syn}^{IF} = 1\,Hz$ a mean-field approximation with a reduction factor of 1000 and a time step of $0.1\,ms$ is 10 times slower than the non-simplified model. On the other hand, the substitution of each spiking neuron by a firing rate unit (that is, $R = 1$), is always inefficient from a computational point of view, independently of network activity. Thus the mean-field approximations are computationally efficient only for large reduction factors and high levels of activity. However, the use of large reduction factors can lead to the loss of fast and / or local dynamics important for the problem under study. Therefore, the two options should be contrasted before implementation using the kind of tools we describe here.

6 Discussion

In this work we show that large neural networks of I & F neurons can be optimally implemented if techniques such as the ones described here are included. Thus large realistic models can be simulated in real time using a personal computer such as a Pentium III 600 MHz. This allows the use of stimuli taken in real time from cameras, microphones, etc. as input to the model.

The implementation of neural models in real time using real world stimuli is important for a proper validation of the model. It also allows a direct comparison to biological experiments: the protocols the biologists do can be entirely replicated and the real stimuli used in the biological experiment (visual stimuli, sound, etc.) can be used in the model setup as "they are". In addition this could reduce the number of animals used in the laboratory (the preliminary experiments can be done in the model setup).

Real time simulations of neural networks also allow the study of animal - model interaction (for example the replacement of a neural circuit in the animal by a modeled one, connecting both systems through micro-electrodes): currently this technique is only possible using FPGAs, which have strong limitations in the number of elements they compute, thus being restricted to small networks. The optimization techniques we describe in this paper make possible the real-time interaction between a large neural circuit implemented in the computer and the animal.

There are other techniques to speed up the simulations that can be used in addition to the techniques here described. For example, the numerical integration could be performed using a variable time-step which is larger in silent periods and smaller in periods of activity. Another technique which can be exploited is the parallelization of the simulation [1, 11]. This is convenient for neural models since they are intrinsically parallel. Finally, plasticity algorithms based on spikes can also be implemented efficiently (for a review see [2]).

Acknowledgements

We wish to thank L. F. Lago-Fernández, P .F. M. J. Verschure, R. Huerta, F. Corbacho and F. Rodríguez for useful discussions and suggestions. This work was supported by grant BFI2000-0157 from MCyT.

References

1. Brettle, D., Niebur, E.: Detailed parallel simulation of a biological neuronal network. IEEE Comp. Sci. Eng. **1** (1994) 31-43
2. Giugliano, M.: Synthesis of generalized algorithms for fast computation of synaptic conductances with markov kinetic models in large network simulations. Neural Computation **12** (2000) 903–31
3. Koch, C., Segev, I. (eds.): Methods in Neuronal Modeling. From Ions to Networks, 2nd edition. MIT Press: Cambridge, Massachusetts (1998)
4. Koch, C.: Biophysics of computation. New York: Oxford UP (1999)
5. Köhn, J., Wörgötter, F.: Employing the Z-Transform to optimize the calculation of the synaptic conductance of NMDA and other synaptic channels in network simulations. Neural Computation **10** (1998) 1639–1651
6. Lago-Fernández, L.F., Sánchez-Montañés, M. A., Corbacho, F.: A biologically inspired visual system for an autonomous robot. Neurocomputing (2001) in press
7. Lambert, J. D.: Computational methods in ordinary differential equations. New York: Wiley (1973)
8. Olshausen, B.: Discrete-time difference equations for simulating convolutions (Tech. Memo). Pasadena: California Institute of Technology (1990)
9. Oppenheim, A. V., Schafer, R. W.: Digital-signal processing. London: Prentice Hall International (1975)
10. Sánchez-Montanés, M. A., König, P., Verschure, P. F. M. J.: Learning in a neural network model in real time using real world stimuli. Neurocomputing (2001) in press
11. Verschure, P.F.M.J.: Xmorph: A software tool for the synthesis and analysis of neural systems. Technical Report ETH-UZ, Institute of Neuroinformatics (1997)

A Neural Network Model of Working Memory
Processing of "What" and "Where" Information

Tetsuto Minami and Toshio Inui

Graduate School of Informatics, Kyoto University, Yoshida-Honmachi, Sakyo-ku, Kyoto, 606-8501, Japan
minami@cog.ist.i.kyoto-u.ac.jp

Abstract. Physiological studies have revealed that the prefrontal cortex (PF) plays an important role in working memory, which retains relevant information on line. Rao et al. (1997) found neurons contributing to both object and spatial working memory. However, their mechanisms are still unknown. In this study, we propose a neural network model of working memory in order to shed light on its mechanism. Our model has two input streams and can cope with a task in which two kinds of information have to be retained at the same time. We simulated some physiological results with this model. As a result, we simulated temporal activity patterns of neurons responding to both object and location information, as shown by Rao et al. (1997). We also considered domain-specificity by constructing three neural network architectures and the physiological results were simulated best by a no-domain specificity model. This result suggests that there is no domain-specificity in PF working memory.

1 Introduction

In higher brain functions such as language processing and thinking, not only parallel processing, but also that of time series plays an important role. Generally, we can interpret information of language as a hierarchical time series having symbolic meaning. However, it is not clear how such a hierarchical time series structure is processed in the brain.

Working memory is a basic process for recognizing a sequential behavior, and elucidation of its structure is very useful in understanding such cognitive functions. This was made clearer by some results of physiological studies in monkeys and fMRI experiments which found that working memory is closely related to PF cortex activity. However, there is still much to learn about the mechanism.

The first problem is whether we should pay attention to only maintenance as a function, or both maintenance and processing. Furthermore, a second problem is whether or not it is "domain specific" that Wilson et al. (1993) suggest as the mechanism in the brain exists. The domain specific theory suggested that visual working memory is organized into two systems within the PF cortex, with spatial working memory supported by the dorsolateral PF cortex and object working memory supported by the ventrolateral PF cortex. Inconsistent results with a segregation model, however, were found in some results from electrophysiological studies of monkeys and fMRI studies on humans (Rao et al., 1997; Rainer et al., 1998; Postle and D'Esposito, 1999).

J. Mira and A. Prieto (Eds.): IWANN 2001, LNCS 2084, pp. 126-133, 2001.

Recently Rao et al. (1997) reported data from a physiology experiment contradicting the domain specific theory. They showed that over half of the PF neurons with delay activity showed both what and where tuning by training monkeys to perform object and spatial delayed response tests within the same trial. Rainer et al. (1998) also found a high degree of integration of "what" and "where" in the principal sulcus of monkeys performing a delayed matching task that required memory for an object in a specific location.

In this study, we simulated the function of working memory using recurrent neural networks to investigate the mechanism. We first review some results of physiological experiments which contradict the domain specific theory. We then propose a PF cortex model in which the processing module is embedded, and simulate the results shown by Rao et al. (1997). The paper concludes with a discussion of the results and future work using the model.

2 Some Physiological Results

Wilson et al. (1993) investigated neural activities in the PF cortex of monkeys doing a task in which they had to remember an object or a location. As a result, most of the neurons which play a role in spatial working memory were found in the dorsolateral PF cortex (area 46 and 9), while object working memory neurons were found in the ventrolateral PF cortex (area 12). These results suggest that visual working memory is organized into two systems within the PF cortex (Goldman-Rakic, 1987; Wilson, O'Scalaidhe, and Goldman-Rakic, 1993).

In previous studies, however, working memory for what and where was examined in two separate tasks: an object task and a spatial task. However, this separation seldom occurs in natural situations. In order to investigate whether object and spatial information is integrated by individual PF neurons, Rao et al. (1997) employed a task involving what and where information being used together. On each trial, while the monkey maintained its gaze on a fixation spot, a sample object was briefly presented at the center of sight. After a delay (a "what" delay), two test objects were briefly presented at two of four possible extrafoveal locations. One of the test objects matched the sample, the other was a nonmatch. After another delay (a "where" delay), the monkey had to make a saccade to the remembered match location. Thus, in this task, the monkey had to memorize both the object and the location simultaneously.

They recorded the activity of 195 neurons from the lateral PF cortex of two monkeys. Some PF neurons showed delay activity that was significantly tuned to either the sample object only or the sample location only, while more than half of the PF neurons (52 %) were tuned to both an object and a position. They were highly selective for both objects and locations. On average, there was a 64 % increase in what delay activity after a good (preferred) sample object over the activity after a poor (nonpreferred) sample object, and a 71 % increase in where delay activity after cueing of a good location compared to the activity after a poor location. Thus, these "what-and-where" cells conveyed object and spatial information during different epochs of the same behavioral trial, and appeared to contribute to both object and spatial working memory. What cells, where cells, and what-and-where cells were distributed equally between the dorsolateral PF cortex and the ventrolateral PF cortex.

In addition, Rainer et al. (1998) recorded the activity of PF neurons in monkeys performing a delayed-match-to-object-and-place task, in which they needed to remember both the identity and location of the sample object. Their results showed that when monkeys need to remember an object and its location, the activity of many lateral PF neurons reflects this combined what and where information.

Furthermore, Postle and D'Esposito (1999) designed some experiments to investigate the neural basis of spatial and nonspatial working memory in the human PF cortex, and throughout the brain, with event-related functional magnetic resonance imaging (fMRI). Their what-then-where delayed-response task was based on Rao et al. (1997). Their experiment failed to find evidence for anatomical segregation of the spatial and object working memory functions in the PF cortex.

These results suggest two things: 1. Prefrontal working memory is not domain specific. 2. "What" and "where" information is conjuncted in the PF cortex.

3 Model

Zipser (1991) showed that the hidden unit activity of a fully recurrent neural network model, trained on a simple memory task, matched the temporal activity patterns of memory-associated neurons in monkeys performing delayed match-to-sample tasks. Zipser's short-term active memory model consists of two linear input units and a number of fully connected sigmoid units representing short-term memory. The input units, which have connections to all memory nodes, are an analogue input to carry an analogue value to be stored, and a gate input that is required because the single neuron firing data show that new values are loaded only at appropriate times during a memory task.

We extended the Zipser (1991) model in order to assess the function of information processing, such as selection, as well as retention. Zipser's model deals with only one kind of information, such as an object or a location. For example, the model explains a delayed response task in which only information of location must be retained.

Consequently, we considered three architectures in our model to process two kinds of information (Figure 1(a)). Model A has an unseparated storage module, which does not presume domain-specificity. Models B and C has two storage modules that independently receive inputs: "what" working memory and "where" working memory. The difference between Model B and C is that the two modules in Model B are mutually connected, but are not in Model C.

The network of Model A is shown in Figure 1(b). It consists of a set of fully interconnected model neurons. In our model, there are two main input streams: one represents "what" information (object-module) and the other "where" information (location-module), each of which has a load-in. The initial input is composed of five sets of four neurons. Each set in the initial input denotes five positions: the center of gaze and four possible extrafoveal locations. The object-module consists of four neurons, one of which is activated. It can code four objects. The location-module also consists of four neurons, one of which is activated. It can code four locations, that is, four extrafoveal

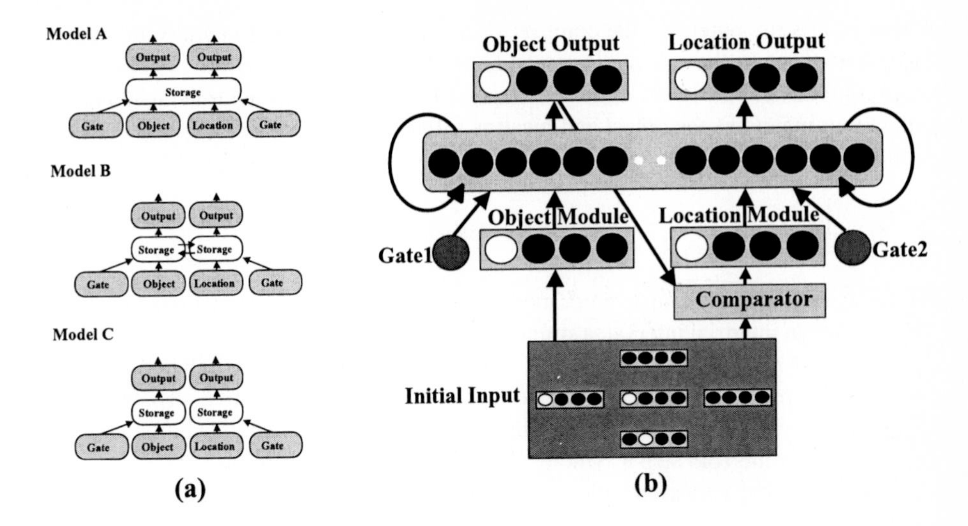

Fig. 1. (a) Three model architectures (b) The input-output structure of the model: Inputs are separated into two modules: one module corresponding to object information and the other corresponding to location module. Each module has a load-in input. The input lines connect all model neurons. There are two output modules.

locations. This coding can be said to be valid considering neurons in the dorsal pathway that show object selectivity equivalent to those shown in the ventral pathway (Sereno and Maunsell, 1998).

Initial inputs to the networks are processed in two modules representing two systems: the ventral and dorsal systems. Hidden units corresponding to PF neurons receive these inputs. There are comparator units between the input layer and the hidden layer. These units play a role in receiving inputs from the location-module and object output unit, and activating the neuron set in the module that coincides with the object output. This feedback means that what's loaded into working memory goes back to sensory processing (Miller et al., 1996).

The output of the ith model unit in the network on a time cycle $t+1$ is given by

$$y_i(t+1) = f(\sum_j w_{ij} y_j(t) - b) \tag{1}$$

where $f(x)$ is the sigmoid function

$$f(x) = \frac{1}{1+e^{-x}} \tag{2}$$

which ranges between 0 and 1 as x ranges from minus infinity to plus infinity.

The output of comparator units is defined as

$$m_{jk} = \sum_i w_i a_{ij} a_{ik} \tag{3}$$

where m_{jk} is the output of the comparator units, w_i is the weight connecting the ith comparator unit to the output (Oppenheim et al., 1996). $a_{ij}a_{ik}$ is the product of the object outputs and the location-module inputs. The hidden units are trained so that the output unit maintains the object and location information when the load-in is activated until the next time the load-in becomes active.

4 Results

At the start of each trial, the network model was set to a resting level by loading in a low value. After an intertrial interval, a stimulus object pattern (representing one object) was loaded in at the 5th step. After a delay, a location information that included the same pattern as the previous object was loaded in at the 15th step. At the 16th step, the object input was reset. The period from the 5th to 15th step represented a "what" delay and the period after the 15th step a "where" delay as in Rao et al. (1997).

In the temporal activation patterns of a typical trained network, the output unit has a moderately stable sustained activity reflecting the stored value. The spectrum of hidden unit activity patterns has sustained activity during the delay, elevated activity only at the time of gating, or a mixture of these two patterns. We compared the characteristics of hidden units (90 units) in each model in networks trained in this way with those of the PF neurons shown in Rao et al. (1997). As a result, in model hidden units there were object-tuned neurons (Figure 4(a)) , location-tuned neurons (Figure 4(b)) and both-tuned neurons (Figure 4(c)). "Good" or "poor" refer to the object or location that elicited the most or least activity, respectively. Basically, a large difference between "good" and "poor" value means that the neuron had selectivity for the stimulus. According to Figure 4(a) and (b) these neurons contributed to the mechanism of retention of object or location information. On the other hand, according to Figure 4(c) a single neuron had selectivity for both object and location and was related to the retention of both types of information. In this way, our model simulated the results of physiological experiments in which processing two kinds of information is needed.

Next, we compared the results of three models. In each model, object-tuned neurons and location-tuned neurons were found, but both-tuned neurons were not found in Model C. Moreover we calculated the ratio of object-tuned units, location-tuned units and both-tuned units among hidden units showing delay activity. Then, we compared the results with those of Rao et al. (Table 1). They showed that the ratio of each neuron in Model A was very similar to the result of Rao et al. (1997). However, the result of Model B does not agree with physiological results that more than half the neurons respond to both object and location.

Considering these results, the architecture of prefrontal working memory is not a domain-specific module that independently receives inputs as in Models B and C, but the domain-free module as in Model A.

5 Discussion

In this article, we presented a PF working memory model to simulate temporal sequences in monkeys that must retain both an object and a location. The model is mainly

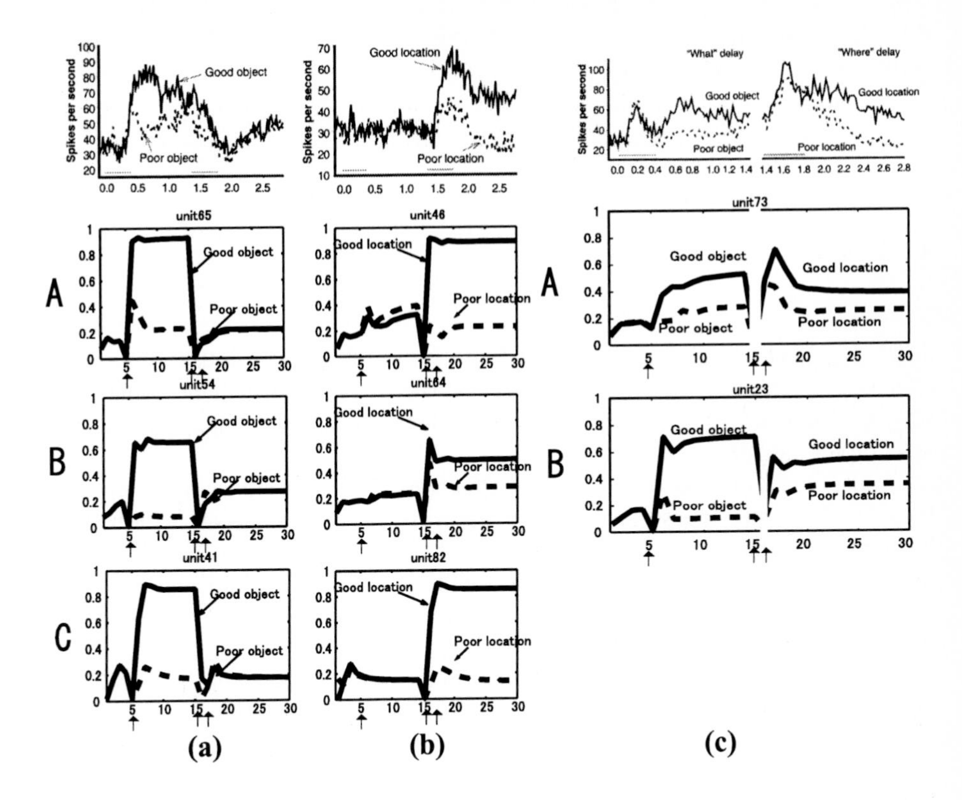

Fig. 2. Comparison of the temporal activity patterns of Rao et al. (1997) and our model simulation: (a) neurons showing either object-tuned (b) location-tuned delay activity (c) neurons showing object-tuned activity in the "what" delay and location-tuned activity in the "where" delay! (The top graph is from Rao et al. (1997) and the others are from the temporal activity patterns of hidden units in our model). "Good" or "poor" refer to the object or location that elicited the most or least activity, respectively. The arrows indicate the time when information was gated in.

based upon the dynamics of recurrent neural networks. The spectrum of firing was similar in the different cortical areas, and its properties indicated that information was being stored as patterns of neural activity. The similarity between the firing patterns observed in different cortical areas suggests that a similar general mechanism may be used for active information storage. The model described here is concerned with the nature of this general mechanism.

In addition to the dynamic activity of recurrent networks, some physiological findings (Sereno and Maunsell, 1998; Miller et al., 1996) form the foundation of the model investigated here. That is, the physiological results of Sereno and Maunsell (1998) were incorporated in the representation of the inputs, and those of Miller et al. (1996) to the architecture of our model.

Sereno and Maunsell (1998) showed that many neurons in a primate posterior parietal cortex (the "where" pathway) had shape selectivities to simple, two-dimensional

Table 1. Comparison of the results of Rao et al. (1997) and those of our model: The ratio of the neurons selective for objects, selective for locations, and selective for both.

	Rao et al.(1997)	Model A	Model B	Model C
Selective for objects	7 %	13 %	39 %	47 %
Selective for locations	41 %	30 %	30 %	53 %
Selective for both	52 %	56 %	30 %	0 %

geometric shapes while the monkey performed a simple fixation task and a delayed match-to-sample task. The selectivities are equivalent to any shown in the "what" pathway. According to this result, we expressed an initial input as a set of neurons equivalent to an object-module.

Miller et al. (1996) suggested that the PF cortex is a major source of the top-down inputs to the posterior association cortex and is heavily interconnected with both the IT cortex and other visual areas. In other words, what's loaded into the PF working memory goes back to sensory processing. Tomita et al. (1999) showed evidence of this top-down signal from the PF cortex. We installed this backward stream in our model as the feedback to the comparator module.

The temporal activity patterns of neurons in the model matched those of prefrontal neurons shown in Rao et al. (1997). This result suggests that the dynamics of recurrent connections play an important role in maintaining the information in the PF cortex. In addition, attractors in the model have an important connection with the retention mechanism. An attractor in a nerve system is thought to be important in storage and maintenance (Zipser, 1991).

The main purpose of this paper was to find neurons that mimicked the temporal activity of those in Rao et al. (1997). However, the neurons found in other physiological studies were found in the hidden units of our model (e.g., Funahashi et al.,1991).

Furthermore, we investigated the problem of domain specificity of working memory using a structural method. We assumed three architectures, considering the structure of the hidden layer equivalent to the PF cortex in our model. We found object-tuned units, location-tuned units and both-tuned units in one moduled hidden layer, but did not find both-tuned units in a two-divided hidden layer. Taking this result into account, the physiological results by Rao et al. (1997) support the possibility that PF working memory has no domain specificity.

In addition, these neurons may contribute to the binding of "what" information to "where" information needed to guide behavior. In other words, we can assume that the information passing through both streams from V1 in the occipital lobe is integrated in a large PF domain of working memory domain, if the same neuron takes roles of both "what" and "where" working memory. Fuster (2000) suggested that PF cortex neurons are part of integrative networks that represent behaviorally meaningful cross modal associations. In this way, the PF cortex is essential for the temporal integration of sensory information in behavioral and linguistic sequences.

We should emphasize that the neuronal models we used are still very simple. It is true that the PF cortex plays an important role in working memory, but working memory is not likely to be supported only by the PF cortex. This is obvious in that activated regions in the task where working memory is needed have a very wide range (a parietal association area, a temporal association area, basal ganglia, the cerebellum, hippocampus, etc.). Therefore, we must study the interaction of the PF cortex and other regions in order to understand the neural mechanism of working memory more profoundly. However, our results are very important in that physiological results are produced by simple neural networks using only recurrent loops in the cortex. In future studies, we will elucidate the mechanism of working memory by simulating more physiological data.

Acknowledgements

This work was supported by the "Research for the Future Program," administered by the Japan Society for the Promotion of Science (Project No. JSPS-RFTF99P01401).

References

1. Baddeley, A. D., 1986. Working Memory. Oxford: Oxford University Press.
2. Funahashi, S., Bruce, C. J., Goldman-Rakic, P. S., 1991. Neuronal activity related to saccadic eye movements in the monkey's dorsolateral prefrontal cortex. Journal of Neurophysiology 65(6), 1464–1483.
3. Fuster, J. M., Bodner, M., Kroger, J. K., 2000. Cross-modal and cross-temporal association in neurons of frontal cortex. Nature 405, 347–351.
4. Miller, E. K., Erickson, C. A., Desimone, R., 1996. Neural mechanisms of visual working memory in prefrontal cortex of the macaque. The Journal of Neuroscience 16, 5154–5167.
5. Postle, B. R., D'Esposito, M., 1999. What-then-where in visual working memory: An event-related fmri study. Journal of Cognitive Neuroscience 11 (6), 585–597.
6. Rainer, G., Asaad, W. F., Miller, E. K., 1998. Memory fields of neurons in the primate prefrontal cortex. Proc. Natl. Acad. Sci. USA 95, 15008–15013.
7. Rao, S. C., Rainer, G., Miller, E. K., 1997. Integration of what and where in the primate prefrontal cortex. Science 276, 821–824.
8. Sereno, A. B., Maunsell, J. H. R., 1998. Shape selectivity in primate lateral intraparietal cortex. Nature 395, 500–503.
9. Tomita, H., Ohbayashi, M., Nakahara, K., Hasegawa, I., Miyashita, Y., 1999. Top-down signal from prefrontal cortex in executive control of memory retrieval. Nature 401, 699–703.
10. Williams, R. J., Zipser, D., 1989. Experimental analysis of the real-time recurrent learning algorithm. Connection Science 1, 87–111.
11. Wilson, F. A. W., O'Scalaidhe, S. P., Goldman-Rakic, P. S., 1993. Dissociation of object and spatial processing domains in primate prefrontal cortex. Science 260, 1955–1958.
12. Zipser, D., 1991. Recurrent network model of the neural mechanism of short-term active memory. Neural Computation 3, 179–193.
13. Zipser, D., Kehoe, B., Littlewort, G., Fuster, J., 1993. A spiking network model of short-term active memory. The Journal of Neuroscience 13 (1), 3406–3420.

Orientation Selectivity of Intracortical Inhibitory Cells in the Striate Visual Cortex: A Computational Theory and a Neural Circuitry

Mehdi N. Shirazi

Osaka Institute of Technology, Faculty of Information Science
1-79-1 Kitayama, Hirakata-shi, Osaka, Japan

Abstract. Intracortical inhibitory cells are thought to play a major role in giving rise to orientation selectivity of simple and complex cells in the striate visual cortex. There is, on the other hand, an ample experimental evidence supporting that the intracortical inhibitory cells are orientation selective as well, thus, giving rise to the so-called bootstraping problem. This article shows how to solve the bootstraping problem in the striate visual cortex by introducing a computational theory that consists of a probabilistic model, a computational goal, a parallel algorithm to achieve the computational goal and a physiologically-plausible neural circuitry to implement the algorithm.

1 Introduction

One of the most distinctive features of simple and complex cells in the striate visual cortex is orientation selectivity [1]. Simple and complex cells lose their orientation selectivity during a drug-induced localized blockade of intracortical inhibition [2], indicating that intracortical inhibitory cells play an important role in shaping the simple and complex cells' orientation selectivity. Records taken from major different types of smooth cells, the basket cells, axo-axonic cells, double bouguet cells and GABAergic cells in layer 1 have revealed that smooth cells have oriented receptive fields [3]. Since smooth cells are thought to be inhibitory, it can be safely concluded that the intracortical inhibitory cells are also orientation selective.

If intracortical inhibitory cells make the simple and complex cells orientation selective—an assumption which is still controversial [4]— , *then what makes the intracortical inhibitory cells orientation selective?*

If we agree that orientation selectivity of cortical cells (simple cells, complex cells as well as the intracortical inhibitory cells) is mediated by oriented intracortical inhibitory cells, then it is necessary to demonstrate how it is possible to generate oriented intracortical inhibitory cells in the striate visual cortex from the non-oriented LGN (lateral geniculate nucleus) afferents. This is the so-called bootstraping problem [5] which is an obstacle to the acceptance of the attractive notion that the orientation selectivity of simple and complex cells is created merely by intracortical inhibition mechanisms.

The goal of this article is to show how the bootstraping problem might be solved by a physiologically plausible neural circuitry in the striate visual cortex.

J. Mira and A. Prieto (Eds.): IWANN 2001, LNCS 2084, pp. 134-141, 2001.

2 Computational scenario

There are one layer of LGN cells and one layer of striate cortical (SC) cells. The LGN and SC cells are in a topographical and one-to-one relation. Each single LGN cell can be either active or less-active; whereas each single SC cell can be either active or silent. An LGN activity pattern might comprises of lines, blobs, and dots corresponding to strings, clusters, and singletons of active LGN cells. Driven by the LGN layer, we assume that the SC layer function as a horizontally-tuned orientation-selective spatial filter.

The question is: *what circuitry is needed between LGN and SC cells to make the SC cells horizontally-tuned?*

To answer this question, I start with a computational scenario. Figure 1 shows a pair of activity patterns of the LGN and SC cells . The SC-layer activity pattern

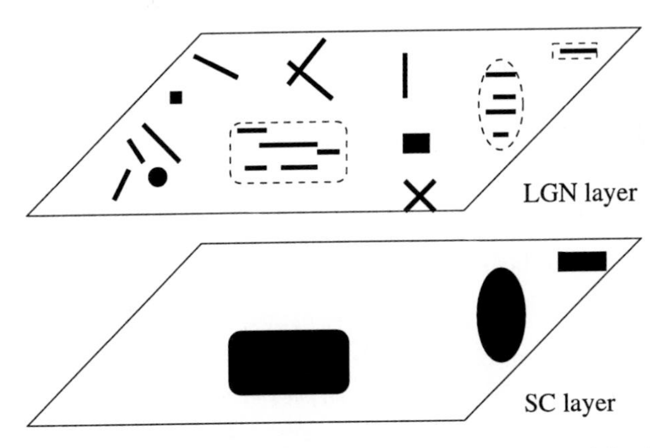

Fig. 1. The relationship assumed between activity patterns of horizontally-tuned SC-cells and the non-oriented LGN-cells.

comprises of active (black) and silent (white) regions. The LGN activity pattern comprises of: (1) horizontal lines of arbitrary lengths scattered randomly within regions topographically corresponding to the SC-layer active (black) regions; and (2) non-horizontal lines, blobs and dots scattered randomly over the remaining region. Given an LGN activity pattern, the computational scenario is to estimate the locations and extensions of the regions containing horizontal lines.

The actual locations and lengths of the horizontal lines within their confining regions and the locations, shapes and sizes of the SC-layer active regions might be different from one LGN activity pattern to another. This might suggest us to think of LGN- and SC-layer activity-pattern pairs as samples of two interactive spatial stochastic processes and thus naturally try to formulate our computational scenario within a probabilistic framework.

3 Problem formulation

It is assumed that the LGN and SC cells are located on two identical rectangular lattices denoted by $\mathcal{L}$ and assumed that the activity patterns of the LGN- and SC-layer are samples from two interactive spatial stochastic processes defined on $\mathcal{L}$ denoted by $Y_{\mathcal{L}} = \{Y_{ij}; (i,j) \in \mathcal{L}\}$ (LGN process) and $X_{\mathcal{L}} = \{X_{ij}; (i,j) \in \mathcal{L}\}$ (SC process), respectively. Furthermore, the composite stochastic process $(X_{\mathcal{L}}, Y_{\mathcal{L}})$ is modeled by a hierarchical Markov random fields (MRFs), ([6], [8]).

3.1 A probabilistic computational goal

The computational goal of the circuitry between the LGN and SC layers is formulated as the Maximum-A-Posteriori (MAP) estimate of the SC-layer activity pattern from a given LGN-layer activity pattern $y_{\mathcal{L}}$, ([7]). In other words, the computational goal is formulated as solving the following optimization problem

$$\hat{x}_{\mathcal{L}} = \arg \max_{x_{\mathcal{L}} \in \Omega_X} P(y_{\mathcal{L}} \mid x_{\mathcal{L}}) P(x_{\mathcal{L}}), \tag{1}$$

where $P(x_{\mathcal{L}})$ denotes the joint discrete density function of $X_{\mathcal{L}}$, $P(y_{\mathcal{L}} \mid x_{\mathcal{L}})$ denotes the conditional joint discrete density function of $Y_{\mathcal{L}}$ given $X_{\mathcal{L}} = x_{\mathcal{L}}$, and Ω_X denotes the $X_{\mathcal{L}}$'s configuration space.

3.2 A parallel algorithm

The following local updating rule gives an approximate solution to the MAP-estimation problem, Eq. (1)(see [7])

$$\begin{aligned}\hat{x}_{i,j}^{(n+1)} = \arg \max_{x_{i,j} \in \{\underline{X}, \overline{X}\}} \{ &- \sum_{p \in \{-1,1\}} \sum_{q \in \{-1,1\}} \mathcal{E}(y_{i,j}, y_{i+p,j+q} \mid x_{i,j}) \\ &- \sum_{p \in \{-1,1\}} \sum_{q \in \{-1,1\}} \mathcal{E}(x_{i,j} \mid \hat{x}_{i+p,j+q}^{(n)}) \\ &- \max_{k \in \{\underline{Y}, \overline{Y}\}} \{ - \sum_{p \in \{-1,1\}} \sum_{q \in \{-1,1\}} \mathcal{E}(k, y_{i+p,j+q} \mid x_{i,j}) \} \, \}. \end{aligned} \tag{2}$$

for all $(i,j) \in \mathcal{L}$. Here, $\overline{Y}$ and $\underline{Y}$ denote the active and less-active states of the LGN cells, whereas $\overline{X}$ and $\underline{X} = 0$ denote the active and silent states of the SC cells respectively, and $\mathcal{E}$s denote the doubleton-clique energies ([8]).

3.3 Doubleton-clique energies

- Doubleton-clique energies generating horizontal lines of active LGN cells at locations where the corresponding SC cells are active

A horizontal doubleton-clique energy is defined as

$$\mathcal{E}(y_{i,j}, y_{i+p,j+q} \mid \overline{X}) = \begin{cases} -\beta & \text{if } \; y_{i,j} = y_{i+p,j+q} \\ \beta & \text{otherwise,} \end{cases} \tag{3}$$

whereas, a non-horizontal one is defined as

$$\mathcal{E}(y_{i,j}, y_{i+p,j+q} \mid \overline{X}) = \begin{cases} \beta & \text{if } y_{i,j} = y_{i+p,j+q} \\ -\beta & \text{otherwise.} \end{cases} \quad (4)$$

- Doubleton-clique energies generating non-horizontal lines, blobs, and dots of active LGN cells at locations where the corresponding SC cells are silent

  A horizontal doubleton-clique energy is defined as

$$\mathcal{E}(y_{i,j}, y_{i+p,j+q} \mid \underline{X}) = \begin{cases} \beta & \text{if } y_{i,j} = y_{i+p,j+q} \\ -\beta & \text{otherwise,} \end{cases} \quad (5)$$

  whereas, a non-horizontal one is defined as

$$\mathcal{E}(y_{i,j}, y_{i+p,j+q} \mid \underline{X}) = \begin{cases} -\beta & \text{if } y_{i,j} = y_{i+p,j+q} \\ \beta & \text{otherwise.} \end{cases} \quad (6)$$

- Doubleton-clique energies generating non-oriented blobs of active SC cells

  Doubleton-clique energies are defined as

$$\mathcal{E}(x_{i,j} \mid x_{i+p,j+q}) = \begin{cases} -\gamma & \text{if } x_{i,j} = x_{i+p,j+q} \\ 0 & \text{otherwise.} \end{cases} \quad (7)$$

The clique strengths, β and γ are assumed positive.

4 Physiologically-imposed requirements

Any computational model aspiring to explain the orientation-selective inhibition mechanism in the striate visual cortex must fulfill—at least— the following physiologically-imposed requirements: (1) it must be implementable in a massively parallel manner; (2) it must converge fast enough — within one or two dozens of iterations at most —; (3) it must be completely LGN-driven; (4) it must be implementable by physiologically-plausible neural circuits; (5) it must tolerate noise — at the LGN level —; (6) it must be insensitive to contrast ([9]); and (7) it must be robust with respect to model-parameter deviations.

The algorithm given by Eq. (2) fulfills the first two requirements. Although the algorithm is LGN-driven, it is not completely LGN-driven. First, the local updating rule , Eq. (2) needs to be initialized because of its second term. This implies that a mechanism is needed for re-initializing whenever the activity pattern of LGN cells change. The need for re-initialization is one of the reasons that makes the algorithm not-completely LGN-driven. The second reason and the most important one is due to the third term in the updating rule, Eq. (2). To calculate the third term, we need to know the activity values of the active and less-active LGN cells, i.e., $\overline{Y}$ and $\underline{Y}$. In real world, these values change all the time both within and between incoming LGN activity patterns.

4.1 Making the algorithm completely LGN-driven

For both $x_{i,j} = \underline{X}$ and $x_{i,j} = \overline{X}$, it can be shown that

$$\max_{k \in \{\overline{Y}, \underline{Y}\}} \{ - \sum_{p \in \{-1,1\}} \sum_{q \in \{-1,1\}} \mathcal{E}(k, y_{i+p,j+q} \mid x_{i,j}) \} =$$
$$\max \{ - \sum_{p \in \{-1,1\}} \sum_{q \in \{-1,1\}} \mathcal{E}(y_{i,j}, y_{i+p,j+q} \mid x_{i,j}),$$
$$- \sum_{p \in \{-1,1\}} \sum_{q \in \{-1,1\}} \overline{\mathcal{E}}(y_{i,j}, y_{i+p,j+q} \mid x_{i,j})\} \} \quad (8)$$

where $\overline{\mathcal{E}}(y_{i,j}, y_{i+p,j+q} \mid x_{i,j}) = -\mathcal{E}(y_{i,j}, y_{i+p,j+q} \mid x_{i,j})$. Replacing the third term in Eq. (2) with the right-hand side of Eq. (8) makes the updating rule not only LGN-driven but also contrast insensitive.

The initialization and re-initialization problem can be solved by starting the algorithm with $\gamma = 0$, and then let γ increase as the algorithm proceeds until it saturates to a pre-assigned value. Note that Eq. (2) generates a Maximum-Likelihood (ML) estimate when $\gamma = 0$, (see [7]).

4.2 Making the algorithm noise tolerant

This property can be achieved simply by replacing the identity requirements, $y_{i,j} = y_{i+p,j+q}$, in (3)-(6) with a similarity requirement. Two activity levels z_1 and z_2 are considered similar if $\mid z_1 - z_2 \mid \leq T_s$. The similarity parameter, $T_s \geq 0$ controls the degree of similarity and thus the algorithm's insensitivity to the LGN-noise, that is, the LGN cells' instantaneous firing.

5 Simulation Result

The modified algorithm was applied to the test image shown in Fig. 2(a). Figure 2 (a) shows the activity pattern of 128×128 LGN-cells. the mean activity-level of the active cells in the central area is 100 spikes/sec, whereas of those in the peripheral area is 80 spikes/sec. The mean activity-level of the background less-active LGN cells is 30 spikes/sec. The activity levels of active and less-active LGN cells are corrupted as it can be seen in Fig. 2 (g). Figure 2 (b) shows the ML-estimate derived by applying the completely LGN-driven version of the MAP-estimation local updating rule with γ set to zero. Figures 2 (c)-(f) show the algorithm's 1st, 2nd, 3rd, and 6th iteration, respectively. $\gamma^{(n)} = 5\tanh(n)$ was used to update γ and similarity parameter, T_s was set to 10 spikes/sec and β to 10.

To see whether the algorithm is robust or not with respect to the doubleton-clique parameters, the values of both excitatory, $+\beta$, and inhibitory, $-\beta$, doubleton cliques were changed randomly at all sites and then the algorithm was applied to the same test image. Exactly the same results shown in Figs. 2 (b)-(f) were obtained.

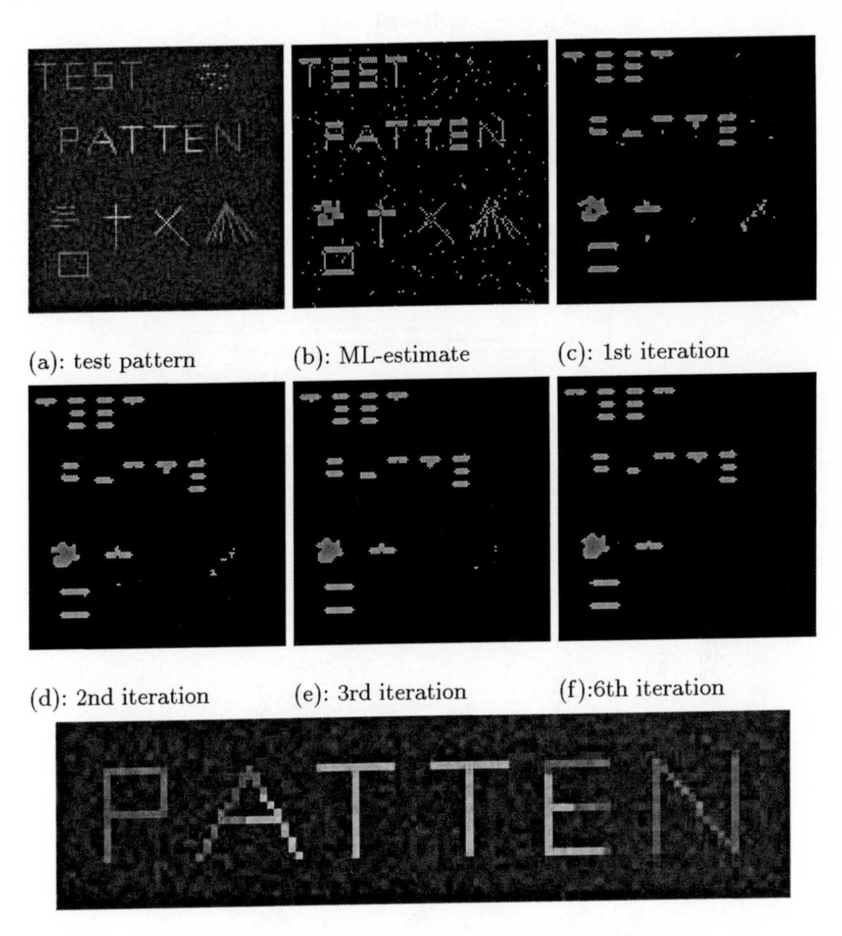

(g): an enlarged patch from the LGN-layer activity pattern

Fig. 2. The activity pattern of the LGN cells and the dynamics of the horizontally-tuned SC cells. Dark-gray in (a) denotes background less-active LGN cells, whereas lighter one denotes active LGN cells. Light-gray in (b)-(f) denotes active SC cells whereas black denotes silent ones.

6 A firing-rate-based neural circuitry

Figure 3 shows a neural module associated with horizontal cliques at ij-site. Micro-circuits 1 and 2 implement the similarity function. For instance in the micro-circuit 1, S is active and $\overline{S}$ is silent when Y_{ij} and Y_{ij-1} are similar and vice versa when they are not similar. Micro-circuits 3 and 4 implement the max operator (third term) in Eq. 2. For the sake of graphical simplicity, only inputs associated with the horizontal cliques have been shown at the inhibitory cells in the lower parts of micro-circuits 3 and 4. Actually these cells receive inputs

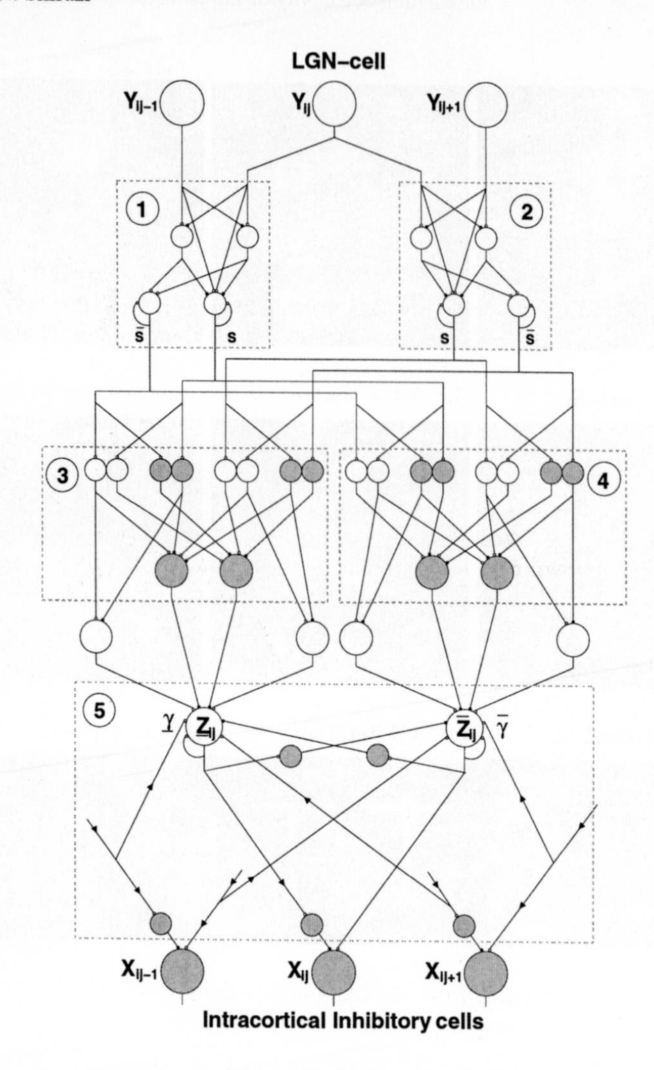

Fig. 3. The neural module associated with the ijth horizontal cliques. A white/gray circle denotes an excitatory/inhibitory interneuron. A small white/black circle denotes an excitatory/inhibitory synapse.

associated with the vertical, west-eastward, east-westward cliques as well. The excitatory cells depicted between 3-4 micro-circuit and micro-circuit 5 need to be incorporated for making the signals arrive at the excitatory cells $\underline{Z}_{ij}$ and $\overline{Z}_{ij}$ at almost the same time. The upper-part of micro-circuit 5 implements the arg-max operator in Eq. 2, whereas its lower-part implements the second term in Eq. 2 by feeding its previous state to the upper-part.

The left half of the module up to the $\underline{Z}_{ij}$-cell implements one of the two terms of the arg-max operator in Eq. 2. It is associated with the model generating non-horizontal patterns, i.e., the model associated with $\underline{X}$. The corresponding right half of the module implements the other term of the arg-max which is associated with the model generating horizontal patterns, i.e., the model associated with $\overline{X}$. The dynamics of the inhibitory feedbacks between the $\overline{Z}_{ij}$- and $\underline{Z}_{ij}$-cells forces a winner-take-all situation. If the $\underline{Z}_{ij}$-cell wins, then its activity will inhibit the X_{ij}-cell, otherwise the $\overline{Z}_{ij}$-cell is the winner which its activity will activate the X_{ij}-cell. The level of activity of the the X_{ij}-cell can be controlled by the activity levels of the $\overline{Z}_{ij}$- and $\underline{Z}_{ij}$-cells and the strengths of the involved inhibitory and excitatory synapses. In our implementation, it is essential for the $\underline{Z}_{ij}$-cell to make the X_{ij}-cell silent when it becomes active.

7 Conclusion

This article introduces a computational model that can account for the orientation selectivity of intracortical inhibitory cells in the striate visual cortex. The model has the following properties: (1) it is implementable in a massively parallel manner; (2) it is strikingly fast, it needs less than a dozen of iterations to converge; (3) it is driven exclusively by geniculate cells; (4) it is implementable by physiologically-plausible neural circuits; (5) it tolerates noise in firing-rates of geniculate cells; (6) it is insensitive to the contrast of visual stimuli; (7) it is insensitive to model-parameter deviations.

References

1. Hubel, D.H., Wiesel, T.N.: Receptive fields, binocular interaction and functional architecture in the cat's visual cortex. J. Phys. **160** (1962) 106–154
2. Sillito, A.M.: Inhibitory circuits and orientation selectivity in the visual cortex. Models of the Visual Cortex, Ed. by Rose, D. and Dobson, V.G., John Wiley & Sons, (1985) 396–407
3. Martin, K.A.C.: From single cells to simple circuits in the cerebral cortex. Quarterly J. Exp. Phys. **73** (1988) 637–702
4. Chung, S., Ferster, D.: Strength and orientation tuning of the thalamic input to simple cells revealed by electrically evoked cortical suppression. Neuron **20** (1998) 1177–1189
5. Gilbet, C.D.: Microcircuitry of the visual cortex. Ann. Rev. Neurosci. **6** (1983) 217–247
6. Besag, J.E.: Spatial interaction and the statistical analysis of lattice systems. J. Roy. Stat. Soc., Ser. B. **36** (1974) 192–226
7. N. Shirazi, M., Nishikawa, Y.: A Computational theory for orientation-selective simple cells based on the MAP estimation principle and Markov random fields. Lecture Notes in Computer Science 1240, Springer, (1997) 248–256
8. Geman, S., Geman, D.: Stochastic relaxation, Gibbs distributions, and the Bayesian restoration of images, IEEE Trans. Patt. Anal. & Mach. Intell. **PAMI-6** (1984) 721–741
9. Sclar, G., Freeman, R.D.: Orientation selectivity in the cat's striate cortex is invariant with stimulus contrast. Exp. Brain Res. **46** (1982) 457–461

Interpreting Neural Networks in the Frame of the Logic of Lukasiewicz*

Claudio Moraga[1] and Luis Salinas[2]

[1] Department of Computer Science, Computer Engineering and Computing Education, University of Dortmund, 44221 Dortmund, Germany. moraga@cs.uni-dortmund.de
[2] Department of Computer Science, Technical University Federico Santa María, Valparaíso, Chile. lsalinas@inf.utfsm.cl

Abstract. Neural networks using a piecewise linear ramp-step activation function may be interpreted as expressions in the propositional logic of Kleene-Lukasiewicz. These expressions even though information preserving may have a high degree of complexity impairing their understandability. The paper discloses a strategy which combines classical logic with the logic of Lukasiewicz to decompose a complex rule into a set of simpler rules that cover the former.

1. Introduction

In [CaT 98] the authors proved that a neural network using a piece-wise linear ramp-step activation function may be given an information preserving interpretation in the logic of Lukasiewicz or, in other words, such neural networks may be interpreted as fuzzy if-then rules evaluated pointwise by using the t-norm and conorm of Lukasiewicz [Gil 76], sometimes called bold t-norm, together with the standard fuzzy negation [Tri 79].

The main result is given by the following

Theorem 1 (T1) [CaT 98]:

Let $\Psi: \mathbb{R} \rightarrow [0,1]$, s.t. $\forall r \in [0,1]$ $\Psi(r)$ is non decreasing, $\Psi(r) = 0$ if $r \leq 0$ and $\Psi(r) = 1$ if $r \geq 1$. Ψ is called a bounded squashing function. Moreover let N_f be a neural network with one hidden layer using Ψ as activation function and one linear output node, realizing a function f: $(\mathbb{R}_c)^n \rightarrow \mathbb{R}_c$ (where $\mathbb{R}_c$ is a compact subset of $\mathbb{R}$). The following holds:

(i) $$f(x) = \sum_{j=1}^{k} \beta_j \Psi\left(b_j + \sum_{i=1}^{n} w_{ij} x_i \right)$$

* Work leading to this paper has been partially supported by the Ministry of Education and Research, Germany, under grant BMBF-CH-99/023 and by the Technical University Federico Santa María, Chile, under grant UTFSM/DGIP-Intelligent Data Mining in Complex Systems/2000-2001.

J. Mira and A. Prieto (Eds.): IWANN 2001, LNCS 2084, pp. 142-149, 2001.

(ii) $f(x) = \sum_{j=1}^{k} \beta_j \Psi\left(P_j(\alpha_{1,j}y_1, \ldots, \alpha_{n,j}y_n, \alpha_{n+1,j})\right)$

where $y_i = x_i + M_i$, $1 \leq i \leq n$, $M_i \in \mathbb{Z}$, $\alpha_{i,j} \in [-1,1]$ s.t. $\alpha_{i,j}(x_i + M_i) \in [0,1]$, $\alpha_{n+1,j} \in [0,1]$ and P_j: $[0,1]^{n+1} \rightarrow [0,1]$ is a truth function of a proposition in the Kleene-Lukasiewicz propositional logic.

T1 is supported by the following

Auxiliary Lemma (A1):

$\forall\, r \in \mathbb{R}$ and $a \in [0,1]$ holds:

$$\phi(r + a) = [\phi(r) \oplus a] \otimes [\phi(-r)]^c,$$

where ϕ is a bounded squashing function s.t. $\phi(a)=a$ (*i.e.* ϕ is a unit ramp function) and $\forall\, x,y \in [0,1]$, $x \oplus y = \min(1, x+y)$, $x \otimes y = \max(0, x+y-1)$, $x^c = 1-x$.

2. Contributions

The present work introduces a new auxiliary Lemma A2 and several corollaries. (*proofs of which are given in the appendix*). Furthermore it analyzes the complexity of using T1 and suggests a hybrid method combining classical and fuzzy logic to improve the understandability of the extracted rules.

Auxiliary Lemma (A2):

$\forall\, r \in \mathbb{R}$ and $a \in [0,1]$ holds:

$$\phi(r - a) = [\phi(r - 1) \oplus a^c] \otimes \phi(r)$$

Corollaries to A1 and A2:

$\forall\, r \in \mathbb{R}$, $a, x \in [0,1]$, $n, q \in \mathbb{N}$ and $K \in \mathbb{Z}$ holds:

C.1: $\phi(r + 1) = [\phi(-r)]^c$
$\phi(-r) = [\phi(r + 1)]^c$

C.2: $\phi(na) = [a \oplus a \oplus \ldots \oplus a]$ (n times)

C.3: $$\phi(Kx + a) = \begin{cases} \bigoplus_{i=1}^{K} x \oplus a & \text{if } K > 0 \\ a \otimes \bigotimes_{i=1}^{|K|} x^c & \text{if } K < 0 \\ a & \text{if } K = 0 \end{cases}$$

$$\textbf{C.4}: \phi(nx + qa) = \bigoplus_{i=1}^{n} x \oplus \bigoplus_{j=1}^{q} a$$

C.5: $\phi(2x - 1) = x \otimes x = (x^c \oplus x^c)^c$

C.6: $\phi(1 - 2x) = x^c \otimes x^c = (x \oplus x)^c$

C.7: $\phi(a - r) = [\langle\phi(r+1)\rangle^c \oplus a] \otimes [\phi(r)]^c$

C.8: $\phi(r + a) = [\phi(r) \oplus a] \otimes \phi(r + 1)$

C.9: $\phi(3x - 1) = [(x \otimes x) \oplus x] \otimes [x \oplus x]$

Complexity analysis:

Consider a most simple neural network with only one elementary processor using ϕ as activation function, two bounded inputs and one output, as shown in figure 1.

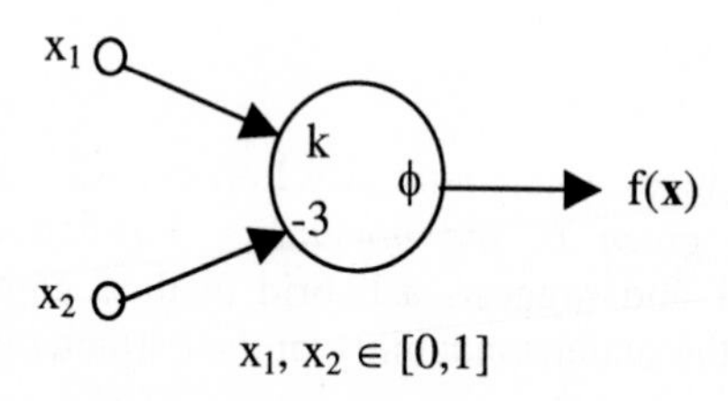

Fig. 1. Test neural network

(i) Let k = 1. Then:

$$f(\mathbf{x}) = \phi(x_1 - 3x_2) =_{A1} [\phi(-3x_2) \oplus x_1] \otimes [\phi(3x_2)]^c = x_1 \otimes [\phi(3x_2)]^c$$

$$=_{C2} x_1 \otimes [x_2 \oplus x_2 \oplus x_2]^c$$

$$f(\mathbf{x}) = \left(p_1 \wedge \neg \bigvee_{i=1}^{3} p_2 \right)(x_1, x_2)$$

(ii) Let k = 2. Then:

$$f(\mathbf{x}) = \phi(2x_1 - 3x_2) = \phi[(x_1 - 3x_2) + x_1] =_{A1} [\phi(x_1 - 3x_2) \oplus x_1] \otimes [\phi(3x_2 - x_1)]^c$$

but

$$\phi(3x_2 - x_1) =_{A2} [\phi(3x_2 - 1) \oplus x_1^c] \otimes \phi(3x_2)$$

and

$$\phi(3x_2 - 1) = \phi((2x_2 - 1) + x_2) =_{C8} [\phi(2x_2 - 1) \oplus x_2] \otimes \phi(2x_2)$$

$$=_{C5} [(x_2 \otimes x_2) \oplus x_2] \otimes (x_2 \oplus x_2)$$

Then

$$\phi(3x_2 - x_1) = [\{[(x_2 \otimes x_2) \oplus x_2] \otimes (x_2 \oplus x_2)\} \oplus x_1^c] \otimes \phi(3x_2)$$

Replacing this expression in the equation for $f(\mathbf{x})$ follows that:

$$f(\mathbf{x}) = [\phi(x_1 - 3x_2) \oplus x_1] \otimes \{[\{[(x_2 \otimes x_2) \oplus x_2] \otimes (x_2 \oplus x_2)\} \oplus x_1^c] \otimes \phi(3x_2)\}^c$$

and with the result of (i)

$$f(\mathbf{x}) = [\{x_1 \otimes [x_2 \oplus x_2 \oplus x_2]^c\} \oplus x_1] \otimes$$

$$\{[\{[(x_2 \otimes x_2) \oplus x_2] \otimes (x_2 \oplus x_2)\} \oplus x_1^c] \otimes \phi(3x_2)\}^c$$

leading to:

$$f(\mathbf{x}) = \left[p_1 \wedge \neg \left(\bigvee_{i=1}^{3} p_2 \right) \vee p_1 \right] \wedge$$

$$\wedge \neg \left[\left\langle \left((p_2 \wedge p_2) \vee p_2 \right) \wedge (p_2 \vee p_2) \vee \neg p_1 \right\rangle \wedge \left(\bigvee_{i=1}^{3} p_2 \right) \right] (x_1, x_2) \qquad (1)$$

It becomes apparent that the simple change of k from 1 to 2 induces a tremendous increment of complexity. Since in the logic of Lukasiewicz there exist no distributivity between $\wedge$ and $\vee$, and furthermore the logic is not idempotent, it is extremely difficult to simplify the above given expression. It is beyond discussion that the interpretability of such an expression as a rule by a non-expert is far from being useful.

In what follows it is shown, based on the same example, that a combination of classical logic with the logic of Lukasiewicz using a *divide and conquer* approach, alleviates the problems discussed above.

(iii) If $x_1 \leq ½$ then:

$$\phi(2x_1 - 3x_2) = \phi((-3x_2) + (2x_1)) =_{A1} [\phi((-3x_2) \oplus 2x_1] \otimes [\phi(3x_2)]^c$$

$$= (2x_1) \otimes (x_2 \oplus x_2 \oplus x_2)^c$$

leading to the rule:

$$\text{if } x_1 \leq \frac{1}{2} \text{ then } f(x) = p_1 \wedge \neg \left(\bigvee_{i=1}^{3} p_2 \right) (2x_1, x_2) \qquad (2a)$$

with the special case

$$\text{if } x_1 = \frac{1}{2} \text{ then } f(x) = \neg\left(\bigvee_{i=1}^{3} p_2\right)(x_2) \tag{2b}$$

(iv) If $x_2 = 1/3$ then

$\phi(2x_1 - 3x_2) = \phi(2x_1 - 1) =_{C.5} x_1 \otimes x_1$

leading to:

$$\text{If } x_2 = 1/3 \text{ then } f(\mathbf{x}) = (p_1 \wedge p_1)(x_1, x_2) \tag{3}$$

(v) If $x_2 < 1/3$ then $0 \leq (3x_2) < 1$

$\phi(2x_1 - 3x_2) =_{A2} [\phi(2x_1 - 1) \oplus (3x_2)^c] \otimes \phi(2x_1)$

$=_{C.5,\, C.2} [x_1 \otimes x_1 \oplus (3x_2)^c] \otimes (x_1 \oplus x_1)$

leading to:

$$\text{If } x_2 < 1/3 \text{ then } f(\mathbf{x}) = \left(\left[\bigwedge_{i=1}^{2} p_1 \vee \neg p_2\right] \wedge \bigvee_{l=1}^{2} p_1\right)(x_1, 3x_2) \tag{4}$$

(vi) If $1 \geq x_1 > 1/2$ and $1 \geq x_2 > 1/3$ then $1 \geq (2x_1 - 1) > 0$ and $(x_1 - 3x_2) \leq 0$.

It follows:

$\phi(2x_1 - 3x_2) = \phi((1-3x_2) + (2x_1-1)) =_{A1} [\phi(1- 3x_2) \oplus (2x_1 - 1)] \otimes [\phi(3x_2 - 1)]^c$

$= (2x_1 - 1) \otimes [\phi(3x_2 - 1)]^c =_{C9} \phi(2x_1 - 1) \otimes [[(x_2 \otimes x_2) \oplus x_2] \otimes (x_2 \oplus x_2)]^c$

$=_{C5} (x_1 \otimes x_1) \otimes [[(x_2 \otimes x_2) \oplus x_2] \otimes (x_2 \oplus x_2)]^c$

This leads to:

If $1 \geq x_1 > 1/2$ and $1 \geq x_2 > 1/3$ then

$$f(\mathbf{x}) = \left(\bigwedge_{i=1}^{2} p_1 \wedge \neg\left[\left(\bigwedge_{j=1}^{2} p_2 \vee p_2\right) \wedge \bigvee_{l=1}^{2} p_2\right]\right)(x_1, x_2) \tag{5}$$

Eqs. (4) and (5) represent the most complex rules. They are however strongly simpler than the rule expressed with eq. (1). The covering relation is shown in figure 2, where the numbers in parentheses refer to the corresponding equations.

f(**x**)	$x_2<1/3$	$x_2=1/3$	$x_2>1/3$
$x_1<1/2$	(2a)(4)	(2a)(3)	(2a)
$x_1=1/2$	(2b)(4)	(2b)(3)	(2b)
$x_1>1/2$	(4)	(3)	(5)

Fig. 2: Covering matrix for f(**x**)

It becomes apparent that the simple observation of the problem when the variables are constrained to contribute less than 1, exactly 1 and more than 1 to the related expressions allows deducing a set of simple rules with the general structure

if <domain restrictions> then <simplified Lukasiewicz expression>

which represents a combination of classical and Lukasiewicz logics and increases the understandability of the rules.

References

[CaT 97] Castro J.L., Trillas E.: The Logic of Neural Networks. *Mathware and Softcomputing*, Vol. V (1), 23-27, (1998)
[Gil 76] Giles R.: Lukasiewicz logic and fuzzy set theory. *Int. Jr. Man-Machine Studies* 8, 313-327, (1976)
[Tri 79] Trillas E.: Sobre funciones de negación en la teoría de conjuntos difusos, *Stochastica* **III-I**, 47-60, (1979)

Appendix

Auxiliary Lemma (A2):

$\forall$ r $\in \mathbb{R}$ and a $\in$ [0,1] holds: $\phi(r - a) = [\phi(r - 1) \oplus a^c] \otimes \phi(r)$

Proof:

$\phi(r - a) = \phi(r - 1 + 1 - a) = \phi(r - 1 + a^c) = [\phi(r - 1) \oplus a^c] \otimes [\phi(1 - r)]^c =$

$[\phi(r - 1) \oplus a^c] \otimes \{[\phi(- r) \oplus 1] \otimes [\phi(r)]^c\}^c = [\phi(r - 1) \oplus a^c] \otimes \phi(r)$

Corollaries

$\forall\, r \in \mathbb{R}$, $a, x \in [0,1]$, $n, q \in \mathbb{N}$ and $K \in \mathbb{Z}$ holds:

C.1: $\phi(r+1) = [\phi(-r)]^c$ or $\phi(-r) = [\phi(r+1)]^c$

Proof:

$$\phi(r+1) =_{A1} [\phi(r) \oplus 1] \otimes [\phi(-r)]^c = 1 \otimes [\phi(-r)]^c = [\phi(-r)]^c$$

C.2: $\phi(na) = [a \oplus a \oplus \ldots \oplus a]$ (n times)

Proof: (Induction)

$$\phi(a) = a$$

Assume

$$\phi(q \cdot a) = [a \oplus a \oplus \ldots \oplus a] \text{ (q times)}$$

Then

$$\phi((q+1)a) = \phi(q \cdot a + a) =_{A1} [\phi(q \cdot a) \oplus a] \otimes [\phi(-q \cdot a)]^c = [\phi(q \cdot a) \oplus a] \otimes [0]^c =$$

$$= [\phi(q \cdot a) \oplus a] \otimes [\phi(-q \cdot a)]^c = [a \oplus a \oplus \ldots \oplus a] \text{ (q +1 times)}$$

The assertion follows.

C.3: $$\phi(Kx + a) = \begin{cases} \bigoplus_{i=1}^{K} x \oplus a & \text{if } K > 0 \\ a \otimes \bigotimes_{i=1}^{|K|} x^c & \text{if } K < 0 \\ a & \text{if } K = 0 \end{cases}$$

Proof:

K>0: $$\phi(Kx + a) =_{A1} [\phi(Kx) \oplus a)] \otimes [\phi(-Kx)]^c = [\phi(Kx) \oplus a)] \otimes 0^c =$$

$$= \phi(Kx) \oplus a =_{C2} x \oplus x \oplus x \oplus \ldots \oplus x \oplus a$$

K<0: $$\phi(Kx + a) =_{A1} [\phi(Kx) \oplus a)] \otimes [\phi(-Kx)]^c = a \otimes [\phi(|K|x)]^c =_{C2}$$

$$= a \otimes x^c \oplus x^c \oplus x^c \oplus \ldots \oplus x^c$$

C.4: $\phi(nx + qa) = \bigoplus_{i=1}^{n} x \oplus \bigoplus_{j=1}^{q} a$

Proof: Apply twice C3 first case.

C.5: $\phi(2x - 1) = x \otimes x = (x^c \oplus x^c)^c$

Proof:

$\phi(2x - 1) = \phi((x - 1)+x) =_{A1} [\phi(x - 1) \oplus x] \otimes [\phi(1-x)]^c = x \otimes [\phi(x^c)]^c = x \otimes x$

C.6: $\phi(1 - 2x) = x^c \otimes x^c = (x \oplus x)^c$

Proof:

$\phi(1 - 2x) = \phi(-(2x - 1)) =_{C1} [\phi(2x)]^c =_{C2} (x \oplus x)^c = x^c \otimes x^c$

C.7: $\phi(a - r) = [\langle\phi(r+1)\rangle^c \oplus a] \otimes [\phi(r)]^c$

Proof: Use A1 and apply C1 to the first term. The assertion follows

C.8: $\phi(r + a) = [\phi(r) \oplus a] \otimes \phi(r + 1)$

Proof: Use A1 and apply C1 to the last term. The assertion follows

C.9: $\phi(3x - 1) = [(x \otimes x) \oplus x] \otimes [x \oplus x]$

Proof:

$\phi(3x - 1) = \phi((2x - 1) + x) =_{A1} [\phi(2x - 1) \oplus x] \otimes [\phi(1 - 2x)]^c$

$=_{C5, C6} [x \otimes x \oplus x] \otimes [x \oplus x]$

Time-Dispersive Effects in the J. Gonzalo's Research on Cerebral Dynamics

Isabel Gonzalo[1] and Miguel A. Porras[2]

[1] Departamento de Optica. Facultad de Ciencias Físicas.
Universidad Complutense de Madrid. Ciudad Universitaria s/n. 28040-Madrid. Spain
igonzalo@eucmax.sim.ucm.es

[2] Departamento de Física Aplicada. ETSIM. Universidad Politécnica de Madrid.
Rios Rosas 21. 28003-Madrid. Spain

Abstract. In the present work, certain aspects, restricted to the visual system, of the J. Gonzalo's research, are interpreted according to a simple linear theory of time-dispersion applied to the dynamic cerebral system, characterized by its excitability and its reaction velocity.

1 Introduction

As exposed in previous work [8]-[12], J. Gonzalo characterized what he termed the *central syndrome* associated to a unilateral lesion in the parieto-occipital cortex, equidistant from the visual, tactile and auditory projection areas (central lesion). The projections paths are untouched while rather unspecific "central" cerebral mass is lost, producing a deficit in the cerebral excitability.

In the central syndrome *all* sensory systems are involved, in all their functions and with *symmetric* bilaterality. As was explained previously [8]-[11], the deficit in the excitability produces a diminution in the reaction (response) velocity of the cerebral system leading to an allometric *dissociation* or *desynchronization* of sensory qualities (normally united in perception) according to their excitability demands. Concerning the visual system, for instance, when the illumination of a vertical white arrow diminishes, the perception of the arrow is at first upright and well-defined; next the arrow is perceived to be more and more rotated, becoming at the same time smaller, and losing its form and colors in a well defined order. The sensorial perception thus splits into components; several functions appear, as the direction function which gives place to the striking phenomenon of the inverted vision: about 160 degrees in patient "M" under low stimulation (systematically studied for the first time by J. Gonzalo [7, 8]).

The importance of this type of involvement (central syndrome) lies in the fact that the change in the nervous excitability can reveal the dynamics of the cerebral cortex, the organization of the sensorial structures and a continuity from the simplest sensorial function to the more complex ones, according to the nervous excitability and following a physiological order. The cerebral system follows the same organization plan and the same physiological laws as in the normal case but in a smaller scale [9]. J. Gonzalo's contribution to the knowledge of the brain

J. Mira and A. Prieto (Eds.): IWANN 2001, LNCS 2084, pp. 150–157, 2001.

is connected with the research of other authors, e.g., [6] [13]-[16] [20]-[23], and was considered [1]-[4] [18] accurate, rich in physiological data and clinical proofs, as well as theoretically elaborated. His research appears to be related to recent approaches in cerebral dynamics in which integrative and adaptative aspects are involved, e.g., [5, 17, 22].

In the present work we deal with the excitability and the reaction velocity of the cerebral system. It is then suitable to study cases with different magnitude of central syndrome since their cerebral systems are slower in different degree from the normal case. We show that the simple mathematical theory of linear time-dispersion can describe some manifestations of the central syndrome. This simple model can also account for the phenomenon of summation [8, 9]. The slowness of the cerebral system in the central syndrome makes the cerebral excitation to dissipate slowly, confering the system with temporal summation capability (iterative aptitude). There is also intersensorial summation or *facilitation*, that is, the perception in a sensory system (e.g., visual) can be improved by stimulating other sensory systems. In particular is very noticeable the *reinforcement* or facilitation obtained by strong muscular contraction (very pronounced in case M). Unlike the summation by iteration, this type of summation modifies the system essentially, and makes the cerebral system more rapid and excitable, i.e., supplies in part the neural mass lost in the central lesion [8, 9]. This effect is greater as the deficit (the lesion) is greater, but is null in a normal case.

2 The Model

In the theory of dynamical systems, a system is said to be *time-dispersive* if its response at a time t depends not only on the stimulus at that time, but also at all previous times, $t' \leq t$. Thus, if we consider an stimulus $S(t')$ acting on the cerebral system, the excitation $E(t)$ produced at time t in the cerebral system is determined by the stimulus at times $t' \leq t$. In the linear case, the excitation can be expressed by

$$E(t) = \int_{-\infty}^{t} \chi(t, t') S(t') dt' = \int_{0}^{\infty} \chi(\tau) S(t - \tau) d\tau, \quad (1)$$

where $\tau = t - t'$, χ is related to the capability of the system to be excited and will be called excitation permeability. The quantity χ can also be interpreted as the excitation produced by a delta pulse stimulus, i.e., $E_\delta(t) = \int_0^\infty \chi(\tau)\delta(t-\tau)d\tau = \chi(t)$. A simple form of χ, which accounts for the presumed dispersive behaviour, is

$$\chi(\tau) = \chi_0 e^{-a\tau}. \quad (2)$$

With this choice, a delta pulse stimulus results in a sudden growth of the excitation up to χ_0 followed by the exponential temporal decay $e^{-a\tau}$, where a is related to the velocity of the system. For Eq. (1), we then have

$$E(t) = \int_{0}^{\infty} \chi_0 e^{-a\tau} S(t - \tau) d\tau. \quad (3)$$

For the case of a constant stimulus S during the time interval [0,t], the excitation at time t can evaluated from Eq. (3), to be

$$E(t) = \chi_0 S \int_0^t e^{-a\tau} d\tau = \frac{\chi_0}{a} S(1 - e^{-at}). \tag{4}$$

3 Threshold Curves of Nervous Excitation

We examine now, in the framework of the above model, some of the basic physiological data of the visual system obtained by J. Gonzalo [8, 9].

(a) We first chose the fundamental threshold curves stimulus-time (also known as strength-duration curves) of the nervous excitation by electrical stimulation of retina with capacitor discharge (cathode on eyelid). As shown in Fig.1(a), the data were taken for the normal case N, patient M (very pronounced central syndrome), the same patient under strong muscular contraction (40 Kg held in each hand), i.e., under *reinforcement*, and for patient T with less intense central syndrome (smaller central lesion). The main source of errors are related to the experimental difficulties to obtain the data [8], especially for patient M, and for long times in all cases. Each point of the curves represents, for a given electrical intensity $I = V/R$ (expressed indirectly in volts V), the needed time $t = RC$ (expressed in microfarads) during which the intensity must act to obtain minimum luminous sensation. V is the potential applied, R is a constant resistance and C is the capacity of the capacitor.

These curves can be explained from the above model if a few reasonable assumptions are made. We assume that the stimulus just on the retina is proportional to the electrical intensity, i.e., $S = G(V/R)$, G being a constant, and that the cerebral excitation threshold E_{th} necessary to have a particular sensorial function (minimum luminous sensation) is the same for the patient with central syndrome (functional depression) and for the normal man. The threshold can be reached by different ways, e.g., by low intensity during long time or higher intensity during shorter time. The corresponding values S and t are related by Eq. (4) for $E = E_{th}$ constant, that is,

$$S = \frac{aE_{th}}{\chi_0} \frac{1}{1 - e^{-at}}. \tag{5}$$

What is expected to be different for patients with central lesions with respect to a normal case is the value of the permeability χ_0 and the reaction velocity a of the cerebral system. To make a direct comparison between Eq. (5) and experimental data, we rewrite Eq. (5) as

$$V = \frac{R}{G} \frac{aE_{th}}{\chi_0} \frac{1}{1 - e^{-aR10^{-6}c}} \equiv \frac{B}{1 - e^{-Ac}} \tag{6}$$

where c is given in microfarads, $A \equiv aR10^{-6}$ and $B \equiv (RE_{th}/G)(a/\chi_0)$. The fitted curves given by Eq. (6) are shown in Fig. 1(a). The obtained values to the best fitting are, $B_N = 2$, $A_N = 2$ for normal case; $B_T = 11.5$, $A_T = 1.3$ for case

T; $B_{M_r} = 14.5$, $A_{M_r} = 0.7$ for case M reinforced; and $B_M = 22$, $A_M = 0.48$ for M inactive. Patient M under reinforcement (that supplies in part the lesion) has an active cerebral system intermediate between M inactive and T. We see that, when the deficit in the active central cerebral mass grows, the reaction velocity a decreases: $a_T/a_N = 0.65$, $a_{Mr}/a_N = 0.35$, $a_M/a_N = 0.24$, and the excitation permeability χ_0 also decreases: $\chi_{0T}/\chi_{0N} = 0.17$, $\chi_{0Mr}/\chi_{0N} = 0.14$, $\chi_{0M}/\chi_{0N} = 0.09$, but the ratio a/χ_0 increases significantly: $(a/\chi_0)_T/(a/\chi_0)_N = 5.91$, $(a/\chi_0)_{M_r}/(a/\chi_0)_N = 7$, $(a/\chi_0)_M/(a/\chi_0)_N = 12$.

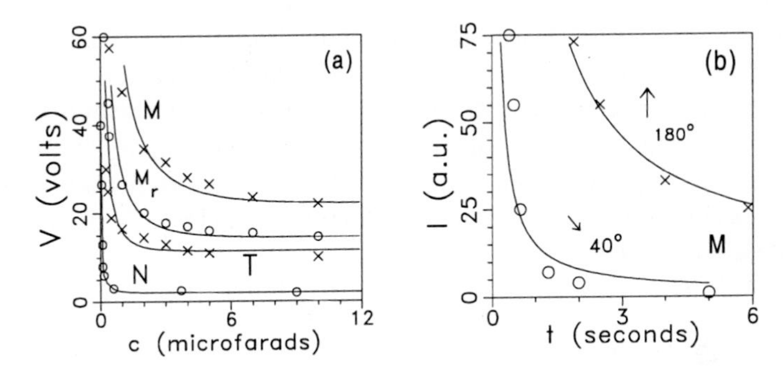

Fig. 1. Threshold curves stimulus-time. **(a)** Electrical stimulation of retina: For a given intensity $I = V/R$, the time necessary $t = R10^{-6}c$ to obtain minimum luminous sensation, for normal case (N) and patients with central lesions (T and M) [8, 9] (see the text). **(b)** Two perception levels (40° and 180°) of the direction function for case M, right eye. For each level, time necessary to illuminate with an intensity I an upright test arrow in order to be perceived as rotated a certain angle [8, 9] (see the text).

(b) Here we consider also threshold curves stimulus-time, but for different perception levels of a particular sensorial function perceived by the same observer (patient M, right eye). In order to illustrate an striking phenomenon we chose the direction function explained in the preceding section. Different perception levels correspond to different degrees of rotation under which a vertical upright white test arrow is perceived, according to the intensity of illumination of the arrow. Under low stimulation (illumination), the arrow is perceived as very turned, almost inverted in patient M, but under high illumination (and then greater cerebral excitation), the image becomes upright. From detailed studies of visual, tactile and auditory inversion, it was proposed [8, 9] that the perception originated in the projection area is inverted and constricted, but is then magnified and reinverted, i.e., elaborated or integrated, in the whole cortex, and particularly in the central zone. The reinversion of the arrow made by the cerebral system was called direction function and reaches its maximum value (180 degrees) when the image is seen upright.

Each point of Fig. 1(b) represents the time necessary to illuminate the test arrow with an intensity I, in order to be perceived as turned 140 degrees ($180 - 140 = 40$ degrees of direction function), or in order to be perceived upright (180 degrees of direction function). We consider that the stimulus S on the cerebral system is $S = KI$, K being a constant factor. From the theoretical Eq. (5) we then obtain,

$$I = \frac{aE_{th}}{\chi_0 K} \frac{1}{1 - e^{-at}} \equiv \frac{D}{1 - e^{-at}}, \tag{7}$$

where $D \equiv aE_{th}/(\chi_0 K)$ and E_{th} is here the threshold or required excitation for a given level of perception (40° or 180°). The fitting of Eq. (7) to the experimental data leads to $D_{40} = 1.3, a_{40} = 0.09$ and $D_{180} = 10.6, a_{180} = 0.088$. It is reasonable to assume that the different values of D are due to different values of E_{th} only, the factor (a/χ_0) being considered to be the same, in a first approximation, for the different sensorial perception levels of a given sensorial function, as it can be seen in the next case (c). We can then assert that for the most "elaborated" perception (maximum of the direction function), the required excitation $E_{th}^{(180)}$ is greater than for the less "elaborated" perception $E_{th}^{(40)}$. In addition, since $a_{180} \approx a_{40}$, we can deduce from the above assumption that $\chi_0^{180} \approx \chi_0^{40}$. This suggests that each sensorial function (direction in this case) is characterized by the values of a and χ_0.

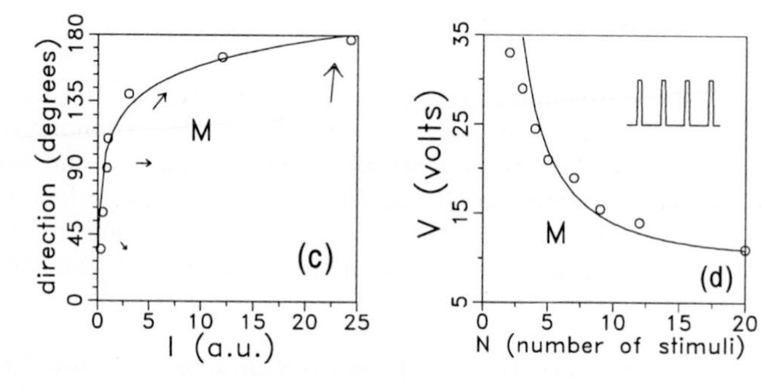

Fig. 2. (c) Perceived direction of the upright test arrow for each value of the intensity I illuminating the arrow for indefinite time (case M, right eye) [8, 9]. **(d)** Threshold curve for patient M in similar experience as in Fig. 1(a): For a given electrical intensity (given indirectly by V), the number of stimuli (each one of duration of the order of milliseconds and spaced 1/12 seconds) necessary to perceive minimum luminous sensation [8, 9].

(c) We continue with patient M, right eye. Figure 2(c) shows the maximum value reached by the direction function (perceived direction of the upright test arrow) for each value of the the intensity I illuminating the arrow for indefinite time, i.e., $t \to \infty$. The experimental data were obtained by switching on the illumination I and waiting for long time enough for the image that the patient sees to become stable. From our model, Eq. (4), we obtain the excitation at $t \to \infty$ as $E_\infty = \frac{\chi_0}{a} S$. If we assume that the perception (sensation) of a sensorial function F (here the direction function) is proportional to the logarithm of the cerebral excitation, we obtain the Fechner law, well-known in normal cases,

$$F = Z \log E_\infty = Z \log \frac{\chi_0 S}{a} = Z \log \frac{\chi_0 Q I}{a} \equiv Z \log(YI), \tag{8}$$

where Z and Q are constants (we consider that S is proportional to the intensity, $S = QI$) and $Y \equiv Q\chi_0/a$ is constant if (χ_0/a) can be considered as a constant. In fact, the fitting of Eq. (8) to the data (curve of Fig. 2(c)) is good enough with the values $Z = 54$, $Y = 90$, to consider that the logarithm law describes satisfactorily the sensorial growth versus the stimulus, and that the ratio χ_0/a can be considered constant for a given sensorial function of a given cerebral system. For other data in the case of patient T for example, or in the case of other sensorial functions, the behaviour is similar.

4 Iterative Aptitude

As shown in Sec. 2, the cerebral excitation for a single short stimulus has an exponential decay. If a second stimulus arrives before the excitation has completely fallen down, there is an effect of summation in the cerebral excitation. The same can be said for a train of impulse stimuli, so that it is possible to achieve the threshold to produce a sensorial perception despite a single one is unable to do it. The iterative aptitude is then more pronounced as the cerebral system is slower, i.e., as the central lesion is greater [8, 9].

In Fig. 2(d) this property was shown [8, 9] in a similar experience as in Fig. 1(a), except that in horizontal axis we have the number of stimuli. Each one has a duration $t_0 = R'C$ of the order of milliseconds ($C = 1.5 \times 10^{-6}$ farads and R' of the order of several thousand ohms), and the time between successive stimuli is $T = 1/12$ s.

Let us now consider that $S(t)$ is a train of stimuli, as illustrate in the inset of Fig. 2(d), which can be expressed as

$$S(t) = S \sum_{n=0}^{\infty} \left[\theta(t - nT) - \theta(t - nT - t_0)\right], \tag{9}$$

where $\theta(t)$ is the Heaviside step function, and S is a constant. Evaluation of Eq. (1) with Eq. (9) yields

$$E(t) = \frac{\chi_0 S}{a} \left[f(t) - g(t)e^{-at}\right], \tag{10}$$

where $f(t) \equiv \sum_{n=0}^{\infty} [\theta(t - nT) - \theta(t - nT - t_0)]$, $g(t) \equiv \sum_{n=0}^{\infty} [\theta(t - nT)e^{anT}$ $-\theta(t - nT - t_0)e^{a(nT+t_0)}]$. The patient has received $N = 1, 2, ...$ complete stimuli at $t = (N-1)T + t_0$. As in the case (a) of Sec. 3, $S = G(V/R')$, and $E(t) = E_{th}$ for each experimental point. Then, from Eq. (10) and $B' \equiv (R'E_{th}/G)(a/\chi_0)$ we can write

$$V = \frac{B'}{f(t) - g(t)e^{-at}} \tag{11}$$

for the relation voltage-number of impulses to have minimum luminous sensation. The fitting to the data of patient M inactive, right eye, yields $B' = 0.55$ and $a = 1.65$. From the available data it is not possible to know the value of R', thus, it is not possible to know the exact value of t_0. However, in contrast to case (a) of Sec. 3, without iteration, here we can neglect t_0 compared to $(N-1)T$ in the expression of t, so that the value of a can be obtained from the fitting.

5 Conclusions

We have shown that it is possible to model the temporal dynamics of simple sensorial functions as a first-order linear time-dispersive system, having, as a first approach, a response $\chi_0 e^{-a\tau}$ to a short impulse stimulus. The reaction velocity a and the permeability to the excitation χ_0 of the cerebral system decrease when the deficit in the active "central" cerebral mass, or central lesion, grows, but the ratio a/χ_0 increases significantly. For a given active cerebral system, the ratio a/χ_0 can be considered to be a constant for each sensorial function. Even we can suggest that a and χ_0 are two constants that characterize the sensorial function. The fact that the sensorial perception is related to the logarithm of the stimulus also in pathological cases was already found by J. Gonzalo, and here it is expressed in the framework of the model employed. It allows to confirm that the perception is proportional to the logarithm of the cerebral excitation, at least in the range of the data analyzed. The model can also describe the iterative aptitude, and the value of a can be deduced from it.

This simple model can then be considered as a first approach to the mathematical description of some aspects of the cerebral dynamics studied by J. Gonzalo. The model can be improved in several ways: the excitation permeability $\chi(\tau)$ of Eq. (2) can be replaced by a more realistic function; the macroscopic magnitudes a and χ_0 should be related to the neural mass and its microscopic structure (quantity of neurons, conexions); changes in the value of (a/χ_0) for different sensorial functions must be analyzed; and intersensorial summation is expected to be described by including spatial dispersion.

References

1. de Ajuriaguerra J. et Hécaen H.: *Le Cortex Cerebral. Etude Neuro-psychopathologique*, Masson, Paris 1949.
2. Bender M.B. and Teuber H.L.:"Neuro-ophthalmology" in: Progress in Neurology and Psychiatry, Ed. Spiegel E.A., **III** , Chap. 8, 163-182 (1948)
3. Critchley Mc.D.: *The Parietal lobes*, Arnold, London 1953.
4. Delgado García A.G.: *Modelos Neurocibernéticos de Dinámica Cerebral*, Ph.D. Thesis. E.T.S. de Ingenieros de Telecomunicación. U.P.M. Madrid 1978.
5. Engel A.K. et al.: "Temporal coding in the visual cortex: new vistas on integration in the nervous system", TINS, **15**, 6, 218-226 (1992).
6. Gelb A. and Goldstein K.: *Psychologische Analysen hirnpathologischer Fälle.* Barth, Leipzig 1920. "Psychologische Analysen hirnpathologischer Fälle auf Grund Untersuchungen Hirnverletzter: VII Ueber Gesichtsfeldbefunde bei abnormer Ermüdbarkeit des Auges (sog. Ringskotome)", Albrecht v. Graefes Arch. Ophthal., **109**, 387-403 (1922).
7. Gonzalo J. "Investigaciones sobre Dinámica Cerebral. La acción dinámica en el sistema nervioso. Estructuras sensoriales por sincronización cerebral", Memory communicated to Consejo Superior de Investigaciones Científicas, Madrid 1941.
8. Gonzalo J.: *Investigaciones sobre la nueva dinámica cerebral. La actividad cerebral en función de las condiciones dinámicas de la excitabilidad nerviosa*, Publ. C.S.I.C., Instituto S. Ramón y Cajal. Vol. **I**: Optic functions, 342 pp., Madrid 1945. Vol. **II**: Tactile functions, 435 pp., Madrid 1950.

9. Gonzalo J.: "Las funciones cerebrales humanas según nuevos datos y bases fisiológicas. Una introducción a los estudios de Dinámica Cerebral", Trabajos del Instituto Cajal de Investigaciones Biológicas, C.S.I.C., Vol. **XLIV**, 95-157 (1952).
10. Gonzalo I. and Gonzalo A.: "Functional gradients in cerebral dynamics: The J. Gonzalo theories of the sensorial cortex" in *Brain Processes, theories, and models. An international conference in honor of W.S. McCulloch 25 years after his death*, 78-87, Moreno-Díaz R. and Mira-Mira J. (Eds.), The MIT Press, Cambridge, Massachusetts 1996.
11. Gonzalo I.: "Allometry in the Justo Gonzalo's Model of the Sensorial Cortex", Lecture Notes in Computer Science, **124**, Proceedings of the International Work-Conferenceon on Artificial and Natural Neural Networks, Lanzarote, Canary Islands, Spain, June 1997, Mira J., Moreno-Díaz R. and Cabestany J. (Eds.), Springer-Verlag, Berlin 1997.
12. Gonzalo I.: "Spatial Inversion and Facilitation in the J. Gonzalo's Research of the Sensorial Cortex", Lecture Notes in Computer Science, **1606**, Proceedings of the International Work-Conferenceon on Artificial and Natural Neural Networks, Alicante, Spain, June 1999, Vol. I, Mira J., and Sánchez-Andrés J.V. (Eds.), Springer-Verlag, Berlin 1999.
13. Köhler W.: *Gestalt Psychology*, Liveright, New York 1929. *Dynamics in Psychology*, Liveright, New York 1940.
14. Lapique L.: *L'excitabilité en Fonction du Temps*, Hermann et Cie., Paris 1926.
15. Lashley K.S.: *Brain mechanisms and intelligence*, Univ. of Chicago Press, Chicago 1929. "Integrative functions of the cerebral cortex", Psychol. Rev., **13**, 1-42 (1933). "Studies of cerebral function in learning", Comp. Psychol. Monogr., **11** (2), 5-40 (1935). "Functional determinants of cerebral localization", Arch. Neurol. Psychiat. (Chicago), **30**, 371-387 (1937). "The problem of cerebral organization in vision", Biol. Symp., **7**, 301-322 (1942).
16. Luria A.R.: *Restoration of Function after Brain Injury*, Pergamon Press, Oxford 1963. *Traumatic Aphasia*, Mouton, Paris 1970.
17. Llinás R.R.: "The intrinsic electrophysiological properties of mammalian neurons: Insights into central nervous system function", Science, **242**, 1654-1664 (1988).
18. Mira J., Delgado A.E. and Moreno-Diaz R.: "The fuzzy paradigm for knowledge representation in cerebral dynamics", Fuzzy Sets and Systems, **23**, 315-330 (1987).
19. Mira J., Delgado A.E., Manjarrés A., Ros S. and Alvarez J.R.: "Cooperative processes at the symbolic level in cerebral dynamics: reliability and fault tolerance" in *Brain Processes, theories, and models. An international conference in honor of W.S. McCulloch 25 years after his death*, 244-255, Moreno-Díaz R. and Mira-Mira J. (Eds.), The MIT Press, Cambridge, Massachusetts 1996.
20. Piéron H.: *La connaissance sensorielle et les problèmes de la vision*, Hermann, Paris 1936. "Physiologie de la vision" in *Traité d'ophtalmologie*, Masson, Paris 1939.
21. Pötzl O.:"Die optish-agnostichen Störungen" in *Handbuch der Psychiatrie*, Aschaffenburg, Wien 1928.
22. Rakic D. and Singer W. (Eds): *Neurobiology of Neocortex*, Wiley, 1988.
23. Ramón y Cajal S.: *Histologia del sistema nervioso del hombre y de los vertebrados*, Vol. **II**, Madrid 1899.

Verifying Properties of Neural Networks

Pedro Rodrigues[1], J. Félix Costa[2], and Hava T. Siegelmann[3]

[1] Departamento de Informática, Faculdade de Ciências da Universidade de Lisboa, Campo Grande, 1749-016 Lisboa, Portugal
pmr@di.fc.ul.pt
[2] Departamento de Matemática, Instituto Superior Técnico, Lisbon University of Technology, Av. Rovisco Pais, 1049-001 Lisboa, Portugal
fgc@math.ist.utl.pt
[3] Faculty of Industrial Engineering and Management, Technion City, Haifa 32 000, Israel
iehava@ie.technion.ac.il

Abstract. In the beginning of nineties, Hava Siegelmann proposed a new computational model, the Artificial Recurrent Neural Network (ARNN), and proved that it could perform hypercomputation. She also established the equivalence between the ARNN and other analog systems that support hypercomputation, launching the foundations of an alternative computational theory. In this paper we contribute to this alternative theory by exploring the use of formal methods in the verification of temporal properties of ARNNs. Based on the work of Bradfield in verification of temporal properties of infinite systems, we simplify his tableau system, keeping its expressive power, and show that it is suitable to the verification of temporal properties of ARNNs.

1 Introduction

During the forties and the beginning of the fifties, the Josiah Macy Jr. Foundation patronized an annual scientific meeting that brought together researchers from disparate areas such as Biology, Physics, Mathematics and Engineering, to discuss possible models of computation machines. They all agreed that a computer should resemble the brain. From this principle, two research programs emerged: the first one, less known, originated the theory and technology of analog computation; the second one, focused on the theory and technology of digital computation.
In 1943, McCulloch and Pitts presented their most influential paper "A logical calculus of the ideas immanent in nervous activity", in which they introduced the idea that processes in the brain can be metaphorically modeled by a finite interconnection of logical devices. Although the logical abstraction of McCulloch and Pitts was controversial, von Neumann developed this idea towards the design of the digital computer and, since then, the digital approach to computation has prevailed.
Recently, researchers have speculated that although the Turing machine, the theoretical model of digital computation, is indeed able to simulate a large class of computations, it does not necessarily provide a complete picture of the computations possible in nature. This speculation emerged with several proposals of different models of computation. In the beginning of nineties, Hava Siegelmann proposes what she called the "Artificial Recurrent Neural Network" (ARNN), an analog model

J. Mira and A. Prieto (Eds.): IWANN 2001, LNCS 2084, pp. 158–165, 2001.

capable of performing hypercomputation. By showing that ARNNs were equivalent to other analog computational models, also able to perform hypercomputation, Siegelmann claims that ARNNs should be taken as *the* analog computational model. Upon this postulate some research has been conducted: foundations [11, 9, 5], languages and compilers [6, 10, 7, 8], etc.
In this paper we will be contributing to this emergent alternative computer science by exploring the use of formal methods in the verification of temporal properties of ARNNs. More specifically, we will be using "model checking". "Model checking" is a way of certifying the agreement between a specification and a system: a system is described as a transition graph and the proof of properties is established by traversing parts of the graph. We will extend the work of Bradfield [1, 2, 3] who has developed a "mechanism" of "model checking" for infinite systems and analyzed its applicability to conventional Petri Nets, by simplifying the "mechanism" and studying its applicability to ARNNs.
In section 2 we will present the language we shall be using to describe properties we wish to verify. In section 3, we will present the tableau system and, in section 4, we will present the ARNN and the applicability of the tableau system to prove properties of ARNNs.

2 Mu-calculus (Syntax and Semantics)

Mu-calculus is a temporal logic with vast expressive power – it subsumes several standard temporal logics [4]. There are several versions of it in the literature. We shall be using the version that Bradfield used in his PhD thesis [1], a version in which the indexation of modal operators was enlarged from labels to sets of labels.
Let us assume two fixed sets: the set of variables $\mathcal{V}$ (countably infinite) and the set of labels $\mathcal{E}$ (finite and non-empty). X, Y and Z will be used to designate variables and, a,b will be used to designate labels.

Definition A mu-formula is inductively defined as follows:

- X is a mu-formula;
- if φ_1 and φ_2 are mu-formulas, then $(\varphi_1 \wedge \varphi_2)$ and $(\varphi_1 \vee \varphi_2)$ are mu-formulas;
- if φ is a mu-formula, then $(\neg\varphi)$, $([K]\varphi)$ and $(\langle K\rangle\varphi)$ are mu-formulas, where K stands for a set of labels;
- if φ is a mu-formula and any free occurrence of X in φ is within the scope of an even number of negations, then $(\nu X.\varphi)$ and $(\mu X.\varphi)$ are mu-formulas. ◊

Notation To minimize the use of parenthesis, we adopt the following convention of precedence (from greater to a smaller precedence): $\neg$, the modal operators [K] and $\langle K\rangle$, the logical operators $\wedge$ and $\vee$, and, finally, the fix-point operators $\nu X.\varphi$ and $\mu X.\varphi$. We also assume that the logical operators associate to the left. ◊

Notation We will be using $\sigma X.\varphi$ to denote either of the two fix-point operators. ◊

Mu-formulas are interpreted in models.

Definition A labeled transition system is a triple $\langle \mathcal{S}, \mathcal{A}ct, \rightarrow \rangle$ where:

- $\mathcal{S}$ is a set whose elements are called states;
- $\mathcal{A}ct$ is a set whose elements are called actions;
- $\rightarrow \subseteq \mathcal{S} \times \mathcal{A}ct \times \mathcal{S}$ is a relation, called the transition relation. ◊

Definition A model $\mathcal{M}$ for the set of mu-formulas is a pair $\langle \mathcal{T}, \mathcal{I} \rangle$ where $\mathcal{T} = \langle \mathcal{S}, \mathcal{E}, \rightarrow \rangle$ is a labeled transition system, with $\mathcal{E}$ as the action set, and $\mathcal{I}: \mathcal{V} \rightarrow 2^{\mathcal{S}}$, an interpretation that assigns a set of states to each variable. ◊

Definition [1] Let $\mathcal{M} = \langle \mathcal{T}, \mathcal{I} \rangle$ be a model for the set of mu-formulas. The denotation $\|\varphi\|^{\mathcal{T}}_{\mathcal{I}}$ of a mu-formula in a model $\mathcal{M}$ is inductively defined as follows:

- $\|X\|^{\mathcal{T}}_{\mathcal{I}} = \mathcal{I}(X)$
- $\|\neg\varphi\|^{\mathcal{T}}_{\mathcal{I}} = \mathcal{S} - \|\varphi\|^{\mathcal{T}}_{\mathcal{I}}$
- $\|\varphi_1 \wedge \varphi_2\|^{\mathcal{T}}_{\mathcal{I}} = \|\varphi_1\|^{\mathcal{T}}_{\mathcal{I}} \cap \|\varphi_2\|^{\mathcal{T}}_{\mathcal{I}}$
- $\|\varphi_1 \vee \varphi_2\|^{\mathcal{T}}_{\mathcal{I}} = \|\varphi_1\|^{\mathcal{T}}_{\mathcal{I}} \cup \|\varphi_2\|^{\mathcal{T}}_{\mathcal{I}}$
- $\|[K]\varphi\|^{\mathcal{T}}_{\mathcal{I}} = \{ s \in \mathcal{S} : \forall s' \in \mathcal{S}\ \forall a \in K\ (s \xrightarrow{a} s' \Rightarrow s' \in \|\varphi\|^{\mathcal{T}}_{\mathcal{I}})\}$
- $\|\langle K\rangle\varphi\|^{\mathcal{T}}_{\mathcal{I}} = \{ s \in \mathcal{S} : \exists s' \in \mathcal{S}\ \exists a \in K\ (s \xrightarrow{a} s' \text{ e } s' \in \|\varphi\|^{\mathcal{T}}_{\mathcal{I}})\}$
- $\|\nu X.\varphi\|^{\mathcal{T}}_{\mathcal{I}} = \cup\{ T \subseteq \mathcal{S} : T \subseteq \|\varphi\|^{\mathcal{T}}_{\mathcal{I}[X:=T]} \}$
- $\|\mu X.\varphi\|^{\mathcal{T}}_{\mathcal{I}} = \cap\{ T \subseteq \mathcal{S} : \|\varphi\|^{\mathcal{T}}_{\mathcal{I}[X:=T]} \subseteq T \}$

  where $\mathcal{I}[X:=T]$ agrees with $\mathcal{I}$ in every variable except, eventually, in X, being $\mathcal{I}[X:=T](X) = T$. ◊

Notation We shall be writing $\|\varphi\|_{\mathcal{I}}$ instead of $\|\varphi\|^{\mathcal{T}}{}_{,\mathcal{I}}$ whenever $\mathcal{T}$ is irrelevant. ◊

The meaning of mu-formulas not involving fix-point operators should be easily understandable from this last definition. For the meaning of more complex mu-formulas cf. [1, 2].

3 The Tableau System

Bradfield [1] presented a tableau system for proving that a set of states satisfies a property described as a mu-formula without determining its full denotation. We shall be using an equivalent tableau system in which the unfolding rule can only be applied once to each subformula whose external operator is a fix-point operator. This small change had already been mentioned by Bradfield [1] and used by him [2] though not considering the indexation of modal operators by sets of states. In our opinion, this small change simplifies tableaux construction.

We shall not present the proofs of correctness and completeness of the tableau system, but they are fairly similar to the ones presented by Bradfield [1].

Let us assume a fixed model $\mathcal{M} = \langle \mathcal{T}, \mathcal{I} \rangle$. We wish to prove that a set of states S satisfies a property φ, i.e., $S \subseteq \|\varphi\|_{\mathcal{I}}$. This is only possible, using the tableau system, if

φ is in the positive normal form: it is proven that every mu-formula has an equivalent mu-formula, which is in the positive normal form (negations are only applied to variables, there are no variables simultaneously free and bounded in the formula and the same variable can not be bounded by two fix-point operators).
A tableau is a proof tree built on sequents of the form $S' \vdash \varphi'$; a sequent represents the goal of showing $S' \subseteq \|\varphi'\|_{\mathcal{V}'}$. If we wish to prove that $S \subseteq \|\varphi\|_{\mathcal{V}}$, we start from the sequent $S \vdash \varphi$ and go on applying the rules of the system, that transform goals into one or two subgoals, until we reach terminal sequents. Once built the tableau, the root sequent goal is proven iff a given success condition is verified.

Definition tableau rules (presented as inverted proof rules):

(and) $\dfrac{S \vdash \varphi_1 \wedge \varphi_2}{S \vdash \varphi_1 \quad S \vdash \varphi_2}$

(or) $\dfrac{S \vdash \varphi_1 \vee \varphi_2}{S_1 \vdash \varphi_1 \quad S_2 \vdash \varphi_2}$ where $S = S_1 \cup S_2$

(box) $\dfrac{S \vdash [K]\varphi}{S' \vdash \varphi}$ where $S' = \{ s' \in \mathcal{S} : \exists s \in S\ \exists a \in K\ (s \xrightarrow{a} s')\}$

(diamond) $\dfrac{S \vdash \langle K \rangle \varphi}{S' \vdash \varphi}$ where S' is the range of an application $f: S \rightarrow \mathcal{S}$

such that, if $f(s) = s'$, then $\exists a \in K\ (s \xrightarrow{a} s')$

(unfolding) $\dfrac{S \vdash \sigma X.\varphi}{S \vdash \varphi}$ (weakening) $\dfrac{S \vdash \varphi}{S' \vdash \varphi}$ where $S' \supseteq S$ ◊

These rules, except for the last two, should be, straightforward. To explain the unfolding rule, let us suppose that the goal represented by $S \vdash \sigma X.\varphi$ is $S \subseteq \|\sigma X.\varphi\|_{\mathcal{V}'}$. Then, the goal represented by $S \vdash \varphi$ should be seen as $S \subseteq \|\varphi\|_{\mathcal{V}'[X:=S]}$. According to the way the semantics is defined, if this latest inclusion is true, then $S \subseteq \|\sigma X.\varphi\|_{\mathcal{V}'}$ as long as $\sigma X.\varphi$ is $\nu X.\varphi$. If this is not the case, then something more has to be proven.
The weakening rule is strictly needed for the completeness of the system. Still, it has only to be applied immediately before the unfolding rule. We shall present an example of this later, in the next section.
Given a root sequent, the tableau is built downwards, using tableau rules, until the only rule applicable to leaf sequents is the weakening rule (these sequents are called terminal nodes).

Definition Let φ be the mu-formula present in the root sequent and $n' = S' \vdash \varphi'$, a terminal node.

- If φ' is X and X is free in φ, then n' is called a free terminal.
- If φ' is $\neg X$ and X is free in φ, then n' is called a negated terminal.
- If φ' is X and X is bounded in φ, then n' is called a σ-terminal.

- If φ' is X and there is a subformula μX.ψ of φ, then n' is called a μ-terminal.
- If φ' is X and there is a subformula νX.ψ of φ, then n' is called a ν-terminal. ◊

For a tableau to be successful, every terminal node must be a successful node.

Definition Let n' = S' |– X be a σ-terminal. Its companion is the lowest node n''= S'' |– σX.φ above it. ◊

Definition A terminal node n' = S' |– φ' is successful iff:

- φ' is ¬X and S' ∩ ℐ(X) = ∅ or
- S' = ∅ or
- n' is a free terminal and S' ⊆ ℐ(φ') or
- n' is a ν-terminal with companion n'' = S'' |– νX.φ'' and S' ⊆ S'' or
- n' is a μ-terminal with companion n'' = S'' |– μX.φ'', S' ⊆ S'' and n'' satisfies the μ-success condition presented below. ◊

Definition A path from a state s at a node n to a state s' at a node n', s@n $\longrightarrow$ s'@n', is a sequence (s,n) = (s_0,n_0), (s_1,n_1), …, (s_k,n_k) = (s', n') of pairs (state, node) such that:

- n_{i+1} is a child of n_i;
- if $n_i = S_i$ |– φ_i, then $s_i \in S_i$;
- if the rule applied to n_i is the box or the diamond rule, then ∃a ∈ K ($s_i \xrightarrow{a} s_{i+1}$), otherwise, s_{i+1} é s_i. ◊

Definition There is an extended path from a state s at a node n to a state s' at a node n', s@n $\longrightarrow$ s'@n', if:

i) s@n $\longrightarrow$ s'@n' or

ii) there exists a node n'' = S'' |– σX.φ, companion of k (k≥1) σ-terminal nodes, n_1, …, n_k, and a finite sequence of states, s_0, …, s_k, such that s@n $\longrightarrow$ s_0@n'', for each i (0≤i<k), s_i@n'' $\longrightarrow$ s_{i+1}@n_{i+1}, and s_k@n'' $\longrightarrow$ s'@n'. ◊

Definition Let n = S |– σX.φ be a companion node. We define the relation $\sqsupset_n$ on the elements of S in the following way: s $\sqsupset_n$ s' iff s@n $\longrightarrow$ s'@n' for some σ-terminal n' = S' |– X. ◊

Definition Let n = S |– μX.φ be a companion node. n satisfies the μ-success condition iff the relation $\sqsupset_n$ is well founded. ◊

We will postpone an example of the use of the tableau system to the next section.

4 Tableau System for Neural Networks

The neural network architecture we will be using is the one used by Siegelmann [11] to study the computational power of neural networks. Nevertheless, this model is pretty similar to other neural network architectures and so the work we will present is easily adaptable to those architectures.

Definition An artificial recurrent neural network of n ($n \in \mathbb{N}$) neurons and u ($u \in \mathbb{N}$) input units is a quadruple $\mathfrak{R}_{n,u} = \langle A, B, C, f\rangle$ where: A is a $n \times n$ matrix of real numbers; B is a $n \times u$ matrix of real numbers; C is a $n \times 1$ matrix of real numbers; f is an application from $\mathbb{R}$ to $\mathbb{R}$. ◊

Definition The dynamics of an ARNN $\mathfrak{R}_{n,u} = \langle A, B, C, f\rangle$ is given by an application F: $\mathbb{R}^n \times \{0,1\}^u \to \mathbb{R}^n$ defined in the following way:

$F(x, u) = x^+$ where, for all i such that $1 \le i \le n$, $x_i^+ = f\left(\sum_{j=1}^{n} a_{ij}x_j + \sum_{j=1}^{u} b_{ij}u_j + c_i\right)$ ◊

The computation of an ARNN is infinite: given an initial state and an infinite succession of stimulus to be presented to the input units, the net will be evolving through a succession of states infinitely.

Definition Let $\mathfrak{R}_{n,u} = \langle A, B, C, f\rangle$ be an ARNN. An input stream for $\mathfrak{R}_{n,u}$ is an application u: $\mathbb{N}_0 \to \{0,1\}^u$. ◊

Definition Let $\mathfrak{R}_{n,u} = \langle A, B, C, f\rangle$ be an ARNN. A computation of $\mathfrak{R}_{n,u}$ is a triple $\langle x_{init}, u, x\rangle$ where: x_{init} is an element of $\mathbb{R}^n$, the initial state; u is an input stream for $\mathfrak{R}_{n,u}$; x is an application from $\mathbb{N}_0$ to $\mathbb{R}^n$ defined in the following way: $x(0) = x_{init}$; for all $t \ge 0$, $x(t+1) = F(x(t), u(t))$ ◊

As we can notice, an ARNN is not a labeled transition system. Nevertheless, its behavior can be simulated by a labeled transition system.

Definition Let $\mathfrak{R}_{n,u}$ be an ARNN. The associated labeled transition system to $\mathfrak{R}_{n,u}$ is a triple $\mathcal{T}_{\mathfrak{R}_{n,u}} = \langle \mathcal{S}, \mathcal{A}ct, \to\rangle$ where $\mathcal{S} = \mathbb{R}^n$, $\mathcal{A}ct = \{0,1\}^u$ and $\forall s, s' \in \mathcal{S}\ \forall a \in \mathcal{A}ct$ ($s \xrightarrow{a} s'$ iff $F(s,a) = s'$). ◊

According to this latest definition, an ARNN and its associated labeled transition system evolve in the same way (a stimulus presented to the input units becomes an action performed on the associated labeled transition system). Therefore, we will say that a set of states S, of an ARNN $\mathfrak{R}_{n,u}$, satisfy a given property φ iff the same set S is in the denotation of φ, considering $\mathcal{T}_{\mathfrak{R}_{n,u}}$ as the labeled transition system.

We will finish this section with an example. Assume an ARNN with 2 neurons and 1 input unit defined as follows:

$$A = \begin{bmatrix} -1 & 1 \\ 1 & 0 \end{bmatrix},\ B = \begin{bmatrix} 1 \\ 0 \end{bmatrix},\ C = \begin{bmatrix} 0 \\ 0 \end{bmatrix} \text{ and f is the signal function, i.e., } f(x) = \begin{cases} 0 & \text{se } x \leq 0 \\ 1 & \text{se } x > 0 \end{cases}$$

Let us prove that, for all states of the ARNN in which both neurons have the same activation, there is a computation in which the state (1,0), the state in which the neuron 1 has activation 1 and the neuron 2 has activation 0, is visited infinitely often.

Translating our purpose to the mu-calculus, we wish to prove that the set $\{(x,x){:}x\in\mathbb{R}\}$ is in the denotation of $\nu Y.\mu X.(P \vee \langle - \rangle X) \wedge \langle - \rangle Y$, assuming $\{(1,0)\}$ as the interpretation for P (we are using $\langle - \rangle$ as an abbreviation for $\langle \{0,1\} \rangle$).

$$\{(x;x){:}x\in\mathbb{R}\} \vdash \nu Y.\mu X.(P \vee \langle - \rangle X) \wedge \langle - \rangle Y$$

$$n_1\ \{(x;x){:}x\in\mathbb{R}\}\cup\{(0,1),(1,0)\} \vdash \nu Y.\mu X.(P \vee \langle - \rangle X) \wedge \langle - \rangle Y$$

$$n_2\ \{(x;x){:}x\in\mathbb{R}\}\cup\{(0,1),(1,0)\} \vdash \mu X.(P \vee \langle - \rangle X) \wedge \langle - \rangle Y$$

$$\{(x;x){:}x\in\mathbb{R}\}\cup\{(0;1);(1,0)\} \vdash (P \vee \langle - \rangle X) \wedge \langle - \rangle Y$$

$$\{(x;x){:}x\in\mathbb{R}\}\cup\{(0;1);(1,0)\} \vdash P \vee \langle - \rangle X \qquad \{(x;x){:}x\in\mathbb{R}\}\cup\{(0;1);(1;0)\} \vdash \langle - \rangle Y\ n_3$$

$$n_4\{(1,0)\} \vdash P \quad \{(x;x){:}x\in\mathbb{R}\}\cup\{(0;1)\} \vdash \langle - \rangle X\ n_5 \qquad \{(0,1),(1,0)\} \vdash Y\ n_6$$

$$n_7\ \{(0,1),(1,0)\} \vdash X$$

As it can be seen, in this case, it would be impossible to prove what we wanted if the weakening rule (the first rule applied) was not available. This rule is strictly essential to satisfy the inclusion condition of the sets associated to σ-terminals and its companions.

Before showing that this is a successful tableau, we must present the functions implicit in the application of the diamond rule:

- Let f_3 denote the function considered in the application of the diamond rule to n_3. This function is defined in the following way: for all $x \leq 0$, $f_3(x,x) = (1,0)$; for all $x > 0$, $f_3(x,x) = (0,1)$; $f_3(0,1) = (1,0)$, $f_3(1,0) = (0,1)$.
- Let f_5 denote the function considered in the application of the diamond rule to n_5. This function is defined in the following way: for all $x \leq 0$, $f_5(x,x) = (1,0)$; for all $x > 0$, $f_5(x,x) = (0,1)$; $f_5(0,1) = (1,0)$.

Let us prove that the tableau is a successful tableau. n_4 and n_6 are clearly successful terminals: n_4 is a successful terminal because $\{(1,0)\}$ is contained in the interpretation of P; n_6 is also a successful node because $\{(0,1), (1,0)\}$ is contained in $\{(x,x){:}x\in\mathbb{R}\}\cup\{(0,1),(1,0)\}$, the set associated to n_1, the companion of n_6. Finally, n_7 is also a successful terminal since $\{(0,1), (1,0)\}$ is contained in $\{(x,x){:}x\in\mathbb{R}\}\cup \{(0,1),(1,0)\}$, the set associated to n_2, the companion of n_7, and the relation $\sqsupset_{n2}$ is well founded: $(0,1) \sqsupset_{n2} (1,0)$; for all $x \leq 0$ $(x,x) \sqsupset_{n2} (1,0)$; for all $x > 0$ $(x,x) \sqsupset_{n2} (0,1)$. Since all terminal nodes are successful, the tableau is a successful tableau.

5 Conclusions

We have presented a system for reasoning about ARNNs by adapting standard techniques of model-checking whose applicability had only been studied on conventional Petri Nets. The system presented is clearly undecidable and so, can not be fully automated. Still, if we confine the elements of the matrices to integers, the ARNN becomes a finite system and, consequently, the tableau system becomes decidable (we have implemented a fully automated tableau system for this subclass of ARNNs). Nevertheless, since we are dealing with a local model-checking technique, we shall be able to automatically prove properties about ARNNs with an infinite state space. How can it be done? We don't have the answer yet and the work on finite ARNNs has not given us much clues. We believe we should recover the work on finite ARNNs and try to improve the results obtained.

References

1. Bradfield, J.: Verifying Temporal Properties of Systems with Applications to Petri Nets, PhD thesis, University of Edinburgh (1991)
2. Bradfield, J.: Proving Temporal Properties of Petri Nets. In Rozenberg, G. (eds.): Advances in Petri Nets 1991. Lecture Notes in Computer Science, Vol. 524. Springer-Verlag, Berlin (1991) 29-47
3. Bradfield, J., Stirling, C.:,Local Model Checking for Infinite State Spaces. Theoretical Computer Science, Vol. 96. (1992) 157-174
4. Emerson, E.A., Lei, C.-L.: Efficient Model Checking in Fragments of the Propositional Mu-calculus. Proceedings of the 1st IEEE Symposium on Logic in Computer Science, (1986) 267-278
5. Gilles, C., Miller, C., Chen, D., Chen, H., Sun, G., Lee, Y.: Learning and Extracting Finite State Automata with Second-Order Recurrent Neural Networks. Neural Computation, Vol. 4 **3** (1992) 393-405
6. Gruau, F., Ratajszczak, J., Wiber, G.: A neural compiler. Theoretical Computer Science, Vol. 141(1-2) (1995) 1-52
7. Neto, J. P., Siegelmann, H. T., Costa, J. F.: On the Implementation of Programming Languages with Neural Nets. In First International Conference on Computing Anticipatory Systems (CASYS'97). CHAOS, **1** (1998) 201-208
8. Neto, J. P., Siegelmann, H. T., Costa, J. F.: Building Neural Net Software. submitted, (1999)
9. Pollack, J.: On Connectionism Models of Natural Language Processing. PhD thesis. University of Illinois, Urbana (1987)
10. Siegelmann, H. T.: On NIL: the Software Constructor of Neural Networks. Parallel Processing Letters. Vol. 6 **4**, World Scientific Publishing Company (1996) 575-582
11. Siegelmann, H. T.: Neural Networks and Analog Computation: beyond the Turing limit. Birkhäuser, Boston (1999)

Algorithms and Implementation Architectures for Hebbian Neural Networks

J. Andrés Berzal and Pedro J. Zufiria

Grupo de Redes Neuronales
Dpto. de Matemática Aplicada a las Tecnologías de la Información
E.T.S. Ingenieros de Telecomunicación, Universidad Politécnica de Madrid
Ciudad Universitaria S/N, E-28040 Madrid, SPAIN

{abf,pzz}@mat.upm.es

Abstract. Systolic architectures for Sanger and Rubner Neural Networks (NNs) are proposed, and the local stability of their learning rules is taken into account based on the indirect Lyapunov method. In addition, these learning rules are improved for applications based on Principal Component Analysis (PCA). The local stability analysis and the systolic architectures for Sanger NN and Rubner NN are presented in a common framework.

1 Introduction

This paper presents a type of systolic architecture for two types of hebbian NNs, Sanger NN and Rubner NN (with linearized hebbian training approach for direct connections) [1, 2, 4]. The proposed NNs and their architectures implement an on-line PCA from a sample sequence of a stationary vector stochastic process, their weights converging to the eigenvectors of the autocorrelation matrix associated with the input process. The main eigenvector, related to the largest eigenvalue, is approximated by the weights of the first neuron, and so on for the remaining eigenvectors. Hence, PCA finds the uncorrelated directions of maximum variance in the data space, as well as providing the optimal linear projection in the least square sense. PCA is a basic procedure for implementing the Karhunen-Loeve Transform (KLT), widely employed in signal processing systems. Since these NNs implement the KLT from the input data directly, they can perform real time signal processing tasks without explicit computation of the autocorrelation matrix, as well its eigenvalues and eigenvectors (image coding, component analysis of multispectral images, etc) [1, 2, 5, 7].

The implementations of hebbian NNs with a single processor have speed limitations, so they are not appropriate for real time applications being very demanding in computation, high volume data, etc. On the other hand, the systolic architectures with two or more processors are an alternative to the single processor implementations providing parallel processing, modular hardware implementation for VLSI, etc [4]. In fact, the architectures introduced by this paper are a practical hardware alternative for the real time adaptive calculation of the PCA.

J. Mira and A. Prieto (Eds.): IWANN 2001, LNCS 2084, pp. 166-173, 2001.

In section 2, Sanger NN and Rubner NN and their training algorithms are presented. In section 3, the local stability of these algorithms is enunciated taking into account the demonstration based on the indirect Lyapunov method. This analysis is appropriate for practical numerical or hardware implementations, where NNs are considered as discrete time dynamical systems. In section 4, these training algorithms are improved extending the PCA to the covariance matrix besides the autocorrelation matrix, Adaptive Learning Rate (ALR), etc. In sections 5 and 6, for both Sanger NN and linearized Rubner NN, some new systolic architectures are presented for hardware implementation. Concluding remarks are indicated in section 7.

2 Sanger NN and Rubner NN Learning Rules and the KLT

The Sanger NN structure and its learning rule, for M neurons with N inputs, are defined by:

$$y_i = \sum_{j=1}^{N} w_{ij} x_j , \qquad 1 \le i \le M,\ 1 \le M \le N , \tag{1}$$

$$w_{ij,n+1} = w_{ij,n} + \eta_i y_i \left(x_j - \sum_{k=1}^{i} y_k w_{kj,n} \right), \qquad \eta_i \in \Re , \tag{2}$$

where y_i, x_j, w_{ij} and η_i are the output, input, weights of the net, and learning rate, respectively. Direct weights are obtained by Sanger learning rule, a modified hebbian learning rule, since it incorporates a second term (Eq. 2). This learning rule leads to the weight vectors to approach the eigenvectors of the autocorrelation matrix of the input samples (patterns), $\mathbf{C}$=E[$\mathbf{xx}^T$].

On the other hand, the Rubner NN structure and its learning rule, for M neurons with N inputs, are defined by:

$$y_i = \sum_{j=1}^{N} w_{ij} x_j + \sum_{j=1}^{j<i} u_{ij} y_j , \quad 1 \le i \le M,\ 1 \le M \le N , \tag{3}$$

$$w_{ij,n+1} = \frac{w_{ij,n} + \eta_i y_{i,n} x_{j,n}}{\left| \mathbf{w}_{ij,n} + \eta_i y_{i,n} \mathbf{x}_{j,n} \right|}, \quad u_{ij,n+1} = u_{ij,n} - \gamma_i y_{i,n} y_{j,n}, \quad \eta_i, \gamma_i \in \Re , \tag{4}$$

where y_i represents the output, x_j is an input, w_{ij} are the direct weights, u_{ij} are the lateral weights between y_j and y_i, and η_i and γ_i are the learning rates for the direct and lateral weights, respectively. The direct weights are adjusted upon normalized hebbian learning and the lateral weights, upon antihebbian learning (Eq. 4). The normalization of the direct weights of the Rubner learning rule can be avoided for small η_i, so that this learning rule for w_{ij} can be approximated by the linear term of the Taylor expansion [2] (Eq. 5). Also, this linear formulation reduces the complexity of the learning rule in terms of operations (processing time).

For Rubner NN and its linearized version, the direct weights converge to the eigenvectors of the autocorrelation matrix of the input samples (patterns), $\mathbf{C}$=E[$\mathbf{xx}^T$], and the lateral weights converge to the null vector, $\mathbf{0}$.

$$w_{ij,n+1} = w_{ij,n} + \eta_i y_{i,n}\left(x_{j,n} - (\mathbf{w}_{i,n}^T \mathbf{x}_n) w_{ij,n}\right), \qquad \eta_i \in \Re . \tag{5}$$

3 Local Stability for Sanger NN and Rubner NN Learning Rules

The local stability of the Sanger and linearized Rubner learning rules is analytically proved via the indirect Lyapunov method [2]. Other studies to prove the local stability of Sanger and Rubner learning rules are based on their associated ODE (Ordinary Differential Equation) [4, 6]. However, in this paper the local stability study used maintains its discrete time evolution. Therefore, it allows to obtain the local stability conditions taking into account the learning rates.

For the indirect Lyapunov method, the Sanger and linearized Rubner learning rules are expressed in terms of the weights, and the mean operator is applied to both sides of the learning equations which are transformed from the deterministic framework to the probabilistic one. Their fixed points are obtained assuming that input samples ($\mathbf{x}$) and weights ($\mathbf{w}_i$ and $\mathbf{u}_i$ only for linearized Rubner learning rule) are statistically independent, and that there are different scales of time for the evolution of each weight. For Sanger, its fixed points take the form of vector $\pm\mathbf{e}_j$ ($i \leq j$) and for Rubner, its fixed points are vector pairs ($\pm\mathbf{e}_j$, $\mathbf{0}$) ($i \leq j$), where $\mathbf{e}_j$ is an eigenvector of $\mathbf{C}$=E[$\mathbf{xx}^T$]. Applying Lyapunov method, for Sanger learning rule, the asymptotically stable fixed points are $\mathbf{w}_i = \pm\mathbf{e}_i$ if $\eta_i < 1/\lambda_i$ and for linearized Rubner learning rule, they are ($\mathbf{w}_i$, $\mathbf{u}_i$)=($\pm\mathbf{e}_i$, $\mathbf{0}$) if $\eta_i < 1/\lambda_i$ (assuming $\gamma_i >> 2\eta_i$).

4 Improvements for Sanger NN and Rubner NN

The performance of Sanger and Rubner learning algorithms can be improved [1, 3]:

- The direct weights can converge to the eigenvectors of the covariance matrix of the input samples. This is accomplished by a dynamic computation of the time mean value of the input samples $\overline{\mathbf{x}}_n$ (Eq. 6) and subtracting this value to the input to the NN. $\overline{\mathbf{x}}_n$ converges to E[$\mathbf{x}$] in probability , assuming $\mathbf{x}_n$ are uniform random variables with finite variance.

$$\overline{\mathbf{x}}_n = \overline{\mathbf{x}}_{n-1} + \frac{\mathbf{x}_n - \overline{\mathbf{x}}_{n-1}}{n}, \qquad \overline{\mathbf{x}}_1 = \mathbf{x}_1 . \tag{6}$$

- The dynamic evaluation of the eigenvalues $\tilde{\lambda}_{in}$ (Eq. 7) of the autocorrelation or covariance matrix of the input samples, whose magnitude and location (neuron i) provides information about the importance and the convergence of the weights (eigenvectors). $\tilde{\lambda}_{i,n}$ converges to E[y_i^2]=λ_i in probability, being y_i^2 uniform random variables with finite variance (assuming the weights of the NN converge to their fixed points).

$$\tilde{\lambda}_{i,n} = \tilde{\lambda}_{i,n-1} + \frac{y_{i,n}^2 - \tilde{\lambda}_{i,n-1}}{n}, \qquad \tilde{\lambda}_{i,n} = y_{i,1}^2, \ \tilde{\lambda}_{i,n} = \overline{y_i^2}_{\ n} . \tag{7}$$

- The Adaptive Learning Rate (ALR) defined by Eq. 8 for each neuron accelerates and synchronises the training phase. This is only useful for the Sanger NN, since for the Rubner NN it amplifies the oscillation in each component of the weights [1]. The experimental simulations of Sanger and Rubner learning algorithms have proved that the ALR does not modify the convergence properties (fixed points and local stability) with respect to their initial algorithms (section 3).

$$\eta_{i,n} = \eta_0 \frac{\tilde{\lambda}_{1,n}}{\tilde{\lambda}_{i,n}}, \qquad \tilde{\lambda}_{i,n} \neq 0 . \tag{8}$$

5 Systolic Architectures for Hebbian Neural Networks

In this section, we propose some new systolic architectures for the Sanger NN and the Rubner NN which implement the retrieving phase (input and output function of the NN) and training phase for both NN. The graphical representation of the systolic architecture is given by the Dependence Graph (DG). It shows the topology of the network, nodes, dependencies (input and output), flow of data, etc, and all of them define the function to implement. The systolic architecture is suitable for hebbian NNs and it is a 2-D array (Fig. 1). The number of rows and columns correspond to the number of output and inputs of the NN respectively. The operations of each node are basically multiply-accumulate and commutation.

The projection of the DGs onto a linear array of processors defines the architecture. The vertical projection for the hebbian NNs, where each processor executes a row of the DG (neuron) is selected in this paper since it gives more simplicity than others (Fig. 2). The number of cycles to process one input sample for the Sanger NN linear arrays for both retrieving and training architectures is $N+M-1$, less than NM cycles required by classical architectures (one processor). For linearized Rubner NN architectures this situation also happens, namely the number of cycles for their linear arrays for both retrieving and training architectures is $N+2M-1$, less than $NM+(M-1)M/2$ cycles required by classical architectures. The interconnection between retrieving and training linear arrays for linerized Rubner NN requires a synchronization for input/output data, unnecessary for Sanger NN.

The processors, flow of data, etc, for these linear arrays are different and mainly they depend on the NN and the function to implement (retrieving or training):

- **Retrieving Linear Array** of **Sanger NN** (Fig. 2). Its processor j has as serial inputs the weights of the neuron j ($\mathbf{w}_j$) stored in a circular register and the components of the input samples ($\mathbf{x}$). After N cycles this processor provides the output of neuron j (y_j). The processor j+1 receives $\mathbf{x}$ in serial with a delay of one cycle; then y_{j+1} is obtained one cycle later than y_j. The cycle for these processors is determined mainly by the time needed to execute one multiplication and one addition in serial.

Fig. 1. Dependence Graphs (DGs) of a systolic architectures for Sanger NN and Rubner NN.

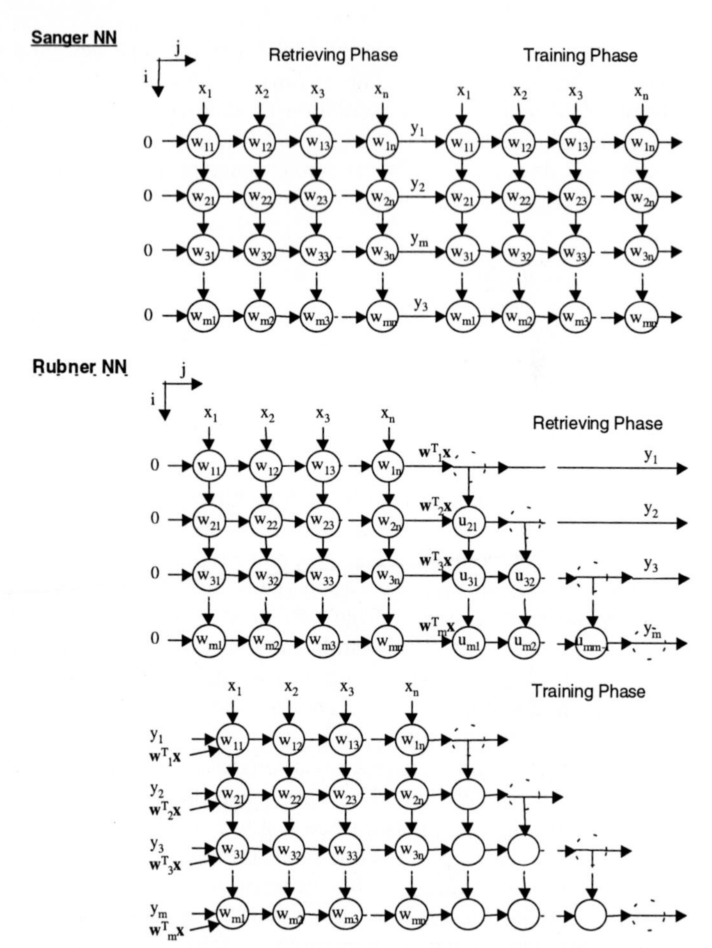

- **Retrieving Linear Array** of **linearized Rubner NN** (Fig. 2). Its processor j differs from the processor j of the retrieving linear array of the Sanger NN in several aspects for the processor j, its computing time to provide y_j is $N+j$-1 cycles, the direct and lateral weights ($\mathbf{w}_j$ and $\mathbf{u}_j$) are stored in a circular register (first $\mathbf{w}_j$), a new output for the scalar product $\mathbf{w}^T_j\mathbf{x}$ is incorporated, and processor j sends to processor $j+1$ $\mathbf{x}$ and y_k ($0<k<j$) in serial. The cycle for this processor is determined mainly by the time needed to execute two multiplications and one addition in serial.

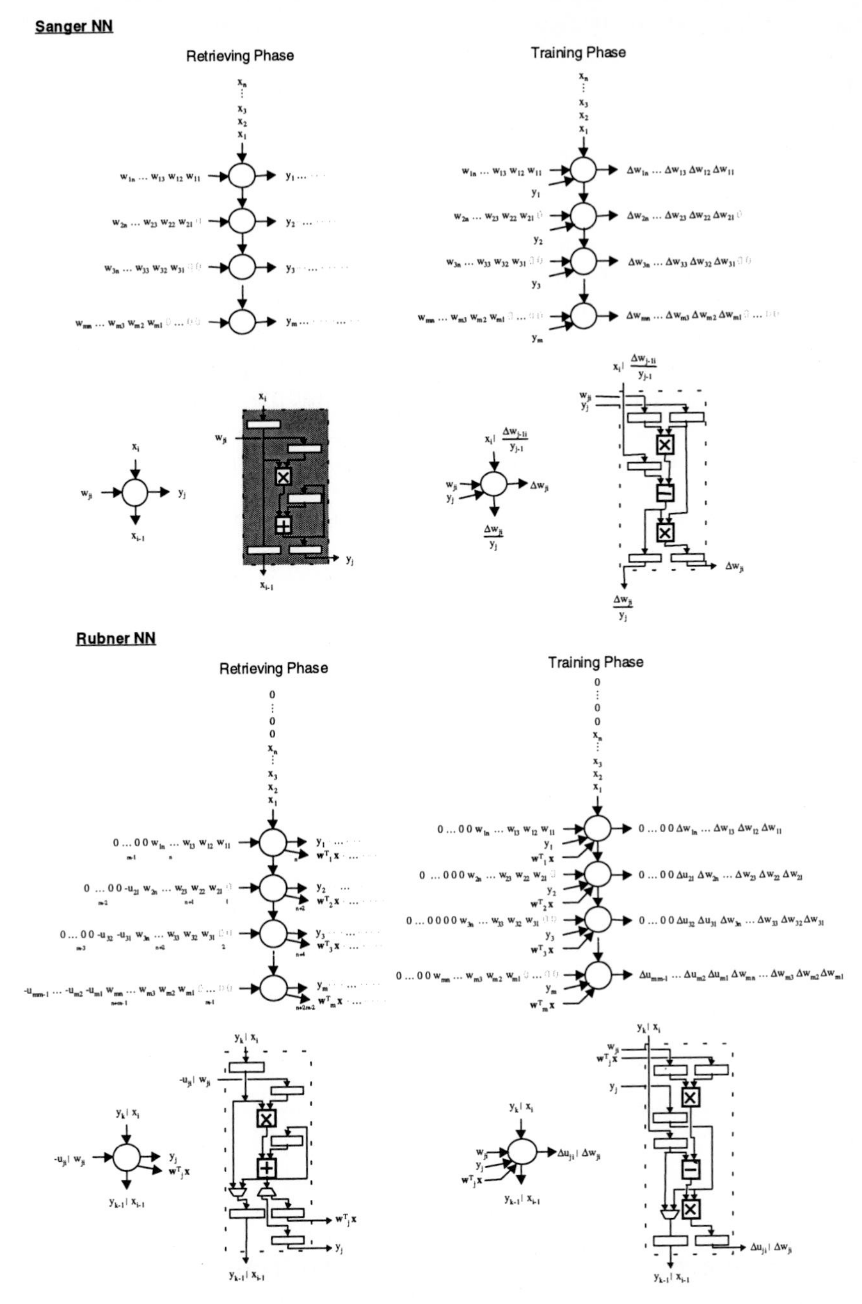

Fig. 2. Linear arrays of processors (vertical projection of the DG) and their processor schemes of a systolic architecture for Sanger NN and Rubner NN (retrieving and training phases).

– **Training Linear Array** of **Sanger NN** (Fig. 2). Its processor j has as inputs the weights of the neuron j ($\mathbf{w}_j$) stored in a circular register, its associated output (y_j)

and the components of the weight increments of the previous neuron normalized by its output ($\Delta\mathbf{w}_{j-1}/y_{j-1}$), except for the first processor where we have the components of the input samples ($\mathbf{x}$). After N cycles this processor provides the weight increment $\Delta\mathbf{w}_j$ (update $\mathbf{w}_j$ next step). The processor j+1 receives $\Delta\mathbf{w}_j/y_j$ in serial with a delay of one cycle then $\Delta\mathbf{w}_{j+1}$ is obtained one cycle later than $\Delta\mathbf{w}_j$. The cycle for these processors is determined mainly by the time needed to execute two multiplications and one subtraction in serial.

- **Training Linear Array** of **linearized Rubner NN** (Fig. 2). Its processor j differs from the processor j of the training linear array of the Sanger NN in the following aspects: the computing time to provide $\Delta\mathbf{w}_j$ and $\Delta\mathbf{u}_j$ for the neuron j are $N+j-1$ (update $\mathbf{w}_j$ and $\mathbf{u}_j$ next step), the direct and null weights ($\mathbf{w}_j$ and $\mathbf{0}$) are stored in a circular register (first $\mathbf{w}_j$), a new input for the scalar product $\mathbf{w}^T_j\mathbf{x}$ is incorporated, and the processor j sends to processor j+1 $\mathbf{x}$ and and y_k ($0<k<j$) in serial. The cycle for this processor is determined mainly by the time needed to execute two multiplications and one subtraction in serial. This linear array is valid for APEX NN [4] if the null weights are replaced by $\mathbf{u}_j$.

For the training phase there is an alternative to the Data Adaptive (DA) scheme described in the previous points, weights being updated before the next input pattern is processed (Fig. 2). This is the Block Adaptive (BA) scheme, the weight update being postponed until the end of each training data block. Note that the BA scheme requires less computation than the DA scheme since the weight updating is done less frequently according to the size of the blocks. On the other hand, the size of the blocks should affect minimally to the convergence speed versus the computing time.

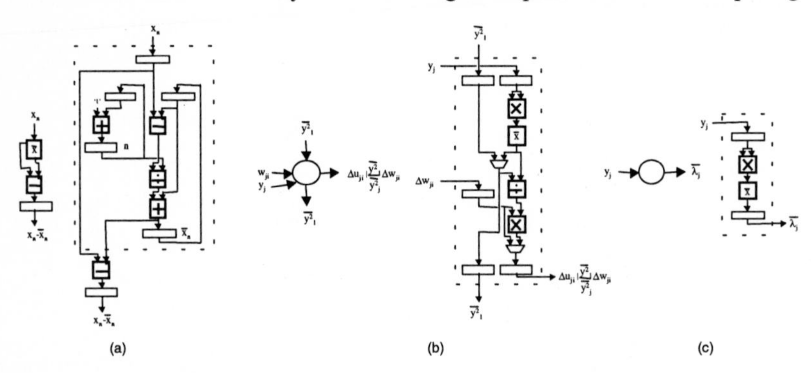

Fig. 3. Improving the features of the processors of the systolic architectures for Sanger NN and Rubner NN: (a) PCA with the covariance matrix, (b) ALR and (c) eigenvalue calculation.

6 Improvements for the Hebbian NN Architectures

The improvements for the Sanger NN and Rubner NN defined in section 4 can also be implemented by the systolic linear array architectures.

These linear architectures can be adapted to support PCA with covariance matrix if the input for input samples incorporates a preprocessing input block. This block should have a dynamic time mean unit (Fig. 3.a) with circular registers (memory for each component) synchronized with the components of the input samples, then the input for the Sanger NN or Rubner NN is the output of the dynamic time mean unit minus the initial input. The cycle of this block is the time needed for two subtractions, one sum and one division in serial. As well, it is possible that these linear array architectures use ALR. Then, the weight increment outputs of each processor are treated by other processor which modulates the increments according to Eq. 8 (Fig. 3.b). The cycle of this processor is the time needed for two multiplications, the time for one step of the dynamic mean value unit and one division in serial. The eigenvalue associated with each neuron can be calculated using a processor for each neuron whose input is the output of its associated neuron (retrieving array processor) and the output is the time mean value of its square input which converges to the eigenvalue of its associated neuron (Fig. 3.c). The cycle of this processor is the time for one multiplication and the time for one step of the dynamic mean value unit.

7 Concluding Remarks

The present work has focused on the study of Sanger NN and Rubner NN, as well as for Rubner NN its linearized formulation, and a new systolic architecture for their implementation. First of all, the presented NN has been introduced from their empirical formulation, some improvements of their algorithms have been addressed and their local stability has been enunciated. Finally, a systolic architecture for these NNs is described using its DG and projected on a linear array of processors including the improvements for their algorithms.

References

1. Berzal, J.A., Zufiria, P.J. and Rodríguez, L.: Implementing the Karhunen-Loeve Transform via Improved Neural Networks. Proceedings of the International Conference on Engineering Applications of Neural Networks, London 15-17 June 1996, 375-378.
2. Berzal, J.A. and Zufiria, P.J.: Linearized Trainning of Hebbian Neural Networks, Aplication to Multispectral Image Processing. Proceedings of the International Conference on Engineering Applications of Neural Networks, Gibraltar 10-12 June 1998, 1-8.
3. Chen, L.H. and Chang, S.: An adaptive Learning Algorithm for Principal Component Analysis. IEEE, Transactions on Neural Networks, September 1995, 1255-1263.
4. Diamantaras K.I. and Kung, S. Y.: Principal Component Neural Networks, Theory and Applications. John Wiley & Sons, Inc, 1994.
5. Dony, R.D. and Haykin, S.: Neural Networks Approaches to Image Compression. Proceedings of the IEEE, February 1995, 288-303.
6. Weingessel, A. and Hornik, K.: Local PCA Algorithms. IEEE, Transactions on Neural Networks, November 2000, 1242-1250.
7. Zufiria, P.J., Berzal, J.A., Martínez, M.A. and Fernández Serdán, J.M.: Neural Network Processing of Satellite Data for the Nowcasting and Very Short Range Forecasting. Proceedings of the International Conference on Engineering Applications of Neural Networks, Warsaw, 13-15 September 1999, 241-246.

The Hierarchical Neuro-Fuzzy BSP Model: An Application in Electric Load Forecasting

Flávio J. de Souza[1], Marley Maria R. Vellasco[2], Marco Aurélio C. Pacheco[3]

[1]DSc Electrical Engineering – PUC-Rio
Department of Computer & Systems Engineering – FEN/UERJ
Rua S. Francisco Xavier 524, CEP 20550-013, Rio de Janeiro, Brazil
fjsouza@eng.uerj.br

[2] Ph.D. University College London, UK
Department of Computer & Systems Engineering – FEN/UERJ
Department of Electrical Engineering – PUC-Rio
Rua Marquês S. Vicente 225, Gávea, CEP 22453-900, Rio de Janeiro, Brazil
marley@ele.puc-rio.br

[3]Ph.D. University College London, UK
Department of Computer & Systems Engineering – FEN/UERJ
Department of Electrical Engineering – PUC-Rio
Rua Marquês S. Vicente 225, Gávea, CEP 22453-900, Rio de Janeiro, Brazil
marco@ele.puc-rio.br

Abstract. This work presents the development of a novel hybrid system called Hierarchical Neuro-Fuzzy BSP (HNFB) and its application in electric load forecasting. The HNFB system is based on the BSP partitioning (Binary Space Partitioning) of the input space and has been developed in order to bypass the traditional drawbacks of neuro-fuzzy systems: the reduced number of allowed inputs and the poor capacity to create their own structure. To test the HNFB system, we have used monthly load data of six electric energy companies. The results are compared with other forecast methods, such as Neural Networks and Box & Jenkins.

1. Introduction

Neuro-Fuzzy Systems (NFS) [1],[3],[8],[9] are hybrid systems, as they use more than one systems identification technique for the solution of a problem. Neuro-Fuzzy Systems match the learning capacity of artificial neural networks (ANNs) [2] with the linguistic interpretation power of the Fuzzy Inference Systems (FISs) [6]. The basic idea of a Neuro-Fuzzy System is to implement a Fuzzy Inference System in a distributed parallel architecture in such a way that the learning paradigms used in the neural networks can be employed in this hybrid architecture.

This work introduces and apply a new Neuro-Fuzzy System, called Hierarchical Neuro-Fuzzy BSP (HNFB) [8],[9] in monthly load forecasting. The HNFB System is

J. Mira and A. Prieto (Eds.): IWANN 2001, LNCS 2084, pp. 174-183, 2001.

based on the BSP partitioning (Binary Space Partitioning) of the input space and was developed in order to bypass the traditional drawbacks of the NFSs.

This article is organized in 5 additional sections. Section 2 briefly describes the BSP partitioning scheme. Section 3 introduces the HNFB model, its basic cell, and its architecture. Section 4 presents the case studies, using the monthly load data of six electric energy companies. Section 5 compares the obtained results with other forecasting systems such as neural networks and Box & Jenkins. Section 6 finishes the article presenting the conclusions.

2. BSP Partitioning

Neuro-Fuzzy Systems and the Fuzzy Systems perform a mapping of fuzzy regions of the input space into fuzzy regions of the output space. This mapping is made through fuzzy rules. The input and output variables of Fuzzy and Neuro-Fuzzy Systems are divided in some linguistic terms (for example: low, high) that are used by the fuzzy rules. The input space partitioning indicates how the fuzzy rules are related in the space. The most common partitioning is the fuzzy grid, which, in spite of being the simplest, limits the number of input variables. The HNFB System [8],[9] uses a recursive partitioning, called BSP, that aims to reduce this limitation, besides having a unlimited capacity to create and to expand its own structure.

The BSP Partitioning divides the space successively, in a recursive way, in two regions. An example of such partitioning, illustrated in figure 1(a), shows that initially the space was divided into two parts in the vertical direction – x_2 (for example high and low).

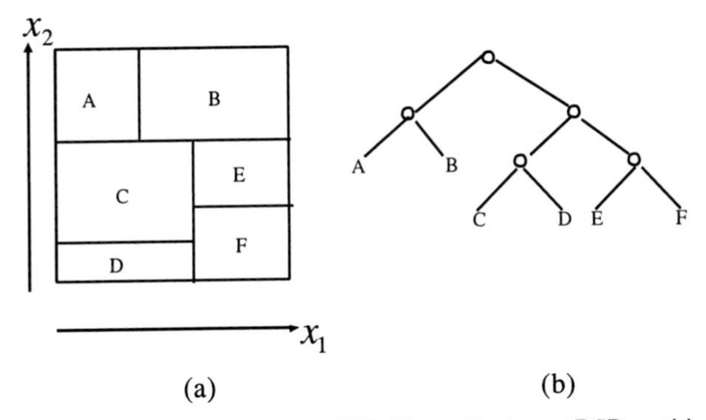

Fig. 1. (a) BSP Partitioning; (b) BSP Tree referring to BSP partitioning.

In this example, the upper partition was subdivided in two new partitions A and B, according to the horizontal direction – x_1. The inferior partition, in turn, was subdivided successively, in the horizontal and vertical route, generating the partitions C, D, E and F. Figure 1(b) shows each final partition represented by letters in the BSP tree. Interior knots represent the existing intermediate partitions.

The BSP partitioning is flexible and minimizes the problem of the exponential growth of rules, since it only creates new rules locally, according to the training set. Its main advantage is to build automatically its own structure. This type of partitioning is considered recursive because it uses recursive processes in its generation, what results in models with hierarchy in their structure and, consequently, hierarchical rules.

3. Hierarchical Neuro-Fuzzy BSP (HNFB) Model

3.1 Basic Neuro-Fuzzy BSP (Binary Space Partitioning) Cell

An HNFB (Hierarchical Neuro-Fuzzy BSP) cell is a neuro-fuzzy mini-system that performs binary fuzzy partitioning of the input space. The HNFB cell generates a precise (crisp) output after a defuzzification process.

Figure 2 (a) illustrates the cell's defuzzification process and the concatenation of the consequents. In this cell, 'x' represents the input variable, while ρ and μ are the membership functions that generate the antecedents of the two fuzzy rules.

Figure 2(b) illustrates the representation of this cell in a simplified manner.

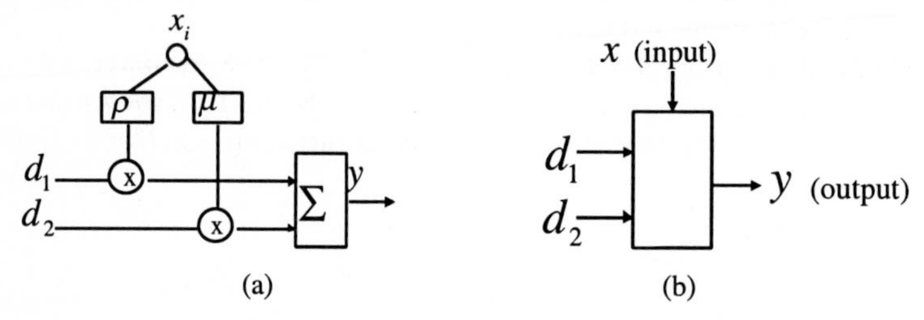

Fig. 2. *(a) Interior of the Neuro-Fuzzy BSP cell. (b)Simplified Neuro-Fuzzy BSP Cell.*

The linguistic interpretation of the mapping implemented by the HNFB cell is given by the following set of rules:

Rule 1: If $x \in \rho$ then $y = d1$.

Rule 2: If $x \in \mu$ then $y = d2$.

Each rule corresponds to one of the two partitions generated by BSP partitioning. When the inputs occur in partition 1, it is rule 1 that has a higher firing level. When they are incident to partition 2, it is rule 2 that has a higher firing level. Each partition can in turn be subdivided into two parts by means of another HNFB cell.

The profiles of membership functions ρ and μ are presented in Figure 3.

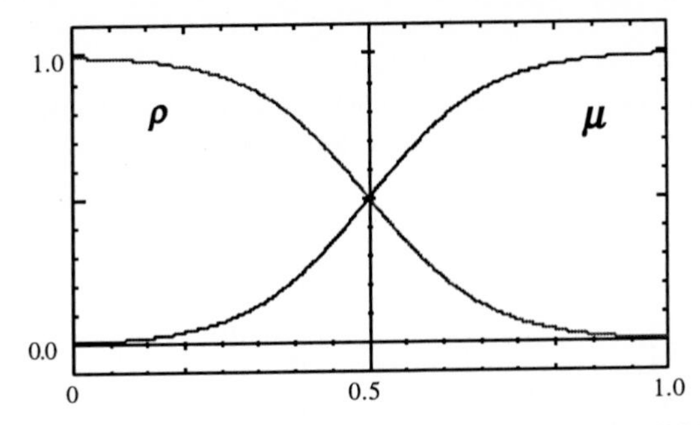

$$\mu(x)=1/(1+\exp(-x))$$
$$\rho(x)=1-\mu(x).$$

Fig. 3. Example of a profile of the BSP cell Membership functions:

The (crisp) output 'y' of an HNFB cell is given by the weighted average shown in equation (1).

$$y=\frac{\rho(x)*d1+\mu(x)*d2}{\rho(x)+\mu(x)}. \qquad (1)$$

Due to the fact that the membership function ρ is the complement to 1 of the membership function μ, in the HNFB cell the following equation applies:

$$\rho(x)+\mu(x)=1. \qquad (2)$$

Therefore, the output equation is simplified as:

$$y=\rho(x)*d1+\mu(x)*d2. \qquad (3)$$

or

$$y=\sum_{i=1}^{2}\alpha_i*di. \qquad (4)$$

where the α_i's symbolize the firing level of the rules and are given by: $\alpha_1=\rho(x)$; $\alpha_2=\mu(x)$.

Each *di* of equation (4) corresponds to one of the three possible consequents below:

- A **singleton**: The case where di = constant.
- A **linear combination** of the inputs: The case where $di=\sum_{k=0}^{n}w_k x_k$.

where: x_k is the system's k-th input; the w_k represent the weights of the linear combination; and 'n' is equal to the total number of inputs. The w_0 weight, with no input, corresponds to a constant value (bias).

- The **output of a stage** of a previous level: The case where $di=y_j$,

where y_j represents the output of a generic cell 'j', whose value is also calculated by equation (4).

3.2 HNFB Architecture

An HNFB model may be described as a system that is made up of interconnections of HNFB cells. Figure 4 illustrates an HNFB system along with the respective partitioning of the input space.

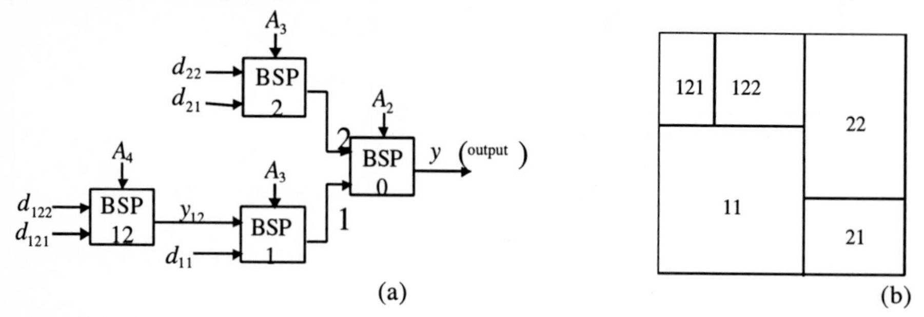

Fig. 4. (a) Example of an HNFB system. (b) Input space Partitioning of the HNFB system.

In this system, the initial partitions 1 and 2 ('BSP 0' cell) have been subdivided; hence, the consequents of its rules are the outputs of subsystem 1 and subsystem 2, respectively. In turn, the latter has, as consequents, values d11, y12, d21 and d22, respectively. Consequent y12 is the output of the 'BSP 12' cell. The output of the system in figure 4 (a) is given by equation 5.

$$y = \alpha_1.(\alpha_{11}.d_{11} + \alpha_{12}.(\alpha_{121}.d_{121} + \alpha_{122}.d_{122})) + \alpha_2.(\alpha_{21}.d_{21} + \alpha_{22}.d_{22}). \quad (5)$$

3.3 Learning Algorithm

The HNFB system has a training algorithm based on the gradient descent method [2][8] for learning the structure of the model and consequently, the linguistic rules. The parameters that define the profiles of the membership functions of the antecedents and consequents are regarded as the fuzzy weights of the neuro-fuzzy system. Thus, the di's and parameters 'a' and 'b' are the fuzzy weights of the model.

In order to prevent the system's structure from growing indefinitely, a parameter, named decomposition rate (δ), was created. It is an adimensional parameter and acts as a limiting factor for the decomposition process. Information on how this algorithm functions may be found in greater detail in [8],[9].

4. Case Studies

This section presents an electric load forecasting application of the proposed HNFB System, based on monthly historical data.

This experiment made use of data related to the monthly power values (electric load) of 6 utilities of the Brazilian electrical energy sector. The data obtained correspond to the period between January 1983 and August 1998. However, in order to compare the results obtained with the available results of other forecasting techniques, only data from the period between January 1983 and December 1994 was considered.

The forecasts were made with a one-step-ahead (one month ahead) prediction horizon. The data sets were divided into two subsets:

- the first, with 132 patterns, corresponding to the period from January 1983 to December 1993, was selected for training;
- the second subset, with 12 patterns, corresponding to the period from January to December 1994, was selected for testing.

4.1 Modeling of HNFB Systems in Load Forecasting

The most important steps in modeling the problem of load series forecasting by an HNF system are:

- *initial study of the load series* – Involves issues related to seasonal aspects of the load series and to whether or not the series is stationary (in terms of average and variance).
- *selection of model's inputs* – Consists of determining which inputs are relevant to a good forecast, such as: past load values; past temperature values and predicted temperature value; information on the day/month/year to be forecast.
- *data sets employed* – Regards the formatting and normalization of the data sets used in training and in the tests to validate the models.
- *development of the HNF models* – Once the inputs to be used in the models have been determined, the latter are trained and thus generate an HNF structure that is capable of making predictions.
- *analysis of the results obtained* – A few metrics are employed in order to evaluate the quality of the forecast generated by the model at hand.

The aspects described above were based on [11],[12] where back-propagation neural net forecasting is considered. Step 1 was not performed because the evaluation addressed the forecasting capability of the HNFB model in nonstationary series. The third step was unnecessary because the learning algorithm of the HNFB model already executes a kind of normalization on the input data.

The windowing technique was also employed in the HNFB model. A window whose maximum size was equal to 12 (referring to the past 12 monthly values) and two additional informative items (outside the window) regarding the month and year of the forecast were used.

4.2 Evaluation of the Performance of the HNFB Model

In order to evaluate the forecasting performance of the HNFB model, a few metrics associated with prediction errors were used: the **MAPE** ("*Mean Absolute Percentage Error*"), the **RMSE** ("*Root Mean Square Error*") and the **Theil's U** coefficient.

The MAPE indicates, as its name says, the mean percentage error value of the predictions over the entire test set.

The RMSE metric penalizes higher errors. Thus, a forecasting technique which presents excellent results in most predictions but contains high errors in one specific prediction will provide a high RMSE.

The Theil's U coefficient measures to what extent the results are better than those of a naive or trivial prediction (i.e., "*the best estimate of the next value is the present value*"). This coefficient makes it possible to analyze the quality of a forecast in the following manner:

- when $U \geq 1$, the error in the model is higher than the naive error.
- when $U < 1$, the error in the model is lower than the naive error (good forecast); the closer this coefficient is to zero, the better the forecast.

4.4 Results Obtained by the HNFB Models

For the purpose of load forecasting, 6 HNFB models, one for each company, were used and their performances are described in table 1 below.

The *Theil's U* coefficients obtained (all below 1.0) show the good quality of the forecasts yielded by the HNFB models that were trained for this task. However, the remaining metrics are more valuable when compared with other results such as the ones in the section 5.

Company	Input Sequence	# conseq.	Theil's U	MAPE	RMSE
COPEL	2,14,11,10,1	36	0.6318	1.17 %	28.89
CEMIG	2,14,11	6	0.7336	1.12 %	58.49
LIGHT	14,2,13,1	25	0.6931	2.22 %	81.74
FURNAS	14, 2, 1, 6, 7, 3, 13	27	0.7981	3.76 %	50.78
CERJ	14,2,1,5,6,10	15	0.5724	1.35 %	15.66
ELETROPAULO	2, 14, 10, 9 , 11, 1	23	0.6339	1.17 %	108.11

Input sequence: sequence of the inputs at each level
 # conseq. : Number of parameters in the consequents

Table 1- Architectures employed and prediction errors obtained by the HNFB model

5. Comparison with Other Techniques

In order to evaluate the performance of the HNFB model, the results were compared with those obtained by the *Backpropagation* algorithm and by other techniques, such as the *Holt-Winters* and *Box&Jenkins* techniques, and with those of the Bayesian Neural Nets (BNN) [10], trained by Gaussian approximation [4],[5] and by the MCMC method [7].

Table 2 below presents the compared performances of the HNFB model and the BNNs. For the sake of better visualization, the shaded cells in the tables indicate the lowest MAPE and RMSE. It can be noticed that the overall performance of the HNFB model was a bit lower, when compared with that of the Bayesian Nets tested in [10]. However, the data used in the HNFB model were not treated in terms of their seasonal aspects, nor were they made stationary as was the case of the Bayesian Neural Nets tested in [10].

COMPANY	BNN (G. Approx.)		BNN (MCMC)		HNFB	
	MAPE	RMSE	MAPE	RMSE	MAPE	RMSE
COPEL	1.45%	17.4	1.16%	16.4	1.17 %	28.89
CEMIG	1.29%	16.7	1.28%	17.1	1.12 %	58.49
LIGHT	1.44%	17.3	2.23%	27.5	2.22 %	81.74
FURNAS	1.33%	17.3	3.85%	56.2	3.76 %	50.78
CERJ	1.50%	21.2	1.33%	19.7	1.35 %	15.66
E.PAULO	0.79%	14.3	0.78%	10.6	1.17 %	108.11

Table 2 – Comparison of the HNFB model with the techniques described in [10].

In addition, it should be emphasized that in the case of the BNNs, according to [10], the training time for forecasting those load series was about 8 hours. This was a much longer period than the time required by the HNFB system to perform the same task, which was of the order of tens to hundreds of seconds, on equipment with a similar performance.

Table 3 below presents the comparisons between the performance of the HNFB models and that of the *BackPropagation*, the *Holt-Winters* and the *Box&Jenkins* forecasting techniques described in [11],[12]. Once more, the shaded cells in the tables indicate the lowest MAPE and RMSE.

COMPANY	HNFB		BackProp.		Box&Jenkins		Holt-Winters	
	MAPE	RMSE	MAPE	RMSE	MAPE	RMSE	MAPE	RMSE
COPEL	1.17 %	28.89	1.57%	26.57	1.63%	30.60	1.96%	31.91
CEMIG	1.12 %	58.49	1.47%	69.23	1.67%	79.93	1.75%	85.54
LIGHT	2.22%	81.74	3.57%	123.1	4.02%	122.2	2.73%	88.13
FURNAS	3.76%	50.78	5.28%	50.32	5.43%	55.29	4.55%	47.51
CERJ	1.35 %	15.66	3.16%	28.46	3.24 %	31.15	2.69%	23.91
E.PAULO	1.17 %	108.11	1.58%	127.5	2.23%	165.8	1.85%	148.61

Table 3 – Comparison of the HNFB model with a neural net, a Box & Jenkins model and with a Holt-Winters forecast.

The HNFB model always performed better than the *Holt-Winters*, *Box&Jenkins* and *BackPropagation* models with regard to the MAPE. As for the RMSE metric, the HNFB performed better in most cases. The few instances in which this did not occur were due to the circumstantial occurrence of some major error (greater than the average).

7. Conclusions

The objective of this paper was to introduce the HNFB model, a new neuro-fuzzy model, and present its application on time series forecasting. The HNFB model was developed with the purpose of improving the weaknesses of neuro-fuzzy systems. The results obtained by the HNFB model showed that it performs well in time series forecasting as well as in other types of which have not been described in this study but are reported in [8],[9]. In addition, this model is able to create its own structure and allows the extraction of knowledge in a fuzzy rule-base format.

An interesting prospect for future work, which is already in the research stage, is the use of this model for classification systems that allow the extraction of knowledge in the format of nonhierarchical fuzzy rules for data mining applications.

References

[1] Halgamuge, S. K., Glesner,M.: Neural Networks in Designing Fuzzy Systems for Real World Applications. Fuzzy Sets and Systems N0.65, pp. 1-12. (1994).

[2] Haykin, S.: Neural Networks – A Comprehensive Foudation. Macmillan College Publishing Company, Inc.(1994).

[3] Kruse, R., Nauck, D.: NEFFCLASS – A Neuro-Fuzzy Approach for the Classification of Data. Proc. Of the 1995 ACM Symposium on Applied Computing, Nashville.

[4] Mackay, D., 'Bayesian Methods for Adaptive Models", Ph.D. thesis, California Institute of Technology, 1992.

[5] Mackay, D., A Practical Bayesian Framework for Backpropagation Networks. Neural Computation, 4 (3):448-472, 1992.

[6] Mendel, J.: Fuzzy Logic Systems for Engineering: A Tutorial. Proceedings of the IEEE, Vol.83, n.3, pp.345-377. (1995).

[7] Neal, R., "Bayesian Learning for Neural Networks", Lecture Notes in Statistics 118, Springer Verlag, 1996.

[8] Souza, Flávio Joaquim de: Modelos Neuro-Fuzzy Hierárquicos. Tese de Doutorado. Departamento de Engenharia Elétrica da Pontifícia Universidade Católica do Rio de Janeiro (1999) (in portuguese).

[9] Souza, F. J., Vellasco, M.M.B.R., Pacheco, M.A.C, *Hierarchical Neuro-Fuzzy QuadTree Models,* Fuzzy Sets & Systems, (to be published), 2001.

[10] Tito, E.H., Zaverucha, G., Vellasco, M.M.B.R., Pacheco, M.A.C, "*Applying Bayesian Neural Networks to Electrical Load Forecasting*", Proceedings of The 6th International Conference on Neural Information Processing (ICONIP'99), Perth, Australia, 16-20 November 1999.

[11] Zebulum, R.S., Vellasco, M.M.B.R., Guedes, K., and Pacheco, M.A.C, *An Intelligent Load Forecasting System*, Proceedings of the IEEE International Conference on Electricity Sector Development and Demand Side Management - ESDDSM'95, pp. 96-103, Kuala Lumpur, Malasia, 21-22 November 1995.

[12] Zebulum, R.S., Guedes, K., Vellasco, M.M.B.R., and Pacheco, M.A.C, *Short-Term Load Forecasting Using Neural Nets*, Lecture Notes in Computer Science 930, From Natural to Artificial Neural Computation, Springer-Verlag, Proceedings of the International Workshop on Artificial Neural Networks (IWANN'95), pp.1001-1008, Torremolinos (Málaga), Spain, 7-9 June 1995.

The Chemical Metaphor in Neural Computation*

J. Barahona da Fonseca[1], I. Barahona da Fonseca[2], C.P. Suárez Araujo[3], J. Simões da Fonseca[4]

[1]Dept. of Electrical Engineering; FCT.UNL. Portugal E-mail: jbfonseca@ip.pt.[2]Faculty of Psychology of Lisbon; University of Lisbon. E-mail: ibf@ fpce.ul.pt. [3]Faculty of Computer Sciences. Dept. of Computer Sciences and Systems, University of Las Palmas de Gran Canaria, E-mail: cpsuarez@dis.ulpgc.es. [4]Faculty of Medicine of Lisbon; Univ. of Lisbon; E-mail: j.s.da.fonseca@ ip.pt

Abstract: Chemical reactions provide new insights concerning information processing in neural networks. Batch reactions are considered in first place and it is presented a model in which computational instructions will be conceptualised as being sent in parallel to multiple operational structures in sites where they are transformed in a way that generates a variety of computational viewpoints associated to a set of data.

Anastomotic computations are considered as pre-processing operations implemented by convolutional weighing functions. Quantum valence theory is used to construct computational models for cognitive processes. The model of isolobal fragments is interpreted in a way that gives rise to classes of equivalence that represent concepts in extenso.

The conscious quality of cognition is derived from the emergence of new macroscopic qualities from the relationships between valence bonds. Binding of representations is considered as dependent on dendro-dendritic relationships homeomorphic with orbital valence frontiers.

1. Introduction

Our aim is to study information processing in neural networks in order to understand both the architecture of cognition and brain processes from micro to macro structures, and to build artificial neural networks which meet more strict neurophysiological requirements than those which can be represented using exclusively cell body and axonal computations.

We define a conceptual framework, which allows the implementation of formal and computational neural information processes adequate for the representation of high level symbolic stages of perception and cognition, and to quantify the results. The general conceptual framework stems from chemical and physical paradigms. Chemical reactions lead to new computational structures in which data are directly associated to computational instructions that are distributed over neuron ensembles. Each neuron modifies the incoming instructions for operation generating a particular component or viewpoint in a global process.

---

* This work was supported by a grant from Bial foundation, Oporto, Portugal.

J. Mira and A. Prieto (Eds.): IWANN 2001, LNCS 2084, pp. 184-195, 2001.

Besides cell body and axonal processes, local dendro-dendritic operations are also considered providing new insights and the use of relevant metaphors.

Chemical processes either involving batch reactors, plug flow reactors or else the model of isolobal fragments in organometallic complexes and their valence bonds. This provides new paradigms for the study of some information processes which are involved in high level symbolic operations performed by dendro-dendritic networks and cell bodies and axons to which they are connected. These models allow the study of the dynamics of the information processing in terms of final equilibrium states and the prediction of convergence towards a specific final state. Velocity of reaction and thermodynamic characteristics of the reaction, parallel reactions as well as multi-stage reactions are also useful for the study of the dynamics of information processing in neural networks. It should be noted that we make a distinction between low level analogic representation of data and events which is performed bottom-up according with genetically prescribed rules and on the other hand a second system of encoding in which significations depend on relationships that are acquired through learning. The concept of isolobal fragments of organometallic complexes together with their quantum valence implications is used as a paradigm for the representation of classes of equivalence that are essential to the definition of concepts at the cognitive level. It also provides a paradigm for the way in which new macroscopic qualities, like consciousness emerge from complex interrelationships between microstructures.

The chemical metaphor contributes to clarify some types of data pre-processing that are not simply serial or else parallel but rather anastomotic as it was proposed by W.S. McCulloch. They may be easily represented taking batch or plug flow reactors as paradigms.

Another relevant consequence of the use of chemical metaphors is the possibility of understanding the effects of changes in synaptic weights not as a set of independent changes but rather as a structured set of complex pattern transformations that implement Gestaltic entities.

The orbital concept of chemical valence leads to the understanding of the binding of neural processes distributed over large and disperse structures forming high level language correlates for cognitive processes.

Besides the usefulness of those chemical paradigms to study the dynamics of neural network processes from the viewpoint of complexity theory, they contribute with relevant metaphors to the construction of neural network models for cognitive processes.

2. Neuronal and Computational Implementation: The Dendro-Dendritic Network Model

2.1 Batch Reactors-A Neurophysiological Interpretation

Within the batch reactor metaphor, processes in a dendro-dendritic complex provoke the formation of 'compound messages' such that the time needed to attain a 'steady state' for a new compound message Ay is

$$t = -\frac{1}{k}\ln\frac{N_{Ay}}{N_{Ay0}} \tag{1}$$

In (1) N_{Ay} stands for the number of messages of a class at any instant, and N_{Ay0} that number at t=0. k is related to the 'reaction rate', something like the readiness of the dendro-dendritic net to produce new messages.

Now, a main feature of the batch reactor analogy is the potentiality to generate parallel processes, besides the serial ones, which lead to some kind of anastomotic computation. For parallel 'reactions' the proportion of different compound messages can be written as

$$S = \frac{\alpha_{Ay/A}}{\alpha_{Ay/A} + \alpha_{By/A}} \tag{2}$$

where $\alpha_{Ay/A}$ is the proportion of compound Ay and $\alpha_{By/A}$ is the proportion of By. In any case these parameters are introduced to illustrate the potentiality of the metaphor.

In case the 'compound message' triggers the adequate patterns of action, the Ascending Reticular System of the Brain Stem and of the Thalamus will energise the rate of production of that 'compound message' performing an action equivalent to a reactors' temperature elevation during an adequate time interval Δt_i. The following classical reactor equations express the idea that 'compound message formation' rate will effectively increase with a raise of temperature, in a reversible endothermic reaction. At equilibrium, $r_A=0$; $x_A=x_A^*$ (x_A^* is the reactant conversion at equilibrium) and substituting into the reaction rate expression we have,

$$\frac{x_A^*}{1-x_A^*} = \frac{k_1}{k_{-1}}\exp\left(\frac{\Delta E_{-1} - \Delta E_1}{RT}\right) \tag{3}$$

In (3) ΔE_1 is the forward activation energy which must be greater than the reverse activation energy, ΔE_{-1}, k_{-1} are defined by

$$r_A = k_1\, C_A - k_{-1}\, C_B \tag{4}$$

In (4) C_A is the concentration of A and C_B is the concentration of B.

In case the 'compound messages' are no longer adequate, the energising action of the Reticular System ceases, and a new ensemble of sets of characteristic dendro-dendritic oscillators will be activated, renewing the cycle of interactions which will now include new dendro-dendritic operators. Note that it is implied that reactants change but not the form of the reactions.

Finally, Work Memory and Executive Attention will activate new areas of the Brain and associate Sub-Cortical structures which will create a similar but distinctly organised functional architecture, until the Intentional action is completed. What changes here in the model is the type of relationships between isolated reactions which become organised in a structured system. To embody Work Memory and Executive Attention Control and Command, a dendro-dendritic network of fully connected multiple centres in the sense that any centre is connected by inputs and outputs to all the remaining ones and to itself, forms a sort of a 'keyboard' that is able to be activated sequentially or in parallel and in turn produce changes in the pattern of relationships between more elementary units that are organised by the 'keyboard'. A switching control makes a temporal plan of activations according with a prescribed set of criteria which act within the context of local time relationships which form an invariant pattern for decision making. This local time and space parameters form the context necessary for an internal model representation of an interaction with the environment using both recursive and hyperincursive goal directed computations which generate behaviours which will be performed according with external time-space parameters and produces a state which represents the interaction from the viewpoint of internal decision making performed according with local time-space parameters completely distinct from the external ones. In our hypothesis the set of decision making rules, the 'keyboard' programming of both plans of action and connected internal representations, internal local time-space relationships and the availability to an self reflexive knowledge about this processes in a summarised manner in which many links are missing, forms the basis for conscious awareness of the environment, of the Self, and of both elementary decisions and complex plans of action and Intentions.

Central planning of action furthermore recruit, at a distance, other specialised operators to produce new processes made from a different viewpoint. These centres which work in a distributed manner make decisions of action and back propagate information about their actual state of activity. They correspond to Tertiary Multisensory Associative Areas, as well as to centres responsible for the use of linguistic competence and idiomotor symbolic representations of commands of structured action.

These high level centres either non linguistic or else linguistic will send encoded messages, that upon being received, play a role in the programming of the commands and representations to which they are contributing. Possibly it is the encoded structure of this high level symbolic messages that performs the double role of carrying a semantic content and simultaneously specific programming of relational instructions.

2.2- Neuronal Implementation - The Dendro-Dendritic Network Model

If we consider neural architectures, dendro-dendritic networks possess the remarkable characteristic of allowing both forward and backward propagation. The complex geometry of these networks permits back propagation of action potentials generated at the axon hillock. Ultimately when such potentials attain fine dendritic ramifications they can eventually fade out. When this is not the case such potentials can circulate along dendro-dendritic closed loops which implement local computations.

Although dendro-dendritic networks are extremely dense anatomical structures with interspersed almost reticulate structures linked by multiple synapses, they may be thought of as an ensemble of computational compartments. Let us consider the case in which information arrives to the dendritic neuropile through axo-dendritic and dendro-dendritic synapses.

If we attribute to each neuronal cell body a set of characteristic dendro-dendritic closed loops we may consider each such set as an ensemble of oscillators with characteristic frequencies. We assume that complex information processing is performed by each neuron due to its closed loops. Similar functions are served by identical sets of closed loops and distinct functions are implemented by different sets of closed loops, A, B,..., Z.

If the same information arises to an ensemble of distinct sets of closed loops each one performing the analysis and processing of the incoming information from a distinct viewpoint, we may assert that an input vector {Y} (coming from the output of a predecessor structure) activates distinct sets A, B, ..., Z of dendro-dendritic closed loop operators.

From the viewpoint of our analysis it may be considered that 'compounds' Ay, By, ..., Zy have been formed and the corresponding information will be transmitted by neuronal cell bodies at a distance to decision making neurons responsible for the generation of coordinated patterns of action. A, B, ..., Z, are closed loop operators and oscillators $y \in \{Y\}$ are a subset of ordered components of {Y} with dimension $\dim.(y) \leq \dim(\{Y\})$.

2.3 Anastomotic Computation

One way of understanding anastomotic computation is to consider a convolution matrix of dimension k x k which acts through a inner product over a data matrix of the same dimension. It is furthermore supposed that the convolution weighing matrix is constant while data do vary.

This is a way to understand preprocessing of data along an active set of information transmission and computational parallel and interactive lines.

Furthermore chemical paradigm allows the creation of systems in which the convolution weighing components are transmitted along together with the data matrix in a computational dynamic flow.

A further generalisation leads to the concept of a dynamic convolution with variable coefficients. This last non-linear paradigm may also contribute to understand the way in which operative instructions are transmitted at a distance and are transformed in characteristic an distinct ways by the neurons that receive them.

Besides the usual concept of data transmission between computational stations we have here a paradigm for the transmission of computational instructions which produce a local transformation at each computational station which receives them.

This computational model is likely to exist in the peripheral and Central Nervous System. The chemical metaphor closer to it is the model of plug flow reactors. Let us remind the basic characteristics of a PFR.

2.4 Plug Flow Reactors (PFRs)

Now that we have 'designed' a simple batch reactor we will take a look at plug-flow reactors. Tubular reactors are used for many large-scale gas reactions. In tubular reactors, there is a steady movement of reagents in a chosen direction. No attempt is made to induce mixing of fluid between different points along the overall direction flow.

A formal material balance over a differential volume element requires that we know (or assume) patterns of fluid behaviour within the reactor, e.g. the velocity profile. The simplest set of assumptions about the fluid behaviour in a tubular reactor is known as the plug flow (or piston flow) assumption. Reactors approximately satisfying this assumption are called plug flow reactors (PFRs). The plug flow assumptions are as follows,

(a) flow rate and fluid properties are uniform over any cross-section normal to fluid motion;
(b) there is negligible axial mixing- due to either diffusion or convection.

The plug flow assumptions tend to hold when there is good radial mixing (achieved at high flow rates $Re > 10^4$) and when axial mixing may be neglected (when the length divided by the diameter > 50 (approximately)). This means that if we consider a differential element within the reactor (with its boundaries normal to the fluid motion), it can be taken to be perfectly mixed and, as it travels along the reactor, it will not exchange any fluid with the element in front of or behind it. In this way, it may be considered to behave as a differential batch reactor. Furthermore, we will make an additional assumption that the reactor is at steady state. In a batch reactor, composition changes from moment to moment. In continuous operation at steady state in a PFR, the composition changes with position, but at a given position there is no change with time.

We are now able to perform a material balance (we now know enough about the

l	length (m)
L	total reactor length (m)
V	reactor volume (m^3)
A	cross - sectional area (m^2)
n_A	molar flow rate of component A (mol s^{-1})
n_T	total molar flow rate (mol s^{-1})
v_T	volumetric flow rate (m^3 s^{-1})
0	denotes inlet conditions (subscript)
e	denotes exit conditions (subscript)

behaviour of the fluid within the reactor). We must choose an appropriate element over which to perform the material balance. The following symbols are used:
The material balance over the differential element has the usual form,

Accumulation = input – output – loss through reaction (5)

We now substitute in mathematical expressions for each of these terms. We have

$$0=n_A\big|_l - n_A\big|_{l+dl} - r_A dV \tag{6}$$

Dividing throughout by dl the differential is obtained

$$0=-\frac{dn_A}{dl}-r_A\frac{dV}{dl} \tag{7}$$

Dividing by the cross-sectional area and remembering that $A\ dl = dV$,

$$r_A=-\frac{dn_A}{dV} \tag{8}$$

Equation (8) is known as the design equation for PFRs.

The PFR rate of conversion may be found by performing the integration of the following design equation

$$\int_0^V dV = V = -\int_{n_{A0}}^{n_A} \frac{dn_A}{r_A} \tag{9}$$

The definition for conversion can be rewritten in terms of molar flow rates

$$x_A = \frac{n_{A0} - n_A}{n_{A0}} \tag{10}$$

Substituting (10) in (9) we have

$$V = \int_0^{x_A} \frac{n_{A0}\, dx_A}{r_A} \tag{11}$$

The reaction rate will be given by

$$r_A = k\, C_A = k \frac{n_A}{v_T} \tag{12}$$

Substituting (12) in (11) and integrating we have

$$V = -\frac{v_T}{k} \ln \frac{n_A}{n_{A0}} = -\frac{v_T}{k} \ln (1 - x_A) \tag{13}$$

$$\tau = \frac{V}{v_T} = \frac{1}{k} \ln \frac{n_{A0}}{n_A} = -\frac{1}{k} \ln (1 - x_A) \tag{14}$$

The residence time is given by

3. Towards an Homeomorphism of Isolobal Analogy

One of the main theories, which explain chemical reactions, is the molecular orbital theory. In organic molecules it is possible to substitute carbon atoms by metallic heteroelements if we use metal ligands and we create fragments with the adequate symmetry, level of energy and adequate valence structure in the bounding surface of the new component fragments.

The valence frontiers of both the organic fraction and the metallic complex do not need more than an approximate match. Hoffmann used a transformed Mendeleiff table to characterise the orbital structure of each element.

These two fragments are called *isolobal* if the number, symmetry properties, approximate energy and shape of the frontier orbital and the number of electrons in them are similar (Hoffmann, 1982).

With the support of this data and using considerations issued from group theory and Quantum Mechanics Perturbation Theory he was able to make exact predictions about sets of equivalent component fragments that matched and reacted with complex organic molecules.

Here we find a paradigm for classes of equivalence of organometallic compounds.

Returning to the proposed homeomorphism between dendro-dendritic closed loops on one hand and orbitals on the other, we can define interactional surfaces that are composed by dendritic computational compartments that are attached to a complex set of neural bodies.

The valence sites are identified as computational compartments of dendro-dendritic closed loops which include input and output synapses. These synapses contribute to an interaction surface through which this computational assembly can be linked to distinct assemblies. This metaphor is convenient to represent some characteristics of cognitive processes.

Namely we are thinking about the knowledge that cognitive operators concerning lower level sensory and motor processes and memory representations.

The inquiry about what is occurring at a sensory and motor level in a transaction with the environment may be compared to the successive linkage of a matrix of organic complex molecule to distinct equivalent isolobal metallic fragments.

A first characteristic of this class of equivalence is that it may represent a concept in extenso. On the other hand as this organometallic complexes are very unstable they may contribute to represent a sequence of inquiry operations in which distinct attributes of a concept are successively scanned.

All of them remain available to a further inquiry but only those which form stable complexes give rise to a conscious experience.

The correlate for the conscious character of cognition may be found in the way microscopic interactions generate the emergence of macroscopic qualities.

Comparative neuroanatomy and neurophysiology help to specify as a further requirement that consciousness must be related to secondary and terciary multisensory and idiomotor areas of the Brain in Primates and Man.

Ultimately the Quantum valence bounding surfaces may be considered putatively as a possible correlate for the type of binding of distinct sources of information which are involved in complex cognitions.

The resemblance between two neuron assemblies together with their characteristic dendro-dendritic closed loops that can participate in neural connections allows them to generate a complex and self-organised structure. In our proposal it can be defined as an isocompartment analogy, which can hold a distributed micro-resemblance defining an isosynaptic structural analogy. We will call two neurones or neural structures *isocompartment* if the number, symmetry properties, approximate energy and shape of the ensemble of the computational compartments corresponding to dendro-dendritic circuits on the neuron and the geometrical properties and number of synapses in them are similar. The isocompartment analogy provides the connection capabilities and the self-organisation of the neural structures needed to represent high level symbolic operations. Finally, this analogy also provides a hypothetical framework to understand abstract cross-modal representations.

4. The relationship of signification in visual cognitive processes

Dendritic computation in the Brain provides very powerful means for information processing which are involved in cognitive operations. If we consider visual cognition the observations we made in preceding sections need to be completed by additional features which are required if we try to represent symbolic processing for example in perceptive phenomena that attain the cognitive level. The basic stages of perceptive processes have been discovered in visual perception in very ingenious experiments first by Lettvin, Maturana, Pitts and McCulloch (1959), Hubel and Wiesel (1965), Blakemore (1974) to mention only the initial works in which the concept of feature detectors was introduced. Results concerning the study of simple cells by Hubel and Wiesel imply a referential level of processing in which elementar mappings in the Brain are directly related to elementary characteristics of stimuli. At this first level neuronal structures and their dendro-dendritic networks are responsible for a referential attribution. This reference to external objects represented by analagous mappings is in itself already very complex as it can be immediately verified if we consider the phenomenon of sensory projection on peripheral receptors or the phenomena of phantom limb.

At this level an analogical relationship between the spatial distribution of the stimuli and spatial configuration of dendro-dendritic and cell body mappings in the Brain is a first mode of iconic and diagrammatic representation of visual information. These simple components of sensory information are combined by dendro-dendritic networks that implement the specification of functions defined over sets of components of the iconic and diagrammatic level of representation. The complex and hipercomplex types of response (detection of segments of straight line with variable lengths and detectors of angles independently of rotation keeping constant vertex) suggest that the relationships necessary to represent data at a higher symbolic level are already present but the representation is still referentially linked to the characteristics of stimuli in an analogous manner. The encoding is still genetically determined, although it suffers the influence of the interaction with the environment along some initial stages of the development process (Blakemore, 1972). The next step introduces the type of encoding which is characteristic for cognition, namely the referential mappings are substituted by abstract relationships defined over them. The relationship is already not analogical but rather belongs to a second level of symbolisation. At this second level the relationship is one of significant-significate in which the significate is derived by specific rules associated to the significant. These rules are formed under the influence of experience and learning and give to knowledge structures an abstract character adequate for cognitive processes. Data concerning possible schemes of motor interaction and predictions about the outcome of transactions with the environment, are integrated in multi-modal sensory and motor secondary and tertiary representations. At this stage the isolobal inquiry matrices scan data selecting those which may contribute to successive stages of cognition. The final choice of attributes is attained and they acquire a stable configuration which gives rise to the cognitive experience. Attributes not essential for immediate cognition are organised as a field which remains available to further scanning whenever knowledge actualisation requires it.

The isolobal matrices are responsible for the implementation of the signification relationship which leads to the characterisation of the concept through its attributes and to the subjective experience which acquires a transient conscious dimension.

Complex forms imply the constitution of predicates and relationships in a visual language largely independent of the linguistic system of reference. A visual pattern can then be represented by a set of complex descriptors **P(X)** and a relationship **R** defined over them. The implementation of recognition of a pattern by a dendritic network might correspond to an input formed by the assembling of descriptors **P(X)** defined over an ensemble of elementary symbols either visual or belonging to other modalities as well as motor data and motor predictions and components of past sensory meanings assembled by isolobal matrices. The final stable result may be mathematically specified by **R** that correspond to a matrix which would produce a new vector of identified pattern components. Characterisation and identification may be represented formally by the comparison between a final vector of attributes and a reference vector. Identification would correspond to the fulfilment of a distance criterion and subjective experience would be due to the formation of a stable complex which would link the inquiry relationships to members of the class of isolobal intermediate level attribute representation. A criterion vector **Ref** would be compared with **C** and the pattern recognition would correspond to the verification that distance between **C** and **Ref** was smaller than a criterion α. This corresponds to

$$[R]\,[X]=[C]$$

dist([C],[Ref]) $< \alpha$

5. Final Comments

It results from the qualitative analysis of experimental data and model representation of concepts and perceptions that the Quantum Mechanics analogy provides an adequate description and new insights relevant for the understanding of self-reflexive cognition and consciousness.

References

Barahona da Fonseca, J., Barahona da Fonseca, I. and Simões da Fonseca, J. (1995): "Visual Information Processing from the Viewpoint of Symbolic Operations", in J. Mira y Mira and F. Sandoval (Eds), *From Natural to Artificial Neural Computation*, 253-260, Springer-Verlag.

Barahona da Fonseca, I., Barahona da Fonseca, J. and Simões da Fonseca, J. (1995): "Dendritic Computation in the Brain", in J. Mira and F. Sandoval (Eds), *From Natural to Artificial Neural Computation*, 260-267, Springer-Verlag.

Blakemore, C. and Cooper, J. F. (1970): "Development of the Brain Depends on the Visual Environment", Nature 228, 477-478.

Blakemore, C. (1974): "Development Factors in the Formation of Feature Extracting Neurons", in Neuroscience Third Research Program, 105-113.

Hoffmann, R. (1982), Building bridges between inorganic and organic chemistry, Angewandte Chemie Vol. 21, Nº 10, 711-724.

Hubel, D.H., Wiesel, T.N. (1965):"Receptive Fields and Functional Architecture in two Non-Striate Visual Areas (18 and 19) of the Cat", J. Neurophysiol.

Lettvin, J. Y., Maturana, H. R., McCulloch, W. S., Pitts, W. H. (1959): "What the Frog's Eye Tells the Frog's Brain", in Proc. Inst. Radio Eng., New York.

Simões da Fonseca, J., Barahona da Fonseca, J. and Barahona da Fonseca, I. (1996): "Cognitive Processes-Representation of Events and Operations from the Viewpoint of Dendritic Computation", in R Moreno-Díaz and J Mira Mira (Eds). *Brain Processes, Theories and Models*, 173-183, MIT Press.

Tolman, E. C (1932): *Purposive Behaviour in Animals and Men*, Century Co., New York.

Tolman, E. C. (1938): "The Determiners of Behaviour at a Choice Point"; Psychol. Rev., 1, 41.

The General Neural-Network Paradigm for Visual Cryptography

Tai-Wen Yue and Suchen Chiang

Department of Computer Science and Engineering, Tatung University
40 Chungshan North Road, Section 3, Taipei 104, Taiwan
twyu@cse.ttu.edu.tw and suchen@comm.cse.ttu.edu.tw

Abstract. This paper proposes the general paradigm to build Q'tron neural networks (NNs) for *visual cryptography*. Given a visual encryption scheme, usually described using an access structure, it was formulated as a optimization problem of integer programming by which the a Q'tron NN with the so-called *integer-programming-type* energy function is, then, built to fulfill that scheme. Remarkably, this type of energy function has the so-called *known-energy* property, which allows us to inject bounded noises persistently into Q'trons in the NN to escape local minima. The so-built Q'tron NN, as a result, will settle down onto a solution state if and only if the instance of the given encryption scheme is realizable.

1 Introduction

Visual cryptography is a cryptographic scheme to achieve visual secret sharing [6,7]. It finds many applications in cryptographic field such as key management, message concealment, authorization, authentication, identification, and entertainment. Given a visually recognizable *target image*, the (n, k) encryption scheme (or *access scheme*) is to cryptologically decompose the image into a set of n shares, called *shadow images*, which are recorded in transparencies, such that the pictorial meaning stored in the original image is recognizable if and only if a viewer is able to acquire at least k shares and stack them together. More complex encryption schemes can be described using the so-called *access structures* [1,2]. Traditionally, a codebook approach was used to produce shares for the instances of an encryption scheme. This implies that different access schemes require different codebooks. For complex ones, the codebooks are, in fact, hardly to be found or, even, inexistent.

The Q'tron NN model [8–10] had been applied to $(2, 2)$ access scheme successfully [11,12]. In this particular scheme, the energy function to build the Q'tron NN that effectively fulfills the scheme can be magically found. However, the same strategy is inapplicable to other access schemes. The $(2, 2)$ contains two plain shares, i.e., each small area in the share almost has the same mean greylevel, and one target image. Some complex access schemes require that both shares and target images carry meaningful informations. Furthermore, the subsets of shares in an access scheme can be described either as being *qualified* or *forbidden*. That is, the stacking of a forbidden subset (Γ_{forb}) of shares will reveal meaningless information, while the stacking of a qualified subset (Γ_{qual}) of shares will reveal the preassigned one.

J. Mira and A. Prieto (Eds.): IWANN 2001, LNCS 2084, pp. 196–206, 2001.

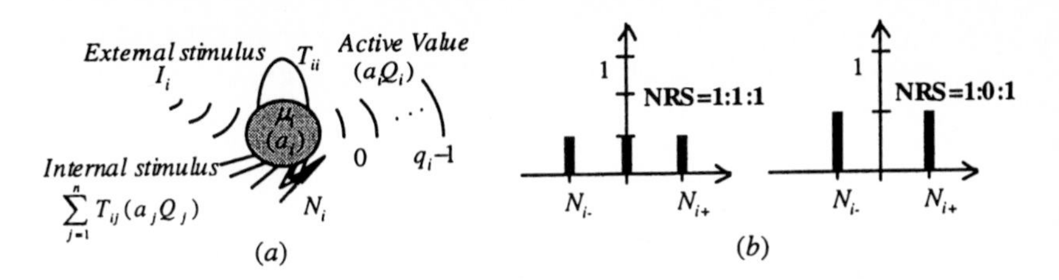

Fig. 1. (a)The Q'tron. (b)Two commonly used noise ratio spectra.

Building a Q'tron NN to function properly for a generalized access scheme requires that its energy function possesses the so-called *known-energy property*. Specifically, the highest value of energy, which corresponds to the solution quality of a problem, of the Q'tron NN that represents a valid or feasible solution has to be prior known. Accordingly, the bounded noise-spectra of Q'trons can be determined to escape the NN from local-minima. By injecting such noises into Q'trons persistently, the NN finally will settle down on a state whose energy is lower than or equal to the specified highest value so long as such a state does exist. Such a noise injecting mechanism is considerably different from traditional one since a schedule, such as the cooling schedule for *simulated annealing*, to control the noise strengths is needless.

The content of the paper is organized as follows: Section 2 gives a brief introduction on the Q'tron NN model and investigates its noise-injecting mechanism. Section 3 discusses Q'tron NN structure and the main energy terms to build Q'tron NNs for visual cryptography. Section 4 shows some experimental results using the proposed Q'tron NN. Some conclusions will be drawn in section 5.

2 Q'tron NN Model

In this model of NN, the basic processing elements are called Q'tron's. The number of output-level of each Q'tron can be greater than two. Specifically, let μ_i represent the ith Q'tron in a Q'tron NN, where the output of μ_i, denoted as Q_i, takes up its value in a finite integer set $\{0, 1, ..., q_i - 1\}$ with q_i (≥ 2) being the number of output-level. In addition, Q_i is also weighted by a specific positive value a_i called *active weight*, which stands for the unit excitation strength of that Q'tron. The term a_iQ_i is thus referred to as the *active value* of μ_i, which represents the total excitation strength of that Q'tron. In a Q'tron NN, for a pair of connected Q'trons μ_i and μ_j, there is only one connection strength, i.e. $T_{ij} = T_{ji}$. In Q'tron NN model, each Q'tron in the NN is allowed to be noise-injected. Thus, *noise-injected stimulus* $\hat{\mathcal{H}}_i$ for the Q'tron μ_i is defined as:

$$\hat{\mathcal{H}}_i = \mathcal{H}_i + \mathcal{N}_i = \sum_{j=1}^{n} T_{ij}(a_jQ_j) + I_i + \mathcal{N}_i, \tag{1}$$

where $\mathcal{H}_i$ denotes the *noise-free net stimulus* of μ_i, which apparently is equal to the sum of *internal stimuli*, namely, $\sum_{j=1}^{n} T_{ij}(a_jQ_j)$, and *external stimulus* I_i. The term $\mathcal{N}_i$ denotes the piece of random noise fed into μ_i, and n denotes the number of Q'trons in the NN. The schematic diagram for such neural network model is shown in Figure 1 (a). A Q'tron NN is said to run in *simple mode* if it is noise-free, i.e., $\mathcal{N}_i = 0$ for all i; otherwise, it is said to run in *full mode*. At each time step only one Q'tron is selected for level transition subject to the following rule:

$$Q_i(t+1) = Q_i(t) + \Delta Q_i(t), \tag{2}$$

with

$$\Delta Q_i(t) = \begin{cases} +1 & \hat{\mathcal{H}}_i(t) > \frac{1}{2}|T_{ii}a_i| \text{ and } Q_i(t) < q_i - 1; \\ -1 & \hat{\mathcal{H}}_i(t) < -\frac{1}{2}|T_{ii}a_i| \text{ and } Q_i(t) > 0; \\ 0 & \text{otherwise,} \end{cases} \tag{3}$$

where we here have assumed that the ith Q'tron is selected at time $t+1$. According to this rule, clearly whenever a Q'tron reaches a stable state, each Q'tron say, μ_i must satisfy one of the following conditions: i) $\hat{\mathcal{H}}_i > \frac{1}{2}|T_{ii}a_i|$ and $Q_i = q_i - 1$, which corresponds to output saturation at the highest level; ii) $\hat{\mathcal{H}}_i < -\frac{1}{2}|T_{ii}a_i|$ and $Q_i = 0$, which corresponds to output saturation at the lowest level; iii) $-\frac{1}{2}|T_{ii}a_i| \leq \hat{\mathcal{H}}_i \leq \frac{1}{2}|T_{ii}a_i|$, which corresponds to output suspension at certain level. From the model description for a Q'tron NN, one can easily seen that if we let each Q'tron, say μ_i have $q_i = 2$, $a_i = 1$ and $T_{ii} = 0$, and let the NN run in simple mode, then the Q'tron NN model is reduced to the original Hopfield model [4, 5].

To make each Q'tron to be able to function either as an input or as an output node, a Q'tron can either be operated in *clamp-mode*, i.e., its output-level is clamped fixed at a particular level, or in *free mode*, i.e., its output-level is allowed to be updated according to the level transition rule specified in Eq. (2).

System Energy – Stability

The system energy $\mathcal{E}$ embedded in a Q'tron NN, called *Liapunov energy*, is defined by the following form:

$$\mathcal{E} = -\frac{1}{2}\sum_{i=1}^{n}\sum_{j=1}^{n}(a_iQ_i)T_{ij}(a_jQ_j) - \sum_{i=1}^{n} I_i(a_iQ_i) + K; \tag{4}$$

where n is total number of Q'trons in the NN, and K can be any suitable constant. It was shown that, in simple mode, the energy $\mathcal{E}$ defined above will monotonically decrease with time. Therefore, if a problem can be mapped into one which minimizes the function $\mathcal{E}$ given in the above form, then the corresponding NN, hopefully, will autonomously solve the problem after $\mathcal{E}$ reaches a global/local minimum.

Noise Spectra for Q'trons

Building a Q'tron NN to solve a problem, there may be enormous number of local minima, which will lead the NN to be stuck on a state representing an unsatisfactory solution. Random noise $\mathcal{N}_i$ injected into the ith Q'tron, then, serves to escape the NN from the unsatisfactory states. In our approach, we let the strength of $\mathcal{N}_i$ always be bounded, i.e., $\mathcal{N}_i \in [\mathcal{N}_{i-}, \mathcal{N}_{i+}]$. One convenient way to generate such a piece of noise is to specify the so-called *noise ratio spectrum* (NRS) of a Q'tron. It is defined as:

$$\mathrm{NRS}_i = P(\mathcal{N}_i = \mathcal{N}_{i-}) : P(\mathcal{N}_i = 0) : P(\mathcal{N}_i = \mathcal{N}_{i+}) \tag{5}$$

with

$$P(\mathcal{N}_i = \mathcal{N}_{i-}) + P(\mathcal{N}_i = 0) + P(\mathcal{N}_i = \mathcal{N}_{i+}) = 1; \tag{6}$$

where NRS_i represents the NRS of the ith Q'tron, and $P(\mathcal{N}_i = x)$ represents the probability that the strength of noise $\mathcal{N}_i = x$ is generated. Figure 1(b) shows some typical NRS that we used in the experiments.

The Known-Energy Systems and Solution Qualifier

We'll use a simple example to highlight the concept. Consider the generalized sum-of-subset problem defined below.

Let $S = \{s_1, s_2, \ldots, s_n\}$ be a set of different nonnegative integers, let M be an arbitrary real number, and let $\frac{\Delta}{2}$ be a nonnegative real number. The problem is to find a subset, say, $S' = \{s_{1'}, s_{2'}, \ldots, s_{k'}\} \subseteq S$ that satisfies

$$-\frac{\Delta}{2} \leq E = M - \sum_{i=1}^{k} s_{i'} \leq \frac{\Delta}{2}, \tag{7}$$

where E denotes the error between M and the subset sum. Apparently, setting $\frac{\Delta}{2} \leq 0.5$, the problem is to find a subset of S whose sum equals to the integer closest to M. Furthermore, if M is an integer, then it is reduced to the typical sum-of-subset problem. Setting $\frac{\Delta}{2} > 0.5$, hence, means that larger errors are tolerable. Because the value of $\frac{\Delta}{2}$ is correlated with the quality of the resulted solution, it is called the *solution qualifier* for the solution space of the problem.

We now build a Q'tron NN to solve the problem. According to the problem nature, let the Q'tron NN contain n 2-level Q'trons, say, $\mu_1, \mu_2, \ldots, \mu_n$, and let μ_i have active weight $a_i = s_i$. Here use $Q_i = 0$ and $Q_i = 1$ to represent $s_i \notin S'$ and $s_i \in S'$, respectively. Consider the following energy function.

$$\mathcal{E} = \frac{1}{2}\left(\sum_{i=0}^{n} a_i Q_i - M\right)^2. \tag{8}$$

Clearly, running the Q'tron NN built using this energy function in simple mode serves to reduce the value of $\mathcal{E}$ and, hence, $|E|$ monotonically. This property implies that, without incorporating with any noise injecting mechanism, the NN may be stuck at a local-minimum state which does not satisfy the condition specified in Eq. (7). From the problem setting, one can easily see that the Q'tron NN reaches a state of satisfactory solution only if the following inequality also holds

$$0 \leq \mathcal{E} \leq \frac{1}{2}\left(\frac{\Delta}{2}\right)^2 \tag{9}$$

This implies that the energy range that the Q'tron NN should reach to solve the problem is prior known. Hence, we call the Q'tron NN built using Eq. (8) as a *known-energy system.*

The Main Theorem – Completeness

The following theorem states, without proving, that the noise spectra for Q'trons in the NN constructed above can be determined to solve the generalized sum-of-subset problem in probabilistically complete sense.

Theorem 1. *Let $\mathcal{S}$ be the Q'tron NN constructed above to solve the generalized sum-of-subset problem. Then, with $\mathcal{N}_{i+} = -\mathcal{N}_{i-} = \frac{1}{2}|T_{ii}a_i| - \Delta/2$ and $NRS_i = a_i : b_i : c_i$, where $a_i \neq 0$ and $c_i \neq 0$, for all i, the following will be true.*

1. *If $M < 0$ or $M > \sum_{i=1}^{n} s_i$, then $\mathcal{S}$ will always settle down with probability one on the state with Q'trons output-level being all 0 or 1, respectively.*
2. *If $M \leq \sum_{i=1}^{n} s_i$, then $\mathcal{S}$ will always settle down with probability one on a state if and only if there exists a vector $(\xi_1, \ldots, \xi_n)^T \in \{0,1\}^n$ such that*

$$-\frac{\Delta}{2} \leq M - \sum_{i=1}^{n} s_i \xi_i \leq \frac{\Delta}{2}. \tag{10}$$

Furthermore, if $\mathcal{S}$ does settle down on a state, say, $(Q_1, \ldots, Q_n)^T$, then it must satisfy

$$-\frac{\Delta}{2} \leq M - \sum_{i=1}^{n} s_i Q_i \leq \frac{\Delta}{2}. \tag{11}$$

3 The Q'tron NN for Visual Cryptography

Without loss of generality, we'll use the *full access scheme* for three shares to describe the general Q'tron NN structure for visual cryptography. In this scheme, the stacking of any combination of shares will form a meaningful image. Hence, there will be seven images in total involved in the scheme. Using the Q'tron NN approach, these images are specified as greylevel images. Setting these images as input to the Q'tron NN, we need the NN to report a set of binary shares that fulfills the scheme in halftone sense when it settles down.

Let S_1, S_2 and S_3 be the three shares, and let G_1, G_2 and G_3 be their representative greylevel images, respectively. In the scheme, we'll use G_{ij}, where $i, j \in 1, 2, 3$ and $i \neq j$, to denote the representative greylevel images when shares S_i and S_j are stacked together, and use G_{123} to denote the one when all shares are stacked together. Because the resulted images of the Q'tron NN are all binary, i.e., halftone images, we will use H_x to denote the halftone version of the corresponding greylevel image G_x described above. In the following discussion, we will use $A = A_1 \cup A_2 \cup A_3$, where $A_1 = \{1, 2, 3\}, A_2 = \{12, 13, 23\}$ and $A_3 = \{123\}$, to denote the set of Q'tron-plane indices for the scheme.

The Q'tron NN Structure

Figure 2 shows the Q'tron NN structure for the full access scheme for three shares. In the figure, Q'tron planes G_x's and H_x's represent the aforementioned images, and Q'tron planes X_y's serve to make the system a known-energy one, which will be discussed shortly.

This Q'tron NN will run in full mode. To perform encryption, clamp all G_x's in the corresponding Q'tron planes, and let all other Q'tron planes be free. Apparently, the energy of the NN will be pulled very high right after G_x's are clamped. Because noises are injected into the NN persistently, and the system is a known-energy one, the NN becomes unstable immediately and, hence, will continuously changes its state until it reaches a state whose energy is lower enough.

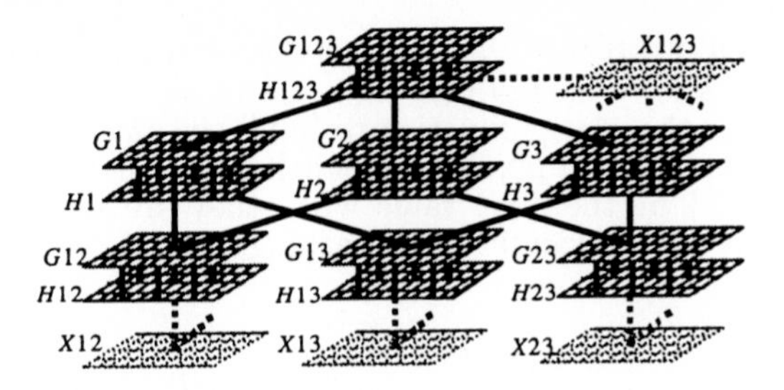

Fig. 2. The Q'tron NN structure for full scheme of three shares.

The Main Energy Items

To make the Q'tron NN shown in Figure 2 workable, its energy function should contain the following terms:

1. Energy functions to convert each greylevel image, say G_x, to a halftone image, say H_x, i.e., H_x is a binary image that looks similar to G_x.
2. Energy functions to satisfy the stacking rules of the access scheme. For example, stacking H_1 and H_2 will produce H_{12} which, of course, looks similar to G_{12}.

A computer usually uses 0 to represent a black pixel (the darkest luminance) and 255 to represent a white pixel (the brightest luminance). However, in this approach, we use the converse, i.e., 0 for white (or, more precisely, the transparent one) and 255 for black. Therefore, as the pixel value, called the *darkness level*, changes from 0 to 255, the luminance of the pixel will transit from the brightest one to the darkest one gradually.

In the following discussion, we assume that the size of images in the scheme is $M \times N$. We will use $Q_{ij}^{Gx} \in \{0, 1, 2, \ldots, 255\}$ to denote the output level for ij^{th} Q'tron in Q'tron plane G_x and use $Q_{ij}^{Hx} \in \{0, 1\}$ to denote the output level for ij^{th} Q'tron in Q'tron plane H_x. Setting $a_{ij}^{Gx} = a^G = 1$ and $a_{ij}^{Hx} = a^H = 255$, the active value of that Q'tron thus stands for the darkness level of its corresponding pixel.

To perform halftoning, one requires that the mean darknesses for each pair of small areas, e.g., 3×3 in the following energy function, in Q'tron planes G_x and H_x are close. Such a criterion can be described using the following energy function, say $\mathcal{E}_{\text{htone}}$.

$$\mathcal{E}_{\text{htone}} = \frac{1}{2} \sum_{x \in A} \sum_{i=1}^{M} \sum_{j=1}^{N} \left(\sum_{k=i-1}^{i+1} \sum_{l=j-1}^{j+1} a^G Q_{kl}^{Gx} - \sum_{k=i-1}^{i+1} \sum_{l=j-1}^{j+1} a^H Q_{kl}^{Hx} \right)^2 \tag{12}$$

For simplicity, in Eq. (12), the energy subterms for boundary pixels have not been specially noticed. Apparently, the minimization of Eq. (12) achieves the goal of image halftoning. Now, consider the following inequality

$$0 \leq E = \left| \sum_{k=i-1}^{i+1} \sum_{l=j-1}^{j+1} a^G Q_{kl}^{Gx} - \sum_{k=i-1}^{i+1} \sum_{l=j-1}^{j+1} a^H Q_{kl}^{Hx} \right| \leq \frac{\Delta}{2}, \tag{13}$$

for all $1 \leq i \leq M$,$1 \leq j \leq N$ and $x \in A$. Clearly, the value of $\frac{\Delta}{2}$ in Eq. (13) is correlated with the quality of halftone images in the access scheme. Considering it to be the solution qualifier of the halftoning process, the energy term $\mathcal{E}_{\text{htone}}$, then, possesses the known-energy property.

We now consider the energy function to fulfill the stacking rules. Let $p_1, \ldots, p_n \in \{0, 1\}$ be the values of n pixels to be stacked together in n different shares, and let $p_{1..n} \in \{0, 1\}$ be the pixel value when they are stacked. The following inequality accounts for the validity of a stacking.

$$0 \le np_{1..n} - \sum_{i=1}^{n} p_i \le n - 1 \tag{14}$$

Note that we have used $p_i = 1$ and $p_i = 0$ to denoted an inked (black) and an uninked (transparent) pixels in a transparency, respectively. With a little investigation, one then see that Eq. (14), in fact, specifies that $p_{1..n}$ is black if and only if at least one of p_i's is also black. However, the inequality nature of Eq. (14) disallows us to formulate a known-energy system. By introducing a surplus variable $x_{1..n} \in \{0, 1, \ldots, n-1\}$, we obtain the following equality, which has the same meaning as Eq. (14).

$$np_{1..n} - \sum_{i=1}^{n} p_i - x_{1..n} = 0 \tag{15}$$

The Q'tron planes X_y's, where $y \in A_2 \cup A_3$, in Figure 2 then serve to make the Q'tron NN to become a know-energy system. Now, consider the following energy term.

$$\mathcal{E}_{\text{stack}} = \frac{1}{2} \sum_{xy \in A_2} \sum_{i=1}^{M} \sum_{j=1}^{N} \left[2a^H Q_{ij}^{Hxy} - (a^H Q_{ij}^{Hx} + a^H Q_{ij}^{Hy}) - a^X Q_{ij}^{Xxy} \right]^2 +$$
$$\frac{1}{2} \sum_{xyz \in A_3} \sum_{i=1}^{M} \sum_{j=1}^{N} \left[3a^H Q_{ij}^{Hxyz} - (a^H Q_{ij}^{Hx} + a^H Q_{ij}^{Hy} + a^H Q_{ij}^{Hz}) - a^X Q_{ij}^{Xxyz} \right]^2 \tag{16}$$

Referring to Eq. (15), it is then clear that we will have $Q_{ij}^{Xxy} \in \{0, 1\}$ for $xy \in A_2$, $Q_{ij}^{Xxyz} \in \{0, 1, 2\}$ for $xyz \in A_3$ and $a^X = a^H = 255$. Furthermore, it is not hard to see that $\mathcal{E}_{\text{stack}}$ reaches zero once the stacking rules for the access scheme are satisfied. This implies $\mathcal{E}_{\text{stack}}$ also possesses the known-energy property.

The Total Energy of the Q'tron NN

To account for both halftoning and stacking, the energy function to build the Q'tron NN for the access scheme now is described as:

$$\mathcal{E} = \lambda_1 \mathcal{E}_{\text{htone}} + \lambda_2 \mathcal{E}_{\text{stack}}, \tag{17}$$

where λ_1 and λ_2 are positive weighting factors. The value λ_2 should be set much larger than the value of λ_1 such that any violation of stacking rules will pull the total energy $\mathcal{E}$ to very high. On computer simulation, we set $\lambda_1 = 1$ and set $\lambda_2 = 4096$.

The Construction of Q'tron NN for Visual Cryptography

The structure of Q'tron NN shown in Figure 2 has the following energy function:

$$\mathcal{E} = -\frac{1}{2}\sum_{x\in P}\sum_{i=1}^{M}\sum_{j=1}^{N}\sum_{y\in P}\sum_{k=1}^{M}\sum_{l=1}^{N}(a^x Q_{ij}^x)T_{ij,kl}^{xy}(a^y Q_{kl}^y) - \sum_{x\in P}\sum_{i=1}^{M}\sum_{j=1}^{N} I_{ij}^x(a^x Q_{ij}^x), \tag{18}$$

where $P = P_H \cup P_G \cup P_X$ with $P_H = \{H_x : x \in A\}$, $P_G = \{G_x : x \in A\}$ and $P_X = \{X_x : x \in A_2 \cup A_3\}$, $T_{ij,kl}^{xy}$ represents the connection strength between the ij^{th} Q'tron in plane x and the kl^{th} Q'tron in plane y, and I_{ij}^x represents the external stimulus fed into the ij^{th} Q'tron in plane x. By mapping Eq. (17) to Eq. (18), one can then determine the values of all connection strengths $T_{ij,kl}^{xy}$'s and all exte rnal stimuli I_{ij}^x's (= 0, in the scheme). With careful investigation, one will see that the Q'tron NN is sparsely connected, i.e., each Q'tron only connected to neighborhood Q'trons, and the connection strengths between neighborhood Q'trons also reveal regular patterns.

The Noise Spectra for Q'tron

The detail discussion on determining the Q'trons' noise spectra of the NN will be skipped in the paper. The noise spectra that we used in experiments for the full access scheme are summarized as follows:

$$\mathcal{N}_{ij}^x \in [-\mathcal{N}_x, \mathcal{N}_x] \tag{19}$$

with

$$\mathcal{N}_x = \begin{cases} \frac{255}{2}(9\lambda_1 + 3\lambda_2) - 9k\frac{\Delta}{2}\lambda_1 & x \in \{H_1, H_2, H_3\} \\ \frac{255}{2}(9\lambda_1 + 4\lambda_2) - 9k\frac{\Delta}{2}\lambda_1 & x \in \{H_{12}, H_{13}, H_{23}\} \\ \frac{255}{2}(9\lambda_1 + 9\lambda_2) - 9k\frac{\Delta}{2}\lambda_1 & x \in \{H_{123}\} \\ 0 & \text{otherwise} \end{cases}, \tag{20}$$

where $\frac{\Delta}{2} = \frac{255}{2}$, and $k \geq 2$.

The Q'tron NNs for General Access Schemes

Although the Q'tron NN built in the above is for the full access scheme with three shares, it can be easily modified to cope with the arbitrary access schemes for three shares. For example, if we change the plane index subset A_2 to $A_2 = \{12, 23\}$, it then implies that the stacking of S_1 and S_3 is a forbidden configuration. The energy subterms corresponding the stacking of these two shares will, then, be withdrawn from Eq. (12) and (16). The Q'tron NN built using this renewed energy function thus deals with a non-full access scheme. This reveals that the Q'tron NN constructing method introduced in this section is, in fact, very general.

4 The Experimental Results

As was mentioned, using this approach, the input to a Q'tron NN will be a set of greylevel images of shares and targets, to be clamped on planes G_x's, for the corresponding access scheme. However, before feeding these images into the NN,

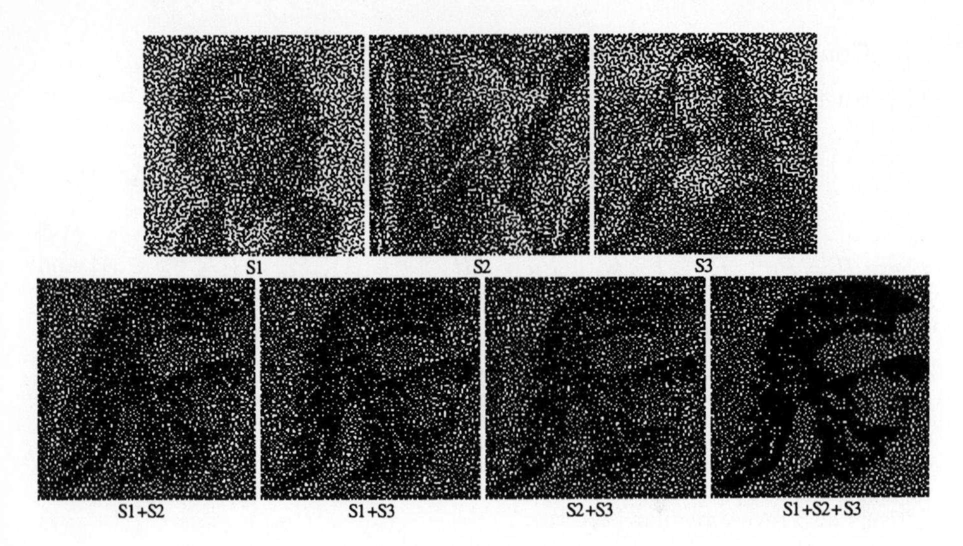

Fig. 3. The simulation result for (3, 2) access scheme.

their histograms of greylevel usually have to be reallocated. For example, it is unreasonable to stack two darker images to obtain a brighter one. We call the problem to determine feasible greylevel ranges for the involving images in an access scheme as the *histogram allocation problem*. This problem is nontrivial while will be skipped in the discussion.

We now present some experimental results running on the Q'tron NN constructed in the above as follows:

1. Figure 3 shows an application for the full access scheme of three shares. In the application, the three shares represent three different images. However, the stacking any combination of them forms the images of the same meaning, i.e., they differ only on lightness. Traditionally, such a application is called $(3, 2)$, i.e., two out of three. Theoretically, it can be verified that there always exists a feasible allocation of histograms for (n, m), i.e., the m out of n. In the experiment, the value of k in Eq. (20) is set to 2.
2. Figure 4 shows another application for the scheme. It differs from the above one only on the assignment of stacked target images. In the application, all images involved are different. It can be verified that a feasible allocation of histograms, in general, is inexistent for this kind of application. However, by guessing reasonable histograms for images and setting k larger (k=3) to tolerate larger downgrading of halftone images, the result is satisfactory.
3. Figure 5 shows the application with one forbidden subset. Although the total number of images in this application is only one less than the full one, it becomes theoretically realizable. In this experiment, we set $k = 2$.

More experimental results and the enlarged version of Figure 3 to 5 now are available in:

http://www.cse.ttu.edu.tw/twyu/vc/gscheme.html

One can also visit the site to obtain extra informations.

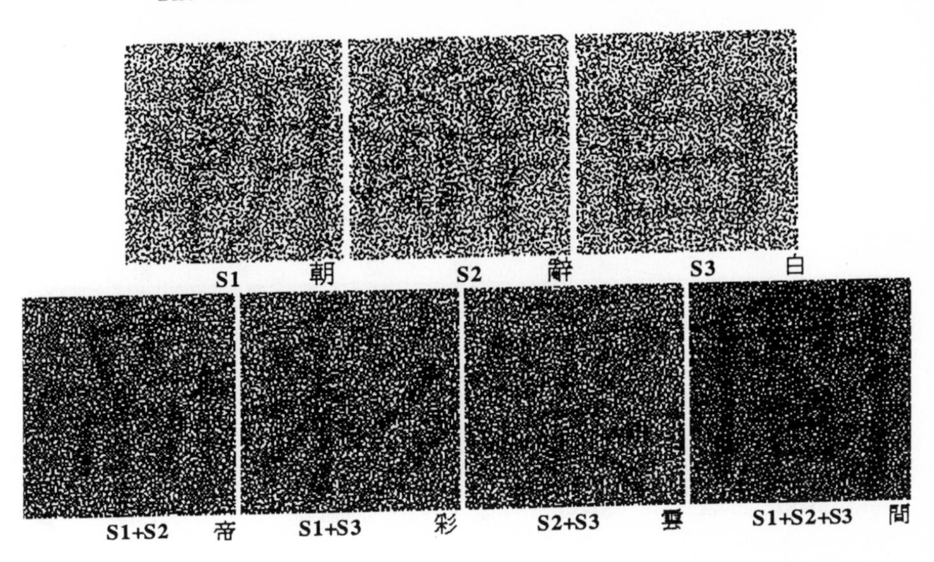

Fig. 4. A simulation result for full access scheme (a 7-character Chinese verse).

5 Conclusions

In the paper, we propose a novel approach for visual cryptography using Q'tron NN model, which is a generalized version of the Hopfield NN model. The remarkable feature for this new model of NN is that it can incorporate with a persistent noise injecting mechanism to escape the local minima corresponding to infeasible solutions. Without such a mechanism, the Q'tron NNs that we built in the article are almost useless. The same strategy, in fact, can be also used to solve other optimization problems, e.g., NP-hard problems [3].

This Q'tron NN approach for visual cryptography is completely different from the traditional approach. Using traditional approach, different codebooks are required for different access schemes. For complex schemes, the codebooks are hardly to be found or, even, inexistent. Furthermore, it requires that the target images involved in an scheme are all binary, and the size of shares produced by the algorithm will be enlarged. Using the Q'tron NN approach, however, a codebook is not required, all images are specified as greylevel images, and the share images produced by the NN have the same size as the input images. Besides, even though an access scheme that is unrealizable using the traditional approach, it can be solved using a Q'tron NN by adjusting the values of the solution qualifier. The computer experiment also shows that the results are surprisingly good.

Acknowledge. This research is supported by National Science Council under the grant NSC89-2218-E-036-003.

References

1. G. Ateniese, C. Blundo, A. D. Santis, D. R. Stinson, "Visual Cryptography for General Access Structures", Information and Computation 129(2), 86-106 (1996)
2. G. Ateniese, C. Blundo, A. D. Santis, and D. R. Stinson, "Extended Scheme for Visual Cryptography," submitted to Theoretical Computer Science, http://cacr.math.uwaterloo.ca/~dstinson/visual.html.

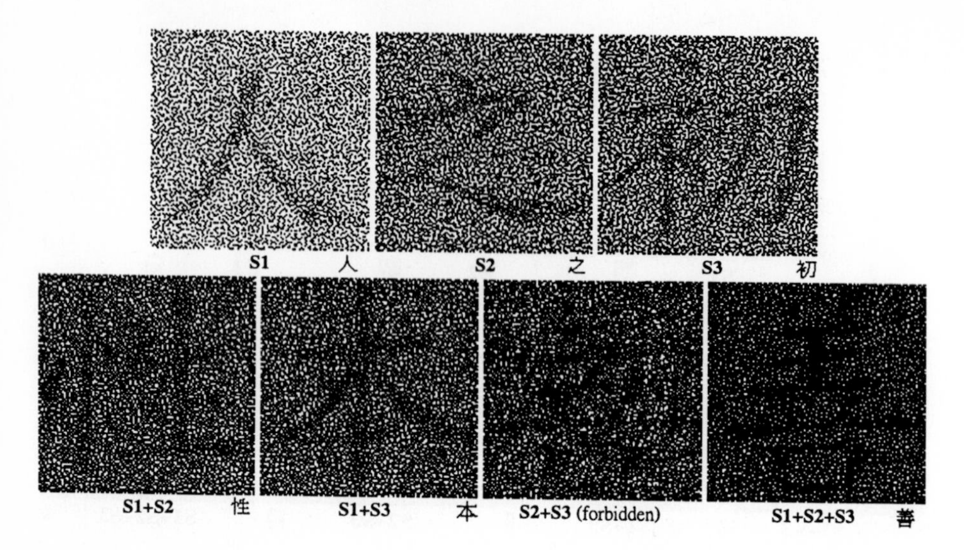

Fig. 5. An access scheme with forbidden subset, i.e., $S1 + S2$ means nothing.

3. Harry R. Lewis, Christos H. Papadimitriou, *Elements of The Theory of Computation*, pp.275-351, Prentice Hall International Editions, 1998.
4. J. J. Hopfield, "Neural Networks and Physical Systems with Emergent Collective Computational Abilities," *Proc. Natl. Acad. Sci. USA,* Vol.79, 2554-2558, April 1982.
5. J. J. Hopfield and D. W. Tank, "Neural Computation of Decisions in Optimization Problems," *Biological Cybernetics*, 52, pp.141-152,1985.
6. M. Naor, B. Pinkas, "Visual Authentication and Identification," *CRYPTO* 1997, pp. 322-336.
7. M. Naor and A. Shamir, "Visual Cryptography," in Advances Cryptology-Eurocrypt '94, A. DeSantis Ed. Vol. 950 of Lecture Notes in Computer Science, Springer-Verlag, Berlin, pp. 1-12, 1995.
8. Tai-Wen Yue, *A Goal-Driven Neural Network Approach for Combinatorial Optimization and Invariant Pattern Recognition*, Phd's Thesis, Department of Computer Engineering, National Taiwan University, Taiwan, 1992.
9. Tai-Wen Yue and Li-Cheng Fu, "A Local-Minima-Free Neural Network Approach for Building A/D Converters and associative Adders", *IEEE IJCNN-91-Seattle*, Vol. 2, pp.976, 1991.
10. Tai-Wen Yue and Li-Cheng Fu, "A Complete Neural Network Approach to Solving a Class of Combinatorial Problems", *IEEE IJCNN-91-Seattle*, Vol. 2, pp.978, 1991.
11. Tai-Wen Yue, Suchen Chiang, "A Neural Network Approach for Visual Cryptography, " Proceedings of the IEEE-INNS-ENNS International Joint Conference on Neural Networks, Vol. 5, pp. 494-499, July 2000.
12. Suchen Chiang, Tai-Wen Yue, "Neural Networks for Visual Cryptography – with Examples for Complex Access Schemes, " Proceedings of the ISCA 13th International Conference on Computer Applications in Industry and Engineering, pp. 281-286, November 2000.

Π–DTB, Discrete Time Backpropagation with Product Units

J. Santos and R. J. Duro

Grupo de Sistemas Autónomos - Departamento de Computación - Univ. de A Coruña
Facultad de Informática - Campus de Elviña - A Coruña 15071 Spain
{santos, richard}@udc.es

Abstract

This paper introduces a neural network that combines the power of two different approaches to obtaining more efficient neural structures for processing complex real signals: the use of trainable temporal delays in the synapses and the inclusion of product terms within the combination function. In addition to the neural network structure itself, the paper presents a new algorithm for training this particular type of networks and provides a set of examples using chaotic series, which compare the results obtained by these networks and training algorithm to other structures.

1. Introduction

Even though the term "high order" is not clearly delimited in the artificial neural network field, it encompasses those networks that incorporate elements with a larger processing capacity than the classical nodes, that use linear or non-linear activation functions acting over a combination that is the addition of the contribution of each input modulated by the corresponding connection weight.

Giles and Maxwell [4] define the order of a neural network through an interpretation of the equation:

$$(1)\quad y_i(x) = S\left[w_0(i) + \sum_j w_1(i,j)x(j) + \sum_j \sum_k w_2(i,j,k)x(j)x(k) + \ldots \right]$$

That is, in addition to the summation term of first order units, the higher order units use both addition and multiplication. Through these processes they define the so called *Sigma-Pi* units. Consequently, and using this definition, a unit that includes terms up to and including degree k will be called a *kth-order* unit. For these networks to be useful, as indicated by Giles and Maxwell [4], they should be matched to the order of the problem, and in this sense, a single *kth-order* unit will solve a *kth-order* problem, as defined by Minski and Papert [8].

Many approaches have been used to obtain non-linearity in the nodes [5], some even using the conditional operator leading to the concept of inferential neural networks studied by Mira [9], where one implementation can be found in [10]. The authors have employed a symbolic formalism (a micro-frame based model) to represent the information processing and learning algorithms of the connectionist models.

Notwithstanding the previous comments the increase in the processing power of a neural network cannot only lay in the capacity of the nodes. The synaptic connections of a neural network are forgotten elements that can also be used to this end. Two

J. Mira and A. Prieto (Eds.): IWANN 2001, LNCS 2084, pp. 207-214, 2001.

examples of this can be found in [2] and [3]. The first one considers neural networks that include a variable delay term in each synapse in addition the connection weight. It represents the time the information requires to go from the origin node to the postsynaptic node and resembles the length of the biological synapses. Like the weights, the time delays between the nodes are trainable. The second case [3] corresponds to substituting the numerical weight term in the synapses by a gaussian function with three free trainable parameters: amplitude, variance and center of the gaussian. This way, the contribution of a connection to the postsynaptic node is not only a function of the connection weight, but also of the current value of the input to that connection. In other words, the gaussian functions act as a filter of the inputs, which is automatically obtained though learning to the best adaptation to the problem. These structures demonstrate their higher processing capabilities over traditional ANN architectures with a reduction in the number of processing elements

Particularizing in the application of this paper, in the realm of signal identification and modelling, the main problem is to obtain a model of the signal that is as accurate as possible and which permits performing short and long term predictions. This is, we are interested in the dynamic reconstruction of the measured signal.

Most neural network based implementations have not taken into account the basic premises on which the modelling of signals must be based, that is, that in order to model a time dependent non linear signal one must be able to define its state space, or an equivalent state space, in an unambiguous manner in order to prevent multivalued sets of coordinates which would lead to ambiguous predictions. In fact, one of the best areas to appreciate this is in the case of chaotic time series, where if the state space is not well chosen the orbits of the series in this space will cross, generating points of ambiguity. In this article we make use of a trainable delay based artificial neural network that directly implements a form of the embedding theorem [6][11], with the advantage of being able to autonomously obtain the embedding dimension and the normalized embedding delay.

As a first step in directly training the delays introduced in a generalized synaptic delay network, using a gradient descent method, we have developed the DTB algorithm [2], that results in easy and precise training of temporal event processing networks straight from the input signals, with no windowing or preprocessing required, and what is more important, the embedding dimension does not have to be defined a priori. In the context of signal processing, and taking into account that many signal related correlations take place in the frequency domain, we have now added the possibility of including product terms in some of the nodes which, combined with the automatic selection of the signal points to be multiplied (through the training of the delay terms), allow for learnable signal correlation operations within the network.

In the following sections we will present the network and algorithm used to train it as well as some examples of its application to chaotic signal modelling.

2. Discrete Time Backpropagation with Π Units

The artificial neural network we consider for training consists of several layers of neurons connected as a Multiple Layer Perceptron (MLP). There are two differences with traditional MLPs. The first one is that the synapses include a delay term in addition to the classical weight term. That is, now the synaptic connections between neurons are described by a pair of values, (W, τ), where W is the weight, representing the ability of the synapse to transmit information, and τ is a delay which in a certain sense provides an indication of the length of the synapses. The longer it is it will take more time for information to traverse it and reach the target neuron. The second one is that some of the nodes implement a product combination function instead of the traditional sum.

We have developed an extension of the backpropagation procedure for the training of both parameters of the connections, and have called it Pi Discrete Time Backpropagation (Π-DTB). This algorithm permits training the network through variations of synaptic delays and weights, in effect changing the length of the synapses and their transmission capacity in order to adapt to the problem in hand. Here we summarize the basis of the algorithm, explained in detail in [2], concentrating in the new equations for the Π units.

If we take into account the description of the network in terms of synaptic weights and synaptic delays, the main assumption during training is that each neuron in a given layer can choose which of the previous outputs of the neurons in the previous layer it wishes to input in a given instant of time. Time is discretized into instants, each one of which corresponds to the period of time between an input to the network and the next input. During this instant of time, each of the neurons of the network computes an output, working its way from the first to the last layer. Thus, for each input, there is an output assigned to it.

In order to choose from the possible inputs to a neuron the ones we are actually going to input in a given instant of time, we add a selection function to the processing of the neuron. This selection function could be something as simple as:

$$(2)\quad \delta_{ij} = \begin{cases} 1 \rightarrow i = j \\ 0 \rightarrow i \neq j \end{cases}$$

so that the output of a traditional (3) and a product (4) node k in an instant of time t is given by:

$$(3)\quad O_{kt} = F\left(\sum_{i=0}^{N}\sum_{j=0}^{t}\delta_{j(t-\tau_{ik})} w_{ik} h_{ij}\right) \qquad (4)\quad O_{kt} = F\left(\prod_{i=0}^{N}\sum_{j=0}^{t}\delta_{j(t-\tau_{ik})} w_{ik} h_{ij}\right)$$

where F is the activation function of the neuron, h_{ij} is the output of neuron i of the previous layer in instant j and w_{ik} is the weight of the synapse between neuron i and neuron k. The first sum (or product) is over all the neurons that reach neuron k (those of the previous layer) and the second one is over all the instants of time we are considering (let´s say since the beginning of time, although in practical applications the necessary time is finite).

The result of this function is the sum or product of the outputs of the hidden neurons in times $t-\tau_{ik}$ (where τ_{ik} is the delay in the corresponding connection) weighed by the corresponding weight values.

Now that we know what the output of each neuron is as a function of the outputs of the neurons in the previous layer and the weights and delays in the synapses that connect these neurons to it, what we need is an algorithm that allows us to modify these weights and delays so that the network may learn to associate a set of inputs to a set of outputs. The basic gradient descent algorithm employed in traditional backpropagation may be used, but we must now take into account the delay terms when computing the gradients of the error with respect to weights and delays.

As shown in [2], these gradient terms are:

$$(5)\ \frac{\partial E_{total}}{\partial w_{jk}} = \Delta_k h_{j(t-\tau_{jk})} \qquad (6)\ \frac{\partial E_{total}}{\partial \tau_{jk}} = \Delta_k w_{jk}\left(h_{j(t-\tau_{jk})} - h_{j(t-\tau_{jk}-1)}\right)$$

being E_{total} the total squared error for all the training vectors and

$$(7)\ \Delta_k = \frac{\partial E_{total}}{\partial O_k}\frac{\partial O_k}{\partial ONet_k} = 2(O_k - T_k)F'(ONet_k)$$

where T_k is the desired output, O_k the one really obtained and $ONet_k$ is the combination of inputs to neuron k, when we consider output neurons. For hidden neurons connected to the input layer, and defining as before

$$(8)\ \Delta_k = \frac{\partial E_{total}}{\partial hNet_k} = F'(hNet_k)\sum_r \Delta_r w_{kr}$$

where index r represents the neuron of the next layer, whether output or hidden, we have the following derivatives for the weights and connections:

$$(9)\ \frac{\partial E_{total}}{\partial w_{jk}} = \frac{\partial E_{total}}{\partial hNet_k}\frac{\partial hNet_k}{\partial w_{jk}} = \Delta_k I_{j(t-\tau_{jk})}$$

$$(10)\ \frac{\partial E_{total}}{\partial \tau_{jk}} = \frac{\partial E_{total}}{\partial hNet_k}\frac{\partial hNet_k}{\partial \tau_{jk}} = \Delta_k w_{jk}\left(I_{j(t-\tau_{jk})} - I_{j(t-\tau_{jk}-1)}\right)$$

where the second derivative in (10) is the result of:

$$(11)\ \frac{\partial hNet_k}{\partial \tau_{jk}} = \frac{\partial\left[\sum_{i=0}^{N}\sum_{n=0}^{t}\delta_{n(t-\tau_{ik})} w_{ik} I_{in}\right]}{\partial \tau_{jk}}$$

when we consider neurons in a hidden layer. It may be observed that the derivative in *hNet* of equation (11) has been discretized in order to obtain (10), implicitly assuming there is a certain continuity in the temporal variation of the outputs of the neurons, which in practice turns out to be a valid assumption.

Regarding the product units in the first hidden layer, we have:

$$(12)\ hNet_k = \prod_{r=0}^{N} w_{rk}\ I_{r(t-\tau_{rk})} \quad \text{and the derivatives are}$$

$$(13)\ \frac{\partial hNet_k}{\partial w_{jk}} = \frac{\prod_{r=0}^{N} w_{rk}\ I_{r(t-\tau_{rk})}}{w_{jk}} \qquad (14)\ \frac{\partial hNet_k}{\partial \tau_{jk}} = \frac{\prod_{r=0}^{N} w_{rk}\ I_{r(t-\tau_{rk})}}{I_{j(t-\tau_{jk})}}\left(I_{j(t-\tau_{jk})} - I_{j(t-\tau_{jk}-1)}\right)$$

where index r denotes the input nodes connected to the corresponding product hidden unit. The importance of the procedure is that now the appropriate delays for these connections, and consequently the temporal values that make up the product terms in the *pi* units are obtained automatically, and are not imposed beforehand.

As we show, by discretizing the time derivative we obtain simple expressions for the modification of the weights and delays of the synapses, in an algorithm that is basically a backpropagation algorithm where we have modified the transfer function of the neuron to permit a choice of inputs between all of the previous outputs of the neurons of the previous layer.

3. Chaotic Time Series Prediction

To test the proposed structure in the realm of time series prediction, we have chosen the Mackey Glass chaotic time series often found in the literature as a benchmark.

$$\frac{\partial x(t)}{\partial t} = a\frac{x(t-\tau)}{1+x(t-\tau)^{10}} - bx(t) \tag{15}$$

The Mackey Glass delay-differential equation was derived as a model of blood production. $x(t)$ represents the concentration of blood at time t, that is, when the blood is produced and $x(t-\tau)$ is the concentration of blood when there is a demand for more blood. In patients with different pathologies, such as leukemia, the time τ may become very large, and the concentration of blood will oscillate, becoming chaotic when $\tau > 17$. We are thus going to choose values for τ in this range so as to test our strategy in the hardest conditions. In order to be as standard as possible in the series we employ, we have chosen values for a and b of 0.2 and 0.1 respectively, which are the values we have found to be most often used in the literature. We have taken as boundary conditions $x(t)=0.8$ for $t<0$. In every case we scaled the signal in order to fit it into the [0,1] interval but did not remove the offset, as some other authors do.

The idea behind the experiments was to be as close as possible to reality, and thus the networks were not trained on segments of the signal but directly on the signal through a single input. In this work we have performed basically three tests regarding the ability of Π-DTB to obtain networks that predict and generate chaotic signals. The first one consists in demonstrating the appropriateness of the networks to predict future values of the signals it learns from past values. The second one was to make these networks become signal generators, which is an extreme case of multistep prediction and, finally, as a third test, we compared the error results obtained to those of other authors.

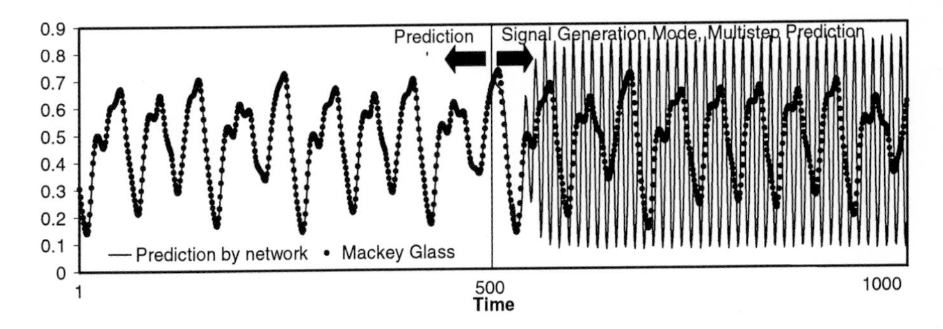

Figure 1: Prediction one instant into the future and signal generation with a delay based neural network (1-14-20-1) without Pi units.

In figures 1 and 2 we display the results corresponding to predictions obtained by a regular DTB trained network. On the left side of the pictures we have the prediction values obtained and in the right hand the values obtained when the input is turned off and the prediction made by the network in a given instant of time is used as input for the next prediction, that is, multistep prediction or signal generation. In the first case, the network which had two hidden layers with 14 and 20 nodes was able to predict the Mackey Glass series quite well, but the model was not good enough for a multistep prediction process not to diverge and that is just what it did after about 50 steps into the future.

In figure 2 we display the results for a similar network, but we have increased the number of neurons in the hidden layers to 20-20. Now the model obtained of the signal is much better. This can be seen in the right part of the figure, where the multistep prediction values obtained are acceptable up to 200 steps into the future and, after that the extreme divergence obtained in the previous case does not occur, and only a phase shift appears. By increasing the number of neurons one can obtain quite good models using these types of networks with delays.

Figure 3 shows the results when Π-DTB is used. In this case the prediction is much better even though we are using far less neurons than in the previous cases, that is, a 10-20 network (consequently only half the number of synapses as the previous case). In this case we can say that the network has learnt an almost perfect model of the

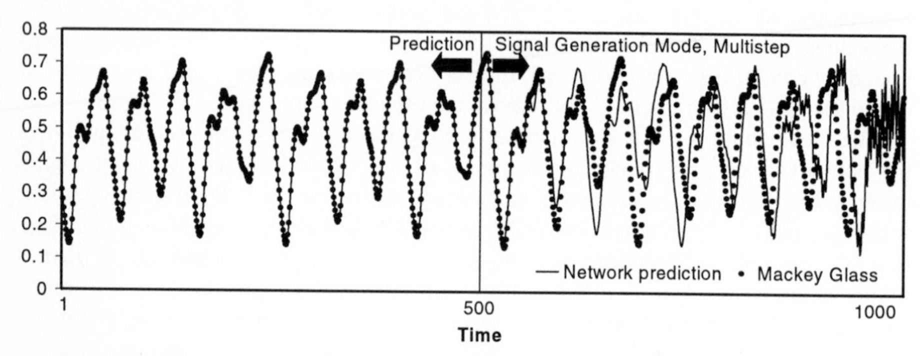

Figure 2: Prediction and reconstruction of the Mackey Glass series with a DTB trained network (1-20-20-1).

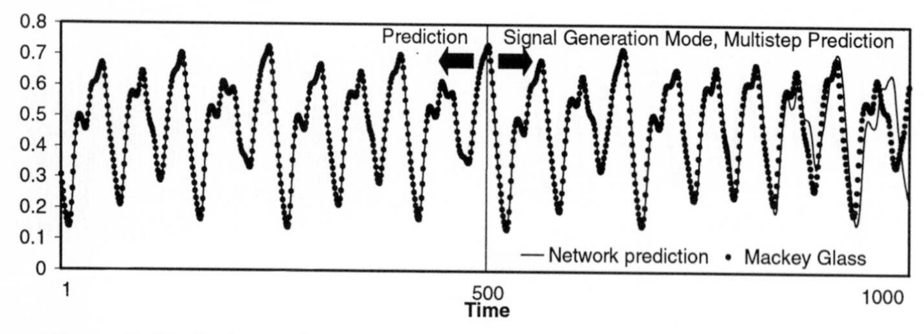

Figure 3: Prediction and reconstruction of the Mackey Glass series with a Π-DTB trained network.

signal thus allowing it to predict accurately more than 400 instants in to the future and reasonably thereafter.

In terms of the Mean Squared Error obtained by the two types of networks, just mention that during training and after the first few iterations the errors obtained by the Π-DTB algorithm are consistently better, reaching up to 80% improvement over DTB, as shown in figure 4. Initially, during the first stages of training, the Π-DTB based networks perform worse due to the higher sensitivity of the product terms with respect to changes, making the problem of reaching stable solutions harder.

The MSE of $9{,}7\mathrm{x}10^{-6}$ obtained for one step ahead prediction in a trained network with Pi units, using a 2000 point test set after training the network on-line on the signal, compares quite favorably with other results, such as the MSE value of 10^{-5} of the DTB network [2] and those obtained by other authors. The best result we have found is the one obtained by McDonnell and Waagen [7], whose predictions are obtained using an evolved recurrent non linear IIR filter network (MSE $=5\mathrm{x}10^{-5,}$ about five times ours), and Chow and Leung [1] who obtained a result of around MSE$=10^{-5}$ in the case of a one step prediction, but after a non-linear preprocessing stage. The differences are more evident in the case of multistep prediction, comparing the error of MSE of $5\mathrm{x}10^{-5}$ obtained through the Π-DTB algorithm in the generation of 10 points into the future and $9\mathrm{x}10^{-5}$ reconstructing 50 points into the future or even $4\mathrm{x}10^{-3}$ when predicting 500 points into the future, to the value of 5.11 of [7], considering only 3 points of signal generation. In fact, most authors do not even show results beyond 10 points into the future.

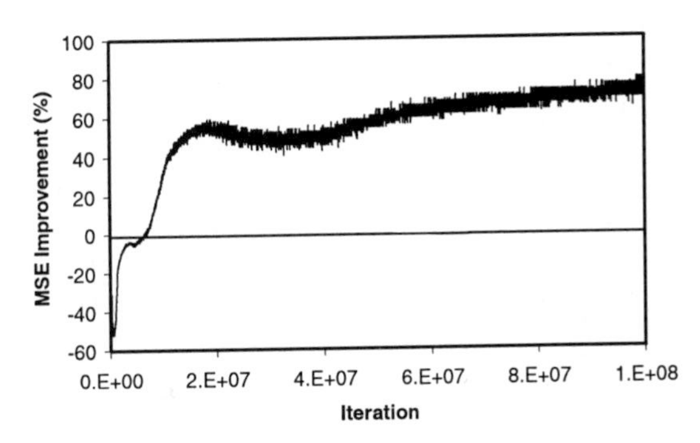

Figure 4: MSE improvement of the network with Pi units with respect to the one without Pi units during on-line training.

4. Conclusions

We have used the Π-DTB algorithm for training a time delay based neural network for the prediction and reconstruction of a chaotic signal. The use of the product units and the training algorithm permits reducing the prediction and especially the reconstruction error with respect to a network without this type of units. At the same time the number of nodes required is reduced, indicating this way the adequateness of the product terms in the signal and reconstruction problems. This can be interpreted by the fact that, as many signal related correlations take place in the frequency domain, the inclusion of the product terms makes it easier for the network to take into account these types of correlations. In addition the Π-DTB algorithm performs an automatic selection of the signal points to be multiplied through the appropriate determination of the delay terms in the synapses. Consequently, when speaking in the language of the dynamic reconstruction of signals, the network automatically obtains the embedded delay and embedded dimension.

Acknowledgements

This work was funded by Xunta de Galicia under project PGIDT99PXI10503A and the MCYT of Spain under project number TIC 2000-0739-C04-04.

References

[1] T. Chow and C.T. Leung, "Performance Enhancement Using Nonlinear Preprocessing", *IEEE Trans on Neural Networks*, V. 7, N.4, pp.1039-1042, 1996.
[2] R.J. Duro and J. Santos, "Discrete Time Backpropagation for Training Synaptic Delay Based Artificial Neural Networks", *IEEE Transactions on Neural Networks,* Vol.10, No. 4, pp. 779-789, 1999.
[3] R.J. Duro, J.L. Crespo, and J. Santos, "Training Higher Order Gaussian Synapses", *Foundations and Tools for Neural Modeling*, J. Mira & J.V. Sánchez-Andrés (Eds.), *Lecture Notes in Computer Science*, pp. 537-545, Vol. 1606, Springer-Verlag, Berlín 1999.
[4] C.L. Giles and T. Maxwell, "Learning, Invariance and Generalization in High Order Neural Networks", *Applied Optics,* Vol. 26, No. 23, pp. 4972-4976, 1987.
[5] E.B. Kosmatopoulos, M.M. Polycarpou, M.A. Christodoulou, and P.A. Ioannou, "High-Order Neural Network Structures for Identification of Dynamical Systems", *IEEE Transactions on Neural Networks,* V. 6, N. 2, pp. 422-431, 1995.
[6] Mañe, R., "On the Dimension of the Compact Invariant Sets of Certain Non-Linear Maps", In D. Rand and L.S. Young (Eds.), *Dynamical Systems and Turbulence, Warwick 1980 Lecture Notes in Mathematics*, Vol. 898, pp. 230, Springer Verlag, 1981.
[7] J.R. McDonnel and D. Waagen, "Evolving Recurrent Perceptrons for Time Series Modeling", *IEEE Trans. Neural Networks*, Vol. 5, No. 1, pp. 24-38, 1994.
[8] M.L. Minski and S. Papert, *Perceptrons*, MIT Press, Cambridge, MA, 1969.
[9] J. Mira, and A.E. Delgado, "Linear and Algorithmic Formulation of Cooperative Computation in Neural Nets*", Computer Aided Systems Theory - EUROCAST 91*, F. Pichler and R. Moreno-Díaz (Eds.), *Lecture Notes in Computer Science*, Vol. 585, pp. 2-20, Springer-Verlag, Berlín, 1991.
[10] J. Santos, R.P. Otero, and J. Mira, "NETTOOL: A Hybrid Connectionist-Symbolic Development Environment", *From Natural to Artificial Neural Computation*, J. Mira & F. Sandoval (Eds), *Lecture Notes in Computer Science*, Vol. 930, pp. 658-665, Springer-Verlag, Berlín, 1995.
[11] Takens, F., "On the Numerical Determination of the Dimension of an Attractor", In D. Rand and L.S. Young (Eds.), *Dynamical Systems and Turbulence, Warwick 1980 Lecture Notes in Mathematics*, V. 898, pp.366-381, Springer Verlag, 1981.

Neocognitron-Type Network for Recognizing Rotated and Shifted Patterns with Reduction of Resources *

Shunji Satoh[1,2,4], Shogo Miyake[2], and Hirotomo Aso[3]

[1] Japan Society for the Promotion of Science
[2] Department of Applied Physics, Graduate School of Engineering, Tohoku University, Aobayama 05, Sendai 980-8579, Japan
[3] Department of Electrical and Communication Engineering, Graduate School of Engineering, Tohoku University, Aobayama 04, Sendai 980-8579 Japan
[4] shun@aso.ecei.tohoku.ac.jp
http://www.aso.ecei.tohoku.ac.jp/~shun/research-e.html

Abstract. A rotation-invariant neocognitron was proposed by authors for recognition of rotated patterns. In this paper, we propose a new network in order to reduce the number of cells for the same purpose. The new network is based on the rotation-invariant neocognitron in its structure and based on an idea of hypothesis and its verification in its process. In the proposed model the following two processes are executed: 1) making a hypothesis of an angular shift of an input supported by an associative recall network and 2) verification of the hypothesis realized by mental rotation of the input. Computer simulations show that 1) the new network needs less cells than the original rotation-invariant neocognitron and 2) the difference of recognition rates between the proposed network and the original network is very little.
Keywords. rotation invariant neocognitron, the number of cells, associative recall, mental rotation, hypothesis and verification,

1 Introduction

Two dimensional patterns like hand-written letters, its appearance is usually spoiled by deformation of the shape, the scaling of size, shift in position, rotation and/or noise. Although many neural networks for recognizing letters have been proposed [1, 2, 3, 4], these are not tolerant for all variations described above. For example, a neocognitron[5], which is a hierarchical neural network for pattern recognition, is tolerant for deformation, the scaling of size, shift in position and noise, but the network can not recognize rotated patterns in large degrees. Some neural networks are tolerant for rotation as well as deformation, size, shift and/or noise [2, 4]. These networks, however, have a combinational explosion problem on the number of neurons when one make the networks possible to be tolerant for all variations. And these networks themselves are not tolerant for all variations.

* This work was supported by JSPS Research Fellowships for Young Scientists.

J. Mira and A. Prieto (Eds.): IWANN 2001, LNCS 2084, pp. 215-222, 2001.

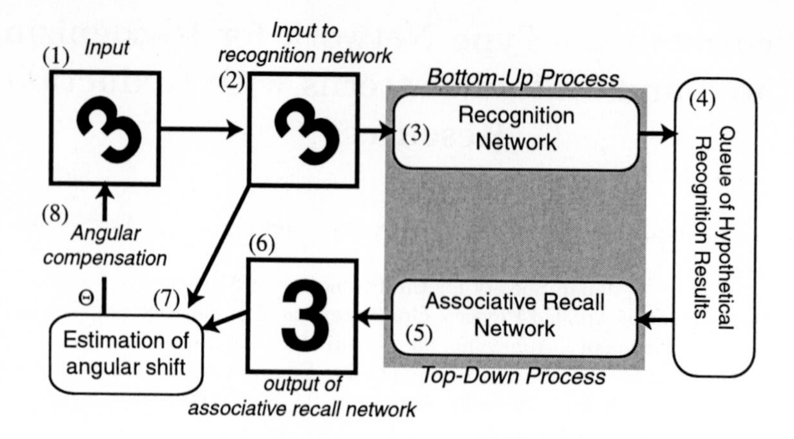

Fig. 1. Overview of the proposed algorithm for recognition of rotated patterns.

For example, in order to recognize rotated patterns by use of the neocognitron only, one should prepare some neocognitrons corresponding to any angles of patterns and train all the neocognitrons using rotated patterns. More concretely, because a neocognitron has rotational tolerant about 30 degrees, 360/30=12 neocognitrons should be prepared and trained.

To solve the problem on the number of resources and training time, authors proposed a rotation-invariant neocognitron [6, 7]. The rotation-invariant neocognitron is a bottom-up type network and a hierarchical neural network same as a neocognitron, and the model can recognize rotated patterns in any angles by learning not-rotated(standard) patterns only. It was shown that the model was tolerant for any variations of patterns by computer simulations.

By the way, a network for recognizing occluded patterns have been proposed and the network was based on an idea of *hypothesis and its verification* [8]. The main process of the network is 1) generating a hypothesis of the position and the category of an occluded pattern in an input and 2) verification the hypothesis. The idea is beneficial for reduction of resources of a network because it is not necessary for the network to be completely tolerant for the shift in positions of target patterns, and for occluded target patterns.

A network based on the idea for recognizing rotated patterns has been proposed by authors from a cognitive psychological point of view [9]. In this paper, we propose a new neural network for the purpose of reducing the number of neurons based on a rotation-invariant neocognitron as its structure and based on the idea of hypothesis and its verification as its process. We will show the effectiveness of the new network in comparison with the rotation-invariant neocognitron from the next three points: 1) the number of neurons needed for recognizing rotated patterns, 2) recognition rate and 3) recognition time.

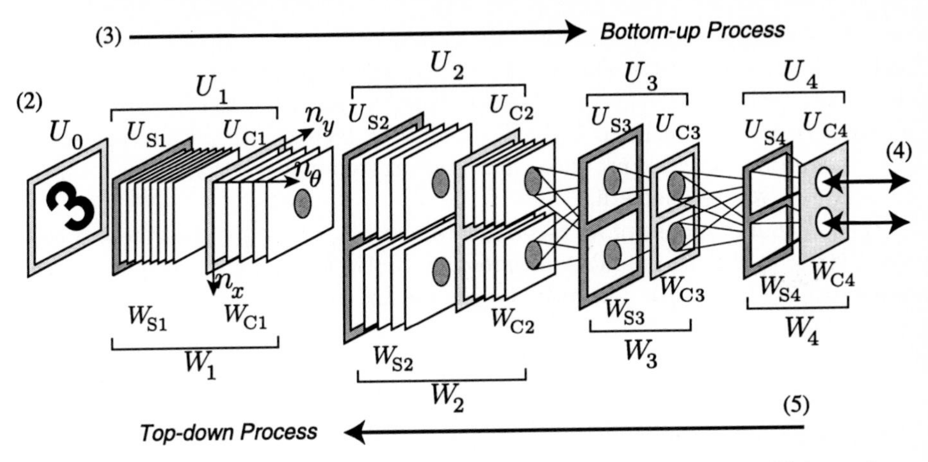

Fig. 2. The structure of a recognition network and an associative recall network.

2 Process and Structure of the New Model

Figure 1 shows the schematic overview of the algorithm. The direction of an arrow in Fig. 1 shows a direction of information flow and numbers in parenthesis represent the order of processing. An input pattern is given in a retina model of a recognition network as shown in Fig. 1-(2) and Fig. 2-(2). The retina model is referred to U_0.

2.1 Recognition by bottom-up process

The recognition network tries to recognize the pattern appeared in retina U_0 (Fig. 1-(3)). The structure of the recognition network is depicted in Fig. 2. The recognition network is composed of modules $U_l\,(l = 1, 2, 3, 4)$. A module U_l is composed of two kinds of layer, U_{Sl} and U_{Cl}, and both layers are named as lth S-layer and lth C-layer respectively.

The lower modules, U_1 and U_2, are composed of the modules of the rotation-invariant neocognitron. A layer in the lower modules is composed of a number of *cell-plane stacks* and different cell-plane stacks detect different features of inputs. A cell-plane stack is composed of a number of cell-planes, and each cell-plane detects a different rotation angle of the local features appeared in training patterns. Each cell in a cell-plane stack is located in a three dimensional space. In the model the angle of a local pattern in an input is represented by a number n_θ assigned to a cell-plane in a cell-plane stack, and positional information of the pattern is represented by coordinates (n_x, n_y) of a firing cell in the cell-plane of n_θ. The higher modules are composed of the modules of the original neocognitron. In other words, the higher modules include *cell-planes*.

Because the recognition network is composed of neocognitron-type layers in the higher modules and consequently the recognition network can not recognize largely rotated patterns, all gnostic cells (grandmother cells) in U_{C4} would make no response if a largely rotated pattern is presented in the retina U_0. At that time, the recognition network executes the following two processes in order to make one or more gnostic cells fire at force; (i): the recognition network decreases the value of selectivity (threshold) of S-cells in U_{S3} and U_{S4}, and (ii): spreads the blurring region of C-cells in U_{C2}, U_{C3} and U_{C4}. By executing these processes, the recognition network produces *hypothetical* (tentative) recognition result(s) for a largely rotated pattern.

The output response of an S-cell located on $\boldsymbol{n} = (n_x, n_y, n_\theta)$ of the kth cell-plane stack in the lth module is denoted by $U_{Sl}(\boldsymbol{n}, k)$, and the output response of a C-cell by $U_{Cl}(\boldsymbol{n}, k)$. The output response $U_{Sl}(\boldsymbol{n}, k)$ is given by

$$u_{Sl}(\boldsymbol{n},k) = r_l \cdot \phi \left[\frac{1 + \sum_{\kappa=1}^{K_{Cl-1}} \sum_{\boldsymbol{\nu} \in A_l} a_l(\boldsymbol{\nu}, \boldsymbol{n}, \kappa, k) \cdot u_{Cl-1}(\boldsymbol{n} \underset{T_{Cl-1}}{\oplus} \boldsymbol{\nu}, \kappa)}{1 + \frac{r_l}{1+r_l} \cdot b_l(k) \cdot u_{Vl}(\boldsymbol{n})} - 1 \right], \quad (1)$$

where the function $\phi(x) = \max(x, 0)$. A binomial operator $\underset{M}{\oplus}$ with M is defined by

$$\begin{cases} n_x \underset{M}{\oplus} \nu_x \overset{\text{def}}{=} n_x + \nu_x, \\ n_y \underset{M}{\oplus} \nu_y \overset{\text{def}}{=} n_y + \nu_y, \\ n_\theta \underset{M}{\oplus} \nu_\theta \overset{\text{def}}{=} (n_\theta + \nu_\theta) \bmod M. \end{cases} \quad (2)$$

Here r_l denotes a threshold value of a cell, $a_l(\boldsymbol{\nu}, \boldsymbol{n}, \kappa, k)$ represents an excitatory connection from C-cells in the anterior layer to an S-cell and $b_l(k)$ an inhibitory connection from a V-cell to an S-cell [1]. Each connection is linked to a restricted number of C-cells in the preceding module, and $A_l (\subset \boldsymbol{Z}^3)$ defines the number of C-cells. K_{Cl} denotes a number of cell-plane stacks in the U_{Cl} layer, and T_{Cl} a number of cell-planes in the U_{Cl} layer [2].

The output response of a V-cell is given by

$$u_{Vl}(\boldsymbol{n}) = \sqrt{\sum_{\kappa=1}^{K_{Cl-1}} \sum_{\boldsymbol{\nu} \in A_l} c_l(\boldsymbol{\nu}) \cdot \left\{ u_{Cl-1}(\boldsymbol{n} \underset{T_{Cl-1}}{\oplus} \boldsymbol{\nu}, \kappa) \right\}^2}, \quad (3)$$

where $c_l(\boldsymbol{\nu})$ is an excitatory connection from a C-cell to a V-cell, which takes a fixed value during a learning. The output response of a C-cell is given by

$$u_{Cl}(\boldsymbol{n},k) = \psi \left[\sum_{\boldsymbol{\nu} \in D_l} d_l(\boldsymbol{\nu}) \cdot u_{Sl}(\boldsymbol{n} \underset{T_{Sl}}{\oplus} \boldsymbol{\nu}, k) \right], \quad (4)$$

[1] V-cells are not depicted in Fig. 2 for simplicity.
[2] The cells in the higher layers are described by setting parameters such that $T_{S3} = T_{C3} = T_{S4} = T_{C4} = 1$ in the equations (1), (3) and (4).

where the function ψ is defined by $\psi(x) = \phi(x)/(\phi(x)+1)$. Here $d_l(\boldsymbol{\nu})$ is an excitatory connection from an S-cell to a C-cell, $D_l(\subset \boldsymbol{Z}^3)$ represents a restricted region of the connection, T_{Sl} a number of cell-planes in the U_{Sl} layer.

While the learning stage, excitatory connections $a_l(\boldsymbol{\nu}, \boldsymbol{n}, \kappa, k)$ and inhibitory connections $b_l(k)$ are modified. A learning method of neocognitron-type models as the rotation-invariant neocognitron is a kind of winner-take-all rule. A winner cell is referred to a seed-cell. If a seed-cell, $u_{Sl}(\hat{\boldsymbol{n}}, \hat{k})$, is selected, the modification rules of plastic connections of the S-cell are denoted by the following equations,

$$\Delta a_l(\boldsymbol{\nu}, \hat{\boldsymbol{n}}, \kappa, \hat{k}) = q_l \cdot c_l(\boldsymbol{\nu}) \cdot u_{Cl-1}(\hat{\boldsymbol{n}} \oplus \boldsymbol{\nu}, \kappa), \tag{5}$$

$$\Delta b_l(\hat{k}) = q_l \cdot u_{Vl}(\hat{\boldsymbol{n}}), \tag{6}$$

where the value q_l is a constant positive value.

2.2 Queue of hypothetical recognition results

The hypothetical recognition result(s) given by the recognition network are queued as shown in Fig. 1-(4) and Fig. 2-(4). Since the hypotheses must not be a correct result and the reliability would be low, a verifying process is executed.

2.3 Recall a standard pattern by top-down process

The structure of the associative recall network is also depicted in Fig. 2. The associative recall network is composed of modules W_l $(l = 4, 3, 2, 1)$ and a module W_l includes two layers, W_{Sl} and W_{Cl}. As shown in Fig. 2, every cell of the associative recall network is identical with a cell of the recognition network. The structure and the values of connections are also the same as the recognition network. The difference is the direction of information flow, that is, from the upper layer to the lower layer (right to left in Fig. 2). The associative recall network recalls a standard not-rotated pattern corresponding to the queued hypothesis (Fig. 1-(5) and Fig. 2-(5)). The result of recall is appeared in W_{S2} (Fig. 1-(6) and Fig. 2-(6)).

The output of a w_{Cl} cell in W_{Cl} and the cell w_{Vl} in W_{Sl} in the backward paths are given by

$$w_{Cl}(\boldsymbol{n}, k) = \psi\left[\alpha_l \cdot \{e' - i'\}\right], \tag{7}$$

$$e' = \sum_{\kappa=1}^{K_{Sl+1}} \sum_{\boldsymbol{\nu} \in A_{l+1}} a_{l+1}(\boldsymbol{\nu}, \boldsymbol{n}, k, \kappa) \cdot w_{Sl+1}(\boldsymbol{n} \underset{T_{Sl+1}}{\ominus} \boldsymbol{\nu}, \kappa), \tag{8}$$

$$i' = \sum_{\boldsymbol{\nu} \in A_{l+1}} c_{l+1}(\boldsymbol{\nu}) \cdot w_{Vl+1}(\boldsymbol{n} \underset{T_{Sl+1}}{\ominus} \boldsymbol{\nu}), \tag{9}$$

$$w_{Vl+1}(\boldsymbol{n}) = \frac{r_{l+1}}{1 + r_{l+1}} \sum_{\kappa=1}^{K_{Sl+1}} b_{l+1}(\kappa) \cdot u_{Sl+1}(\boldsymbol{n}, \kappa), \tag{10}$$

where α_l is a positive constant. The output of a w_{Sl} cell in W_{Sl} is given by

$$w_{Sl}(\boldsymbol{n}, k) = \beta_l \cdot \sum_{\boldsymbol{\nu} \in D_l} d_l(\boldsymbol{\nu}) \cdot w_{Cl}(\boldsymbol{n} \underset{T_{Cl}}{\ominus} \boldsymbol{\nu}, k), \tag{11}$$

where β_l is a positive constant.

2.4 Estimation of angular shift

An angular shift of the input, Θ, is estimated by calculating the shift between the firing patterns appeared in U_{S2} and W_{S2} (Fig. 1-(7)). The angular shift is given by

$$\Theta = \underset{\Delta\theta}{\operatorname{argmax}} \left\{ \sum_{n_x, n_y \in N_{S2}} u_{S2}(\boldsymbol{n}, k) \cdot \sum_{n_x, n_y \in N_{S2}} w_{S2}\left(\boldsymbol{n} \underset{T_{S2}}{\oplus} (0, 0, \Delta\theta), k \right) \right\}, \tag{12}$$

where $\Delta\theta = 0 \pm 1, \cdots, \pm T_{S2} - 1$.

2.5 Mental rotation and verification of hypothesis

A largely rotated pattern is compensated in the angle Θ in order to verify the hypothesis (this process is referred to mental rotation, Fig. 1-(8)), and the rotationally corrected pattern is recognize by the recognition network again. If the recognition result given by the second trial by the recognition network is equal to the hypothesis, the hypothesis is adopted as the final recognition result because the hypothesis is proofed. On the other hand, if the result given by the second trial is not equal to the hypothesis, the hypothesis is dismissed and other hypothesis in the queue will be verified.

2.6 Learning Process and the Number of Cells

The major learning rule for neocognitron-type models is an unsupervised learning with a kind of winner-take-all process, and the rule is adopted to S-cells (feature-extracting cells) in the layer U_{S2}, U_{S3} and U_{S4} The learning of a module begins when learning of the preceding modules has been completely finished. The rule is summarized as the following: (i) A new cell-plane or cell-plane stack is generated, if a training pattern is presented and there are no S-cells which detect a local pattern in the training pattern. Then the generated S-cells in the plane or stack are so reinforced that the S-cells detects the local pattern. (ii) If there are already S-cells that detects the local pattern, a new cell-plane or cell-plane stack is not generated.

Therefore, the number of cells tends to increase with increase of the number of training patterns because local patterns which are not learned yet may appear in new training patterns.

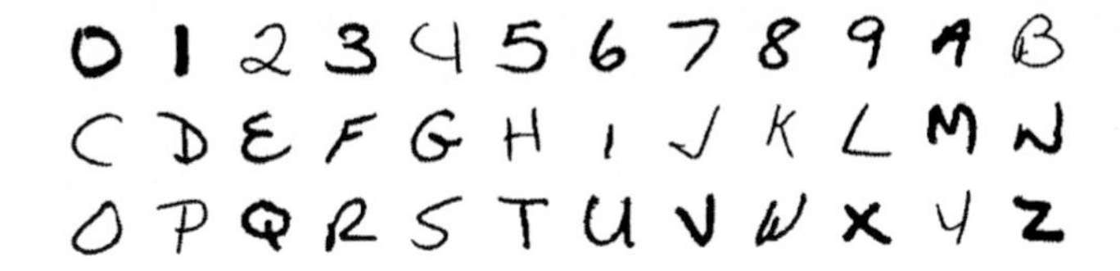

Fig. 3. Examples of training patterns given in CEDER database.

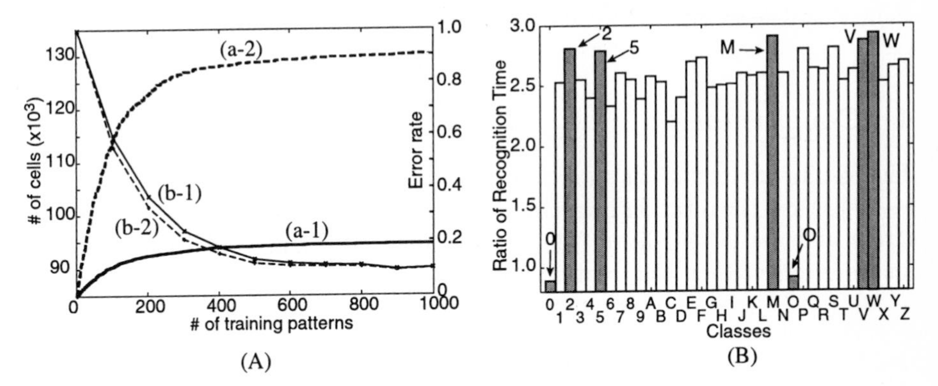

Fig. 4. (A): the number of cells and the error rate for test pattern. (B): recognition time for each letter.

3 Computer Simulations

The past examinations on a rotation-invariant neocognitron showed that the network was completely tolerant for rotations, so the model is trained so as to recognize point symmetrical patterns, "6"–"9"and "M"–"W," as identical ones in the recognition layer U_{C4}. On the other hand, the proposed network correctly distinguishes these patterns. In both models we regard the class "O" as the same class with "zero". Parameters for the number of cells are same as in [9].

We compare the number of cells needed in the rotation-invariant neocognitron with the number in the proposed model using training patterns. a set of handwritten alphabets and numeral figures given by CEDER database are shown in Fig. 3. After training the module U_2 using $1,000$ training patterns, we prepare ten different initial values of plastic connections in the third module for the both models, and train the U_3 for each model using $1 \sim 1,000$ training patterns. The Fig 4(A) shows the change of the averaged number of cells over 10 initial values with the increase of the number in training patterns. A line marked by (a-1) in Fig 4(A) and a line (a-2) are the number of cells of the proposed network and the rotation-invariant neocognitron respectively. A line (b-1) in Fig 4(A) and a line (b-2) are error rates for test patterns, which are rotated in any angles, by the proposed network and the rotation-invariant neocognitron respectively. This figure shows that the proposed model needs less cells than the rotation-invariant neocognitron, though the difference of error rates between the both networks is very little.

The ratio of averaged recognition time for each class of patterns by the proposed network to that by the rotation-invariant neocognitron are shown in Fig. 4(B). We can observe that each recognition time for "zero" and "O" by proposed model is less than that by the rotation-invariant neocognitron Reasons for this are; (i) the proposed model is composed of less cells than the rotation-invariant neocognitron and (ii) a recall process by the top-down process and an angle estimation are not needed for these patterns. On the contrary, very similar patterns for rotations, *e.g.* "2"–"5," "M"–"W"–"V", take more execution time than other classes since there are many hypothetical recognition results for these patterns and a verifying process for each hypothesis should be done.

4 Conclusion

The results of simulations show that the proposed model works efficiently with less cells than the rotation-invariant neocognitron and the difference of the recognition abilities is very little. The recognition time for rotated patterns in any angles is longer than the rotation-invariant neocognitron. But the longer recognition time of the proposed model is not a serious problem because it is rare case that all letters are rotated in any angles in letter recognition. The rotation-invariant neocognitron is effective for recognition of coin images or chip set images.

References

[1] M. Fukumi, S. Omatsu and Y. Nishikawa: "Rotation-invariant neural pattern recognition system estimating a rotation angle," IEEE Trans. Neural Network, **8**, 3, 569–581, 1997.
[2] B. Widrow, R. G. Winter and R. A. Baxter, "Layered neural nets for pattern recognition," IEEE Trans. Acoust, Speech, Signal Processing, **ASSP-36**, 1109–1118, 1988.
[3] S. Kageyu, N. Ohnishi, and N. Sugie, "Augmented multilayer perceptron for rotation-and-scale invariant hand written numeral recognition," Proc. Int. J. Conf. Neural Networks, **1**, 54–59, 1991.
[4] M. B. Reid, L. Spirkovska and E. Ochoa, "Rapid training of higher-oder neural networks for invariant pattern recognition," Proc. Int. J. Conf. Neural Networks, **1**, 689–692, 1989.
[5] K. Fukushima: "Neocognitron: A hierarchical neural network capable of visual pattern recognition," Neural Networks, **6**, 119–130, 1988.
[6] S. Satoh, J. Kuroiwa, H. Aso and S. Miyake, "A rotation-invariant neocognitron," Systems and computers in Japan, **30**, 4, 31–40, 1999.
[7] S. Satoh, J. Kuroiwa, H. Aso and S. Miyake: "Recognition of rotated patterns using neocognitron," Proc. of the fourth inter. conf. neural information processing, **1**, 112–116, 1997.
[8] J. Basak and S. K. Pal: "Psycop —a psychologically motivated connectionist system for object perception," IEEE Trans. Neural Networks, **6**, 6, 1337–1354, 1995.
[9] S. Satoh, H. Aso, S. Miyake and J. Kuroiwa, "Pattern recognition system with top-down process of mental rotation," Proc. of the fifth Intern. Work-Conf. Artificial and Natural Neural networks, **1**, 816–825, 1999.

Classification with Synaptic Radial Basis Units

J. David Buldain

Dept. Ingeniería Electrónica & Comunicaciones
Centro Politécnico Superior. C/ Maria de Luna 3.
50015 - Zaragoza. Spain
buldain@posta.unizar.es

Abstract. A new type of Multilayer network including certain class of Radial Basis Units (RBU), whose kernels are implemented at the synaptic level, is compared through simulations with the Multi-Layer Perceptron (MLP) in a classification problem with a high interference of class distributions. The simulations show that the new network gives error rates in the classification near those of the Optimum Bayesian Classifier (OBC), while MLP presents an inherent weakness for these classification tasks.

1 Introduction

Although there are theorems that prove the existence of a three-layer perceptron capable of approximating a given real valued and continuous function to any desired degree of accuracy [9][11], the back-propagation (BP) algorithm does not seem to perform adequately as a general technique to find the weights of the network. There are numerous reports in the literature or ways to improve the standard BP method, trying to cope with diverse types of difficulties. As noted in [18], some of the difficulties of the BP algorithm [17] applied to a Multi-Layer Perceptron (MLP) may be related to the fact that MLP do not have the best approximation property, as opposed to regularization networks [5]. These networks contain units implementing localized functions, while perceptrons generate hyperplanes that have not local properties. Some solutions are proposed in [18] and [13] with the use of combinations of sigmoids and local bumps in the construction of MLP trying to introduce local properties in the MLP network for function approximation.

In this work, MLP is faced with a type of classification problem with a high superposition between the class distributions that the non-local decision regions of the perceptron cannot process adequately. It is shown how a multi-layer network with diverse types of Radial Basis Units (RBU) can manage this problem with error rates that approximate those obtained with the Optimum Bayesian Classifier (OBC).

Section 2 describes briefly the proposed model of RBU. Section 3 defines a classification problem with a high level of class superposition to simulate the network models. Section 4 compares the characteristics of both types of neural networks. Finally section 5 exposes the performance of the different classifiers.

J. Mira and A. Prieto (Eds.): IWANN 2001, LNCS 2084, pp. 223–234, 2001.

2 Dendritic and Synaptic Radial Basis Units

The Radial Basis Units (RBU) generate bounded responses within closed regions of input-spaces and are mainly applied in the Radial Basis Functions (RBF) model [14]. To facilitate the exposition we call dendritic level of a RBU to the set of synaptic connections coming from a predendritic layer, and the dendritic fascicle of certain layer is the set of dendrites of its RBUs from the predendritic layer. The local kernel that the RBU recognizes can be defined by two parameters: the kernel position in the input space, given by the centroid and the kernel extension, imposed by the width. Centroids are always synaptic adaptive parameters, but widths can be defined at dendritic or synaptic level (for a more detailed exposition see [3] [4]).

When the width is a dendritic parameter, the RBU will be denominated Dendritic RBU (DRBU) and, if the widths are defined at synaptic level, the RBU will be denominated Synaptic RBU (SRBU). The SRBU model includes as a particular case the DRBU model and provides better processing capabilities, since a SRBU can process each synaptic input separately, but with the disadvantage of doubling the number of synaptic parameters and therefore memory requirements.

The multilayer network of RBUs simulated in this work presents hidden layers with DRBUs and output layers with SRBUs. The more precise stimulus adjustment of SRBUs are preferred at the output level and the broad stimulus adjustment of hidden DRBUs are good enough in this processing stage and reduces the memory size of the network.

Diverse authors treat with learning of RBU, mainly in RBF model. The preferred method is competitive dynamics like in [8]; also there are K-clustering methods [14], strategies derived from gradient descent methods [15], and Genetic Algorithms [19]. In this work, the hidden DRBUs are trained with a competitive K-Winner Takes All algorithm (K-WTA) considering only one unit as winner of the layer. The algorithm defines certain competitive factor, similar to the factor defined in FSCL algorithm [1], trying to avoid under-utilization of neurons [6]. The winner unit receives a value one in its activation while the rest of units in the layer receive zeros. Those units, whose activation have value one, update their centroids and widths with equations well known in competitive learning [12].

The SRBUs in the output layer do not compete; they are supervised by assigning their activations with the supervisory labels of the samples: a value one means that the sample belongs to the class associated to the SRBU and a value zero means the contrary. Only the SRBU with the value one in its supervised activation adjust its parameters. Therefore the SRBU learn to recognize the distribution of samples of its class and it is not influenced by the data samples of other class distributions. This property its crucial if the data of several classes present interference, since the interference zone is belonging to several decision regions of the units at the same time.

The Synaptic or Dendritic RBU functionality in recall phase can be resumed as follows: calculate the square of Euclidean distance between the synaptic or dendritic input stimulus (the synaptic input component in SRBU or the input vector in DRBU) and the centroid. This distance is divided by the synaptic/dendritic kernel width and this ratio is subtracted the value one to obtain the synaptic/dendritic excitation. In the

SRBU the synaptic excitations are summed and averaged by the number of synaptic components to obtain the unit excitation, which is passed through the output function. Many output functions are possible and some authors [7][16] demonstrate that this choice has minor importance for the processing executed by the first stage of the RBF model. The main characteristic of the output function is that it generates an elevated response in the kernel. In this work, it is used a certain normalized sigmoid function.

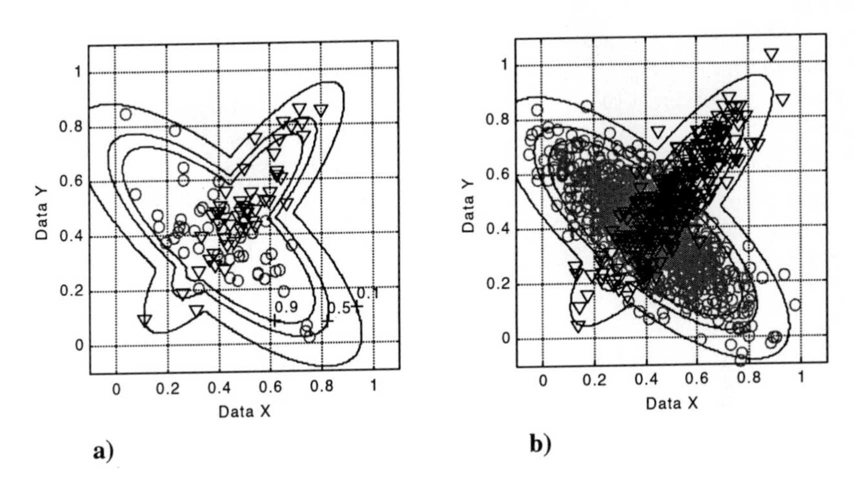

Fig. 1. Both figures represent three contours of the Joint Probability Distribution Function (PDF) in values: 0.1, 0.5 and 0.9. Figure **a)** shows the learning set (50 samples for each class) using triangles for the samples of class 1 and circles for the samples of class 2. Figure **b)** presents the test set (300 samples in class and 900 samples in class 2)

3 The Classification Problem

It was chosen the simplest classification task: a two-class problem generated with two gaussian distributions with strong interference. Both class distributions are of elliptical form bounded in the data area 1x1 with both class means very close, therefore a strong interference zone appears in the middle of both means that we call the interference zone. To facilitate the visualization of the resulting decision regions of the classifiers, the gaussian distributions present two data components: X and Y.

The resulting Probability Distribution Function (PDF) and both sets of patterns are represented in the Fig.1. The PDF is represented with three contours associated to the values: 0.1, 0.5 and 0.9. The samples of class 1 are depicted with triangles and circles for class 2.

The classes have different a priori probabilities: class 2 is three times more probable than class 1. Therefore the test set has 1200 samples: 300 of class 1 and 900 of class 2. The number of samples in the test set was chosen following the results in

[10]. The learning set only contains 50 samples of each class. This situation simulate the situation where the a priori probabilities of the classes are not known and the training set has to include the same number of samples for all the classes.

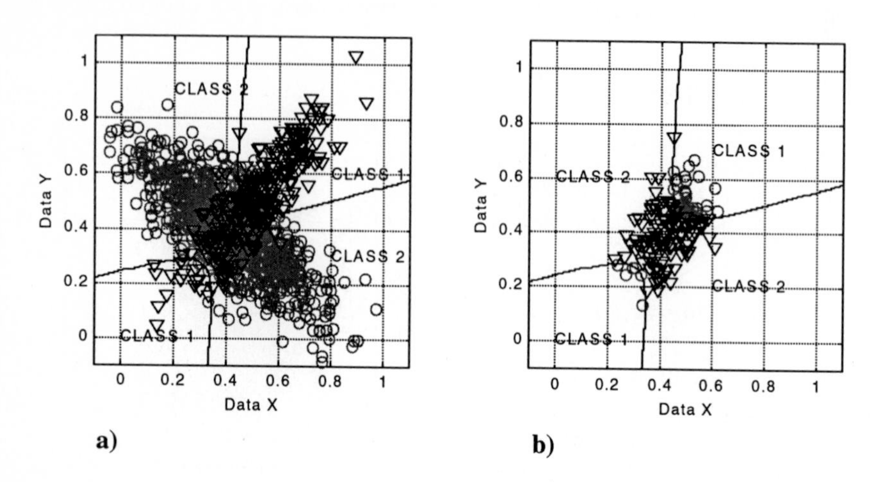

Fig. 2. Both figures show the decision region of the Optimum Bayesian Classifier. Figure **a)** shows the decision region compared with the test set and the figure **b)** shows the samples of the test set confused by OBC: 119 of class 1 and 40 of class 2

3.1 Optimum Bayesian Classifier

Since the gaussian distributions and the a priori probabilities are known, we can obtain the Optimum Bayesian Classifier (OBC), which from the mathematical point of view is the best classifier. The covariance matrixes are:

$$\sigma^2(u_1) \quad \begin{matrix} 115.03 & 113.25 \\ 113.25 & 154.97 \end{matrix} \;;\; \sigma^2(u_2) \quad \begin{matrix} 80.24 & 53.34 \\ 53.34 & 61.43 \end{matrix} \;; \tag{1}$$

The probability of generating the sample **x**, $\Pr(u_i|\mathbf{x})$, is given by the PDFs where K=1,2; is the class index and $m_i(u_k)$ are the mean components:

$$\varphi_K(x) \quad \frac{\left|\sigma^2_{ij}(u_K)\right|^{1/2}}{2\pi} \quad \exp\left(\frac{1}{2}\left(\sum_{i,j} \sigma^2_{ij}(u_K)\ (x_i \quad m_i(u_K))\ (x_j \quad m_j(u_K))\right)\right); \tag{2}$$

The a priori probabilities are: $\Pr(u_1)=1/4$ for class 1 and $\Pr(u_2)=3/4$ for class 2. Therefore the Bayesian decision region can be obtained by applying the maximization rule to the class PDFs weighted by their a priori probabilities. The resulting frontier is depicted in Fig.2: where figure a) represents the test set with 1200 samples and the figure b) shows the samples confused by the OBC. It confuses 159 samples: 119 of class 1 (triangles) and 40 of class 2 (circles). Therefore its classification error rate is 13,25%.

4 The Neural Classifiers

Two types of multilayer networks are simulated to learn the classification problem: the Multi-Layer Perceptron (MLP) and a Multi-Layer of Radial Basis Units (MLRBU). Both models are implemented with one hidden layer and one output layer. Each exemplar of a network architecture and model was simulated independently five times and their averaged classification results are shown in section 5.

4.1 Multi-Layer Perceptron

The simulated exemplars of the MLPs present a variable number of hidden units, while the output layer contains two units for all the exemplars. Thought one output unit is sufficient for this problem, but two output units were included to facilitate the comparison with the MLRBU. The output function of the perceptrons is the bipolar sigmoid that varies in the interval (-1, 1) instead of the unipolar sigmoid varying in the interval (0, 1). The second sigmoid was rejected because it gave reiteratively poorer results than the first one in the initial simulations. The MLP is able to learn this problem with as few hidden units as 4, but the best classification rates appeared when the number of hidden units was over 6. However, the increasing number of hidden units did not reduce the error rate significantly, therefore they were simulated three networks exemplars including 6, 9 and 16 units.

4.2 Multi-Layer Radial Basis Units

The MLRBU networks exemplars include 9 or 16 hidden DRBUs and the output layer has two SRBUs for each classification output. The hidden layer representation was simulated to generate two different forms of representations: the adaptive representation and the sparse map representation.

The adaptive representation is generated in the same training process of the class units with the class data. The competition process forces all the hidden DRBUs to cover the zones of the input space where the class data of the learning set appear.

The map representation is generated by training the hidden layer with data from a uniform distribution in the same input space of the class data. The hidden DRBUs compete to cover the input space (in this case the area 1x1 of X-Y space) and generate a sparse map representation over the class data space. The term sparse map is taken from [2], indicating that neighbor units in the map are not physically neighbor units like in the topological maps of [12]. This map representation of RBUs is "frozen" during the training of the class SRBUs with the class data.

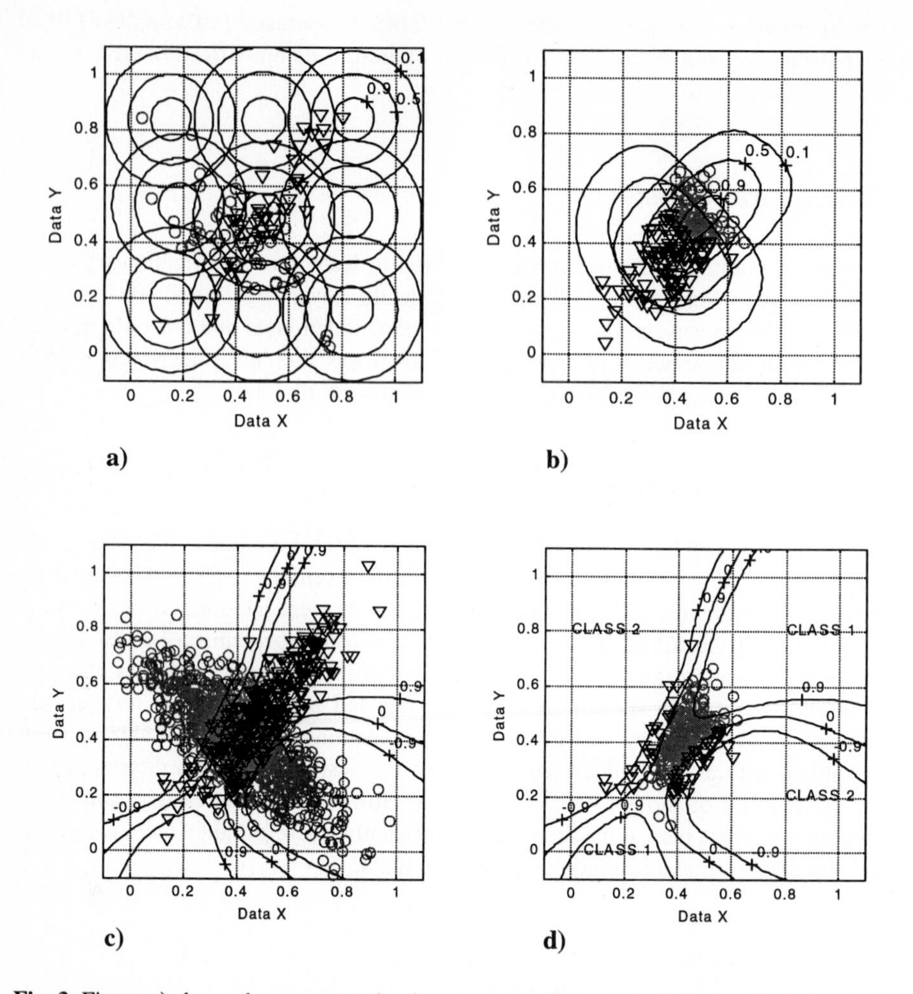

Fig. 3. Figure **a)** shows the representation in sparse map form of the 9 hidden RBUs by depicting three contours in values: 0.1, 0.5 and 0.9. The learning set is superimposed as comparison. Figure **b)** shows the classification decision regions of the SRBUs with three contours and the confused samples of the test set are superimposed. Figure **c)** represents the test set and the decision region of class unit 1 with three contours in values: -0.9, 0 and 0.9, found in an exemplar of MLP. Figure **d)** shows the same unit decision region including the samples of test set confused in the classification

Both forms of representations present pros and cons. The adaptive representation uses all the hidden DRBUs in the specialization on the particular samples of the learning set, while the map representation is covering all the input space with no particularization on any class data. The map representation is a universal representation that can be reused if data from new classes are introduced in the trained network (since adding a new class only needs a new output unit that specializes in the new class data). This is not possible with the adaptive representation, because its DRBUs are specialized over the previous class data. In the map representation is possible to prune those DRBUs that are not being activated during the training with the class data, to reduce the number of hidden DRBUs only to those covering the class data distributions. The main disadvantage of the map representation is that it degrades the performance of the network if incomplete data is used, while the adaptive representation can cope with this situation [3].

4.3 Comparison of Decision Regions

The main difference of both models comes from the processing nature of the decision regions that their units can generate. The perceptrons generate hyperplanes that separate two zones of their input space. The decision region is an open decision frontier that has no local properties. In the case of the RBUs, the decision region is a closed zone of the input space. This difference influences in the form of the classification regions that each model generates when class interference appears.

When an output perceptron is trained in a classification task, it needs to be supervised to generate the response associated to the recognition of a sample from its associated class (for example response in value 1), and to generate the opposite response (for example –1) with samples of other classes. This kind of supervision forces the perceptron to learn and forget the same data patterns in the zones of class interference.

The locality of the SRBUs makes their supervision easier than the supervision of perceptrons. The SRBUs only need to be trained to give the high response (for example 1), because the low response (for example 0) appears for zones of the output space out of their local kernel. Therefore the SRBUs only learn to recognize the zones where the samples of their associated class are presented, but do not try to avoid samples of other classes.

Fig. 3 and Fig.4 depict the output responses of units in both models to compare the kind of decision regions they can generate. The Fig.3a shows the map representation of 9 hidden DRBUs by depicting three contours in their response values: 0.1, 0.5 and 0.9; and the learning set is superimposed with triangles for samples of class 1 and circles for class 2. The Fig.3b shows, with three contours in the same response values, the two classification decision regions of the SRBUs in the output layer. The confused samples of the test set are superimposed. Observe that both SRBU´s regions are of elliptical form and cross like the class distributions do, without applying any previous calculation to generate it as in [5].

The Fig.3c represents the decision region of the output perceptron for class 1 found in an exemplar of MLP. Its decision region is depicted with three contours in values: -0.9, 0 and 0.9. The test set is superimposed to appreciate how the unit region is adapted to include the samples of class 1 and tries to avoid samples of class 2. The Fig.3d shows the same region including the confused samples of the test set. Compare the confused samples of the OBC in Fig.2b with those in Fig.3b and Fig.3d, and see the similitude of the confused set in the OBC and the MLRBU. The confused set of the MLP has a different structure as the result of the distorted decision regions that the output perceptrons generate in the interference zone.

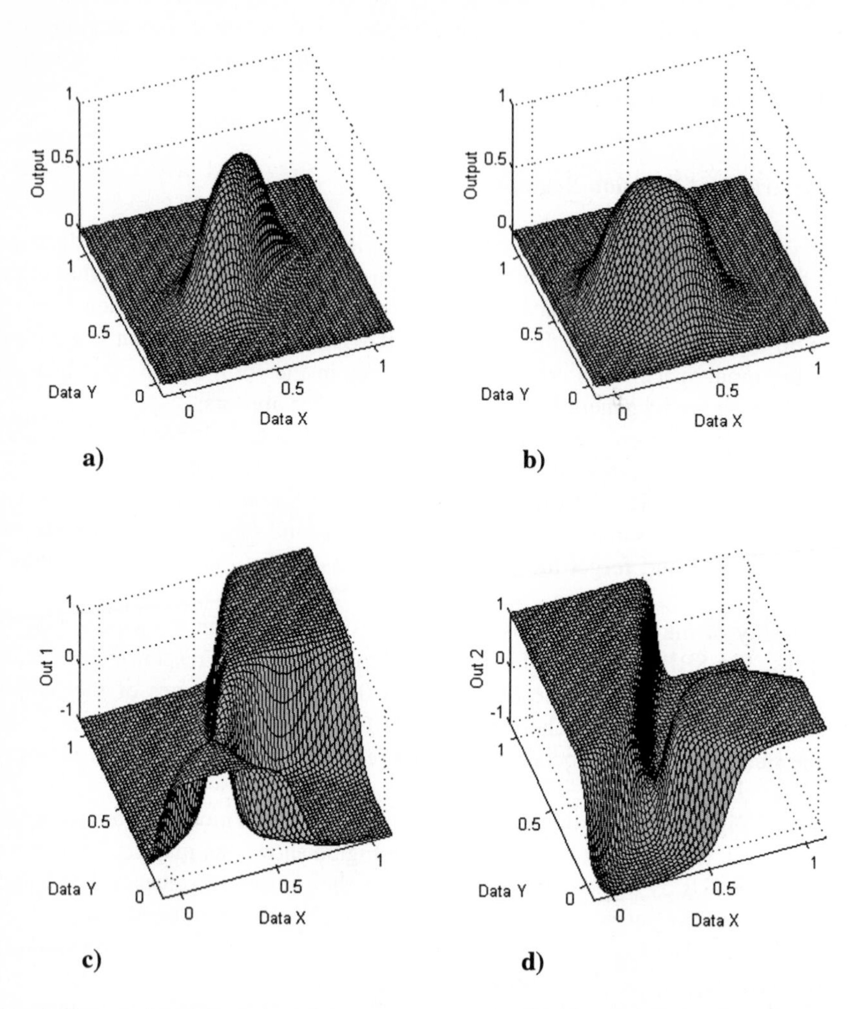

Fig. 4. Figures **a)** and **b)** present the response surface of SRBUs for class 1 and 2 respectively, and figures **c)** and **d)** present the response surfaces of output units in MLP for class 1 and 2

The 3D response surfaces of the classification units are shown in Fig.4. Figures 4a and 4b correspond to the SRBUs responses, while figures 4c and 4d are the response of the output perceptrons. The responses of SRBUs have bump forms, but the responses of the perceptrons present a distortion like a fold of their regions in the interference zone, where they have to learn and to avoid similar samples. The response interval of bipolar sigmoid is able to manage with this problem because the averaged value in the interference zone becomes zero and the scalar product annuls this contribution, but the perceptrons with unipolar sigmoid had severe problems to process this situation as their average value is 0.5 in the interference zone. This is the reason of the bad performance of the MLPs implementing the unipolar sigmoid.

5 Classification Results

The classification results of the neural networks were obtained in five independent exemplars of each network type and averaged. The number of networks exemplars is as low as five because the results were very similar in all the simulations. The three MLPs differ in the number of hidden units: 16, 9 and 6. While the four networks with RBUs differ in the number of hidden units: 16 and 9, and in the type of representation generated in their hidden layer: map or adaptive representation.

The classification performance is resumed by the average error percent committed in each class and the total error. These errors have been depicted in graphic form in Fig.5 and with numeric values in the Table 1, where the class errors includes the class samples not correctly assigned to their class-unit. Fig.5 presents in the horizontal axis the different classifiers: three MLP networks, the OBC and four MLRBUs, using three bars in each classifier for each error rate.

Table 1. Errors rates in the classifiers (presented in graphic form in the Fig.4)

Classifier	Total_Error	Class_1_Error	Class_2_Error
MLP 16 H.Units	23,03	18,13	24,67
MLP 9 H.Units	23,18	18,47	24,76
MLP 6 H.Units	23,03	18,13	24,67
OBC	13,25	39,67	4,44
Map 16 DRBUs	16,13	35,93	9,53
Map 9 DRBUs	16,83	34,20	11,04
Adpt.16 DRBUs	18,42	31,27	14,13
Adpt. 9 DRBUs	19,07	21,07	18,40

The classification rates obtained in the three MLPs are quite the same; they do not get better with increasing number of hidden units and present an average total error of value 23%. The class errors are slightly greater than 18% in class 1 and close to 25% in class 2. Note that the OBC presents a very different results: the total error is 13,25% and the class errors are: 39,66% in class 1 and 4,44% in class 2. Clearly the OBC generates biased class error rates by assigning almost all the samples in the interference zone to the class 2, since the largest a priori probability is assigned to the class 2.

The MLRBUs present very diverse classification results. The best classifier is the network with 16 hidden DRBUs with a map representation giving a total error rate of value 16,13%, and class errors of values: 35,94% in class 1 and 9,54% in class 2. The biased error rates shown by the OBC are almost replicated by this network. The network with 9 hidden DRBUs with a map representation presents a worse classification results, but better than those obtained with MLPs. The total error is 16,83% and the class errors are: 34,2% and 11,05%. Finally, the MLRBUs with adaptive hidden representations obtain classification rates better than the MLPs, but far of those obtained in the MLRBUs with hidden map representations.

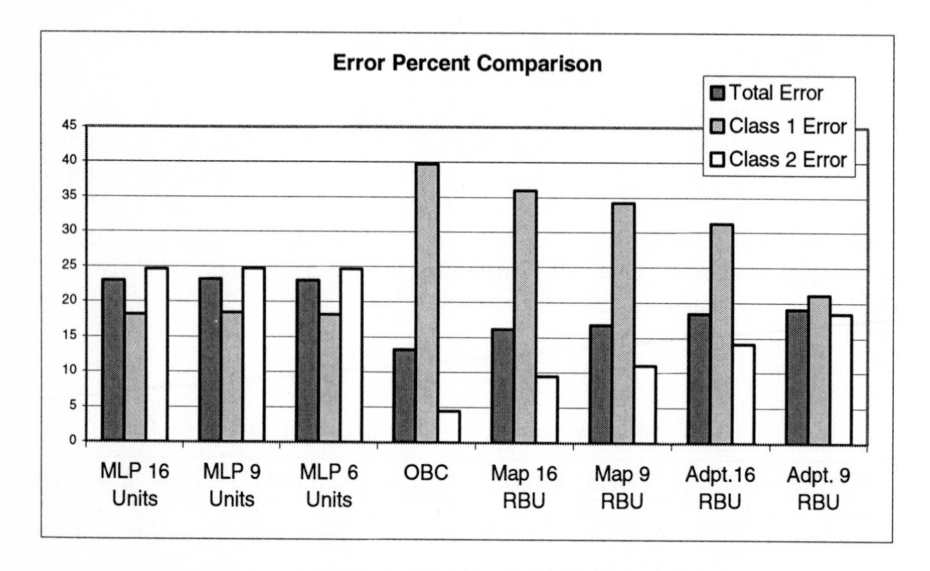

Fig. 5. Comparison of the mean errors committed by five independent exemplars of each type of network and the result of the OBC. Three different MLPs: with .6, 9 and 16 hidden units; and four different MLRBU: with 16 and 9 hidden units in sparse map representation and with 16 and 9 hidden units in adaptive representation

6 Conclusions

The Multi-Layer Perceptron has some disadvantages dealing with classification tasks where class distributions interfere. This weakness stems from the non-local property of the hyperplanes generated by perceptrons. Multi-Layer networks with Radial Basis Units generate error rates close to those obtained by the Optimum Bayesian Classifier; quite better than the Multi-Layer Perceptron trained with back-propagation algorithm.

To compare several possibilities of network implementations the hidden representation of Radial Basis Units is established in two types of representation: the sparse map representation and the adaptive representation.

The map representation is not specialized in any data of the class problem and merely provides a vector quantization of the input space, so it can be reused in the network when new classes are added to the problem. The adaptive representation can resist better the incompletion of patterns. In general, the map representation gives place to a better performance of the network. Than the adaptive hidden representation.

The robustness of both types of representation can be combined in the same network by including two hidden layers for each type of representation. The corresponding output layer should have two separated dendritic connections that can be modulated to process the data by the dendritic path more suited to the changing data.

References

1. Ahalt, S.C., Krishnamurthy, A.K., Chen, P., & Melton, D.E. (1990) Competitive learning algorithms for Vector Quantization. Neural Networks. Vol. 3, pp277-290.
2. Banzhaf, W & Haken, H. (1990) Learning in a Competitive Network. Neural Networks Vol. 3, pp423-435
3. Buldain, J.D. (1998) Doctoral Thesis: Modelo Neuronal Artificial Multi-Dendritico de Unidades con Campo Receptivo. Dept. Ingeniería Electrónica & Comunicaciones. Universidad de Zaragoza.
4. Buldain, J.D. & Roy, A. (2000) Multi-Dendritic Neural Networks with Radial Basis Units, In revision in Neural Networks.
5. Chen,S. & Cowan, C.F.N. & Grant, P.M. (1991) Orthogonal Least Squares Learning Algorithm for Radial Basis Function Network. IEEE Trans.on Neural Networks. Vol. 2, pp302-309.
6. DeSieno, D. (1988) Adding a conscience to competitive learning. Proc. Int. Conf. on Neural Networks. pp117-124. IEEE Press, New York.
7. Epanechnik, V.A. (1969) Nonparametric Estimation of a Multidimensional Probability Density. Theory of Probability Application. Vol. 14, pp153-158.
8. Firenze, F. & Morasso, P. (1994) The "Capture Effect " a new self-organizing network for adaptive resolution clustering in changing environments. Proc. Int. Conference on Neural Networks. pp653-658.
9. Funahashi, K. (1989) On the Approximate realization of Continuous Mapping by Neural Networks. Neural Networks, Vol. 2, pp183-192.
10. Guyon, I.& Makhoul, J. & Schwartz, R. & Vapnik, V. (1998) What Size Test Set Gives Good Error Rate Estimates ? IEEE Transactions on Pattern Analysis and Machine Intelligence, 20, pp52-64.
11. Hornik, K. & Stinchcombe, M. & White, H. (1989) Multilayer Feedforward Networks are Universal Approximators. Neural Networks, Vol.2, pp359-366.
12. Kohonen, T. (1982) Self-organized formation of topologically correct feature maps. Biological Cybernetics, 43, pp59-69.
13. Lapedes, A. & Farber, R. (1988) How Neural Networks Work. Neural Information processing Systems, pp442-456, D.Z.Anderson,Ed New York: American Institute of Physics

14. Moody, J. & Darken, C. (1989) Fast learning in networks of locally-tuned processing units. Neural Computation. Vol.1 nº2 pp281-294.
15. Poggio, T. & Girosi, F. (1990) Networks for approximation and learning. Proceedings of the IEEE, Special Issue on Neural Networks vol.I, Sept.90, pp1481-1497.
16. Powell, M.J.D. (1987) Radial Basis Function for Multivariable Interpolation: a Review. Algorithms for Approximation. J.C.Mason and M.G.Cox, Eds, Oxford, UK: Clarendon, pp143-167
17. Rumelhart, D.E. & McClelland, J.L. (1986) Parallel Distributed Processing. Vol. 1: Foundations. MIT Press.
18. Geva, S. & Sitte, J. (1992) A Constructive Method for Multivariate Function Approximation by Multilayer perceptrons. IEEE Trans. on Neural Networks. Vol. 3, nº 4, pp621-624.
19. Whitehead, B.A. & Choate, T.D. (1996) Cooperative-Competitive Genetic Evolution of Radial Basis Function Centers and Widths for Time Series Prediction. IEEE Trans. on Neural Networks. Vol. 7, nº 4, pp869-881.

A Randomized Hypercolumn Model and Gesture Recognition

Naoyuki Tsuruta[1], Yuichiro Yoshiki[1], and Tarek El. Tobely[2]

[1] Department of Electronics Enginieering and Computer Science, Fukuoka University
8-19-1, Nanakuma, Jonan, Fukuoka 814-0180 JAPAN
{tsuruta@, yoshiki@mdmail.}tl.fukuoka-u.ac.jp
http://www.tl.fukuoka-u.ac.jp

[2] Department of Information Science and Electrical Engneering, Kyushu University
6-1, Kasuga-Koen, Kasuga, Fukuoka 816-8580 JAPAN
tobely@al.is.kyushu-u.ac.jp
http://www.al.is.kyushu-u.ac.jp

Abstract. Gesture recognition is an appealing tool for natural interface with computers especially for physically impaired persons. In this paper, it is proposed to use Hypercolumn model (HCM), which is constructed by hierarchically piling up Self-organizing maps (SOM), as an image recognition system for gesture recognition, since the HCM allows alleviating many difficulties associated with gesture recognition. It is, however, required for on-line systems to reduce the recognition time to the range of normal video camera rates. To achieve this, the Randomized HCM (RHCM), which is derived from HCM by replacing SOM with randomized SOM, is introduced. With RHCM algorithm, the recognition time is drastically reduced without accuracy deterioration. The experimental results to recognize hand gestures using RHCM are presented.

1 Introduction

The goal of gesture understanding research is to redefine the way people interact with computers. By providing computers with the ability to understand gestures, speech, and facial expressions, it is possible to bring human-computer interaction closer to human-human interaction. Researches on gesture recognition can be divided into image-based systems [1],[2] and instrument glove-based systems[3]. The image-based systems enable natural interaction, while the user cannot help wearing glove-like instrument in the glove-based systems.

In image-based systems, the recognition process should be divided into two stages (see Fig. 1). The first stage converts an input image sequence into a discrete posture state sequence. The second stage applies a pattern-matching algorithm to the input and gesture models. Finally, the gesture recognition systems answer the category of the best match gesture model. The first stage must achieve the following four goals.

1. The first stage must output the discrete states, which are useful for stochastic analysis of the second stage. From this point of view, vector quantization techniques are suitable.

J. Mira and A. Prieto (Eds.): IWANN 2001, LNCS 2084, pp. 235-242, 2001.

2. The first stage must output the discrete states, which are independent of camera position, individuals and light source condition. Therefore, invariant image recognition techniques are suitable.
3. The first stage must finish the conversion within the range of normal video camera rates to realize real-time interaction.
4. The first stage must perform well under scenes of complex background, and trace the target object.

For the second stage, the hidden markov model is commonly used. This is not mentioned in this paper.

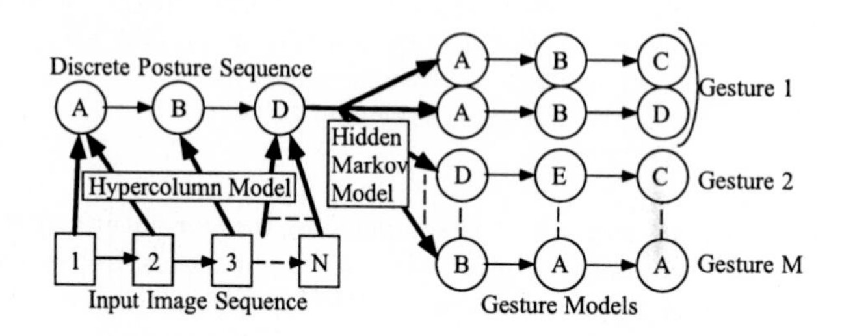

Fig. 1. A proposed gesture recognition system.

Figure 1 shows a proposed gesture recognition system. HCM is used for the first stage. HCM is a SOM base image recognition system, and can learn discrete states, whose a priori probability is a good approximation of distribution density of the learning data [4]. With this characteristic, HCM can achieve the goal (1). Furthermore, HCM is an improved model of SOM by combining Neocognitron, and can recognize complex objects with shift, rotation and distortion. With this characteristic, HCM can achieve the goal (2). HCM also has a mechanism of selective attention to overcome goal (4)[5]. Then, if the recognition time of HCM can be reduced within the range of normal video camera rates, HCM achieves all four goals.

In this paper, a randomize technique is proposed for HCM to reduce the execution time. The randomized HCM is applied to hand-gesture recognition for Jan-Ken-Pon game. In section 2, normal SOM is applied to Jan-Ken-Pon game to be compared with randomized HCM. Then, HCM is briefly mentioned in section 3. Finally, randomize technique is introduced in section 4, and compared with normal HCM.

2 Self-Organizing Maps (SOM)

Kohonen's SOM is a biological model of the hypercolumn, and a vector quantization algorithm for pattern recognition [6]. Figure 2 shows the basic structure of the SOM. All neurons are presented the same input data I. Each neuron has a weight vector W_u, that is, a codebook of the set of I. The codebooks, which are organized by SOM learning algorithm, have two remarkable characteristics:

1. The distribution density of the codebooks is a good approximation of the distribution density of the learning data. Therefore, from the point of view of information theory, this method is equivalent to a histogram equalization and provides good stochastic information for the gesture pattern matching.
2. The topographical order of the learning data is preserved in the codebooks, even if the dimensionality of the SOM is smaller than that of the learning data. This characteristic of topographic mapping is important for randomization technique mentioned later.

To beginning with, a SOM was applied to the Jan-Ken-Pon game. The game includes three postures for three-hand positions. These postures are called "Goo", "Choki", and "Par" as shown in Fig. 3, respectively. Fig. 3 shows three

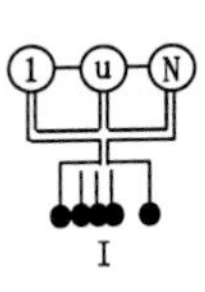

Fig. 2. Structure of SOM.

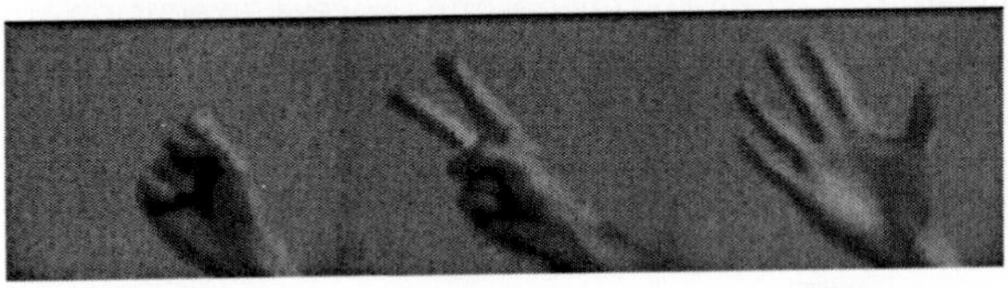

Goo Choki Par

Fig. 3. Three learning images for the postures of Jan-Ken-Pon game.

examples of learning data. The image size was 160×120 pixels. The images were collected from five persons under the same lighting condition. Fig. 4 shows a feature map trained using 872 images. In the feature map, codebooks for "Goo" and "Par" form continuous codebook cluster, respectively, according to the topographic mapping characteristic. Codebooks for "Choki", however, are divided into two codebook clusters. In addition, there are mixtured pattern of two categories near to category boundaries. These are due to lacking in ability of shift or rotate invariant recognition.

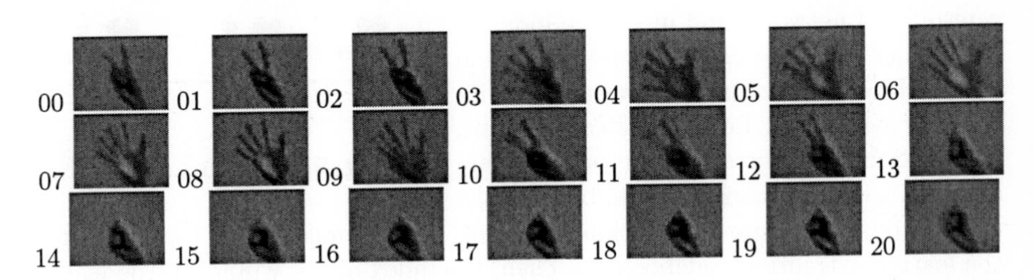

Fig. 4. A feature map trained using 872 images.

The recognition process of SOM consists of three stages. In the first stage, an input image is presented to all neuron. In the next stage, each neuron calculates a distance between the input and its codebook, respectively. For example, in a simplified model using the Euclidean distance as the measure, neuron c is selected, where

$$||I - W_c|| = \min_u (||I - W_u||) . \quad (1)$$

Finally, in the third stage, the best match neuron of the input is selected as the winner, and the category name of the winner is replied. In the experiment, recognition accuracy is 72.2% for test data collected from another three parsons.

SOM is too pure to recognize shifted, rotated and distorted input data. To overcome this, two additional processes are necessary. The first is off-line process. By it, sub-images, which include "hand", are selected from codebooks. The second is on-line process. In the second process, calculation of distance between the input and the selected sub-image is applied on multiple positions and multiple angles. According to this on-line process, execution time should be extremely longer. For example, when 10×10 positions were used for the SOM to achieve shift invariant recognition, the execution time for one video frame was 12.4 second using Alpha 21164A/600MHz processor.

3 Hypercolumn Model

HCM is constructed by pyramidally piling up hierarchical SOMs (Fig. 5.) The hierarchical SOM (HSOM) has two layers. The lower layer is called feature extraction layer. The recognition process is the same as the shift invariant recognition process using SOM. Input fields of the each SOMs, which are denoted in Fig. 5 as from I_1^0 to I_k^0, are slightly shifted by one pixel or one neuron. To introduce shift-invariant recognition, all SOMs have the same codebooks. The upper layer is called feature integration layer. This layer is also SOM network, and input the index of the winner from the lower feature extraction layer. Therefore, feature integration layer input the same data for all shifted patterns. In addition, slightly distorted patterns are also identified as a same pattern, because the number of the feature integration codebook M is smaller than the number of the feature extraction codebook N. Then, the index of the winner in the feature integration layer is presented to the next feature extraction layer. These local feature recognitions are hierarchically piling up. Finally, top layer achieve global shift, rotation and distortion invariant recognition.

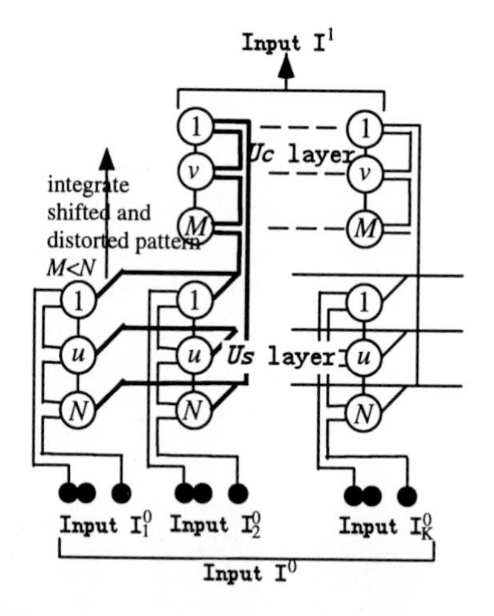

Fig. 5. Structure of HCM.

For example, Table 1 shows a correspondence between the feature map of the SOM mentioned in section 2 and a feature map of HCM top layer. The HCM

Table 1. A correspondence between the feature map of SOM and one of HCM.

SOM	0	1	2	3	4	5	6	7	8	9	10	11	12	13	14	15	16	17	18	19	20
Category	C	C	C	P	P	P	P	P	P	P	C	C	C	C	G	G	G	G	G	G	G
HCM	1	1	1	2	1	1	2	2	2	2	1	1	1	1	1	0	0	0	0	0	0

had three layers of HSOM, and three codebooks in the top layer. The HCM was trained using the same data used in section 2. When a codebook of the SOM was presented to the HCM as an input image, the winner of the HCM was defined as a correspondence of the input. From table 1, it is clear that the SOM and the HCM have very similar feature map according to topographic mapping. Moreover, the codebooks of the "Choki", which are divided into two different codebook clusters (0–2, 10–13) in the SOM, are integrated into one neuron (1) in the HCM.

In HCM, multiple codebook images are effectively encoded into the hierarchical structure sharing local features. Therefore, for complex image recognition, execution time of HCM is shorter than one of SOM. For example, in this experiment, the HSOMs of the lowest layer and the second layer integrated 5×5 and 2×2 shifted patterns into one pattern, respectively. This means that the HCM totally recognized 10×10 shifted patterns and certain distorted patterns. In addition, recognition accuracy 66.7% was about the same as the recognition accuracy of the SOM 72.2%. Therefore, execution time of the HCM can be compared with the SOM mentioned in section 2. To perform as well as the SOM, HCM could reduce the number of neurons, and the execution time was 7.87 second for one frame. (The execution time of the SOM was 12.4 second.) It is, however, far from the range of normal video rate yet.

4 Randomized Hypercolumn Model

In this section, RSOM[7] is introduced as a competition algorithm for the HCM. Since HCM is a hierarchical structure of SOM, so with RSOM algorithm its recognition time can be reduced.

4.1 Randomized Self-Organizing Maps (RSOM)

Algorithm In general, the normal competition algorithm of SOM is applied between all feature map neurons based on the entire input space. In this case, the required computation to select the winner increases as the numbers of input neurons and feature map neurons increase.

In RSOM algorithm, the winner computation is less depending on the network size and mainly depends on the statistics of the input data and network codebook. The competitions in RSOM are done in two phases: The first phase uses subset of the input image to estimate the position of the winner on the feature map; the winner in this phase is called the winner candidates, and its competition runs as follow:

1. Select random samples of pixels $\{S\}$ from the input image with size k and apply to the network.
2. With any competition scheme, apply the competition between the pixels in $\{S\}$ and the corresponding codebooks of all feature map's neurons. If shift invariant recognition is needed, apply all positions.
3. The winner selected from this competition is called the winner candidates.

The ratio k/N, where N is the size of input image, is called pixel usage ratio, and is denoted by PUR. The execution time of this phase B_1 is represented by $B_1 = B \times PUR$, where B denotes the execution time for the competition of normal SOM.

In the second phase, the entire input image pixels are used to search for the winner in the set of feature map neurons neighbor to the winner candidate. The winner selected from this phase is considered as the final SOM winner. The competition in this phase runs as follow:

1. Input all the image pixels to the network.
2. With the same competition scheme used in the first phase, apply the competition between the set of feature map's neurons neighbor to the winner candidate. For shift invariant type, positions are also restricted near to the winning position.
3. The winner selected from this competition is the final SOM winner.

The number of neighborhood neurons is called neighborhood range, and is denoted by NR. The execution time of this phase B2 is represented by $B_2 = B \times NR/M$, where M is the size of the feature map.

Now, there are two parameters PUR and NR. To reduce the execution times, both of PUR and NR should be small. There is, however, a tradeoff point. When a PUR is small, accuracy of the first stage deteriorates, and a large NR is required. The PUR depends on the probability so that variation of the difference between an input image and codebooks is within the NR. The probability may be estimated by the standard deviation of pixel values of codebooks. Formalization of this estimation method is one of future works.

Experimental Result At first, the recognition using normal SOM was applied to show the correct correspondence between the input images and codebooks. After that, the same gesture images were used again to estimate the performance of the RSOM algorithm. The experiments also applied on Alpha 21164A / 600 MHz processor. Fig. 6 shows the recognition time of 10×10 positions using the normal SOM and the RSOM algorithm with different values of PUR and NR.

The recognition accuracy of the RSOM was considered as the rate of selecting the same winner selected by the normal SOM competition algorithm. The recognition accuracy of the experiments in Fig. 6 is shown in Fig. 7.

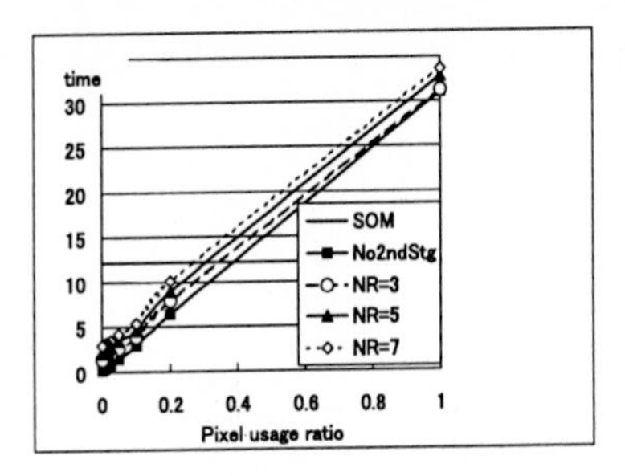

Fig. 6. The recognition time (in second) of 10 × 10 positions using SOM and RSOM with different values of PUR and NR.

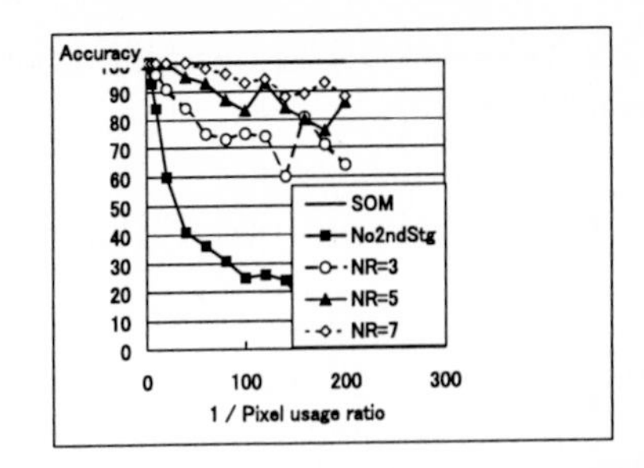

Fig. 7. The recognition accuracy of SOM and RSOM with different values of PUR and NR.

The recognition time of the normal SOM competition algorithm was constant and equal 12.4 second. Table 2 shows the minimum recognition time with maximum accuracy for different values of PUR and NR. As shown in table 2, the recognition time of RSOM with 100% accuracy was 3.4 second. If less recognition accuracy can be accepted, the recognition time can reach 2.3 second.

Table 2. The recognition time (in second) and accuracy of RSOM algorithm with different values of PUR and NR

Recognition Accuracy(%)	Minimum Recognition time(sec)	PUR	NR	$\frac{RSOM}{SOM}$(%)
100	3.4	0.050	5	27.4
96	2.3	0.008	5	18.5

4.2 Randomized Hypercolumn Model

RHCM is derived from HCM by simply replacing SOM for feature extraction with RSOM. The replacement of the lowest layer is especially effective, because of hierarchical structure of HCM. For example, the HCM used in the experiment had 660 HSOM in the lowest layer, and 60 HSOM, one HSOM in the second layer and the top layer, respectively. Therefore, the lowest layer took 91.7% of execution time.

Fig. 8 shows the recognition time of one image using the normal HCM competition algorithm and the RHCM algorithm with different values of PUR and NR. All recognition accuracy of the experiments in Fig. 8 was 100%. The recognition time of one image using the normal HCM is constant and equal 7.87 second. The minimum recognition time with 100% accuracy is 0.336 second, when the values of PUR and NR are 0.0083 and 3, respectively. This means that the recognition of 10×10 positions and certain distorted patterns on more than 3 images per second is feasible [1].

5 Conclusions and Future Works

Gesture recognition based on images should be divided into two stages, image recognition stage and model matching stage. In this paper, to use Hypercolumn model as an image recognition stage was proposed. Though Hypercolumn model allows alleviating difficulties associated with gesture recognition, to reduce the recognition time is required for on-line process. To achieve this, a randomization technique was introduced. With this randomized HCM, the recognition time was drastically reduced without accuracy deterioration. The experimental results to apply the randomized HCM to Jan-Ken-Pon game were presented.

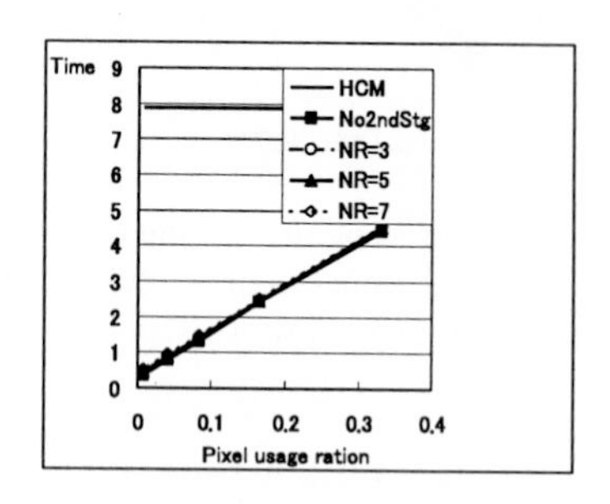

Fig. 8. The recognition time (in second) using HCM and RHCM with different values of PUR and NR.

In the randomization technique, there are two parameters, pixel usage ratio and neighborhood range. In this paper, optimal values of these parameters were estimated experimentally. In future, method to estimate those value in advance must be formalized. In addition, performance of randomized HCM for more complex gesture recognition problem must be estimated.

References

1. Davis J. and Shah M.: Recognizing Hand Gestures, ECCV'94, (1994) 331–340
2. Kameda Y., Minoh M. and Ikeda K.: Three Dimension Pose Estimation Of An Articulated Object From Its Silhouette Image, ACCV'93 (1993) 612–615
3. Freeman W. and Roth M.: Orientation Histgrams For Hand Gesture Recognition, Int. Workshop on Automatic Face- and Gesture- Recognition, IEEE Computer Society (1995)
4. Tsuruta N., Taniguchi R. and Amamiya M.: Hypercolumn Model: A Combination Model of Hierarchical Self-Organizing Maps and Neocognitron for Image Recognition, System and Computer in Japan, Vol. 31, No. 2 (2000) 49–61
5. Tsuruta N., Taniguchi R. and Amamiya M.: Hypercolumn Model: A Modified Model of Neocognitron Using Hierarchical Self-Organizing Maps, IWANN'99, Vol. 1 (1999) 840–849
6. Kohonen T.: Self-organizing maps, Springer Series in Information Sciences (1995)
7. Tobely T. El., Yoshiki Y., Tsuda R., Tsuruta N. and Amamiya M., Randomized Self-Organizing Maps and Its Application, 6th Int. Conf. on Soft Computing, IIZUKA2000 (2000) 207–214

[1] If gcc optimizer is used, the time is 0.15 second. Recent pentium processor and optimizer are much faster than it.

Heterogeneous Kohonen Networks

Sergio Negri[1] and Lluís A. Belanche[2]

[1] Politecnico di Torino
Corso Duca degli Abruzzi, 16
Torino, Italy
sernegri@tin.it

[2] Dept. de Llenguatges i Sistemes Informàtics
Universitat Politècnica de Catalunya
c/Jordi Girona Salgado, 1-3
08034 Barcelona, Spain
belanche@lsi.upc.es

Abstract. A large number of practical problems involves elements that are described as a mixture of qualitative and quantitative information, and whose description is probably incomplete. The self-organizing map is an effective tool for visualization of high-dimensional continuous data. In this work, we extend the network and training algorithm to cope with heterogeneous information, as well as missing values. The classification performance on a collection of benchmarking data sets is compared in different configurations. Various visualization methods are suggested to aid users interpret post-training results.

1 Introduction

Kohonen networks (also known as self-organizing maps) [4, 7] were born to emulate the human brain characteristic of topological and geometrical organization of information. The training algorithm aims at finding analogies between *similar* incoming data in a non-supervised process.

The algorithm places the weight vectors such that geometrically close vectors (in weight space) are also topologically close in the grid represented by the network. In other words, for similar incoming vectors (in input space), the neurons responding more vigorously should also be similar (in terms of their weight vectors) and located in nearby positions in the network grid.

The interest in using these networks is not limited to the discovery of regularities or to tracking the input data density. Once the training process has ended, the result can be *visualized*. Furthermore, it the class labels are available, they can be superimposed to the discovered regularities. Much information can be extracted from such graphical plots, although it has to be identified with care.

1.1 Data heterogeneity

Real-world data come from many different sources, described by mixtures of numeric and qualitative variables. These variables include continuous or

J. Mira and A. Prieto (Eds.): IWANN 2001, LNCS 2084, pp. 243-252, 2001.

discrete numerical processes, symbolic information, etc. In particular, qualitative variables might have a different nature. Some are *ordinal* in the usual statistical sense (i.e., with a discrete domain composed by k categories, but totally ordered w.r.t a given relation) or *nominal* (discrete but without an ordering relation). The data also come with their own peculiarities (vagueness, uncertainty, incompleteness), and thus may require completely different treatments.

In the neural network paradigm, this *heterogeneity* is traditionally handled, if at all, by *preparing* the data using a number of coding methods, so that all variables are treated as real quantities. However, this pre-processing is not part of the original task and may have deep consequences in the structure of the problem. These consequences range from a change in input distribution to an increase in dimension, which results in a growth in the number of weights the network is forced to learn, an added difficulty in their interpretation, an increase in training time, and so on. The choice of representation (if any) should be as faithful as possible, in the sense that the relations between the represented entities should correspond to meaningful relations on the original data items.

1.2 Euclidean geometry

The common assumption of artificial neural models about the Euclidean nature of the input space leads naturally to the use of a scalar product or a distance metric as the standard neuron models (usually followed by a non-linear activation function). In particular, this space is taken to be $\mathbb{R}^n$, with the customary definition of scalar product. This assumption not only means that the features of the problem at hand can be expressed in terms of vectors of real quantities, but also that scalar product and Euclidean distance are adequate ways of measuring the similarity between elements in the space.

In consequence, in order to determine which neuron has a weight vector that is more similar to the input, there are two possibilities: choose the neuron which maximizes the scalar product or choose that which minimizes the distance. These methods may or may not be meaningful for a particular problem, and in principle are only appropriate for real-valued vectors. What to do in cases where input patterns contain heterogeneous information? In this case, a *similarity* index can be used, as a more flexible way to measure likeness.

1.3 Aims and structure

The aim of this work is to extend the main ideas of Kohonen networks in such a way that they can work in generic heterogeneous spaces, making the neurons compute a similarity measure among the elements of the space. This is to be done in stages, which we call *network configurations*, in order to appreciate the effect of each decision. The resulting networks are studied regarding two aspects: classification accuracy and ability to express meaningful information in a visual way. From the point of view of classification accuracy it is shown how the consideration of heterogeneous and/or incomplete information without the need of a coding scheme results in significantly better classifiers. On the other hand, it is illustrated how a solution can be visually analyzed.

The paper is organized as follows. In section (2) we introduce the data characteristics considered in this work. Sections (3) and (4) briefly review the Kohonen algorithm and the basics of a similarity measure. Section (5) describes the proposed extension of the algorithm. The last two sections present practical matters, about classification ability in benchmarking data sets –section (6)– and visualization of the results –section (7).

2 Data heterogeneity

We consider in this work the following types of variables, for which corresponding similarity measures are to be defined.

Nominal (categorical) : non-numerical variable on which no order relation has been defined. It thus can be seen as having a *set* of values (finite or not).

Ordinal : variable (numerical or not) for which a linear order relation has been defined on a finite number of values, where each value has a precise sense.

Continuous : numerical and crisp variable for which a linear order relation has been defined on a continuum of values.

Linguistic : variable whose values are expressing uncertainty in the form of *vagueness* (e.g. *cool, fast, young*).

The values of the last type can be obtained —where appropriate— by converting an existing set of values (ordinal or continuous) into fuzzy quantities. In all cases, we assume some values may be missing in a particular data set.

3 The Kohonen network and algorithm

The classical Kohonen network [4,7] assumes a set of laterally interacting adaptive neurons, usually arranged as a two-dimensional sheet (a rectangular grid of neurons). Each neuron r is represented by an n-dimensional prototype vector $\boldsymbol{w}_r$, where n is the dimension of the input space. On each training step t, a data sample $\boldsymbol{x}(t)$ is presented to the network and the unit $\boldsymbol{w}_s$ *most similar* to $\boldsymbol{x}(t)$ is identified (the Best Matching Unit or BMU). The adaptation step shifts $\boldsymbol{w}_s$ (and those $\boldsymbol{w}_r$ corresponding to neighbouring units r to s) towards $\boldsymbol{x}(t)$:

$$\boldsymbol{w}_r(t+1) = \boldsymbol{w}_r(t) + \epsilon(t)\, h_{rs}(t)\, (\boldsymbol{x}(t) - \boldsymbol{w}_r(t)), \qquad \text{for all units } r \text{ in the grid.} \quad (1)$$

In this formula, $h_{rs}(t)$ establishes the scope and amount of the changes, centered on the BMU s, at time t. In other words, it represents the (varying) neighbourhood. The factor $\epsilon(t)$ acts as a learning ratio, controlling the size of the adaptive steps towards the input vector at time t. Both are decreasing functions of time. In this work, we use the following commonly found formulas:

$$h_{rs}(t) = exp\left\{-\frac{d(r,s)^2}{\sigma(t)^2}\right\} \text{ with } \sigma(t) = \sigma_{in}\left(\frac{\sigma_{fin}}{\sigma_{in}}\right)^{\frac{t}{t_{max}}} \tag{2}$$

$$\epsilon(t) = \begin{cases} \epsilon_0 + \frac{t}{t_1}(\epsilon_{t_1} - \epsilon_0) & \text{if } t \leq t_1 \\ \epsilon_{t_1} + \frac{t-t_1}{t_{max}-t_1}\epsilon_{t_1} & \text{if } t \geq t_1 \end{cases}, \quad \text{for all } t \leq t_{max}. \tag{3}$$

In formula (2), $d(r,s)$ is the (topological) distance between units in the network structure. In case this structure is a rectangular grid, the standard Euclidean distance $d(r,s) = ||r - s||$ can be used. Note also that formula (1) ensures that the BMU would be the same in the hypothetical case $\boldsymbol{x}(t+1) = \boldsymbol{x}(t)$. The initial vectors $\boldsymbol{w}_r(0)$ are set to small random values.

As stated in (1.2), in order to determine the BMU, there are two basic possibilities: pick the neuron with the greatest scalar product to the input vector, or pick that with the smallest (Euclidean) distance. The first choice usually involves vector normalization (both input and weight). The Euclidean distance is a more general criterion (which reduces to scalar product for normalized vectors) and is usually adopted to compute similarity (closeness, in this case) in input space. In this work, all real-valued variables are standardized (to zero-mean, unit standard deviation) so that all of them have a uniform influence in the computation of distance.

4 Similarity measures

4.1 Definition

Let us represent input patterns belonging to a space $X \neq \emptyset$ (about which the only assumption is the existence of an equality relation) as vectors $\boldsymbol{x}_i$ of n components, where each component x_{ij} represents the value of a particular feature (descriptive variable) a_j for object i, from a predefined set of features $A = \{a_1, a_2, \ldots, a_n\}$, judged by the investigator as relevant to the problem. A *similarity measure* is a unique number expressing how "like" two given objects are, given only the features in A [2]. Let us denote by s_{ij} the similarity between $\boldsymbol{x}_i$ and $\boldsymbol{x}_j$, that is, $s : X \times X \to \mathbb{R}^+ \cup \{0\}$ and $s_{ij} = s(\boldsymbol{x}_i, \boldsymbol{x}_j)$.

Definition 11 *A similarity measure in X fulfills the following properties:*

1. Non-negativity. $s_{ij} \geq 0 \quad \forall \boldsymbol{x}_i, \boldsymbol{x}_j \in X$
2. Symmetry. $s_{ij} = s_{ji} \quad \forall \boldsymbol{x}_i, \boldsymbol{x}_j \in X$
3. Boundedness. *There is a maximum attained similarity (that of an object with itself):* $\exists s_{max} \in \mathbb{R}^+ : \; s_{ij} \leq s_{max} \quad \forall \boldsymbol{x}_i, \boldsymbol{x}_j \in X$.
4. Minimality $s_{ij} = s_{max} \Leftrightarrow \boldsymbol{x}_i = \boldsymbol{x}_j \quad \forall \boldsymbol{x}_i, \boldsymbol{x}_j \in X$
5. Semantics. *The meaning of* $s_{ij} > s_{ik}$ *is that object* i *is more similar to object* j *than is to object* k.

4.2 Heterogeneous measures

In this section we define similarity measures (with $s_{max} = 1$) for the types of variables mentioned in (§2). Let $\boldsymbol{x}, \boldsymbol{y}$ be two heterogeneous vectors.

- For **nominal** variables:

$$s(x_k, y_k) = \begin{cases} 1 & \text{if } x_k = y_k \\ 0 & \text{if } x_k \neq y_k \end{cases} \tag{4}$$

- For **ordinal** variables:

$$s(x_k, y_k) = \frac{1}{1 + \frac{|x_k - y_k|}{rank(k)}} \tag{5}$$

  where $rank(k)$ is the number of values the variable k can take.
- For **continuous** variables:

$$s(x_k, y_k) = \frac{1}{1 + |x_k - y_k|} \tag{6}$$

  where x_k, y_k are standardized continuous variables.
- For **linguistic** variables (where their values are fuzzy intervals of the LR-type [9]) the problem is a bit more complex. We started using trapezoids, but we found that better results were in general achieved with the following bell-shaped function:

$$\mu_{\tilde{A}}(x) = \frac{1}{1 + (a(x - m))^2} \tag{7}$$

  where m is the mean and a controls the fuzziness. Let $\mathbb{F}(X)$ be the (crisp) set of all such fuzzy intervals in X. Given $\tilde{A}, \tilde{B} \in \mathbb{F}(\Delta)$, with Δ a real interval, and respective support sets $\Delta_{\tilde{A}}, \Delta_{\tilde{B}} \subset \Delta$, we define [1]:

$$I(\tilde{A}, \tilde{B}) = \int_{\Delta_{\tilde{A}} \cup \Delta_{\tilde{B}}} \mu_{\tilde{A} \cap \tilde{B}}(u) du; \qquad U(\tilde{A}, \tilde{B}) = \int_{\Delta_{\tilde{A}} \cup \Delta_{\tilde{B}}} \mu_{\tilde{A} \cup \tilde{B}}(u) du \tag{8}$$

  and then

$$s(x_k, y_k) = \frac{I(x_k, y_k)}{U(x_k, y_k)} \tag{9}$$

4.3 Gower's similarity index

A basic but very useful similarity-based neuron can be devised using a Gower-like similarity index, well-known in the literature on multivariate data analysis [3]. For any two objects $\boldsymbol{x}_i$, $\boldsymbol{x}_j$ of cardinality n, this index is given by the expression:

$$s_G(\boldsymbol{x}_i, \boldsymbol{x}_j) = \frac{\sum_{k=1}^{n} s_k(x_{ik}, x_{jk}) \, \delta_{ijk}}{\sum_{k=1}^{n} \delta_{ijk}} \tag{10}$$

where s_k is the partial similarity index according to variable k, and δ_{ijk} is a binary function expressing whether the objects are *comparable* or not according to variable k. Let $\mathcal{X}$ represent a missing value, then:

$$\delta_{ijk} = \begin{cases} 1 & \text{if } x_{ik} \neq \mathcal{X} \wedge x_{jk} \neq \mathcal{X} \\ 0 & \text{otherwise} \end{cases} \tag{11}$$

5 Heterogeneous algorithm for the Kohonen network

The weight propagation in equation (1) can only be applied to real values. For other variable types, the following propagation formulas are used. Let $w_{r,i}(t)$ represent the i-th component of weight vector $\boldsymbol{w}_r(t)$.

- For **ordinal** variables:

$$w_{r,i}(t+1) = \lfloor w_{r,i}(t) + \epsilon(t)\, h_{rs}(t)\,(x_i(t) - w_{r,i}(t)) + 0.5 \rfloor \tag{12}$$

- For **linguistic** variables, formula (1) is applied to both m and a in (7).
- For **nominal** variables, a deeper explanation is needed. In all of the previous cases, there is a linear order (continuous or discrete). Hence the notion of "getting closer" to the input vector makes sense and the basic formula (1) can be applied. In absence of an order, there is no shift, but a *change* in value (which can be regarded as a more general concept). In addition, though the intuition behind the product $\epsilon(t)\, h_{rs}(t)$ must be kept, its practical role has to be different. We take it as the *probability* of such a change.
  In essence, a random number $\xi \in [0,1]$ with uniform probability is generated. Then, the updating rule is given by:

$$w_{r,i}(t+1) = \begin{cases} x_i(t) & \text{if } \epsilon(t)\, h_{rs}(t) > \xi \\ w_{r,i}(t) & \text{otherwise} \end{cases} \tag{13}$$

  This scheme is intuitively pleasing but has a serious drawback: it ignores the past. We hence propose an updating rule in which the probability of changing the weight vector component increases proportionally to the number of times this change was already attempted, relative to the number of attempts of the current value of the weight and the rest of possible values, and influenced by the distance $d(r,s)$, where s is the BMU for $\boldsymbol{x}(t)$ (see the Appendix for details).

6 An experimental comparison

A number of experiments are carried out to illustrate the validity of the approach, using several benchmarking problems. These are selected as representatives because of the diversity in the kind of problem and richness in data heterogeneity.

6.1 Problem description

A total of ten learning tasks are worked out, taken from [6] and [5], and altogether representative of the kinds of variables typically found in real problems, while displaying different degrees of missing information (from 0% to 26%). Their main characteristics are displayed in Table 1.

Table 1. Basic features of the data sets. Missing refers to the *percentage* of missing values. R (real), N (nominal), I (ordinal) and L (linguistic).

Name	Cases	Heterogeneity	Classes	Class distribution	Missing
Credit Card	690	6R-9N-0I-0L	2	55.5%/44.5%	0.52%
Heart Disease	303	5R-7N-1I-0L	2	54.1%/45.9%	0.00%
Horse Colic	364	7R-4N-4I-5L	2	61.5%/24.2%/14.3%	26.1%
Solar Flares	244	0R-6N-4I-0L	2	70.9%/29.1%	0.00%
Gene Sequences	3175	0R-60N-0I-0L	3	25%/25%/50%	0.00%
Mushroom	8124	0R-22N-0I-0L	2	51.8%/49.2%	1.26%
Cylinder bands	540	20R-15N-0I-0L	2	42.2%/57.8%	5.29%
Meta data	528	14R-2N-5I-0L	3	32.4%/37.1%/30.5%	0.00%
Servo data	167	2R-0N-2I-0L	2	47.3%/52.7%	0.00%
Annealing data	898	6R-10N-3I-0L	6	1.75%/10.0%/75.5% 0.25%/7.75%/4.75%	12.2%

6.2 Different network configurations

The extended algorithm has been tested in seven different configurations. In all cases, formula (10) is used, as follows:

1. Kohonen network, treating all attributes as real-valued, with the usual 1-out-of-k coding for nominal ones (with k the number of values) and an extra attribute signaling a missing value for those attributes that can be missing.
2. As 1., but coding nominals as 0,1,2...
3. As 2., without the coding for missing values –thus treating them directly by using formula (10).
4. As 3, using formula (12) for ordinal attributes, (considering all linguistic attributes as ordinals) and (13) for nominals.
5. and 6., as 4. but using two different ways of enhancing (13) for nominals (see the Appendix).
7. As 5., considering linguistic attributes as indicated.

In all cases, the interest is not in building a classifier *per se*, but in assessing the relative differences between network configurations. Hence, all the information is used to train the network (without the class labels). After this process has ended, the network is *colored* using the weighted global method – see section (7)– and its classification accuracy computed. The results are shown graphically in Fig. (1).

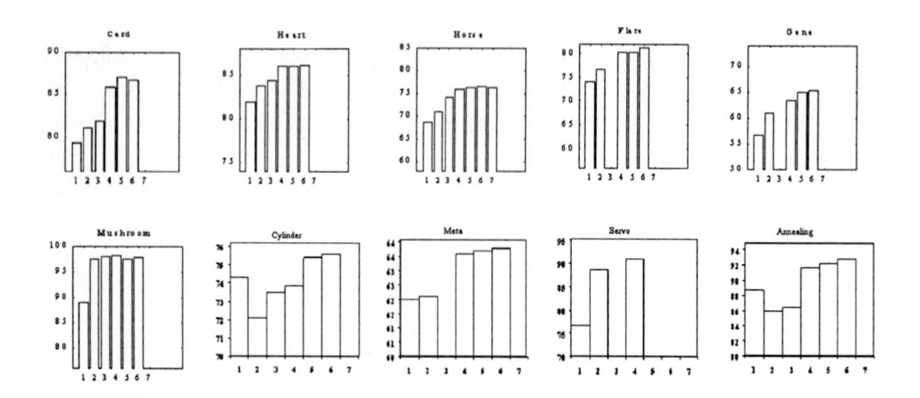

Fig. 1. Graphical impression of the results. The horizontal axis stands for the network configuration. The vertical axis shows classification accuracy. Each plotted value is the average of forty runs.

7 Visualization methods

In order to help users understand the post-training result of the network we use two different visualization methods.

7.1 Weighted global BMU

At the end of the training process we assign a class to each neuron of the grid (that is, we *color* the grid). For each neuron r, we compute its *receptive field* $F(r)$, composed of all those input vectors for which its BMU is r. A basic criterion is then to assign to r the majority class in $F(r)$. This criterion does not take into account the relevance of each vector in $F(r)$. A simple way to do this is to weight each vector in $F(r)$ by its similarity to $\boldsymbol{w}_r$. Afterwards, the class with greater weighted similarity to $\boldsymbol{w}_r$ is chosen to label r. If a neuron is never a BMU, it is not assigned to any class, and is painted "neutral" –*white* in Fig. 2 (left).

7.2 U-matrix

As shown in [8] the U-matrix is useful to visualize similarity among contiguous map units. The *minimum* similarity between each grid unit and its four adjacent neighbours, is computed, and displayed using grey shade –Fig. 2 (center). We have also found very useful displaying the *contour plot* of the same U-matrix –Fig. 2 (right)– and compare it to the weighted global BMU.

8 Conclusions

An extension of Kohonen's self-organizing network and the corresponding training algorithm has been introduced that works in heterogeneous spaces (for

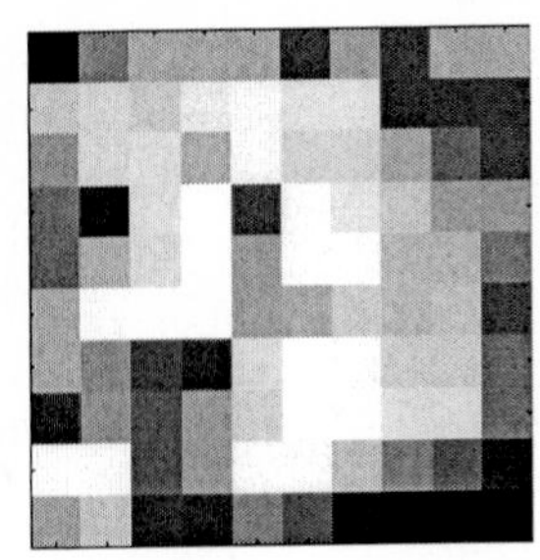
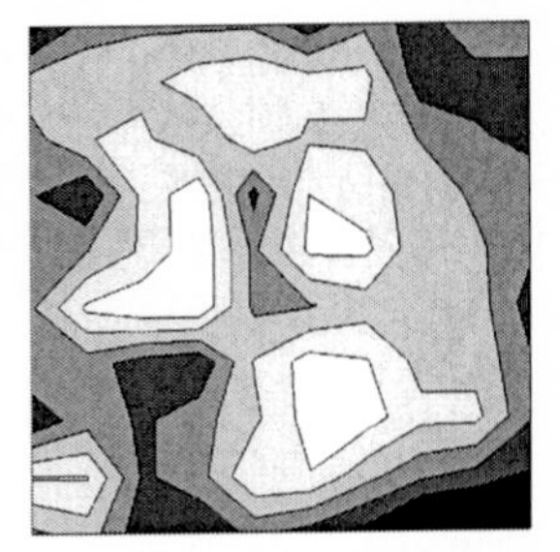

Fig. 2. Visualization of the results in a two-class example. Left: *Global BMU*. Center: *U-matrix*. Right: *U-matrix contour plot*.

both input and weight vectors). The inner workings of a neuron are grounded on a similarity measure, which allows for more flexibility and a built-in treatment of heterogeneous or incomplete data. It is also shown how to visualize the result of the training process. Some results that have been obtained for illustrative purposes in several benchmarking data sets indicate a superior performance of the extended algorithm.

Acknowledgements:
This work is supported by the Spanish CICYT grant TAP99-0747.

References

1. Belanche, Ll. Similarity-based Heterogeneous Neuron Models. In Procs. of ECAI'2000: European Conf. on Artificial Intelligence. IOS Press, 2000.
2. Chandon, J., Pinson, S. *Analyse Typologique. Théorie et Applications.* Masson, 1981.
3. Gower, J.C. A General Coefficient of Similarity and some of its Properties. *Biometrics*, 27: 857-871, 1971.
4. Kohonen, T. *Self-Organization and Associative Memory*. Springer-Verlag, 1988.
5. Murphy, P.M., Aha, D. UCI Repository of machine learning databases. UCI Dept. of Information and Computer Science, 1991.
6. Prechelt, L. Proben1: A set of Neural Network Benchmark Problems and Benchmarking Rules. Fac. für Informatik, Univ. Karlsruhe. Tech. Report 21/94, 1994.
7. Ritter, H., Kohonen, T. Self-Organizing Semantic Maps. *Biological Cybernetics*, 61: 241-254, 1989.
8. Vesanto, J. SOM-based data visualization methods. *Intel. data analysis* 3(2), 1999.
9. Zimmermann, H.J. *Fuzzy set theory and its applications.* Kluwer Acad. Publ., 1992.

A Appendix

During weight propagation, for nominal variables, a random number $\xi \in [0,1]$ with uniform probability is generated. Then, $w_{r,i}(t+1)$ is updated to $x_i(t)$ if $p(t) > \xi$, with $p(t) = \epsilon(t)\, h_{rs}(t)$, otherwise it is left unchanged.

Let n_α be the number of times $w_{r,i}(t)$ has been proposed to be the value of $w_{r,i}$ up to time t, n_β the number of times $x_i(t)$ has been proposed, and let n_γ generally denote the number of times *any other* possible value of nominal variable i has been proposed to be $w_{r,i}$, up to time t. If there are no more possible values (that is, if $rank(i) = 2$), then n_γ is undefined. Define then:

$$f(t) = \begin{cases} \frac{n_\beta^2}{n_\alpha} \sqrt{\frac{rank(i)-2}{\sum_{\gamma=1,\gamma\neq\alpha,\gamma\neq\beta}^{rank(i)} n_\gamma^2}} & \text{if } rank(i) > 2 \\ \frac{n_\beta^2}{n_\alpha} & \text{if } rank(i) = 2 \end{cases} \tag{14}$$

In these conditions, the new probability of change is defined as:

$$p'(t) = \left(\epsilon(t)\, h_{rs}(t)\right)^{\frac{1}{ln(e-1+f(t))}} \tag{15}$$

Whereas a simpler possibility is:

$$p''(t) = \left(\epsilon(t)\, h_{rs}(t)\right)^{\frac{1}{f(t)}} \tag{16}$$

Network configuration 4. uses $p(t)$, configuration 5. uses $p'(t)$ and configuration 6. uses $p''(t)$.

Divided-Data Analysis in a Financial Case Classification with Multi-dendritic Neural Networks

J. David Buldain

Dept. Ingeniería Electrónica & Comunicaciones
Centro Politécnico Superior. C/ Maria de Luna 3.
50015 - Zaragoza. Spain
buldain@posta.unizar.es

Abstract. A dendritic description of multilayer networks, with Radial Basis Units, is applied to a real classification problem of financial data. Simulations demonstrate that the dendritic description of networks is suited for classification where input data is divided in subspaces of similar information content. The input subspaces with reduced dimensions are processed separately in the hidden stages of the network and combined by an associative stage in the output. This strategy allows the network to process any combination of the input subspaces, even with partial data patterns. The division of data also permits to deal with many input components by generating a set of data subspaces whose dimensions have a manageable size.

1 Introduction

Many classification problems present a certain informational structure in their data patterns, containing partial information, or that must be changed regularly to include or exclude data that the changing environment or testing methods generate.

An informational structure in the problem means that the input pattern can be separated in several data subspaces, each one containing input components with similar conceptual information. This data division is an analogy of the perceptual system of animals, where the separated sensorial stimulations are processed by independent senses and finally integrated in higher stages of the nervous system.

Models that can utilize partial data are necessary when the collected data is incomplete to the degree that the adequate generalization cannot be achieved on the basis of the complete examples, as noted in [23]. In many application domains, the larger the input space, the more likely that a training example will contain some missing components. For example, a large process control system may rely on hundreds of remote sensors and, at any time, some fraction of these may not be operational for various reasons. The strategy of data division [24] also attacks the problem of large dimensionality in some data patterns denominated the curse of dimensionality [3].

In this work, we used for the evaluation of the proposed neural model a classification problem where the solvency or bankrupt of banks is determined based on several

J. Mira and A. Prieto (Eds.): IWANN 2001, LNCS 2084, pp. 253–268, 2001.

ratios obtained from the data that they publish. There are many empirical studies made with accounting data on assessment of bankrupt risk. Usually the analysis of the financial situation is made from several ratios obtained from the accounting data published by the companies, to which conventional methods based on univariant or multivariant analysis are applied. Beaver carried out the first study in 1966 [2].

In Spain, Laffarga et al. [11,12] and Pina [23] carried out some interesting studies about the Spanish banking crisis of 1977-1985. These methods achieved a high percent of success in the classification of the banks, but they have been criticized because of several problems, and the use of neural systems has been suggested to overcome them [9]. An example is [17] where a Multi-Layer Perceptron (MLP) was compared with conventional methods for prediction of bankruptcy.

1.1 Neural Networks Structure in Automatic Diagnosis Systems

The situation where a data pattern structure is continuously modified can be understood if we describe as an example a medical diagnosis system. In the medical field, the information of analytic tests is being incessantly refined but, at the same time, is more complex to interpret, since new testing methods are discovered and old ones abandoned. Physicians have to specialize in the understanding of the new diagnostic tests and forget methods that become obsolete. The mastering of such a changing data world implicates a lot of effort and is time consuming.

Neural networks are good candidates to implement adaptive classifiers, but an appropriate neural system to implement a diagnostic system for changing or incomplete data should include at least two processing stages. The first stage would contain several primary networks processing independently the results of the analytic tests (or input subspaces). The second stage, whose output is the result of the whole system, would associate the partial results of the primary networks to generate a global response of the system.

In this structure, each primary network processes the data from one analytic test and could be trained with self-organization or with supervised methods depending on the existence of supervisory data. The second stage should implement an associative mechanism capable of giving coherent responses with any combination of the diverse results from the primary networks. This system structure has some advantages:

1. The system is modular and admits modifications easily. The addition of a new analytic test means to include one or several new primary networks and to train them and the corresponding communication channels to the associative stage, without modifying the rest of the system. The elimination of an obsolete test consists in the opposite strategy.
2. It is possible the modulation of the system to process only data from specific or available tests, by enabling or disabling the communication channels of the associative stage from the primary networks. Modulation also permits to develop the diagnostic process step by step, introducing in the system the available tests data to generate intermediate diagnostic results that indicate whether more diagnostic tests

are necessary or not. This step-by-step analysis can provide savings in time decision, patient suffering and economic costs.

3. The system is capable of processing the particular clinic history of the patient by adding a set of primary networks trained with the relevant patient characteristics or medical data that could be important for the pathology.
4. Finally, the pathologies would have their corresponding diagnostic neural systems and the analysis of the patient's symptoms and tests would be introduced in many of these neural systems working in parallel.

As we will see in the next section, multidendritic networks can implement easily such a system.

2 Multidendritic Network with Radial Basis Units

The multidendritic network description defines a new processing level in the usual multilayer structure: the dendritic connections among layers. A multidendritic structure description was exposed in [13] and applied to the Multilayer Perceptron (MLP). In this work, the multidendritic networks contain processing units that generate spherical closed decisions in their dendritic connections like those of the Radial Basis Units (RBU) implemented in the Radial Basis Functions (RBF) networks [16][20]. These units are trained in a competitive dynamic briefly described in the Appendix 1.

The neural system structure proposed in the previous section (see an example in Fig.1) consists in the combination of a primary stage where several hidden layers are connected to each input subspace. These groups of layers, containing a diverse number of units with different processing parameters, provide an intermediate stage where input data are processed in a vector quantization form. Some layers present a few units with broad decision regions and others layers present many units with small decision regions. The combination of both types of representations is a redundant representation with generalization and particularization properties.

The output stage consists in another group of layers that receive dendritic connections from all the hidden layers and from the input subspace dedicated to the supervision labels. These layers associate the different responses of the hidden layers with the supervised inputs using a competitive training strategy that we call associative supervision [4].

2.1 Simulated Network Architectures

One of the simulated network architectures is depicted in Fig.1, where arrows represent the dendritic fascicles between layers instead of depicting all the synaptic connections among the units. The number of units in each layer is represented between parentheses following the label used to enumerate the layer.

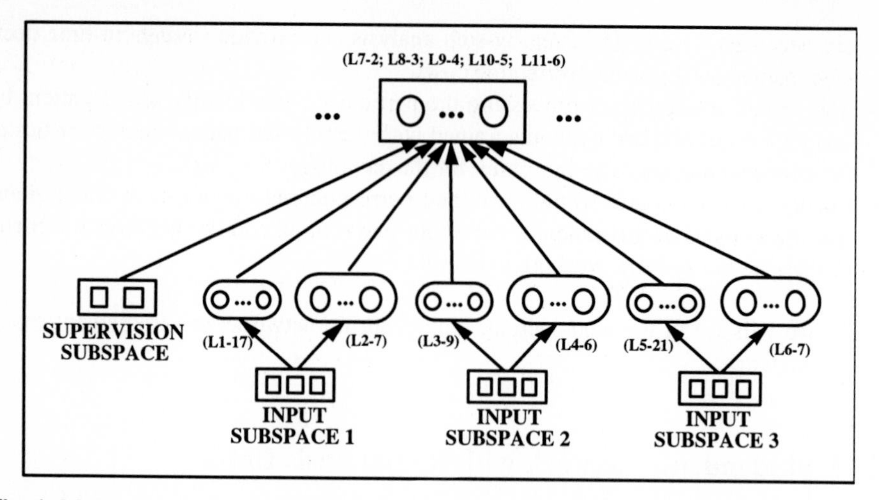

Fig. 1. Modular network architecture with 11 layers and 4 input subspaces. The numbers of units are presented close the icons of layers. Layers L1-L6 are hidden layers and layers L7 - L11 (only one is represented) are associative layers. Three input subspaces introduce the 9 data components and the fourth subspace introduces the two-class supervision labels.

The number of units in the hidden layers was chosen by measuring the entropies [26] of several layers with different number of units (from 2 units to 30) trained competitively with the corresponding data of the subspaces. The layer with few units (generalization) and the layer with large number of units (particularization) contain the numbers of units that generated local minima in the curves of the entropies that are not shown for the sake of brevity.

The network presents four input subspaces (see section 3 for detailed comments about the data division): three of the subspaces with three input components for the financial data and the fourth with two components for the supervision labels.

Each input subspace is connected to two hidden layers and these are connected to the associative layers. Several associative layers, with number of units varying from 2 to 6, will show how the number of associative units modifies the classification performance. These associative layers are the output of the network or can be used as intermediate result for a layer with linear units to configure a complex RBF network

The second network architecture consists in two layers with 2 and 6 DRBUs connected to the same set of subspaces. This simple network gives as good results in classification as the previous one (see section 4.2), but their unit responses are less sharp and, therefore, prone to higher indetermination. This second network is included to point out the possibility of its implementation as the hidden layer in a RBF network and is compared with a Multi-Layer Perceptron network.

2.2 Modulation of Dendritic Channels

The dendritic channels can be enabled or disabled during network operation. This control is carried out by modulating the values of parameters assigned to the dendritic fascicles of each layer, that we call dendritic gains. If we assign a value zero to any dendritic gain (see equation 4 of the Appendix) its corresponding dendritic excitation is omitted in the resulting unit excitation, but if the value of certain dendritic gain is the largest of all dendritic fascicles of the layer, then the corresponding dendritic excitation has the most important effect in the unit responses. Therefore, by modulating the dendritic gains we can select those dendritic channels that the layer will process and strengthen or weaken the effect of any dendritic fascicle respect to the others.

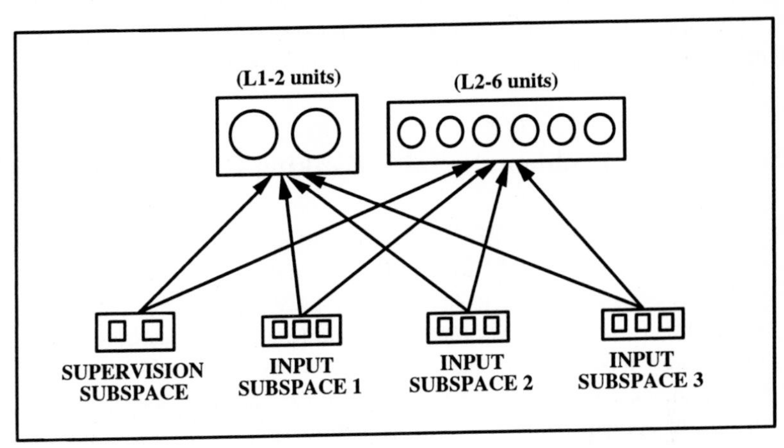

Fig. 2. Second network architecture with 2 layers whose numbers of units are 2 and 6, connected to the same 4 input subspaces of the Fig.1.

This control modulation is also applied to graduate the effect of the supervisory inputs along the training process. At the beginning of the learning phase, the supervisory dendritic gain is very large (value 10) compared with the gains of dendritic fascicles from hidden or input layers (value 1), but it decays with an exponential expression to the value 1 in the final training cycles.

This strategy of supervisory-gain decrement forces the units in each associative layer to compete initially for the supervisory labels. The result is a separation of the units in class groups. The decay of the supervisory gains makes the units in each class-group finally to compete among them for the examples of their class, which makes their kernels to spread over the class distribution.

Other strategies where proved: a) when we maintain a large supervisory gain during all the training phase, then the units in each class-group do not spread over the class distribution, but they crowd near the mean of the class distribution and, b) if we assign a low supervisory-gain value (for example 1) during all the training phase, the separation of units in class-groups is not forced and the supervision effect is weak.

Notice that the supervisory data are not imposed to the units, since all units in the network follow a competitive process during their learning phase. This strategy allows the units to learn mainly the interesting examples and not to give importance to particular examples with irrelevant or corrupted data that are a continuous source of error in other supervised models.

Once trained the network, the response of the associative units must be analyzed to recognize in which class they had specialized and, then, we assign each test input sample to the class containing the unit with maximum response. This method permits to include any number of associative units in the layer that collectively give a better approximation to the class decision boundary

Of course, during a recall phase, the supervisory gains must be annulled to omit the influence of supervisory labels. Following the same gain control, the processing channels of any input subspace can be disabled or enhanced in the test evaluation. In such a form, the gain control makes possible to evaluate the response of the network for all the possible combinations of enabled/disabled input subspaces.

3 The Financial Problem: Spanish Banking Crisis of 1977-1985

The Spanish banking crisis of 1977-1985, when no fewer than 58 out of 108 Spanish banks were in crisis involving the 27% of the external resources of the Spanish banking system, has been compared to the crash of 1929 in the USA. The data is taken from the *Anuario Estadístico de la Banca Privada* (Statistical Yearbook of the Private Banks), published in Spain. The data contains information from 66 banks (29 of them bankrupt). In the case of bankrupt banks, we use the data from one year before their crisis, whilst for solvent banks we used data from 1982. The data was published in [25] and has not been included for the sake of brevity.

The Table 1 shows the ratios that were also used in [18][14][15]. Every input component is normalized to mean zero and variance 1, which keeps the values in a similar range. The ratios R1, R2 and R3 have meaning of liquidity. The ratios R5, R6 and R7 have meaning of profitability. The ratios R4, R8 and R9 have different concepts.

Table 1. Ratios used in the bankrupt classification

Symbol	Ratio
R1	Current Assets/Total Assets
R2	(Current Assets-Cash and Banks)/Total Assets
R3	Current Assets/Loans
R4	Reserves/Loans
R5	Net Income/Total Assets
R6	Net Income/Total Equity Capital
R7	Net Income/Loans
R8	Cost of Sales/Sales
R9	Cash-Flow/Loans

As we seek for a separation of input variables based on their meaning, this three groups form a good election: the liquidity, the profitability and the Cost-Cash-Reserves (CCR) subspaces. The supervisory data consist in two inputs implementing a complementary two-class decision: the label of solvency and the label of bankrupt taking values 0/1 (while the supervisory output for the MLP takes values –1/1). Each one of these subspaces is introduced in a separated input layer in the networks (see Fig. 1 and 2), while the MLP networks receive the 9 input data in the same input layer.

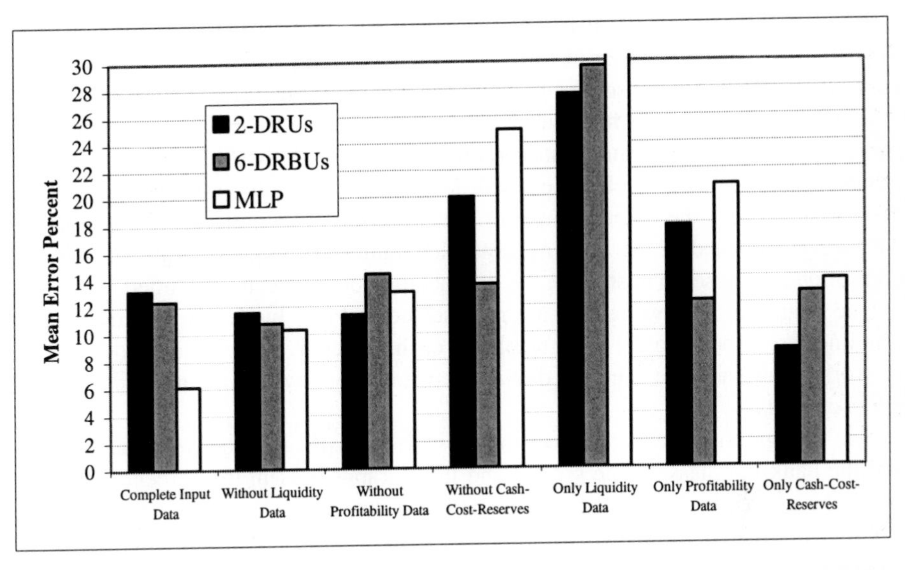

Fig. 3. Bar graph with the mean error percentages committed by the two layers of DRBUs, with 2 and 6 units, and the MLP. The error of the MLP in the combination with only the liquidity inputs is 51%.

4 Classification Results

The data examples were separated in six different pairs of training and test sets, containing 42 samples in the training set and 14 samples in the corresponding test set (6 of bankrupt class and 8 in the solvent class). In the section 4.1, we will comment the classification results of the first network (Fig. 1) also evaluating classes for indetermination and null responses. The second network (Fig.2) is analyzed in the section 4.2 and is compared with the results of the MLP networks.

4.1 Classification with an Associative Network of DRBUs

Six network exemplars were trained (200 cycles) and evaluated with each pair of training and test sets. After the recognition of the class for each associative unit, the test examples were classified by the associative layers to determine its assigned class, and if the sample corresponds to an outlayer (a sample far from the center of its class distribution) or a frontier sample. This analysis can be inferred by defining two thresholds: the indetermination-class threshold and the null-class threshold, as in [8].

The indetermination threshold was set at value 0.2 and detects if two or more units from different classes give output values that are closer than this threshold. This means that the example appears in a frontier between unit regions from different classes. The null-class threshold was set at value 0.1 and detects if the responses of all the units in a layer are closer to zero than this value. This means that all unit kernels are far from this example, so it can be inferred that the sample is an outlayer

These error evaluations have sense if the cost of an incorrect classification is high, for example in a medical diagnostic problem. The samples that belong to the indetermination-class cannot be classified with safe; therefore more clinical tests must be generated to break the indetermination. The detection of an outlayer, if this situation has rarely happened in the learning examples, indicates the possibility of some data corruption or that some mistake in the data acquisition has been committed.

The results of the evaluations are represented in Fig.4 by depicting three graphs. The lines correspond to the error results of each associative layer averaged in the six network exemplars (the standard deviations were very low and are not presented for the sake of clarity).

The first graph presents the error means in the classification between the solvency and the bankrupt classes. The second and third graphs present the percentages of the samples assigned to the indetermination and null classes.

In the graph, the horizontal axis is labeled with all combinations of input subspaces processed by the networks. The third position corresponds to the evaluations without disabling any input subspace (except the supervisory ones) or hidden layer.

In the first combination the networks process the inputs from all subspaces, but only enabling the hidden layers with many units. The second combination only enables hidden layers with few units. These combinations show how the associative layers can process the input data through dendritic channels with different levels of specialization: the many-units representations providing particularization and the few-units representations providing a higher level of generalization.

The rest of combinations are obtained by disabling the input subspaces that we do not want the associative layers to process.

Notice in the upper graph that, in general, the higher the number of associative units, the better the classification results. The layers with 6 and 5 units present in almost all the modulator combinations the lowest error rates, whilst the layers with 2 and 3 units have the worst ones. This effect was expected, as increasing the number of associative units reduces the kernels of the units and, consequently, the frontier of the decision region among classes becomes more detailed.

The combination where networks process only the liquidity subspace is the worst situation, with the error rates varying between 35% and 40%. These particular combination rates are not shown to reduce the range of the graphs. It is clear that the liquidity inputs contain poor information about the solvency of the banks.

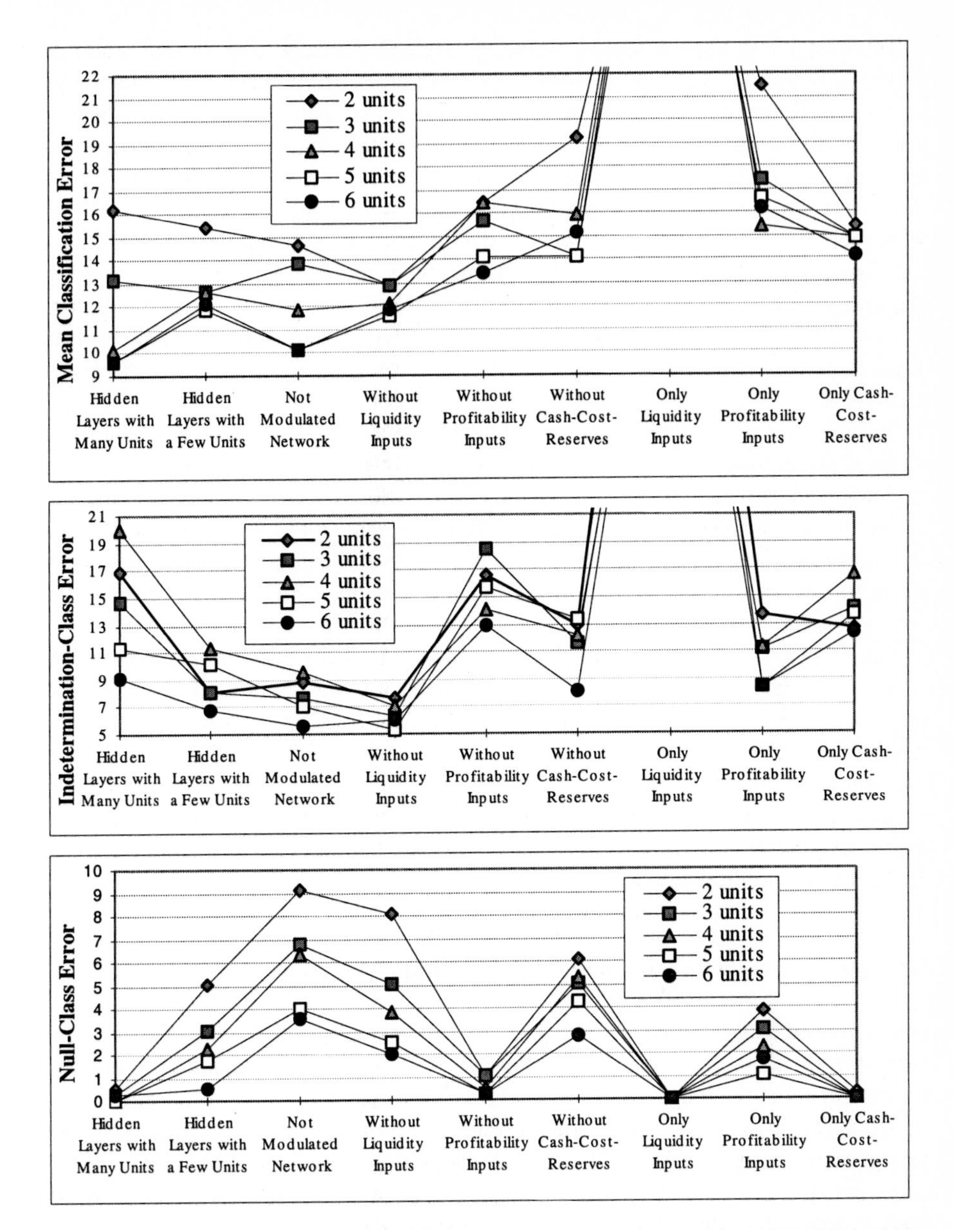

Fig. 4. The graphs represent the mean classification-error percentages (upper graph), the indetermination-class error (centered graph) and the null-class error (lower graph). Each line corresponds to the evaluations in each associative layer of the six averaged network exemplars. The legend denotes the number of units of each associative layer. In the horizontal axis, the modulator combinations that were evaluated, to process the different input subspaces and the alternative hidden channels, are briefly described.

This result was also found in [14] using a Self-Organizing Map. In that work, the profitability and the CCR ratios were good predictors of the bank's solvency. Here, we appreciate that the combination where liquidity inputs are disabled gives a similar result for all the associative layers close to the error value at 12% and is the best of all the combinations with any disabled subspace. The combinations where only profitability or CCR subspaces are enabled have a similar performance.

Respect to the modulation between hidden layers, it results that the layers with 4, 5 and 6 associative units give better discriminations if the dendritic channels from hidden layers with few units are disabled. When these layers only see the channels from the few units their classification errors increase. The combination of both dendritic channels is better for the associative layer with 2 units, whilst the layer with 3 units gets worse.

However, all the layers in the not-modulated network combination get better in if we look at the graph with the indetermination-class errors. The redundancies of the hidden layers provide a better discrimination than those obtained in the combinations with only one enabled hidden layer. The opposite behavior is generated in the null-class errors depicted in the bottom graph. It is clear that the indetermination and null errors follow an inverse relation. When indetermination is high the null error is low and vice versa.

The graphs of indetermination-class and null-class errors provide an insight of the data characteristics of each input subspace comparing the errors with only one subspace enabled. A high null-class error indicates that the class distributions in this subspace are sparse or that the samples appear far from the unit kernels. Therefore increasing the number of units or the corresponding dendritic widths would get better performance. A high indetermination-class error indicates that the data distributions are too close in this subspace and reducing the corresponding dendritic widths would give a better discrimination.

4.2 Classification with Layers of DRBUs and MLP

The results of classification with the second network architecture (Fig.2) have a simpler analysis than those of the previous section, as the indetermination and null classes are not included.

Six network exemplars were trained during 100 cycles and evaluated with each pair of training and test sets. The MLP architecture presents a hidden layer with 6 units and one output unit using the bipolar sigmoid. The MLPs were trained with Back-Propagation Algorithm [22] during 5000 cycles. The responses of the MLP in the combinations where an input subspace is omitted are obtained by annulling the corresponding input variables.

The resulting mean class errors percentages for both networks are shown in the Fig.3 in a bar graph. The first bar (2-DRBUs) represents results of the 2-units layer, the second bar (6-DRBUs) for the 6-units layer and the third bar (MLP) for the multilayer perceptrons. The measured standard deviations were less than 5 percent.

In the graph, horizontal axis is labeled with the different combinations of input subspaces processed by the networks. The first position corresponds to the evaluations with all the input data subspaces enabled (except the supervisory ones). The other combinations are obtained by annulling the corresponding dendritic gains to disable the input subspaces that we do not want the networks to process.

A remarkable result is the coincidence of the three classifiers in the error rates on the different input combinations. The graph also shows a profile of error rates similar to the ones obtained in the networks of the previous section.

In the 2-units layer, the class error average is 13.2% and the performance of the two units gets better if we omit the data from liquidity (11,6%) or profitability subspaces (11,4%). Again, the combination where networks process only the liquidity subspace is the worst situation. The classification with only the data from CCR subspace is the best (8.6%) of this layer.

The 6-units layer also generates its best classification when liquidity inputs are omitted (10.8%) but, even with more units, it doesn't give a better classification than the 2-units layer. However, in the combination where the CCR data is omitted (13,6%) or when only data of profitability is processed (12,2%), the 6-units layer gives lower errors than the 2-units layer and the MLPs. The 6-units layer shows the steadiest performance of the three networks in all the combinations, except when networks process only liquidity ratios (29,6%).

The errors of the MLP are quite similar to those of the 2-units layer, but it is the best when complete input data is processed (6,13%). The MLP also gives good error rates in other combinations, especially in the one where the subspace of liquidity is disabled (10,32%) but, when CCR subspace is omitted, it is the network with the worst results.

The good error rates of the MLP have to do with the fact that the input variables are normalized to mean zero. Since the omission of input variables consists in their substitutions with zeros, this strategy is equivalent to the presentation of the means of the input variables. The performance of the MLP possibly would not be as good using other kind of normalization.

The performance of the DRBUs in this problem is quite good if we notice that a competitive algorithm trains the DRBUs and their regions are the result of combinations of simple spherical dendritic kernels. But if we see their performance in the indetermination-class error (not shown for the sake of brevity) it is clear that the networks of the previous section has a better performance as the responses of their units are more sharp and precise.

5 Conclusions

The situations where data are incomplete are common in problems of high dimensionality, where many inputs components can be absent or corrupted, or in medical diagnosis, where data from clinical tests are generated in a step-by-step procedure. The simulations using a financial classification problem demonstrate that the multi-dendritic description of network architectures permits the a priori design of processing structures implementing the division of input data components in separated groups of similar information content. The architecture modulates easily the communication channels in its internal structure providing multiple combinations of input evaluations even with incomplete or absent data.

Appendix: Functionality of Dendritic Radial Basis Units

The functionality of Radial Basis Units is extended to deal with separated dendritic kernels. Each dendritic kernel is defined with a dendritic centroid vector that localizes its center in the dendritic subspace and the dendritic width that establish the extension of the kernel. Lets consider four processing levels in the mathematical description of network magnitudes: layers, units, dendrites and synapses, specified by numeric labeling. Labels for the layer and dendrites will appear as super-indices, while labels for units and synapses will appear as sub-indices. For example, a weight represented as:

$$w_{ij}^{(md}(t) \quad (1)$$

is associated to the synapse 'j' (with the presynaptic unit 'j') included in the dendrite 'd' of the postsynaptic unit 'i' belonging to the postdendritic layer 'm'.

The recall phase of a certain unit 'i' in the layer 'm' begins by calculating the square of Euclidean distance between the dendritic input vector and the dendritic centroid in all of its dendrites:

$$D_i^{(md}(t) \quad \sum_{j\ 1}^{N^{(d}} (x_j^{(d}(t) \quad w_{ij}^{(md}(t))^2 \quad \left| \boldsymbol{x}^{(d}(t) \quad \boldsymbol{w}_i^{(md}(t) \right|^2; \quad (2)$$

The resulting dendritic distance is compared with the corresponding dendritic-width to obtain the dendritic excitation with the expression:

$$\mathrm{E_D}_i^{(md}(t) = \frac{D_i^{(md}(t)}{\sigma\mathrm{D}_i^{(md}(t)} - 1 \ ; \text{ resulting } \mathrm{E_D}_i^{(md}(t) \quad [-1, \ ). \quad (3)$$

The dendritic excitation is negative if the input pattern appears in the dendritic kernel and is positive in other cases. Next, we obtain the unit excitation as the weighted sum by certain dendritic parameters called dendritic-gains that are assigned to the dendritic fascicles of the layers:

$$\mathrm{E_U}_i^{(m}(t) = \sum_{d=1}^{M} g^{(md} \ \mathrm{E_D}_i^{(md}(t) = \boldsymbol{g}^{(m} \ \boldsymbol{E_D}_i^{(m}(t); \quad (4)$$

$$\text{Resulting} \quad \mathrm{E_U}_i^{(m}(t) \quad [-G^{(m}, \ ). \quad (5)$$

Finally, the unit-excitation is passed by the output function to obtain the unit response. Many functions can be used as output functions and some authors [6][21] demonstrate that this choice has minor importance for the processing executed by the first stage of the RBF model. The main characteristic of the output function is that it generates an elevated response in the kernel.

We define the Multi-Dendritic Sigmoid (MD-Sigmoid) like an extension of the sigmoid capable to deal with multi-dendritic operations:

$$o_i^{(m}(t) \quad \frac{1 \quad \exp \quad G_T^{(m}}{1 \quad \exp \; Eu_i^{(m}(t)}; \tag{6}$$

$$\text{resulting} \quad o_i^{(m}(t) \quad 0{,}1 \qquad i=1, ..., N^{(m}; \quad m=1, ..., M; \tag{7}$$

The numerator expression is a normalizing factor. The quantity $G_T^{(m}$ is the total gain, obtained as the sum of dendritic-gains assigned to all fascicles of the layer.

The dendritic-gains are not represented with time dependency as they are used as modulator parameters to enable or disable their corresponding dendritic excitations. The gains also vary the slope of the sigmoid, so modifying the sharpness of the unit response in the frontiers of the kernel from linear response with low gain to hard limited with high gain. This effect is used in the Multi-Layer Perceptron by [10][19] and is used to vary the kernel widths of RBU in [7].

A.1 Unit Functionality in Learning Phase

Many authors deal with RBU learning, mainly in the RBF model. There are K-clustering methods [16], strategies derived from gradient descent methods [20], Genetic Algorithms [27] and competitive dynamics [7].

The chosen method in this study is a K-Winner Takes All algorithm with one unit as the winner of the competition in each layer.

The winner unit receives a value one in its internal activation while the rest of units receive zeros. In the next step, the units accumulate their activations in an internal variable called activation level that must be set at 1 in the beginning of the training process. These activation levels are used as conscience parameters to avoid the under-utilization problem [5]. This effect is obtained by calculating the competitive output as the unit response divided by its activation level [1].

A.2 Adjustment of Centroids and Widths

The centroids and widths of the winner unit are adjusted in each dendrite separately with its particular dendritic learning factor, whose expression is explained in the next subsection. Each component of the centroid has an initial value assigned at random (close to 0.5), and is updated following the equations:

$$w_{ij}^{(md}(t) = act_i^{(m}(t)\ \ \eta_i^{(md}(t)\ \ o_j^{(d}(t)\ \ w_{ij}^{(md}(t-1)\ ; \tag{8}$$

$$w_{ij}^{(md}(t) = w_{ij}^{(md}(t-1) + \ \ w_{ij}^{(md}(t); \tag{9}$$

The dendritic widths are initialized with high values, so the initial kernels of the units can cover the whole input space. During the training process, each dendritic width decreases to cover only the dendritic input region where the captured samples have appeared. The dendritic-width updating has a formulation like the one used for centroid components:

$$\sigma D_i^{(md}(t) = act_i^{(m}(t)\ \ \eta_i^{(md}(t)\ \ D_i^{(md}(t)\ \ \ \sigma D_i^{(md}(t-1)\ ; \tag{20}$$

$$\sigma D_i^{(md}(t) = \sigma D_i^{(md}(t-1) + \ \ \sigma D_i^{(md}(t); \tag{31}$$

This equation does not limit the value of the widths that should not reach a null value, so a certain limitation is needed. The updated width is compared with two limits: the minimum dendritic width, $\sigma D_{min}^{(md}$, and the maximum dendritic-width, $\sigma D_{max}^{(md}$. If the adjusted dendritic width exceeds any of both, it is equaled to the nearest limit value.

A.3 Dendritic Learning Factors

As dendritic data are independent in their mathematical values the adaptation of each dendritic fascicle needs independent factorization. Each dendritic learning factor depends on the corresponding dendritic width value, as the dendritic width is wider, the factor becomes higher, so facilitating the plasticity of the unit in the dendrite, but if the width is the minimum allowed, then the factor becomes small and stability is provided:

$$\eta_i^{(md}(t) = \eta^{(md}\ \ \frac{\sigma D_i^{(md}(t)}{\sigma D_{max}^{(md}}; \tag{42}$$

The constant value $\eta^{(md}$ is the external dendritic learning factor imposed on the dendritic fascicle 'd'; the same for all units in layer 'm'.

The external dendritic learning factor values were between 0.1 and 0.8 in the hidden layers, without resulting significant differences in the generated representations, except in training velocity, faster when the factor is higher. The layers in higher processing stages receive smaller factor values to ensure the stability of the multi-layer structure (Fukushima, 1980).

References

1. Ahalt, S.C., Krishnamurthy, A.K., Chen, P., & Melton, D.E. (1990) Competitive Learning Algorithms for Vector Quantization. Neural Networks. Vol. 3, pp277-290.
2. Beaver,W (1966). Financial Ratios as Predictors of failure. J. Account Research, Suppl. Empirical Research in Accounting: Selected Studies, V, pp71-111.
3. Bellman, R. (1961) Adaptive Control Processes: a Guided Tour. Princeton University press.
4. Buldain, J.D. & Roy, A. (1999) Association with Multi-Dendritic Radial Basis Units. Proc. Of the Int. Work-Conference on Artificial Neural Networks,Vol. I, pp573-581.
5. DeSieno, D. (1988) Adding a conscience to competitive learning. Proc. Int. Conf. on Neural Networks. pp117-124. IEEE Press, New York.
6. Epanechnik, V.A. (1969) Nonparametric Estimation of a Multidimensional Probability Density. Theory of Probability Application. Vol. 14, pp153-158.
7. Firenze, F. & Morasso, P. (1994) The "Capture Effect " a new self-organizing network for adaptive resolution clustering in changing environments. Proc. Int. Conference on Neural Networks. pp653-658.
8. Guyon, I. (1991) Applications of Neural Networks to Character Recognition. Int. J. Pattern recognition Artificial Intell. vol.5, pp 353-382.
9. Hartvigsen, G (1992) Limitations of Knowledge-Based Systems for Financial Análisis in Banking. Expert Systems Application, vol 4, pp19-32.
10. Kruschke, J.K. & Movellan, J.R. (1991) Benefits of Gain: Speeded Learning and Minimal Hidden Layers in Back-Propagation Networks. IEEE Trans. on Systems, Man and Cybernetics. Vol. 21, n°1, pp273-280.
11. Laffarga, J & Martín, JL & Vázquez, MJ (1985) El análisis de la solvencia en las instituciones bancarias: propuesta de una metodología y aplicaciones de la banca española. Esic-Market, 48, April-June.
12. Laffarga, J & Martín, JL & Vázquez, MJ (1986) El pronóstico a corto plazo del fracaso en las instituciones bancarias: metodología y aplicaciones al caso español. Esic-Market, 53, July-September.
13. Li, S., Chen, Y. & Leiss, E.L. (1994) GMDP: a Novel Unified Neuron for Multilayer Feed-forward Neural Networks. IEEE Int.Conf. on NN' 94. pp107-112.
14. Martín-del-Brío, B. & Serrano-Cinca, C. (1993) Self-organizing Neural Networks for the Análisis and Representation of Data: Some Financial Cases. Neural Computing & Applications, Vol.1, pp193-206.
15. Martín-del-Brío, B. & Serrano-Cinca, C. (1995) Self-organizing neural networks for analysis and representation of data: The financial state of Spanish companies. Neural Networks in the Capital Markets. A.N. Refenes (Ed.) John Wiley and Sons, Cap. 23, pp 341-357
16*14. Moody, J. & Darken, C. (1989) Fast learning in networks of locally-tuned processing units. Neural Computation. Vol.1 n°2 pp281-294.
17*91. Odom,M.D. & Sharda, R. (1990) A Neural Network Model for Bankrupcy Prediction. Proc. Third Int. Conf. On Neural networks (San diego, CA), vol. II, pp163-168.
18*23. Pina, V (1989) Estudio empírico de la crisis bancaria. Rev. Española Financ. Y Contab., XVIII (58).
19*93. Plant, D.C., Nowlan, S.J. & Hinton, G.E. (1986) Experiments of Learning by Back-Propagation. Tech. Report Carnegie Mellon University-Computer Science, June86.
20*15. Poggio, T. & Girosi, F. (1990) Networks for approximation and learning. Proceedings of the IEEE, Special Issue on Neural Networks vol.I, Sept.90, pp1481-1497.

21*16. Powell, M.J.D. (1987) Radial Basis Function for Multivariable Interpolation: a Review. Algorithms for Approximation. J.C.Mason and M.G.Cox, Eds, Oxford, UK: Clarendon, pp143-167

22*17. Rumelhart, D.E. & McClelland, J.L. (1986) Parallel Distributed Processing. Vol. 1: Foundations. MIT Press.

23*60. Samad, T. & Harp, S.A. (1992) Self organization with partial data. Network 3, pp205-212.

24*80. Sanger, T.D. (1992) A tree-structured adaptive network for function approximation in high-dimensional spaces. IEEE Trans. on NN vol 2, nº4, pp624-627.

25*9. Serrano-Cinca, C. & Martín-del-Brío, B. (1993) Predicción de la quiebra bancaria mediante el empleo de Redes Neuronales Artificiales. Rev. Española Financ. y Contab. 74, pp153-176.

26*40. Wan, E.A. (1990) Neural Network Clasification: a Bayesian interpretation. IEEE Trans.on NN vol.1, pp303-305.

27*19. Whitehead, B.A. & Choate, T.D. (1996) Cooperative-Competitive Genetic Evolution of Radial Basis Function Centers and Widths for Time Series Prediction. IEEE Trans. on Neural Networks. Vol. 7, nº 4, pp869-881.

Neuro Fuzzy Systems: State-of-the-Art Modeling Techniques

Ajith Abraham

School of Computing & Information Technology
Monash University, Churchill 3842, Australia
http://ajith.softcomputing.net
Email: ajith.abraham@infotech.monash.edu.au

Abstract: Fusion of Artificial Neural Networks (ANN) and Fuzzy Inference Systems (FIS) have attracted the growing interest of researchers in various scientific and engineering areas due to the growing need of adaptive intelligent systems to solve the real world problems ANN learns from scratch by adjusting the interconnections between layers. FIS is a popular computing framework based on the concept of fuzzy set theory, fuzzy if-then rules, and fuzzy reasoning. The advantages of a combination of ANN and FIS are obvious. There are several approaches to integrate ANN and FIS and very often it depends on the application. We broadly classify the integration of ANN and FIS into three categories namely concurrent model, cooperative model and fully fused model. This paper starts with a discussion of the features of each model and generalize the advantages and deficiencies of each model. We further focus the review on the different types of fused neuro-fuzzy systems and citing the advantages and disadvantages of each model.

1. Introduction

Neuro Fuzzy (NF) computing is a popular framework for solving complex problems. If we have knowledge expressed in linguistic rules, we can build a FIS, and if we have data, or can learn from a simulation (training) then we can use ANNs. For building a FIS, we have to specify the fuzzy sets, fuzzy operators and the knowledge base. Similarly for constructing an ANN for an application the user needs to specify the architecture and learning algorithm. An analysis reveals that the drawbacks pertaining to these approaches seem complementary and therefore it is natural to consider building an integrated system combining the concepts. While the learning capability is an advantage from the viewpoint of FIS, the formation of linguistic rule base will be advantage from the viewpoint of ANN. In section 2 we present cooperative NF system and concurrent NF system followed by the different fused NF models in section 3. Some discussions and conclusions are provided towards the end.

2. Cooperative and Concurrent Neuro-Fuzzy Systems

In the simplest way, a cooperative model can be considered as a preprocessor wherein ANN learning mechanism determines the FIS membership functions or fuzzy rules from the training data. Once the FIS parameters are determined, ANN goes to the

J. Mira and A. Prieto (Eds.): IWANN 2001, LNCS 2084, pp. 269-276, 2001.

background. The rule based is usually determined by a clustering approach (self organizing maps) or fuzzy clustering algorithms. Membership functions are usually approximated by neural network from the training data.

In a concurrent model, ANN assists the FIS continuously to determine the required parameters especially if the input variables of the controller cannot be measured directly. In some cases the FIS outputs might not be directly applicable to the process. In that case ANN can act as a postproces sor of FIS outputs. Figures 1 and 2 depict the cooperative and concurrent NF models.

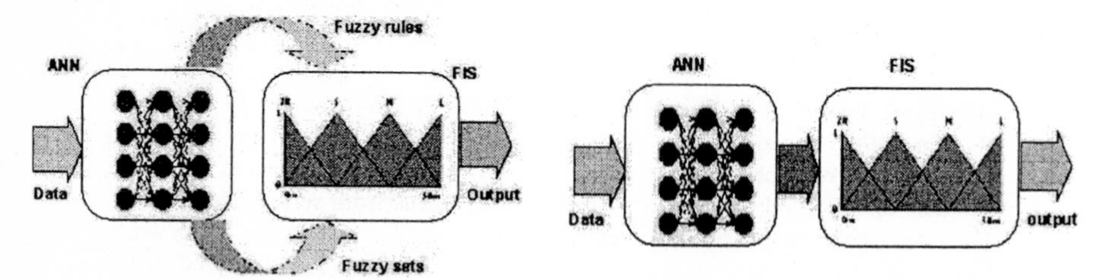

Figure 1. Cooperative NF model

Figure 2. Concurrent NF model

3. Fused Neuro Fuzzy Systems

In a fused NF architecture, ANN learning algorithms are used to determine the parameters of FIS. Fused NF systems share data structures and knowledge representations. A common way to apply a learning algorithm to a fuzzy system is to represent it in a special ANN like architecture. However the conventional ANN learning algorithms (gradient descent) cannot be applied directly to such a system as the functions used in the inference process are usually non differentiable. This problem can be tackled by using differentiable functions in the inference system or by not using the standard neural learning algorithm. Some of the major woks in this area are GARIC [9], FALCON [8], ANFIS [1], NEFCON [7], FUN [3], SONFIN [2], FINEST [4], EFuNN [5], dmEFuNN[5], evolutionary design of neuro fuzzy systems [10], and many others.

- **Fuzzy Adaptive learning Control Network (FALCON)**

FALCON [8] has a five-layered architecture as shown in Figure 3. There are two linguistic nodes for each output variable. One is for training data (desired output) and the other is for the actual output of FALCON. The first hidden layer is responsible for the fuzzification of each input variable. Each node can be a single node representing a simple membership function (MF) or composed of multilayer nodes that compute a complex MF. The Second hidden layer defines the preconditions of the rule followed by rule consequents in the third hidden layer. FALCON uses a hybrid-learning algorithm comprising of unsupervised learning to locate initial membership functions/ rule base and a gradient descent learning to optimally adjust the parameters of the MF to produce the desired outputs.

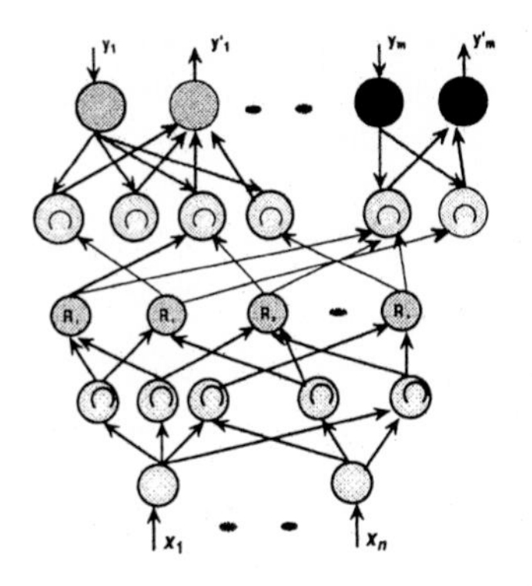

Figure 3. Architecture of FALCON

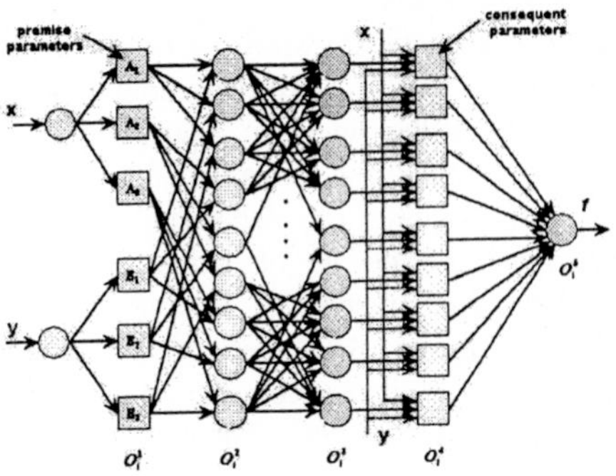

Figure 4. Structure of ANFIS

- **Adaptive Neuro Fuzzy Inference System (ANFIS)**

ANFIS [1] implements a Takagi Sugeno FIS and has a five layered architecture as shown in Figure 2. The first hidden layer is for fuzzification of the input variables and T-norm operators are deployed in the second hidden layer to compute the rule antecedent part. The third hidden layer normalizes the rule strengths followed by the fourth hidden layer where the consequent parameters of the rule are determined. Output layer computes the overall input as the summation of all incoming signals. ANFIS uses backpropagation learning to determine premise parameters (to learn the parameters related to membership functions) and least mean square estimation to determine the consequent parameters. A step in the learning procedure has got two parts: In the first part the input patterns are propagated, and the optimal consequent parameters are estimated by an iterative least mean square procedure, while the premise parameters are assumed to be fixed for the current cycle through the training set. In the second part the patterns are propagated again, and in this epoch, backpropagation is used to modify the premise parameters, while the consequent parameters remain fixed. This procedure is then iterated.

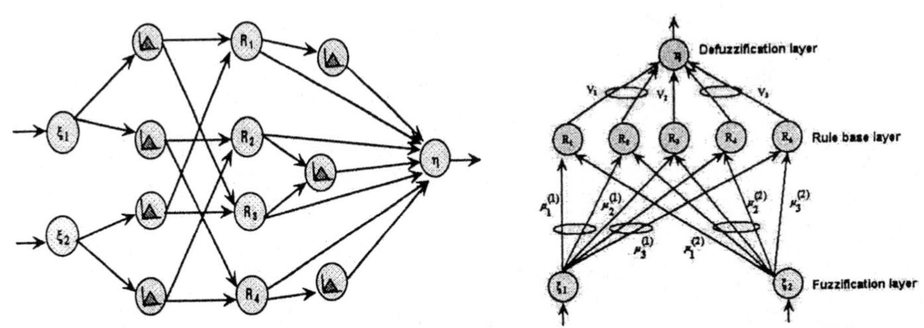

Figure 5. ASN of GARIC

Figure 6. Architecture of NEFCON

- **Generalized Approximate Reasoning based Intelligent Control (GARIC)**

GARIC [9] implements a neuro-fuzzy controller by using two neural network modules, the ASN (Action Selection Network) and the AEN (Action State Evaluation

Network). The AEN is an adaptive critic that evaluates the actions of the ASN. ASN of GARIC is feedforward network with five layers. Figure 5 illustrates the structure of GARIC – ASN. The connections between layers are not weighted. The first hidden layer stores the linguistic values of all the input variables. Each input unit is only connected to those units of the first hidden layer, which represent its associated linguistic values. The second hidden layer represents the fuzzy rules nodes, which determine the degree of fulfillment of a rule using a *softmin* operation. The third hidden layer represents the linguistic values of the control output variable ?. Conclusions of the rule are computed depending on the strength of the rule antecedents computed by the rule node layer. GARIC makes use of local mean-of-maximum method for computing the rule outputs. This method needs a crisp output value from each rule. Therefore the conclusions must be defuzzified before they are accumulated to the final output value of the controller. GARIC uses a mixture of gradient descent and reinforcement learning to fine-tune the node parameters.

- **Neuro-Fuzzy Control (NEFCON)**

NEFCON [7] is designed to implement Mamdani type FIS and is illustrated in Figure 6. Connections in NEFCON are weighted with fuzzy sets and rules (μ_r, $?_r$ are the fuzzy sets describing the antecedents and consequents) with the same antecedent use so-called shared weights, which are represented by ellipses drawn around the connections. They ensure the integrity of the rule base. The input units assume the task of fuzzification interface, the inference logic is represented by the propagation functions, and the output unit is the defuzzification interface. The learning process of the NEFCON model is based on a mixture of reinforcement and backpropagation learning. NEFCON can be used to learn an initial rule base, if no prior knowledge about the system is available or even to optimize a manually defined rule base. NEFCON has two variants: NEFPROX (for function approximation) and NEFCLASS (for classification tasks) [7].

- **Fuzzy Inference and Neural Network in Fuzzy Inference Software (FINEST)**

FINEST [4] is capable of two kinds of tuning process, the tuning of fuzzy predicates, combination functions and the tuning of an implication function. The generalized modus ponens is improved in the following four ways (1) Aggregation operators that have synergy and cancellation nature (2) A parameterized implication function (3) A combination function that can reduce fuzziness (4) Backward chaining based on generalized modus ponens. FINEST make use of a backpropagation algorithm for the fine-tuning of the parameters. Figure 7 shows the layered architecture of FINEST and the calculation process of the fuzzy inference. FINEST provides a framework to tune any parameter, which appears in the nodes of the network representing the calculation process of the fuzzy data if the derivative function with respect to the parameters is given.

- **FUzzy Net (FUN)**

In FUN [3], the neurons in the first hidden layer contain the membership functions and this performs a fuzzification of the input values. In the second hidden layer, the conjunctions (fuzzy-*AND*) are calculated. Membership functions of the output variables are stored in the third hidden layer. Their activation function is a fuzzy-*OR*. Finally the output neuron performs the defuzzification. The network is initialized with a fuzzy rule base and the corresponding membership functions and there after uses a stochastic learning technique that randomly changes parameters of membership functions and connections within the network structure. The learning process is driven by a cost function, which is evaluated after the random modification. If the modification resulted in an improved performance the modification is kept, otherwise it is undone.

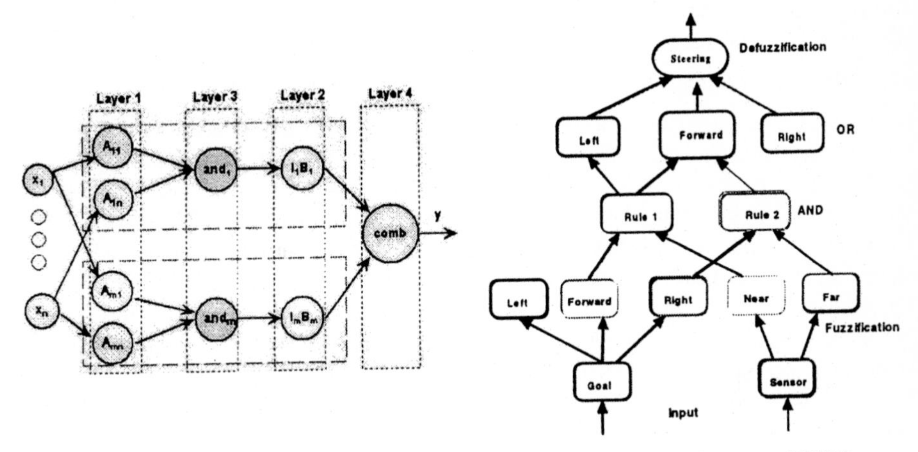

Figure 7. Architecture of FINEST

Figure 8. Architecture of FUN

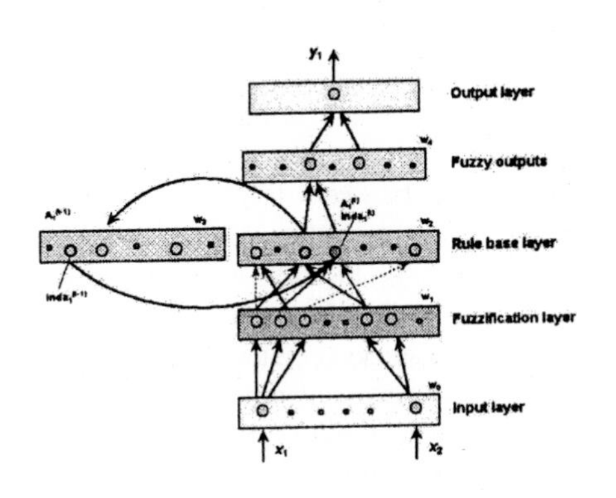

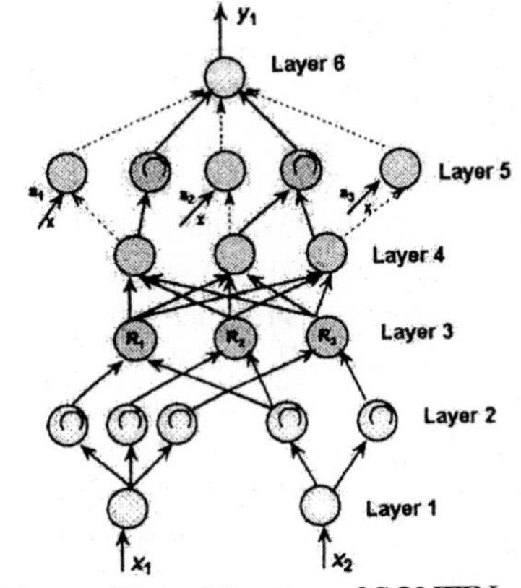

Figure 9. Architecture of EFuNN

Figure 10. Architecture of SONFIN

- **Evolving Fuzzy Neural Network (EFuNN)**

In EFuNN [5] all nodes are created during learning. The input layer passes the data to the second layer, which calculates the fuzzy membership degrees to which the input values belong to predefined fuzzy membership functions. The third layer contains fuzzy rule nodes representing prototypes of input-output data as an association of hyper-spheres from the fuzzy input and fuzzy output spaces. Each rule node is defined by 2 vectors of connection weights, which are adjusted through the hybrid learning technique. The fourth layer calculates the degrees to which output membership functions are matched by the input data, and the fifth layer does defuzzification and calculates exact values for the output variables. Dynamic Evolving Fuzzy Neural Network (dmEFuNN) [5] is a modified version of EFuNN with the idea that not just the winning rule node's activation is propagated but a group of rule nodes is dynamically selected for every new input vector and their activation values are used to calculate the dynamical parameters of the output function. While EFuNN implements fuzzy rules of Mamdani type, dmEFuNN estimates the Takagi-Sugeno fuzzy rules based on a least squares algorithm.

- **Self Constructing Neural Fuzzy Inference Network (SONFIN)**

SONFIN [2] implements a modified Takagi-Sugeno FIS and is illustrated in Figure 10. In the structure identification of the precondition part, the input space is partitioned in a flexible way according to an aligned clustering based algorithm. As to the structure identification of the consequent part, only a singleton value selected by a clustering method is assigned to each rule initially. Afterwards, some additional significant terms (input variables) selected via a projection-based correlation measure for each rule are added to the consequent part (forming a linear equation of input variables) incrementally as learning proceeds. For parameter identification, the consequent parameters are tuned optimally by either least mean squares or recursive least squares algorithms and the precondition parameters are tuned by backpropagation algorithm.

- **Evolutionary Design of Neuro-Fuzzy Systems**

In the evolutionary design of NF systems [10], the node functions, architecture and learning parameters are adapted according to a five-tier hierarchical evolutionary search procedure as shown in Figure 11(b). The evolving NF model can adapt to Mamdani or Takagi Sugeno type FIS. Only the layers are defined in the basic architecture as shown in Figure 11(a). The evolutionary search process will decide the optimal type and quantity of nodes and connections between layers. Fuzzification layer and the rule antecedent layer functions similarly to other NF models. The consequent part of rule will be determined according to the inference system depending on the problem type, which will be adapted accordingly by the evolutionary search mechanism. Defuzzification/ aggregation operators will also be adapted according to the FIS chosen by the evolutionary algorithm. Figure 11(b) illustrates the computational framework and interaction of various evolutionary search procedures. For every learning parameter, there is the global search of inference mechanisms that proceeds on a faster time scale in an environment decided by the inference system and the problem. For every inference mechanism there is the global search of fuzzy rules (architecture) that proceeds on a faster time scale in an environment decided by the learning parameters, inference system and the problem. Similarly, for every architecture, evolution of membership function parameters proceeds at a faster time scale in an environment decided by the architecture, inference mechanism, learning rule, type of inference system and the problem. Hierarchy of the different adaptation procedures will rely on the prior knowledge. For example, if there is more prior knowledge about the architecture than the inference mechanism then it is better to implementthe architecture at a higher level.

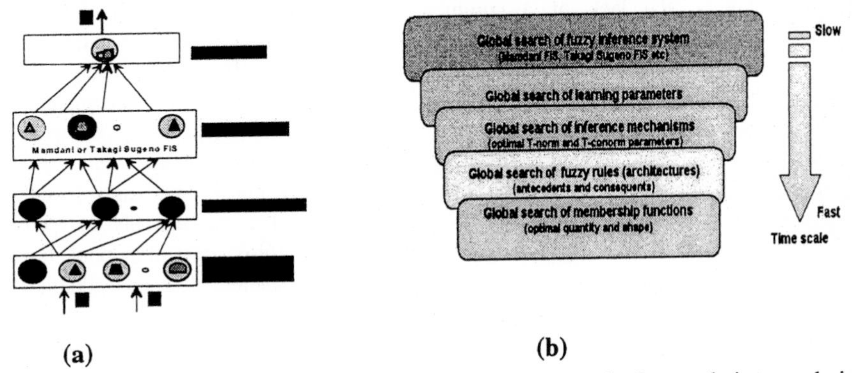

(a) **(b)**

Figure 11(a). Architecture and **(b)** computational framework for evolutionary design of neuro fuzzy systems

4. Discussions

As evident, both cooperative and concurrent models are not fully interpretable due to the presence of ANN (black box concept). Whereas a fused NF model is interpretable and capable of learning in a supervised mode. In FALCON, GARIC, ANFIS, NEFCON, SONFIN, FINEST and FUN the learning process is only concerned with parameter level adaptation within fixed structures. For large-scale problems, it will be too complicated to determine the optimal premise-consequent structures, rule numbers etc. User has to provide the architecture details (type and quantity of MF's for input and output variables), type of fuzzy operators etc. FINEST provides a mechanism based on the improved generalized modus ponens for fine tuning of fuzzy predicates & combination functions and tuning of an implication function. An important feature of EFuNN and dmEFuNN is the one pass (epoch) training, which is highly capable for online learning. Since FUN system uses a stochastic learning procedure, it is questionable to call FUN a NFy system. As the problem become more complicated manual definition of NF architecture/parameters becomes complicated. Especially for tasks requiring an optimal NF system, evolutionary design approach might be the best solution. Table 1 provides a comparative performance [11] of some neuro fuzzy systems for predicting the Mackey-Glass chaotic time series [6]. Training was done using 500 data sets and NF models were tested with another 500 data sets.

Table 1. Performance of NF systems and ANN

System	Epochs	RMSE
ANFIS	75	0.0017
NEFPROX	216	0.0332
EFuNN	1	0.0140
dmEFuNN	1	0.0042
SONFIN	-	0.0180

5. Conclusions

In this paper we have presented the state of art modeling of different neuro-fuzzy systems. Due to the lack of a common framework it remains often difficult to compare the different neuro-fuzzy models conceptually and evaluate their performance comparatively. In terms of RMSE error, NF models using Takagi Sugeno FIS performs better than Mamdani FIS even though it is computational expensive. As a guideline, for NF systems to be highly intelligent some of the major requirements are fast learning (memory based - efficient storage and retrieval capacities), on-line adaptability (accommodating new features like inputs, outputs, nodes, connections etc), achieve a global error rate and computationally inexpensive. The data acquisition and preprocessing training data is also quite important for the success of neuro-fuzzy systems. All the NF models use gradient descent techniques to learn the membership function parameters. For faster learning and convergence, it will be interesting to explore other efficient neural network learning algorithms (e.g. conjugate gradient search) instead of backpropagation.

References

[1] Jang R, *Neuro-Fuzzy Modeling: Architectures, Analyses and Applications*, PhD Thesis, University of California, Berkeley, July 1992.

[2] Juang Chia Feng, Lin Chin Teng, *An Online Self Constructing Neural Fuzzy Inference Network and its Applications*, IEEE Transactions on Fuzzy Systems, Vol 6, No.1, pp. 12-32, 1998.

[3] Sulzberger SM, Tschicholg-Gurman NN, Vestli SJ, *FUN: Optimization of Fuzzy Rule Based Systems Using Neural Networks*, In Proceedings of IEEE Conference on Neural Networks, San Francisco, pp 312-316, March 1993.

[4] Tano S, Oyama T, Arnould T, *Deep combination of Fuzzy Inference and Neural Network in Fuzzy Inference*, Fuzzy Sets and Systems, 82(2) pp. 151-160, 1996.

[5] Kasabov N and Qun Song, *Dynamic Evolving Fuzzy Neural Networks with 'm-out-of-n' Activation Nodes for On-line Adaptive Systems*, Technical Report TR99/04, Department of information science, University of Otago, 1999.

[6] Mackey MC, Glass L, *Oscillation and Chaos in Physiological Control Systems*, Science Vol 197, pp.287-289, 1977.

[7] Nauck D, Kruse R, *Neuro-Fuzzy Systems for Function Approximation,* 4th International Workshop Fuzzy-Neuro Sy stems, 1997.

[8] Lin C T & Lee C S G, Neural Network based Fuzzy Logic Control and Decision System, IEEE Transactions on Comput. (40(12): pp. 1320-1336, 1991.

[9] Bherenji H R and Khedkar P, Learning and Tuning Fuzzy Logic Controllers through Reinforcements, IEEE Transactions on Neural Networks, Vol (3), pp. 724-740, 1992.

[10] Abraham A & Nath B, Evolutionary Design of Neuro-Fuzzy Systems – A Generic Framework, In Proceedings of the 4th Japan – Australia joint Workshop on Intelligent and Evolutionary Systems, Japan, November 2000.

[11] Abraham A & Nath B, Designing Optimal Neuro-Fuzzy Systems for Intelligent Control, In proceedings of the Sixth International Conference on Control Automation Robotics Computer Vision, (ICARCV 2000), Singapore, December 2000.

Generating Linear Regression Rules from Neural Networks Using Local Least Squares Approximation

Rudy Setiono

School of Computing
National University of Singapore
Singapore 117543

Abstract. Neural networks are often selected as the tool for solving regression problem because of their capability to approximate any continuous function with arbitrary accuracy. A major drawback of neural networks is their complex mapping which is not easily understood by a user. This paper describes a method that generates decision rules from trained neural networks for regression problems. The networks have a single layer of hidden units with hyperbolic tangent activation function and a single output unit with linear activation function. The crucial step in this method is the approximation of the hidden unit activation function by a 3-piece linear function. This linear function is obtained by minimizing the sum of squared deviations between the hidden unit activation values of data samples and their linearized approximations. Once the activation function of the hidden units have have been linearized, the rules are generated. The conditions of the rules divide the input space of the data into subspaces, while the consequence of each rule is a linear regression function. Our experimental results indicate that the method generates more accurate rules than those from similar methods.

1 Introduction

While neural networks are widely used as the tool for solving regression problems, they are not without drawbacks. One of the more often mentioned disadvantages of neural networks is their complex input-to-output mapping which is difficult for a human user to comprehend. For some applications, linear functions that relate the inputs and outputs of the data are more desirable. Compared to the nonlinear mapping of the networks, linear functions are much more comprehensible. The coefficients of a linear regression function indicate the expected change in the output for a unit change in the corresponding inputs.

In the machine learning community, many decision tree generating algorithms have been developed for classification. Among the most well-known of these algorithms is C4.5 of Quinlan [9]. In order to apply this robust classification algorithm to regression problems, some researchers proposed methods which transform a regression task into a classification problem by discretizing the continuous target variable. The RECLA system [12] allows the user to choose from one of the three methods for dividing the interval of the target variable into subintervals:

J. Mira and A. Prieto (Eds.): IWANN 2001, LNCS 2084, pp. 277-284, 2001.

(1) equally probable intervals, where each subinterval contains the same number of data samples; (2) equal width interval; and (3) k-means clustering, where the sum of the distances of all data samples in an interval to its gravity center is minimized.

A more recent method for discretizing the continuous attributes of a data set is the Relative Unsupervised Discretization (RUDE) algorithm [8]. The algorithm discretizes not only the continuous target variable, but also all continuous input variables. A key component of this algorithm is a clustering algorithm which groups values of the target variables into subintervals that are characterized by more or less similar values of some input attributes. Once the variables have been discretized, C4.5 can be applied for solving the original regression problem. The experimental results on five benchmark problems show that when compared to the trees from the datasets discretized using the equal width interval and the k-mean clustering methods, the decision trees generated using RUDE-discretized data sets have fewer nodes but lower predictive accuracy.

As noted by the developers of RECLA and RUDE, the difficulty in casting a regression problem as a classification problem lies in the splitting of the continuous target variable. Splitting the interval into many fine subintervals leads to smaller deviations of the predictions within each subinterval. However, many subintervals translates into many classes in the resulting classification problem which would degrade the prediction accuracy of the classification tree. The wrapper approach [6] has been used to overcome this difficulty in RECLA. It is an iterative approach that tries varying number of intervals to find one that gives the best estimated accuracy.

This paper describes a method to generate linear regression rules from trained neural networks for regression problems. Unlike RECLA and RUDE, our proposed method does not require discretization of the target variable, and hence, the difficulty associated with its discretization can be avoided.

Neural networks, capable of approximating any continuous function with arbitrary precision [2, 5], would be a natural first choice for solving regression problems. In practice, they have also been shown to be able to predict with good accuracy for this type of problems [10]. The method that we propose here attempts to extract comprehensible explanation in terms of linear functions from neural networks that have been trained for regression The networks have a single layer of hidden units with hyperbolic tangent activation function and a single output unit with linear activation function. The conditions of the rules generated by the proposed method divide the input space of the data into subspaces, while the consequence of each rule is given as a linear regression function. The crucial step in this method is the approximation of the hidden unit activation function by a 3-piece linear function. This linear function is obtained by minimizing the sum of squared deviations between the hidden unit activation value of a data sample and its linearized approximation. Preliminary experimental results indicate that the method generates compact set of rules that are higher in accuracy than other methods that generate decision trees for regression.

The next section describes our neural network training and pruning. In Section 3 we describe how the hidden unit activation function of the network is approximated by a 3-piece linear function such that the sum of the squared errors of the approximation is minimized. In Section 4 we present our method to generate a set of regression rules from a neural network. In Section 5 we present our results and compare them with those from other methods for regression. Finally, in Section 6 we conclude the paper.

2 Network training and pruning

We divide the available data samples $(\mathbf{x}^i, y^i), i = 1, 2, \ldots$, where $\mathbf{x}^i \in \mathrm{I\!R}^N$ and $y^i \in \mathrm{I\!R}$, randomly into the training set, the cross-validation set and the test set. Using the training data set, a network with a single hidden layer consisting of H units is trained so as to minimize the sum of squared errors $E(\mathbf{w}, \mathbf{v})$ augmented with a penalty term $P(\mathbf{w}, \mathbf{v})$:

$$E(\mathbf{w}, \mathbf{v}) = \sum_{i=1}^{K} \left(\tilde{y}^i - y^i\right)^2 + P(\mathbf{w}, \mathbf{v}) \tag{1}$$

$$P(\mathbf{w}, \mathbf{v}) = \epsilon_1 \left(\sum_{m=1}^{H} \sum_{\ell=1}^{N} \frac{\eta w_{m\ell}^2}{1 + \eta w_{m\ell}^2} + \sum_{m=1}^{H} \frac{\eta v_m^2}{1 + \eta v_m^2} \right) + \epsilon_2 \left(\sum_{m=1}^{H} \sum_{\ell=1}^{N} w_{m\ell}^2 + \sum_{m=1}^{H} v_m^2 \right) \tag{2}$$

where K is the number of samples in the training data set, $\epsilon_1, \epsilon_2, \eta$ are positive penalty parameters, and $\tilde{y}^i$ is the predicted function value for input sample $\mathbf{x}^i$

$$\tilde{y}^i = \sum_{m=1}^{H} \tanh\left((\mathbf{x}^i)^T \mathbf{w}_m\right) v_m + \tau, \tag{3}$$

$\mathbf{w}_m \in \mathrm{I\!R}^N$ is the vector of network weights from the input units to hidden unit m, $w_{m\ell}$ is its ℓ-th component, $v_m \in \mathrm{I\!R}$ is the network weight from hidden unit m to the output unit, $\tanh(\xi)$ is the hyperbolic tangent function $(e^{\xi} - e^{-\xi})/(e^{\xi} + e^{-\xi})$, and $(\mathbf{x}^i)^T \mathbf{w}_m$ is the scalar product of $\mathbf{x}^i$ and $\mathbf{w}_m$, and τ is the output unit's bias.

We obtain a local minimum of the error function $E(\mathbf{w}, \mathbf{v})$ by applying the BFGS method [3, 4] due to its faster convergence rate than the gradient descent method. Once the network has been trained, irrelevant and redundant hidden units and input units are removed from the network by applying the algorithm N2PFA (Neural Network Pruning for Function Approximation) [10]. This is an iterative algorithm which checks the accuracy of the network on the cross-validation samples to determine if a unit can be removed. If the removal of an input or a hidden unit does not cause a deterioration in the network's accuracy on this set of samples, then the unit will be removed. The benefit of network pruning is two-fold. First, pruned networks are less likely to overfit the training data samples, and hence they would generalize better than fully connected networks. Second, we can expect to generate more compact sets of rules from the less complex pruned networks.

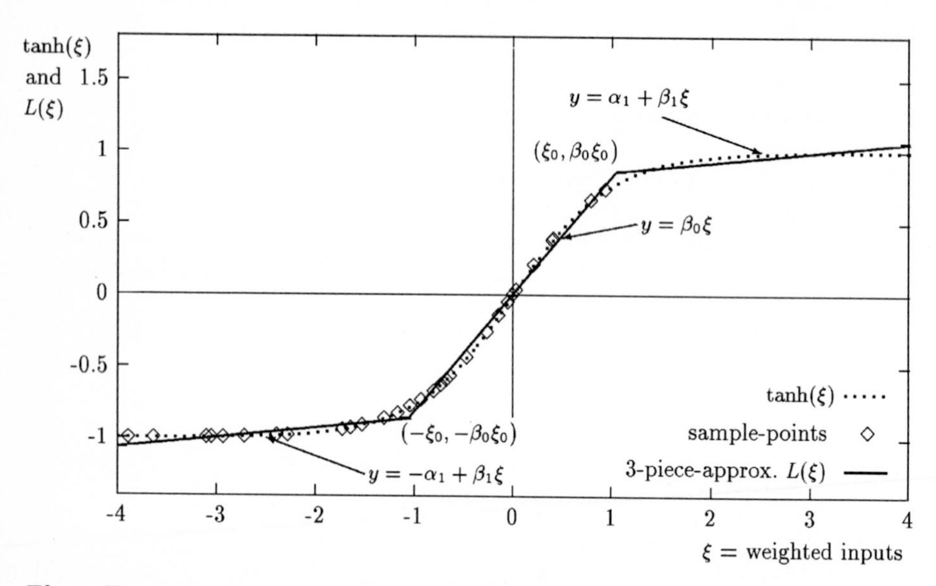

Fig. 1. The 3-piece linear approximation of the hidden unit activation function tanh(x) given 30 training samples (◇).

3 A local least squares approximation of tanh(x)

The method which we develop to approximate the hyperbolic tangent activation function entails finding the cut-off points ξ_0 and $-\xi_0$, the slope and the intersection of each of the three line segments of the piece-wise linear function $L(\xi)$ (see Figure 1). We minimize the sum of the squared deviations to obtain the values of these parameters:

$$\min_{\xi_0,\beta_0,\beta_1} \sum_{i=1}^{K} \left(\tanh(\xi_i) - L(\xi_i)\right)^2 , \tag{4}$$

where $\xi_i = (\mathbf{x}^i)^T \mathbf{w}$, the weighted input of sample $i, i = 1, 2, \ldots K$ and

$$L(\xi) = \begin{cases} -\alpha_1 + \beta_1 \xi & \text{if} \quad \xi < -\xi_0 \\ \beta_0 \xi & \text{if} \quad -\xi_0 \le \xi \le \xi_0 \\ \alpha_1 + \beta_1 \xi & \text{if} \quad \xi > \xi_0 \end{cases} \tag{5}$$

Since $\tanh(0) = 0$, we also require $L(0) = 0$, hence, for the middle line segment we need to determine only its slope β_0. For continuity and symmetry, we require that the first line segment passes through $(-\xi_0, -\beta_0\,\xi_0)$ and the third line segment $(\xi_0, \beta_0\,\xi_0)$. The slopes of the line segments that minimize the total sum of the squared deviations are as follows:

$$\beta_0 = \frac{\sum_{|\xi_i| \leq \xi_0} \xi_i \, \tanh(\xi_i)}{\sum_{|\xi_i| \leq \xi_0} \xi_i^2} \tag{6}$$

$$\beta_1 = \frac{\sum_{|\xi_i| > \xi_0} (\tanh(\xi_i) - \tanh(\xi_0))(\xi_i - \xi_0)}{\sum_{|\xi_i| > \xi_0} (\xi_i - \xi_0)^2} \tag{7}$$

While the intercept α_1 is given by:

$$\alpha_1 = (\beta_0 - \beta_1)\,\xi_0 \tag{8}$$

The weighted inputs of all samples ξ_i are checked as a possible optimal value for ξ_0 starting from the one which has the smallest magnitude.

4 Regression rules generation

A set of linear regression rules can be generated from a pruned network once the network hidden unit activation function $\tanh(\xi)$ has been approximated by 3-piece linear function as described in the previous section. The steps are as follows:

1. For each hidden unit $m = 1, 2, \ldots, H$, generate the 3-piece linear approximation $L_m(\xi)$ (Eg. 5).
2. Using the pair of points $-\xi_{m0}$ and ξ_{m0} from function $L_m(x)$, divide the input space into 3^H subregions.
3. For each non-empty subregion, generate a rule as follows:
   (a) Define a linear equation that approximates the network's output for input sample i in this subregion as the *consequence* of the rule:

$$\hat{y}_i = \sum_{m=1}^{H} v_m L_m(\xi_{mi}) + \tau \tag{9}$$

$$\xi_{mi} = (\mathbf{x}^i)^T \mathbf{w}_m \tag{10}$$

   (b) Generate the rule *condition*: $(\mathcal{C}_1$ and $\mathcal{C}_2$ and $\cdots \mathcal{C}_H)$, where $\mathcal{C}_m$ is either $\xi_{mi} < -\xi_{m0}$, $-\xi_{m0} \leq \xi_{mi} \leq \xi_{m0}$, or $\xi_{mi} > \xi_{m0}$.
4. (Optional step) Apply C4.5 [9] to simplify the rule conditions.

The rule condition (C_1 and C_2 and ... and C_H) defines intersections of half-spaces in the input space. As the boundaries of the region defined by such intersections involve the network weights, C4.5 may be applied to remove these weights from the rule conditions. The result will be a set of rules that can be much easier to understand and analyze.

5 Experimental results

The proposed method has been tested on benchmark approximation problems from various domains. The data sets (Table 1) are downloaded from Luis Torgo's home page `http://www.ncc.up.pt/~ltorgo/Research/`. They are also available from the UCI repository [1].

Table 1. Test data sets. D = discrete attribute, C = continuous attribute

Name	#samples	Attributes	NN inputs	Prediction task
Abalone	4177	1D, 7C	9	age of abalone specimens
Auto-mpg	392	3D, 4C	25	car fuel consumption
Housing	506	1D, 12C	13	median value of homes in Boston suburbs
Machine	209	6C	6	relative CPU performance
Servo	167	4D	19	response time of a servo mechanism

A ten-fold cross validation evaluation was conducted on each data set. The data were randomly divided into ten subsets of equal size. Eight subsets were used for network training, one subset for deciding when network pruning should terminate, and one subset for measuring the predictive accuracy of the pruned network and the rules. This procedure was repeated ten times so that each subset was tested once.

The same experimental settings were used for all problems. The networks started with eight hidden units and the penalty parameter ϵ was set to 0.5. Network pruning was terminated if removal of a hidden unit or an input unit caused the accuracy of the resulting network on the cross validation set to drop by more than 10% [10]. The coding scheme for the input data was as follows. One input unit was assigned to each continuous attribute in the data set. The values of the continuous attributes were normalized so that they range in the interval $[0, 1]$. Discrete attributes were binary-coded. A discrete attribute with D possible values was assigned D network inputs, except when $D = 2$, where one input unit was sufficient.

The average accuracy of the rules on the test samples in Table 2 was measured in terms of the Mean Absolute Deviation (MAE) and the Mean Absolute Percentage Error (MAPE). They were computed as follows

$$\text{MAE} = \frac{1}{\hat{K}} \sum_{i=1}^{\hat{K}} |\hat{y}^i - y^i| \tag{11}$$

$$\text{MAPE} = \frac{1}{\hat{K}} \sum_{i=1}^{\hat{K}} 100 \times |(\hat{y}^i - y_i)/y^i| \tag{12}$$

Table 2. Summary of the rules extracted from neural networks.

Data set	MAE	MAPE	#Rules	#Attributes
Abalone	1.57 ± 0.06	15.88 ± 0.86	4.10 ± 1.45	4.90 ± 2.64
Auto-mpg	1.96 ± 0.32	8.56 ± 1.40	7.50 ± 5.08	9.20 ± 6.07
Housing	2.53 ± 0.46	13.36 ± 3.51	25.30 ± 17.13	9.20 ± 2.70
Machine	20.99 ± 11.38	32.27 ± 13.23	3.00 ± 3.00	4.70 ± 0.67
Servo	0.34 ± 0.08	33.94 ± 7.96	4.70 ± 2.31	10.50 ± 4.45

In Table 3 we compare the performance of our proposed method to that of other regression methods. The predictive accuracy of three variants of a regression tree generating method called HTL[11] are shown under the columns KRTrees, kNNTrees and LinearTrees. HTL grows a binary regression tree by adding nodes to minimize the mean squared errors of the patterns in the leaf nodes. The prediction error for a training sample is computed as the difference between the actual target value and the average target value of all training samples in the same leaf node. Once the tree is generated, predictions for new data are made using different techniques. The KR method employs kernel regression with a gaussian kernel function to compute the weights to be assigned to selected samples in a leaf node. The kNN prediction is computed as the average values of its k nearest neighbors. Each leaf node of a Linear Tree is associated with a linear regression function which is used for prediction. The last column of the table shows the results from RUDE [8].

It is clear that regression rules extracted from neural networks by our method gives more accurate predictions than the other methods for all the five problems tested. For one problem, Housing, the decrease in average mean absolute error is as much as 30% from the next lowest average error obtained by KRTrees. In addition to achieving higher accuracy, our method also generates fewer rules. The mean size of RUDE trees ranges from 21.5 for Servo to 65.8 for Housing. For these two data sets, the number of rules from neural networks are 10.5 and 9.2, respectively.

Table 3. MAEs of NN rules and those of other regression methods.

Data set	NN rules	KRTrees	kNNTrees	LinearTrees	RUDE
Abalone	1.57 ± 0.06	1.7 ± 0.1	1.7 ± 0.1	1.8 ± 0.1	2.13 ± 0.09
Auto-mpg	1.96 ± 0.32	2.4 ± 0.4	2.3 ± 0.4	18.0 ± 5.6	3.96 ± 0.34
Housing	2.53 ± 0.46	2.8 ± 0.5	2.9 ± 0.4	3.9 ± 2.7	4.07 ± 0.34
Machine	20.99 ± 11.38	31.2 ± 15.1	31.5 ± 14.7	35.7 ± 11.7	51.49 ± 16.25
Servo	0.34 ± 0.08	0.4 ± 0.2	0.4 ± 0.2	0.9 ± 0.2	0.44 ± 0.16

6 Conclusion

We have presented a method that generates a set of linear equations from a neural network trained for regression. Linear equations that provide predictions of the continuous target values of data samples are obtained by approximating each hidden unit activation function by a 3-piece linear function. A piece-wise linear function is computed for each hidden unit such that it minimizes the sum of squared deviations between the actual activation values of the training samples and their approximated values.

Using the relative mean absolute error as the performance measure, the rules generated from the neural networks achieve higher accuracy than those from the other methods for regression which first discretize the continuous target values and then build classification trees.

By converting the nonlinear mapping of a neural network into a set of linear regression equations, it is hoped that the method be able to provide its user better understanding about the problem being solved.

References

1. Blake, C., Keogh, E. and Merz, C.J. (1998) UCI Repository of Machine Learning Databases, Dept. of Information and Computer Science, University of California, Irvine. `http://www.ics.uci.edu/ ~mlearn/MLRepository.html`.
2. Cybenko, G. (1989) Approximation by superpositions of a sigmoidal function. *Mathematics of Control, Signals, and Systems*, 2, 303-314.
3. Dennis Jr. J.E. and Schnabel, R.E. (1983) Numerical methods for unconstrained optimization and nonlinear equations. Englewood Cliffs, New Jersey: Prentice Halls.
4. Setiono, R. and Hui, L.C.K. (1995) Use of quasi-Newton method in a feedforward neural network construction algorithm, *IEEE Trans. on Neural Networks*, 6(1), 273-277.
5. Hornik, K. (1991) Approximation capabilities of multilayer feedforward networks. *Neural Networks*, 4, 251-257.
6. John, G., Kohavi, R. and Pfleger, K. (1994) Irrelevant features and the subset selection problem. In *Proc. of the 11th ICML Learning*, Morgan Kaufman, San Mateo, 121-129.
7. Khattree, W. and Naik, D.N. (1999) Applied multivariate statistics with SAS software. SAS Institute, Carey, N.C.
8. Ludl, M-C. and Widmer, G. (2000) Relative unsupervised discretization for regression problems. In *Proc. of the 11th ECML, ECML 2000*, Lecture Notes in AI 1810, Springer, R.A. Mantaras and E. Plaza (Eds), 246-253, Barcelona.
9. Quinlan, R. (1993) C4.5: Programs for machine learning. Morgan Kaufmann, San Meteo, CA.
10. Setiono, R. and Leow, W.K. (2000) Pruned neural networks for regression. In *Proc. of the 6th Pacific Rim Conference on Artificial Intelligence, PRICAI 2000*, Lecture Notes in AI 1886, Springer, R. Mizoguchi and J. Slaney (Eds), 500-509, Melbourne.
11. Torgo, L. (1997) Functional models for regression tree leaves. In *Proc. of the ICML, ICML-97*, Fisher, D. (Ed), Morgan Kaufman, San Mateo, CA.
12. Torgo, L. and Gama, J. (1997) Search-based class discretization. In *Proc. of the 9th European Conference on Machine Learning, ECML-97*, Lecture Notes in AI 1224, Springer, M. van Someren and G. Widmer (Eds), 266-273, Prague.

Speech Recognition Using Fuzzy Second-Order Recurrent Neural Networks

A. Blanco, M. Delgado, M.C. Pegalajar*, and I. Requena

Department of Computer Science and Artificial Intelligence
E.T.S.I. Informática. University of Granada
Avenida de Andalucía, 38
18071 Granada (Spain)
*Corresponding author e-mail : mcarmen@decsai.ugr.es

Abstract. The use of Recurrent Neural Networks is not so widely extended as Feedforward Neural Networks. In recent years, these neural networks are being very studied and they are giving better results than feedforward neural networks in problems related on control, pattern recognition, etc.
In this paper, we present a neural model for speech recognition, whose main characteristic is the little number of neurons that uses and the good results that obtains.

1 Introduction

A lot of systems in real world are non-linear systems or systems whose behaviour depends on its current state. The Artificial Neural Networks that have given the best results in identification problems related with this kind of systems are the Recurrent Neural Networks (RNN). In recent years, a great number of works have been focused to study the capabilities and limitations of RNNs applied to subjects associated to pattern recognition and control.

It has been shown that the feedforward neural networks are not able to learn space-time patterns (their elements are not simultaneously but sequentially considered in time), and therefore, with this kind of neural networks it is not possible to extract the rules that can define the structure of patterns of variable length.

However, these recurrent neural networks can treat and store information that depends on the time and, therefore, they can learn space-time relationships. This ability is attributed to recurrent connections, since when processing a symbol they are keeping the activation values of the neurons in the previous instant of time. An example of this kind of problem is the grammatical inference using neural networks from examples.

On the other hand, a problem that is related on pattern recognition and Artificial Intelligence is speech recognition. Speech recognition can be considered as a problem whose examples are space-time patterns, and therefore, it seems reasonable to apply methods as the Recurrent Neural Networks for its learning.

J. Mira and A. Prieto (Eds.): IWANN 2001, LNCS 2084, pp. 285-292, 2001.

This paper provides a brief introduction to speech recognition in section 1.1. In section 2, the neural network used and a modified version of real-time recurrent learning algorithm is given.

Finally, the model to recognize a group of words is shown in the section 4.

1.1 Automatic recognition of the speech

The main goal of automatic recognition of speech is the conception and realization of automatic systems which are able to interpret the vocal sign coming from some human speakers in terms of linguistic categories of a given universe [5].

From user point of view and without taking into account the complexity of the system needed to carry on, the applications of oral systems can be classified in four different groups:

1. Classical applications. These are those tasks in that operators interact with machine through speech. Some typical examples of these kind of applications are: quality control and inspection, automatic management and classification of packages, pieces, and so on, machine-tools control.
2. Applications oriented to fully replace to the human operator in routine tasks. Typical examples are: automatic phone answer or information, documental services from a database, distribution of phone calls through switchboard, and so on.
3. Help systems to physically handicapped people of special relevance due its social interest: wheelchair controlable by speech, systems for speech learning of deafmutes, sinthetic word prosthesis for helping to people with speech discorders, and so on.
4. Systems of oral communications with computers. In a near future it is foreseen these systems will be part of standard peripherals: alphanumeric screens (and/or graphicals) with sinthetic speech, oral input data devices, oral order languages, and so on.

Preprocess and segmentation As previously commented, speech recognition is a problem associated to Pattern Recognition. Generally, when recognition pattern is treated, the first step is to obtain a representation of those objects that constitute the universe to study. This representation is ussually obtained from, using an appropiate sensor, an specific physical magnitude (those we wish recognizing) into an electrical one that it can be subsequently processed. This step is named preprocess and it is needed to obtain an specific numerical representation of the signal converted by the sensor. Also, segmentation is defined as the process of specifying the different subobjects in which the researched object can be divided or, more precisely, to the differentiation between the environment and the object considered.

Next, each process will be separately analyzed.

- **Preprocess.**
  Code and filtrate are characteristic tasks of the preprocess. Frequently, encoded signal can become noisy and, therefore, the quality of the representation of the studied object can be affected. To overcome this drawback is needed the use of filtrate techniques, eliminating certain frequency components due to noise using low-pass filters or averaging in each point by nearest points.
- **Segmentation.**
  This interpretative process is usually needed when in the (preprocessed) representation of the external object several subsets for recognizing exist.These objects can be structurally related to each other. The goal of the segmentation
  consists in breaking down into parts the representation of the object, so that each part can be individually recognized later.

2 Neural Model For Carrying Out Speech Recognition

The fuzzy neural model [1–4] (Fig. 1) we propose consists of:

- N hidden recurrent neurons labeled S_j^t, $j = 0..N-1$
- L input neurons labeled I_k^t, $k = 0..L-1$
- $N^2 \times L$ weights labeled w_{ijk}
- One linear output neuron, O, connected to the hidden recurrent neurons by N weights labeled u_j, $j = 0..N-1$,

L being the number of symbols in the input alphabet.

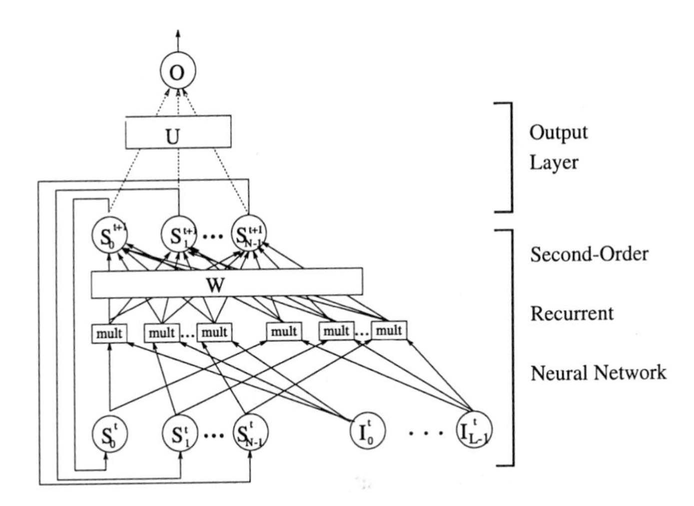

Fig. 1. Fuzzy Recurrent Neural Model

This neural network accepts an input sequence ordered on time. Each symbol belonging to a sequence to be processed is sequentially encoded in the input

neurons at each step in time t. Let us assume the alphabet is the symbol set $\{a_0, ...a_{L-1}\}$. If the $t-th$ character in the sequence to be processed is a_i, then it will be encoded in the input neurons exactly in time t by: $I_i^t = 1$, $I_j^t = 0$ $\forall j = 0..L-1$, where
$j \neq i$.

The membership degree is computed by the linear output neuron once the input sequence has been fully processed by the SORNN (Second-Order Recurrent Neural Network).

The dynamics of the neural network may be summarized into the following steps:

1. The initial values on ($t = 0$) of the recurrent hidden neurons are fixed to be $S_0^0 = 1$ and $S_i^0 = 0$ $\forall i \neq 0$.
2. On any instant given an input sequence, Equation (1) is evaluated for each hidden recurrent neuron, obtaining the values they will take at the instant $t+1$, S_i^{t+1}

$$S_i^{t+1} = g(\theta_i^t) \tag{1}$$

$$\theta_i^t = \sum_{j=0}^{N-1} \sum_{k=0}^{L-1} I_k^t . S_j^t . w_{ijk}, \tag{2}$$

g being the sigmoidal function:

$$g(x) = \frac{1}{1 + e^{(-x)}}. \tag{3}$$

Each element in a sequence to be processed is sequentially encoded in the input neurons at each step in time t by means of Kronecker's delta.
3. Once the recurrent neural network has processed the whole input sequence, the value of the output neuron, O, is obtained from the values S_i^m, $i = 0..N-1$(m being the sequence length) , Equation (4). The output neuron given us the membership degree to the class we want to identify.

$$O = \sum_{i=0}^{N-1} u_i S_i^m. \tag{4}$$

2.1 A learning based on the error gradient

The development of learning algorithms for RNNs has centered on using gradient descent algorithms, of which there are two basic types:

- Real Time Recurrent Learning (RTRL) [7,9]. The main drawback of this algorithm is its high computational cost for each iteration.
- Backpropagation through time (BTT) [6,8]. The main limitation of this algorithm is that the length of the input sequence.

Although these algorithms are based on the backpropagation algorithm [8], they are computationally much more hard than it when used for feedforward networks.

A variation of the RTRL algorithm to train our neural network is presented below.

A new version of RTRL Algorithm The error function in the output neuron is

$$E = \frac{1}{2}(T - O)^2, \tag{5}$$

where T is the desired value for and O is the actually obtained one.

The training algorithm updates the weights at the end of the each sequence presentation when $E > \epsilon$, ϵ being an error tolerance bound. The modification of the weights is given by

$$\Delta u_i = -\alpha \frac{\partial E}{\partial u_i} \tag{6}$$

$$\Delta w_{lon} = -\alpha \frac{\partial E}{\partial w_{lon}}, \tag{7}$$

where α is the learning rate. The partial derivative in (6) can be directly obtained as:

$$\frac{\partial E}{\partial u_i} = (T - O).S_i^m, \tag{8}$$

where S_i^m is the final value of neuron i once the network has processed the whole sequence. The derivatives associated with w_{lon} are calculated by

$$\frac{\partial E}{\partial w_{lon}} = (T - O). \sum_{i=0}^{N-1} u_i.g'(\theta_i^m).[\delta_{il}.S_o^{m-1}.I_n^{m-1} + \sum_{j=0}^{N-1} \sum_{k=0}^{L-1} w_{ijk}.I_j^{m-1}.\frac{\partial S_j^{m-1}}{\partial w_{lon}}]. \tag{9}$$

Since obtaining $\frac{\partial E}{\partial w_{lon}}$ requires a recursive computation, we fix the initial value of:

$$\frac{\partial S_i^0}{\partial w_{lon}} = 0. \tag{10}$$

3 The Speech Recognizer

To illustrate the use of our SORNN to speech recognition we study an specific example: to recognize the first ten numerals in spanish, i.e. {*cero*, *uno*, *dos*, *tres*, *cuatro*, *cinco*, *seis*, *siete*, *ocho*, *nueve*} using ten fuzzy recurrent neural networks (topology 2), see figure 1, that they act as classifiers, each specialized in recongnizing an specific digit.

We have a set of examples: ten different speakers, female and male, having pronounced each spanish numeral-word ten times obtaining in this way 1000 example sequences.

Data were acquired at 8533Hz sampled frequency and the parameter set represent spectral coeficients of the output of a Mel's scale filter bank (D-FFT). These parameters was measured at 100Hz as subsampled frequency. A vectorial quantization was later carried out and a process of labelling to a 16 symbols $\{0, 1, 2, 3, 4, 5, 6, 7, 8, 9, A, B, C, D, E, F\}$.

Next, we present some examples from the files of examples we used. An example associated to word "CERO" ("ZERO" in English) is the following: $2222256188881D888CC99444$

An example associated to word "CUATRO" ("FOUR" in English) is the following: $4944BBBBBBCC222266166CCCCC99944$

Notice that the sequences have different length and, thus, feedforward networks can not be used.

Training neural networks to recognize each numerals, we have built a training set including 100 positive examples associated to the corresponding numerals, together with 90 negative ones. Among these 90 examples, 1/9 belong to each one of the remaining classes and they are ramdomly chosen. Neural networks that recognizes each one of the digits have been obtained by trial-error, starting from those which have two neurons in the hidden layer. The speech recognizer system has the following form (Figure 2). Given an associated sequence to a digit is processed by each one of the neural recognizers and each one of them give as output the memebership degree to the learned digit.

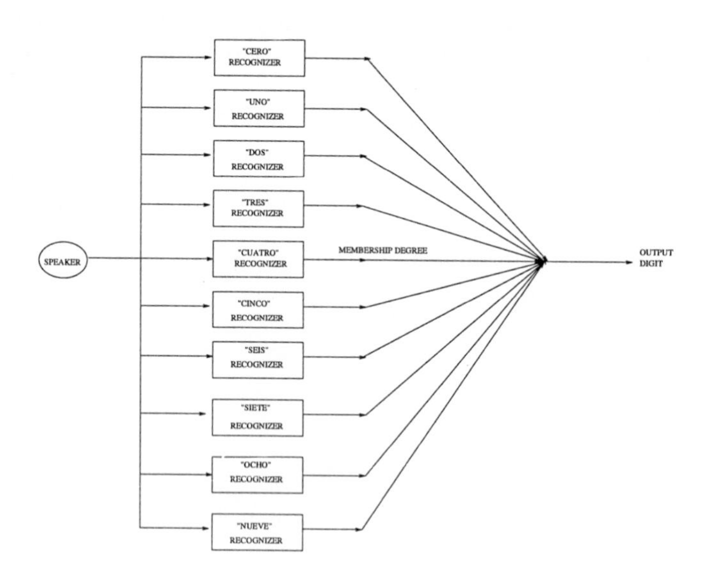

Fig. 2. Recognizer system of the 10 spanish digits.

3.1 Neural topologies associated to each recognizer

Table 1, describes the main parameters of the neural networks we have found to recognize each digit. Table includes the following for each recognizer: number of recurrent neurons used, learning rate, α, momentum, μ, answer tolerance, ϵ, training cycles, % of learned examples, average error obtained when network has been trained.

Recognizer	N	α	μ	ϵ	Cycles	%	Error
CERO	2	0.1	0.0	0.1	2544	100%	0.00354
UNO	2	0.1	0.0	0.1	2188	100%	0.00409
DOS	2	0.2	0.9	0.1	3100	95%	0.01793
TRES	3	0.1	0.0	0.1	1907	93%	0.012
CUATRO	4	0.1	0.0	0.1	2476	91%	0.0102092
CINCO	3	0.1	0.0	0.1	2029	92%	0.023
SEIS	3	0.1	0.0	0.1	3100	99.5 %	0.0025
SIETE	3	0.1	0.9	0.1	1657	95%	0.0060446
OCHO	3	0.1	0.0	0.1	2126	94%	0.05
NUEVE	3	0.1	0.0	0.1	3143	91.05%	0.01042

Tabla 1. The parameters used and the results obtained

The recognizer associated to digit CUATRO has been the most difficult to find because of it is recquired a higher number of simulations.

Figure 3 shows an example about the behaviour of the recognizer system when the following string $2222256188881D888CC99444$ is introduced. As we can observed the membership degrees given as output by recognizers are near to 0, except output of the recognizer associated to digit CERO that is 0.96. This input sequence would be classified, hence, as belonging to CERO class.

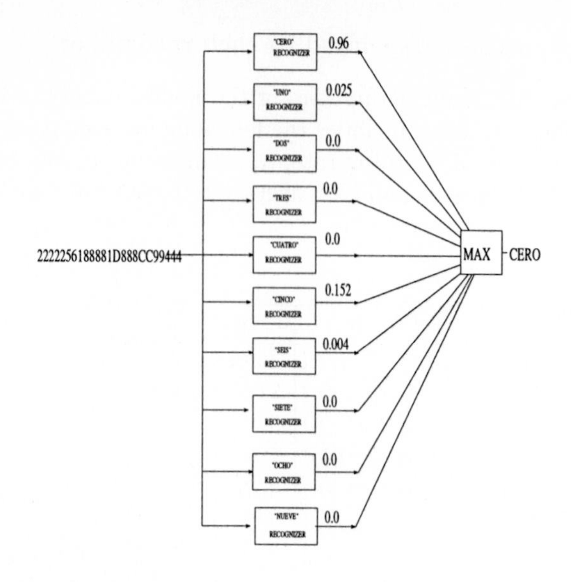

Fig. 3. Example of recognizer output when input is 2222256188881D888CC99444

References

[1] Blanco A., Delgado M. and Pegalajar M.C: Extracting Rules from a (Fuzzy/Crisp) Recurrent Neural Networks using a Self-Organizing Map. International Journal of Intelligent Systems **15** (2000) 595–621

[2] Blanco A., Delgado M. and Pegalajar M.C: A Real-Coded Genetic Algorithm for Training Recurrent Neural Networks. Neural Network **14** (2001) 93–105.

[3] Blanco A., Delgado M. and Pegalajar M.C: A Genetic Algorithm to obtain the Optimal Recurrent Neural Network. International Journal of Approximate Reasoning, **23** (2000) 67–83.

[4] Blanco A., Delgado M. and Pegalajar M.C: Fuzzy Automaton Induction using Neural Network. International Journal of Approximate Reasoning. (In press)

[5] Casacuberta F., Vidal E: Reconocimiento automático del habla". Ed. Marcombo, 1987

[6] Pearlmutter, B: Learning state space trajectories in recurrent neural networks. Neural Computation **1** (1989) 263–269.

[7] Pineda, F: Dynamics and architecture for neural computation. Journal of Complexity, **4** (1988) 216–245.

[8] Rumelhart, D., abd McClelland J. Parallel Distributed Processing: Explorations in the Microstructure of Cognition: Foundations. **1** (1986) MIT Press, Cambridge, MA.

[9] Williams, R., and Zipser D: A learning algorithm for continually running fully recurrent neural networks. Neural Computation, **1** (1989) 270–277.

A Measure of Noise Immunity for Functional Networks

Enrique Castillo[1], Oscar Fontenla-Romero[2], Bertha Guijarro-Berdiñas[2], and Amparo Alonso-Betanzos[2]

[1] Department of Applied Mathematics and Computational Sciences, University of Cantabria and University of Castilla-La Mancha, Spain
castie@unican.es

[2] Laboratory for Research and Development in Artificial Intelligence, Department of Computer Science, University of A Coruña, Campus de Elviña s/n, 15071, A Coruña, Spain
oscarfon@mail2.udc.es, {ciamparo,cibertha}@udc.es
http://www.dc.fi.udc.es/lidia

Abstract. In this paper a study of the influence of input perturbations on a functional network is presented. A quantitative measure, related to the mean squared error degradation in presence of input noise, is introduced. This measure, based on statistical sensitivity, provides an estimation of the generalization ability and noise immunity of functional networks and lets to predict the performance degradation of a functional network. The experimental results corroborated the validity of the proposed model.

1 Introduction

In the area of neural networks several studies have been done to verify the influence of input perturbations in these systems [3][4][9][10] . Furthermore, several measures have been proposed to estimate their tolerances to this kind of noise. Choi et al. [8] proposed statistical sensitivity as a measure of noise immunity or fault tolerance for Multilayer Perceptrons (MLPs) with only one output. Also, in [1][2] the authors presented a measure to estimate the noise immunity to input deviations in a MLP. They showed the existing relation between the mean squared error (MSE) and the statistical sensitivity introducing a new measure, termed mean squared sensitivity, that predicts accurately the degradation of the MSE when the inputs or weights are distorted. These measures can be used as a selection criterion for choosing different networks and sets of weights that present a similar performance but differ with respect to input noise immunity. However, this kind of studies have not been applied in the field of functional networks. In this paper a measure, based on statistical sensitivity and mean squared sensitivity, is proposed for functional networks. This measure generalizes the one proposed by the authors of [1][2], in the same way as functional networks generalize neural networks, and is an accurate estimator of the generalization ability and noise immunity of functional networks.

J. Mira and A. Prieto (Eds.): IWANN 2001, LNCS 2084, pp. 293-300, 2001.

2 Statistical Sensitivity of a Functional Network

Consider the generalized functional network in Figure 1. This network is composed of M layers, where each layer m has N_m functional neurons. Each functional neuron i in layer m is connected to the N_{m+1} neurons of the $m+1$ layer. The set $\{y_1^{(0)}, \ldots, y_{N_0}^{(0)}\}$ is the input of the network and the output of functional neuron m is defined as

$$y_i^{(m)} = f_i^{(m)}(y_1^{(m-1)}, \ldots, y_{N_{m-1}}^{(m-1)}); \; i = 1, \ldots N_m; \; m = 1, \ldots, M, \tag{1}$$

where each $f_i^{(m)}$ is defined as the following functions composition:

$$\begin{gathered} f_i^{(m)}(y_1^{(m-1)}, \ldots, y_{N_{m-1}}^{(m-1)}) = g_i^{(m)}(h_{i1}^{(m)}(y_1^{(m-1)}), \ldots, h_{iN_{m-1}}^{(m)}(y_{N_{m-1}}^{(m-1)})); \\ i = 1, \ldots N_m; \; m = 1, \ldots, M \; . \end{gathered} \tag{2}$$

The $g_i^{(m)}$ functions are known and fixed during the training process and each $h_{ij}^{(m)}$ function is considered to be a linear combination of known functions. Thus, we have

$$\begin{gathered} h_{ij}^{(m)}(y_j^{(m-1)}) = \sum_{z=1}^{n_{ij}^{(m)}} a_{ijz}^{(m)} \phi_{ijz}^{(m)}(y_j^{(m-1)}); \\ i = 1, \ldots N_m; \; j = 1, \ldots N_{m-1}; \; m = 1, \ldots, M, \end{gathered} \tag{3}$$

where $n_{ij}^{(m)}$ is the number of kernel functions and the coefficients $a_{ijz}^{(m)}$ are the parameters of the functional network. The functional network described and showed in Figure 1 is a generalized model of those proposed by Castillo et al. [5][6][7].

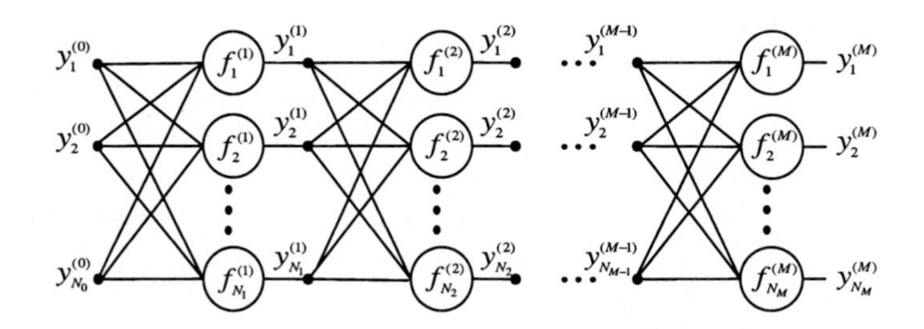

Fig. 1. A Generalized Functional Network

The output of a functional neuron, $y_i^{(m)}$, will change if there are variations in the values of either the $a_{ijz}^{(m)}$ parameters or the neuron inputs ($y_j^{(m-1)}$; $j = 1, \ldots, N_{m-1}$). If these changes are small, then the corresponding deviation can be approximated as

$$\Delta y_i^{(m)} \approx \sum_{r=1}^{N_{m-1}} \left(\frac{\partial y_i^{(m)}}{\partial y_r^{(m-1)}} \Delta y_r^{(m-1)} + \sum_{s=1}^{n_{ir}^{(m)}} \frac{\partial y_i^{(m)}}{\partial a_{irs}^{(m)}} \Delta a_{irs}^{(m)} \right) \; . \tag{4}$$

Let $u_{ij}^{(m)} = h_{ij}^{(m)}(y_j^{(m-1)})$; $i = 1, \dots N_m$; $j = 1, \dots N_{m-1}$; $m = 1, \dots, M$, then the partial derivatives in equation 4 are

$$\frac{\partial y_i^{(m)}}{\partial y_r^{(m-1)}} = \frac{\partial y_i^{(m)}}{\partial u_{ir}^{(m)}} \frac{\partial u_{ir}^{(m)}}{\partial y_r^{(m-1)}} = \frac{\partial y_i^{(m)}}{\partial u_{ir}^{(m)}} \sum_{z=1}^{n_{ir}^{(m)}} a_{irz}^{(m)} \frac{d\phi_{irz}^{(m)}(y_r^{(m-1)})}{dy_r^{(m-1)}} \tag{5}$$

and

$$\frac{\partial y_i^{(m)}}{\partial a_{irs}^{(m)}} = \frac{\partial y_i^{(m)}}{\partial u_{ir}^{(m)}} \frac{\partial u_{ir}^{(m)}}{\partial a_{irs}^{(m)}} = \frac{\partial y_i^{(m)}}{\partial u_{ir}^{(m)}} \phi_{irs}^{(m)}(y_r^{(m-1)}) \; . \tag{6}$$

In [8] Choi et al. proposed statistical sensitivity as a measure of tolerance to perturbations in MLPs. Statistical sensitivity, which measures the alterations of the output when the values of the parameters change, is defined as

$$S_i^{(m)} = \lim_{\sigma \to 0} \frac{\sqrt{var(\Delta y_i^{(m)})}}{\sigma} \; , \tag{7}$$

where σ is the standard deviation of the changes and $var(\Delta y_i^{(m)})$ is the variance of the deviation in the output due to these changes.

In this work, an additive model for the input deviations is used. Also, the noise distribution is considered to have zero mean and to be uncorrelated between inputs. Thus, the following conditions are satisfied

- (c.1) $E[\Delta y_i^{(0)}] = 0; \; i = 1, \dots, N_0$,
- (c.2) $E[\Delta y_i^{(0)} \Delta y_j^{(0)}] = \sigma^2 \delta_{ij}; \; i, j = 1, \dots, N_0$,

where the parameter σ represents the standard deviation of the noise distribution, δ is the Kronecker delta and $E[\cdot]$ is the expectation operator.

Theorem 1. $E[\Delta y_i^{(m)}] = 0; \; \forall i = 1, \dots, N_m; \; \forall m = 1, \dots, M$ *if* $E[\Delta y_i^{(0)}] = 0; \; \forall r = 1, \dots, N_0$

Proof: It will be shown by induction over m *and assuming that the parameters of the network are free of errors, i.e.,* $\Delta a_{irs}^{(m)} = 0; \; \forall i, r, s, m$.
For $m = 1$

$$E[\Delta y_i^{(1)}] = E\left[\sum_{r=1}^{N_0}\left(\frac{\partial y_i^{(1)}}{\partial y_r^{(0)}}\Delta y_r^{(0)} + \sum_{s=1}^{n_{ir}^{(1)}} \frac{\partial y_i^{(1)}}{\partial a_{irs}^{(1)}} \Delta a_{irs}^{(1)}\right)\right]$$

$$= \sum_{r=1}^{N_0}\left(\frac{\partial y_i^{(1)}}{\partial y_r^{(0)}} E[\Delta y_r^{(0)}] + \sum_{s=1}^{n_{ir}^{(1)}} \frac{\partial y_i^{(1)}}{\partial a_{irs}^{(1)}} E[\Delta a_{irs}^{(1)}]\right) = 0 \; .$$

Suppose that result holds for $m-1$, *i.e.* $E[\Delta y_i^{(m-1)}] = 0 \; \forall i = 1, \dots, N_{m-1}$, *then for layer* m:

$$E[\Delta y_i^{(m)}] = E\left[\sum_{r=1}^{N_{m-1}}\left(\frac{\partial y_i^{(m)}}{\partial y_r^{(m-1)}}\Delta y_r^{(m-1)} + \sum_{s=1}^{n_{ir}^{(m)}} \frac{\partial y_i^{(m)}}{\partial a_{irs}^{(m)}} \Delta a_{irs}^{(m)}\right)\right]$$

$$= \sum_{r=1}^{N_{m-1}} \left(\frac{\partial y_i^{(m)}}{\partial y_r^{(m-1)}} E[\Delta y_r^{(m-1)}] + \sum_{s=1}^{n_{ir}^{(m)}} \frac{\partial y_i^{(m)}}{\partial a_{irs}^{(m)}} E[\Delta a_{irs}^{(m)}] \right) = 0 \ .$$

Theorem 2. *The statistical sensitivity to input noise can be expressed as*

$$S_i^{(m)} = \sqrt{\sum_{r=1}^{N_{m-1}} \sum_{z=1}^{N_{m-1}} \frac{\partial y_i^{(m)}}{\partial y_r^{(m-1)}} \frac{\partial y_i^{(m)}}{\partial y_z^{(m-1)}} C_{rz}^{(m-1)}}; \ \forall i = 1, \ldots, N_m; \ \forall m = 1, \ldots, M \tag{8}$$

where the terms $C_{ik}^{(m)}$ *are recursively calculated as*

$$C_{ik}^{(m)} = \sum_{r=1}^{N_{m-1}} \sum_{z=1}^{N_{m-1}} \frac{\partial y_i^{(m)}}{\partial y_r^{(m-1)}} \frac{\partial y_k^{(m)}}{\partial y_z^{(m-1)}} C_{rz}^{(m-1)} \ . \tag{9}$$

Proof:

$$\begin{aligned}
&E[\Delta y_i^{(m)} \Delta y_k^{(m)}] = \\
&= E\left[\left(\sum_{r=1}^{N_{m-1}} \frac{\partial y_i^{(m)}}{\partial y_r^{(m-1)}} \Delta y_r^{(m-1)} + \sum_{s=1}^{n_{ir}^{(m)}} \frac{\partial y_i^{(m)}}{\partial a_{irs}^{(m)}} \Delta a_{irs}^{(m)} \right) \right. \\
&\qquad \left. \left(\sum_{z=1}^{N_{m-1}} \frac{\partial y_k^{(m)}}{\partial y_z^{(m-1)}} \Delta y_z^{(m-1)} + \sum_{t=1}^{n_{kz}^{(m)}} \frac{\partial y_k^{(m)}}{\partial a_{kzt}^{(m)}} \Delta a_{kzt}^{(m)} \right) \right] = \\
&= \sum_{r=1}^{N_{m-1}} \sum_{z=1}^{N_{m-1}} \frac{\partial y_i^{(m)}}{\partial y_r^{(m-1)}} \frac{\partial y_k^{(m)}}{\partial y_z^{(m-1)}} E[\Delta y_r^{(m-1)} \Delta y_z^{(m-1)}] + \\
&+ \frac{\partial y_i^{(m)}}{\partial y_r^{(m-1)}} \sum_{t=1}^{n_{kz}^{(m)}} \frac{\partial y_k^{(m)}}{\partial a_{kzt}^{(m)}} E[\Delta y_r^{(m-1)} \Delta a_{kzt}^{(m)}] + \frac{\partial y_k^{(m)}}{\partial y_z^{(m-1)}} \sum_{s=1}^{n_{ir}^{(m)}} \frac{\partial y_i^{(m)}}{\partial a_{irs}^{(m)}} E[\Delta a_{irs}^{(m)} \Delta y_z^{(m-1)}] + \\
&+ \sum_{s=1}^{n_{ir}^{(m)}} \sum_{t=1}^{n_{kz}^{(m)}} \frac{\partial y_i^{(m)}}{\partial a_{irs}^{(m)}} \frac{\partial y_k^{(m)}}{\partial a_{kzt}^{(m)}} E[\Delta a_{irs}^{(m)} \Delta a_{kzt}^{(m)}] = \\
&= \sum_{r=1}^{N_{m-1}} \sum_{z=1}^{N_{m-1}} \frac{\partial y_i^{(m)}}{\partial y_r^{(m-1)}} \frac{\partial y_k^{(m)}}{\partial y_z^{(m-1)}} E[\Delta y_r^{(m-1)} \Delta y_z^{(m-1)}] \ .
\end{aligned} \tag{10}$$

If $C_{ik}^{(m)}$ *is defined as* $C_{ik}^{(m)} \equiv \frac{E[\Delta y_i^{(m)} \Delta y_k^{(m)}]}{\sigma^2}$ *then*

$$E[\Delta y_i^{(m)} \Delta y_k^{(m)}] = \sigma^2 \sum_{r=1}^{N_{m-1}} \sum_{z=1}^{N_{m-1}} \frac{\partial y_i^{(m)}}{\partial y_r^{(m-1)}} \frac{\partial y_k^{(m)}}{\partial y_z^{(m-1)}} C_{rz}^{(m-1)} \tag{11}$$

where

$$C_{ik}^{(m)} = \sum_{r=1}^{N_{m-1}} \sum_{z=1}^{N_{m-1}} \frac{\partial y_i^{(m)}}{\partial y_r^{(m-1)}} \frac{\partial y_k^{(m)}}{\partial y_z^{(m-1)}} C_{rz}^{(m-1)} \ . \tag{12}$$

It is straightforward to obtain the initial condition for equation 12, $C_{ik}^{(0)} = \delta_{ik}$*, using the* $C_{ik}^{(m)}$ *definition and (c.2).*
Using $var(\Delta y_i^{(m)}) = E[(\Delta y_i^{(m)})^2] - (E[\Delta y_i^{(m)}])^2$ *and Theorem 1 in equation 7 we have:*

$$S_i^{(m)} = \frac{\sqrt{E[(\Delta y_i^{(m)})^2]}}{\sigma} . \tag{13}$$

Substituting $E[(\Delta y_i^{(m)})^2] = \sigma^2 C_{ii}^{(m)}$ *in equation 13 we get:*

$$S_i^{(m)} = \frac{\sqrt{\sigma^2 C_{ii}^{(m)}}}{\sigma} = \sqrt{\sum_{r=1}^{N_{m-1}} \sum_{z=1}^{N_{m-1}} \frac{\partial y_i^{(m)}}{\partial y_r^{(m-1)}} \frac{\partial y_i^{(m)}}{\partial y_z^{(m-1)}} C_{rz}^{(m-1)}} . \tag{14}$$

3 Mean Squared Sensitivity

Mean squared error (MSE) is usually employed in the learning of functional networks to measure the error between the real and the desired output, and therefore to evaluate the performance of the system. Formally, the MSE is defined as

$$MSE = \frac{1}{T} \sum_{t=1}^{T} \sum_{i=1}^{N_M} \epsilon_i^2(t) = \frac{1}{T} \sum_{t=1}^{T} \sum_{i=1}^{N_M} (d_i(t) - y_i^{(M)}(t))^2 \tag{15}$$

where T is the number of input patterns and $d_i(t)$ the desired output.

If the inputs are affected by deviations, the MSE is altered and thus considering the Taylor expansion of equation 15 the following equation is obtained:

$$\begin{aligned} MSE' = MSE_0 &+ \frac{1}{T} \sum_{t=1}^{T} \sum_{i=1}^{N_M} \frac{\partial \epsilon_i^2(t)}{\partial y_i^{(M)}} \Delta y_i^{(M)}(t) \\ &+ \frac{1}{2T} \sum_{t=1}^{T} \sum_{i=1}^{N_M} \sum_{j=1}^{N_M} \frac{\partial^2 \epsilon_i^2(t)}{\partial y_i^{(M)} \partial y_j^{(M)}} \Delta y_i^{(M)}(t) \Delta y_j^{(M)}(t) + 0 \\ = MSE_0 &+ \frac{2}{T} \sum_{t=1}^{T} \sum_{i=1}^{N_M} (y_i^{(M)}(t) - d_i(t)) \Delta y_i^{(M)}(t) + \frac{1}{T} \sum_{t=1}^{T} \sum_{i=1}^{N_M} (\Delta y_i^{(M)}(t))^2 \end{aligned} \tag{16}$$

where MSE_0 is the nominal MSE obtained after the learning process. Taking expectations in equation 16 and using equation 13 we get:

$$E[MSE'] = MSE_0 + \frac{\sigma^2}{T} \sum_{t=1}^{T} \sum_{i=1}^{N_M} (S_i^{(M)}(t))^2 . \tag{17}$$

The second term in the right hand side of the above equation was termed by Bernier et al. [1][2] the Mean Squared Sensitivity (MSS), due to the analogy with the definition of the MSE. Thus, we have:

$$E[MSE'] = MSE_0 + \sigma^2 MSS \tag{18}$$

where

$$MSS = \frac{1}{T}\sum_{t=1}^{T}\sum_{i=1}^{N_M}(S_i^{(M)}(t))^2 \ . \tag{19}$$

As it can be observed in equation 18, a lower value of MSS implies a lesser degradation of the MSE.

4 Experimental Results

The measure presented in the previous section (equation 18) was evaluated with the generalized associativity functional network in Figure 2. The time series used

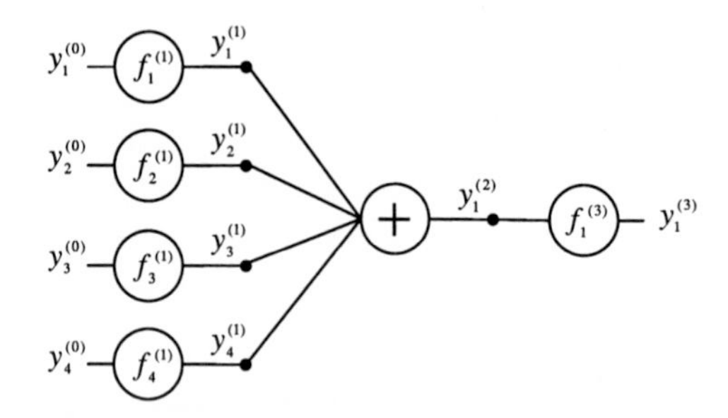

Fig. 2. The generalized associativity functional network employed in simulation

in our simulation was generated by the chaotic Mackey-Glass differential delay equation [11] defined as

$$\frac{dx(t)}{dt} = -0.1x(t) + \frac{0.2x(t-\tau)}{1+x^{10}(t-\tau)} \ . \tag{20}$$

We used the data available from CMU Learning Benchmark Archive[1], which was generated with $\tau = 17$ and using a second order Runge-Kutta method with a stepsize of 0.1. Training data is formed by 3000 samples ($200 \leq t < 3200$) while test data contains 500 instances ($5000 \leq t < 5500$). The goal of the functional network is to use known values of the chaotic time series (t-18, t-12, t-6 and t) to predict the value at some point in the future (t+85).

Table 1 shows the MSE estimated by equation 18 after the training and the averaged values obtained experimentally over 100 simulations of 500 test samples. In each simulation all the inputs were distorted by adding a random variable with zero mean and standard deviation equal to σ (from 0.01 to 0.2). The confidence interval, for a confidence level of 0.95, is also presented in Table 1.

Figure 3 is a graph of the values of Table 1. In addition, the confidence interval is plotted for each point.

[1] http://legend.gwydion.cs.cmu.edu/neural-bench/benchmark/mackey-glass.html

Table 1. Predicted values of MSE (P) and MSE obtained by simulation (S)

Standard deviation (σ)	P ($\times$ 1e − 2)	S ($\times$ 1e − 2)
0.01	1.0959	$1.0959 \pm 1.1813e-3$
0.02	1.1087	$1.1086 \pm 2.3459e-3$
0.03	1.1301	$1.1298 \pm 3.4990e-3$
0.04	1.1599	$1.1594 \pm 4.6455e-3$
0.05	1.1984	$1.1973 \pm 5.7912e-3$
0.06	1.2453	$1.2434 \pm 6.9416e-3$
0.07	1.3008	$1.2977 \pm 8.1027e-3$
0.08	1.3648	$1.3601 \pm 9.2805e-3$
0.09	1.4374	$1.4304 \pm 1.0481e-2$
0.10	1.5185	$1.5084 \pm 1.1710e-2$
0.11	1.6082	$1.5941 \pm 1.2975e-2$
0.12	1.7064	$1.6872 \pm 1.4279e-2$
0.13	1.8131	$1.7876 \pm 1.5630e-2$
0.14	1.9284	$1.8951 \pm 1.7033e-2$
0.15	2.0522	$2.0094 \pm 1.8493e-2$
0.16	2.1845	$2.1304 \pm 2.0015e-2$
0.17	2.3254	$2.2578 \pm 2.1603e-2$
0.18	2.4748	$2.3915 \pm 2.3262e-2$
0.19	2.6327	$2.5311 \pm 2.4996e-2$
0.20	2.7992	$2.6765 \pm 2.6808e-2$

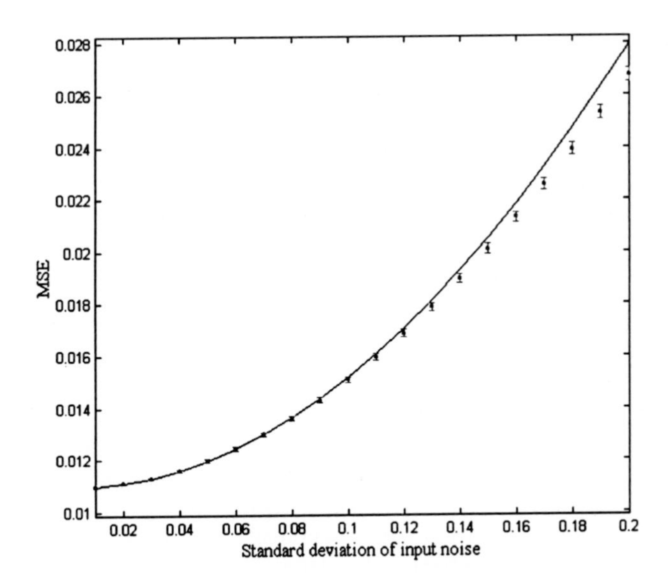

Fig. 3. Estimated and experimental MSE.

5 Conclusions

In this work, a measure for mean squared error degradation of a functional network in presence of input noise is presented. It provides an estimation of the smoothness of the error surface with respect to input perturbations. Therefore, it can be used as an alternative model selection criterion for noise immunity and generalization ability. Also, as the statistical sensitivity of a particular functional neuron can be computed independently, various lines of research are open to analyze the influence of each particular element in the global performance of the system. As future work, new learning methods, based on new cost functions that includes the mean squared sensitivity, will be developed in order to maximize the noise immunity and to enhance generalization capability while maintaining the learning performance.

Acknowledgements

This research has been funded in part by PGIDT99COM10501, the Pre-Doctoral Grants Programme 2000/01 of the Xunta de Galicia and CICYT project PB98-0421.

References

1. Bernier, J.L., Ortega, J., Ros, E., Rojas, I., Prieto, A.: A Quantitative Study of Fault Tolerance, Noise Immunity, and Generalization Ability of MLPs. Neural Computation, **12** (2000) 2941–2964
2. Bernier, J.L., Ortega, J., Ros, E., Rojas, I., Prieto, A.: A New Measurement of Noise Immunity and Generalization Ability for MLPs. International Journal of Neural Systems, **9**(6) (1999) 511–522
3. Bishop, C.M.: Neural Netwoks for Pattern Recognition. Oxford University Press, New York (1995)
4. Bishop, C.M.: Training with Noise is Equivalent to Tikhonov Regularization. Neural Computation **7**(1) (1995) 108–116
5. Castillo, E.: Functional Networks. Neural Processing Letters, **7**(3) (1998) 151–159
6. Castillo, E., Cobo, A., Gutierrez, J.M., Pruneda, R.E.: Functional Networks with Applications. A Neural-Based Paradigm. Kluwer Academic Publishers, Dordrecht (1998)
7. Castillo, E., Gutierrez, J.M.: A Comparison of Functional Networks and Neural Networks. Proc. of the IASTED Int. Conference on Artificial Intelligence and Soft Computing (1998) 439–442
8. Choi, J.Y., Choi, C.: Sensitivity Analysis of Multilayer Perceptron with Diffenciable Activation Functions. IEEE Trans. on Neural Networks **3**(1) (1992) 101–107
9. Edwards, P.J., Murray, A.F.: Modelling Weight and Input-Noise in MLP learning". Proc. Int. Conference on Neural Networks **1** (1996) 78–83
10. Minnix, J.I.: Fault Tolerance of the Backpropagation Neural Network Trained on Noisy inputs". Int. Joint Conference on Neural Networks **1** (1992) 847–852
11. Mackey, M.C., Glass, L.: Oscillation and chaos is physiological control systems. Science, Vol. 197 (1977) 287-289

A Functional-Neural Network for Post-Nonlinear Independent Component Analysis

Oscar Fontenla Romero, Bertha Guijarro Berdiñas, and Amparo Alonso Betanzos

Laboratory for Research and Development in Artificial Intelligence,
Department of Computer Science, University of A Coruña,
Campus de Elviña s/n, 15071, A Coruña, Spain
oscarfon@mail2.udc.es, {ciamparo,cibertha}@udc.es
http://www.dc.fi.udc.es/lidia

Abstract. In this paper a hybrid approach, based on a functional network and a neural network, for post-nonlinear independent component analysis is presented. In order to obtain the independence among the outputs, it was used as cost function a measure based on Renyi's quadratic entropy and Cauchy-Schwartz inequality. Also, the Kernel method was used for nonparametric estimation of the probability density function. The experimental results corroborated the soundness of the approach and a comparative study with a neural network showed its superior performance.

1 Introduction

In the field of independent component analysis (ICA) [4][7] most of the neural network approaches deal with the linear case. However, in some real applications the linear model is not suitable and it can not be used. Hence, it is interesting to extend the problem to the nonlinear case. Recently, several researchers have started addressing the ICA formulation to nonlinear mixing models [1][6][8]. It was proved in [6][14] that in the general case it is not possible to separate the sources without nonlinear distortion. Therefore, we address our work on particular nonlinear mixtures known as post-nonlinear mixtures, as denoted by Taleb and Jutten [13]. In this model, which has been proved by these authors to be a special separable problem, nonlinear mixing models are added to the linear model and the goal is to find the inverse of both. Formally, let us denote by $\mathbf{s}(i) = [s_1(i), \ldots, s_n(i)]^T$ a zero-mean n-dimensional random variable, where $i = 1, \ldots, m$ is a time index, and $\mathbf{z}(i) = [z_1(i), \ldots, z_n(i)]^T$ its n-dimensional transform defined as

$$\mathbf{z}(i) = \mathbf{f}(\mathbf{v}(i)) \tag{1}$$

where $\mathbf{v}(i) = \mathbf{A}\mathbf{s}(i)$, $\mathbf{A}$ is a $n \times n$ mixture matrix and $\mathbf{f} = [f(v_1(i)), \ldots, f(v_n(i))]^T$ is an unknown invertible derivable nonlinear function. The aim of post-nonlinear ICA is to find a mapping $\mathbf{h} = [h(z_1(i)), \ldots, h(z_n(i))]^T$ and a matrix, $\mathbf{W}$, which estimates the original variable as

$$\mathbf{y}(i) = \mathbf{W}\mathbf{h}(\mathbf{z}(i)) \tag{2}$$

J. Mira and A. Prieto (Eds.): IWANN 2001, LNCS 2084, pp. 301-307, 2001.

The sources $\mathbf{s}(i)$ are recovered if $\mathbf{h}$ and $\mathbf{W}$ are the inverse function for $\mathbf{f}$ and $\mathbf{A}$ respectively.

This work presents a functional-neural network (FN-ANN) approach for the post-nonlinear independent component analysis model presented in equation 1. This unsupervised network solves the problem by minimizing the Cauchy-Schwartz independence measure [15].

2 Cauchy-Schwartz independence measure

Xu, Principe and Fisher [9] have shown that the use of Renyi's entropy, instead of Shannon's, can lead to expressions of mutual information with meaningful computational savings. They proposed in [15] a novel independence measure based on the Cauchy-Schwartz inequality and Renyi's quadratic entropy defined, this last, as

$$H_{R_2} = -log \sum_{k=1}^{m} f(\mathbf{y})^2 \tag{3}$$

where $f(\mathbf{y})$ is the probability density function (pdf) of $\mathbf{y}$. Moreover, the non-parametric kernel-based method [12] was employed to estimate this pdf:

$$\hat{f}(\mathbf{y}) = \frac{1}{m} \sum_{i=1}^{m} K(\mathbf{y} - \mathbf{y}(i)) \tag{4}$$

where $K(\cdot)$ is a kernel function. In this work we used the Gaussian kernel, due to its differentiability properties, $K(\mathbf{y}) = G(\mathbf{y}, 2\sigma^2\mathbf{I})$ where σ^2 is the smoothing parameter or kernel size, $\mathbf{I} \in \Re^{n\times n}$ is the identity matrix and

$$G(\mathbf{y}, \mathbf{\Sigma}) = \frac{1}{(2\pi)^{\frac{n}{2}} |\mathbf{\Sigma}|^{\frac{n}{2}}} \exp\left(-\frac{1}{2}\mathbf{y}^T \Sigma^{-1} \mathbf{y}\right) . \tag{5}$$

Let $\{\mathbf{y}(i), i = 1, \ldots, m\}$ a set of observed data then the final expression for the Cauchy-Schwartz independence measure (Ic), generalized for n variables, is shown in equation 6:

$$Ic = \log \frac{P(\{\mathbf{y}(i)\}) \prod_{r=1}^{n} P_r(\{\mathbf{y}(i)\})}{P_c(\{\mathbf{y}(i)\})^2} \tag{6}$$

where

$$P(\{\mathbf{y}(i)\}) = \frac{1}{m^2} \sum_{i=1}^{m} \sum_{j=1}^{m} G(\mathbf{y}(i) - \mathbf{y}(j), 2\sigma^2\mathbf{I}) , \tag{7}$$

$$P_r(\{\mathbf{y}(i)\}) = \frac{1}{m} \sum_{j=1}^{m} P_r(j, \{\mathbf{y}(i)\}) , \tag{8}$$

$$P_r(j, \{\mathbf{y}(i)\}) = \frac{1}{m} \sum_{i=1}^{m} G(y_r(j) - y_r(i), 2\sigma^2 I_r) , \tag{9}$$

$$P_c(\{\mathbf{y}(i)\}) = \frac{1}{m}\sum_{j=1}^{m}\prod_{z=1}^{n} P_z(j, \{\mathbf{y}(i)\}) \tag{10}$$

and I_r the rth diagonal element of $\mathbf{I}$.

It can be proved that Ic is non-negative and equal to zero if and only if the variables $\{y_1, \ldots, y_n\}$ are statistically independent.

3 Learning algorithm

Figure 1 shows the architecture of the employed hybrid system. The network is composed by two subsystems. The first one is a functional network [2, 3], in which the associated functions (f) with each of the computing units are not fixed but have to be learned from data. These neural functions are approximated by finite combinations of known functions from a given family. In this work the polynomial functional family ($x_j = f(z_j) = \sum_{k=0}^{m} a_k z_j^k$) was used. Therefore, the weights can be suppressed because they are subsumed by the functions themselves. The second part of the system is a feed-forward neural network with tanh nonlinearities ($g_1, \ldots, g_n$) where $y_i = g_i(u_i)$ and $u_i = \sum_{j=1}^{n} w_{ij} x_j$.

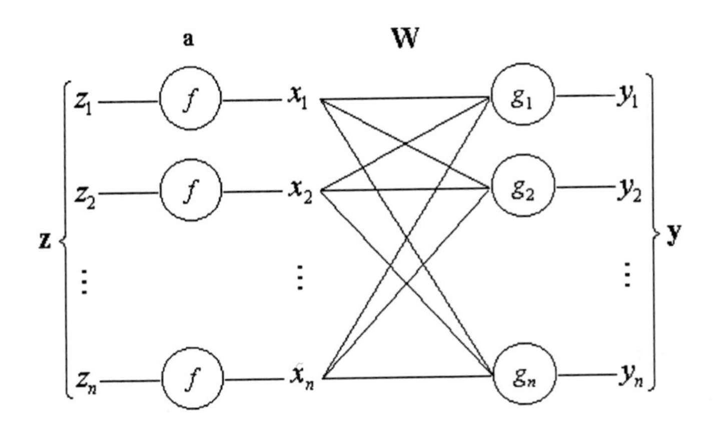

Fig. 1. Functional-Neural Network (FN-ANN) architecture.

The independence measure presented in equation 6 was employed as the objective function to minimize. The gradient descent rule was used to update the weights ($\mathbf{W}$) in the instant $l + 1$:

$$w_{kp}(l+1) = w_{kp}(l) - \eta\frac{\partial Ic}{\partial w_{kp}} = w_{kp}(l) - \eta\sum_{t=1}^{m}\frac{\partial Ic}{\partial\, y_k(t)}\frac{\partial y_k(t)}{\partial u_k(t)}\frac{\partial u_k(t)}{\partial w_{kp}} \tag{11}$$

where η is the step size. The two last partial derivatives are easy to calculate and their value is $\partial y_k(t)/\partial u_k(t) = 1 - y_k^2(t)$ and $\partial u_k(t)/\partial w_{kp} = x_p(t)$. The term

$\partial Ic/\partial y_k(t)$ in equation 11 can be calculated as

$$\frac{\partial Ic}{\partial y_k(t)} = \frac{1}{m^2\sigma^2}\left(\frac{1}{P(\{\mathbf{y}(i)\})}\gamma_1 + \frac{1}{P_k(\{\mathbf{y}(i)\})}\gamma_2 - \frac{1}{P_c(\{\mathbf{y}(i)\})}\gamma_3\right) \tag{12}$$

where

$$\gamma_1 = \frac{\partial P(\{\mathbf{y}(i)\})}{\partial y_k(t)} = \sum_{j=1,j\neq t}^{m} G(\mathbf{y}(j)-\mathbf{y}(t), 2\sigma^2\mathbf{I})(y_k(j)-y_k(t)) \tag{13}$$

$$\gamma_2 = \frac{\partial(\prod_{r=1}^{n} P_r(\{\mathbf{y}(i)\}))}{\partial y_k(t)} =$$

$$= \sum_{j=1,j\neq t}^{m} G(y_k(j)-y_k(t), 2\sigma^2 I_k)(y_k(j)-y_k(t)) \tag{14}$$

$$\gamma_3 = \frac{\partial P_c(\{\mathbf{y}(i)\})}{\partial y_k(t)} =$$

$$= \sum_{j=1}^{m} G(y_k(j)-y_k(t), 2\sigma^2 I_k)(y_k(j)-y_k(t)) \prod_{r=1,r\neq k}^{n} P_r(j,\{\mathbf{y}(i)\}) \tag{15}$$

Finally, the learning rule for the parameters of the functional units (**a**) was also obtained using the chain rule and the gradient descent on Ic as follows

$$a_k(l+1) = a_k(l) - \mu\frac{\partial Ic}{\partial a_k} = a_k(l) - \mu\sum_{t=1}^{m}\left(\sum_{i=1}^{n}\frac{\partial Ic}{\partial y_i(t)}\frac{\partial y_i(t)}{\partial u_i(t)}\frac{\partial u_i(t)}{\partial a_k}\right) \tag{16}$$

$$a_k(l+1) = a_k(l) - \mu\sum_{t=1}^{m}\left(\sum_{i=1}^{n}\frac{\partial Ic}{\partial y_i(t)}\frac{\partial y_i(t)}{\partial u_i(t)}\sum_{j=1}^{n}\frac{\partial u_i(t)}{\partial x_j(t)}\frac{\partial x_j(t)}{\partial a_k}\right) \tag{17}$$

where μ is the step size, $\partial x_j/\partial a_k = z_j^k$ and $\partial u_i/\partial x_j = w_{ij}$. The result of $\partial Ic/\partial y_i$ and $\partial y_i/\partial u_i$ is the same of those obtained in equation 11.

4 Simulation results

The proposed post-nonlinear independent component analysis method based on the FN-ANN system was evaluated using the test signals, $\mathbf{s}(i) = [s_1(i), s_2(i)]^T$, in figure 4. Firstly, the signals were linearly mixed, $\mathbf{v}(i) = \mathbf{A}s(i)$, where the mixture matrix was randomly chosen as

$$\mathbf{A} = \begin{bmatrix} 0.3891 & -0.4906 \\ 0.2909 & 0.1509 \end{bmatrix} .$$

Afterwards, a nonlinear function, $f(x)$, was applied individually to each component $\mathbf{z}(i) = [f(v_1(i)), \ldots, f(v_n(i))]^T$. In order to assess the behaviour of the system and estimate the quality of the outputs we used the performance measure

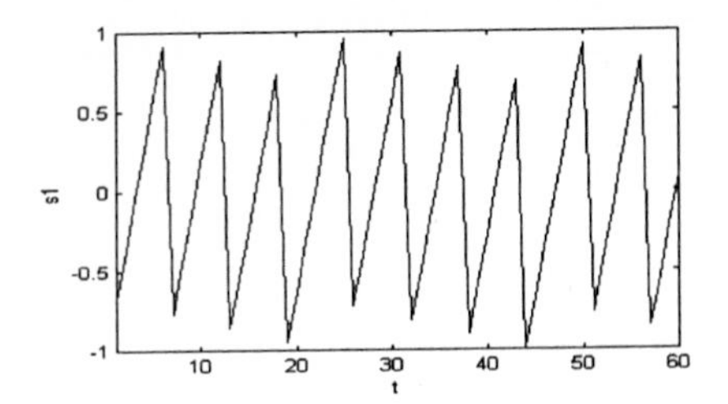

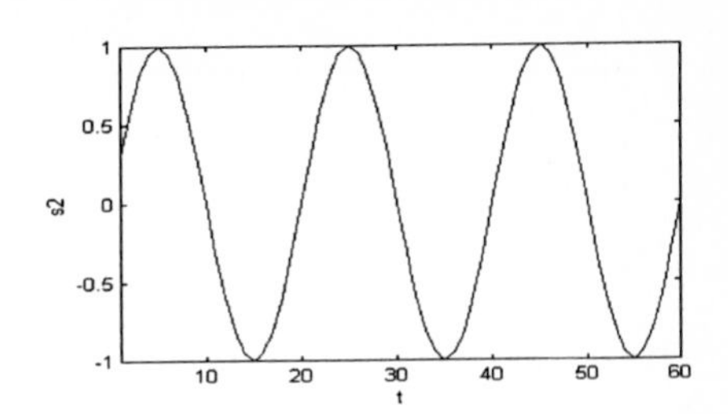

Fig. 2. Source signals.

proposed by Schobben et al. [11]. Several experiments were carried out employing different nonlinearities. Figure 3 shows the results for $f(x) = c_1x^5 + c_2x^3 + c_3x$, where $c_1 = 1.5, c_2 = -1.1$ and $c_3 = 0.8$ were chosen randomly. The signals on top are the nonlinear mixtures, the middle ones are the outputs of the FN-ANN network and finally the training curves at bottom show the performance measure for each of the outputs. The step sizes were initialized to 0.5 and the value of μ was decreased during the training, to avoid stabilization problems, employing the exponential law $\mu(n) = 0.995\mu(n-1)$. Also, the smoothing parameter (σ^2) of the kernel function was held to $1/\sqrt{m}$. This value was obtained empirically as a modification of that proposed in [5].

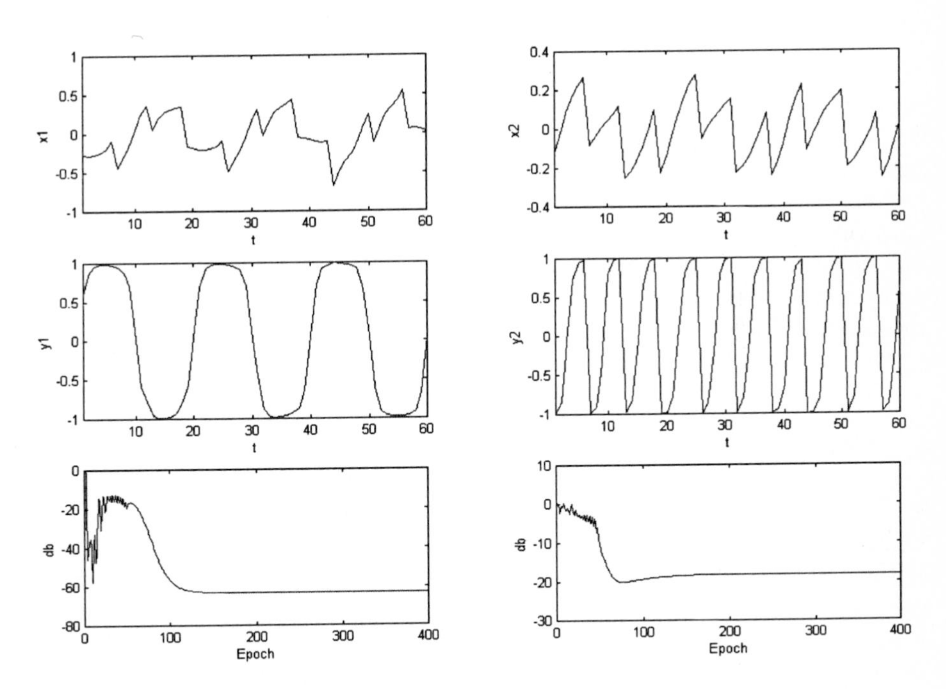

Fig. 3. Results employing the nonlinear function $f(x) = c_1x^5 - c_2x^3 + c_3x$.

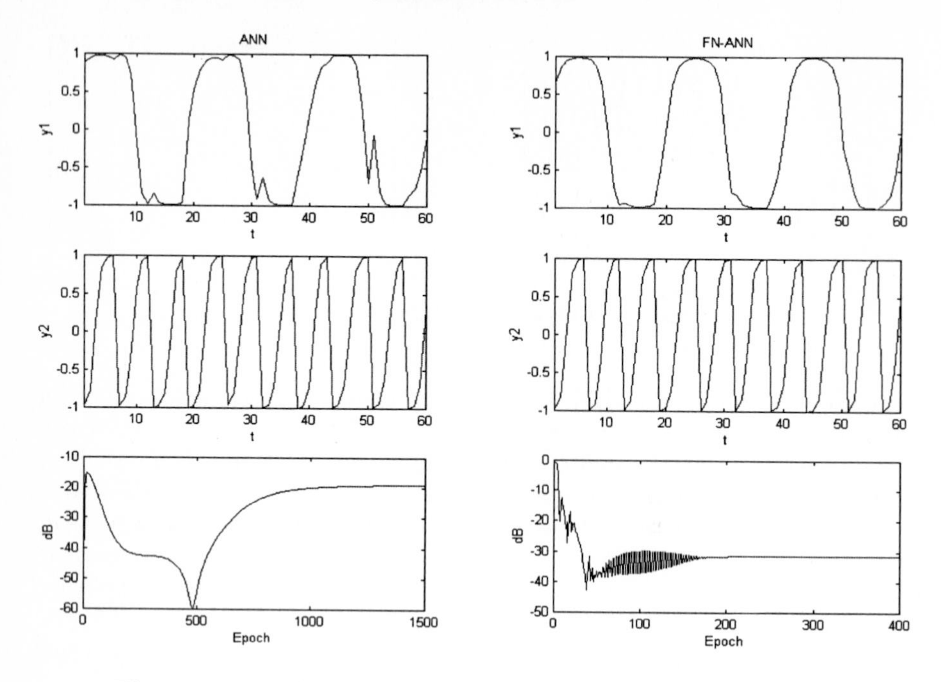

Fig. 4. Results employing the nonlinear function by $f(x) = tanh(x)$.

Moreover, we compared the performance of the FN-ANN system with the same architecture (figure 1) but without the first layer of functional units. This new system, that is only an artificial neural network (ANN), was proposed in [10]. Again, several simulations were accomplished with different nonlinear functions and in all cases the FN-ANN obtained better results than the ANN. Figure 5 contains the results of both systems using the distortion function $f(x) = tanh(x)$. The signals on the left correspond to the two outputs of the ANN and the mean performance curve obtained during the training, while signals on the right correspond to the FN-ANN. It can be observed that the FN-ANN exhibits superior performance (-31.33 dB) than the ANN (-18.93 dB) at convergence.

5 Conclusion

In this paper, a method for post-nonlinear component analysis is presented. The proposed approach combines two different paradigms, functional networks and neural networks, to recover the original data. Moreover, mutual information measure according to Renyi's entropy was applied to avoid computational complexities that arise from using Shannon's definition. As it was showed, the proposed hybrid network yields goods results in test examples and exhibits better performance than a one-layer feedforward neural network also trained with a gradient descent learning and the same cost function. However, it is well-known that the choice of the smoothing parameter is of crucial importance in density estimation and therefore in the global performance of the system. Hence, and as a future work, techniques to select this parameter as a function of the data will be analysed.

References

1. Burel, G.: A non-linear neural algorithm. Meural Networks **5** (1992) 937–947
2. Castillo, E.: Functional Networks. Neural Processing Letters, **7**(3) (1998) 151–159
3. Castillo, E., Cobo, A., Gutierrez, J.M., Pruneda, R.E.: Functional Networks with Applications. A Neural-Based Paradigm. Kluwer Academic Publishers, Dordrecht (1998)
4. Comon, P.: Independent Component Analysis–a new concept?. Signal Processing **36** (1994) 287–314
5. Duda, R.O., Hart, P.E.: Pattern Classification and Scene Analysis. John Wiley & Sons, New York (1973)
6. Hyvarinen, A., Pajunen, P.: Nonlinear Independent Component Analysis: Existence and Uniqueness results. Neural Networks **12**(3) (1999) 429–439
7. Jutten, C., Herault, J.: Blind separation of sources, part I: An adaptive algorithm based on neuromimetic architecture. Signal Processing **24** (1991) 1–10
8. Pajunen, P.: Nonlinear independent component analysis by self-organizing maps. Proceedings ICANN (1996) 815–819
9. Principe, J.C., Fisher III, J., Xu, D.: Information Theoretic Learning. In S. Haykin (Ed.): Unsupervised Adaptive Filtering, Vol. I. John Wiley & Sons, New York (2000)
10. Principe, J.C., Xu, D.: Information Theoretic Learning using Renyi's Quadratic Entropy. Proceedings ICA'99 (1999) 407–412
11. Schobben, D.W.E., Torkkola, K., Smaragdis, P.: Evaluation of Blind Signal Separation Methods. Proceedings ICA'99 (1999) 261–266
12. Silverman, B.W.: Density estimation for statistics and data analysis. Chapman & Hall, London (1986)
13. Taleb, A., Jutten, C.: Nonlinear source separation: The post-nonlinear mixtures. Proceedings ESANN (1997) 279–284
14. Taleb, A., Jutten, C.: Source separation in post-nonlinear mixtures. IEEE Trans. on Signal Processing **47**(10) (1999) 2807–2820
15. Xu, D., Principe, J.C., Fisher III, J., Wu, H.-C.: A Novel Measure for Independent Component Analysis (ICA). Proceedings ICASSP, Vol. II (1998) 1161–1164

Optimal Modular Feedforward Neural Nets Based on Functional Network Architectures

A.S. Cofiño and José M. Gutiérrez

Dept. of Applied Mathematics and Computer Science
Universidad de Cantabria, Santander 39005, Spain
{cofinoa, gutierjm}@unican.es
http://personales.unican.es/gutierjm

Abstract. Functional networks combine both domain and data knowledge to develop optimal network architectures for some types of interesting problems. The topology of the network is obtained from qualitative domain knowledge, and data is used to fit the processing functions appearing in the network; these functions are supposed to be linear combinations of known functions from appropriate families. In this paper we show that these functions can also be estimated using feedforward neural nets, making no assumption about the families of functions involved in the problem. The resulting models are optimal modular network architectures for the corresponding problems. Several examples from nonlinear time series prediction are used to illustrate the performance of these models when compared with standard functional and neural networks.

1 Introduction

In the last decades an enormous research activity has been devoted to several computing techniques inspired in different biologic mechanisms. Among these biological metaphors, feedforward neural networks are simple computing techniques which have proven their efficiency in many practical problems. Multilayer perceptrons (MLPs) are layered feedforward networks with sigmoidal functional units connected by weights [1]. Once a standard network architecture is chosen, data from the problem of interest is used to fit the weights using backpropagation- like algorithms. Their main advantage is that they are easy to use and can approximate any input/output map [2]. The key disadvantage is their rigid structure of fully connected layers with many degrees of freedom that may overfit the data, train slowly, or converge to local minima.

1.1 Modular Neural Networks

Several attempts to solve the above problems deal with more flexible feedforward network topologies based on modules [3–5]; modules are groups of unconnected neurons, each connected to the same set of nodes. The concept of modularity is linked to the notion of local computation, in the sense that each module is an independent system and interacts with others in a whole architecture, in order

J. Mira and A. Prieto (Eds.): IWANN 2001, LNCS 2084, pp. 308–315, 2001.

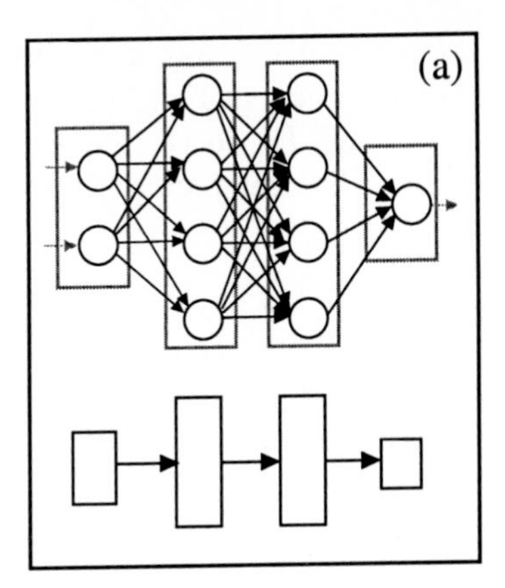

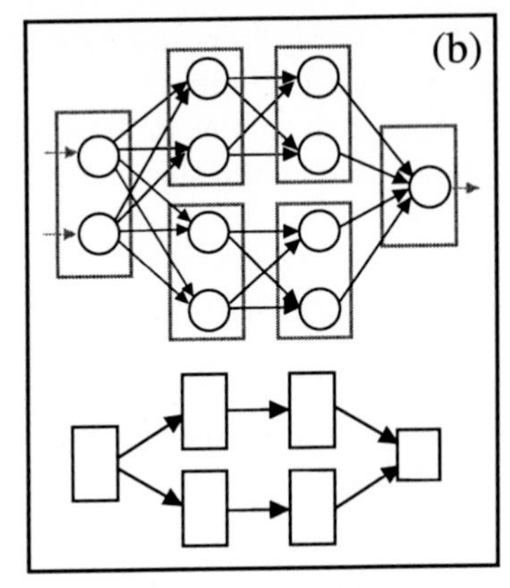

Fig. 1. (a) A fully connected feedforward $2:4:4:1$ MLP; (b) a modular network.

to perform a given task. The local structure and the complexity reduction of modular networks have shown to overcome some of the problems of fully connected feedforward nets [6]. However, in order to have meaningful and efficient models, each module has to perform an *interpretable* and *relevant* function according to the mathematical or physical properties of the system. Unfortunately, it is unclear how to best design the modular topology based on the data.

For instance, given the trivial modular network shown in Fig. 1(a) (a MLP consisting of four modules: an input layer, two hidden layers and an output layer, respectively), several nontrivial modular networks can be easily obtained by splitting up some layers into sub-layers, hereby reducing the number of weights (one of such modular networks is shown in Fig. 1(b)). However, there is no general procedure to design optimal modular structures according to some available domain knowledge. This problems has been partially addressed with functional networks.

1.2 Functional Networks

Functional networks are a soundness and efficient generalization of neural networks which allow combining both domain and data knowledge to overcome the above limitation of modular neural nets, leading to optimal functional structures for several problems [7]. For this purpose, the topology of the network (functional units and connections) is obtained using qualitative knowledge of the problem at hand (symmetries or other functional properties) which are analyzed and simplified using functional equations [8]. Then, the resulting functional units are fitted to data as linear combinations of a predetermined set of appropriate functions (polynomial, trigonometric Fourier functions, etc.) [9, 10]. These models have been applied to several practical problems outperforming standard neural networks (see [11] for an overview of functional network applications).
Apart from the input, output and hidden layers of processing units, a functional network also includes a layer of *intermediate units*, which do not perform any

computation, but only store intermediate information and may force the outputs of different processing units to be equal. For instance, in Fig. 2(a) the output unit u is connected to the processing units I and F_3, so their outputs must be coincident. Therefore, this network is the graphical representation of the functional equation:

$$F_4(x,y) = F_3\left(F_1(x,z), F_2(y,z)\right), \ \forall x, y, z. \tag{1}$$

This property characterizes the so-called general associativity models, which generalize the class of models which combine the separate contribution of independent variables.

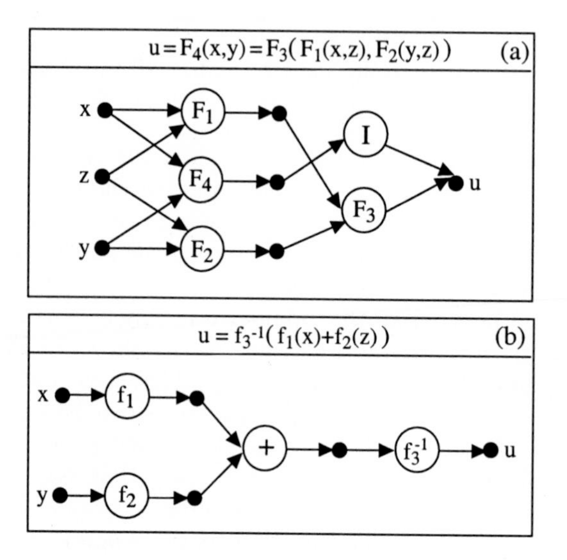

Fig. 2. (a) Functional network with two neurons connected to the same intermediate unit (dots represent intermediate units, circles represent the processing units, and I denotes the identity function); (b) simplified functional network.

The above equation imposes some constraints in the neuron functions $F_1, \ldots, F_4$, allowing the simplification of the functional structure. The solution to this functional equation is given by the following theorem [8].

Theorem 1. *The general solution continuous on a real rectangle of the functional equation (1) with F_2 invertible in the first argument, F_1 invertible in both arguments, F_3 invertible in the first argument and F_4 invertible in the second, is*

$$\begin{array}{ll} F_4(x,y) = f_3^{-1}\left(f_1(x) + f_2(y)\right), & F_3(x,y) = f_3^{-1}\left(f_4(x) + f_5(y)\right), \\ F_1(x,y) = f_4^{-1}\left(f_1(x) + f_6(y)\right), & F_2(x,y) = f_5^{-1}\left(f_2(x) - f_6(y)\right), \end{array} \tag{2}$$

where $f_1, \ldots, f_6$ are arbitrary continuous and strictly monotonic functions. □

Replacing (2) in (1) we obtain the simplest functional expression of general associativity models models

$$u = F_4(x, y) = F_3 (F_1(x, z), F_2(y, z)) = f_3^{-1} (f_1(x) + f_2(y)), \quad (3)$$

defining the functional network in Figure 2(b).

The problem of learning the above functional network reduces to estimating the neuron functions f_1, f_2 and f_3 from the available data of the form (x_i, y_i, u_i), $i = 1, \ldots, n$. To this aim each of the functions f_s in (3) is supposed to be a linear combination of known functions from given families $\phi_s = \{\phi_{s1}, \ldots, \phi_{sm_s}\}$, $s = 1, 2, 3$, i.e.,

$$\hat{f}_s(x) = \sum_{i=1}^{m_s} a_{si} \phi_{si}(x); \; s = 1, 2, 3,$$

where the coefficients a_{si} are the parameters of the functional network (see [9, 10] for implementation details).

1.3 An Illustrative Example

In order to illustrate the performance of the above methodologies, we have implemented several simple experiments using the Matlab Neural Network Toolbox (`http://www.mathworks.com`) and some Matlab code for functional networks developed by the authors. First, we generated some data by randomly simulating a hundred of points from the model

$$u_i^3 = x_i + x_i^2 + y_i^2 + y_i^3 + \epsilon_i, \; i = 1, \ldots, 100, \quad (4)$$

where ϵ is a noise term given by a uniform random variable in $(-0.01, 0.01)$. The dots in Fig. 3 show the resulting data used for training several modular and functional networks; the 25×25 noiseless points in the regular grid were used for validating the models:

1. *Feedforward networks:* As a first experiment, the network in Fig. 1(a) was used as a black-box method for this problem, fitting the weights to data using backpropagation algorithm; we performed ten experiments with networks consisting of two neurons in each of the hidden layers (2 : 2 : 2 : 1 MPL with 15 parameters) and random initial weight configurations. The average Root Mean Squared validation Error (RMSE) obtained in these experiments was 0.0074. The same experiment was repeated considering three neurons in each of the hidden layers (2 : 3 : 3 : 1 with 25 parameters) obtaining a significant improvement (RMSE=0.0031).
2. *Modular networks:* The same experiment was performed using the modular network shown in Fig. 1(b) (29 parameters) obtaining similar results (RMSE=0.0033). No improvement was achieved due to the wrong selection of the modular structure for this problem.

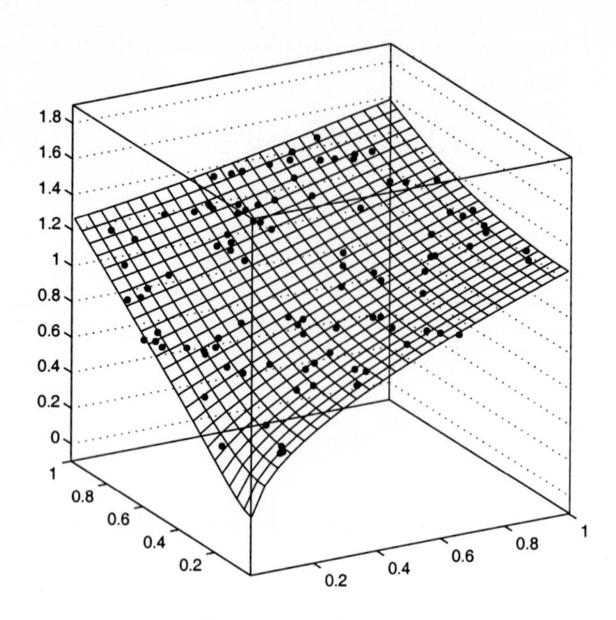

Fig. 3. Training data (dots) and validation grid.

3. *Functional networks:* Suppose we know that the underlying model satisfies the generalized associative property and also, that the functions involved are polynomials (domain knowledge). Then we can use the functional network shown in Fig. 2(b) selecting an appropriate functional family, say $\{1, x, x^2, x^3\}$, obtaining the model:

$$\begin{aligned} f_1(x_1) &= 0.3350 + 0.322x_1 + 0.352x_1^2 - 0.0095x_1^3, \\ f_2(x_2) &= 0.334 - 0.00372x_2 + 0.331x_2^2 + 0.339x_2^3, \\ f_3(y) &= 0.659 + 0.0322y - 0.037y^2 + 0.346y^3, \end{aligned}$$

with a RMSE $= 0.0027$ for the validation data. Note how the error is substantially reduced, even though the functional model consists only of 12 parameters.

The efficiency of the functional network in the above example is due to the knowledge of the optimal network structure and the optimal family of functions to be chosen. However, in general it may not be possible to specify such an optimal family of functions and the functional network may not improve fully connected MLPs. In these cases, it would be desirable to learn the neuron functions using some standard non-parametric technique, such as neural networks.

2 Functional Networks as Modular Neural Networks

Figure 4 illustrates the architecture resulting from the functional network in Fig. 2(b) when using MLPs for approximating each of the processing units in

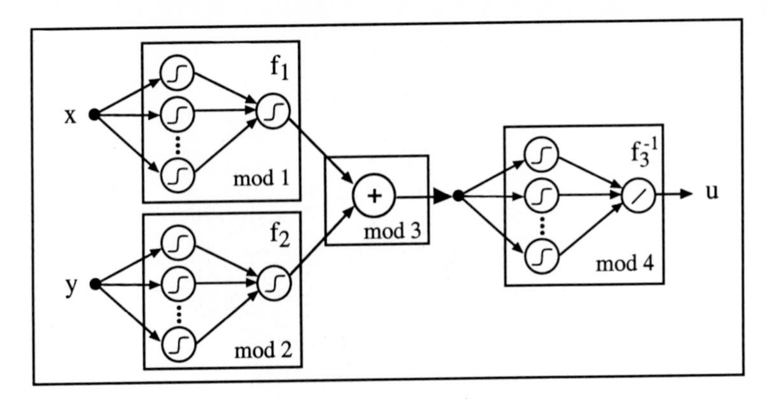

Fig. 4. Modular neural network resulting form the functional network in Fig. 2(b).

the functional network. The resulting model can be considered a particular type of modular neural network, where some of the processing units may be known (such as the sum operator in module 3) and other may be unknown (modules 1, 2 and 4); then, the whole structure can be estimated from data using modular backpropagation [5]. Therefore, the resulting network is an optimal modular neural network architecture for this problem.

Using this model for the problem in Section 1.3 and considering MLPs with single hidden layer of two neurons for each of the neural modules ($7+7+6=20$ parameters) we obtained a RMSE=0.003. This result is similar to the one reported for functional networks and outperforms the one obtained with non-optimal modular neural nets. In the following section we illustrate the performance of these models using a more complex example from time series forecasting.

3 Nonlinear Time Series Prediction

A challenging problem for time series analysis techniques is modeling nonlinear chaotic systems from observed data. Feedforward neural networks [12] and functional networks [11] have been efficiently applied to this problem inferring deterministic nonlinear models from data. In this section we show that the modular functional/neural networks introduced this paper outperform standard functional and neural networks for this task.
We analyze the special case of 2D delayed maps of the form $x_n = F(x_{n-1}, x_{n-2})$; one of the most illustrative and widely studied of these models is the non-differentiable Lozi map [13]:

$$x_n = 1 - 1.7\,|x_{n-1}| + 0.5\,x_{n-2}. \tag{5}$$

We generated a time series $\{x_n\}$, $n = 1, \ldots, 500$, corresponding to the initial conditions $x_0 = 0.5$ and $x_1 = 0.7$. This time series is represented by the dots in Fig. 5, which also shows the surface (5) embedding the time series and two slices

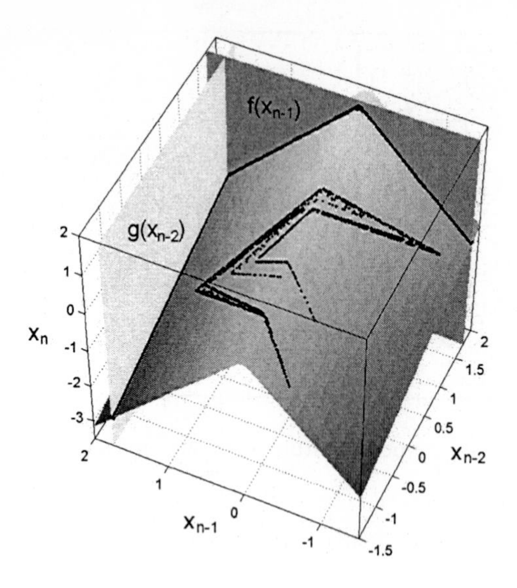

Fig. 5. Lozi time series (dots) over the surface $x_n = 1-1.7\,|x_{n-1}|+0.5\,x_{n-2}$. The planes show the slices $f(x_{n-1})$ and $g(x_{n-2})$ corresponding to fixed values of $x_{n-2} = 1.75$ and $x_{n-1} = 1.75$, respectively.

of the surface corresponding to fixed values of each of the variables x_{n-1} and x_{n-2}, respectively. From this figure we can see that the rest of slices keep the same functional structure (a linear function, or a tent-like function); therefore we can suppose the model satisfies the general associativity property (in this particular case the function $f_3^{-1}(x)$ is simply the identity function) and use the functional network in Fig. 4 without module number 4 for this problem (this particular case of associative functional network is known as *separable network*). Separable functional networks were applied to this problem using polynomial and Fourier families for the processing units f_1 and f_2 [14]. The results obtained when using a Fourier trigonometric family $\{sin(x), \ldots, sin(mx), cos(x), \ldots, cos(mx)\}$ for both f_1 and f_2 are shown in Table 1 under the label FN (the model contains $4m$ parameters); this table also includes the training errors obtained with two different modular neural/functional networks. In the first case (MFNN1), we trained a symmetric modular network with a m-neuron single hidden layer MLP for each of the processing units f_1 and f_2 ($6m$ parameters). The second modular network (MFNN2) is an asymmetric one with $2m-2$ and 2 neurons for units f_1 and f_2, respectively (in this case, we give more flexibility to the functional term x_{n-1} of the time series, keeping the number of parameters).

From Table 1 we can see how both modular functional/neural nets outperform the results obtained using a predetermined functional family based on Fourier terms (similar results are also obtained using polynomial functions); this shows that, in this case, there is no information to choose a convenient functional family for the problem and, therefore, it is more efficient to consider general input/output approximators, such as neural networks.

Table 1. Performance of several Fourier functional networks (FN) modular symmetric and asymmetric functional/neural networks (MFNN1 and MFNN2, respectively) for the Lozi time series.

FN		MFNN1		MFNN2	
parameters	RMSE	parameters	RMSE	parameters	RMSE
$m = 6$ (24 par)	0.0214	$m = 4$ (24 par)	$9.0\,10^{-3}$	$m = 4$ (24 par)	$2.7\,10^{-3}$
$m = 8$ (32 par)	0.0121	$m = 5$ (30 par)	$4.9\,10^{-3}$	$m = 5$ (30 par)	$8.6\,10^{-4}$
$m = 9$ (36 par)	$9.2\,10^{-3}$	$m = 6$ (36 par)	$2.5\,10^{-3}$	$m = 6$ (36 par)	$5.7\,10^{-4}$
$m = 11$ (44 par)	$5.3\,10^{-3}$	$m = 7$ (42 par)	$1.5\,10^{-3}$	$m = 7$ (42 par)	$4.1\,10^{-4}$

From the above example we note the convenience of distributing the neurons among the different modules, giving more flexibility to those functions performing harder tasks (the best results are obtained with an asymmetric modular network which gives more flexibility to one of the modules). We started investigating this problem using genetic algorithms for obtaining the optimal configuration (layers and number of neurons) used for each of the modules; this study will be the scope of a future paper.

References

1. Hertz, J., Krogh, A. and Palmer, R.G.: Introduction to the Theory of Neural Computation. Addison Wesley, 1991.
2. Cybenko, G.: Approximation by supperpositions of a sigmoidal function. Mathematics of Control, Signals, and Systems **2**, 303-314, 1989.
3. Wermter, S. and Sun, R. editors: Hybrid Neural Systems. Springer-Verlag (Lecture Notes in Computer Science, Vol. 1778), 2000.
4. Happel, B. and Murre, J.: Design and evolution of modular neural network architectures. Neural Networks **7**, 985-1004, 1994.
5. Ballard, D.: Modular learning in neural networks. Sixth National Conference on AI **1**, 13-17, 1987.
6. Boers, E.J.W., Kuiper, H., Happel, B.L.M., and Sprinkhuizen- Kuyper, I.G.: Designing modular artificial neural networks. In H.A. Wijshoff, editor: Proceedings of Computing Science in The Netherlands, 87-96, 1993.
7. Castillo, E., Cobo, A., Gutiérrez, J.M. and Pruneda, E.: An Introduction to Functional Networks with Applications. Kluwer Academic Publishers, 1999.
8. Aczél, J.: Lectures on Functional Equations and Their Applications. Academic Press, 1966.
9. Castillo. E.: Functional networks. Neural Processing Letters **7**, 151-159, 1998.
10. Castillo, E. and Gutiérrez, J.M.: A minimax method for learning functional networks. Neural Processing Letters **11**, 39-49, 2000.
11. Castillo, E., Gutiérrez, J.M., Hadi, A.S. and Lacruz, B.: Some applications of functional networks in statistics and engineering. Technometrics **43**, 10-24, 2001.
12. Stern, H.S.: Neural networks in applied statistics. Technometrics **38**, 205-214, 1993.
13. Misiurewicz, M.: The Lozi map has a strange attractor. In R. Hellemann, editor: Nonlinear Dynamics, Annals of the NY Academy of Science **357**, 348-358, 1980.
14. Castillo, E. and Gutiérrez, J.M.: Nonlinear time series modeling and prediction using functional networks. Physics Letters A **244**, 71-84, 1998.

Optimal Transformations in Multiple Linear Regression Using Functional Networks

Enrique Castillo[1], Ali S. Hadi[2], and Beatriz Lacruz[3]

[1] Department of Applied Mathematics, University of Cantabria, Spain,
[2] Department of Statistical Sciences, Cornell University, USA and Department of Mathematics, The American University in Cairo, Egypt,
[3] Department of Statistical Methods, University of Zaragoza, Spain.

Abstract. Functional networks are used to determine the optimal transformations to be applied to the response and the predictor variables in linear regression. The main steps required to build the functional network: selection of the initial topology, simplification of the initial functional network, uniqueness of representation, and learning the parameters are discussed, and illustrated with some examples.

1 Introduction

Functional networks are a very useful general framework for solving a wide range of problems in probability and statistics such as characterization of univariate and bivariate distributions, finding conjugate families of distributions, obtaining reproductive families and stable families with respect to maxima operations, modeling fatigue problems, computing convenient posterior probability distributions, conditional specification of statistical models, time series and regression modeling, etc. (see [5], [9], and [7]) for further details of the above mentioned applications).

In this paper we are concerned with the problem of discovering nonlinear transformations of the variables in multiple linear regression models. A multiple linear regression model expresses the relationship between a dependent or response variable Y and k predictor or explanatory variables, $X_1, X_2, \ldots, X_k$. The relationship is formulated as

$$Y = \beta_0 + \beta_1 X_1 + \beta_2 X_2 + \ldots + \beta_k X_k + \varepsilon, \tag{1}$$

where $\beta_0, \beta_1, \beta_2, \ldots, \beta_k$ are the parameters of the model and ε is a random disturbance or error that measures the discrepancy in the approximation. Parameters are estimated from data consisting of n observations on the response and the explanatory variables. Each observation can be written as

$$y_i = \beta_0 + \beta_1 x_{1i} + \beta_2 x_{2i} + \ldots + \beta_k x_{ki} + \varepsilon_i, \tag{2}$$

for $i = 1, 2, \ldots, n$. The ε's are assumed to be independent normal random variables with zero mean and common variance σ^2.

J. Mira and A. Prieto (Eds.): IWANN 2001, LNCS 2084, pp. 316-324, 2001.

The evidence as to whether a variable transformation might be helpful is based on theoretical considerations about the true relationship among the variables; empirical indications such as a nonnegative response, a response with large range or the presence of outliers; and the observation of systematic departures from the assumptions after examining the residuals from the fit.

Furthermore, the usefulness of the linear model lies on the possibility of converting nonlinear into linear relationships, satisfying the hypothesis above, by transforming the variables properly. Transformations can be required in the response variable, in some or all of the predictors, or in both. But, the appropriate transformations are usually selected by the exploration of their effects on the model. Then, several possibilities can be tried before deciding which one is the best. See [2], [4], and [10] for further details. Functional networks provide a new method for selecting these transformations automatically.

In the remainder of the paper we assume that our data come in a form that is not immediately suitable for analysis, and transformations are the more appropriate solution to ensure linearity, to achieve normality, and/or to stabilize the variance. It is a matter of empirical observation that all three criteria are often satisfied by the same transformation.

2 Identifying Transformations with Functional Networks

Working with functional networks requires several steps:

1. Selecting the initial topology of the functional network.
2. Simplifying the initial functional network.
3. Solving the uniqueness of representation to avoid estimation problems.
4. Learning the parameters of the neural functions.

2.1 Selection of the Initial Topology

The selection of the initial topology of the functional network is based on the characteristics of the problem at hand, which usually lead to a single clear network structure. We are concerned with the problem of discovering the structure of the transformations f (sometimes assumed to be invertible) and h in

$$f(Y) = h(X_1, X_2, \ldots, X_k), \tag{3}$$

that lead to a linear model in the parameters, as that specified in (1).

To approximate the right hand side of this equation we only consider two possible functional networks known as separable models. The first one (Model I) is expressed by the functional relation

$$f(Y) = \sum_{r_1=1}^{q_1} \sum_{r_2=1}^{q_2} \cdots \sum_{r_k=1}^{q_k} c_{r_1 r_2 \ldots r_k} \phi_{r_1}(X_1) \phi_{r_2}(X_2) \ldots \phi_{r_k}(X_k), \tag{4}$$

where $c_{r_1 r_2 \ldots r_k}$ are some of the parameters in the model, $\Phi_s = \{\phi_{r_s}(X_s),\ r_s = 1, 2, \ldots, q_s\}$ are families of linearly independent functions, and q_s is the number of elements in Φ_s, $s = 1, 2, \ldots, k$. This model can be represented by the functional network shown in Figure 1, where we have considered two variables and the same number of functions in both families.

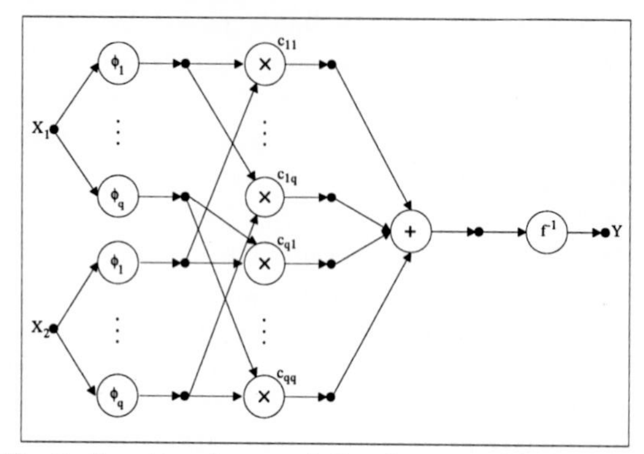

Fig. 1. Functional network for the separable model I.

If, for illustrative purposes, we consider $k = 2$ explanatory variables, and we use the polynomial families $\Phi_s = \{1, X_s, X_s^2, \ldots, X_s^q\}$, where $s = 1, 2$ and $q_s = q + 1$, the right hand side of (4) can be written as:

$$f(Y) = h_1(X_1) + h_2(X_2) + h_3(X_1, X_2), \tag{5}$$

where, without loss of generality, the constant term is included in $h_1(X_1)$, and the structure of $h_1(X_1), h_2(X_2)$ and $h_3(X_1, X_2)$ is given below.

$$\begin{array}{rr|llll|l}
\hline
 & c_{00} \cdot 1 \cdot 1 & +c_{01} \cdot 1 \cdot X_2 & +c_{02} \cdot 1 \cdot X_2^2 & +\ldots & +c_{0q} \cdot 1 \cdot X_2^q & \rightarrow h_2(X_2) \\
\cline{3-6}
 & +c_{10} \cdot X_1 \cdot 1 & +c_{11} \cdot X_1 \cdot X_2 & +c_{12} \cdot X_1 \cdot X_2^2 & +\ldots & +c_{1q} \cdot X_1 \cdot X_2^q & \\
h_1(X_1) \leftarrow & +c_{20} \cdot X_1^2 \cdot 1 & +c_{21} \cdot X_1^2 \cdot X_2 & +c_{22} \cdot X_1^2 \cdot X_2^2 & +\ldots & +c_{2q} \cdot X_1^2 \cdot X_2^q & \\
 & +\ldots & +\ldots & +\ldots & +\ldots & +\ldots & \rightarrow h_3(X_1, X_2) \\
 & +c_{q0} \cdot X_1^q \cdot 1 & +c_{q1} \cdot X_1^q \cdot X_2 & +c_{q2} \cdot X_1^q \cdot X_2^2 & +\ldots & +c_{qq} \cdot X_1^q \cdot X_2^q & \\
\hline
\end{array}$$

The second model (Model II) is expressed by the functional relation

$$f(Y) = \sum_{r=1}^{|J|} \prod_{j \in J_r} g_{rj}(X_j), \tag{6}$$

where $J = \{J_1, \ldots J_{|J|}\}$, $J_r \subseteq \{1, \ldots, k\}$, and $|J|$ is the number of subsets in J.

For example, if $k = 2$, $J = \{\{1\}, \{2\}, \{1, 2\}\}$, and $|J| = 3$, the model in equation (6) becomes

$$f(Y) = \sum_{r=1}^{|J|} g_{r1}(X_1) g_{r2}(X_2) = g_{11}(X_1) + g_{22}(X_2) + g_{31}(X_1) g_{32}(X_2), \tag{7}$$

where $g_{12}(X_2) = g_{21}(X_1) = 1$. Note that this is a particular case of the model in (5), where $h_3(X_1, X_2)$ is now the product $g_{31}(X_1)g_{32}(X_2)$.

This model can be represented by the functional network shown in Figure 2. These models correspond to the linear (in the parameters) model that

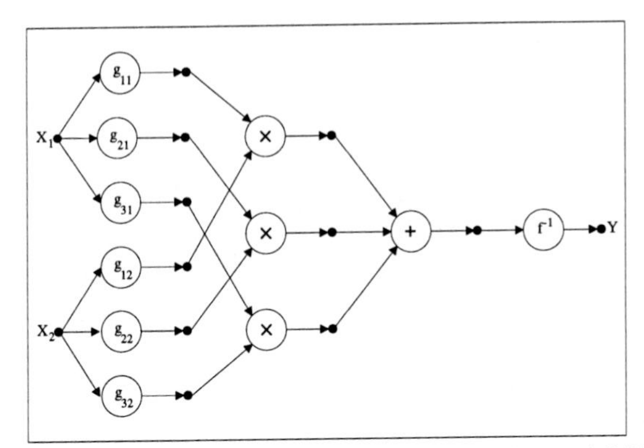

Fig. 2. Functional network for the separable model II.

includes polynomial terms in both sides, and interactions of the explanatory variables. The additive model is a particular case of both. For example, if $J = \{\{1\}, \{2\}, \ldots, \{k\}\}$, the additive model without interactions resulting from (6) becomes

$$f(Y) = \sum_{r=1}^{|J|} \prod_{j \in J_r} g_{rj}(X_j) = g_{11}(X_1) + g_{22}(X_2) + \ldots + g_{kk}(X_k), \tag{8}$$

i.e., the linear model in equation (1) after transforming both sides. For simplicity, in the rest of this paper we work with this model and consider only two ($k = 2$) explanatory variables. Figure 3 shows the functional network for this additive model, an extension of the well-known associative model.

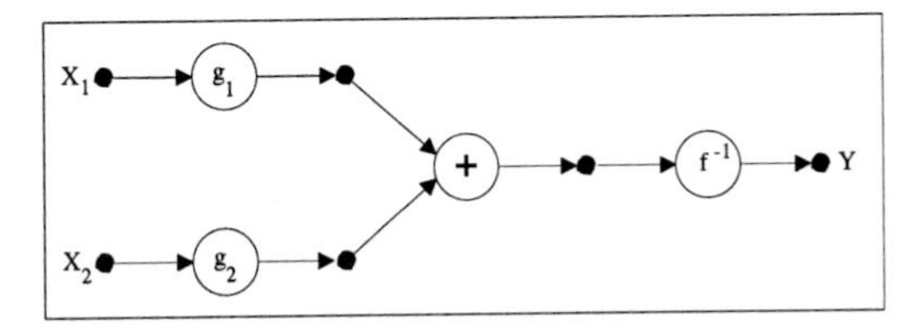

Fig. 3. Functional network for the additive model.

2.2 Simplifying Functional Networks

The simplification of the initial functional network is done using functional equations. For a general introduction to functional equations and methods to solve them see [1] or [8].

In the three cases introduced above, since there are no interconnections among neuron outputs, simplification does not apply.

2.3 Uniqueness of Representation

To avoid estimation problems, the uniqueness of representation of the network must be analyzed before learning the neural functions. Non-uniqueness occurs when there are several sets of functions leading to the same output for any input. For example, in the additive model (8), for two explanatory variables, when there exist two different triplets of functions $\{g_1, g_2, f\}$ and $\{g_1^\star, g_2^\star, f^\star\}$ such that

$$f^{-1}[g_1(X_1) + g_2(X_2)] = f^{\star-1}[g_1^\star(X_1) + g_2^\star(X_2)], \ \forall X_1, X_2.$$

This functional equation has as general solution (see [5]):

$$g_1(X) = a\, g_1^\star(X) + b; \ \ g_2(X) = a g_2^\star(X) + c; \ \ f(X) = a\, f^\star(X) + b + c, \tag{9}$$

where a, b, and c are arbitrary constants.

Thus, to have uniqueness of solution we need to fix each of the functions f, g_1, and g_2 at a point. Appart from some points, that lead to no solution, different points lead to different but equivalent models, i.e., they have no influence on the final models, as it will be shown.

2.4 Parametric Learning

When dealing with functional networks, there are two types of parametric learning:

- *Exact learning.* It consists of identifying the functions that are solutions of the functional equation associated with the functional network, and
- *Approximate learning.* It consists of estimating the neural functions based on given data.

Since a data set is always available in multiple linear regression modeling, approximate learning is always possible. The learning algorithm consists of minimizing a function of the errors to obtain $\hat{f}$, $\hat{g}_1$, and $\hat{g}_2$.

Let $D = \{y_i, x_{1i}, x_{2i}; i = 1, 2, \ldots, n\}$ be a set of observations of the response and the explanatory variables. It is assumed that, in the range of the observations studied, the additive model provides an acceptable approximation to the true relation between the response and the predictor variables. Then for each observation in D, we can consider the residuals

$$e_i = \hat{f}(y_i) - \hat{g}_1(x_{1i}) - \hat{g}_2(x_{2i}), \ i = 1, \ldots, n,$$

and use several (loss) functions of the errors such as the least squares method, the minimax method or the minimum description length method ([6], and [7]). Here, we have chosen the least squares method because it is the most commonly used estimation method in multiple linear regression. Thus, we minimize the sum of squared errors

$$\sum_{i=1}^{n} e_i^2 = \sum_{i=1}^{n} \left(\sum_{j=1}^{q_0} a_j \phi_{0j}(y_i) - \sum_{j=1}^{q_1} b_j \phi_{1j}(x_{1i}) - \sum_{j=1}^{q_2} c_j \phi_{2j}(x_{2i}) \right)^2 \tag{10}$$

subject to

$$\sum_{j=1}^{q_0} a_j \phi_{0j}(y_0) = \alpha_0, \qquad \sum_{j=1}^{q_1} b_j \phi_{1j}(x_{10}) = \alpha_1, \qquad \sum_{i=1}^{q_2} c_j \phi_{2j}(x_{20}) = \alpha_2. \tag{11}$$

where $\hat{f}$, $\hat{g}_1$, and $\hat{g}_2$ have been approximated by a linear combination of linearly independent functions $\Phi_s = \{\phi_{sj}, j = 1, 2, \ldots, q_s\}$, $s = 0, 1, 2$. The Lagrange multipliers technique leads to a system of linear equations.

One of the most appropriate family of functions, for transformations in linear regression, is the polynomial family $\Phi_s = \{1, t, t^2, \ldots, t^q\}$, $\forall s$. In this case, the resulting system has a unique solution if y_0, x_{10}, and x_{20} are not equal to 0. In addition, α_0, α_1, and α_2 cannot be simultaneously equal to 0, because this leads to an homogeneous system with a trivial solution. Furthermore, if one of the α's is equal to 0, the corresponding estimated function is the null polynomial when the set of approximation functions includes only one term.

For selecting the best model a search procedure is needed based on a goodness-of-fit measure. We have chosen the exhaustive method which calculates this measure for all possible models and selects the one leading to the optimum value. Even though, this method usually requires more computational power than other alternatives, a C++ program running in a PC solve all the linear equations in a few seconds.

The adjusted R_a^2 was used as the goodness-of-fit measure:

$$R_a^2 = 1 - \frac{\sum_{i=1}^{n} e_i^2/(n-p)}{\sum_{i=1}^{n} (\hat{f}(y_i) - \overline{\hat{f}(y_i)})^2/(n-1)}, \tag{12}$$

where p is the number of parameters in the model and $\overline{\hat{f}(y)} = \frac{1}{n}\sum_{i=1}^{n} \hat{f}(y_i)$. These measures allow us comparing models when different families of polynomials are used.

3 Examples of Applications

To illustrate the proposed technique, sets of sample data of size 200 for several models have been simulated. Figure 4(a) shows the scatter plot of the linear model $Y = X_1 + X_2 + \varepsilon$, where X_1 and X_2 are $U[0,5]$, and ε is $N[0,1]$. Figure

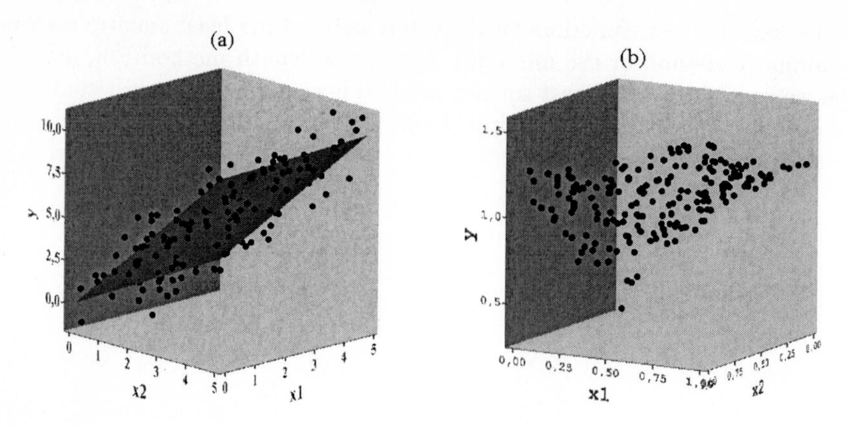

Fig. 4. Scatter plot for (a) the linear model and (b) the polynomial model.

4(b) shows the scatter plot of the polynomial model $Y^3 = X_1 + X_1^2 + X_2^2 + X_2^3 + \varepsilon$, where X_1 and X_2 are $U[0,1]$ and ε is $N[0,0.1]$. A classical linear regression analysis for the relationship among variables leads to $R_a^2 = 0.831$ for the first model, and to $R_a^2 = 0.986$ for the second. In both cases all the coefficients are significantly different from 0, except for the constant term, that can be assumed to be 0. The coefficients estimates are all close to 1.

We have also simulated the model $\ln Y = X_1 + X_2 + \varepsilon$, where X_1 and X_2 are $U[0,1]$ and ε is $N[0,0.1]$. Figure 5 shows the scatter plot for this model.

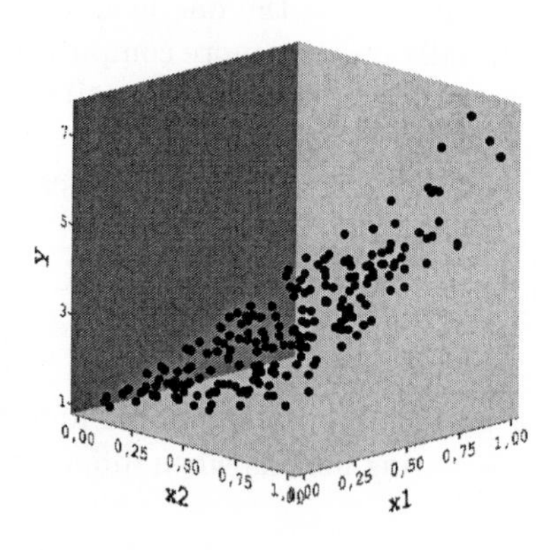

Fig. 5. Scatter plot for the exponential model.

A classical linear regression analysis for $\ln Y$ over X_1 and X_2 leads to $R_a^2 = 0.942$. All the coefficients are significantly different from 0, except for the constant term, and the estimations for the coefficients are all close to 1.

Note that for the linear model no transformation is required. However, transformations for the response and the predictor variables are looked for in the case of the polynomial model.

The exhaustive method selects the true model in the linear case with corrected $R_a^2 = 0.8303$ when using $\Phi_s = \{1, t, t^2\}$, $\forall s$. In the polynomial case, the selected model, when using $\Phi_s = \{1, t, t^2, t^3\}$, $\forall s$, is

$$Y^3 = 0.1826 + 0.7169\, X_1 + 1.2670\, X_1^2 - 0.0468\, X_2 + 2.9393\, X_2^2.$$

with $R_a^2 = 0.9839$.

Figure 6(a) shows the very good fit of the estimated polynomial for the explanatory variables and the true relationship against Y. In the exponential case using $\Phi_s = \{1, t, t^2, t^3\}$, $\forall s$, the maximum adjusted R^2 value is 0.9367 that corresponds to the model

$$-0.596813 + 0.784923\, Y - 0.11358\, Y^2 + 0.0063452\, Y^3 = 0.827367\, X_1 + 0.808534\, X_2.$$

The left hand side of the equation approximates very well the true transformation $\ln Y$, as it is shown in Figure 6(b).

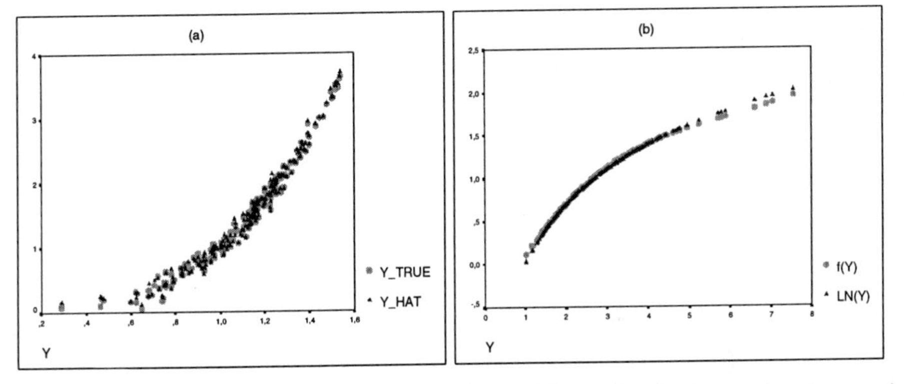

Fig. 6. Scatter plot for (a) the estimated polynomial function in the explanatory variables and the true relationship against Y, (b) the estimated polynomial function for the response variable and $\ln Y$ against Y.

4 Conclusions

Functional networks provide a new method for automatically selecting transformations in linear regression. Breiman and Friedman ([3]) have proposed also Alternate Conditional Expectations (ACE) as a method for discovering both side transformations based in computing iteratively bivariate conditional expectations. The disadvantage of their technique is that the transformations must be deduced from the graphical representation of those conditional expectations for every function $f, g_1, \ldots, g_k$ versus the $Y, X_1, \ldots, X_k$, respectively. Contrary, our approximation gives the functional form of the transformation.

More complex models than those given in Section 2.1 are also possible. Thus, functional networks can also be used for including interactions among explanatory variables and for more general and complex nonlinear regression models.

References

[1] Aczél, J.: Lectures on Functional Equations and Their Applications, Vol. **19**. Mathematics in Science and Engineering, Academic Press, New York, (1966)
[2] Atkinson, A. C.: Plots, Transformations and Regression. An Introduction to Graphical Methods of Diagnostic Regression Analysis. Clarendon Press, Oxford, (1985)
[3] Breiman, L., Friedman, J. H.: Estimating Optimal Transformations for Multiple Regression and Correlation (with comments). Journal of the American Statistical Association, Vol. **80**, No. 391, (1985) 580–619
[4] Carroll, R. J., Ruppert, D.: Transformations and Weigthing in Regression. Chapman and Hall, (1988)
[5] Castillo, E., Cobo, A., Gutiérrez, J. M., Pruneda, E.: An Introduction to Functional Networks with Applications. Kluwer Academic Publishers: New York, (1998)
[6] Castillo, E., Gutiérrez, J. M., Cobo, A., Castillo, C.: A Minimax Method for Learning Functional Networks. Neural Processing Letters, **11**,1, (2000) 39–49
[7] Castillo, E., Gutiérrez, J. M., Hadi, A. S., Lacruz, B.: Some Aplications of Functional Networks in Statistics and Engineering. Technometrics, (in press), (2001)
[8] Castillo, E., Ruiz-Cobo, R.: Functional Equations in Science and Engineering. Marcel Dekker: New York, (1992)
[9] Castillo, E. and Gutiérrez, J. M.: Nonlinear Time Series Modeling and Prediction Using Functional Networks. Extracting Information Masked by Chaos. Physics Letters A, Vol. **244**, (1998) 71–84
[10] Chaterjee, S., Hadi, A. S., Price, B.: Regression Analysis by Example. Third Edition. John Wiley and Sons, Inc., (2000)

Generalization Error and Training Error at Singularities of Multilayer Perceptrons

Shun-ichi Amari, Tomoko Ozeki and Hyeyoung Park

Laboratory for Mathematical Neuroscience
RIKEN Brain Science Institute
2-1 Hirosawa, Wako, Saitama, 351-0198, Japan
{amari,tomoko,hypark}@brain.riken.go.jp

Abstract. The neuromanifold or the parameter space of multilayer perceptrons includes complex singularities at which the Fisher information matrix degenerates. The parameters are unidentifiable at singularities, and this causes serious difficulties in learning, known as plateaus in the cost function. The natural or adaptive natural gradient method is proposed for overcoming this difficulty. It is important to study the relation between the generalization error and and the training error at the singularities, because the generalization error is estimated in terms of the training error. The generalization error is studied both for the maximum likelihood estimator (mle) and the Bayesian predictive distribution estimator in terms of the Gaussian random field, by using a simple model. This elucidates the strange behaviors of learning dynamics around singularities.

1 Introduction

Multilayer perceptrons are basic models of neural networks which are nonlinear and adaptive, having capabilities of universal function approximators. Learning takes place in the parameter space of perceptrons, which is called a neuromanifold, and it is important to study the geometrical properties of the space in order to understand the behaviors of learning dynamics.

We use the statistical approach where the input-output behavior of a perceptron is represented by a conditional probability distribution of the output y conditioned on input $\boldsymbol{x}$. The probability distributions are parameterized by the modifiable connection weights of the network. Because of the symmetric hierarchical structure of multilayer perceptrons, singularities exist in a neuromanifold, at which the parameters are redundant and unidentifiable. It has been shown that such singularities give rise to plateaus in learning, which make the learning behaviors so slow. Natural gradient learning and adaptive natural gradient learning (Amari[1], Amari, Park and Fukumizu[5], Park, Amari and Fukumizu[12]) have been proposed to overcome such difficulties in learning, where the geometrical structures of singularities are taken into account.

From the point of view of statistical inference, the conventional paradigm of the Cramér-Rao theorem does not hold at singularities. Therefore, we need to

J. Mira and A. Prieto (Eds.): IWANN 2001, LNCS 2084, pp. 325-332, 2001.

establish a different theory for analyzing the accuracy of estimation or of learning at around singularities. Abnormal phenomena at singularities were studied by Hagiwara et al[9]. Watanabe[14, 15] used the algebraic-geometrical framework to analyze the behavior in generalization error of the Bayesian predictive distribution. Fukumizu[7, 8] also analyzed the behaviors of the maximum likelihood estimator in such cases. Amari and Ozeki[4] used a simple toy model to show how singularities affect the behavior of learning.

The generalization error is an important factor for evaluating the capability of a model. There is a nice relation between the generalization error and the training error which latter can be calculated from the data. When the error is measured by the negative of the log probability, which is equivalent to the least square error in the case of Gaussian noise, the generalization error is larger than the noise entropy by a fixed term, and the training error is smaller than the noise entropy by the same fixed term. This duality was shown by Amari and Murata[2] and many others. The fixed term is the number of the parameters devided by the number of examples, as is well known in statistics, and this fact gives the AIC for model selection.

However, when one compares two models of different numbers of hidden units, their intersections are at singularities of the larger model. Therefore, the conventional Cramér-Rao paradigm which justifies the AIC (or MDL) does not hold. Hence, it is important to study the relation between the generalization error and the training error at singularities.

The present paper analyzes such a relation in the case of the mle (least square error estimator), and also in the case of the Bayesian predictive estimator. The gap between the generalization error and the training error is larger in the case of the mle than that in the regular case, but is much smaller in the case of the Bayesian predictive distribution. To this end, we use the framework of the Gaussian random field, used by Dacunha-Castelle and Gassiat[6], Fukumizu[7] and Hartigan[10].

2 Singularities in Multilayer Perceptrons

Let us consider a one hidden-layer multilayer perceptron (MLP) consisting of k hidden units and one output unit. Given an input vector $\boldsymbol{x}$, the ith hidden unit calculates a nonlinear function of a weighted sum of inputs, $\varphi(\boldsymbol{w}_i \cdot \boldsymbol{x})$, and the output unit again calculates their weighted sum $\sum v_i \varphi(\boldsymbol{w}_i \cdot \boldsymbol{x})$. The total output is disturbed by a Gaussian noise, so that its input-output behavior is written as

$$y = \sum v_i \varphi(\boldsymbol{w}_i \cdot \boldsymbol{x}) + n. \tag{1}$$

Here, $\boldsymbol{w}_i$ is the weight vector of the ith hidden unit, v_i is its weight to the output unit, φ is the sigmoid function given by tanh, and n is an additive noise subject to $N(0, 1)$.

The input-output behavior is hence represented by the conditional distribution of y given input $\boldsymbol{x}$,

$$p(y|\boldsymbol{x}) = \frac{1}{\sqrt{2\pi}} \exp\left\{-\frac{1}{2}\left(y - \sum v_i \varphi\left(\boldsymbol{w}_i \cdot \boldsymbol{x}\right)\right)^2\right\}. \tag{2}$$

Let us summarize the set of all parameters by a vector $\boldsymbol{\theta} = (\boldsymbol{w}_1, \cdots, \boldsymbol{w}_k; v_1, \cdots, v_k)$. Then, the set of all such perceptrons forms a manifold of probability distributions, where $\boldsymbol{\theta}$ is a coordinate system in it. Since it is a space of (conditional) probability distributions, information geometry(Amari and Nagaoka[3]) can be applied to it, giving a Riemannian metric tensor in terms of the Fisher information matrix.

However, the manifold of MLPs has lots of singularities, where the Fisher information matrix degenerates and the parameters are not identifiable. For example, when $v_i = 0$, the behavior of MLP is the same whatever value $\boldsymbol{w}_i$ has. Similarly, when $\boldsymbol{w}_i = \boldsymbol{w}_j$, the two neurons may be merged into one. It is known that this type of singularities give rise to plateaus in the cost function of MLP, where learning becomes extremely slow. The natural gradient and adaptive natural gradient learning rules (Amari[1], Amari, Park and Fukumizu[5], Park, Amari and Fukumizu[12]) are proposed to overcome such difficulties, and have been proved to work well (Rattray, Saad and Amari[13]).

It is important to analyze the behavior of learning and estimation of the parameters near singularities, because model selection methods in general concern about such behaviors. In the following, we study the interesting relations between the generalization error and training error, when the true model lies at a singularity, for the maximum likelihood estimator (the least square error estimator) and the Bayes predictive distribution. Their behaviors are quite different from those at regular points. We use a simple model of one hidden neuron, but we can generalize the results.

3 Gaussian Random Fields

Let us consider a simplest singular model of multilayer perceptrons, whose input-output relation is written as

$$y = \xi \varphi(\boldsymbol{w} \cdot \boldsymbol{x}) + n, \tag{3}$$

where ξ is the parameter along the cone and $\boldsymbol{w}$ is the parameter which is not identifiable when $\xi = 0$. This may be regarded as one hidden-unit perceptron, where v is replaced by ξ, but a more general case of singularities has such a cone structure. The conditional probability function of the output y is given by

$$p(y|\boldsymbol{x}, \xi, \boldsymbol{w}) = c \exp\left[-\frac{1}{2}\left\{y - \xi\varphi(\boldsymbol{w} \cdot \boldsymbol{x})\right\}^2\right]. \tag{4}$$

Let

$$D = \left\{(\boldsymbol{x}_1, y_1), \cdots, (\boldsymbol{x}_n, y_n)\right\} \tag{5}$$

be a set of n examples (training data), which are generated from the true perceptron of $\xi = 0$. This implies that y_i are independent random variable in this case subject to the standard Gaussian distribution $N(0,1)$.

The log likelihood of D is written as

$$L(\xi, \boldsymbol{w}|D) = \sum_{i=1}^{n} \log p\left(y_i|\boldsymbol{x}_i, \xi, \boldsymbol{w}\right) = -\frac{1}{2}\sum_i \left\{y_i - \xi\varphi\left(\boldsymbol{w}\cdot\boldsymbol{x}_i\right)\right\}^2. \tag{6}$$

The maximum likelihood estimator is the one that maximizes $L(\xi, \boldsymbol{w})$. However, $\partial L/\partial \boldsymbol{w} = 0$ at $\xi = 0$, so that we cannot analyze the behaviors of the mle by the Taylor expansion of the log likelihood in this case. Following Hartigan (see also Fukumizu[7] and Hagiwara et al[9]), we first fix $\boldsymbol{w}$ and search for the ξ that maximizes L. This is easy since L is a quadratic function of ξ. The maximum $\hat{\xi}$ is given by

$$\hat{\xi}(\boldsymbol{w}) = \frac{\sum y_i\varphi(\boldsymbol{w}\cdot\boldsymbol{x}_i)}{\sum \varphi^2\left(\boldsymbol{w}\cdot\boldsymbol{x}_i\right)}. \tag{7}$$

By substituting $\hat{\xi}(\boldsymbol{w})$ in (6), the log likelihood function becomes

$$\begin{aligned}\hat{L}(\boldsymbol{w}) &= L\left(\hat{\xi}(\boldsymbol{w}), \boldsymbol{w}\right)\\ &= -\frac{1}{2}\sum y_i^2 + \frac{1}{2}\frac{\left\{\sum y_i\varphi\left(\boldsymbol{w}_i\cdot\boldsymbol{x}\right)\right\}^2}{\sum\varphi^2\left(\boldsymbol{w}\cdot\boldsymbol{x}_i\right)}.\end{aligned} \tag{8}$$

Therefore, the mle $\hat{\boldsymbol{w}}$ is given by the maximizer of $\hat{L}(\boldsymbol{w})$,

$$\hat{\boldsymbol{w}} = \arg\max \hat{L}(\boldsymbol{w}) \tag{9}$$

and

$$\hat{\xi} = \hat{\xi}\left(\hat{\boldsymbol{w}}\right). \tag{10}$$

Here, we define two quantities which depend on D and $\boldsymbol{w}$,

$$Y(\boldsymbol{w}) = \frac{1}{\sqrt{n}}\sum y_i\varphi\left(\boldsymbol{w}\cdot\boldsymbol{x}_i\right), \tag{11}$$

$$Z(\boldsymbol{w}) = \frac{1}{n}\sum\varphi^2\left(\boldsymbol{w}\cdot\boldsymbol{x}_i\right). \tag{12}$$

Since y_i are subject to $N(0,1)$, we have

$$E\left[y_i\varphi\left(\boldsymbol{w}\cdot\boldsymbol{x}_i\right)\right] = 0, \tag{13}$$

$$V\left[y_i\varphi\left(\boldsymbol{w}\cdot\boldsymbol{x}_i\right)\right] = E\left[y_i^2\varphi^2\left(\boldsymbol{w}\cdot\boldsymbol{x}_i\right)\right] = P(\boldsymbol{w}). \tag{14}$$

Hence, by the central limit theorem, Y is subject to a Gaussian random variable whose mean is 0 and whose variance is $P(\boldsymbol{w})$. That is, $Y(\boldsymbol{w})$ is a Gaussian random variable defined over the parameter space of $W = \{\boldsymbol{w}\}$. Such a random function is called a random field (process) over W. We define the normalized Gaussian field by

$$S(\boldsymbol{w}) = \frac{Y(\boldsymbol{w})}{\sqrt{P(\boldsymbol{w})}}. \tag{15}$$

The covariance function of the normalized random field $S(\boldsymbol{w})$ is given by

$$V(\boldsymbol{w}, \boldsymbol{w}') = E\left[S(\boldsymbol{w})S(\boldsymbol{w}')\right] = \frac{E\left[\varphi(\boldsymbol{w}\cdot\boldsymbol{x})\varphi\left(\boldsymbol{w}'\cdot\boldsymbol{x}\right)\right]}{\sqrt{P(\boldsymbol{w})P(\boldsymbol{w}')}}, \tag{16}$$

where $V(\boldsymbol{w}, \boldsymbol{w}) = 1$.

Under certain regularity conditions (e.g., W is a compact set) concerning the convergence of the above stochastic process (see Dacunha-Castelle and Gassiat[6], Fukumizu[7]), we see that the mle is given by the maximizer

$$\hat{\boldsymbol{w}} = \arg\max_{\boldsymbol{w}} \left|S^2(\boldsymbol{w})\right| \tag{17}$$

of the Gaussian process.

It should be noted that the estimated probability density $p\left(y|\boldsymbol{x}, \hat{\xi}, \hat{\boldsymbol{w}}\right)$ is consistent in the sense that it converges to the true one as n goes to infinity. The estimated $\hat{\xi}$ also converges to 0 in the order of $1/\sqrt{n}$. However, different from the regular case, neither $\hat{\xi}$ nor $\hat{\boldsymbol{w}}$ are subject to the Gaussian distribution. We cannot apply the central limit theorem. This implies that the Cramér-Rao paradigm fails here.

4 Generalization Error and Training Error of MLE

In order to define the generalization error, we use the Kullback-Leibler divergence $K[p:q]$ between the two conditional probability distributions p and q,

$$K[p:q] = E_p\left[\log\frac{p}{q}\right]. \tag{18}$$

When p and q are close, this reduces to a half of the squared Riemannian distance of the two distributions. The generalization error of the mle is defined by

$$E_{\text{gen}} = E_D E\left[\frac{1}{2}\left\{y - \hat{\xi}\varphi\left(\hat{\boldsymbol{w}}\cdot\boldsymbol{x}\right)\right\}^2\right], \tag{19}$$

where E and E_D denote, respectively, expectation which respect to a new data $(\boldsymbol{x}, y)$ and the training data D. Then it is seen that the generalization error is related to the KL-divergence by

$$E_{\text{gen}} = H_0 + E_D\left[K\left[p_0 : \hat{p}\right]\right] \tag{20}$$

where H_0 is the entropy of the noise term n, $\hat{p} = p(y, \boldsymbol{x}, \hat{\xi}, \hat{\boldsymbol{w}})$.

When the distribution $\hat{p}$ given by the mle is evaluated by using the training data D itself, instead of a new data, we have the training error,

$$E_{\text{train}} = E_D\left[\frac{1}{2n}\sum_i\left\{y_i - \hat{\xi}\varphi\left(\hat{\boldsymbol{w}}\cdot\boldsymbol{x}_i\right)\right\}^2\right]. \tag{21}$$

The learning curves are given by the generalization error and the training error as functions of the number of the training examples n.

Let us define

$$\alpha = E_D \left[\max_{\boldsymbol{w}} S^2(\boldsymbol{w}) \right] \tag{22}$$

which is the expectation of the maximum of the squared random field. This α is a characteristic number of the field.

Theorem 1. The generalization error is given asymptotically by

$$E_{\text{gen}} = H_0 + \frac{\alpha}{2n}. \tag{23}$$

Theorem 2. The training error is given by

$$E_{\text{train}} = H_0 - \frac{\alpha}{2n}. \tag{24}$$

The proofs are omitted because of the limitation of space, but they are not so difficult, once they are formulated in the above way.

Remarks:

1. We have assumed that the parameter space W is compact, that is, there exists a constant M such that

$$|\boldsymbol{w}| < M. \tag{25}$$

When this does not hold, it is known that α diverges to infinity. Hartigan[10] proved this for the model of Gaussian mixtures, Hagiwara et al[9] in the case of a family of Gaussian radial basis functions, Kitahara et al[11] in the case of hard-limiter neural networks, and Fukumizu[7] in the case of multilayer perceptrons. In many cases, α diverges to infinity in the order of $\log n$ or of $\log\log n$.

2. It is in general difficult to calculate α in a closed form. In the regular case where the true distribution is not at a singular point, the ordinary statistical theory holds and $\alpha = d$, where d is the number of the parameters. In general, $\alpha > d$ at singularities.

5 Generalization Error of Bayes Predictive Distribution

The Bayes paradigm assumes a prior distribution on the parameter space. Let $\pi(\xi, \boldsymbol{w})$ be a prior distribution of the parameters ξ and $\boldsymbol{w}$. One assumes a Gaussian distribution, uniform distribution, or the Jeffrey noninformative distribution, some of which might be improper. As long as the prior is a smooth function, there is no difference in the first order asymptotic theory in the regular case. Here, we assume a uniform prior over ξ and a Gaussian prior for $\boldsymbol{w}$, for simplicity's sake. We can analyze the Jeffrey prior in a similar way, but the result is different.

The posterior distribution of the parameters ξ and $\boldsymbol{w}$ based on the observation D is written as

$$\begin{aligned} p(\xi, \boldsymbol{w}|D) &= c(D)\pi(\xi, \boldsymbol{w}) \prod_i p\left(y_i|\boldsymbol{x}_i, \xi, \boldsymbol{w}\right) q\left(\boldsymbol{x}_i\right) \\ &= c(D)\pi(\xi, \boldsymbol{w}) \exp\left\{L(\xi, \boldsymbol{w}|D)\right\}, \end{aligned} \tag{26}$$

where $c(D)$ is the normalization factor depending only on D. The Bayesian predictive distribution is obtained by averaging $p(y|\boldsymbol{x}, \xi, \boldsymbol{w})$ with respect to the posterior distribution $p(\xi, \boldsymbol{w}|D)$, and is given by

$$p(y|\boldsymbol{x}, D) = \int p(y|\boldsymbol{x}, \xi, \boldsymbol{w}) p(\xi, \boldsymbol{w}|D) d\xi d\boldsymbol{w}. \tag{27}$$

The predictive distribution is a distribution of y for a new input $\boldsymbol{x}$. It is a weighted mixture of probability distributions in the model, so that it does not belong to the model.

Theorem 3. The generalization error of the predictive distribution is given by

$$E_{\text{gen}}^{\text{Bayes}} = H_0 + \frac{1}{2n} E\left[V^2\left(\boldsymbol{w}, \boldsymbol{w}'\right)\right], \tag{28}$$

where V is the covariance function of the related Gaussian random field and the expectation is taken with respect to the prior distribution $\pi(\boldsymbol{w})$ and $\pi\left(\boldsymbol{w}'\right)$.

Proof is more complicated in the present case.

Remarks:

1. When the true network parameters are not singular, it is known that the Bayes predictive distribution, the Bayes maximum posterior estimator and the mle have the same first order asymptotics. However, the Bayes predictive estimator has much smaller generalization error than that of mle (Watanabe[14]). We have shown this in terms of the related Gaussian field.
2. The generalization error depends on the Bayes prior. It should be noted that the Jeffrey noninformative prior is singular at singularities, as the Fisher information is. It is not easy to calculate the generalization error in the closed form except for toy models.
3. It is interesting to see how the generalization error is related to the number of the parameters. Becuase of $V^2\left(\boldsymbol{w}, \boldsymbol{w}'\right) \leq 1$, $E_{\text{gen}}^{\text{Bayes}}$ does not depend on the number of parameters or dimensions of $\boldsymbol{x}$, in this simple cone model (3), provided the uniform prior is used.

6 Conclusions

We have analyzed the generalization error and training error of multilayer perceptrons, when the true or teacher perceptron has a smaller number of hidden units. This is important for model selection or deciding the number of hidden units of neural networks. The classic paradigm of statistical inference based on the Cramér-Rao theorem does not hold in such a singular case, and we need a new theory. We have used the Gaussian random field to derive the generalization and training errors for the maximum likelihood estimator and for the Bayes predictive distribution.

References

1. Amari, S. : Natural gradient works efficiently in learning, *Neural Computation,* 10, 251-276, 1998.
2. Amari, S. and Murata, N. : Statistical theory of learning curves under entropic loss criterion *Neural Computation,* 5, 140-153, 1993.
3. Amari S. and Nagaoka, H. : *Information Geometry*, AMS and Oxford University Press, 2000.
4. Amari, S. and Ozeki, T. : Differential and algebraic geometry of multilayer perceptrons, *IEICE Transactions on Fundamentals of Electronics, Communications and Computer System,* E84-A, 31-38, 2001.
5. Amari, S., Park, H., and Fukumizu, F. : Adaptive method of realizing natural gradient learning for multilayer perceptrons, *Neural Computation*, 12, 1399-1409, 2000.
6. Dacunha-Castelle, D. and Gassiat, E. : Testing in locally conic models, and application to mixture models, *Probability and Statistics,* 1, 285-317, 1997.
7. Fukumizu, K. : Statistical analysis of unidentifiable models and its application to multilayer neural networks, *Memo at Post-Conference of the Bernoulli-RIKEN BSI 2000 Symposium on Neural Networks and Learning*, October 2000.
8. Fukumizu, K. : Likelihood Ratio of Unidentifiable Models and Multilayer Neural Networks, *Research Memorandum*, 780, Inst. of Statitical Mathematics, 2001.
9. Hagiwara, k., Kuno, K. and Usui, S. : On the problem in model selection of neural network regression in overrealizable scenario, *Proceeding of International Joint Conference of Neural Networks,* 2000.
10. Hartigan, J. A. : A failure of likelihood asymptotics for normal mixtures, *Proceedings of Berkeley Conference in Honor of J. Neyman and J. Kiefer,* 2, 807-810, 1985.
11. Kitahara, M., Hayasaka, T., Toda, N. and Usui, S. : On the probability distribution of estimators of regression model using 3-layered neural networks (in Japanese), *Workshop on Information-Based Induction Sciences (IBIS 2000)*, 21-26, July, 2000
12. Park, H., Amari, S. and Fukumizu, F. : Adaptive natural gradient learning algorithms for various stochastic models, *Neural Networks*, 13, 755-764, 2000.
13. Rattray, M., Saad, D. and Amari S. : Natural Gradient Descent for On-line Learning, *Physical Review Letters*, 81, 5461-5464, 1998.
14. Watanabe, S.: Algebraic analysis for non-identifiable learning machines, *Neural Computation*, to appear.
15. Watanabe, S.: Training and generalization errors of learning machines with algebraic singularities (in Japanease), *The Trans. of IEICE A*, J84-A, 99-108,2001.

Bistable Gradient Neural Networks: Their Computational Properties

Vladimir Chinarov and Michael Menzinger

Department of Chemistry, University of Toronto, M5S 3H6
{vhinaro; mmenzing@chem.utoronto.ca}

Abstract: A novel type of gradient attractor neural network is described that is characterized by bistable dynamics of nodes and their linear coupling. In contrast to the traditional perceptron-based neural networks which are plagued by spurious states it is found that this Bistable Gradient Network (BGN) is *virtually free from spurious states.* The consequences of this – greatly *enhanced memory capacity, high speed of training and perfect recall* – are illustrated by and compared with a small Hopfield network.

1. Introduction

Attractor neural networks [1] are dissipative dynamical systems composed of N formal neurons or nodes that are mutually connected according to certain network topologies allowing for massive feedback. By adjusting the (synaptic) coupling strengths w_{ij} between the nodes, using an appropriate learning algorithm, the network is able to memorize a set of input or memory vectors $\{\rho^k;\ k= 1,\ldots P\}$. In doing so, its fixed point attractors are made to coincide with the input vectors. Amit [1] has reviewed the computational, error correcting and associative recall abilities of attractor networks.

A classical example of attractor networks is the fully connected, symmetrically coupled Hopfield network [1,5] which shares the following essential node dynamics with most other perceptron-based NNs [4]. A caricature of the network dynamics is

$$dx/dt = -x_i + f((\,\Sigma\, w_{ij}\, x_j\,) - \theta_i),\ \ (i,j = 1,2\ldots N) \tag{1}$$

where x_i is the state variable of the i-th node, and $f(.)$ is generally a nonlinear (step or sigmoid) function of the coupling terms $w_{ij}x_j$ and firing threshold θ_i . Thus each node is monostable and linear and the nonlinearity lies entirely in the coupling terms. Although ANNs and other NNs based on a node structure similar to (1) are widely used in engineering applications [4], their principal drawback is an abundance of *spurious states* – attractors that arise as a byproduct of the training process without having been memorized. These spurious states occupy valuable phase space, reducing the memory capacity, the rate of learning and above all the accuracy of pattern recall.

J. Mira and A. Prieto (Eds.): IWANN 2001, LNCS 2084, pp. 333-338, 2001.

We describe here a novel ANN that overcomes these deficiencies [2]. We refer to it as the Bistable Gradient (neural-like) Network (BGN). Rather than basing it on the early physiological views of induced neuronal firing as summarized by Eq.(1), BGN is based on the well established principles of many body physics and nonlinear dynamics. Its most striking property is the absence of spurious states. This frees up valuable phase space for storage of memory patterns and results in enhanced memory capacity. Due to the gradient nature of the system, trajectories converge rapidly to their attractor from within its basin of attraction, making the recall of patterns fast. Fault tolerance is high and recall from within a pattern's basin of attraction is perfect.

These properties are illustrated for a N=5 BGN and are compared with those of an equivalent Hopfield net.

2. The Bistable Gradient Network

The gradient network is fully connected and is composed of N nodes:

$$dx_i/dt = -\partial V/\partial x_i, \qquad (i=1,\ldots,N) \tag{2}$$

where the total potential $V(x_1, x_2, \ldots x_N)$ is defined as

$$V = V_o + V_{int} \tag{3}$$

the sum of the total local potential V_o and the interaction term V_{int}. V_o is the sum, over all nodes of local double-well potentials

$$V_o(x_1,\ldots x_N) = \Sigma_i V_i(x_i) \tag{4}$$

$$V_i(x_i) = -x_i^2/2 + x_i^4/4, \tag{5}$$

which have their local minima at $x_i=\pm 1$. The interaction term V_{int} is

$$V_{int} = ½\, \alpha\, \Sigma_i \Sigma_j\, w_{ij} x_i x_j, \tag{6}$$

where w_{ij} are the symmetric coupling coefficients and α is a coupling strength parameter. With these definitions, the full gradient system (1) is characterized by nonlinear, bistable local dynamics and by a linear coupling of nodes:

$$dx_i/dt = g(x_i) + \alpha\, \Sigma_j\, w_{ij} x_j, \quad \text{where} \tag{7.a}$$

$$g(x_i) = -\partial V_o/\partial x_i = x_i - x_i^3. \tag{7.b}$$

Hoppenstaedt and Izhikievich [6] have studied weakly connected NNs of this type with $\alpha \ll 1$. Here we concentrate here on the case $\alpha=1$.

Each network element is thus an overdamped nonlinear oscillator moving in a double-well potential. All limit configurations of the net are contained within its set of fixed point attractors. For a given coupling matrix w_{ij}, the network rapidly converges towards such limit configurations, starting from any input configuration of the bistable elements, as long as it lies within the basin of attraction of this particular limit configuration (fixed point).

The coupling matrix w_{ij} may be readily constructed using Hebb's learning rule [3,4]

$$w_{ij} = 1/P\Sigma_k^P\beta_k(\rho^k)_i(\rho^k)_j\,, \tag{8}$$

where ρ^k $(k=1,...,P)$ are the P vector configurations or patterns submitted to the network for memorization and β_k is a weight parameter. In traditional applications $\beta_k=1$ for all k.. Note that neither the components of the input patterns ρ^k nor the coordinates of the fixed points or memory states of (7) are integer binary digits (±1) as they are in Hopfield NNs, but they are real valued. Even for integer (±1) input patterns, the network attractors become real valued due to the coupling interaction, but the sign structure (see Fig.1) is preserved. Note furthermore that perfect training is very efficiently achieved by Hebb's rule (8) and that the slow, iterative gradient descent learning [4] that constitutes the bottleneck in training of many nonlinearly coupled NNs is not required for the BGN.

3. Examples

We illustrate the associative performance of a small BGN with N=5 nodes by first training it, using rule (8), with sets $\{\rho^k, k=1,...P\}$ of real valued input patterns whose sign structure is shown in Fig.1. Black and white cells denote negative and positive elements, i.e. left and right wells of the local potential. Two sets $\{\rho^k, k=1,...P\}$ of P=5 and P=8 patterns were memorized. In the first case, patterns ρ^1-ρ^5, in the second ρ^1-ρ^8 were submitted to the network using Equ.(8) with $\beta_k=1$.

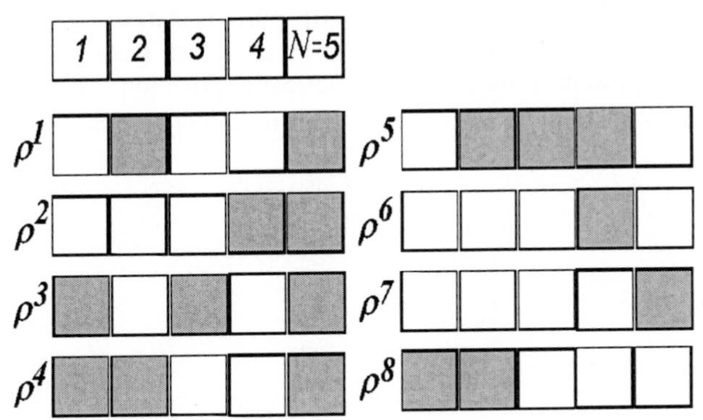

Fig 1. Patterns submitted as memory states to the BGN. Black and white cells represent elements with negative and positive signs.

Test patterns were generated from these input patterns by inverting (corrupting) the signs of one or more of their elements.

The performance results at the retrieval phase are shown in Fig.2 for the *P=5* memory patterns in the top row and for *P=8* in the bottom row. The left panels illustrate the limit configurations to which the network converges from randomly chosen initial conditions in the *E*-$\langle X \rangle$ phase plane. Here *E* is the total net energy defined as

$$E = V_o + 1/2\alpha \, \Sigma_i \Sigma_j w_{ij} x_i x_j \tag{9}$$

and $\langle X \rangle = 1/N \Sigma_i^N x_i$ is the mean activity of the network.

The righthand panels are retrieval diagrams obtained by binning the outcomes of a large number of trajectories with randomly chosen initial conditions. The top left panel shows only the original *P=5* memory patterns, while the bottom left *(P=8)* includes the isoenergetic "mirror states" with reversed signs at all nodes. In addition to the attractors the top left *P=5* panel displays the total network energy (9) as a function of the mean node activity $\langle X \rangle$. In panel (c), black and white circles denote attractors that correspond to the memorized patterns and to their mirror

The retrieval histograms (Fig.2.b,d), which measure the relative probabilities of reconstructing the patterns $\{\rho^k, k=1,...P\}$, were obtained from a large number of trajectories with randomly chosen initial conditions. As such these probabilities represent also the relative volumes of the attractor basins and they measure the resilience to corruption of patterns. Comparison of panels (a,b) and (c,d) show that attractors with lower total energy (9) tend to have larger basins of attraction.

Note that all of the P=5,8 memory patterns could be successfully stored and recalled. Hence the memory capacity of the BGN is of order K~N. Similar results were obtained for larger nets.

For comparison, we attempted to recall the same set of five patterns $\{\rho_1 ... \rho_5\}$ from an equivalently built Hopfield network (Eq.1) with N=5. In that case only one memory pattern (two if one includes the mirror state), namely ρ_5, was perfectly retrieved from the network This number coincides, of course, with the theoretical estimate of storage capacity K=0.14N giving perfect recall for Hopfield nets [1,4]. At the same time the Hopfield net generates 5 (10) spurious states.

Unequal retrieval probabilities for the stored patterns, as shown in Figs.2.b,d, may represent a drawback in practical applications. This may be readily circumvented by adjusting the weights β_k in learning rule (8) accordingly.

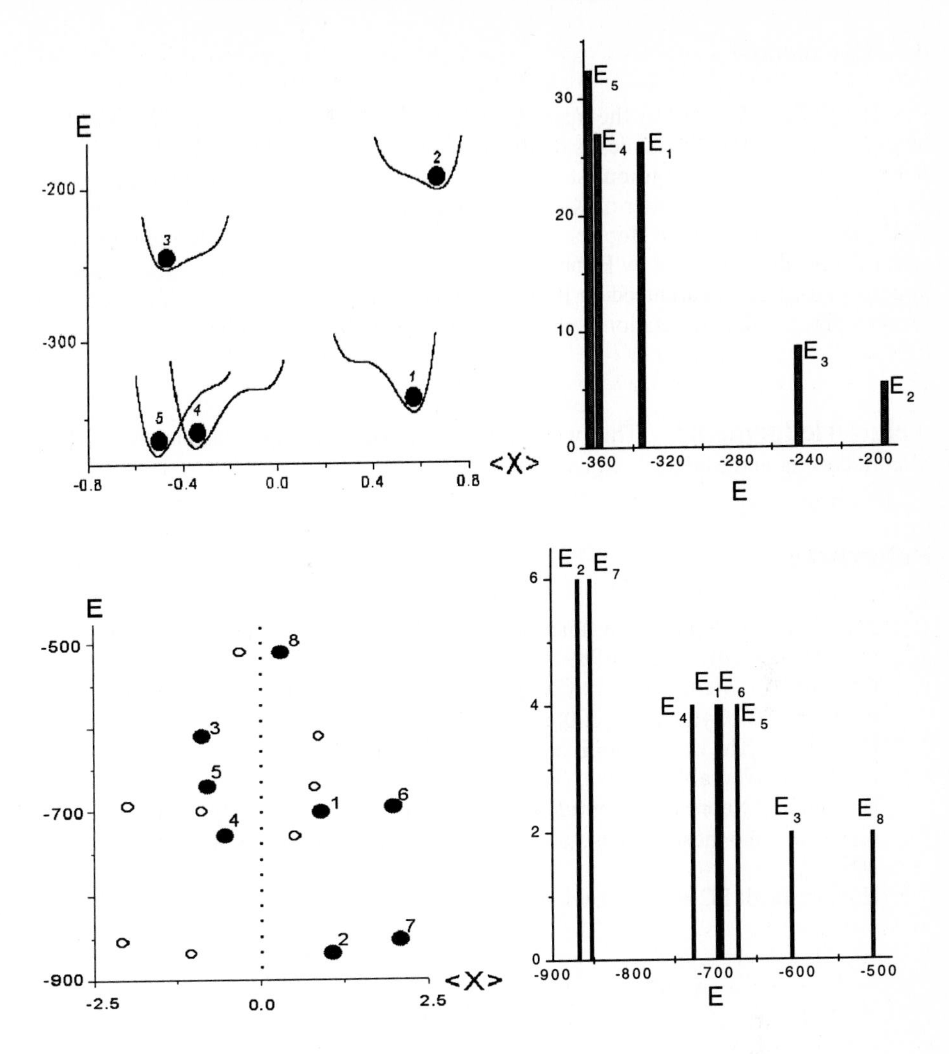

Fig 2. Retrieval performance of GBN with N=5 for P=5 (top panels) and P=8 (bottom) memorized patterns. The left panels (a,c) show the E-<X> phase planes with the final states to which the network relaxes from a random set of initial conditions. Black circles: memorized patterns; white circles: mirror states. The right hand panels (b,d) are retrieval histograms.

4. Discussion

The BGN is inspired by the many-body physics of spin glasses and the Ising model. As a result, its dynamics is relatively simple and clean. Its outstanding feature – the virtual absence of spurious states – frees valuable phase space and increases the memory capacity to the order of K~N. This is a substantial improvement over the K~0.14N limit [1,4] of the Hopfield network. BGN also stands out with regard to its ease of "one shot" training by Hebb's rule. Fast learning and fast convergence to the system's attractors, guaranteed by its gradient dynamics, make the BGN a powerful tol with promising computational properties in a wide range of applications.

Acknowledgement: This work was performed with the financial support of Manufacturing and Materials Ontario (MMO).

References

1. Amit D.J.: Modeling brain function. The world of attractor neural networks. (Cambridge Univ. Press, 1989)
2. Chinarov V., Menzinger M.: Computational dynamics of gradient bistable networks , BioSytems **55** (2000), 137-142.
3. Hebb D.O.: The organization of behaviou*r* (Wiley, 1949)
4. Haykin S.: Neural Networks, (Prentice Hall, N.Y. 1999)
5. Hopfield J.: Neurons with graded response have collective computational properties like those of two-state neurons. Proc.Natl.Acad.Sci. **81** (1984), 3088-3092.
6. Hoppenstaedt FC and Izhikevich (199?),

Inductive Bias in Recurrent Neural Networks

S. Snyders[1] and C.W. Omlin[2]

[1] University of Stellenbosch, Department of Computer Science, Stellenbosch 7600, South Africa
snyders@cs.sun.ac.za

[2] University of Western Cape, Department of Computer Science, Bellville 7535, South Africa
comlin@uwc.ac.za

Abstract. The use of prior knowledge to train neural networks for better performance has attracted increased attention. Initial domain theories exists for many machine learning applications. In both, feed forward and recurrent neural networks, algortihms for encoding prior knwoledge has been constructed. We propose a heuristic for determining the strength of the prior knowledge (inductive bias) for recurrent neural networks encoded with a DFA as initial domain knowledge. Our heuristic uses gradient information in weight space in the direction of the prior knowledge to enhance performance. Tests on known benchmark problems demonstrate that our heuristic reduces training time, on average, by 30% compared to a random choice of the strength of the inductive bias. It also achieves, on average, near perfect generalization for that specific choice of the inductive bias.

1 Introduction

Prior knowledge about a certain domain is available for many machine learning problems. These domain theories can be encoded into neural networks prior to training. The use of prior knowledge reduces the learning complexity and can improve both training and generalization performance [2]. Hybrid methods of combining inductive methods, e.g. neural networks, with symbolic knowledge reduces the complexity of the learning task [1] and introduces the means of revising the initial information [3]. Prior knowledge allows the model to be chosen that limits the variance of the model, but is not too restictive to prohibit the learning of new dependencies. The initial model thus created by the prior information greatly helps to reduce the bias/variance dilemma [4].

The prior information encoded in the network is used to derive an initial hypothesis, from which the search for a solution can start. The prior knowledge presumably defines a good starting point in weight-space and leads to faster convergence; it provides an inductive bias which focuses the network's attention on relevant input features or favors a desirable connectionist representation. The inductive bias is determined by the value of the encoded weights in the network.

J. Mira and A. Prieto (Eds.): IWANN 2001, LNCS 2084, pp. 339-346, 2001.

Encoding of prior knowledge in feed-forward networks (Knowledge-Based Neural Networks)[8] has been predominantly done using Horn clauses. In recurrent neural networks, prior knowledge in the form of DFAs has been encoded[9]. A network is then trained using either real-time recurrent learning (RTRL)[7] or back propagation trough time (BPTT)[10]. The advantages to using prior knowledge are thus as follows: (1) The learning performance may lead to a faster convergence to a solution, (2) networks that are trained with hints may generalize better on future unseen examples (3) revision of current prior knowledge and addition of new information about the specific domain and (4) better rules may be extracted from trained networks.

How strong must this inductive bias be? How much emphasis must be placed on the prior knowledge for a specific domain? This paper proposes a heuristic for determining this inductive bias for recurrent neural networks; this has been done for feed forward networks in [5]. The method uses gradient information of the error function in the direction of the programmed weights. The network starts its search for a solution in weight space where the gradient is maximal, thus speeding up convergence.

2 Neural Network Architecture

2.1 Second Order Recurrent Neural Network

The continuous network dynamics are described by the following equations:

$$S_i(t+1) = \sigma(\text{net}_i(t)) = \frac{1}{1+e^{-\text{net}_i(t)}} \tag{1}$$

$$\text{net}_i(t) = b_i + \sum_{j,k} w_{ijk} S_j(t) I_k(t), \tag{2}$$

where S_i is the activation of the hidden recurrent state neurons, I_k is the k-input, w_{ijk} is the corresponding weight and b_i is the bias for neuron i. The product $S_j(t)I_k(t)$ directly corresponds to the state transition of the automaton: $\delta(q_j, a_k) = q_i$ where a_k is the kth symbol, represented by the input $\boldsymbol{I}$ using unary encoding, i.e. $I_k(t) \in \{0,1\}$. A special neuron S_0 represents the output and decides whether or not the string is accepted.

The weight updates for a specific string is done according to the quadratic error function

$$E = \frac{1}{2}(\tau_0 - S_0(f))^2 \tag{3}$$

where τ is the desired output for the string. The weight updates are computed by

$$\Delta w_{lmn} = \alpha(T_0 - S_0(f))\frac{\partial S_0(f)}{\partial w_{lmn}} + \eta \Delta w'_{lmn} \tag{4}$$

where α is the learning rate, η the momentum rate and $\Delta w'_{lmn}$ the previous weight update. $\frac{\partial S_0(f)}{\partial w_{lmn}}$ can be computed recursively with equations

$$\frac{\partial S_i(t+1)}{\partial w_{lmn}} = net'_i(t) \left(\delta_{il} S_m(t) I_n(t) + \sum_{jk} w_{ijk} \frac{\partial S_j(t)}{\partial w_{lmn}} I_k(t) \right) \tag{5}$$

$$\frac{\partial S_i(t+1)}{\partial b_l} = net'_i(t) \left(\delta_{il} + \sum_{jk} w_{ijk} \frac{\partial S_j(t)}{\partial b_l} I_k(t) \right) \tag{6}$$

where $net'_i(t)$ is the derivative of the sigmoidal discriminate function and δ_{il} is the Kronecker-delta. δ_{il} is equal to 1 if $i = l$, 0 otherwise.

An aspect of the second order recurrent neural network is that the product $S_j(t)I_k(t)$ in the recurrent network directly corresponds to the state transition $\delta(q_j, a_k) = q_i$ in the DFA. After a string has been processed, the output of a designated neuron S_0 decides whether the network accepts or rejects a string.

2.2 DFA Encoding Algorithm

The DFA encoding algorithm follows directly from the similarity of state transitions in a DFA and the dynamics of a recurrent neural network: Consider a state transition $\delta(q_j, a_k) = q_i$. We arbitrarily identify DFA states q_j and q_i with state neurons S_j and S_i, respectively. One method of representing this transition is to have state neuron S_i have a high output ≈ 1 and state neuron S_j have a low output ≈ 0 after the input symbol a_k has entered the network via input neuron I_k. One implementation is to adjust the weights w_{jjk} and w_{ijk} accordingly: setting w_{ijk} to a large positive value will ensure that $S_i(t+1)$ will be high and setting w_{jjk} to a large negative value will guarantee that the output $S_j(t+1)$ will be low. All other weights are set to small random values. In addition to the encoding of the known DFA states, we also need to program the response neuron, indicating whether or not a DFA state is an accepting state. We program the weight w_{0jk} as follows: If state q_i is an accepting state, then we set the weight w_{0jk} to a large positive value; otherwise, we will initialize the weight w_{0jk} to a large negative value. We define the values for the programmed weights as a rational number H, and let large *programmed* weight values be $+H$ and small values $-H$. We will refer to H as the *strength* of a rule. We set the value of the biases b_i of state neurons than have been assigned to known DFA states to $-H/2$. This ensures that all state neurons which do not correspond to the the previous or the current DFA state have a low output. Thus, the rule insertion algorithm defines a nearly *orthonormal internal representation* of all known DFA states. We assume that the DFA generated the example strings starting in its initial state. Therefore, we can arbitrarily select the output of one of the state neurons to be 1 and set the output of all other state neurons initially to zero.

The initial state $\boldsymbol{S}(0)$ of the network is, thus

$$\boldsymbol{S}(0) = (S_0(0), 1, 0, 0, ..., 0),$$

where initial value of the response neuron $S_0(0)$ is 1 if the DFA's initial state q_0 is an accepting state and 0, otherwise.

3 Strength of Inductive Bias

An indiscriminant choice of the inductive bias has two major drawbacks: (1) It is conceivable that different applications require different choices of the inductive bias H which leads to fast convergence, and (2) it does not provide a mechanism for dealing with uncertainty about the initial domain theory. This section proposes a method for choosing the strength of the inductive bias which takes the two mentioned shortcomings into account, i.e. the choice of H depends on the application represented by the initial domain theory, the network architecture, and the training data, and it adjusts its confidence into the prior knowledge according to the amount and the quality of the available prior knowledge.

Consider an error function E used to train a network. The idea for determining a good value for the inductive bias H is to start the search for solution in a point in weight space where the gradient $|\partial E/\partial H|$ is maximal, i.e. we choose H such that the search starts in a point where the error function in the direction of the inductive bias is steep; this avoids the need for determining H through trial-and-error or traversing flat regions of the weight space during the initial training phase. Furthermore, the value H which gives good training performance depends on the prior knowledge and the training data. The function $\partial E/\partial H$ takes both these dependencies into consideration. The more prior knowledge that is available and the more accurate that knowledge is, the more the function $\partial E/\partial H$ influences the gradient-descent search for a solution in weight space.

We will now derive a recursive procedure for evaluating the gradient $\partial E(H)/\partial H$ prior to training which is similar to the derivation of the real-time recurrent learning algorithm[1].

Consider the quadratic error function in Equation 3. Notice that $S_0(f)$ depends on the particular choice of H Then, the derivative $\partial E/\partial H$ for a specific string[2] is given by

$$\frac{\partial E}{\partial H} = -(\tau_0 - S_0(f))\frac{\partial S_0(f)}{\partial H} \tag{7}$$

[1] The value of the error function E depends on the particular choice of H, thus $E(H)$. For simplicity, we omit the argument H in the equations for the computation of $\partial E(H)/\partial H$.

[2] $\partial E(H)/\partial H$ is calculated for a specific string. Normalization according to the number of strings in the training set is neccessary for a comparable value.

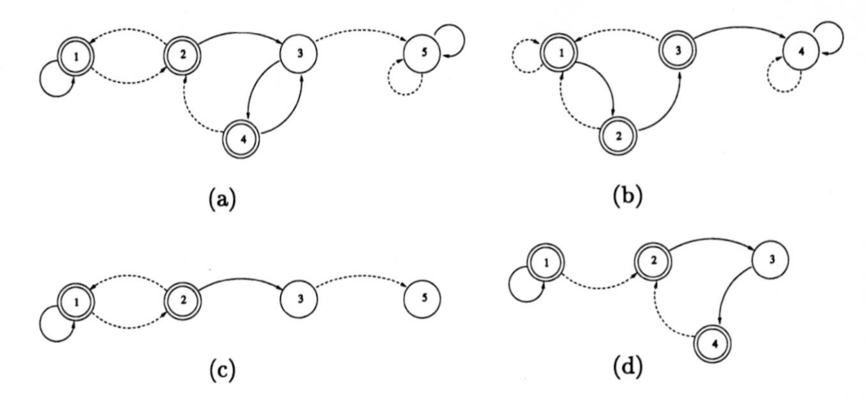

Fig. 1. DFAs used for prior knowledge: (a) Full DFA of Tomita's 3rd language (b) Full DFA of Tomita's 4th language (c) Partial DFA of Tomita's 3rd language (d) Another partial DFA of Tomita's 3rd language. Initial state is numbered 1, solid arcs and dashed arcs represent 0 and 1 transitions, respectively.

We can compute $\partial S_i(t)/\partial H$ recursively as follows:

$$\frac{\partial S_i(t)}{\partial H} = net_i'(t)\left[\frac{\partial b_i}{\partial H} + \sum_{jk}\frac{\partial w_{ijk}}{\partial H}S_j(t-1)I_k(t-1) + \sum_{jk} w_{ijk}I_k(t-1)\frac{\partial S_j(t-1)}{\partial H}\right] \tag{8}$$

where

$$\frac{\partial W_{ijk}}{\partial H} = \begin{cases} +1 & \text{if } w_{ijk} = +H \\ -1 & \text{if } w_{ijk} = -H \\ 0 & \text{otherwise} \end{cases} \quad and \quad \frac{\partial b_i}{\partial H} = \begin{cases} +1/2 & \text{if } b_i = +H/2 \\ -1/2 & \text{if } b_i = -H/2 \\ 0 & \text{otherwise} \end{cases} \tag{9}$$

When $t = 0$, $\boldsymbol{S}$ does not depend on H since it is the initial state of the network, thus

$$\frac{\partial S_i(t)}{\partial H} = 0. \tag{10}$$

4 Results

The heuristic was tested on two languages as described by Tomita[6] (Figures 1 (a) and (b)). The DFAs describing these languages were encoded into the neural networks. For full DFA encoding, the encoding was done as described in Section 2.2. For partial prior knowledge encoding, only the weights corresponding to known DFA transitions were encoded, and the rest of the weights were random initialized to values $[-0.1, 0.1]$.

All networks encoded with Tomita's third DFA were trained using a learning rate $\alpha = 0.5$ and a momentum rate $\eta = 0.1$, for Tomita's fourth DFA, $\alpha = 0.3$

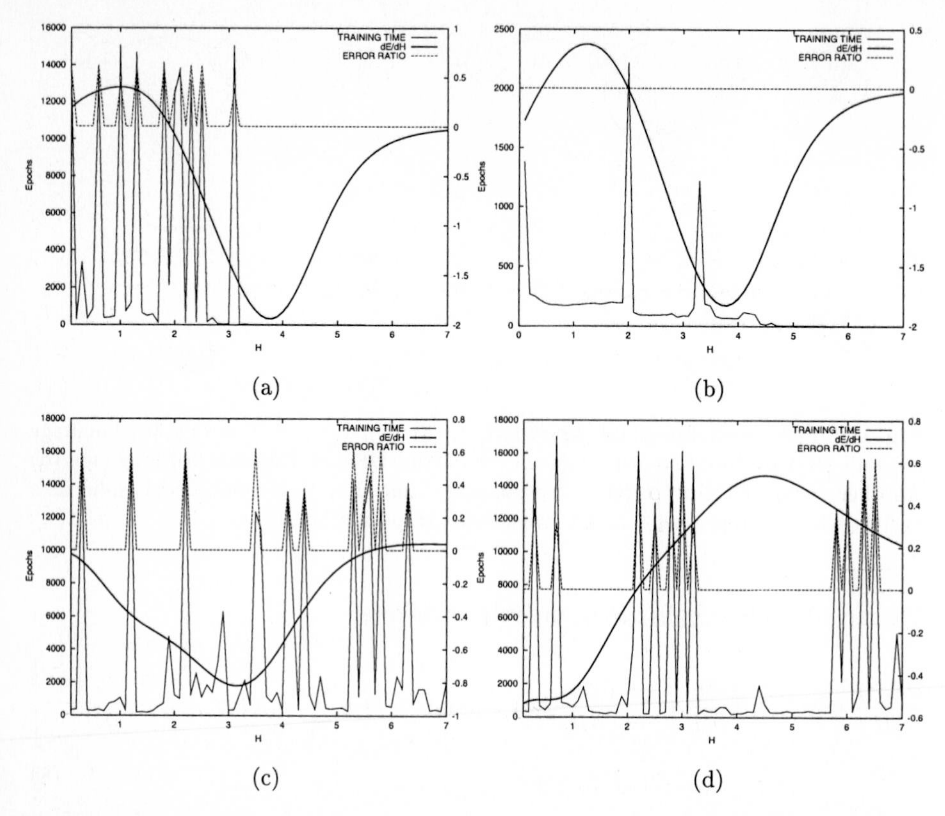

Fig. 2. Training Time and Generalization Performance: (a) Full DFA encoding of Tomita's 3rd language (b) Full DFA encoding of Tomita's 4th language (c) and (d) Partial DFA encodings of Tomita's 3rd language according to Figures 1 (c) and 1 (d), respectively.

and $\eta = 0.1$[3]. The training set consisted of a 1000 strings up to length 10. The networks were not trained *wholesale* on all the strings, but incrementally; the initial working set contained 30 strings. Training was subdivided into cycles. In each cycle the network was trained on the working set up to a maximum of 1000 epochs or until all strings in the working set were correctly classified[4]. After such a cycle the network is evaluated on the *whole* training set. If all the strings were correctly classified the training stops otherwise the first 10 misclassified strings, not already in the working set, is added to the working set and a new training cycle then started. Training was also stopped when the working set grew to a size bigger than 200 strings.

[3] These learning and momentum rate values are not necessarily optimal for the tested networks.

[4] Strings were misclassified, for training, to within $\psi = 0.2$ of desired output. Testing of networks were performed on $\psi = 0.5$.

Figures 2 (a) and 2 (b) show the training and generalization performance of full DFA encoding for different values of H ranging from 0.1 to 7.0 in increments of 0.1. It can clearly be seen that after a certain H value, no training is needed. Our heuristic for choosing the H value where the function $|\frac{\partial E}{\partial H}|$ is maximal, proves to find values of H where the training and generalization performance is closely correlated to these positions.

Figures 2 (c) and 2 (d) show typical performances for partial DFA encoding of Tomita's third language according to the partial DFAs in Figures 1 (c) and 1 (d), respectively. Statistically significant results for these experiments are shown in Table 1. It shows the average and standard deviation of the training and generalization performance, comparing values with the fastest training time, training time and generalization performance obtained using our heuristic, and the average performance.

We observe that our heuristic of choosing the inductive bias H, also yields good performance for partial prior knowledge. The value of $H_{|\partial E/\partial H|}$ seems independant of the random initialized weights, but dependant on the specific prior information (DFA). For partial knowlegde encoded according to Figure 1 (d), the graph of $\partial E/\partial H$ has two very close values for the maximal gradient. The maximum gradient is at $H = 4.5$ and the other is close to zero ($H = 0.1$). The value close to zero may seem contrary to intuition, as partial *correct* knowledge was encoded. We speculate that the partial knowledge encoded does not significantly explain the training data. The heuristic's confidence in the partial knowledge, to help learn the specific data, is almost lacking.

In all figures, the value of the function $\frac{\partial E}{\partial H}$ approaches the x-axis for bigger values of H. This can be interpreted that at such high values, little error information is available to propagate back through the network, resulting in a small gradient. The graphs show very erratic behaviour in training performance. This behaviour can be due to the error surface containing alot of local minima's, which, when the network gets stuck in such a local minima, results in bad generalization performance.

Inductive Bias H	Figure 1 (c)				Figure 1 (d)			
	training time		generalization error		training time		generalization error	
	μ	σ	μ	σ	μ	σ	μ	σ
$H_{optimal}$	160	16	0%	0%	152	14	0%	0%
$H_{\|\partial E/\partial H\|}$	2114	1983	0%	0%	1991	3881	5.6%	1.7%
$H_{average}$	2931	466	6.6%	1.8%	3063	459	7.9%	1.6%

Table 1. Results for Partial encoding of Tomita's third language: The table shows average and standard deviation for the optimal training time, the training performance achieved with our heuristic for determining the strength of the inductive bias, and the average training time of all runs.

5 Conclusions

We have seen that training time and generalization performance varies with different values of H. Our results have demonstrated that, although the error surface is very non-linear, our heuristic for determining the inductive bias H yields fairly good performance. We have performed preliminary experiments on known benchmark problems and have shown that choosing the correct inductive bias is very important, especially when partial prior knowledge is used, as this attributes to an even more complex error surface to traverse.

Our method takes the prior information, the network architecture and the training data into consideration. Out heuristic achieves, on average, 30% better training performance than a random choice of H. Generalization performance is optimal in most cases. It would be interesting to analyze the generalization performance of rules that are extracted form networks trained using the inductive bias proposed by our hueristic, compared to any other value of the inductive bias. Synergistically combining inductive and symbolic knowledge remains an open research problem.

References

1. Abu-Mostafa, Y.S.: Learning from Hints in Neural Networks. Journal of Complexity **6** (1990) 192
2. Fu, L.: Learning Capacity and Sample Complexity on Expert Networks. IEEE Transactions on Neural Networks **7** no. **6** (1996) 1517–1520
3. Omlin, C.W., Giles, C.: Rule revision with recurrent neural networks. IEEE Transactions on Knowledge and Data Engineering **8** no. **1** (1996) 183–188
4. Geman, S., Bienenstock, E., Doursat, R.: Neural Networks and the Bias/Variance Dilemma. Neural Computation **4** (1992) 1–58
5. Snyders, S., Omlin, C.W.: What Inductive Bias Gives Good Neural Network Training Performace? Proceedings of the IEEE-INNS-ENNS International Joint Conference on Neural Nerworks **3** (2000) 445–450
6. Tomita, M.: Dynamic construction of finite-state automata from examples, using hill-climbing. Proceedings of the Fourth Annual Cognitive Science Conference (1982) 105–108
7. Williams, R., Zipser, D.: A learning algorithm for continually running fully recurrent neural networks. Neural Computation **1** (1989) 270–280
8. Towell, G., Shavlik, J.: Knowledge-based artificial neural networks. Artificial Intelligence **70** no. **1-2** (1994) 119–165
9. Omlin, C.W., Giles, C.: Constructing Deterministic Finite-State Automata in Recurrent Neural Networks. Journal of the ACM **43** no. **6** (1996) 937–972
10. Rumelhart, D., Hinton, G., Williams, R.: Learning internal representations by error propagation. Parallel Distributed Processing chp. **8** (1986)

Accelerating the Convergence of EM-Based Training Algorithms for RBF Networks

Marcelino Lázaro, Ignacio Santamaría, Carlos Pantaleón *

Dpto. Ing. Comunicaciones, ETSII y Telecom. Universidad de Cantabria
Avda. Los Castros, 39005 Santander, Spain
marce@gtas.dicom.unican.es,
WWW home page: http://gtas.dicom.unican.es

Abstract In this paper, we propose a new Expectation-Maximization (EM) algorithm which speeds up the training of feedforward networks with local activation functions such as the Radial Basis Function (RBF) network. The core of the conventional EM algorithm for supervised learning of feedforward networks consists of decomposing the observations into their individual output units and then estimating the parameters of each unit separately. In previously proposed approaches, at each E-step the residual is decomposed equally among the units or proportionally to the weights of the output layer. However, this approach tends to slow down the training of networks with local activation units. To overcome this drawback in this paper we use a new E-step which applies a soft decomposition of the residual among the units. In particular, the residual is decomposed according to the probability of each RBF unit given each input-output pattern. It is shown that this variant not only speeds up the training in comparison with other EM-type algorithms, but also provides better results than a global gradient-descent technique since it has the capability of avoiding some unwanted minima of the cost function.

1 Introduction

Radial Basis Function (RBF) networks have become one of the most popular feedforward neural networks with applications in regression, classification and function approximation problems [1,2]. The RBF network approximate nonlinear mappings by weighted sums of Gaussian kernels. Therefore, an RBF learning algorithm must estimate the centers of the units, their variances and the weights of the output layer. Typically, the learning process is separated into two steps: first, a nonlinear optimization procedure to select the centers and the variances and, second, a linear optimization step to fix the output weights. To simplify the nonlinear optimization step, the variances are usually fixed in advance and the centers are selected at random [3] or applying a clustering algorithm [4].

Other approaches try to solve the global nonlinear optimization problem using supervised (gradient-based) procedures to estimate the network parameters,

* This work has been supported by the European Community and the Spanish Government through FEDER project 1FD97-1863-C02-01.

J. Mira and A. Prieto (Eds.): IWANN 2001, LNCS 2084, pp. 347-354, 2001.

which minimize the mean square error (MSE) between the desired output and the output of the network [5,6,7]. However, gradient descent techniques tend to be computationally complex and suffer from local minima.

As an alternative to global optimization procedures, a general and powerful method such as the Expectation-Maximization algorithm [8] can be applied to obtain maximum likelihood (ML) estimates of the network parameters. In the neural networks literature, the EM algorithm has been applied in a number of problems: supervised/non-supervised learning, classification/function approximation, etc. Here we concentrate on its application to supervised learning in function approximation problems. In this context, Jordan and Jacobs proposed to use the EM algorithm to train the mixture of experts architecture for regression problems [9]. The EM algorithm has been also applied to estimate the input/output joint pdf, modeled through a Gaussian mixture model, and then estimating the regressor as the conditional pdf [10]. In both cases the hidden variables select the most likely member of the mixture given the observations, and then each member is trained independently.

More recently, the EM algorithm has been applied for efficient training of feedforward and recurrent networks [11,12]. The work in [11] connects to the previous work of Feder and Weinstein for estimating superimposed signals in noise [13]. In both methods, the complete set is the set of individual neuron outputs (signal components in [13]), and the E-step reduces to decompose at each iteration the total residual into N components (N being the number of neurons). In [13] it is shown that the variables used to decompose the residual can be arbitrary as long as they sum one. Different alternatives are then possible: for instance, in [13] the residual is decomposed into N equal components, whereas in [11] it is decomposed proportionally to the weights of the output layer. Both approaches work well for feedforward networks with global activation functions such as the MLP, but tend to be rather slow for networks with local activation functions since each individual unit is forced to approximate regions far away from the domain of the activation unit.

To overcome this drawback in this paper we propose a new EM algorithm, specific for RBF networks, which aims to accelerate its convergence. We perform a soft decomposition of the residual using a sequence of posterior probabilities, in this way we take into account the locality of the basis functions. Different examples show that this modification speeds up the convergence in comparison with previous EM approaches. Moreover, in comparison with global gradient-descent optimization procedures, the proposed approach has the capability of escaping from some local minima.

2 EM-based training of feedforward networks

In this section we introduce the notation and describe previous work on training two-layer feedforward networks using EM-based approaches. Without loss of generality, let us consider an RBF network with I units, which approximates an one-dimensional mapping, $g(x) : \mathcal{R} \to \mathcal{R}$, as follows

$$\widetilde{g}(x) = \sum_{i=1}^{I} \lambda_i o_i(x) \quad (1)$$

where i indexes the RBF units, λ_i is the amplitude, and $o_i(x)$ is the activation function of each unit, which is given by

$$o_i(x) = \exp -\left(\frac{(x-\mu_i)^2}{2\sigma_i^2}\right). \quad (2)$$

Our training problem consists in estimating the amplitudes, λ_i, centers, μ_i, and variances, σ_i^2, of an RBF model given a set of inputs and the corresponding noisy outputs, $\{x_k, y_k\}$. The noisy observations may be characterized using the following model

$$y_k = \sum_i g_i(x_k) + e_k \quad (3)$$

where $g_i(x_k) = \lambda_i o_i(x_k)$ and, as usually, we assume that e_k is a zero-mean white Gaussian noise of variance σ^2. Then, the log-likelihood of the parameters is given by

$$L(\mathbf{G}; \mathbf{x}, \mathbf{y}) = K - \frac{1}{2\sigma^2} \sum_k \left(y_k - \sum_i g_i(x_k)\right)^2 \quad (4)$$

where K is a constant which can be neglected in the optimization process, $\mathbf{G} = (G_1, \cdots, G_I)$ and $G_i = (\lambda_i, \mu_i, \sigma_i)$.

From (4) we see that, under the Gaussian noise assumption, to obtain ML estimates reduces to minimize the conventional MSE. This multiparameter nonlinear optimization process can be accomplished through a global gradient descent algorithm [5,6,7]: its shortcomings have been already mentioned.

A computationally more efficient procedure to obtain ML estimates is based on the EM algorithm. At each E-step the algorithm computes the expectation of some unobserved (hidden) data using the current parameter estimates. Then, in the M-step new estimates are obtained and the process is iterated. Each EM iteration increases the likelihood, so the algorithm is guaranteed to converge to a maximum (not necessarily the global one) of the likelihood function [8].

The important point regarding the EM algorithm is that a good choice of the hidden variables can help to simplify the maximization step. A particularly useful selection of hidden variables for this problem was proposed in [13]: each observation is decomposed into I hidden variables according to

$$y_{k,i} = g_i(\mathbf{x}_k) + e_{k,i} \qquad i = 1, \cdots, I \quad (5)$$

where the residuals $e_{k,i}$ are also obtained by decomposing the total residual $e_k = y_k - \sum_i g_i(\mathbf{x}_k)$ into I components, i.e.,

$$e_{k,i} = t_{k,i} e_k \qquad i = 1, \cdots, I. \quad (6)$$

In [13], it was shown that the decoupling variables $t_{k,i}$ are arbitrary (obviously they are constrained to sum 1). Different alternatives are then possible: in [13] the residual is decomposed equally among all the neurons

$$t_{k,i} = \frac{1}{I}, \qquad i = 1, \cdots, I; \tag{7}$$

while in [11] it is decomposed proportionally to the weights of the output layer

$$t_{k,i} = \frac{\lambda_i^2}{\sum_j \lambda_j^2}, \qquad i = 1, \cdots, I. \tag{8}$$

Finally, using whether (7) or (8) to decompose the residual, the EM algorithm for training two-layer feedforward networks can be summarized as follows

E-step: for $i = 1, \cdots, I$ compute

$$y_{k,i} = g_i(\mathbf{x}_k) + t_{k,i}\left(y_k - \sum_j g_j(\mathbf{x}_k)\right) \tag{9}$$

M-step: for $i = 1, \cdots, I$ compute

$$(\lambda_i, \mu_i, \sigma_i^2) = \min_{G_i} \sum_k \left(y_{k,i} - g_i(\mathbf{x}_k)\right)^2 \tag{10}$$

where the index denoting iteration has been omitted. Note that the problem of globally training an RBF network with I neurons has been decoupled into I simpler problems of training a single neuron.

3 Fast EM training of RBF networks

The decoupling variables (7) or (8), which are constant over the whole input space, have shown to be very effective in feedforward neural networks with non-local activation functions, such as the Multilayer Perceptron (MLP). However, they are not well suited for networks with local activation functions, such as the RBF. For this type of networks its convergence is slow due to the fact that, using the previous decoupling variables, at each M-step we are trying to fit a Gaussian to a very large region of the input space. Intuitively, we could make a better job if we localize somehow the error associated to each RBF unit. This is the idea that we exploit in this paper to obtain a faster convergence and also to avoid some unwanted local minima of the cost function. Specifically, we propose a modification, which consists in using as decoupling variables the following posterior probabilities

$$t_{k,i} = P(G_i | x_k, y_k), \tag{11}$$

i.e., the probability of the ith RBF unit given the kth input-output pattern. Using (11), the algorithm performs a soft adaptive decomposition of the residual taking into account the local nature of the activation functions. We denote this modification as soft-EM as opposed to the classical EM versions using constant decoupling variables [11,13].

Now we consider the estimation of (11): applying Bayes, the posterior probabilities can be estimated as

$$t_{k,i} = \frac{P(x_k, y_k|G_i)}{\sum_j P(x_k, y_k|G_j)}, \tag{12}$$

and the probabilities $P(x_k, y_k|G_j)$ can be obtained through

$$P(x_k, y_k|G_j) = P(y_k|G_j, x_k)P(x_k|G_j). \tag{13}$$

It is interesting to remark that the probabilities $P(x_k|G_j)$ are the key variables responsible for introducing the local character of the RBF units, since they can be estimated as

$$P(x_k|G_s) = \frac{o_s(x_k)}{\sum_j o_j(x_k)}. \tag{14}$$

In order to estimate the output given the input sample and the RBF unit, $P(y_k|G_j, x_k)$, we have several possibilities depending on the assumed model for the data. Here, we start by remembering that y_k can be decomposed into a set of hidden variables as follows

$$y_k = \sum_i y_{k,i}, \tag{15}$$

and then considering that

$$P(y_k|G_j, x_k) \propto \sum_i P(y_{k,i}|G_j, x_k). \tag{16}$$

Finally, we assume for simplicity the following Gaussian model for $P(y_{k,i}|G_j, x_k)$

$$P(y_{k,i}|G_j, x_k) \propto \frac{1}{\sqrt{2\pi}\sigma_{ei}} \exp -\frac{(y_{k,i} - g_j(x_k))^2}{2\sigma_{ei}^2}, \tag{17}$$

where the variance σ_{ei}^2 can be estimated at each iteration as

$$\sigma_{ei}^2 = \frac{\sum_k t_{k,i} e_k^2}{\sum_k t_{k,i}}. \tag{18}$$

Actually, Gaussianity is a rather crude approximation for the pdf of the residuals associated to each RBF, however it seems to work very well in practice.

To summarize, using (17), (16) and (14) we estimate the decoupling variables as (12): these estimates are then used in the E-step.

In the maximization step, new RBF parameters need to be obtained through minimizing (10) for each component. The amplitude of each RBF unit can be obtained solving a linear least squares problem, while its center and variance can be updated using a gradient descent procedure [7].

Let us point out that each RBF unit is adapted separately therefore simplifying the global optimization problem and allowing an easy parallelization. The extension to multidimensional input spaces is straightforward.

4 Experimental Results

4.1 1-D Simulation Example

In this example we consider a set of four 1-D functions, each one composed by the superposition of slow sinusoids with random amplitudes, phases and frequencies. The experiments are carried out generating 100 samples of the signals. We compare the performance of the proposed soft-EM approach with the classical EM alternatives [13] and [11], denoted as EM-1 and EM-2, respectively. In all the experiments, an RBF network with 5 units is initialized using the solution provided by the Orthogonal Least Squares (OLS) algorithm [14], with several different initial values of σ^2. Figure 1 shows the evolution of the MSE with the number of iterations. It can be seen that the soft-EM approach provides a faster convergence than the conventional non-localized EM alternatives.

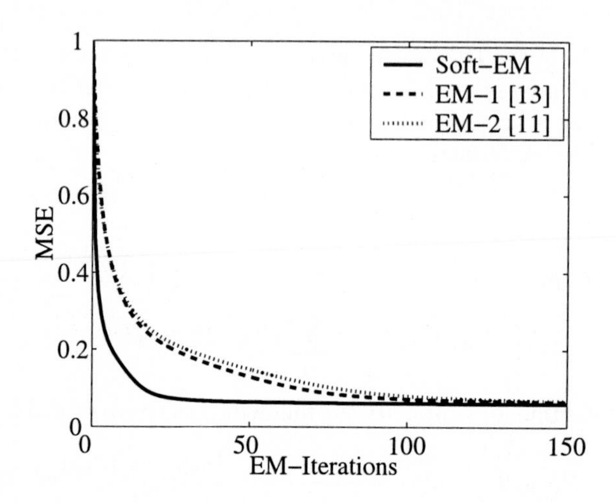

Figure 1. MSE evolution of the diferent EM alternatives (1-D example)

4.2 2-D Simulation Example

In this experiment we consider the set of eight 2-D functions used in [15] (see Table 1).Now we use a Generalized Radial Basis Function (GRBF) allowing a different variance along each input dimension. In each experiment, a GRBF network with 10 neurons is initialized using a different initial value of σ^2 and applying the OLS algorithm. When compared with the conventional EM approaches, again an important improvement in convergence speed is observed (see Fig. 2). Moreover, it also presents a lower sensitivity to get trapped into local minima.

This lower sensitivity to local minima is more noticeable when we compare the performance of the soft-EM with a global gradient descent approach [7]. A GRBF network with 10 neurons is employed and 500 iterations of both the gradient method and the soft-EM approach are carried out. To reduce the computational cost of the soft-EM algorithm, we apply a single gradient iteration at each M-step. In this way both alternatives have a similar computational cost. Table 2 shows the mean signal to error ratio (SER) in dB obtained for each function with both alternatives.

It can be seen that, for most of the functions, the soft-EM provides better results than the global gradient approach. The reason is that the gradient approach usually is trapped into local minima, while the soft-EM is frequently able to avoid these local minima, thus providing a better approximation.

Name	Function	Domain
Fun 1	$y = sin(x_1 x_2)$	[-2,2]
Fun 2	$y = exp(x_1 sin(\pi x_2))$	[-1,1]
Fun 3	$y = \frac{40 * exp(8((x_1-0.5)^2+(x_2-0.5)^2))}{exp(8((x_1-0.2)^2+(x_2-0.7)^2))+exp(8((x_1-0.7^2+(x_2-0.2)^2))}$	[0,1]
Fun 4	$y = (1 + sin(2x_1 + 3x_2))/(3.5 + sin(x_1 - x_2))$	[-2,2]
Fun 5	$y = 42.659(0.1 + x_1(0.05 + x_1^4 - 10x_1^2 x_2^2 + 5x_2^4))$	[-0.5,0.5]
Fun 6	$y = 1.3356[1.5(1 - x_1) + exp(2x_1 - 1)sin(3\pi(x_1 - 0.6)^2)$ $+exp(3(x_2 - 0.5))sin(4\pi(x_2 - 0.9)^2)]$	[0,1]
Fun 7	$y = 1.9[1.35 + exp(x_1)sin(13(x_1 - 0.6)^2)$ $+exp(3(x_2 - 0.5))sin(4\pi(x_2 - 0.9)^2)]$	[0,1]
Fun 8	$y = sin(2\pi\sqrt{x_1^2 + x_2^2})$	[-1,1]

Table 1. Functions used to generate the data sets

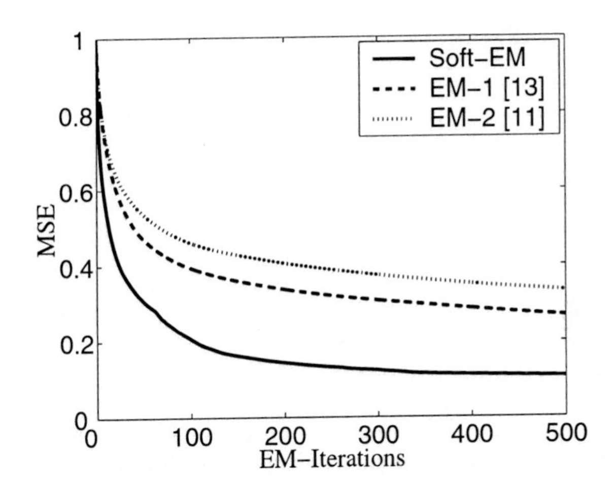

Figure 2. MSE evolution of the diferent EM alternatives (2-D example)

5 Conclusions

The decoupling variables used in the E-step of EM-based learning algorithms can be selected to control the rate of convergence of the algorithm and also to reduce its sensitivity to local minima. We have studied in this paper a suitable selection of these variables for feedforward networks with local activation functions (mainly, RBF networks). Specifically, these variables are estimated as the posterior probability of each RBF unit given each input-output pattern. By means of several simulation examples, it has been shown that this modification accelerates the convergence of the algorithm and also avoids some unwanted minima of the cost function.

	Fun 1	Fun 2	Fun 3	Fun 4	Fun 5	Fun 6	Fun 7	Fun 8
Gradient	18.0	23.5	26.7	5.6	11.1	19.7	12.8	10.4
Soft-EM	19.2	23.9	32.3	7.3	14.9	22.5	17.8	10.3

Table 2. Mean SER (in dB) for the different test functions

References

1. S. Haykin, *Neural Networks: A Comprehensive Foundation*, Macmillan, New York, 1994.
2. C. Bishop, *Neural Networks for Pattern Recognition*, Clarendon Press, Oxford, 1997.
3. D. S. Broomhead, D. Lowe, " Multivariable functional interpolation and adaptive networks", *Complex Systems*, vol. 2, pp. 321-355, 1988.
4. J. E. Moody, C. J. Darken, "Fast learning in networks of locally-tuned processing units", *Neural Computation*, vol. 1, pp. 281-294, 1989.
5. N. B. Karayanis, "Gradient descent learning of radial basis neural network", *Proc. 1997 IEEE Int. Joint Conf. Neural Networks*, Houston, TX, pp. 1825-1820, 1997.
6. D. Lowe, "Adaptive radial basis function function nonlinearities and the problem of generalization, *1st IEE Conf. on Artificial Neural Networks*, London, UK, pp. 171-175, 1989.
7. I. Santamaría, *et al* "A nonlinear MESFET model for intermodulation analysis using a GRBF Network", *Neurocomputing*, vol. 25, pp. 1-18, 1999.
8. A. P. Dempster, N. M. Laird, D. B. Rubin, "Maximum likelihood from incomplete data via the EM algorithm", *J. Roy. Statisc. Soc B*, vol. 39, pp. 1-38, 1977.
9. M. I. Jordan, R. A. Jacobs, "Hierarchical mixtures of experts and the EM algorithm", *Neural Computation*, vol. 6, pp. 181-214, 1994.
10. Z. Ghahramani, M. I. Jordan, "Supervised learning from incomplete data via an EM approach", in *Advances in NIPS VI*, J. D. Cowan, G. Tesauro and J. Alspector, Eds., San Mateo, CA: Morgan Kaufmann, 1994.
11. S. Ma, C. Ji, J. Farmer "An efficient EM-based training algorithm for feedforward neural networks", *Neural Networks*, vol. 10, pp. 243-256, 1997.
12. S. Ma, C. Ji, "Fast training of recurrent networks based on the EM algorithm", *IEEE Trans. on Neural Networks*, vol. 9, pp. 11-26, 1998.
13. M. Feder, E. Weinstein, "Parameter estimation of superimposed signals using the EM algorithm", *IEEE Trans. Acoust., Speech, Signal Processing*, vol. 36, pp. 477-489, 1988.
14. S. Chen, C.F.N. Cowan, P. M. Grant, "Orthogonal least squares learning algorithm for radial basis functions networks", *IEEE Trans. on Neural Networks*, vol. 2, pp.302-309, 1991.
15. V. Cherkassky, D. Gehring, F. Mulier, "Comparison of adaptive methods for function estimation from samples", *IEEE Trans. on Neural Networks*, vol. 7, no. 4, pp.969-984, 1996.

Expansive and Competitive Neural Networks

J.A. Gomez-Ruiz, J. Muñoz-Perez, E. Lopez-Rubio, and M.A. Garcia-Bernal

Dept. of Computer Science and Artificial Intelligence
University of Malaga, Campus de Teatinos s/n
29071 Malaga, Spain.
{janto, munozp, ezeqlr, magb}@lcc.uma.es

Abstract. In this paper, we develop a necessary and sufficient condition for a local minimum to be a global minimum to the vector quantization problem and present a competitive learning algorithm based on this condition which has two learning terms; the first term regulates the force of attraction between the synaptic weight vectors and the input patterns in order to reach a local minimum while the second term regulates the repulsion between the synaptic weight vectors and the input's gravity center, favoring convergence to the global minimum. This algorithm leads to optimal or near optimal solutions and it allows the network to escape from local minima during trainning. Experimental results in image compression demonstrate that it outperforms the simple competitive learning algorithm, giving better codebooks.

1 Introduction

Vector quantization (VQ) is a block coding method for approximating a multidimensional signal space with a finite number of representative vectors. A vector quantizer statistically encodes data vectors in order to quantize and compress the data. The design of vector quantizers is a multidimensional optimization problem in which a distortion function is minimized. Vector quantization has been widely used for image compression and speech signals. In the same way, vector quantization is an approach to data clustering of a data set by combinatorial optimization which divides the data into clusters according to a suitable cost function.

According to Shannon´s rate distortion theory, vector quantization can always achieve better compression performance than any conventional coding technique based on the encoding of scalar quantities [1]. However, practical use of VQ techniques has been limited because of the prohibitive amount of computation associated with existing encoding algorithms. Linde, Buzo and Gray [2] proposed an algorithm (LBG), with no differentiation, for VQ design which is the standard approach to compute the codebook. The LBG algorithm converges to a local minimum, but is not guaranteed to reach the global minimum.

The aim of competitive neural networks is to cluster or categorize the input data and it can be used for data coding and compression through vector quantization. Competitive learning is an appropriate algorithm for VQ of unlabelled data. Ahalt,

J. Mira and A. Prieto (Eds.): IWANN 2001, LNCS 2084, pp. 355-362, 2001.

Krishnamurthy and Chen [3] discussed the application of competitive learning neural networks to VQ and developed a new training algorithm for designing VQ codebooks which yields near-optimal results and can be used to develop adaptive vector quantizers. Pal, Bezdek and Tsao [4] proposed a generalization of learning vector quantization for clustering which avoids the necessity of defining an update neighbourhood scheme and the final centroids do not seem sensitive to initialization. Xu, Krzyzak and Oja [5] developed a new algorithm called rival penalized competitive learning which for each input not only the winner unit is modified to adapt itself to the input, but also its rival delearns with a smaller learning rate. Ueda and Nakano [6] presented a new competitive learning algorithm with a selection mechanism based on the equidistortion principle for designing optimal vector quantizers. The selection mechanism enables the system to escape from local minima. Uchiyama and Arbib [7] showed the relationship between clustering and vector quantization and presented a competitive learning algorithm which generates units where the density of input vectors is high and showed its efficiency as a tool for clustering color space in color image segmentation based on the least sum of squares criterion.

We propose a new expansive and competitive algorithm. Its main contribution is an additional term in the weight updating rule that implicitly leads the synaptic weights (code vectors) toward a global minimum of the distortion function. It represent a heuristic strategy to escape from local minima and reach a global minimum.

2 Vector Quantization and Clustering

A vector quantizer of dimension N and size K is a mapping Q from the N-dimensional Euclidean space $\Re^N$ into a finite subset $C=\{\mathbf{y}_1,\mathbf{y}_2,\ldots,\mathbf{y}_K\}$ of $\Re^N$ containing K output or representative vectors, called code vectors, reference vectors, reproduction vectors, prototypes or *codewords*. The collection of all possible reproduction vectors is called the *reproduction alphabet* or more commonly *the codebook.* Hence, the input vector space, $\Re^N$, is divided into K disjoint and exhaustive regions, $S_1, S_2,\ldots,S_K$, where

$$S_i = \{\mathbf{x}\in\Re^N : Q(\mathbf{x}) = \mathbf{y}_i\}, \quad i=1,2,\ldots,K. \tag{1}$$

All inputs vectors in S_i are approximated by $\mathbf{y}_i$. The cost introduced by this approximation is given by a nonnegative distortion measure $d(\mathbf{x},\mathbf{y}_i)$. The most common distortion measure is the *squared error distortion measure* defined by the squared Euclidean distance between the two vectors:

$$d(\mathbf{x},\mathbf{y}) = \|\mathbf{x}-\mathbf{y}\|^2 = \sum_{j=1}^{N}(x_j - y_j)^2 \tag{2}$$

The distortion measure can be viewed as the cost of representing $\mathbf{x}$ by $\mathbf{y}_i$. Note that for every reproduction set C, $Q(\mathbf{x}) = \mathbf{y}_i$ if and only if $d(\mathbf{x},\mathbf{y}_i) \le d(\mathbf{x},\mathbf{y}_k)$, $\forall \mathbf{y}_k \in C$.

When we have the number K of reproduction vectors, and a stochastic input vector X distributed according to a time invariant probability distribution $f_X(x)$, the goal is to select a set C that minimizes the average distortion,

$$D(C) = \mathrm{E}[d(X,Q(X))] = \sum_{i=1}^{K} \int_{S_i} d(\mathbf{x}, \mathbf{y}_i) f_X(\mathbf{x}) d\mathbf{x} \tag{3}$$

where E[·] represents the expectation operator with respect to underlying probability distribution of X. To obtain an optimal vector quantizer, the expected distortion should be minimized with respect to the partition $\{S_1, S_2, ..., S_K\}$ and the codebook C. A vector quantizator is said to be optimal if no other quantizer has a smaller distortion cost for the same source X, vector dimension, and codebook with size K.

At present, no general analytical solution for this minimization problem exists because the underlying distribution $f_X(x)$ is in general unknown. In other words, we have to solve this minimization problem with a *learning* procedure over a given set of training vectors. For a finite training set, $T = \{\mathbf{x}_i \in \Re^N \mid i=1,2,...,n\}$, the expected value is replaced by the sample mean, and the vector quantization is a combinatorial problem that attempts to represent T (with large information contents) by a reduced set of reproduction vectors C. In other words, the goal is to select a set C such that the sample distortion function

$$D(\mathbf{y}_1, \mathbf{y}_2, ..., \mathbf{y}_K) = \frac{1}{n} \sum_{i=1}^{K} \sum_{j \in S_i} \left\| \mathbf{x}_j - \mathbf{y}_i \right\|^2 \tag{4}$$

is minimized. Finally, we could state the problem as follows:

$$\min\ D(\mathbf{y}_1,\mathbf{y}_2,...,\mathbf{y}_K) = \frac{1}{n} \sum_{j=1}^{n} \sum_{i=1}^{K} \left\| \mathbf{x}_j - \mathbf{y}_i \right\|^2 \delta_i(\mathbf{x}_j, \mathbf{y}_1, ..., \mathbf{y}_K) \tag{5}$$

where

$$\delta_i(\mathbf{x}, \mathbf{y}_1, ..., \mathbf{y}_K) = \begin{cases} 1 & \text{if } \left\| \mathbf{x} - \mathbf{y}_i \right\|^2 = \min_r \left\| \mathbf{x} - \mathbf{y}_r \right\|^2 \\ 0 & \text{otherwise} \end{cases}, \quad i=1,2,...,K. \tag{6}$$

This problem can be interpreted as a cluster analysis problem. When T and the number of clusters K is given, the clustering based on the least sum squares is the partition $S = \{S_1, S_2,...,S_K\}$ that minimizes the within-cluster sum of squares. Uchiyama and Arbib [7] have shown how the problem of selecting the clustering S becomes the problem of selecting $C=\{\mathbf{y}_1,\mathbf{y}_2,...,\mathbf{y}_K\}$.

The above distortion function is generally not convex and may contain multiple local minima. The resulting assignment problem is known to be a NP-hard combinatorial optimization problem. The two necessary conditions for minimization of the problem in Eq. (5) are well-known [1]:

C1. Voronoi partition (competitiveness): For fixed representative vectors, the optimal partition should be constructed in such a manner that:

$$Q(\mathbf{x}) = \mathbf{y}_i \Leftrightarrow d(\mathbf{x},\mathbf{y}_i) \leq d(\mathbf{x},\mathbf{y}_k),\ \forall \mathbf{y}_k \in C. \tag{7}$$

This implies a partition of the space $\Re^N$ into the following disjoint regions:

$$S_i = \{\mathbf{x} \in \Re^N : \|\mathbf{x}-\mathbf{y}_i\| < \|\mathbf{x}-\mathbf{y}_j\|, \forall j \neq i\} \tag{8}$$

C2. Generalized Centroid Condition: For a given partition $\{S_1,\dots,S_K\}$, each codeword $\mathbf{y}_i$ is the centroid of the region S_i, given by $\mathbf{y}_i = E[\mathbf{x} \mid \mathbf{x} \in S_i]$. For a finite training set, the expected value is replaced by the sample mean of the i-th region. That is,

$$\mathbf{y}_i = \frac{\sum_{j \in S_i} \mathbf{x}_j}{|S_i|} \tag{9}$$

where $|S_i|$ is the cardinality of S_i.

Note that the two above conditions are sufficient for a local minimum of the distortion function, but are only necessary for a global minimum. In proposition 1, we propose a necessary and sufficient condition for a local minimum codebook to be optimal.

At present the most widely used techniques to design VQ codebooks are the LBG algorithm and competitive neural networks. These algorithms guarantee only local optimality and their performances strongly depend on the initial values of the representative vectors. Furthermore, the LBG algorithm is slow since it requires an exhaustive search through the entire codebook on each iteration.

3 Expansive and competitive Learning

The simple competitive learning (SCL) rule [8] can be derived as a stochastic gradient descent of the distortion function $D(\mathbf{y}_1,\mathbf{y}_2,\dots,\mathbf{y}_k)$. It can sometimes escape from unstable local minima with the help of its stochastic characteristics, whereas the LBG algorithm cannot [6].

We consider a single layer of K units (neurons or processing elements), where o_i denotes the output of unit i and $\mathbf{w}_i=(w_{i1},w_{i2},\dots,w_{iN})$ the synaptic weight vector of unit i, for $i=1,2,\dots,K$. For a particular input, $\mathbf{x}=(x_1,x_2,\dots,x_N) \in \Re^N$, its output is $\mathbf{o}=(o_1,o_2,\dots,o_K)$, where

$$o_i = \begin{cases} 1 & \text{if} \quad h_i = \underset{j}{Max}\{h_j\} \\ 0 & \text{otherwise} \end{cases}, \qquad h_i = \sum_{j=1}^{N} w_{ij}x_j - \theta_i \quad \text{and} \quad \theta_i = \frac{1}{2}\sum_{j=1}^{N} w_{ij}^2 \tag{10}$$

The SCL rule [8] is given by

$$\Delta\mathbf{w}_i(k+1) = \begin{cases} \alpha_i(k)(\mathbf{x}(k)-\mathbf{w}_i(k)) & \text{if } o_i = 1 \\ 0 & \text{otherwise.} \end{cases} \tag{11}$$

where $\alpha_i(k) \in [0,1]$ is the learning rate for unit i. Thus, in practice only the *winning unit* or *best matching unit* is updated. Note that the synaptic weights of these neural networks provide the prototypes to the vector quantization problem.

Since the distortion function is in generally not convex and may contain multiple local minima [9], competitive learning often produces non optimal codebooks. For this reason, we need to include in the learning rule some specific property to search a global minimum. Thereby, we now show that when the prototypes (centroids) are a global minimum solution to the problem in Eq. (5) then they are the centroids with larger variance.

Proposition 1: Let $\{\mathbf{w}_1^*,\mathbf{w}_2^*,\ldots,\mathbf{w}_K^*\}$ be an optimal solution (global minimum) to the problem in Eq. (5). Then we have

$$\sum_{i=1}^{K} n_i^* \left\|\mathbf{w}_i^* - \overline{\mathbf{x}}\right\|^2 \geq \sum_{i=1}^{K} n_i \left\|\mathbf{w}_i - \overline{\mathbf{x}}\right\|^2 \tag{12}$$

for any local optimal solution $\{\mathbf{w}_1,\mathbf{w}_2,\ldots,\mathbf{w}_K\}$ to the problem in Eq. (5), where

$$\mathbf{w}_i = \frac{\sum_{j\in C_i} \mathbf{x}_j}{n_i} \quad \text{and} \quad \overline{\mathbf{x}} = \frac{\sum_{j=1}^{n} \mathbf{x}_j}{n} \tag{13}$$

This result implies that we have to search the local minimum with larger dispersion since

$$\overline{\mathbf{x}} = \frac{\sum_{j=1}^{n} \mathbf{x}_j}{n} = \frac{\sum_{i=1}^{K} n_i \mathbf{w}_i}{n} \tag{14}$$

Proof of proposition: Since $\{\mathbf{w}_1^*,\mathbf{w}_2^*,\ldots,\mathbf{w}_K^*\}$ and $\{\mathbf{w}_1,\mathbf{w}_2,\ldots,\mathbf{w}_K\}$ are local minima then

$$\mathbf{w}_i^* = \frac{\sum_{j\in C_i^*} \mathbf{x}_j}{n_i^*}, \quad \mathbf{w}_i = \frac{\sum_{j\in C_i} \mathbf{x}_j}{n_i}, \quad i=1,2,\ldots,K \quad \text{and} \quad \overline{\mathbf{x}} = \frac{\sum_{i=1}^{K} n_i \mathbf{w}_i}{n} \tag{15}$$

It is well know [10] that

$$\sum_{j=1}^{n} \left\|\mathbf{x}_j - \overline{\mathbf{x}}\right\|^2 = \sum_{i=1}^{K} \sum_{j\in C_i^*} \left\|\mathbf{x}_j - \mathbf{w}_i^*\right\|^2 + \sum_{i=1}^{K} n_i^* \left\|\mathbf{w}_i^* - \overline{\mathbf{x}}\right\|^2 \tag{16}$$

$$\sum_{j=1}^{n} \left\|\mathbf{x}_j - \overline{\mathbf{x}}\right\|^2 = \sum_{i=1}^{K} \sum_{j\in C_i} \left\|\mathbf{x}_j - \mathbf{w}_i\right\|^2 + \sum_{i=1}^{K} n_i \left\|\mathbf{w}_i - \overline{\mathbf{x}}\right\|^2 \tag{17}$$

and thus

$$\sum_{i=1}^{K} n_i^* \left\|\mathbf{w}_i^* - \overline{\mathbf{x}}\right\|^2 \geq \sum_{i=1}^{K} n_i \left\|\mathbf{w}_i - \overline{\mathbf{x}}\right\|^2 \tag{18}$$

The SCL algorithm updates the synaptic weights of the winner unit directly towards $\mathbf{x}_j$ while this property suggests that the synaptic weights should be moved away from $\overline{\mathbf{x}}$. We propose a new version of the competitive learning algorithm that puts together these two features by considering two learning parameters in the SCL rule, α and β. The expansive and competitive learning (ECL) rule is given by

$$\Delta \mathbf{w}_i(k+1) = \begin{cases} \alpha_i(k)(\mathbf{x}(k) - \mathbf{w}_i(k)) - \beta_i(k)(\bar{\mathbf{x}} - \mathbf{w}_i(k)) & \text{if } o_i = 1 \\ 0 & \text{otherwise} \end{cases} \quad \textbf{(19)}$$

where $\alpha_i(k)$ is the learning parameter that controls the effect of moving the synaptic weight vector $\mathbf{w}_i$ of the winning unit i toward the input pattern $\mathbf{x}(k)$; on the other hand, $\beta_i(k)$ controls the effect of pushing the synaptic weight vector $\mathbf{w}_i$ away from $\bar{\mathbf{x}}$. The parameter $\alpha_i(k)$ will be called *parameter of attraction* while the parameter $\beta_i(k)$ will be called *parameter of repulsion*. Hence, the synaptic weights are influenced by two forces; an attraction toward the input patterns, and a repulsion from the gravity center of the input patterns. At the beginning of the learning process the attraction force must be larger than the repulsion, and it must decrease slower than the repulsion force to reach local minima. The learning parameters $\alpha_i(k)$ and $\beta_i(k)$ are assumed to be a monotonically decreasing function of the number of iterations k. Moreover, $\beta_i(k)$ has to decrease fast while $\alpha_i(k)$ has to decrease slow since a goal of this algorithm is to find cluster centroids (local minima). On the other hand, if $\beta_i(k)$ decreases too slow then some synaptic weight vector can stay out of reach, that is, it stays far from any input vector and so it may never wins and never learns (dead units). Hence, $\alpha_i(k)$ can be taken as

$$\alpha_i(k) = \alpha_0(1\text{-}k/T), \quad k=1,2,\ldots,T \quad \textbf{(20)}$$

where T is the maximum number of iterations that the learning process is allowed to execute and $\alpha_0 \in [0,1]$ is the initial value of the learning parameter while $\beta_i(k)$ can be taken as

$$\beta_i(k) = \begin{cases} \beta_0(1 - k/T_0), & k = 1,2,\ldots,T_0 \\ 0 & k > T_0 \end{cases} \quad \textbf{(21)}$$

where $\beta_0 \in [0,1]$, $\beta_0 < \alpha_0$ and $T_0 << T$.

Therefore, the synaptic weights are expanded from their initial values at the same time they are attracted by the input data until they are trapped on a local minimum (attractor) of the distortion function. The codebook given by the algorithm can be used as initial values for the synaptic weights in a new iteration, because the second term of the learning rule enables to escape from local minima.

Note that

$$\sum_{i=1}^{K} \left\| \mathbf{w}_i - \bar{\mathbf{w}} \right\|^2 = \frac{1}{K} \sum_{i<j} \left\| \mathbf{w}_i - \mathbf{w}_j \right\|^2 \quad \textbf{(22)}$$

and so if the left term is increased then the right term is also increased. Thus, when $\mathbf{w}_i$ is moved away from $\bar{\mathbf{x}}$ then the overall distance between the synaptic weight vectors is increased.

(a) (b)

Fig. 1. (a) Original Image. (b) Compressed image with mean squared error equal to 52.99.

Table 1. Mean squared error of compressed images for ten simulations with β=0 and β=013.

β=0	54.91	53.58	52.99	52.56	52.56	53.14	55.06	55.20	52.99	53.44
β=0.13	48.58	52.99	53.73	48.44	52.85	52.85	45.70	54.46	47.33	52.99

4 Experiments Results

In this section we illustrate the effectiveness of proposed approach in image compression. Next it is explained the procedure to perform image compression with unsupervised learning. First of all, the input sample is set up by subdividing the monochrome image into square subimages named *windows*. Hence, if the image size is $M \times N$ pixels and the window size is $k \times k$ pixels, we have $(M{\times}N)$ / $(k \times k)$ windows. These windows are our input patterns with $k \times k$ components (these patterns are obtained by arranging the pixel values row by row from top to bottom). The compression process consists in selecting a reduced set of K representative windows and replacing every window of the original image with the "closest" prototype (representative window), the given by the neural network after the learning process. In this experiment we have considered a window size of 3x3 pixels and K=32 representative windows. Thus, the neural network has 32 output neurons. We use the image shown in Fig. 1(a) with a resolution of 399x276 = 110124 pixels (108,65 Kbytes in BMP format) and 256 grey values. Then, we have 12236 input patterns and 32 representative windows.

The network's initial weights were chosen randomly from all input patterns. $\alpha(k)$ and $\beta(k)$ decay linearly with initial values $\alpha_0=0.5$, $\beta_0=0.13$ chosen empirically. The input patterns have been introduced ten times and the parameter β operates only in the first time, so we have taken T=122360 and T_0=12236. Table 1 shows the mean squared error in ten simulations with standard competitive learning (β=0) and with expansive and competitive learning (β= 0.13).

The best solution given by the standard competitive learning is 52.56, whereas the best solution given by the expansive and competitive learning is 45.70.

An example of this compression is shown in Fig. 1(b) with a size of 7.76 Kbytes. The original monochrome image with 256 grey levels requires 8 bits per pixel. In this example the window has a size of 9 pixels, that is, we need 72 bits to store it and since there are only 32 different classes, only 5 bits are necessary to represent the windows. In this way we obtain a compression rate of 72 to 5, that is 14 to 1.

5 Conclusions

The competitive neural network paradigm has been applied to solve the vector quantization problem, yielding better results than the LBG algorithm. However, it converges to a local minimum and it is not guaranteed to reach the global minimum. Moreover, it is sensitive to initialization. We have proposed a new competitive learning rule with two learning parameter, α and β; the parameter α regulates the force of attraction between the synaptic weight vectors and the input patterns in order to get a local minimum while the parameter of repulsion β regulates the force of repulsion between the synaptic weight vectors and the center of gravity of the input patterns in order to get a global minimum. Moreover, the parameter β enables the system to escape from local minima. This algorithm leads to better results than the simple competitive algorithm to solve problems of vector quantization with many local minima as the compress image problem. We have shown that the algorithm leads to a substantial reduction of the mean squared error.

References

1. Gray, R.M.: Vector quantization. IEEE ASSP Magazine, Vol. 1 (1980) 4-29.
2. Linde, Y., Buzo A., Gray, R.M.: An Algorithm for Vector Quantizer Design. IEEE Trans. on Communications, Vol. 28, nº 1 (1980) 84-95.
3. Ahalt, S.C., Krishnamurphy, A.K., Chen P., Melton, D.E.: Competitive Learning Algorithms for Vector Quantization. Neural Networks, Vol. 3 (1990) 277-290.
4. Pal, N.R., Bezdek, J.C., Tsao, E.C.: Generalized Clustering Networks and Kohonen's Self-Organizing Scheme. IEEE Trans. Neural Networks, Vol. 4, nº 4 (1993) 549-557.
5. Xu, L., Krzyzak, A., Oja, E.: Rival Penalized Competitive Learning for Clustering Analysis. IEEE Trans. Neural Networks, Vol. 4, nº 4 (1993) 636-649.
6. Ueda, N., Nakano, R.: A New Competitive Learning Approach Based on an Equidistortion Principle for Designing Optimal Vector Quantizers. Neural Networks, Vol. 7, nº 8 (1994) 1211-1227.
7. Uchiyama, T., Arbib, M.A.: Color image segmentation using competitive learning. IEEE Trans. on Pattern Analysis and Machine Intelligence, Vol. 16, nº 12 (1994) 1197-1206.
8. Hertz, J., Krogh, A., Palmer, R.G.: Introduction to the Theory of Neural Computation. Addison-Wesley, New York (1991).
9. Gray, R.M., Karnim, E.D.: Multiple local optima in vector quantizers. IEEE Trans. Inform. Theory, Vol. IT-28, nº 2 (1982) 256-261.
10. Spath, H.: Cluster Analysis Algorithms for Data Reduction and Classification of Objects. Wiley, New York (1980).

Fast Function Approximation with Hierarchial Neural Networks and Their Application to a Reinforcement Learning Agent

Joern Fischer[1], Ralph Breithaupt[1], and Mathias Bode[2]

[1] GMD (AIS.ARC) National Research Center, Autonomous Intelligent Systems, D-53754 Sankt Augustin, Germany
{joern.fischer, ralph.breithaupt}@gmd.de

[2] Cortologic AG Berlin & University of Muenster, Department of Applied Physics, D-48149 Muenster, Germany,
bode@cortologic.de

Abstract. Current function approximators especially neural networks are often limited in several directions: most of the architectures can hardly be extended with more "informational" capacity, often neural networks with high capacity are too costly in calculation time (especially for an implementation on a microcontroller of a real world robot) and functions with high gradients can hardly be learned. The following approach shows, that these limitations can be overcome by using an adaptive hierarchical vector quantizing algorithm. With this algorithm the calculation time of a classification can decrease down to O(log(n)), where n is the number of implemented prototypes. If a given number of prototypes can not carry the "information" of the function which has to be approximated, the "informational" capacity can be increased by adding prototypes. Proposed in this article the algorithm is tested in a Reinforcement Learning task.

1 Introduction

Reinforcement Learning (RL) [1] [2] is a learning method, where a value-function, which contains all necessary information, is step by step built on the basis of external reward signals and state transitions. This value-function can be approximated by a neural network. Combining RL algorithms with Neural Networks [3] seems to be a promising method to extend RL with generalizing capabilities. This was impressively demonstrated in the TD-Gammon example [4], where a Reinforcement-Backpropagation combination learned to play Backgammon at master level.

But there are still various difficulties to overcome when such a combination of a RL algorithm and a neural network is constructed. Plenty of questions appear like: "How many neurons or prototypes should the network have to be capable to solve a given problem?", "Which architecture is advisable?" or if the network is big "Is the computer fast enough for realtime actions?".

Our investigation leads to an algorithm which is easy to implement, which finds

J. Mira and A. Prieto (Eds.): IWANN 2001, LNCS 2084, pp. 363-369, 2001.

on its own the number of necessary prototypes and last but not least is extremly fast in calculation time. Only O(log(n)) operations are needed to calculate a function value for n prototypes, which makes this algorithm fit even into simple microcontrollers.

2 Classifying methods

One of the most famous hierarchical classifying systems is quadtree [5]. It is often used in picture compression and seems useful especially in low dimensional spaces. In two dimensions a rectangle is divided in horizontal and vertical direction, Fig 1a. The four resulting rectangles may be recursively divided in the same way until the areas that have to be seperated are resolved with acceptable accuracy.

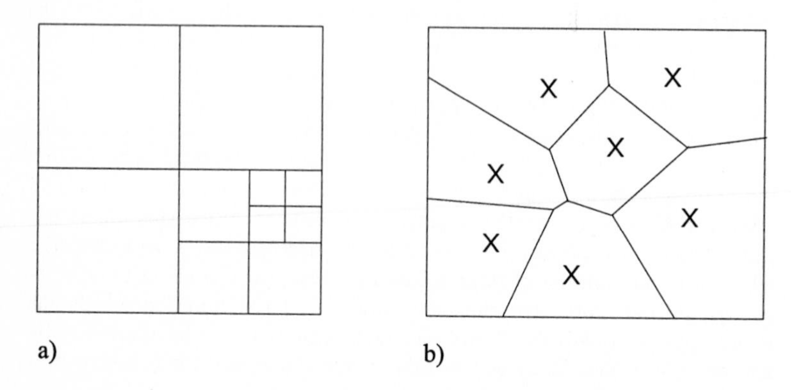

Fig. 1. a) Quadtree: The area is always divided in horizontal and vertical direction to enhance the resolution. b) Prototypes of a vector-quantizer in a 2-dimensional input space (crosses) and their respective Voronoi-cells.

The disadvantage of this algorithm is that there are always $2^{dimensions}$ divisions for every step to enhance the resolution, which is too much in high dimensional spaces. In general many of these divisions separate areas belonging to the same class. These divisions are not at all useful, but slow down learning and blow up the memory usage. Moreover each division is parallel to the axes of the space but one, a property which is not satisfying in many cases (e.g. to classify regions separated by an oblique hyper-plane).

A way to divide even high-dimensional (e.g. m-dimensions) state spaces is to use prototypes. A prototype is defined by its m-dimensional vector. In recent non-hierarchical vector-quantizing [6] [7] the Euclidean distance from a represented state vector to all prototype-vectors is calculated, to determine the prototype which is closest to the state. This rule implies that each prototype defines an area called Voronoi-cell Fig. 1b.

This algorithm is often used, but suffers from a lack of efficiency. The calculation time rises with O(n), with n being the number of prototypes. A more promising method is to combine the advantages of these two algorithms and to use only few prototypes in a first hierarchy level to classify first roughly and to enhance the resolution by classifying in several hierarchy levels. In the following section an efficient hierarchical vector quantizer is introduced, which classifies much more specifically than quadtree and which is much faster than recent vector quantizers.

3 A hierarchical vector-quantizer

Classifying vectors in a vector quantizer with multiple level architecture is much more efficient than using only one layer of prototypes. In multiple layer vector quantizing [8] the prototypes are organized as nodes in a tree. Every prototype has some prototypes of the next layer it points to. To minimize calculation time we choose a binary tree [9].

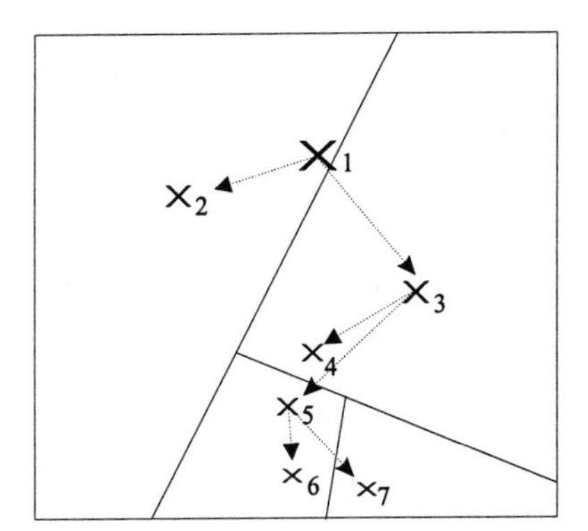

Fig. 2. Prototypes in a 2-dimensional input space and their respective areas. Classifying a pattern vector, we begin with the first prototype and pursue the arrows to the prototypes which are the nearest to the pattern vector until we get to the "winning" prototype.

To classify a given state vector S, first we scan the primary layer to select the prototype with the shortest Euclidean distance from S, then we take the nearest one of the two successors of the above prototypes to obtain a second layer result, then the nearest of the successors of this successor etc.. We go on until we come to a leaf-prototype without successor. This one is our "winning" prototype. The area it represents does not look like the classical Voronoi-cell, but like a splinter of broken glass Fig.2.

The classification is one part of the neural algorithm. Even more care should be invested to construct such a tree of prototypes. According to problem requirements there exist different methods to organize predefined prototypes like in clustering [10] or to generate new ones by dividing prototypes. In our investigation we start from an online learning procedure, that is initialized with one prototype. Prototype areas are divided if necessary by adding prototypes.

4 A hierarchical vector-quantizer as function approximator

To change a classifying system into a simple function approximator every prototype has to store a function value, which here is the mean value of all presented values in its area. We now introduced the following rule: A prototype area, with a function-value variance higher than a predefined variance-limit, has to be divided. The responsible prototype (the winning one) has to point to the two new successor prototypes which should reduce the function value variance by learning new function mean-values of their areas.

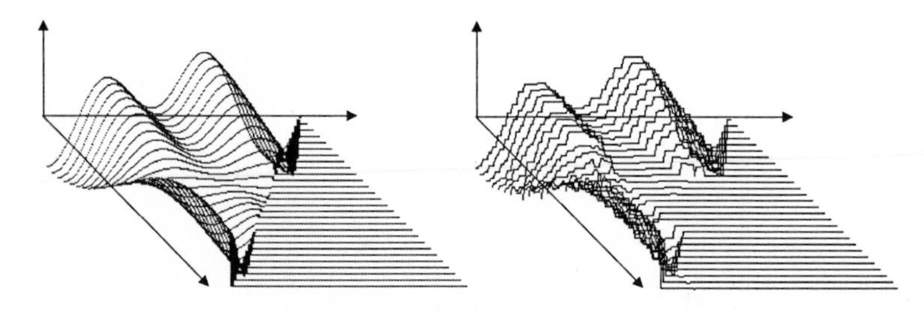

Fig. 3. A variance dependent function approximation with 837 Prototypes (left: the original function, right: the approximated one).

Because of the influence of every prototype on its neighbors, adding a new one might cause more trouble than it could solve. Therefore we need a possibility to estimate the ideal position before we divide. To benefit from the experience made by the old, unsophisticated prototype during the learning process we extend every prototype by two virtual prototypes, a "pos-prototype" and a "neg-prototype". If the function value of the presented vector is higher than the mean value of the prototype, the virtual "pos-prototype" moves towards the presented state-vector. If it is lower than the mean value, the virtual "neg-prototype" moves towards the presented vector. This implies that the virtual prototypes divide the area in a higher- and a lower-value part.

Now, if a prototype's function value variance is too high, two successors are generated and placed just at the position of the virtual prototypes. These new prototypes now get two new virtual prototypes each. Both successor prototypes learn the function value of their area. Fig. 3 shows a function approximation with 837 used prototypes.

5 A Reinforcement Learning task

We used this method in a Q-Learning algorithm with eligibility trace to approximate the Q-function [2]. The problem the agent had to learn was to play a squash game simulation. There is a racket which can be moved by the agent into 10 different positions by the actions "one step left", "one step right" or "stand still" and a ball with a constant speed reflecting in 45 degree angles from the wall.

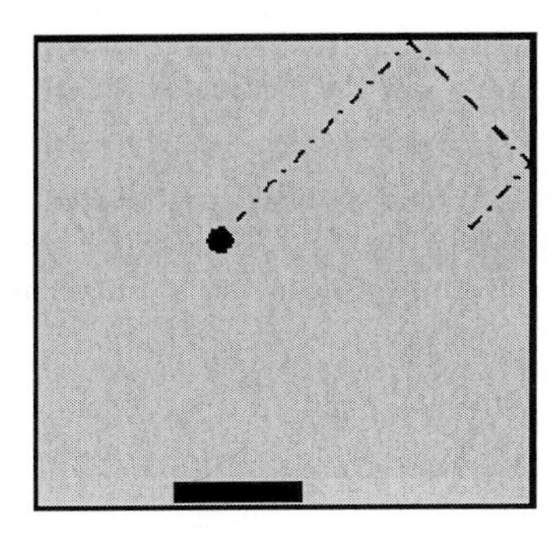

Fig. 4. Squash game environment. The racket can move to the left, to the right or can stand still. If the ball touches the ground the reward is (r = -1), if it is rejected into the field it is (r = +1).

If the ball touches the ground, the reward signal is r=-1, if the racket rejects the ball into the field, the reward is r=+1. The states the agent can perceive comprise the ball position (10 possible x-positions and 9 y-positions), the 4 possible directions of the ball and the 10 different positions of the racket, Fig. 4.

Fig. 5 shows the success of the algorithm determined in 100 games. Without predefining a discretization of the state space the algorithm places between 2 and 10 prototypes dependent on the initial state. Mostly the goal could be learned accurately. A normal Reinforcement algorithm which could perceive all 3600 states needed about the same number of episodes to converge. For detailed description see [11] [12] [13].

6 Conclusions

The capability to hierarchically discretizise the state space in an efficient manner based on variance control, allows us to construct fast RL algorithms which can even be implemented on microcontroller architectures. In our model the prototypes are placed nearly exactly where they are needed in contrast to existing models of state space quantization [14] [15]. The introduction of virtual prototypes makes this method work also properly in high dimensional state spaces. It is still possible to implement Reinforcement expansions like Suttons Dyna-Q

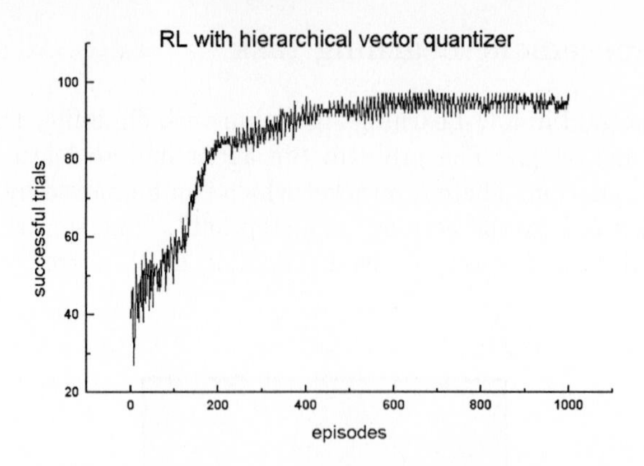

Fig. 5. Total success of 100 agents in a squash game simulation. While the algorithm discretizises the state space on its own the Q-Learning part evaluates the Q-values. The result is always a compromise between the economical use of resources and the permitted error-limit.

[2], priorized sweeping [16] or Multi-Time Models [17], because the state space is discrete.

Further research will be focused on the self-discretization of the full state-action space and on building an adaptive world model, in which far reaching concepts of planning are possible.

References

1. Watkins, C.J.C.H. "Learning with delayed rewards", PhD thesis, Cambridge University Psychology Dep., 1989.
2. Sutton R S, Barto A G, "Reinforcement Learning: An Introduction", MIT Press Cambridge, Massachusetts, London, England, 1998
3. Herz, J., Krogh, A., Palmer, "Introduction to the Theory of Neural Computation." Addison-Wesley, 1991
4. Tesauro G, "Temporal Difference Learning of Backgammon Strategy", Machine Learning, ed. by D. Sleeman and P. Edwards. San Mateo, Morgan Kaufmann Pub Inc., pp. 451-57., 1992
5. Samet,H., "Region representation", quadtrees from boundary codes; University of Maryland, Computer Science Center, 1979
6. Kohonen T, "Self-Organization and Associative Memory", Springer Series in Information Sciences, Springer, Third edition, 1989
7. Okabe A, Boots B, Sugihara K, Chiu S N, "Concepts and Applications of Voronoi Diagrams", (second edition) , Chichester: John Wiley, 2000
8. Fischer J.: "Verfahren zur Erkennung von Mustern" submitted to Deutsches Patent- und Markenamt, 80297 Mnchen, Germany
9. Knuth, Donald Ervin: "The Art of Computer Programming", Vol. 2, Seminumerical Algorithms; Prentice Hall, Muenchen, 1998
10. Vesanto J., Alhoniemi E.,"Clustering of the self-organizing map",IEEE Transactions on Neural Networks, Volume 11, Issue 3,Pages 586-600, 2000

11. Fischer J, Breithaupt R, Bode M, "Adaptive state space quantization for Reinforcement Learning Agents", submitted
12. Fischer J, "Strategiebildung mit neuronalen Netzen" (german), diploma thesis in the department of applied physics, WWU-Muenster, 1999.
13. Breithaupt R, "Adaptive und kooperative Automaten in statischen und dynamischen Umgebungen" (german), diploma thesis in the department of applied physics, WWU-Muenster, 1999.
14. Kroese Ben J A, van Dam J W, "Adaptive space quantisation for reinforcement learning of collision-free navigation", Faculty of mathematics and computer Science, University of Amsterdam Kruislan 403, NL-1098 SJ Amsterdam, 1992
15. Samejima K, Omori T, "Adaptive internal state space construction method for reinforcement learning of a real world agent", Faculty of Engeneering, Tokyo University of Agriculture and Technology, Nakachi 2 24-26 Koganei, Tokyo, in Neural Networks, Vol.12,No.7-8,p1143-1156,1999
16. Moore A.: "Memory based Reinforcement Learning: efficient computation by priorized sweeping". In SJ Hanson, CL Giles, JD Cowan, Neural Information Processing Systems 5, San Mateo, CA, Morgan Kaufmann, 1993.
17. Precup D., Sutton R.S.: "Multi-Time Models for Reinforcement Learning"; University of Massachusetts Amherst. In Proceedings of the ICML'97 Workshop on Modelling in Reinforcement Learning, 1997

Two Dimensional Evaluation Reinforcement Learning

Hiroyuki Okada[1] ,Hiroshi Yamakawa[1] and Takashi Omori[2]

[1] Real World Computing Partnership,
9-3, Nakase 1-Chome, Mihama-ku, Chiba City, Chiba 261-8588, Japan,
HiroyukiOkada@jp.fujitsu.com,
WWW home page: http://www.fujitsu.rwcp.or.jp/okada/index.html
[2] Hokkaido University
Nishi 8 Chome, Kita 13 Jo, Kita-ku, Sapporo 060-8628, Japan

Abstract. To solve the problem of tradeoff between exploration and exploitation actions in reinforcement learning, the authors have proposed two-dimensional evaluation reinforcement learning, which distinguishes between reward and punishment evaluation forecasts. The proposed method uses the difference between reward evaluation and punishment evaluation as a factor for determining the action and the sum as a parameter for determining the ratio of exploration to exploitation. In this paper we described an experiment with a mobile robot searching for a path and the subsequent conflict between exploration and exploitation actions. The results of the experiment prove that using the proposed method of reinforcement learning using the two dimensions of reward and punishment can generate a better path than using the conventional reinforcement learning method.

1 Introduction

Reinforcement learning refers to general learning to obtain appropriate action strategies by trial and error without an explicit target system model; instead, learning is accomplished using rewards obtained in the execution environment as the result of self-generated actions . This learning method is now being actively studied as a framework for autonomous learning because actions can be learned using only scalar evaluation values and without explicit training[1].

The purpose of reinforcement learning is to maximize the total rewards depending on the present and future the environment. This kind of learning has two properties. One is optimality (exploration), that is, to ultimately obtain as many rewards as possible. The other is efficiency (exploitation), which is to obtain rewards even in the middle of the learning process. These properties are in a tradeoff relationship. If exploration is overemphasized, convergence into an optimum policy is much longer as the environment becomes more complicated. Furthermore, only small rewards can be obtained in the learning process. Conversely, if exploitation is always emphasized, the learning results decrease to

J. Mira and A. Prieto (Eds.): IWANN 2001, LNCS 2084, pp. 370-377, 2001.

the local minimum and no optimum policies may be available at the end of the learning process.

Knowledge obtained from rats and monkeys about operand-conditioned subjects [2] [3] and from humans having damaged brains [4] indicates that distinguishing between the evaluations of successes and failures has a tremendous effect on action learning [5] [6]. With this in mind, the authors propose reinforcement learning by a two-dimensional evaluation. This evaluation involves an evaluation function based on the dimensions of reward and punishment. An evaluation immediately after an action is called a reward evaluation if its purpose is to obtain a favorable result after repeated attempts to learn an action, or punishment evaluation if its purpose is to suppress an action.

2 Reinforcement Learning Based on Two Dimensions of Reward and Punishment

2.1 Search by *Interest* and resource allocation

Two-dimensional reinforcement learning basically consists of two aspects. One is to distinguish between reward evaluation and punishment evaluation forecasts. The other is to determine an action according to the combined index of positive and negative reward forecasts.

The conventional reinforcement learning method uses only the difference (*Utility*) between reward and punishment reinforcement signals in an evaluation to determine an action. In comparison, the proposed method determines the sum (*Interest*) of reward and punishment evaluation signals and considers it as a kind of criticality.

2.2 Distinction of the time discount ratio of forecast reward

In reinforcement learning, a forecast reward is discounted more if it's more likely to be received in the future. This discount ratio is called the time discount ratio (hereinafter called γ) of the forecast reward. The value of γ ranges from 0 to 1.0.

In many practical problems, a reward reinforcement signal is related to the method used to move toward a goal and a forecast reward signal is used for learning a series of actions to reach the goal. To consider the effect of a goal that is far away, the γ setting must therefore be large. Meanwhile, if a punishment reinforcement signal for avoiding a risk has an effect too far away from the risk, an avoidance action may be generated in many input states. In turn, the search range of the operating subject is reduced, thereby lowering the performance of the subject. Therefore, to generate a punishment reinforcement signal for initiating an action to avoid an obstacle only when the obstacle is immediately ahead, the value of γ must be small in the signal.

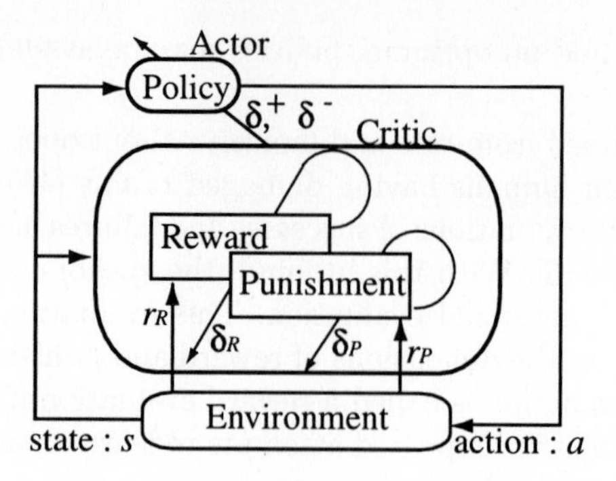

Fig. 1. Actor Critic model based on reward and punishment replusive evaluation.

2.3 Actor-Critic Architecture Based on Two-dimensional Evaluation

Actor-Critic architecture Figure.1 shows the actor-critic architecture [7] based on the proposed two-dimensional evaluation. Critic consists of a Reward section for reward evaluation and a Punishment section for punishment evaluation. Each section receives a state (s), a reward evaluation (r_R), and a punishment evaluation (r_P) according to the environment, and each section learns the forecast values. Both r_R and r_P are positive values and *Interest* ((δ^+)) and *Utility* ((δ^-)) are defined according to the TD differences (δ_R, δ_P) related to the forecasts of reward and punishment evaluations, which are shown below:

$$\delta^- = \delta_R - \delta_P \qquad : \textit{Utility} \tag{1}$$

$$\delta^+ = |\delta_R| + |\delta_P| \qquad : \textit{Interest} \tag{2}$$

Actor learns an action strategy using δ^-(*Utility*) as a de facto reinforcement signal and δ^+(*Interest*) to determine the ratio of exploitation action to environmental search action.

State evaluation (Critic) The reward evaluation, $V_R(s(t))$, and punishment evaluation, $V_P(s(t))$, to state $s(t)$ at time t are defined as follows:

$$\begin{aligned} V_R(s(t)) &= r_R(t) + \gamma_R r_R(t+1) + \gamma_R^2 r_R(t+2) + \dots \\ &= \sum_{i=t}^{\infty} \gamma_R^{i-t} r_R(i) \end{aligned} \tag{3}$$

$$\begin{aligned} V_P(s(t)) &= r_P(t) + \gamma_P r_P(t+1) + \gamma_P^2 r_P(t+2) + \dots \\ &= \sum_{i=t}^{\infty} \gamma_P^{i-t} r_P(i) \end{aligned} \tag{4}$$

where $r_R(t)$ and $r_P(t)$ are the reward and punishment evaluation values (positive), respectively, at time t. $r_R(t)$ represents the time discount ratio of the reward evaluation forecast and $r_P(t)$ represents the time discount ratio of the punishment evaluation forecast.

Based on equations 3 and 4, the following relationship can be established between the evaluation forecast value $\hat{V}(s(t))$ at the current time and $\hat{V}(s(t+1))$ at the subsequent time:

$$\hat{V_R}(s(t)) = r_R(t) + \gamma_R \hat{V_R}(s(t+1)) \tag{5}$$
$$\hat{V_P}(s(t)) = r_P(t) + \gamma_P \hat{V_P}(s(t+1)) \tag{6}$$

By learning to approximate forecast errors ($\delta_R(t)$, $\delta_P(t)$) to 0, status evaluations can be accurately forecast.

$$\delta_R(t) = r_R(t) + \gamma_R \hat{V_R}(s(t+1)) - \hat{V_R}(s(t)) \tag{7}$$
$$\delta_P(t) = r_P(t) + \gamma_P \hat{V_P}(s(t+1)) - \hat{V_P}(s(t)) \tag{8}$$

where $\delta_R(t)$ represents forecast errors related to reward evaluations and $\delta_P(t)$ represents those related to punishment evaluations.

Determination of action (Actor) Actor is used to develop an action strategy that maximizes the reward forecast by Critic. The proposed method determines action strategy $\pi(s,a)$ for taking action $a(t)$ in state $s(t)$ at time t as follows:

$$\begin{aligned} \pi(s(t),a(t)) &= \Pr\{a(t) = a|s(t) = s\} \\ &= \frac{\exp(\frac{p(s(t),a(t))}{\delta^+(t)})}{\sum_b \exp(\frac{p(s(t),b(t))}{\delta^+(t)})} \end{aligned} \tag{9}$$

where $p(s(t),a(t))$ indicates whether it is preferable to take action $a(t)$ in state $s(t)$ at time t. *Interest*($\delta^+(t)$) realizes a search function for large errors in TD learning. This function is equivalent to the automatic temperature control in Boltzmann selection that determines the ratio of exploitation to environmental search. As *Interest* becomes greater, actions become more random with a greater priority placed on the search. If *Interest* is small, a slight difference in $p(s(t),a(t))$ has a great effect on action selection. This difference is corrected using *Utility*($\delta^-(t)$) as expressed below, where positive constant β represents the learning rate:

$$p(s(t),a(t)) \leftarrow p(s(t),a(t)) + \beta\delta^-(t) \tag{10}$$

3 Searching for a Goal in Environment Containing Many Obstacles

3.1 Experimental Settings

To verify the effectiveness of the proposed method for reinforcement learning by two-dimensional evaluation, the authors conducted an experiment where a subject searches for a goal in an environment that has many obstacles. In a 20 x 20 grid world, a robot moves to the front, back, right, or left while searching for a path from the start position at the lower left (0,0) to the goal at the upper right (19,19). The robot has distance sensors facing eight different directions to measure distances to the walls or obstacles. The robot advances one step at a time in one of the four directions.

The robot receives a reward evaluation after reaching the goal. Punishment evaluations are given to the robot for collisions against several obstacles or wall. The robot can go over obstacles. It turns back if it collides against the wall. If the robot collides against an obstacle or wall more than 10 times or if it moves more than 80 steps, the trial terminates as a single trial failure.

To the robot, the following evaluation (r_R, r_P) is given:

$$r_R = 5.0 \qquad \text{reward(reaching the goal)} \tag{11}$$

$$r_P = 0.05 \qquad \text{punishment(collision)} \tag{12}$$

The time discount rates for reward and punishment evaluation forecasts in the experiment were set at $\gamma_R = 0.9$ and $\gamma_P = 0.1$, respectively. Since the robot is able to go over obstacles, the number of steps before it reaches the goal increases if the utmost efforts are made to avoid obstacles. Despite collisions against obstacles, generating a path to reach the goal results in many rewards if the number of collisions is kept within the limit (10 times).

For comparison, the authors conducted a similar experiment using conventional reinforcement learning based on the Actor-Critic architecture that does not distinguish between reward and punishment. For the conventional method, the punishment evaluation to a collision against an obstacle or wall was set to a negative value of -0.05. For evaluation by the conventional method, a preliminary experiment was conducted with the time discount rate varied from 0.1 to 0.9 in increments of 0.1 and set to 0.9 for when the goal reaching probability was greatest.

The experiment consists of learning and test trials. With several obstacles placed on the grid world at random, the learning trial was repeated 1000 times at first. Then, the test trial was repeated 100 times by changing the random number. A trial was classified as a success if the robot reaches the goal in 80 or fewer steps and with 10 collisions or fewer against obstacles. The probability of reaching the goal was then calculated.

3.2 Experimental Results

Comparison with the conventional method Figure.2 shows the experimental results. In the figure, the bold line indicates the probability of reaching the goal when number of obstacles were increased, the solid line indicates the proposed method, and the dotted line indicates the conventional method.

According to the proposed method, the robot can surely learn a path to the goal even if obstacles are placed on half of the grid (number of obstacles: 200).

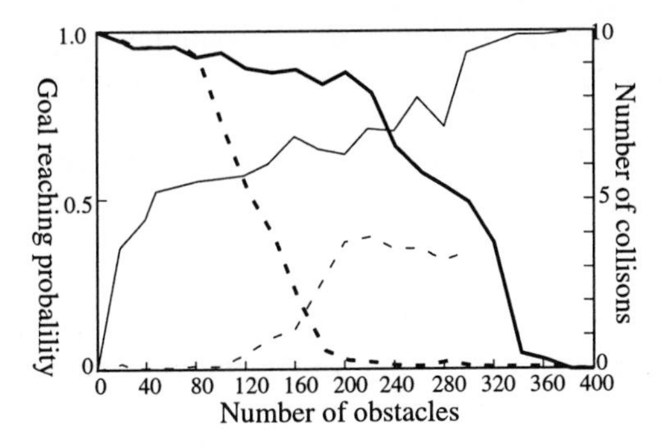

Fig. 2. Performace of proposed method (solid line) and traditional actor-critic method (dotted line) as a effeciency of number of obstacles. See text for the detail.

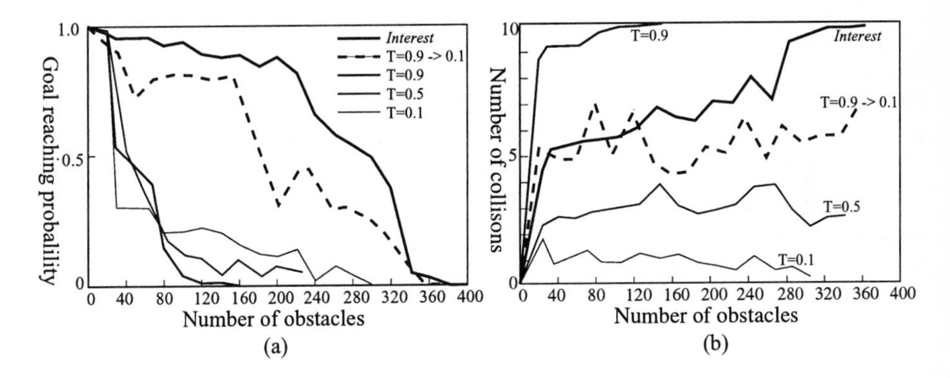

Fig. 3. Goal reaching probability for *Interest* and various values of T (a) and Goal reaching probability for *Interest* and various values of T (b)

In contrast, the probability of reaching the goal using the conventional method showed a quick drop when the number of obstacles exceeded 100.

In Figure.2, the thin line indicates the number of obstacles that the robot collided against before reaching the goal in successful trials. The solid line indicates the results using the proposed method, and the dotted line indicates the results using the conventional method. Use of the proposed method resulted in five or more collisions immediately after the start of experiment. When the number of obstacles was over 300, the robot collided against 10 obstacles (maximum allowed) in some trials. Use of the conventional method resulted in four collisions or less against obstacles and generated a path to reach the goal with minimal collisions in successful trials.

Effect of Interest The authors proposed *Interest* as a value for determining the ratio of exploration to exploitation. The effect of this value was confirmed.

At $\delta^{+}(t) = T$ in formula (9), the proposed technique was compared by setting T at 0.9, 0.5, or 0.1, and by gradually reducing the value of T from 0.9 to 0.1 during 1000 learning trials.

Figure.3(a) shows the probability of reaching the goal and Figure.3(b) shows the number of collisions against obstacles in successful trials.

When T was set at 0.9, the probability of generating a random action grew and the trial ended in a failure after collisions with more than 10 obstacles. When T was set at 0.1, the actions became definite. In many unsuccessful trials, the robot was kept trying to avoid an obstacle immediately ahead and could not reach the goal within the limit of 80 steps.

When value of the T was gradually reduced, the curve is similar to that when *Interest* was used. However, the proposed technique showed better probabilities of reaching the goal even when the number of obstacles were increased.

Compared with using the conventional method, the number of collisions against obstacles was not always small when the proposed method was used, as shown in Figure.3(b). However, the objective is not to minimize the number of collisions but to consider both the number of collisions and the length of the path to reach the goal. When the conventional method was applied in the experiment, the robot was instructed to only avoid obstacles. This instruction minimized the number of collisions but extended the path to reach the goal, thereby not allowing the robot to reach the goal within the specified number of steps (80 steps) in many trials. To specify the tradeoff, an effective method is considered to be using a large temperature parameter at the initial stage of the learning process before gradually reducing it. However, the proposed method was confirmed to have even better performance.

Discussion of Experimental Results Compared with the conventional method, using the proposed method resulted in more collisions against obstacles in successful trials but also stable generation of paths to reach the goal even when the number of obstacles were increased in the environment. Using the conventional method resulted in few collisions but the probability of reaching the goal decreased as the number of obstacles were increased in the environment.

This means that obstacle avoidance and reward approach are not balanced in the conventional method because the robot devoted more attempts to avoid obstacles immediately ahead rather than achieve the substantial purpose by searching for the goal. In the proposed method, however, a practical solution of reaching the goal within 10 collisions was selected by defining a time discount rate of punishment stimulus forecast that is smaller than that of reward stimulus forecast.

In an experiment to confirm the effect of *Interest*, the number of collisions and the probability of reaching the goal decreased in successful trials as its effect became small. This is because a slight difference in $p(s,a)$ affects action selection as the effect of *Interest* becomes small. In turn, the probability of avoiding an obstacle immediately ahead increases, thereby making it difficult to reach the goal. In comparison, when *Interest* was used, the robot randomly searched for exploration while moving toward the goal. A good balance of search and execution was thus confirmed.

4 Conclusion

To solve the problem of tradeoff between exploration and exploitation actions in reinforcement learning, the authors have proposed two-dimensional evaluation reinforcement learning, which distinguishes between reward and punishment evaluation forecasts.

In the proposed method of reinforcement learning using the two dimensions of reward and punishment, a reinforcement signal dependent on the environment is distinguished into reward evaluation after successful action and punishment evaluation after an unsuccessful action. The proposed method uses the difference between reward evaluation and punishment evaluation (*Utility*) as a factor for determining the action and the sum (*Interest*) as a parameter for determining the ratio of exploration to exploitation.

This paper has described an experiment with a mobile robot searching for a path and the subsequent conflict between exploration and exploitation actions. The results of the experiment prove that using the proposed method of reinforcement learning using the two dimensions of reward and punishment can generate a better path than using the conventional reinforcement learning method. MEMORABLE enables the proposed method to be effective for actual robots searching a path in a physical environment.

5 Acknowledgement

This study was conducted as part of the Real World Computing Program .

References

[1] L.P.Kaelbling, K.L.Littman and A.W.Moore: Reinforcement learning :A survey, Journal of Artificial Intelligence Research, vol.4, 237–285(1996).
[2] N.E.Miller:Liberalization of basic S-R concepts:extensions to conflict behavior, motivation and social learning,in Koch.S(Ed), Psychology:A Study of a Science, study 1 vol.2 ,196–292, New York:McFraw-Hill(1959).
[3] J.R.Ison and A.J.Rosen:The effect of amobarbital sodium on differential instrumental conditioning and subsequent extinction, Psyhopharmacologia,vol.10, 417–425(1967).
[4] B.Milner:Effects of different brain lesions on card sorting, Archives of Neurology,vol.9, 10–100(1963).
[5] H.Okada and H.Yamakawa:Neuralnetowrk model for attention and reinforcement learning,SIG-CII-9710, 4–14(1997).
[6] H.Okada, H.Yamakawa and T.Omori:Neural Network model for the preservation behavior of frontal lobe injured patients,ICONIP'98, 1465–1469(1998).
[7] A.G.Barto, R.S.Suttond and C.W.Anderson:Neuronlike Adaptive Elements That Can Solve Difficut Learning Control Problems,IEEE Transaction on Systems, Man and Cybernetics, vol.13, no.5, 834–846(1983).

Comparing the Learning Processes of Cognitive Distance Learning and Search Based Agent

Hiroshi Yamakawa[1], Yuji Miyamoto[2] and Hiroyuki Okada[1]

[1] Real World Computing Partnership,
9-3, Nakase 1-Chome, Mihama-ku, Chiba City, Chiba 261-8588, Japan,
yamakawa@flab.fujitsu.co.jp,
WWW home page: http://www.fujitsu.rwcp.or.jp/yamakawa/HomePage-e.html
[2] FUJITSU BROAD SOLUTION & CONSULTING Inc.,
15-33, Shibaura 4-Chome, Minato-ku, Tokyo 108-8531, Japan

Abstract. Our proposed cognitive distance learning agent generates sequence of actions from a start state to goal state in problem state space. This agent learns cognitive distance (path cost) of arbitrary combination of two states. The action generation at each state is selection of next state that has minimum cognitive distance to the goal.
In this paper, we investigate a learning process of the agent by a computer simulation in a tile world state space. An average search cost is more reduced more the prior learning term is long and our problem solver is familiar to the environment. After enough learning process, an average search cost of proposed method is reduced to 1/20 from that of conventional search method.

1 Introduction

Search algorithms, such as general problem solvers (GPSs) [1], are widely known as agents that generate a series of actions required to move from an start state to a goal state if both states are given to a state space on the problems space. These agents include a forward model, which is also called the "world model." As shown in Fig.1(a), a forward model predicts state $s(t+1)$, which will be moved from state $s(t)$ if action $a(t+1)$ [1] is taken in state $s(t)$. (t indicates time in these parameters.) A series of actions are taken at each state from the start to goal states to form a plan and then to execute the plan. A fundamental drawback of these algorithms is that the search costs required to form a plan are rather high.

Reactive planning can be used for systems that require real-time operation. As shown in Fig.1(b), reactive planning creates action $a(t+1)$ directly from state $s(t)$ [2]. A drawback of this algorithm is that it lacks flexibility because the goal state cannot be changed.

[1] Many papers use $a(t)$ to represent an action corresponding to state $s(t)$ at time t. This paper uses $a(t+1)$ because it enables us to better explain our proposed method later in this paper.

J. Mira and A. Prieto (Eds.): IWANN 2001, LNCS 2084, pp. 378-385, 2001.

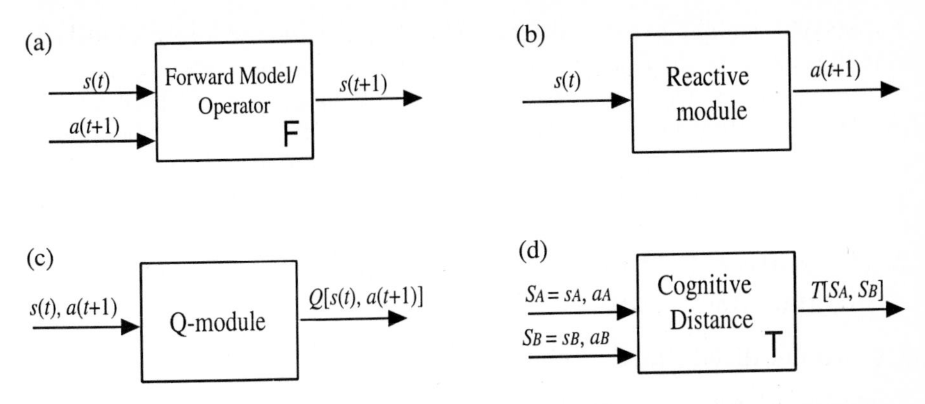

Fig. 1. Modules for agent: (a)Forward model or operator module, (b)Reactive planning module, (c)Example of reinforcement learning (Q-learning module), (d)Cognitive distance module (proposing now): The cognitive distance module outputs distance $T[S_A, S_B]$, the distance from S_A to S_B, where S_A is a start state and S_B is an end state in general cases. In contrast to Q-learning, s_A and a_A are a state and an action that are recognized at the same time.

To overcome this drawback, reinforcement learning has been proposed [3],[4]. Reinforcement learning can change the goal state of reactive planning by expressing the goal state as a reward and learns a predicted reward in the middle of a series of actions. Q-learning, a kind of reinforcement learning, uses a Q module shown in Fig.1(c) that uses an immediate reward to conduct learning. The Q-learning module can estimate $Q(s(t), a(t+1))$, which is the Q value as a predicted reward for action $a(t+1)$ at state $s(t)$. By preferentially selecting an action that produces higher Q values, Q-learning ensures the rational selection of actions. As described above, reinforcement learning, such as the Q-learning model, can quickly determine an action. However, since this method requires a large amount of learning cost to estimate a reward for a specific goal state, it is not easy to change the goal state. With reinforcement learning, if the location of the goal moves, the learning results are essentially no longer usable.

Another major problem in both search algorithms and reinforcement learning is agent cannot determine that it can reach a given goal by itself. Search algorithms require almost the same amount of processing cost as for forming a plan to judge that it can reach a given goal. Reinforcement learning cannot judge this at any moment.

Our propsed cognitive distance module [5] that learns cognitive distance $T[S_A, S_B]$, a cost for moving from start state S_A to end state S_B, as shown in Fig.1(d) and Fig.2. In many cases, this cost is expressed as the moving distance of a state. (In the example in Fig.1(d), S_A is a generalized state created using the concatenation of state s_A and action a_A, which are both recognized at the same time step t.) The purpose of our proposed algorithm is to develop a agent that can reduce calculation costs during execution of a plan while being able to respond to the change of the goal as flexibly as can GPS.

The proposed algorithm is better than search algorithms and reinforcement learning algorithms because its calculation cost to determine an action and to calculate the probability of reaching a goal given to a state is lower than the above-mentioned algorithms.

2 Cognitive Distance Learning Agents that handles Generalized States

2.1 Generalized States

A generalized state concatenates $s(t)$, sensor input at t, and $a(t)$, an action reflecting the output of an action at the preceding time, without separately handling them; they are separately handled by Markov processes. (In the balance of this paper, capital letters are used to indicate generalized states.)

For an example, Fig.2 shows generalized state S_A, which is a state at time t if a moving agent exists in a problem space. S_A is a state in which $s(t) = s_A$, sensor input for that location, and $a(t) = a_A$, action that moved the state from left to right, are concatenated.

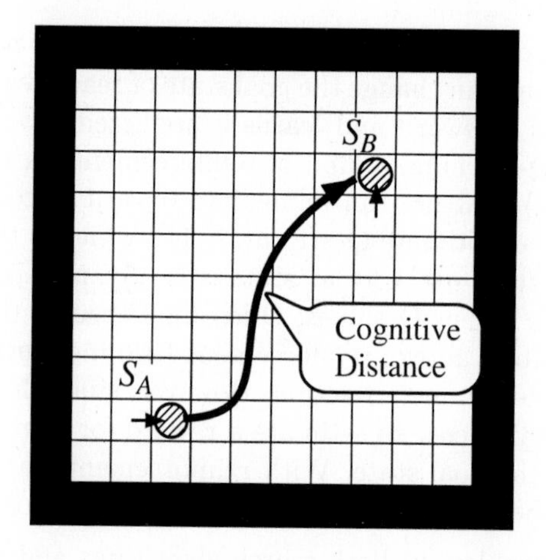

Fig. 2. Generalized states and cognitive distance on a problem state space

The action of the agent during the formulation of generalized state $S(t)$ is determined by generalized state $D(t)$, which is the desired at the next time. (In the balance of this paper, $D(t)$ is called a "subgoal.") The environment is defined as a transition network in which a state can move. If subgoal $D(t)$ received from an agent is acceptable, generalized state $S(t)$, which was given to the agent at the next time, is identical to $D(t)$. However, if subgoal $D(t)$ is received but is unacceptable, the environment transits to a state $S(t)$ that is different from $D(t)$.

2.2 Cognitive Distance Learning Agent

Our proposed agent shown in Fig.3 combines cognitive distance module T with forward model F to determine an action, and uses state history H to learn the cognitive distance T and forward model F.

Cognitive distance module T holds $T[S_A, S_B]$, which is the cognitive distance between S_A and S_B. S_A is a generalized start state and S_B is a generalized end state in a problem space. State history H is a queue that holds a generalized state $S(t)$ that has lasted for a certain period. If X is input as a generalized state, forward model F outputs a list of candidate states that can be directly moved from X. This specific forward module is created by modifying the conventional forward model (Fig.1(a)), to support the framework of generalized states.

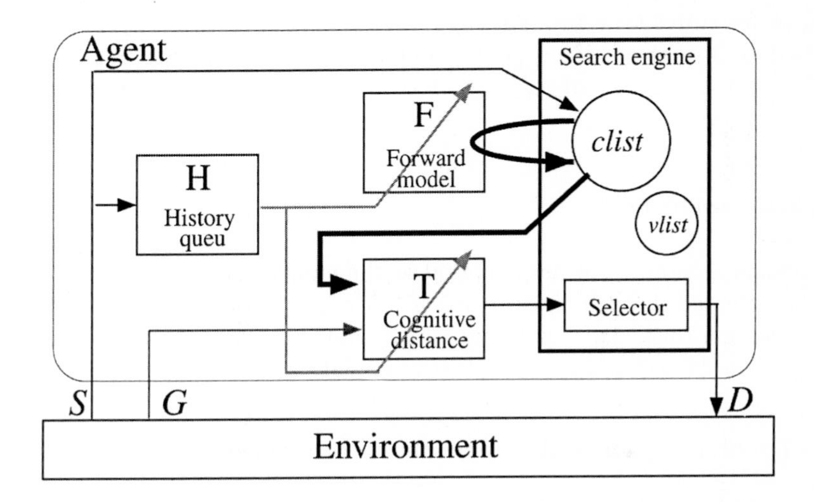

Fig. 3. Agent that uses generalized cognitive distance

During each cycle, the agent obtains state S and goal G from the environment and outputs subgoal D, all of which are generalized states.

2.3 Learning Processes

First, latest state S is added to state history H, which is a queue. Cognitive distance module T then checks cognitive distance $T[H[i], H[0]]$ between $H[i]$, the state at i time-step earlier, and latest state $S(= H[0])$ to see if $T[H[i], H[0]]$ has any value. If the value of $T[H[i], H[0]]$ is null or more than i, the value will be registered as $T[H[i], H[0]] \leftarrow i$. The same registration and update processing as those applied to $S(= H[0])$ will be applied to some of the states that are obtained by applying forward model F to latest state S. If $H[0]$ has not been registered in $F[H[1]]$, which was a transition state list one time-step earlier, the subsequent learning of forward model F registers $H[0]$ to $F[H[1]]$.

2.4 Execution Procedures

The execution process outputs subgoal D, which is required to reach goal G. First, the execution process obtaines state list *clist* that was obtained by forward model $F[S]$ from the current state S. An element of *clist* is expressed as "X." If *clist* contains a state X that contains the cognitive distance to goal G, the agent outputs state X that can obtain the minimum $T[X, G]$ as subgoal D. If *clist* does not contain a state X that contains the cognitive distance to goal G, the agent calculates forward model $F[X]$ for each state X in the current *clist* to create a new *clist*. This same process is repeated to reach *maxdepth*, which is the deepest search depth. If cognitive distance $T[X, G]$ cannot be obtained in the deepest search depth *maxdepth*, the search fails. To prevent a repeat of a search of the same state, *vlist*, which is a list of states already searched, is used to check for states that have already been searched.

As described above, if any goal G is given at any curent state S, an agent can immediately generated subgoal D by learning cognitive distance T in a problem space in advance.

3 Simulation

Our computer simulation revealed that as the number of obtained cognitive distances increased through an extended advance (prior) learning period before the execution of a task, the average search cost during task execution was reduced. This section describes the results of our simulation.

3.1 Problem Space and Experiment Procedures

For the experimental environment, we used a 10-by-10-matrix tile world, as shown in Fig.2. Sensor input $s(t)$ corresponding to the agent at time t is given absolute coordinates x and y. Action output $a(t)$ has values 0, 1, 2, and 3 in the one-dimensional coordinate, the values of which indicate the upward, downward, left, and right directions. A three-dimensional vector that concatenates all of these is expressed as generalized state $S(t)$.

Agents can be placed on 100 points on a map. Since generalized states are handled with actions included, four states that correspond to four actions can be observed. However, since only three states are possible at each of the four corners, N, the maximum number of observable states, is 396 ($= (10^2 \times 4) - 4$). Therefore, n_{cd}, the cognitive distance size obtained when all of cognitive distances have been learned, is N^2 = 156,316.

The experiment consisted of a prior walking phase and an execution phase.

- Prior walking phase
  In this phase, w-step random walks were applied in a problem space. The values of w were 500, 1,000, 1,500, 2,000, 3,000, 4,000, 5,000, 7,000, 10,000, and 15,000.

- Execution phase
  In this phase, the probability of reaching a goal is evaluated. For this phase, we repeated a sequence 100 times. In this sequence, a goal, which is the target of the move, is randomly output. If the goal is reached, a new goal is output. The results of the 100 sequences are then averaged.

We conducted the experiment using cognitive distance algorithms and search algorithms.

3.2 Results of Experiment

All 396 states were visited at w = 3,000, by the random actions series used for the prior walking steps in this experiment, Fig.4, Fig.5, and Fig.6 show the change in performance in the execution phase relative to the change of w, the number of prior walking steps.

Fig.4 shows that the goal reaching probability and the minimum number of steps do not depend on algorithms because all the algorithms use forward model F to search for the goal until it is found. However, since not all searches can find the goal in an early learning stage, the goal reaching probability will decline. If a state has never reached the goal or if it is impossible to search for the goal in some cases, the goal reaching probability will also decline. Therefore, the goal reaching probability is less than the ratio of nodes that are already visited during the prior walking phase. For example, 60% (237 states) of all states (396 states) have been visited after the completion of 500 learning steps, whereas the goal reaching probability stays 36%.

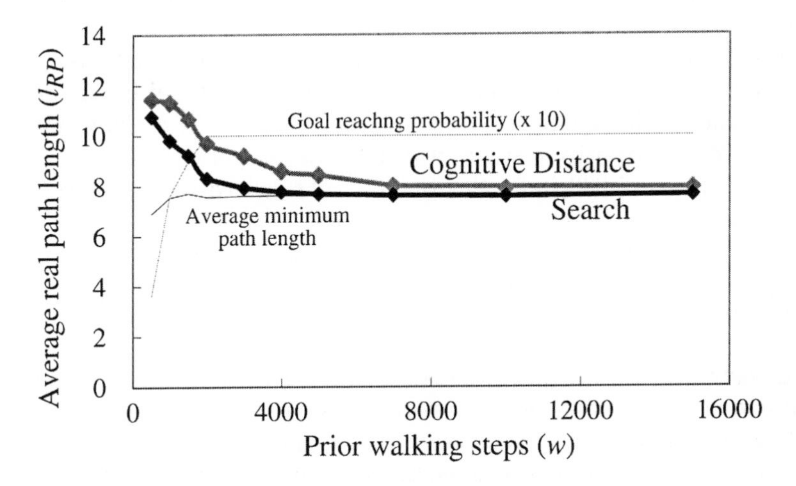

Fig. 4. Average real path length and goal reaching probability

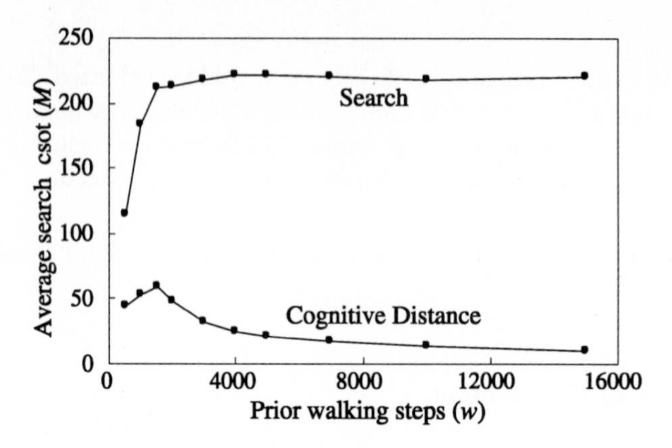

Fig. 5. Comparing average search costs through learning

Fig.4 shows the changes in l_{RP}, real path length in relation to the number of w, prior walking steps for both algorithms. This figure shows that use of cognitive distance slightly increased l_{RP}, real path length. We speculate that this is because searches that use cognitive distance T probably make more detours than searches that use forward model F.

Fig.5 shows the relationships between w, the number of prior walking steps, and M, the average search cost. In the search algorithms, since the function of forward model F in memory improves as w, the number of prior walking steps, increases,M reaches its plateau around w = 4,000. If cognitive distance is used, the search cost increases if w is small, but if w exceeds 1,500, the search cost decreases as more cognitive distances are obtained. If cognitive distance is used after w exceeds 15,000, the search cost$M$ is reduced to about one twentieth of that of search algorithms.

As Fig.6 shows, cognitive distance size n_{cd} increases monotonously with w, the number of prior walking steps. The maximum size of cognitive distance that can be learned is the square of N, (N^2 = 156,316), where N is the total number of states,. When cognitive distance was used, the size of cognitive distance n_{cd} was 99,786 at w = 15,000, indicating that 64% of cognitive distance was obtained from learning.

4 Conclusion

Our already proposed cognitive distance learning agent generates sequence of actions from a start state to goal state. This agent learns cognitive distance of arbitrary combination of two states. The action generation at each state is selection of next state that has minimum cognitive distance to the goal.

In this paper, we compare the performance of proposed algorithm and traditional search algorithm by a computer simulation that uses a tile world. As

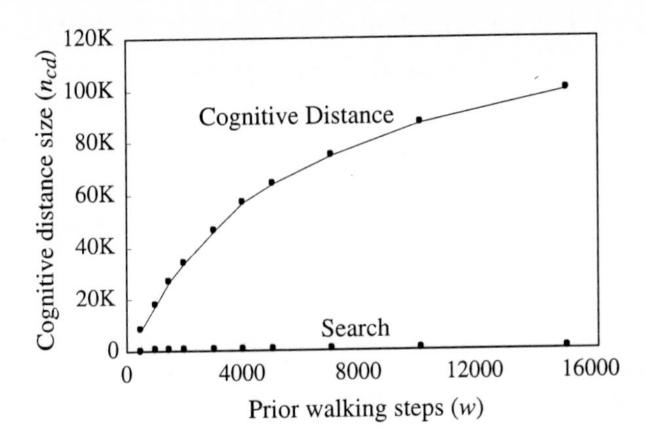

Fig. 6. Comparing cognitive distance sizes through learning

indicated by the experiment results, the proposed algorithm can produce a goal reaching probability as high as that of conventional search algorithms, but at a low average search cost necessary to determine actions in an early learning stage for which w is relatively small. After enough learning process, an average search cost of proposed method is reduced to about one twentieth of that of conventional search method. Therefore, this algorithm is effective for real-time evaluations because search costs further decrease as learning progresses.

Compared to conventional reinforcement learning, the proposed algorithm has much strength. For example, the proposed algorithm can use the results of learning to flexibly respond to changes in the goal state. Because the proposed algorithm combines search algorithms, another strength is that it can obtain a high goal reaching probability from the early learning stage.

One problem is that, if cognitive distance size becomes considerably large, the maximum memory required increases to the square of N. (N is the number of states in the space.) In the future, we intend to solve this problem by applying a layered structure to cognitive distance learning agent.

References

[1] A. Newell and H. A. Simon: GPS, a program that simulates human thought, In H. Billing (Ed.), Lernede Automaten, 109–124 (1961).

[2] Seiji Yamada: Reactive Planning, Japanese J. Artificial Intelligence, **8**, **6**, 729–735 (1993).

[3] Richard S. Sutton and Andrew G. Barto: Reinforcement Learning: An Introduction, Adaptive Computation and Machine Learning. MIT Press (1988).

[4] L. P. Kaelbling and et al: Reinforcement Learning: A Survey. J. Artificial Intelligence Research. **4** 237–285 (1996).

[5] H. Yamakawa and et al : Proposing Problem Solver using Cognitive Distance, Proc. MACC2000, (2000).

Selective Learning for Multilayer Feedforward Neural Networks

AP Engelbrecht

Department of Computer Science, University of Pretoria, South Africa
engel@driesie.cs.up.ac.za

Abstract. Selective learning is an active learning strategy where the neural network selects during training the most informative patterns. This paper investigates a selective learning strategy where the informativeness of a pattern is measured as the sensitivity of the network output to perturbations in that pattern. The sensitivity approach to selective learning is then compared with an error selection approach where pattern informativeness is defined as the approximation error.

1 Introduction

"What can be done with fewer is done in vain with more" - William of Ockham (1285-1349). Much research has been done to show that Ockham's statement is true when considering the architecture of neural networks (NN). Several approaches have been developed to optimize the architecture of NNs, including network pruning [?,6, 10, 18], network construction [5, 7], and regularization [9, 16]. Ockham's principle also applies to the training set, in which case the objective is to make optimal use of the training data through selection of the most informative patterns - a process referred to as *active learning*.

In active learning the learner is given some control over which information to use for training. The network uses its current knowledge, as encapsulated in the weights, to select the most informative patterns. Rather than passively accepting training examples from a teacher, the learner guides the search for the most informative patterns.

With careful dynamic selection of training examples, shorter training times and better generalization may be obtained. Provided that the added complexity of the example selection method does not exceed the reduction in training computations (due to a reduction in the number of training patterns), training time will be reduced [8, 15, 17]. Generalization can potentially be improved, provided that selected examples contain enough information to learn the task. Cohn shows through average case analyses that the expected generalization performance of active learning is significantly better than passive learning [1]. Seung, Opper and Sompolinsky [14], Sung and Niyogi [15] and Zhang [17] report similar improvements. Results presented by Seung, Opper and Sompolinsky indicate that generalization error decreases more rapidly for active learning than for passive learning [14].

Active learning strategies can be grouped into two classes:

J. Mira and A. Prieto (Eds.): IWANN 2001, LNCS 2084, pp. 386-393, 2001.

- **Incremental learning**, where training starts on an initial subset of a candidate training set. During training, at specified selection intervals (e.g. after a specified number of epochs, or when the error on the current training subset no longer decreases), further subsets are selected from the candidate examples using some criteria or heuristics, and added to the training set. The training set consists of the union of all previously selected subsets, while examples in selected subsets are removed from the candidate set. Thus, as training progresses, the size of the candidate set decreases while the size of the actual training set grows. Examples of incremental learning algorithms are optimal experiment design [1], information-based objective functions [11], integrated squared bias minimization [12], error selection [13, 17], and sensitivity analysis incremental learning [3].
- **Selective learning**, where the network selects at each selection interval a new training subset from the original candidate set. Selected patterns are not removed from the candidate set. At each selection interval, all candidate patterns have a chance to be selected. The subset is selected and used for training until some convergence criteria on the subset is met (e.g. a specified error limit on the subset is reached, the error decrease per iteration is too small, the maximum number of epochs allowed on the subset is exceeded). A new training subset is then selected for the next training period. This process repeats until the NN is trained to satisfaction. Examples of selective learning algorithms are selective updating [8], sensitivity analysis selective learning [2], and selective learning for time series approximation [4].

The advantages of active learning, and selective learning specifically, have been illustrated extensively in the references cited above. The objective of this paper is to further investigate the sensitivity analysis selective learning algorithm (SASLA) developed by Engelbrecht *et al* [2]. The influence of the subset selection constant is investigated, and the performance of SASLA is compared to an approach where patterns are selected based on prediction error. For this purpose the paper considers only classification problems.

Section 2 presents an overview of the two main approaches to selective learning, namely (1) pattern selection based on sensitivity information, and (2) based on error selection. Results are presented and discussed in section 3.

2 Selective Learning Approaches

The success of selective learning is based on the measure of pattern informativeness. Two main approaches to selective learning can be defined based on the pattern informativeness measure:

- **Output Perturbation Methods:** Engelbrecht *et al* developed a selective learning algorithm based on sensitivity analysis of the NN output to input perturbations [2]. The partial derivatives $S_{o,k}^{(p)} = \frac{\partial o_k}{\partial z_i^{(p)}}$ is calculated for each input z_i and output o_k for each pattern p. The informativeness of a pattern is then measured as the influence small perturbations in input parameters have on the output of the network, using $\Phi_\infty^{(p)} = \max_{k=1,\cdots,K}\{|S_{o,k}^{(p)}|\}$ where K is the total number of output units. All the patterns from the candidate training set for which $\Phi_\infty^{(p)} > (1-\beta)\overline{\Phi}_\infty$ are selected for training, where $\overline{\Phi}_\infty = \frac{\sum_{p=1}^{P_C} \Phi_\infty^{(p)}}{P_C}$, with P_C the total number of patterns in the candidate training set. The selection of a training subset occurs at each epoch.

The effect of this selective learning approach is that those patterns that lie close to a decision boundary is selected for training. The subset selection constant β is crucial to the efficacy of the algorithm. This selection constant, which lies in the range $[0, 1]$, is used to control the region around decision boundaries within which patterns will be considered as informative. The larger the value of β, the more patterns will be selected. If β is too small, only a few patterns will be selected which may not include enough information to form boundaries, with a consequent reduction in generalization performance. Low values for β will however mean less computational costs. A conservative choice of β close to 1 improves the chances of selecting patterns representing enough information about the target concept, ensuring most of the candidate patterns to be included in the initial training subset. A conservative value for β does, however, not mean a small reduction in training set size. As training progresses, more and more patterns become uninformative, resulting in larger reductions in training set size. If $\beta = 1$, SASLA simply generalizes to normal fixed set learning. The effects of different values for β are investigated in section 3.

- **Error Selection (ES) Methods:** Another, simpler, approach to selective learning is simply to discard those patterns that have been classified correctly. For this purpose the following pattern selection scheme is used: if there exists an output unit such that $(t_k^{(p)} = 0.9)\&\&(o_k^{(p)} < 0.7)$ or $(t_k^{(p)} = 0.1)\&\&(o_k^{(p)} > 0.3)$, then select pattern p for training ($t_k^{(p)}$ is the target value associated with output $o_k^{(p)}$). The effect of such an approach is that the training set will include those patterns that lie close to decision boundaries. A comparison of SASLA with ES is presented in section 3.

3 Experimental Results

This section investigates the effect of the selection constant, and compares SASLA with ES. For this purpose the following performance criteria were used: training error and generalization, computational complexity and the convergence characteristics. The *wine* and *glass* classification problems from the UCI repository were used. For each problem and selection method 50 simulations were executed. The results reported in this section are averages over these 50 simulations, together with 95% confidence intervals as obtained from the t-distribution.

3.1 Selection Constant's Effects

The effect of the selection constant for SASLA was tested for $\beta = 0.1, 0.3, 0.5, 0.7, 0.9$. The results are summarized in tables 1, 2, 5 and 7 for the *glass* problem (after 2000 epochs), and tables 3, 4, 6 and 7 for the *wine* problem (after 500 epochs). Training errors and generalization are expressed as the percentage of correctly classified patterns, saving in computational costs are computed as the difference in the total number of calculations between the selective learning algorithm and training on a fixed set using gradient descent, and convergence is expresses as the percentage of simulations that converged to different generalization levels.

Tables 1 and 3 show that the lower the value of β, the worse the accuracy of SASLA: The average training error $\overline{\mathcal{E}}_T$ and generalization $\overline{\mathcal{E}}_G$ over the 50 simulations deteriorated for both problems, while the best generalization $\overline{\mathcal{E}}_G^{best}$ (the best generalization performance over the 50 simulations) also decreased for the *glass* problem. Also note that higher β values achieved generalization accuracies larger than the training accuracy. As the value of β becomes smaller, the difference between generalization and training accuracy becomes smaller. Lower β values do, however, save on computations, due to the fact that less patterns are selected for the candidate set. However, the savings in computational cost using low β values do not justify the loss in accuracy for the *glass* problem, while the *iris* problem's decrease in accuracy is acceptable. Tables 5 and 6 illustrate for the *glass* and *wine* problems how lower β values decreased the training set size for selected epochs. Table 7 shows that the smaller the value of β the more simulations do not converge to generalization levels specified in the table, indicating that the lower β values do not select enough information to achieve the higher generalization accuracies.

What can be concluded from the above experiments is that the best value for β is problem dependent. In selecting a value for β the trade-off between computational cost savings and decreased performance needs to be considered. It is suggested that future research includes an investigation into adaptive β values, where the value of β increases when performance deteriorates, and decreases when performance improves.

3.2 Error Selection

This section compares SASLA with the ES approach to selective learning. The comparison is done for $\beta = 0.9$, since SASLA produced the best results for this selection constant. Table 8 summarizes the error performance results for the ES approach in comparison with SASLA for $\beta = 0.9$, while table 9 and table 10 respectively summarizes the complexity and convergence results. For all problems SASLA showed to have significantly better generalization performance than ES. From table 9 it is evident that ES is computationally less intensive than SASLA, using substantially less training patterns than SASLA. This very large reduction in training set size by ES compared to SASLA is a cause of the worse ES generalization performance, since the reduced training set may not contain sufficient patterns to refine the boundaries. SASLA showed to have better convergence characteristics than ES. Table 10 illustrates that ES had substantially more simulations that did not converge to the different generalization levels.

In conclusion, although ES is computationally more efficient than SASLA, the better convergence and generalization results of SASLA make it the preferred selective learning algorithm.

4 Conclusions

This paper investigated the effect of the subset selection constant β of sensitivity analysis selective learning (SASLA). The value of the selection constant is crucial. While smaller β values do reduce computational effort substantially, the reduction is at the cost of degraded generalization performance. A comparison between SASLA and error selection (ES) showed SASLA to have significantly better generalization and convergence results, while ES did save on computations. Future research will include developing a scheme for adaptive β values, and an investigation into the effects of outliers on SASLA and ES.

References

1. Cohn, D.A.: Neural Network Exploration using Optimal Experiment Design. AI Memo No 1491, Artificial Intelligence Laboratory, Massachusetts Institute of Technology (1994)
2. Engelbrecht, A.P., Cloete, I.: Selective Learning using Sensitivity Analysis. IEEE World Congress on Computational Intelligence, International Joint Conference on Neural Networks, Anchorage, Alaska (1998) 1150-1155
3. Engelbrecht, A.P., Cloete, I.: Incremental Learning using Sensitivity Analysis. IEEE International Joint Conference on Neural Networks, Washington DC, USA (1999) paper 380
4. Engelbrecht, A.P., Adejumo, A.: A New Selective Learning Algorithm for Time Series Approximation using Feedforward Neural Networks. In: Development and Practice of Artificial Intelligence Techniques, VB Bajić, D Sha (eds). Proceedings of the International Conference on Artificial Intelligence, Durban, South Africa (1999) 29-31
5. Fritzke, B.: Incremental Learning of Local Linear Mappings. International Conference on Artificial Neural Networks, Paris (1995)
6. Hassibi, B., Stork, D.G., Wolff, G.: *Optimal Brain Surgeon: Extensions and Performance Comparisons*, JD Cowan, G Tesauro, J Alspector (eds), Advances in Neural Information Processing Systems. Vol. 6. (1994) 263-270
7. Hirose, Y., Yamashita, K., Hijiya, S.: Back-Propagation Algorithm which Varies the Number of Hidden Units. Neural Networks. **4** (1991) 61-66
8. Hunt, S.D., Deller, J.R. (jr): Selective Training of Feedforward Artificial Neural Networks using Matrix Perturbation Theory. Neural Networks. **8**(6) (1995) 931-944
9. Kamimura, R., Nakanishi, S.: Weight Decay as a Process of Redundancy Reduction. World Congress on Neural Networks. Vol. 3. (1994) 486-491

10. Le Cun, Y., Denker, J.S., Solla, S.A.: Optimal Brain Damage. In D Touretzky (ed), Advances in Neural Information Processing systems. Vol. 2. (1990) 598-605
11. MacKay, D.J.C.: Information-Based Objective Functions for Active Data Selection. Neural Computation. **4** (1992) 590-604
12. Plutowski, M., White, H.: Selecting Concise Training Sets from Clean Data. IEEE Transactions on Neural Networks. **4**(2) (1993) 305-318
13. Röbel, A.: The Dynamic Pattern Selection Algorithm: Effective Training and Controlled Generalization of Backpropagation Neural Networks. Technical Report, Institute für Angewandte Informatik, Technische Universität Berlin. (1994) 497-500
14. Seung, H.S., Opper, M., Sompolinsky, H.: Query by Committee. Proceedings of the 5th Annual ACM Workshop on Computational Learning Theory. (1992) 287-299
15. Sung, K.K., Niyogi, P.: A Formulation for Active Learning with Applications to Object Detection. AI Memo No 1438, Artificial Intelligence Laboratory, Massachusetts Institute of Technology. (1996)
16. Weigend, A.S., Rumelhart, D.E., Huberman, B.A.: Generalization by Weight-Elimination with Application to Forecasting. In R Lippmann, J Moody, DS Touretzky (eds), Advances in Neural Information Processing Systems. Vol. 3. (1991) 875-882
17. Zhang, B.-T.: Accelerated Learning by Active Example Selection. International Journal of Neural Systems. **5**(1) (1994) 67-75
18. Zurada, J.M., Malinowski, A., Usui, S.: Perturbation Method for Deleting Redundant Inputs of Perceptron Networks. Neurocomputing. **14** (1997) 177-193

Table 1. Comparison of error performance measures for different β values for the *glass* problem

β	$\overline{\mathcal{E}}_T$	$\overline{\mathcal{E}}_G$	$\overline{\mathcal{E}}_G^{best}$	**Presentation**	$\overline{\rho}$
0.9	$69.06 \pm 1.07\%$	$70.90 \pm 1.83\%$	88.10%	281485	1.029 ± 0.030
0.7	$66.43 \pm 1.37\%$	$68.10 \pm 1.82\%$	85.71%	248760	1.031 ± 0.036
0.5	$62.73 \pm 1.76\%$	$64.38 \pm 1.70\%$	83.33%	200510	1.032 ± 0.036
0.3	$59.95 \pm 1.41\%$	$60.90 \pm 2.11\%$	80.95%	177310	1.023 ± 0.045
0.1	$57.26 \pm 1.19\%$	$57.10 \pm 1.99\%$	78.57%	121385	0.999 ± 0.035

Table 2. Comparison of computational complexity for different β values for the *glass* problem

β	$\overline{P}_T$	**Total Pattern Presentations**	**Cost Saving** $(\times 10^6)$
0.9	131.42 ± 1.67	$287\,906.74 \pm 1\,847.39$	176.860996 ± 49.938683
0.7	111.60 ± 1.75	$252\,744.90 \pm 2\,124.33$	$-773.633863 \pm 57.424777$
0.5	94.48 ± 1.98	$214\,170.60 \pm 2\,405.99$	$-1\,816.374341 \pm 65.038596$
0.3	81.30 ± 1.94	$180\,981.50 \pm 2\,468.39$	$-2\,713.542092 \pm 66.725490$
0.1	69.20 ± 3.03	$152\,859.90 \pm 2\,698.49$	$-3\,473.725183 \pm 72.945519$

Table 3. Comparison of error performance measures for different β values for the *wine* problem

β	$\overline{\mathcal{E}}_T$	$\overline{\mathcal{E}}_G$	$\overline{\mathcal{E}}_G^{best}$	**Presentation**	$\overline{\rho}$
0.9	$98.33 \pm 0.33\%$	$98.89 \pm 0.64\%$	100.0%	21317	1.006 ± 0.007
0.7	$97.56 \pm 0.36\%$	$97.97 \pm 0.86\%$	100.0%	9823	1.004 ± 0.011
0.5	$97.01 \pm 0.46\%$	$97.42 \pm 0.91\%$	100.0%	9956	1.004 ± 0.012
0.3	$96.95 \pm 0.41\%$	$97.14 \pm 0.94\%$	100.0%	8958	1.002 ± 0.012
0.1	$96.64 \pm 0.45\%$	$96.22 \pm 0.95\%$	100.0%	9597	0.996 ± 0.013

Table 4. Comparison of computational complexity for different β values for the *wine* problem

β	$\overline{P}_T$	**Total Pattern Presentations**	**Cost Saving** ($\times 10^6$)
0.9	78.45 ± 2.88	$45\,117.77 \pm 1\,068.71$	-32.810974 ± 4.913917
0.7	60.42 ± 1.94	$34\,624.29 \pm 749.06$	-81.060013 ± 3.444165
0.5	51.39 ± 1.35	$28\,384.00 \pm 1\,138.55$	-109.752868 ± 5.235048
0.3	45.48 ± 1.35	$22\,697.76 \pm 1\,311.81$	-125.124575 ± 2.123308
0.1	40.97 ± 1.15	$22\,112.00 \pm 821.32$	-138.591524 ± 3.776447

Table 5. Training set reduction for different β values for the *glass* problem

	Reduction at epoch				
β	**50**	**100**	**500**	**1000**	**2000**
0.9	$1.14 \pm 0.30\%$	$2.97 \pm 0.43\%$	$7.76 \pm 0.51\%$	$12.86 \pm 1.06\%$	$18.36 \pm 1.15\%$
0.7	$7.03 \pm 0.62\%$	$9.80 \pm 0.84\%$	$21.66 \pm 1.14\%$	$28.35 \pm 1.27\%$	$50.55 \pm 1.02\%$
0.5	$16.69 \pm 1.46\%$	$21.84 \pm 1.06\%$	$37.24 \pm 1.31\%$	$39.15 \pm 1.61\%$	$45.07 \pm 1.15\%$
0.3	$27.66 \pm 1.47\%$	$33.29 \pm 0.84\%$	$47.84 \pm 0.95\%$	$51.21 \pm 1.49\%$	$52.73 \pm 1.13\%$
0.1	$37.30 \pm 2.06\%$	$41.13 \pm 1.43\%$	$54.95 \pm 1.66\%$	$59.91 \pm 1.47\%$	$59.77 \pm 1.76\%$

Table 6. Training set reduction for different β values for the *wine* problem

	Reduction at epoch			
β	**25**	**50**	**100**	**500**
0.9	$19.21 \pm 1.57\%$	$27.37 \pm 1.42\%$	$32.44 \pm 1.62\%$	$45.07 \pm 2.11\%$
0.7	$37.89 \pm 1.54\%$	$43.98 \pm 1.48\%$	$48.64 \pm 1.19\%$	$57.97 \pm 1.35\%$
0.5	$48.62 \pm 1.04\%$	$54.02 \pm 1.20\%$	$57.95 \pm 0.97\%$	$63.81 \pm 0.95\%$
0.3	$56.47 \pm 0.71\%$	$61.22 \pm 0.79\%$	$64.63 \pm 0.75\%$	$68.36 \pm 0.81\%$
0.1	$60.97 \pm 0.65\%$	$65.58 \pm 0.67\%$	$68.04 \pm 0.73\%$	$71.15 \pm 0.81\%$

Table 7. Comparison of convergence results for different β values

β	Generalization Levels for *glass*							Generalization Levels for *wine*			
	60%	**65%**	**70%**	**75%**	**80%**	**85%**	**90%**	**94%**	**96%**	**98%**	**100%**
0.9	0%	0%	10%	29%	65%	90%	98%	0%	4%	12%	12%
0.7	2%	6%	18%	46%	80%	96%	100%	0%	6%	38%	38%
0.5	2%	20%	38%	70%	94%	100%	100%	0%	10%	36%	36%
0.3	8%	28%	76%	88%	98%	100%	100%	4%	12%	42%	42%
0.1	16%	52%	84%	94%	100%	100%	100%	4%	12%	38%	38%

Table 8. Comparison of SASLA ($\beta = 0.9$) and ES error performance measures

Problem		$\overline{\mathcal{E}}_T$	$\overline{\mathcal{E}}_G$	$\overline{\mathcal{E}}_G^{best}$	Presentation	$\overline{\rho}$
glass:	*SASLA*	$69.06 \pm 1.07\%$	$70.90 \pm 1.83\%$	88.10%	281485	1.029 ± 0.030
	ES	$59.28 \pm 1.15\%$	$60.76 \pm 2.27\%$	78.60%	192125	1.029 ± 0.043
wine:	*SASLA*	$98.33 \pm 0.33\%$	$98.89 \pm 0.64\%$	100.0%	21317	1.006 ± 0.007
	ES	$92.29 \pm 0.02\%$	$90.97 \pm 0.03\%$	100.0%	5704	0.985 ± 0.020

Table 9. Comparison of SASLA ($\beta = 0.9$) and ES computational complexity

Problem		$\overline{P}_T$	Total Pattern Presentations	Cost Saving ($\times 10^6$)
glass:	*SASLA*	131.42 ± 1.67	$287\,906.74 \pm 1\,847.39$	176.860996 ± 49.938683
	ES	67.92 ± 1.88	$183\,540.80 \pm 2\,037.27$	$-4\,333.405094 \pm 55.071407$
wine:	*SASLA*	78.00 ± 2.88	$45\,117.77 \pm 1\,068.71$	-32.810974 ± 4.913917
	ES	11.16 ± 3.46	$7\,766.29 \pm 486.73$	-290.322597 ± 2.237983

Table 10. Comparison of SASLA ($\beta = 0.9$) and ES convergence results

Problem		Generalization levels						
glass		**60%**	**65%**	**70%**	**75%**	**80%**	**85%**	**90%**
	SASLA	0%	0%	10%	30%	64%	90%	98%
	ES	10%	18%	62%	90%	100%	100%	100%
wine		**90%**	**95%**	**96%**	**97%**	**98%**	**99%**	**100%**
	SASLA	0%	0%	4%	4%	12%	12%	12%
	ES	10%	48%	48%	78%	78%	78%	78%

Connectionist Models of Cortico-Basal Ganglia Adaptive Neural Networks During Learning of Motor Sequential Procedures

J. Molina Vilaplana, J. Feliú Batlle, J. López Coronado

Departamento Ingeniería de Sistemas y Automática. Universidad Politécnica de Cartagena. Campus Muralla del Mar. C/ Doctor Fleming S/N Cartagena 30202. Murcia. Spain
Javi.Molina@upct.es, Jorge.Feliu@upct.es, JLCoronado@upct.es.

Abstract. In this paper two neural models of basal ganglia function during motor sequential behaviour are presented. Two connectionist models of neuron – like elements that mimic some aspects of anatomy and physiology of cortico – basal ganglia – thalamo - cortical loops have been developed. The aim of this work is to report a new computational model of motor sequence learning guided by reinforcement signals from neuronal systems that evaluate behaviours. The models are partially recurrent neural networks known as Jordan networks trained under a reinforcement learning paradigm. To validate these models, experimental findings of Tanji and Shima [5] on monkeys have been reviewed. The hypothesis that cortico – basal ganglionic loops learn and perform sequences successfully driven by reinforcement signals has been demonstrated in computer simulations of the models presented in this paper.

1 Introduction

In the absence of a comprehensive, testable theory of basal ganglia operations, neurobiological techniques alone are unlikely to provide a complete answer to the question of what functions are subserved by the basal ganglia. What is generally lacking are adaptive network models that capture essential aspects of basal ganglia able to mimic some of the behavioural functions attributed to these structures. The advent of such models is likely to prove crucial for a full understanding of how the basal ganglia contribute to behavioural processing. It appears that is essential to develop functional models of basal ganglia circuitry that capture essential aspects of the known anatomy and physiology and were able to simulate some of the behavioural functions that have been imputed to these structures.

In this paper two neural network models of basal ganglia function during motor sequential behaviour are presented. It has been developed two connectionist models of neuron – like elements that mimic some aspects of anatomy and physiology of

J. Mira and A. Prieto (Eds.): IWANN 2001, LNCS 2084, pp. 394-401, 2001.

cortico – basal ganglia – thalamo - cortical loops. The aim of this work is to present a computational model of motor sequence learning guided by the modification of synapses modulated by non-specific signals of neuronal systems that reflect the global evaluation of recent behaviour. Neuron responses in the model are subject to an amplification carried out by a local Hebbian synaptic rule that strengths or weaks connections between active neurons. This selective amplification of neuronal responses is modulated by a system that establishes if the movements carried out in response to a sensory and motor context is the correct movement within a specific sequence. This system release non-specific signals measuring the value of motor actions in the environment, these signals act in the local synaptic rule in a diffuse way, resembling the kind of modulation over neural activity and plasticity that dopaminergic neurons do through vast areas of cortex and striatum [1].

To validate these models, we have reviewed experimental findings of Tanji and Shima [5], on monkeys like Doya and Bapi [2] have reported in previous works. We use the experimental paradigm of Tanji and Shima to simulate the learning process and posterior performance of the models.

The paper is organised in the following form. In section 2 a review of some aspects of basal ganglia anatomy and physiology is shown. In section 3 we review the experimental findings of Tanji and Shima that have guided the design of the models and the simulations. Section 4 describes the neural network models of sequence learning and execution. In section 5 simulation results are reported and section 6 finishes with a discussion of results and future works and developments on this work.

2 Cortico – Basal Ganglionic – Thalamocortical loops

The basal ganglia are adaptive subcortical neural circuits whose outputs terminate in thalamic regions that in turn have access to a broad region of the frontal lobe of the cerebral cortex. This connectivity allows the basal ganglia to participate in the selection, initiation, execution, modulation and learning, of sequential movements. It's hypothesised that the basal ganglia acts by opening normally closed gates via selective phasic removal of tonic inhibition that output stages of basal ganglia, carry out on their thalamic targets. The selective opening of thalamic pathways to cortical areas occurs in relation to expected or current tasks requirements.

The basal ganglia anatomy and physiology have been characterised in great detail. The input stage, the striatum, receives a diverse input from virtually the entire neocortex. The striatal projection neurons, which are GABAergic and inhibitory, project to the globus pallidus, which itself is comprised of an internal and external segment. The STN also receives an excitatory cortical input. The STN projects via diffuse excitatory neurons to the internal segment of the globus pallidus (GPi). Thus the GPi receives an inhibitory projection from the striatum and an excitatory projection from the STN. These opposing influences in the GPi will be central to mechanisms of selection of actions by the basal ganglia. The GPi, which is the output stage of the basal ganglia, projects via GABAergic neurons to the ventrolateral thalamus, which in turn projects back to the cortex closing the loop.

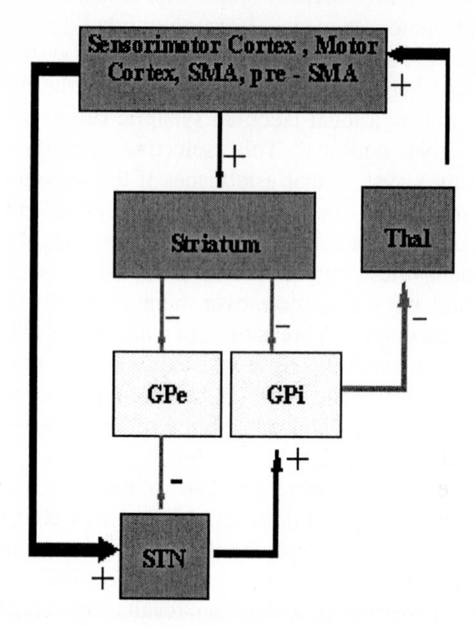

Fig. 1. Schematic diagram of the proposed action selection function of basal ganglia structures. Sensory and motor maps from the cortex project in a convergent manner to the striatum. The sensory representation roughly represents the parietal cortex and sends a processed representation of the environment to the striatum. The striatum sends two projections: a highly convergent one to the globus pallidus externus (GPe) and a moderately convergent one to the globus pallidus internus (GPi). Striatal neurons that reach firing threshold inhibits its target in the GPi. Inhibition of the GPi neuron leads to disinhibition of the corresponding thalamic neuron (Thal) and consequent gating of ascending information to the corresponding cortical motor neurons. Simultaneously, inhibition of the GPe neuron leads to disinhibition of the subthalamic (STN) neuron, which then diffusely excites the other GPi neurons. The diffuse excitation prevents another action from being selected.

In this paper new models have been developed that show how the circuitry of the basal ganglia can select the correct action at any given moment. The output stage of the basal ganglia, the internal segment of the globus pallidus (GPi), is known to be almost wholly GABAergic and tonically active. GPi tonically inhibits the target thalamic nuclei. Because these nuclei also gate ascending information to their cortical targets (motor, supplementary motor, prefrontal cortex), it is reasonable that they should be tonically inhibited until the ascending information is required for action. Given that the GPi neurons must be tonically active, in order for one information stream to be selected, one GPi channel must be turned off, thus allowing for disinhibition of the corresponding thalamic target. The supplementary motor area (SMA) assumes a prominent role in the models we present in this article. In these models the SMA contains a representation of previous actions. Because of the SMA-striatal projection the selection of action function, incorporates information about previously selected actions and thus can lead to the production of sequences of motor actions. In the model we present below positive or negative reward signals acts in the striatum to change the cortico-striatal mapping during learning of sequential procedures [1], [2], [4] .

3 Tanji & Shima experiments

Tanji and colleagues [3], [5] designed experiments where they made single cells recordings in monkeys performing sequential arm movements. The recordings were made in several brain areas such supplementary motor area (SMA), pre – supplementary motor area (pre –SMA), sensorimotor cortex (SMC), primary motor cortex (MI) and basal ganglia which comprises striatum (STR) and globus pallidus (GP).

Monkeys were initially trained to respond with a push, pull or turn movement on a manipulandum when a red, yellow or green light were presented respectively, on a display panel. After establishing initial colour to action mapping, monkeys were required to sequence these basic actions by following a succession of colour lights . The trials guided by colour lights were called Visual Trials (VT). There were another kind of trials called Memory Trials (MT). In these trials the monkeys were required to carry out a specific sequence of the three movements without visual guidance. In a block of trials, monkeys experienced five Visual Trials followed by six Memory Trials. The order of appearance of different sequences varied randomly and each block used only one sequence. The end of block (the end of sequence and the beginning of a new block and a new sequence) was indicated by flashing of all lights.

Many studies have shown that the SMA is related to sequential movements. SMA recordings carried out by Tanji and colleagues [5] revealed a significant activation showing three main properties: First, a sequence specific activity in some cells that fire specifically before the performance of a concrete order of movements (example Push – Pull –Turn and not Turn – Pull – Push). Second a transition specific activity showed by others cells that fired tonically after performing a specific movement within a sequence and before executing next movement in the sequence, (example Push followed by Turn). Third they found cells with movement related activity, say, some cells fired specifically when a specific movement was executed. This kind of activity (selective to single movements but not sequence specific) was recorded also in primary motor cortex (MI). SMA cells were found active mainly during MT phase, while SMC cells were active only during VT phase and MI cells showed activity during all phases of the trial. Cells in preSMA were active only during first VT trial in each block likely pointing the start of a new sequence and acting like sequence switching mechanism. The activity of pre-SMA neurons may be related with the acquisition of new sequences. In this way, the pre-SMA would control, rather than execute, motor programs. Such a control mechanism would allow efficient learning of a sequential procedures because the performance of the procedure initially is dependent on sensory information but eventually becomes automatic [3].

In relation with basal ganglia recordings, Schultz [4] reviewed motor sequence specific activity in striatum/putamen neurons. Putamen constitutes the major input stage to basal ganglia and some neurons here showed activity mainly during MT phase and specifically during particular transitions in sequences or specifically to some sequences.

4 Description of the models

In this paper two neural network models of cortico – basal ganglia – thalamo – cortical interactions during learning and performing motor sequences are reported. The models are partially recurrent neural networks known as Jordan networks. The first model is a Jordan network with a hidden layer and, an output layer with no modifiable associate synapses. The second model is a Jordan network with one hidden layer and, an output stage with modifiable synapses. Neuron responses in the model are subject to an amplification carried out by a local Hebbian synaptic rule based on the associative reward - penalty [1], [2], [4] that strengths or weaks connections between active neurons. This selective amplification of neuronal responses is modulated by a system that establishes if the movements carried out in response to a motor context is the correct movement within a specific sequence. This system release non-specific signals measuring the correctness of motor actions, these signals act in the local synaptic rule in a diffuse way, resembling the kind of modulation over neural activity and plasticity that dopaminergic neurons in substantia nigra pars compacta do through vast areas of cortex and striatum. This system generates a positive signal (+1) if the movement that has been made is correct. A penalty signal (-0.1) is released if the movement is wrong. The networks learn during VT trials and recall sequences during MT trials. In our model the SMA neurons are identified with state units in Jordan networks so our SMA module contains an explicit representation of previous actions acting like short term context of the ongoing sequence. Figure 2 shows the architecture of the first model. The input layer in this model consists of two modules and a set of state units resembling the function of SMA neurons. In the pre – SMA module the sequence number is encoded. Neurons in this module are active specifically to some sequences, acting as a long term context during sequence performance. Sensorimotor cortex module (SMC) encode the visual stimulus during Visual Trials. During Memory Trials there is no activity in SMC and there is activity in pre – SMA only during first movement in sequence. For the rest of sequence generation, short term context of the ongoing sequence encoded in state units/SMA is used by the neural network.

The first model is designed under the idea that striatum (STR) which is topologically connected to basal ganglia output structure, the globus pallidus (GP) selects the next action to be executed through the disinhibitory mechanism explained in section 2, resembling the functionality of direct pathway within basal ganglia circuitry. The STN module receives the same inputs from cortex that striatum does. Diffuse excitatory connections from STN to GP prevents the selection of other actions enhancing the selection made by the our *direct pathway*.

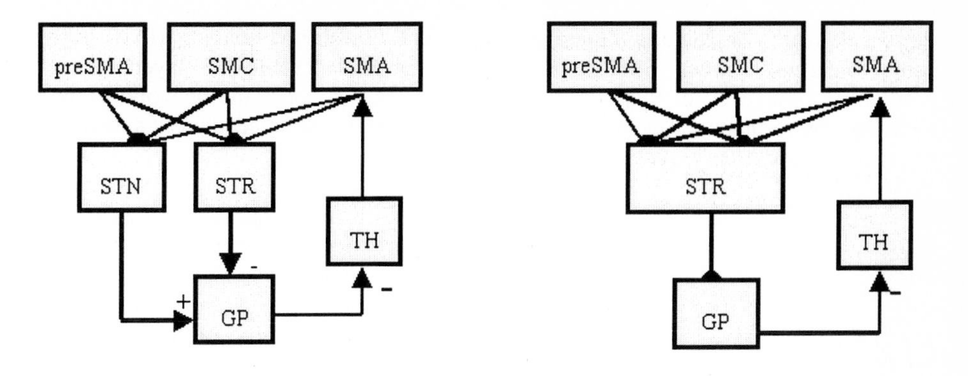

Fig. 2. Neural network models of basal ganglia function during motor sequential behaviour. The first model (left) is a Jordan network with a hidden layer and, an output layer with no modifiable associate synapses. The second model (right) is a Jordan network with one hidden layer and, an output stage with modifiable synapses. Neuron responses in the model are subject to an amplification carried out by a local Hebbian synaptic rule based on the associative reward - penalty scheme of Barto and Jordan with a fixed critic (not shown in figures). Arrows finishing in semicircle represents adaptive connections. Arrows finishing in triangles are fixed and non – adaptive connections between different neural populations.

Modifiable synapses between cortex and STR and STN change during learning process. The second architecture is shown in Figure 2. This neural network has a hidden layer and an output module (GP) with modifiable synapses. We have suppress the STN module so the new architecture is a classical Jordan network that learns guided by a diffuse modulatory system called critic in reinforcement learning paradigms. Neurons in both models are classical McCulloch & Pitts neuron – like adaptive elements interconnected in a Jordan partially recurrent architecture. Elements in hidden and output layer have sigmoid transfer function. The local synaptic learning rule that has been used is the classical Hebbian learning rule extended with Oja`s rule in preventing unbounded growth of synaptic efficacies. Learning is modulated by evaluative signals of critic resembling dopaminergic modulation of synaptic plasticity in wide brain areas. The change in synaptic efficacy between neuron i and neuron j is

$$\Delta w_{ij} = \eta \, \delta s_i \, (s_j - s_i \, w_{ij}) \tag{1}$$

where w_{ij} is the synaptic efficacy between neuron i and neuron j, s_i the activity of neuron i, δ the non- specific modulatory reinforcement signal and η is the learning rate. In the following section results of the simulations of the neural models are presented, in the context of the experimental paradigm designed by Tanji & Shima.

5 Simulation results

The models are initially trained on single colour inputs and made to learn the corresponding action. These trials named, stimulus / response (SR) trials, are given before sequence training and a good performance is established in about 20 SR trials. Initial weights between SMC-to-STR/STN layer and STR-to-GP layer are set randomly. As a result of this initial training, during VT phase the networks shown very high performance. The aim of the networks is to generate sequences accurately in the MT phase when visual guidance is absent.

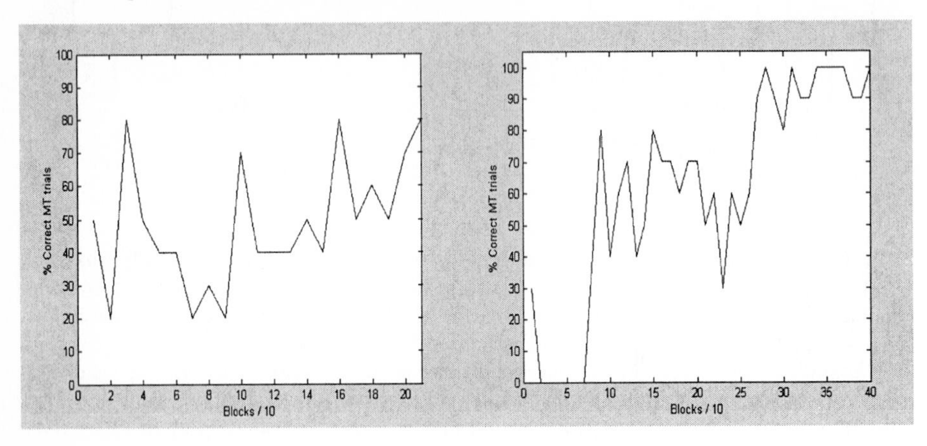

Fig. 3. . Learning profiles of two models during 200 blocks (first model ,left part of figure) and 400 blocks (second model, right part of figure). Performance of the models is taken as the percentage of correct MT trials carry out by the system in each block of training.

In simulations of both models, first a set of two sequences (Seql, Seq3) was chosen from a pool of the four sequences used in Tanji 's experiment, Seql (Push – Pull – Turn), Seq2 (Push – Turn – Pull), Seq3 (Turn – Push – Pull) , and Seq4(Turn – Pull – Push). Networks were trained with one sequence per block for 200 blocks for first neural network and 400 for second neural network. In each block a sequence is chosen randomly from the two sequence set. This set is fixed during the complete simulation. Each block consisted of five VT followed by six MT. During each VT, colour inputs are given and the networks are required to choose the appropriate action. Colour input is encoded as a 3-bit binary vector, with each bit representing one of the three colours (red, yellow, or green). Motor output is encoded as 3 component vector in thalamic module that drives this information to another 3 component vector representing neurons in motor cortex , with each component representing one of the three actions (push, pull, or turn). During the MT, sensory input is absent and the networks are required to generate appropriate actions at each time step in the correct order. Figure 3 shows the learning profiles of the networks during the simulation with sequences (Seql, Seq3). Each data point is obtained by calculating the correct number of MT in the preceding ten blocks of trials. As we have said above, during VT phase, when sensory inputs from sensorimotor cortex (SMC) are available, the networks shown a performance of 100 % due to initial training in SR trials.

In the first neural network model, performance on MT increases very slowly and reaches 80% level in 200 blocks of simulation. Performance on MT in second model increases during learning and reaches 95% level in 400 blocks of simulation.

6 Discussion and Future works

In this paper it has been hypothesised that adaptive neural networks in basal ganglia due to training, learn sequential procedures in a sensory guided mode and, after training process, permit them to internally generate those sequences learned previously. Simulations have shown how two connectionist neural network models with anatomical correspondence to known cortico – basal ganglionic recurrent loops learn by trial and error sequences in a process driven by internal reinforcement signals. The hipothesis that cortico – basal ganglionic recurrent loops learn and perform sequences successfully driven by reinforcement signals has been demonstrated in simulations. The chaining of successive movements is enabled by short term context inputs from SMA structures. In simulations, a fixed critic that generates reinforcement signals has been used, an adaptive critic implementation that will allow the system to be driven under a Temporal Difference (TD) [4] learning paradigm, in which reinforcement signals are adaptively computed as predictions of future rewards, is left to future works.

References

1. Aosaki, T., Graybiel, A. M., Kimura, M. Effect of nigrostriatal dopamine system on acquired neural responses in the striatum of behaving monkeys. *Science Wash. DC* 265: 412-415, (1994)

2. Bapi, R.S, Doya. K. Sequence representation in animals and networks: Study of a recurrent network trained with reinforcement learning. NIPS*98, Denver, CO, USA Nov 30 – Dec 5, (1998)

3. Shima, *K.,* Mushiake, H., Saito, N., & Tanji, J. Role for cells in the pre supplementary motor area in updating motor plans, *Proc. Natl. Acad. Sci. USA,* 93, pp.8694-8698 (1996).

4. Schultz, W. Predictive reward signal of dopamine neurons. J. Neurophysiology 80: 1 – 27 (1998).

5. Tanji. J, Shima, K. Role for supplementary motor area cells in planning several movements ahead, *Nature,* 371, 29, pp.413-416 (1994).

Practical Consideration on Generalization Property of Natural Gradient Learning

Hyeyoung Park

Laboratory for Mathematical Neuroscience
RIKEN Brain Science Institute
2-1 Hirosawa, Wako, Saitama, 351-0198, Japan
hypark@brain.riken.go.jp

Abstract. Natural gradient learning is known to resolve the plateau problem, which is the main cause of slow learning speed of neural networks. The adaptive natural gradient learning, which is an adaptive method of realizing the natural gradient learning for neural networks, has also been developed and its practical advantage has been confirmed. In this paper, we consider the generalization property of the natural gradient method. Theoretically, the standard gradient method and the natural gradient method has the same minimum in the error surface, thus the generalization performance should also be the same. However, in the practical sense, it is feasible that the natural gradient method gives smaller training error when the standard method stops learning in a plateau. In this case, the solutions that are practically obtained are different from each other, and their generalization performances also come to be different. Since these situations are very often in practical problems, it is necessary to compare the generalization property of the natural gradient learning method with the standard method. In this paper, we show a case that the practical generalization performance of the natural gradient learning is poorer than the standard gradient method, and try to solve the problem by including a regularization term in the natural gradient learning.

1 Introduction

Since the slow learning speed of neural networks is one of the biggest problem in practical applications, there have been a lot of researches to solve the problem. The slow convergence is konwn to be mainly caused by plateaus. A plateau is a period in the whole learning process in which the decrease of learning error is extremely slow, taking very long time to escape from it. Recently, there have been several theoretical studies on the plateau problem. Statistical-mechanical analysis on the dynamics of feedforward neural networks has made it clear that plateaus are ubiquitous in the gradient descent learning, and take dominant parts in the whole learning process[4, 9]. By investigating the learning dynamics, Saad and Solla[9] said that in the early stages of learning, all weight parameters of

J. Mira and A. Prieto (Eds.): IWANN 2001, LNCS 2084, pp. 402-409, 2001.

the hidden nodes try to learn the same values so that the network gets trapped in the symmetric phase in which there is no differentiation among the weight values of hidden nodes. It takes dominant time to break the symmetric phase, which makes plateaus in learning. On the other hand, Fukumizu and Amari[5] investigated the geometrical structure of the parameter space of neural networks. They showed that a critical point of a smaller network forms a one-dimensional critical subspace in the parameter space of a larger network in which the points can be local minima or saddles of the larger network according to a condition. They also suggested that the saddle points on the critical subspace is a main cause of making plateaus. This subset of the critical points of a smaller network, the set of the critical subspace in the larger network, includes the symmetric phase of the network mentioned in [9].

The natural gradient method can solve the plateau problem efficiently. Since Amari[1] proposed the concept of natural gradient learning and proved that the natural gradient achieves Fisher efficiency, Park and Amari[6] suggested that the natural gradient algorithm has possibility of avoiding or alleviating plateaus. This possibility has been theoretically confirmed by statistical-mechanical analysis [8]. In order to implement the concept of natural gradient in learning of feedforward neural networks, Amari, Park, and Fukumizu[2] proposed an adaptive method of obtaining an estimate of the natural gradient. This method is called adaptive natural gradient method, and Park, Amari and Fukumizu[7] extended it to various stochastic neural network models. Through a number of computational experiments on well-known benchmark problems, they confirmed that the adaptive natural gradient learning can be applied to various practical applications successfully, and it can avoid or alleviate plateaus.

On the other hands, beside the problem of learning speed, the generalization performance is also another important problem in neural networks. The ultimate purpose of learning is not to give correct representation of the training data, but to give accurate predictions of output values for new inputs. If a learning system can give good predictions, then the system is said to have a good generalization performance. This generalization problem is closely related to the optimization problem of the complexity of the model, such as structure optimization in neural networks[3].

In this paper, we consider the generalization property of the natural gradient learning. Actually, the learning algorithm itself does not make much influence on the generalization performance. Theoretically, when the network structure and the error function are fixed, all learning algorithms based on the gradient descent method have the same equilibrium points in the error surface. Thus, in theoretical sense, it is hard to say that the generalization performance of the natural gradient method is much different from the standard gradient method. In practical sense, however, the results may come to be different. Let us assume the following usual practical situations. First, we do not know the optimal number of hidden nodes for a given problem, and thus we use a sufficient number of hidden nodes that is not optimal. Second, since we do not know the minimum error that can be achieved by the network, we stop learning when the decrement of training error comes to be very small for a while, and no improvement by learning is expected. In this practical situation, the solution obtained by the natural gradient method can be different from that by the standard method, because the standard gradient learning is subject to be trapped in a long plateau and easily be misunderstood that the learning process is over. In this case, it is obvious that the generalization performances of the methods are different. Therefore, it is possible that the generalization performance of the natural gradient learning is worse than that of the standard method due to overfitting. In this paper, we investigate this situation, and give some solutions for this problem using a regularization term.

2 Natural Gradient Learning

Let us begin with a brief introduction of the natural gradient method. Since the natural gradient learning is a kind of the stochastic gradient descent learning methods, we consider a stochastic neural network model defined by

$$y = f(\boldsymbol{x}, \boldsymbol{\theta}) = \sum_{i=1}^{m} v_i \varphi(\boldsymbol{w}_i \cdot \boldsymbol{x}) + \xi, \tag{1}$$

where $\boldsymbol{x}$ is an n-dimensional input subject to a probability distribution $q(\boldsymbol{x})$; y is an output from a linear output unit; $\boldsymbol{w}_i$ is an n-dimensional connection weight vector from the input to the i-th hidden unit($i = 1, \cdots, m$); v_i is the connection weight from the i-th hidden unit to the output unit; and ξ is an additional random noise subject to $N(0, \sigma^2)$. The parameters $\{\boldsymbol{w}_1, \cdots, \boldsymbol{w}_m; \boldsymbol{v}\}$ can be represented by a single vector $\boldsymbol{\theta}$, which plays the role of a coordinate system in the space S of the multilayer perceptrons. We use here a basic network model for simplicity, but the natural gradient learning algorithm for general models has been developed (see Park, Amari and Fukumizu[7] for details).

This network can be considered as the conditional probability distribution of output y conditioned on input $\boldsymbol{x}$,

$$p(y|\boldsymbol{x}, \boldsymbol{\theta}) = \frac{1}{\sqrt{2\pi}\sigma} \exp\left[-\frac{1}{2\sigma^2}\{y - f(\boldsymbol{x}, \boldsymbol{\theta})\}^2\right]. \tag{2}$$

Then we can consider a space of the probability density functions $\{p(\boldsymbol{x}, \boldsymbol{y}; \boldsymbol{\theta}) | \boldsymbol{\theta} \in \Re^M\}$, and define an appropriate error function $E(\boldsymbol{\theta})$ on the space. The typical error function is the loss function defined by the negative logarithm of the likelihood function which is of the form,

$$E(\boldsymbol{\theta}) = -\log p(\boldsymbol{y}^* | \boldsymbol{x}; \boldsymbol{\theta}), \tag{3}$$

where $p(\boldsymbol{y}|\boldsymbol{x}; \boldsymbol{\theta})$ is the conditional pdf of $\boldsymbol{y}$ conditioned on $\boldsymbol{x}$. Learning is a process of finding an optimal point in the space of the probability density functions, which minimizes the value of the error function.

The natural gradient learning method is based on the fact that the space of $p(\boldsymbol{x},\boldsymbol{y};\boldsymbol{\theta})$ is a Riemannian space in which the metric tensor is given by the Fisher information matrix $G(\boldsymbol{\theta})$ defined by

$$G(\boldsymbol{\theta}) = \iint \frac{\partial \log p}{\partial \boldsymbol{\theta}} (\frac{\partial \log p}{\partial \boldsymbol{\theta}})^T p(\boldsymbol{y}|\boldsymbol{x},\boldsymbol{\theta}) q(\boldsymbol{x})\, d\boldsymbol{y}\, d\boldsymbol{x} \tag{4}$$

$$= E_{\boldsymbol{x}} \left[E_{\boldsymbol{y}|\boldsymbol{x};\boldsymbol{\theta}} \left[\frac{\partial \log p(\boldsymbol{y}|\boldsymbol{x};\boldsymbol{\theta})}{\partial \boldsymbol{\theta}} (\frac{\partial \log p(\boldsymbol{y}|\boldsymbol{x};\boldsymbol{\theta})}{\partial \boldsymbol{\theta}})^T \right] \right] \tag{5}$$

where $E_{\boldsymbol{x}}[\cdot]$ and $E_{\boldsymbol{y}|\boldsymbol{x};\boldsymbol{\theta}}[\cdot]$ denote the expectation with respect to $q(\boldsymbol{x})$ and $p(\boldsymbol{y}|\boldsymbol{x};\boldsymbol{\theta})$, respectively, and T denotes the transposition. Using the Fisher information matrix of Eq. 5, we can obtain the natural gradient $\tilde{\nabla} E$ and its learning algorithm for the stochastic systems;

$$\tilde{\nabla} E(\boldsymbol{\theta}) = G^{-1}(\boldsymbol{\theta}) \nabla E(\boldsymbol{\theta}), \tag{6}$$

$$\boldsymbol{\theta}_{t+1} = \boldsymbol{\theta}_t - \eta_t \tilde{\nabla} E(\boldsymbol{\theta}_t) = \boldsymbol{\theta}_t - \eta_t G^{-1}(\boldsymbol{\theta}_t) \nabla E(\boldsymbol{\theta}_t). \tag{7}$$

Since the practical implementation of the Fisher information matrix $G(\boldsymbol{\theta})$ and its inverse is almost impossible, we need an approximation method. The adaptive natural gradient[7] is a method of estimation of the inverse of Fisher information matrix adaptively, which is written by

$$\hat{G}_{t+1}^{-1} = \frac{1}{1-\varepsilon_t} \hat{G}_t^{-1} \tag{8}$$

$$- \frac{\varepsilon_t}{(1-\varepsilon_t)} \hat{G}_t^{-1} \nabla F_t \left((1-\varepsilon_t) I + \varepsilon_t \nabla F_t^T \hat{G}_t^{-1} \nabla F_t \right)^{-1} \nabla F_t^T \hat{G}_t^{-1} \tag{9}$$

where

$$\nabla F_t = (\nabla f_1(\boldsymbol{x}_t, \boldsymbol{\theta}_t) \ldots, \nabla f_L(\boldsymbol{x}_t, \boldsymbol{\theta}_t)). \tag{10}$$

In the case of batch learning, if the number of parameter is small, then the sample mean can be used instead of the real expectation of Eq. 5, and its direct inversion can be used with reasonable computational cost.

3 Generalization Property of Natural Gradient Learning

Before investigating the generalization performance, it would be helpful to consider the difference of the natural gradient learning from the standard method. From the Eq. 7, we can see that the natural gradient learning is using a search direction which is obtained by multiplying a matrix $G^{-1}(\boldsymbol{\theta})$ to the standard search direction. This simple difference can get rid of or alleviate the plateaus in the learning curve. Some intuitive reason why it is possible can give useful hints about the generalization performance of the natural gradient learning. Firstly, the Fisher information matrix $G(\boldsymbol{\theta})$ comes to be singular in the plateau area where more than two hidden nodes have the same weight $\boldsymbol{w}$. Thus if the learning parameter approaches to a plateau, the update size come to be very big and escapes from it. Secondly, by multiplying a matrix to the standard search direction ∇E, it comes to be possible to force hidden nodes not to approach to the same values. Thus, the symmetry phenomena hardly occur in the natural gradient learning. In other words, natural gradient are trying to use all hidden nodes effectively by making them to have different parameter $\boldsymbol{w}$. For this reason, in practical applications, if we use more than the optimal number of hidden nodes, the natural gradient learning may be easily overfitted.

To check this intuitive guess, we conducted computer simulations with a toy problem. The task is to approximate a function defined by

$$y = \frac{(x-1)(2x+1)}{(1+x^2)}. \tag{11}$$

The shape of the function is illustrated in Fig. 1. We generated 21 training data

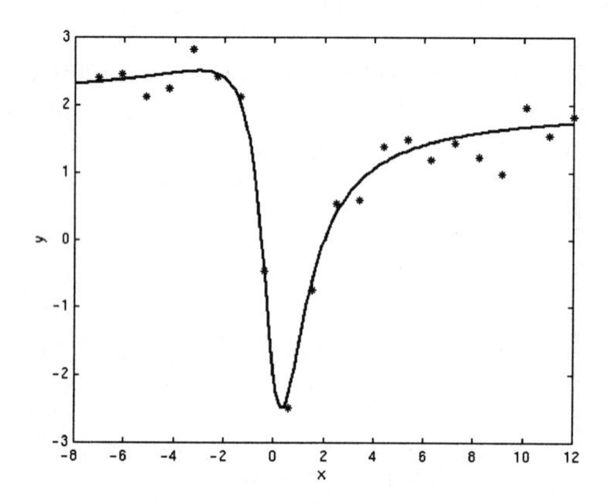

Fig. 1. True function and noisy data used for learning.

with additional noise subject to $N(0, 0.3^2)$. The training data are also plotted in Fig. 1. The learning network has 1 input and 1 output, and we tried 4 cases of the number of hidden nodes; two, three, six, and ten. From the shape of the true function, we can guess that two or three hidden nodes are optimal.

We compared two learning method; the standard gradient descent method and the natural gradient method. Since the network size is small, we approximated the Fisher information matrix with the sample mean and directly inverse it to get the natural gradient. For both of the learning methods, we stopped learning if the decrement of training error during one hundred learning cycles comes to be smaller than 10^{-5}. The result is shown in Tab. 1 and Fig. 2. From the results, we can confirm our guess about the generalization property of natural gradient learning. The natural gradient can utilize all hidden nodes and converges to optimal point rapidly, while the standard gradient learning is trapped in a long plateau. Considering this kind of noisy data problem which is usual in practical applications, the generalization problem of natural gradient learning should be treated more carefully than that of the standard gradient method.

Table 1. Learning results of the standard gradient method and the natural gradient method

Hidden	Standard Gradient		Natural Gradient	
Nodes	Learning Cycle	Training Error	Learning Cycle	Training Error
2	19400	0.0645	750	0.0630
3	15400	0.0533	800	0.0502
6	7600	0.0500	25200	0.0112
10	9600	0.0499	9800	0.0115

4 Learning with Reglarization Term

One simple solution for the overfitting to noisy data is to add a regularization term to the usual error function. In this paper, we add the most basic regularization term to the standard error function and use the new cost function defined by

$$C(\boldsymbol{\theta}) = E(\boldsymbol{\theta}) + \nu \frac{1}{2} ||\boldsymbol{\theta}||^2 \tag{12}$$

instead of the standard error function $E(\boldsymbol{\theta})$, where ν is a user-defined regularization parameter. Since the definition of the Fisher information matrix does not depend on the cost function, the natural gradient learning algorithm with regularization term can be written by

$$\boldsymbol{\theta}_{t+1} = \boldsymbol{\theta}_t - \eta_t G^{-1}(\boldsymbol{\theta}_t) \nabla C(\boldsymbol{\theta}_t) \tag{13}$$

$$= \boldsymbol{\theta}_t - \eta_t G^{-1}(\boldsymbol{\theta}_t)(\nabla E(\boldsymbol{\theta}_t) + \nu \boldsymbol{\theta}_t). \tag{14}$$

The same experiments on the function approximation problem was done, and the results are shown in Tab. 2 and Fig. 3. We used $\nu = 0.02$ which is empirically obtained. Form the results, we can confirm that the generalization performance is successfully improved using the regularization method.

Table 2. Learning results of the natural gradient learning with regularization term.

Hidden	Natural Gradient with Regularization	
Nodes	Learning Cycle	Training Error
2	410	0.0674
3	380	0.0533
6	400	0.0532
10	380	0.0504

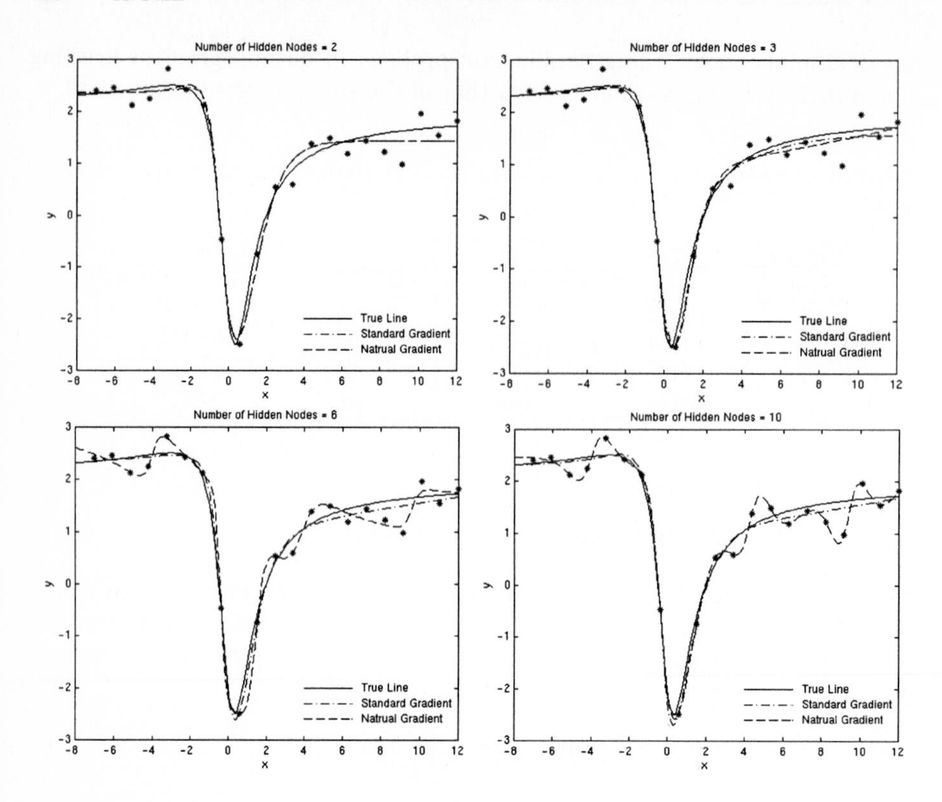

Fig. 2. Approximation results of the standard gradient method and the natural gradient method

5 Conclusions and Discussions

In this paper, we investigated the generalization problem of the natural gradient learning, which can occur in practical situations. Due to the ability of avoiding plateau of natural gradient method, the generalization problem of the natural gradient learning could be more serious than that of the standard gradient method. To solve the problem, we presented a natural gradient learning method with regularization term, and through computational experiments, we showed that it can improve the generalization performance.

This paper is a preliminary work for the combination of the natural gradient learning and the structure optimization, which is another important problem in the field of neural network. As future works, it is necessary to develop an appropriate method for deciding the regularization parameter. Furthermore, an efficient and proper structure optimization method such as pruning or growing needs to be developed and combined with the natural gradient method.

Acknowledgements

The author would like to thank prof. Shun-ichi Amari for helpful discussions.

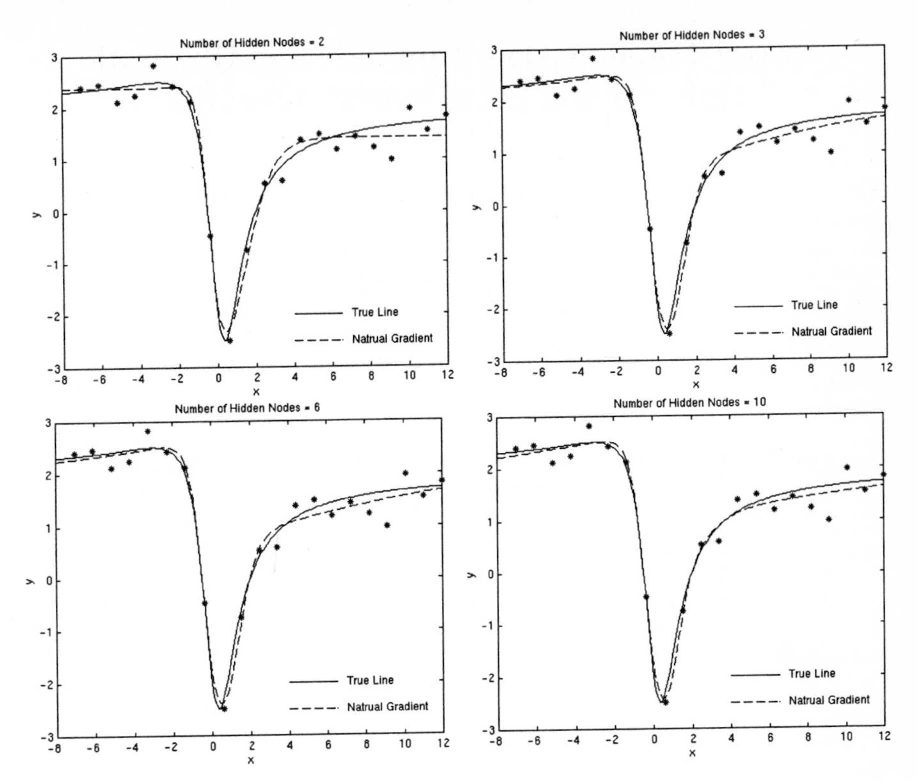

Fig. 3. Approximation results of the natural gradient learning with regularization term.

References

1. S. Amari : Natural Gradient Works Efficiently in Learning, *Neural Computation,* 10, 251-276, 1998.
2. Amari, S., Park, H., and Fukumizu, F. : Adaptive method of realizing natural gradient learning for multilayer perceptrons, *Neural Computation*, 12, 1399-1409, 2000.
3. Bishop, C. : *Neural Networks for Pattern Recognition*, Oxford University Press, 1995.
4. Biehl, W., Riegler, P. and Wöhler, C. : Transient Dynamics of On-line Learning in Two-layered Neural Networks, *Journal of Physics, A*, 29, 4769-4780, 1996.
5. Fukumizu, K. and Amari, S. : Local Minima and Plateaus in Hierarchical Structures of Multilayer Perceptrons, in preparation, 1999.
6. Park, H. and Amari, S. : Escaping from Plateaus of Multilayer Perceptron Learning by Natural Gradient, *The 2nd RIEC International Symposium on Design and Architecture of Information Processing Systems Based on the Brain Information Principles*, 189-192, 1998.
7. Park, H., Amari, S. and Fukumizu, K. : Adaptive natural gradient learning algorithms for various stochastic models, *Neural Networks,* 13, 755-764, 2000.
8. Rattray, M., D. Saad, and S. Amari : Natural Gradient Descent for On-line Learning, *Physical Review Letters*, 81, 5461-5464, 1998.
9. Saad, D. and Solla, S. A. : On-line Learning in Soft Committee Machines, *Physical Review E*, 52, 4225-4243, 1995.

Novel Training Algorithm Based on Quadratic Optimisation Using Neural Networks

Ganesh Arulampalam, Abdesselam Bouzerdoum

School of Engineering & Mathematics,
Edith Cowan University,
100, Joondalup Drive, Joondalup,
WA 6027, Australia
{g.arulampalam,a.bouzerdoum}@ecu.edu.au

Abstract. In this paper we present a novel algorithm for training feedforward neural networks based on the use of recurrent neural networks for bound constrained quadratic optimisation. Instead of trying to invert the Hessian matrix or its approximation, as done in other second-order algorithms, a recurrent equation that emulates a recurrent neural network determines the optimal weight update. The development of this algorithm is presented, along with its performance under ideal conditions as well as results from training multilayer perceptrons. The results show that the algorithm is capable of achieving results with less errors than other methods for a variety of problems.

1 Introduction

The training of feedforward neural networks is based on the minimisation of an objective function related to the output error. The general strategy for supervised learning is based on combining a quickly convergent local method with a globally convergent one [1]. The local methods are based on local models of the generally complex error surface. Most algorithms are based on a linear (first order) model or quadratic (second order) model. Quadratic methods tend to have faster convergence, though they occasionally get trapped in local minima.

Second order methods rely on minimising a quadratic approximation to the error function, $E(\mathbf{w})$, that uses the first three terms of the Taylor-series expansion about the current point, $\mathbf{w}_o$, given by

$$E(\mathbf{w}_o + \Delta\mathbf{w}) \approx E(\mathbf{w}_o) + \mathbf{g}^T \Delta\mathbf{w} + \tfrac{1}{2}\Delta\mathbf{w}^T \mathbf{H} \Delta\mathbf{w} \tag{1}$$

where $\Delta\mathbf{w}$ is the weight change, $\mathbf{g}$ is the gradient vector and $\mathbf{H}$ is the Hessian matrix.

Solving this equation yields the optimal change in the weight matrix, $\Delta\mathbf{w}_{opt} = \mathbf{H}^{-1}\mathbf{g}$. However, the calculation of the Hessian $\mathbf{H}$ and its inverse is computationally prohibitive, thereby leading to approximation methods being investigated. There are also the problems where the Hessian is not positive definite or is singular or ill-conditioned [1]. The matter is further complicated if constraints are imposed on the solution, as is the case for Shunting Inhibitory Artificial Neural Networks (SIANNs) where certain weights need to be constrained [2].

J. Mira and A. Prieto (Eds.): IWANN 2001, LNCS 2084, pp. 410-417, 2001.

The *Quadratic Neural Network* (QNN) algorithm is a novel second order method that uses a recurrent "neural network" to determine the minimum point of the objective function to be minimised. It is based on work using recurrent neural networks for bound constrained quadratic minimisation done by Bouzerdoum and Pattison [3, 4]. In the actual implementation of the algorithm, the recurrent neural network is simulated by a recurrent equation that is iterated.

The following section outlines the development of the algorithm and modifications made. In section 3, the speed of convergence of this method under ideal conditions is demonstrated. Section 4 details the practical implementation of the algorithm for training feedforward neural networks, followed by experimental results comparing the performance of the QNN algorithm with other algorithms in Section 5 and conclusions in Section 6.

2 The Quadratic Neural Network (QNN) Algorithm

2.1 Development of the Algorithm

Bouzerdoum and Pattison's method uses a recurrent neural network to perform bound constrained quadratic optimisation such as

$$\min\{E(\mathbf{w}_o + \Delta\mathbf{w}) : \mu \leq \Delta\mathbf{w} \leq \upsilon\} \tag{2}$$

with $\mu, \upsilon \in \mathbf{R}^n$ for $\mathbf{w} \in \mathbf{R}^n$.

In order to ensure the constraints are always satisfied, let

$$\Delta\mathbf{w} = \mathbf{p}(\mathbf{u}) \equiv \mathbf{B}\mathbf{f}(\mathbf{u}) \tag{3}$$

where $\mathbf{f} : \mathbf{R}^n \rightarrow \mathbf{R}^n$ is defined as

$$f_i(\mathbf{u}) \equiv f_i(u_i) = \begin{cases} \varsigma_i & u_i < \varsigma_i \\ u_i & u_i \in [\varsigma_i, \xi_i] \\ \xi_i & u_i > \xi_i \end{cases} \tag{4}$$

The n-dimensional vector $\mathbf{u}$ is permitted to vary without constraint, $\mathbf{B}$ is an n-by-n positive diagonal matrix that serves as a *preconditioner*, and $\varsigma, \xi \in \mathbf{R}^n$ are the constraints μ, υ on $\Delta\mathbf{w}$ mapped onto corresponding values of $\mathbf{u}$ such that $\varsigma = \mathbf{B}^{-1}\mu$ and $\xi = \mathbf{B}^{-1}\upsilon$. The function f is a piecewise-linear activation function. By identifying $\mathbf{p}(\mathbf{u})$ such that $\Delta\mathbf{w}$ is confined to the constraint region, the problem now becomes an *unconstrained* minimisation of $M(\mathbf{u})$ over $\mathbf{u}$ where

$$M(\mathbf{u}) = \mathbf{g}^T\mathbf{p}(\mathbf{u}) + \tfrac{1}{2}\mathbf{p}(\mathbf{u})^T\mathbf{H}\mathbf{p}(\mathbf{u}) \tag{5}$$

Consider now the single-layered recurrent neural network whose state vector is defined by the differential equation

$$\frac{d\mathbf{u}}{dt} = -\mathbf{g} - \mathbf{A}\mathbf{u} - \mathbf{C}\mathbf{f}(\mathbf{u}) \tag{6}$$

where $\mathbf{g}$ is the external input, $\mathbf{f}(\mathbf{u})$ is the network output, $\mathbf{C}$ is the lateral feedback matrix with zero diagonal entries, and $\mathbf{A}$ is a positive diagonal matrix representing the passive decay rate of the state vector.

To map the constrained quadratic problem onto the neural network, set

$$\mathbf{A} = \text{diag}(\mathbf{HB}) \tag{7}$$

$$\mathbf{C} = \mathbf{HB} - \mathbf{A} \tag{8}$$

where diag(.) selects the diagonal elements of its matrix argument.

The desired output $\Delta\mathbf{w} = \mathbf{Bf}(\mathbf{u})$ is obtained from the network output $\mathbf{f}(\mathbf{u})$ through multiplication with the diagonal preconditioner $\mathbf{B}$. Bouzerdoum & Pattison [4] showed that, provided the matrix $\mathbf{H}$ is positive definite, the neural network in (6) has a unique equilibrium point $\mathbf{u}^*$ which is mapped by $\mathbf{p}$ onto $\Delta\mathbf{w}^*$, the optimal constrained step to the minimum of $E(\mathbf{w})$. They have also shown that the network is globally convergent to this equilibrium point.

The *spectral condition number* of a matrix is defined as the ratio of the maximum to the minimum singular values of the matrix. If the state-feedback matrix has a large condition number, then it is susceptible to round-off errors and errors in the weights of the state-feedback matrix. Preconditioning is used to keep the condition number small. It has also been shown that preconditioning speeds up convergence [3].

For the system in question, a simple choice for the preconditioner matrix $\mathbf{B}$ is

$$b_{ii} = \frac{\alpha}{h_{ii}} \qquad \alpha > 0 \tag{9}$$

where b_{ii} and h_{ii} are the diagonal elements of $\mathbf{B}$ and $\mathbf{H}$ respectively. The choice of preconditioning has the added advantage of simplifying the expression for matrix $\mathbf{A}$, which then simply becomes

$$\mathbf{A} = \alpha\mathbf{I} \tag{10}$$

where $\mathbf{I}$ is the identity matrix.

The matrix $\mathbf{C} = \mathbf{HB} - \mathbf{A}$ can then be defined by

$$c_{ij} = \begin{cases} \alpha \dfrac{h_{ij}}{h_{jj}} & i \neq j \\ 0 & i = j \end{cases} \qquad i, j \in 1,\ldots,n \tag{11}$$

In this training algorithm, we approximate the operation of the recurrent network for the quadratic minimisation by a discrete time recurrence equation. At each iteration of the training, we "construct" a recurrent neural network with constraints based on the state of the network being trained at that point. The "recurrent network" modelled by the recurrent equation will return the optimal weight update for that iteration and the network being trained will have its parameters updated using (3).

The differential equation (6) can be approximated by

$$\frac{\mathbf{u}(k+1)-\mathbf{u}(k)}{d}=-\mathbf{g}-\mathbf{A}\mathbf{u}(k)-\mathbf{C}\mathbf{f}(\mathbf{u}(k)) \tag{12}$$

$$\therefore \mathbf{u}(k+1)=\mathbf{u}(k)-d\,(\mathbf{g}+\mathbf{A}\mathbf{u}(k)+\mathbf{C}\,\mathbf{f}(\mathbf{u}(k))) \tag{13}$$

where d is the discrete 'time-step'.

The recurrent equation (13) will be iterated a finite number of times to obtain an approximate optimal value of $\mathbf{u}$, then calculate the update of the weights, $\Delta\mathbf{w}$, from (3). The recurrence equation can be iterated a fixed number of times, for even if the weight update is sub-optimal, the overall effect of any error can be ignored since the process will be repeated for each epoch.

2.2 Applying the QNN Algorithm without Constraints

When applying the QNN algorithm to neural networks such as multilayer perceptrons (MLPs), the fact that the weights of MLPs are unconstrained can be exploited. This means that the problem is an unconstrained minimisation already, so (4) simply becomes

$$\mathbf{f}(\mathbf{u})=\mathbf{u} \tag{14}$$

This then simplifies the recurrence equation given in (13) to

$$\mathbf{u}(k+1)=\mathbf{u}(k)-d(\mathbf{g}+(\mathbf{A}+\mathbf{C})\,\mathbf{u}(k)) \tag{15}$$

$$\Rightarrow \mathbf{u}(k+1)=\mathbf{u}(k)-d\,(\mathbf{g}+\mathbf{HB}\,\mathbf{u}(k)) \tag{16}$$

since by definition $\mathbf{A}+\mathbf{C}=\mathbf{HB}$ from (8).

2.3 Adaptive Determination of the Parameters for the Algorithm

The variables that affect the algorithm are α, d and i, the number of iterations to update the recurrence equation. One important observation that was made during these experiments was that the product αd could be taken as one parameter for the algorithm since parameters α and d had inverse effects, For example, setting $\alpha = 1$ and $d = 0.1$ produces exactly the same results as $\alpha = 2$ and $d = 0.05$. As such, the term α was fixed at 1 and only the d and i parameters were varied.

To further reduce the number of parameters to be specified, a method was developed to adaptively determine i, the number of iterations for $\mathbf{u}(k)$. The rationale was that the iterations could be stopped when the percentage change in $\mathbf{u}(k)$ dropped below a certain limit. The change, $\delta\mathbf{u}(k)$, is given by

$$\delta\mathbf{u}(k)=\frac{\mathrm{norm}(\mathbf{u}(k+1)-\mathbf{u}(k))}{\mathrm{norm}(\mathbf{u}(k+1))} \tag{17}$$

The procedure is to iterate $\mathbf{u}(k)$ until $\delta\mathbf{u}(k) < 0.01$ (1%) or until a maximum number is reached, typically 50.

A number of methods have been tried to determine the parameter d adaptively. The most successful to date is where the starting value at each epoch, d_o, is increased by a factor based on the epoch number

$$d_0(\text{current epoch}) = d_0(1) * \left(1 + \frac{\text{current epoch}}{\text{total epochs}}\right). \tag{18}$$

The value of d is set to d_o then reduced as iterations go on, depending on the change in the objective function $M(\mathbf{u}(k))$

$$d(k+1) = \begin{cases} 0.90d(k) & M(\mathbf{u}(k)) \leq 1.1\, M(\mathbf{u}(k-1)) \\ 0.75d(k) & M(\mathbf{u}(k)) > 1.1\, M(\mathbf{u}(k-1)) \end{cases} \tag{19}$$

3 QNN Algorithm Performance Under Ideal Conditions

In order to compare the performance of the QNN algorithm with other algorithms under ideal conditions, the algorithm was used to solve the minimisation of the Powell function [5],

$$f(x_1, x_2, x_3, x_4) = (x_1 + 10x_2)^2 + 5(x_3 - x_4)^2 + (x_2 - 2x_3)^4 + 10(x_1 - x_4)^4 \tag{20}$$

and Rosenbrock's function [6],

$$f(x_1, x_2) = 100(x_2 - x_1{}^2)^2 + (1 - x_1)^2 \tag{21}$$

The exact expressions for the gradient vectors and the Hessian matrices were derived from (20) and (21) above, and used in the evaluation of the problem instead of using approximations. The performance of the QNN algorithm was compared to the Conjugate-Gradient (CG) algorithm. The results are shown in Fig. 1.

3.1 Constraining the QNN Update

A modification made to the QNN algorithm was to use the ability of the algorithm to handle constraints by imposing a constraint on $\mathbf{u}$ such that it is bounded by the function $\mathbf{f}$ to the hypercube defined by 100 times the components of the gradient vector. The rationale behind this is that it would keep the updates in the general gradient descent quadrant, thereby reducing the possibility of instability. The performance of the constrained QNN algorithm is shown by the dotted line in Fig. 1.

Figure 1 illustrates the general trend where both versions of the QNN algorithm outperform the CG algorithm. The constrained QNN algorithm is slightly slower than the unconstrained version initially but is generally able to achieve better final results.

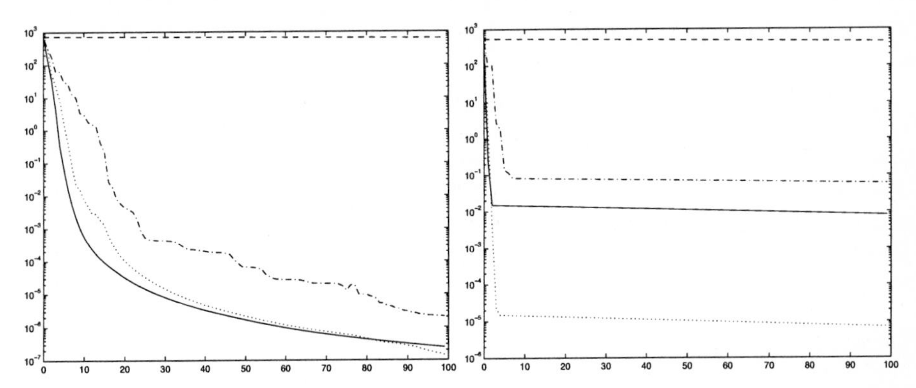

Fig. 1. Output vs. iterations using LM (*dashed*), CG (*dot-dash*), constrained QNN (*dotted*) and unconstrained QNN (*solid*) for minimisation of (a) the Powell function and (b) Rosenbrock's function

4 Applying the QNN Algorithm to Neural Networks

In practical implementations, the computational cost of calculating the Hessian matrix is too high, so approximations of the Hessian are used. The QNN algorithm has been implemented for MLPs based on the Levenberg-Marquadt (LM) algorithm [7]. The same approximations for the gradient and Hessian based on the Jacobian as in LM have been used:

$$\mathbf{g} \approx \mathbf{J}^T(\mathbf{w})\mathbf{e}(\mathbf{w}) \tag{22}$$

$$\mathbf{H} \approx \mathbf{J}^T(\mathbf{w})\mathbf{J}(\mathbf{w}) + \mu\mathbf{I} \tag{23}$$

where $\mathbf{J}(\mathbf{w})$ is the Jacobian matrix, $\mathbf{e}(\mathbf{w})$ is the vector of residuals (errors) for the training set, $\mathbf{I}$ is the identity matrix and μ is a variable parameter that determines the trust region.

This approximation of the Hessian has been used instead of the Gauss-Newton approximation ($\mathbf{H} \approx \mathbf{J}^T(\mathbf{w})\mathbf{J}(\mathbf{w})$) to overcome the problems of rank deficiency, since neural network training problems are intrinsically ill-conditioned [8, p212], as well as the requirement of the QNN algorithm that the Hessian be positive definite. Tests with the Gauss-Newton approximation for the QNN algorithm resulted in non-convergence due to the above-mentioned problems.

The only difference between the implementation of the LM and QNN algorithms is that the step where the change in weights is calculated with the matrix inversion in LM has been replaced with the "recurrent neural network" ie. the recurrent equations given in (13) and (16) for the constrained and unconstrained versions respectively.

To adaptively vary the initial 'time-step' term for each epoch, d_o, the parameter μ was used as a measure of how close to quadratic the objective function is during training. If μ decreases, the quadratic approximation is improving, therefore d_o is increased by a factor of 1.1, otherwise it is decreased by a factor of 0.9.

5 Experimental Results

The QNN algorithm was tested on MLPs for a number of different problems. Ten networks with suitable structures and random weights were generated for each problem and trained, with results then averaged over the ten networks. The problems used were the 5-bit parity problem with 2-bit output (even and odd parity), Monks-1 and Spiral problems. The same networks were also trained using the LM, Conjugate-Gradient with Fletcher-Reeves update (CGF) and Conjugate-Gradient with Powell restarts (CGP) algorithms for comparison. Summarised results are given in Table 1.

Table 1. Results of comparison between the unconstrained QNN (QNN-U), constrained QNN (QNN-C) and LM, CGP and CGF algorithms for 5-bit parity, Monks-1 and Spiral problems

		No. of nets (out of 10)			Average Epochs		Average error	
Algorithm	Error Goal	Reached goal	All correct	Conv-erged	Reached goal	All Correct	% Err	(Std. dev)
5-bit parity using 5-5-2 MLP								
QNN-U	1e-6	8	8	10	120	120	1.88	(4.93)
QNN-C	1e-6	7	7	8	737	737	0.78	(2.21)
LM	1e-6	6	6	9	177	177	3.47	(5.51)
CGP	1e-6	5	5	10	284	284	3.44	(4.53)
CGF	1e-6	2	2	8	472	472	9.77	(6.98)
Monks-1 problem using 6-9-1 MLP								
QNN-U	1e-5	8	7	8	17.8	16.9	0.64	(1.80)
QNN-C	1e-5	10	9	10	22.9	22.9	0.49	(1.54)
LM	1e-5	10	8	10	11.1	9.5	1.99	(4.31)
CGP	1e-5	10	3	10	24.6	21.3	2.20	(2.32)
CGF	1e-5	10	1	10	37	24	3.24	(2.74)
Spiral-1 problem using 2-36-12-1 MLP								
QNN-U	1e-5	8	0	10	3152	*	12.1	(1.84)
QNN-C	1e-5	7	0	10	7046	*	11.7	(1.52)
LM	1e-5	10	0	10	135	*	14.7	(3.81)
CGP	1e-5	10	0	10	198	*	13.1	(2.34)
CGF	1e-5	5	0	10	2030	*	14.6	(3.16)

The results show that for the Monks and 5-bit parity problems the unconstrained QNN algorithm is faster than the CG algorithms in terms of number of epochs but slower than LM. However it is able to lower error than these other algorithms. The constrained QNN algorithm is able to achieve the lowest error rates of all but at the cost of longer training.

Overall the QNN algorithm is able to achieve better results in terms of generalisation (average error) and number of networks able to get all correct classifications.

6 Conclusion

In this paper we have shown how the idea of using a recurrent neural network for bound constrained quadratic optimisation can be developed into a training algorithm for feedforward neural networks. The Quadratic Neural Network (QNN) algorithm is a second order algorithm that avoids the need to invert the Hessian matrix by using a recurrent equation that simulates a recurrent neural network. It has been shown that under ideal conditions this is can produce better results than the Conjugate-Gradient algorithm.

The QNN algorithm has been successfully applied to training MLPs for a number of standard problems, and the results show that this algorithm is able to train the networks to achieve better results than the LM and CG algorithms.

Overall the QNN algorithm has been shown to be a viable training algorithm that is capable of producing good results. It has the added advantage of being able to readily incorporate constraints that may need to be imposed during training.

Further investigations will be carried out in order to speed up training, particularly into the methods for adaptively modifying the algorithm's parameters

References

1. R. Battiti, "First- and Second-Order Methods for Learning: Between Steepest Descent and Newton's Method," *Neural Computation*, vol. 4, pp. 141-166, 1992.
2. G. Arulampalam and A. Bouzerdoum, "Training Shunting Inhibitory Artificial Neural Networks as Classifiers," *Neural Network World*, vol. 10, pp. 333-350, 2000.
3. A. Bouzerdoum and T. R. Pattison, "Neural Network for Quadratic Optimization with Bound Constraints," *IEEE Transactions on Neural Networks*, vol. 4, pp. 293-304, 1993.
4. A. Bouzerdoum and T. R. Pattison, "Constrained Quadratic Optimisation using Neural Networks," presented at 4th ACNN, Melbourne, 1993.
5. G. V. Rekliatis, A. Ravindran, and K. M. Ragsdell, *Engineering Optimization: Methods and Applications*. New York: John Wiley & Sons, 1983.
6. H. H. Rosenbrock, "An automated method for finding the greatest of least value of a function," *Computer J.*, vol. 3, pp. 175-184, 1960.
7. M. T. Hagan and M. B. Menhaj, "Training Feedforward Networks with the Marquadt Algorithm," *IEEE Transactions on Neural Networks*, vol. 5, pp. 989 - 993, 1994.
8. S. Haykin, *Neural Networks: A Comprehensive Foundation*. New York: Macmillan, 1994.

Non-symmetric Support Vector Machines

Jianfeng Feng

Sussex University, Brighton BN1 9QH, UK
http://www.cogs.susx.ac.uk/users/jianfeng

Abstract. A novel approach to calculate the generalization error of the support vector machines and a new support vector machine–non-symmatic support vector machine–is proposed here. Our results are based upon the extreme value theory and both the mean and variance of the generalization error are exactly ontained.

1 Introduction

Multilayer perceptrons, radial-basis function networks and support vector machines are three approaches widely used in pattern recognition. In comparison with multilayer perceptrons and radial-basis function networks, the support vector machine optimizes its margin of separation and ensures the uniqueness of the final result. It seems support vector machines have become established as a powerful technique for solving a variety of classification, regression and density estimation tasks[2]. In practical applications, it is also recently reported that the SVM outperforms conventional learning algorithms[1]. How much does the SVM improve a machine's capability of generalization? There are a few authors [2, 9, 3] who have carried out a detailed analysis on the performance of the SVM. Nevertheless, the exact behaviour of the SVM on generalization remains elusive: all results obtained up to date are upper bounds of the mean of the generalization error (see next section for definition).

Here we propose a novel approach in terms of the extremal value theory [8, 4, 5] to exactly calculate the generalization error of a SVM. Although we confine ourselves to the case of one dimension, the conclusions obtained are illuminating. Firstly the mean and variance (or distribution) of the generalization error are exactly calculated. In the literature, as we mentioned above, only upper bounds of the mean are estimated. Secondly our approach enables us to go a step further to compare different learning algorithms. We assert that the support vector machine does improve the generalization error, both mean and variance, by a factor of a constant. Thirdly we then further propose a new version of the SVM, called non-symmetric support vector machine, which could, in some circumstances, further reduce the mean of the generalization. The basic idea of the non-symmetric support vector machine is that to employ not only the support vectors which are the only information used in the SVM, but also the mean of samples. In fact the advantage of non-symmetric support machine could be easily understood. The essence of the SVM is to rely only on the set of samples

J. Mira and A. Prieto (Eds.): IWANN 2001, LNCS 2084, pp. 418–426, 2001.

which take extremal values, the so-called support vectors. From the statistics of the extremal values, we know that the disadvantage of such an approach is that the information contained in most samples (no extremal values) is simply lost and is bound to be less efficient than an algorithm taking into the lost information. We refer the reader to [6] for more details.

2 Models

The models we are going to consider are the support vector machine and the worst learning machine. For the former, the basic assumption is that the learning is only dependent on the support vectors, and a maximization of the separation margin is fulfilled. For the later,we assume that after learning, the machine is only able to correctly recognize learned samples (see Fig. 1). Very different from most approaches in the literature, where the learning machines with high dimensional inputs are considered, here we consider only one dimensional case due to the following reasons. Firstly in the one dimensional case, we are able to carry out a rigorous calculation of the mean and variance of the generalization error. Secondly in the one dimensional case, we could fully understand why and how the support vector machine outperforms the worst learning machine and gain insights onto how to further improve the generalization capability of a learning machine (see Discussion).

Let us first introduce the model here. Suppose that the target function is $x = 0$, i.e. the correct separation function (target hyperplane, is $\text{sign}(x)$. After learning t examples from $A_1 = \{x(i) > 0, i = 1, \cdots, t\}$ and $A_2 = \{y(i) < 0, i = 1, \cdots, t\}$, a new coming signal $\xi(t+1)$ is sampled from $U(0,1)$, the uniform distribution over $[0,1]$, with probability 1/2 and $U(-1,0)$ with probability 1/2. The generalization error is defined by

$$\epsilon(t) = P(0 \leq \xi(t+1) \leq x_0|\mathcal{F}_t)I_{\{x_0>0\}} + P(0 \geq \xi(t+1) \geq x_0|\mathcal{F}_t)I_{\{x_0<0\}} \quad (1)$$

where $\mathcal{F}_t$ is the sigma-algebra generated by $x(i), y(i), i = 1, \cdots, t$ and I_A is the indication function of the set A, i.e. $I_A(x) = 1$ if $x \in A$ and 0 otherwise.

Denote

$$x(tt) = \min\{x(i), i = 1, \cdots t\} \qquad y(tt) = \max\{y(i), i = 1, \cdots, t\}$$

for the SVM the separation hyperplane is given by

$$x_0 = \frac{x(tt) + y(tt)}{2}$$

for the worst learning machine the separation hyperplane is given by

$$x_0 = x(tt)$$

In the literature the expectation of $\epsilon(t)$ is called the generalization error. Here since we are able to calculate not only the mean of $\epsilon(t)$, but also the variance etc., we prefer to call $\epsilon(t)$ the generalization error, which is a random variable.

3 Generalization Errors: Symmetric Cases

The basic idea is to apply the extremal value theory in statistics to estimating the generalization error. To this end, we first introduce a lemma here[1].

Lemma 1. *Suppose that $x(i) \sim U(0,1)$, the uniform distribution over $[0,1]$, is identically and independently distributed for $i = 1, \cdots, t$. When $t \to \infty$ we have*

$$P(x(tt) \geq \frac{x}{t}) = \exp(-x), \qquad x > 0 \tag{2}$$

$$\langle x^k(tt)\exp(-\alpha x(tt))\rangle = \frac{k!}{(\alpha+1)^{k+1}t^k}, \qquad \alpha > 0, k = 1, 2, \cdots, \tag{3}$$

where $\langle \cdot \rangle$ denotes the expectation. In other words, the distribution density of $x(tt)$ is $h(x) = t\exp(-tx)$.

Proof. From example 1.7.9 in[8] we know that $P(\eta(tt) \leq 1 - x/t) = \exp(-x)$ for $\eta(tt)$ representing the largest maximum of $x(i)$. Then Eq. (2) is a simple consequence of the symmetry between 1 and 0 of the uniform distribution. In terms of Eq. (2) we have

$$\langle x^k(tt))\exp(-\alpha x(tt))\rangle = \int x^k t \exp(-(\alpha+1)tx)dx = \frac{k!}{(\alpha+1)^{k+1}t^k}$$

Lemma 1 simply tells us the asymptotic distribution of $x(tt)$ when t is large enough. Let $\alpha = 0$ in Eq. (3), we see that $\langle x^k(tt)\rangle$ conveges to zero with a rate of $1/t^k$. For a given random sequence $x(i)$, we could calculate its exact distribution rather than its asymptotic distribution, which will provide us with further information on its behaviour with small samples[7].

From now on we assume that both $x(i)$ and $y(i)$ are uniformly distributed random variables and will report further work in [7]. The generalization error of the SVM defined in the previous section can now be rewritten as a function of extremal values

$$\begin{aligned}\epsilon(t) &= \frac{1}{2}P(0 \leq \xi(t+1) \leq \frac{x(tt)+y(tt)}{2}I_{\{x(tt)+y(tt)>0\}}|\mathcal{F}_t) \\ &\quad + \frac{1}{2}P(0 > \xi(t+1) \geq \frac{x(tt)+y(tt)}{2}I_{\{x(tt)+y(tt)<0\}}|\mathcal{F}_t) \\ &= \frac{1}{2}\frac{x(tt)+y(tt)}{2}I_{\{x(tt)+y(tt)>0\}} - \frac{1}{2}\frac{x(tt)+y(tt)}{2}I_{\{x(tt)+y(tt)<0\}}\end{aligned}$$

Here we have used the fact that $x(i)$ is uniformly distributed. Due to the symmetry between $x(i)$ and $y(i)$ we further conclude that

$$\langle \epsilon^k(t)\rangle = \langle [\frac{x(tt)}{2}I_{\{x(tt)+y(tt)>0\}} + \frac{y(tt)}{2}I_{\{x(tt)+y(tt)>0\}}]^k\rangle \qquad k = 1, 2, \cdots \tag{4}$$

[1] In the sequence, we take the convention that all terms of order $O(\exp(-t))$ in an equality are omitted

We first consider the mean of the generalization error by calculating the mean of each term in the equation above. The first term is

$$\begin{aligned}\langle\frac{x(tt)}{2}I_{\{x(tt)+y(tt)>0\}}\rangle &= \langle\frac{x(tt)}{2}\int_{-x(tt)}^{0} t\exp(ty)dy\rangle \\ &= \langle\frac{x(tt)}{2}(1-\exp(-tx(tt)))\rangle \\ &= [\frac{1}{2t}-\frac{1}{8t}] = \frac{3}{8t}\end{aligned}$$

The second term turns out to be

$$\begin{aligned}\langle\frac{y(tt)}{2}I_{\{x(tt)+y(tt)>0\}}\rangle &= \langle\frac{y(tt)}{2}\int_{-y(tt)}^{1} t\exp(-tx)dx\rangle \\ &= \langle\frac{y(tt)}{2}[\exp(ty(tt))-\exp(-t)]\rangle \\ &= -[\frac{1}{8t}-\frac{1}{2t}\exp(-t)]\end{aligned}$$

Therefore we have the following conclusion.

Theorem 1. *The mean of the generalization error of the support vector machine is*

$$\langle\epsilon(t)\rangle = \frac{1}{4t} \tag{5}$$

Although the proof of Theorem 1 is almost straightforward, it is very interesting to see the implications of its conclusion. In the literature, different upper bounds for the mean of the generalization error of the support vector machine have been found (see for example [9]). However, it seems the result of Theorem 1 is the first rigorous, and exact value of the mean. It is generally believed that the generalization error of the support vector machine is improved, in comparison with other conventional learning rules. How much does it exactly improve? We answer it in the following theorem.

Theorem 2. *For the worst learning machine, the mean of the generalization error is given by*

$$\langle\epsilon(t)\rangle = \frac{1}{2t}$$

Proof. Now the generalization error is simply given by

$$\epsilon(t) = x(tt)/2$$

which, combining with Lemma 1, implies the conclusion of the theorem.

It is well known in the literature that the mean of the generalization error of a learning machine decays at a rate of $O(1/t)$, independent of the distribution of input samples. The mean of the generalization error of both the support vector machine and the worst learning machine is of order $1/t$, as we could expect. The illuminating fact here is that the support vector machine improves the mean of the generalization error by a factor of $1/2$, in comparison with the worst learning machine. We want to emphasize here that the conclusion in Theorem 2 is independent of distributions, i.e. *universally* for the worst learning machine its generalization error is $1/t$ (see Lemma 3 in [5]for a proof). Nevertheless, for the support vector machine, the conclusions in Theorem 1 are obtained in terms of the assumptions of the uniform distribution of input samples. For any given distribution, we could calculate, as we developed in Theorem 1, its mean generalization error. The key and most challenging question is that whether the obtained conclusion is univeral, i.e. independent of input distribution, or not. A detailed analysis is outside the scope of the present letter and we will report it in [7].

In the literature the generalization error of the support vector machine is expressed in terms of the separation margin. We could easily do it here as well. Denote $d = x(tt) - y(tt)$ as the separation margin.

Theorem 3. *The mean of the generalization error of the support vector machine is*

$$\langle \epsilon(t) \rangle = \frac{\langle d \rangle}{8} \tag{6}$$

Proof. Since $\langle x(tt) \rangle = 1/t$, the conclusion follows.

So far we have known that in terms of the mean of the generalization error, the support vector machine improves the performance. How is the variance of the generalization error of the support vector machine, in comparison with conventional learning rules? To the best of our knowledge, there is no report in the literature to successfully calculate the variance of the generalization error. Due to Lemma 1, we have the following conclusions.

Theorem 4. *For the worst learning machine its variance of the generalization error is*

$$var(\epsilon(t)) = \frac{1}{4t^2}$$

For the support vector machine we have

$$var(\epsilon(t)) = \frac{1}{16t^2}$$

Proof. We only need to calculate $\langle \epsilon^2(t) \rangle$ of the support vector machine. For Eq. (4) we have

$$\begin{aligned} \langle x^2(tt) I_{\{x(tt)+y(tt)>0\}} \rangle &= \langle x^2(tt) \int_{-x(tt)}^{0} yt \exp(ty) dy \rangle \\ &= \langle x^2(tt)[1 - \exp(-tx(tt))] \rangle \\ &= [\frac{2}{t^2} - \frac{1}{4t^2}] = \frac{7}{4t^2} \end{aligned}$$

and

$$
\begin{aligned}
\langle y^2(tt)I_{\{x(tt)+y(tt)>0\}}\rangle &= \langle y^2(tt)\int_{-y(tt)}^{1} t\exp(-tx)dx\rangle \\
&= \langle y^2(tt)(\exp(ty(tt)) - \exp(-t))\rangle \\
&= [\frac{1}{4t^2} - \frac{2}{t^2}\exp(-t)]
\end{aligned}
$$

Furthermore

$$
\begin{aligned}
\langle x(tt)y(tt)I_{\{x(tt)+y(tt)>0\}}\rangle &= \langle x(tt)\int_{-x(tt)}^{0} yt\exp(ty)dy\rangle \\
&= \langle x(tt)[x(tt)\exp(-tx(tt)) - \int_{-x(tt)}^{0}\exp(ty)dy]\rangle \\
&= \langle x(tt)[x(tt)\exp(-tx(tt)) - \frac{1}{t}(1-\exp(-tx(tt)))]\rangle \\
&= [\frac{1}{4t^2} - \frac{1}{t^2} + \frac{1}{4t^2}]
\end{aligned}
$$

which gives the desired results.

In words, the support vector machine also improves the standard deviation of the generalization error by a factor of 1/2, comparing with the worst leaning machine. As aformentioned it seems results on $var(\epsilon(t))$ have not been reported in the literature.

We could go further to estimate the distribution density of the generalization error. However, from the fact that the mean and the standard deviation of the generalization error are equal to each other, we could guess that the distribution density of the generalization error is negatively distributed with the parameter $\langle\epsilon(t)\rangle$, for both the worst learning machine and the support vector machine, a conclusion which is proved in [7].

In summary, under some assumptions on its input distributions (see [7] as well), we grasp a complete picture of the generalization behaviour of the one diminsional support vector machine.

4 Generalization Error: Non-symmetric Cases

In the previous sections we have considered the support vector machine with symmetric input distributions. Certainly we do not expect that $x(i)$ and $y(i)$ are identically distributed in problems arising from practical applications. In this section we assume that $yL \sim U(-L, 0)$ and the generalization error is then

$$
\epsilon(t) = \frac{1}{2}\frac{x(tt)+Ly(tt)}{2}I_{\{x(tt)+Ly(tt)>0\}} - \frac{1}{2L}\frac{x(tt)+Ly(tt)}{2}I_{\{x(tt)+Ly(tt)<0\}} \tag{7}
$$

where $L > 0$ is a constant. The first term in Eq. (7) is

$$
\begin{aligned}
\langle\frac{x(tt)}{2}I_{\{x(tt)+Ly(tt)>0\}}\rangle &= \langle\frac{x(tt)}{2}\int_{-x(tt)/L}^{0} t\exp(ty)dy\rangle \\
&= \langle\frac{x(tt)}{2}(1-\exp(-tx(tt)/L))\rangle \\
&= [\frac{1}{2t} - \frac{L^2}{2(L+1)^2 t}]
\end{aligned}
$$

and the second term

$$\begin{aligned}\langle\frac{Ly(tt)}{2}I_{\{x(tt)+Ly(tt)>0\}}\rangle &= \langle\frac{Ly(tt)}{2}\int_{-Ly(tt)}^{1} t\exp(-tx)dx\rangle \\ &= \langle\frac{Ly(tt)}{2}(\exp(tLy(tt)) - \exp(-t))\rangle \\ &= -[\frac{L}{2(L+1)^2t} - \frac{L}{2t}\exp(-t)]\end{aligned}$$

Similarly for the third and the fourth term we have

$$\begin{aligned}\langle\frac{x(tt)}{2}I_{\{x(tt)+Ly(tt)<0\}}\rangle &= \langle\frac{x(tt)}{2}\int_{-1}^{-x(tt)/L} t\exp(ty)dy\rangle \\ &= \langle\frac{x(tt)}{2}(\exp(-tx(tt)/L) - \exp(-t))\rangle \\ &= \frac{L^2}{2(L+1)^2t}\end{aligned}$$

and

$$\begin{aligned}\langle\frac{Ly(tt)}{2}I_{\{x(tt)+Ly(tt)<0\}}\rangle &= \langle\frac{Ly(tt)}{2}\int_{0}^{-Ly(tt)} t\exp(-tx)dx\rangle \\ &= \langle\frac{Ly(tt)}{2}(1 - \exp(tLy(tt)))\rangle \\ &= -[\frac{L}{2t} - \frac{L}{2(1+L)^2t}]\end{aligned}$$

Summing together we finally obtain

$$\langle\epsilon(t)\rangle = \frac{L}{4(1+L)t} + \frac{1}{4(1+L)t} = \frac{1}{4t} = \frac{\langle d\rangle}{4(1+L)} \tag{8}$$

where $d = x(tt) - Ly(tt)$.

Eq. (8) tells us that the mean of the generalization error is the same as the symmatic case and is proportional to $1/(1+L)$, where $1+L$ is conventionllly thought of as the gap between support vectors. It is somewhat surprising to note that the mean of the generalization error of the support vector machine is independent of the scaling of the input distribution. From the proof of Eq. (8) we see that the summation of the first and second term of Eq. (7) is not equal to the summation of the third and the fourth term, much as $\langle\epsilon(t)\rangle$ is independent of L. This reveals one of the difficulties to prove a general conclusion as developed in [5]. We have obtained a general conclusion and we will report it in [7].

5 Non-symmetric SVM

For the nonsymmetric case considered in the previous section, if the separation hyperplane is again $[x(tt) + y(tt)]/2$ then the generalization error is

$$\epsilon(t) = \frac{1}{2}\frac{x(tt)+y(tt)}{2}I_{\{x(tt)+y(tt)>0\}} - \frac{1}{2L}\frac{x(tt)+y(tt)}{2}I_{\{x(tt)+y(tt)<0\}}$$

From Theorem 1 we know that

$$\langle \epsilon(t) \rangle = \frac{1}{8t} + \frac{1}{8tL}$$

When L is large we then have

$$\langle \epsilon(t) \rangle = \frac{1}{8t}$$

a further reduction of the mean of the generalization error is achieved. The similar result is true for the variance of the generalization error.

The idea above can be implememted in the following way. We assume that A_1 and A_2 are the data set to be learnt.

1. According to the support vector machine algorithm, we obtain the separation hyperplane s_1.
2. Calculate the distance beween the mean of A_1, A_2 and s_1, denoting as d_1 and d_2 respectively. When L is large, we have $d_1 = 1/2 - [x(tt) + Ly(tt)]/2 \sim 1/2$ and $d_2 = L/2 + [x(tt) + Ly(tt)]/2 \sim L/2$.
3. In parallel with the hyperplane s_1, we find a new hyperplane s_2 so that

$$\frac{c_1}{d_1} = \frac{c_2}{d_2} \tag{9}$$

where c_1 and c_2 are the distance between s_2 and A_1 and A_2 respectively. We have $c_1 = x(tt) - [x(tt) + y(tt)]/2 = [x(tt) - y(tt)]/2$ and $c_2 = [x(tt) + y(tt)] - Ly(tt) \sim -Ly(tt)$. Hence when $s_2 = [x(tt) + y(tt)]/2$, we have Eq. (9).

Since in general the obtained separation hyperplane is not symmetric about the support vectors (maximization of the separation margin), we call s_2 the separation hyperplane of a non-symmetric support vector machine. The fact that the non-symmetric support vector machine improves the mean of the generalization error of s_1 could be easily understood. The support vector machine use only the information contained in the support vectors, while the non-symmetric support vector machine explore the information of the whole data set since the mean of the data is also taken into account.

6 Discussion

By virtue of the extremal value theory, we present here a novel approach to calculate the mean and variance of the generalization error of the support vector machine and the worst learning machine. The exact mean and variance of the generalization error are obtained. To estimate upper bounds for the support vector machine is currently a very active topic. Our results reveal, for the first time in ther literature, that how much the SVM improves the generalization error, comparing with other learning algorithms. Much as we consider a very simple case here, our results could also be used as a criteria to check how tight an estimated upper bound of general cases is (see [9, 3] and references therein). The extremal value theory is somewhat similar to the central limit theorem: a powerful and universal theorem and is almost independent of the sample distributions (see [8, 4, 5] for details). We hope the new techniques introduced here could help us to clarify some issues related to the SVM. Some of them we would like to further pursue in future publications are the following.

- The support vector machine improves the mean of the generalization error by a factor of a constant. From the calculations presented in the paper, we see that the next term in the mean of the generalization error is of order $\exp(-t)$. To find a learning algorithm with the mean of the generalization of order $\exp(-t)$–an exponential machine– would be a real breakthrough in the field. The approach presented here provides us with such a possibility.
- We only consider the case of one dimension. Certainly it is interesting to consider the models of high dimension. We will report it elsewhere[7].
- The extremal values are more sensitive to perturbations than other statistical quantities such as the mean or median of samples. It is therefore interesting to carry out a study on how the SVM relies on perturbations.

Acknowledgement. The work was partially supported by BBSRC and a grant of the Royal Society.

References

1. Brown M., Grundy W., Lin D., Cristianini N., Sugnet C., Furey T., Ares Jr. M., and, Haussler D.(1999) Knowledge-based Analysis of Microarray Gene Expression Data using Support Vector Machines. *Proceedings of the National Academy of Sciences* **97** 262-267.
2. Cristianini, N., and, Shawe-Taylor J. (2000) *An introduction to support vector machines* Cambridge University Press: Cambridge UK.
3. Dietrick R., Opper M., Sompolinsky H. (1999) Statistical mechanics of support vector networks *Phys. Rev. Letts*, **82** 2975-2978.
4. Feng, J.(1997) Behaviours of spike output jitter in the integrate-and-fire model. *Phys. Rev. Lett.* **79** 4505-4508.
5. Feng, J.(1998) Generalization errors of the simple perceptron. *J. Phys. A.* **31**, 4037-4048.
6. Feng, J., and Williams P. (2001) Calculation of the generalization error for various support vector machines *IEEE T. on Neural Networks* (in press).
7. Feng, J., and Williams P. (2001) Support vector machines-a theoretical and numerical study (in Prepartion)
8. Leadbetter, M.R., Lindgren, G. & Rootzén, H.(1983) *Extremes and Related Properties of Random Sequences and Processes,*Springer-Verlag, New York, Heidelberg, Berlin.
9. Vapnik V., and, Chapelle O.(2000) Bounds on error expectation for support vector machines *Neural computation* **12** 2013-2036.

Natural Gradient Learning in NLDA Networks

José R. Dorronsoro, Ana González, Carlos Santa Cruz *

Depto. de Ingeniería Informática and Instituto de Ingeniería del Conocimiento
Universidad Autónoma de Madrid, 28049 Madrid, Spain

1 Introduction

Neural network training is usually formulated as a problem in function minimization. More precisely, if W are the weights defining a network's architecture and $e(W)$ is the weight depending error function, its gradient $\nabla e(W)$ is usually employed to arrive at the optimal weight set W^*. There may be several ways of exploting this information and the simplest is just plain gradient descent, which assumes an "euclidean" structure in the underlying space of the W weights. Although very natural, this may result sometimes in quite slow network learning in some problems, both in batch and, especially, on line error minimization, where the global error function $e(W)$ is replaced by an individual, Z pattern depending error function $e(Z, W)$. Several procedures such as adaptive learning rates or the addition of momentum terms have been proposed [6]. A different approach is suggested by the fact that in some instances, there may be metrics other than the euclidean one better suited to describe weight space. This has been shown to be the case for a related problem, likelihood estimates for parametric probability models [1, 4], for which a Riemannian structure can be defined in weight space. The same reasoning can be applied for a concrete network model, Multilayer Perceptrons (MLPs). When used in regression problems, that is, when the MLP tries to establish a relationship between an input X and output y for each pattern $Z = (X, y)$, a probability model $p(Z; W) = p(X, y; W)$ can be defined in pattern space so that the on line MLP error function $e(Z, W) = e(X, y, W) = (y - F(X, W)^2/2$ is seen as the log–likelihood of $p(Z; W)$; here $F(X, W)$ denotes the network's transfer function. This allows one to recast network learning as the likelihood estimation of a certain semi–parametric probability density $p(X, y, W)$. In this setting, there is [2] a natural Riemannian metric on the space $\{p(X, y; W) : W\}$ of these densities, determined by a metric tensor given by the matrix

$$G(W) = E\left[(\nabla_W \log p)(\nabla_W \log p)^t\right] = \int\int \frac{\partial \log p}{\partial W}\left(\frac{\partial \log p}{\partial W}\right)^t p(X, y; W) dX dy.$$

$G(W)$ is also known as the Fisher Information matrix, as it gives the variance of Cramer–Rao bound for the optimal parameter estimator. This suggests to use the

* All authors partially supported by Spain's CICyT, grant TIC 98–965

J. Mira and A. Prieto (Eds.): IWANN 2001, LNCS 2084, pp. 427-434, 2001.

"natural" gradient in the Riemannian setting, that is, $G(W)^{-1}\nabla_W e(X, y; W)$, instead of the ordinary euclidean gradient $\nabla_W e(X, y; W)$.

This approach has received recently considerable attention in the literature for MLPs in regression and classification problems, and in other settings such as Independent Component Analysis and Systems Space [2]. Among other results, it has been experimentally checked that the natural gradient can noticeably speed up the convergence of on line MLP training, in the sense that many less training epochs (i.e., presentations of individual patterns) are needed (see [9] for an alternative, statistical mechanics analysis of convergence in this case). This suggests that, in other settings, gradient descent network learning can also be improved by an adequate modification of gradient values. Its principled application would require the definition of an appropriate Riemannian structure on the weight space of the concrete network transfer function to be used, which would need first the definition of a certain distance between two network transfer functions [3] and then the computation of the appropriate metric. This may be quite difficult in many instances and alternative ways of introducing natural gradients have appeared in the literature.

For instance, if the network's error function is given by a distance $d(W) = d(f(X, W), t)$ between the network's transfer function $f(X, W)$ and targets t, a distance $D(W, W') = d(f(X, W), f(X, W'))$ is introduced in [8] between the transfer functions corresponding to different weights W, W'. It is then shown that the appropriate weight update formula is $\Delta W = -\eta H^{-1}\nabla d(W)$, where $H = H(W)$ is the hessian of the distance D. In general, H will not coincide with a true Fisher information matrix, although it will when the distance d is given as a log–likelihood, but, on the other hand, the same learning accelerating effect has been reported. Another way of introducing a kind of Fisher matrix (the one we will follow here) is possible when the gradient $\nabla_W J$ of a network's error function J can be expressed as the expectation $\nabla_W J = E[Z(X, W)]$ of a certain random vector $Z = Z(X, W)$ defined in terms of the network's inputs and weights. A natural extension of the Fisher information matrix can be then defined as the covariance

$$\mathcal{I} = E[(Z(X, W) - \overline{Z})(Z(X, W) - \overline{Z})^t]$$

of the $Z(X, W)$ function. Again, this will coincide with the usual Fisher information matrix in the log–likelihood network error case, but it has the advantage that it can be defined even when network training is not formulated in terms of errors with respect to some targets.

In this work we shall consider one such case, the training of Non Linear Discriminant Analysis Networks (NLDA). In them a new feature set is extracted from D–dimensional patterns X using also a Multilayer Perceptron (MLP) architecture, but the optimal network weights are selected as those minimizing not the usual MLP square error but a Fisher discriminant criterion function instead. More precisely, consider the simplest possible MLP architecture for a two class problem. It will have D input units, a single hidden layer with H units and a single, linear output unit; recall that $C-1$ outputs are used in Fisher's linear discriminants for a C class problem. We denote network inputs as $X = (x_1, \ldots, x_D)^t$

(we may assume that $x_D \equiv 1$ for bias effects), the weights connecting these D inputs with the hidden unit h as $W_h^H = (w_{1h}^H, \ldots, w_{Dh}^H)^t$ and the weights connecting the hidden layer to the single output by $W^O = (w_1^O, \ldots, w_H^O)^t$. As in an standard MLP, the network transfer function $F(X, W)$ is thus given by $y = F(X, W) = (W^O)^t O$, with $O = O(X, W) = (o_1, \ldots, o_H)^t$ the outputs of the hidden layer, that is, $o_h = f((W_h^H)^t X)$; we shall use here as f the sigmoidal function. The selection of the optimal weights $W_* = (W_*^O, W_*^H)$ is done in terms of the minimization of the criterion function $J(W) = s_T/s_B$. Here $s_T = E[(y - \overline{y})^2]$ is the total covariance of the output y and $s_B = \pi_1(\overline{y}_1 - \overline{y})^2 + \pi_2(\overline{y}_2 - \overline{y})^2$ the between class covariance. By $\overline{y} = E[y]$ we denote the overall output mean, $\overline{y}_c = E[y|c]$ denotes the output class conditional means and π_c the class prior probabilities. Notice that no targets are used in defining $J(W)$.

The relationship between MLPs and classical linear Fisher analysis has been pointed out by several authors. For instance, it is proven in [11] that if the vectors $(0, \ldots, 1, \ldots, 0)^t$ are used as targets, the optimal weights of an MLP with linear output connections maximize the quantity $\mathrm{tr}((S_T^O)^{-1} \tilde{S}_B^O)$. Here $S_T^O = E[(O - \overline{O})(O - \overline{O})^t]$ is the total covariance of the outputs O of a network's last hidden layer, and $\tilde{S}_B^O = \sum_c \pi_c^2 (\overline{O}_c - \overline{O})(\overline{O}_c - \overline{O})^t$ is a weighted version (notice the π_c^2 factors) of the between class covariance. For two class problems $\tilde{S}_B^O$ essentially coincides with the ordinary between class covariance S_B^O. Hence, in this concrete setting the optimal weights of a 2 hidden layer MLP with a $D \times H \times 1 \times 2$ architecture, linear connections from the first hidden layer and targets $(1, 0)$, $(0, 1)$, define at the single unit of the second hidden layer, the same transformation than a $D \times H \times 1$ NLDA network. However, besides having a simpler architecture, NLDA networks allow a more general and flexible use of any other of the criterion functions customarily employed in linear Fisher analysis. In [7, 10] other properties of NLDA networks are studied, and it is shown that the features they provide can be more robust than those obtained through MLPs in problems where there is a large class overlapping and where class sample sizes are markedly unequal.

As defined above, the minimization of an NLDA criterion function is a problem of batch learning, in the sense that second order moment information is needed. Thus global optimization procedures such as conjugate gradients or quasi–Newton methods can be used. However, the weight space geometry of the learning surface of an NLDA network can be quite complicated, which may result in numerical difficulties for some components of these procedures, such as line minimizations. Simple gradient descent learning can overcome some of these problems, although perhaps adding some of its own, particularly very long converging times. This makes worthwhile to investigate the possibility of accelerating gradient descent, something we shall do here in terms of a Fisher–like matrix. The paper is organized as follows. In the next section we shall briefly review gradient descent learning for NLDA networks. We shall derive the required gradient computations and write them as random variable expectations, from which we shall derive the Fisher matrix and natural gradient descent. The third section illustrates the use of these procedures and the paper ends with a short discusion.

2 Natural gradient descent in NLDA networks

As mentioned before, two weight sets are involved in our simplest NLDA network: the hidden weights W^H connecting inputs with the single hidden layer, and the output weights W^O connecting the hidden layer with the single output unit. Once the hidden layer unit outputs O are computed, the W^O can be simply obtained performing standard Fisher's linear discriminant analysis. More concretely, we compute first for each class the averages $\overline{O}_1, \overline{O}_2$ of these hidden layer outputs; the optimal Fisher weights are then given by $W^O = (S_T^O)^{-1}(\overline{O}_1 - \overline{O}_2)$, with S_T^O the total covariance matrix of the O outputs. As it is well known, the Fisher weights are unique up to dilations, and we shall normalize them to unit norm in what follows. These considerations show that in some sense, the weights W^H determine the values of the W^O weights: in fact, they establish for each input X the values $o_h = f((W_h^H)^t X)$ of the hidden layer outputs that, in turn, determine the Fisher weights. This suggests the following general setting for standard gradient descent in NLDA networks: starting with a given hidden weight set W_t^H at time t, we update it to W_{t+1}^H as follows:

1. Compute the hidden output O values and then the associated Fisher weights W_{t+1}^O by the previous procedure.
2. Compute the new weights W_{t+1}^H as

$$W_{t+1}^H = W_t^H - \eta_t \nabla_{W^H} J_t^H(W_t^H), \tag{1}$$

   where $J_t^H(W^H)$ represents the general criterion function $J(W^H, W^O) = s_t/s_B$ when the weights W^O are clamped at the W_{t+1}^O values. Here η_t denote a possibly time varying learning rate.

To apply this gradient descent procedure we need to compute the gradient $\nabla_W J_t^H(W)$, something we do next. To simplify the subsequent notations, we shall drop the t reference and the H superscript, writing with a slight abuse $J(W)$ instead of $J_t^H(W) = J(W, W_{t+1}^O)$.

The above computation of the output weights W^O show that they are invariant with respect to translations in the last hidden layer. In particular, we can center around zero the hidden outputs without affecting the Fisher weights W^O or the value of J. We will thus assume that at the last hidden layer, $\overline{O} = 0$, which implies that the network outputs y will have zero mean as well, for $\overline{y} = (W^O)^t \overline{O}$. Under these simplifications, we therefore have $s_T = E[y^2]$, $s_B = \pi_1(\overline{y}_1)^2 + \pi_2(\overline{y}_2)^2$. It is easy now to see that

$$\frac{\partial s_T}{\partial w_{kl}} = 2E\left[y \frac{\partial y}{\partial w_{kl}}\right] = 2w_l E\left[y f'(a_l) x_k\right];$$

$$\frac{\partial s_B}{\partial w_{kl}} = 2\sum_c \pi_c \overline{y}_c E_c\left[\frac{\partial y}{\partial w_{kl}}\right] = 2w_l \sum_c \pi_c \overline{y}_c E_c\left[f'(a_l) x_k\right],$$

where we recall that $E_c[z] = E[z|c]$ denotes class conditional expectations. Therefore, we can compute the gradient of J as follows:

$$\frac{\partial J}{\partial w_{kl}} = \frac{1}{s_B}\left(\frac{\partial s_T}{\partial w_{kl}} - J\frac{\partial s_B}{\partial w_{kl}}\right)$$
$$= \frac{2w_l}{s_B}\left(E\left[yf'(a_l)x_k\right] - J\sum \pi_c \overline{y}_c E_c\left[f'(a_l)x_k\right]\right). \quad (2)$$

Let us now define $\Psi = (\Psi_{kl})$ as

$$\Psi_{kl}(X, W) = \frac{2w_l}{s_B} yf'(a_l)x_k - \lambda_{kl}(W) = z_{kl}(X, W) - \lambda_{kl}(W),$$

where the nonrandom term $\lambda_{kl}(W)$ is given by

$$\lambda_{kl}(W) = \frac{2w_l}{s_B} J(W) \sum_c \pi_c \overline{y}_c(X, W) E_c[f'(a_l(X, W))x_k].$$

Now (2) shows that ∇J can be written as the expectation of the Ψ_{kl}:

$$\frac{\partial J}{\partial w_{kl}}(W) = E[\Psi_{kl}(X, W)] = E[z_{kl}] - \lambda_{kl}.$$

In particular, we can define the covariance matrix $\mathcal{I}$ of the Ψ_{kl} as

$$\mathcal{I}_{(kl)(mn)} = (\mathcal{I})_{(kl)(mn)}(W) = E[(\Psi_{kl} - \overline{\Psi}_{kl})(\Psi_{mn} - \overline{\Psi}_{mn})]$$
$$= E[(z_{kl} - \overline{z}_{kl})(z_{mn} - \overline{z}_{mn})], \quad (3)$$

and therefore $(\mathcal{I})_{(kl)(mn)} = cov(z_{kl}z_{mn})$. We shall use the matrix $\mathcal{I}$ as the Fisher matrix for NLDA natural gradient descent. Besides its definition as the covariance of the gradient of the error function, it has some other points in common with more usual appearances of the Fisher information matrix. For instance, it is shown in [5] that it if X_n is a sequence of i.i.d. random vectors distributed as X, the distribution of the weights $\widetilde{W}_N$ that solve the minimization equations

$$\frac{1}{N}\sum_{j=1}^{N}(z_{kl}(X_j, W) - \lambda_{kl}(W)) = \frac{1}{N}\sum_{j=1}^{N}\left(\frac{2w_l}{s_b} y^j f'(a_l^j)x_k^j - \lambda_{kl}(W)\right) = 0, \quad (4)$$

converges in probability to a minimum W^* of the criterion function. Moreover, $\sqrt{N}(\widetilde{W}_N - W^*)$ converges in distribution to a multivariate normal $N(0, \mathcal{C}^*)$, where $\mathcal{C}^* = (\mathcal{H}^*)^{-1}\mathcal{I}^*(\mathcal{H}^*)^{-1}$. Here $\mathcal{I}^*$ denotes the Fisher matrix evaluated at the minimum W^* and $\mathcal{H}^*$ is the Hessian of J.

We can now define natural gradient descent for NLDA networks simply changing the usual gradient descent formula (1) to

$$W_{t+1}^H = W_t^H - \eta_t \mathcal{I}^{-1} \nabla_W J_t(W_t^H). \quad (5)$$

We shall numerically illustrate its use in the next section.

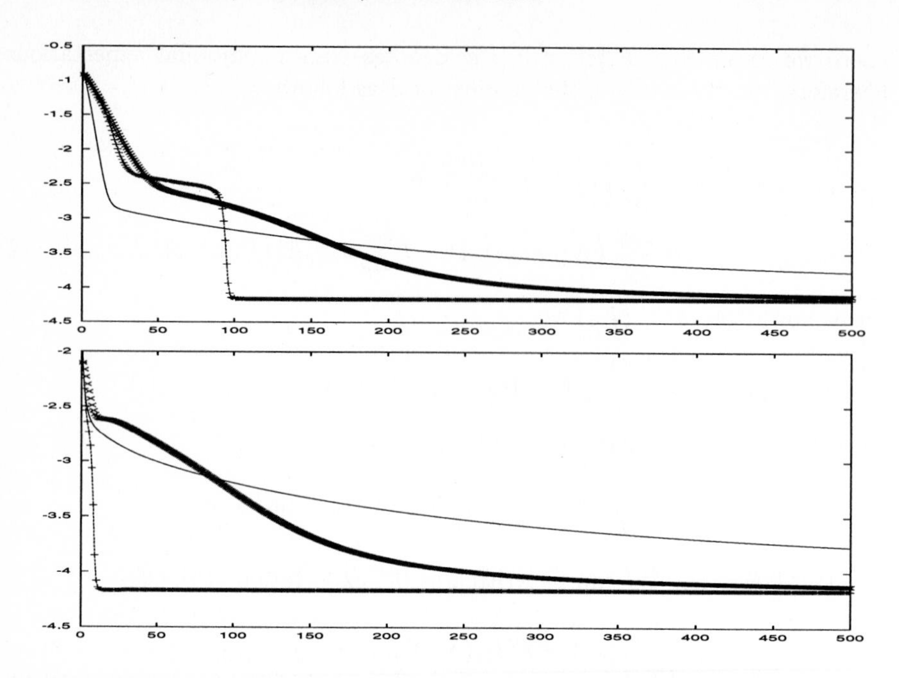

Fig. 1. Two different evolutions of the logarithm $\log(J-1)$ of the criterion function J for standard gradient (continuous line), and natural (crosses and lines) and simplified natural gradient (crosses). Only the first 500 epochs are shown. Convergence of natural gradient is somewhat slow (100 epochs) in the first figure although 5 times faster than that of its simplified version and 15 times faster than that of standard gradient. It is however much faster in the second figure (just 15 epochs).

3 A numerical illustration

We will apply the preceding techniques to a toy synthetic problem with 2 unidimensional classes. The first one, C_0, follows a $N(0, 0.5)$ distribution and the other one, C_1, is given by a mixture of two gaussians, $N(-2, 0.5)$ and $N(2, 0.5)$. The prior probabilities of these two gaussians are 0.5, which is also the prior probability for both classes C_0 and C_1. Sample sizes will be 1000 patterns for each class. As a classification problem, it is an "easy" one, for the mean error probability (MEP) of the optimal Bayes classifier has a rather low value of about 0.48 %. On the other hand, the class distributions are not unimodal, and neither linearly separable. NLDA networks will thus realize feature enhancing rather than feature extraction. It is easily seen that the optimal architecture for this problem just needs 2 hidden units. In fact, the optimal weight structure is quite easy to guess here: the weights associated to each one of the two hidden units will perform a pairing of patterns of the central, 0–centered class with each one of the ± 2 centered classes. Typical values of the hidden outputs center at $(0, 0)$ for the 0–centered gaussian, and at either $(1, 0)$ or $(0, 1)$ for each one of the

± 2–center gaussians. In this case, the value of the unit norm Fisher weights is about $(\pm 0.71, \mp 0.71)$. After training, the MEP of the δ_{NLDA} classifier coincides with the optimal Bayes value of 0.48.

To compute the gradient and the Fisher matrix we have used the sample versions of formulae (2) and (3) for ∇J and $\mathcal{I}$ respectively. That is, denoting by the superscript ^{cj} values corresponding to the j–th element of class c, and assuming N_c patterns in class c and $N = \sum N_c$, we have

$$\frac{\partial J}{\partial w_{kl}} \simeq \frac{2\hat{w}_l}{\hat{s}_B} \frac{1}{N} \sum_{c,j} y^{cj} f'(a_l^{cj}) x_k^{cj} - \hat{J} \frac{1}{N} \sum_c \frac{1}{N_c} \left(\sum_{j=1}^{N_c} y_c^{cj} \right) \left(\sum_{j=1}^{N_c} f'(a_l^{cj}) x_k^{cj} \right);$$

$$\hat{\mathcal{I}}_{(kl)(mn)} \simeq \frac{4\hat{w}_l \hat{w}_n}{(\hat{s}_B)^2} \frac{1}{N} \sum_{c,j} (y^{cj})^2 f'(a_l^{cj}) x_k^{cj} f'(a_m^{cj}) x_n^{cj} -$$
$$\frac{4\hat{w}_l \hat{w}_n}{(\hat{s}_B)^2} \frac{1}{N^2} \left(\sum_{c,j} (y^{cj}) f'(a_l^{cj}) x_k^{cj} \right) \left(\sum_{c,j} (y^{cj}) f'(a_m^{cj}) x_n^{cj} \right).$$

We denote by $\hat{w}_h$, $\hat{w}_{dh}$ the sample derived weights. These formulae show that the cost of a single gradient computation is $O(NDH)$. In the natural gradient case, the cost of computing the $DH \times DH$ Fisher matrix has also to be considered. This basically requires the computation of D^2H^2 values $\hat{\mathcal{I}}_{(kl)(mn)}$, each with an individual cost of $O(N)$. The cost of matrix inversion has to be added, which is $O(D^3H^3)$. Thus the total overhead of natural gradient with respect to standard gradient is $O(ND^2H^2 + D^3H^3)$. In practice $N >> DH$ is to be hoped, which brings the actual overhead to $O(ND^2H^2)$. A simple way to lower this cost is to approximate $\mathcal{I}$ by its diagonal $\tilde{\mathcal{I}} = \mathrm{diag}(\mathcal{I}_{(\mathrm{kl})(\mathrm{kl})})$, which reduces the natural gradient weight update to

$$W_{t+1}^H = W_t^H - \eta_t \frac{1}{\mathcal{I}_{(kl)(kl)}} \frac{\partial J_t}{\partial w_{kl}} (W_t^H). \tag{6}$$

The cost of using (6) becomes now $O(NDH)$ as only DH different terms have to be computed, which in practice is about twice the cost of the plain gradient updates.

In our experiments we have used the three updates (1), (5) and (6). Figure 1 presents the first 500 epochs of two evolutions of the three gradients. All of them converge to the optimal value -4.1 of the natural logarithm $\log(J-1)$. Since always $J \geq 1$, large negative values of $\log(J-1)$ mean that $J \simeq 1$ and that a good convergence has been achieved. In the first one, full natural gradient is somewhat "slow" requiring 100 epochs, while its diagonal version needs 500. On the other hand, it takes about 1500 epochs for the standard gradient to converge, despite the fact that its learning rate was 0.1 and that of the natural gradients was 0.01. In this case $D = 1$ and $H = 2$, so that the cost of the full natural gradient is about twice that of the simplified one. Nevertheless, this larger cost is not significant, given that the simplified natural gradient took 5 times as many epochs to converge. In any case, much faster convergence of the full natural gradient can be expected, as the second figure shows. In it, a different random initial weight set was chosen (uniform random values in $[-1, 1]$ were used in both figures). The full natural gradient converges after just 15 epochs, while its simplified version needs again 500 (about 30 times more) and standard gradient about 1500 epochs (100 times more).

4 Conclusions

In this work we have shown a possible way to define natural gradients for NLDA networks using as a Fisher matrix the covariance of the random variable that defines the gradient of an NLDA criterion function. Although very simple, the proposed approach shows very good convergence acceleration properties and suggests the interest of further studying the applicability of natural gradient learning for NLDA networks. Among the topics that will receive future attention will be more precise definitions of the Fisher matrix, for which the asymptotic distribution theory of optimal weights may suggest worthwhile alternatives. Another question is the appropriate choice of learning rates. Here fixed values have been used, but adaptative choices may result in still faster convergence.

References

1. S. Amari, **Differential Geometric Methods in Statistics**, Lecture Notes in Statistics 28, 1985.
2. H. Park, S. Amari, K. Fukumizu, "Adaptive Natural Gradient Learning Algorithms for Various Stochastic Models", Neural Networks 13 (2000), 755–764.
3. S. Amari, "Natural Gradient Works Efficiently in Learning", Neural Computation 10 (1998), 251–276.
4. M. Murray, J.W. Rice, **Differential Geometry and Statistics**, Chapman & Hall, 1993.
5. J.R. Dorronsoro, A. González, C. Santa Cruz, "Arquitecture selection in NLDA networks", submitted to ICANN 2001, Vienna, August 2001.
6. R.O. Duda, P.E. Hart, D.G. Stork, "Pattern classification" (second edition), Wiley, 2000.
7. J.R. Dorronsoro, F. Ginel, C. Sánchez, C. Santa Cruz, "Neural Fraud Detection in Credit Card Operations", IEEE Trans. in Neural Networks 8 (1997), 827–834.
8. T. Heskes, "On natural Learning and pruning in multilayered perceptrons", Neural Computation 12 (2000), 1037–1057.
9. M. Rattray, D. Saad, S. Amari, "Natural gradient descent for on–line learning", Physical Review Letters 81 (1998), 5461–5464.
10. C. Santa Cruz, J.R. Dorronsoro, "A nonlinear discriminant algorithm for feature extraction and data classification", IEEE Transactions in Neural Networks 9 (1998), 1370–1376.
11. A.R. Webb, D. Lowe, "Optimized Feature Extraction and the Bayes Decision in Feed–Forward Classifier Networks", IEEE Trans. on Pattern Analysis and Machine Intelligence 13 (1991), 355–364.

AUTOWISARD: Unsupervised Modes for the WISARD

Iuri Wickert and Felipe M. G. França

Computer Systems Engineering Program - COPPE, UFRJ
Caixa Postal 68511, 21945-970, Rio de Janeiro, RJ, Brazil
{iuri, felipe}@cos.ufrj.br

Abstract. This work introduces two new unsupervised learning algorithms based on the WISARD weightless neural classifier model. The first one, the standard AUTOWISARD model, is able to perform fast one-shot learning of unsorted sets of input patterns. The second one is a recursive version of AUTOWISARD which not only keeps the good features of the basic model, agility, plasticity and stability, but also produces hierarchically structured tree-like classifications. Although the standard AUTOWISARD model exhibits good classification skills when exposed to symbolic patterns (by producing only few classes containing patterns of different meanings), its recursive version produces even less confusing classes. The stability of both learning algorithms is also demonstrated.

1 Introduction

The WISARD is a well-known weightless neural network model, based on a very simple neuron model (RAM) and architecture (class discriminators) [1], [2]. Despite its simplicity, this model provides good generalisation and is able to perform fast, one-shot learning of binary input patterns. The main purpose of this work is to introduce new unsupervised learning algorithms for this neural network model, with the intention to bring together the good features of both WISARD and ART models [3].

WIS-ART, a hybrid approach where both models coexist, was firstly introduced by Fulcher [4]. The new AUTOWISARD models introduced here do not include the ART model explicitly, like in WIS-ART, but have embedded the essence of its behaviour within the WISARD model. Moreover, as it will be shown later in this text, the ability of the ART model on representing classes of patterns through vector quantisation (stylisation) is generalised in the AUTOWISARD models in the sense that each class can be defined by a family of vectors. This may lead to a greater richness in terms of the topology of the frontiers among the different resulting classes.

The following is how the remainder of this paper is organised. A quick overview of the WISARD and ART models is given in the next section. Section 3 introduces the basic AUTOWISARD model, including the proof of its stability, what is true for both AUTOWISARD models. Next, in Section 4, the

J. Mira and A. Prieto (Eds.): IWANN 2001, LNCS 2084, pp. 435-441, 2001.

recursive version of AUTOWISARD is described. The natural generation of tree-like structured classifications over binary input patterns is demonstrated and its performance analysed. The experiments carried out are presented in Section 5. Section 6 includes a discussion over the obtained results. The conclusions are finally presented in Section 7.

2 The WISARD Neural Network

WISARD (Wilkie, Stonham and Aleksander's Recognition Device) [1], [2] is a multi-discriminator weightless neural paradigm originally designed to perform image pattern recognition. Each discriminator is composed by multiple n-input RAM-type memory units (n-tuples or simply *neurons*) having their outputs connected to a summing device. Before any kind of training occurs, all positions of all tuples of all discriminators are set to 0.

The training phase of a discriminator starts by presenting randomly chosen subsets of the pixels of the entire target binary input image to all n-tuples in such a way that the whole input image is covered. The mapping of the connection of all pixel subsets to the n-tuples will be kept for further training and recognition phases. As illustrated by Fig. [1], each subset of pixels of the input image forms an address defining the n-tuple position to be set to 1.

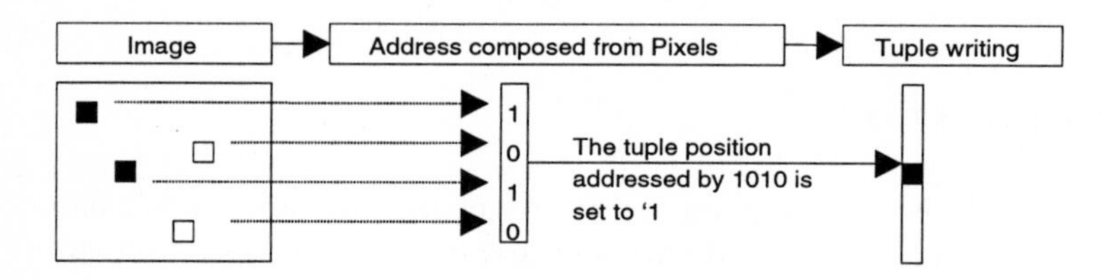

Fig. 1. A Boolean neuron.

The training phase of a discriminator is finished upon the presentation of all different input images that will define its representative class during the recognition phase. After the training of all descriminators, the recognition phase can be started by simultaneously presenting a target input image to all discriminators and comparing their responses r, i.e., the sums of all neurons' outputs belonging to each discriminator. Fig. [2] illustrates the recognition process. Observe that d, the difference between the two best responses, can be used to define a positive class identification. For instance, the relative *confidence* $c = d/r_{\text{best}}$ is usually a good practical choice.

3 The AUTOWISARD Model

The standard AUTOWISARD model is an unsupervised learning algorithm for the WISARD network, which allows a single-pass clustering of unsorted and

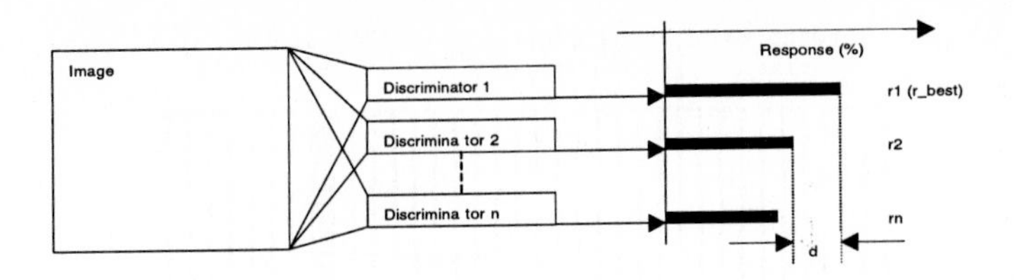

Fig. 2. The WISARD multi-discriminator system.

un-normalized, raw binary data. The main concepts behind AUTOWISARD are the *learning window*, *partial learning* and *semi-probabilistic allocation of classes* (discriminators).

At the beginning, AUTOWISARD can be seen as a WISARD which trains the winner class with a well-matching input sample and allocates dynamically a new class when an input sample couldn't be acceptably recognised by the existing ones. Although this is a workable stategy, one can see that it leads to saturation of the discriminators and doesn't reach stability. Otherwise, if the learning clause above was removed, it can be thought as a simple vector quantization model.

To ensure that AUTOWISARD would reach stability, the allocation of new classes is ruled by a learning window policy. It consist of a recognition interval given by w_{min} and w_{max}, $0 \leq w_{\text{min}} \leq w_{\text{max}} \leq r_{\text{max}}$, r_{max} being the number of neurons (RAM nodes) of a discriminator (Fig.[3]).

The learning window policy operates as follows: given r_{best} the WISARD's best recognition value for an input sample, if

- $0 \leq r_{\text{best}} \leq w_{\text{min}}$, a new discriminator is allocated and trained with that sample;
- $w_{\text{min}} < r_{\text{best}} < w_{\text{max}}$, it can either allocate a new class, or submit the winner class to a partial training, probabilistically;
- $w_{\text{max}} \leq r_{\text{best}}$, nothing is done.

The learning window ensures that no well-fitted sample will disturb its class with any extra training, and that ill-fitted samples will get new classes to represent themselves. The cases "in between" (inside the learning window) uses a probability function of the relative position of the recognition inside the window, to choose to allocate a class or to train an existing one. If a r_{best} falls inside the window, but nearer to w_{min}, the probability of allocating a class is greater than to train the existing winner; on the other side, if r_{best} is nearer to w_{max}, then the probability of the winner class be trained is greater. This probabilistic feature was thought to allow a sample that (wrongfully) felt inside the window to have a chance to create a new class to itself and not to contribute to the best class' representation. It was called "semi-probabilistic" because the probabilistic behavior shows only when $w_{\text{min}} < r_{\text{best}} < w_{\text{max}}$. In an attempt to control, or to minimize the effects of saturation on the discriminators, a partial learning method was developed. It consists of training a discriminator just enough to make it able to recognise that sample ($r_{\text{best}} = w_{\text{max}}$). The method estimates the number of

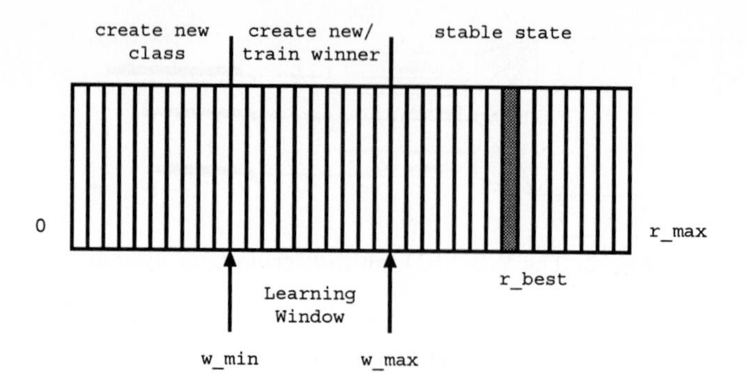

Fig. 3. The Learning Window.

neurons to be taught, $n = w_{\max} - r_{\text{best}}$, randomly selects n neurons from a list of mismatching neurons and trains them according to the input sample.

The learning window policy, associated with the partial training method and the monotonicity of the discriminator's recognition function, leads to a network that stabilizes with an order-less, single-pass training: for a set of training input samples, once the network was trained (in a single pass), it is guaranteed that any of them will have a minimum recognition of $w_{\max}$.

4 The Hierarchical AUTOWISARD

The hierarchical AUTOWISARD is a recursive, tree-like structure using AUTOWISARDs as nodes. Its aim is to recursively find a clustering hierarchy inside the data set, creating sucessive levels of even more specialised (discriminative) AUTOWISARDs, using a *grouping metric* to determine when a sufficiently fine clusterization was achieved. The grouping metric used is the recognition interval of the samples assigned to a discriminator: it is assumed that when a discriminator has a recognition interval above an arbitrary threshold, it is very likely to contain more than one data cluster.

The recursive method to create the hierarchical structure is to link a new AUTOWISARD network to each discriminator that has a large recognition interval, until the recognition intervals of all the discriminators (at any level) are below that threshold or a maximum number of levels is reached. Each network is trained without supervision only with the set of input samples assigned to the parent discriminator. It is desirable to have a mechanism to control the (free) growing of the network, as the creation of deep, super-specialised AUTOWISARDs, tend to present a vector-quantization behavior.

5 Experiments

The example application for the classification abilities of the AUTOWISARD models is the classification of handwritten digits, represented as binary images

(32x32 pixels), whose symbolic classes are previously known. The digits database employed is the training set of "Optical Recognition of Handwritten Digits" database [5], in its un-normalized version, freely available.

The input image set has the following digits distribution:

Digit	0	1	2	3	4	5	6	7	8	9
Qty.	189	198	195	199	186	187	195	201	180	204

The experiments consists of training a set of standard AUTOWISARD networks with the unsorted images and to compare the clusters created inside each network for *symbolic and quantitative confusion*, and *overfitting.*

Network	Classes	Classes with multiple symbols				Winners' recognition	Overfit-
		2	3	4	5+	average.(%)	ting (%)
1	126	30	7	7	0	80.53	75.40
2	140	29	14	6	1	75.27	80.00
3	131	21	13	4	2	77.49	76.34
4	136	28	11	4	3	79.84	75.00
5	133	35	5	2	2	80.81	76.69
6	135	27	5	5	1	79.98	77.04
7	116	25	6	6	4	78.91	77.59
8	139	30	13	5	0	79.82	74.82
9	133	33	10	6	0	80.26	72.18
10	123	28	7	3	1	80.25	78.05
11	117	27	11	2	1	80.38	73.50
12	133	29	8	7	1	79.28	75.19
13	130	26	15	4	1	80.31	76.15
14	144	28	12	5	1	77.49	79.17
15	134	27	5	5	1	79.01	79.85
16	128	23	14	5	0	81.07	67.97
17	126	33	11	5	0	76.47	73.81
18	122	37	12	4	0	78.78	74.59
19	138	29	7	4	2	76.96	77.54
20	122	30	8	5	3	82.38	71.31
AVG.	130.3	28.75	9.70	4.70	1.20	79.26	75.61
STDEV	7.49	3.70	3.24	1.34	1.12	1.70	2.96

Table 1. The standard AUTOWISARD results.

The symbolic and quantitative confusion are measured as the percentile of classes that contains multiple symbols (digit labels) and, for that classes, the mean percentile of the winner symbol's stances, respectively, and were meant to characterize the quality of (symbolic) classification and to search for classification "trends" inside the ambiguous discriminators. The overfitting measure is the percentile of classes that represents a smaller number of stances than an overfitting threshold, giving feedback of the network's generalisation performance.

The hierarchical AUTOWISARD test consists of a single training phase with a limited maximum number of levels. For matters of space, only a branch of the hierarchy will be shown.

The experiments were realized with fixed set of parameters (tuple size, learning window parameters and probability distribution function), because the analysis of the AUTOWISARD sensibility to this parameters was beyond the scope of this paper. The experiments were thought to bring to the readers a feeling of the models capacities rather than to be an authoritative, extensive models benchmarking

6 Results and Analysis

A set of 20 AUTOWISARDs were trained with the same set of randomly ordered input images, with 6-bits tuples, $w_{\mathrm{min}} = 30\%$, $w_{\mathrm{max}} = 40\%$. The overfitting threshold is 1%: an overfitted class represents 19 or less images, from a set of 1924 images. The results are shown in Table [1]. It also shows the mean and standard deviation of these results, illustrating that, in spite of training the networks with random ordered samples and using different input mappings, the networks shows classification profiles qualitatively similar to each other's.

The low deviation of the winner's average on the classes with more than one symbol indicates that, even when there is confusion inside a class, still exists a clear winner digit and that these discriminators wasn't driven into saturation (when it wouldn't be possible to point a winner digit).

The hierarchical AUTOWISARD was trained with the same set of images used above, with a maximum of 2 layers (using tuples of 4 and 8 bits, respectively), $w_{\mathrm{min}} = 30\%$, $w_{\mathrm{max}} = 50\%$, and a subclassing threshold of 35%. A selected branch is shown in Fig. [4]. Observe that, as the sub-networks were trained only with their parents' pre-classified images, it is easier to classify them into a more specialised way.

7 Conclusions

The new AUTOWISARD unsupervised learning approaches, based on a simple weightless neural model, were introduced and their classification potentialities identified. An interesting feature offered by AUTOWISARD lies in its implicit recognition phase ($w_{\mathrm{max}} < r_{\mathrm{best}}$) during training. A potential advantage of that is the greater spacial stability of the clusters created by AUTOWISARD's constructive approach, thanks to the (recognition) monotonicity provided by the basic WISARD model.

The experimental results have shown a strong consistency on the formation of classes, as indicated by (i) the number of hierarchical clusters generated under different trials, and; (ii) that inside the few classes formed by patterns of different

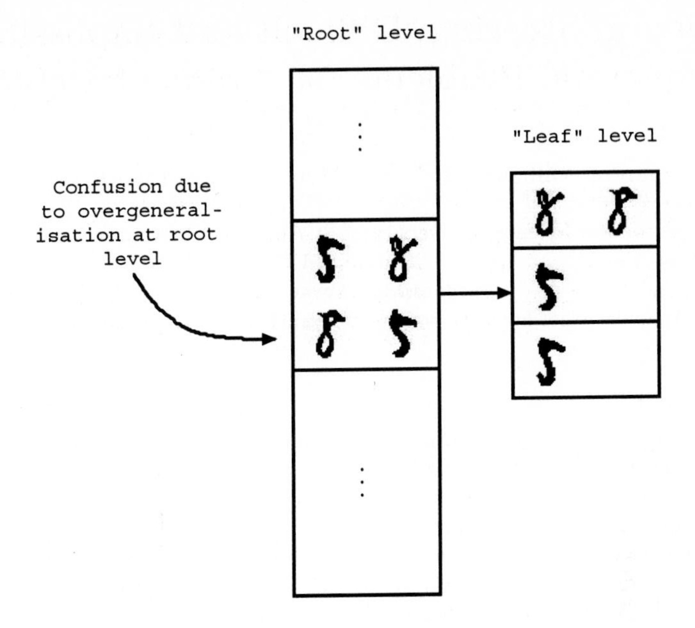

Fig. 4. Better classification through (hierarchical) specialization

symbolic meanings, it is easy to evidentiate a winner, i.e., wrongly classified patterns are a minority inside such classes.

8 Acknowledgements

The authors would like to thank Sergio Exel for the helpful comments, and the grant support from CNPq - Brazilian Research Council.

References

[1] Aleksander, I., Thomas, W., Bowden, P.: WISARD, a radical step forward in image recognition. Sensor Ver., (1984) 120–124

[2] Aleksander, I., Morton, H..: An Introduction to Neural Computing. Chapman and Hall, London (1990)

[3] Grossberg, S.: Competitive learning: From interactive activation to adaptive resonance. Cognitive Science **11** (1987) 23–63

[4] Fulcher, E.: WIS-ART: Unsupervised clustering with RAM discriminators. Intl. J. of Neural Systems, **3**, n. 1 (1992) 57–63

[5] Alpaydin, E., Kaynak, C.: Bogazici University, Turkey. ftp://ftp.ics.uci.edu/pub/ machine-learning-databases/optdigits/ (1998)

Neural Steering: Difficult and Impossible Sequential Problems for Gradient Descent

Gordon Milligan, Michael K Weir, and Jonathan P Lewis

School of Computer Science, University of St Andrews, St Andrews, Fife KY16 9SS, Scotland, UK
gordonm@dcs.st-and.ac.uk,
WWW home page: http://www.gordon-milligan.pwp.blueyonder.co.uk/

Abstract. A common underlying design feature of neural network methods is the approximation of a fixed goal through a single I/O characterisation. By contrast, the problems investigated here are those which contain sequences of multiple goals. Such sequences may occur when the environmental or task requirements alter unpredictably to vary the outputs appropriate to the given inputs, or where completion of a task requires a sequence of modes of behaviour.
Some of these problems prove impossible for training based on gradient descent due to the goal sequence. Other problems may be found to be increasingly difficult as the sequence extends.
A Neural Steering mechanism is presented to improve training which chains subgoals so that actual and subgoal excitations are *zipped* together towards the goal. The results show that the steering method is able to solve such problems feasibly and robustly.

1 Introduction

A great deal of mainstream effort has been put into neural learning techniques for static pattern recognition and association. A unifying theme is that feedforward architectures are commonly used with the trained weight state attempting to capture the spatial structure of the desired I/O mapping.

Approaches to sequential problem solving by neural networks on the other hand are varied and not as unified as the mainstream approaches. In part, this may be because sequences can be seen to pose greater fundamental difficulties. For example, the sequences may be more complex in being large or unbounded, of variable length, and involving temporal as well as spatial consideration.

One approach to learning sequences has involved representing the sequences as part of spatial mappings. A well known example is that of back-propagation through time [RHW86] which unfolds moments in time as spatial layers within the network. The large amount of data that accumulates over time in many tasks poses problems for such a technique. Another approach is to try to handle the increased amount of data better by selecting out the novel parts of the sequences [Sch91a]. Recurrent networks have also been used to control sequence learning by

J. Mira and A. Prieto (Eds.): IWANN 2001, LNCS 2084, pp. 442-449, 2001.

using the network's internal states to act as memories for suitable tasks [Elm90] [Pea89] [Moz94].

Ulbricht [Ulb96] has addressed the variable length of sequences for autonomous agent tasks by getting the network to change its state slowly in response to sequential changes. Schmidhuber [Sch91b] has used subgoals to provide decompositional solutions for action sequences where each subgoal is tackled using a sub-program which is known to attain the subgoal. The subgoal programs act as stepping stones to enable action sequences to attain their goal.

Despite their disparate nature, a common design feature in the above is to use neural networks that meet a fixed goal through a single I/O relation.

By contrast, the sequences of concern to the present paper are those which require multiple goals and I/O mappings. Such sequences may occur when the environmental or task requirements alter unpredictably to vary the outputs appropriate to the given inputs, or where completion of a task requires a sequence of modes of behaviour.

We focus here on the applicability of mainstream training in this area. In particular, we investigate the suitability of common gradient descent techniques for neural networks in tackling *multi-goal sequential problems*. As the problems will not involve internal states, we restrict ourselves to feedforward networks.

Our definition of a multi-goal sequential problem for a feedforward network is one where a sequence of I/O mappings is required. We present a number of such problems where each individual goal problem can be attempted by nets using standard gradient descent. One aim is to show that some of these problems are impossible for such descent due to the sequence the goals occur in. A second aim will be to show that other problems become increasingly difficult as the sequence is extended.

2 Multi-goal sequential problems

Two types of multi-goal sequential problem are described here, which are either impossible or difficult for steepest gradient descent. These types indicate that steepest gradient descent is likely to give poor performance generally on sequential problems. The problems can nevertheless be solved feasibly by a steering approach to search which is described in the next section.

The impossibility of the first type of sequential problem for steepest gradient descent can be readily seen. The problem type consists of a sequence of two individual problems where the global minimum basin for the first problem lies in the local minimum basin for the second problem as depicted in Fig. 1 **(a)**.

The other type of sequential problem is difficult as opposed to impossible for steepest gradient descent to solve. These problems are made up from a sequence of individual problems where each individual has a significant likelihood of failure using random initialisation and steepest gradient descent. As the sequence is extended, the chances of success using steepest gradient descent diminish.

The general method for creating sequential problems for this type is to include individual problems where travel to at least one goal takes too long due to poor

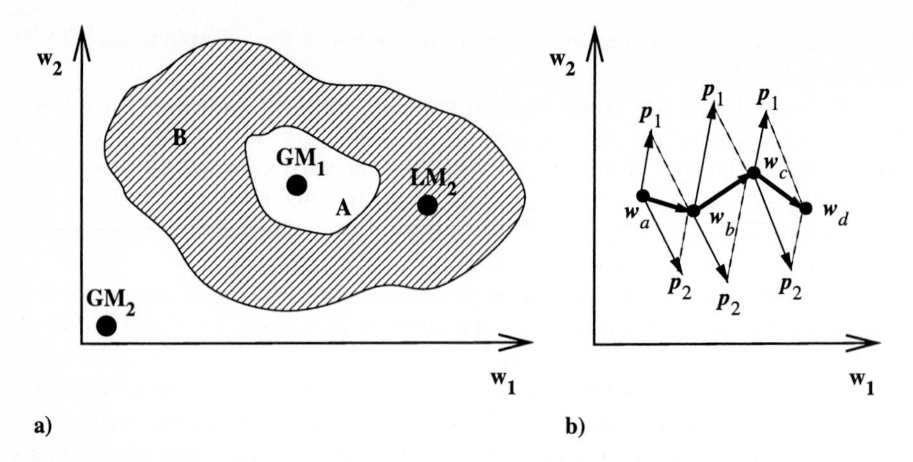

Fig. 1. **(a)** shows the basin **A** of the global minimum $\mathbf{GM_1}$ for problem 1 lying in the local minimum basin **B** (with local minimum $\mathbf{LM_2}$) for problem 2. **(b)** Weight state transitions from $\boldsymbol{w}_a$ to $\boldsymbol{w}_d$ formed through summation of gradient vectors $\boldsymbol{p}_i$ for pattern i are weakly directed and may form sequences containing nearly opposing directions. This often occurs when distant from the goal.

travel direction. The probability of such a goal being set increases as the goal sequence extends.

The quality of travel direction is influenced by the goal distance. If the goal is nearby, the error-weight surface is likely to be bowl-shaped, yielding relatively good direction for gradient descent techniques. Where the goal occurs further away, direction is more influenced by local undulations, so that the gradient summations are more arbitrary and unsteady. This situation is depicted in Fig. 1 **(b)**.

3 Steering

The approach taken in this paper is that sequential problems generally, and the ones described above in particular, can be solved feasibly using steering. Steering here means taking action in line with setting and resetting intermediate targets or subgoals as a means of achieving the desired ends or goals. In neural terms, instead of setting the desired goal immediately as the single fixed target, an adaptive sequence of subgoals leading to the goal, a *subgoal chain*, will be continually shaped to progress the behaviour towards each goal.

The motivation and justification for such an approach is that the difficulties standard steepest gradient descent runs into are seen as being caused by the use of desired goals as targets which are distant from the initial state. The travel surface between the initial and goal states is full of local undulations which make for poor direction as indicated in Fig. 1 **(b)** and hence infeasible travel by gradient descent. Sequential problems compound the difficulty by requiring a sequence of goals to be met without being able to re-initialise the state during the sequence. Once the sequence has been started, learning has to succeed for each goal on a *first-time* basis.

The subgoal chaining method attempts such first-time learning by instead trying to *zip* the actual and subgoal states together. That is, transitions are continually made in such a way that keeps the process from becoming stuck, and actual and target states are placed close to one another to promote smooth steady travel.

In the method described below, the outputs are converted at various stages into equivalent excitations so that linear scaling on these excitation transitions may be performed. The scaling of the excitations is achieved by scaling the neural weights.

The initial subgoal chain is set to be a simple linear path between the initial state and the first goal state. That is, we may use a parameterised training set $S(\boldsymbol{\lambda})$, where each setting of the vector $\boldsymbol{\lambda}$ corresponds to a subgoal state setting, and

$$S(\boldsymbol{\lambda}) = \{\boldsymbol{in}_p, T_p(\lambda_p)\} = \{\boldsymbol{in}_p, ex_p + \lambda_p(t_p - ex_p)\} \tag{1}$$

where λ_p, the elements of $\boldsymbol{\lambda}$, are increased uniformly for all P training patterns in discrete steps from 0 to 1, such that $\lambda_p = \lambda_q$, $\forall p \neq q$, $\boldsymbol{in}_p$ is the input vector for training pattern p, $T_p(\lambda_p)$ is the desired target response for pattern p as a function of λ_p, ex_p is the initial excitation pattern response to in_p and t_p is the goal target response for pattern p.

To keep the process from getting stuck, a different technique from descent in LMS error is used. An extrapolation of a small regular amount is made from the current weight state. This is done using the last cubic in a spline running through the weight states realised so far. The resulting transition, called *the unsteered transition*, is then steered by correcting its direction to be closer to the subgoal direction if possible. The resulting *steered transition* is thereby designed to always keep the process on the move in a locally progressive direction.

The specifics of this steering are described as follows. Each output unit has a directional measure Q defined in terms of transitions in its output, specifically the angle Θ between the *unsteered* transition vector $\boldsymbol{o}_U$ and the *subgoal* transition vector $\boldsymbol{o}_{SG}$. The origin of these vectors is taken to be the current output state $\boldsymbol{o}_c$. We now describe the learning equations for improving Θ.

$$Q = \cos\Theta = \frac{\boldsymbol{o}_U \cdot \boldsymbol{o}_{SG}}{\|\boldsymbol{o}_U\| \cdot \|\boldsymbol{o}_{SG}\|} \tag{2}$$

so that

$$Q = \sum_{p=1}^{P} Q_p = \sum_{p=1}^{P} \frac{o_{Up}}{\|\boldsymbol{o}_U\|} \cdot \frac{o_{SGp}}{\|\boldsymbol{o}_{SG}\|} \tag{3}$$

where o_{Up} and o_{SGp} are the pth components of $\boldsymbol{o}_U$ and $\boldsymbol{o}_{SG}$ respectively for P training patterns.

Q varies between -1 and 1 for each output unit with the optimal value being 1. To compute an optimal direction for improving Q, the directional-weight gradients may be used to perform gradient ascent. For each pattern p, one component

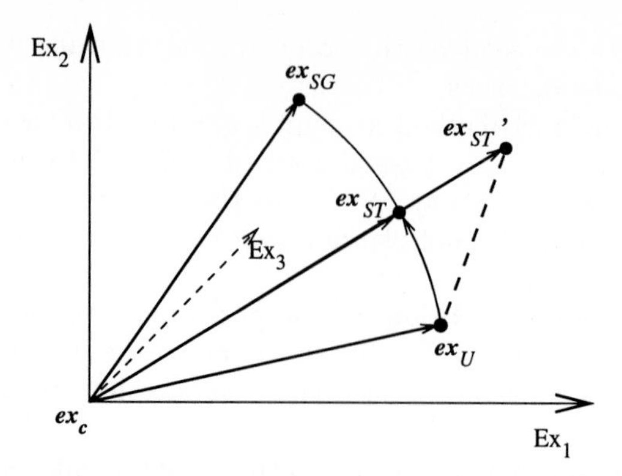

Fig. 2. Directed correction of the unsteered transition $\boldsymbol{ex}_U$ through increasing Q in (2) provides a steered transition $\boldsymbol{ex}_{ST}{}'$ which is then scaled to the length of the subgoal excitation $\boldsymbol{ex}_{SG}$ and results in $\boldsymbol{ex}_{ST}$.

of the gradient is given by

$$\frac{\partial Q_p}{\partial o_{Up}} = \frac{o_{SGp}}{\|\boldsymbol{o}_{SG}\| \cdot \|\boldsymbol{o}_U\|} - \frac{o_{SGp} \cdot {o_{Up}}^2}{\|\boldsymbol{o}_{SG}\| \cdot \|\boldsymbol{o}_U\|^3} \tag{4}$$

We may then define Δw_{Uij} to be the change in weight for link ij from its unsteered value w_{Uij} so that

$$\Delta w_{Uij} = \varepsilon \cdot \frac{\partial Q}{\partial w_{Uij}} = \varepsilon \cdot \sum_{p=1}^{P} \frac{\partial Q_p}{\partial w_{Uij}} = \varepsilon \cdot \sum_{p=1}^{P} \frac{\partial Q_p}{\partial o_{Ujp}} \cdot \frac{\partial o_{Ujp}}{\partial ex_{Ujp}} \cdot \frac{\partial ex_{Ujp}}{\partial w_{Uij}} \tag{5}$$

where ε is a learning rate. When j is an output unit we have $\frac{\partial Q_p}{\partial o_{Ujp}}$ defined as for $\frac{\partial Q_p}{\partial o_{Up}}$ in (4) and the 2 rightmost component terms in (5) have the same form as their equivalents in back-propagation [RHW86]. When j is a hidden unit connected to output units, we have $\frac{\partial Q_p}{\partial o_{Ujp}}$ defined through a further chain rule:

$$\frac{\partial Q_p}{\partial o_{Ujp}} = \sum_{k=1}^{K} \frac{\partial Q_p}{\partial o_{Ukp}} \cdot \frac{\partial o_{Ukp}}{\partial ex_{Ukp}} \cdot \frac{\partial ex_{Ukp}}{\partial o_{Ujp}} \tag{6}$$

where k are the K output units connected to hidden unit j and o_{Ujp} is the transition from the current output for unit j to the output after the unsteered transition. (Changes in weights along links to hidden units connected to other hidden unit layers may be recursively defined in terms of (6)).

At this stage, the steering has found a transition $\boldsymbol{o}_{ST}{}'$ in output closer to the desired subgoal in direction. The distance to the subgoal needs to be made reasonably close as well. We do a scaling on $\boldsymbol{ex}_{ST}{}'$, the excitation equivalent of $\boldsymbol{o}_{ST}{}'$, to achieve this.

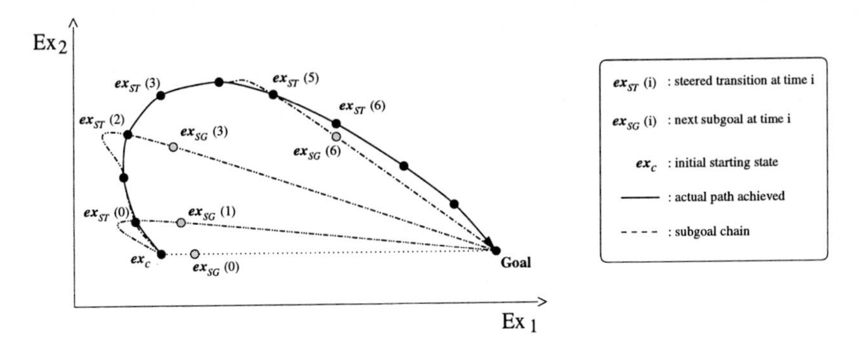

Fig. 3. An adapting subgoal chain in excitation space zipping together the desired and actual path (not all subgoals or actual states are shown).

In order to make the local progress connect to the goal, a subgoal chain is created in the form of a (cubic) spline. This spline is made to run through excitation states corresponding to the steered transitions $\boldsymbol{ex}_{ST}$ made so far and the goal excitation. The next subgoal is set at a small distance $\boldsymbol{\lambda}$ along the spline between the last steered excitation realised and the goal. Once a steered transition has been made, the spline is updated. Initially the spline takes the linear form defined in (1). As excitation states occur which depart from the linear form, the λ_p are set to reflect the non-linear forms so that $\lambda_p \neq \lambda_q$ for all training patterns $p \neq q$. As more transitions occur and the distance to the goal diminishes, the spline is intended to give increasingly accurate interpolations of the next states that are realisable as depicted in Fig. 3. In this way, the next subgoals become increasingly realisable and so are zipped closer to the actual output.

4 Problems

4.1 A dual minimum problem

The topology and data set here has been previously used to demonstrate another subgoal chaining technique [LW00]. Fig. 4 illustrates the data set design for a 2-1 net for which the local minimum is a separation near the line $\mathbf{H_{LM}}$ and the global minimum near the line $\mathbf{H_{GM}}$, with the arrows indicating the side with output greater than 0.5. The sequential problem set here has 2 goals where the first goal has targets corresponding to the outputs of the local minimum and the second goal has the targets of the original data set. Hence the problem consists of a sequence of two goals where the global minimum basin for the first goal lies in the local minimum basin for the second goal. This presents an effectively impossible sequence for BP to achieve as can be seen from its extremely low success rate, whereas NS manages to achieve 100% success (see IS in Table 1).

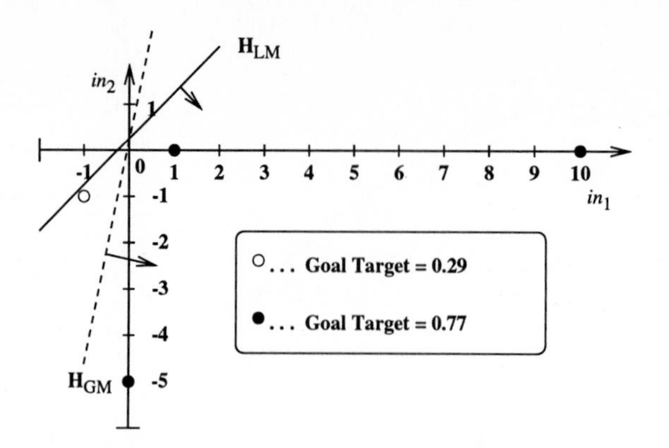

Fig. 4. Input space with axes in_1 and in_2 showing a data set for a 2-1 net consisting of 4 training patterns. This data set creates a local and global minimum.

4.2 Linear sequential problems

In this type of problem, a sequence of goals is positioned along a linear excitation path. If the goals are strung out far enough apart there should be a degree of poor travel direction for gradient descent which may cause failure within the given time limit for at least one of the goals if the problem is hard enough.

Data sets derived from two well known problems, AND and XOR, were used as multi-goal problems. All the generated goals have the same input set, namely $\{(0,0),(0,1),(1,0),(1,1)\}$. An initial and final goal was set for each problem, where the final goal is the goal of the well known problem. For the AND-related problem the outputs corresponding to the input set for the two goal states were $(0.8, 0.8, 0.8, 0.2)$ and $(0.2, 0.2, 0.2, 0.8)$. For the XOR-related problem they were $(0.5, 0.5, 0.8, 0.2)$ and $(0.2, 0.8, 0.8, 0.2)$. The target states for an additional 2 intermediate goals are set equally spaced along the line between the initial and final goal's equivalent target excitation states. A suitably large and indicative range of learning parameter values were explored for both BP and NS for all problems.

The problem based on AND was first investigated using a 2-1 net to give an indicator of the relative lack of smoothness between gradient descent and the NS method. The results show that the NS method was roughly 3 times better than gradient descent in average angle deviation per cycle between the same number of weight state transitions (see average weight deviations in Table 1 for DS1). They also show that if superior speed rather than smoothness is important, NS can be faster than BP, again by a factor of roughly 3 (average cycles in Table 1 for DS2).

The problem based on XOR was then investigated using a 2-2-1 net. Despite the greater success of NS relative to BP in terms of success and the average cycles (see Table 1 for DS3) the results here show that the NS method requires further development to encourage smoothness in multi-layer training (see avg. weight deviation in Table 1 for DS3).

5 Conclusion

In conclusion, the results show that the Neural Steering method is able to solve problems, feasibly and robustly, which are found by gradient descent to be either difficult or impossible.

Table 1. Table of Results, where BP and NS refer to back-propagation and neural steering respectively. For all BP problems a momentum of 0.9 was used and a more generous timeout was allowed for BP due to the higher computational cost per iteration for NS. All results are based on 100 random initialisations for each problem.
Key: IS – Impossible Sequence; DSi – Difficult Sequence i.

Problem	Topology	no. goals set	learning rate		no. goals achieved		Success %		avg. weight deviation		average cycles	
			BP	NS	BP	NS	BP	NS	BP	NS	BP	NS
IS	2-1	2	0.67	0.1	0.24	2	4	100	26.6	82.2	77.8	52.5
DS 1	2-1	2	0.84	10^{-4}	2	2	100	100	9.3	3.2	76.0	75.9
DS 2	2-1	2	0.39	0.01	2	2	100	100	8.2	11.5	63.9	21.3
DS 3	2-2-1	4	0.99	0.38	1.86	3.8	16	91	9.6	65.7	92.6	36.9

References

[Elm90] J.L. Elman. Finding structure in time. *Cognitive Science*, 14:179–212, 1990.

[LW00] J.P. Lewis and M.K. Weir. Using subgoal chaining to address the local minimum problem. In *Proceedings of Second International ICSC Symposium on Neural Computation*, 2000.

[Moz94] Michael C. Mozer. *Neural Net Architectures for Temporal Sequence Processing*, pages 243–264. Addison Wesley Publishing, Redwood City CA, 1994.

[Pea89] B.A. Pearlmutter. Learning state space trajectories in recurrent neural networks. *Neural Computation*, 1:263–269, 1989.

[RHW86] D.E. Rumelhart, G.E. Hinton, and R.J. Williams. *Parallel distributed processing: Explorations in the microstructure of cognition*, volume 1, chapter 8, pages 318–362. MIT Press/Bradford Books, Cambridge MA, 1986.

[Sch91a] J. Schmidhuber. Adaptive history compression for learning to divide and conquer. In *International Joint Conference on Neural Networks*, volume 2, pages 1130–1135, 1991.

[Sch91b] J. Schmidhuber. Learning to generate subgoals for action sequences. In T. Kohonen, K. Makisara, O. Simula, and J. Kangas, editors, *Artificial Neural Networks*, pages 967–972. Amsterdam: Elsevier Publishers B.V., North-Holland, 1991. Proceedings of the 1991 International Conference on Artificial Neural Networks (ICANN '91) Espoo, Finland.

[Ulb96] C. Ulbricht. Handling time-warped sequences with neural networks. In P. Maes, M.J. Mataric, J-A. Meyer, J. Pollack, and S.W. Wilson, editors, *Proceedings of the Fourth International Conference on Simulation of Adaptive Behavior, From Animals to Animats 4*, pages 180–189, Cape Cod, Massachussetts, September 9th–13th 1996. MIT Press, Cambridge, MA.

Analysis of Scaling Exponents of Waken and Sleeping Stage in EEG

Jong-Min Lee, Dae-Jin Kim, In-Young Kim, and Sun I. Kim

Department of Biomedical Engineering, College of Medicine, Hanyang University
Haengdang-dong, Seongdong-ku, Seoul, 133-791, Korea
sunkim@email.hanyang.ac.kr
WWW home page: http://bme.hanyang.ac.kr

Abstract. A classic problem in physics is the analysis of highly nonstationary time series that typically exhibit long-range correlations. Here we tested the hypothesis that the scaling exponents of the dynamics of sleeping EEG have more stable pattern than those of waken EEG by analyzing its fluctuations. We calculated the modified fluctuations of EEG stage with detrended fluctuation analysis(DFA). DFA is very useful to detect a long-range correlation in the time-series. We found a scaling exponent of sleeping stage is larger than that of waken. ...

1 Introduction

Many kinds of biological time series analysis may be used to identify disclosed important dynamical properties. But the statistical characteristics of biological signals often change with time in many cases, so such analysis is complicated by the fact that biological signals are typically both highly irregular and non-stationary [1] [2]. Human brain activities, for example, normally fluctuate in a complex manner. These fluctuations may arise from a complex nonlinear dynamical system rather than being an epiphenomenona of environmental stimuli [3].

In recent years long-range power-law correlations have been discovered in a remarkably wide variety of systems [4] [5] [6]. Such long-range power-law correlations are a physical fact that in turn gives rise to the increasingly appreciated fractal geometry of nature. Since as soon as we find power-law correlations we can quantify them with a critical exponent, recognizing the ubiquity of long-range power-law correlations can help us in our effort to understand electroencephalogram (EEG) properties [7]. Under healthy conditions, many physiological time series exhibit long-range power-law correlations [1]. We apply a detrended fluctuation analysis (DFA) to the EEG for quantifying correlation property in non-stationary physiological time series. Cause this method permits the detection of long-range correlations embedded in a seemingly non-stationary time series, it may be of use in distinguishing waken stage data from sleep data sets [3].

The purpose of the present study is to analyze the EEG with different sleep stages using detrended fluctuation analysis and to discriminate the waken stage from the sleeping stage. We used the digitized EEG over very long time series recorded in the MIT/BIH Polysomnography data.

J. Mira and A. Prieto (Eds.): IWANN 2001, LNCS 2084, pp. 450-456, 2001.

2 Methods & Materials

2.1 Scaling Exponent

There are prominent fluctuations on the exponent characterizing long-range correlations in the finite-length sequences [8]. This scaling exponent is important since we can find correlation properties of the time series from it.

Time series is said to be self-similar when the statistical property of original sequence is identical to rescaled version of subset. In detail, we can see this fact in mathematical terms with

$$y \overset{d}{\equiv} a^{\alpha} y(\frac{t}{a}) \tag{1}$$

where $\overset{d}{\equiv}$ means that the statistical properties of both sides of the equations are same. The exponent α is called the self-similarity parameter or the scaling exponent. Generally scaling exponent α is calculated by

$$\alpha = \frac{\ln M_y}{\ln M_x} = \frac{\ln s_2 - \ln s_1}{\ln n_2 - \ln n_1} \tag{2}$$

where M_y is magnification factor of y-axis and M_x is of x-axis, n is a data size and s is a standard deviation of each signal respectively.

Now consider the case of $\alpha = 0$, this means that standard deviations of two signals are the same. For example we can see that white noise has the same standard deviations regardless of signal size. But it has not a long-term correlation but random property. To overcome this problem, we use the modified version of analysis. This is a detrended fluctuation analysis.

2.2 Detrended Fluctuation Analysis (DFA)

DFA is to calculate the root-mean-square fluctuation of integrated and detrended time series. In detail, each of time series with the number of N samples is integrated first as

$$y(k) = \sum_{t=1}^{k} \{X(t) - X_{ave}\} \tag{3}$$

Here $X(t)$ is the sequence at time t, and X_{ave} is the average of entire time series. Next $y(k)$, integrated time series is divided into sub sequences of equal length, n. And in each box, the y-coordinate of a least-square line which fits to the data is denoted by $y_n(k)$. Finally the average fluctuation as a function of box size, n is given by

$$F(n) = \sqrt{\frac{1}{N} \sum_{t=1}^{N} [y(k) - y_n(k)]^2} \tag{4}$$

Then the $log - log$ plot of $F(n)$ vs. n is drawn. The slope of this graph is a scaling exponent α which is the characteristic of fluctuations. [3]

It is known that white noise where the value at one instant is not correlated with any previous value has uncorrelated integrated series $y(k)$ such like a random walk and scaling exponent α is equal to 0.5 [9]. The case of $\alpha = 1$ is a special one, the time series corresponds to $1/f$ noise [10]. And $\alpha = 1.5$ indicates Brownian noise which is the integration of white noise [7]. When $0 < \alpha < 0.5$, power-law anti-correlations are present such that large values are more likely to be followed by small values and vice versa [11]. If α is greater than 0.5 and less than or equal to 1.0, then there is a long-range power-law correlations in that we are interested [3].

We used three kinds of time series to verify the DFA algorithm that we developed. White noise, $1/f$ noise, which is computer-generated fractional Gaussian noise with long-range correlations, and the scaling exponent is approximately 0.8, and Brownian noise. The results are shown in Fig. 1.

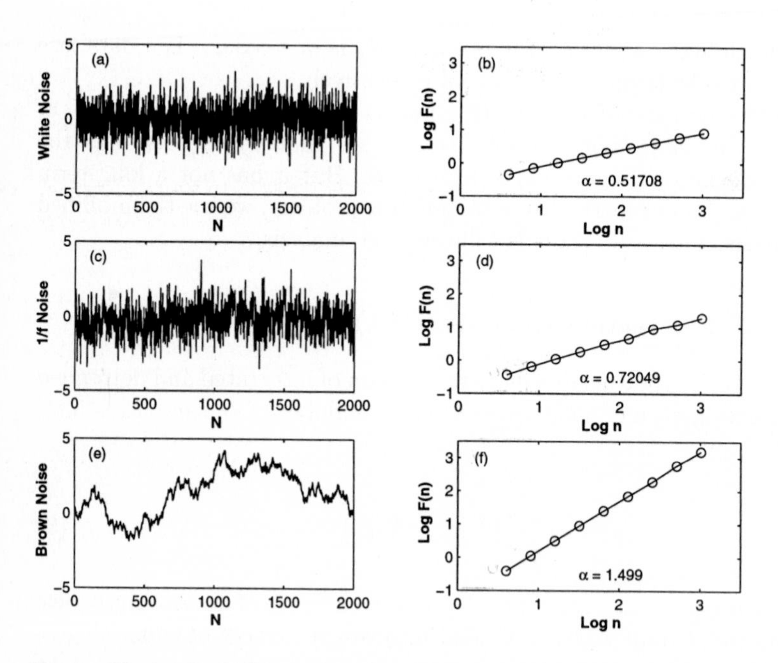

Fig. 1. Slope of scaling exponents of white noise in (a) is approximately 0.5 in (b). In the case of $1/f$ noise in (c), average slope is 0.7 in (d), set to approximately 0.8 by computer algorithm in advance. Brown noise in (e) is the integration of white noise, and its slope in (f) is near to 1.5.

3 Results

Our analysis is based on a set of 16 sections of the digitized electroencephalograms recorded in the MIT/BIH Polysomnography data. Section is derived from EEG part of each data after reading annotation file. The property of each section is listed in Table 1. Among the data, we select a sleep EEG in SLP01A of stage 4 with slope 1.3785 and a waken EEG in SLP01B with slope 1.0354, for example and plot them in Fig. 2. We truncate each time series to 10^5 points (400 seconds) of sleeping or non-sleeping stages for the convenience. We find that the slope of scaling exponent of EEG with waken stage is larger than with sleeping stage in many cases like Fig. 2.

Table 1. Polysomnography data in MIT/BIH used in this study.

Data Name	Section	Samp. Num.	SR (Hz)	Electrode Pos.	Stage	Ave. α
SLP01A	217501-285001	67500	250	C4-A1	4	1.3706
SLP01A	292501-315001	22500	250	C4-A1	3	1.3834
SLP01A	855001-990001	135000	250	C4-A1	4	1.3785
SLP01B	75001-1057501	982500	250	C4-A1	W	0.9880
SLP01B	1470001-1725001	255000	250	C4-A1	W	1.0354
SLP01B	2437501-2460001	22500	250	C4-A1	W	1.0816
SLP02A	525000-540000	15000	250	O2-A1	2	1.2974
SLP02A	547500-562500	15000	250	O2-A1	3	1.3148
SLP02A	570000-585000	15000	250	O2-A1	2	1.3276
SLP02A	907500-1245000	337500	250	O2-A1	2	1.2721
SLP02A	1252500-1522500	270000	250	O2-A1	W	1.1592
SLP02A	1537500-1552500	15000	250	O2-A1	W	1.0018
SLP02B	7500-172500	165000	250	O2-A1	2	1.3291
SLP02B	472500-495000	22500	250	O2-A1	W	1.2242
SLP02B	877500-1027500	150000	250	O2-A1	W	1.1618
SLP02B	1485000-2017500	532500	250	O2-A1	W	1.1110

Using 16 sections of EEG, we calculate mean and standard deviation of sleeping and waken stage. Plot of them is drawn in Fig. 3. Since mean of sleeping EEG is near to approximately 1.5, we may think that sleeping EEG has a property of Brownian noise. In fact, if we may think that there are less brain activities and physical movements in sleeping stage, it is reasonable for EEG to be smooth-pattern. In the case of waken stage, average slope of scaling exponents is approximately 1, that is $1/f$ noise-like. Since EEG is a recording of the electrical activity of the brain, it may have not a random property but a long-range correlation such like $1/f$ noise.

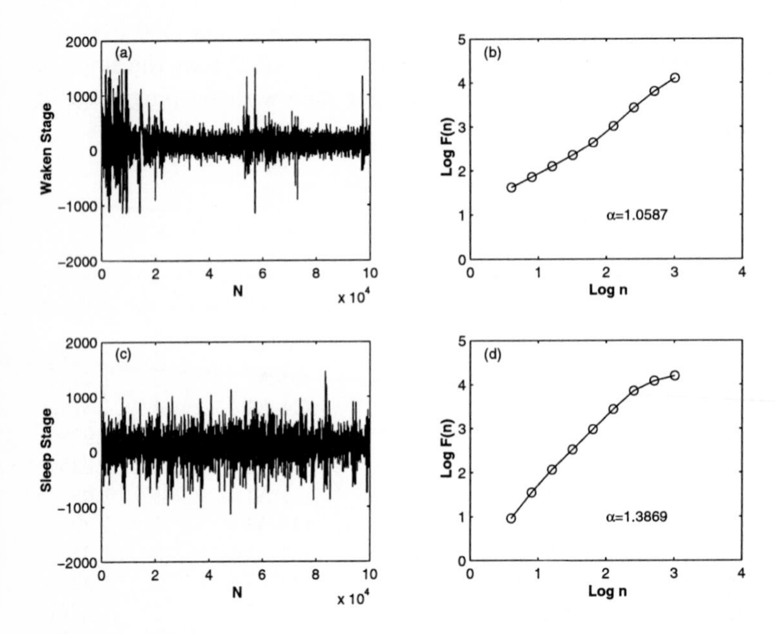

Fig. 2. Waken and sleeping EEG representation in (a) and (c), and double log plot of DFA computation results in (b) and (d). We truncate each time series to 10^5 points (400 seconds) of sleeping or non-sleeping stages. Slope is changed by a little bit because of truncated data size. (a) Time series is captured at waken stage in SLP01B of MIT/BIH Polysomnography. (b) Average slope of DFA results of (a) is approximately 1, this means that waken EEG have $1/f$ noise-like characteristics. (c) Time series is captured at stage 4 in SLP01A of MIT/BIH Polysomnography. (d) Slope is steeper than (b), and closer to Brownian noise. In fact, sleeping EEG is smoother than waken EEG. It is a property of Brownian noise typically.

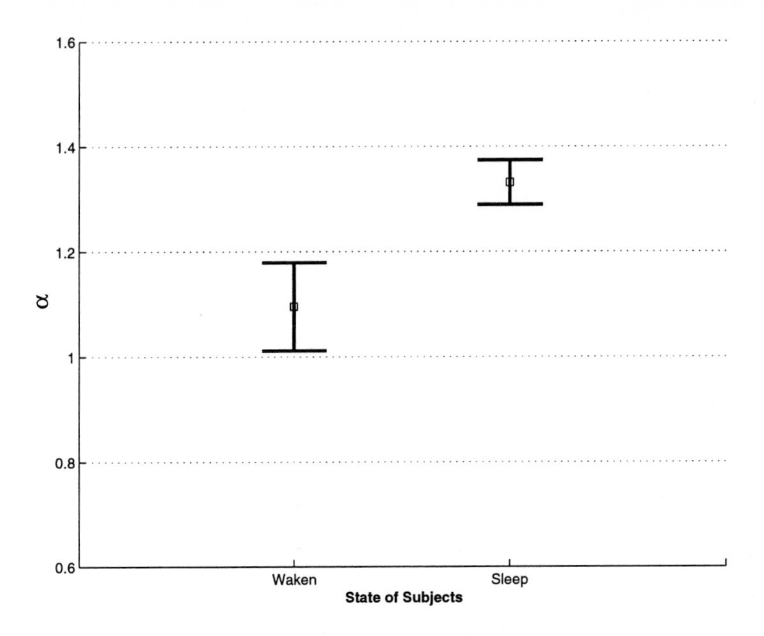

Fig. 3. Plot of means and standard deviations for each stage. Data is based on Table 1. Mean and standard deviation of waken EEGs are 1.0954 and 0.0840 respectively. In the sleeping stage, mean is 1.3317 and standard deviation is 0.042. We are able to see that slope of DFA results is larger at sleeping stages.

4 Discussion & Conclusion

In summary, we applied a modified fluctuation analysis to the nonstationary EEG time series. Our results showed that the sleeping and the waken EEG are different from each other in the scaling exponents. The larger scaling exponent found for sleeping states suggests that the dynamics of sleeping EEG is more likely to Brownian noise than waken EEG. It means that the fluctuations in sleeping stage are in a more stable fashion than in waken. Our results also show that the dynamics of EEG has long-range correlation, and is different from random noise. This finding is of interest since it motivates new modeling approaches to account for the brain mechanisms. From a practical point of view, these properties of scaling exponents may have useful applications for brain modeling and bedside monitoring. Though we don't determine how to stage in the sleeping EEG yet, in the future work, we will classify sleeping EEG into more details, e.g. REM, stage 1, 2, 3, and 4 respectively. Moreover we will obtain meaningful results after the tests of more clinical data.

Acknowlegements This work was supported by a grant NO .(HMP-98-G-1-012) of the 98 Highly Advanced National Projects on the development of Biomedical Engineering and Technology, Ministry of Health and Welfare, R.O.K.

References

1. Gandhimohan M. Viswanathan, C.-K. Peng, H. Eugene Stanley, and A.L. Goldberger: Deviations from uniform power law scaling in nonstationary time series. Phys. Rev. **E 55** (1997) 845–849
2. Plamen Ch. Ivanov, Michael G. Rosenblum, C.-K. Peng, Joseph Mietus, Shlomo Havlin, H. Eugene Stanley, and A.L. Goldberger: Scaling behavior of heartbeat intervals obtained by wavelet-based time-series analysis. Nature **383**. (1996) 323–327
3. C.-K. Peng, Havlin S, H.E. Stanley, A.L. Goldberger: Quantification of scaling exponents and crossover phenomena in nonstationary heartbeat time series. Chaos. **5** (1995) 82–87
4. Mandelbrot, B.B.: The Fractal Geometry of Nature. San Francisco, CA, Freeman. (1982)
5. Keshner, M.S.: 1/f Noise. Proc. IEEE, **70** (1982) 212–218
6. Pentland, A.P.: Fractal-Based Description of Natural Scenes. IEEE Trans. Pattern Anal. Machine Intell. **6**, (1984) 661–674
7. S.V. Buldyrev, A.L. Goldberger, S. Havlin, C.-K. Peng, and H.E. Stanley: in Fractals in Science, edited by A. Bunde and S. Havlin (Springer-Verlag, Berlin, 1994) 48–87
8. C.-K. Peng, S.V. Buldyrev: Finite-size efforts on long-range correlations: Implications for analyzing DNA sequences. Phys. Rev. **E 47** (1993) 3730–3733
9. E.W. Montroll, M.F. Shlesinger: in Nonequilibrium Phenomena II. From Stochastics to Hydrodynamics, edited by J. L. Lebowitz and E. W. Montroll (North-Holland, Amsterdam, 1984). 1–121.
10. P. Bak, C. Tang, and K. Wiesenfeld: Phys. Rev. Lett. **59** (1987) 381–384
11. J. Beran: Statistics for Long-Memory Processes. Mew York: Chapman & Hall. (1994)

Model Based Predictive Control Using Genetic Algorithms. Application to Greenhouses Climate Control*

Xavier Blasco, Miguel Martínez, Juan Senent and Javier Sanchis

Predictive Control and Heuristic Optimization Group
Department of Systems Engineering and Control
Universidad Politecnica de Valencia (Spain)
http://www.isa.upv.es/~jsalcedo/WebCPOH/index.html

Abstract. Solving multivariables and non-linear problems with constrains is usual when dealing with control problems. The classical way to solve this was through the decomposition into less complex problems: sub-problems with less variables and through the use of linear approximated models. These methodologies can present good results, but for some, only a suboptimal solution with a poor quality can be reached.
The aim of this work is to combine Model Based Predictive Control (MBPC), a powerful control technique, with Genetic Algorithms, a powerful optimization technique. This combination can overcome limitations when approaching very complex problems in an integral way. This work extends this application to Multi Inputs Multi Outputs modeled with state space representation (a general way to include a wide range of non-linearities) and shows its application to Greenhouse Climate Control.

1 Introduction

Model Based Predictive Control (MBPC) is one of the most intuitive and powerful control techniques. This methodology can be summarized in a few words:

> With a process model and its past behaviour, it is possible to produce predictions of the process dynamic evolution for different control laws. If we could set a cost for each one of these predictions, it would be possible to select the best control law to achieve a fixed objective.

This easy and intuitive way to describe how MBPC works has been the basis of its success in industry [15], [7], [13], [14], [2], [16], [6]. On the other hand, several research works and industrial applications have shown its control capabilities. This means that it is a very interesting methodology for process control.

When described in more detail, all of the controllers with this methodology, have three fundamental elements:

* This work has been partially financed by European FEDER funds, project 1FD97-0974-C02-02

J. Mira and A. Prieto (Eds.): IWANN 2001, LNCS 2084, pp. 457-465, 2001.

1. A **Predictor** that supplies controlled variable predictions for different manipulated variable combinations (control law). These predictions are based on process information (model and variable mesures).
2. A **Cost function** that assigns a cost to each prediction depending on previous fixed objectives.
3. An **Optimization technique** to search for the best control law.

Usually the bottleneck of this methodology is the Optimization Technique. Accurate models commonly have to include non-linearities, even if linear models are accurate enough, a realistic cost function could introduce non-linearities. All these aspects generally produce fairly difficult optimization problems. A Genetic Algorithm (GA) is a competitive way to solve difficult optimization problems when the computing time is not a problem.

Therefore, combining MBPC with GA is a promising alternative to solve complex control problems. This alternative was already proposed and analyzed for SISO (single-input single-output) transfer function model with additional non-linearities such as saturation, dead-zone and backlash [11]. This work extends MBPC with GA to MIMO (multi-input multi-output) processes using a state space representation as a general way to model non-linear proceses.

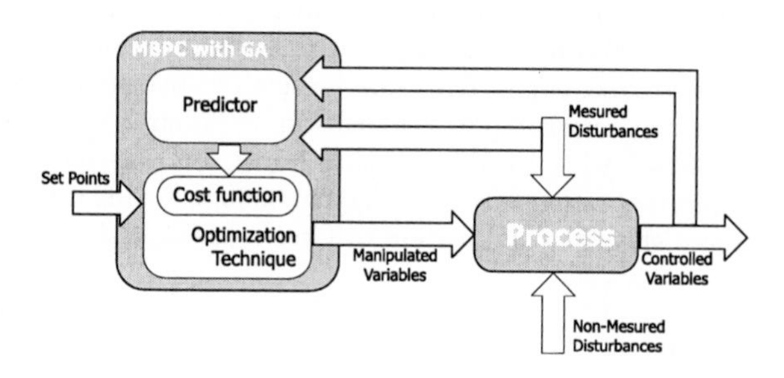

Fig. 1. MIMO Control structure for Model Base Predictive Control with GA.

2 MBPC using GA for MIMO models

MBPC control structure is similar in SISO (see [11]) and MIMO models (Fig. 1). In this case, MBPC elements: predictor, cost function and optimization technique are defined in the following way.

2.1 Predictor

The state space representation of a process is a widely used way to model linear and non-linear processes:

$$\begin{aligned} \dot{X}(t) &= g(X(t), U(t), D(t)) \\ Y_u(t) &= f(X(t), U(t), D(t)) \end{aligned} \tag{1}$$

where:

- $Y_u(t) = [y_{u1}(t), y_{u2}(t), \ldots, y_{ur}(t)]^T$, array of controlled variables.
- $U(t) = [u_1(t), u_2(t), \ldots, u_e(t)]^T$, array of manipulated variables.
- $D(t) = [d_1(t), d_2(t), \ldots, d_p(t)]^T$, array of disturbances.
- $X(t) = [x_1(t), x_2(t), \ldots, x_q(t)]^T$, array of state variables.

To increase robustness in MBPC, meaning better behaviour under uncertainty or non-modeled dynamic or noisy measures, the model is usually corrected by means of an additional term:

$$Y(t) = Y_u(t) + N(t) \tag{2}$$

- $N(t) = [n_1(t), n_2(t), \ldots, n_r(t)]^T$, an array of correction variables.
- $Y(t) = [y_1(t), y_2(t), \ldots, y_r(t)]^T$, array of corrected controlled variables.

An alternative with good performances, to define correction variables ([3], [4]) is to use an ARIMA model:

$$n_i(t) = \frac{T_i(z^{-1})}{\Delta A_i(z^{-1})} \xi_i(t) \tag{3}$$

Where: $T_i(z^{-1})$ is a polynomial designed to increase robustness against noisy mesures (default one is $T_i(z^{-1}) = 1$), $A_i(z^{-1})$ is a polynomial related to $y_i(t)$ dynamic (it could be set to $A_i(z^{-1}) = 1$), $\Delta = (1 - z^{-1})$ to improve steady state behaviour and $\xi_i(t)$ is a white noise.

With this structure, predictions for time '$t+j$' with measures till time 't' are calculated by:

$$Y(t+j|t) = Y_u(t+j|t) + N(t+j|t) \tag{4}$$

$y_{ui}(t+j|t)$ are obtained from the state space model and $n_i(t+j|t)$ are obtained from the following original development:

$$n_i^f(t) = n_i(t)/T_i(z^{-1}) = \xi_i(t)/\Delta A_i(z^{-1}) \tag{5}$$

Then:

$$n_i^f(t+j) = (1 - \Delta A_i(z^{-1}))n_i^f(t+j) + \xi_i(t+j)$$

As $\xi_i(t)$ is a white noise, the best predictions are: $\xi_i(t+j|t) = 0 \ \ \forall j > 0$

$$\begin{aligned} n_i^f(t+j|t) = (1 - \Delta A_i(z^{-1}))n_i^f(t+j|t) &= -\hat{a}_{i1} n_i^f(t+j-1|t) - \\ -\hat{a}_{i2} n_i^f(t+j-2|t) - \ldots &- \hat{a}_{i,j-1} n_i^f(t+1|t) - \hat{a}_{ij} n_i^f(t) - \\ -\hat{a}_{i,j+1} n_i^f(t-1) - \ldots &- \hat{a}_{i,n_a+1} n_i^f(t+j-n_a-1) \end{aligned} \tag{6}$$

Where:

$$\Delta A_i(z^{-1}) = 1 + \hat{a}_{i1} z^{-1} + \hat{a}_{i2} z^{-2} + \cdots + \hat{a}_{i,n_a+1} z^{-(n_a+1)}$$

The best prediction of $n_i(t+j|t)$:

$$n_i(t+j|t) = T_i(z^{-1}) n_i^f(t+j|t) \tag{7}$$

Past data of $n_i(t)$ needed in the calculus are obtained from the difference between measure controlled variables and the controlled variables the from model:

$$n_i(t-j) = y_i(t-j) - y_{ui}(t-j) \tag{8}$$

2.2 Cost function

The next MBPC element to be defined is the cost function, an alternative related to the IAE indicator (a usual way to evaluated control performances [10]) is:

$$J = \sum_{i=1}^{r} \left(\sum_{j=N_{i1}}^{N_{i2}} \alpha_{ij} \left| y_i(t+j|t) - w_i(t+j) \right| \right) \tag{9}$$

Where:

- $y_i(t+j|t)$, controlled variable prediction.
- $w_i(t+j)$, set point for variable $y_i(t+j)$.
- N_{i1} and N_{i2}, prediction horizon limits for variable y_i.
- α_{ij} weighting factors.
- J is a function of manipulated variables:

  $$U = [u_1, u_2, \dots, u_e]^T$$

  As usual in all MBPC, the evolution of control laws is restricted to the control horizon, meaning that each manipulated variable can only change during $[t, t+N_{uk}]$:

  $$\begin{aligned} u_1 &= [u_1(t), \dots, u_1(t+N_{u1}-1)] \\ u_2 &= [u_2(t), \dots, u_2(t+N_{u2}-1)] \\ \vdots \quad & \quad \vdots \\ u_e &= [u_e(t), \dots, u_e(t+N_{ue}-1)] \end{aligned}$$

  N_{uk} are called control horizon.

A correct selection of prediction and control horizon and weighting factors are crucial to achieve good control performances. A default selection would have to be:

- A Prediction horizon has to include the complete transient evolution of controlled variables.
- A Low control horizon produces a more conservative response, meaning less aggressive control actions. With a high control horizon control, actions are higher and obtaining better performances is possible. A conservative selection would be $N_{uk} = 1$, a more aggressive selection would be the number of state variables.
- Weighting factors have to normalized the variables involved in cost function, a natural way to do it would be:

$$\alpha_{ij} = \frac{1}{|w_i(t+j)|} \tag{10}$$

The objective is to normalize the predicted error when compared to a set-point. If increasing the contribution of a variable becomes necessary, it can be done by changing the weighting factor in α'_{ij} (see [5]):

$$\alpha_{ij} = \frac{\alpha'_{ij}}{|w_i(t+j)|} \tag{11}$$

Changing α'_{ij}, it is possible to change the priority of control for each variable in the cost function.

2.3 Optimization technique

The choice of optimization technique depends on the type of problem to be solved. The analytical solution is the best option if possible: it is the most exact one and the least computer time consuming because optimization is done off-line. But for complex models and indexes, the optimization problem is so difficult to solve that an analytical solution is impossible and not even classical numerical solutions are available, as most of them require convexes problems. For these optimization problems, Genetic Algorithms are a good alternative and they have demonstrated very good performances.

Genetic Algorithms ([8], [12], [9]) are optimization techniques based on the simulated evolution of species. Problem solution is obtained from the evolution of several generations in a population formed by a set of possible solutions. Evolution is performed following rules represented by genetic operators: selection, crossover and mutation. Differences between GA are marked by: Chromosome codification and Genetic operators. A good GA choice for MBPC is [1]:

- Real value codification.
- Linear ranking.
- Selection operator: Stochastic Universal Sampling.
- Crossover operator: Linear combination.
- Mutation operator: Oriented mutation.

3 Greenhouse climate control

This section shows an application of MBPC with GA for greenhouse climate control. This greenhouse is devoted to rose crop[1]. Process model can be obtained from the mass and energy balance, including the biological behaviour of the plants (see [1] for more details).

[1] This greenhouse is located at the Instituto Valenciano de Investigaciones Agrarias (IVIA) in Valencia (Spain).

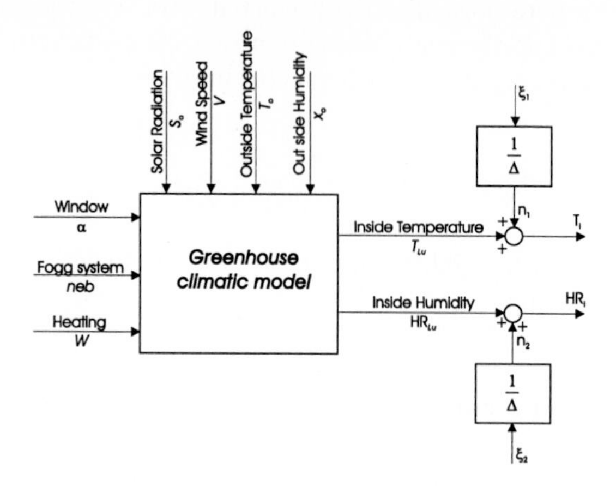

Fig. 2. Climatic greenhouse model used for prediction.

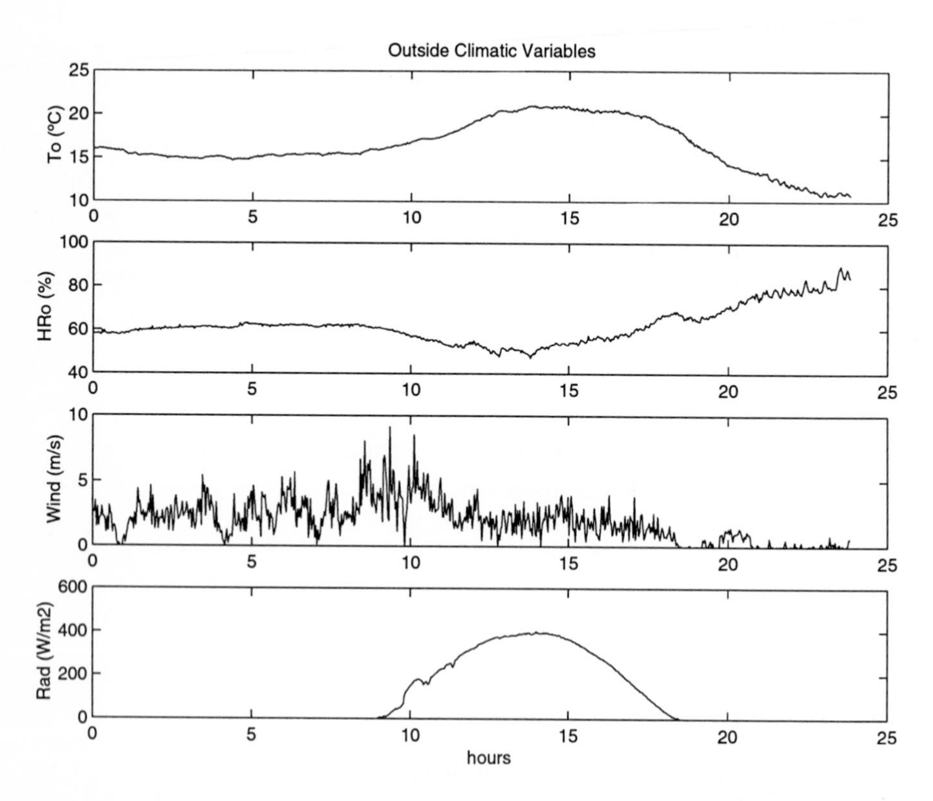

Fig. 3. External perturbation measured on the 28th November 1998. Temperature (^oC), Relative Humidity (%), Wind Speed (m/s) and Solar Radiation (W/m^2).

Controlled variables are:

- Inside Humidity (HR_i in %).
- Inside Temperature (T_i in ^oC).

Manipulated variables are:

- Aperture of roof vents (α).
- Heating power (W).
- Water rate of the fog system (neb).

Disturbances are:

- Solar radiation (S_o in $W \cdot m^{-2}$).
- Outside temperature (T_o in ^oC).
- Outside Humidity (HR_o in %).
- Wind speed (V in m/s).

The model has important non-linearities that are described in detail on [1], in addition, it is a complex process with multiple disturbances (some of them measurable). Therefore, MBPC with GA can constitute a valid alternative. A limiting factor would be computing time, but this process has a slow dynamic and allows for a high sample time (2 minutes).

Predictions are obtained from the model and an additional term designed to increase robustness, which for this particular example is set to $T(z^{-1}) = 1$. The model used to generate the predictions is shown on figure 2.

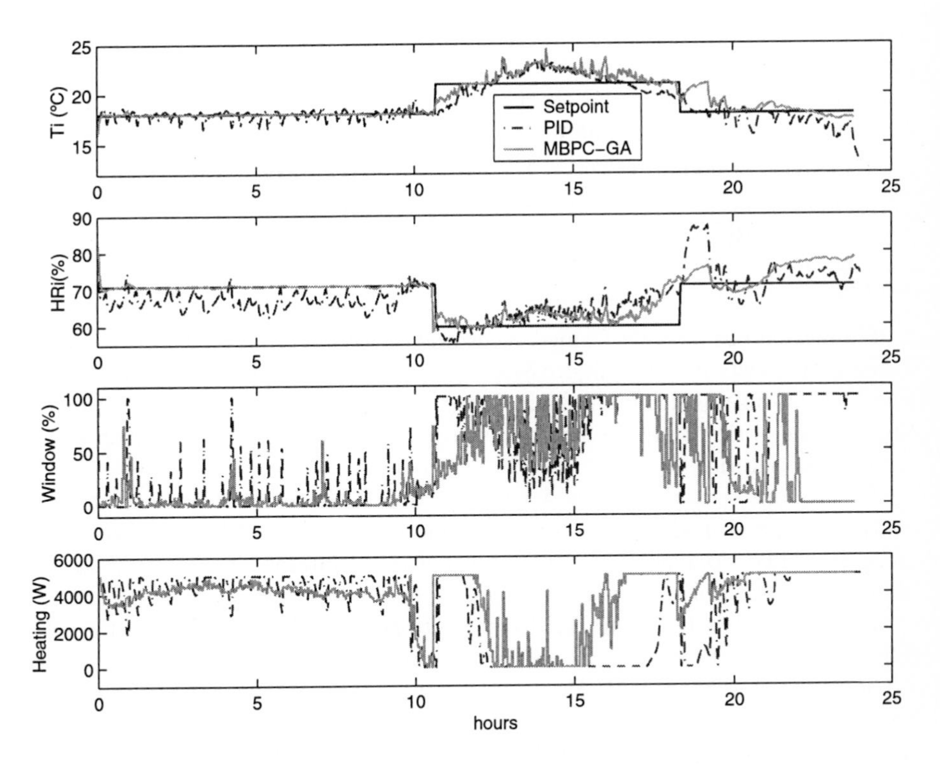

Fig. 4. Temperature and Humidity MBPC-GA vs PID control.

The cost function used in this application is:

$$J = \sum_{j=1}^{N_2} \alpha_{j1} \left| HR_i(t+j|t) - HR_{io}(t) \right| + \sum_{j=1}^{N_2} \alpha_{j2} \left| T_i(t+j|t) - T_{io}(t) \right|$$

Where: $HR_{io}(t)$ and $T_{io}(t)$ are relative humidity and temperature setpoints.

$$\alpha_{j1} = \frac{1}{|HR_{io}(t)|} \; ; \; \alpha_{j2} = \frac{1}{|T_{io}(t)|}$$

The cost function does not include future setpoints, it is calculated based on setpoint in time 't'. If following an established setpoint trajectory where needed, the capability could be improved by introducing future setpoints in the cost function.

The prediction and control horizon are set to: $N_2 = 5$ and $N_u = 1$. With $N_2 = 5$ (meaning 10 min), the measurable perturbations could be considered constant in all the prediction horizon without inducing to too much error (perturbation doesn't show any dramatic changes during those 10 min, with the exception of wind speed); and this time would be enough to be able to accurately watch the transient response of the controlled variables. If N_2 is set to a greater value, a perturbation estimation would be necessary to obtain reliable predictions.

$N_u > 1$ is a possible setting which would produce a more aggressive control but as it can be seen, the manipulated variables are working near their saturation value and therefore increasing the computational cost is unnecessary if this more aggressive control law can not be applied.

Optimization technique is the GA describe in previous section, setting parameters to (these settings allows a realtime execution):

- Crossover probability $P_c = 0.7$, weighting factor in linear combination α is randomly set for each individual.
- Mutation probability $P_m = 0.1$.
- Number of individuals in each generation $NIND = 25$.
- Maximum number of generations $MAXGEN = 15$.

Simulation has been performed according to the perturbations corresponding to the 28th november 1998 (figure 3). An additional restriction set by greenhouse users is that during autumn and winter time, the fog system is turned off (during summer and spring time the heater is turned off), these restrictions respond to economical criteria.

Results are shown in figure 4 where it is possible to see conventional PID control results (tuned using a linearized model). MBPC with GA have better control quality: controlled variables are better maintained near setpoints and manipulated variables are not so aggressive compared to PID control.

During the first ten hours MBPC is quite good, controlled variables exactly follows setpoints, PID control is less accurate. During the day both control have similar performances but MBPC is a bit less aggressive. For the period of time between 18h00 and 24h00 in both controls manipulated variables are usually saturated meaning it is not possible to reach setpoints, still MBPC seems more accurate than PID control.

4 Conclusions

This work shows how to combine Predictive Control with Genetic Algorithms in order to overcome complex control problems (non linear, restrictions, disturbances, etc.). An application for Greenhouse climate control is presented showing an improved performance compared to conventional PID control, even without using all the possibilities of MBPC, such as noise rejection with polinomial $T(z^{-1})$.

For this technique to be useful, it requires a reasonably accurate non linear model. Nevertheless, non linear techniques require and therefore it is not presented as a disadvantage. If this model can be obtained, control becomes quite easy and intuitive. With the only real limitation being computational cost.

References

1. F.X. Blasco. *Model based predictive control using heuristic optimization techniques. Application to non-linear and multivariables proceses.* PhD thesis, Universidad Politécnica de Valencia, Valencia, 1999 (In Spanish).
2. E.F. Camacho and C. Bourdons. *Model Predictive Control in the Process Industry.* Springer, 1995.
3. D.W. Clarke, C. Mohtadi, and P.S. Tuffs. Generalized Predictive Control-Part I. *Automatica*, 23(2):137–148, 1987.
4. D.W. Clarke, C. Mohtadi, and P.S. Tuffs. Generalized Predictive Control-Part II. Extensions and Interpretations. *Automatica*, 23(2):149–160, 1987.
5. J.L. Cohon. *Multiobjective programming and plannig.* Academic Press, UK, 1978.
6. C.R. Cutler and D.L. Ramaker. Dynamic matrix control - a computer control algorithm. In *Proceedings of the joint automatic control conference (JACC)*, San Francisco, CA, 1980.
7. J.B. Froisy. Model predictive control: past, present and future. *ISA Transactions*, 33:235–243, 1994.
8. D.E. Goldberg. *Genetic Algorithms in search, optimization and machine learning.* Addison-Wesley, 1989.
9. J.H. Holland. *Adaptation in natural and artificial systems.* Ann Arbor: The University of Michigan Press, 1975.
10. T. E. Marlin. *Process Control. Designing Processes and Control Systems for Dynamic Performance.* Mc Graw-Hill, 1995.
11. M. Martínez, J.S. Senent, and F.X. Blasco. Generalized predictive control using genetic algorithms (GAGPC). *Engineering applications of artificial intelligence*, 11(3):355–368, 1998.
12. Z. Michalewicz. *Genetic Algorithms + Data Structures = Evolution Programs.* Springer, 1996.
13. M. Morari and J. H. Lee. Model predictive control: past, present and future. In *PSE'97-ESCAPE-7 symposium*, Trondheim, Norway, 1997.
14. K.R. Muske and J.B. Rawlings. Model predictive control with linear models. *AIChE Journal*, 39(2):262–287, 1993.
15. S.J. Qin. An overview of industrial model predictive control technology. In *Proceedings AIchE symposium serie 316*, volume 93, pages 232–256, 1996.
16. J. Richalet, A. Rault, J. L. Testud, and J. Papon. Model predictive heuristic control: Applications to industrial processes. *Automatica*, 14:413–428, 1978.

Nonlinear Parametric Model Identification with Genetic Algorithms. Application to a Thermal Process*

X. Blasco, J.M. Herrero, M. Martínez, J. Senent

Predictive Control and Heuristic Optimization Group
Department of Systems Engineering and Control
Universidad Politecnica de Valencia (Spain)
http://ctl-predictivo.upv.es

Abstract. One of the first steps taken in any technological area is building a mathematical model. In fact, in the case of process control, modelling is a crucial aspect that influences quality control. Building a nonlinear model is a traditional problem. This paper illustrates how to built an accurate nonlinear model combining first principle modelling and a parametric identification, using Genetic Algorithms. All the experiments presented in this paper are designed for a thermal process.

1 Introduction

Nonlinear models are difficult to obtain because of the high degree of complexity presented by both the structure determination and the parameter estimation [4].

We can divide mathematical models into two types [2]: Phenomenological and behavioural models. The first ones are obtained from first principles and result in a state space representation (a set of differential equations). In these models, parameters have a meaning, which could be useful to validate the model as it means that the use of prior information is viable. Behavioural models try to approximate process evolution without prior information, for instance, with a polynomial, a neural network, a fuzzy set, etc. The selection among these types of structures is not simple. This paper will be focused on exploiting prior knowledge, so the model will be obtained from first principles.

A second aspect to consider is parameter estimation. There is a well established set of identification techniques [4],[2] for linear models, although the same can not be said about nonlinear ones. Most of these techniques are based on an optimization of a cost function. When the cost function is convex, the local optimizer supplies the solution, but in the case of a nonconvex function, a global optimization becomes necessary. In the case of nonlinear models, obtaining a nonconvex function is easy and in this context, Genetic Algorithms [3],[1] offers a good solution for off-line optimization.

* This work has been partially financed by European FEDER funds, project 1FD97-0974-C02-02

J. Mira and A. Prieto (Eds.): IWANN 2001, LNCS 2084, pp. 466-473, 2001.

2 Parameters Identification by Means of Genetic Algorithms

The technique is based on the acceptance of an initial structure of the model, whose parameters are unknown (or at least part of them), the main goal is determining the parameters of the model. Habitually, the model of the process (linear or nonlinear) is represented by means of a set of first order differential equations that can be obtained from the first physical principles.

$$\begin{aligned} \dot{x} &= f(x, u, \theta) \\ \hat{y} &= g(x, u, \theta) \end{aligned} \tag{1}$$

where:

- f, g: model's structure.
- θ: parameters of the model to identify.
- u: inputs to the model (m).
- $\hat{y}$ outputs from the model (l).
- x: state variables (n).

The main goal is to obtain a model behaviour as similar as possible to the real process.

The process behaviour can be obtained by means of experiments and in the case of nonlinear processes, the experiments covering the whole set of behaviours of the model becomes even more important. The input signals do not need to be white noises (as in other identification techniques); usually, a set of inputs of step shape along the inputs' suitable space is considered enough.

The model's behaviour can be obtained from its simulation, applying the same input signals used in the experiment. Generally, the models are continuous and so, it is necessary to use numerical integration methods such as Runge-Kutta's.

Mathematically, this goal is achieved through minimizing the cost function where the differences presented along the experiment, among the process' output and the model, are penalized. For example:

$$J(\theta) = \sum_{j=1}^{te} \sum_{i=1}^{l} k_{ij} |y_i(j) - \hat{y}_i(j)| \tag{2}$$

where:

- te: experiment samples.
- l: number of outputs.
- k_{ij}: output pondering coefficient i, for sample j.
- $y_i(j)$: sample j of the real process' output i.
- $\hat{y}_i(j)$: simulation of the model's output i for the instant of time corresponding to sample j.

Using different pondering values for k_{ij} has the following goals in mind:

- Escalating the rank of the different outputs.
- Giving more importance to certain samples of outputs taken in certain instants of time, as strategic points in the process' answer (over-oscillation zones,etc.)

The objective has been focused on finding parameters to minimize the $J(\theta)$. It is here where Genetic Algorithms play their role. Genetic Algorithms (GA) are optimization techniques based on simulating the phenomena that take place in the evolution of species and adapting it to an optimization problem. These techniques imply applying the laws of natural selection onto the population to achieve individuals that are better adjusted to their environment.

The population is nothing more than a set of points in the search space. Each individual of the population represents a point in that space by means of his chromosome (identifying code). The adaptation degree of the individual is given by the objective function.

Applying genetic operators to an initial population simulates the evolution mechanism of individuals. The most usual operators are as follows:

- Selection: Its main goal is selecting the chromosomes with the finest qualities to integrate the next population
- Crossbreeding: Combining the chromosomes of two individuals, new ones are generated and integrated into the society.
- Mutation: Aleatory variations of parts of the chromosome of an individual in the population generate new individuals.

A good identification could result from the optimizing phase, in which case, validating the model for a different set of data would be the next step. On the other hand, if the identification results unsatisfactory, the initial structure of the model must be restated or some of the GA's parameters (search space, cross operators, selection and mutation) must be modified if a local minimum found by the optimizer is suspected.

3 Application to the Identification of a Thermal Process

3.1 Description of the Process

The process consists of a scale model of a furnace in which the typical behaviours of thermal processes are studied. The energy contribution inside the process is due to the power dissipated by the resistance placed inside. A ventilator continually and constantly introducing air from the outside produces the air circulation inside the process.

The actuator is constituted by a tension source controlled by tension, the input rank of the actuator is $0 \div 100\%$. Two thermopairs are used to measure the resistance temperature and the temperature inside the furnace, in the rank of $-50 \div 250°C$. The purpose is modelling the dynamic of the resistance temperature.

3.2 Problems Concerning the Obtention of the Model

It is well known that the behaviour of the process is nonlinear with respect to the input and depends on its inside temperature (perturbation) and on the air renovation flow (assumed constant). Non-linearity can be observed on figure 1, where after introducing two step inputs to the process and identifying (through traditional methods) the answer of the first step by means of linear model 3, this does not accurately fit the dynamics of the second step.

$$\frac{T(s)}{u(s)} = \frac{0.935}{(210s+1)(8s+1)} \frac{^oC}{\%} \tag{3}$$

These linear models are only suitable around a working point. When the rank of work is broad, to achieve a solution it is necessary to obtain linear multimodels (usually unfeasible when the dynamics of the process depend on too many parameters, or when non-linearity is not too strong) or to set the structure of a nonlinear model according to physical laws and adjusting its parameters.

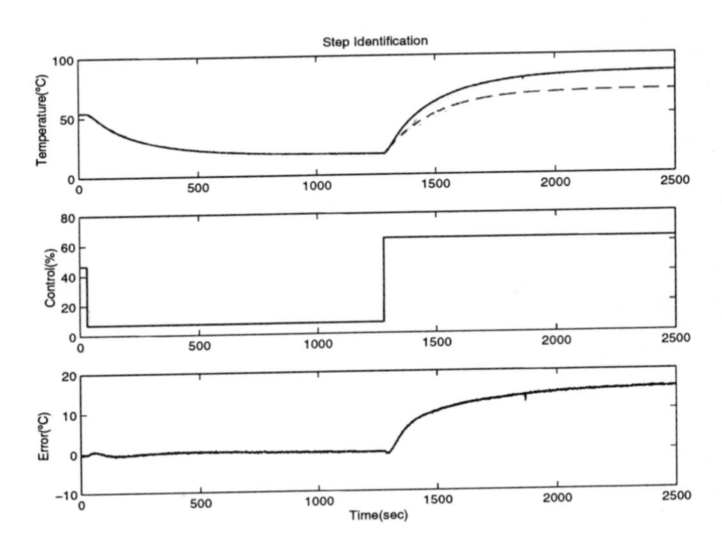

Fig. 1. Identification of the process by means of a lineal model. The continuous line represents the answer of the process, the discontinuous line represents the model's. Underneath, the input to the process and the modelling error are illustrated.

3.3 Structure of the Nonlinear Model. Identification of Its Parameters

According to thermodynamics' physic laws [6] the initial structure of the model can be defined by means of the following differential equations, in which convection and conduction heat losses are modelled.

$$\dot{x_1} = \frac{k_1}{k_2}u^2 - \frac{1}{k_2}(x_1 - T_i)$$
$$\dot{x_2} = \frac{1}{k_3}(x_1 - x_2)$$
$$\hat{y} = x_2$$

where:

- $\hat{y}$: temperature of the resistance (°C)
- u: tension of input to the actuator(%).
- T_i: temperature in the resistance's periphery (°C).
- $k_{1..3}$: parameters of the model to be identified.

The cost function employed for the identification is as follows:

$$J(k_1, k_2, k_3) = \sum_{j=1}^{te=2500} |y(j) - \hat{y}(j)| \tag{4}$$

The Genetic Algorithm employed displays the following characteristics:

codification	Real
n^o of individuals in the population	400
type of selection operator	Stochastic Universal Sampling with ranking
type of cross operator	Lineal recombination. Pc=0.9
type of mutation operator	Aleatory with normal distribution with 50% variance of the search space of each parameter. Pm=0.2

The implementation of Genetic Algorithms has been performed in C and with the mathematical libraries NAG [5].

The search spaces for each parameter are as follows:

$$k_1 \in [0, 0.1] \quad k_2 \in [100, 300] \quad k_3 \in [3, 12]$$

Figure 2 illustrates the result of the identification, the value of the parameters obtained by the GA are as follows:

$$k_1 = 0.014778 \quad k_2 = 186.5761 \quad k_3 = 6.5918$$

The cost function represents a value of 2607, considering an experiment of 2500 samples (with 1-sec period) has been performed, the mean error is 1.05 [grades/sample]. The improvement of the model's identification could be considered by resetting the initial structure of the model presented in (5) in adding

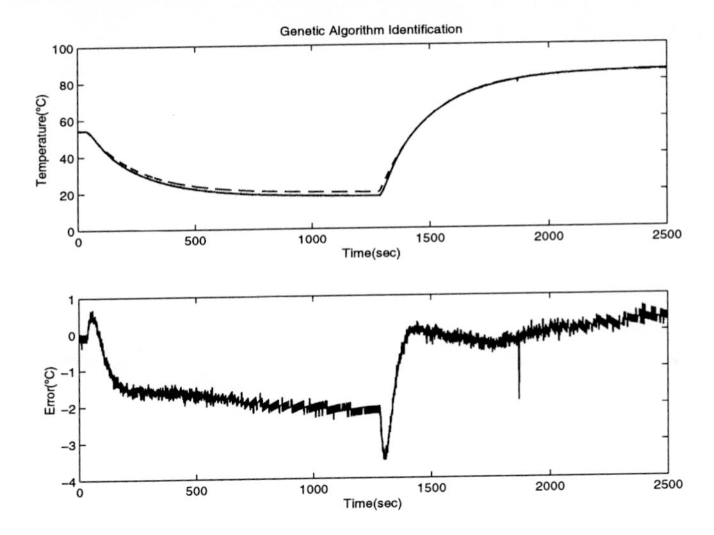

Fig. 2. Identification of the nonlinear model 5 by means of Genetic Algorithms. The continuous line represents the answer of the process, the discontinuous line represents the model's. Underneath, the modelling error is illustrated.

radiation losses to the model.

$$\dot{x_1} = \frac{k_1}{k_2}u^2 - \frac{1}{k_2}(x_1 - T_i) - k_3(\frac{273 + x_1}{100})^4$$
$$\dot{x_2} = \frac{1}{k_3}(x_1 - x_2) \tag{5}$$
$$\hat{y} = x_2$$

The identification process is repeated adding a new search dimension (the one corresponding to parameter k_4), maintaining the cost function $J(k_1, k_2, k_3, k_4)$ and increasing to 600 the number of individuals used by the optimizer. The rank used for k_4 has been $\in [0, 0.001]$.

Figure 3 illustrates the result of the identification, the values of the parameters obtained by the GA are as follows:

$$k_1 = 0.0160 \quad k_2 = 196.550 \quad k_3 = 5.0523 \quad k_4 = 0.000153$$

Figure 4 illustrates the evolution of the parameters in each optimizer iteration. Once the model is completed, it has to be validated, figure 5 illustrates the answer of the model with respect to the answer of the process obtained for a different experiment; the model can be observed to be fitting the answer of the process and therefore being validated. The cost function presents a value of 549.6, so the mean error is 0.22 [grades/sample].The mean error obtained in this last experiment, considering it carries 5300 samples and the cost function is

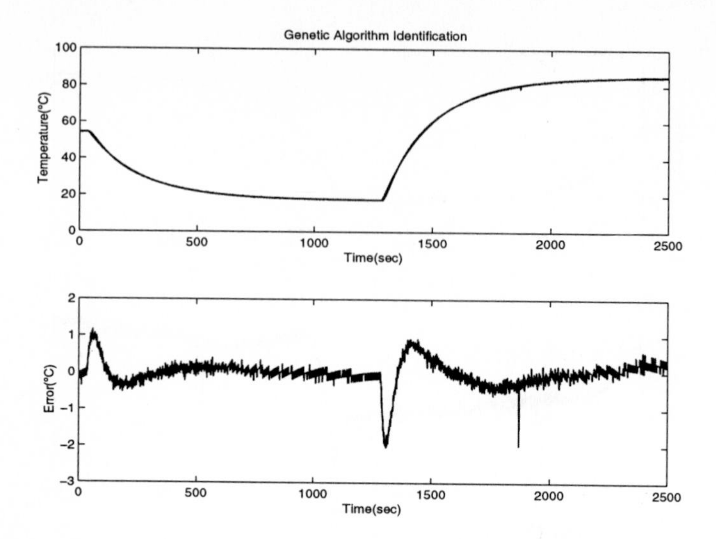

Fig. 3. Identification of the nonlinear model 7 by means of Genetic Algorithms. The continuous line represents the answer of the process, the discontinuous line represents the model's. Underneath, the modelling error is illustrated.

2076, amounts to 0.39 [grades/samples], which can be considered adequate taking into account that the quantification error when sampling the signal equals 0.15 grades.

4 Conclusions

The potential of GA as global optimizers permits undertaking complex optimization problems and therefore allows for greater degrees of freedom in the selection of the model's structure.

Genetic Algorithms have proven to be an alternative for the (off line) identification of parameters in models, especially in nonlinear models, allowing the a priori use of the physical knowledge of the process. Therefore, this technique is an attractive alternative to those methods based on neural network or fuzzy logic.

The use of GA is considered as a future line of work for the optimal fitting (off line) of linear controllers such as PID, so if a reliable nonlinar model is given, the optimum PID can be fitted for a given cost function. In conclusion, it is a similar idea to that of the nonlinear identification stated in this paper.

As a final point, even though the example used in this paper corresponds to a MISO process, the translation of the method to the identification of MIMO processes is direct. Nevertheless, the complexity presented by the identification of the model's parameters is partly determined by the number of existing parameters.

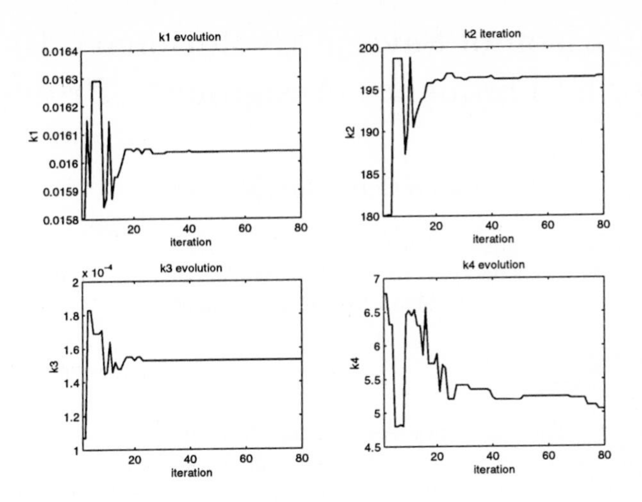

Fig. 4. Values of the parameters obtained by the GA in each iteration.

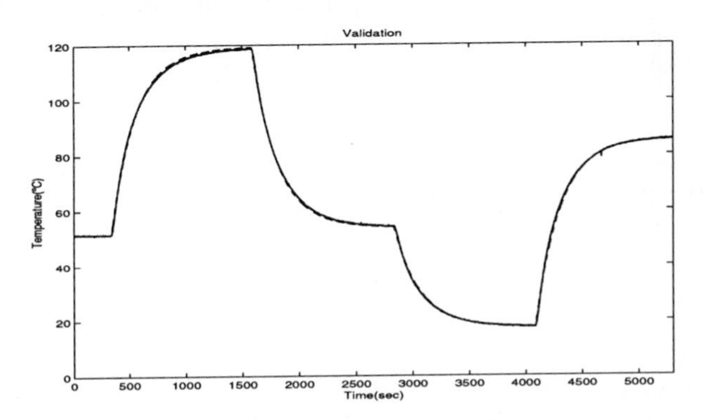

Fig. 5. Validation of the nonlinear model 7. The continuous line represents the answer of the process, the discontinuous one represents the model's.

References

1. F.X. Blasco. *Model based predictive control using heuristic optimization techniques. Application to non-linear and multivariables proceses.* PhD thesis, Universidad Politécnica de Valencia, Valencia, 1999 (In Spanish).
2. L. Pronzalo E. Walter. *Identification of parametric models from experimental data.* Springe Verlang, 1997.
3. D.E. Goldberg. *Genetic Algorithms in search, optimization and machine learning.* Addison-Wesley, 1989.
4. R. johansson. *System modeling identification.* Prentice Hall, 1993.
5. Numerical Algorithms Group NAG. NAG Fortran Library. Introductory guide. Mark 18. Technical report, 1997.
6. M. Zamora. *A study of the thermodynamic systems.* Universidad de Sevilla, 1998 (In Spanish).

A Comparison of Several Evolutionary Heuristics for the Frequency Assignment Problem

Carlos Cotta, José M. Troya

Dpto. de Lenguajes y Ciencias de la Computación, Univ. de Málaga
Campus de Teatinos (3.2.49), 29071 - Málaga - Spain
{ccottap, troya}@lcc.uma.es

Abstract. The Frequency Assignment Problem (FAP) is a very important problem of practical interest. This work compares several evolutionary approaches to this problem, based both in the *forma analysis* framework, and in the *decoder* paradigm. All approaches are studied from the point of view of two different quality measures of assignments: the number of distinct frequencies, and the frequency span. It is shown that using decoders as embedded heuristics is more adequate than performing a direct search in the feasible solution space. Furthermore, despite the apparent symmetry of the problem, a recombination operator based on multiple-emitter-to-frequency preservation performs better than focusing on multiple-frequency-to-emitter preservation.

1 Introduction

The term Frequency Assignment Problem (FAP) comprises a number of optimization problems of great difficulty (NP-hard in general). Although presented under different flavors, all FAPs essentially consist of finding an assignment of a set of frequencies to a set of emitters fulfilling some specific constraints (e.g., avoiding interference between closely located emitters). The actual proliferation of cellular phone networks, local television stations, etc. clearly underpins the practical interest of these problems.

The above mentioned NP-hardness of most FAPs imply that exact techniques are inherently limited for solving these problems. For this reason heuristic techniques such as tabu search, simulated annealing or genetic algorithms (GAs) are frequently used for the resolution of FAPs [3, 4, 8]. This work focuses on the application of GAs for this purpose. To be precise, we compare two evolutionary approaches to this problem, a direct search in feasible space via specifically designed operators, and an indirect search via permutation decoders.

The remainder of the paper is organized as follows. First, Section 2 provides a formal definition of the FAPs considered in this work. Next, the two approaches are presented in Section 3, describing some different variants of each one. Subsequently, empirical results are reported in Section 4. Finally, some conclusions are extracted and future work is outlined in Section 5.

J. Mira and A. Prieto (Eds.): IWANN 2001, LNCS 2084, pp. 474-481, 2001.

2 Frequency Assignment Problems

As mentioned in the previous section, there exist a number of FAP variants, so it is necessary to give a precise definition of the particular FAPs considered in this work. As with any optimization problem, three elements must be given in order to define a FAP: a characterization of problem instances, a characterization of problem solutions, and a quality measure.

Definition 1 (FAP Instance). *An instance of the FAP is a tuple FAP*$(\mathcal{E}, \mathcal{F}, \mathcal{D}, \mathcal{R}, \mathcal{I})$ *where*

- $\mathcal{E} = \{e_1, \cdots, e_n\}$ *is a set of* emitters.
- $\mathcal{F} = \{f_1, \cdots, f_m\}$ *is a set of available* frequencies.
- $\mathcal{D} : \mathcal{E} \times \mathcal{E} \to \mathbb{R}$ *is a function such that* $D(e, e')$ *is the distance between emitters* e *and* e'.
- $\mathcal{R} : \mathcal{E} \to \mathbb{N}$ *is a function such that* $\mathcal{R}(e)$ *is the number of frequencies required by emitter* e.
- $\mathcal{I} : \mathbb{R} \to \mathbb{N}$ *is a function such that* $\mathcal{I}(d)$ *is the frequency separation needed to avoid interference between two emitters separated by a distance* d.

According to this definition, it is easy to see that there exist two central constraints in a FAP instance that must be satisfied: the number of frequencies assigned to an emitter must be equal to the number of frequencies it demands, and these frequencies must not interfere with frequencies assigned to other emitters. This is formalized below:

Definition 2 (FAP Solution). *A solution for a FAP instance* $I(\mathcal{E}, \mathcal{F}, \mathcal{D}, \mathcal{R}, \mathcal{I})$ *is a vector* $\alpha = \langle \alpha_1, \cdots, \alpha_n \rangle \in [2^{\mathcal{F}}]^n$ *such that*

- $|\alpha_e| = \mathcal{R}(e)$, *i.e., each emitter is assigned the number of different frequencies it demands.*
- $\forall e, e' \in \mathcal{E} \nexists f, f' \in \mathcal{F} \; : f \in \alpha_e, \; f' \in \alpha_{e'}, \; |f - f'| < \mathcal{I}(\mathcal{D}(e, e'))$, *i.e., no interfering frequencies are assigned to two emitters.*

For the purposes of this work, we will consider that $|\mathcal{F}|$ is high enough to allow the existence of FAP solutions as shown in the previous definition. An upper bound for the cardinality of $\mathcal{F}$ is thus $\mathcal{I}(0) \cdot \sum_{e \in \mathcal{E}} \mathcal{R}(e)$, assuming that $\mathcal{I}$ is monotonically decreasing, as usual.

A quality function must be defined now, in order to quantify the goodness of a particular FAP solution. In this work, we will consider two different quality measures. The first one is termed the *frequency span*, and is defined below.

Definition 3 (Frequency Span). *The frequency span* $F(\alpha)$ *of a FAP solution* α *is*

$$F(\alpha) = \max_{e, e' \in \mathcal{E}} \left[\max_{f \in \alpha_e, f' \in \alpha_{e'}} (|f - f'|) \right] , \tag{1}$$

i.e., the maximum separation between assigned frequencies.

Thus, the optimal solution with respect to this quality measure is the one that satisfies the problem constraints within the smallest frequency interval. This is important in situations in which the frequency spectrum is partitioned into disjoint compact sets (e.g., a set of frequencies per city or province), and we require to fit the frequency demand of a group of emitters within one of these sets. A related, but generally different measure of FAP solution is its *size*:

Definition 4 (Assignment Size). *The size S of a FAP solution α is*

$$S(\alpha) = |\cup_{e\in\mathcal{E}}\alpha_e| \ , \tag{2}$$

i.e., the number of different frequencies assigned to emitters in $\mathcal{E}$.

Hence, the above quality measure tries to promote frequency re-utilization (notice that this re-utilization does not necessarily result in lower frequency spans).

3 Two Evolutionary Approaches for the Frequency Assignment Problem

This section will describe two different approaches for tackling FAP instances. Both mechanisms are based on restricting the search to the feasible space, but differ in the way they achieve this. On one hand, FAP solutions can be directly manipulated during recombination and mutation. On the other hand, this manipulation can be done indirectly via a construction heuristic. These two approaches are discussed below.

3.1 Direct Manipulation in Feasible-Space

As mentioned above, the first approach consists of directly manipulating frequency assignments. It is thus necessary to define the information units that will be subject to this manipulation. Let us consider the set of equivalence relations $\Psi = \{\psi_{ef} \mid e \in \mathcal{E},\ f \in \mathcal{F}\}$, where $\psi_{ef}(\alpha, \alpha') = \texttt{TRUE}$ if, and only if, frequency f is assigned to emitter e both in α and α' or in neither of them. Subsequently, each equivalence relation ψ_{ef} induces two equivalence classes, respectively comprising solutions assigning f to e (ψ^1_{ef}) or not (ψ^0_{ef}). Each of these equivalence classes is termed a *basic forma* [6].

Ψ can be shown to be an independent set covering the feasible space, so it can be used to induce a representation of solutions, i.e., $\alpha = \langle\alpha_1, \cdots, \alpha_n\rangle \equiv \bigcap_{i=1}^{n}\bigcap_{f\in\alpha_i}\psi^1_{e_if}$. On the basis of this representation, three different information units can be processed: partial emitter assignments (i.e., formae ϕ^F_e, $e \in \mathcal{E}$, $F \in 2^{\mathcal{F}}$, defined as $\phi^F_e = \bigcap_{f\in F}\psi^1_{ef}$), partial frequency assignments (i.e., formae η^E_f, $f \in \mathcal{F}$, $E \in 2^{\mathcal{E}}$ defined as $\eta^E_f = \bigcap_{e\in E}\psi^1_{ef}$), and single emitter-frequency assignments (i.e., single formae ψ^1_{ef}, $e \in \mathcal{E}$, $f \in \mathcal{F}$).

In order to manipulate these information units, it must be taken into account that none of them are orthogonal, although they are separable. This is formally established below.

Proposition 1. *Single emitter-frequency assignments are not orthogonal.*

Proof. The proof is straightforward. Given two formae ψ^1_{ef} and $\psi^1_{e'f'}$, their intersection is empty if $|f - f'| < \mathcal{I}(\mathcal{D}(e, e'))$. This is true for many values of e, e', f, and f' unless $\mathcal{R}(d) = 0$ for all d (a trivial situation without interest from an optimization point of view). □

Since partial emitter-assignments and partial frequency-assignments are defined as the intersection of single emitter-frequency assignments, it follows as a corollary that none of these units is orthogonal.

Proposition 2. *Single emitter-frequency assignments are separable.*

Proof. It must be shown that given ψ^1_{ef} and $\psi^1_{e'f'}$ $(\psi^1_{ef} \cap \psi^1_{e'f'} \neq \emptyset)$, no $\alpha \in \psi^1_{ef}$, $\alpha' \in \psi^1_{e'f'}$ exist such that $\psi^1_{ef} \cap \psi^1_{e'f'} \cap \Omega = \emptyset$, where Ω is the intersection of all basic formae common to α and α'. For this latter intersection to be empty, it must be that either f or f' (or both) interfere with some frequency assignments in Ω. But this is impossible because the existence of α and α' respectively implies that $\psi^1_{ef} \cap \Omega \neq \emptyset$ and $\psi^1_{e'f'} \cap \Omega \neq \emptyset$. □

This separability result implies that the information common to two assignments α and α' can be respected and, simultaneously, compatible information can be assorted. Nevertheless, this is generally incompatible with forma transmission, i.e., it may be necessary to introduce some exogenous information -not present either in α or α'- in the assignment produced during recombination. This new information could be selected at random, or by means of a heuristic. In this work we consider the latter approach, which falls within the patching model termed *locally optimal completion* [7] (the precise heuristics used for this purpose will be described in the next subsection).

3.2 The Decoder Approach

The decoder approach is a completely different way of carrying out the search in the feasible region. In this case, assignments are not directly manipulated. On the contrary, some external structures are processed, being a so-called *decoder* used to translate these structures into feasible assignments in order to perform evaluation. A very typical situation is the use of permutation decoders, due to the fact that permutations are a well-known structure for which different reproductive operators are available (a good survey can be found in [2]). This is the approach considered in this work.

Before defining the particular decoders considered in this work, notice that FAPs are closely related to coloring problems, as pointed out in [5]. A FAP instance can be represented as a labeled graph $G(V, E)$, where $V \equiv \mathcal{E}$, and $(e, e', \delta) \in E \Leftrightarrow \mathcal{I}(\mathcal{D}(e, e')) = \delta$ $(\delta > 0)$. A FAP solution would then be a multicoloring of the graph, such that each vertex is assigned as many colors as frequencies demands, and the colors assigned to adjacent vertices satisfy the separation constraint δ of the edge connecting them. An algorithm for obtaining such a coloring of the graph is shown in Fig. 1.

First-Available-Frequency Heuristic

1. Let $P = \langle e_{i_1}, e_{i_2}, \cdots, e_{i_n} \rangle$ be a permutation of the vertices in V.
2. For all $j \in \{1, \cdots, n\}$ do $\mathcal{A}_j \leftarrow \mathcal{F}$.
3. For all $j \in \{1, \cdots, n\}$ do
   (a) Let $\alpha_{e_{i_j}} \leftarrow \emptyset$.
   (b) For all $k \in \{1, \cdots, \mathcal{R}(e_{i_j})\}$ do
      i. Let $f \leftarrow \min_{f' \in \mathcal{A}_j} f'$.
      ii. Let $\alpha_{e_{i_j}} \leftarrow \alpha_{e_{i_j}} \cup \{f\}$.
      iii. For all $j' \in \{j, \cdots, n\}$, $(e_{i_j}, e_{i_{j'}}, \delta) \in E$ do
      $\mathcal{A}_{e_{j'}} \leftarrow \mathcal{A}_{e_{j'}} - \{f' \mid \delta > |f - f'|\}$.

Fig. 1. Pseudocode of the First-Available-Frequency heuristic.

Notice now that associated to the mentioned labeled graph, there exists a dual graph in which the vertices are frequencies, and edges are labeled with subsets of $\mathcal{E}$. In this dual graph, the edge (f, f', σ) means that frequencies f and f' cannot be simultaneously assigned to nodes $e, e' \in \sigma$. Hence, a FAP solution can be also obtained by multicoloring this graph. This can be done using the algorithm depicted in Fig. 2.

First-Available-Emitter Heuristic

1. Let $P = \langle e_{i_1}, e_{i_2}, \cdots, e_{i_n} \rangle$ be a permutation of the vertices in V.
2. For all $j \in \{1, \cdots, n\}$ do
   (a) Let $\alpha_{e_{i_j}} \leftarrow \emptyset$.
   (b) Let $\mathcal{A}_j \leftarrow \mathcal{F}$.
3. Let $T \leftarrow \sum_{e \in \mathcal{E}} \mathcal{R}(e)$.
4. Let $f \leftarrow \min_{f' \in \mathcal{F}} f'$.
5. While $T > 0$ do
   (a) For all $j \in \{1, \cdots, n\}$ do
   If $\left[|\alpha_{e_{i_j}}| < \mathcal{R}(e_{i_j})\right] \wedge (f \in \mathcal{A}_{e_j})$ then
      i. Let $\alpha_{e_{i_j}} \leftarrow \alpha_{e_{i_j}} \cup \{f\}$.
      ii. For all $(e_{i_j}, e_{i_{j'}}, \delta) \in E$ do
      $\mathcal{A}_{e_{j'}} \leftarrow \mathcal{A}_{e_{j'}} - \{f' \mid \delta > |f - f'|\}$.
      iii. Let $T \leftarrow T - 1$.
   (b) Let $f \leftarrow f + 1$.

Fig. 2. Pseudocode of the First-Available-Emitter heuristic.

Both algorithms can be used as a decoders in a permutation-based GA. This allows the utilization of classical recombination/mutation operators during the reproductive stage.

4 Experimental Results

The test suite used in this work is composed of eight 21-emitter FAP instances. These eight instances correspond to the combination of two emitter layouts and four frequency-demand vectors. The first layout is a random distribution of emitters within a 6×6 plane, and the second one is the well-known Philadelphia layout [1], based on a cellular-phone network. Both frequency-demand vectors and interference constraints are taken from [8].

The first experiments consist of a fitness-variance analysis. The goal of these experiments is estimating which of the two views of the problem (emitter-based or frequency-based) carries more significant fitness information. This is important from the perspective of both the direct approach and the decoder approach. The results of this analysis are shown in Fig. 3. As it can be seen, the fitness variance is lower (and hence the fitness information is more significant) when processing frequency-based units (i.e., manipulating partial frequency assignments, or using the First-Available-Emitter heuristic).

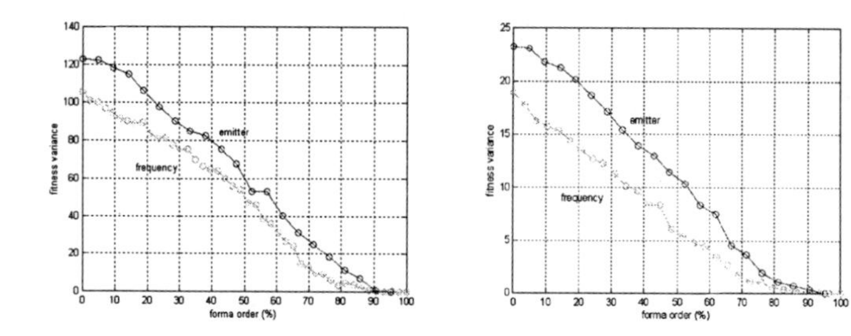

Fig. 3. Fitness variance for emitter-based units, and for frequency-based units. (Left) Frequency span (Right) Number of frequencies.

To confirm these results, the four random-layout instances are used. To be precise, the two decoder variants have been tested, using an elitist generational genetic algorithms ($popsize$ = 100, p_c = .9, p_m = 0.013, $maxevals$ = 100.000) utilizing ranking selection (η^+ = 2.0). Three different recombination operators have been used: cycle crossover (CX), order crossover (OX), and partially mapped crossover (PMX). In all cases, mutation is done via the *swap* operator. The results are shown in Table 1.

As it can be seen, the First-Available-Emitter heuristic is globally better that the First-Available-Frequency heuristic, confirming the hypothesis extracted from the fitness-variance analysis. Hence, the former heuristic will be used as the patching algorithm in subsequent experiments. These are done on the Philadelphia instances, using the same experimental setup mentioned above. The results are shown in Table 2.

The obtained results are conclusive. On one hand, FX (frequency crossover) performs better than EX (emitter crossover) or EFX (emitter-frequency crossover).

Table 1. Comparison of different genetic operators in the decoder approach (random-layout instances). All results correspond to series of twenty runs.

	IR1			IR2			IR3			IR4		
	CX	OX	PMX	CX	OX	PMX	CX	OX	PMX	CX	OX	PMX
	Frequency Span (mean)											
FAF	421.0	421.0	421.0	295.8	295.8	295.6	288.8	288.6	289.6	846.0	846.0	846.0
FAE	421.0	421.0	421.0	295.0	295.0	295.0	219.1	219.0	219.1	846.0	846.0	846.0
	Number of Frequencies (mean)											
FAF	301.2	299.7	300.8	261.0	258.4	259.1	220.4	220.3	220.1	594.1	595.0	596.1
FAE	295.8	296.2	298.1	258.9	260.8	263.7	220.0	220.0	220.0	590.1	594.2	593.8

This is in accordance with the previous fitness-variance analysis, and with the fact that EFX is an operator performing a strong mixture of information taken from the parents (similar to UX in binary representations). This is a detrimental property in such a constrained problem, in which assignment values closely interact. On the other hand, the decoder approach yields the overall best results. Actually, the optimal solution is found for three of the problem instances (IP1, IP3, and IP4), since the GA reaches the lower bounds given in [8]. This good performance can be partially explained by the fact that the decoder provides locally optimal solutions of high quality, difficult to achieve by random recombination. Clearly, this is a specific property of the particular decoding algorithm used in this work. In this sense, the First-Available-Emitter heuristic seems to be very appropriate to introduce problem-specific knowledge in the GA.

5 Conclusions

This work has compared two different approaches for the resolution of frequency assignment problems. The obtained results have confirmed the goodness of fitness-variance estimations in order to predict GA performance. It has been shown that, despite the apparent emitter/frequency symmetry of the problem, manipulating partial frequency assignments is more adequate that manipulating partial emitter assignments. This results also holds for two different construction heuristics used as decoders in a permutation-based GA. Furthermore, there seems to be a good interplay between the GA and the First-Available-Emitter heuristic, resulting in much better solutions than those obtained by means of *blind* recombination operators.

Future work will be directed to study other construction heuristics, as well as tackling different variants of FAPs. Over-constrained instances in which the goal is minimizing unfeasibility rather than optimizing feasibility are a specifically interesting line of future work.

Acknowledgments. This work is partially supported by CICYT under grant TIC99-0754-C03-03.

Table 2. Comparison of different genetic operators on the Philadelphia test-suite. Patching is done via the First-Available-Emitter heuristic. All results correspond to series of twenty runs.

Frequency Span												
	IP1			IP2			IP3			IP4		
Operator	best	mean	σ	best	mean	σ	best	mean	σ	best	mean	σ
EX	434	459.40	12.95	287	308.25	11.73	243	287.70	12.08	875	919.50	25.89
FX	434	451.10	8.41	269	275.65	4.40	240	244.65	1.49	871	907.55	19.55
EFX	435	459.25	13.27	296	308.00	8.14	243	270.60	8.79	885	920.50	24.07
OX	426	426.00	0.00	273	281.75	4.94	239	239.00	0.00	855	855.90	0.94
PMX	426	426.00	0.00	271	285.45	5.32	239	239.00	0.00	855	855.65	0.57
CX	426	426.25	0.43	282	288.25	3.16	239	239.00	0.00	855	857.10	1.45
Number of Frequencies												
	IP1			IP2			IP3			IP4		
Operator	best	mean	σ	best	mean	σ	best	mean	σ	best	mean	σ
EX	360	360.10	0.30	271	275.25	3.39	240	241.95	4.34	720	720.70	0.95
FX	360	360.00	0.00	270	270.00	0.00	240	240.00	0.00	720	720.00	0.00
EFX	360	363.70	2.33	273	282.95	4.66	240	249.25	9.03	720	727.40	4.86
OX	360	360.00	0.00	270	270.00	0.00	240	240.00	0.00	720	720.00	0.00
PMX	360	360.00	0.00	270	270.00	0.00	240	240.00	0.00	720	720.00	0.00
CX	360	360.00	0.00	270	270.00	0.00	240	240.00	0.00	720	720.00	0.00

References

1. L.G. Anderson. A simulation study of some dynamic channel assignment algorithms in a high capacity mobile telecommunications system. *IEEE Transactions on Communications*, COM-21:1294–1301, 1973.
2. C. Cotta and J.M. Troya. Genetic forma recombination in permutation flowshop problems. *Evolutionary Computation*, 6(1):25–44, 1998.
3. C. Crisan and H. Mühlenbein. The breeder genetic algorithm for frequency assignment. In A.E. Eiben et al., editors, *Parallel Problem Solving From Nature V - LNCS 1498*, pages 897–906. Springer-Verlag, Berlin, 1998.
4. S. Hurley, D.J. Smith, and S.U. Thiel. FASoft: a system for discrete channel frequency assignment. *Radio Science*, 32:1921–1939, 1997.
5. R.A. Murphey, P.M. Pardalos, and M.G.C. Resende. Frequency assignment problems. In D.-Z. Du and P. M. Pardalos, editors, *Handbook of combinatorial optimization*, volume 3. Kluwer Academic Publishers, 1999.
6. N.J. Radcliffe. Equivalence class analysis of genetic algorithms. *Complex Systems*, 5:183–205, 1991.
7. N.J. Radcliffe and P.D. Surry. Fitness variance of formae and performance prediction. In L.D. Whitley and M.D. Vose, editors, *Foundations of Genetic Algorithms III*, pages 51–72, San Mateo CA, 1994. Morgan Kauffman.
8. C. Valenzuela, S. Hurley, and D. Smith. A permutation based genetic algorithm for minimum span frequency assignment. In A.E. Eiben et al., editors, *Parallel Problem Solving From Nature V - LNCS 1498*, pages 907–916. Springer-Verlag, Berlin, 1998.

GA Techniques Applied to Contour Search in Images of Bovine Livestock

Horacio M. González Velasco, Carlos J. García Orellana, Miguel Macías Macías, and M. Isabel Acevedo Sotoca

Departamento de Electrónica e Ingeniería Electromecánica
Universidad de Extremadura
Av. de Elvas, s/n. 06071 Badajoz - SPAIN
horacio@nernet.unex.es

Abstract. In this work a system based on genetic algorithms is presented that generates valid initializations for deformable models methods. Following a systematics similar to that used by other authors, a model of the shape we are looking for (cows in lateral position) is constructed using PDM, and later the search within the image is made based on instances of that model, and using genetic algorithms techniques. Since we have color images, several objective functions are suggested that take advantage of this information, which are tested later over a database of 309 animal images taken directly in the field.

1 Introduction

A very important task related with the control and the conservation of the purity in certain breeds of bovine livestock is the morphological evaluation. This process consists of scoring a series of defined characteristics of the animal [12] which mainly involve the morphology of it, and are summed later with weights (previously defined) to obtain the final score. Usually this process should be carried out by highly qualified staff, that requires a great experience in the task, and therefore the number of them is very small. This, together with the need of uniformity in criterions, lead us to consider the utility of a semiautomatic system to help morphological evaluation, as it has already been discussed in [11] and in [5].

In the publications on the topic [12, 11] and in the information received from the consulted experts it is suggested that a great part of the characteristics can be simply evaluated analyzing three pictures corresponding to three positions of the animal, and fundamentally this information is contained in their profiles. In this work we try to present a technique based on genetic algorithms [4] which allows us locating the animal into the image, understanding as localization an approximate position it that could be used later as initialization [1, 8] for deformable model methods, as the ASM [2], which are quite sensitive to the initial position. Particularly we will use the approach designed by Hill et al. [7, 1] that has already been applied with great success to another type of images (usually

J. Mira and A. Prieto (Eds.): IWANN 2001, LNCS 2084, pp. 482-489, 2001.

grey-scale medical images [6, 3]), which we will adapt to be used with the colour photographs directly obtained by means of a digital camera in the field. In order to carry it out we propose a first processing of the images which takes advantage of the colour information in an appropriate way, so the new image (that we will call *potencial*) together with an appropriate objective function (that we will also define, and we will try to maximize) allows us to carry out an efficient search using genetic algorithms.

With this work outline, in section 2 we will describe in detail the technique used to model and parameterize the shape we want to search in the images. Later on in section 3 the objective functions which we will try to maximize with genetic algorithms are described. The exposition of the trials we have made on our database of images as well as the results can be found in section 5, whereas conclusions and possible improvements of our work will be presented, finally, in section 5.

2 Model building

In order to achieve our aim, the first step consists in representing appropriately the shape we want to search as well as its possible variations. In the works of Hill et al. [6, 1, 10], as well as other authors [3] they use a general systematic known as PDM [2], which consist of deformable models that represent the contours by means of ordered groups of points (fig. 1,b) located at specific positions of the object, which are constructed statistically, based on a set of examples. As a result we will have modelled the *average shape* of our object, as well as the allowed deviations from this average shape, based on the set used in the training process. Through an appropriate election of the elements used for the training, with this technique we have the advantage of not needing any heuristic assumption about which contours have an acceptable shape and which not.

Furthermore, another fundamental advantage consist in that the technique provides us a manner of parameterizing the contours. As can be consulted in [2, 1], each contour will be represented mathematically by a vector $\boldsymbol{x}$ such that

$$\boldsymbol{x} = \boldsymbol{x}_m + \mathbf{P} \cdot \boldsymbol{b} \tag{1}$$

where $\boldsymbol{x}_m$ is the average shape, $\mathbf{P}$ is the matrix of eigenvectors of the covariance matrix and $\boldsymbol{b}$ a vector containing the weights for each eigenvector and is which properly defines the contour in our description. Fortunately, considering only few eigenvectors corresponding to the largest eigenvalues of the covariance matrix, we will be able to describe practically all the variations that take place in the training set.

This representation (model) of the contour is, however, in a normalized space. To project *instances* of this model to the space of our image we will need a transformation that conserves the form, which will consist in general in an translation $\boldsymbol{t}$, a rotation θ and a scale s.

In our specific case we have centred in one of the three kinds of images referred in introduction, the lateral image, since the method we are using is general and

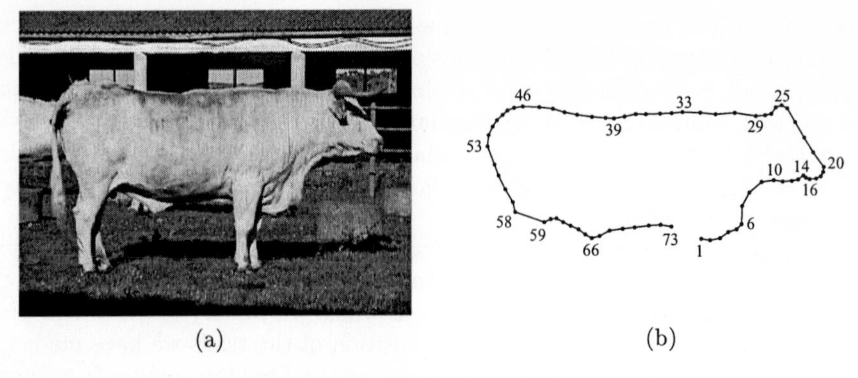

(a) (b)

Fig. 1. In figure (a) a representative example is shown of the images we are dealing with. The model of the cow contour is plotted in figure (b).

can be equally applied to the other types. We have described the cow's contour, without limbs, using a set of 73 points (fig. 1,b), with 16 significative ones, which are numbered in the figure. The model has been constructed using a training set of 20 photographs, distributed evenly over the breeds present in the database, trying to cover as much as posible the small variations in position that were found on it. So, with a set of 12 parameters (the 8 coefficients of the eigenvectors, and the four needed for the transformation) we will have described any instance of our model in the image.

It is necessary to emphasize that in [1, 10, 3], besides modelling the objects' shape, they also model the structure of grey levels (the *grey-level appearance*) around the different points, under the hypothesis that, since each point corresponds with a certain part of the object, the grey-level patterns around these points in images of different examples should be similar. However, in our case we are dealing with colour images, and due to the characteristics of the object we are looking for (animals of different breeds and colours) and the origin of the images (taken directly in the field, and therefore with a great variety of different backgrounds), we have opted to use only the shape information for our model's construction. Later on we will take advantage of the colour information to define an objective function that allows us an effective search.

3 Image search using genetic algorithms

The method used to carry out the search inside the image of one of our shapes (modelled according to the technique presented in the previous section) it is described in [7] and used later, with small changes, in [6, 10, 3]. This method consists of defining a population of instances of our model, on which we will calculate a certain objective function that evaluates if the instance matches an object inside the image or not. As we have already stressed, the model that we previously built has allowed us the parametrización of the shapes as well as the allowed deviations, and so we are sure that any set $\{\boldsymbol{b}, \boldsymbol{t}, \theta, s\}$ that defines one instance, has an allowed shape of searched object.

Therefore the main task to use this systematic is the definition of the objective function. In the consulted works we found two fundamental ideas to define this function. On one hand, in [7, 6] the function is built so it becomes minimum when strong edges (great difference of grey-level), of similar magnitude, are located in the image near the points of a certain instance of the model. On the other hand, in [1, 10, 3] an objective function is used based on the similarities between the grey-level profiles found around the points of the instance, and those stored during the model's construction.

In our case we intend to use a similar approach to the first one of those mentioned above, but with a fundamental change regarding the methodology used in the previous works: our instances of the model will not be compared with the own image, but with the result of a processing of it that we will call *potential*, and will consist in a grey-level image with light values in those positions to which we want our points to tend to (strong edges between objects), and darker pixels in those points which are not interesting.

3.1 Potential image

To generate the potential image, we have tried to take full advantage of the fact that we are dealing with colour images. In order to use that information, instead of applying a conventional edge detection method on the luminance coordinate, we worked with the three colour components, in the system proposed by the CIE, $L^*a^*b^*$ [9]. This colour space has the advantage that is perceptually uniform, i.e. a small perturbation to a component value is approximately equally perceptible across the whole range of that value. As our goal it is to base our edges on colour differences between areas, the perceptual uniformity allows us to treat the three components of colour *linearly*, i.e., to extract edges in the images corresponding to the colour coordinates by a conventional method (particularly we have used the Sobel one [9]) and subsequently to put them together using a linear combination. Figure 2 shows a comparison of the results using only the luminance and using the method we propose.

3.2 Objective functions and genetic algorithms

To carry out the search with genetic algorithms we have used three approaches to reach the aims of [7, 6] mentioned above, with the difference that we will try to maximize the function, not to minimize it. From now on we will call $\boldsymbol{x} = (x_0, y_0; x_1, y_1; \ldots; x_{n-1}, y_{n-1})$ the set of points that form one instance of our model, whereas we will name $V(x, y)$ our potential function, i.e., the grey-levels in our potential image. Also, we will call $O(x, y)$ one function which returns 1 if the point (x, y) is within the image, and returns 0 if it is not.

We are going to identify each objective function with a name:

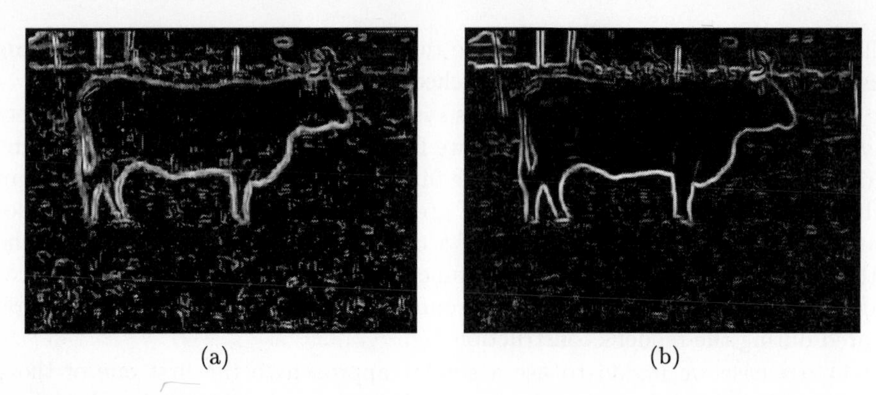

(a) (b)

Fig. 2. Representation of the final edge determined from the colour information, as against the edge of the coordinate L^* in which only the luminance coordinate has been used.

Function POINTS: This is the simplest function that we will use. It is defined in the following way:

$$f_P(\boldsymbol{x}) = \sum_{j=0}^{n-1} O_j \cdot V(x_j, y_j) \tag{2}$$

where $O_j = O(x_j, y_j)$. With this function we only try to achieve that the points of our instance lie on strong edges. Though the approach is very simple, it has the advantage of requiring very few calculations for its evaluation. Parameter O_j is included for not to consider those points that are outside our image.

Function LINES: With this function we intend not only to search strong edges in the points of our instance, but also in all the points that form our contour. With this aim we define $\boldsymbol{r_j}$ as the set of points that form the straight line that joins (x_j, y_j) and (x_{j+1}, y_{j+1}), and K_j as the number of points that such a set contains. The function is defined as

$$f_L(\boldsymbol{x}) = \left(\sum_{j=0}^{n-2} O_{rj} \cdot K_j \right)^{-1} \cdot \sum_{j=0}^{n-2} \left(O_{rj} \cdot \sum_{(x,y) \in \boldsymbol{r_j}} V(x, y) \right) \tag{3}$$

where $O_{rj} = O(x_j, y_j) \cdot O(x_{j+1}, y_{j+1})$. As we can see, the result is the average intensity over the whole set of points of the contour for which the potential function is computed. Now we must divide by the number of points in the contour, since not every contour has the same number of points.

Function UNIFORM: The definition of this function will be based in the previous one, but it is going to include another term:

$$f_U(\boldsymbol{x}) = \frac{f_L(\boldsymbol{x})}{\sigma(\boldsymbol{x})} \text{ and } \sigma(\boldsymbol{x}) = \sqrt{\sum_{j=0}^{n-2} \left(O_{rj} \cdot \sum_{(x,y) \in \boldsymbol{r_j}} \left(V(x,y) - f_L(\boldsymbol{x})\right)^2 \right)} \tag{4}$$

This way, the function will have a larger value when $f_L(\boldsymbol{x})$ increases (strong edges), but also it will be larger when the edge is more uniform over the whole contour, since in that case differences $V(x,y) - f_L(\boldsymbol{x})$ will be small. The main drawback of this function is the great quantity of calculations that it involves for its evaluation.

4 Experiments and results

The results that we present next have been obtained using a GA with *standard* parameters [6]: rate of crossover=0.6, using the two-point version of crossover operator; rate of mutation=0.005; roulette selection method, uniform replacement method and elitism. Trying to achieve the maximum simplicity in chromosomes, each of our 12 parameters has been coded using 8-bit-length gray-code binary integers. To establish the maximum range for parameters we have used *a priori* knowledge about the images. This way, we restrict the rotation to values $\theta \in [-1.137, -0.963]$ radians (we know that the cow will be more or less horizontal in the image), while the scale is limited to $s \in [55, 95]$ (the searched shape will not be arbitrarily small, nor larger than the own image). Limits in translation $\boldsymbol{t}$ have been established in such a way that even with the smallest scale it would be possible to reach all the zones of the image, resulting $x_c \in [225, 175]$ y $y_c \in [275, 125]$. Finally, for parameters b_i the criterion presented in [2] has been applied: $b_i \in [-3\sqrt{\lambda_i}, 3\sqrt{\lambda_i}]$.

With these parameters we have carried out the search for each one of the 309 images in the database, using the three suggested objective functions and three different population sizes successively, with the aim of discovering the influence of this parameter in our real problem. Concretely we have used the standard value of 50 referred in [6], and values of 100 and 500 individuals.

Table 1. Summary of the results, showing the number of successes (percentage) and the average time needed to do the search per image in a Pentium III 600 system

Population	*POINTS*	*LINES*	*UNIFORM*
Pop. 50	92 (29.7 %)	86 (27.8 %)	102 (33.0 %)
	t=5.2 s	t=17.5 s	t=23.7 s
Pop. 100	120 (38.8 %)	161 (52.1 %)	163 (52.7 %)
	t=11.0 s	t=33.6 s	t=46.5 s
Pop. 500	209 (67.6 %)	204 (66.1 %)	225 (72.9 %)
	t=59.4 s	t=168.8 s	t=238.6 s

For each of our experiments we have checked all the results visually, so in table 1 the number of successes for each case is presented (over the 309 images)

as well as the average time needed for that trial. It should be noticed again that the aim of our search is the initialization of a deformable model, so we have considered *success* all those cases where the final contour is placed near enough to the real profile of the animal, though they are not completely matched. Considering the relative subjectivity of the success concept, the visual recount was carried out by two different people, obtaining very similar results, which are averaged in the table 1.

In that table of results there are several aspects which attract our attention:

- On the first place, we should stress the relatively low rate of successes for the 50 individuals population. However, the great difficulty that our images present must be considered, since they have been taken directly in the field, and then their backgrounds contains a great quantity of edges (for example, see fig. 1,a). This will make that, to be able to find a global maximum, a big population it is needed which explores the search space efficiently.
- In second place, as it should be expected, the increase of the successes percentage with the population size is remarkable (though not lineal, evidently). However, the problem is to be found in the processing time, which does grow lineally with population size. Taking into account that, in our case, time is not a very important parameter (provided that we are dealing with reasonable times, of the order of minutes), we can even think about the use of larger populations.
- In third place, and maybe the more surprising fact, is that the rate of successes for the three functions turns out to be quite similar, fundamentally in the case of the 500 individuals population. This leads us to think that, with our approach, the most important thing consists of trying to place the points in the strongest edges as possible, without caring in excess about the uniformity of these. In a great extent, that is due to the wide diversity of backgrounds that can appear behind a particular searched object in an image, as we can see in fig. 1,a, and in edges of fig. 2,a, where we observe that, due to the background, the edges of our object has different intensities depending on the zone, and then the global contour obtained for it is not uniform, so looking for uniform edges does not improve the search in many cases.

5 Conclusions and future research

In light of the presented results, we could stress several aspects of our present approach that also allow us to define several future investigation lines. In the first place, as we have discussed above, the approach that searches the uniformity in the contours does not improve significantly the performance, and produces a great increase in calculation time. So we must expect that it is better to use the extra time in a larger population size than in a complicated objective function. In this respect, we are now making experiments with larger populations and with parallel GAs. Also, the increase in time can be neutralized by the use of a parallel simulation system, as the one we have operative in our laboratory (Beowulf cluster with 24 nodes).

On the other hand we should try to not only exploit the color information for the elaboration of the potential image, but also to define another kind of objective function which performs better. We should notice that, in most of the images, what defines the entity of the object that we are looking for is, more or less, its uniform color, provided that we eliminate the luminance information.

Finally we must notice that, in spite of all we have explained before, the results described are not bad at all, since with a very simple approach, and over a set of quite heterogeneous images, we have achieved a satisfactory initialization for our snake in approx. a 70% of the cases.

Acknowledgements

This work has been supported by the Junta de Extremadura through IPR 98B037. We also wish to express our gratitude to the CENSYRA for helping us with everything related to cattle.

References

1. Cootes, T.F., Hill, A., Taylor, C.J., Haslam, J.: The Use of Active Shape Models For Locating Structures in Medical Images. Image and Vision Computing, vol 12, nº 6, pp 335-366. July 1994.
2. Cootes, T.F., Taylor, C.J., Cooper D.H., Graham, J.: Active Shape models – Their Training and Application. Computer Vision and Image Understanding, vol 61, nº 1, pp 38-59. Jan. 1995.
3. Felkel, P., Mrázek, P., Sýkora, L., Zára, J., Jezek, B.: On Segmentation for Medical Data Visualization. Proc. of EWVSC'96, pp 189-198. April 1996.
4. Goldberg, D.E.: Genetic Algorithms in Search Optimization and Machine Learning. Addison-Wesley, 1989
5. González, H.M., López, F.J., García, C.J., Macías, M., Acevedo, M.I.: Application of ANN Techniques to Automated Identification of Bovine Livestock . Proc. of IWANN'99, pp 422-431. June 1999.
6. Hill, A., Cootes, T.F., Taylor, C.J., Lindley, K.: Medical Image Interpretation: A Generic Approach using Deformable Templates. Journal of Medical Informatics, vol 19, nº 1, pp 47-59. Jan. 1994.
7. Hill, A., Taylor, C.J.: Model-Based Image Interpretation Using Genetic Algorithms. Image and Vision Computing, vol. 10, nº 5, pp 295-300. June 1992.
8. Hug, J., Brechbühler, C., Székely, G.: Model-based Initialisation for Segmentation. Proc. of ECCV 2000, pp 290-306. June 2000.
9. Jain, A.K.: Fundamentals of Digital Image Processing. Prentice Hall, 1989
10. Lanitis, A., Hill, A., Cootes, T.F., Taylor, C.J.: Locating Facial Features Using Genetic Algorithms. Procs. of ICDSP'95 , pp.520-525. 1995.
11. López, S., Goyache, F., Quevedo, J.R. et al.: Un Sistema Inteligente para Calificar Morfológicamente a Bovinos de la Raza Asturiana de los Valles, Revista Iberoamericana de Inteligencia Artificial, nº 10, pp 5-17. 2000.
12. Sánchez-Belda, A.: Razas Bovinas Españolas. Manual Técnico, Ministerio de Agricultura, Pesca y Alimentación. España, 1984.

Richer Network Dynamics of Intrinsically Non-regular Neurons Measured through Mutual Information

F.B. Rodriguez[1,2], P. Varona[1,2], R. Huerta[1,2],
M. I. Rabinovich[1], Henry D. I. Abarbanel[1,3]

[1] GNB. E.T.S. Ingeniería Informática, Universidad Autonóma de Madrid,
Cra. Colmenar Viejo, Km. 15, 28049 Madrid, Spain.
[2] Institute for Nonlinear Science, University of California,
San Diego La Jolla, CA 92093-0402, USA.
[3] Marine Physical Laboratory, Scripps Institution of Oceanography,
Department of Physics, University of California, San Diego
La Jolla, CA 92093-0402, USA.

Abstract. Central Pattern Generators (CPGs) are assemblies of neurons that act cooperatively to produce regular signals to motor systems. The individual behavior of some members of the CPGs has often been observed as highly variable spiking-bursting activity. In spite of this fact, the collective behavior of the intact CPG produces always regular rhythmic activity. In this paper we show that simple networks built out of intrinsically non-regular units can display modes of regular collective behavior not observed in networks composed of intrinsically regular neurons. Using a measure of mutual information we characterize several patterns of activity observed by changing the coupling strength and the network topology. We show that the cooperative behavior of these neurons can display a rich variety of information transfer while maintaining the regularity of the rhythms.

1 Introduction

A Central Pattern Generator is a group of neurons that control motor activity. CPGs are responsible for activities like chewing, walking and swimming [1–3]. The inner properties of every neuron in the CPG together with the connection topology of the network and the modulatory inputs determine the shape and phase relationship of the electrical activity. A group of neurons can generate many different patterns of activity that control a variety of motor movements. An essential property for these neural assemblies is the presence of robust and regular rhythms in the membrane potentials of their member neurons within the CPG. However, some of these cells can display highly irregular spiking-bursting activity when they are isolated from the other members of the CPG. In this paper we show that intrinsically irregular model neurons can cooperatively produce robust and regular oscillations which can be characterized by a measure of mutual information. We do this using reduced networks of simple spiking-bursting model

J. Mira and A. Prieto (Eds.): IWANN 2001, LNCS 2084, pp. 490-497, 2001.

neurons. In the following sections, we will discuss the fact that some regular modes of operation of networks built out of irregular spiking-bursting cells are absent in similar configurations of intrinsically regular neurons.

2 Dynamics of the neural model

We have built networks of three electrically coupled Hindmarsh-Rose (HR) neurons [4]. The single neuron model is able to produce chaotic spiking-bursting activity similar to the one observed in isolated stomatogastric CPG neurons [5, 6]. The dynamics of the model can be described by the following system of equations:

$$\frac{dx_i(t)}{dt} = y_i(t) + 3x_i^2(t) - x_i^3(t) - z_i(t) + e_i - \sum_{j \neq i} g_{ij}(x_i(t) - x_j(t)),$$

$$\frac{dy_i(t)}{dt} = 1 - 5x_i^2(t) - y_i(t), \quad \frac{1}{\mu}\frac{dz_i(t)}{dt} = -z_i(t) + S\left[x_i(t) + 1.6\right], \tag{1}$$

where the last term of the first equation represents the total electric current arriving to unit i, and g_{ij} is the electrical coupling conductance between units i and j. The sum in (1) extends over the neurons j that are connected to neuron i. The model parameters were chosen to set the single units in the regular regime ($e_i = 3.0$, $\mu = 0.0021$ and $S = 4$), or in the non-regular regime ($e_i = 3.281$, $\mu = 0.0021$ and $S = 4$).

The isolated dynamics of a single HR neuron with this selection of parameters produces the characteristic regular or irregular spiking-bursting activity shown in Figure 1. Here x_i can be considered as the membrane potential of the neuron. This dynamics is characterized by two different time scales. The subsystem (x_i, y_i) is responsible for the generation of fast spikes, and the slow subsystem (x_i, z_i) is responsible for the slow waves over which the spikes appear. The fast subsystem is placed nearby the homoclinic bifurcation and, under the influence of the slow subsystem can produce a chaotic spiking-bursting behavior depending on the value of parameter e_i[7].

3 Neural Information Measure

To measure information exchange among the neurons we used Shannon's concept of mutual information [8]. The mutual information between the activities r_i and s_i of two neurons, drawn from sets $R = \{r_i\}$ and $S = \{s_i\}$ of possible measurements, is given in bits [1]by the expression:

$$\log_2 \frac{P_{RS}(r_i, s_i)}{P_R(r_i)P_S(s_i)}, \tag{2}$$

where $P_{RS}(r_i, s_i)$ is the joint probability density and $P_R(r_i)$ and $P_S(s_i)$ represent the individual probabilities for measurements R and S. If variables r_i and s_i

[1] Henceforth all measures of information or entropy will be given in bits.

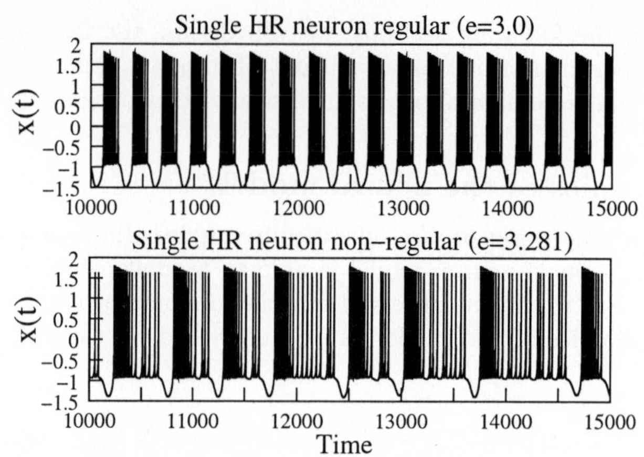

Fig. 1. *Time series of regular ($e = 3.0$) and non-regular ($e = 3.282$) activity (x_i) of isolated HR neurons. Units are dimensionless.*

are completely independent, then the mutual information between these two variables is zero.

The average over all measurements is called average mutual information [8, 9], and it is defined by:

$$I_{RS} = \sum_{r_i, s_i} P_{RS}(r_i, s_i) \log_2 \frac{P_{RS}(r_i, s_i)}{P_R(r_i) P_S(s_i)} = H(R) - H(R|S) \geq 0 \quad (3)$$

where $H(R)$ is the entropy for the measurement R and $H(R|S)$ is the conditional entropy for the measurement R given the measurement S. The entropy and the conditional entropy are defined, respectively, as:

$$H(R) = -\sum_{r_i} P_R(r_i) \log_2 P_R(r_i),$$

$$H(R|S) = -\sum_{r_i} P_S(s_i) \sum_{r_i} P_R(r_i|s_i) \log_2 P_R(r_i|s_i). \quad (4)$$

What does it mean that the average mutual information between sets R and S of all possible measurements is large or small? The entropy is basically a measure of disorder, it gives an idea of the uncertainty in a set of possible measurements. Let us suppose that we have measured $I_{RS} = a$ and $I_{R\hat{S}} = \hat{a}$, and we have obtained the result that $a >> \hat{a}$. This means that the uncertainty of the set R given $\hat{S}$ is bigger than the uncertainty of R given S. The variability for the set R of all possible measurements is further reduced through the interaction with S.

Signal and Coding Space for Information Measure

We assume that the information in the network is completely contained in the temporal evolution of the membrane potential of the neurons. In this paper we will use a discrete treatment in the neural signal. There is experimental evidence

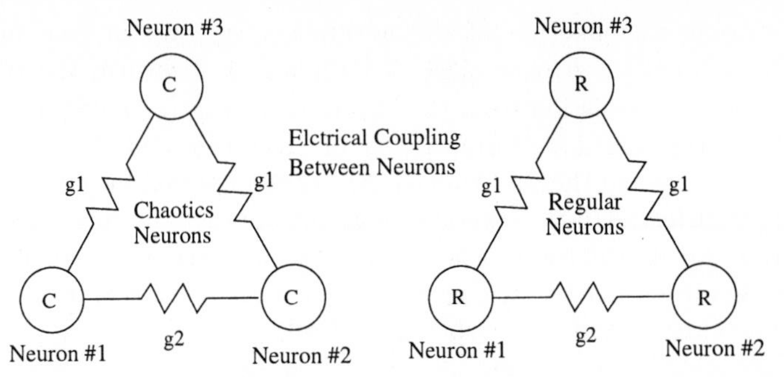

Fig. 2. *Architecture of the networks. Electrical couplings are a predominant type of interconnection in many CPGs. The neurons can be tuned in the chaotic or in the regular regime. Here* $g_{12} = g_2$ *and* $g_{23} = g_{31} = g_1$. *In some simulations we externally inhibited the connection from neuron* 3 ($g_1 = 0$).

which shows that information processing in some neural systems is governed by discrete events such as spikes [10] and bursts [11, 12]. Several authors have used this discrete treatment in the temporal evolution of the membrane potential in order to measure the average of mutual information among neurons [13, 14]. In order to discretize the temporal evolution of the membrane potential we divide our time space in N time windows of size Δt. We map the membrane potential in time series of N samples. We assign the value 1 to each bin Δt when an event is present and 0 otherwise. We will assume (see neural model and Figure 1) that the main informational event in our time series is the occurrence of a burst. There are several experimental evidences showing that this kind of event is the main carrier of information in some real neural systems. For example in the Central Pattern Generator of the lobster's stomatogastric nervous system, the bursts frequency is strongly related to the frequency of the mussels of the stomach or pylorus [11, 12]. To specify the occurrence of this event, we search for the regions of hyperpolarization in the time series of membrane potential. Consequently, if we find a maximum hyperpolarization in a bin Δt we assign the code 1 and 0 otherwise. Now we have a discrete temporal sequence $\{e_t, t = 1 \ldots N\}$ where e_t can be 0 or 1. We define a word W_t^L of size L at time t as a sequence of symbols $W_t^L = \{e_t, e_{t+1} \ldots, e_{t+L}\}$. Thus, there are $N - L + 1$ words of L bits in each time series. We have already defined our coding space in order to measure what kind of information is transfered among neurons. In the next section we will study how the information depends on the parameters that define the coding space, Δt, L and the rules to identify an event.

Information Measure

We consider one neuron in the network as the output neuron (R, response neuron) and another neuron as the input neuron (S, stimulus neuron). Then we calculate $P_{RS}(r_i, s_i)$ to estimate the average of mutual information for a given sequence of words for the stimulus and the response. We calculate the joint probabilities making use of empirical probabilities for the stimulus and the response in the simulation time series.

We are going to make use of the normalized average mutual information given by the expression $E_{RS} = \frac{I_{RS}}{H(S)}$. This quantity measures the efficiency of the information transmission from the stimulus neuron to the response neuron. Normalized average mutual information is dimensionless and the variation range is $0 \leq E_{RS} \leq 1$. Due to the fact that $H(S)$ is the maximal information amount that can be transferred to the response neuron from the stimulus neuron, $E_{RS} = 0$ means that all information is lost (response and stimulus are independent quantities). On the other hand, $E_{RS} = 1$ means that there is a perfect matching between the neurons (complete synchronization).

4 Results

We run several simulations using networks of three HR neuron with the connection topology shown in Figure 2. The networks were composed of three identical HR neurons with electrical couplings in an all to all connectivity. We tuned the neurons in the chaotic or in the regular regime by selecting the value of the parameter e_i. In some simulations, connection from neuron 3 was inhibited so that the effective coupling g_1 become zero. We have chosen six representative simulations from a large number of trials. The parameters used in these simulations are given in Table 1. The resulting time series of membrane potential activity are given in Figure 3.

Table 1. *Parameters used in the six simulations discussed in the text. The architecture of the networks is shown in Figure 2. When $g_1 = 0$ connection from neuron 3 was externally inhibited. The column labeled by $(C-C)_{max}$ represents the maximum cross–correlation between the time series of neuron 1 and neuron 2.*

Sim.	g_1	g_2	Regime	$(C-C)_{max}$	Delay	Sim.	g_1	g_2	Regime	$(C-C)_{max}$	Delay
A	≈ 0	≈ 0	Chaotic	39761.6	-165373	D	0.0	0.02	Chaotic	959205.9	-1763
B	0.01	0.01	Chaotic	209641.4	67	E	0.01	0.01	Regular	924600.7	-106
C	0.03	0.03	Chaotic	458572.6	-7	F	0.0	0.02	Regular	967294.5	113

Panels A, B, C and D of Figure 3 show simulations using intrinsically irregular neurons (parameters were chosen to tune the isolated neurons in the chaotic region, see Table 1). Panels E and F correspond to the activity of the networks with intrinsically regular neurons. A small coupling among the neurons is not enough to regularize the activity of the chaotic cells as shown in panel A. However, a larger conductance for the electrical coupling induces more regular behavior as shown in panel B. In this case, we can sometimes find the neurons in phase, although this behavior is transient. A further increase in the strength of the coupling (see panel C in Figure 3) makes the neurons oscillate almost in phase. When we inhibit the connection from neuron 3 (effective coupling $g_1 = 0$), a singular behavior between neuron 1 and neuron 2 is observed: these two neurons oscillate in *anti–phase* with regularized bursts (see panel D in Figure 3). The panels E and F correspond to different synchronization levels when the neurons have intrinsically regular activity. One can see in these panels that the

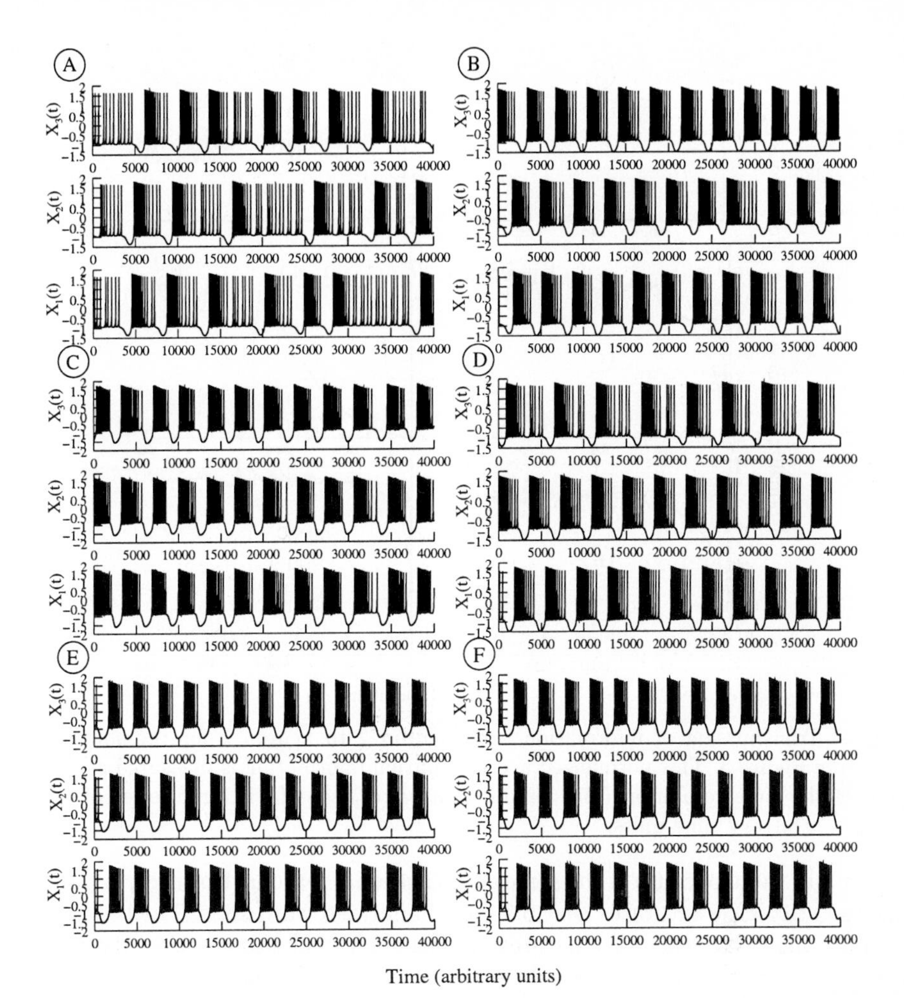

Fig. 3. Membrane potential time series of the three neurons within the network. Parameters used are shown in Table 1. Network architecture is shown in Figure 2. We used a Runge-Kutta 6(5) method with variable time step to solve the coupled differential equations of system 1.

synchronization in these cases is almost perfect. However anti–phase regularized activity is not observed when connection from neuron 3 is inhibited. Figure 4 shows several measurements of the normalized average mutual information E_{RS} as a function of the time resolution (Δt) and the word size (L). We have analyzed the transmission of information from neuron 2 to neuron 1 for the different times series shown in Figure 3 using E_{RS}. The time series of simulation A are basically independent, and the quantity E_{RS} should be close to zero. In fact, $E_{RS} \approx 0$ when $L = 2$. However, for $L > 2$ the number of samples needed to estimate the probability distribution is unpracticable and an artifact appears. In panel B the quantity E_{RS} saturates near 0.15 for word sizes $L = 2$ and $L = 6$. For $L = 10$ an $L = 14$ the normalized average mutual information does not reach a stable

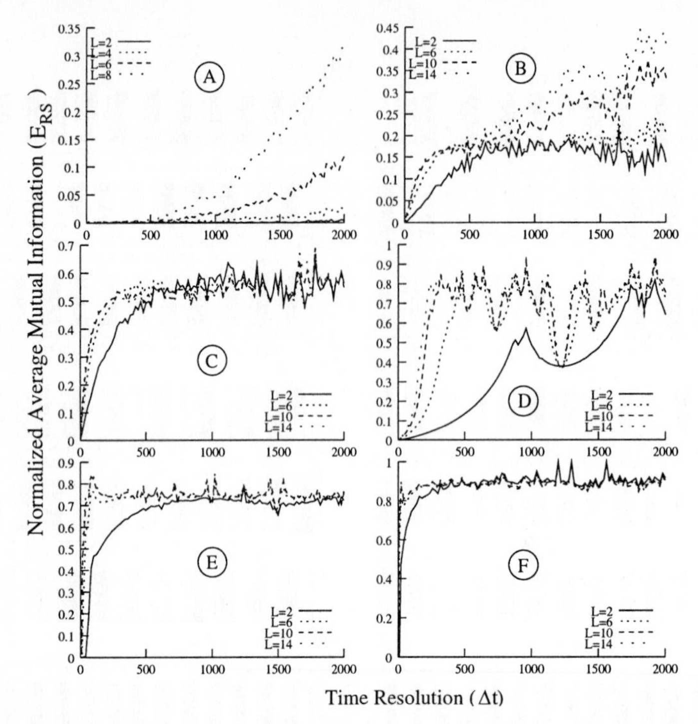

Fig. 4. Transmission of information from neuron 2 to neuron 1 for the simulations shown in Figure 3. This normalized average mutual information is measured as a function of the time resolution (Δt) for different word sizes (L) in time series of length 2×10^6 arbitrary time units.

maximum value because of the same computational reasons. However, in panel C the quantity E_{RS} saturates near 0.5 for all analyzed word sizes. The oscillations around this value reflect the non-regular behavior of the time series. Panel D in the same figure shows a strong dependence of the information transmission on time resolution. This fact is due to the singular pattern of anti–phase regularized behavior between neuron 1 and 2. However, one can see that the normalized average of mutual information sometimes reaches values very close to 1. Finally, in panels E and F the neurons are in the regular regime and the quantity E_{RS} is close to the maximum value 1 due to the existing synchronization. The saturation levels of information in the different time series can be compared with the maximum cross–correlation,$(C-C)_{max}$, between neuron 1 and neuron 2 shown in the last columns of Table 1.

5 Discussion

Several experiments have established that the oscillations of isolated neurons from the pyloric central pattern generator of the stomatogastric ganglion of the lobster are chaotic [15]. However, these neurons exhibit a robust regular spiking-bursting activity when operating within the intact CPG. In this paper we have pointed out several modes of oscillations in networks of three spiking-bursting neurons where the behavior is different depending on whether the neurons are intrinsically regular or intrinsically chaotic. Those modes that produced regular spiking-bursting activity within the network are of particular interest. We have given a measure of information that clearly distinguish these modes of operation and characterizes the information exchange. This measure can also be useful when analyzing the response of complex networks of synaptically coupled cells to well-identified inputs. The role of chaos in neural systems has been discussed widely in recent papers [16–18]. The organization of chaotic bursting neurons in assemblies where their synchronization drives regular adaptive and reliable activity can be a general enough principle used by nature to process information and accomplish critical functionality.

6 Acknowledgments

This work has been supported partially by CICyT grant n^o TIC98-0247-C02-02, grant from the vicerrectorado de Investigación de la Universidad Autónoma de Madrid and MCT grant n^o BFI2000-0157.

References

1. Selverston, A.: What invertebrate circuits have taught us about the brain. Brain Research Bulletin **50**(5-6) (1999) 439–40
2. Selverston, A.: General principles of rhythmic motor pattern generation derived from invertebrate CPGs. Progress in Brain Research **123** (1999) 247–57
3. Selverston, A., Elson, R., Rabinovich, M., Huerta, R., Abarbanel, H.: Basic principles for generating motor output in the stomatogastric ganglion. Ann. N.Y. Acad. Sci. **860** (1998) 35–50
4. Hindmarsh, J.L., Rose, R.M.: A Model of Neuronal Bursting Using Tree Coupled First Order Differential Equations. Philos. Trans. Royal Soc. London. **B221** (1984) 87–102
5. Rabinovich, M.I., Abarbanel, H.D.I., Huerta, R., Elson, R., and others. Self-regularization of chaos in neural systems: Experimental and theoretical results. IEEE Transactions on Circuits and Systems I-Fundamental Theory and Applications **44** (1997) 997–1005
6. Szucs, A., Varona, P., Volkovskii, A.R., Abarbanel, H.D.I., Rabinovich, M.I., and Selverston, A.I.: Interacting biological and electronic neurons generate realistic oscillatory rhythms. NeuroReport, **11** (2000) 563-569
7. Bazhenov, M., Huerta, R., Rabinovich, M.I, Sejnowski, T.: Cooperative behavior of a chain of synaptically coupled chaotic neurons. Physica **D116** (1998) 392-400
8. Shannon, C.E.: A Mathematical Theory of Communication. Bell Sys. Tech. J. **27** (1948) 379–423 623–656
9. Cover, T.M., Thomas, J.A.: Elements of Information Theory. Wiley and Sons (1991)
10. Rieke, F., Warland, D., de Ruyter van Steveninck, R., Bialek, W.: Spikes: Exploring the Neuronal Code. A Bradford Book. MIT Press Cambridge. Massachusetts, London, England (1997)
11. Selverston, A.I., Moulis, M.: The Crustaceam Stomatogastric System: a Model for the Study of Central Nervous Systems. Berlin; New York: Springer-Verlag. (1987)
12. Harris–Warrick, R.M.: Dynamic Biological Network: The Stomatogastric Nervous System. Canbridge, Mass.: MIT Press. (1992)
13. Strong, S.P., Koberle, R., de Ruyter van Steveninck, R., Bialek, W.: Entropy and Information in Neural Spike Train. Physical Review Letters **80** 1 (1998) 197–200
14. Eguia, M.C., Rabinovich, M.I., Abarbanel, H.D.I.: Information Transmission and Recovery in Neural Communications Channels. Phys. Rev. E **62** (2000) 7111-7122.
15. Abarbanel, H.D.I., Huerta, R., Rabinovich, M.I., Rulkov, N.F., Rowat, P., Selverston, A.I: Synchronized action of synaptically coupled chaotic model neurons. Neural Computation **8** (1996) 1567–1602
16. Rabinovich, M.I., Abarbanel, H.D.I.: The Role of Chaos in Neural Systems. Neuroscience **87** 1 (1998) 5–14
17. Richardson, K.A., Imhoff, T.T., Grigg, P., Collins, J.J.: Encoding Chaos in Neural Spike Trains. Physical Review Letters **80** 11 (1998) 2485–2488
18. Huerta, R., Varona, P., Rabinovich, M.I., Abarbanel. H.D.I.: Topology selection by chaotic neurons of a pyloric central pattern generator. Biological Cybernetics **84** (2001) L1–L8

RBF Neural Networks, Multiobjective Optimization and Time Series Forecasting

J. González, I. Rojas, H. Pomares and J. Ortega

Department of Computer Architecture and Computer Technology
University of Granada
Campus de Fuentenueva
E. 18071 Granada (Spain)

Abstract. This paper presents the problem of optimizing a radial basis function neural network from training examples as a multiobjective problem and proposes an evolutionary algorithm to solve it properly. This algorithm incorporates some heuristics in the mutation operators to better guide the search towards good solutions. An application to the Mackey-Glass chaotic time series is presented. The prediction accuracy of the proposed method is compared with that of other approaches in terms of the root mean squared error.

1 Introduction

The automatic optimization of a Radial Basis Function Neural Network (RBFNN) from training data [10, 11] is a problem in which two clearly competing objectives must be satisfied. The model's prediction error must be minimized in order to achieve a well fitted model, while the number of Radial Basis Functions (RBFs) should be as low as possible to obtain a reliable interpolator. The problem here is how to minimize both objectives simultaneously. Improving one of them will probably cause a worse behavior in the remaining one. This kind of problems is known as Multi-Objective Problems (MOPs), and their solutions are usually sub-optimal for each objective in particular, but "acceptable" taking all the objectives into account, where "acceptable" is totally subjective and problem dependent.

The algorithms proposed in the literature to construct RBFNNs from examples try to find a unique model with a compromise between its complexity and its prediction error. This is not an adequate approach. In MOPs there are usually more than one alternative optimal solutions (having different compromises between their multiple objectives) that should be considered equivalent. Thus, conventional optimization techniques have great difficulty to be adapted to solve MOPs because they weren't designed to deal with more than one solution simultaneously. Nevertheless, Evolutionary Algorithms (EAs) maintain a population of potential solutions for the problem, thus making easier their adaption to solve MOPs [2]. In particular, the fitness of the individuals must be adapted to comprise all the objectives to be satisfied and some new mutation operators must be designed to alter the structure of RBFNNs.

J. Mira and A. Prieto (Eds.): IWANN 2001, LNCS 2084, pp. 498-505, 2001.

2 Multiobjective evolution

The great difference between single objective and multiple objective problems is that the set of solutions is not completely ordered for MOPs. For two objective vectors, the following relations can be defined:

$$\begin{array}{ll} f(\iota_1) = f(\iota_2) \Longleftrightarrow f_i(\iota_1) = f_i(\iota_2) & \forall i \in 1,2,...,n_{obj} \\ f(\iota_1) \leq f(\iota_2) \Longleftrightarrow f_i(\iota_1) \leq f_i(\iota_2) & \forall i \in 1,2,...,n_{obj} \\ f(\iota_1) < f(\iota_2) \Longleftrightarrow (f(\iota_1) \leq f(\iota_2)) \wedge (f(\iota_1) \neq f(\iota_2)) & \end{array} \tag{1}$$

where n_{obj} is the number of competing objectives. Taking into account the above relations the Pareto-dominance criterion can be defined as:

$$\begin{array}{lll} \iota_1 \prec \iota_2 \quad (\iota_1 \text{ dominates } \iota_2) & \Longleftrightarrow & f(\iota_1) < f(\iota_2) \\ \iota_1 \preceq \iota_2 \; (\iota_1 \text{ weakly dominates } \iota_2) & \Longleftrightarrow & f(\iota_1) \leq f(\iota_2) \\ \iota_1 \sim \iota_2 \quad (\iota_1 \text{ is indifferent to } \iota_2) & \Longleftrightarrow & f(\iota_1) \not\leq f(\iota_2) \wedge f(\iota_2) \not\leq f(\iota_1) \end{array} \tag{2}$$

A Pareto-optimum solution is defined as an individual that cannot be dominated by any one in the solution set:

$$\nexists \iota_i \in D : \iota_i \prec \iota_j \tag{3}$$

A good multiobjective algorithm should find as many Pareto-optimum solutions as possible, to provide the final user the possibility of choosing the right solution following his own criteria.

There are several ways of adapting EAs to solve MOPs. In this paper is used the approach proposed in [2], which estimates an scalar dummy fitness for each individual in the current population based in its rank. The rank of each individual is defined as:

$$\text{rank}(\iota_j^t) = 1 + dom_j^t \tag{4}$$

where dom_j^t represents the number of individuals dominating ι_j^t in the current population. Once that each individual is assigned a rank, a dummy fitness is obtained for all the individuals by interpolating between the maximum and minimum rank in the population. This simple modification allows a generic EA [8] to solve a MOP transparently, that is, without changing any other of its components.

Multi-Objective Evolutionary Algorithms (MOEAs) are a robust optimization technique that have been successfully applied to several optimization problems. Its strength is based on their simplicity and easy implementation. However, MOEAs are a generic optimization technique whose results can be improved if some expert knowledge about the problem to be solved is incorporated in them. These changes produce algorithms that hybridizes the robustness and strength of EAs and the expertness of some heuristics for the problem. This adapted

MOEAs obtain better results than the generic ones because they can guide the search towards solutions that the expert knowledge expects to be superior. The easiest way of incorporating this expert knowledge of the problem in an EA is to construct some mutation operators specific for the problem to be solved. These operators differ from the original ones in the sense that they do not apply blind changes to the individuals they affect. They try to improve the mutated individual by analyzing and altering it using some heuristics. A complete description of the MOEA used in this paper can be found in [3], here we will focus only en the expert mutation operators designed for this problem and on how they can guide the search towards better solutions.

3 Proposed mutation operators

The purpose of the MOEA described in this paper is to search for RBFNNs with different compromises of complexity and prediction error, so that the final user can choose among diverse possibilities the model that better solve his requirements. Thus, some mutation operators have to be designed to alter the structure of the net, adding or deleting RBFs until good nets are obtained. These mutation operators incorporate some heuristics to alter the RBFNNs "cleverly", detecting the RBFs that less contribute to the net output and eliminating them, or, on the other hand, estimating the input regions that increase the prediction error and adding there more RBFs to obtain a better model.

3.1 RBF Splitting

This mutation operator is based in a splitting mechanism for RBFs proposed in [6], but with some modifications because the original mechanism was oriented to classification problems and the proposed algorithm will be used to forecasting time series. The basic functioning of this operator is as follows:

- detect that RBF producing the highest error increment in the prediction error, and
- split it into two RBFs to obtain a net that covers better the input space.

The prediction error is caused by some RBFs that become highly activated by some input vectors and produce an output value different of the expected one. The RBFs that are not activated by an input vector don't contribute to the prediction error for that input vector. Thus, the prediction error can be proportionally divided between all the RBFs that are activated by an input example as:

$$e^j = \sum_{k=1}^{n} \frac{\phi_j(\boldsymbol{x}^k)}{\sum_{i=1}^{m} \phi_i(\boldsymbol{x}^k)} \left|\mathcal{F}(\boldsymbol{x}^k) - y^k\right|, \quad j = 1, ..., m \tag{5}$$

where n is the number of input examples, m is the number of RBFs in the net, ϕ_j is the j-th RBF, $\mathcal{F}(\boldsymbol{x}^k)$ is the prediction of the net for the input vector $\boldsymbol{x}^k$, and y^k is its expected prediction. A high value of e^j shows that the j-th RBF is not sufficient to learn the input examples that activate it, thus the bigger e^j, the higher probability of division for ϕ_j. So, each RBF ϕ_h is assigned a division probability inversely proportional to e^j and a basis function is randomly selected according to these probabilities.

Once one RBF ϕ_j has been selected, the 2-means algorithm is run with the input examples that are closer to the center of ϕ_j than to any other RBF center, obtaining two new positions for two new RBFs, ϕ_{j1} and ϕ_{j2}, that will substitute ϕ_j in the affected net. The radii for ϕ_{j1} and ϕ_{j2} are calculated using the CIV heuristic [6] and the optimum weights for all the weights of the new net are obtained using the Cholesky method [9].

3.2 OLS based pruning

Orthogonal Least Squares (OLS) [1] is one of the most widely used method to prune the less relevant RBFs of a net. Basically, this method calculates a vector of error reduction ratios $\boldsymbol{err}$ where each one of its components $[err]_j$ gives an idea of the output variance explained by its associated RBF ϕ_j. The lower $[err]_j$, the less relevant is ϕ_j in the RBFNN output. OLS allows this mutation operator to assign a prune probability to each RBF based in its associated error reduction ratio. Less important RBFs will have more likelihood to be deleted, while more relevant RBFs will be more deletion-protected.

Once that an RBF is selected and deleted, the weights of the remaining basis functions are optimally recalculated using the Cholesky method.

3.3 SVD based pruning

Another good heuristic to prune RBFs is the Singular Value Decomposition (SVD) of the activation matrix P of the RBFNN [5]. This orthogonal transformation provides a vector $\boldsymbol{\sigma}$ of Singular Values (SVs), each one of them concerning one of the RBFs in the net. These SVs reveal the degree of linear independence of the columns of P. Columns with a low (nearly null) singular value are almost linearly dependent and may cause a singular activation matrix. Thus, if these columns are identified and deleted, the system will become simpler and more robust.

With the aforementioned idea in mind, this mutation operator assigns a pruning probability to each RBF inversely proportional to its SV and deletes a basis function randomly selected, having less relevant RBFs more likelihood to be pruned than those more important ones. After the deletion, the weights of the remaining RBFs are optimally obtained using the Cholesky method.

4 Results

The algorithm presented above have been tested with the time series generated by the Mackey-Glass time-delay differential equation [7]:

$$\frac{ds(t)}{dt} = \alpha \cdot \frac{s(t-\tau)}{1+s^{10}(t-\tau)} - \beta s(t) \tag{6}$$

Following previous studies [12], the parameters have been fixed to $\alpha = 0.2$, $\beta = 0.1$, obtaining a chaotic time series without a clearly defined period; it will not converge or diverge, and it is very sensitive to initial conditions.

As in [4], to obtain the time series value at integer points, we applied the fourth-order Runge-Kutta method to find the numerical solution for the above equation. The values $s(0) = 1.2$, $\tau = 17$, and $s(t) = 0$ for $t < 0$ are assumed. This data set can be found in the file `mgdata.dat` belonging to the FUZZY LOGIC TOOLBOX OF MATLAB 5.

Following the conventional settings for predicting these time series, we will predict the value $s(t+6)$ from the current value $s(t)$ and the past values $s(t-6)$, $s(t-12)$, and $s(t-18)$, thus, the training vectors for the model will have the following format:

$$[s(t-18), s(t-12), s(t-6), s(t), s(t+6)] \tag{7}$$

The first 500 input vectors have been used to train the model and the next 500 vectors have been used to test the RBFNNs obtained. The proposed algorithm has been run 5 times with a population of 25 individual during 1000 generations, and the best solutions found have been applied the Levenberg-Marquard minimization algorithm to fine-tune their parameters. Table 1 shows minimum, mean and standard deviation of the best RMSEs obtained in five different runs of the algorithm using a population of 25 individuals.

Table 2 shows the best solutions found by other approaches. The comparison of these results with those obtained by the proposed algorithm (see table 1)

Num.	Training RMSE			Test RMSE		
RBFs	Min.	Mean	St. Dev.	Min.	Mean	St. Dev.
8	0.0065	0.0099	0.0034	0.0057	0.0095	0.0035
9	0.0061	0.0085	0.0024	0.0061	0.0085	0.0025
10	0.0055	0.0071	0.0018	0.0054	0.0070	0.0017
11	0.0055	0.0070	0.0019	0.0056	0.0086	0.0043
12	0.0061	0.0069	0.0014	0.0060	0.0061	0.0002
13	0.0057	0.0062	0.0010	0.0056	0.0063	0.0011
14	0.0037	0.0042	0.0008	0.0039	0.0043	0.0007

Table 1. Minimum, mean and standard deviation of the training and test errors for all the different RBFNN structures.

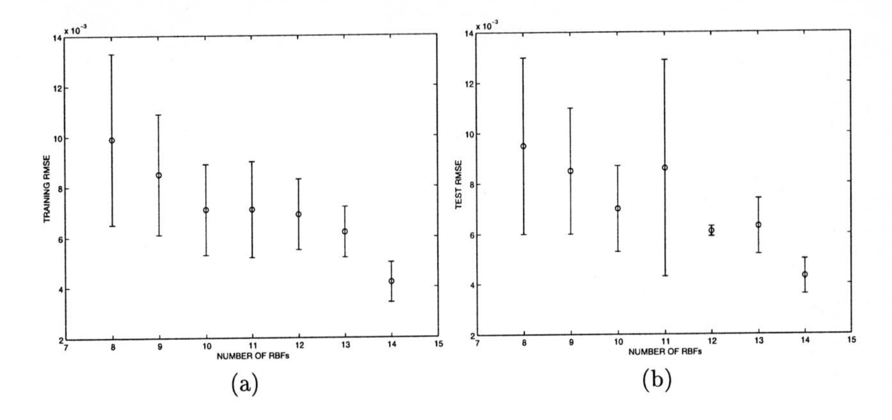

Fig. 1. Mean and standard deviation of the RMSE in five runs of the algorithm.

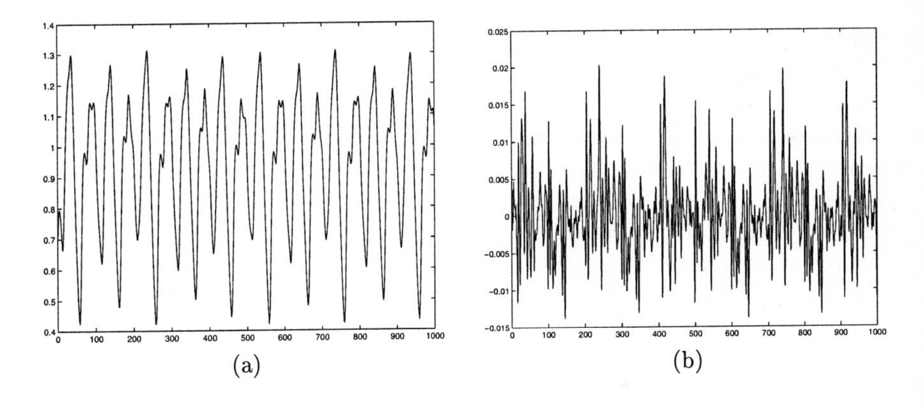

Fig. 2. (a) Original time series (solid line) and predicted values (dashed line) for training and test data. (b) Training and test residuals.

reveals that the use of expert mutation operators enhances significantly the search results. All minimum performance indices are superior to those obtained by other approaches, while the standard deviation from the mean RMSE are small enough to conclude the robustness of our approach. As an example, figure 2 shows the original series (solid line), and the predicted values (dashed line) for the training and test data (a), and the training and error residuals (b) for the best RBFNN of 10 basis functions found.

5 Conclusions

Identifying the adequate structure for a RBFNN created from training data is not an easy problem. There exist several optimal solutions depending on the chosen compromise between the prediction error and the complexity of the model.

Approach		Test RMSE
Linear Predictive Method		0.55
Auto Regressive Model		0.19
L-X. Wang	T-Norm: Prod.	0.0907
	T-Norm: Min.	0.0904
Cascade Correlation ANN		0.06
6th-order Polynomial		0.04
D. Kim & C. Kim	5 MFs/var.	0.0492
(Genetic Algorithm	7 MFs/var.	0.0423
+ Fuzzy System)	9 MFs/var.	0.0379
Retropropagation ANN		0.02
ANFIS (ANN + Fuzzy Logic)		0.007

Table 2. RMSE of other approaches

MOEAs are able to find a sufficient number of Pareto-optimal solutions, providing the final user a wide variety of possibilities, in order he can choose the solution that better satisfies his requirements.

MOEAs can also be specialized by means of expert mutation operators that improve the search results. These expert mutation operators can incorporate some heuristics for the problem, such as OLS and SVD for the pruning mechanism, or an error sharing scheme to identify RBFs producing bigger errors, to perform "clever changes" when altering an individual. Particularly, the proposed algorithm has found very good RBFNNs for the prediction of the Mackey-Glass time series.

Acknowledgement

This work has been partially supported by the CICYT Spanish projects TAP97-1166 and TIC2000-1348.

References

1. S. Chen, C. F. N. Cowan, and P. M. Grant. Orthogonal least squares learning algorithm for radial basis function networks. *IEEE Trans. Neural Networks*, 2:302–309, 1991.
2. C. M. Fonseca and P. J. Fleming. Genetic algorithms for multiobjective optimization: Formulation, discussion and generalization. In S. Forrest, editor, *Proceedings of the Fifth International Conference on Genetic Algorithms*, pages 416–423. Morgan Kaufmann, 1993.
3. J. González, I. Rojas, H. Pomares, M. Salmerón, and A. Prieto. Evolutive identification of fuzzy systems for time series prediction. In A. Ollero, S. Sánchez, B. Arrue, and I. Baturone, editors, *Actas del X Congreso Español sobre Tecnologías y Lógica Difusa, ESTYLF 2000*, pages 403–409, Sevilla, Spain, Sept. 2000.

4. J. S. R. Jang. Anfis: Adaptive network-based fuzzy inference system. *IEEE Trans. Syst., Man, Cybern.*, 23:665–685, May 1993.
5. P. P. Kanjilal and D. N. Banerjee. On the application of orthogonal transformation for the design and analysis of feed-forward networks. *IEEE Trans. Neural Networks*, 6(5):1061–1070, 1995.
6. N. B. Karayiannis and G. W. Mi. Growing radial basis neural networks: Merging supervised and unsupervised learning with network growth techniques. *IEEE Trans. Neural Networks*, 8(6):1492–1506, Nov. 1997.
7. M. C. Mackey and L. Glass. Oscillation and chaos in physiological control systems. *Science*, 197(4300):287–289, 1977.
8. Z. Michalewicz. *Genetic Algorithms + Data Structures = Evolution Programs.* Springer-Verlag, 3rd edition, 1996.
9. H. Pomares, I. Rojas, J. Ortega, J. González, and A. Prieto. A systematic approach to a self-generating fuzzy rule-table for function approximation. *IEEE Trans. Syst., Man, Cyber. Part B*, 30(3):431–447, June 2000.
10. I. Rojas, J. González, A. Cañas, A. F. Díaz, F. J. Rojas, and M. Rodriguez. Short-term prediction of chaotic time series by using rbf network with regression weights. *Int. Journal of Neural Systems*, 10(5):353–364, 2000.
11. I. Rojas, H. Pomares, J. González, J. L. Bernier, E. Ros, F. J. Pelayo, and A. Prieto. Analysis of the functional block involved in the design of radial basis function networks. *Neural Processing Letters*, 12(1):1–17, Aug. 2000.
12. B. A. Whitehead and T. D. Choate. Cooperative-competitive genetic evolution of radial basis function centers and widths for time series prediction. *IEEE Trans. Neural Networks*, 7(4):869–880, July 1996.

Evolving RBF Neural Networks

V.M. Rivas[1], P.A. Castillo[2], and J.J. Merelo[2]

[1] Dpto. Informática, Univ. de Jaén,
E.P.S., Avda. de Madrid, 35, E.23071, Jaén(Spain)
vrivas@ujaen.es, http://wwwdi.ujaen.es/~vrivas
[2] Dpto. de Arquitectura y Tecnología de Computadores, Univ. de Granada,
Fac. de Ciencias, Campus Fuentenueva, S/N. E.18071, Granada (Spain)
todos@geneura.ugr.es, http://geneura.ugr.es

Abstract. Determining the parameters of a Radial Basis Function Neural Network (number of neurons, and their respective centers and radii) is often done by hand, or based in methods highly dependent on initial values. In this work, Evolutionary Algorithms are used to automatically build a RBF NN that solves a specified problem. The evolutionary algorithms are implemented using a new evolutionary computation framework called EO, which allows direct evolution of problem solutions, so that no internal representation is needed, and specific solution domain knowledge can be used to construct evolutionary operators, as well as cost or fitness functions. Results show that this new approach finds nets with good generalization power, while maintaining a reasonable size.
Keywords: RBF, evolutionary algorithms, EO, functional estimation, time series forecasting.

1 Introduction

A *Radial Basis Function (RBF)* can be characterized by a point of the input space, c, and a radius or width, r, such that the RBF can behave in two different ways: one, it reaches its maximum value when applied to c, and decreases to its minimum value when applied to points far from c; and two, it reaches its minimum value in c and increases as distance does. The radius, r, controls how distance affects that increment or decrement. Because of this, mathematicians have used groups of RBF to successfully interpolate data. Typical examples of RBF are the Gaussian function and the multiquadric function, although there are many others.

Radial basis function neural networks (RBF NNs) were introduced in [3], being their main application function approximation and time series forecasting. A RBF NN is a two-layer, feed-forward network (see fig. 1) in which the activation functions of the neurons of the hidden layer are RBF. Gaussian functions are commonly used. Each hidden neuron computes the distance from its input to the neuron's central point, c, and applies the RBF to that distance. The neurons of the output layer perform a weighted sum between the outputs of the hidden layer and the weights of the links that connect both output and hidden layer neurons, i.e.,

J. Mira and A. Prieto (Eds.): IWANN 2001, LNCS 2084, pp. 506-513, 2001.

$$h_i(x) = \phi(||x - c_i||^2/r_i^2) \quad (1)$$

$$o_j(x) = \sum w_{ij} h_i(x) + w_0 \quad (2)$$

where x is the input, ϕ is the RBF, c_i is the center of the *i-th* hidden neuron, r_i is its radius, w_{ij} is the weight of the links that connects hidden neuron number i and output neuron number j, and w_0 is a bias for the output neuron.

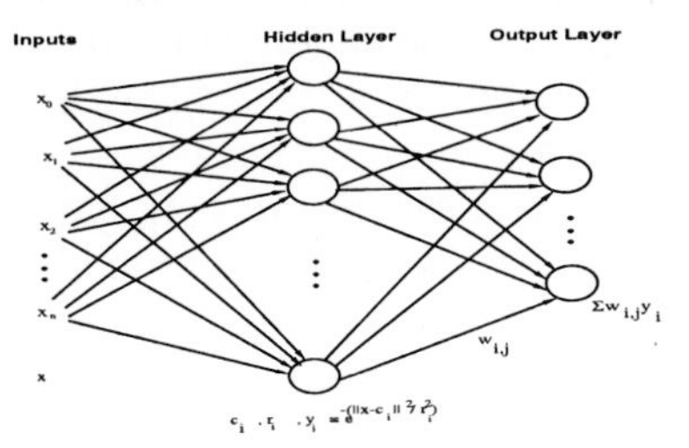

Fig. 1. RBF Neural Network

It has been proved ([8]) that when enough units are provided, a RBF NN can approximate any multivariate continuous function as much as desired. This is possible because once the centers and the radii have been fixed, the weights of the links between the hidden and the outputs layers can be calculated analytically using Singular Value Decomposition [15] or any algorithm suitable to solve lineal algebraic equations, making unnecessary the use of training algorithms, as those used in other kinds of neural networks such as multilayer perceptrons. Thus, the main problem in RBF NNs design concerns establishing the number of hidden neurons to use and their centers and radii.

The need of automatic mechanisms to build RBF NNs is already present in Lowe and Broomhead's work [3], where they showed that one of the parameters that critically affects the performance of a RBF NN is the number of hidden neurons. If its number is not sufficient, the approximation offered by the net is not good enough; in the other hand, nets with many hidden neurons will approximate very well those points used to calculate the connection weights, while having a very poor predictive power, this is the so-called overfitting problem [2]. Thus, establishing the number of neurons (that is, the number of centers and values related to them) that solves a given problem is one of the most important tasks researchers have faced.

In this paper an evolutionary algorithm is used to find the best components of the RBF NNs for approximating a set of functions.

The rest of the paper is organized as follows: section 2 describes some of the methods used to solve the cited problems; section 3 shows our proposed solution, and describes the evolutionary computation paradigm we use (EO). Next section (4) shows some functional approximation experiments; and finally, our conclusions and proposals of future work can be read in section 5.

2 State of the Art

There are several papers that address the problem of automatic RBF NN design. The paper by Leonardis and Bischof [7] offers a good overview. Methods can be divided in three classes.

1. Methods in which **the number of radial basis functions must be given a priori**; after this, computing the values for the centers and the widths is done choosing points randomly from the training set or performing any kind of clustering method with them [11].
2. Methods that automatically **search for the number of RBF and their values.** Most of these methods grow the final net from an initially empty one to which hidden neurons are added until a given condition is reached. One of such methods is Orthogonal Least Squares [4], based on a mechanism called *forward selection*, that allows the addition of hidden neurons to an initially empty net until the approximation error is under a prespecified tolerance value or threshold. An improvement over this algorithm is Regularised OLS [5], based in *regularised forward selection*, in which high hidden-to-output weight values are penalyzed using a penalty term, a process termed regularization, [2]. Later, in [12], Orr newly used forward selection as the growing mechanism, and delete-1 (or leave-one-out) and generalized cross-validation as the methods to stop adding neurons; in any of it forms, cross-validation is used to compute the generalization error of many different nets in competition one with each other, halting the algorithm when that error reaches a minimun. Other methods of the same characteristics are Resource Allocation Networks [14], and Growing Cell Structures [6], which try to estimate the centers using techniques based on gradient descent, making them liable to fall in local optima.
3. There are also **pruning methods** that start with an overspecified RBF NN and, iteratively, remove hidden neurons and/or weights. Leonardis and Bischof's algorithm, [7], can be classified as a pruning method, in which the Minimum Description Length (MDL) measure is used to achieve a reduction in the complexity of the RBF NN. According to the MDL principle, a set of RBFs describing the training data with the shortest possible encoding are selected among a larger set. The main drawback of pruning methods is that they tend to be very restrictive, and find suboptimal networks.
4. **The values for the radii are also important parameters**. Very narrow radii will lead to overfitting, and radii wider than necessary can give even worst results. Thus, in order to obtain a good performance, and taking into account that each component of the points in the input space can have ranges with very different limits, many radii with many different values must be fixed. There are some algorithms intended to estimate the radii; Orr in [13] chooses a set of radii, and after this, the best of those radii is found (as efficiently as possible) by means of 1) constructing the RBFs, 2) using them, and 3) regarding the performance of the nets. It is obviously necessary to start with a good set of radii in order to succeed. Furthermore, it could

happen that a good value for the radii were not considered because a value close to it, and present in the initial set, had produced bad results.

5. The use of **Evolutionary Algorithms** to construct neural nets is also well known in the literature. Some reviews can be found in [1] and [19]. All of them try to optimize only one of the parameters of the net (number of neurons, topology, learning rate, and so on), leaving the other parameters fixed. A different case can be found in the work by Castillo et al. [16,17], in which evolutionary algorithms are used to create good Multilayer Perceptrons (MLPs), optimizing all the parameters at the same time, and obtaining very good results. A similar and previous example can be found in Merelo's [9], where a neural network called Learning Vector Quantization (LVQ) is genetically optimized. Nevertheless, both Castillo and Merelo's methods can not be applied directly to the construction of RBF NNs since their genetic operators are geared towards their specific network architecture.

3 EvRBF

The method introduced in this work uses an evolutionary algorithm, **EvRBF**, to build RBF NNs, optimizing their predictive power by finding the number of neurons in the hidden layer, and their centers and radii.

The evolutionary algorithm itself has been programmed using the new evolutionary computation framework, **EO** (**E**volving **O**bjects), current version is 0.9.2 [18]. It can be found at `http://eodev.sourceforge.net`

This new framework is the result of the cooperative work carried out by several research teams in Europe. The main advantage of this new paradigm is that what can be evolved is not necessarily a sequence of genes (bitstring or floating point values), but anything to which a fitness or cost function can be assigned. In this sense, everything one can imagine can be converted into something evolvable, and consequently, something that EO can evolve, optimizing it.

EO allows to define operators that increase or decrease the diversity in the population, that is, crossover-like operators and mutation-like operators. Some of these operators are generic and can be applied to many classes of evolvable objects, but new operators can also be specifically defined for each object; taking into account that, using EO, evolutionary algorithms deal with the solution of the problem being optimized itself, instead of a representation of that solution, problem domain knowledge can be inserted in the operators, obtaining a better behaviour. This is also in accordance with Michalewicz's ideas published in [10].

In the present research, an standard genetic algorithm has been used with specific operators. Fixed size population, tournament selection, and elitist replacement policy have been used, and two kind of operators, mutation-like and crossover-like, have been created. The algorithm finishes when a previously specified number of generations has been reached.

Binary crossover operator

The binary operator is applied to two RBF NNs; it takes a random uniformly chosen number of consecutive hidden neurons from the first network, and another sequence from the second, and interchanges the two sequences, no matter if both sequences have different number of neurons.

Unary operators

Given that in a RBF NN the weights from hidden neurons to output ones can be easily computed, the unary or mutation-like operators affect the hidden neuron components, centers and radii, in their quantities and values, but not to the hidden-to-output weights.

Centers mutator. This operator changes a given percentage of the components of the center of every hidden neuron. Each component of the center is modified by adding or subtracting a random value, which is chosen following a Gaussian probability function with mean 0 and standard deviation 0.1.

Radii mutator. This operator is applied to one of the existing nets, but affects to the radii of the RBF. Radii are modified using a Gaussian function as defined previously. Radii are different for each center, and for each component of the center.

Hidden neuron adder. This operator duplicates every hidden neuron with a given probability (the same for all the neurons). Once duplicated, the values for the centers and radii of the new neuron are modified using a Gaussian function.

Hidden neuron deleter. Erases a given percentage of hidden neurons in every net. Neurons are randomly chosen.

The breeder

Every new generation is created in the following way:

1. Select a few individuals using tournament selection of fixed size.
2. Delete the rest of the population.
3. Generate all new individuals using crossover operator (the parents remain unchanged).
4. Apply the unary operators to the new individuals.
5. Set the weights of the new individuals using SVD.
6. Remove the non-sense neurons, i.e., those whose weights to output neurons are very close to 0.
7. Evaluate the net, and set its fitness.

Once tournament selection, performed with a parametric number of individuals, has finished, the crossover operator is applied as many times as needed until the population reaches its fixed size. Later, mutation operators are applied with a given fixed probability (not necessarily the same for all of them) to each neuron of each RBF NN, so that if one operator is used with a probability of 0.01, it will modify 1% of the RBF on average.

Nets' fitness are calculated as 1 divided by the root mean square error:

$$fitness = \frac{1}{\sqrt{\frac{\sum_{i=0}^{n-1}(y_i - o(x_i))^2}{n}}} \tag{3}$$

where y_i is the expected output, $o(x_i)$ is the output calculated by the net, and n is the number of input-output pairs in the validation set; none of these points has been seen previously by the nets, because they are not used to compute the weights.

4 Experiments and results

This new approach has been tested on the task of function approximation. Two functions, borrowed from literature, have been used. For any of them, three different sets of input-output pairs have been created: one for the training process (with 188 input-output pairs), another for validation (with 62 pairs), and the last to test the generalization power of the best RBF NN found (composed of 250 input-output pairs). The training set is used by the SVD algorithm to compute the weights of the new NN; the validation set is used to compute the fitness of the individuals.

For each function, the EA have been ran 3 times, and average errors with their respective standard deviation are shown. The values in the tables belong to the errors obtained by the best nets in the task of approximating the validation set and the generalization set.

These functions are borrowed from [3], and are shown using solid lines in figure 2. Consider T_d as an ordered sequence of iterations of the doubling map: $x_{n+1} = 2x_n$; and T_q a sequence of iterations of the quadratic map: $x_{n+1} = 4x_n(1 - x_n)$. These maps are known to be chaotic in the interval $[0, 1]$.

Figure 2 also shows the predictions (filled triangles) made by the best RBF NN found.

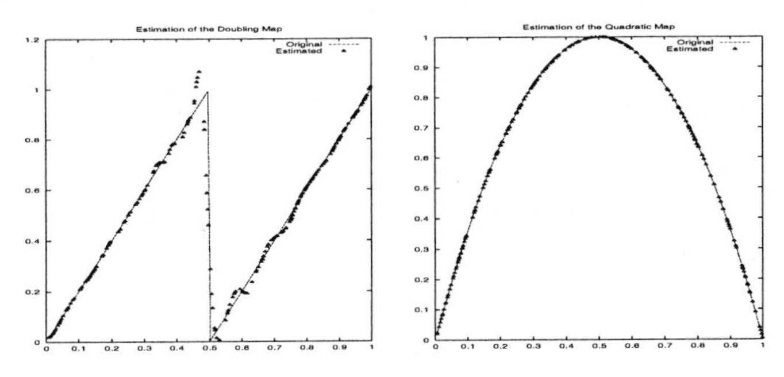

Fig. 2. Double (left) and quadratic (right) Time Series Maps. The function to approximate is shown as a dashed line, the RBFNN approximation, with triangles.

We used 10 generations and 50 individuals in the population in every run in order to avoid long simulation times. After an exhaustive test, we have considered to use a tournament size of 3 and to apply genetic operators using a crossover rate of 1 and a mutation rate of 0.01.

Tables 1 and 2 show, on logarithmic scale, the normalized error (i.e., the error divided by its standard deviation) with respect to the number of hidden neurons for the generalization process. It can be clearly seen that EvRBF can find smaller RBF NNs while obtaining a best generalization power.

Lowe and Broomhead					EvRBF
Neurons	50	100	150	200	***36.33 ± 2.31***
Generalization	*-0.85*	*-0.7*	*-0.9*	*-1.15*	***-1.47 ± 0.06***

Table 1. Comparison between results obtained by Lowe, and results obtained by EvRBF for the doubling map problem. Generalization error is given as $\log_{10}(Normalised\, error)$, thus lower negative values are better.

Lowe and Broomhead					EvRBF
Neurons	50	100	150	200	***16.33 ± 7.23***
Generalization	*-8*	*-8.7*	*-8.1*	*-8*	***-9.61 ± 0.43***

Table 2. Comparison between results obtained by Lowe, and results obtained by EvRBF for quadratic map problem. Generalization error is given as $\log_{10}(Normalised\, error)$, thus lower negative values are better.

5 Conclusions

Creating the best RBF NN that solves a given problem is a difficult task because many parameters have to be set at the same time: number of hidden neurons, and centers and radii for them. Evolutionary algorithms can help find the optimal values for those parameters, obtaining RBF NNs with a good power of generalization.

In order to use specific problem knowledge, the RBF NNs are evolved as such, without the use of a representation and a decoder. This is possible thanks to the new evolutionary computation framework, EO, implemented as a C++ class library, because it allows to evolve any object to which a fitness or cost function can be assigned.

The results obtained by our algorithm, EvRBF, show that good RBF NNs are found when it is applied to function approximation.

Acknowledgment

This work has been supported in part by FEDER I+D project 1FD97-0439-TEL1, CICYT TIC99-0550 and INTAS 97-30950.

References

1. A.V. Adamopoulos, E.F. Georgopoulos, S.D. Likothanassis, and P.A. Anninos. Forecasting the MagnetoEncephaloGram (MEG) of Epilectic Patients Using Genetically Optimized Neural Networks. In *Proceedings of the genetic and Evolutionary Computation Conference, GECCO'99*, volume 2, pages 1457–1462. Morgan-Kaufmann Publ., July 1999.
2. C.M Bishop. *Neural Networks for Pattern Recognition.* Oxford University Press, 1995. ISBN 0-19-853849-9 (hardback) or 0-19-853864-2 (paperback).
3. D.S. Broomhead and D. Lowe. Multivariable Functional Interpolation and Adaptative Networks. *Complex Systems*, 11:321–355, 1988.
4. S. Chen et al. Orthogonal Least Squares algorithm for constructing Radial Basis Function Networks. *IEEE Transactions on Neural Networks*, 2(2):302–309, 1991.
5. S. Chen et al. Regularised Orthogonal Least Squares Learning for Radial basis function Networks. Submitted to International Journal Control, 1995.
6. B. Fritzke. Supervised learning with growing cell structures. In J.D. Cowan, G. Tesauro, and J. Aspector, editors, *Advances in Neural Information Processing Systems*, volume 6, pages 255–262. Morgan Kaufmann, 1994.
7. A. Leonardis and H. Bischof. And efficient MDL-based construction of RBF networks. *Neural Networks*, (11):963–973, 1998.
8. W.A. Light. Some aspects of Radial Basis Function approximation. *Approximation Theory, Spline Functions and Applications*, 356:163–190, 1992.
9. J.J. Merelo and A. Prieto. G-LVQ a combination of genetic algorithms and LVQ. In D.W. Pearson, N.C. Steele, and R.F. Albrecht, editors, *Artificial Neural Nets and Genetic Algorithms.* Springer-Verlag, 1995.
10. Zbigniew Michalewicz. *Genetic algorithms + data structures = evolution programs.* Springer-Verlag, NewYork USA, 3 edition, 1999.
11. J. E. Moody and C. Darken. Fast learning in networks of locally tuned processing units. *Neural Computation*, 2(1):281–294, 1989.
12. M.J.L. Orr. Regularisation in the Selection of Radial Basis Function Centres. *Neural Computation*, 7(3):606–623, 1995.
13. M.J.L. Orr. Optimising the Widths of Radial Basis Functions. V Brazilian Congress on Neural Networks, 1998.
14. J. Platt. A resource-allocating network for function interpolation. *Neural Computation*, 3(2):213–225, 1991.
15. W.H. Press, S.A. Teukolsky, W.T. Vetterling, and B.P. Flannery. *Numerical Recipes in C.* Cambridge University Press, 2nd edition, 1992.
16. P.A. Castillo; J. Carpio; J.J. Merelo; V. Rivas; G. Romero; A. Prieto. Evolving Multilayer Perceptrons. *Neural Processing Letters, vol. 12, no. 2, pp.115-127. October*, 2000.
17. P.A. Castillo; J.J. Merelo; V. Rivas; G. Romero; A. Prieto. G-Prop: Global Optimization of Multilayer Perceptrons using GAs. *Neurocomputing, Vol.35/1-4, pp.149-163*, 2000.
18. J.J. Merelo; M. G. Arenas; J. Carpio; P.A. Castillo; V. M. Rivas; G. Romero; M. Schoenauer. Evolving objects. *In M. Graña, editor, FEA2000 (Frontiers of Evolutionary Algorithms) proceedings. Proc. JCIS'2000. P.P. Wang(ed). Editorial Association for Intelligent Machinery. ISBN:0-9643456-9-2. Vol I, pp.1083-1086. Atlantic City, NJ, Feb. 27-March 3.*, 2000.
19. D. White and P. Ligomenides. GANNet: a genetic algorithm for optimizing topology and weights in neural networks design. In *Lectures Notes in Computer Science*, volume 686, pages 322–327, 1993. Proceedings of IWANN'93.

Evolutionary Cellular Configurations for Designing Feed-Forward Neural Networks Architectures

G. Gutiérrez[1], P. Isasi[2], J.M. Molina[2], A. Sanchís[3], and I. M. Galván[3]

Departamento de Informática, Universidad Carlos III de Madrid,
Avenida de la Universidad 30, 28911, Leganés, Madrid.
[1]ggutierr@inf.uc3m.es, [2]{isasi,molina}@ia.uc3m.es,
[3]{masm,igalvan}@inf.uc3m.es

Abstract. In the recent years, the interest to develop automatic methods to determine appropriate architectures of feed-forward neural networks has increased. Most of the methods are based on evolutionary computation paradigms. Some of the designed methods are based on direct representations of the parameters of the network. These representations do not allow scalability, so to represent large architectures, very large structures are required. An alternative more interesting are the indirect schemes. They codify a compact representation of the neural network. In this work, an indirect constructive encoding scheme is presented. This scheme is based on cellular automata representations in order to increase the scalability of the method.

1 Introduction

The design of Neural Network (NN) architectures is crucial in the successful application of the NN because the architecture may strongly drive the neural network's information processing abilities. In most of the cases exist a large number of architectures of feed-forward neural networks set suitable to solve an approximation problem. The design of the NN architecture can be seen as a search problem within the space of architectures, where each point represents an architecture. Evidently, this search space is huge and the task of finding the simplest network that solves a given problem is a tedious and long task.

In the last years, many works have been centred toward the automatic resolution of the design of neural network architecture [1, 2, 3, 4, 5, 6]. Two main representation approaches exist to find the optimum net architecture using Genetic Algorithms (GA): one based on the complete representation of all possible connections and other based on an indirect representation of the architecture. The first one is called Direct Encoding Method, and is based on the codification of the complete network (connections matrix) into the chromosome of the GA [7, 8, 9, 10, 11]. The direct representation is relatively simple and straightforward to implement but requires much larger chromosomes. This could end in a too huge space search that could make the method impossible in practice.

In order to reduce the length of the genotype, indirect encoding scheme has been proposed in the last years. These methods consists on codifying, not the complete network, but a compact representation of it [1], avoiding the scalability problem. One

J. Mira and A. Prieto (Eds.): IWANN 2001, LNCS 2084, pp. 514-521, 2001.

of those previous works was from Kitano [12], introducing a constructive scheme based on grammars. The solution proposed by Kitano was to encode networks as grammars. An extension of the of Kitano's method can be found in [13]. Other works have considered fractal representations [14], arguing that this kind of representation is more related with biological ideas than constructive algorithms.
In this work, an indirect constructive encoding scheme, based on Cellular Automata (CA) [15], is proposed to find automatically an appropriate feed-forward NN architecture. In this scheme, positions of several seeds in a two-dimensional grid are represented (codified) in the chromosome. The seeds are defined through two co-ordinates that indicate their positions in the grid and they are used as the initial configuration of a cellular automaton. That initial configuration is evolved using some cellular automata rules. The rules allow the convergence of the automata toward a final configuration depending on the initial configuration and it is translated into a feed-forward NN. The automata rules have been chosen such that a wide variety of feed-forward NN architecture can be obtained, from full-connected NNs to architectures with few connections.
The main interest of this paper is to show, experimentally, that cellular configurations allow obtaining appropriate architectures, as in domains relatively simple as in domains in which big architectures are required. The motivation of this approach is based on the idea that few seeds can produce big architectures. Thus, the chromosome length is reduced and the scalability of the method is increased.

2 Cellular Approach

Three different modules compose the global system proposed in this paper: the Genetic Algorithm, the Cellular Automaton and the module responsible of NN training, as is shown in Figure 1.

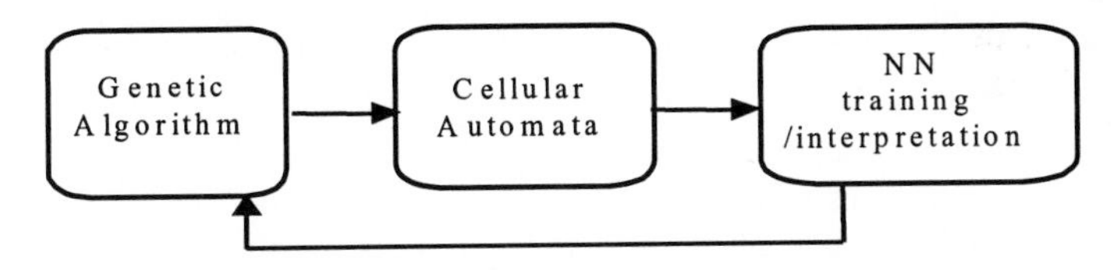

Fig.1. System's architecture and modules relationship

The GA module takes charge of generating initial configurations of the cellular automata, i.e. seed positions and to optimise these configurations from the information obtained from the training module. The cellular automaton takes the initial configuration and generates a final configuration corresponding to a particular NN architecture. Finally, the generated architecture has to be trained and evaluated for a particular problem and relevant information about the NN (error, size, etc.) is used as the fitness value for the GA. The GA carries out the architecture optimisation. However, the cellular automaton is used as a constructive way of generating the architectures. In the next subsections, the description of the different modules is presented in detail.

2.1 Genetic Algorithm Module

This module works with a population of chromosomes that codifies the seeds positions in a two-dimension grid. The GA module operates to maximise a fitness function provided by the NN module, which evaluates the efficiency of the NN to solve the considered problem.
The size of chromosomes in the GA corresponds with the number of seeds, and it codifies all the possible locations of seeds in the grid.
Chromosomes have been codified in base b, where b is the number of rows in the grid and is given through the number of inputs plus the number of outputs. Each seed is determined by a co-ordinate (x,y). A unique gene, indicating the row in which the seed is located, represents the first co-ordinate x. The second co-ordinate y will require more than one gene, if, as usual, the maximal number of hidden neurones is bigger than b. In this particular case, two genes have been used to codify the y co-ordinate, what allows a maximum of b*b hidden neurones. For instance, if there are 3 inputs and 2 outputs, the maximum size of the hidden layer is 25 (4x5+5 = 25). This could be a good estimation of the maximum number of neurones in the hidden layer, but any other consideration could be taken into account without modifying the proposed method. Hence, the chromosome will have 3 genes for each seed to be placed in the grid.

2.2 Cellular Automata Module

For generating neural networks architectures, a two-dimension CA has been used. The size of the two-dimension grid, Dimx*Dimy, is defined as follows: Dimx (rows) is equal to the number of input neurones plus the number of output neurones; Dimy (columns) corresponds with the maximum number of hidden neurones to be considered.
Each cell in the grid could be in three different states: active (occupied by a seed) or inactive. Two different kinds of seeds have been introduced: growing seeds and decreasing seeds. The first kind allows making connections and the second one removing connections. Each seed type corresponds with a different type of automata rule, so there are two rules called growing rule and decreasing rule respectively. The rules determine the evolution of the grid configuration and they have been designed allowing the reproduction of growing and decreasing seeds. In the description of the rules s is a specific growing seed, d is a decreasing seed, i is an inactive state for the cell and a means that the cell could be in any state or contains any type of seed (even a decreasing seed).
Growing rules: They reproduce a particular growing seed when there are at least three identical growing seeds in its neighbourhood. There are different configurations, growing seeds located in: rows, columns, or in a corner of the neighbourhood. In table 1(a) some of those rules are shown (the others are symmetrical). The growing rules allow obtaining feed-forward NN with a large number of connections.
Decreasing rules: They remove connections in the network deactivating a cell in the grid when the cell has a seed and a cell of its neighbourhood contains also a decreasing seed. One situation in which the decreasing rules can be applied is shown in table 1(b); the others can be obtained symmetrically.

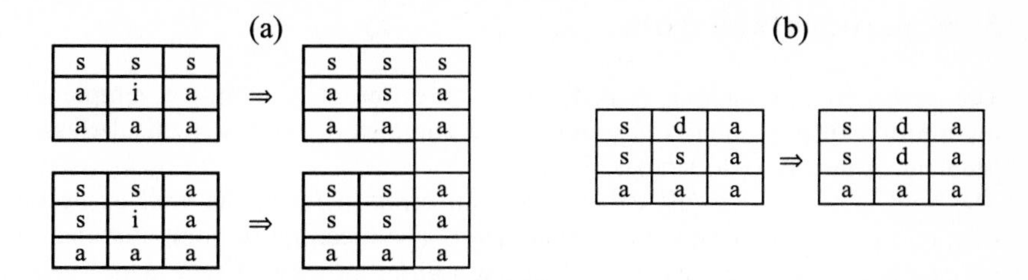

Table 1. Some Automata Rules some configuration of seeds in the neighbourhood of a particular cell. (a) Growing rules. (b) Decreasing Rules

The mechanism of expanding the CA is as follows:

1) The growing seeds are located in the grid.
2) An expansion of the growing seeds takes place. This expansion consists on replicating each seed in turns, over its quadratic neighbourhood, in such a way that if a new seed has to be placed in a position previously occupied by another seed, the first one is replaced.
3) The growing rules are applied until no more rules could be fired.
4) The decreasing seeds are placed in the grid. If there are some other seeds in those places, they are replaced.
5) The decreasing rules are applied until the final configuration is reached.
6) The final configuration of the CA is obtained replacing the growing seeds by a 1 and the decreasing seeds or inactive cells by a 0.

2.3 Neural Network Module

To relate the final configuration of the CA with an architecture of a NN, the following meaning for a cell in the (x,y) grid is defined: if x(n, with n the number of input neurons, (x,y) represents a connection between the x-th input neuron and the y-th hidden neuron; if x>n, (x,y) represents a connection between the y-th hidden neuron and the (x-n)-th output neuron.
In the final configuration a 1 is interpreted as a connection, and a 0 as the absence of connection. Thus, the rows and columns in the matrix with values 0 are removed. A new and shorter binary matrix (M) is obtained. If Mij =1 then a connection between the i-th input neuron and the j-th hidden neuron is created, or between the j-th hidden neuron and the (i-n)-th output neuron, as is previously described. If Mij=0, there do not exist connection between that neurons.
The neural network obtained is trained to solve the particular problem considered. Weights of the NN are randomly initialised, and learned using the back propagation learning method. A value measuring the efficiency of the architecture is computed after the learning phase of the network. This value is used as the fitness function of the chromosome.

3 Experimental Results

The proposed approach is tested with two different domains, a simple one: the minimum coding problem [16], and a more complex domain: a medical classification problem, Ann-Thyroid-Database [17].
The goal is to get networks that, given an error to reach (as well as needed), its training means a lower computational effort. The meaning of computational effort is the number of weights changed along the training process. Then, neural nets with a minimal number of hidden nodes, and reaching an error as soon as possible along training, are looked for.
In the Neural Network Module, when the final matrix connection is obtained from the final configuration of CA there are some special cases take into account, following this steps:

- If there is a node in hidden layer without any connection to output, this node is eliminated from the net.
- When a hidden node has no connection from input, but it's connected to output layer, two chances have been considerate: penalizes the net and don't train it, or eliminate that node and is training.
- If an output node has no connection from hidden layer, the net is penalized and is not trained.

As it was previously mentioned, the aim is to find NN architectures requiring the least computational cost to reach an appropriate level of error. Therefore, the fitness function used in this work is defined as follows: a maximum number of learning cycles and a level of error are defined. If the network reaches that error, the training is stopped and the fitness function is evaluated as:

$$\text{Fitness} = \frac{1}{(c * tc)};\text{c number of weights, tc training cycles} \tag{1}$$

where c*tc measures the computational cost of network. When NN architectures obtained by the CA is penalized or the training of the NN end without reaching the level of error defined, maximum computational cost is assigned. Using this fitness function, the indirect encoding approach presented in this paper has been tested. The performance of this approach has been compared with direct encoding methods.

3.1 Minimum Encoding Problem

In this domain there are four inputs and two outputs. The first two inputs are just noise, and the relation between the other two inputs and the outputs is the Gray coding for integer represented by this two relevant inputs.
In this case, the direct codification is implemented as follows: the matrix of connections, previously indicated in NN Module, is codified in the chromosome. As every possible connection of a network, with 36 hidden nodes, is indicated in the chromosome, then the size of chromosome is 216. Using that codification 400 generations have been carried out and a NN with five hidden nodes is obtained (see figure 2(a)) .
In the proposed method, the position of growing and decreasing seeds in the grid of CA are codified into the genotype. The length of chromosome is 30, 3 genes for each growing or decreasing seed. Again, 400 generations are realized and the NN obtained

is shown in figure 2(b). For both approaches, the evolutions of average fitness along 400 generations are shown in figure 3. As it is possible to observe in figures 2 and 3, the indirect cellular encoding is able to provide more optimal architectures than direct encoding. 100 generations are required by the indirect encoding to find each.

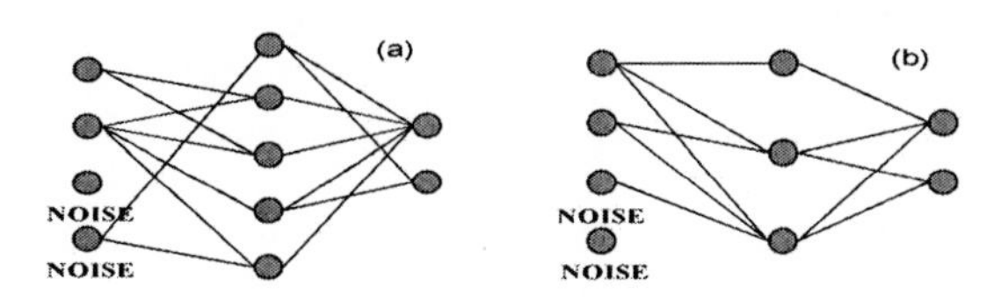

Fig. 2. Architectures obtained by direct encoding (a) and indirect encoding (b)

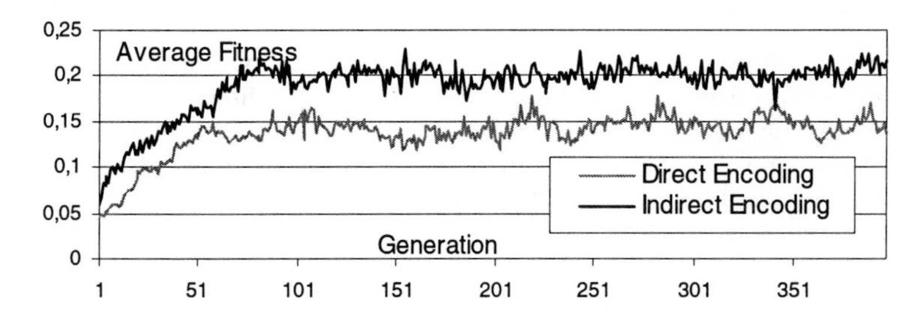

Fig. 3. Average fitness for minimum encoding problem.

3.2 Ann-Thyroid-Database

The thyroid data are measurements of the thyroid gland. Each pattern has 21 attributes, and can be assigned to any of three classes, hyper-, hypo-, and normal function of thyroid gland. This classification problem is hardly to solve for neural nets.

Experiments with a direct encoding have been done in [18]. The authors have obtained architectures with different number of hidden nodes and the connectivity percentage is calculated. One of the best architecture has 10 hidden nodes with 86 % of connectivy.

Several experiments have been developed with indirect encoding. Initially with 30 genes in the chromosome, 5 growing and 5 decreasing seeds. After, the number of seeds has been reduced to 5-3 and 3-3.Thus the length of chromosome is reduced. In all of them similar architectures have been found. Networks with 10 hidden nodes and about 72 % of connectivity are obtained. The indirect cellular encoding find smaller architectures than direct encoding. The evolution of average fitness is shown in figure 4. As it is observed in that figure few generations have been needed. With the direct encoding a large number of generation has been realized (around 1000) [18].

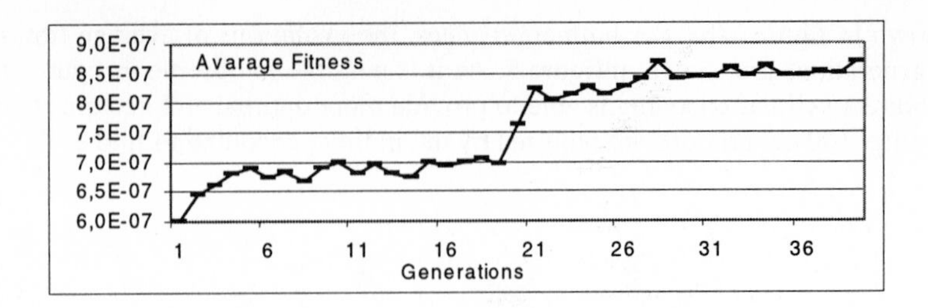

Fig. 4. Average fitness for Ann-Thyroid-Database

4 Conclusions and Future Works

The election of good neural network architectures is an important step in many problems where there is few knowledge about the problem itself. Evolutionary computation techniques are good approaches for automatically generate those good architectures. However the codification of the network is a crucial point in the success of the method. Direct codification's become inefficient from a practical point of view, making bigger and redundant the search space. To solve this problem an indirect encoding has to be used.

Indirect encoding is driven to reduce the search space in such a way that similar solutions are eliminated and represented by the only one representative. In these cases, the codification makes the method able to find better architectures.

CA are good candidates for non-direct codification's. The constructive representation introduced in this work solves some of the problems for non-direct codification's. The final representation has a reduced size and could be controlled by the number of seeds used.

The results shown that the indirect encoding approach presented in this paper is able to find appropriate NN architectures as a simple domain, as a large one. In additions the number of generations over the population is less when the indirect encoding approach is used.

Since the final configuration of the two-dimensional grid will represent a feed-forward NN connection matrix, the rules of the automata used in this first approach have been design such that a wide variety of architectures may be obtained. However, the influence of the rules in the CA evolution and the capability of the rules to generate a complete space of NN architectures must be still studied. Besides, some issues about Neural Network Module and fitness function used, i.e. how punish the nets to increase the search, will be studied in future works.

References

[1] S. Harp, Samad T. and Guha A. Towards the Genetic Synthesis of Neural Networks. Proceedings of the Third International Conference on Genetic Algorithms and their applications, pp 360-369, San Mateo, CA, USA, 1989.

[2] G.F. Miller, P.M. Todd and S.U. Hegde. Designing neural networks using genetic algorithms. In Proc. of the third international conference on genetic algorithms and their applications, pp 379-384, San Mateo, CA, USA, 1989.

[3] S. Harp, Samad T. and Guha A. Designing Application-Specific Neural Networks using the Genetic Algorithm, Advances in Neural Information Processing Systems, vol2, 447-454, 1990.

[4] F. Gruau. Genetic Synthesis of Boolean Neural Networks with a Cell Rewriting Developmental Process. Proc. of COGANN-92 International Workshop on Combinations of Genetic Algorithms and Neural Networks, pp. 55-74, IEEE Computer Society Press, 1990.

[5] F. Gruau. Automatic Definition of Modular Neural Networks. Adaptive Behaviour, vol. 2, 3, 151-183, 1995.

[6] P.A. Castillo, J. González, J.J. Merelo, V. Rivas, G. Romero and A. Prieto. Optimization of Multilayer Perceptron Parameters using Simulated Anneling. Lectures Notes in Computer Science, Vol 1606, pp 661-670, Springer-Verlang, 1998.

[7] T. Ash. Dynamic Node Creation in Backpropagation Networks ICS Report 8901, The Institute for Cognitive Science, University of California, San Diego (Saiensu-sh, 1988), 1988.

[8] D.B. Fogel, Fogel L.J. and Porto V.W. Evolving Neural Network, Biological Cybernetics, 63, 487-493, 1990.

[9] T.P. Caudell and Dolan C.P. Parametric Connectivity: Training of Constrained Networks using Genetic Algorithms, Proc. of the third International Conference on Genetic Algorithms and their Applications, 370-374. Morgan Kaufman, 1989.

[10] J.D. Schaffer, R.A. Caruana and L.J. Eshelman. Using genetic search to exploit the emergent behaviour of neural networks. Physica D, 42, pp 244-248, 1990.

[11] E. Alba, J.F. Aldana and J.M. Troya. Fully automatic ANN design: A genetic approach. In Proc. of International workshop on artificial neural networks, pp 179-184, 1993.

[12] H. Kitano. Designing Neural Networks using Genetic Algorithms with Graph Generation System, Complex Systems, 4, 461-476, 1990.

[13] Molina, J. M., Torresano, A., Galván, I., Isasi, P., Sanchis, A. (2000) Evolution of Context-free Grammars for Designing Optimal Neural Networks Architectures , GECCO 2000, Workshop on Evolutionary Computation in the Development of ANN. USA.

[14] J.W.L. Merril and R.F. Port. Fractally configured Neural Networks. Neural Networks, 4, 53-60, 1991.

[15] S. Wolfram. Theory and applications of cellular automata. World Scientific, Singapore, 1988.

[16] D.H. Ackley, G.E. Hinton and T.J. Sejnowski. A learning algorithm for Boltzmann machines. Cognitive Science, 9, 147-169, 1985

[17] Blake, C.L. & Merz, C.J. (1998). UCI Repository of machine learning databases [http://www.ics.uci.edu/~mlearn/MLRepository.html]. Irvine, CA: University of California, Department of Information and Computer Science.

[18] W. Schiffmann, M. Joost, and R. Werner. Synthesis and performance analysis of multilayer neural network architectures. Technical Report 16, University of Koblenz, Institute of Physics, 1992.

A Recurrent Multivalued Neural Network for the N-Queens Problem

Enrique Mérida[1], José Muñoz[2], and Rafaela Benítez[2]

[1] Departamento de Matemática Aplicada,
[2] Departamento de Lenguajes y Ciencias de la Computación,
Universidad de Málaga
Complejo Tecnológico, Campus Universitario de Teatinos s/n, 29079 Málaga
merida@ctima.uma.es, munozp@lcc.uma.es, benitez@lcc.uma.es

Abstract. This paper presents a multivalued Hopfield-type neural network as a method for solving combinatorial optimization problems with a formulation free of fine-tuning parameters.
As benchmark of the performance of the network we have used N-Queen problems. Computer simulations confirm that this network obtains good results when is compared with other neural networks.
It is shown also that different dynamics are easily formulated for the network leading to obtain more sophisticated algorithms with better performance.

1 Introduction

Optimization problems arise in a wide variety of scientific and engineering applications including signal processing, system identification, filter design, function approximation, regression analysis, and so on. Since Hopfield and Tank introduced neural networks for solving combinatorial optimization problems in 1985 ([2], [3]), many attempts have been made to propose alternative neural networks for solving linear and non-linear programming problems. Recurrent stable networks represent an interesting alternative to classical algorithms for the solution of optimization problems because of the parallelism inherent in the neural approach and their recogninized behavior, that is, the fact that they are capable of converging on a finite number of solutions that are as close as possible to optimal solutions.

A neural network with a good computational performance should satisfy some considerations. First, the convergence of the neural network with an arbitrary given initial state should be guaranteed. Second, the network design preferably contains no variable parameter. Third, the equilibrium points of the network should correspond to the solutions. From a mathematical point of view, these characteristics are relevant to the optimization techniques employed for deriving optimization neural networks models.

The Hopfield network has demonstrated that a distributed system of simple processing elements can collectively solve optimization problems. However, the original Hopfield network generates poor quality and/or invalid solutions and so

J. Mira and A. Prieto (Eds.): IWANN 2001, LNCS 2084, pp. 522-529, 2001.

a lot of different approaches have been proposed. These techniques have demonstrated significant improvement over the generic Hopfield network. It is crucial to incorporate as many problem-specific constraints as possible into the structure of the neural network to improve scalability by limiting the search space. In this way Takefuji et al. [8] and Lee et al. [5] handle a class of NP-complete optimization problems, which is used to be hard to solve by a neural network. This model has shown to provide powerful approaches for combinatorial optimization problems ([2], [5], [6], and [9]) and for polygonal approximation [1].

In this paper, we present a new neural network model called multivalued recurrent model inspired in Hopfield network and apply it to a constraint satisfaction problem. As benchmark problem, the N-queens problem is selected because several neural network approaches have been presented for this problem. In Funabiki et al. ([2]) the maximum neural model is applied to constraint satisfaction problems. They selected the N-queens problem to show its performance. They demonstrated that their network showed the far better performance among the existing neural networks. However in order to avoid local minimum convergence they had to apply heuristic methods. One of mayor problem with the heuristics is the lack of rigorous guidelines in selecting appropriate values of the parameters used in them. Recently, Galán-Marín et al. [3] presented an application of a new maximum neuron model, the OCHOM (Optimal Competitive Hopfield Model), to the N-queens model. They showed by simulation results that this model without help of heuristic methods performs better than the best known neural network of Funabiki et al [2] in terms of solution quality and computation time. Therefore, we have selected both neural networks, by Funabiki and by Galán-Marín for the performance comparison with our multivalued neural network.

2 Mutivalued Recurrent Network

We propose a neural network with single layer formed by N neurons fully interconnected. The state of neuron i is defined by its output V_i for $i \in I$ and it takes values in a discrete set $\mathcal{M} = \{1, 2, \ldots, M\}$. The state of the network at time k is given by a N-dimensional vector which represents the output of N neurons $V(t) = (V_1(t), V_2(t), \ldots, V_N(t))$.

Like in Hopfield model, in this network there exists an energy function characterizing its dynamics. The energy function depends on the outputs and indexes from each neuron:

$$E = -\frac{1}{2}\sum_{i=1}^{N}\sum_{j=1}^{N} w_{i,j} f(V_i, V_j, i, j) \tag{1}$$

where $i, j \in \mathcal{I}$, y $f(V_i, V_j, i, j)$ measures the interactions between ith and jth neurons.

This energy function is more general than original Hopfield energy:

$$E = -\frac{1}{2}\sum_{i=1}^{N}\sum_{j=1}^{N} w_{i,j}V_iV_j$$

since we can take $f(V_i, V_j, i, j) = V_iV_j$.

We introduce now the notion of group update, that is, the possibility of selecting more than a single neuron for update considering discrete-time dynamic. A group G containing a number n of neurons is previously selected at time t. Then it is computed the difference in the energy, $\Delta E_G = E_G(t+1) - E_G(t)$, that results if only the states of the neurons in the group G are altered. More precisely, we can consider that neurons in G at time $t+1$ only can take some states in $\mathcal{M}$, so in this application we only consider the possibility of permuting their outputs.

3 Representation of N-Queens Problem with a Type-Hopfield Network

We present several formulations by other authors for the N-Queen problem.

a) Jacet Mandziuk [7] used a Hopfield Network with following energy function:

$$2E = A\sum_{i=1}^{N}\sum_{j=1}^{N}\left[\left(\sum_{\substack{k=1\\k\neq j}}^{N} V_{ik}\right)V_{ij} + \left(\sum_{\substack{k=1\\k\neq i}}^{N} V_{kj}\right)V_{ij}\right] +$$

$$+B\sum_{i=2}^{N}\sum_{j=1}^{i-1}\left[\left(\sum_{\substack{k=i-j+1\\k\neq i}}^{N} V_{k,k-i+j}\right)V_{ij}\right] + B\sum_{i=1}^{N}\sum_{j=1}^{N}\left[\left(\sum_{\substack{k=1\\k\neq i}}^{N+i-j} V_{k,k-i+j}\right)V_{ij}\right]$$

$$+B\sum_{i=1}^{N}\sum_{j=n-i+1}^{N}\left[\left(\sum_{\substack{k=i+j-N\\k\neq i}}^{N} V_{k,i+j-k}\right)V_{ij}\right] + B\sum_{i=1}^{N-1}\sum_{j=1}^{N-i}\left[\left(\sum_{\substack{k=1\\k\neq i}}^{i+j-1} V_{k,i+j-k}\right)V_{ij}\right]$$

$$+C\left(\sum_{i=1}^{N}\sum_{j=1}^{N} V_{ij} - (n+\sigma)\right)^2$$

b) Nobou Funabiki el al. [2] proposed:

$$E = \frac{A}{2}\sum_{j=1}^{N}\left(\sum_{k=1}^{N} V_{kj} - 1\right)^2 + \frac{B}{2}\sum_{i=1}^{N}\sum_{j=1}^{N} V_{ij} \times$$

$$\times \left(\sum_{\substack{1 \leq i+k, j+k \leq N \\ k \neq 0}} V_{i+k,j+k} + \sum_{\substack{1 \leq i+k, j-k \leq N \\ k \neq 0}} V_{i+k,j-k} \right)$$

c) Recently, Gloria Galán et al. ([3], [4]) applied the next energy function:

$$E = \frac{1}{2}\sum_{i=1}^{N}\left(\sum_{j=1}^{N} V_{ij} - 1\right)^2 + \frac{1}{2}\sum_{j=1}^{N}\left(\sum_{i=1}^{N} V_{ij} - 1\right)^2 + \frac{1}{2}\sum_{i=1}^{N}\sum_{j=1}^{N} V_{ij} \times$$

$$\times \left[\sum_{\substack{1 \leq i-k \\ 1 \leq j-k \\ k > 0}} V_{i-k,j-k} + \sum_{\substack{1 \leq i+k \\ 1 \leq j+k \\ k > 0}} V_{i+k,j+k} + \sum_{\substack{1 \leq i+k \\ 1 \leq j-k \\ k > 0}} V_{i-k,j-k} + \sum_{\substack{1 \leq i-k \\ 1 \leq j-k \\ k > 0}} V_{i-k,j+k} \right]$$

4 Representation of N-Queens Problem with a Multivalued Recurrent Network

In order to solve the problem with a multivalued network, we consider N neurons with outputs $V_i \in \{1, \ldots, N\}$. The index i of a neuron indicates the ith row in the chessboard (only a queen in each row) and the output V_i means the column where the queen in this row is placed. That is, $V_i = k$ specifies that a queen is located in the cell (i, k) so the state space of the network θ has cardinality N^N.

With this representation we can avoid allocating more than one queen in the same row. The states of the network with no interferences in columns are characterized by vectors V which components are permutations of the indexes $1, 2, \ldots, N$. The transitions between states can be limited in order this constraint will not be violated, therefore the possible configurations of the network will be one of $N!$ states formed by all permutations of indexes associated to the neurons. These states expresses that queens have no interferences per row and per column and they are called quasi-feasible states.

The network is initialized in a quasi-feasible state V and all transitions will keep on the network in a quasi-feasible state.

4.1 Energy Function

We have proposed a representation where one and only one queen is located per row and per column, now it must be considered a new constraint; no more than one queen must be located in any diagonal line. This is carried out choosing an adequate expression of energy function.

Property 1 (Condition of interference between queens in a diagonal line): Two queens which positions (r_1, c_1) and (r_2, c_2), respectively are placed in the same diagonal line if and only if $|r_1 - r_2| = |c_1 - c_2|$

Proof: Given (r_1, c_1) and (r_2, c_2), two cells that correspond to the position of two queens in the diagonal line with direction $(1,1)$ to (N,N), it can be observed that **r-c=constant**. Then two queens with positions (r_1,c_1) y (r_2,c_2) in the same diagonal line will verify $r_1 - c_1 = cte = r_2 - c_2 \Rightarrow r_1 - r_2 = c_1 - c_2$.

On the other hand, two cells (r_1, c_1) and (r_2, c_2), in the diagonal placed in the direction $(1, N)$ to $(N, 1)$ will satisfy that **r+c=constant**, then when two queens are located in the considered diagonal it means that $r_1 + c_1 = cte = r_2 + c_2 \Rightarrow r_1 - r_2 = -(c_2 - c_1)$. □

Next, we propose the analytic expression of the energy function where the number of interferences between queens in a diagonal line is included by considering *Property 1*. In this expression we use the following function:

$$g(x, y) = (x \equiv y) = \begin{cases} 0 & x \neq y \\ 1 & x = y \end{cases}$$

Finally the energy function of the neural network is given by

$$E(V) = \sum_{i=1}^{n-1} \sum_{j>i}^{n} (|V_i - V_j| \equiv |i - j|) \tag{2}$$

This expression can be obtained from equation (1) taking:

$$W = (w_{i,j}) = \begin{cases} -2 & i < j \\ 0 & i \geq j \end{cases} \quad \text{and} \quad f(V_i, V_j, i, j) = |V_i - V_j| \equiv |i - j|$$

4.2 Dynamics of the Network

We sequentially select a group with two neurons $G = \{a, b\}$. Let $V_a(t) = k$ and $V_b(t) = l$ be the state of these neurons.

The next state of the network will be: $V_a(t+1) = l$, $V_b(t+1) = k$, $V_i(t+1) = V_i(t), i \notin \{a, b\}$ if $E(V(t+1)) \geq E(V(t)$. Otherwise, $V_i(t+1) = V_i(t), \forall i$.

That is because the objective is to get a state with minimum energy so it is interesting to achieve a search wide enough, with no stop until a feasible configuration is reached.

We compute the energy increment directly by expression:

$$\Delta E_{(a,b),(k,l)} = \sum_{i \neq a} (|V_i - k| == |i - a|) + \sum_{i \neq b \wedge i \neq a} (|V_i - l| == |i - b|) \tag{3}$$

and on the analogy of above expression:

$$\Delta E_{(a,b),(l,k)} = \sum_{i \neq a} (|V_i - l| == |i - a|) + \sum_{i \neq b \wedge i \neq a} (|V_i - k| == |i - b|)$$

The computation dynamic (sequential permutation of two neurons) can be expressed as:

$$V_i(t+1) = \begin{cases} V_i(t) & (i \neq a) \wedge (i \neq b) \\ k & (i = a) \wedge (\Delta E_{(a,b),(k,l)} < \Delta E_{(a,b),(l,k)}) \\ l & (i = a) \wedge (\Delta E_{(a,b),(k,l)} \geq \Delta E_{(a,b),(l,k)}) \\ l & (i = b) \wedge (\Delta E_{(a,b),(k,l)} < \Delta E_{(a,b),(l,k)}) \\ k & (i = b) \wedge (\Delta E_{(a,b),(k,l)} \geq \Delta E_{(a,b),(l,k)}) \end{cases}$$

The energy function in the new state is:

$$E(t+1) = E(t) - \Delta E_{(a,b),(k,l)} + \Delta E_{(a,b),(l,k)}$$

Good results are obtained with this dynamic (percentage of successes, time and number of iterations). However, this algorithm is stabilized in local minimum with low values of energy when deals with small size problems ($n < 25$). This is a common problem that arises if other networks model are applied to the same problem. In order to avoid it we have modified the update schedule. First of all a iteration with sequential update is made where all combinations of two neurons are updated then several iterations with randomly selected neurons to be updated are considered. The objective is to keep the network away from non optimal configurations.

Some updating schedules have been tested; when a group of two neurons is selected randomly several times the results are better than in a sequential mode but an optimal solution is not guaranteed. Then we randomly select a group with three neurons, $G\{a,b,c\}$. The possible states at time $t+1$ are

$$\begin{array}{llll} V_a(t+1) = k, & V_b(t+1) = l, & V_c(t+1) = m, & \text{(actual state)} \\ V_a(t+1) = m, & V_b(t+1) = k, & V_c(t+1) = l, & \text{(change to right)} \\ V_a(t+1) = l, & V_b(t+1) = m, & V_c(t+1) = k, & \text{(change to left)} \end{array}$$

It is easy to show that the energy difference resulting in any case is:

$$\Delta E_{(a,b,c),(k,l,m)} =$$

$$= \sum_{i \neq a} |V_i - k| \equiv |i-a| + \sum_{i \notin \{a,b\}} |V_i - l| \equiv |i-b| + \sum_{i \notin \{a,b,c\}} |V_i - m| \equiv |i-c|$$

Among the three possible states the state with minimum energy increment will be the state of the network at time $t+1$. So the energy in $t+1$ is

$$E(t+1) = E(t) - \Delta E_{S(t)} + \Delta E_{S(t+1)}$$

in this way the energy will never increases.

5 Simulation Results

In this section we illustrate, through simulations the theoretical analysis derived above. Simulations are made with 100 iterations for problems with $N > 100$ and 500 iterations for $N \leq 100$ with a initial configuration randomly selected. Table-1 shows the results with the multivalued network, and they are compared with the performance of the networks included in Funabiki et al. [2] and Galán et al. [3]. The comparison is done in terms of convergence rate and number of iterations Matlab and run on a Pentium II computer with a 233 MHz processor.

Table 1. Simulations results.

	Funabiki's Network		OCHOM Network		Proposed Network		
N	Conv.	Iter.	Conv.	Iter.	Conv.	Iter.	Time
8	96%	76.3	90%	4.92	100%	1.5637	0.12
10	96%	123.5	91%	8.77	100%	3.4680	0.47
20	100%	122.7	100%	8.60	100%	1.8099	1.77
50	99%	64.8	100%	9.65	100%	1.0889	15.91
100	100%	47.3	100%	9.96	100%	0.7397	78.37
200	100%	44.6	100%	9.96	100%	0.5420	489.56
300	100%	45.6	100%	13.58	100%	0.3906	1256.42
400	100%	46.7	100%	13.88	100%	0.3628	2536.23
500	100%	48.7	100%	16.58	100%	0.2850	3954.81

As it can be observed in Table-1, the number of iterations[1] required to reach the optimal solution in the multivalued network is smaller than in Funabiki and OCHON networks, and the convergence rate is better.

There exists problems with convergence in Funabiki and OCHON networks for small sizes N as it can be tested in Table-1, however this is no problem for our network with the proposed dynamic. That it is not true if we consider only sequential permutations in groups of two neurons, in this case we find the above problem in smaller sizes (for $N \leq 15$ the convergence rate is bigger than 98% but for $N = 6$ is about 91.5%).

In this paper we have used the same sizes of N than in [4], where we have applied two computation dynamics; a sequential permutation of two neurons and a random selection of three neurons to be updated only 0.2N times.

Simulation results described in table-1 show that in the application of a multivalued model the number of iterations is reduced when the size of the problem is augmented.

[1] In all models, an iteration needs N(N-1) evaluations of the increments of the energy function. But our expressions have a lower complexity.

6 Conclusions

In this paper a multivalued recurrent network is introduced for the N-queens problem. This new model expresses in a natural and simple way the constraints of this classical combinatorial problem so that the energy function does not contain parameters to be fine-tuned.

The expression of the energy function associated to our model reflects the number of collisions between the queens and a measures of similarity between outputs of neurons is included. The result is a generalization of Hopfield energy function.

We have proposed a simple computational dynamic that guarantee the descent of the energy function when a group of neurons is updated.

The experimental results have shown that our network obtains better results than the Takefuji and OCHOM models.

References

[1] P.C. Chung, C.T. Tsai, E.L. Chen and Y.N. Sun. Polygonal approximation using a Competitive Hopfield Neural Network. Pattern Recognition. Vol 27, n^0 11, pp 1505-1512, (1994).

[2] Funabiki, N., Takenaka Y. & Nishikawa, S. A maximum neural network approach for N-queens problem. Biological Cybernetics, vol. 76, pp. 251-255, (1997).

[3] Galán Marín, Gloria Tesis Doctoral. Universidad de Málaga, (2000) (in Spanish).

[4] Galán Marín, Gloria & Muñoz Perez, José. IEEE Transactions on Neural Network, Vol 12, n^0 2 pp 1-11, (2001).

[5] K.C. Lee, N. Funabiki and Y. Takefuji. A Parallel Improvement Algorithm for the Bipartite Subgraph Problem. Neural Networks, Vol 3,n^0 1, pp 139-145, (1992).

[6] K.C. Lee and Y. Takefuji. A generalized maximum neural network for the module orientation problem. Int. J. Electronics, Vol 72, n^0 3 pp 331-355, (1992).

[7] Jacek Mandziuk Solving the N-Queens problem with a binary Hopfield-type network. Biological Cybernetics, vol 72 439-445, (1995).

[8] Y. Takefuji, K.C. Lee and H. Aiso. An artificial maximum neural network: a winner-take-all neuron model forcing the state of the system in a solution domain. Biological Cybernetics, vol 67, pp. 243-251, (1992).

[9] Y. Takefuji. Neural Network Parallel Computing. Kluwer Academic Publ. (1992).

A Novel Approach to Self-Adaptation of Neuro-Fuzzy Controllers in Real Time

Héctor Pomares, Ignacio Rojas, Jesús González, Miguel Damas

Dept. of Computer Architecture and Computer Technology,
University of Granada, Granada, Spain
{hpomares, irojas, jgonzalez, mdamas}@atc.ugr.es

Abstract. In this paper, we present a novel approach to achieve global adaptation in neuro-fuzzy controllers. The adaptation process is achieved by means of two auxiliary systems: the first one is responsible for adapting the consequents of the main controller's rules with the target of minimizing the error arising at the plant output. The second auxiliary system compiles real input/output data obtained from the plant, which are then used in real time taking into account, not the current state of the plant but rather the global identification performed. Simulation results show that this approach leads to an enhanced control policy avoiding overfitting.

1 Introduction

Implementing self-adapting systems in real time is an important issue in intelligent control. The ability of man-made systems to learn from experience and, based on that experience, improve performance is the focus of machine learning [1]. As is well known, Fuzzy Logic and ANNs have proved successful in a number of applications where no analytical model of the plant to be controlled is available. Among the different ways of implementing a neuro-fuzzy controller, adaptive and/or self-learning controllers are, at least in principle, able to deal with unpredictable or unmodelled behaviour, which enables them to outperform non-adaptive control policies when the real implementation is accomplished.

The first adaptive fuzzy controller, called the 'linguistic self-organizing controller' (SOC), was introduced by Procyk and Mamdani in 1979 [2]; they proposed a learning algorithm capable of generating and modifying control rules by assigning a credit or reward value to the individual control actions(s) that make a major contribution to the present performance.

In more recent approaches, hybrid adaptive neuro-fuzzy systems have focused on merging concepts and techniques from conventional adaptive systems into a fuzzy systems framework. Most notable is the work of Wang [3] where the author deals with plants governed by a certain class of differential equations whose bounds must be known. The algorithms proposed by Wang also need off-line pretraining before working in real time.

J. Mira and A. Prieto (Eds.): IWANN 2001, LNCS 2084, pp. 530-537, 2001.

In [4], [5], algorithms are presented in which the rules are adapted in real time based on the actual error committed at the plant output. All such rules are based on the consideration that the current state of the plant is the responsibility of the rules which are activated a certain number of instants before (depending on the delay of the plant). Thus, the reward or penalty resulting from this actual error is used to modify the rules in proportion to their degree of activation at that particular instant. Nevertheless, although they function well, these algorithms are only capable of modifying the consequences of the rules, not the parameters that define the membership functions.

In [6], however, based on the work of [7], the premises of the rules can also be tuned without the knowledge of the plant equations. In this interesting approach, the controller output error is used to provide real input/output data concerning the system to be controlled. Nevertheless, this method has the disadvantage of requiring the system to have been previously controlled by another controller, as it does not attempt to reduce the plant output error directly.

The approach described in this paper, which is an improved version of [8], synergically incorporates the advantages of the above methods and attempts to resolve their disadvantages. Requiring practically no knowledge of the system to be controlled and by means of two auxiliary systems, the algorithm is able to exploit virtually all the information that can be obtained during the on-line control of the plant and manages to control the system in real time with no off-line pretraining. After this introduction, Section 2 states the mathematical basis of the problem to be solved. In Section 3 the strategy of the proposed approach is presented, which is then explained in detail in Sections 3.1 and 3.2. A simple example to illustrate the main characteristics of the proposed method are presented in Section 4. Finally, some conclusions are drawn in Section 5.

2 Problem Specification

Our objective is to achieve real time control of a system which, in general, may be non-linear and variable in time. Furthermore, both its equations and the optimization of this control are unknown. In the proposed approach, we attempt to optimize the controller's rules and the parameters defining it in order to translate the state of the plant to the desired value in the shortest possible time.

The system or plant to be controlled is usually expressed in the form of its differential equations or, equivalently, by its difference equations, providing these are obtained from the former with the use of a short enough sampling period. In mathematical terms:

$$y(k+d) = f(y(k),...,y(k-p),u(k),...,u(k-q)) \tag{1}$$

where d is the delay of the plant and f is a continuous function.

The restriction usually imposed on plants is that they must be controllable, i.e., that there always exists a control policy capable of translating the output to the desired value (within the operating range). This means that there must not be any state in which the output variable does not depend on the control input. Therefore, and as the plants are continuous with respect to all their variables, we can assume there exists a function F such that the control signal given by:

$$u(k) = F(\vec{x}(k)) \tag{2}$$

with $\vec{x}(k)$ =(r(k),y(k),...,y(k-p),u(k-1),...,u(k-q)), and where r(k) is the desired output at instant k; then y(k+d) = r(k)

In the proposed algorithm, no information is needed on the equations determining the plant, although it is necessary to determine its monotony, delay and which are the inputs that have a significant influence on the plant output.

This paper, as usual in neuro-fuzzy controllers, takes the product as the T-norm and the sum as the T-conorm, with the consequences of the rules being scalar values. Thus, the output of our neuro-fuzzy controller is given by:

$$\mathrm{u(k)} = \hat{\mathrm{F}}(\vec{x}(k)) = \frac{\sum_{i=1}^{n_{rules}} R_i \cdot \mu_i(\vec{x}(k))}{\sum_{i=1}^{n_{rules}} \mu_i(\vec{x}(k))} \tag{3}$$

The value of this depends on the rule set and membership functions defining it.

It should be noted that, according to (3), the control field could be deemed as a problem of function approximation [9], when I/O data of the true inverse plant function are available. Nevertheless, tasks such as those discussed in this paper, i.e. real time control starting from no knowledge, are much more complex due to the fact that the approximation of (3) to the real inverse plant function must be done while working in real time and attempting to direct the plant output to the target set point at every instant.

3 Control Strategy

Fig. 1 depicts the block diagram of the proposed control architecture. The larger block is the main neuro-fuzzy controller, where special emphasis has been laid on the correspondence between rule premises and rule consequents and the different layers of the network. Coupled to the main controller, there exist two auxiliary systems, which are in charge of finding the suitable parameters of the neuro-fuzzy controller from the control evolution:

- The 'Adaptation Block' (A-Block) is capable of a coarse tuning of the rule consequents, using information such as the monotony of the plant output with respect to the control signal and the delay of the plant. This block is necessary in the first iterations of the control process when no initial parameter values are available.
- The 'Global Learning Block' (GL-Block) is responsible for the fine-tuning of the membership functions and the rule consequents, using I/O data collected from the plant evolution.

Finally, the input selection block provides the input values needed by every block and collects I/O data in real time; this is subsequently used by the learning process, as we will show in section 3.2.

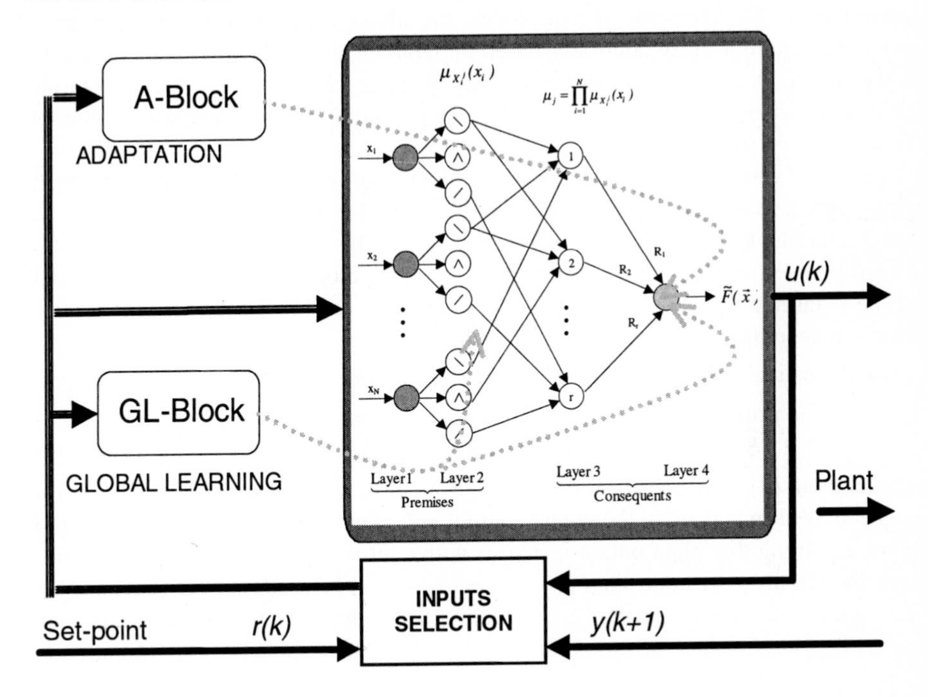

Fig. 1. Block diagram showing the main neuro-fuzzy controller and its interaction with the plant to be controlled and the auxiliary systems in charge of the adaptation and learning process.

3.1 Adaptation Block Procedure

The main problem when real time control problems must be faced lies in the fact that, as the internal functioning of the system to be controlled is unknown, we are unaware of how to modify the controller's parameters. To use a gradient-based algorithm, we would have to compute $\partial y/\partial u$, an unknown derivative. Moreover, in the case of long sampling periods, such a derivative cannot be approximated by $\Delta y/\Delta u$.

Nevertheless, as stated above, we do have the information regarding the monotony of the plant, which allows us to obtain the right direction in which to move the consequents of our rules. This is the basis of the SOC proposed by Procyk and Mamdani [2] and has the advantage of needing neither a model of the plant nor the desired control output at each instant of time. Common approaches based on SOC use a fuzzy auxiliary system in charge of this modification of the consequents of the fuzzy rules [4], [5].

In order to accurately build a self-organizing controller, the auxiliary system should possess information on how the plant output varies with respect to the control signal for every possible operation region. This entails knowing the Jacobian matrix of the plant function but, unfortunately, this information is normally unavailable in a control task. To overcome this problem, the above-mentioned authors use as an auxiliary system a fuzzy controller based on heuristically-built metarules constructed using the plant output error and error rate as inputs. Nevertheless, these metarules assume that step functions are required as set points to the plant and are not suitable when these are time variable.

From the above, it is evident that with the kind of information available from the plant, only a relatively coarse control can be applied to the system. In this paper, the adaptation block shown in Fig. 1 is responsible for this coarse control, implementing a modified self-organizing controller (MSOC). This block evaluates the current state of the plant and proposes the correction of the rules responsible for the existence of such a state, either as a reward or as a penalty.

Mathematically, if the output of the adaptation block is denoted by F_{AB}, the rule modification proposed at instant *k* (see Eq. (3)) is given by:

$$\Delta R_i(k) = \mu_i(k-d) \cdot F_{AB}(k) \tag{4}$$

where, as in [4], this modification is proportional to the degree with which the rule was activated in achieving the control output *u(k-d)* now being evaluated at instant *k*. Now, assuming that plant output increases with control input, if this output is smaller than the required set point then *u(k-d)* would have to be made bigger (alternatively, we should have used a smaller control signal if the monotony were of the opposite sign). This implies that the output of the adaptation block must be proportional to the current plant error:

$$F_{AB}(k) = C \cdot e_y(k) = C \cdot \left(r(k-d) - y(k) \right) \tag{5}$$

C being negative when the plant output decreases with the control sign. In the above expression, *r(k-d)* is the set point required of the plant output at instant *k-d* and *y(k)* is the current plant output. Note that it would be incorrect to use *r(k)* in (5), as the rules that are activated at instant *k-d* serve to achieve the desired value *r(k-d)* and not *r(k)*.

The next task is to assign a proper value to parameter *C*. As is apparent from the proposed control architecture in Fig. 1, both 'Adaptation Block' and 'Global Learning Block' are in charge of tuning the consequents of the fuzzy rules. In order to make them work harmoniously, we must concentrate on what we expect of the adaptation block. As stated above, with the relatively small amount of information available from the plant, the adaptation block is only capable of a coarse tuning of the fuzzy rules, necessary in the first steps of the control evolution. As the process advances, the adaptation block must give way to the global learning block since only this is capable of fine tuning the control actions. Thus, the influence of the adaptation block must decrease with time in the following form:

$$C = C(k) = C_0 \cdot exp(-k/\beta) \tag{6}$$

where β should not be too small, to let the rule consequents take suitable values, nor too big, which would make the learning process too slow. A typical value of β might be in the range 1000-5000 epochs. C_0 is a scale factor that is necessary to avoid out of range modifications in the fuzzy rules. This factor can be off-line determined by $C_0=\Delta u/\Delta y$, where Δy is the range in which the plant output is going to operate and Δu is the range of the controller's actuator.

3.2 Global Learning Block

In this section, we show how it is possible to use the gradient descent methodology, based on the error in the control output instead of that in the plant output, in order to achieve a fine-tuning of the main controller parameters. For this purpose, we base our approach on a modified version of the algorithm proposed by Andersen et al [6]. The main characteristic of the methodology presented in this section is that, analogously to the adaptation algorithm of the previous section, it does not rely on a plant model or need to know its differential equations or require a reference model.

When the controller provides a control signal at instant k, $u(k)$ and the output is evaluated d sampling periods later $y(k+d)$, the error committed at the plant output is not the only information that may be obtained. Regardless of whether or not this was the intended response, we now know that, if the same transition from the same initial conditions but now with $r(k) = y(k+d)$ is ever required again, the optimal control signal is precisely $u(k)$. Therefore, at every sampling time, we do get an exact value of the true inverse function of the plant.

One way to use this on-line information concerning the true inverse plant function is to store the recently obtained data values in a memory M. As this memory has a finite capacity, the data being received must be filtered in some way. The best way to do this is to define a grid in the input space and to store the most recent datum belonging to each of the hypercubes defined by such a grid, substituting a pre-existing datum in the hypercube. By these means, the memory contains a uniform representation of the inverse function of the plant, and with every step of the gradient descent algorithm, the learning process is performed in a global way, taking into consideration the whole input space and thus eliminating overfitting. Since some operation regions are more important than others, a weight parameter is assigned to each of the hypercubes, indicating the number of times a datum belonging to that hypercube is collected. Thus, the more important dynamic regions will have a greater influence on the fine-tuning of the main controller parameters.

In mathematical terms, the control signal exerted at the plant at instant k is given by (see Eq. (3)):

$$u(k) = \hat{F}\left(\vec{x}(k);\Theta(k)\right) \tag{7}$$

where $\Theta(k)$ represents the set of parameters that define the controller at instant k (rules plus membership functions). After d iterations, we obtain at the plant output the value $y(k+d)$. If we now replace the input vector $\vec{x}(k)$ by:

$$\hat{x}(k) \equiv (y(k+d), y(k), \ldots, y(k-p), u(k-1), \ldots, u(k-q)) \tag{8}$$

an expression that only differs from $\vec{x}(k)$ in the first element, where $y(k+d)$ replaces $r(k)$, we obtain the following datum belonging to the actual inverse plant function:

$$u(k) = F\left(\hat{x}(k)\right) \tag{9}$$

this datum is stored in the corresponding position of M, increasing its associated weight.

To perform the global learning process on the basis of the data stored in M, we must evaluate, for each datum, the output $\hat{u}$ given by the controller for each possible input, thus obtaining an error signal in the output of the controller:

$$e_u(m) = u(m) - \hat{u}(m) \tag{10}$$

where $u(m)$ and $\hat{u}(m)$ are the desired output and that obtained by the current parameters of the main fuzzy controller, respectively. Index m runs through the size of M. It is important to note that, although $\hat{u}(m)$ is produced by the controller, it is not applied to the plant. Its only purpose is to calculate $e_u(m)$.

In each iteration k it is necessary to compute the error in the output of the controller for each of the valid data stored, where the magnitude to be minimized is given by:

$$J(k) \equiv \frac{1}{2}\sum_{m=1}^{M} w_m \cdot e_u^2(m) = \frac{1}{2}\sum_{m=1}^{M} w_m \cdot \left[u(m) - \hat{F}(\hat{x}(m);\Theta(k))\right]^2 \tag{11}$$

Therefore, the parameters of the main fuzzy controller are optimized in each iteration in the following way:

$$\Delta\vec{\Theta}(k) = -\eta \cdot \nabla_{\Theta} J(k) \tag{12}$$

4 An Example of the Proposed Procedure

This section presents a simple example showing the functioning of the above-described algorithm. Consider the system described by the following difference equation:

$$y(k+1) = -0.3 \cdot sin(y(k)) + u(k) + u^3(k) \tag{13}$$

for which we will use random set points in the range [-4,4].

As input variables, we use the desired plant output $r(k)$ and its current output $y(k)$ using a total of 25 rules. All rule consequents are initially taken as zero. For this example, C_0 can be roughly taken as 0.5. This is a positive value since the plant output increases with respect to the control signal (positive monotony). Finally, we use $\eta = 0.1$, $\beta = 2000$ and a memory of 100 data.

To see how the control process develops, the initial control evolution is plotted in Fig. 2a. The dots represent the desired plant values and with solid lines, the plant output obtained. When the learning process has practically reached a steady condition, the control evolution is as shown in Fig. 2b, from which it can be concluded that the control actions are now excellent.

5 Conclusions

This paper proposes a new methodology for the automatic implementation of real time fuzzy controllers to control non-linear time-variable systems with one input and one output. The principal feature of the algorithm is that it requires practically no knowledge of the system to be controlled. By means of combining sinergically an auxiliary system that attempts to minimize the actual output error and a second learning system that uses real data of the true inverse function of the plant, it is possible to exploit virtually all the information that can be obtained during the on-line control of the plant. Finally, the validity of the proposed algorithm was tested by applying it to a simple example.

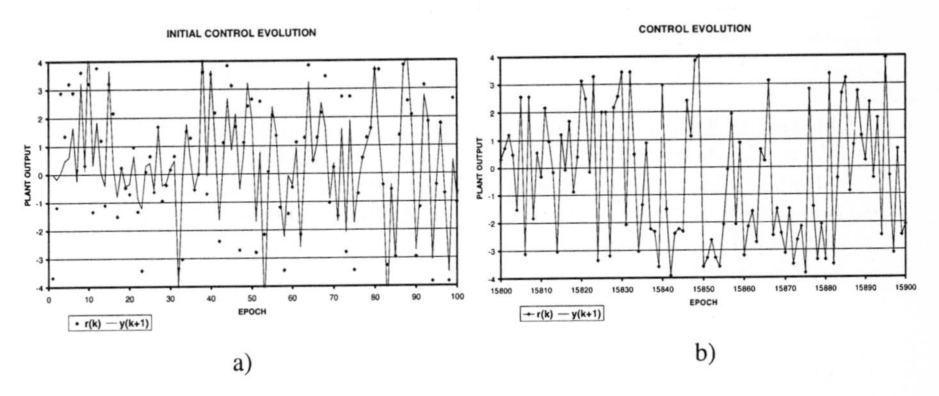

Fig. 2. Control evolution a) First epochs b) When learning process is successfully committed.

References

[1] Antsaklis, P.: Intelligent learning control, IEEE Control Systems, vol. 14, no. 4, (1995) 5-7.

[2] Procyk, T., Mamdani, E.: A linguistic self-organizing process controller, Automatica, vol.15, no.1, (1979) 15-30.

[3] Wang, L.X.: Adaptive fuzzy systems and control. Design and stability analysis, Prentice Hall, (1994).

[4] Rojas, I., Pomares, H., Pelayo, F.J., Anguita, M., Ros, E., Prieto, A.: New methodology for the development of adaptive and self-learning fuzzy controllers in real time, Int. J. Approximate Reasoning, vol.21, (1999) 109-136.

[5] Singh, Y.P.: A modified self-organizing controller for real-time process control applications, Fuzzy Sets and Systems, vol.96, (1998) 147-159.

[6] Andersen, H.C., Lotfi, A., Tsoi, A.C.: A new approach to adaptive fuzzy control: The Controller Output Error Method, IEEE Trans. Syst. Man and Cyber.- Part B, vol.27, no.4, (1997) 686-691.

[7] Albus, J.S.: Data storage in the cerebellar model articulation controller (CMAC), Trans. ASME, J. Dyn. Syst., Meas., Contr., vol.97, (1975) 228-233.

[8] Pomares, H., Rojas, I., Fernández, F.J., Anguita, M., Ros, E., Prieto, A.: A New Approach for the Design of Fuzzy Controllers in Real Time, Proc. 8th Int. Conf. Fuzzy Systems, Seoul, Korea, (1999) 522-526.

[9] Rojas, I., Pomares, H., Ortega, J., Prieto, A.: Self-Organized Fuzzy System Generation from Training Examples, IEEE Trans. Fuzzy Systems, vol.8, no.1, (2000) 23-36.

[10] Pomares, H., Rojas, I., Ortega, J., Gonzalez, J., Prieto, A.: A Systematic Approach to a Self-generating Fuzzy Rule-table for Function Approximation, IEEE Trans Syst., Man and Cyber., vol.30, no.3, (2000) 431-447.

Expert Mutation Operators for the Evolution of Radial Basis Function Neural Networks

J. González, I. Rojas, H. Pomares and M. Salmerón

Department of Computer Architecture and Computer Technology
University of Granada
Campus de Fuentenueva
E. 18071 Granada (Spain)

Abstract. This paper compares some mutation operators containing expert knowledge about the problem of optimizing the parameters of a Radial Basis Function Neural Network. It is shown that the expert knowledge is not always able to improve the results obtained by a blind evolutionary algorithm, and that the final results depend strongly on how the expert knowledge is utilized.

1 Introduction

Evolutionary Algorithms (EAs) are a robust optimization technique that have been successfully applied to several optimization problems [8]. Its strength is based on their simplicity and easy implementation. However, EAs can be considered a generic optimization technique whose results can be improved if some expert knowledge about the problem to be solved is incorporated in them. These changes produce algorithms that hybridizes the robustness and strength of EAs and the expertness of some heuristics for the problem. This adapted EAs obtain better results than the generic ones because they can guide the search towards solutions that the expert knowledge expects to be superior. The easiest way of incorporating this expert knowledge of the problem in an EA is to construct some mutation operators specific for the problem to be solved. These operators differ from the original ones in the sense that they do not apply blind changes to the individuals they affect. They try to improve the mutated individual by analyzing and altering it using some heuristics. Specifically, in the problem of optimizing the parameters that define a Radial Basis Neural Network (RBFNN) [10, 11] from training samples, there exist two good heuristics to determine the relevance of a basis function in the net output: Orthogonal Least Squares (OLS) [1] and Singular Value Decomposition (SVD) [6].

This paper presents a brief review of the orthogonal transformations OLS and SVD in section 2 and describes some mutation operators based on these orthogonal methods and specifically designed for the optimization of RBFNNs in section 3. These operators are compared with a random (blind) type mutation in section 4, and some interesting conclusions are presented in section 5.

J. Mira and A. Prieto (Eds.): IWANN 2001, LNCS 2084, pp. 538-545, 2001.

2 Orthogonal transformations

Once that the parameters concerning the RBFs have been fixed, their associated weights can be optimally calculated using a linear systems solver [9]. An RBFNN can be seen as an special case of linear regression:

$$\boldsymbol{y} = P\boldsymbol{\omega} + \boldsymbol{e} \tag{1}$$

where $\boldsymbol{y} = [y^1, ..., y^n]^T \in \Re^n$ is the column vector containing all the expected outputs, $P = [\boldsymbol{p_1}, ..., \boldsymbol{p_m}] \in \Re^{n \times m}$ is a matrix whose columns $\boldsymbol{p_j} = [p_{1j}, ..., p_{nj}]^T \in \Re^n$ represent the output of the j-th basis function for all the input vectors, $\boldsymbol{e} = [e^1, ..., e^n]^T \in \Re^n$ is a vector having all the errors committed by the model, and $\boldsymbol{\omega}$ is a vector containing the net weights. Usually, the number of training samples n is greater than the number of RBFs m. So we only have to solve the following linear system:

$$\boldsymbol{y} = P\boldsymbol{\omega} \tag{2}$$

to minimize the approximation error of the net and find the optimal weights (in the least squares sense). There are several ways to solve this linear system. In the following sections, two of the most commonly used methods are presented: OLS and SVD.

2.1 Orthogonal Least Squares (OLS)

This method was originally employed by Chen *et al.* in [1] to calculate the optimum weights of an RBFNN. It also estimates the relevance of each RBF ϕ_j in the output of the net by assigning it an error reduction ratio $[err]_j$. OLS transforms the columns of the activation matrix P in a set of orthogonal vectors $\boldsymbol{u_j}$. This transformation is performed applying the Gram-Schmidt orthogonalization method [3] and produces:

$$P = UR \tag{3}$$

where $\boldsymbol{g} = R\boldsymbol{\omega}$. As $\boldsymbol{u_j}$ y $\boldsymbol{u_l}$ are orthogonal $\forall j \neq l$, the sum of squares of y^k can be written as:

$$\boldsymbol{y}^T\boldsymbol{y} = \sum_{j=1}^{m} g_j^2 \boldsymbol{u_j}^T \boldsymbol{u_j} + \boldsymbol{e}^T\boldsymbol{e} \tag{4}$$

Dividing both sides of equation (4) by n, it can be seen how the model variance is decomposed in explained and residual variances:

$$\frac{\boldsymbol{y}^T\boldsymbol{y}}{n} = \frac{\sum_{j=1}^{m} g_j^2 \boldsymbol{u_j}^T \boldsymbol{u_j}}{n} + \frac{\boldsymbol{e}^T\boldsymbol{e}}{n} \tag{5}$$

So, as $g_j^2 \boldsymbol{u_j}^T \boldsymbol{u_j}/n$ is the contribution of $\boldsymbol{u_j}$ to the total output variance, we can define the *error reduction ratio* of $\boldsymbol{u_j}$ as [1]:

$$[err]_j = \frac{g_j^2 \boldsymbol{u_j}^T \boldsymbol{u_j}}{\boldsymbol{y}^T\boldsymbol{y}} \qquad \forall j = 1, ..., m \tag{6}$$

This ratio can be used to select a subset of important RBFs in a direct way. If we want to keep the r more relevant RBFs ($r < m$), we will select the r basis functions having higher error reduction ratios. This method can be used to prune the least relevant RBFs in the net, obtaining a simpler model whose approximation error is as close as possible to the error of the original net. A more detailed discussion of the OLS transformation can be found in [1].

2.2 Singular Value Decomposition

The Singular Value Decomposition (SVD) of the activation matrix P produces:

$$P = U\Sigma V^T \tag{7}$$

where $U \in \Re^{n\times m}$ is a matrix with orthogonal columns, $\Sigma \in \Re^{m\times m}$ is a diagonal matrix whose elements σ_j are positive or zero and are called singular values of P, and $V \in \Re^{m\times m}$ is an orthogonal matrix.

Each of the columns $\boldsymbol{u_j}$ in U is related with an RBF of the net, and has an associated singular value σ_j that estimates its "linear independence degree" in the system. The bigger σ_j is, the more linearly independent is $\boldsymbol{u_j}$ in U. So, if we only want to select the r more relevant basis functions of the net ($r < m$), we only have to maintain the r RBFs with higher associated singular values [3]:

$$P_r = U_r\Sigma_r V_r^T \tag{8}$$

where Σ_r is a diagonal matrix containing the higher r singular values, and U_r and V_r keep the columns of U and V associated with the singular values in Σ_r. This is an easy method to discard the $m - r$ less relevant RBFs in the net.

The SVD decomposition is a complicated orthogonal transformation and its implementation is out of the scope of this paper. Here, we will only mention that there are two possible implementations, the fastest is SVD-QRCP, proposed in [3]. This implementation uses Householder transformations to get the SVD faster, but obtains the columns of U with an altered order, so it is necessary to apply the QR algorithm with Column Pivoting (QR-CP) to get an estimation of their true order. The other method is the Kogbetliantz algorithm [7], which maintains the original order of the columns, but is much slower, although easy to parallelize [12].

3 Proposed mutation operators

This section presents some new mutation operators specifically designed for the problem of optimizing the parameters of an RBFNN. All these new operators apply random changes to the individuals they affect to maintain the diversity in the population and to provide mechanisms to escape from local minima [5], but they apply different criteria.

We will present the Locally Random Mutation (LRM), a blind operator, and some heuristically guided operators such as the Local OLS based Mutation (LOLSM), the Local SVD based Mutation (LSVDM), the Global OLS based Mutation (GOLSM), and the Global SVD based Mutation (GSVDM). As can be seen, an important difference is the scope of the changes they perform, that can be local to an RBF (LRM, LOLSM and LSVDM), or global for the whole net (GOLSM and GSVDM).

3.1 LRM: Locally Random Mutation

This operator is a direct adaptation of the classical blind genetic mutation operator [5]. All the parameters concerning the RBFs of the net have the same probability of being altered, and once that a parameter is selected, it will suffer a locally random change. Due to the big amount of parameters defining an RBFNN, there are several possible implementations for this operator. The chosen one will alter only the centers and radii of the net, because the weights can be optimally calculated once those parameters have been fixed, as described in section 2. Basically, LRM performs the following steps:

- Select randomly an RBF to be altered. All the RBFs have the same probability of being chosen.
- Decide whether altering its radius or its center. This decision is made randomly, having equal probability each one of the possibilities.
- Perform a local alteration of the center or the radius [4] and obtain the optimum weights for the new net.

3.2 LSVDM: Local SVD based Mutation

One of the orthogonal transformations described in section 2 to obtain a measure of the relevance of each RBF, is the SVD of the net's activation matrix P. This decomposition provides a set of singular values, one per RBF. High singular values identify important RBFs, while low singular values detect less relevant basis functions.

The set of singular values give us an idea of the sensitivity of each RBF to a random displacement. If we move an RBF having a high singular value, we probably obtain a worse fitted net. On the other hand, RBFs whose singular values are nearly zero are not making any significant contribution to the net's output, so they can be altered freely without increasing the net's error.

Having this idea in mind, this mutation operator will select a basis function to be locally altered with a probability inversely proportional to its associated singular value. Less relevant RBFs will have more probability of being altered while more important RBFs will be more change-protected [4].

3.3 LOLSM: Local OLS based Mutation

The other orthogonal transformation described in section 2 is OLS. This method also assigns a relevance value to each RBF, but with an important difference: OLS takes into account the expected output for each input vector in the training set. So the relevance of each RBF is closely related to its contribution to the reduction of the training error. Those RBFs making a bigger contribution to the training error reduction will be more sensitive to a random displacement than others making a smaller contribution.

OLS calculates a vector of error reduction ratios $\boldsymbol{err}$ (6) in which there will be an $[err]_j$ for each RBF ϕ_j in the net. The bigger $[err]_j$ is, the more sensitive ϕ_j is to a random change. So, this mutation operator defines a distribution where each ϕ_j has a probability of being altered inversely proportional to its associated error reduction ratio $[err]_j$ [4]. An RBF is chosen using this distribution and a local change is applied to its center or radius.

3.4 GSVDM: Global SVD based Mutation

Another way of hybridation between SVD and a mutation operator is to select the RBF to be altered uniformly (all the RBFs have the same likelihood to be chosen) and apply a random displacement according to its associated singular value. Basis functions with small singular values will suffer big movements, while sensitive RBFs will be applied only small perturbations.

This behavior can be implemented applying a random shift to the center or the radius of the selected RBF whose modulus varies inversely with the magnitude of its singular value.

3.5 GOLSM: Global OLS based Mutation

This evolutive operator performs exactly the same steps than the above one, but using OLS to detect the relevance of the RBFs instead of SVD. All the basis functions have the same likelihood to be altered, and when one RBF ϕ_j is chosen, its center or its radius suffers a random displacement inversely proportional to its error reduction ratio $[err]_j$ (6).

4 Comparison

The purpose of this section is to analyze a simple experiment to give an insight about each mutation operator behavior. This experiment is related with the approximation of the target function:

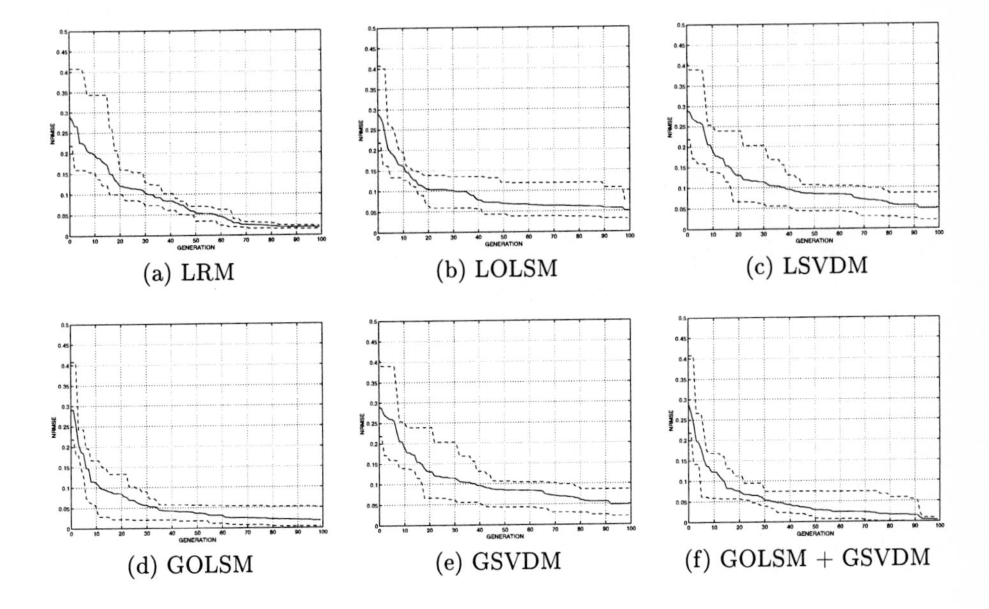

(a) LRM (b) LOLSM (c) LSVDM

(d) GOLSM (e) GSVDM (f) GOLSM + GSVDM

Fig. 1. Effect of the proposed operators

$$f(x) = 3x(x-1)(x-1.9)(x+0.7)(x+1.8), \quad x \in [-2.1, 2.1] \tag{9}$$

proposed in [2] from a set of 100 equidistributed samples. This function has been approximated with several EAs which were exactly the same in all their features except in the mutation operator. This makes a total of 5 EAs, one per mutation operator. All of these EAs have been used with 5 different initial populations and all the executions have started with the same random seed, to favor a fair competition. Particularly, all EAs have used a population composed by 30 individuals with 7 RBFs each one, and all of them have been run for 100 generations. These parameters have been fixed arbitrarily, but as they are identical for all the EAs, provide a fair comparison between the mutation operators.

Figure 1 shows the evolution of the different EAs. The solid line represents the mean of the best individuals in each generation, and the dashed line indicates the minimum and maximum of the best individuals found in every generation. These figures suggest that LRM achieves a very good solution, even better than LOLSM and LSVDM. This detail discovers that the expert knowledge can stuck the search in local minima if it is not correctly used. As LOLSM and LSVDM only perform local changes to the individuals, they avoid less relevant RBFs to move freely in the input space. Although less important RBFs are given more chances to be altered, as the random changes are always made in a local way, they only affect contiguous regions in the input space. These local alterations don't allow an RBF which makes little contribution to the net's output to move to a region in the input space where it could reduce the approximation error,

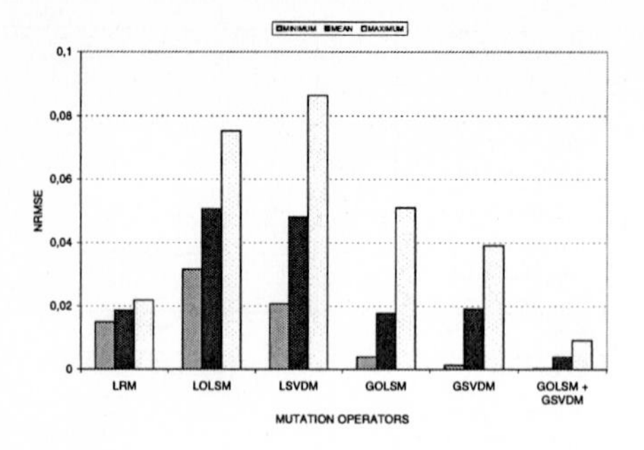

Operator	Mean	Min.	Max.
LRM	0.0186	0.0150	0.0219
LOLSM	0.0507	0.0316	0.0753
LSVDM	0.0481	0.0208	0.0864
GOLSM	0.0178	0.0039	0.0510
GSVDM	0.0192	0.0013	0.0392
GOLSM+ GSVDM	0.0039	0.0002	0.0093

Fig. 2. Mean, minimum and maximum of the best solutions found for each mutation operator

if this region is not close to the RBF. So, such two mutation operators tend to trap the population in local minima.

On the other hand, the operators GOLSM and GSVDM, obtain much better results. This is because GOLSM and GSVDM perform global changes to the individuals using displacements inversely proportional to the relevance of the RBF they affect. Thus, important basis functions will only suffer local perturbations and less relevant RBFs will be able to move freely. Figures 1 (d) and 1 (e) show that GOLSM and GSVDM accelerate the EA's convergence more than LRM. Starting in the same initial populations, GOLSM and GSVDM are able to discover solutions with an approximation error that is half of the approximation error that reaches the EA using LRM in the same generation. This behavior justifies the selection of GOLSM and GSVDM as the more appropriated mutation operators for this problem.

If we now pay attention to the last generations, LRM, GOLSM and GSVDM reach a similar mean approximation errors, although the best solution is always better for GOLSM and GSVDM. This comparison is graphically shown in figure 2.

Finally, as we can add as many mutation operators as we want in an EA, the same experiments have been run with an EA combining GOLSM and GSVDM. When an individual is chosen to be altered, one of these two operators will be applied. The probability of application has been 0.5 for both. This last EA shows an easy way of combining the effects of several operators. Figures 1 (f) and 2 show that this combination produces even better solutions than each operator applied separately.

5 Conclusions

The final conclusion in this paper is that the incorporation of expert knowledge and heuristics in the mutation operators does not always produce a good EA.

This paper has shown two clear examples (LOLSM and LSVDM) where expert mutation operators don't favor the convergence to good local optima.

The objective of mutation operators is to add diversity to the population and to provide mechanisms to favor the exploration of the search space. If the expert knowledge limits in any way this objective, the EA will not search properly and will tend to get stuck in a local optimum. On the other hand, if the heuristics cleverly guide the alterations towards better solutions, an EA using such operators can achieve better solutions than a blind one.

Acknowledgement

This work has been partially supported by the CICYT Spanish projects TAP97-1166 and TIC2000-1348.

References

1. S. Chen, C. F. N. Cowan, and P. M. Grant. Orthogonal least squares learning algorithm for radial basis function networks. *IEEE Trans. Neural Networks*, 2:302–309, 1991.
2. J. A. Dickerson and B. Kosko. Fuzzy function approximation with ellipsoidal rules. *IEEE Trans. Syst. Man and Cyber. - Part B*, 26(4):542–560, August 1996.
3. G. H. Golub and C. F. Van Loan. *Matrix Computations.* Johns Hopkins University Press, Baltimore, 3rd edition, 1996.
4. J. González, I. Rojas, H. Pomares, M. Salmerón, and A. Prieto. Evolution of fuzzy patches for function approximation. In A. Ollero, S. Sánchez, B. Arrue, and I. Baturone, editors, *Actas del X Congreso Español sobre Tecnologías y Lógica Difusa, ESTYLF 2000*, pages 489–495, Sevilla, Spain, Sept. 2000.
5. J. J. Holland. *Adaption in Natural and Artificial Systems.* University of Michigan Press, 1975.
6. P. P. Kanjilal and D. N. Banerjee. On the application of orthogonal transformation for the design and analysis of feed-forward networks. *IEEE Trans. Neural Networks*, 6(5):1061–1070, 1995.
7. E. G. Kogbetliantz. Solution of linear equations by diagonalization of coefficients matrix. *Quart. Appl. Math.*, 13:123–132, 1955.
8. Z. Michalewicz. *Genetic Algorithms + Data Structures = Evolution Programs.* Springer-Verlag, 3rd edition, 1996.
9. H. Pomares, I. Rojas, J. Ortega, J. González, and A. Prieto. A systematic approach to a self-generating fuzzy rule-table for function approximation. *IEEE Trans. Syst., Man, Cyber. Part B*, 30(3):431–447, June 2000.
10. I. Rojas, J. González, A. Cañas, A. F. Díaz, F. J. Rojas, and M. Rodriguez. Short-term prediction of chaotic time series by using rbf network with regression weights. *Int. Journal of Neural Systems*, 10(5):353–364, 2000.
11. I. Rojas, H. Pomares, J. González, J. L. Bernier, E. Ros, F. J. Pelayo, and A. Prieto. Analysis of the functional block involved in the design of radial basis function networks. *Neural Processing Letters*, 12(1):1–17, Aug. 2000.
12. M. Salmerón, J. González, J. Ortega, I. Rojas, and C. G. Puntonet. Métodos ortogonales paralelos para predicción de series. In J. Ortega, editor, *Perspectivas en Paralelismo de Computadores. Actas de las XI Jornadas de Paralelismo*, pages 277–282, Granada, España, Sept. 2000. Universidad de Granada.

Studying Neural Networks of Bifurcating Recursive Processing Elements – Quantitative Methods for Architecture Design and Performance Analysis

Emilio Del Moral Hernandez

Escola Politecnica da Universidade de Sao Paulo
Departamento de Eng. de Sistemas Eletronicos
CEP 05508-900 - Sao Paulo - SP - Brazil
edmoral @ usp.br

Abstract. This paper addresses quantitative techniques for the design and characterization of artificial neural networks based on Chaotic Neural Nodes, Recursive Processing Elements, and Bifurcation Neurons. Such architectures can be programmed to store cyclic patterns, having as important applications spatio temporal processing and computation with non fixed-point attractors. The paper also addresses the performance measurement of associative memories based on Recursive Processing Elements, considering situations of analog and digital noise in the prompting patterns, and evaluating how this noise reflects in the Hamming distance between the desired stored pattern and the answer pattern produced by the neural network.

Keywords: associative memories, chaotic neural networks, design of neural architectures, parametric recursive processing elements, non-linear dynamics, bifurcation and chaos.

1 Introduction

The research addressing recursive processing elements (RPEs), chaotic neural networks (CNNs) and similar model neurons can be found in an increasing number of works [1,2,3,4,5,6]. As discussed in [3] and [6], while in traditional model neurons the functionality of nodes is described by a non linear activation function, which frequently assumes the form of a sigmoid or a radial basis function [7], in RPEs the functionality of the nodes is more properly characterized by a first order parametric recursion acting on a variable state of discrete time, as generically represented by $\boldsymbol{x}_{n+1} = \boldsymbol{R}_p(\boldsymbol{x}_n)$. In this recursive equation, $\boldsymbol{n}$ represents the discrete time, $\boldsymbol{x}_n$ and $\boldsymbol{x}_{n+1}$ represent consecutive values of the state variable $\boldsymbol{x}$, and $\boldsymbol{R}_p$ represents the parametric recursion (or parametric map) that relates $\boldsymbol{x}_n$ and $\boldsymbol{x}_{n+1}$. Another complementary and important characterization of the network node functionality is provided by its bifurcation diagram [8].

In [6] we used architectures based on RPEs to implement associative memories of binary patterns, and we developed the idea of encoding zeros and ones by two of the many modalities of dynamic behavior that can be exhibited by bifurcating recursions. Fig. 1 shows an example of a distorted prompting pattern leading to the proper pattern

J. Mira and A. Prieto (Eds.): IWANN 2001, LNCS 2084, pp. 546–553, 2001.

previously stored in the associative memory. Such results were obtained for nodes defined by the logistic recursion, where $R_p(x_n) = p.x_n.(1- x_n)$, the state variable x can assume continuous values in the range 0 to 1, and the bifurcation parameter p can range from 0 to 4, although in this work we concentrate our attention in the sub-range from 3 to 4. In the previous work, we demonstrated the validity of such associative architecture through results obtained for a prototypical network of 20 RPEs. In this paper, we expand those results to larger networks (100 nodes) and we present and discuss quantitative methods developed for the architecture design and network performance analysis.

2 Designing Networks of Recursive Processing Elements

One of the main concerns during the design of architectures based on bifurcating RPEs nodes is to make an appropriate choice for the operating point of the bifurcation parameters. When we say "operating point", we are referring to the sub-range of values that the bifurcation parameters p will be allowed to exercise. During network operation, such p parameters start from a minimum value named here ***minp***, possibly increase with time as needed for the search of a stored pattern, reach a maximum value named here ***maxp*** (which promotes chaotic searches), and then decrease and return to ***minp***, after the search is finished. Notice that the evolution of the p parameters over the range [***minp***, ***maxp***] is driven by the state of the network and the auto correlation coupling matrix **Wij**, calculated from the stored patterns [6,7].

In order to define the best value for the constant ***minp***, we used as our first source of information the bifurcation diagram of the recursive node, so we could locate the range of p values for which the period-2 dynamics behavior is observed. Such a periodicity is necessary for the temporal encoding of binary quantities described in [6] and briefly summarized in Section 1. According to the bifurcation diagram of the logistic recursion presented in Fig. 2 (lower part of the figure), as the parameter p (horizontal axis) ranges from 3 to 4, we observe that the long-term behavior of the state variable x_n correspond to attractors with increasing periods (period 2, then 4, 8 and so on), and eventually we reach chaotic attractors as p ranges from 3.56 to 4. It is clear from the diagram that the value of ***minp*** has to be larger than 3 to guarantee period-2 cycling; in addition, it cannot be larger than 3.4, otherwise we risk the bifurcation to period-4 cycling modalities.

Within the range 3 to 3.4, we then guide our choice of the ***minp*** value by using two complementary quantitative techniques described ahead. In particular, we assume that the choice of its value, in addition to promote the desired period-2 behavior, has also to promote maximal immunity to noise and maximal network speed of convergence. Based on these goals, we have developed methods for the automatic measurement of the set of values visited by the long term periodic trajectories performed by x_n (we call these values "the limit set" of the cycle). We also developed mechanisms for the automatic calculation of two associated measures: 1 - the smallest distance between the values that compose the final cyclic attractor; 2 - the measure of dispersion observed for each value that compose the limit set of the cyclic attractor. These measurements are used to generate two plots, respectively named "Limit Set Separation Curve" and "Limit Set Dispersion Curve". We obtained such plots for the range of p values between 3 and 4, where the periodicity of cycling ranges from 2 to infinite. These two curves appear in Fig. 2, together with the bifurcation diagram of the logistic recursion.

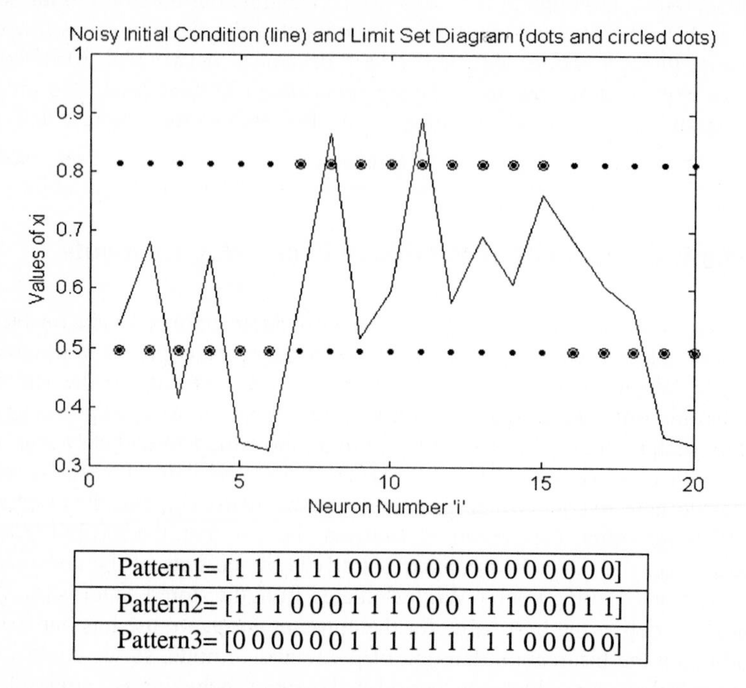

Pattern1= [1 1 1 1 1 1 0 0 0 0 0 0 0 0 0 0 0 0 0 0]
Pattern2= [1 1 1 0 0 0 1 1 1 0 0 0 1 1 1 0 0 0 1 1]
Pattern3= [0 0 0 0 0 0 1 1 1 1 1 1 1 1 1 0 0 0 0 0]

Fig. 1. The lower part of the figure lists three binary patterns (20 bits wide) which were stored in an associative memory employing RPEs and coding based on the phase of period-2 cycles [6]. The upper part of the figure shows the recovery of one of the stored patterns (Pattern 3) from a distorted version of it. The solid line represents the distorted (or noisy) initial condition, the horizontal axis represents the 20 neurons, the pairs of dots right above a given neuron number represent the period-2 cyclic pattern performed by that neuron, and the circled dots indicate the phase of the period-2 cycling. We name this type of diagram showing the long-term behavior of the network "Limit Set Diagram" (or simply LSD) [3,5,6].

Based on the two curves presented in Fig. 2, we can optimize the choice of the value ***minp*** and achieve maximal tolerance to noise in the prompting patterns. In our networks we have chosen ***minp*** = 3.25. For this value, we have a relatively high value in the Limit Set Separation Curve and a clear deep in the logarithmic plot of the Limit Set Dispersion Curve. These two facts are favorable conditions to noise immunity and fast convergence to attractors.

Notice that similar procedures to the ones described above for the definition of the ***minp*** can be used to choose the appropriate value of ***maxp***, the maximal value to be reached by the bifurcation parameters p during pattern search and recovery. In this case, the appropriate choice of ***maxp*** has to imply high entropy for the x_n variable and high coverage of its state space [3,8].

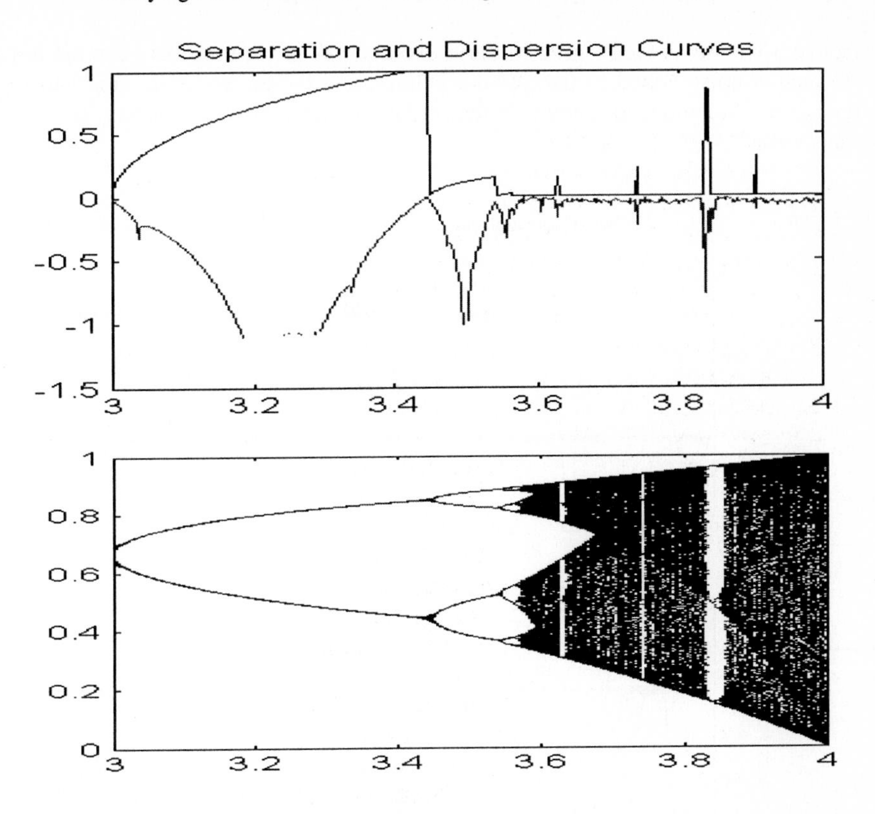

Fig. 2. In the upper part of the figure, the first curve measures the distance between the values of the limit set of a cyclic attractor (this curve is named "Limit Set Separation Curve" and ranges from 0 to 1 in its normalized form), and the second curve measures the spread observed for each of the values of the limit set (this curve is named "Limit Set Dispersion Curve" and ranges from 0 to –1.5, in the logarithmic form presented here). The lower part of the figure is the traditional Bifurcation Diagram for the logistic recursion $\boldsymbol{x_{n+1} = p.x_n.(1- x_n)}$: the horizontal axis represents the bifurcation parameter $\boldsymbol{p}$ and the vertical axis represents the set of values visited by the $\boldsymbol{x_n}$ state variable in the long term. All the plots have the same horizontal scale, allowing us to correlate qualitative and quantitative information related to specific values of the bifurcation parameter $\boldsymbol{p}$, as it ranges from 3 to 4.

3. Measures of Immunity to Analog Noise

In this section we present characterization results for associative memories based on RPEs that show high immunity to noise (or distortion) in the initial prompting condition. In Fig. 3 we present two illustrative experiments showing the operation of an associative network when submitted to specific levels of noise / distortion in the initial state. The two presented Limit Set Diagrams (LSDs) show the long term dynamic behavior of a network with 100 nodes and initial conditions (prompting patterns) with analog noise factors of 0.4 in the first experiment and 0.3 in the second one (we present latter a plot showing the results for generic levels of analog noise factors). The noise factor used in these experiments is the ratio between the level of the analog noise added to the initial condition and the amount 0.5. This value (0.5) represents a rough measure of the separation between x_n values in period-2 trajectories.

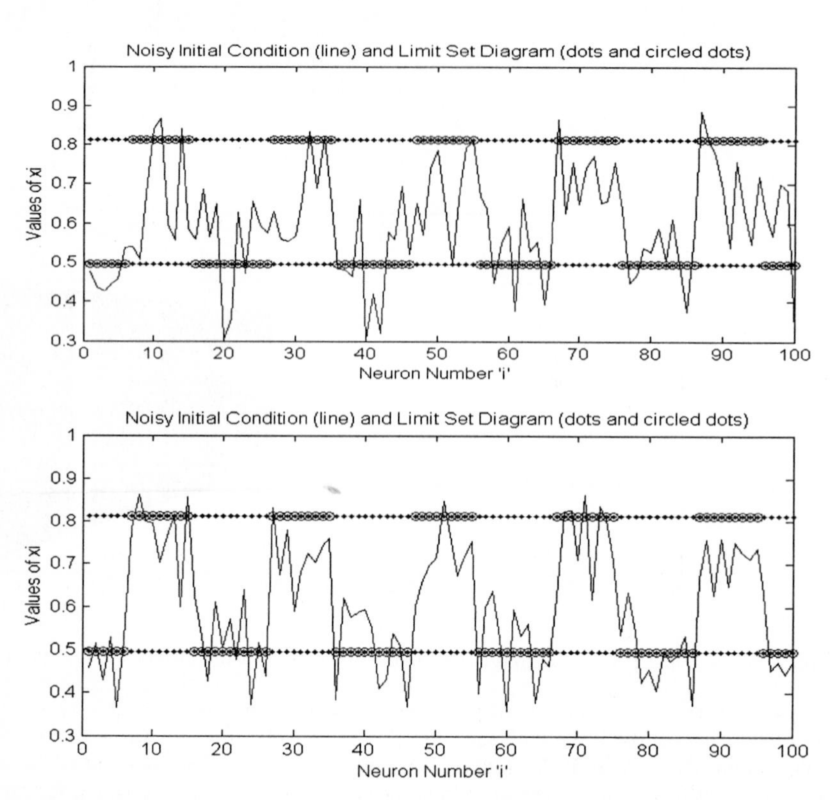

Fig 3. Limit Set Diagrams observed in networks prompted by noisy / distorted initial conditions. As in Fig. 1, the solid lines represent the initial condition, the horizontal axis represent 100 neurons of the network, and the pairs of dots and circled dots represent the final cyclic state of the network. In the first figure, the experiment was performed with level of analog noise 0.4, while in the second experiment, the level was 0.3.

In both experiments represented in Fig. 3 we observe a perfect recovery of the stored pattern, although the prompting initial condition is highly noisy. Notice that the patterns stored in this larger network (100 nodes) were obtained by five spatial repetitions of the patterns previously listed in the table of Fig. 1, which were used for the tests with networks of 20 nodes. Because of the perfect pattern recovery observed in the figure, we say that the "Hamming error" for these two experiments is zero, meaning that in both experiments represented in Fig. 3 the Hamming distance between the recovered pattern and the stored pattern is zero bits. A more general result for arbitrary levels of analog noise factors is presented in Fig. 4, which shows a plot relating the Hamming error and the level of analog noise introduced in the initial condition.

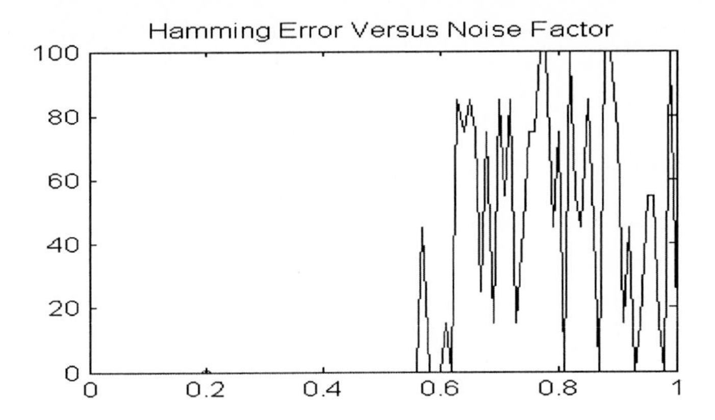

Fig. 4. Plot of the Hamming error (vertical axis) for arbitrary levels of analog noise in the initial condition (horizontal axis). We observe virtually no error in the recovery of patterns as long as the analog noise factor is limited to values below 0.55 (the main text describes in detail the way we defined this factor).

4 Network Immunity with Respect to Digital Noise

In the previous section we studied the network immunity to analog noise and distortion in the prompting pattern. This is a relevant issue in applications where the prompting patterns come from analog sensory systems. Another context in which the study of analog noise is important is when we consider hardware implementation of neuro-like architectures: then the analog nature of variables and the issue of noise cannot be ignored. (see [3] for hardware implementation of recursive processing elements). In certain classes of applications though, the prompting pattern is naturally digital and the neuro-computing system can be assumed to be free of noise (in software implementation of neural networks for example, this last assumption usually applies). Because of these classes of application, we also studied the robustness of the network for the case of noisy / distorted prompting patterns that are purely digital, have no analog noise added, but have a certain percentage of bits flipped. With this in mind, we did the study of the dependence of the recovery Hamming error with the level of digital noise, obtaining plots as we did for the case of analog noise. From the results of such study presented in Fig. 5 we can observe that as long as the number of flipped bit does not exceed 25%, there are no errors in the recovery of stored patterns. Beyond that value, we start to have some recovery problems, but the Hamming error is still moderate (around 15%). This situation changes drastically when we have more than 50% of the bits flipped. At this point, the Hamming error of the recovered patterns grows fast towards 100%. According to our interpretation, this indicates the recovery of mirrored patterns: since the digitally distorted prompting patterns have more than 50% of their bits flipped, they become more similar to the mirror versions of the prompting patterns than to the prompting patterns themselves. This in the same phenomena observed in Hopfield networks, for which the storage of a given pattern implies the storage of its mirror image as a spurious state [7].

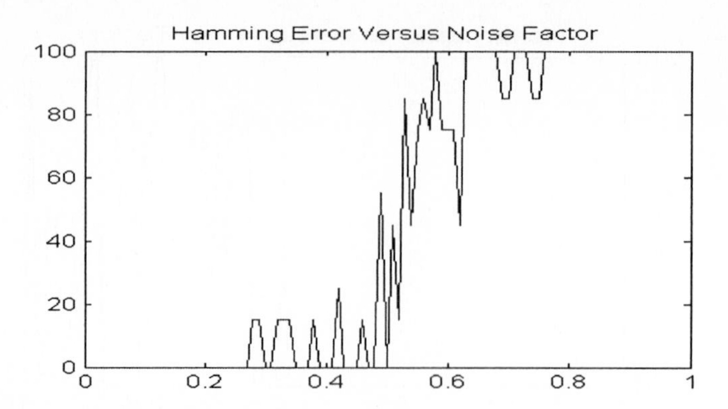

Fig. 5. Plot showing how the Hamming error (vertical axis) changes with different levels of digital noise in the initial condition (horizontal axis).

5 Monitoring the Values of the *p* Parameters

As we performed experiments with increasing levels of distortion in the prompting condition and experiments with increasing number of M stored patterns, we observed that the network had increasing problems of long transients, and non-convergence of the LSDs. This performance degradation with noise and number of stored patterns made us introduce an additional monitoring mechanism, to evaluate the success or the failure of the network in reaching some stable answer to a prompting pattern: we decided to record the final state of the ***p*** variables, at the end of each experiment. With that, we could evaluate how far these parameters were from their final goal value (the amount ***minp*** discussed in section 2). We observed that the recording of the final values of ***p*** could be useful in many different ways. First, the monitoring of the ***p*** variables can help us to detect the end of network operation. Second, the fact that after long time the ***p*** variables do not decrease to ***minp*** for a given prompting pattern can be used as an indicator of non-recognition of such pattern as one of the stored in the network. Finally, such a situation of final ***p*** larger than ***minp*** can also be used to indicate the saturation of the memory, as the number of M stored patterns increases.

Fig. 6 presents two Limit Set Diagrams for networks with 20 nodes for which this type of monitoring of the final values of the ***p*** parameter of each node was done. One of them is an LSD for a situation in which we start to observe saturation of the memory, with M=6 stored patterns: the final state is moderately stable and most of the ***p*** values reached ***minp*** (just one neuron – neuron #19 - in the network operates with high entropy cycling). The second LSD shows a more critical situation, where M=10, the final state is very unstable and many of the ***p*** variables have not reached low values, being stuck at the value ***maxp***.

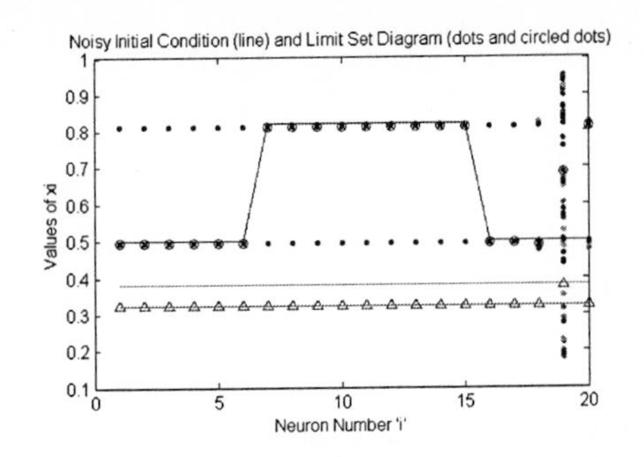

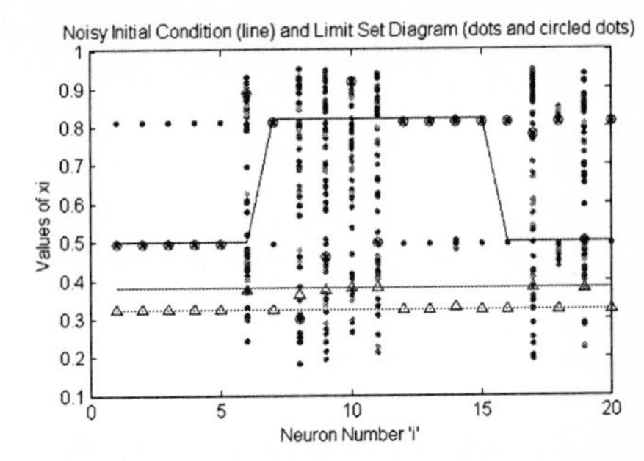

Fig. 6. Two Limit Set Diagrams including the p parameters monitoring technique: in addition to the values visited by the x variables, the diagrams represent, through small triangles, the final values of the p parameters of all nodes. The horizontal lines, on top of which the triangles are, represent the values $minp = 3.25$ and $maxp = 3.8$ (these p values appear here divided by 10, so we could use the same vertical scale to represent values of x and values of p). In the first LSD we have a moderate saturation of the memory. In the second LSD we have a clear situation of network instability.

6 Conclusions

The results presented here show that ANNs based on Recursive Processing Elements can implement effective associative memories. More important, the paper describes quantitative methods for the design and characterization of such networks. As a final comment, we want to mention that such methods can also be used in the design of architetures in which the nodes have more complex dynamics than the period-2 cycling used here.

References

1. K. Aihara et al. Chaotic Neural Networks, Physica Letters A, v.144 p.333, 1990
2. K. Kaneko, The Coupled Map Lattice: Introduction, Phenomenology, Lyapunov Analysis, Thermodynamics and Applications, in Theory and Applications of Coupled Map Lattices, John Wiley Sons, 1993
3. E. Del Moral Hernandez, Artificial Neural Networks Based on Bifurcating Recursive Processing Elements, UMI Dissertation Services, Ann Arbor, MI, 1998
4. N. H. Farhat et al. Complexity and Chaotic Dynamics in Spiking Neuron Embodiment, in Adaptive Computing, Mathematics, Electronics, and Optics, SPIE Critical Review, v.CR55 p.77, 1994
5. N. H. Farhat and E. Del Moral Hernandez, Logistic Networks with DNA-Like Encoding and Interactions, in Lecture Notes in Computer Science, v.930 p.214, Springer Verlag, 1995
6. E. Del Moral Hernandez, Bifurcating Pulsed Neural Networks, Chaotic Neural Networks and Parametric Recursions, IJCNN2000, IEEE, 2000
7. S. Haykin, Neural Networks: a Comprehensive Foundation, Prentice Hall, 1999
8. R. C. Hilborn, Chaos and Nonlinear Dynamics, Oxford U. Press, 1994

Topology-Preserving Elastic Nets

Valery Tereshko

Mogilev State Technical University, Prospect Mira 43, 212005 Mogilev, Belarus
tereshko@phys.belpak.mogilev.by

Abstract. We have developed a topology-preserving elastic net which combines both lateral and synaptic interactions to obtain topologically ordered representations (receptive fields) of an external stimulus. Existing neural models that preserve the topology by utilizing lateral interactions, such as the Kohonen map and Goodhill's mapping, and by utilizing synaptic interactions, such as cortical mapping and elastic net, appear as limiting cases of this model.

The development of neural receptive fields in a way that they mimic stimulus distribution and become ordered is biologically meaningful [1]. Competitive learning neural nets that utilize lateral interactions to perform a mapping from the stimulus space to the response space with preserving neighbourhood relations are called topology-preserving, or simply Kohonen, maps [2]. Goodhill applied the batch mode of Kohonen mapping with a special lateral interaction function to modelling the topography and ocular dominance of the visual cortex [3]. On the another hand, the cortex can be modeled as a dimension-reducing map from a high-dimensional stimulus space to its two-dimensional surface [5]. Utilizing the soft competition mechanism and synaptic interactions, the dimension-reducing mapping pulls the cortical sheet towards the stimula but, at the same time, keeps the cortical area as minimal as possible. In the batch mode, one obtains the elastic net model which was first applied to solve the travelling salesman problem (TSP) [4]. Researchers have paid attention to similarities in topology-preserving maps and elastic nets [2, 3, 6, 7], but up to date these approaches are developing independently. We propose a unified model combining lateral and synaptic interactions, the topology-preserving elastic net, so that all the above models appear as limiting cases.

Let us consider a one-dimensional net of n stochastic neurons trained by N patterns. The "distortion energy" of a one-dimensional net of n stochastic neurons, for a given input pattern, is defined as

$$E_i = \frac{1}{2}\sum_{j=1}^{n} h(i,j)|\boldsymbol{x}^{\mu} - \boldsymbol{w}_j|^2, \tag{1}$$

where $\boldsymbol{x}^{\mu}$ is a given sample pattern, $\boldsymbol{w}_j$ are the weight vectors, and $h(i,j)$ is the neighbourhood function.

The distortion energy represents the composition of local distortions of weight vectors $\boldsymbol{w}_j$ from a given stimulus $\boldsymbol{x}^{\mu}$ weighted by the neighbourhood function

J. Mira and A. Prieto (Eds.): IWANN 2001, LNCS 2084, pp. 554-560, 2001.

$h(i,j)$ which is a representation of lateral interactions in the net. Typically, the neighbourhood function falls off with distance from a chosen neuron. One can refer to (1) as topographic distortions.

The "tension energy" which minimizes the difference between neighbouring weight vectors is

$$E_{tens} = \frac{\gamma}{2} \sum_{i=1}^{n-1} |\boldsymbol{w}_{i+1} - \boldsymbol{w}_i|^2, \tag{2}$$

where γ is the elasticity parameter.

Instead of the "hard" assignment of Kohonen's original algorithm [1] with a unique winner, we assume a "soft" assignment [8, 9] where every i-th neuron is assigned to a given μ-th pattern with a probability $p(i,\mu)$, with $\sum_i p(i,\mu) = 1$.

Then, the *free energy*, composed of the averaged energy (that is a composite of the topographic distortions weighted by their assignment probabilities and the elastic synaptic tensions) and thermal noise energy, is

$$F = \sum_{i=1}^{n} p(i,\mu) E_i + E_{tens} - TS, \tag{3}$$

where $p(i,\mu)$, $\sum_i p(i,\mu) = 1$, are assignment probabilities,
$S = -\sum_i p(i,\mu) \ln p(i,\mu)$ is the entropy, and $T = 1/\beta$ is the "temperature" of the system.

For a given input pattern and set of weight vectors, the assignment probabilities minimising the free energy (3) are

$$p(i,\mu) = \frac{\mathrm{e}^{-\beta E_i}}{\sum_{k=1}^{n} \mathrm{e}^{-\beta E_k}}, \tag{4}$$

and, respectively, the minimal free energy is

$$F = -\frac{1}{\beta} \ln \Big(\sum_{i=1}^{n} \mathrm{e}^{-\beta E_i} \Big) + \frac{\gamma}{2} \sum_{i=1}^{n-1} |\boldsymbol{w}_{i+1} - \boldsymbol{w}_i|^2 \tag{5}$$

Incremental learning strategies (on-line learning and cyclic learning [10]) are derived through a steepest descent minimization of this free energy function. The dynamics follows the gradient of free energy (5):

$$\Delta \boldsymbol{w}_j = -\eta \frac{\partial F}{\partial \boldsymbol{w}_j} = \eta \Big(\sum_{i=1}^{n} p(i,\mu) h(i,j) (\boldsymbol{x}^\mu - \boldsymbol{w}_j) + \gamma \Theta(\boldsymbol{w}) \Big), \tag{6}$$

where $\Delta \boldsymbol{w}_j = \boldsymbol{w}_j(t+1) - \boldsymbol{w}_j(t)$, $\Theta(\boldsymbol{w}) = \boldsymbol{w}_{j+1}(t) - 2\boldsymbol{w}_j(t) + \boldsymbol{w}_{j-1}(t)$, and η is the learning rate.

This is a generalized soft *topology-preserving* mapping. Soft mapping is based on soft competition which allows all neurons to adjust their weights with probabilities proportional to their topographic distortion. This makes the weights move more gradually to the presented patterns. The strength of the competition is adjusted by a "temperature". The underlying mechanism, deterministic annealing, is derived from statistical physics — it mimics an ordering process during a system's cooling.

The limit of no synaptic interactions ($\gamma \to 0$) results in a soft topology-preserving mapping[8, 9]:

$$\Delta \boldsymbol{w}_j = \eta \sum_{i=1}^{n} p(i,\mu) h(i,j)(\boldsymbol{x}^\mu - \boldsymbol{w}_j) \tag{7}$$

At low temperatures ($\beta \to \infty$), equation (7) reduces to *Kohonen's* map [1]:

$$\Delta \boldsymbol{w}_j = \eta h(j,j^*)(\boldsymbol{x}^\mu - \boldsymbol{w}_j), \tag{8}$$

where j^* is the winning unit.

Kohonen's topology-preserving, or self-organizing, map has become a standard unsupervized learning algorithm [2]. It has been successfully applied to describe the development of the visual cortex — simultaneous formation of retinotopy, i.e. a continuous topographic mapping from the retina to the visual cortex, ocular dominance, and orientation preference [11]. Applied to the TSP, the Kohonen algorithm produces slightly longer tours than the elastic net, but the former is computationally faster [15, 16]. The generalization in the definition of the winner, when topographic distortions instead of Euclidean distances are applied to select a winner, leads to another promising application of the above algorithm – data compression using the so-called topology-preserving vector quantization where a set of data is encoded by a reduced set of reference vectors in a way that captures the essential spatial relations of the data set [12]. The deterministic annealing version of the above scheme is called soft topology-preserving vector quantization [13, 8].

Alternatively, in the limit of no lateral interactions ($h(i,j) \to \delta(i-j)$, where $\delta(i-j)$ is a Dirac delta function), equation (6) tends to a *cortical* mapping [5]:

$$\Delta \boldsymbol{w}_j = \eta \Big(\tilde{p}(j,\mu)(\boldsymbol{x}^\mu - \boldsymbol{w}_j) + \gamma \Theta(\boldsymbol{w}) \Big), \tag{9}$$

where $\tilde{p}(j,\mu) = \exp(-0.5\beta|\boldsymbol{x}^\mu - \boldsymbol{w}_j|^2) / \sum_{k=1}^{n} \exp(-0.5\beta|\boldsymbol{x}^\mu - \boldsymbol{w}_k|^2)$.

The idea of the cortex as a dimension-reducing map from high-dimensional stimulus space to its two-dimensional surface has proved to be fruitful [7, 5]. The backward projection of each position on the cortex sheet to the position in stimulus space is a convenient way to consider cortex self-organization — the way in which it fills stimulus space defines the receptive field properties. The learning algorithm (equation (9)) performing such a mapping induces two conflicting tendencies: (1) the cortical surface should pass through the representative points in stimulus space; (2) the area of the sheet should be kept a minimum. This ensures the formation of smooth receptive fields and, hence, the minimal "wiring" interconnecting the cortical cells, which, in turn, ensures the closeness of the cortical cells representing similar stimuli. The stripes and patches seen within cortical areas have been argued to be adaptations that allow the efficient wiring by such structures [17].

Figure 1 illustrates the development of a one-dimensional cortical chain embedded into a two-dimensional stimulus space. The simulations are performed for 90 cortical neurons with an initial uniform random distribution of the weight vectors. The training is cyclic with a fixed sequence. The 64 stimulus are sites on a 8×8 regular square. The learning rate (with $\eta_0 = 1$) and the elasticity parameter (with $\gamma_0 = 0.01$) are linearly decreasing functions of time. The annealing schedule corresponds to linearly lowering the temperature from $\beta = 4$ to $\beta = 140$ in steps of 0.025. One can see the evolution of the cortex from a random configuration to an optimally ordered configuration that covers all points in stimulus space, and, at the same time, is smoothly mapped (i.e. continuous and without intersections) and is of a minimal length. If one fixes the boundaries of the chain at the start and end, such an algorithm can, obviously, be used to find shortest routes providing that all intermediate "stations" are visited. Remarkably, Kohonen mapping is found to minimize a measure that is closely related to wirelength [6].

In the *batch* mode, the updating rule is averaged over the set of training patterns before changing the weights. Minimization of the free energy function (equation (5)) in this mode results in a generalized elastic net algorithm:

$$\Delta \boldsymbol{w}_j = -\eta \frac{1}{N} \sum_{\mu=1}^{N} \frac{\partial F}{\partial \boldsymbol{w}_j} = \hat{\eta} \Big(\sum_{\mu=1}^{N} \sum_{i=1}^{n} p(j,\mu) h(i,j)(\boldsymbol{x}^{\mu} - \boldsymbol{w}_j) + \gamma \Theta(\boldsymbol{w}) \Big), \quad (10)$$

where $\hat{\eta} = \eta / N$, and N is the number of training patterns.

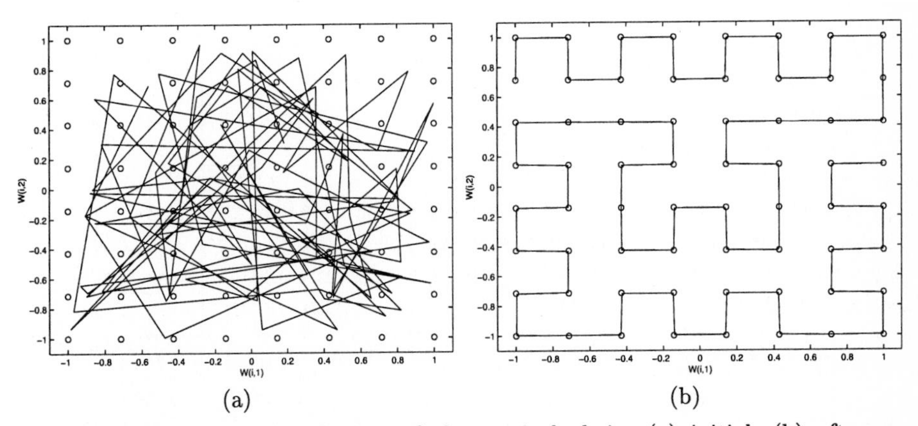

(a) (b)

Fig. 1. Weight vector distribution of the cortical chain: (a) initial, (b) after applying cortical mapping (Eq. 9, see details in the text). Stimula and weight vectors are marked by open and filled circles, respectively.

In the limit of no lateral interactions ($h(i,j) \to \delta(i-j)$), equation (10) results to the original elastic net algorithm [4]:

$$\Delta \boldsymbol{w}_j = \hat{\eta}\Big(\sum_{\mu=1}^{N} \tilde{p}(j,\mu)(\boldsymbol{x}^{\mu} - \boldsymbol{w}_j) + \gamma\Theta(\boldsymbol{w})\Big) \tag{11}$$

Indeed, with a definition $\beta \equiv 1/\sigma^2$, where σ is the variance of the Gaussian function, equation (11) is the exact form of Durbin-Willshaw's elastic net [4]. Shrinking the variance σ in Durbin-Willshaw's algorithm is thus similar to lowering the temperature. This clearly shows how an annealing procedure is incorporated into the system.

Using both synaptic and lateral interactions can improve the system performance. Let us demonstrate this with theTSP example. Figure 2 illustrates the formation of an optimal tour. The simulations are performed for a 90 point elastic ring. Initially, the weight vectors are distributed equidistantly on the unit radius circle. The 64 "cities" are sites on a 8×8 regular square. The lateral interactions are nearest-neighbours and fixed with very weak strengths (($h(i,j)$ $= 1$, 0.01, and 0 for $i = j$, $|i-j| = 1$, and $|i-j| \geqslant 2$ respectively). The learning rate (with $\eta_0 = 1.5$) and the elasticity parameter (with $\gamma_0 = 0.005$) are linearly decreasing functions with time. The annealing schedule corresponds to linearly lowering the temperature from $\beta = 4$ to $\beta = 230$ in steps of 0.25. With the same annealing schedule, the original elastic net algorithm (11) leads to optimal tours for larger initial elasticity. The use of both lateral and elastic interactions is, thus, a necessity when the strength of any or both types of interactions is limited. Together with finding the shortest tours of the TSP, the elastic net algorithm can be successfully applied to modelling retinotopic maps from two eyes onto the visual cortical surface [18, 19]. The model accounts for the development

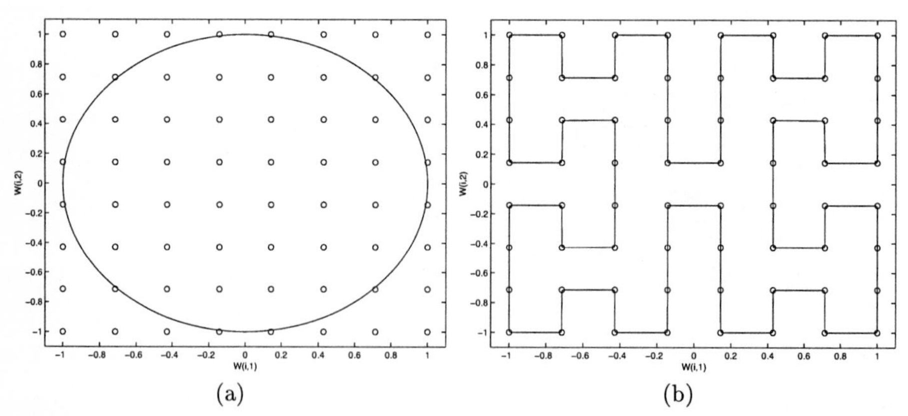

(a) (b)

FIG. 2. Weight vector distribution of the elastic ring: (a) initial, (b) after applying the elastic net algorithm (Eq. 10, see details in the text). Stimula and weight vectors are marked by open and filled circles, respectively.

of both topography and ocular dominance; forming the stripes similar to that seen in biological investigations.

Alternatively, with no synaptic interactions ($\gamma \to 0$), equation (10) reduces to the *batch* version of the soft *topology-preserving* map [8, 9]:

$$\Delta \boldsymbol{w}_j = \hat{\eta} \sum_{\mu=1}^{N} \sum_{i=1}^{n} p(i,\mu) h(i,j)(\boldsymbol{x}^\mu - \boldsymbol{w}_j), \tag{12}$$

At low temperatures ($\beta \to \infty$), equation (10) tends to the *batch* mode of the *Kohonen* map:

$$\Delta \boldsymbol{w}_j = \hat{\eta} \sum_{\mu=1}^{N} h(j,j^*)(\boldsymbol{x}^\mu - \boldsymbol{w}_j), \tag{13}$$

If one defines the lateral interaction function $h(j,j^*) = \exp(-0.5\beta|j-j^*|^2)/\sum_{k=1}^{n} \exp(-0.5\beta|j-j^*|^2)$, equation (13) is merely Goodhill's map [3] which has been applied to the formation of topography and ocular dominance in the visual cortex. The elastic net appears to slightly outperform this map [3]. Another limitation of Goodhill's mapping is the use of a shrinking lateral interaction function, which is not biologically justified (this is pointed by Goodhill himself [14]).

We applied soft mapping (equation (13)) with fixed nearest-neighbour lateral interactions to the problem of *ocular dominance*. The simulations are performed for 32 cortical neurons with initial uniform random distribution of the weight vectors within $[-0.0667, 0.0667] \times [-1, 1]$ rectangular. The stimula are placed regularly within the two columns of 16 units each at the left and right boundary of the rectangular, which represent left and right "eyes" respectively. The ratio

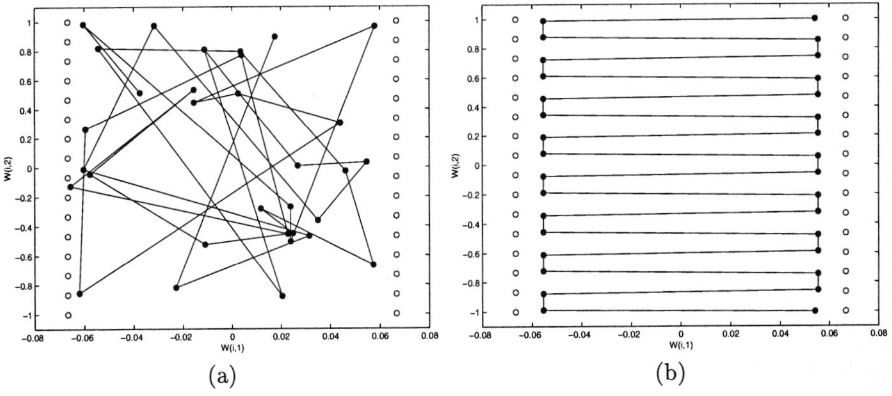

FIG. 3. Weight vector distribution of the cortical chain: (a) initial, (b) after applying the batch mode of soft topology-preserving mapping (Eq. 12) with fixed nearest-neighbour lateral interactions (see details in the text). Stimula and weight vectors are marked by open and filled circles, respectively.

of the separation units between retinae to the separation of neighbouring units within a retina defines the correlation of retinal units, which is ≈ 1. The lateral interactions are nearest-neighbours and fixed with strengths as in Fig. 1. The learning rate (with $\eta_0 = 1$) is a linearly decreasing function of time. The annealing schedule corresponds to linearly lowering the temperature from $\beta = 4$ to $\beta = 1000$ in steps of 0.025. One can see the simultaneous formation of retinotopy and ocular dominance stripes where regular clusters of cortical cells are associated with either left or right eyes (Fig. 3).

The considered models, exploiting lateral and/or synaptic interactions, have common characteristics such as competive activation, clustering, smoothing, and fine tuning through the annealing or annealing-like processes. Owing to a similarity in mechanisms, they produce very similar results when applied to the same problem. The use of both lateral and synaptic interactions can be beneficial, and it is a necessity when the strength of any or both types of interaction is limited. Fixed nearest-neighbour lateral interactions we utilized can be implemented in artificial neural nets much easier than the shrinking lateral interaction function.

References

1. T. Kohonen, Biol. Cybern. **43**, 59 (1982).
2. J. Hertz, A. Krogh, and R.G. Palmer, *Introduction to the Theory of Neural Computation* (Addison-Wesley, Reading, 1991).
3. G.J. Goodhill, Cognitive Science Research Paper 226 (University of Succex, UK, 1992).
4. R. Durbin and D. Willshaw, Nature **326**, 689 (1987).
5. R. Durbin and G. Mitchison Nature **343**, 644 (1990).
6. G. Mitchison, Neural Comput. **7**, 25 (1995).
7. N.V. Swindale, N. V. Network: Comput. Neural Syst. **7**, 161 (1996).
8. T. Graepel, M. Burger, and K. Obermayer, Phys. Rev. E **56**, 3876 (1997).
9. T. Heskes, in *Kohonen Maps*, edited by E. Oja and S. Kaski (Elsevier, Amsterdam, 1999) p. 303.
10. T. Heskes and W. Wiegerinck, IEEE Trans. Neural Networks **7**, 919 (1996).
11. K. Obermayer, G.G. Blasden, and K. Schulten, Phys. Rev. A **45**, 7568 (1992).
12. S.P. Luttrell, IEEE Trans. Neural Network **1** 229 (1990).
13. K. Rose, E. Gurewitz, and G.C. Fox, IEEE Trans. Inform. Theory **38**, 1249 (1992).
14. G.J. Goodhill, Biol. Cybern. **69**, 109 (1993).
15. B. Angéniol, C. Vaubois, and J.Y. Le Texier, Neural Networks **1**, 289 (1988).
16. Fort, J. C., Biol. Cybern. **59**, 33 (1988).
17. G. Mitchison, Proc. Roy. Soc. B **245**, 151 (1991).
18. G.J. Goodhill and D.J. Willshaw, Network: Comput. Neural Syst. **1**, 41 (1990).
19. G.J. Goodhill and D.J. Willshaw, Neural Comput. **6**, 615-621 (1994).

Optimization with Linear Constraints in the Neural Network

Mikiya Oota, Naohiro Ishii, Kouichiro Yamauchi, and Mayumi Nakamura

Department of Intelligence and Computer Science
Nagoya Institute of Technology
Gokiso-cho, Showa-ku, Nagoya, 466-8555,
Japan
ishii@ics.nitech.ac.jp

Abstract. The solution method of optimization problem using the Hopfield model is equal to the optimization of the objective quadratic function without constraints. The conventional penalty method is difficult in the optimization problem with constraints. In this study, a contimuous gradient projection method is applied to the problem with linear constraints in the neural network. It is shown that the optimum solution of the constrained problem is obtained effectively by this method.

1 Introduction

Hopfield et al. [1] showed neural network properties which minimize an energy function which is decided by weight and threshold in the network [1]. The neural network is assumed to have symmetrical connections. Using this property, it was shown that the traveling salesman problem was possible to solve. It has been tried that various optimization problems are solved using the Hopfield models [2], [4–6].

There is a formulation by the penalty method as the method for handling the optimization problem with constraint conditions in the Hopfield model [2]. This method is included in the objective function which minimizes energy function using penalty method, and it is converted into the optimization problem without constraints. However, the parameters in the penalty method have to be experientially decided, since there is no objective method for the parameters stimation.

In this study, for the purpose of obtaining the solution which satisfies constraints, a design method of energy function for the optimization problem with linear constraints is proposed.

2 Formulation by Penalty Method

The Hopfield model is decided by the weight W and the threshold $\boldsymbol{\theta}$ of the neuronal circuit which minimizes energy function shown in the following. The

J. Mira and A. Prieto (Eds.): IWANN 2001, LNCS 2084, pp. 561-569, 2001.

weight W is symmetrical here, and the diagonal component is assumed 0. The solution is based on the quadratic function E in the equation (1).

$$E = -\frac{1}{2}\boldsymbol{x}^T W \boldsymbol{x} + \boldsymbol{\theta}^T \boldsymbol{x} \tag{1}$$

Therefore, in the penalty method, it is necessary to convert the problem into the case in which optimization problem with constraints is transformed to the optimization problem in which there are no constraints.

The optimization problem is formulated using the penalty method as shown in the following equation (2).

$$\begin{array}{ll} \min & f(\boldsymbol{x}) \\ \text{subject to} & g_j(\boldsymbol{x}) = 0 \quad (j = 1, \cdots, m) \\ & 0 \leq x_i \leq 1 \quad (i = 1, \cdots, n) \end{array} \tag{2}$$

x is the n-dimensional variable vector, and constraints $g_j(\boldsymbol{x}) = 0 \ \ (j = 1, \cdots, m)$ show linear constraints. The function which takes the minimum value, when constraints are satisfied for $g_j(\boldsymbol{x}) = 0$, is defined in order to convert into the optimization problem without constraints, as shown in the following equation (3).

$$\begin{array}{ll} \min & f(\boldsymbol{x}) + \sum_j \lambda_j \left(g_j(\boldsymbol{x})\right)^2 \\ & \qquad (j = 1, \cdots, m) \\ \text{subject to} & 0 \leq x_i \leq 1 \quad (i = 1, \cdots, n) \end{array} \tag{3}$$

This function is minimized as an objective function. Here, $\lambda_j \ \ (j = 1, \cdots, m)$ is penalty parameters. Next, change of variables is carried out using the sigmoid function. Supreme and lower-limit constraint of variable x is eliminated by this variable change. Then equation (4) is derived.

$$\begin{array}{l} \min f(\boldsymbol{y}) + \sum_j \lambda_j \left(g_j(\boldsymbol{y})\right)^2 \quad (j = 1, \cdots, m) \\ y_i = \frac{1}{1+\exp(-x_i)} \quad (i = 1, \cdots, n) \end{array} \tag{4}$$

This becomes an optimum solution of the equation (2), in optimization of the equation (4) without constraints, when x minimizes the objective function.

The objective function does not necessarily decrease, even if constraints are satisfied, when the value of the parameter is too great. The form of the energy function must be considered to take an appropriate balance, when there are several constraints.

Therefore, in this paper, we develop a new design method for the energy function in the optimization problem with linear constraints.

3 Proposed Method

3.1 Continuous Gradient Projection Method

In this chapter, a proposed method is based on Tanabe's Continuous gradient projection method [3], in the following.

$$\begin{array}{ll} \min & f(\boldsymbol{x}) \\ \text{subject to} & g_j(\boldsymbol{x}) = 0 \quad (j = 1, \cdots, m) \end{array} \tag{5}$$

The optimization problem is described in the following.

$$L(\boldsymbol{x}, \boldsymbol{\lambda}) \equiv f(\boldsymbol{x}) - \boldsymbol{\lambda}^T g(\boldsymbol{x}) \tag{6}$$

Lagrange function of this optimization problem is made to be $L(\boldsymbol{x}, \boldsymbol{\lambda}) \equiv f(\boldsymbol{x}) - \boldsymbol{\lambda}^T g(\boldsymbol{x})$. Here, $\boldsymbol{\lambda} \equiv (\lambda_1, \lambda, \cdots, \lambda_m)^t \in \boldsymbol{R}^m$ is the Lagrange multiplier. $\boldsymbol{\lambda}^* \in \boldsymbol{R}^m$ exists for the optimum solution $\boldsymbol{x}^*$, and $(\boldsymbol{x}^*, \boldsymbol{\lambda}^*)$ satisfies the Karus-Kuhn-Tucker condition, namely the simultaneous nonlinear equation on variable $\boldsymbol{x}, \boldsymbol{\lambda}$ as shown in the following.

$$\begin{cases} \nabla_{\boldsymbol{x}} L(\boldsymbol{x}, \boldsymbol{\lambda}) = \nabla_{\boldsymbol{x}} f(\boldsymbol{x}) - J(\boldsymbol{x})^T \boldsymbol{\lambda} = 0 \\ \nabla_{\boldsymbol{\lambda}} L(\boldsymbol{x}, \boldsymbol{\lambda}) = g(\boldsymbol{x}) \qquad\qquad\qquad = 0 \end{cases} \tag{7}$$

$J(\boldsymbol{x})$ is a Jacobian matrix of $m \times n$ of $g(\boldsymbol{x})$ in the following, and the rank is m.

$$J(\boldsymbol{x}) = \nabla g(\boldsymbol{x}) = \begin{pmatrix} \frac{\partial g_1(\boldsymbol{x})}{\partial \boldsymbol{x}_1} & \cdots & \frac{\partial g_1(\boldsymbol{x})}{\partial \boldsymbol{x}_n} \\ \vdots & \ddots & \vdots \\ \frac{\partial g_m(\boldsymbol{x})}{\partial \boldsymbol{x}_1} & \cdots & \frac{\partial g_m(\boldsymbol{x})}{\partial \boldsymbol{x}_n} \end{pmatrix} \tag{8}$$

In short, it is necessary to solve this simultaneous equation (7) in order to solve the optimization problem with constraints. The simultaneous equation (9) is obtained, when the Neutonian method is applied in this simultaneous equation (7).

$$\begin{bmatrix} I_n & -J(\boldsymbol{x}_k)^T \\ J(\boldsymbol{x}_k) & 0 \end{bmatrix} \begin{bmatrix} d\boldsymbol{x}_k \\ \boldsymbol{\lambda}_{k+1} \end{bmatrix} = - \begin{bmatrix} \nabla_{\boldsymbol{x}} f(\boldsymbol{x}) \\ g(\boldsymbol{x}) \end{bmatrix} \tag{9}$$

$d\boldsymbol{x}_k$, $\boldsymbol{\lambda}_{k+1}$ will be computed as following.

$$d\boldsymbol{x}_k = -(I_n - J_k^+ J(\boldsymbol{x}_k)) \nabla f(\boldsymbol{x}_k) - J_k^+ g(\boldsymbol{x}_k) \tag{10}$$

$$\boldsymbol{\lambda}_{k+1} = (J_k^+)^T \nabla f(\boldsymbol{x}_k) - (J(\boldsymbol{x}_k) J(\boldsymbol{x}_k)^T)^{-1} g(\boldsymbol{x}_k) \tag{11}$$

, where, $J_k^+ \equiv J(\boldsymbol{x}_k)^T (J(\boldsymbol{x}_k) J(\boldsymbol{x}_k)^T)^{-1}$ is the Moore-Penrose pseudo-inverse of $J(\boldsymbol{x}_k)$. It can be expected that $\boldsymbol{x}_k$, $\boldsymbol{\lambda}_k$ is converged to the optimization values by equations (9) and (10).

3.2 Application to the Neural Network

In this chapter, the solution to the optimization problem with linear constraints applied to the Hopfield model on the basis of equation (10), is explained. The optimization problem with linear constraints is shown in the equation (12).

$$\begin{array}{ll} \min & f(\boldsymbol{x}) \\ \text{subject to} & A\boldsymbol{x} - \boldsymbol{b} = 0 \\ & 0 \le x_i \le 1 \quad (i = 1, \cdots, n) \end{array} \tag{12}$$

,where $\boldsymbol{x}$ shows n-dimensional variable vector, A on $m \times n$ coefficient matrix, $\boldsymbol{b}$ with the m dimension constant vector. From (10) and (12), the following equation (13) is obtained.

$$d\boldsymbol{x} = -(I_n - A^+ A) \nabla f(\boldsymbol{x}) - A^+ (A\boldsymbol{x} - \boldsymbol{b}) \tag{13}$$

$A^+ = A^T(AA^T)^{-1}$ is Moore-Penrose pseudo-inverse of matrix A. In this equation, a function P is introduced so that the value of variable $\boldsymbol{x}$ may not exceed region $[0,1]^n$.

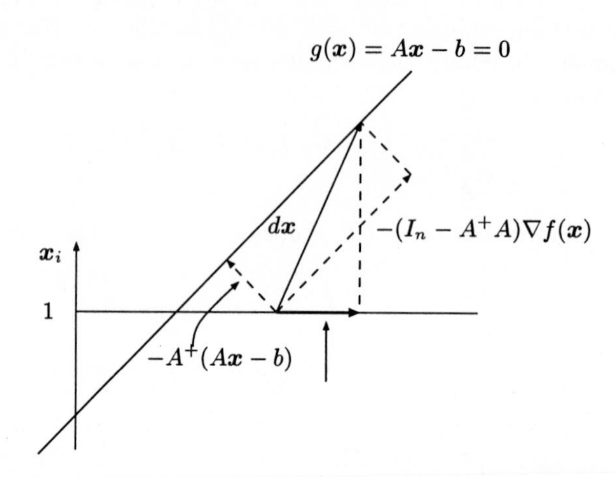

Fig. 1. Drawback of Eq(14). Eq(14) can not find the optimal solution on $[0,1]^n$.

$$d\boldsymbol{x} = P\left\{-(I_n - A^+A)\nabla f(\boldsymbol{x}) - A^+(A\boldsymbol{x} - \boldsymbol{b})\right\} \tag{14}$$

Here, matrix P is a diagonal matrix of equation (15).

$$P = \begin{pmatrix} x_1(1-x_1) & & 0 \\ & \ddots & \\ 0 & & x_n(1-x_n) \end{pmatrix} \tag{15}$$

It is possible to divide equation (13) into two following equations by the difference between the search direction, to prevent $[0,1]^n$ failure on the boundary.

$$d_{grad} = -(I_n - A^+A)\nabla f(\boldsymbol{x}) \tag{16}$$
$$d_{const} = \alpha\{-A^+(A\boldsymbol{x} - \boldsymbol{b})\} \tag{17}$$

,where a control parameter α is introduced to seek a better direction for the solution. Steepest descent direction $\nabla f(\boldsymbol{x})$ of the objective function was projected so that equation (16) may become parallel plane $A\boldsymbol{x} - \boldsymbol{b}$. Equation (17) is search direction of the Newtonian method to the plane $A\boldsymbol{x} - \boldsymbol{b}$ direction which satisfies constraints.

4 Computer Experiments

4.1 Two Variables Problem

In the simple two variable problem (18), whether the proposed method obtains the solution which satisfies constraints, is confirmed.

$$\begin{aligned} \min \quad & -\frac{1}{2}(x_1^2 + x_2^2 - 0.2x_1x_2) \\ & \quad +0.64x_1 + 0.53x_2 \\ \text{subject to} \; & -2x_1 + x_2 + 0.2 = 0 \\ & 0 \le x_i \le 1 \;\; (i = 1, 2) \end{aligned} \tag{18}$$

Objective function $-\frac{1}{2}(x_1^2 + x_2^2 - 0.2x_1x_2) + 0.64x_1 + 0.53x_2$ is a convex function which has the maximum value at $\boldsymbol{x} = (0.7, 0.6)$ (Fig. 2). Then, it has

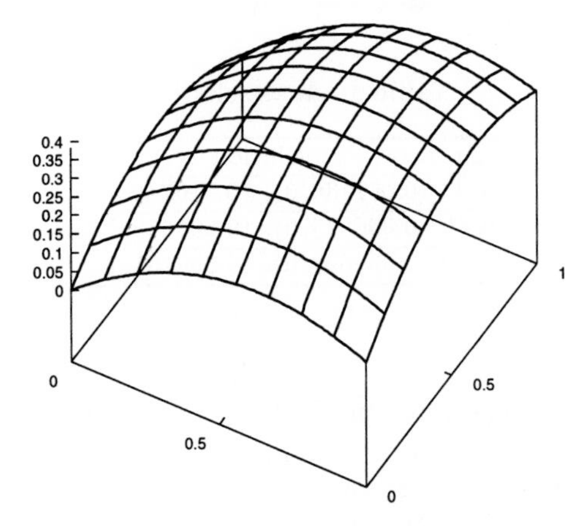

Fig. 2. 3D view of the objective function.

two local optimum solutions of $\boldsymbol{x} = (0.1, 0)$ and $(0.6, 1)$, and their values are 0.059, 1.254 respectively. For the experiment, the Gaussian method was used, and the time width was made to be $\Delta t = 0.001$ in the neuronal change of state. The experiment was carried out for 5 initial values. The initial values of neuronal circuit were set at the center $\boldsymbol{x} = (0.7, 0.6)$ with 0.1 radius, as shown in Fig. 3. The locus of an output of each element is shown in Fig. 4. It is proven to converge on the solution which satisfies constraints, even if it is started from every initial value as shown in the figure.

The value of parameter α was introduced in order to clarify the effect of amount of change d_{const} to d_{grad} on the direction which satisfies constraints.

To begin with, the values obtained at $\Delta t = 0.1$, 0.01, and 0.001 of Δt are compared with two local optimum solutions of $(0.1, 0)$ and $(0.6, 1)$. The results

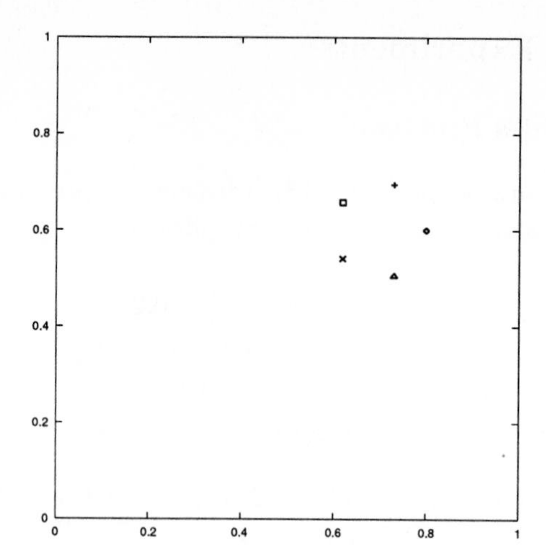

Fig. 3. Initial value of $\boldsymbol{x}$.

Table 1. Δt vs solution.

Δt	solution
0.1	$(0.062136, 0)$, $(0.616740, 1)$
0.01	$(0.096223, 0)$, $(0.601652, 1)$
0.001	$(0.099623, 0)$, $(0.600165, 1)$

are shown in Table 1. The error increases in proportion to the size of Δt, as shown in the table.

Next, it computer simulation was tried to clarify results by the fixation to $\Delta t = 0.001$ with $\alpha = 1$, 10, 100, 1000 ($= 1/\Delta t$), and 10000. The results are shown in Table 2. The accuracy of the solution has improved, as the value of α

Table 2. α vs solution.

α	solution
1	$(0, 0)$, $(0.791558, 1)$
10	$(0, 0)$, $(0.999892, 1)$
100	$(0.096224, 0)$, $(0.601652, 1)$
1000	$(0.099623, 0)$, $(0.600165, 1)$
10000	$(0.099962, 0)$, $(0.600017, 1)$

is increased. However, since there is no an objective method for the decision of the value of α, it seems to be appropriate be $\alpha = 1/\Delta t$ as one standard views from the results.

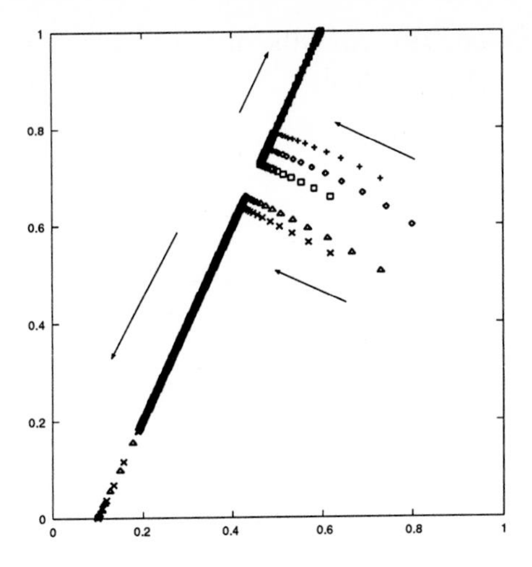

Fig. 4. Loci of $\boldsymbol{x}$.

4.2 Shortest Path Problem

In this section, we apply the proposed method to the shortest path problem. The shortest path problem, here, is to find the path ways with minimum costs, where the path from the node i to the node j, has the cost C_{ij} in the graph with the total nodes N. To attack this problem, some methods were proposed by using Hopfield models [4–6]. However, they assumed the energy functions in the penalty method, thus the optimum path ways cannot be always found. The shortest problem is formulated is the following.

$$\begin{aligned}
\min \quad & \sum_{i=1}^{N} \sum_{\substack{j=1 \\ j \neq i}}^{N} C_{ij} x_{ij} \\
\text{subject to} \quad & \sum_{j=1}^{N} x_{ij} - \sum_{j=1}^{N} x_{ji} = 0 \\
& \qquad\qquad (i = 1, \cdots, N) \\
& x_{ds} = 1 \\
& x_{ij} \in \{0, 1\} \quad (i, j = 1, \cdots, N)
\end{aligned} \tag{19}$$

, where C_{ij} shows the cost of the edge from the node i to the node j. When there is no edge, the cost becomes $C_{ij} = \infty$. The variable x_{ij} shows whether the connected path exists from the node i to the node j. Thus, the variable x_{ij} is defined in the following,

$$x_{ij} = \begin{cases} 1 & \text{if the edge from the node } i \text{ to the node } j, \\ & \text{is included in the path way.} \\ 0 & \text{otherwise.} \end{cases} \tag{20}$$

The constraints equation (19) imply the first condition that the incoming edges to the node i, is equal to the number of the outcoming edges from the node i, as shown in Fig. 5.

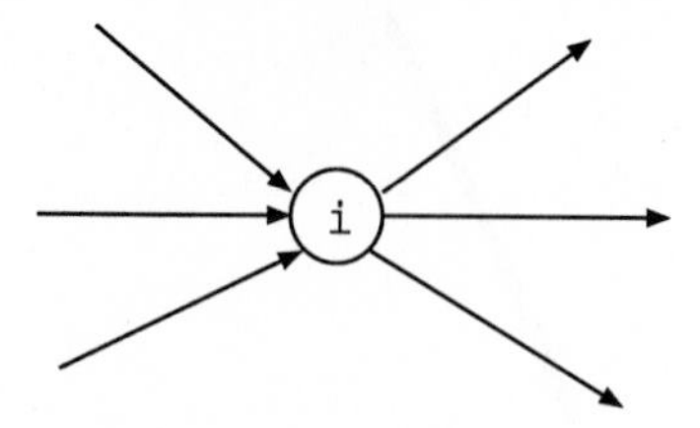

Fig. 5. Constraint Number of input edges must be the same as that of output edgees.

The second condition is that the edge from the end node d to the initial node s, must be included in the path way. To satisfy this condition, a fictitious edge between s and d must be added. Thus, the first condition is satisfied in the problem.

The third condition is that the variable x_{ij} should be 0 or 1. The shortest path can be found by the minimization of the object function under the above constraint conditions in the neural network. The output x_{ij} of network is the same as defined in the above equation (20).

Computer experiments are carried out to show the effectiveness of the proposed method in the problem of the shortest path from the node 2 to the node 7 in Fig. 6.

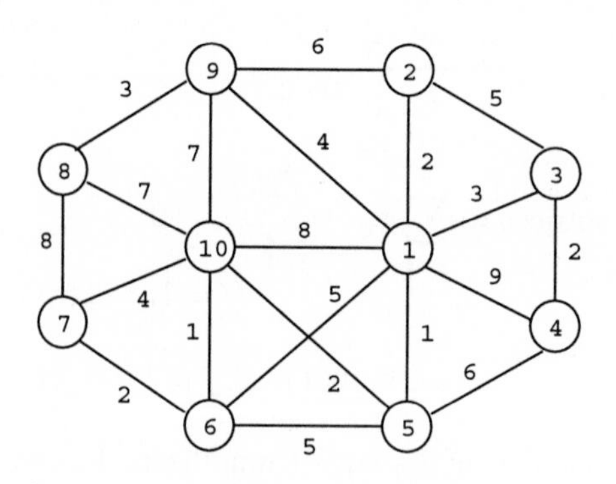

Fig. 6. Graph used in the experiment.

To attack this problem, parameters in the neural network, were set as the time interval $\Delta t = 0.001$, the coefficient $\tau = 100$. The initial values, were set randomly in the interval $0.4 \leq x_i \leq 0.6$ $(i = 1, \cdots, 100)$. We assigned the value 1 to the output of the neuron being more than the value 0.5. Also, the value 0 is assigned to the output of the neuron being less than the value 0.5. However, the values were more than 0.9 or less than 0.05 in the output of the neurons, in the experiments. Experiments were carried out 100 times in Fig. 6 problem. Then, the optimum shortest route, $2-1-5-10-6-7$, was obtained in all 100 experiments, while the conventional penalty method got 12 optimum shortest route, 23 other routes and 65 unsolved solutions.

5 Conclusion

In this paper, an optimization method which satisfies Karush-Kuhn-Tacker conditions, is proposed by using Neural Network model. The method has the linear constraints in the optimization problem. This method converges to the optimum values in the problem. Computer experiments show that the proposed method is effective to the optimization problems in the neural network.

References

1. J. J. Hopfield and D. W. Tank, ""Neural" Computation of Decision in Optimization Problems," Biological Cybernetics, Vol. 52, pp. 141-152, 1985.
2. Shigeo Abe, Andrew H. Gee, "Global Convergence of the Hopfield Neural Network with Nonzero Diagnal Elements," IEEE Transactions on Circuits and Systems II: Analog and Digital Signal Processing, Vol. 42, No. 1, pp. 39-45, 1995.
3. K. Tanabe, Geometric Method in Nonliner Programming, "Journal of Optimization Theoryand Applications," Vol. 30, No. 2, pp. 181-210, 1980.
4. Sun-Gi Hong, Sung-Woo Kim and Ju-Jang Lee, "The Minimum Cost Path Finding Algorithm Using a Hopfield Type Neural Network," Proceedings of the 1995 IEEE International Conference on Fuzzy Systems, Part 4, pp. 1719-1726, 1995.
5. Trenton Haines and Juraj V. Medanic, "A Neural Network Shortest Path Algorithm," Proceedings of the 1994 IEEE International Symposium on Intelligent Control, pp. 382-387, 1994.
6. S. Cavalieri, A. Di Stefano and O. Mirabella, "Optimal Path Determination in a Graph by Hopfield Neural Network," Neural Networks, Vol 7, No. 2, pp. 397-404, 1994.

Optimizing RBF Networks with Cooperative/Competitive Evolution of Units and Fuzzy Rules

A.J. Rivera[1], J. Ortega[2], I. Rojas[2], A. Prieto[2]

[1]Departamento de Informática. Universidad de Jaén.
arivera@ujaen.es
[2]Departamento de Arquitectura y Tecnología de Computadores. Universidad de Granada
{julio,ignacio,aprieto}@atc.ugr.es

Abstract. This paper presents a new evolutionary method to design optimal networks of Radial Basis Functions (RBFs). The main characteristics of this method lie in the estimation of the fitness of a neurone in the population and in the choice of the operator to apply; for this latter objective, a set of fuzzy rules is used. Thus, the estimation of the fitness considered here, is done by considering three main factors: the weight of the neuron in the RBF Network, the overlapping among neurons, and the distances from neurons to the points where the approximation is worst. These factors allow us to define a fitness function in which concepts such as cooperation, speciation, and niching are taken into account. These three factors are also used as linguistic variables in a fuzzy logic system to choose the operator to apply. The proposed method has been tested with the Mackey-Glass series.

Introduction

This paper deals with the design of neural networks, more specifically networks of Radial Basis Functions [2,3,4], by using evolutionary algorithms. A Radial Basis Function Network (RBFN) implements the mapping $f(x)$ from R^n onto R:

$$f(x) \approx f'(x) = w_0 + \sum_{i=1}^{m} w_i \phi_i(x) \quad (1)$$

where $x \in R^n$, and the m RBF functions ϕ_i have the form $\phi_i = \phi(|x - c_i|/ d_i)$ with a scaling factor $d_i \in R$ and a centre $c_i \in R^n$. Among the possible choices for ϕ, in this paper we consider gaussian RBFs, $\phi(r)=exp(r^2/2)$, where r is the scaled radius, $|x - c|/ d$, and $|x - c|$ is a euclidean norm on R^n.

RBFs were originally proposed as an interpolation method. In this context, if the value of f is known at p data points $x_1, \ldots, x_p$, a RBF can be centred on one data point. In the context of neural networks, on the other hand, it is commonly assumed that there are significantly fewer basis functions than data points. The central problem is the placement of the centres and the determination of the widths d_i and weights, to achieve the best approximation and generalization performance. The optimization of a RBFN can be considered as a multicriteria optimization problem [12] in which the

J. Mira and A. Prieto (Eds.): IWANN 2001, LNCS 2084, pp. 570-578, 2001.

two criteria are the complexity of the network, measured as the number of parameters, and its performance evaluated as the approximation error. The research work done in the field of evolutionary computation to optimize neural networks can be classified according to the following major trends:

- *Competition among neural networks.*The population is a set of whole neural networks that compete by using a fitness function corresponding to the performance of each network. Alternatives within this general procedure include: (a) to evolve only the structural specification of the untrained networks, and use conventional non-evolutionary learning algorithms to train the networks [7], and (b) to use evolutionary algorithms to determine both the weights and the structure of the network [6]. In this case, no additional training is required but the search space of possible neural networks is much larger. Other methods use a compromise between the two previous approaches [13].
- *Incremental/Decremental algorithms.* These are based on adding [8] and/or deleting neurones [9] in the hidden layer one by one. These methods, by optimizing one unit at a time, can become trapped in local optima.

Evolving cooperating and competing units. Each individual in the population is a RBF of the network instead of a population of competing networks, and the neurons reproduce or die depending on their performance [5]. Thus, the neurons of the population work together to determine a neural network that performs better with successive generations. Two problems must be solved in order to reach this goal. The first is the credit apportionment problem [10], which involves determining a performance measure for each individual in the population that reflects the performance of the whole population; this is the measure that has to be optimized, and which can be evaluated. The second problem is that of niching [14], which implies maintaining a population of neurons evolving to different parts of the space of solutions. Thus, the individuals in the population have to both cooperate and compete. In [5], the credit assigned to each RBF is based on the contribution of this RBF to the overall performance of the network, and niche creation is implicitly obtained [11] by changing the intensity of competition between neurons according to the degree of overlap in their activations.

The procedure proposed here can be included in this latter group of methods. It uses a population of Radial Basis Functions that cooperate to build a neural network that performs sufficiently well for the application considered, and which compete in order to obtain a set of RBFs distributed across the whole input domain of the application. The procedure selects the group among the worst neurons (RBFs) and applies them an operator chosen from a given set of operators by using fuzzy rules to determine if the RBFs are created, deleted or adjusted. These transformations can be applied simultaneously, and several neurons can appear and/or disappear at the same time, in contrast to the incremental/decremental methods. The procedure differs from other algorithms of the same type previously proposed in the set of transformations that are used to accomplish the evolution in the set of RBFs. Moreover, the procedure has been devised with the goal of an efficient parallel and distributed implementation.

This algorithm is an improvement on the algorithm we propose in [15]. In the present procedure we have simplified the way of selecting the operators to apply to the current solution are selected and better results are achieved.

In the next Section a description of the procedure is provided. Then Section 3 describes the experiments that have been implemented and the experimental results obtained, and Section 4 gives the conclusions of the paper.

2. Description of the Algorithm

In this Section, the algorithm used to determine an optimal network of Radial Basis Functions (RBFN) is presented. The steps of the algorithm are presented in Figure 1.

```
1. Initialize the RBFN
2. Train the RBFN
3. Evaluate fitness of RBFs
4. Select the r worst RBFs
5. Apply operators to RBFs
6. Add RBFs to worst isolated points
7. Train the RBFN
8. If End condition is not verified go 3
   else terminate
```

Fig. 1. RBF network optimization algorithm

Step 1
In the first step, the algorithm builds an RBFN. This is done by initializing a population of r_{init} RBFs with centres allocated to randomly selected points and widths set to a given initial value.

Step 2 (and 7)
After determining the population of RBFs, a given number of training iterations are applied by using the LMS algorithm, to adapt the weights of the RBFN. The value of α in the LMS algorithm is modified during the training iterations by decreasing it after applying the training set n times.

Step 3
The next step of the algorithm described in Figure 1 corresponds to the evaluation of the fitness of each RBF. This fitness is determined by considering three characteristics of the corresponding RBF: its weight in the present network of RBFs, its distance from badly predicted points, and the distances from other RBFs in the network.

The effect, e_i, of the value of the weight associated to a RBF is taken into account by using a function of this weight, w_i, as follows

$$e_i = f_u(|w_i|) \tag{2}$$

where

$$f_u(x) = \begin{cases} x & \text{if } x \leq u \\ u & \text{otherwise} \end{cases} \tag{3}$$

The influence on the fitness of the closeness of the neurone i to the p worst approximated points of the training set is quantified by the parameter m_i , defined by

$$m_i = 1 + \sum_{j}^{p} m_{ij} \tag{4}$$

where m_{ij} is related to the closeness of ϕ_i and the point j (pt_j) by the expression

$$m_{ij} = \begin{cases} (1-(D(\phi_i, pt_j)/d_i)) & \text{if } D(\phi_i, pt_j) < d_i \\ 0 & \text{otherwise} \end{cases} \quad (5)$$

The overlapping of RBFs is quantified by a function s_i which is assigned to the RBF i and is defined as:

$$s_i = 1 + \sum_j^m s_{ij} \quad (6)$$

where m is the number of RBFs and s_{ij} measures the overlapping between the RBFs i and j, (ϕ_i and ϕ_j):

$$s_{ij} = \begin{cases} (1-(D(\phi_i, \phi_j)/d_i)) & \text{if } D(\phi_i, \phi_j) < d_i \\ 0 & \text{otherwise} \end{cases} \quad (7)$$

Finally, by using these three parameters the fitness associated to each RBF ϕ_i is:

$$fitness(\phi_i) = e_i/(s_i * m_i) \quad (8)$$

Step 4
Before applying these operators it is necessary to select the RBFs that are going to be modified. The set of candidate RBFs is built with the r RBFs having the worst *fitness*. The cardinal, r, of the candidate RBFs is taken as a small value, to keep a parsimonious evolution of the RBF network in order to maintain the main characteristics of the network behaviour. The parameter r is also adaptive and changes with the number or iterations of the algorithm. The p worst approximated points are also determined in this step.

Step 5
A set of operators is applied. These operators are the following:
- *OP1 (RBF creation)*: A new RBF is added in a given zone where there is one of the p worst approximated points. First, the width, of the new RBF, is set to *er*. The centre of the new RBF is set to the coordinates of the selected point and is modified by applying a randomly selected perturbation lower than *pc1*(a given small percentage of *er*). To determine the final width of the RBF, *er* is modified by applying another random perturbation less than *pc1*, taking into account the closeness of the rest of the RBFs in the variable c.

$$c = 1 + \sum_j^m c_j \quad (9)$$

where c_j is the distance between the new RBF ϕ_n and the RBF j and m is the number of RBFs

$$c_j = \begin{cases} (1-(D(\phi_n,\phi_j)/d_i))/g & \text{if } D(\phi_n,\phi_j) < d_i \\ 0 & \text{otherwise} \end{cases} \quad \mathbf{(10)}$$

where g is set to a small value (for example 1.5 or 2). The final width d_n is set to:

$$d_n = er / c \quad \mathbf{(11)}$$

- *OP2 (RBF elimination)*: A selected RBF is pruned.
- *OP3 (RBF small adjust)*: The centre and the width of a selected RBF are adjusted by a small percentage of the width in the following way. For each point of the training set inside the width of the RBF, the error *err* (difference between *f(x)* and *f'(x)*) is calculated, and the rules of Table 1 are applied to adjust the centre and the width of the RBF.

Table 1. Rules used to adjust a RBF

	$err > 0$	$err < 0$
$w_i>0$	Close the RBF to the point and increase the width	Move the RBF away from the point and decrease the width
$w_i<0$	Move the RBF away from the point and decrease the width	Close the RBF to the point and increase the width

Finally, the centre and width are changed by adding a randomly selected amount less than the percentage *pc2* of the initial width.
To choose the operator to apply, a set of fuzzy rules are applied by considering three parameters for each RBF ϕ_i (e_i, m_i, s_i), and three linguistic variables, one for each parameter (*ve* for *e*, *vm* for *m*, and *vs* for *s*). The membership functions for the linguistic variable are represented in Figure 2.

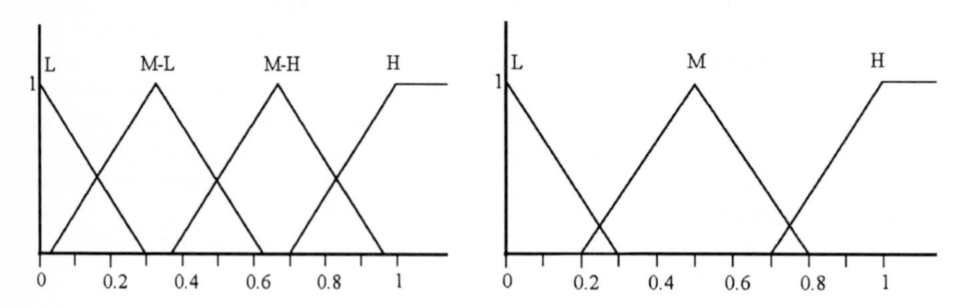

Fig. 2. Membership functions for *ve* (left) and for vm and vs (right)

The consequent variables of the rules are *vop1*, *vop2* and *vop3,* and have the same membership functions as *ve* (Figure 2, left). The fuzzy rules are given in Table 2.

Table 2. Set of rules of the fuzzy inference system

Antecedents			Consequences			Antecedents			Consequences		
ve	*Vm*	*vs*	*vop1*	*vop2*	*vop3*	*ve*	*vm*	*vs*	*vop1*	*vop2*	*vop3*
L	L	L	M-H	L	M-L	M-H	L	L	L	L	H
L	L	M	M-H	M-L	M-L	M-H	L	M	L	M-L	H
L	L	H	M-H	L	M-L	M-H	L	H	L	M-L	H
L	M	L	M-H	L	M-L	M-H	M	L	L	L	H
L	M	M	M-H	M-L	M-L	M-H	M	M	L	M-L	H
L	M	H	M-H	L	M-L	M-H	M	H	L	M-L	H
L	H	L	H	L	M-L	M-H	H	L	L	L	H
L	H	M	H	L	M-L	M-H	H	M	L	L	H
L	H	H	H	L	M-L	M-H	H	H	L	L	H
M-L	L	L	M-L	L	M-H	H	L	L	L	L	H
M-L	L	M	M-L	M-L	M-H	H	L	M	L	L	H
M-L	L	H	M-L	M-L	M-H	H	L	H	L	M-L	H
M-L	M	L	M-L	L	M-H	H	M	L	L	L	H
M-L	M	M	M-L	M-L	M-H	H	M	M	L	L	H
M-L	M	H	M-L	M-L	M-H	H	M	H	L	M-L	H
M-L	H	L	M-H	L	M-H	H	H	L	M-H	L	H
M-L	H	M	M-H	L	M-H	H	H	M	M-H	L	H
M-L	H	H	M-H	L	M-H	H	H	H	M-H	L	H

The reasoning mechanism used to derive a reasonable output or conclusion from the rules and a given condition is the Mamdani Fuzzy Model. In this type of fuzzy inference system, min and max are chosen as fuzzy AND and OR operators. To extract a crisp value from a fuzzy set, the centroid area is used.

From this fuzzy inference system three crisp values are obtained that correspond to the probabilities for adding a new RBF, deleting an RBF and modifying an RBF.

Step 6

In this step, less than *ni* RBFs are randomly created centred on points that are outside the width/1.5 zone of any RBF (we call them *isolated points*). The parameter *ni* is adaptive and depends on the number of RBFs

3. Experimental Results

The algorithm has been applied to a time series prediction problem. The time series used is the well-known and frequently used Mackey-Glass series, which is generated by the following equation:

$$\frac{dx(t)}{dt} = -bx(t) + \frac{ax(t-\tau)}{1+x(t-\tau)^{10}} \tag{12}$$

where $a = 0.2$, $b = -0.1$ y $\tau = 17$. As mentioned by Martinez et al.[1] this series is quasi-periodic and chaotic for the above parameters. The data for the Mackey-Glass time series was obtained by applying the integration Runge-Kutta method, with the initial condition x(t)=0.3 for $0<t<\tau$. The inputs to the network of RBFs consist of four lagged data points, x(t), x(t-6), x(t-12), and x(t-18) and the output is x(t+85). The training data were extracted from the points 4000 to 4500, and the following 500 data points were used as testing data.

The normalized root-mean-square (RMS) error E was used to evaluate the performance of the algorithm, which is determined by the RMS value of the absolute prediction error divided by the standard deviation of $x(t)$. We ran 100 iterations of our algorithm over this benchmark with the values for the static parameters provided in Table 3.

Table 3. Values for static parameters

Parameter	Value
p	Initial number of RBFs
er	0.25
pc1	15%
g	1.5

The adaptive parameters *r* and *pc2* follow a lineal progression. The number of worst RBFs, *r*, is calculated as:

$$r = number\_RBFs / dn \tag{13}$$

where *dn* is also adaptive and depends on the number of iterations. Thus, *dn* is 4 for 0 iterations and increases linearly up to 7 for 100 iterations. The value of *pc2* is set to 1 when the RMSE of the RBFN is 0.05 and rises to 10 when the RMSE is 0.5. The algorithm was executed several times with initial population sizes of 15, 20 and 25 RBFs. As can be seen, the results improve on those provided in [15].

Table 4. Parameters and results

	[15]		Here Proposed	
RBFs	Best Fitness	Av. Fitness	Best Fitness	Av. Fitness
10	0.410	0.465	0.423	0.476
12	0.375	0.446	0.360	0.408
15	0.280	0.376	0.304	0.351
18	0.256	0.331	0.271	0.316
20	0.262	0.312	0.257	0.294
22	0.261	0.296	0.246	0.280
25	0.252	0.285	0.222	0.261
28	0.251	0.268	0.215	0.246
30	0.259	0.266	0.209	0.240

The procedure was compared with other procedures using the same benchmark [5]. The results for the normalized RMS provided by two of these algorithms are given in Table 5. As can be seen, the procedure proposed here improves on the performance of both methods.

Table 5. Results for other algorithms

Algorithm	RBFs	Normalized RMS
Fitness sharing [5]	25	0.29
K-means clustering (taken from [5])	25	0.53

4. Conclusions

This paper presents a new evolutionary procedure to design networks of radial basis functions. The procedure describes a new method to calculate the fitness of a given RBF according to its contribution to the final performance of the neural network, thus proposing a solution to the credit assignment problem. Here, the fitness depends on three factors: the weight of the RBF, the closeness to other RBFs and the closeness to badly predicted points. Other characteristics and elements of the procedure are a set of rules which control the application of some transformations to the RBFs, the dynamic-adaptive character of the parameters, and the parsimonious evolution of the behaviour of the network as the procedure advances.

Good experimental results were obtained, and the method performs better than other, more mature methods, previously proposed. Future research aims are the relations between the major parameters of this algorithm, i.e. the parameters that calculate the fitness and the design of the set of rules. Moreover, as the algorithm has a distributed character, we consider that a parallel version could significantly improve the speed of the procedure.

Acknowledgements. This paper has been supported by project TIC2000-1348 of the Spanish *Ministerio de Ciencia y Tecnología*.

References

[1] M. C. Mackey and L. Glass, "Oscillation and chaos in physiological control system" Sci vol. 197, pp. 287-289, 1.977.

[2] J. Platt, "A resource-allocating network for function interpolation", Neural Computation 3, 213-225. 1.991

[3] L. Yingwei, N. Sundararajan, P. Saratchandran, "A sequential learning scheme for function approximation using minimal radial basis function neural networks." Neural Computation 9, 461-478. 1.997

[4] J. Moody and C. Darken, "Fast learning networks of locally-tuned processing units," Neural Computation, vol. 3 n. 4 pp.579-588, 1.991

[5] Whitehead, B.A.; Choate, T.D.:"Cooperative-competitive genetic evolution of Radial Basis Function centers and widths for time series prediction". IEEE Trans. on Neural Networks, Vol.7, No.4, pp.869-880. July, 1996.

[6] Angeline, P.J.; Saunders, G.M.; Pollack, J.B.:"An evolutionary algorithm that constructs recurrent neural networks". IEEE Trans. on Neural Networks, Vol.5, No.1, pp.54-65. January, 1994.

[7] Maniezzo, V.:"Genetic evolution of the topology and weight distribution of neural networks". IEEE Trans. on Neural Networks, Vol.5, No.1, pp.39-53. January, 1994.

[8] Fahlman, S.E.; Lebiere, C.:"The cascade-correlation learning architecture". In *Advances in Neural Information Processing Systems, 2,* Lippmann, P.; Moody, J.E.; and Touretzky, D.S. (Eds.), Morgan Kaufmann, pp.524-532, 1991.
[9] Hwang, J.-N.; Lay, S.-R.; Maechler, M.; Martin, R.D.; Schiemert, J.:"Regression modeling in backpropagation and projection pursuit learning". IEEE Trans. on Neural Networks, Vol.5, No.3, pp.342-353. March, 1994.
[10] Smalz, R.; Conrad, M.:"Combining evolution with credit apportionment: a new learning algorithm for neural nets". Neural Netwroks, Vol.7, No.2, pp.341-351, 1994.
[11] Horn, J.; Goldberg, D.E.; Deb, K.:"Implicit niching in learning classifier system: Nature's way". Evolutionary Computation, Vol.2, No.1, pp.37-66, 1994.
[12] Coello, C.C.:"An Updated Survey of Evolutionary Multiobjective Optimization Techniques: State of the art and future trends". Congress on Evolutionary Computation, CEC'99, pp.3-13, 1999.
[13] Whitehead, B.A.; Choate, T.D.:"Evolving space-filling curves to distribute radial basis functions over an input space". IEEE Trans. on Neural Networks, Vol.5, No.1, pp.15-23. January, 1994.
[14] Sareni, B.; Krähenbühl, L.:"Fitness sharing and niching methods revisited". IEEE Trans. on Evolutionary Computation, Vol.2, No.3, pp.97-106. September, 1998.
[15] Rivera, A.;Ortega, J.; Prieto A.:"Design of RBF networks by cooperative/competitive evolution of units" International Conference on Artificial Neural Networks and Genetic Algorithms, ICANNGA 2001. April 2001.

Study of Chaos in a Simple Discrete Recurrence Neural Network

Jose D. Piñeiro, Roberto L. Marichal, Lorenzo Moreno, Jose F. Sigut, and Evelio J. González

{jpineiro, rlmarpla, lmoreno, jfsigut, ejgonzal}@ull.es
Department of Applied Physics. University of La Laguna.
La Laguna 38271, Tenerife, Spain.
Tel: +34 922 318286 / Fax: +34 922 319085

Abstract. A simple class of discrete recurrent neural network is analyzed to establish its possible chaotic dynamics. A two-neuron network is selected for simplicity and two cases of chaotic dynamics are distinguished, one degenerate (one-dimensional) and a more general situation with a two-dimensional attractor. The robustness of the found configurations is assessed through evaluation of Lyapunov exponents in a range of parameter values. In every situation there is a lack of robustness, as suggested in a conjecture of Barreto et al.

1 Introduction

The objective of this work is to present some results in order to analyze the chaotic dynamics of a discrete recurrent neural network. The particular model of network in which we are interested is the Williams-Zipser network, also known as Discrete-Time Recurrent Neural Network (DTRNN) in [1]. Its state evolution equation is

$$y_i(k+1) = f\left(\sum_{n=1}^{N} w_{in} y_n(k) + \sum_{m=1}^{M} w'_{im} u_m(k) + w''_i\right) \tag{1}$$

where
$y_i(k)$ is the ith neuron output
$u_m(k)$ is the mth input of the network
w_{in}, w'_{im} are the weight factors of the neuron outputs, network inputs and w''_i is a bias weight.
$f(\cdot)$ is a continuous, bounded, monotonically increasing function such as the hyperbolic tangent.

The problem of chaos in neural networks has received much attention lately. In particular, see [2] for a study of chaos "robustness" on neural networks and recent references on properties of these systems, such as studies of routes to chaos and properties of chaotic attractors.

Our motivation for studying this kind of systems is the possibility of using them as temporal pattern detectors in biomedical systems [3]. More generally, the applications

J. Mira and A. Prieto (Eds.): IWANN 2001, LNCS 2084, pp. 579-585, 2001.

of dynamic neural network systems cover a wide range of possibilities. They have been used, for example, in modeling dynamic processes, speech recognition, in signal processing applications or even as a source of pseudorandom numbers (of great interest in cryptography).

Although a dynamic network of this kind can successfully represent complex temporal patterns, several drawbacks have been pointed out in the literature. These are related to the training process. One of them is the problem of learning long-term dependencies in data [4]. Other problems will be related to the presence of chaos and the sensibility with initial conditions and numerical errors that it induces. In other paper [5] we have studied the capabilities of these systems, characterising the possible dynamics of a simple case of this kind of network. For this objective the strategy was to determine different classes of equivalent dynamics. A system will be equivalent to another if its trajectories have the same qualitative behaviour. This is made precise mathematically in the definition of topological equivalence [6]. The simplest trajectories are those who are points of equilibrium or fixed points that do not change in time. Their character or stability is given by the local behaviour of nearby trajectories. A fixed point can attract (sink), repel (source) or have directions of attraction and repulsion (saddle) of close trajectories [7]. Next in complexity are periodic trajectories, quasiperiodic trajectories or even chaotic sets, which are the main objective of this work. All this features are similar in a class of topologically equivalent systems. When a system parameter is varied the system can reach a critical point in which there are qualitative changes in its properties. This is called a bifurcation, and the system will exhibit new behaviors. The study of how these changes can be carried out is a powerful tool in the previously mentioned analysis.

A single neuron of this kind is not capable of having chaotic dynamics. It is necessary al least two neurons to show interesting dynamics. In [8], chaos is found in a simple two-neuron network in a specific (degenerate) weight configuration by demonstrating its equivalence with a 1-dimensional chaotic system (the logistic map). We will begin by reviewing these results for a two-neuron system in the degenerate case (Jacobian determinant equals zero) and continue with the general situation, identifying the conditions in which chaos is found and assessing its robustness by means of Lyapunov exponents. In the following, we will ignore the input in the model proposed and will consider for simplicity zero bias weigths.

2 Chaos in Two-Neuron Network: Degenerate Case

We will consider the case in which the transformation defining the dynamical system has a Jacobian determinant zero. To be more concrete, suppose a weight matrix like

$$W = \begin{bmatrix} a & ka \\ b & kb \end{bmatrix} \tag{2}$$

With this weights, it can be easily shown that the mapping (1) of two neurons is equivalent to a one-dimensional map f on a interval $[z_0, z_1]$ which is defined as [8]

$$f(z_{i+1}) = tanh(\mu a z_i) + k tanh(\mu b z_i) \tag{3}$$

where $z_0 = -1 - |k|$ and $z_1 = 1 + |k|$

It can be demostrated that this last mapping is topologically equivalent to a class of well-known unimodal maps such as the logistic map [7]. With this equivalency, it can be expected to find similar behaviors and, in particular, similar ways of finding chaotic dynamics ("routes to chaos") by varying its parameters.

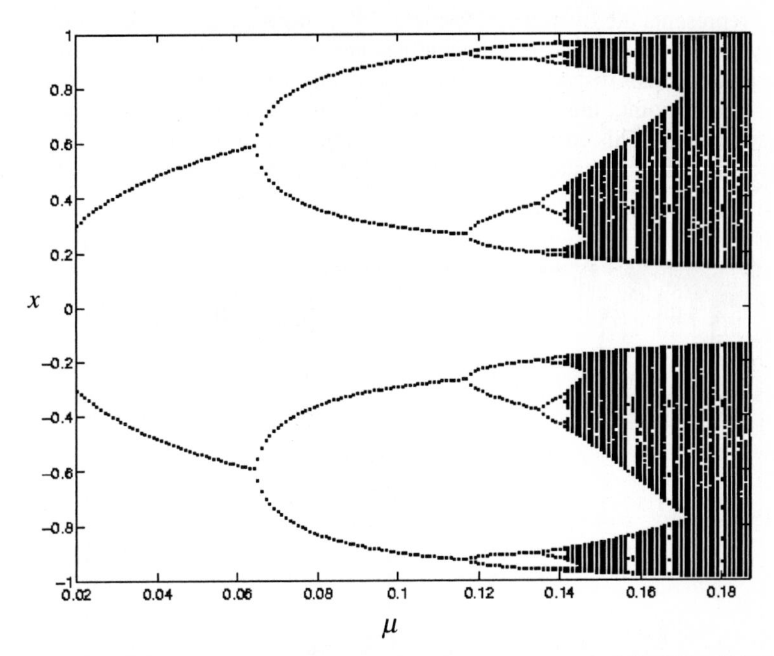

Fig. 1. Period-doubling route to chaos. A cascade of successive period-doubling bifurcations is observed as the parameter is changed, ending with a continuum of equilibria which denotes the chaotic dynamics. Interspersed within the region of chaotic dynamics, there are parameter values with give dynamics with periodic trajectories (windows)

Fig. 1 shows one of these routes to chaos (period-doubling cascade). This diagram is obtained by plotting the final points of many simulations versus a parameter value. It can be observed that the equilibrium points of the trajectories seem to bifurcate (actually, a periodic point is substituted by two new points of double periodicity) until a continuum is reached. This is the chaotic attractor, which is a much more complex equilibrium set. There are "windows" of periodicity intermixed with chaotic parameter values. In many cases, both dynamics are densely intertwined: small changes to the parameter can destroy the chaotic dynamics to give place to a periodic trajectory. On the contrary, when small changes in the parameter cannot alter substantially the chaotic dynamics, the chaos is said to be *robust*.

In [2], a similar case is studied in a somewhat more particular situation. A new neuron capable of displaying chaos is defined that corresponds with the mapping (3) if the parameters a y b are made equal.

One of the hallmarks of chaotic dynamics is the sensitive dependence with initial conditions. The Lyapunov exponents are magnitudes that measure how two close initial states diverge with time. The definition and value for the one-dimensional case are given in (4). For a chaotic trajectory, this number is positive. If it is negative, it implies a stable trajectory towards which nearby points tend to converge.

$$L = \lim_{n\to\infty} \frac{1}{n}\ln\sum_{j=1}^{n}\left|f'(z_j)\right| = \lim_{n\to\infty} \frac{1}{n}\ln\left(\sum_{j=1}^{n}\left|a(1-\tanh(az_j)^2)+kb(1-\tanh(bz_j)^2)\right|\right) \tag{4}$$

Fig. 2 represents the Lyapunov exponent versus the *a* parameter. It can be seen that for many parameter values, the exponent becomes negative, indicating stable periodic trajectories and chaos dissapearing. For certain values, those in which the derivative is zero for some point, the Lyapunov exponent is undefined (represented as large downward spikes). This corresponds to superstable trajectories. The specific values of parameters in which this situation occurs are called a *spine locus* in parameter space.

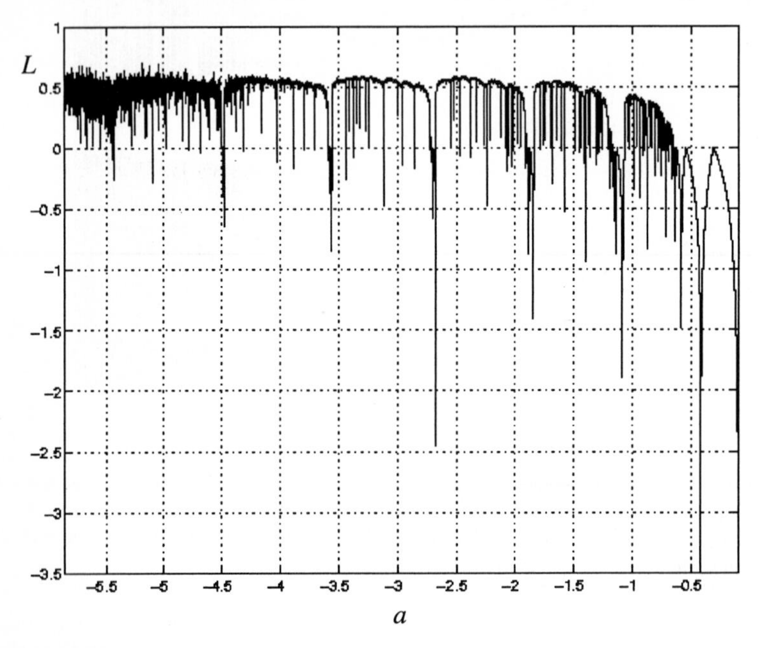

Fig. 2. Lyapunov exponent (L) as the parameter a is varied. It can be observed many downward spikes reaching negative values denoting non-chaotic dynamics mixed with regions with positive values

This experimental result is coherent with a conjecture made by Barreto et al. in [9]. According to it, if a chaotic attractor has k positive Lyapunov exponents system and there are n parameters that can be varied, then typically a slight variation of n parameters can destroy the chaos if $n \geq k$. If $n < k$ then the chaos typically cannot be so destroyed. For this one-dimensional system, with a single Lyapunov exponent, the chaos will typically be destroyed by varying a single parameter like Fig. 2 shows.

3 Chaos in Two Neuron Network: Non-degenerate Case

In this case we have a true two-dimensional system (it cannot be reduced to an equivalent monodimensional system) because its associated Jacobian is typically full rank. This situation can be derived from the former by perturbing the weight matrix slightly to make it nonsingular. Fig. 3 shows the shape of the attractor in two-dimensional state space. There are three saddle equilibrium points, marked by circles in the figure. The straight lines are the stable manifolds of each saddle. These are the points in state space which eventually will converge to the corresponding saddle point. The curved lines are the unstable manifolds of the lower and upper saddle points (lower half and upper half respectively) and of the saddle at the origin (both halves). Conversely, the unstable manifolds are points that will converge to the corresponding saddle by the inverse mapping. These manifolds play an important part to establish the existence of chaotic attractors. Since these manifolds appear to cross transversally, it can be expected that this be a Horseshoe-type of chaos, with both homoclinic and heteroclinic crossings, see [7]. For this dynamic, the unstable manifold constitutes the chaotic attractor.

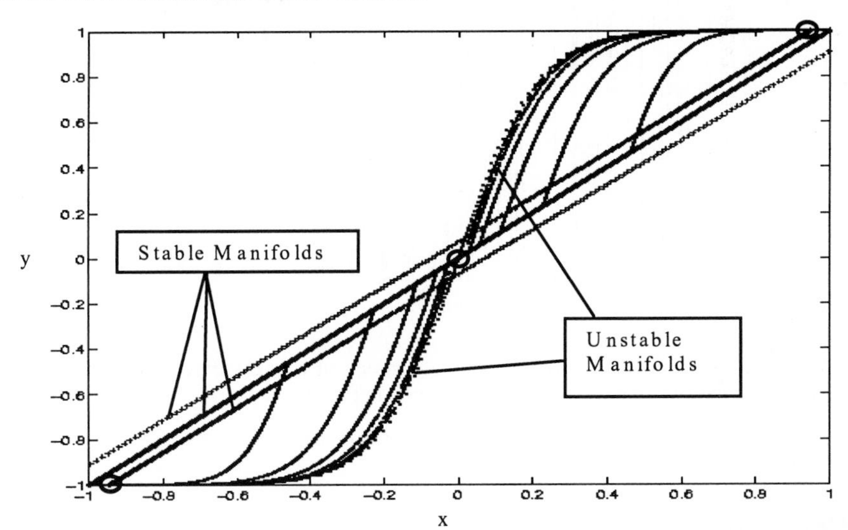

Fig. 3. Chaotic configuration with parameters (w_{11}=- 25 ,w_{12}= 25.5 ,w_{21}= -124.5, w_{22}= 125). Circles represent the saddle points with their stable and unstable manifolds

In Fig. 3 it can be seen that the extreme saddle fixed points are situated in a state space point that makes the Jacobian determinant zero.

$$|J(x_p, y_p)| = |W|\left(1 - tanh(w_{11}x_p + w_{12}y_p)^2\right)\left(1 - tanh(w_{21}x_p + w_{22}y_p)^2\right) = 0 \quad (5)$$

where x_p, y_p are the saddle point coordinates and $|W|$ the weight matrix determinant.

In principle, the weight matrix determinant is not zero and no finite values of weights and state space points can cancel out the other factors. In practice, the finite numerical accuracy in the determination of the hyperbolic tangent function makes zero one of these terms and this is indeed now the case for the IEEE double precision arithmetic used. This has the implication of making the mapping non-invertible and gives birth to the multiple branches of the unstable manifold (chaotic attractor).

The conditions for the onset of chaos are multiple: it is necessary the existence of three saddles and the contact condition of the two extreme saddles for non-invertibility. The determination of the exponents of Lyapunov in two dimensions is more involved. In this work we have used the QR decomposition algorithm [10]. One of the Lyapunov exponents is always negative. Fig. 4 shows the other exponent, demonstrating the lack of robustness of the attractor.

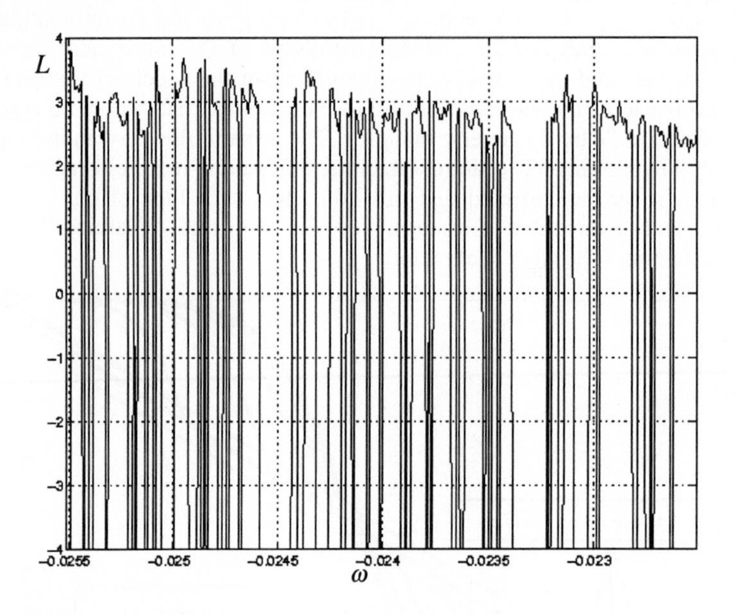

Fig. 4. Lyapunov exponent (*L*) as the weights are varied in a linear trajectory in parameter space. Here the stability windows are wider than those in Fig. 3

4 Conclusion

The chaotic dynamic of a simple model of a recurrent discrete neural network has been analyzed. Two different chaotic behaviors have been identified. In both cases there is a lack of robustness, that is, slight perturbations of the parameters take the system out from the chaotic situation. The two-neuron network discussed are quite simple, but they are potentially useful since the complexity found in these simple cases might be carried over to larger discrete recurrent neural networks. There exists the possibility of generalising some of these results to higher dimensions and use them to design training algorithms that avoid the problems associated with the learning process. In particular, the results on robustness can be of utility when traversing the parameter space in the process of training.

References

1. Hush, R., Horne, G.B.: Progress in supervised Neural Networks. IEEE Signal Processing Magazine (1993) 8-39
2. Potapov, A., Ali, M.K.:Robust Chaos in Neural Networks. Physics Letters A 277 (2000) 310-322
3. Piñeiro, J. D., Marichal, R.L., Moreno, L., Sigut, J., Estévez, I., Aguilar, R., Sánchez, J.L., Merino J.: Evoked Potential Feature Detection with Recurrent Dynamic Neural Networks. International ICSC/IFAC Symposium on Neural Computation 98 (1998)
4. Bengio, Y., Simard, P. Frasconi, P.: Learning Long-Term Dependences with Gradient Descent is Difficult. IEEE Trans. on Neural Networks 5 (1994) 157-166
5. Piñeiro, J. D., Marichal, R.L., Moreno, L., Sigut, J., Estévez, I., Aguilar, R.: Dynamics of a Small Discrete Recurrent Neural Network. 2nd ICSC Symposium on Neural Computation 2000 (2000)
6. Kuznetsov, Y.A.: Elements of Applied Bifurcation Theory. Applied Mathematical Sciences, Vol. 112. 2nd Ed. Springer-Verlag, New York (1998)
7. Robinson, C.: Dynamical Systems: Stability, Symbolic Dynamics, and Chaos. CRC Press (1995)
8. Wang, X.: Period-Doublings to Chaos in a Simple Neural Network: An Analytical Proof. Complex Systems, 5. (1991) 425-441
9. Barreto, E., Hunt, B.R., Grebogi, C., Yorke, J.A.: From High Dimensional Chaos to Stable Periodic Orbits: The Structure of Parameter Space. Phys. Rev. Lett. 78 (1997) 4561-4564
10. Eckmann, J.P., Ruelle, D.: Ergodic Theory of Chaos. Rev. Mod. Phys. Vol. 57 (1985) 617-655

Genetic Algorithm versus Scatter Search and Solving Hard MAX-W-SAT Problems

Habiba Drias

USTHB, Electrical and Computer Engineering Faculty,
Computer Science Department,
BP 32 El-Alia Bab-Ezzouar, 16111 Alger
Drias@wissal.dz

Abstract. The genetic algorithm approach is well mastered nowadays. It has been applied to a diverse array of classical and real world problems such as the time-tabling and the traveling salesman problems. The scatter search, yet another evolutionary approach has been developed recently. Some known versions have been tested on problems such as the quadratic assignment and the linear ordering problems.
In this paper, both approaches are studied for the NP-hard satisfiability problem namely SAT and its optimization version MAX-W-SAT. Empirical tests are performed on Johnson benchmarks and the numerical results show a slight performance difference in favor of the scatter search.

1 Introduction

The satisfiability problem or SAT for short has known an imposing interest in research these last decades. It has been studied on both the theoretical and practical viewpoints, in computational complexity, in problem solving and in first order, modal and temporal logics settings. SAT has also known applications to concrete problems. Detection and correction of anomalies appearing in a data or knowledge base can be evoked as an example.
The evolutionary population based approaches have a significant number of similarities. They are qualified in the literature as Iterative Population Improvement Methods (IPI). The main functioning of these methods consists in improving a population in each generation by combining its individuals or solutions to create new ones and replacing less adapted individuals by better ones. Genetic algorithms and scatter search, which are part of the subject of this work, belong to the IPI method family.
Several genetic algorithms have been conceived for solving the satisfiability problem [3,4,6]. The fitness function, the genetic operators and sometimes the solution representation are designed differently from one study to another and thus lead to distinct performances.
Scatter search is a recent evolutionary approach although it argues its principle from a formulation that dates from 1977 [5]. Scatter search belongs to the population based metaheuristics family. As in genetic algorithms, it is concerned by defining components such as the solution representation, the initial population, the fitness function and the combination operator. The major difference with the genetic algorithms resides in the way of combining solutions to create new ones. In genetic algorithms, the operator namely crossover respects somehow the principle of the

J. Mira and A. Prieto (Eds.): IWANN 2001, LNCS 2084, pp. 586-593, 2001.

natural genetic reproduction of chromosomes. Often, it is based on probabilistic permutation of chromosomes of the parents solutions. In scatter search, the combination operator is defined on a unified principle and does not resort to randomization. It uses combination of solutions vectors inspired from the surrogate constraint method where decision rules and constraints are combined. Some scatter search methodologies have been developed for problems such as the linear ordering problem [1,2].
In this study, two evolutionary procedures corresponding respectively to the genetic algorithms and the scatter search approaches are designed and implemented for the maximum weighted satisfiability problem. Extensive experiments are performed on a benchmark available on a web site. Comparison is done on the empirical results that specify the performance of the procedures.

1.1 SAT and MAX-W-SAT Problems

In the propositional setting, the problem SAT is theoretically well defined. It consists in specifying whether a logical formula is satisfied by at least one truth interpretation of its variables. For a simplification concern, the formula is usually presented in a conjunctive normal form, that is as a logical conjunction of clauses, a clause being a disjunction of literals. A literal is a variable that can be negated. A solution to the problem is a truth assignment of variables that satisfies all clauses. If such a solution does not exists, one can take an interest in searching the maximum number of clauses that can be satisfied. The definition of the optimization version of SAT is related to this problem. Formally, the decision problem is stated as a pair of a set of clauses and a parameter k and a question asking whether there exists at least k clauses that are simultaneously satisfied.

2 The Genetic Algorithm

As indicated by its name, the genetic algorithm or GA for short traces its framework from the evolution mechanism of biological organisms population. The strongest and the most adapted organisms impose themselves in the population according to the natural selection principles. An organism is represented by its chromosomes set. A GA is an IPI method where the population is made of solutions called individuals. A solution to a given problem is coded upon a scheme, which is an important factor of the GA efficiency. In the SAT problem, the obvious way to represent a solution is by its binary code, a solution is a chain of bits called also chromosomes that correspond to the variable values. Some authors have considered another representation and have shown the impact of the coding in the solution quality[6].
The GA starts by creating an initial population often at random and sometimes by means of a heuristic. The population size is a determining parameter set at the experiments step.
The objective or fitness function measures the adaptation level of the individual. In SAT, this function counts the number of clauses satisfied by the solution or the weights sum of the clauses satisfied by the solution in MAX-W-SAT. This function, which is the same for scatter search, makes the only difference in the development of the GA for the problems SAT and MAX-W-SAT.

The selection of individuals to be reproduced translates the probability of reproduction of the individuals. Technically, it might be a random process based on the wheel method, where each individual occupies a sector whose angle is proportional to its fitness function value.
The selected individuals are then paired off ; a good quality individual generally associated with another one drawn at random. The crossover operator consists in exchanging parts of chromosomes of the individuals at the reproduction process. Three types of crossover are used in the SAT problem: the one point, the two point and the crossover with a mask. The choice of the type is tuned by the experiments.
The mutation viewed usually as a background operator consists in modifying one or many chromosomes chosen at random with a probability of mutation called mutation rate. For MAX-W-SAT, this operator is merely the flipping of a bit drawn randomly. In order to prevent the problem of premature convergence due to the similarity of individuals in the population, we apply an adaptive mutation; we decrease the mutation rate during the first iterations of the algorithm where the individuals are likely to be different, so as to speed the process and we increase the mutation rate at the last iterations so as to promote diversification of individuals and escape from local optima.
The final step of the GA is the replacement of the bad individuals of the population by the fittest ones of the offspring generating this way a new population. The whole process is iterated until finding an optimal solution or reaching a maximum number of iterations imposed by the computational limits.

2.1 A Hybrid Genetic Algorithm for SAT and MAX-W-SAT

In order to increase the performance of the GA, the offspring solutions are improved by means of a heuristic called G-bit, which is an optimization method. The latter consists in modifying the bits of the solutions that enhance their quality. G-bit is applied to good solutions, each of them is scanned bit by bit and the current bit is complemented with another one drawn at random. the solution obtained is saved if it is improved otherwise it is rejected. The hybrid genetic algorithm called GA-SAT is written as:

```
Algorithm GA-SAT
Begin   generate at random the initial population;
        while (the maximum number of generations is not reached and
                 the optimal solution is not found) do
                 begin   repeat  select two individuals;
                                 generate at random a number Rc from [0,100];
                                 if Rc < crossover rate then apply the crossover;
                                 generate at random Rm from [0,100];
                                 while Rm < mutation rate do
                                   begin  choose at random a chromosome from
                                          the individual obtained by the crossover
                                          and flip it;
                                          generate at random Rm from [0,100];
                                  end;
                                 evaluate the new individual;
```

```
                    endrepeat;
                    replace the bad individuals of the population by the fittest
                    new ones;
                    G-bit();
          end;
end;
```

3 The Scatter Search

We distinguish three major operators in the scatter search or SS in abbreviation: a dispersion operator, a combination operator and an optimization operator. The dispersion operator is first used to build the initial population in order to yield solutions well distributed in the search space. It is also employed inside the iteration process to retain a certain diversification degree in order to palliate to the premature convergence of the process. The reference set is a subset of the population individuals to be considered in the combination process. In SS, more than two solutions can be combined to create new ones. Often, the SS uses as a combination operator, the linear combination based on generalized path constructions in the Euclidean and the neighborhood spaces, which generally preserve diversification. However other operator can be considered. The new created solutions are transformed in admissible solutions and then improved by an optimization operator. The latter may use a local search or a simple taboo search as in our problem resolution. The replacement mechanism is elitist that is we substitute the best solutions to the less interesting ones.
Although it belongs to the population-based procedures family, scatter search differs mainly from genetic algorithms by its dynamic aspect that does not involve randomization at all. By combining a large number of solutions, different sub-regions of the search space are implicated to build a solution. Besides, the reference set is modified each time a good solution is encountered and not at the combination process termination. Furthermore, since this process considers at least all pairs of solutions in the reference set, there is a practical need for keeping the cardinality of the set small (<=20).

The utility of the reference set called *RefSet* consists in maintaining the b best solutions found in terms of quality and diversity, where b is an empirical parameter. *RefSet* is partitioned into $RefSet_1$ and $RefSet_2$, where $RefSet_1$ contains the b_1 best solutions and $RefSet_2$ contains the b_2 solutions chosen to augment the diversity. The distance between two solutions is defined to measure the solutions diversity. We compute the solution that is not currently in the reference set and that maximizes the distance to all the solutions currently in this set.

Scatter search generates new solutions by combining solutions of *RefSet*. A new solution replaces the worst one in $RefSet_1$ if its quality is better. On the contrary, the distances between the new solution and the solutions in *RefSet* are computed. If diversification is improved, the new solution replaces the element of $RefSet_2$ that has the smallest distance. Otherwise, it is discarded.
The solution combination procedure starts by constituting subsets from the reference set that have useful properties, while avoiding the duplication of subsets previously generated. The approach for doing this, consists in constructing four different

collections of subsets, with the following characteristics: a subset-Type = 1 of all 2-element subsets, a subset-Type = 2 of all 3-element subsets derived from the 2-element subsets by augmenting each 2-element subset to include the best solution not in this subset, a subset-Type = 3 of all 4-element subsets derived from the 3-element subsets by augmenting each 3-element subset to include the best solution not in this subset and a subset-Type = 4 of the subsets consisting of the best i elements, for i=5 to b.

3.1 Scatter Search for SAT and MAX-W-SAT

The implementation of the meta-heuristic requires the design of the following components ; the initial population, the improvement method, the diversification generator and the solution combination method .
The generation of scattered solutions depends only on the search space. An appropriate generator is described as follows. We choose a parameter h <=n-1, where n is the number of variables and an initial solution equal to (0,0,…,0). For each value smaller than h, two solutions are generated. The first one is obtained using the following formula:
$x'_1 = 1 - x_1$
$x'_{1+h*k} = 1 - x_{1+h*k}$ (k=1,…, n/h)
The second solution is the complement of the first one. This process is repeated until we obtain *p-size* solutions where *p-size* is the cardinality of the population.
Many solutions improvement methods exist in the literature and have been used in meta- heuristics like taboo search and GSAT[7] for instance. The improvement method is context-dependent and is applied to the initial solutions and the generated solutions to enhance their quality. Our improvement method consists in choosing the best solution in the neighborhood of the current solution. The neighborhood is obtained by flipping the bit, which evaluates the best value for the objective function. This operation is repeated until no improvement in the evaluation function is observed.
A function distance d(x,y) between two solutions x and y is calculated to measure the solutions diversity. For MAX-W-SAT, we define this distance to be the sum of the absolute values of the differences between the corresponding bits of the solutions. For example, let us consider an instance of SAT with six variables and two instantiations x_1 and x_2 given as : x_1 = (010110) and x_2 = (100010) then $d(x_1,x_2) = 1+1+0+1+0+0= 3$.
Combining two or more solutions consists in determining for each variable, a value that takes into account the corresponding solutions values and of course the fitness function. It is easy to see that for MAX-W-SAT, the value to attribute to a variable is that one of the solution that maximizes the weights sum of the satisfied clauses. In case of conflict, that is when for a same variable, the opposite values (0 and 1) yield the same result for the objective function, an arbitrary choice is made between both values. The scatter search procedure or SS-SAT can be summarized as follows:

Procedure SS-SAT(var seed: solution);
Begin
    For (Iter = 1 To maxIter) **do**
    begin

```
        P = div-gen(seed, pop-size); { generate a diversified population }
        RefSet(P,b1,b2); { construct the reference set }
        NewElements = TRUE;
        While (NewElements) do
        Begin   NewElements = FALSE
                Generate-subsets(RefSet);
                Suppress all the subsets that do not contain at least one element;
                { apply combination operators }
                Combine_type1(x1,x2) for all (x1,x2) ∈ Type 1;
                Combine_type2(x1,x2,x3) for all (x1,x2) ∈ Type 2;
                Combine_type3(x1,x2, x3,x4) for all (x1,x2, x3,x4) ∈ Type 3;
                Combine_type4(x1, ...,xi) for all i=5 to b;
                For all obtained solution xs do
                Begin   xs* = improve(xs);
                        If (xs* ∉ RefSet and ∃ x∈ RefSet1 / OV(xs*) > OV(x))
                                { OV represents the evaluation function }
                                then    begin   replace x by xs*;
                                                NewElements = TRUE;
                                        End else
                                        if (xs* ∉ RefSet and ∃ x∈ RefSet2 /
                                        dmin(xs*)>dmin(x)) then
                                        Begin   replace x by xs*  { dmin(x) is
                                                the smaller };
                                                NewElements = TRUE;
                                        End
                End
        End;
        If (iter < MaxIter) then { maxIter is the maximum number of iterations }
                Seed = the best solution of RefSet1;
    End;
End;
```

The procedure stops as in many meta-heuristics, when during a small number of iterations no improvement in solutions quality is recorded or when we reach a certain number of iterations limited by physical constraints.

4 Computational Experiments

The procedures GA-SAT and SS-SAT have been implemented in C on a Pentium personal computer and numerical tests have been performed on benchmark instances available on a web site given below. These hard problems have been converted from the Johnson class of the second DIMACS implementations Challenge. The weights of the clauses have been drawn from 1 to 1000 and assigned to clauses at random, the number of clauses being ranging from 800 to 950. The solutions quality has been considered as a performance criterion.

---

http://www.research.att.com/~mgcr/data/index.html

4.1 Parameters Setting

Preliminary tests have been carried out in order to fix the key parameters of both algorithms. Table 1 and Table 2 summarize the parameters values obtained respectively for the genetic algorithm and the scatter search after extensive experiments on the benchmark instances.

Table 1. Empirical parameters for GA-SAT

P-size	<400
Conservation rate	[20,25]
Mutation rate	[10,18]
Crossover rate	[80,85]
Max-iter	200

Table 2. Empirical parameters for SS-SAT

P-size	150
Max-iter	30
b_1	5
b_2	5

Table 3. Experimental results comparing GA-SAT and SS-SAT

Instance	Optimal solution	GA-SAT	SS-SAT
Jnh01	420925	420825	420892
Jnh10	420840	420469	420479
Jnh11	420753	420240	420141
Jhn12	420925	420925	420701
Jnh13	420816	420698	420716
Jhn14	420824	420401	420491
Jhn15	420719	420382	420632
Jhn16	420919	420911	420889
Jhn17	420925	420725	420794
Jhn18	420795	420341	420404
Jhn19	420759	420284	420330
Jnh201	394238	394238	394238
Jnh202	394170	394170	393924
Jnh203	393881	394199	393875
Jnh205	394063	394238	394060
Jnh207	394238	394215	394228
Jnh208	394159	393752	393771
Jnh209	394238	394185	394238
Jnh210	394238	394238	394067
Jhn211	393979	393663	393742
Jhn212	394238	394011	394082
Jhn214	394163	394227	394152
Jhn215	394150	394019	393942
Jhn216	394226	393844	393806
Jhn217	394238	394238	394238
Jhn218	394238	394138	394189

4.2 Performance Comparison

Table 3 contains for each problem, the optimal solution and the best solutions found respectively by GA-SAT and SS-SAT. These results show that SS-SAT is slightly more efficient than GA-SAT in general.

5 Conclusion

In this study, the well known genetic algorithm and the recent meta-heuristic called scatter search have been put in competition for solving the maximum weighted satisfiability problem. These two evolutionary approaches present a significant number of similarities since they belong together to the IPI methods family. The main difference in their frameworks resides in the way of combining solutions. In the GA, only two individuals at a time may be reproduced using an operator that resorts to randomization. However in the SS, more than two solutions may be combined according to the integer programming context. Extensive experimental tests have been carried out on the designed procedures called respectively GA-SAT and SS-SAT in order to get a statement on the procedures performances. It appears that SS-SAT is slightly more effective than GA-SAT.

References

1. Campos V., Glover F., Laguna M., Marti R.: An Experimental Evaluation of a Scatter Search for the Linear Ordering Problem, (1999)

2. Cung V.D., Mautor T., Michelon P. , Tavares A.:A Scatter Search based Approach for the Quadratic Assignment Problem, research Report N° 96/037 Versailles university, (1996)

3. Drias H.: Randomness in Heuristics : an Experimental Investigation for the Maximum Satisfiability Problem, LNCS 1606 proceeding of IWANN'99, Alicante Spain, (1999) 700-708

4. Frank J. : A Study of genetic algorithms to find Approximate Solutions to Hard 3CNF Problems, in Golden West International Conference on Artificial Intelligence, (1994)

5. Glover F.: Heuristics for Integer Programminig Using Surrogate Constraints, Decision Sciences, vol 8, N° 1, (1977) 156-166

6. Gottlieb J., Voss N.: Representations, fitness functions and genetic operators for the satisfiability problem, Springer Verlag Heidelberg (1998)

7. Selman B., Kautz H.A., Cohen B.: Local Search Strategies for Satisfiability Testing, second DIMACS implem challenge, vol 26, D.S. Johnson and M.A. Tricks eds., (1996) 521-531

A New Approach to Evolutionary Computation: Segregative Genetic Algorithms (SEGA)

Michael Affenzeller

Institute of Systems Science
Systems Theory and Information Technology
Johannes Kepler University
Altenbergerstrasse 69
A-4040 Linz - Austria
ma@cast.uni-linz.ac.at

Abstract. This paper looks upon the standard genetic algorithm as an artificial self-organizing process. With the purpose to provide concepts that make the algorithm more open for scalability on the one hand, and that fight premature convergence on the other hand, this paper presents two extensions of the standard genetic algorithm without introducing any problem specific knowledge, as done in many problem specific heuristics on the basis of genetic algorithms. In contrast to contributions in the field of genetic algorithms that introduce new coding standards and operators for certain problems, the introduced approach should be considered as a heuristic appliable to multiple problems of combinatorial optimization, using exactly the same coding standards and operators for crossover and mutation, as done when treating a certain problem with a standard genetic algorithm. The additional aspects introduced within the scope of segregative genetic algorithms (SEGA) are inspired from optimization as well as from the views of bionics. In the present paper the new algorithm is discussed for the travelling salesman problem (TSP) as a well documented instance of a multimodal combinatorial optimization problem. In contrast to all other evolutionary heuristics that do not use any additional problem specific knowledge, we obtain solutions close to the best known solution for all considered benchmark problems (symmetric as well as asymmetric benchmarks) which represents a new attainment when applying evolutionary computation to the TSP.

1 Introduction

A genetic algorithm (GA) may be described as a mechanism that imitates the genetic evolution of a species. The underlying principles of GAs were first presented by Holland [5]. An overview about GAs and their implementation in various fields was given by Goldberg [4] or Michalewicz [7].
Evolutionary Strategies, the second major representative of evolutionary computation, were introduced by Rechenberg [8]. Applied to problems of combinatorial optimization, evolutionary strategies tend to find local optima quite efficiently. But in the case of multimodal test functions, global optima can only be detected

J. Mira and A. Prieto (Eds.): IWANN 2001, LNCS 2084, pp. 594-601, 2001.

by evolutionary strategies if one of the start values is located in the narrower range of a global optimum. Nevertheless, the concept how evolutionary strategies handle the selective pressure has turned out to be very useful in the context of the segregative genetic algorithm (SEGA) presented in this paper.
Furthermore, we have borrowed the cooling mechanism from simulated annealing (SA), introduced by Kirkpatrick [6] in order to obtain a variable selective pressure. This will mainly be needed when segregating and reunifying the population as described in subsection 2.2.
The aim of dividing the whole population into a certain number of subpopulations (segregation) that grow together in case of stagnating fitness within those subpopulations is to combat premature convergence which is the source of GA-difficulties. This segregation and reunification approach is a new idea to overcome premature convergence. The principle idea is to divide the whole population into a certain number of subpopulations at the beginning of the evolutionary process. These subpopulations evolve independently from each other until the fitness increase stagnates because of too similar individuals within the subpopulations. Then a reunification from n to (n-1) subpopulations is done. By this approach of width-search, building blocks in different regions of the search space are evolved at the beginning and during the evolutionary process which would disappear early in case of standard genetic algorithms and whose genetic information could not be provided at a later date of evolution when the search for global optima is of paramount importance. In this context the above mentioned variable selective pressure is especially important at the time of joining some residents of another village to a certain village in order to steer the genetic diversity.
Experimental results on some symmetric and asymmetric benchmark problems of the TSP indicate the supremacy of the introduced concept for locating global minima compared to a standard genetic algorithm. Furthermore, the evaluation shows, that the results of the segregative genetic algorithm (SEGA) are comparable for symmetric benchmark problems and even superior for asymmetric benchmark problems when being compared to the results of the cooperative simulated annealing technique (COSA) [13] which has to be considered as a problem specific heuristic for routing problems. This represents a major difference to SEGA that uses exactly the same operators as a corresponding GA and can, therefore, be applied to a huge number of problems - namely all problems GAs can be applied to.
Moreover it should be pointed out that a corresponding GA is unrestricted included in the SEGA when the number of subpopulations (villages) and the cooling temperature are both set to 1 at the beginning of the evolutionary process.

The rest of the paper is organized as follows: In section 2, we first introduce the concept of a variable selective pressure and the concept of segregation and reunification in detail to end up with the SEGA. In section 3, we discuss the performance of SEGA compared to GA and COSA based on standard benchmarks of the TSP. Finally, section 4 summarizes the main results of this contribution.

2 Introducing the New Concepts

In principle, the new SEGA introduces tw o enhancemerts to the basic concept of genetic algorithms. The first is to bring in a variable selectiv epressure, as described in subsection 2.1, in order to control the diversity of the evolving population. The second concept introduces a separation of the population to increase the broadness of the search process and joins the subpopulation after their evolution in order to end up with a population including all genetic information sufficient for locating the region of a global optimum (subsection 2.2). The incorporation of that measures with a usual genetic algorithm is schematically illustrated in subsection 2.3.

2.1 V ariable Selectiv Pressure

The handling of selective pressure in our context is mainly motivated by evolutionary strategies where μ parents produce λ descendants from which the best μ survive. Within the framework of ev olutionary strategies, the selective pres sure is defined as $s = \frac{\mu}{\lambda}$, where a small value of s indicates a high selectiv e pressure and vice versa (for details see for instance [12]). Applied to the new genetic algorithm this means that from $|POP|$ (population size) number of parents $|POP| * T$ ((size of virtual population) $> |POP|$, i.e. T>1) descendants are generated by crossover and mutation from which the best $|POP|$ survive. Obviously we define the selective pressure as $s = \frac{|POP|}{|POP|*T} = \frac{1}{T}$ where a small value of s, i.e. a great value of T stands for a high selective pressure. When processing SEGA, the temperature slowly decreases from a certain value down to 1 in the ev olution process of each subpopulation.

2.2 Segregation and Reunification

It appears that the GA prematurely conv erges to v ery different regions of the solution space when repeatedly running a GA. Moreover it is known from GA-theory that depending on the problem-type and the problem-dimension there is a certain population size, where exceeding this population size doesn't effect any more improvements in the quality of the solution.
Motivated by that observations, w ehave dev elopedan extended approach to genetic algorithms where the total population is split into a certain number of subpopulations or villages, all evolving independently from each other (segregation) until a certain stage of stagnation in the fitness of those subpopulations is reac hed. Then, in order to bring some new genetic information into each village, the number of villages is reduced by one which causes new overlapping-points of the villages. Fig. 1 shows a schematic diagram of the described process. This process is repeated until all villages are growing together ending up in one town (reunification). The variable selective pressure (subsection 2.1) is of particular importance, if the number of subpopulations is reduced by one because this ev ent brings new diversification into the population. In this case a higher selective pressure is reasonable, i.e. if reunifying members of neighbouring villages,

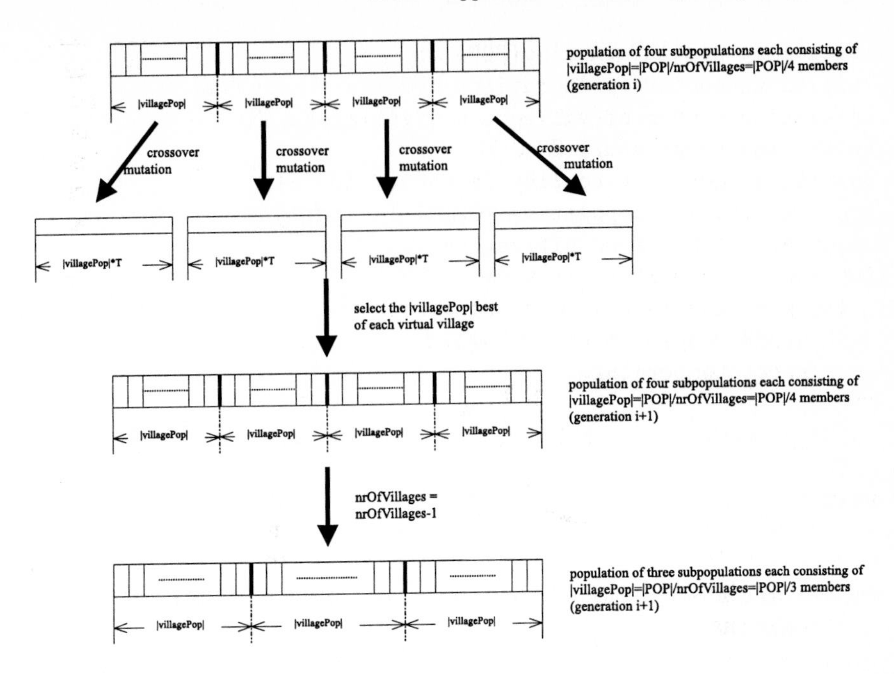

Fig. 1. Evolution of a new population for the instance that four subpopulations are merged to three.

the temperature is reset to a higher lever in order to cool down to 1 as the new system of subpopulations evolves. As the number of villages decreases during evolution, it is recommended to reset the selective pressure to a higher level, because the genetic diversity of the emerging greater subpopulations is growing. For fighting premature convergence, the main advantage of the described strategy is, that building-blocks in different regions of the search space are evolved independently from each other at the beginning of the evolutionary process. The aim is that the best building-blocks survive during the recombination phase, yielding in a final population (if the number of villages is 1) containing all essential building-blocks for the detection of a global optimum. In case of ordinary GAs building blocks that disappear early and which may be important at a later stage of the evolutionary process, when the search for global optima is of paramount importance, can hardly ever be reproduced (premature convergence).

2.3 The New Algorithm

With all strategies described above, finally the new genetic algorithm is stated as follows:

The segregative genetic algorithm

```
procedure SEGA
```

```
initialize population size|POP|
initialize number of iterations nrOfIterations ∈ ℕ
initialize number of villages nrOfVillages ∈ ℕ
initialize temperature T* ∈ ℝ+
initialize adaptive cooling factor α ∈ [0, 1]
generate initial population POP_0 = (I_1, ...., I_|POP|)
calculate dates of reunification
for i:=1 to nrOfIterations do
  if (i = dateOfReunification) then
     nrOfVillages:=nrOfVillags -1
     reset temperature
  end if
  POP_i:=calcNextGeneration(POP_{i-1}, T, nrOfVillages, |POP|)
  T:=T* * α
next i
```

Function "calcNextGeneration" implements the evolution of the next generation of subpopulations.

```
function calcNextGeneration: (POP_{i-1}, T, nrOfVillages, |POP|)
  villagePopulation=|POP| / nrOfVillages
  for i:=(0 to (nrOfVillages-1)) do
    for j:=(i*villagePopulation) to ((i+1)*villagePopulation) do
      calculate fitness_j of each member of the village population
      (like in standard GA).
    next j
    |virtualPopulation|=|villagPop|*T
    for k:=0 to |villagePopulation| do
      generate individuals of virtual population I_k ∈ S
      from the members of the village.
      I_{i*|villagePopulation|} ... I_{(i+1)*|villagePopulation|} ∈ POP_{i-1}
      due to their fitnesses by crossover and mutation
    next k
    select the best |villagePopulation| from the virtual population
    in order to achieve the village population
    I_{i*|villagePopulation|} ... I_{(i+1)*|villagePopulation|} ∈ POP_i
    of the next generation.
  next i
```

SEGA, as described above, uses a fixed number of iterations for termination. Depending on this total number of iterations and the initial number of subpopulations (villages), the dates of reunification are calculated at the beginning of the evolutionary process. Further improvements, particularly in the sense of running time, are possible, if, in order to determine the dates of reunification, a dynamic criterion for the detection of stagnating genetic diversity within the subpopulations is used. A further aspect worth mentioning is the choice of the temperature and the cooling-factor when merging certain subpopulations. Experimental research has indicated that the best results can be achieved resetting the temperature within a range of 0.2 to 2.0 and choosing a cooling-factor α such that the selective pressure converges to one as the subpopulation's genetic diversity is stagnating. Convenient choice of the GA specific parameters can be found in [1], [2], or [3].

3 Experimental Results

In our experiment, all computations are performed on a Pentium III PC with 256 megabytes of main memory. The programs are written in the Java programming language.
We have tested SEGA on a selection of symmetric as well as asymmetric TSP benchmark problem instances taken from the TSPLIB [9] using updated results for the best, or at least the best known, solutions taken from [10]. In doing so, we have performed a comparison of SEGA with a GA using exactly the same operators for crossover and mutation and the same parameter settings and with the COSA-algorithm as an established and successful ambassador of a heuristic especially developed for routing problems.
Fig. 2 shows the experimental results for the problem ch130 (130 city problem) as an example of a symmetric TSP benchmark. This example demonstrates the

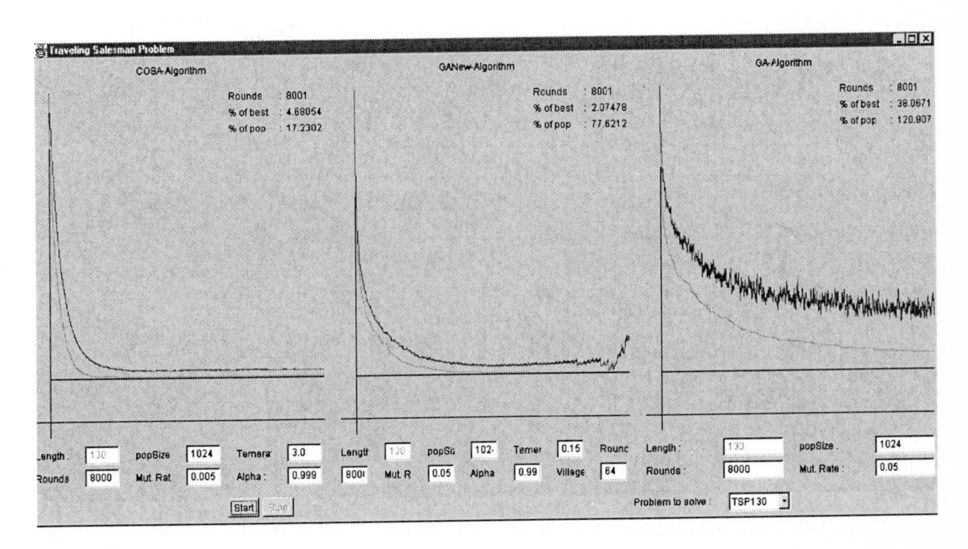

Fig. 2. Comparison of COSA, SEGA, and GA on the basis of the ch130 benchmark problem: For each algorithm, the average fitness and the fitness of the best member of the population is diagrammed relatively to the best known solution represented by the horizontal line.

predominance of the new SEGA compared to the standard-GA. Moreover it even shows the competitiveness of SEGA when compared to the problem specific COSA heuristic. The preeminence of SEGA becomes even more evident, if asymmetric benchmark problems are considered. For the tests the parameters of COSA are set as suggested by the author in [13]. Both, GA and SEGA use a mutation probability of 0.05 and a combination of OX-crossover [7] and ERX-crossover [11] combined with the golden-cage population model (e.g. [13]), i.e. the entire population is replaced with the exception that the best member of the old population survives until the new population generates a better one (wild-card strategy). Within SEGA, the described strategies are applied to each subpopulation. The results of a test presented in the present paper start with 64 villages (subpopulations), each consisting of 8 individuals, i.e. the total population size is set to 1024 for SEGA (as well as for COSA and GA).
Table 1 shows the the experimental results of SEGA, COSA, and GA concerning various types of problems in the TSPLIB. For each problem, the algorithms were run ten times. The efficiency for each algorithm is quantified in terms of the relative difference of the best's individual fitness after a given number or iterations to the best or best-known solution. In this experiment, the relative difference is defined as relativeDifference $= (\frac{Fitness}{Optimal} - 1) * 100\%$.

Table 1. Experimental results of COSA, SEGA and GA.

Problem	Number of Iterations	Average difference(%)		
		COSA	SEGA	GA
eil76 (symmetric)	2000	3.22	1.55	11.21
ch130 (symmetric)	8000	4.76	1.84	35.44
kroA150 (symmetric)	8000	7.90	2.21	40.97
kroA200 (symmetric)	10000	8.54	5.21	45.11
br17 (asymmetric)	100	0.00	0.00	0.00
ftv55 (asymmetric)	2000	44.22	0.76	33.92
kro124p (asymmetric)	10000	26.78	2.61	37.49
ftv170 (asymmetric)	15000	202.33	4.13	131.61

4 Conclusion

The two main changes having been made within the segregative genetic algorithm (SEGA), the introduction of a variable selective pressure and the segregation of the population, have a direct influence on two of the major evolution criteria: speed and convergence. First of all, a far better convergence is noticeable for all tested problems, which is the primary objective of improving the algorithm. Concerning the speed of SEGA, it has to be pointed out that the superior performance concerning convergence requires a higher running time, mainly because of the the greater population size $|POP|$ required. This should allow to transfer already developed GA-concepts to increasingly powerful computer systems in order to achieve better results. Using simultaneous computers seems especially suited to increase the performance of SEGA. Anyway, the corresponding GA is fully included within SEGA if starting with one subpopulation (village) and a temperature constantly set to one, achieving a performance only marginally worse than the performance of the equivalent GA. In other words, SEGAs can be interpreted as a superstructure to GA or as a technique upwards compatible to GAs. Therefore, an implementation of SEGA for a certain problem should be quite easy to do, presumed that the corresponding GA (coding, operators) is known. Even though, because of better comparability, no additional hybrid techniques like commonly used hill-climbing or certain other pre- or post-optimization techniques have been considered in the examples presented in this paper, there absolutely exists no objection of doing so in order to improve the convergence of SEGA.

References

1. Chambers, L. (ed.): Practical Handbook of Genetic Algorithms Volume I. CRC Press. (1995)
2. Chambers, L. (ed.): Practical Handbook of Genetic Algorithms Volume II. CRC Press. (1995)
3. Chambers, L. (ed.): Practical Handbook of Genetic Algorithms Volume III. CRC Press. (1998)
4. Goldberg, D. E.: Genetic Alogorithms in Search, Optimization and Machine Learning. Addison Wesley Longman (1989)
5. Holland, J. H.: Adaption in Natural and Artificial Systems. 1st MIT Press ed. (1992)
6. Kirkpatrick, S., Gelatt Jr., C.D., Vecchi, M.P.: Optimization by Simulated Annealing. Science 220 (1983) 671–680
7. Michalewicz, Z.: Genetic Algorithms + Data Structures = Evolution Programs. 3rd edn. Springer-Verlag, Berlin Heidelberg New York (1996)
8. Rechenberg, I.: Evolutionsstrategie. Friedrich Frommann Verlag (1973)
9. Reinelt, G.: TSPLIB - A Traveling Salesman Problem Library. ORSA Journal on Computing 3 (1991) 376-384
10. Reinelt, G.: TSPLIB - A Traveling Salesman Problem Library. ftp://ftp.zib.de/pub/Packages/mp-testdata/tsp/tsplib/index.html (1997)
11. Schaffer J.: Proceedings of the Third International Conference od Genetic Algorithms. San Mateo (1989) 133–140
12. Schneburg, E., Heinzmann, F., Feddersen, S.: Genetische Algorithmen und Evolutionsstrategien. Addison-Wesley (1994)
13. Wendt, O.: Tourenplanung durch Einsatz naturanaloger Verfahren. Deutscher Universitätsverlag (1995)

Evolution of Firms in Complex Worlds Generalized *NK* Models

Nadia Jacoby

University Paris I Panthéon – Sorbonne, ISYS-MATISSE,
106-112 bd de l'Hôpital, 75013 Paris, France
00.33.1.44.07.81.75, jacoby@univ-paris1.fr

1 Introduction

Evolution is a method of searching among an enormous number of possible "solutions", [10]. If, in biology, this set of possible solutions is the set of possible genetic sequences, evolution can also be seen as a « method for designing innovative solutions to complex problems », [10, p. 5]. The environment changes and thus, it is necessary to search continually new sets of possible solutions. Even if the evolution rules seem to be very complex, in particular because they are responsible for variety and complexity of species, in fact, there are rather simple. « Species evolve by means of random variation, followed by natural selection in which the fittest tend to survive and reproduce, thus propagating their genetic material to future generations », [10, p. 5].

What we want to focus on, in this contribution, concerns firms' evolution. How does the order emerge ? How can we explain it ? To analyze this phenomenon we need to refer to biological concepts such as adaptation and selection. But the context in which we have to do so is the one of natural selection to such an extent that Darwin's view is an essential framework for these questions even if it cannot answer all. Thus, in industrial economics, evolutionary approaches of the firm are particularly relevant especially with their emphasis on the knowledge content of organizational routines, [11]. In particular, these approaches allow to enter the *black-box* and to specify how it works. The firm is a set of routines and core competencies evolving under the influence of both mechanisms of variation and selection. Variation mechanism allows mutation of routines whereas selection mechanism orders them. Through variation process, either routines mutate or they adapt according to the evolution of their environment. Thus, variation concerns either mutation or adaptation. So, adaptation and selection appear fundamental in this theoretical framework we adopt.

The first part of the paper will be devoted to a general approach of interdependency models we want to use for industrial economic applications. After focusing on important concepts mainly borrowed from biology and particularly relevant and useful for our development, we will present the Kauffman's *NK* model of rugged fitness landscape. In the second part, we will implement and state different *NK* applications.

J. Mira and A. Prieto (Eds.): IWANN 2001, LNCS 2084, pp. 602-611, 2001.

2 Interdependency Models

2.1 A General Overview

An useful starting point to analyze these processes is to consider a mapping representation of the population. The concept of *"fitness landscape"* developed by Wright [1931] is defined as a representation of the space of all possible genotypes along with their fitness. The genotype is the set of genes that encode a trait. The different possible setting for a trait are called *alleles* and each gene is located in a particular locus on the chromosome. Finally, the genotype represents different possible configurations of the population. Each genotype has a fitness and the distribution of fitness values over the space of genotypes constitutes the *"fitness landscape"*[1]. In biology, the fitness of an organism is the probability that the organism will live to reproduce, it is a function of the number of offspring the organism has. In our economic purpose, the genotype represents a population of individuals acting in an economic area. The chromosome, composed of genes, refers to a candidate solution for a specific problem and genes encode each particular elements of the candidate solution. *Alleles* are either 0 or 1.

However, those methodologies can be really achieved only with the current development of computer programs. For most of the computational tools, the basic construction consists in designing algorithms to solve specific problems[2]. Thus, in economics, the formulation of this issue has to be presented as a problem to be solved. Problem solving can be straightforwardly understood as a combination of elementary acts leading to a feasible outcome.

Each individual of the population is represented by a dimension of the space associated to a fitness value which depends on how well that individual solves the problem at hand. As described above, the overall distribution of the individual's fitness values over the population constitutes the fitness landscape. For the sake of simplicity, we consider that each population is a bit string of length l and that the distance between two populations is called *"hamming distance"*. It represents the number of locations at which corresponding bits differ. So the fitness landscape can be represented in an $(l+1)$-dimensional plot. But the main interest of this kind of representation is to show how individuals can move all over the landscape, i.e. how to represent evolution process.

According to Wright, evolution causes populations to move along landscapes in particular ways, especially local search processes. Indeed, each population component searches a better position on the landscape, i.e. a position which gives it a higher fitness value. However as the economic rationality of agents is bounded and because they cannot observe all the possible position over the landscape, an accurate representation has to consider a limited space within which components can move. That is the reason why the search process is local. In this context, the concept of

[1] Kauffman [1993] notes that « often, Wright [1931, 1932] thought of the fitness of a given gene or genotype as a function of its frequency in the population ». Like Kauffman in his book [1993], we consider, here, the simpler idea that each genotype can be assigned a fitness.

[2] Genetic algorithms have a particularity. Holland's original goal was to study the phenomenon of adaptation as it occurs in nature and to develop ways in which the mechanisms of natural adaptation might be imported into computer systems.

"neighborhood search", developed by March and Simon [1958], can be compared to the *"local hill climbing"* Holland's [1975] notion. Adaptation can be seen as a moving of population towards local peaks, [14].
Considering a fitness landscape, we have also to consider its relative smoothness, i.e. "the change in fitness value with a change in particular organizational attribute", [6]. If a particular attribute of the population has a strong impact on the performance[3], the landscape cannot be smooth. The landscape is thus more or less rugged. Be that as it may, even if the fitness landscape is smooth with respect to some dimensions of the population, it can be rugged with respect to some others.
In a three-dimension representation, which is commonly accepted for obvious practical reasons, each point has a different height representing its fitness value. In this framework, we can just as easily observe a very rugged landscape or a very smooth one and adaptation evolution is a hill-climbing process. The population can be concentrated on a small part of the landscape or, at the opposite, be spread all over the space. The ultimate goal for the population is to enhance the global fitness value through the increase of components' one. So the population form will be modified.

2.2 NK Model

Among several kind of analytical models available, "*NK* model" proposed by Kauffman [1993] seems to be very interesting to study. It is a simple formal model of rugged fitness landscape that demonstrates that the topology of the fitness landscape is determined by the degree of interdependence of the fitness contribution of the various attributes[4] of the organism. These interactions refer to epistatic effects[5], [12]. N represents the number of parts of the system, i.e. the number of genes of the genotype, the number of individuals of the population. An entity is composed of N attributes where each one can take on two possible values, 0 or 1[6]. Thus, the fitness space consists of 2^N possible configurations. Furthermore, the contribution of each element to the fitness landscape depends upon its own fitness level and upon K other elements among N. Thus, each attribute can take on 2^{K+1} different values depending on the value of the attribute itself and on the value of the K other attributes with which it interacts. K measures the richness of epistatic interactions among the components of the system and influences the relative smoothness or ruggedness of the landscape. K is necessarily inferior to N but can assume all the values between 0 and N-1. Thus, if K=0, the contribution of each element to the global entity is independent of all other attributes and the landscape is smooth, there is a single peak. Conversely, if K is maximum, i.e. if K=N-1, the fitness value of each attribute depends on the value of all other attributes and the landscape is much more rugged, the number of peaks increases. K=N-1 reflects the maximum degree of interdependence that can exist in the landscape.

[3] By the use of the word "performance", we mean fitness value.
[4] Attributes of the organism are genes but in our purpose, they are individuals of the population.
[5] Epistatic effects refer to the action of a specific gene on another one non allele.
[6] Describing this kind of model, Levinthal [1997] notes that « the model can be extended to an arbitrary finite number of possible values of an attribute ».

But how does *NK* model work ? Suppose *N*=10, *K*=3 and that the organization is specified by the string *(1,0,0,0,1,1,1,0,1,0)*. The organization is composed of 10 elements[7] and the value of each element in the string depends on the value of the *K*=3 successive elements of this string[8]. In this way, if the value of the second, the third and/or the fourth elements change, therefore the value of the first has to change also, because of the interdependence effect. Moreover, a random number, generated from an uniform distribution ranging from 0 to 1, is assigned to constitute the fitness contribution of a *1* in the first locus when the values of the second, the third and the fourth elements of the string are *0*. When the value of one of the *K* elements changes, a distinct random number is assigned to constitute the fitness contribution of the element at stake. This assignment is repeated for the *N* elements of the string and the organization overall fitness is the average for the *N* attributes that compose it.
Solving the problem at hand consists in discovering an attractive peak on the rugged fitness landscape, that is a peak with a higher fitness value. One process modeled in *NK* model is a local search process. If the search process reflects any intelligence, it should be sentitive to the fitness value of alternative locations in the space of possible solutions. In this kind of model, the search process takes the form of examining alternatives in the immediate neighborhood of the current proposed solution. But, surprisingly, instead of using the steepest gradient to reach the peak, the search process seeks the first neighbor with a higher fitness level. Search is a step by step process which implies a moving only if the new fitness value is superior. If it's not the case, there is no change and the search process continues. At the end, the search mechanism stops when the local optimum is reached. During the search process and after a few iterations, the number of solutions, yet in touch, declines radically, [5], [6]. It clearly reflects the fact that many initial starting points share the same local optimum. Kauffman [1993] explains that those points belong to a common *"basin of attraction"*. The more rugged the landscape, the higher the number of local optima. The local search process results in a "walk" to the optimum from all the starting points. However, search efforts can be "trapped" at a sub-optimal local peak. Indeed, search process is path-dependent and reveals here its limited nature.
However we can also identify another kind of search process focused on the adoption of alternatives far removed from the organization's current mode of operation. Kauffman [1993] calls, this kind of radical changes, "long-jump". Consequently, the organization adopts this new alternative if its fitness value is higher. Thus the likelihood of a radical organizational change is sensitive to the organization's current performance. Indeed, the more efficient the organization, the less the likelihood for a radical change.
Local search and adaptation lead to the emergence of some local peaks in the landscape. These are local optima that each individual in the population want to reach. Interaction effects preserve diversity among organizational forms[9] even if they coexist within the same market niche. If organizations don't operate in the same market

[7] Elements that compose the organization can be structures, strategies,...
[8] We consider the string as circular, i.e., in this case, the value of the last element of the string depends on the value of the first, the second and the third elements of the same string. More generally, when the string is circular, the neighbor of the last one is the first one.
[9] We can, without any problem, consider that a specific kind of population we want to study are organizational forms.

niche, the diversity persist. However, if all of them operate in the same market niche, a dominant organizational form will emerge through selection process. In the landscape, this dominant form will be the global optima that all landscape's components want to reach[10].

The analytical structure developed by Kauffman which models interaction effects allows to represent epistatic interactions. The degree of epistatic interactions within the organization reflects the relative impact of adaptation and selection processes in changing environments. This degree of interdependence influences the relative smoothness of the fitness landscape and, consequently, the number of peaks in the fitness space. The higher the degree of interdependence, the more rugged the fitness landscape. That is to say, if global fitness is highly interactive, i.e. if the value of specific features depend on a variety of other features, the landscape will tend to be quite rugged. This high interdependence degree implies that the global fitness value changes even if only one component value is changed.

3 Adaptation on Rugged Landscapes

3.1 Standard Application of NK

The model proposed by Levinthal [1997] is based on the Kauffman's *NK* model. It focuses on the interrelationship between processes of firm level change and population level selection pressures.
The first step of the simulation initializes both the fitness landscape and the initial population of organizations. The fitness landscape is initialized by the specification of the fitness value of the 2^N possible organizational forms. For each possible form, each element of its N-length string may take on 2^{K+1} values depending on the value of the K other elements with which it interacts. For each combination, a random number, drawn from an uniform distribution ranging from 0 to 1, is assigned and constitutes its fitness level on the fitness landscape. Once specified, the fitness landscape is fixed. The initial population is initialized by defining each of the N attributes at random with an equal probability. Each attributes takes either the value *0* or the value *1*. The simulation process rests on three main steps repeated for the number of time periods of the simulation. First, organizations which survive from the prior time period are identified. Second, those surviving organizations engage in search processes, both local and distant search. If some of them discover a superior organizational form, i.e. with a higher fitness value, they adopt it. Finally, new organizations are specified to replace those that didn't survive from the prior period. The number of organizations is kept constant over time and, for this simulation, it is set at 100. So the model Levinthal proposes sets N=10, number of organizations=100, number of runs=100 and runs the simulation for K=0, K=1, K=5. The results reflect the average behavior.
RESULTS : The first set of runs used to study emergence of order rests only on local adaptation - without ″long-jump″ nor selection - and is characterized by a high diversity degree. Quite rapidly, the number of organizational forms diminishes

[10] In this perspective, Levinthal [2000] presents the examples of United Parcel Service and Federal Express which share the same market niche even though their activity is identical.

whatever the value of K is. The search process continues till all the organizations reach a local optimum. While the organizations are initially randomly distributed on the fitness landscape, these results show that many of them share the same local optimum. In a second set of simulation, there is no diversity and all the organizations are assigned to the same organizational form. It is assumed that organizations are engaged in local and distant search processes. In the first few periods, roughly half of the organizations can identify attractive forms and adopt them. But quite rapidly, local search process leads to the decline in the number of distinct organizational forms in the population. A third set of simulation postulates a pure selection process without adaptation. Changes occur only by birth and death of organizational forms which are not sensitive to the value of K. The landscape is driven towards the fittest organizational form. Contrary to adaptation processes, the rate of organizational forms differentiation is slower. However, selection process drives the population towards the existence of an unique form while adaptation leads to a set of organizational forms over which selection occurs. But what happens if, now, we consider changing environments. What is the challenge for incumbents in a Schumpeterian environment ? How do adaptation and selection interact in a changing external environment ? To explore these issues, after 25 periods of time, the simulation process re-specifies the fitness landscape by changing one dimension in the fitness contribution. When the degree of epistatic interaction equals zero, all the incumbents survive. But the higher the value of K, the less share of surviving incumbents. With high levels of interaction, the local adaptation is not an efficient response to changes in the fitness landscape. As a result, in a changing environment, survivals are more dependent of a long-jump process result.
The diversity of organizational forms seems to be explained thanks to the diversity of environments in which organizations act. But this kind of argumentation requires the existence of a well-defined scheme of interrelationship between environment and organizations. The degree of epistatic interactions appears as an interesting factor to explain both organizational diversity and persistence of organizations in changing environment, [6]. Even if variation in organizational form results from local search and adaptation, the impact of initial imprinting persists even when organizations are engaged in adaptation.

3.2 Reduction of Complexity through the Decomposition Algorithm

Marengo, Dosi, Legrenzi and Pasquali [2000] propose an alternative representation for problem solving based on the basic structure developped by Kauffman [1993]. On the one hand, economic problems are more and more complex, on the other hand, rational and computational problem solvers' capacities are limited. Therefore, faced with the necessity to reduce the complexity of problems, Marengo and his colleagues introduce a decomposition algorithm. The interdependent elements of the basic problem are decomposed into different less complex sub-problems. They construct a *"near-decomposition"* that improves search times and attains a higher degree of decentralization at the expenses of optimality of the solutions. The problem representation is exogenously given to agents who don't know the "objective" problem but only some imperfect representation of it.

To characterize a problem to solve, they describe several elements ; given :

- the set of components : $S=\{s_1, s_2, ..., s_N\}$, where $s_i \in \{0, 1\}$
- a string for possible solution : $x_i = s_1 s_2 ... s_N$
- the set of possible configurations : $X = \{x_1, x_2, ..., x_p\}$ where $p = 2^N$

Thus, a problem is defined by the couple (X, ≥), where the ordering ≥ allows to identify the prefered strings over the set of possible configurations[11].

The decomposition algorithm considered is climbing and mutational. It moves from a state to another only if the second has a higher fitness value, otherwise it doesn't move. The algorithm can be characterized in terms of sets of bits it can mutate.

Each set of possible configurations is decomposed into non-empty "d_i" blocks which themselves composed the decomposition scheme D : $D = \{d_1, d_2, ..., d_k\}$ where d_i is a subset of components s_i and $d_i \in \{2^I \backslash \varnothing\}$.

The decomposition scheme is therefore a decomposition of the *N* dimensional space of configurations into sub-spaces of smaller dimension, whose union returns the entire problem.

Given a configuration x_j and a block d_k, the block-configuration $x_j(d_k)$ is the sub-string of length $|d_k|$ containing the components of configuration x_j belonging to the block d_k.

Thus, decomposition allows to reduce the complexity of problems. The decomposition scheme $D = \{\{1\}, \{2\}, \{3\}, ..., \{N\}\}$[12] has reduced the problem to the union of N sub-problems of minimum complexity while the scheme $D = \{\{1, 2, 3, ..., N\}\}$ describes a problem of maximum complexity. In the first case, the problem can be solved in linear time, in N steps, while a non-decomposed problem can only be solved in exponential time, in 2^N steps. However, the simplicity due to the decomposition doesn't automatically assure an optimal solution. It is constrained by some path-dependent conditions which become more and more restrictive as the size of the decomposition scheme decreases. For sake of simplicity, Marengo et al. [2000] assume that coordination among blocks takes place through selection mechanisms which selects at no cost and without any friction over alternative block-configurations. The bounded rationality of agents leads to the use of local search process in directions given by the decomposition scheme. As with the standard *NK* model, local search leads to local optima through search-path[13].

The simulation is based on a population of agents characterized by a set of decomposition strategies. The environment in which they leave is defined by some simple rules of selection according to which the worst are eliminated and replaced with "clone" of the best scoring agent. New agents inherit the parent's decomposition scheme but explore the landscape starting from a randomly assigned point. Each agent generates new points by choosing randomly one of the blocks and mutating its bits (one, several or all of them) randomly. After mutation, the fitness value of the new point is observed. If it is higher than the current position, the agent moves to it, otherwise the agent remains where it is. Only search strategy differentiates agents.

[11] We assume that the ordering ≥ is anti-symmetric.

[12] In such a case, *K*=0, there's no interactions between elements ; whereas in the second case, *K* is maximum (=*N*-1) and all the components interact with each others.

[13] In order to economize on space, we don't develop here the specific notations and complements used by Marengo et al. [2000]. For more details on these precise points, see Marengo et al. [2000].

They set six possible decomposition schemes and six types of agents[14]. To avoid the effect of lucky initial conditions, the first simulation implemented shakes the population every 1000 iterations whereas the second one ranks the agents every 10 mutations, removes the 30 worst and replaced them by copies of survivors.
RESULTS : We observe average fitness. Strategies using decompositions not including the one corresponding to the minimum size decomposition scheme are bound to be trapped in local optima. At the opposite, those including the decomposition of minimum size do always reach the global optimum but do so slowly. Moreover they outperform every other strategies. With the second simulation, we observe that, for landscapes of types 2, 3, 4 and 6, agents whose decomposition perfectly reproduces the structure of the landscape tend to dominate the population.

3.3 Adjacency Matrix: A Graph-Theoretic Generalization of the NK

Ghemawat and Levinthal [2000] start with the fact that few problems in business are purely intra-departmental. Most of the time a business problem concerns several entities within the firm. Problems are interdependent. So, how to model inter-temporal linkages in choices ? Thus to better represent the internal working of the firm, we need to introduce, in NK model, other interactivity parameters to complete what K tries to explain. However, *NK* model presents one glaring defect : all choices are assumed to be equally important. In this perspective, Ghemawat and Levinthal [2000] propose a graph-theoretic generalization of the NK simulation-based approach, introducing adjacency matrix which specifies how different choices are linked. They replace interactivity parameter K with this adjacency matrix. In such a matrix, choice variable *j*'s effect on other variables is represented by the set of *0*s and *1*s in column *j*. The value *1* indicates that the variable in the considered row is contingent on variable *j* ; if the value is *0* the two variables are independent. Obviously, the matrix can be read starting across row *i* and explaining the influence on variable *j*. In addition to such direct effects on value contributions, variables may be indirectly related through other variables. Necessarily, the principal diagonal of the matrix is only composed with *1*s, but the matrix need not to be symmetric. When a central choice influences the payoffs of N-1 other choices but there is no other linkages among choices, the corresponding adjacency matrix is composed with *1*s in the first column and in the principal diagonal but *0*s everywhere else.
Representing basic structure of NK model in terms of adjacency matrix reveals a symmetric matrix in which, excepted the elements of the principal diagonal, there are K *1*s in each row and each column but randomly distributed. In the particular case in which K = N-1 (i.e. K is maximum), the graph is totally connected and the matrix is only composed with 1s.
In fact, adjacency matrices mix up different types of effects ; Ghemawat and Levinthal [2000] identify three types : influence, contingency and autonomy. A variable can influence more or less strongly the payoffs of other variables. Their direct payoff is only dependent on their own setting ; no other variable influences it. In the matrix, it is represented by the prevalence of *1*s in the relevant column. Conversely, contingence is measured through the number of *1*s in the relevant row.

[14] For each type of landscapes, we let 180 artificial agents equally distributed between the different types.

The higher this number, the more contingent this variable. As for autonomy, it is characterized by variables that are neither influential nor contingent.
Finally, this model allows to show how the structure of choice landscapes affects organizational fitness levels[15].

4 Conclusion

The evolution of the firm is a complex problem. All the interdependency models proposed here seem to be useful for its explanation. The standard application of *NK* allows to explain not only the diversity of the firms but also their interdependence with the diversity of environments. The evolution of the firms partly depends on the environment in which they are. The more changing the environments, the more diverse the firms. But the complexity of the firms evolution question needs to resort to specific decomposition algorithms. A very precise decomposition algorithm allows the model to represent the path the firm uses to reach the global optimum. So, this kind of model permits, thanks to decomposition, to explain how firms reach the highest point on the landscape. Without this decomposition, most of the time it is not possible, through models, to show how the global optimum is systematically reached. Finally to complete the analysis of firm evolution, we refer to adjacency matrix that allows to represent inter-temporal linkages in choices.
Thus, this representation of these mechanisms through three kinds of interdependency model, all inspired from Kauffman's *NK* model, allows to better explain the evolution of the firm. All those representations can be implemented with computational platform as, for instance, Lsd[16]. A modelization, mainly derived from Marengo et al., and for the moment still in progress, should produce some interesting results on internal evolution of the firm.

References

1. Cohen M., R. Burkhart, G. Dosi, M. Egidi, L. Marengo, M. Warglien, S. Winter, B. Coriat: « Routines and other recurring action patterns of organizations : Contemporary Research Issues », *Industrial and Corporate Change*, vol. 5, (1995) pp. 653-98.
2. Dosi G., R. Nelson, S.G. Winter (eds.): *The Nature and Dynamics of Organisational Capabilities,* Oxford, Oxford University Press (2000).
3. Ghemawat P., D. Levinthal: « Choice structures, business strategy and performance : a generalized NK-simulation approach », *The Wharton School, University of Pennsylvia*, WP 00-05 (2000).
4. Holland J.H.: *Adaptation in natural and artificial systems*, University of Michigan Press, Ann Arbor, MI (1975).
5. Kauffman S.A.: *The Origins of Order*, Oxford, Oxford University Press (1993).

[15] Ghemawat and Levinthal explore three stylized structures of interaction we cannot develop here. However, for more details you can refer to their paper.
[16] Lsd means Laboratory for Simulation Development and has been developped by Marco Valente in Aalborg.

6. Levinthal D.: « Adaptation on rugged landscapes », *Management Science*, vol. 43, pp. 934-950 (1997).
7. Levinthal D.: « Organizational capabilities in complex worlds », in Dosi et al. (eds) (2000)
8. March J., H. Simon: *Organizations*, John Wiley & Co., New York, (1958).
9. Marengo L., G. Dosi, P. Legrenzi, C. Pasquali: « The structure of problem-solving knowledge and the structure of organizations », *Industrial and Corporate Change*, vol.9, n°4, special issue December, (2000), pp. 757-788
10. Mitchell M.: *An introduction to genetic algorithm*, MIT Press (1998).
11. Nelson R.R., S.G. Winter: *An Evolutionary Theory of Economic Change*, Cambridge MA, Harvard University Press (1982).
12. Smith J.M.: *Evolutionary Genetics*, Oxford University Press, New York (1989).
13. Valente M.: Laboratory for simulation development–Lsd, www.business.auc.dk/~mv/, (1998)
14. Wright S.: « Evolution in Mendelian populations », *Genetics*, 16 : 97-159, (1931).

Learning Adaptive Parameters with Restricted Genetic Optimization Method

Santiago Garrido and Luis Moreno

Universidad Carlos III de Madrid, Leganés 28911, Madrid (Spain)

Abstract. Mechanisms for adapting models, filters, regulators and so on to changing properties of a system are of fundamental importance in many modern identification, estimation and control algorithms. This paper presents a new method based on Genetic Algorithms to improve the results of other classic methods such as the extended least squares method or the Kalman method. This method simulates the gradient mechanism without using derivatives and for this reason, it is robust in presence of noise.

1 Introduction

Tracking is the key factor in adaptive algorithms of all kinds. In the problems of adaptive control, it is necessary to adapt the control law on-line. This adaptation can be done by using recursive rules (like MIT rule) or by using an on-line identification of the system (usually a least-squares-based method).

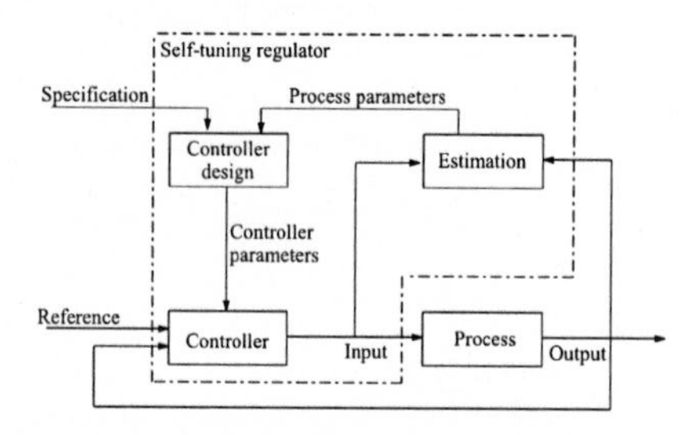

Fig. 1. Block Diagram of a Self-Tuning Regulator.

A similar problem is the estimation of the state vector of a stochastic system.

When trying to estimate the state of a system at the current time-step k provided, there are: knowledge about the initial state, all the measurements up to the current time, the system and the observation models. Because the system model and the observation model are corrupted with noise, some kind of state estimation method must to be used.

$$\begin{cases} x(k) = g(k, x(k-1), \varepsilon(k)) \\ z(k) = h(k, x(k), \eta(k)) \end{cases}$$

J. Mira and A. Prieto (Eds.): IWANN 2001, LNCS 2084, pp. 612–620, 2001.

This estimate lets us integrate all the previous knowledge about the system with the new data observed by the sensors in order to obtain a precise state estimation of the system real state. Due to the stochastical nature introduced by the noise at the sensors and the system it is necessary to use a probabilistic formulation of the estimators to reach reliable values of the state estimate. This estimation problem can be formulated as a Bayesian filtering problem, where the construction of the posterior density $p\{x_k|Z^k\}$ of the current state conditioned on all measurements up to the current time is intended.

Efficient function optimization algorithms are generally limited to regular and unimodal functions like those originated by the assumption of Gaussian noise at sensors measurement and system. However, many functions are multimodal, discontinuous and non differentiable.

In order to optimize this kind of functions, some stochastic methods have been used, like those methods based on the Monte-Carlo technique, which requires a high computational effort. These computational requirements strongly limit the possibility of being used as an on-line adaptive parameters algorithm.

Traditional optimization techniques use the problem characteristic as a way of determining the next parameter adaptation point by using some kind of local slope to determine the best direction, (i.e. gradients, Hessians,etc .) which somehow requires linearity, continuity and differenciability at that point. The stochastic search techniques use some kind of sampling rules and the stochastic decision in order to determine the next adaptation point.

Genetic algorithms are a special and efficient kind of stochastic optimization method, which has been used for the resolution of difficult problems where objective functions do not have good mathematical properties [Davis [1], Goldberg [5], Holland [6], Michalewicz[8]]. These algorithms make their search by using a complete population of possible solutions for the problem and they implement a 'best adapted survival' strategy as a way of searching better solutions.

An interesting type of problems with poor mathematical properties are the problems of optimization where functions are time varying, non-linear, discontinuous and have non Gaussian noise. In this kind of problems, it is almost impossible to use gradient methods because the functions are not differentiable. This kind of problems is frequent in Identification and Control Theory.

2 Introduction to the Restricted Genetic Optimization

In literature, the use of Genetic Algorithms as an stochastic optimization method is traditionally carried out off-line because the computing time is usually quite long. This high computational effort is due to two main reasons: the first one is that Genetic Algorithms are sampling-based methods and the second one is the difficulty of covering a global solution space with a limited number of samples.

The technique proposed in this paper tries to imitate the Nature: it works on-line. When Nature uses Genetic Optimization, it uses it locally, that is: at a given moment and to adapt to determined environmental conditions. For this reason, it is possible to get a fast rate adaptation to changing conditions. We have demonstrated [[2], [3], [4]], that Genetic Algorithms operating in restricted areas of the solution space can be a fast optimization method for time-varying, non linear and non differentiable functions. That is why, the technique proposed is called Restricted Genetic Optimization(RGO).

Usually, GAs are used as a parallel, global search technique. It evaluates many points simultaneously, improving the probability of finding the global optimum.

In Dynamic Optimization, finding the global optimum is useful for the first generations to find the correct basin of attraction. However, it consumes a large computation time. Therefore, a fast semi-local optimization method, such as RGO, is better.

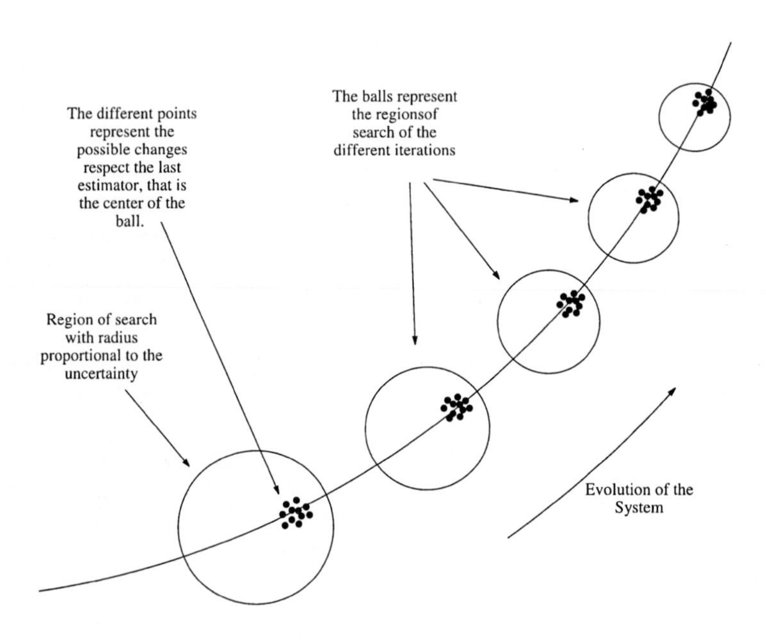

Fig. 2. Tracking mechanism of the RGO

The search of this solution is stochastically made by using a genetic search technique, which has the advantage of being a non gradient-based optimization method (the genetic optimization techniques constitute a probabilistic search method which imitates the natural selection process based on genetic laws).

The proposed method consists of carrying out the search in the point neighborhood, and it takes the best adapted point of the new generation as the search center.

The set of solutions (the population) is modified according to the natural evolution mechanism: selection, crossover and mutation, in a recursive loop. Each loop iteration is called generation, and represents the set of solutions (population) at that moment.

The selection operator tries to improve the medium quality of the solution set by giving a higher probability to be copied to next generation. This operator has a substantial significance because it focuses of the search of best solutions on the most promising regions in the space. The quality of an individual solution is measured by means of the fitness function.

We get new generations oriented in the direction of the steepest slope of the cost function, and with a distance to the center as close as possible to the correct one. This distance corresponds to the velocity at which the system is changing.

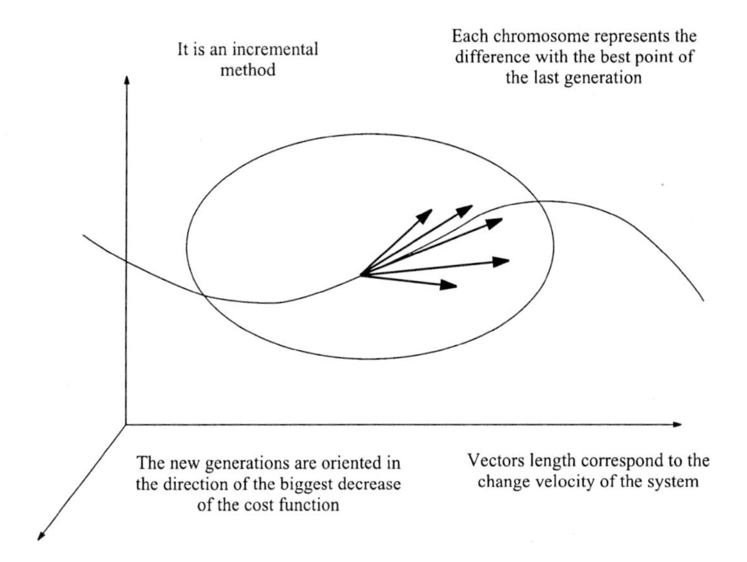

Fig. 3. Way of working of RGO

This behaviour simulates the gradient method without using derivatives and can be used when signals are very noisy. If these signals are not very noisy, a gradient-based method can be used.

We can carry out a search in a big neighborhood in the beginning and then reduce its radius (the radius is taken as proportional to uncertainty). In fact, the method makes a global search at the beginning and a semi-local search at the end. This method reduces the probability of finding a local minimum.

2.1 RGO

In the RGO method the search is done in a neighborhood of the point that corresponds to the last identified model. The best adapted point is taken as the center of the search neighborhood of the new generation and the uncertainty is taken as the radius. This process is repeated for each generation.

Each chromosome corresponds to the difference vector of the point with respect to the center of the neighborhood, in order to get a search algorithm that works incrementally. The center is also introduced to the next generation (it is the zero chain). The best point of each generation that will become the neighborhood center of the next generation is saved apart of the generation and represents absolute coordinates.

In order to evaluate the fitness function, the coordinates of the center are added to the decoded chromosome of each point. This way, the method works incrementally. This behavior simulates the gradient method, but without using derivatives and can be used when signals are noisy.

The radius is proportional to the uncertainty of the estimation with upper and lower extremes. The way of calculating the uncertainty depends of the application of the optimization. For example, in the case of state estimation, the Mahalanobis distance is used:

$$d = (\hat{x}(k|k-1) - \hat{x}(k-1|k-1))' P^{-1}(k|k)(\hat{x}(k|k-1) - \hat{x}(k-1|k-1)) \quad (1)$$

It measures the uncertainty of the estimation $\hat{x}(k)$. In the case of systems identification, the radius is taken as proportional of the MDL criterium (Minimum Description Length) with upper and lower extremes. Other criteria such as AIC (Akaike's Information theoretic Criterion) or FPE (Final Prediction-Error criterion) can be used (Ljung[7]).

In the proposed algorithm, the next operators have been implemented: reproduction, cross, mutation, elitism, immigration, ranking y restricted search.

The ranking mechanism has been used to regulate the number of offspring that a chromosome can have, because if it has a very high fitness, it can have many descendants and the genetic diversity can become very low.

This mechanism has been implemented using $F_n(i) = \frac{1}{c+V_n(i)}$ as fitness function, where $V_n(i) = \sum_{k=1}^{buff} (y_{n-k} - \hat{y}n - k)^2$ is the loss function and c is a constant.

If the system is changing quickly with time, it is possible to include the re-scale mechanism, trialelic dominance and inmigration.

It is important to tell that the RGO method realizes a preferent search in the direction of the steepest slope and it has a preferential distance (that corresponds to the change velocity of the system). This fact permits to reduce the number of dimensions of the search ball. For this reason, this method is comparable to the gradient method but it is applicable when expressions are not differentiable and signals contain noise.

3 Algorithm

1. A first data set is obtained.
2. The initial population is made in order to obtain the type and the characteristics of the system.
3. The individual fitness is evaluated.

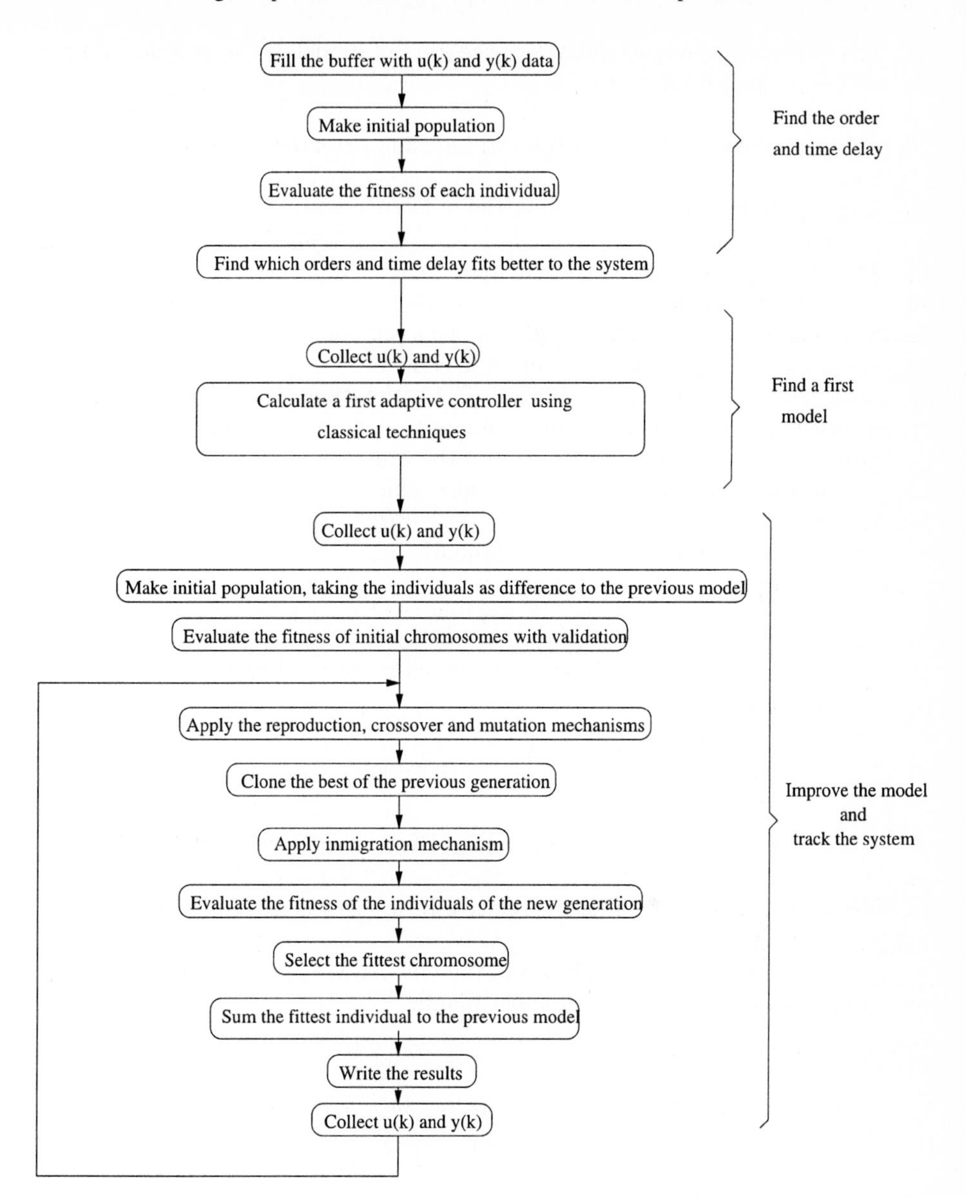

Fig. 4. Flowcart of the Genetic Restricted Optimization (RGO).

4. The order and the time delay and the kind of model of the system that fits better are evaluated.
5. Data u(k), y(k) are collected.
6. A first model estimation is obtained by using extended least squares.
7. $k = 1$
8. A first population is made, using the difference with the previous model as phenotype.
9. The fitness of individuals is evaluated.
10. Begining of the loop.
11. Selection, cross and mutation mechanisms are applied.
12. The best of the previous generation is cloned.
13. Inmigration mechanism is applied. The estimation obtained by using Extended Least Squares (or other technique) is introduced. This way, the algorithm makes sure the improvement of the previous method.
14. The fitness of the individuals of the new generation.
15. The fittest chromosome respect to the desired criterium is selected to be added to the center of the next neighborhood of search.

$$\hat{\theta}^k = \arg\min_{\theta} V_N(\theta). \tag{2}$$

16. The uncertainty of the model is calculated to be used as new radius of the search region.
17. $u(k)$, $y(k)$ data are collected.
18. $k = k + 1$ is done and steps 10-19 are repeated.

4 Comparison with other identification methods

In order to contrast the results, the relative RMS error is used as a performance index.

The ARMAX block of Simulink and the OE, ARMAX and NARMAX methods were applied to the next non-linear plants:

1. Plant 1:

$$\begin{aligned} y(k) - .6y(k-1) + .4y(k-2) = (u(k) - .3u(k-1) + .05u(k-2))sin(.01k) \\ + e(k) - .6e(k-1) + .4e(k-2) \end{aligned}$$

, with zero initial conditions.
This plant is the serial connection of a linear block and a sinusoidal oscillation in the gain.
2. Plant 2: This plant consists of a linear block:

$$\begin{aligned} y(k) - .6y(k-1) + .4y(k-2) = u(k) - .3u(k-1) + .05u(k-2) \\ + e(k) - .6e(k-1) + .4e(k-2) \end{aligned}$$

in serial connection of a backslash of 0.1 by 0.1 (a ten per cent of input signal), with zero initial conditions.

3. Plant 3:

$$y(k) = A_1 + .02\sin(.1k)y(k-1) + A_2y(k-2) + A_3y(k-3)$$
$$+ B_1u(k-1) + B_2u(k-2) + B_3u(k-3)$$

, with $A_1 = 2.62$, $A_2 = -2.33$, $A_3 = 0.69$, $B_1 = 0.01$, $B_2 = -0.03$ and $B_3 = 0.01$.

4. Plant 4:

$$y(k) = (0.8 - 0.5e^{-y^(k-1)})y(k-1) - (0.3 + 0.9e^{y^2(k-1)})y(k-2)$$
$$+ 0.1\sin(\pi y(k-1)) + u(k)$$

with zero initial conditions.

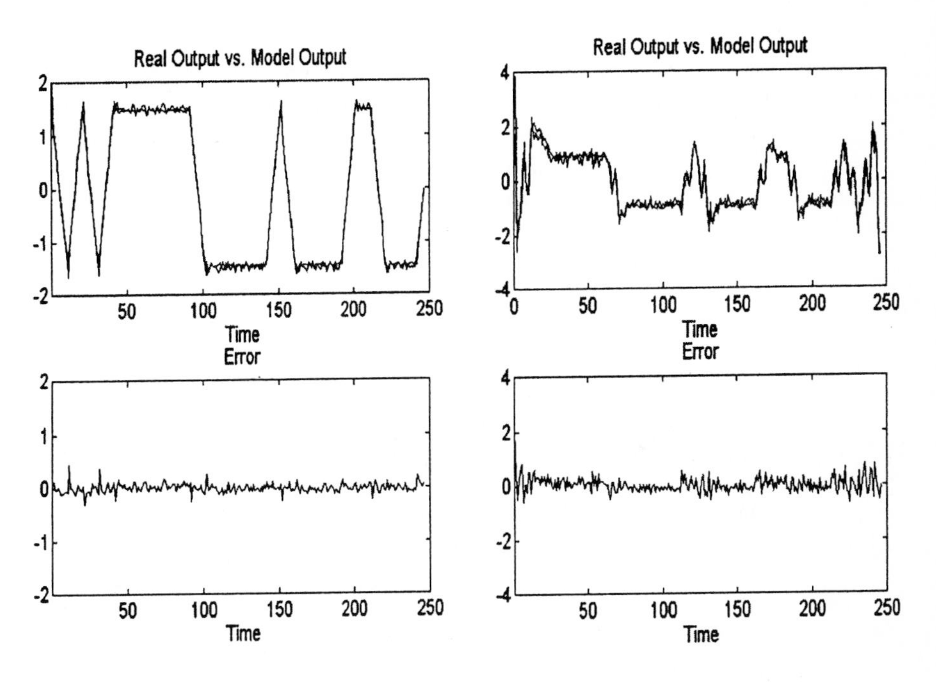

Fig. 5. Results of identification of Plants 1 and 4 with RGO method. Top: Model output vs. system output. Bottom: Error.

The assembly for all the plants is the same. The input signal is a "prbns" signal of amplitude 1. A coloured noise was added to the output of the main block. This noise was produced by filtering a pseudo-white noise of 0.1 of peak with the filter $\frac{z^2+.7z+.2}{z^2-.6+.4}$ The "prbns" input and the output of the global system are carried to the identification block, which gives the coefficients of the OE, ARMAX and NARMAX models, the RMS error and other information.

The obtained results of relative RMS errors can be summarized in the next table :

	1st Plant	2nd Plant	3rd Plant	4th Plant
ARMAX	0.05000	1.33000	209.20000	10.78000
OE RGO	0.05000	0.03000	0.00900	0.05000
ARMAX RGO	0.01800	0.01000	0.00080	0.05000
NARMAX RGO	0.01700	0.00300	0.00030	0.04000

5 Conclusions

The results obtained proved that genetic algorithms can be used to improve the results obtained with other optimization methods that approach a function or a system. For example, least-squares methods need a linear system in the parameters and a non-coloured output signal. When these conditions are not true, the estimate is not good but can be used as a seed to initialize the RGO method for a faster convergence.

References

1. L. Davis, *Handbook of genetic algorithms*, Van Nostrand Reinhold, 1991.
2. S. Garrido, L. E. Moreno, and C. Balaguer, *State estimation for nonlinear systems using restricted genetic optimization*, 11th International Conference on Industrial and Engineering Applications of Artificial Intelligence and Expert Systems, IEA-AIE, 1998.
3. S. Garrido, L. E. Moreno, and M. A. Salichs, *Nonlinear on-line identification of dynamic systems with restricted genetic optimization*, 6th European Congress on Intelligent Techniques and Soft Computing (Aachen, Germany), EUFIT, 1998.
4. S. Garrido, *Identificación, estimación y control de sistemas no-lineales con rgo*, Ph.D. thesis, University Carlos III of Madrid, Spain, 1999.
5. D. Goldberg, *Genetic algorithms in search, optimization, and machine learning*, Addison-Wesley, Reading, MA, 1993.
6. J. H. Holland, *Adaptation in natural and artificial systems*, The University of Michigan Press, Ann Arbor, 1975.
7. L. Ljung, *System identification: Theory for the user*, Prentice-Hall, Englewood Cliffs, N. J., 1987.
8. Z. Michalewicz, *Genetic algorithms + data structures = evolution programs*, AI Series, Springer Verlag, New-York, 1994.

Solving NP-Complete Problems with Networks of Evolutionary Processors

Juan Castellanos[1], Carlos Martín-Vide[2], Victor Mitrana[3], and Jose M. Sempere[4]

[1] Dept. Inteligencia Artificial - Facultad de Informática, Universidad Politécnica de Madrid - Campus de Montegancedo, Boadilla del Monte, 28660 Madrid, Spain - jcastellanos@fi.upm.es - http://www.dia.fi.upm.es

[2] Research Group in Mathematical Linguistics - Rovira i Virgili University - Pça. Imperial Tàrraco 1, 43005 Tarragona, Spain - cmv@correu.urv.es

[3] Faculty of Mathematics, University of Bucharest[‡] - Str. Academiei 14, 70109 Bucharest, Romania - mitrana@funinf.math.unibuc.ro

[4] Department of Information Systems and Computation - Polytechnical University of Valencia, - Valencia 46071, Spain - jsempere@dsic.upv.es

Abstract. We propose a computational device based on evolutionary rules and communication within a network, similar to that introduced in [4], called network of evolutionary processors. An NP-complete problem is solved by networks of evolutionary processors of linear size in linear time. Some furher directions of research are finally discussed.

1 Introduction

A basic architecture for parallel and distributed symbolic processing, related to the Connection Machine [8] as well as the Logic Flow paradigm [5], consists of several processors, each of them being placed in a node of a virtual complete graph, which are able to handle data associated with the respective node. Each node processor acts on the local data in accordance with some predefined rules, and, then local data becomes a mobile agent which can navigate in the network following a given protocol. Only such data can be communicated which can pass a filtering process. This filtering process may require to satisfy some conditions imposed by the sending processor, by the reveiving processor or by both of them. All the nodes send simultaneously their data and the receiving nodes handle also simultaneously all the arriving messages according to some strategies, see, e.g., [6, 8].

Starting from the premise that data can be given in the form of strings, [4] introduces a concept called network of parallel language processors in the aim of investigating this concept in terms of formal grammars and languages. Networks of language processors are closely related to grammar systems, more specifically to parallel communicating grammar systems [3]. The main idea is that one can

‡ Research supported by the Dirección General de Enseñanza Superior e Investigación Cientifica, SB 97–00110508

J. Mira and A. Prieto (Eds.): IWANN 2001, LNCS 2084, pp. 621-628, 2001.

place a language generating device (grammar, Lindenmayer system, etc.) in any node of an underlying graph which rewrite the strings existing in the node, then the strings are communicated to the other nodes. Strings can be successfully communicated if they pass some output and input filter.

In the present paper, we modify this concept in the following way inspired from cell biology. Each processor placed in a node is a very simple processor, an evolutionary processor. By an evolutionary processor we mean a processor which is able to perform very simple ooperations, namely point mutations in a DNA sequence (insertion, deletion or substitution of a pair of nucleotides). More generally, each node may be viewed as a cell having a genetic information encoded in DNA sequences which may evolve by local evolutionary events, that is point mutations. Each node is specialized just for one of these evolutionary operations. Furthermore, the data is each node is organized in the form of multisets, each copy being processed in parallel such that all the possible evolutions events that can take place do actually take place.

These networks may be used as language (macroset) generating devices or as computational ones. Here, we consider them as computational mechanisms and show how an NP-complete problem can be solved in linear time.

It is worth mentioning here the similarity of this model to that of a P system, a new computing model inspired by the hierarchical and modularized cell structure recently proposed in [11].

2 Preliminaries

We start by summarizing the notions used throughout the paper. An *alphabet* is a finite and nonempty set of symbols. Any sequence of symbols from an alphabet V is called *string (word)* over V. The set of all strings over V is denoted by V^* and the empty string is denoted by ε. The length of a string x is denoted by $|x|$.

A *multiset* over a set X is a mappingg $M : X \longrightarrow \mathbf{N} \cup \{\infty\}$. The number $M(x)$ expresses the number of copies of $x \in X$ in the multiset M. When $M(x) = \infty$, then x appears arbitrarily many times in M. The set *supp*(M) is the support of M, i.e., *supp*$(M) = \{x \in X \mid M(x) > 0\}$. For two multisets M_1 and M_2 over X we define their union by $(M_1 \cup M_2)(x) = M_1(x) + M_2(x)$. For other operations on multisets the reader may consult [1].

A *network of evolutionary processors* (NEP for short) of size n is a construct

$$\Gamma = (V, N_1, N_2, \ldots, N_n),$$

where:

- V is an alphabet,
- for each $1 \leq i \leq n$, $N_i = (M_i, A_i, PI_i, FI_i, PO_i, FO_i)$ is the i-th evolutionary node processor of the network. The parameters of every processor are:
  - M_i is a finite set of evolution rules of one of the following forms only
    - $a \to b$, $a, b \in V$ (substitution rules),
    - $a \to \varepsilon$, $a \in V$ (deletion rules),

- $\varepsilon \to a$, $a \in V$ (insertion rules),

More clearly, the set of evolution rules of any processor contains either substitution or deletion or insertion rules.

- A_i is a finite set of strings over V. The set A_i is the set of initial strings in the i-th node. Actually, in what follows, we consider that each string appearing in a node of the net at any step has an arbitrarily large number of copies in that node, so that we shall identify multisets by their supports.
- PI_i and FI_i are subsets of V representing the input filter. This filter, as well as the output filter, is defined by random context conditions, PI_i forms the permitting context condition and FI_i forms the forbidding context condition. A string $w \in V^*$ can pass the input filter of the node processor i, if w contains each element of PI_i but no element of FI_i. Note that any of the random context conditions may be empty, in this case the corresponding context check is omitted. We write $\rho_i(w) = \underline{true}$, if w can pass the input filter of the node processor i and $\rho_i(w) = \underline{false}$, otherwise.
- PO_i and FO_i are subsets of V representing the output filter. Analogously, a string can pass the output filter of a node processor if it satisfies the random context conditions associated with that node. Similarly, we write $\tau_i(w) = \underline{true}$, if w can pass the input filter of the node processor i and $\tau_i(w) = \underline{false}$, otherwise.

By a configuration (state) of an NLP as above we mean an n-tuple $C = (L_1, L_2, \ldots, L_n)$, with $L_i \subseteq V^*$ for all $1 \leq i \leq n$. A configuration represents the sets of strings (remember that each string appears in an arbitrarily large number of copies) which are present in any node at a given moment; clearly the initial configuration of the network is $C_0 = (A_1, A_2, \ldots, A_n)$. A configuration can change either by an evolutionary step or by a communicating step. When changing by a evolutionary step, each component L_i of the configuration is changed in accordance with the evolutionary rules associated with the node i.
Formally, we say that the configuration $C_1 = (L_1, L_2, \ldots, L_n)$ directly changes for the configuration $C_2 = (L'_1, L'_2, \ldots, L'_n)$ by a evolutionary step, written as

$$C_1 \Longrightarrow C_2$$

if L'_i is the set of strings obtained by applying the rules of R_i to the strings in L_i as follows:

- If the same substitiution rule may replace different occurrences of the same symbol within a string, all these occurrences must be replaced within different copies of that string. The result is the multiset in which every string that can be obtained appears in an arbitrarily large number of copies.
- Unlike their common use, deletion and insertion rules are applied only to the end of the string. Thus, a deletion rule $a \to \varepsilon$ can be applied only to a string which ends by a, say wa, leading to the string w, and an insertion rule $\varepsilon \to a$ applied to a string x consists of adding the symbol a to the end of x, obtaining xa.

– If more than one rule, no matter its type, applies to a string, all of them must be used for different copies of that string.

More precisely, since an arbitrarily large number of copies of each string is available in every node, after a evolutionary step in each node one gets an arbitrarily large number of copies of any string which can be obtained by using any rule in the set of evolution rules associated with that node. By definition, if L_i is empty for some $1 \leq i \leq n$, then L'_i is empty as well.

When changing by a communication step, each node processor sends all copies of the strings it has which are able to pass its output filter to all the other node processors and receives all copies of the strings sent by any node processor providing that they can pass its input filter.
Formally, we say that the configuration $C_1 = (L_1, L_2, \ldots, L_n)$ directly changes for the configuration $C_2 = (L'_1, L'_2, \ldots, L'_n)$ by a communication step, written as

$$C_1 \vdash C_2$$

if for every $1 \leq i \leq n$,

$$L'_i = L_i \setminus \{w \in L_i \mid \tau_i(w) = \underline{true}\} \cup \bigcup_{j=1, j\neq i}^{n} \{x \in L_j \mid \tau_j(x) = \underline{true} \text{ and } \rho_i(x) = \underline{true}\}.$$

Let $\Gamma = (V, N_1, N_2, \ldots, N_n)$ be an NEP. By a computation in Γ we mean a sequence of configurations $C_0, C_1, C_2, \ldots$, where C_0 is the initial configuration, $C_{2i} \Longrightarrow C_{2i+1}$ and $C_{2i+1} \vdash C_{2i+2}$ for all $i \geq 0$.

If the sequence is finite, we have a finite computation. The result of any finite computation is collected in a designated node called the output (master) node of the network. If one considers the output node of the network as being the node k, and if $C_0, C_1, \ldots, C_t$ is a computation, then the set of strings existing in the node k at the last step - the k-th component of C_t - is the result of this computation. The time complexity of the above computation is the number of steps, that is t.

3 Solving NP-Complete Problems

In this section we attack one problem known to be NP-complete, namely the Bounded Post Correspondence Problem (BPCP) [2, 7] which is a variant of a much celebrated computer science problem, the Post Correspondence Problem (PCP) known to be unsolvable [9] in the unbounded case, and construct a NEP for solving it. Furthermore, the proposed NEP computes all solutions.

An instance of the PCP consists of an alphabet V and two lists of strings over V

$$u = (u_1, u_2, \ldots, u_n) \quad \text{and} \quad v = (v_1, v_2, \ldots, v_n).$$

The problem asks whether or not a sequence $i_1, i_2, \ldots, i_k$ of positive integers exists, each between 1 and n, such that

$$u_{i_1} u_{i_2} \ldots u_{i_k} = v_{i_1} v_{i_2} \ldots v_{i_k}.$$

The problem is undecidable when no upper bound is given for k and NP-complete when k is bounded by a constant $K \leq n$. A DNA-based solution to the bounded PCP is proposed in [10].

Theorem 1 *The bounded PCP can be solved by an NEP in size and time linearly bounded by the product of K and the length of the longest string of the two Post lists.*

Proof. Let $u = (u_1, u_2, \ldots, u_n)$ and $v = (v_1, v_2, \ldots, v_n)$ be two Post lists over the alphabet $V = \{a_1, a_2, \ldots, a_m\}$ and $K \geq n$. Let

$$s = K \cdot \max\ (\{|u_j| \mid 1 \leq j \leq n\} \cup \{|u_j| \mid 1 \leq j \leq n\}).$$

Consider a new alphabet

$$U = \bigcup_{i=1}^{m} \{a_i^{(1)}, a_i^{(2)}, \ldots, a_i^{(s)}\} = \{b_1, b_2, \ldots, b_{sm}\}.$$

For each $x = i_1 i_2 \ldots i_j \in \{1, 2, \ldots, n\}^{\leq K}$ (the set of all sequences of length at most K formed by integers between 1 and n), we define the string

$$u^{(x)} = u_{i_1} u_{i_2} \ldots u_{i_j} = a_{t_1} a_{t_2} \ldots a_{t_{p(x)}}.$$

We now define a one-to-one mapping $\alpha : V^* \longrightarrow U^*$ such that for each sequence x as above $\alpha(u^{(x)})$ does not contain two occurrences of the same symbol from U. We may take

$$\alpha(u^{(x)}) = a_{t_1}^{(1)} a_{t_2}^{(2)} \ldots a_{t_{p(x)}}^{(p(x))}.$$

The same construction applies to the strings in the second Post list v. We define

$$F = \{\alpha(u^{(x)}\alpha(v^{(x)}) \mid x \in \{1, 2, \ldots, n\}^{\leq K}\} = \{z_1, z_2, \ldots, z_l\}$$

and assume that $z_j = b_{j,1} b_{j,2} \ldots b_{j,r_j}$, $1 \leq j \leq l$, where $|z_j| = r_j$. By the construction of F, no letter from U appears within any string in F for more than two times. Furthermore, if each letter of $z = \alpha(u^{(x)})\alpha(v^{(x)})$ appears twice within z, then x is a solution of the given instance.

We are now ready to define the NEP which computes all the solutions of the given instance. It is a NEP of size $2sm + 1$

$$\Gamma = (U \cup \bar{U} \cup \hat{U} \cup \tilde{U} \cup \{X\} \cup \{X_d^{(c)} \mid 1 \leq c \leq n, 2 \leq d \leq |z|_c\}, N_1, N_2, \ldots N_{2sm+1}),$$

where

$$\bar{U} = \{\bar{b} \mid b \in U\}$$
(the other sets, namely $\hat{U}$ and $\tilde{U}$, which form the NEP alphabet are defined similarly)

$$M_f = \{\varepsilon \to b_f\},$$
$$A_f = \emptyset,$$
$$FI_f = \{X_d^{(c)} \mid 2 \le d \le |z|_c, 1 \le c \le l \text{ such that } b_f \ne b_{c,d}\} \cup \bar{U} \cup \hat{U} \cup \tilde{U},$$
$$PI_f = FO_f = PO_f = \emptyset,$$

for all $1 \le f \le sm$,

$$M_{sm+1} = \{X \to X_2^{(c)} \mid 1 \le c \le l\} \cup \{X_{|z|_c} \to b_{c,1} \cup \{b_d \to \bar{b}_d \mid 1 \le d \le sm\}$$
$$\cup \{X_d^{(c)} \to X_{d+1}^{(c)} \mid 1 \le c \le l, 2 \le d \le |z|_c - 1\},$$
$$A_{sm+1} = \{X\},$$
$$FI_{sm+1} = PI_{sm+1} = FO_{sm+1} = PO_{sm+1} = \emptyset,$$

and

$$M_{sm+d+1} = \{b_d \to \tilde{b}_d, \bar{b}_d \to \hat{b}_d\},$$
$$A_{sm+d+1} = \emptyset,$$
$$FI_{sm+d+1} = (\bar{U} \setminus \{\bar{b}_d\}) \cup \{X_g^{(c)} \mid 2 \le g \le |z|_c, 1 \le c \le l\},$$
$$PI_{sm+d+1} = FO_{sm+d+1} = \emptyset,$$
$$PO_{sm+d+1} = \{\tilde{b}_d, \hat{b}_d\},$$

for all $1 \le d \le sm$.

Here are some informal considerations about the computing mode of this NEP. It is easy to note that in the first stage of a computation only the processors $1, 2, \dots, sm+1$ are active. Since the input filter of the others contains all symbols of the form $X_g^{(c)}$, they remain inactive until one strings from F is produced in the node $sm + 1$.

First let us explain how an arbitrary string $z_j = b_{j,1} b_{j,2} \dots b_{j,r_j}$ from F can be obtained in the node $sm + 1$. One starts by applying the rules $X \to X_2^{(j)}$, $1 \le j \le l$, in the node $sm + 1$. The strings $X_2^{(j)}$, $1 \le j \le l$, obtained in the node $sm + 1$ as an effect of a evolutionary step are sent to all the other processors, but for each of these strings there is only one processor which can receive it. For instance the string $X_2^{(c)}$ is accepted only by the node processor f, $1 \le f \le sm$, with $b_f = b_{c,2}$. In the next evolutionary step, the symbol $b_{j,2}$ is added to the right hand end of the string $X_2^{(j)}$ for all $1 \le j \le l$. Now, a communication step is to be done. All the strings $X_2^{(j)} b_{j,2}$ can pass the output filters of the nodes processors where they were obtained but the node processor $sm+1$ is the only one which can receive them. Here the lower subscripts of the symbol X are increased by one and the process from above is resumed in the aim of adjoining a new letter. This process does not apply to a string $X_r^{(j)} b_{j,2} \ldots b_{j,r}$ anymore, if and only if $r = |z_j|$, when $X_r^{(j)}$ is replaced by $b_{j,1}$ resulting in the string z_j. By these considerations, we infer that all the strings from F are produced in the node $sm + 1$ in $2s$ steps.

Another stage of the computation checks the number of occurrences of any letter within any string obtained in the node $sm + 1$, as soon the string contains only letters in U. This is done as follows. By the way of applying the substitution rules aforementioned, each occurrence of any letter is replaced by its barred version in the node $sm + 1$. Let us consider a string produced by such an evolutionary step. Such a string has only one occurrence of a symbol $\bar{b}_d$, for some $1 \leq d \leq sm$, the other symbols being from $U \cup \hat{U} \cup \tilde{U}$. It can pass the input filter of the processor $sm + d + 1$ only, where it remains for three steps (two evolutionary steps and one comunication one) or forever. The string can leave the node $sm + d + 1$, only if it has an ooccurrence of the symbol b_d. By replacing this occurrence with $\tilde{b}_d$ and $\bar{b}_d$ with $\hat{b}_d$, the string can pass the output filter of the node processor $sm + d + 1$ and goes to the node $sm + 1$. In this way, one checked whether or not the original string had have two occurrences of the letter b_d. After $6s$ steps the computation stops and the node $sm + 1$ has only strings which were produced from those strings in F having two occurrences of any letter. As we have seen, these strings encode all the solutions of the given instance of BPCP. □

4 Concluding Remarks

We have proposed a computational model whose underlying architecture is a complete graph having evolutionary processors placed in its nodes. Being a bio-inspired system, a natural question arises: How far is this model from the biological reality and engineering possibilities? More precisely, is it possible exchange biological material between nodes? Can the input/output filter conditions of the node processors be biologically implemented? What about a technological implementation? We hope that at least some answers to these questions are affirmative.

We have presented a linear algorithm based on this model which provide all solutions of an NP-complete problem.

Further, one can go to different directions of research. In our variant, the underlying graph is the complete graph. In the theory of networks some other types of graphs are common, e.g., rings, grids, star, etc. It appears of interest to study the networks of evolutionary processors where the underlying graphs have these special forms.

A natural question concerns the computational power of this model. Is it computationally complete? However, our belief is that those variants of the model which are "specialized" in solving a few classes of problems have better chances to get implemented, at least in the near future.

References

1. J. P. Banâtre, A. Coutant, D. Le Metayer, A parallel machine for multiset transformation and its programming style, *Future Generation Computer Systems*, 4 (1988), 133–144.
2. R. Constable, H. Hunt, S. Sahni, On the computational complexity of scheme equivalence, *Technical Report* No. 74-201, Dept. of Computer Science, Cornell University, Ithaca, NY, 1974.
3. E. Csuhaj - Varju, J. Dassow, J. Kelemen, Gh. Paun - *Grammar Systems*, Gordon and Breach, 1993.
4. E. Csuhaj-Varjú, Networks of parallel language processors. In *New Trends in Formal Languages* (Gh. Păun, A. Salomaa, eds.), LNCS 1218, Springer Verlag, 1997, 299–318
5. L. Errico, C. Jesshope, Towards a new architecture for symbolic processing. In *Artificial Intelligence and Information-Control Systems of Robots '94* (I. Plander, ed.), World Sci. Publ., Singapore, 1994, 31–40.
6. S. E. Fahlman, G. E. Hinton, T. J. Seijnowski, Massively parallel architectures for AI: NETL, THISTLE and Boltzmann machines. In *Proc. AAAI National Conf. on AI*, William Kaufman, Los Altos, 1983, 109–113.
7. M. Garey, D. Johnson, *Computers and Intractability. A Guide to the Theory of NP-completeness*, Freeman, San Francisco, CA, 1979.
8. W. D. Hillis, *The Connection Machine*, MIT Press, Cambridge, 1985.
9. J. Hopcroft, J. Ulmann, *Formal Languages and Their Relation to Automata*, Addison-Wesley, Reading, MA, 1969.
10. L. Kari, G. Gloor, S. Yu, Using DNA to solve the Bounded Correspondence Problem, *Theoret. Comput. Sci.*, 231 (2000), 193–203.
11. Gh. Păun, Computing with membranes, *J. Comput. Syst. Sci.* 61(2000). (see also *TUCS Research Report* No. 208, November 1998, http://www.tucs.fi.)
12. Gh. Păun, G. Rozenberg, A. Salomaa, *DNA Computing. New Computing Paradigms*, Springer-Verlag, Berlin, 1998.

Using SOM for Neural Network Visualization

G. Romero[1], P.A. Castillo[1], J.J. Merelo[1] and A. Prieto[1]

Department of Architecture and Computer Technology
University of Granada
Campus de Fuentenueva
E. 18071 Granada (Spain)
e-mail: todos@geneura.ugr.es URL: http://www.geneura.org

Abstract. Software visualization is an area of computer science devoted to supporting the understanding and effective use of algorithms. The application of software visualization to Evolutionary Computation has been receiving increasing attention during the last few years. In this paper we apply visualization technique to an evolutionary algorithm for multilayer perceptron training. Our goal is to better understand its internal behavior in order to improve the evolutionary part of the method. The effect of several genetic operators are compare and also the difference with a fitness sharing version of the algorithm.

1 Introduction

Evolutionary Algorithms (EA) work in an algorithmically simple way but produce a vast amount of data. Apart from simple convergence toward the solution, the extraction of useful information to get further insight into the state and course of the algorithm is a non-trivial task. Understanding their behavior is difficult due to the fact that EAs adopt an interactive stochastic approach to searching large problem spaces.

EA users usually examine how the quality of the solutions found changes over time by using a graph of fitness versus generation number. Although such a graph illustrates the improvements in the quality of the solutions considered during the algorithm run, it does not illustrate the structure of the solutions being considered, or the regions of the search space being explored. Visualization is proposed as a useful method for solving this problem; only by understanding the search behavior of their algorithms can EA users be confident about their individual algorithm components and parameter settings.

2 Multidimensional visualization

Most techniques for visualization are limited to representing data depending on one or two variables. This is due to the fact that human vision is limited to three dimensions but there are two possible extensions to go beyond this limitation: using color for the fourth dimension and time as the fifth dimension. Neither

J. Mira and A. Prieto (Eds.): IWANN 2001, LNCS 2084, pp. 629-636, 2001.

possibility is very common and requires practice, especially if time is used for visualizing the fifth dimension. However, if the problem incorporates more than five dimensions a method for visualizing arbitrarily high dimensions must be used.

For the visualization of multidimensional data, a method to transform multidimensional data to a lower dimension is needed, preferably to 2 or 3 dimensions. This transformation should provide a lower-dimensional picture where the dissimilarities between the data points of the multidimensional domain correspond to the dissimilarities of the lower-dimensional domain. These transformation methods are referred to as *multidimensional scaling* ([1, 2]).

To measure the dissimilarity, the distances between pairs of data points is used. These distances can be genuine distance in the high-dimensional domain, for instance the euclidean distance.

One of the best known methods for multidimensional scaling is *Sammon Maps* [3]. Sammon Maps use a Newton method (steepest gradient descent). However this method is not very robust and diverges without special interaction [2]. Besides, it needs to be trained again for every new point that needs to be projected.

Other multidimensional scaling methods are *BFGS* [4] and *RPROP* [5]. BFGS is a standard optimization method included in Matlab. It uses a Quasi-Newton method with a mixed linear quadratic and cubic line search procedure. RPROP is a more robust search method. The RPROP algorithm uses just the change of the sign of the gradient for step size control. It is widely used in the field of neural networks. Both algorithms produce good results. However, when using the BFGS method, multiple runs must be performed and then results must be valued by the user. The RPROP algorithm is slower but produces more consistent results. Like Sammon Maps, they have to be trained again in order to project new points.

Another method is Kohonen's *Self-Organizing Maps* (SOM) [6]. SOM is an artificial neural net specially adapted for the task in question. With other methods, once the data points are transformed from high-dimensional space to the lower-dimensional one, no new points can be projected without repeating the whole process. SOM has the advantage of being able to transform new points between spaces very easily once it has been trained.

3 Visualization of Neural Networks Evolution

A neural net can be seen as a real number vector formed by all its weights. If the net has 2 or 3 weights then we need a 2 or 3 dimensional space to plot neural nets on top. This is easy. The problem arises when the net has more than 3 weights, the most common case. Here is were some multidimensional scaling method is needed. The idea is to use a SOM to reduce from a high dimensional space to a 2 dimensional one easily viewed by anybody.

In a previous work [7], the SOM was trained with all the individuals generated during several experimental runs. This way we were able of inspecting the search

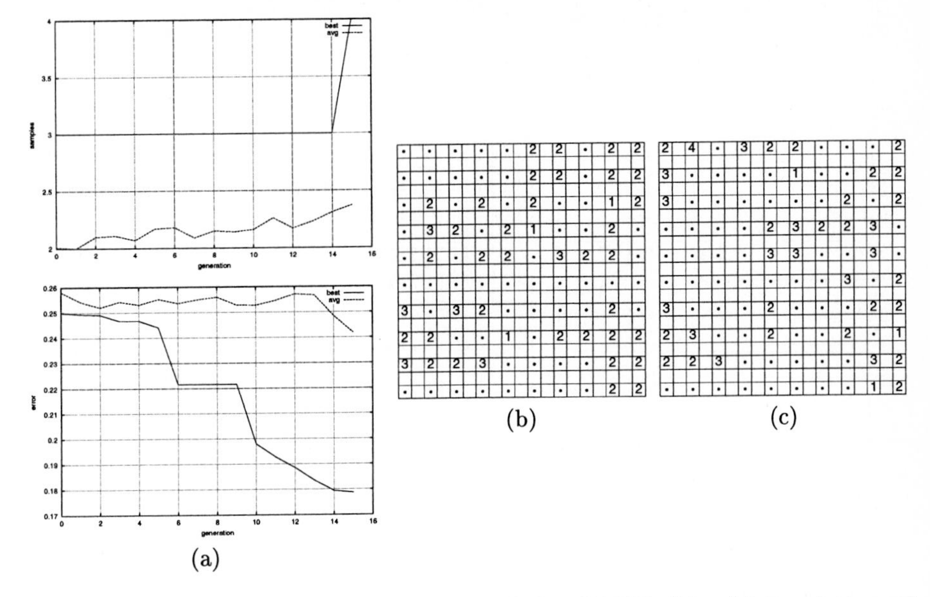

Fig. 1. (a) Number of patterns correctly classified and MSE. (b) y (c) Populations of generations 7 and 15 plotted over a SOM for the XOR problem. Numbers indicate how many patterns have been correctly classified.

space area crossed by the genetic algorithm during its search. Now we want to see all the search space, or, at least, as much as possible. For this reason, this time the SOM was trained with a grid of a n-dimensional space, were n is the length of the vector which represent a neural net. A different map is needed for every kind of problem with a different number of dimensions or a different width inside that n-dimensional space.

As a example, if we work with a neural net with 6 weights, then we need a grid over the 6 dimensional real space to train the SOM. Besides we must limit the width (distance from origin or $< 0, 0, 0, 0, 0, 0 >$) to a certain value easily recoverable from a experimental run of the genetic algorithm, usually between 1 and 5. Another option would be to choose a certain number of random points from the 6-dimensional space.

4 Genetic Operators Effects

The first experiment studies the evolution of MLP's (MultiLayer Perceptrons) with 3 neurons to learn the XOR problem. As every has 2 weights and a bias, the resulting real vector who represent a neural net is of dimension 9. The SOM for this problem was 10x10 neurons and was trained with a grid of with $4^9 =$ 262144 samples.

The first question we tray to see was how the different genetic operator affect the evolutionary process. Usually you know how an operator affects a population,

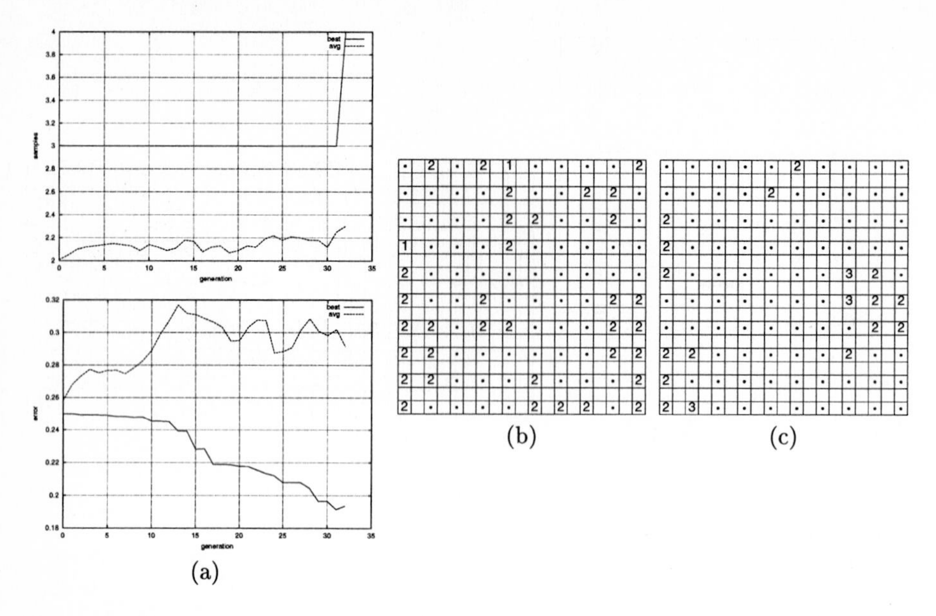

Fig. 2. (a) Number of patterns correctly classified and MSE. (b) y (c) Populations of generations 16 and 32 plotted over a SOM for the XOR problem. Numbers indicate how many patterns have been correctly classified.

but the idea is to be able of seeing it. For this experiment 3 different operators were used: uniform mutation, 1-point crossover and quickprop.

As the problem is very easy, the algorithm always find the solution in a few generations. In typical run of the algorithm, see Figure 1, the solution is found between generations 10 and 20.

To evaluate the effect of the different genetic operators we apply only one of then in the next executions of the genetic algorithm. In Figure 2 we can see the effect of applying only mutation.

The most interesting thing is to watch to evolution of the population over the SOM. This way we observed that mutation operator creates a lot of diversity as new individuals appear in areas that were empty before. As can be seen in Figure 3, this behavior is the opposite of the quickprop operator. Quickprop modifies neural nets making them more similar every generation, it diminishes diversity. The effect is clear if we see all the pictures for every generation. After 25 generations (Figure 3b) there are only a few areas with individuals, and only a couple at the end (Figure 3c).

There is another way of comparing the degree of diversity created, or destroyed, by this operators. In Figure 4 there is a mapping of all the fitnesses obtained over all the generations for the GA using mutation or quickprop. As can be seen the GA using only quickprop leaves some areas unexplored. Here we use a bigger SOM of 20x20 neurons to make the difference more evident.

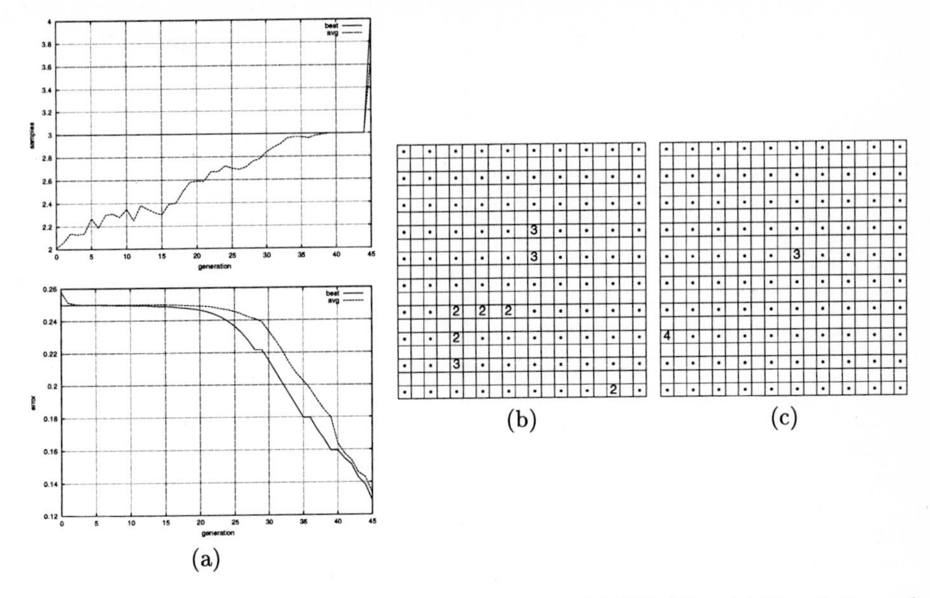

Fig. 3. (a) Number of patterns correctly classified and MSE. (b) y (c) Populations of generations 25 and 45 plotted over a SOM for the XOR problem. Numbers indicate how many patterns have been correctly classified.

No pictures from early generations is shown here because they always look the same. As the initial population is randomly chosen from the search space, the usually cover between 90% and95% of the SOM surface if the population size is big enough.

About the latest generations the opposite can be said. At the end of the evolutionary process the diversity decreases very quickly and only all the individuals are mapped on a small number of SOM neurons. To avoid this common problem we used a fitness sharing technique. The results of the applying of sharing are shown in the next section.

5 Sharing Effects

To study the effect of sharing we choose a more difficult problem proposed by Gorman and Sejnowski [?] called sonar. The problem has with 7 inputs and 2 outputs.

In first place we will show the results of the genetic algorithm, without sharing, just to see how it develops. Then we will see the results of the modified algorithm, now with sharing.

As expected, with a more difficult problem the genetic algorithm quickly lose diversity and after a few generations (see Figure 4b) almost all the neural nets lay in a few cells of the SOM. We sometimes face this problem when using

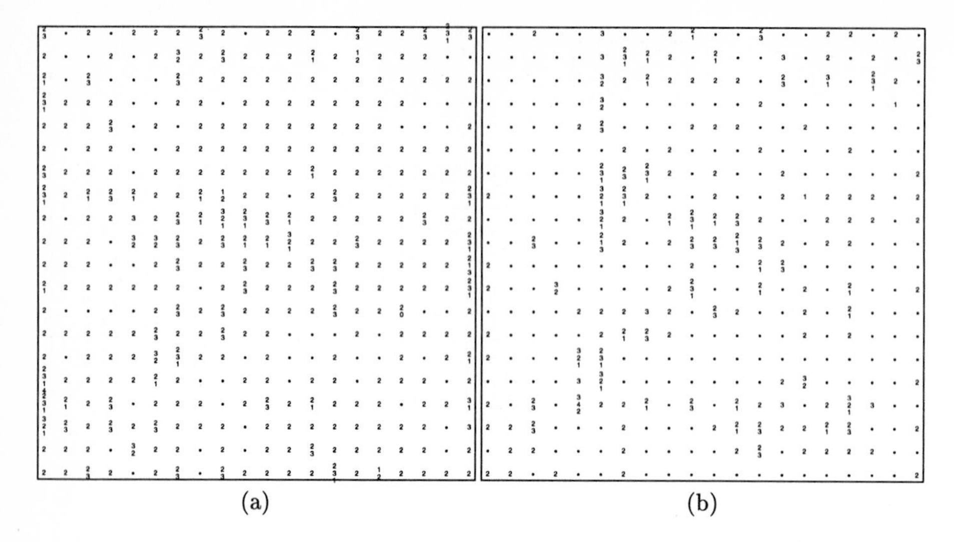

Fig. 4. (a) SOM with nets from GA using mutation. (b) SOM with nets from GA using quickprop. A black dot inside a cell means it is an unexplored region of the search space.

any local search method as operator inside of the genetic algorithm. For this reason the algorithm needs more generations and sometimes it didn't find the solution within a certain time restriction, 1000 generations in this case. In the example show here the process start with in a good random distribution over the SOM, but hardly in 20 generations the search is concentrated in only a couple of small areas. This way a big number of potentially good solutions will never be explored. The situation becomes patent if you look at the sequence formed by all the pictures from every generation. This is a useful visual method to discover the degree of diversity of our population in a look.

Lets see now the effect of the sharing in Figure 6. Again the expected result can be easily extracted from the mapping of the population over the SOM. Now many more areas of the SOM are occupied by individuals during many more generations.

Sharing parameters should be carefully chosen to get the desired rate of exploration and exploitation. Here we implement it as shown by [8]. To explore too many areas can be very time and resource consumpting but it is necessary. The same with exploitation, good individuals should be allow to reproduce and get similar, and hopefuly better, offsprings, but if too much you will be again wasting resources. For this experiment we obtein the best results with alpha = 0.5 and sigma = 2. A rude explanation for this parameter can be: sigma is the distance between neighbors and alpha is how the increasing number of neighbors affect its fitness.

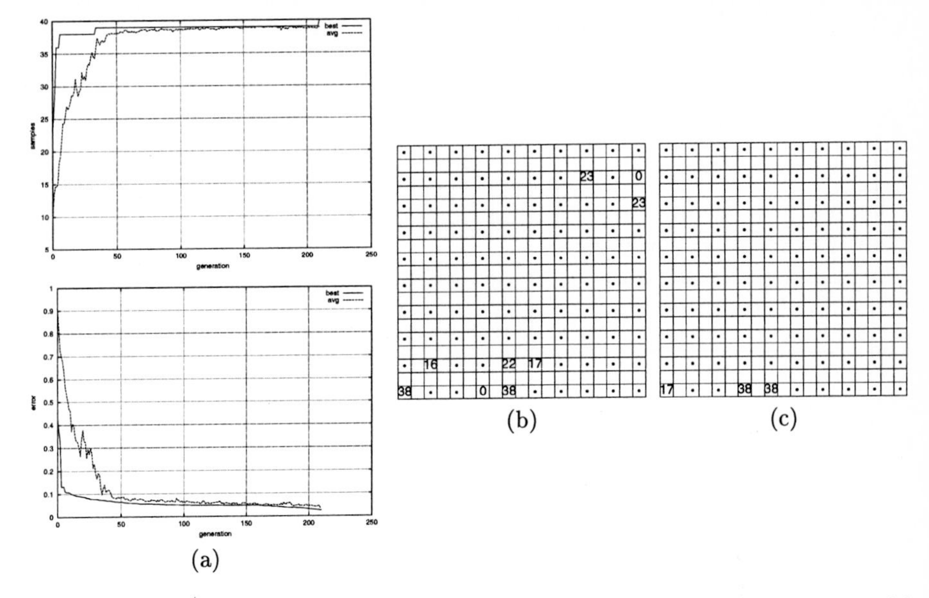

Fig. 5. The basic version of the genetic algorithm lose diversity in a few generations. (a) Number of patterns correctly classified and MSE. (b) y (c) Populations of generations 15 and 40 plotted over a SOM for pattern recognition problem. Numbers indicate how many patterns have been correctly classified.

6 Conclusions and future work

The use of visualization techniques provided a better knowledge about the evolutionary algorithm and gave us some clues about how it can be improved. The following conclusions were drawn from the application of visualization techniques to G-PROP:

- Local search methods, as Quickprop, must be used with care because they can improve the GA search but at the same time they are a diversity killers.
- Fitness sharing techniques are only worth with difficult problems when the GA is more likely to get trap in local solutions. For small or easy problems it is not worth, being only a waste of time and resorces.

Some future lines of work are the evaluation of only potentially good individuals (th ones mapped onto good regions by SOM) and the writting of an application to gain some kind of interactivity during the evolutionary process to change parameters dynamically and add or remove individuals to/from any region of the search space using SOM neurons as MLPs.

References

1. T.F. Cox and M.A.A. Cox. *Multidimensional Scaling.* London: Chapman & Hall, 1994.

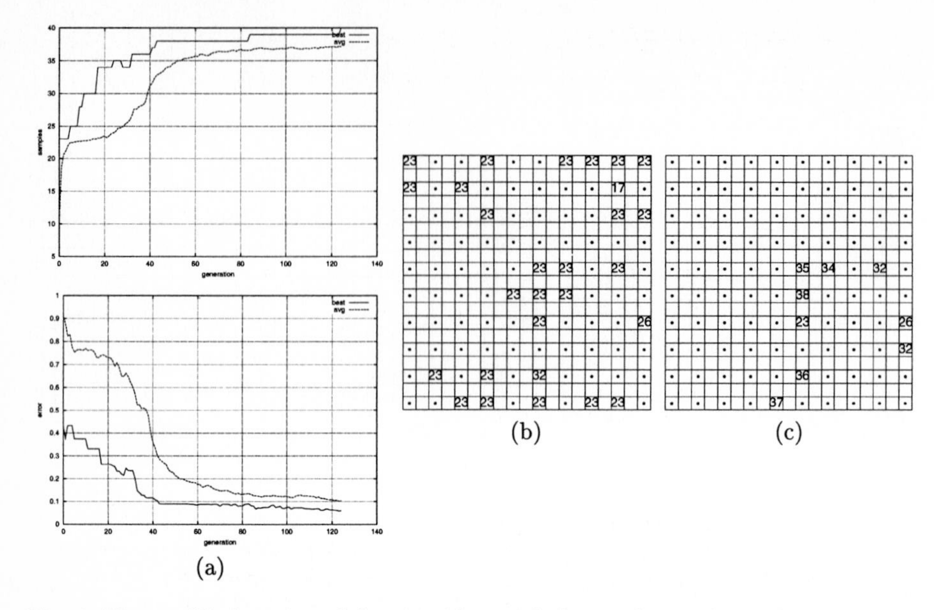

Fig. 6. The modified version of the algorithm with fitness sharing keeps diversity more time. (a) Number of patterns correctly classified and MSE. (b) y (c) Populations of generations 30 and 60 plotted over a SOM for a pattern recognition problem. Numbers indicate how many patterns have been correctly classified.

2. B.D. Ripley. *Pattern Recognition and Neural Networks.* Cambridge, GB: Cambridge University Press, 1996.
3. J.W. Sammon Jr. A nonlinear mapping for data structure analysis. *IEEE Transactions on Computers*, pages 401–409, 1969.
4. Matlab - User Guide. Natick, Mass: The Mathworks, Inc, 1994-1996.
5. M. Riedmiller and H. Braun. A direct adapatative method for faster backpropagation learning: The RPROP algorithm. In H. Ruspini, editor, *Proceedings of the IEEE International Conference on Neural Networks (ICNN)*, pages 586–591, 1993.
6. T. Kohonen. The Self-Organizing Map. In *Proceedings of the IEEE*, volume 78, pages 1464–1480, 1990.
7. G.Romero; P.A. Castillo; M.G. Arenas; J.G. Castellano; J.Carpio; J.J. Merelo; A. Prieto; V. Rivas. Evolutionary Computation Visualization: Application to GPROP. In Schoenauer et al., editor, *Proceeding of the PPSN*, pages 902–912, 2000.
8. Z. Michalewicz. *Genetic Algorithm + Data Structure = Evolution Programs.* Springer Verlag, 1992.

Acknowledgments

This work has been supported in part by FEDER I+D project 1FD97-0439-TEL1, CICYT TIC99-0550 and INTAS 97-30950.

Comparison of Supervised Self-Organizing Maps Using Euclidian or Mahalanobis Distance in Classification Context

F. Fessant[1], P. Aknin[1], L. Oukhellou[2], S. Midenet[1],

[1] INRETS, 2 av. Général Malleret Joinville,
F-94114 Arcueil, France
{fessant, aknin, midenet}@inrets.fr
[2] LETIEF, Université Paris 12, 61 av. Général de Gaulle,
F-94010 Créteil, France
oukhellou@univ-paris12.fr

Abstract. The supervised self-organizing map consists in associating output vectors to input vectors through a map, after self-organizing it on the basis of both input and desired output given altogether. This paper compares the use of Euclidian distance and Mahalanobis distance for this model. The distance comparison is made on a data classification application with either global approach or partitioning approach. The Mahalanobis distance in conjunction with the partitioning approach leads to interesting classification results.

1 Introduction

The self-organizing map -or SOM- is a well-known and quite widely used model that belongs to the unsupervised neural network category concerned with classification processes. The LASSO model (Learning ASsociations by Self-Organization) used in this work, can be considered as an extension of self-organizing maps and allows the classification process in a supervised way [1]. The LASSO model had been tested on pattern recognition tasks [1,2] and it has been shown that the encoding and use of supervision data during the learning phase improve the well-classification rate compared to the standard SOM results.

This paper focuses on the metric choice for the prototype-to-observation distance estimation required during the self-organization and exploitation phases. The distance most commonly used in SOM is the Euclidian distance that considers each observation dimension with the same significance whatever the observation distribution inside classes. Obviously, if the data set variances are not uniformly shared out among the input dimensions, the use of Mahalanobis distance becomes interesting.

After the presentation of supervised SOM and metrics, the article introduces two different classification strategies : a global approach and a partitioning approach. We

J. Mira and A. Prieto (Eds.): IWANN 2001, LNCS 2084, pp. 637–644, 2001.

give classification performances for the two metrics on a data classification application concerning non-destructive evaluation of rail. We show that the choice of a Mahalanobis metric can clearly improve the classification results, specially in the context of partitioning approach.

2 The supervised self-organizing map

2.1 Standard SOM

A SOM is a neural network model made up of a set of prototypes (or nodes) organized into a 2-dimension grid. Each prototype j has fixed coordinates in the map and adaptive coordinates $W(j)$ (or weights) in the input space. The input space is relative to the variables setting up the observations. Two distances are defined, one in the original input space and one in the 2-dimension space of the map for which Euclidian distance and integer map coordinates are always used. The definition choice of first distance will be discussed in section 3. The self-organizing process slightly moves the prototype coordinates in the data definition space -i.e. adjusts W- according to the data distribution. The chosen learning algorithm uses a learning rate and a smooth neighbouring function continuously decreasing with iterations [3].

The SOM is traditionally used for classification purpose. The exploitation phase consists in associating an observation with its closest prototype, called the image-node. The SOMs have the well known ability [4] that the observation projection on the map preserves the proximities : close observations in the original input space are associated with close prototypes in the map.

2.2 LASSO model

Midenet and Grumbach [1] proposed the exploitation of SOM in a supervised way through the LASSO model. As SOM, the LASSO model associates a 2-dimension map with a communication layer. However, 2 sets of nodes are distinguished on the communication layer : observation nodes and class coding nodes. During the learning phase, the whole communication layer (observation nodes and class nodes) is used as input layer (fig. 1). This phase is done as in the classical Kohonen model. The LASSO map self-organizes with observations and associated classes being presented altogether at the network input.

On the other hand, during the exploitation phase, only the observation nodes are used as inputs. The map is used to associate missing data (class estimation) with partial information (observation) in the following way :

1. presentation of an observation X on the input layer

2. selection of the image-node by minimizing the distance between observation and prototypes. The class coding dimensions are simply ignored during the selection
3. the estimation of the class is computed from the weights of the image-node in the class coding dimensions (fig. 1)

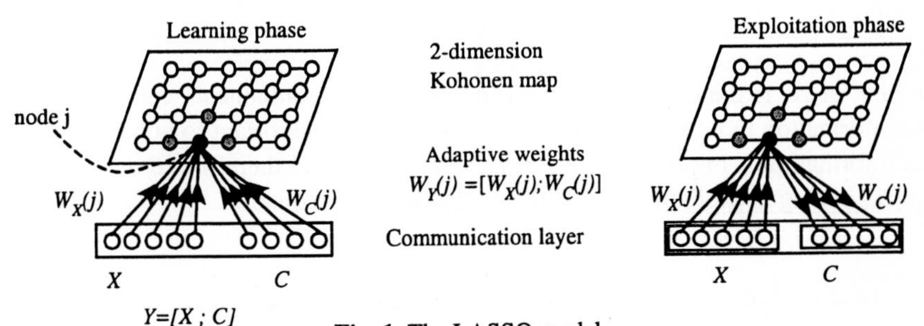

Fig. 1. The LASSO model

3 Metric choice

The original contribution of this paper lies in the use of a non-euclidian metric in learning phase as well as exploitation phase. The distance d_N in a N-dimension space between 2 patterns R and S is given in its generalized form with the following expression :

$$d_N^2(R,S) = (R-S)^T \mathbf{A}^{-1}(R-S) \quad . \tag{1}$$

with **A** being a N by N positive normalisation matrix. If **A** is the identity matrix, the distance is Euclidian. If **A** is the variance-covariance matrix, this distance is refereed as the Mahalanobis distance. In a multi-classes problem, the usual choice of **A** is given by the average of inner class variance matrices [5].

$$\mathbf{A} = \sum_{i=1}^{K} n_{\Omega_i} \mathbf{V}_{\Omega_i} \quad . \tag{2}$$

with n_{Ω_i} the observation number in class Ω_i, $\mathbf{V}_{\Omega_i}$ the variance-covariance matrix of class Ω_i and K the class number.

3.1 Use of Mahalanobis distance in LASSO learning phase

The distance d_N in the N-dimension input space between a complete sample Y (observation X + associated class C) and a prototype j is given by :

$$d_N^2(Y,j) = (Y-W_Y(j))^T \mathbf{A}_Y^{-1}(Y-W_Y(j)) \quad . \tag{3}$$

$\mathbf{A}_Y$ (dimension N x N) being estimated from learning set with $Y=[X\ ;\ C]$. C has N_C dimensions. According to equation 2, $\mathbf{A}_Y$ can be written as :

$$\mathbf{A}_Y = \begin{bmatrix} [\mathbf{A}_X] & 0\cdots0 \\ \ddots & \ddots \\ 0\ \cdots & 0\cdots0 \end{bmatrix} . \quad (4)$$

The rank of $\mathbf{A}_Y$ is the dimension of X (i.e. $N-N_C$). The components corresponding to class information in $\mathbf{A}_Y$ are equal to zero. However, it is important to point out that the class information is implicitly taken into account in $\mathbf{A}_X$ and therefore in d_N. Otherwise the class coding is explicitly used in the weight updating equation.

3.2 Use of Mahalanobis distance in LASSO exploitation phase

During the exploitation phase, the distance is only computed in the observation dimensions ($N-N_C$ dimensions). The class coding dimensions are ignored.

$$d^2_{N\text{-}Nc}(X,j) = (X - W_X(j))^T \mathbf{A}_X^{-1}(X - W_X(j)) \ . \quad (5)$$

Despite the dimension reduction, the part of **A** that is used in equation 5 is the same as in learning phase. Distances used in learning and in exploitation phases are consistent.

4 The classification task

4.1 Global classification approach

A single LASSO model is designed for the simultaneous discrimination of the whole set of classes. Inducing by a disjunctive coding, there are as many class nodes as classes to be discriminated. Each output delivers a level that can be considered as an estimation of class belonging probability of the observation (fig. 2 presents the architecture used to solve a global K=4 classes problem).

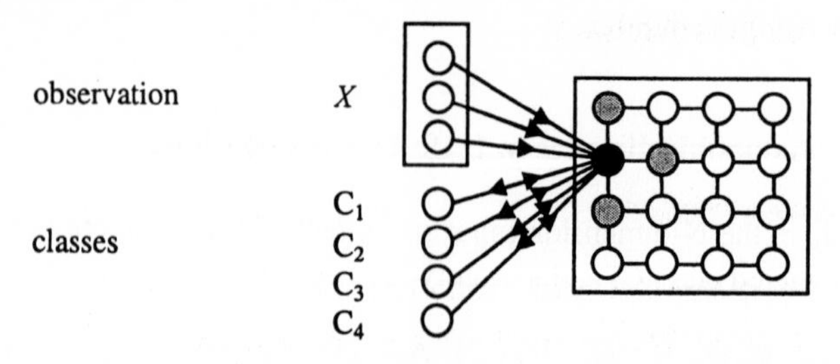

Fig. 2. A single LASSO model for the simultaneous discrimination of 4 classes

The decision rule for the global classifier is simple : during the exploitation phase, the assigned class corresponds to the class node component with the highest weight level.

4.2 Partitioning classification approach

The classification task is now split into several elementary tasks : K sub-problems are generated ; each of them is dedicated to the separation of one class among the others. Each sub-problem is solved with a specific sub-LASSO model. Fig. 3 presents the architecture used to solve a 4 classes classification problem, with 4 sub-networks which deliver some kind of estimation of posterior probability class membership. The decision rule for the final classification corresponds to the maximum posterior probability estimation.

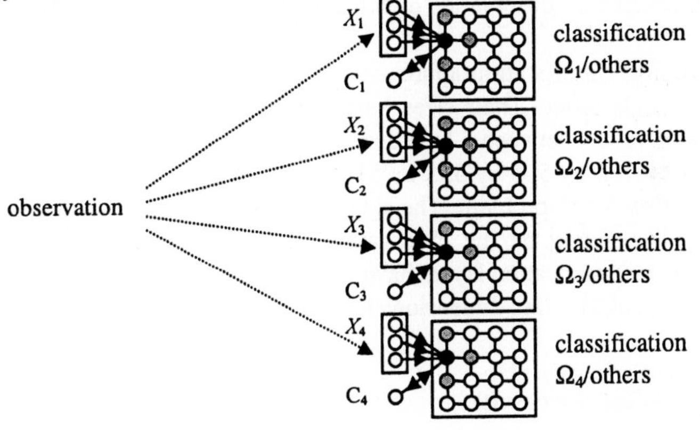

Fig. 3. Complete architecture for a 1 by 4 partitioning approach

The partitioning approach allows us to independently design each sub-network and the input vectors can be adjusted in order to be more relevant for the separability of each particular class ; sub-network input dimension can differ one from another. Moreover, the learning phase of each sub-classifier is independent, which gives us a better control than a global learning.

5 Rail defect classification

5.1 Application description

The application concerns the classification of rail defect signatures measured by a specific eddy current sensor [6]. After a raw signal parametrisation and a reduction of the number of descriptors [7], a classification procedure is carried out in order to assign each defect to one of the 4 defined classes of defects or singularities : external crack, welded joint, joint before crossing, shelling. A database of 140 defects is available for the design and the evaluation of the classifiers (64 from class 1, 39 from class 2, 16 from class 3 and 21 from class 4).

5.2 Practical settings

For the global classification approach, a square map has been used. Its size has been experimentally determined by testing a whole range of values between 3 by 3 and 10 by 10. The best size was found to be 8 by 8. We adopted the disjunctive coding scheme for the class information : all the components of the 4-dimension class vectors are worth zero except the one associated with the represented class. This component is set to one. The observation variables are normalised. The communication layer has the following characteristics : observation dimension=22, class dimension=4.

For the partitioning classification approach, 4 square maps have been implemented, one for each sub-classifier. Their sizes have been determined for each of them by testing a range of values between 3 by 3 and 10 by 10. The best sizes were found to be respectively {7 by 7, 5 by 5, 5 by 5, 5 by 5}. These sizes correspond to the best well-classified rate of each sub-problem. The single class node of each sub-classifier is set to one or zero depending on the observation class. The observation variables are normalised. The communication layers have the following characteristics : observation dimensions={15,15,8,9}, class dimension=1.

5.3 Global classification results

Fig. 4 depicts the four class nodes weight maps for the LASSO with Mahalanobis metric, after self-organization (observation nodes weight maps are not represented). On the figure, a weight value is represented by a dot in gray scale : dark dot means low value whereas light dot means high value.

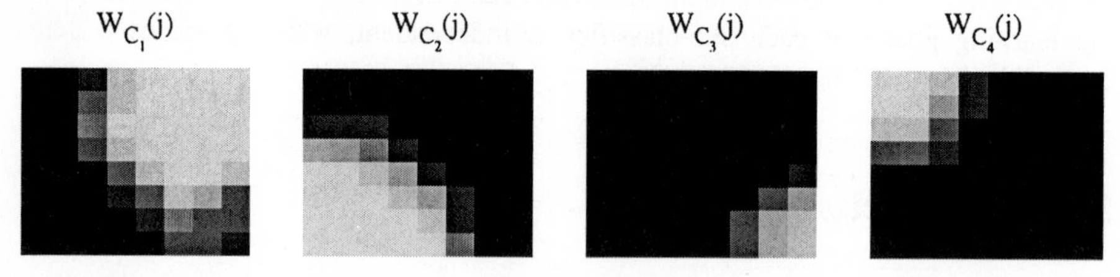

Fig. 4. Class node weight maps of one global classifier with Mahalanobis metric

The prototypes that specialize in one particular class recognition cluster in the map into a related neighbourhood. The areas are compact and do not overlap. The area associated to one class is more or less important in the map according to its occurrence frequency in the learning set ; we can observe that defects of classes 3 and 4 are treated with very few prototypes. This is a consequence of their low representation in the learning set. The class node map configurations reflect some similarity between classes ; for instance class 1 (crack) and 3 (joint before crossing) are lightly close. On the contrary, class 3 and class 4 are far away, indeed they stand in the opposite corners of the map.

The classification performances are summarized table 1. Because of the small number of observations, a "leave-one-out procedure" was applied for the classifiers evaluation [5] : one observation is removed from the complete data base ; then the remaining observations are used for training and the selected observation is used for testing. The procedure is repeated until all the observations of the initial base are listed. The classification rates have been obtained from the average of 20 results given by 20 complete leave-one-out procedures with different initial weights. Standard deviations σ are also pointed out table 1.

Table 1. Global classification performances (in percentage)

	Euclidian metric	Mahalanobis metric
global classifier	**92.8% (σ=1.6)**	**94.1% (σ=1.6)**

The use of the Mahalanobis metric lightly improves the classification results. We can assume that this benefit is due to the processing operated by means of the variance-covariance matrix, which enables to take into account the variances and correlation among components.

5.4 Partitioning classification results

Table 2 summarizes the classifications rates for each sub-classifier studied independently, as well as the classification performance for the complete classifier.

Table 2. Classification performances for partitioning approach (in percentage)

	Euclidian metric	Mahalanobis metric
Class 1/others	89.7%	93.3%
Class 2/others	96.4%	97.9%
Class 3/others	93.7%	94.5%
Class 4/others	98.2%	97.8%
Complete classifier	**90% (σ=1.5)**	**95% (σ=1)**

A partitioning approach in conjunction with the Mahalanobis metric leads to improved classification performances. We can notice that some sub-problems remain rather difficult to solve, like the discrimination of class 1 and 3. The benefit of the Mahalanobis metric is about 5% in partitioning context. This result is quite worthwhile compared to the 1.3% in global context.

6 Conclusion

This paper proposed the comparison of supervised self-organizing maps designed with different distance measures : an Euclidian distance and a Mahalanobis one. This comparison is achieved on a classification task, a particular non destructive evaluation problem involving the discrimination of four classes of defects. Classification performances are given for two classification strategies : a global approach in which the whole classes are discriminated simultaneously and a partitioning one in which the multiclass problem is split into 2-class sub-problems.

Concerning our classification problem, the Mahalanobis distance turns out to be more effective because of the large range of data components variations. In fact, the giving up of Euclidian distance is advisable when the variance of input vector components are highly different.

The best classification results are obtained for a LASSO model using a Mahalanobis metric in conjunction with the partitioning approach. This approach is more efficient than the global one essentially for the reason that the preprocessings and network architecture can be better adapted to each sub-problem.

These results compare with those achieved on the same database with supervised neural networks like multilayer perceptrons or radial basis functions [8].

Acknowledgments. This research is supported by the French Research Ministry within the framework of the PREDIT program (Research program and Innovation in Transportation Systems). The RATP company coordinates this project.

References

1. Midenet, S., Grumbach, A.: Learning Associations by Self-Organization : the LASSO model. Neurocomputing, vol. 6. Elsevier, (1994) 343-361
2. Fessant, F., Aknin, P.: Classification de défauts de rail à l'aide de cartes auto-organisatrices supervisées. NSI'2000, Dinard (2000) 75-79
3. Ritter, H., Martinetz, T., Schulten, K.: Topology conserving maps for learning visuo motor coordination. Neural networks, vol. 2. (1989) 159-168
4. Kohonen, T.: Self-Organizing Maps. Springer, Heidelberg (1995)
5. Bishop, C. M.: Neural Networks for pattern recognition. Clarendon Press, Oxford (1997)
6. Oukhellou, L., Aknin, P., Perrin, J.P.: Dedicated sensor and classifier of rail head defects for railway systems. 8th IFAC Int. Symp. On Transportation Systems, Chania (1997)
7. Oukhellou, L., Aknin, P., Stoppiglia, H., Dreyfus, G.: A new decision criterion for feature selection. Application to the classification of non destructive testing signatures. EUSIPCO, vol. 1 Rhodes (1998) 411-414
8. Oukhellou, L., Aknin, P.: Hybrid training of radial basis function networks in a partitioning context of classification. Neurocomputing, vol. 28 (1999) 165-175

Introducing Multi-objective Optimization in Cooperative Coevolution of Neural Networks*

N. García-Pedrajas, E. Sanz-Tapia, D. Ortiz-Boyer, and C. Hervás-Martínez

University of Córdoba, Dept. of Computing and Numerical Analysis
Campus of Rabanales, C2 Building
14071 Córdoba (Spain)
npedrajas@uco.es, ersanz@uco.es, ma1orbod@uco.es, ma1hemac@uco.es

Abstract. This paper presents *MONet (Multi-Objective coevolutive NETwork)*, a cooperative coevolutionary model for evolving artificial neural networks that introduces concepts taken from multi-objective optimization. This model is based on the idea of coevolving subnetworks that must cooperate to form a solution for a specific problem, instead of evolving complete networks. The fitness of each member of the subpopulations of subnetworks is evaluated using an evolutionary multi-objective optimization algorithm. This idea has not been used before in the area of evolutionary artificial neural networks. The use of a multiobjective evolutionary algorithm allows the definition of as many objectives as could be interesting for our problem and the optimization of these objectives in a natural way.

1 Introduction

In the area of neural networks design one of the main problems is finding suitable architectures for solving specific problems. The problem of finding a suitable architecture and the corresponding weights of the network is a very complex task (for a very interesting review, see [9]). Modular systems are often used in machine learning as an approach for solving these complex problems.

Evolutionary computation is a set of global optimization techniques that have been widely used in late years for training and automatically designing neural networks.

Cooperative coevolution[5] is a recent paradigm in the area of evolutionary computation focused on the evolution of coadapted subcomponents without external interaction. In cooperative coevolution a number of species are evolved together. The cooperation among the individuals is encouraged by rewarding the individuals based on how well they cooperate to solve a target problem. However, the assignment of fitness to the subcomponents is very complex, and remains mainly an open task, as it is difficult to measure how much credit must a subcomponent receive from the performance of the individual where it is a constitutive part.

* This work was supported in part by the Project ALI98-0676-CO2-02 of the Spanish Comisión Interministerial de Ciencia y Tecnología

J. Mira and A. Prieto (Eds.): IWANN 2001, LNCS 2084, pp. 645–652, 2001.

This paper describes a new cooperative coevolutionary model called *MONet (Multi Objective coevolutive NETwork)*, based in a previous work in coevolution not using multi-objective concepts. This model was called *CONet(Cooperative cOevolutive NETwork)*[2][3][1]. The evaluation of the fitness of the coadapted subcomponents in coevolutionary algorithms implies the meeting of many objectives, so the application of multi-objective methods for obtaining the fitness of the individuals could be an interesting technique.

So, the new concept in MONet is the introduction of multi-objective evolutionary programming for the fitness evaluation of the individuals of the subpopulations. MONet develops a method for measuring the fitness of cooperative subcomponents in a coevolutionary model. This method, based on some different criteria that are evaluated using a multi-objective method, could also be applied to other cooperative coevolutionary models not related to the evolution of neural networks.

2 MONet: Multi-objective Cooperative Coevolutionary Network Model

MONet is the *sequel* of CONet, a cooperative coevolutionary model. In such a model, several species are coevolved together. Each species is a subnetwork that constitutes a partial solution of a problem; the combination of several individuals from different species that coevolve together makes up the network that must be applied to the specific problem. The population of subnetworks, that are called *nodules* in CONet, is divided into several subpopulations that evolve independently.

As CONet has been explained elsewhere[2] here we will expose its more important features briefly.

Definition 1. (**Nodule and network**) *A nodule (see Figure 1) is a subnetwork formed by: a set of nodes with free interconnection among them, the connections to these nodes from the input and the connections of the nodes to the output. A nodule cannot have connections with any node belonging to another nodule. A network is a combination of N nodules one from every subpopulation of the nodule population.*

As there is no restriction in the connectivity of the nodule the transmission of the impulse along the connections must be defined in a way that avoids recurrence. The transmission has been defined in three steps: First, each node generates its output as a function of only the inputs of the nodule, these partial outputs are propagated along the connections. Then, each node generates its output as a function of all its inputs. Finally, the output layer of the nodule generates its output.

As the nodules must coevolve to develop different behaviors we have N_P independent subpopulations of nodules that evolve separately. The network will

[1] The original name of CONet was Symbiont, so in these prevoius papers is termed Symbiont.

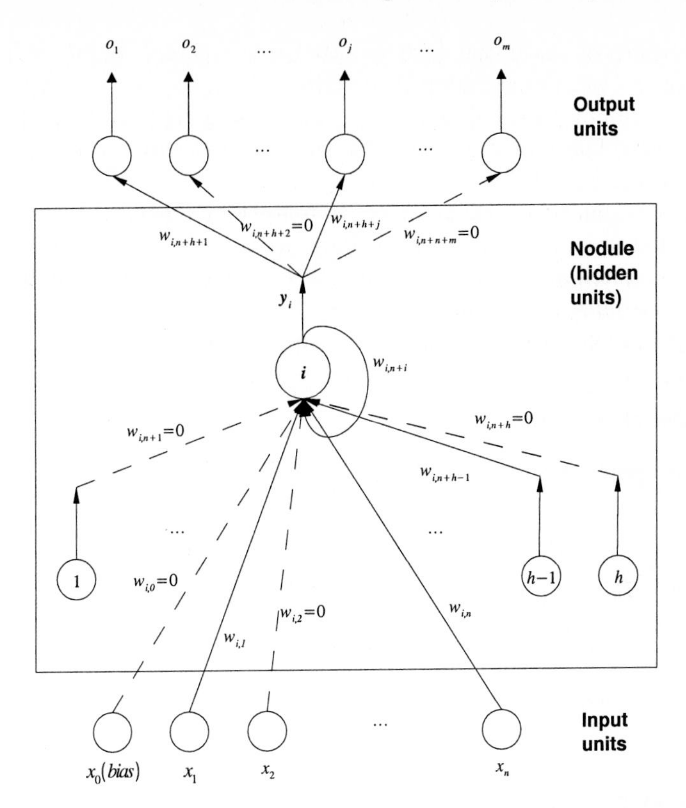

Fig. 1. Model of a nodule. As a node has only connections to some nodes of the nodule, the connections that are missing are represented with dashed lines. The nodule is composed by the hidden nodes and the connections of these nodes from the input and to the output.

always have N_P nodules, each one from a different subpopulation of nodules. Our task is not only to develop cooperative nodules but also to obtain the best combinations. For that reason we have also a population of networks. This population keeps track of the best combinations of nodules and evolves as the population of nodules evolves.

2.1 Nodule Population

The nodule population is formed by N_P subpopulations. Each subpopulation consists of a fixed number of nodules codified directly as subnetworks, that is, we evolve the genotype of Figure 1 that is a one-to-one mapping to the phenotype that is a two-hidden layer network. The population is subject to the operations of replication and mutation.

The algorithm for the generation of a new population is the following:

- The nodules of the initial subpopulation are created randomly. The initial value of the weights is uniformly distributed in the interval $[w_{min}, w_{max}]$.
- The new subpopulation is generated by replicating the best $P\%$ of the former population. The remaining $(1 - P)\%$ is removed and replaced by mutated copies of the best $P\%$.
- The new population is then subject to mutation. There are two types of mutation: parametric and structural. Parametric mutation consists of carrying out a simulated annealing algorithm for a few steps. There are four different structural mutations: Addition of a node, deletion of a node, addition of a connection, and deletion of a connection.

2.2 Network Population

The network population is formed by a fixed number of networks. Each network is the combination of one nodule of each subpopulation of nodules. Crossover is made at nodule level: the parents exchange some of their nodules to generate their offspring. Mutation is also carried out a nodule level. When a network is mutated one of its nodules is selected and it is substituted by another nodule of the same subpopulation selected by means of a roulette algorithm. The network population is evolved using the *steady-state* genetic algorithm[8].

The relationship between the populations of networks and nodules is shown in Figure 2.

2.3 End of Evolutionary Process

In MONet we have evolved the population with and without a validation set. In both cases the evolution of the system depends only on the fitness of the network population. The system is evolved until *the average fitness of the network population stops growing.* The fitness is measured over the training set, unless a validation set is used; in this case the fitness is measured over this validation set.

The best network in terms of training error (or in terms of validation error if a validation set is used) is chosen as the result of the evolution. In case of a tie the smallest network is preferred; if all of them are of equal size, one of them is chosen at random.

3 Multi-objective Fitness Assignment to Nodules

Assigning fitness to the nodules is a complex problem. In fact, the assignment of fitness to the individuals that form a solution in cooperative evolution is one of its key topics. Our credit assignment must fulfill the following requirements to be useful: It must enforce competition among the subpopulations to avoid two subpopulations developing similar responses to the same features of the data, it must enforce cooperation to develop complementary features that together can solve the problem, and it must measure the contribution of a nodule to the

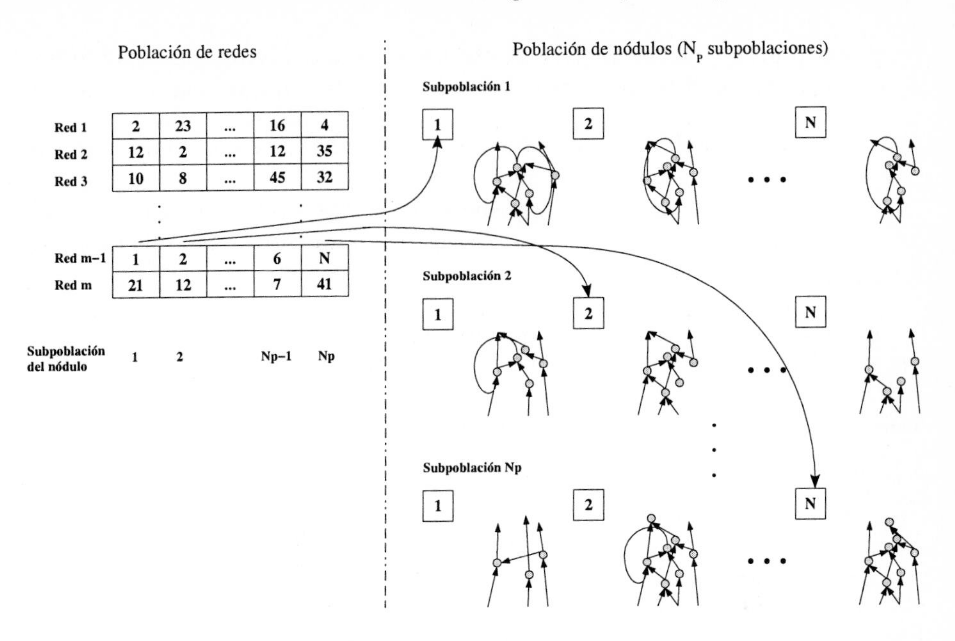

Fig. 2. Network and nodule populations. Each element of the network is a label of an individual of the corresponding nodule subpopulation.

fitness of the network, and not only the performance of the networks where the nodule is present.

With these objectives we could define many different criteria. A good nodule must have the best possible value on every criteria. So the approach of using a multi-objective method is quite natural. These criteria, for obtaining the fitness of a nodule ν in a subpopulation π, are:

1. Substitution. k networks are selected using an elitist method, that is, the best k networks of the population. In these networks the nodule of subpopulation π is replaced by the nodule ν. The fitness of the network with the nodule of the population π substituted by ν is measured. The fitness assigned to the nodule is the averaged difference in the fitness of the networks with the original nodule and with the nodule replaced by ν.
2. Difference. The nodule is removed from all the networks where it is present. The fitness is measured as the difference in performance of these networks.
3. Averaged network fitness. The fitness is the mean of the fitness values of the networks where the nodule ν is present.
4. Number of nodes. Smaller nodules are preferred to bigger ones, so this criterion is used in its negative value.
5. Count. The number of network where the nodule participates.
6. Isolated. The fitness of the nodule if taken as a network alone.

In our previous model, CONet, the fitness of each subcomponent is obtained as a weighted sum of these criteria. The problem with this approach is twofold:

1. It is a very complex task to measure the relevance of each criterion and to assign a weight in function of such relevance.
2. The range of the criteria is very different, so the weights assigned to each one must not only measure its relevance but also correct the differences in range.

For the reasons stated above we decided that these six criteria could be optimized by means of a multi-objective evolutionary programming algorithm based on Srinivas and Deb's NSGA[7]. The original algorithm of Srinivas and Deb is intended for genetic algorithms, so we modified it slightly to be applied to evolutionary programming. The algorithm consists of the following steps:

1. The non-dominated individuals of the population are found. These individuals are assigned a dummy fitness.
2. The non-dominated individuals found in the previous step share their fitness in the standard way.
3. These individuals are removed from the population and the previous two steps are repeated until there are no individuals left in the population.

Many other methods could be used as well as the one we have applied[2]. However, the aim of this work is to ascertain the applicability of multi-objective optimization techniques for the fitness function of and individual in a coevolutive cooperative model, and not to make an study of the different methods for solving multi-objective tasks.

4 Experimental Results

We have applied MONet to the problem of glass classification of the proben1[6] benchmarks. We have also used CONet and a standard back-propagation neural network to solve the same problem, in order to compare the results with those obtained with MONet. The problem is the classification of 6 different kinds of glasses. It is very difficult because there are few patterns, 214, and the number of patterns in each class is very unbalanced. We have chosen this problem because it is extremely difficult[3].

We executed our experiments 30 runs for every method. The data set was divided into 3 sets: 50% for learning, 25% for validating, and 25% for testing the generalization of the networks. In both CONet and MONet, the learning and

[2] Other methods that we have considered for our problem are Zitzler and Theile's Strength Pareto Evolutionary Algorithm (SPEA)[10], Horn, Nafplotis, and Goldberg's niched Pareto genetic Algorithm (NPGA)[4], and Fonseca and Fleming's Multi-Objective Genetic Algorithm (MOGA)[1].

[3] Many papers dealing with this data set reduce the problem to a two class problem that is an easier task.

validation sets were used in the evolutionary process. The results are shown on Table 1. The measure of the error is $E = \frac{1}{P}\sum_{i=1}^{P} e_i$, where P is the number of patterns and e_i is 0 if pattern i is correctly classified, and 1 otherwise. The table shows that MONet clearly outperformed CONet and is far better than the multilayer perceptron.

Table 1. Error rates for Glass data set. For each model we show the averaged error, the variance of the error, and the best and the worst results for the 30 repetitions

Model	Training				Generalization			
	Mean	StD	Best	Worst	Mean	StD	Best	Worst
MONet	0.3584	0.0171	0.3354	0.3913	0.3226	0.0314	0.2830	0.3585
CONet	0.4006	0.0696	0.3106	0.5280	0.4321	0.1120	0.3208	0.6038
MLP	0.4206	0.0897	0.3471	0.5887	0.5101	0.1100	0.4678	0.5906

In order to asses the advantages of MONet we performed another experiment with a different data set. This second experiment was made using the heart disease problem of the same proben1 database. This problem is a bit easier as we have only two classes and more training patterns, 270. The experimental setup was the same that in the previous experiments. The results are shown on Table 2. We can see that in spite of the different are narrower in this problem, MONet still performs better than CONet and MLP.

Table 2. Error rates for Heart data set. For each model we show the averaged error, the variance of the error, and the best and the worst results for the 30 repetitions

Model	Training				Generalization			
	Mean	StD	Best	Worst	Mean	StD	Best	Worst
MONet	0.1244	0.0073	0.1139	0.1386	0.1495	0.0319	0.0588	0.2206
CONet	0.1130	0.0087	0.0941	0.1287	0.1480	0.0284	0.0882	0.2059
MLP	0.1988	0.0050	0.1866	0.2015	0.2230	0.0740	0.1176	0.3088

5 Conclusions

We have proven that multi-objective optimization methods can be applied to the evaluation of fitness of modules in a cooperative coevolutionary model and that their use improves the performance of the model.

Further research is needed in order to ascertain the applicability of this multi-objective method in the approximation of the fitness of the individual. Nevertheless, there are some interesting features of this method, the more important are:

1. The method can tackle with any number of objectives that we could consider interesting for our fitness function.
2. There is no need to adjust weights for each objective. This task is burdensome and the performance of the model may depend highly on the values of these weights.
3. Many different algorithms, not just the one used in our experiments, could be used depending on the special features of the problem at hand.

References

1. C. M. Fonseca and P. J. Flemming. Genetic algorithms for multiobjective optimization: Formulation, discussion, and generalization. In *Proceedings of the Fifth International Conference on Genetic Algorithms*, pages 416–423, 1993.
2. N. García-Pedrajas, C. Hervás-Martínez, and J. Muñoz-Pérez. Symbiont: A cooperative evolutionary model for evolving artificial neural networks for classification. In *8th Interbational Conference on Information Processing and Management of Uncertainty in Knowledge Based Systems*, pages 298–305, Madrid, July 2000.
3. N. García-Pedrajas, C. Hervás-Martínez, and J. Muñoz-Pérez. Symbiont: A cooperarive evolutionary model for evolving artificial neural networks for classification. In B. Bouchon-Meunier, J. Gutiérrez-Ríos, L. Magdalena, and R. R. Yager, editors, *Technologies for Constructing Intelligent Systems*. Springer – Verlag, 2001. (in press).
4. J. Horn, D. E. Goldberg, and K. Deb. Implicit niching in a learning classifier system: Natures's way. *Evolutionary Computation*, 2(1):37 – 66, 1994.
5. M. A. Potter and K. A. de Jong. Cooperative coevolution: An architecture for evolving coadapted subcomponents. *Evolutionary Computation*, 8(1):1–29, 2000.
6. L. Prechelt. Proben1 – a set of neural network benchmark problems and benchmarking rules. Technical Report 21/94, Fakultät für Informatik, Universität Karlsruhe, Karlsruhe, Germany, September 1994.
7. N. Srinivas and K. Deb. Multi-objective function optimization using nondominated sorting genetic algorithms. *Evolutionary Computation*, 2(3):221–248, 1994.
8. D. Whitley and J. Kauth. Genitor: a different genetic algorithm. In *Proceedings of the Rocky Mountain Conference on Artificial Intelligence*, pages 118–130, Denver, CO, 1988.
9. X. Yao. Evolving artificial neural networks. *Proceedings of the IEEE*, 9(87):1423–1447, 1999.
10. E. Zitzler and L. Thiele. Multiobjective optimization using evolutionary algorithms – a comprative case study. *Parallel Problem Solving from Nature*, V:292–301, 1998.

STAR – Sparsity through Automated Rejection

Robert Burbidge, Matthew Trotter, Bernard Buxton, and Sean Holden

University College London, Gower Street, London WC1E 6BT

Abstract. Heuristic methods for the rejection of noisy training examples in the support vector machine (SVM) are introduced. Rejection of training errors, either offline or online, results in a sparser model that is less affected by noisy data. A simple offline heuristic provides sparser models with similar generalization performance to the standard SVM, at the expense of longer training times. An online approximation of this heuristic reduces training time and provides a sparser model than the SVM with a slight decrease in generalization performance.

1 Introduction

The learning of decision functions in noisy domains is one of the main aims of machine learning and has many applications to data mining, pattern recognition, and design of autonomous intelligent agents [1]. The support vector machine (SVM) [2,3] is a recent addition to the toolbox of machine learning algorithms that has shown improved performance over standard techniques in many domains both for classification [4] and regression [2,3]. The advantages of the SVM over existing techniques are often cited as the existence of a global optimum, improved generalization to unseen examples, and sparsity of the model [5]. The sparsity of the model arises as the decision function is an expansion on a subset of the available data. These points are known as the support vectors (SVs) and are often a fraction of the original data. The SVs form a compression scheme for the data [6], hence a model that fits the data with fewer SVs should lead to better generalization.

When learning complex decision functions, and in noisy domains, the number of support vectors can be very high. This leads to a complex model that is difficult to interpret and can be 'abysmally slow in test phase' [3]. When the data are noisy the SVM solution is a trade-off between smoothness of the decision function and training error. The resultant set of SVs contains *all* training errors, and these points make the largest contribution to the decision function. This counterintuitive result is an indirect result of the way that the optimization problem is specified. Heuristics for constructing a decision function to which the training errors contribute less can be incorporated into the SVM to provide a sparser solution with little or no loss in generalization performance.

The remainder of this paper is comprised as follows. In the next section we describe the problem of classification of labelled data, concentrating on the two-class case. The SVM for classification is briefly described, and its advantages and

J. Mira and A. Prieto (Eds.): IWANN 2001, LNCS 2084, pp. 653-660, 2001.

disadvantages discussed with reference to noisy data. In section 3 we describe two heuristics for data cleaning: RaR (Reject and Re-train) and STAR (Sparsity Through Automated Rejection). In section 4 we describe the data sets used and the experimental set-up, followed by performance results and a discussion. Finally, we present some conclusions and ideas for future work.

2 Classification

Consider learning a binary decision function on the basis of a set of training data drawn i.i.d. from some unknown distribution $p(\mathbf{x}, y)$,

$$\{(\mathbf{x}_1, y_1), \ldots, (\mathbf{x}_l, y_l)\},\ \mathbf{x}_i \in \mathcal{R}^d,\ y_i \in \{-1, +1\}\ . \tag{1}$$

One common approach [7] is to search for a pair $(\mathbf{w}, b)$ such that the decision function is given by

$$f(\mathbf{x}) = \mathrm{sgn}(\mathbf{w}^T\mathbf{x} + b)\ . \tag{2}$$

If we assume that the misclassification costs are equal, the pair $(\mathbf{w}, b)$ should be chosen so as to minimize the future probability of misclassification. Minimizing the misclassification rate on the training set, however, does not necessarily minimize future misclassification rate. For example, the function f that takes $f(\mathbf{x}_i) = y_i$ on the training data and is random elsewhere has zero training error but does not generalize to unseen examples. This is an artificial but important illustration of the problem of overfitting. One approach to avoiding overfitting is to use regularization [8], that is, one attempts to minimize training error whilst controlling the expressiveness of the hypothesis space. Roughly speaking, if the data can be classified with low error by a 'smooth' function then that function is likely to generalize well to unseen examples. This is a variation on the principle of Ockham's Razor [2].

2.1 Support Vector Machines

The SVM method avoids overfitting by minimizing the 2-norm of the weight vector, $||\mathbf{w}||_2^2 = \mathbf{w}^T\mathbf{w}$. This is equivalent to maximizing the *margin* between the two classes of the training data, i.e. maximizing the distance between the separating hyperplane and the data on either side of it. If the classes cannot be separated with a linear hyperplane, a Mercer kernel is used to map them into a higher-dimensional feature space where they can be. The standard SVM optimization for the non-linear separation of two classes (once represented in the Lagrangian dual formulation) is

$$\text{Maximize } \frac{1}{2}\alpha^T(\mathbf{y}^T K\mathbf{y} + \mathbf{y}^T\mathbf{y})\alpha - 1^T\alpha\ , \tag{3}$$

$$\text{Subject to } 0 \leq \alpha_i \leq C\ ,$$

where α is a vector of Lagrange multipliers, $\mathbf{y} = (y_1, \ldots, y_l)$ and K is the $l \times l$ matrix with entries $K_{ij} = K(\mathbf{x}_i, \mathbf{x}_j)$. C is a regularization parameter that controls the trade-off between minimizing training error and minimizing $||\mathbf{w}||$. The decision function, resulting from the above optimization is

$$f(\mathbf{x}_i) = \text{sgn}\left(\sum_{i=1}^{N_{SV}} \alpha_i^* y_i K(\mathbf{x}, \mathbf{x}_i) + b^*\right) \tag{4}$$

where N_{SV} is the number of SVs and $b = \alpha^T \mathbf{y}$ is the bias. Note that this formulation corresponds to the case where the hyperplane passes through the origin in feature space. This removes the need to enforce the constraint $\alpha^T \mathbf{y} = 0$. The unbiased hyperplane has slightly worse thoeretical properties [5], but has the same generalization performance as the biased hyperplane on many real-world data sets [9]. A common method of solving the optimization problem 1 is to decompose it into a series of smaller subproblems. The optimization remains convex and has a global optimum. We use the subproblem selection method of Hsu and Lin [9].

2.2 Problems with Noisy Training Data

Noisy training data do not cause as many problems for an SVM as for many other machine learning techniques. Training errors appear in the expansion 4 since $\alpha_i = C$ for points $\mathbf{x}_i$ that are misclassified. Intuitively, we would prefer a decision function that did not depend on noisy training examples. Also, if the data are noisy, or the classes overlap, the SVM solution will have many SVs. This removes one of the oft-cited advantages of SVMs — sparsity of the model. A sparse model is preferable for two reasons: it is easier to interpret and quicker to evaluate.

3 Data Cleaning

We present two heuristics, one offline, the other online, designed to remove outliers and reduce their effect on the SVM solution. The first involves examination of the standard SVM solution, to dictate which points should be removed before retraining on a reduced training set. The second is similar to that of Joachims [10], and allows the SVM to ignore noisy data points during training. We shall see later that each heuristic provides separate benefits to the SVM.

3.1 RaR

A simple method of excluding noisy points from the decision function is to run a standard SVM on the training data, to obtain an intitial solution. Training errors are identified as noisy points and removed. The algorithm is subsequently retrained on the reduced data. We call this method RaR (Remove and Re-train). This method will take longer to train than a standard SVM, but should provide a sparser solution.

3.2 STAR

An obvious extension is to repeat the above, i.e. remove errors after the second pass, retrain, and so on, until convergence. This could take prohibitively long and is not guaranteed to converge. An online heuristic to approximate this is as follows. If a point is consecutively misclassified for a set number, h, of sub-problem iterations, then it is likely to be a training error at the global optimum. Such points are removed during training. We call this algorithm STAR (Sparsity Through Automated Rejection). This method is similar to the shrinking heuristic of Joachims [10]. In shrinking, points that have been at bound for the last h iterations are fixed at bound. Points upper bounded at $\alpha_i = C$, however, are still included in the final solution. In the following section, we compare STAR and RaR with a standard SVM on publicly available, real-world data sets.

4 Results and Discussion

We test the performance of STAR and RaR compared to the standard SVM, on four publicly available datasets. The data sets, available from the UCI Machine Learning Data Repository [11], are as follows. The breast cancer Wisconsin data set has 699 examples in nine dimensions and is 'noise-free', one feature has 16 missing values which are replaced with the feature mean. The ionosphere data set has 351 examples in 33 dimensions and is slightly noisy. The heart data set has 270 examples in 13 dimensions. The Pima Indians diabetes data set has 768 examples in eight dimensions. These last two data sets have a high degree of overlap which leads to a dense model for the standard SVM as many training errors contribute to the solution.

The range of each data set attribute is scaled to $[-1, +1]$. RBF kernels, $K(\mathbf{x}, \mathbf{z}) = \exp(-||\mathbf{x} - \mathbf{z}||^2/d)$, are used, where d is the dimensionality of the input space. The regularization parameter C was varied in $\{1, 10, 100\}$ and chosen as that which minimized error rate. The size of the working set for the decomposition algorithm was set at 10.

The generalization error rate was estimated by ten-fold cross-validation. All significance tests refer to the paired McNemar statistic between two solutions at 95%. The cross-validated error rates and mean sparsity (N_{SV}/l) for the four data sets are shown in figures 1–4. The figures illustrate the effect of the parameter h on the quality of the solution. As h increases the STAR solution approaches the SVM solution. STAR does at least no worse than the SVM on the 'noise-free' breast cancer data. On the slightly noisy ionosphere data STAR provides a sparser solution that has slightly higher error than the SVM (significant except for $h = 80$). On the heart data STAR provides a solution with the same or lower error rate and has sparsity as for the SVM (except for $h = 20$). The results are similar for the diabetes data, although the STAR error rates are significantly higher than for the SVM.

RaR provides the sparsest solution, except for ionosphere where the STAR solution is sparsest. RaR is not significantly worse than the SVM on any data set. RaR error rate is not significantly different from STAR error rate on breast cancer and ionosphere (except for $h = 20$). RaR is significantly better than STAR on diabetes (except for $h = 100$). On the heart data RaR has lower error than STAR for $h = 20$, higher error for $h = 40$, and no significant difference for the other values.

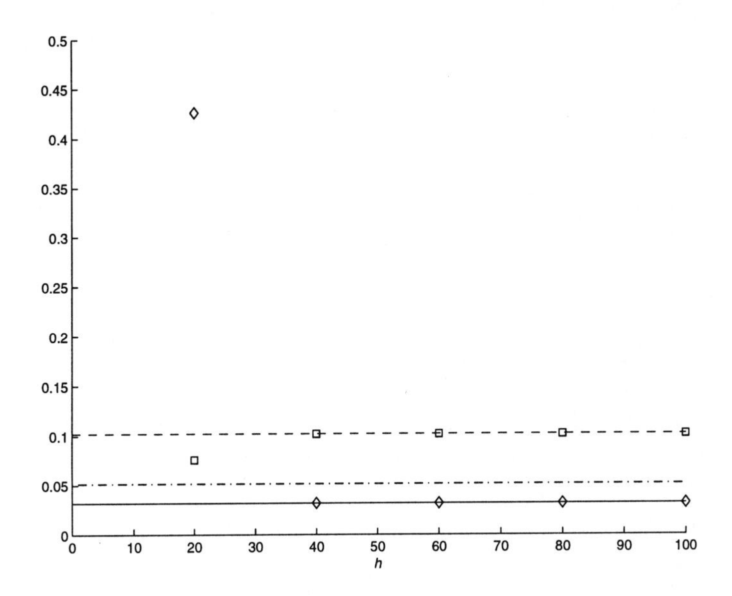

Fig. 1. Cancer data, $C = 1$. STAR error rate (*diamonds*) is the same as the RaR and SVM error rates (*solid line*) except for $h = 20$, when too many points are rejected. STAR sparsity (*squares*) is the same as SVM sparsity (*dashed line*). RaR provides the sparsest model (*dash-dotted line*)

Table 1 shows the training times for each algorithm, for the four datasets and three values of C. RaR has the longest training times over all problems. STAR is faster than the other two techniques, particularly for high values of C.

In short, RaR provides sparser models than the SVM with little or no loss in generalization ability, at the expense of longer training times. STAR is not counter-productive on a 'noise-free' data set. STAR offers the greatest benefit on a slightly noisy data set. On the two more complex data sets STAR trades off accuracy with sparsity, although the solution obtained is generally not as good as that given by the simpler RaR algorithm.

5 Conclusions

An offine heuristic, RaR, has been presented that provides a sparser model than the standard SVM algorithm. RaR does not perform significantly worse than the

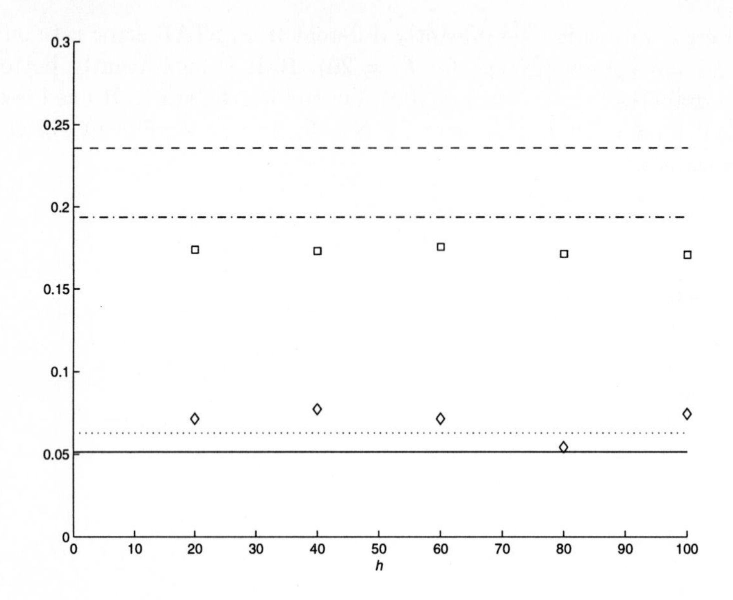

Fig. 2. Ionosphere data, $C = 10$. STAR error rate (*diamonds*) is slightly worse than RaR (*dotted line*) and the SVM (*solid line*). STAR provides a sparser solution (*squares*) than RaR (*dash-dotted line*) and the SVM (*dashed line*)

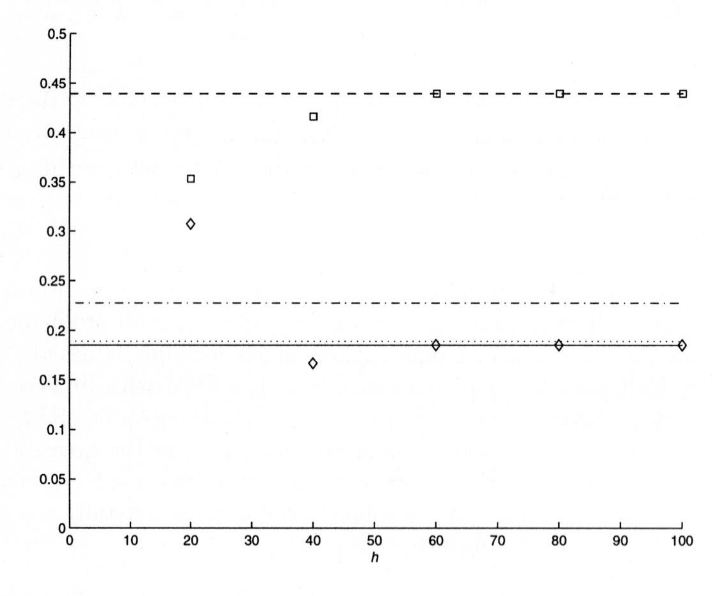

Fig. 3. Heart data, $C = 1$. STAR error rate (*diamonds*) and sparsity (*squares*) approach those of the SVM (*solid and dashed lines respectively*) as h increases. RaR error rate (*dotted line*) is not significantly different from that of the SVM whilst providing a sparser model (*dash-dotted line*)

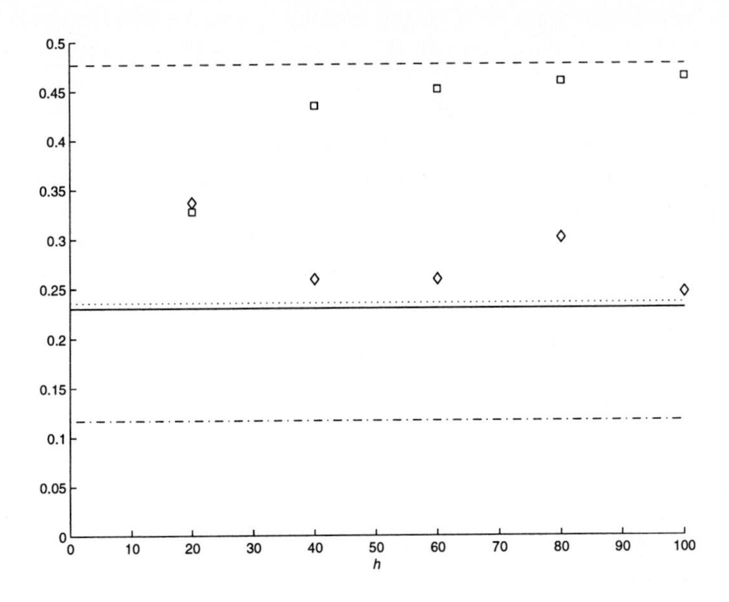

Fig. 4. Diabetes data, $C = 10$. STAR error rate (*diamonds*) and sparsity (*squares*) approach those of the SVM (*solid and dashed lines respectively*) as h increases. RaR error rate (*dotted line*) is not significantly different from that of the SVM whilst providing a sparser model (*dash-dotted line*)

Table 1. Training times, in seconds, for the standard SVM, RaR, and STAR for five values of h. RaR has the longest training times, as it is necessary to run the standard SVM twice to obtain the solution. STAR has the same training times as the SVM for $C = 1$, when the solutions are generally similar. STAR has the shortest training times for higher values of C over all values of h. As h increases so does the training time — STAR is behaving more like the standard SVM.

Data Set	C	SVM	RaR	STAR				
				20	40	60	80	100
Cancer	1	28	41	27	27	27	27	27
	10	67	89	26	35	40	48	54
	100	171	272	57	63	66	77	83
Ionosphere	1	35	56	26	32	35	35	35
	10	66	111	21	27	33	39	44
	100	162	250	18	25	32	37	43
Heart	1	24	36	20	24	24	24	24
	10	67	104	19	26	31	36	42
	100	271	431	34	39	44	50	57
Diabetes	1	138	196	122	137	139	139	139
	10	186	247	94	128	137	148	155
	100	852	1075	144	181	192	200	250

SVM, but requires roughly 50% longer training times. An online approximation to RaR, STAR, has been presented. The trade-off between quality and sparsity of the solution are controlled by the parameter h. STAR is quicker to train than both RaR and the standard SVM.

Further developments to these algorithms include finding a 'rule-of-thumb' for selecting h, and modifying STAR to provide a solution as sparse as that of RaR, without increasing error rate or runtime. Possibilities include removing not only training errors, but also bounded support vectors. These are the points that lie in the margin and may also be regarded as 'noisy' data. Further experiments include investigating the effect of noise level on the error rates and sparsity of RaR and STAR compared to the SVM.

Acknowledgements

This research has been undertaken within the INTErSECT Faraday Partnership managed by Sira Ltd and the National Physical Laboratory, and has been supported by the Engineering and Physical Sciences Research Council (EPSRC), Glaxo SmithKline and Sira Ltd.

Robert Burbidge is an associate of the Postgraduate Training Partnership established between Sira Ltd and University College London. Postgraduate Training Partnerships are a joint initiative of the Department of Trade and Industry and EPSRC.

References

1. Mitchell, T.: Machine Learning. McGraw-Hill International (1997)
2. Vapnik, V.: Statistical Learning Theory. John Wiley & Sons (1998)
3. Burges, C.J.C.: A tutorial on support vector machines for pattern recognition. Data Mining and Knowledge Discovery **2** (1998) 1–47
4. Schölkopf, B., Sung, K.K., Burges, C., Girosi, F., Niyogi, P., Poggio, T., Vapnik, V.: Comparing support vector machines with Gaussian kernels to radial basis function classifiers. IEEE Transactions on Signal Processing **45** (1997)
5. Cristianini, N., Shawe-Taylor, J.: Support Vector Machines. Cambridge University Press (2000)
6. Littlestone, N., Warmuth, M.: Relating data compression and learnability. Technical report, University of California, Santa Cruz (1986)
7. Rosenblatt, F.: The perceptron: a probabilistic model for information storage and organization in the brain. Psychological Review **65** (1959) 386–408
8. Bishop, C.: Neural Networks for Pattern Recognition. Clarendon Press (1995)
9. Hsu, C.W., Lin, C.J.: A simple decomposition method for support vector machines. Machine Learning (2001) To appear.
10. Joachims, T.: Making large-scale SVM learning practical. In Schölkopf, B., Burges, C., Smola, A., eds.: Advances in Kernel Methods: Support Vector Learning. The MIT Press (1999)
11. Blake, C.L., Merz, C.J.: UCI repository of machine learning databases. (1998)

Ordinal Regression with K-SVCR Machines

Cecilio Angulo[1] and Andreu Català[2]

[1] Universitat Politècnica de Catalunya, Systems Engineering Dept.
Vilanova i la Geltrú 08800, Spain. cecilio.angulo@upc.es
http://www-esaii.upc.es/users/cangulo/

[2] Associated European Lab on Intelligent Systems & Advanced Control
Vilanova i la Geltrú 08800, Spain. andreu.catala@upc.es

Abstract. The ordinal regression problem or ordination have mixed features of both, the classification and the regression problem, so it can be seen as an independent problem class. The particular behaviour of this sort of problem should be explicitly considered by the learning machines working on it. In this paper the ordination problem is fomulated from the viewpoint of a recently defined learning architecture based on support vectors, the K-SVCR learning machine, specially developed to treat with multiple classes. In this study its definition is compared to other existing results in the literature.

1 Introduction

Ordinal regression problems arise from many real-world tasks: works on a day are classified according some importance rank; results from a web information search should be showed following the user preference; suppliers or clients are grouped in order to illustrated its satisfaction degree or results prospect; a continuous measure can be partitioned attending some intervalar criterion. So, even if the matter are the classes, a finite number of elements, these items have a ordination in some sense.

Let $\mathcal{T} = \{(\mathbf{x}_p, y_p)\}_{p=1}^{\ell} \subset \mathcal{X} \times \mathcal{Y} \sim P_{\mathcal{X}\mathcal{Y}}^{\ell}$ be an independent identically distributed (i.i.d.) training set containing the entry vectors and the outputs of an unknown system. Usually have been considered that $\mathcal{Y}$ can be a finite unordered set, so the classification problem can be treated with the $0-1$ cost function, or $\mathcal{Y}$ can be a metric space, for example $\mathcal{Y} \subseteq \mathbb{R}$, a regression estimation problem is been considered and it is necessary to employ a cost function having in consideration the metric structure, for example the 2-norm into the real space. In an ordinal regression problem the learning algorithm use an ordination, as in the metric regression case, however the set $\mathcal{Y}$ is finite, as in the classification task, $\mathcal{Y} = \{\theta_1, \ldots, \theta_K\}$ with $\theta_K \succ_{\mathcal{Y}} \theta_{K\text{-}1} \succ_{\mathcal{Y}} \cdots \succ_{\mathcal{Y}} \theta_1$. The existence of this ordinal scale on the output set will allow to make comparisons between the entry vectors.

The treatment of this sort of problems will be made by modifying the original algorithm of the K-SVCR learning machines [AC00]. This new technique use the

J. Mira and A. Prieto (Eds.): IWANN 2001, LNCS 2084, pp. 661-668, 2001.

tri-classification machine to establish an ordination ± 1 for pairs of elements belonging to different classes and assigning zero label if the compared elements are in the same class. Compared to related works like [HGO99] and [P99], the proposed algorithm allows to work on a more reduced number of learning patterns and to make positive and negative comparisons.

2 Ordinal Regression Formulation

To distinguish the ordinal regression problem formulation from those made for the classification and the metric regression problems, it will be named *rank* each element belonging to the output space $\mathcal{Y}$, remarking the order relation between these components.

Conjecture 1. Let

$$\mathcal{T} = \{(\mathbf{x}_p, y_p)\}_{p=1}^{\ell} \subset \mathcal{X} \times \mathcal{Y} = \mathcal{X} \times \{r_1, \ldots, r_K\} \sim P_{\mathcal{X}\mathcal{Y}}^{\ell}, \tag{1}$$

be the i.i.d. training set for a learning machine in a ordination problem, with the order defined by

$$r_K \succ_{\mathcal{Y}} r_{K\text{-}1} \succ_{\mathcal{Y}} \cdots \succ_{\mathcal{Y}} r_1 , \tag{2}$$

where symbol $\succ_{\mathcal{Y}}$ could be translated "*be preferred to*". It will be considered that each mapping f of an approximation space $\mathcal{LM}$ infers an order $\succ_{\mathcal{X}}$ on the entry vectors following the rule

$$\mathbf{x}_i \succ_{\mathcal{X}} \mathbf{x}_j \Leftrightarrow f(\mathbf{x}_i, \omega) \succ_{\mathcal{Y}} f(\mathbf{x}_j, \omega) . \tag{3}$$

It follows from the notation that a *learning method for ordinal regression* should be capable to find the mapping $f(\mathbf{x}, \omega) \in \mathcal{LM}$, inferring an order on the entry space $\mathcal{X}$, with the minimum number of order inversions on the object pairs $(\mathbf{x}_1, \mathbf{x}_2)$ [S90].

The probability to make an order inversion is defined from the *ordinal risk functional* like

$$R^{pref}(\omega) = \mathbf{E}\left[L_{pref}\left(f(\mathbf{x}_1, \omega), f(\mathbf{x}_2, \omega), y_1, y_2\right)\right] , \tag{4}$$

using the *ordinal cost function* defined by

$$L_{pref}(\widehat{y}_1, \widehat{y}_2, y_1, y_2) = \begin{cases} 1 & \text{if } y_1 \succ_{\mathcal{Y}} y_2 \wedge \neg(\widehat{y}_1 \succ_{\mathcal{Y}} \widehat{y}_2) \\ 1 & \text{if } y_2 \succ_{\mathcal{Y}} y_1 \wedge \neg(\widehat{y}_2 \succ_{\mathcal{Y}} \widehat{y}_1) \\ 0 & \text{otherwise} \end{cases} . \tag{5}$$

Applying the empirical risk minimization (ERM) principle, the solution mapping must to minimize the *empirical ordinal risk functional*

$$R_{emp}^{pref}(\omega) = \frac{1}{\ell^2} \sum_{i=1}^{\ell} \sum_{j=1}^{\ell} L_{pref}\left(f(\mathbf{x}_i, \omega), f(\mathbf{x}_j, \omega), y_i, y_j\right) , \tag{6}$$

being an approximation for the unknown risk functional (4).

It can be observed that the former definitions are made on object pairs, so the existence of an ordination on the output ranks is showed by comparing elements. Now, let $\mathbf{x}^{(1)}$ and $\mathbf{x}^{(2)}$ be abbreviations to note the first and the second object in a pair, and let $\mathcal{T}$ be the original training set (1) for a learning machine in an ordination problem. If the operation *rank difference*, noted $\ominus$, is defined like

$$y_1 \ominus y_2 = j - k \quad \text{si } y_1 = r_j \,,\, y_2 = r_k \in \mathcal{Y} \,, \tag{7}$$

then it can be defined the *modified training set* $\tilde{\mathcal{T}}$ by the union of two disjoint subsets

$$\tilde{\mathcal{T}} = \mathcal{T}^{(\approx_\mathcal{Y})} \cup \mathcal{T}^{(\succ_\mathcal{Y}, \prec_\mathcal{Y})} \,, \tag{8}$$

being

$$\begin{array}{l} \mathcal{T}^{(\approx_\mathcal{Y})} = \left\{ \left(\mathbf{x}_i^{(1)}, \mathbf{x}_i^{(2)} \right), z_i \right\}_{i=1}^{\ell^{(1)}} \in \mathcal{X} \times \mathcal{X} \times \{0\} \\ \mathcal{T}^{(\succ_\mathcal{Y}, \prec_\mathcal{Y})} = \left\{ \left(\mathbf{x}_i^{(1)}, \mathbf{x}_i^{(2)} \right), z_i \right\}_{i=1}^{\ell^{(2)}} \in \mathcal{X} \times \mathcal{X} \times \{-1, +1\} \end{array} \,, \tag{9}$$

with

$$z_i = sign \left(y_1 \ominus y_2 \right) \,, \tag{10}$$

each subset having

$$\ell^{(1)} = \#\mathcal{T}^{(\approx_\mathcal{Y})} = \sum_{i=1}^{K} \ell_i^2 \quad , \quad \ell^{(2)} = \#\mathcal{T}^{(\succ_\mathcal{Y}, \prec_\mathcal{Y})} = 2 \sum_{i=1}^{K-1} \sum_{j=i+1}^{K} \ell_i \ell_j \tag{11}$$

training patterns, being ℓ_i the number of patterns for the rank r_i.

In [HGO99] and [P99] have been used a similar formulation to apply support vector machines (SVM) [B98] to ordination problems, however a standard SVM for classification works on dichotomies $+1/-1$, so the authors have considered only the training subset $\mathcal{T}^{(\succ_\mathcal{Y}, \prec_\mathcal{Y})}$. Cardinality of this subset is greater than the refused subset $\mathcal{T}^{(\approx_\mathcal{Y})}$, so the learning process suffers a weak information lost, but the new training set on the pairs is very large, showing that the new formulation will be practical only when the original training set be small. The novel approach based on K-SVCRs machines starts with a wider data set that the former, $\#\tilde{\mathcal{T}} = \ell^{(1)} + \ell^{(2)}$, but finally it will be developed on a more reduced and significant training patterns set like it will be demonstrated.

3 *K*-SVCR Ordinal Regression

Using definitions of Section 2 and the K-SVCR theory will be derived a new algorithm for ordinal regression.

Conjecture 2. It will be assumed that for an approximation set $\mathcal{LM}$, defined on maps inferring an order (3) on the entry space imported of the ordination on the output space $\mathcal{Y}$, there exists a set $\mathcal{U}$ of linear functions on the feature set[1] $\mathcal{F}$ defined from the space $\mathcal{X}$ to $\mathbb{R}$ so that for all mapping $f \in \mathcal{LM}$ there exists a function $U \in \mathcal{U}$ (and vice versa), named usually *utility function*, accomplishing that

$$f(\mathbf{x}, \omega) = r_i \Longleftrightarrow U(\mathbf{x}) \in [\theta(r_{i-1}), \theta(r_i)] \tag{12}$$

where

$$U(\mathbf{x}) = \langle \omega, \mathbf{x} \rangle_{\mathcal{F}} \tag{13}$$

with $\theta(r_0) = -\infty$ and $\theta(r_K) = \infty$.

It is easily demonstrated that the utility function make no error on the i-th example in $\tilde{\mathcal{T}}$ if, and only if,

$$\begin{array}{l} z_i \cdot U\left(\mathbf{x}_i^{(1)}\right) > z_i \cdot U\left(\mathbf{x}_i^{(2)}\right) \Longleftrightarrow z_i \cdot U\left(\mathbf{x}_i^{(1)} - \mathbf{x}_i^{(2)}\right) > 0 \\ z_i \cdot U\left(\mathbf{x}_i^{(1)}\right) = z_i \cdot U\left(\mathbf{x}_i^{(2)}\right) \Longleftrightarrow z_i \cdot U\left(\mathbf{x}_i^{(1)} - \mathbf{x}_i^{(2)}\right) = 0 \end{array} . \tag{14}$$

because U is linear in the feature space[2] $\mathcal{F}$.

Vector pairs $\left(\mathbf{x}_i^{(1)}, \mathbf{x}_i^{(2)}\right)$ with different rank, $z_i = sign(y_1 \ominus y_2) = \pm 1$, are constrained to hold separated on the utility function line, meanwhile pairs with the same rank are constrained into the same interval. The QP problem to be solved will be subject to

$$z_i \cdot \left\langle \omega, \mathbf{x}_i^{(1)} - \mathbf{x}_i^{(2)} \right\rangle_{\mathcal{F}} \geq 1 \;\; i = 1, \ldots, \ell^{(2)}, \tag{15}$$

$$-\delta \leq \left\langle \omega, \mathbf{x}_i^{(1)} - \mathbf{x}_i^{(2)} \right\rangle_{\mathcal{F}} \leq \delta \;\; i = \ell^{(2)} + 1, \ldots, \ell^{(2)} + \ell^{(1)}. \tag{16}$$

To avoid intersection between intervals it is necessary to restrict the insensitivity parameter δ for the $K - SVCR$ formulation. For example, it can be considered $\delta < 0.5$ to assure $|[\theta(r_{i-1}), \theta(r_i)]| < 1$ Having in mind this consideration, it is possible to define the optimization problem associated to the K-SVCR method for the ordinal regression general problem like: for $0 \leq \delta < 0.5$ chosen *a priori*,

$$\arg\min R_{K-SVCR}^{Ord}\left(\omega, \xi, \varphi^{(*)}\right) = \frac{1}{2}\|\omega\|_{\mathcal{F}}^2 + C\sum_{i=1}^{\ell^{(2)}} \xi_i + D\sum_{i=\ell^{(2)}+1}^{\ell^{(2)}+\ell^{(1)}} (\varphi_i + \varphi_i^*) , \tag{17}$$

subject to the constraints (15)–(16).

[1] This space is created according the kernel used in the 'kernel trick' to extend the SVM to the no linear case, $\phi : \mathbb{R}_d \to \mathcal{F}$.

[2] Usually U will be no linear in the original entry space $\mathbb{R}^d$ because the insertion ϕ.

To solve this QP problem a lagrangian formulation is employed

$$\begin{aligned}\arg\min L_p(\omega) = \frac{1}{2}\|\omega\|_{\mathcal{F}}^2 - \\ -\sum_{i=1}^{\ell^{(2)}} \alpha_i \left[z_i \cdot \left\langle \omega, \mathbf{x}_i^{(1)} - \mathbf{x}_i^{(2)} \right\rangle_{\mathcal{F}} - 1 \right] + \\ + \sum_{i=\ell^{(2)}+1}^{\ell^{(2)}+\ell^{(1)}} \beta_i \left[\left\langle \omega, \mathbf{x}_i^{(1)} - \mathbf{x}_i^{(2)} \right\rangle_{\mathcal{F}} - \delta \right] - \\ - \sum_{i=\ell^{(2)}+1}^{\ell^{(2)}+\ell^{(1)}} \beta_i^* \left[\left\langle \omega, \mathbf{x}_i^{(1)} - \mathbf{x}_i^{(2)} \right\rangle_{\mathcal{F}} + \delta \right]\end{aligned}, \tag{18}$$

subject to

$$\begin{aligned}\frac{\partial L_p}{\partial \alpha_i} = 0 \quad ; \quad \alpha_i \geq 0 \qquad i = 1, \ldots, \ell^{(2)} \\ \frac{\partial L_p}{\partial \beta_i^{(*)}} = 0 \quad ; \quad \beta_i^{(*)} \geq 0 \qquad i = \ell^{(2)}+1, \ldots, \ell^{(2)}+\ell^{(1)}\end{aligned}, \tag{19}$$

that in the dual space and noting

$$\gamma_i = \alpha_i \cdot z_i \quad i = 1, \ldots, \ell^{(2)} \tag{20}$$

gives the equivalent restricted problem

$$\begin{aligned}&\arg\min L_p(\gamma) = \\ &\frac{1}{2}\left\langle \left[\sum_{i=1}^{\ell^{(2)}} \gamma_i \cdot \left(\mathbf{x}_i^{(1)} - \mathbf{x}_i^{(2)}\right) - \sum_{i=\ell^{(2)}+1}^{\ell^{(2)}+\ell^{(1)}} (\beta_i - \beta_i^*) \cdot \left(\mathbf{x}_i^{(1)} - \mathbf{x}_i^{(2)}\right) \right], \right. \\ &\left. \left[\sum_{j=1}^{\ell^{(2)}} \gamma_j \cdot \left(\mathbf{x}_j^{(1)} - \mathbf{x}_j^{(2)}\right) - \sum_{j=\ell^{(2)}+1}^{\ell^{(2)}+\ell^{(1)}} (\beta_j - \beta_j^*) \cdot \left(\mathbf{x}_j^{(1)} - \mathbf{x}_j^{(2)}\right) \right] \right\rangle_{\mathcal{F}} \\ &-\sum_{i=1}^{\ell^{(2)}} \frac{\gamma_i}{z_i} + \delta \sum_{i=\ell^{(2)}+1}^{\ell^{(2)}+\ell^{(1)}} (\beta_i + \beta_i^*)\end{aligned}, \tag{21}$$

with

$$\begin{aligned}\frac{\gamma_i}{z_i} \geq 0 \quad i = 1, \ldots, \ell^{(2)} \\ \beta_i, \beta_i^* \geq 0 \quad i = \ell^{(2)}+1, \ldots, \ell^{(2)}+\ell^{(1)}\end{aligned}. \tag{22}$$

Finally, making the substitutions

$$\begin{aligned}\gamma_i = \beta_i \quad i = \ell^{(2)}+1, \ldots, \ell^{(2)}+\ell^{(1)} \\ \gamma_i = \beta_{i-\ell^{(1)}} \quad i = \ell^{(2)}+\ell^{(1)}+1, \ldots, \ell^{(2)}+2\ell^{(1)}\end{aligned} \tag{23}$$

is enunciated the Wolfe optimization problem: for $0 \leq \delta < 0.5$ chosen *a priori*,

$$\arg\min W(\gamma) = \frac{1}{2}\gamma^{\top} \cdot \mathbf{H} \cdot \gamma + \mathbf{c}^{\top} \cdot \gamma, \tag{24}$$

with

$$\gamma^{\top} = \left(\gamma_1, \ldots, \gamma_{\ell}, \gamma_{\ell^{(2)}+\ell^{(1)}}, \ldots, \gamma_{\ell^{(2)}+2\ell^{(1)}}\right) \in \mathbb{R}^{\ell^{(2)}+\ell^{(1)}+\ell^{(1)}}, \tag{25}$$

$$\mathbf{c}^{\top} = \left(\frac{-1}{z_1}, \cdots, \frac{-1}{z_{\ell^{(2)}}}, \delta, \ldots, \delta\right) \in \mathbb{R}^{\ell^{(2)}+\ell^{(1)}+\ell^{(1)}}, \tag{26}$$

$$\mathbf{H} = \begin{pmatrix} (k_{i,j}) & -(k_{i,j}) & (k_{i,j}) \\ -(k_{i,j}) & (k_{i,j}) & -(k_{i,j}) \\ (k_{i,j}) & -(k_{i,j}) & (k_{i,j}) \end{pmatrix}, \tag{27}$$

being

$$(k_{i,j}) = \left\langle \mathbf{x}_i^{(1)} - \mathbf{x}_i^{(2)}, \mathbf{x}_j^{(1)} - \mathbf{x}_j^{(2)} \right\rangle_{\mathcal{F}} \in \mathbb{R}^{\ell^{(a)}} \times \mathbb{R}^{\ell^{(b)}}, \tag{28}$$

$$\mathbf{H} = \mathbf{H}^{\top} \in \mathcal{M}\left(\mathbb{R}^{\ell^{(2)}+\ell^{(1)}+\ell^{(1)}}, \mathbb{R}^{\ell^{(2)}+\ell^{(1)}+\ell^{(1)}}\right), \tag{29}$$

subject to

$$\begin{array}{ll} \gamma_i \cdot z_i \geq 0 & i = 1, \ldots, \ell^{(2)} \\ \gamma_i \geq 0 & i = \ell^{(2)}+1, \ldots, \ell^{(2)}+2\ell^{(1)} \end{array} . \tag{30}$$

Solution mapping f is found from the utility function U by using its definition property (12)

$$U(\mathbf{x}) = \left\langle \sum_{i=1}^{\ell^{(2)}} \gamma_i \cdot \left(\mathbf{x}_i^{(1)} - \mathbf{x}_i^{(2)}\right) - \sum_{i=\ell^{(2)}+1}^{\ell^{(2)}+\ell^{(1)}} \left(\gamma_i - \gamma_{i+\ell^{(1)}}\right) \cdot \left(\mathbf{x}_i^{(1)} - \mathbf{x}_i^{(2)}\right), \mathbf{x} \right\rangle_{\mathcal{F}} \tag{31}$$

that it can be rewrite as

$$U(\mathbf{x}) = \left\langle \sum_{i=1}^{SV} \nu_i \cdot \left(\mathbf{x}_i^{(1)} - \mathbf{x}_i^{(2)}\right), \mathbf{x} \right\rangle_{\mathcal{F}}, \tag{32}$$

if it is noted

$$\begin{array}{ll} \nu_i = \gamma_i & i = 1, \ldots, \ell^{(2)} \\ \nu_i = -\gamma_i + \gamma_{i+\ell^{(1)}} & i = \ell^{(2)}+1, \ldots, \ell^{(2)}+\ell^{(1)} \end{array} . \tag{33}$$

With this new formulation of the ordination problem, the training phase for the learning machine is focused on the pairs belonging the same rank, output 0, and each rank is inserted an bounded interval. In [HGO99] and [P99] are only considered pairs belonging different ranks, so is not used all the available information.

Constraints (16) in (17) assure that each rank is into a bounded ball, meanwhile constraints (15) are only used to distribute these intervals on the line in $\mathcal{F}$. Hence, it is necessary to consider only a pair of patterns for each different pair of ranks and the number of constraints is reduced to $2\ell^{(1)} + K(K-1)$[3].

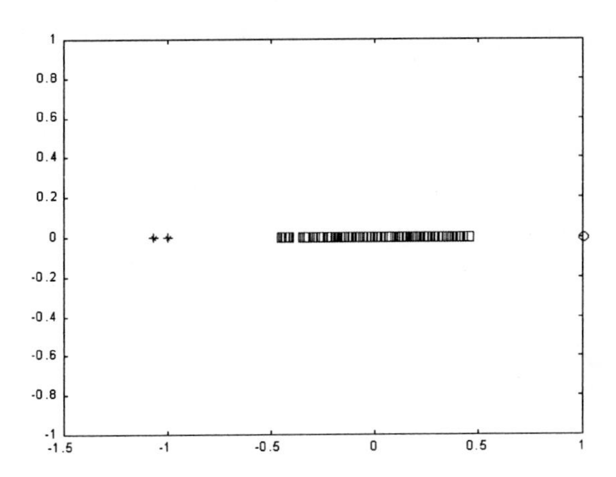

Fig. 1. Constraints in the optimization problem are accomplished for the patterns $\mathbf{x}_i^{(1)} - \mathbf{x}_i^{(2)}$. Learning is focused on pairs of patterns belonging the same rank, output in $[-\delta, \delta]$.

The estimation of the bounds $\theta(r_i)$ is done maximizing the margin

$$\theta(r_k) = \frac{U(\mathbf{x}_1) + U(\mathbf{x}_2)}{2} = \frac{\langle \omega, \mathbf{x}_1 \rangle_{\mathcal{F}} + \langle \omega, \mathbf{x}_2 \rangle_{\mathcal{F}}}{2}$$
$$= \frac{\left\langle \sum_{i=1}^{SV} \nu_i \cdot \left(\mathbf{x}_i^{(1)} - \mathbf{x}_i^{(2)}\right), \mathbf{x}_1 \right\rangle_{\mathcal{F}} + \left\langle \sum_{i=1}^{SV} \nu_i \cdot \left(\mathbf{x}_i^{(1)} - \mathbf{x}_i^{(2)}\right), \mathbf{x}_2 \right\rangle_{\mathcal{F}}}{2}, \quad (34)$$

being $\mathbf{x}_1$ and $\mathbf{x}_2$ so that

$$(\mathbf{x}_1, \mathbf{x}_2) = \arg\min_{(\mathbf{x}_i, \mathbf{x}_j) \in \mathcal{X}\mathcal{X}(k)} \left[U(\mathbf{x}_i) - U(\mathbf{x}_j)\right], \quad (35)$$

with

$$\mathcal{X}\mathcal{X}(k) = \left\{ \left(\mathbf{x}_i^{(1)} - \mathbf{x}_i^{(2)}\right) | y_i^{(1)} = r_k \wedge y_i^{(2)} = r_{k+1} \right\}. \quad (36)$$

New elements will be labeled using the expression (12). The performance of this new learning of preferences is illustrated in Figure 1 and Figure 2.

[3] Moreover, in a practical sense, constraints in (16) accomplishing $\mathbf{x}_i^{(1)} = \mathbf{x}_i^{(2)}$ are identical: $-\delta \le \langle \omega, 0 \rangle_{\mathcal{F}} \le \delta \qquad i = \ell^{(2)} + 1, \ldots, \ell^{(2)} + \ell^{(1)}$.

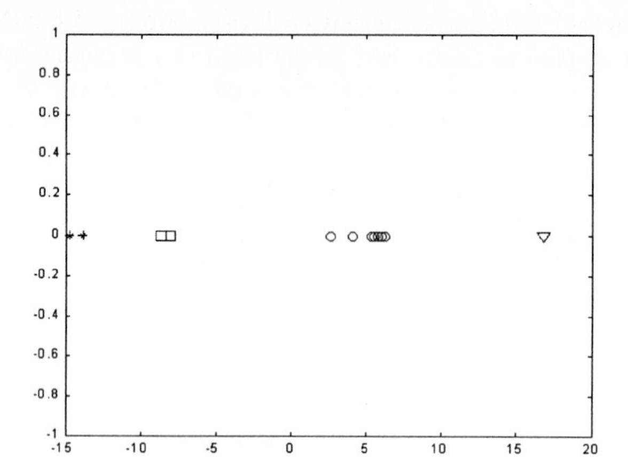

Fig. 2. Entries $\mathbf{x}$ for a 4-ranks ordination problem are ordered into the real line in bounded intervals by using the utility function $U(\mathbf{x})$.

4 Conclusions

Some categorization problems have ordination on the labeled classes, inferring some preference order on the entry space. Learning machines working on them should consider this preference. The new tri-class SV learning machine K-SVCR have a specially useful design to treat with preferences. It is showed by using a ordinal cost function (4) and a modified training set considering pairs of patterns that it is possible to define a utility function introducing the training patterns into bounded intervals of a line living in the feature space.

The new methodology works on fewer patterns that another approaches, however this number is still large compared to the cardinality of the original training set, so it is specially devoted to problems formulated on a reduced number of patterns.

References

[AC00] Angulo, C., Català, A.: K-SVCR. A multi-class support vector machine. Proc. ECML'00, Lecture Notes in Artificial Intelligence **1810** (2000) 31–38.

[B98] Burges, C.J.C.:A Tutorial on support vector machines for pattern recognition. Data Mining And Knowledge Discovery **2** (1998) 1–47.

[HGO99] Herbrich, R., Graepel, T., Obermayer, K.:Support vector learning for ordinal regression. Proc. ICANN'99 (1999) 97–102.

[P99] Phillips, P.J.: Support vector machines applied to face recognition. In Advances in NIPS **11** (1999).

[S90] Sobel, M.:Complete ranking procedures with appropriate loss functions. Communications in Statistics – Theory and Methods **19**(12) (1990) 4525–4544.

Large Margin Nearest Neighbor Classifiers

Sergio Bermejo and Joan Cabestany

Department of Electronic Engineering,
Universitat Politècnica de Catalunya (UPC),
Gran Capità s/n, C4 building, 08034 Barcelona, Spain
e-mail: sbermejo@eel.upc.es

Abstract. Large margin classifiers are computed to assign patterns to a class with high confidence. This strategy helps controlling the capacity of the learning device so good generalization is presumably achieved. Two recent examples of large margin classifiers are *support vector learning machines* (SVM) [12] and *boosting classifiers* [10]. In this paper we show that it is possible to compute large-margin maximum classifiers using a gradient-based learning based on a cost function directly connected with their average margin. We also prove that the use of this procedure in nearest-neighbor (NN) classifiers induce solutions closely related to support vectors.

1.Introduction

In learning pattern recognition, a classifier is constructed for assigning future observations to one of the existing classes based on some knowledge about the problem given by the training set. Typically, these learning machines use the number of misclassifications in the training set as a cost measure to be optimized during training. From a theoretical point of view, this kind of optimization process ensures a good generalization of the learning device once its capacity (i.e. a measure that accounts the complexity of the learning machine) is controlled.

However, recently it has been shown that, we should also take into account the confidence or *margin* of the classifications in order to guarantee a low generalization error. Therefore, the training samples must be assigned to the correct class with high confidence on average during training so the classifier can attain a large margin distribution. The reason why a large margin distribution allows achieving a better control of the capacity of the learning device is related to the constraints imposed in the solution that ensure a higher degree of stabilization.

Two recent examples of large margin classifiers are *support vector learning machines* (SVM) [12] and *boosting classifiers* [10]. However, other learning machines like multilayer perceptrons (MLPs) also belong to this category because the minimisation of the mean squared error (MSE) leads in practice to maximising the margin, as [1] suggests.

In this paper we show that it is possible to define another kind of loss functions, which also maximize the margin and is closely related. Besides, we also demonstrate that the solution achieved in the context of nearest neighbor classifiers (i.e. prototypes) is connected to support vectors.

J. Mira and A. Prieto (Eds.): IWANN 2001, LNCS 2084, pp. 669-676, 2001.

In the next section we introduce large margin classification. Section 3 shows that adaptive soft k-NN classifiers [2][3] and the learn1NN algorithm [4] are large margin classifiers related with support vectors. Finally some conclusions are given.

2. Large Margin Classification

Let $g(\mathbf{x})$ be a classifier that assigns an input pattern $\mathbf{x} \in \Re^p$ to one of the c existing classes. If all the classes have the same risk, the performance of g can be measured with the probability of classification error defined as

$$L(g) = P\{g(\mathbf{x}) \neq y\} \tag{1}$$

where P is the probability distribution on Xx{1,...,c} and y indicates the class label of the pattern $\mathbf{x}$. The best possible classifier g^*, which is called the Bayes classifier, minimises L(g). Since P is usually unknown, an empirical estimator of g, $g_N(\mathbf{x})$, is created by the estimate of L(g), defined as

$$\hat{L}(g_N; D_N) = \frac{1}{N}\sum_{i=1}^{N} 1\{g_N(\mathbf{x}_i) \neq y_i\} \tag{2}$$

where 1(u) is the indicator function which is 1 if u is true and 0 otherwise, and $D_N = \{(\mathbf{x}_1, y_1), \ldots, (\mathbf{x}_N, y_N)\}$ is a set of examples called the training set.

Suppose that g_N is a maximum classifier that uses c discriminant functions $\{d_i(\mathbf{x}), i=1,\ldots,c\}$, subject to the constraint $\sum_{i=1}^{c} d_i(\mathbf{x}) = 1$, which determine the confidence of g_N in each class. Then, g_N discriminates using the following rule:

$$g_N(\mathbf{x}) = j \Leftrightarrow d_j(\mathbf{x}) = \max_{i=1,\ldots,c} d_i(\mathbf{x}) \tag{3}$$

The margin m of g_N given a training sample $(\mathbf{x}_i, y_i)$ can be defined as the difference between the value of the discriminant function of the class which $\mathbf{x}$ belongs to and the maximal value of the other ones [10], i.e.

$$m(g_N, \mathbf{x}_i, y_i) = \sum_{j=1}^{c} 1(y_i = j)\left\{d_j(\mathbf{x}_i) - \max_{\substack{j'=1,\ldots,c \\ j' \neq j}} d_{j'}(\mathbf{x}_i)\right\} \tag{4}$$

Clearly, m is on the interval [-1,1] and only positive values denote correct classifications. As $m \rightarrow 1$, g_N classifies with higher confidence. The margin m is a random variable and consequently can be analyzed in terms of its cumulative distribution function, which is called the margin error estimate [1],

$$\hat{me}(g_N, \gamma; D_N) = \frac{\sum_{i=1}^{N} 1\{m(g_N, \mathbf{x}_i, y_i) < \gamma\}}{N} \tag{5}$$

Note that the margin error estimate includes the misclassification error in the training set because $\hat{me}(g_N, 0; D_N) = \hat{L}(g_N; D_N)$. Since we are interested in measuring the margin over the whole training set, we can simply obtain the average margin of g_N on D_N as

$$\overline{m}(g_N;D_N)=\frac{1}{N}\sum_{i=1}^{N} m(g_N,\mathbf{x}_i,y_i) \tag{6}$$

Equation (6) can be used as the cost function to be optimized in the training phase of large margin classifiers. However note that, in the case of maximum classifiers, the inclusion of the term $-\sum_{i=1}^{N}\sum_{j=1}^{c} 1(y_i = j)\max d_{j'}(\mathbf{x}_i)/N$ term then is not necessary since once we force the right discriminant function to one, the others automatically are forced to zero due to $\sum_{i=1}^{c} d_i(\mathbf{x})=1$. Therefore, the minimization of

$$\hat{L}_m(g_N;D_N)=-\frac{1}{N}\sum_{i=1}^{N}\sum_{j=1}^{c} 1(y_i = j)d_j(\mathbf{x}_i) \tag{7}$$

is equivalent to the maximization of Equation (6). Figure 1 shows the evolution of $\hat{me}(g_N,\gamma;D_N)$ in test set on the Pima database for the adaptive soft k-NN classifier [2][3] which minimizes Equation (7). As the learning time increases, test samples are classified with greater margin so $\overline{m}(g_N;D_N)$ is maximised while the test classification error $\hat{me}(g_N,0;D_N)$ decrease.

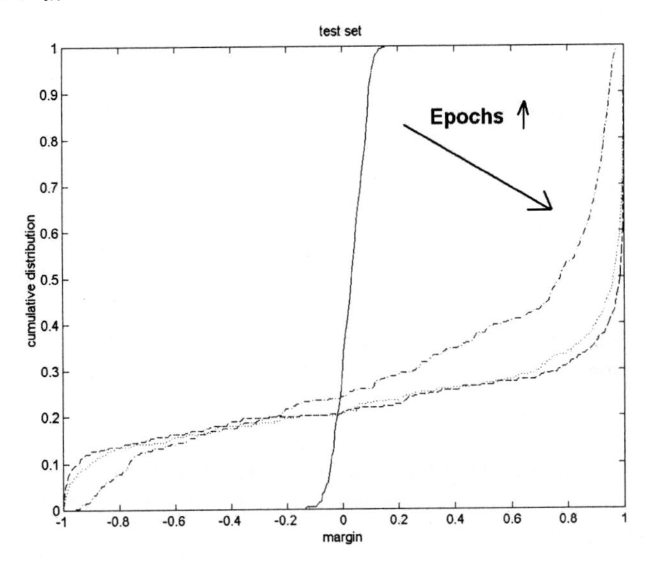

Fig.1. Evolution of the cumulative distributions for the Pima test set as the number of epochs augments.

According to [1], the generalization error of learning machines that maximize $\overline{m}(g_N;D_N)$ during learning is bounded with probability 1-δ by

$$L(g_N)<\hat{me}(g_N,\gamma;D_N)+r\{N,\delta,\mathrm{fat}_G(\gamma)\} \tag{8}$$

where r is a complexity term and fat_G is a scale-sensitive version of the VC dimension called fat-shattering. Typically, the complexity term r augments as the fat_G does. The generalization error is minimized when both terms of the right-hand side of Equation (8) are simultaneously minimized. Hence, the minimization of $\hat{me}(g_N, \gamma; D_N)$ for a given γ, achieved through the maximization of $\overline{m}(g_N; D_N)$ by the learner, does not guarantee a low generalization error since the complexity term r can be arbitrarily large due to the increase of the capacity measure fat_G. Therefore, fat_G must be controlled.

3. Large Margin in NN Classification and Support Vectors

Suppose we have the linear-separable 2-class problem of figure 2. It is possible to compute many linear classifiers that solve the problem. Figure 2 also shows ten solutions computed with the LVQ1 algorithm [7].

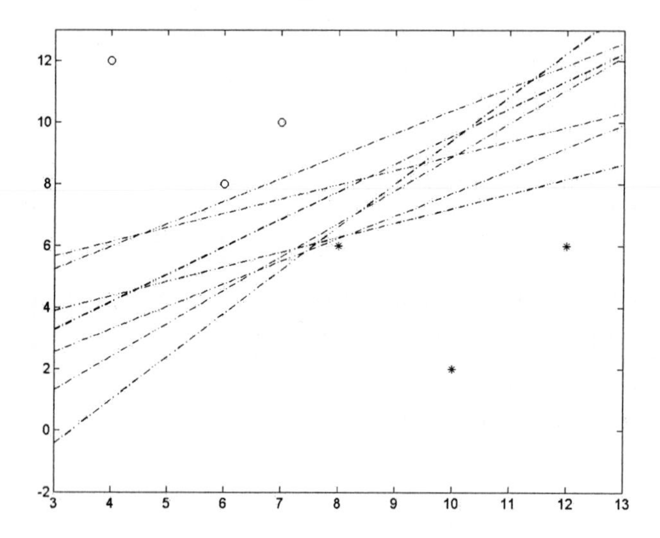

Fig.2. A toy two-class problem that is linear separable. We show the class border of ten linear classifiers (i.e. a 1-nearest-neighbour classifier with 2 prototypes) computed with the LVQ1 algorithm.

Now, we impose a slope of 45° to the linear classifier. Again, many solutions coexist. Nevertheless, it has been observed [12] that the linear classifier with a slope of 45° that achieves a maximal separation (or margin) between the extreme data points (or support vectors) of each class controls effectively its capacity (i.e. its fat_G) and consequently is highly generalizable even if the input space has a high dimensionality (see figure 3). Note that the optimal margin (OH) hyperplane is more robust with respect training patterns and parameters: a slightly variation on a test pattern or on the value of the line will presumably not affect the classification accuracy [11]. Consequently, OH is more "reliable" since it has the largest margin and then can achieve better generalization performance.

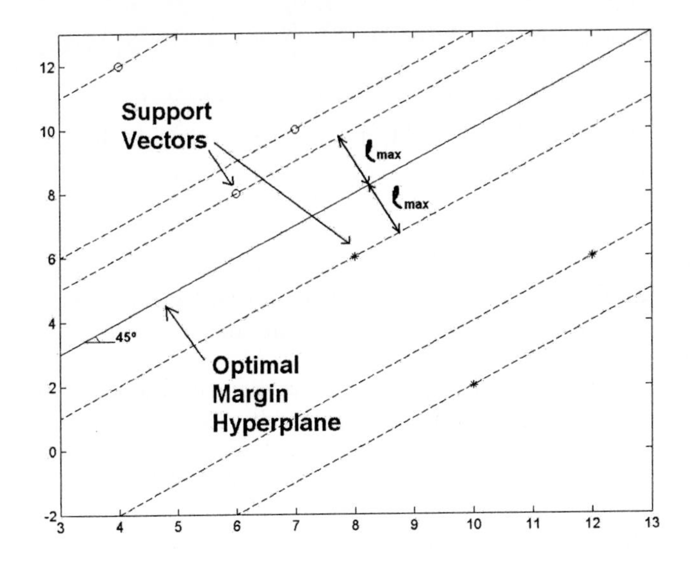

Fig.3. The optimal margin hyperplane for the toy problem (figure 2). This line has a slope of 45° and achieves a maximal separation between the extreme data points of each class. These extreme points are also known as support vectors.

Figure 4 shows the OH for a separable two-class pattern recognition problem. The application of the **Learn1NN algorithm** [4], which uses gaussian kernels and the Euclidean distance metric, is also shown in this problem and converges to OH. Learn1NN transforms 1-nearest-neighbour classifiers as maximum classifiers whose discriminant functions are based on a local mixture model, which can be derived as the following simplification of Parzen windows:

$$d_i(\mathbf{x}) \approx \frac{K(\mathbf{x}-\mathbf{m}_W^i;\gamma)}{\sum_{i'=1}^{C} K(\mathbf{x}-\mathbf{m}_W^{i'};\gamma)} \quad i=1,\ldots,c \tag{9}$$

where $\mathbf{m}_W^i$ is the nearest centre of the mixture model to **x** that belong to class i using the distance metric d and kernel $K(\mathbf{x}-\mathbf{m}_W^i;\gamma)=K(d(\mathbf{x},\mathbf{m}_W^i);\gamma)$. These centres are in fact the prototypes of the 1-NN classifier and are computed through the minimization of Equation (7) using a gradient-descent algorithm. The reason why the linear classifier computed with learn1NN converges to the OH can be explained computing the prototypes that minimises Equation (7). When we only have one prototype for each class $\{\mathbf{m}^j, j=1,\ldots,c\}$, solving $\nabla_{\mathbf{w}^j}\hat{L}_m(g_N;D_N)=0$ for gaussian kernels yields

$$\mathbf{m}^j = \frac{\sum_{i=0}^{N-1} 1(y_i = j) d_j(\mathbf{x}_i)(1 - d_j(\mathbf{x}_i))\mathbf{x}_i - \sum_{i=0}^{N-1}\sum_{\substack{k=1 \\ k \neq j}}^{C} 1(y_i = k) d_j(\mathbf{x}_i) d_k(\mathbf{x}_i)\mathbf{x}_i}{\sum_{i=0}^{N-1} 1(y_i = j) d_j(\mathbf{x}_i)(1 - d_j(\mathbf{x}_i)) - \sum_{i=0}^{N-1}\sum_{\substack{k=1 \\ k \neq j}}^{C} 1(y_i = k) d_j(\mathbf{x}_i) d_k(\mathbf{x}_i)} \quad (10)$$

$j = 1, \ldots, c$

According to Equation (10), prototypes depend on few training samples: only those training points that have a significant activation of their corresponding weight function contribute to form prototypes. Each prototype $\mathbf{m}^j$ are computed with the following subset of training data:

- Samples belonging to the class j which are near the class border since the weight function of these samples is $d_j(1-d_j)$ and reaches its maximum for $d_j=0.5$.
- Samples belonging to any other class which are near the border of class j since the weight function of these samples is $d_j d_k$ and reaches its maximum for $d_j=0.5$ and $d_k=0.5$.

Since the minimum points of $\hat{L}_m(g_N; D_N)$ tend to ensure a minimum number of misclassifications, the set of prototypes that solve $\nabla_{\mathbf{w}^j} \hat{L}_m(g_N; D_N) = 0$ are formed with the sub-set of training samples near class borders, that are hard to classify, that is *hard boundary points* [5] which are in fact the support vectors of each class.

When there is only one prototype for each class learn1NN is equivalent to adaptive soft 1-NN classifiers so both learning machines yield the same solution for the problem in figure 4. The so-called **adaptive soft k-NN classifier** [2][3] is *a soft k-NN rule + a gradient-descent learning algorithm* based on minimizing Equation (7). The soft k-NN rule converts k-NN methods as maximum classifiers whose discriminant functions are a direct extension of the crisp k-NN estimates based on the following use of kernels:

$$d_j(\mathbf{x}; \gamma) = \frac{\sum_{i=1}^{L} 1(y_i^m = j) 1_{kNN}(\mathbf{m}_i, \mathbf{x}) K(\mathbf{m}_i - \mathbf{x}; \gamma)}{\sum_{i'=1}^{L} 1(y_{i'}^m = j) 1_{kNN}(\mathbf{m}_{i'}, \mathbf{x}) K(\mathbf{m}_{i'} - \mathbf{x}; \gamma)}, \quad j = 1, \ldots, c \quad (11)$$

where $\{\mathbf{m}_i\}$ are the prototypes of the classifier, $\{y_i^m\}$ are the labels associated with the prototypes, L is the number of prototypes, $K(\mathbf{u}; \gamma)$ is a bounded and even function on X that is peaked around $\mathbf{0}$ with a locality parameter γ (e.g. $K(\mathbf{u}; \gamma) = \exp(-\|\mathbf{u}\|^2 / 2\gamma)$), and $1_{kNN}(\mathbf{m}_i, \mathbf{x})$ is a function that takes 1 if $\mathbf{m}_i$ is one of the k-nearest-neighbors to $\mathbf{x}$ using the distance metric $D(\mathbf{m}_i, \mathbf{x})$ (e.g. the Euclidean distance) and 0 otherwise.

If the training set D_N is used as the set prototypes, the above discriminant function estimates the posterior class probabilities using a combination of k-NN and Parzen estimations [3]. However, the reduced set of prototypes computed by minimizing Equation (7) typically exhibits better generalization performance. As Figure 1 shows, the margin is maximized during learning so the discriminant functions typically assign data with high confidence to one of the classes, i.e. it assigns values near 1 or

0. But there are some training data near class borders that the classifier assigns with a smaller margin. Among them, we might find the support vectors.

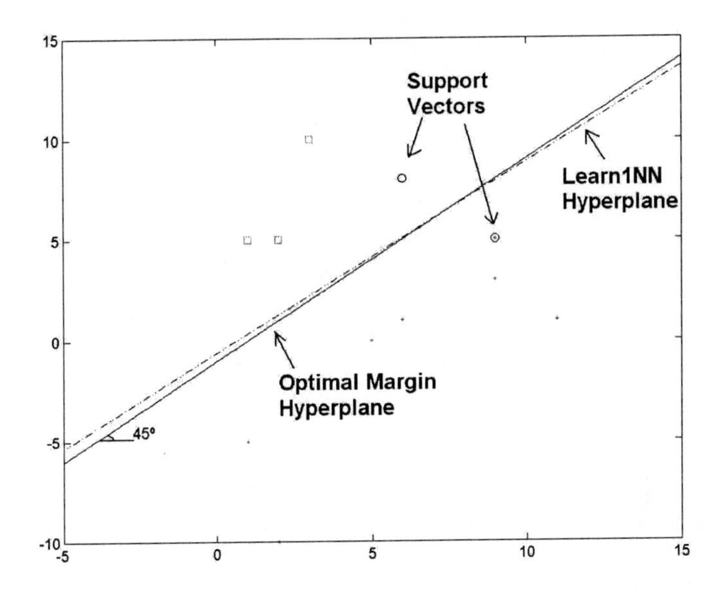

Fig.4. Optimal Margin Hyperplane (OH) for the toy problem that is linearly separable. We also show the class border of the NN classifier computed with learn1NN. Observe that it converges to OH.

Another additional benefit of the gradient-based approach for the computing of large-margin classifiers is related to its poor behavior as an optimizer since gradient-based algorithms are stacked at local minima. The under-computation of gradient descent algorithms prevents over-fitting [6] so capacity can be better controlled. See for instance MLPs [8]. Figure 5 shows an example using the Ripley's problem [9] in which an over-parameterized 1-NN classifier computed with learn1NN does not over-fit training data. The application of learn1NN and adaptive soft k-NN classifiers to real data (e.g. hand-written character recognition) is addressed in [2][3][4].

4. Conclusions

In this paper, we show how to compute large-margin maximum classifiers using a gradient-based learning based on a cost function directly related with the average margin of the classifier on a training set. Besides, we have established a connection between support vectors and large margin nearest-neighbor classifiers. However, further work on this latter topic is needed in order to determine how close both systems really are.

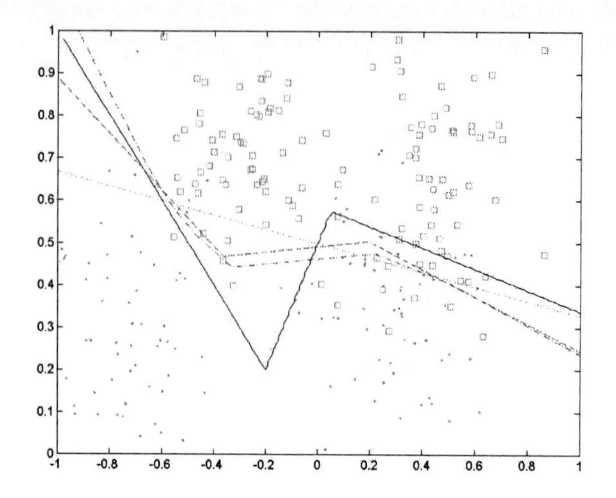

Fig.4. Ripley's synthetic training set with the Bayes border (solid line) and the class borders computed with lear1NN for 2 (dotted line), 6 (dashed line) and 32 (dashdot line) prototypes. The test error for these classifiers was 10.7%, 9.4% and 8.4 % respectively. Note that the NN classifier with 32 prototypes does not over-fit training data.

References

[1] Barlett, P. L. (1998). The Sample Complexity of Pattern Classification with Neural Networks: The Size of the Weights is More Important than the Size of the Network, IEEE Transaction on Information Theory, 44, 525-536.

[2] Bermejo, S., & Cabestany, J. (1999). Adaptive soft k-nearest neighbour classifiers. *Pattern Recognition*, Brief communication, **32**, 2077-2079.

[3] Bermejo, S., & Cabestany, J. (2000a). Adaptive soft k-nearest neighbour classifiers. *Pattern Recognition*, full-length paper, **33**, 1999-2005.

[4] Bermejo, S., & Cabestany, J. (2000b). Learning with nearest neighbour classifiers. To Appear in *Neural Processing Letters*, **13**.

[5] Breiman, L. (1998). Half-&-Half Bagging and Hard Boundary Points, Technical Report No.534, Berkley: University of California, Department of Statistics.

[6] Dietterich, T. (1997). Machine Learning Research: Four Current Directions. AI Magazine, 18, 97-136.

[7] Kohonen, T. (1996). Self-organizing Maps, 2nd Edition, Berlin: Springer-Verlag.

[8] Lawrence, S., Giles, C. L. & Tsoi, A. C. (1997). Lessons in neural network training: overfitting may be harder than expected. *Proceedings of* AAAI-97, 540-545, Menlo Park, CA: AAAI Press.

[9] Ripley, D. (1994). Neural Networks and Methods for Classification, Journal of the Royal Statistical Society, Series B, 56, p. 409-456.

[10] Schapire, R.E., Freund, Y., Bartlett, P. & Lee, W.S. (1998). Boosting the margin: A new explanation for the effectiveness of voting methods. *The Annals of Statistics*, **26**, 1651-1686.

[11] Smola, A. et al. (1999). Introduction to Large Margin Classifiers, in Smola, A. et al. (Eds.) Advances in Large Margin Classifiers Boston, MA: MIT Press.

[12] Vapnik, V. (1998). Statistical Learning Theory, New York: Wiley-Interscience.

Reduced Support Vector Selection by Linear Programs

Winfried A. Fellenz

King's Coll., Dept. of Math., Strand, London WC2R 2LS, UK
fellenz@mth.kcl.ac.uk

Abstract: The problem of selecting a minimal number of data points required to completely specify a nonlinear separating hyperplane classifier is formulated as a concave minimization problem and solved using a linear program. A comparison of the prediction errors for several rule extraction methods shows a good compromise between complexity of the classifier and the errors.

1. Introduction

The support vector machine (SVM) is a powerful technique for classification and regression problems, theoretically motivated by Statistical Learning Theory [13]. For pattern recognition, the problem is to estimate a function $f:\Re^n \rightarrow \{\pm 1\}$ using labeled training data $\{x_i, y_i\}, i=1,\ldots,l; x_i \in \Re^n, y_i \in \{-1,1\}$ such that f will correctly classify new examples generated from the same probability distribution $P(x, y)$ as the training data. If no restriction is put on the class of functions used to estimate f, the estimated function need not generalize well to new examples, even if the training error (the empirical risk),

$$R_{emp}(f)=\frac{1}{l}\sum_{i=1}^{l}\frac{1}{2}\left|f(x_i)-y_i\right|,$$

is minimized for all training examples. To achieve a small test error (the risk) averaged over test examples drawn from the same distribution $P(x, y)$,

$$R(f)=\int\frac{1}{2}\left|f(x)-y\right|dP(x,y),$$

it is necessary to restrict the capacity of f by choosing a function that is suitable for the amount of available training data. Statistical learning theory, or VC (Vapnik-Chervonenkis) theory, provides bounds on the expected test error, which depends both on the empirical risk and the capacity of the function class, called the structural risk. To measure the capacity of a given function class, the VC dimension has been introduced, which is defined as the largest number h of points that can be separated in all possible ways. For a given set of l samples and a VC dimension $h<l$, with probability of at least $1-\delta$ the VC upper bound

$$R(f)\leq 2R_{emp}(f)+\frac{1}{l}\left(4\log\left(\frac{4}{\delta}\right)+4h\log\left(\frac{2el}{h}\right)\right)$$

J. Mira and A. Prieto (Eds.): IWANN 2001, LNCS 2084, pp. 677–684, 2001.

states the expected generalization error of the function class. For a VC dimension h that is similar to the sample size l, the second term will be large, thereby predicting a large test error. To achieve acceptable generalization performance, the capacity (VC dimension) should be small compared to the amount of training data, but large enough to produce a small training error. In practice, the upper bounds appear to be overly pessimistic, possibly caused by distributions generating the real-world data, which do not behave like the worst case distributions used to prove the bounds. A related result estimates the VC lower bound [3]. Given a hypothesis space H with finite VC dimension $h \geq 1$, for any learning algorithm there exist distributions such that with probability at least δ over l random examples, the error is at least

$$\max\left(\frac{h-1}{32l}, \frac{1}{l}\log\left(\frac{1}{\delta}\right)\right)$$

The VC dimension is also central to the probably approximate correct (PAC) model of learning [11]. In the PAC model, an unknown target function $t(x)$ and an arbitrary probability distribution λ over the example space are assumed, and the goal of a learning algorithm is to infer a hypothesis $h \in H$ that closely approximates t. The learning algorithm $L(C, H)$ accepts training examples $s \in S(t)$ for a concept $c \in C$, a recursive subset of $X, c: X \rightarrow \{0,1\}$, and outputs hypotheses $h \in H$. The error of a learned hypothesis is defined as: $err_{\lambda}(h,t) = \lambda\ \{x \in X \mid h(x) \neq t(x)\}$, that is the probability that the hypothesis disagrees with the target function.

Definition: L is a probably approximate correct (PAC) learning algorithm for the hypothesis space H if, given the confidence parameter δ $(0 < \delta < 1/2)$, and the accuracy parameter ε $(0 < \varepsilon < 1/2)$, then there is a positive integer $m_0 = m_0(\delta, \varepsilon)$, such that for any $t \in H$ and for any probability distribution μ on X: Whenever $m \geq m_0$, $\lambda^m\{s \in S(m,t) \mid err_{\mu}(L(s)) < \varepsilon\} > 1 - \delta$.

The accuracy parameter reflects the fact that, without complete knowledge of the data distribution, the learning algorithm is likely to produce a hypothesis that is only approximately correct. The confidence parameter reflects the small chance that the algorithm will draw an unrepresentative sample, and thus the hypothesis will not be as correct as expected. In other words, provided that the size m of the training sample s is large enough, the input-output mapping computed by the learning algorithm is approximately correct with accuracy $1-\varepsilon$ and confidence $1-\delta$.

Of particular interest for practical applications is the issue of sample complexity, or how many random examples must be presented to the learning algorithm for it to acquire sufficient information to learn the target concept. An important result [3] states that a sufficient size of the training set S for a PAC learning algorithm is

$$m = \frac{K}{\varepsilon}\left(h\log\left(\frac{1}{\varepsilon}\right) + \log\left(\frac{1}{\delta}\right)\right),$$

where K is a constant. This general result allows the estimation of the required minimal training sample size m, given the error parameters ε and δ, and the VC dimension h of the used function class, for any PAC learning algorithm.

2. Linear Support Vector Machines

The simplest case for binary classification is the use of a parameterized real-valued function $f : X \subseteq \Re^n \rightarrow \Re$, for which the training data $\{x_i, y_i\}, i = 1,\ldots,l;$ $x_i \in \Re^N, y_i \in \{-1,1\}$ is assigned to the positive class if $f(x) \geq 0$, and otherwise to the negative class. The parameters $(w,b) \in \Re^n \times \Re$ of the linear function f are the weight vector and the bias, and must be learned from the training data. The geometrical interpretation for this classification scheme is the separation of the input space into two parts by the hyperplane defined by the equation $(w \cdot x) + b = 0$ (see Fig 1a). The vector w defines the normal direction to the hyperplane, the value $|b|/||w||$ is the perpendicular distance from the hyperplane to the origin, and $||w||$ is the Euclidian norm of w. Several methods have been proposed to find the parameters of this simple classifier in the statistics and neural network literature, called the linear discriminant or the perceptron. However, here we will motivate a related approach, based on optimization theory to find the optimal separating hyperplane for this classifier.

Let $d_+(d_-)$ be the shortest distance from the separating hyperplane to the closest positive (negative) example, and define the margin to be $d_+ + d_-$. If the data is linearly separable, the optimal separating hyperplane can be found by choosing the hyperplane with the largest margin. Since after appropriate rescaling all positive data points satisfy the constraint $x_i \cdot w + b \geq +1$, and all negative points satisfy $x_i \cdot w + b \leq -1$ (see Fig 1a), it is sufficient to only consider those points, which lie on the two parallel hyperplanes $H_1 : x_i \cdot w + b = +1$ and $H_2 : x_i \cdot w + b = -1$. These points, which are called the support vectors and are marked by circles in Fig 1, completely define the optimal separating hyperplane halfway between the hyperplanes H_1 and H_2. Thus, we can find the pair of hyperplanes that give maximum margin with width $2/||w||$ by minimizing $||w||^2$ subject to the constraints $y_i(x_i \cdot w + b) - 1 \geq 0 \ \forall i$.

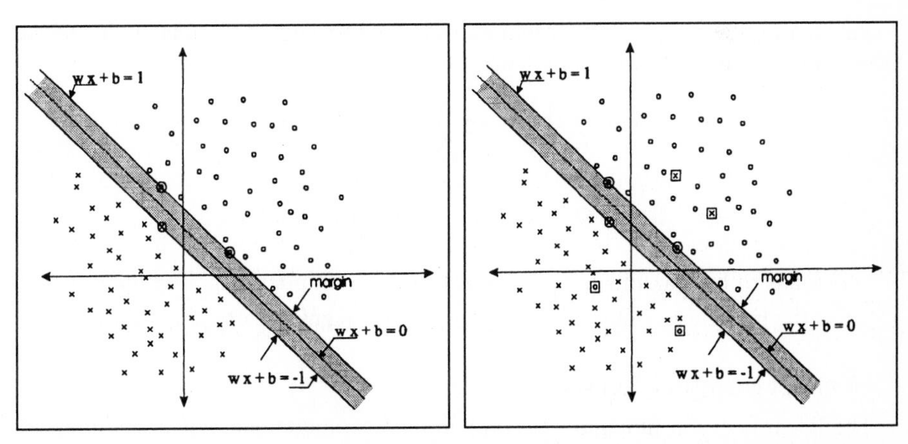

Fig. 1. The largest margin hyperplane for linear separable (a) and nonseparable (b) data.

In order to solve the problem stated above, it can be transformed into a constrained convex optimization problem and solved using a quadratic programming solver

$$\min_{(w,b)} \quad \frac{1}{2}\|w\|^2$$

$$\text{s.t.} \quad y_i((x_i \cdot w)+b) \geq 1 \quad i=1,\ldots,l\,.$$

The silent assumption is that the separating hyperplane and the function f actually exists, or in other words, that the convex optimization problem is feasible. This may not be the case in general, and one wants to allow some errors in the training data (the data points marked by boxes in Fig. 1b, thus trading off errors and flatness of the estimated weight vector. To overcome this sensitivity to noisy training patterns, a standard approach is to allow for the possibility of examples violating the constraints by introducing slack variables ξ_i [2,13] along with relaxed constraints:

$$\min_{(w,b,\xi)} \quad C\sum_{i=1}^{l}\xi_i + \frac{1}{2}\|w\|^2$$

$$\text{s.t.} \quad y_i((x_i \cdot w)+b)+\xi_i \geq 1, \quad \xi_i \geq 0, \quad i=1,\ldots,l$$

where the constant C determines the trade-off. This constraint optimization problem can be solved by introducing Lagrange multipliers $\alpha_i \geq 0$ and a Lagrangian

$$L(w,b,\alpha)=\frac{1}{2}\|w\|^2 - \sum_{i=1}^{l}\alpha_i\left(y_i\left((x_i \cdot w)+b\right)-1\right),$$

which has to be minimized with respect to the primal variables w and b, and maximized with respect to the dual variables α_i. At the corresponding saddle point the gradient of L vanishes with respect to the primary variables, which leads to the conditions:

$$w=\sum_i \alpha_i y_i x_i \text{ and } \sum_i \alpha_i y_i = 0\,.$$

Substituting these equality constraints into the Lagrangian eliminates the primal variables and constructs the Wolfe dual of the optimisation problem:

$$W(\alpha)=\sum_{i=1}^{l}\alpha_i - \frac{1}{2}\sum_{i,j=1}^{l}\alpha_i\alpha_j y_i y_j \left(x_i \cdot x_j\right)$$

$$\text{s.t.} \quad 0 \leq \alpha_i \leq C, \; i=1,\ldots,l \text{ and } \sum_{i=1}^{l}\alpha_i y_i = 0,$$

which has to be maximized with respect to the multipliers α_i. The upper bound C limits the influence of outliers by restricting the magnitude of the multipliers.

3. Kernel Support Vector Machines

The linear decision function f has only limited computational power and may not be well suited for most real-world classification tasks. However, a straightforward extension of the support vector machine allows the construction of nonlinear decision functions in a higher dimensional feature-space. This is achieved by mapping the data into a possibly infinite dimensional Euclidian (Hilbert) space $\aleph$ by a function $\Phi : \Re^n \rightarrow \aleph$, and the calculation of the optimal separating hyperplane in this space. Since both the construction of the hyperplane and the evaluation of the corresponding decision function

$$f(x) = \operatorname{sgn}\left(\sum_{i=1}^{l} y_i \alpha_i \cdot (x \cdot x_i) + b \right)$$

only require the evaluation of dot products $(\Phi(x) \cdot \Phi(y))$, and never the mapped patterns $\Phi(x)$ in explicit form, the dot products can be replaced by any kernel function $k(x, y) = (\Phi(x) \cdot \Phi(y))$. From this it follows, that we do not need to know the underlying feature map in order to learn in the feature space, and furthermore do so in roughly the same time as for the unmapped data. The only assumption on the kernel is that it must be positive definite, which is

$$\sum_{i,j=1}^{l} K(x_i, x_j) c_i c_j \geq 0 \quad \forall c_i, c_j \in \Re .$$

Equivalently, the kernel must satisfy Mercer's condition by proving the finiteness of the integral $\int K(x, y) g(x) g(y) dx dy \geq 0 \ \forall g \in L_2(C)$, where C is a compact subset of $\Re^N$. Examples of kernels satisfying Mercer's condition are the polynomial kernel of degree d: $k(x, y) = ((x \cdot y) + 1)^d$, the radial basis function with widths $\gamma : k(x, y) = \exp\left(-\gamma |x - y|^2\right)$, or the sigmoid kernel with gain κ and threshold Θ, equivalent to a two layer neural network $k(x, y) = \tanh(\kappa (x \cdot y) + \Theta)$.

4. The Generalized Support Vector Machine

The nonlinear separating hyperplane induced by the kernel $k(x, y) \in \Re^m$ and defined by some linear parameters $u \in \Re^m$ and $b \in \Re$ is determined by solving a mathematical program

$$\min_{(u,b,\xi)} \qquad C \sum_{i=1}^{l} \xi_i + \theta(u)$$

$$\text{s.t.} \quad y_i (y_j k(x_i \cdot x_j) u + b) + \xi_i \geq 1 \quad \xi_i \geq 0, \ i = 1, \ldots, l$$

where θ is a convex function on $\Re^m$, typically some norm or seminorm, which suppresses the parameter u [6]. The choice of θ leads to various support vector machines, like the conventional SVM if θ is a quadratic function generated by the kernel defining the nonlinear surface. If θ is chosen to be a piecewise linear convex function, the resulting support vector machine can be solved by a linear program, without assumptions on the kernel. For example, constraining the weight vector w to lie on the unit l_1 sphere instead of the unit l_2 sphere is equivalent to finding the weight vector and threshold that maximizes the minimum l_∞ distance between the training patterns and the decision hyperplane. The most obvious choice for θ is the 1-norm of u, which leads to the following linear program:

$$\min_{(u,b,\xi,t)} \quad C\sum_{i=1}^{l}\xi_i + \sum_{i=1}^{l} t_i$$

$$\text{s.t.} \quad y_i(y_j k(x_i \cdot x_j)u_i + b) + \xi_i \geq 1 \quad i = 1,\ldots,l$$

$$t_i \geq u_i \geq -t_i, \quad \xi_i \geq 0,$$

The advantages of this formulation compared to the conventional quadratic support vector machine if the 2-norm is used, is the possible use of a linear program to solve the convex optimisation problem, which is usually faster than a quadratic program, and the additional suppression of support vectors.

Figures 2 - 4 show results of various linear and nonlinear surfaces, which are extracted by a conventional support vector machine and the generalized support vector machine using radial basis kernels with widths 5.0. As can be evaluated from the figures, the GSVM usually selects viewer support vectors than the SVM, to describe the separating hyperplane. The advantage is most obvious in Fig. 2, where the GSVM only selects about 15 percent of the support vectors compared to the conventional support vector machine.

5. Results

The iris data set was used to compare the performance of several rule extraction techniques. The methods can be divided into combinations of a feed-forward neural network and a rule-extraction procedure (Rulex and Trepan) [1,5], decision tree inducers (C4.5 and CN2) [10,4], support vector classifiers (SMO and GSVM) [9,6], and fuzzy perceptron / fuzzy clustering methods (Nefclass, fcluster) [8,14]. The data set was split into 75 training and test examples, and the prediction performance was evaluated for the test set. While the class Iris Setosa can be correctly predicted by all methods, the second two classes, which are not linear separable, produce between two and six wrong predictions or rejections. The extracted rules are simple if-then-rules, which relate the input dimensions magnitude to the class prediction using a decision-tree, m-of-n rules, which in addition group subsets of input dimensions into conjunctive rules, and functional / trapezoidal rules. The numbers of extracted rules range from three to 12, with an average of six rules. The number of antecedents varies from only four for C4.5 to the 75 support vectors of the multi-class SMO method. To compare the performance of all methods, the number of classification errors and rejections have to be weighted differently, with weights two and one, respectively. The resulting error measure (Table 1) was plotted against the number of rules (Fig. 4a) and the number of antecedents (Fig 4b) with a linear regression line fitted to the measurements. As can be evaluated from the graphs, the GSVM rule extraction method, together with Rulex and CN2 show a good compromise between the number of errors on the test set and the number of and complexity of the extracted rules, placing them in the lower left quadrant of the graphs.

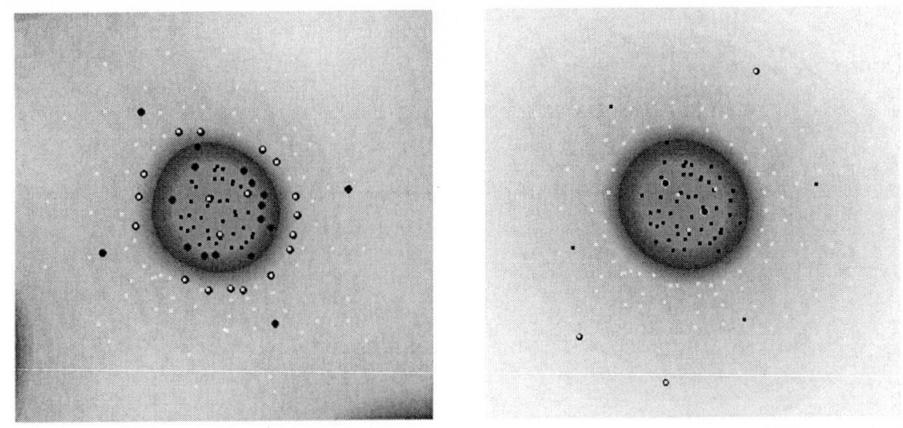

Fig 2. left: the Rbf-Svm selects 36 support vectors of 154; right: the Gsvm selects 5.

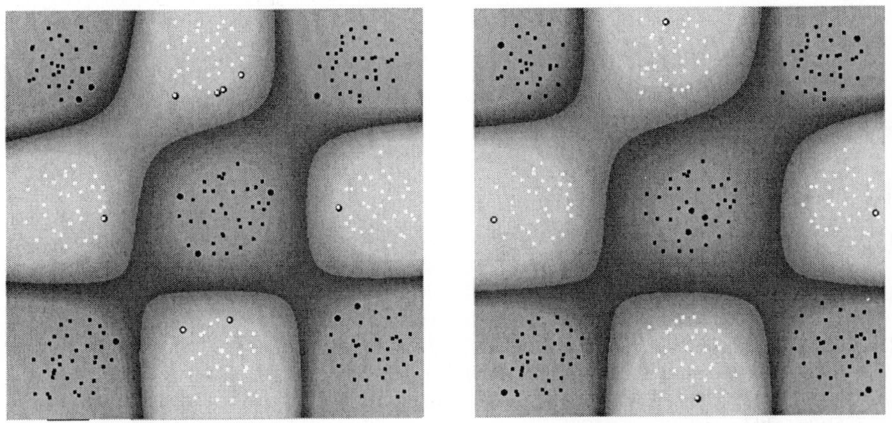

Fig 3. left: the Rbf-Svm selects 17 support vectors of 296; right: the Gsvm selects 11.

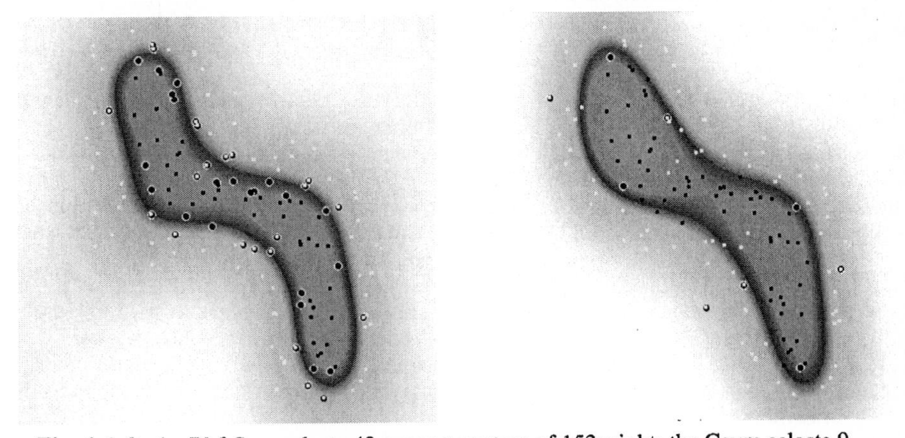

Fig. 4. left: the Rbf-Svm selects 43 support vectors of 153; right: the Gsvm selects 9.

Method:	Rulex	Trepan	C4.5	CN2	SMO	SVM	Nefclass	fcluster
Errors:	6	12	8	7	6	7	4	8
Rules:	6	6	2	4	3	3	12	3
Complexity:	17	12	4	8	75	18	48	6

Table 1. Number of errors, number of rules, and rule complexity for the eight methods.

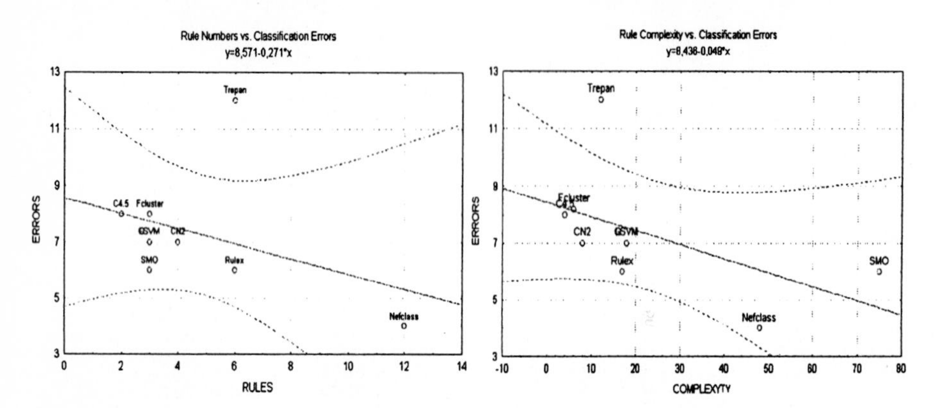

Fig. 5. Number of rules/complexity vs. classification errors for the 8 classification methods.

References

[1]Andrews, R. and Geva, S. (1995). Rulex & Cebp Networks as the Basis for a rule refinement system, In: Hybrid Problems, Hybrid Solutions, J. Hallam (Ed.), pages. 1-12, IOS Press

[2]Bennet, K. P. and Mangasarian, O. L. (1992). Robust linear programming discrimination of two linearly inseparable sets. Optimization Methods and Software, 1:23-34.

[3]Blumer, A., Ehrenfeucht, A., Haussler, D., and Warmuth, M. K. (1989). Learnability and the Vapnik-Chervonenkis dimension. Communications of the ACM, 36(4): 929-965.

[4]Clark, P. and Niblett, T. (1989). The CN2 Induction Algorithm. Machine Learning Journal, 3(4):261-283.

[5]Craven, M. and Shavlik, J. (1999). Rule extraction: where do we go from here?, TR-CS 99-1, University of Wisconsin, Madison.

[6]Mangasarian, O. L. (2000). Generalized support vector machines. In A. J. Smola, P. L. Bartlett, B. Schoelkopf, and D. Schuurmans (eds.) Advances in Large Margin Classifiers, Pages 135-146, Cambridge, MA: MIT Press

[7]Michie, D., Spiegelhalter, D. J., and Taylor, C. C. (Eds.)(1994). Machine Learning, Neural and Statistical Classification, New York: Ellis Horwood.

[8]Nauck, D. and Kruse, R. (1995). NEFCLASS - A neuro-fuzzy approach for the classification of data, Symposium on applied computing.

[9]Platt, J. (1998). Fast training of support vector machines using sequential minimal optimization, In B. Schoellkopf, C.J.C. Burges, and A.J. Smola, (Eds.) Advances in Kernel Methods Support Vector Learning, Pages 185-208, Cambridge, MA: MIT Press.

[10]Quinlan, J. R. (1986). Induction of Decision Trees. Machine Learning 1:81-106.

[11]Valiant, L. G. (1984). A theory of the learnable. Communications of the ACM, 27:1134-42.

[12]Vanderbei, R. J. (2001). Linear Programming: Foundations and Extensions. Kluwer.

[13]Vapnik, V. (1995), The Nature of Statistical Learning Theory, New York: Springer-Verlag.

[14]Wagner, T. and Wagner, O. (1995). Datenanalyse mittels Fuzzy-Clustering, Diplomarbeit, TU Braunschweig.

Edge Detection in Noisy Images Using the Support Vector Machines

Hilario Gómez-Moreno, Saturnino Maldonado-Bascón, Francisco López-Ferreras

Signal Theory and Communications Department. University of Alcalá
Crta. Madrid-Barcelona km. 33,600 D.P. 28871
Alcalá de Henares - Madrid (Spain)
{hilario.gomez, saturnino.maldonado, francisco.lopez}@uah.es

Abstract. In this paper, a new method for edge detection in presence of impulsive noise based into the use of Support Vector Machines (SVM) is presented. This method shows how the SVM can detect edge in an efficient way. The noisy images are processed in two ways, first reducing the noise by using the SVM regression and then performing the classification using the SVM classification. The results presented show that this method is better than the classical ones when the images are affected by impulsive noise and, besides, it is well suited when the images are not noisy.

1 Introduction

The edge detection methods are an important issue for a complete understanding of the image. The most usual classical methods search for several ways to perform an approximation to the local derivatives and they mark the edges by searching the maximum of these derivatives. The Sobel, Prewitt or Roberts filters are some of these approximations [1]. Other approaches are based in a previously smoothing action in order to improve the edge detection and to reduce the effect of noise (basically Gaussian noise) [2]. This pre-filtering process seems to be not adequate when the noise is impulsive and most edge detectors are very sensitive to the pulses of noise.

In this paper we present a way to detect the edges by using a new point of view. We do not try to approximate the derivative or to use other mathematics methods. The main idea in this work is to train the computer to recognize the presence of edges into an image. In order to perform this idea we use the Support Vector Machines (SVM) tool that is given good results in other classification problems.

The training is performed using a few own-created images that represents a clearly defined edges with the edges easily located.

The noise reduction process is performed using the SVM. In this case we use the information in a 3x3 neighborhood of each pixel to make a regression process and then replace the pixel value with the regressed value.

The results obtained with no noise are similar to those from previous methods and clearly superior when compared to some classical methods in noisy images.

J. Mira and A. Prieto (Eds.): IWANN 2001, LNCS 2084, pp. 685-692, 2001.

2 SVM Classification and Regression

There are several ways to classify, Bayesian decision, neural networks or support vector machines, for example. In this work we use the SVM classifier since this method provides good results with a reduced set of data and then we do not require an intensive training like another methods. Thus the SVM gives us a simple way to obtain good classification results with a reduced knowledge of the problem. The principles of SVM have been developed by Vapnik [3] and they are very simple.

In the decision problem we have a number of vectors divided into two sets, and we must find the optimal decision frontier to divide the sets. The frontier chosen may be anyone that divides the sets but only one is the optimal election. This optimal election will be the one that maximizes the distance from the frontier to the data. In the two dimensional case, the frontier will be a line, in a multidimensional space the frontier will be an hyperplane. The decision function that we are searching has the next form,

$$f(\mathbf{x}) = \langle \mathbf{w} \cdot \mathbf{x} \rangle + b = \sum_{i=1}^{n} w_i x_i + b \ . \tag{1}$$

In (1), $\mathbf{x}$ is a vector with n components that must be classified. We must find the vector $\mathbf{w}$ and the constant b that makes optimal the decision frontier. The basic classification process is made by obtaining the sign of the decision function applied to the given vector, a positive value represents the assignment to one class and a negative one represents the assignment to the another class.

Normally, we use another form of this decision function that includes the training input and output vectors information [4]. This new form is,

$$f(\mathbf{x}) = \sum_{i=1}^{l} \alpha_i y_i \langle \mathbf{x}_i \cdot \mathbf{x} \rangle + b \ . \tag{2}$$

The y values that appear into this expression are +1 for positive classification training vectors and –1 for the negative training vectors. Besides, the inner product is performed between each training input and the vector that must be classified. Thus, we need a set of training data (**x**,**y**) in order to find the classification function and the α values that makes it optimal. The l value will be the number of vectors that in the training process contribute in a high quantity to form the decision frontier. The election of these vectors is made by looking at the α values, if the value is low the vector is not significant. The vectors elected are known as support vectors.

Normally the data are not linearly separable and this scheme can not be used directly. To avoid this problem, the SVM can map the input data into a high dimensional feature space. The SVM constructs an optimal hyperplane in the high dimensional space and then returns to the original space transforming this hyperplane in a non-linear decision frontier. The non-linear expression for the classification function is given in (3) where K is the non-linear mapping function.

$$f(\mathbf{x}) = \sum_{i=1}^{l} \alpha_i y_i K(\mathbf{x}_i \cdot \mathbf{x}) + b \ . \tag{3}$$

The choice of this non-linear mapping function or kernel is very important in the performance of the SVM. The SVM applied uses the radial basis function to perform the mapping, since the others proved do not work appropriately. This function has the expression given in (4).

$$K(x, y) = \exp\left(-\gamma (x - y)^2\right). \tag{4}$$

The γ parameter in (4) must be chosen to reflect the degree of generalization that is applied to the data used. The more data is obtained the less generalization needed in the SVM. A little γ reflects more generalization and a big one represents less generalization. Besides, when the input data is not normalized, this parameter performs a normalization task.

When some data into the sets can not be separated, the SVM can include a penalty term (C) that makes more or less important the mismatch classification. The more little is this parameter the more important is the misclassification error. This term and the kernel are the only parameters that must be chosen to obtain the SVM.

The classification scheme may be easily extended to the case of regression. In this case the idea is to train the SVM by using y values different from +1 and –1. The values used are the values known from the function to be obtained. Then, we search for an approximation function that fits approximately the known values [4].

3 Noise Reduction Training

In this section, we explain how the SVM regression may be used to reduce the impulse noise into the images in order to apply an edge detection algorithm without noise interferences.

The main idea is to replace the pixels into noisy image with a new value avoiding the noise. In order to obtain the new pixel value we use the SVM regression. The regression training process is shown in Fig. 1.

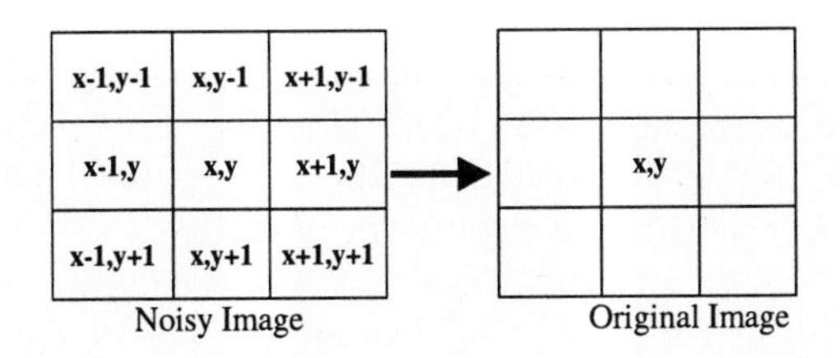

Fig. 1. Pixel regression

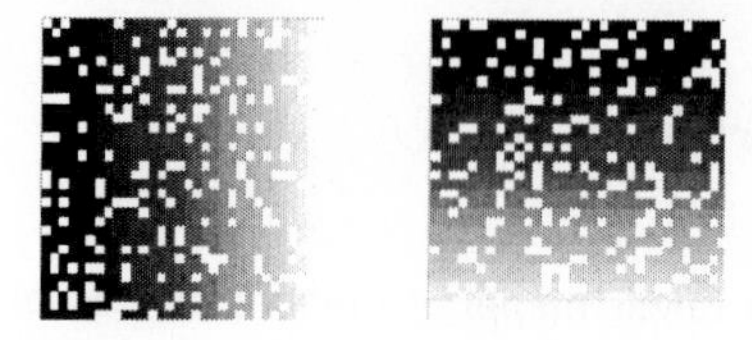

Fig. 2. Example of noisy training images

For each pixel (x,y) of the noisy image we form a vector containing the values of the pixels around it (including the pixel itself). This vector will be the input to the regression function. The output will be the value of the pixel (x,y) of the original image.

The next step is to find the images to be used into the training process. The images elected must be simple to avoid an excessive training time but significant enough to give good results. The option elected was to make controlled images given a gray scale and with random noise in a known position. In Fig. 2 an example is presented.

In this training images we must control the size and the noise ratio. The images in Fig. 2 have a 32x32 size and a 20 percent noise (20 percent pixels of the image are noise). If we use these two images in the training process the number of vectors processed will be 1953. The number of support vectors after the training process is 634 in this case.

4 Edge Detection Training

In this section we present a way to detect edges by using the SVM classification. In this case, the decision needed is between "the pixel is part of an edge" or "the pixel is not part of an edge". In order to obtain this decision we must extract the information needed from the images. In this work a vector is formed for each pixel, given the difference between this one and the pixels in a 3x3 neighborhood around it. This way an eight components vector is calculated at each pixel except for the border of the image, because in this case the differences can not be calculated. The vectors formed are used as inputs to the SVM training.

Fig. 3. Training images for edge detection

The images used to train the SVM are shown in Fig. 3. They are images created by trying to obtain a good model for the detection. The only edges used in the training are vertical, horizontal and diagonal ones and we expect that the other edges will be generalized by the SVM. The pixels considered as edges are those into each image that are in the border between bright and dark zones, i.e. the points in the dark zone near the bright one and vice versa.

The dark and bright zones are not homogenous but the intensity at each pixel is a random value (gaussian values). The values at each zone never reach those of the other zone. By using these random values we try to simulate the no homogenous surface into a real image. The random nature of the training images makes random the SVM training, for example, the number of support vector may change every time we use different images. But the results obtained are very similar in all cases.

A value that must be set in the training process is the mean difference between dark and bright zones, since this parameter controls the sensibility of the edge detector. A little difference makes the detector more sensible and a greater one reduces this sensibility.

5 Edge detection method

When we apply the trained SVM (3) to an image, a value for each pixel into the image is obtained. This value must be (ideally) a value positive or negative near 1. We can use the sign of these values to say when a pixel is an edge or not but, this way, a lost of information is produced. It is better to use the values obtained and say that there is a gradual change between "no edge" and "edge". Then, the value obtained indicates a probability of being an edge or not.

After the process above we must decide the pixels that are considered as edges. This task can be simplified when we have the edges separated into vertical and horizontal directions. In this case we search for the pixels that are local maximum in both directions and the edge image is obtained. The method proposed is able to obtain these vertical and horizontal edges by using vectors formed only with the horizontal and vertical differences as input to the SVM (Fig. 4).

x-1,y-1	x,y-1	x+1,y-1
x-1,y	x,y	x+1,y
x-1,y+1	x,y+1	x+1,y+1

Fig. 4. Vectors for vertical and horizontal detection

Fig. 5 shows and example of the process. The original image "house" is a 256x256 eight bits gray-scaled image. The edge images presented are gray-scaled images where the values from SVM have been translated to gray values, 255 represents edge detection and 0 represents no edge detection. The intermediate values show a gradation between these extreme values. The parameters set for this edge detection were γ = 8e-4, C = 1000, and the mean difference between dark and bright zones in the training images was 50. Besides, in order to reduces the isolated points considered as edges we use only the values above a given threshold. In the examples the threshold is set at 32 in an eight bits gray scale.

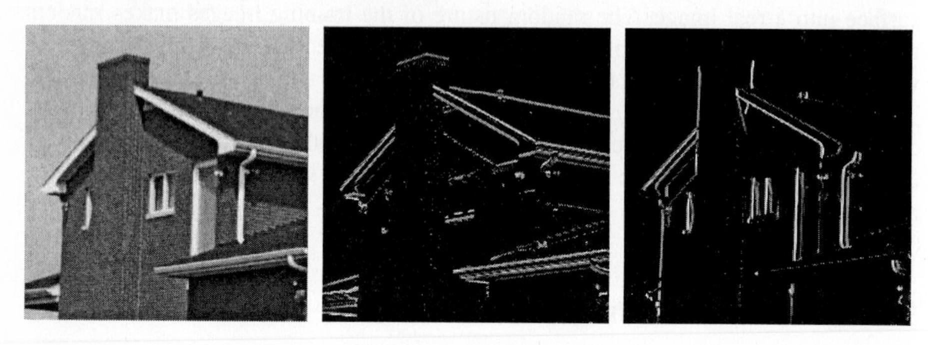

Fig. 5. Horizontal and vertical edge detection example.

Fig. 6 shows a comparison between the Canny edge detector and the detector proposed here. The image used as example presents a complex texture in the walls due to bricks. Thus, the parameters of the Canny detector have been set trying to reduce the false edge detection of this bricks.

We can see how the performance of the method proposed is similar to the Canny edge detector (considered as an standard) although the SVM method is not using any blurring filter like in the Canny edge detector.

Fig. 6. Comparison of edge detection. (a) Canny edge detector. (b) SVM edge detector.

Fig. 7. Example of noise reduction. (a) Noisy image (b) No noise image

Fig. 8. Comparison of edge detectors. (a) SVM detector (b) Canny edge detector.

6 Results

The results presented here have been obtained by using the LIBSVM [5] as implementation for the SVM. The programs used were written in C++ and compiled using the Visual C++ 6.0 compiler. The computer used has a Pentium III processor with 128 Mbytes RAM.

First we show how the noise reduction process performs. The training process uses images like in Fig. 2, 32x32 gray-scaled images with a 20 % impulsive noise. In this case the impulsive noise has white pulses only but the method may be applied in a similar way to "salt&pepper" noise. The γ parameter of the kernel is elected to obtain the best results, in this case the value elected is γ = 3e-5. The C parameter must be elected to obtain accurate results with a reduced number of support vectors. In this work the parameter has a value C = 1000. A greater value increases the training time and a more little value decreases the accuracy of the process. The training process takes about 10 minutes in the case presented. The noise reduction process can be viewed in Fig. 7. Fig. 7-a shows a corrupted image with impulsive noise of 10 % and Fig. 7-b shows the image after the noise reduction process explained here.

The example shows that this reduction process is not well suited for denoising since the recovered image is blurred. However, the impulsive noise disappears from the image and the edge detection process can be performed. Besides, the blurring is useful in the edge detection since the textures into the images are more uniform after blurring and the false detection is reduced.

The next step is to perform the edge detection like previously explained. The kernel parameter in this case was γ = 8e-4 and the parameter C = 1000 like in the previous case. The threshold used in this case is 16. In Fig. 8 is shown a comparison between the SVM method and the Canny edge detector. The parameters of the Canny edge detector have been set in order to reduce to the maximum the detection of the impulsive noise like edges. It is clearly that the SVM method is superior in this case. The only drawback of the method proposed is that the execution takes over 15 seconds while the Canny method takes 1 second.

7 Conclusions

This work shows how the SVM performs the edge detection in presence of impulsive noise in an efficient way and given good results.

The comparison between the edge detection method proposed here with other known methods shows that the SVM method is superior when the impulsive noise is present and it is similar (although slightly inferior) when the image is not noisy.

We think that the method may be improved in execution speed and performance by optimizing the C++ programs and by using more differences between adjacent pixels to make edge detection, e.g. using 5x5 windows around a pixel.

References

1. A.K. Jain. "Fundamentals of Digital Image Processing". Englewood Cliffs, NJ, Prentice Hall, 1989.
2. J.F. Canny. "A computational Approach to Edge Detection". IEEE Trans. On Pattern Analysis and Machine Intelligence, vol. 8, pp. 679-698. 1986.
3. V. Vapnik. "The Nature of Statistical Learning Theory". New York, Springer-Verlag, 1995.
4. N. Cristianini and J. Shawe-Taylor. "An introduction to Support Vector Machines and other kernel-based methods". Cambridge, Cambridge University Press, 2000.
5. C.C. Chang and C.J. Lin. "Libsvm: Introduction and benchmarks". http://www.csie.ntu.edu.tw/~cjlin/papers.

Initialization in Genetic Algorithms for Constraint Satisfaction Problems

Camino R. Vela, Ramiro Varela, Jorge Puente

Centro de Inteligencia Artificial.
Universidad de Oviedo. Campus deViesques. E-33271 Gijón. Spain.
Tel. +34-8-5182032. FAX +34-8-5182125.
e-mail: {camino, ramiro, puente}@aic.uniovi.es
http:\\www.aic.uniovi.es

Abstract. In this paper we propose a strategy to incorporate heuristic knowledge into the initial population of a Genetic Algorithm to solve Job Shop Scheduling problems. This is a generalization of strategy we proposed in a previous work. The experimental results reported confirm that the new strategy improves the former one. In particular, a higher diversity in achieved among the heuristic individuals, and at the same time the mean fitness is improved. Moreover, these improvements translate into a better convergence of the GA.

1 Introduction

In this paper we build on some previous works [5], [6] where we have confronted the problem of solving CSP problems by means of Genetic Algorithms (GA). Our aim is to improve the performance of GAs by means of specific knowledge. GAs were successfully used in solving hard CSP problems, many times by means of an intensive search and an elitist strategy where, for example, only the best individuals obtained are maintained in the population and an offspring does not replace their parents unless it improves their fitness. Here, we address the problem from another point of view: instead of using a "brute force" approach, we start from a conventional GA and try to improve it by means of domain specific knowledge. In particular we introduce this knowledge into a number of the initial chromosomes to obtain a heuristic initial population as an alternative to the random initial population of the conventional GAs. The idea of introducing domain specific knowledge was exploited, for example, in [2].

The remainder of this paper is organized as follows. In section 2 we introduce the problem to be faced: the Job Shop Scheduling (JSS) problem. In section 3 we summarize the heuristic strategy we have exploited to obtain domain knowledge: the variable and value ordering heuristics proposed in [4]. In section 4 we describe our proposed strategy to obtain heuristic chromosomes. In section 5 we present some experimental results over a well know JSS example. And, finally, in section 6 we present the main conclusions.

J. Mira and A. Prieto (Eds.): IWANN 2001, LNCS 2084, pp. 693–700, 2001.

2 The Job Shop Scheduling Problem

The JSS requires scheduling a set of jobs $\{J_1,...,J_n\}$ on a set of physical resources or machines $\{R_1,...,R_q\}$. Each job Ji consists of a set of tasks or operations $\{t_{i1},...,t_{imi}\}$ to be sequentially scheduled. Each task having a single resource requirement and a fixed duration or processing time du_{il} and a start time st_{il} whose value should be determined. We assume that there is a release date and a due date between which all the tasks have to be performed.

Furthermore, there are two non-unary constraints of the problem: *precedence constraints* and *capacity constraints*. Precedence constraints defined by the sequential routings of the tasks within a job translate into linear inequalities of the type: $st_{il} + du_{il} \leq st_{il+1}$ (i.e. st_{il} before st_{il+1}). Capacity constraints that restrict the use of each resource to only one task at a time translate into disjunctive constraints of the form: $st_{il} + du_{il} \leq st_{jk} \vee st_{jk} + du_{jk} \leq st_{il}$ (two tasks that use the same resource can not overlap). The most widely used objective is to come up with a feasible schedule such that the completion time of the whole set of tasks, i.e. the makespan, is minimized.

In the following a problem instance will be represented by a directed graph $G = (V, A \cup E)$ [3]. Each node of the set V represents a task of the problem, with the exception of the dummy nodes *start* and *end* which represents tasks with processing time 0. The set of arcs A represent the precedence constraints and the set of arcs E represent the capacity constraints. The set E is decomposed into subsets E_i with $E=\cup_{i=1..m}E_i$, such that there is one E_i for each resource R_i. The subset E_i includes an arc (v,w) for each pair of tasks requiring the resource R_i. Figure 1 depicts an example with three jobs $\{J_0,J_1,J_2\}$ and three physical resources $\{R_0,R_1,R_2\}$. Solid arcs represent the elements of the set A; whereas dotted arcs represent the elements of the set E. The arcs are weighed with the processing time of the of the task at the source node. The dummy task *start* is connected to the first task of each job; and the last operation of each job is connected to the node *end*. The domain of start time values, constrained by the release date, due date, precedence constraints and the processing time of the tasks, are represented as intervals. For instance [0,7] represents all start times between time 0

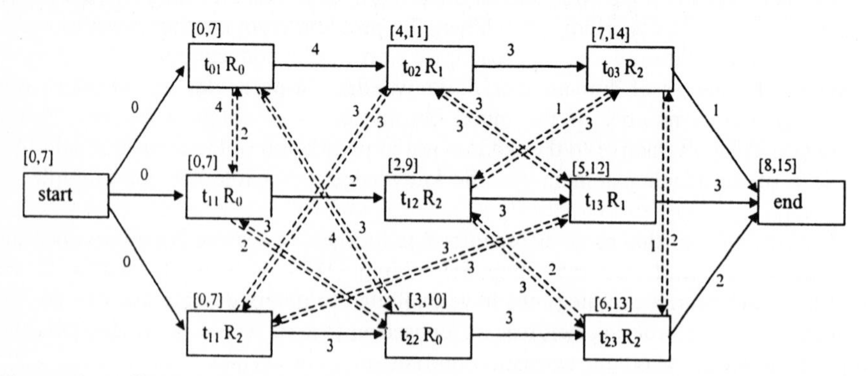

Figure 1. A directed graph representation of a *JSS* problem instance with three jobs. The release date is 0 and the due date is 15. The resource requirement of every task is indicated within the boxes. Arcs are weighted with the processing time of the task at the outcoming node.

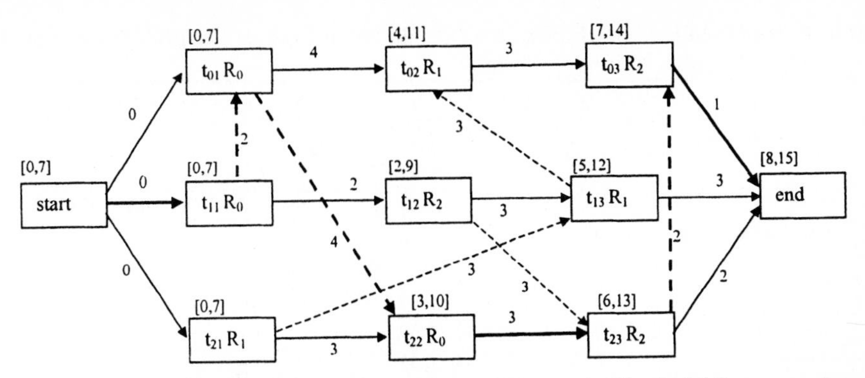

Figure 2. A feasible schedule to the problem of Figure 1. The bold face arcs shown the critical path whose length, that is the makespan, is 12. Hence it is actually a solution to the problem because this value is less that 15, the due date.

and time 5, as allowed by the time granularity, namely {0,1,2,3,4,5,6,7}.
A feasible schedule is represented by an acyclic subgraph *Gs* of *G*, $Gs=(V,A\cup H)$, where $H=\cup_{i=1..m}H_i$, H_i being a hamiltonian selection of E_i. The makespan of the schedule is the cost of a critical path. A critical path is a longest path from node *start* to node *end*. When this value is less than or equal to the due date, the schedule is a solution to the problem. Therefore, to find out a solution can be reduced to discover compatible hamiltonian selections, i.e. orderings for the tasks requiring the same resource or partial schedules, that translates into a solution graph *Gs* without cycles whose critical path does not exceeds the due date. Figure 2 shows a graph representing a feasible solution to the problem of Figure 1.
In this paper we consider the partial schedules as basic building blocks to obtain solutions. In particular, our aim is to build a set of potential solutions from promising partial schedules to seed the initial population of the GA. These partial schedules will be obtained by means of a well known heuristic method: the variable and value ordering heuristics [4]. Of course, two or more partial schedules might result not compatible each one to the others. To solve this problem we have envisaged a method to reorganize one or more of the partial schedules to ensure compatibility.

3 The variable and value ordering heuristics

These heuristics were proposed in [4] and are based on a probabilistic model of the search space. A framework is introduced that accounts for the chance that a given value will be assigned to a variable and the chances that values assigned to different variables conflict with each other. These chances are evaluated from the profile demands of the tasks for the resources. In particular the individual demand and the aggregate demand values are considered. The individual demand $D_{ij}(R_p,\tau)$ of a task t_{ij} for a resource R_p at time interval $T\leq\tau<T+1$ is simply computed by adding the probabilities $\sigma_{ij}(\rho)$ of the resource R_p is demanded by the task t_{ij} at some time ρ within the interval $T-du_{ij}<\rho\leq T$. The individual demand is an estimation of the reliance of a task on the availability of a resource. Consider, for example, the initial state of the

Table 1. Survivabilities of all nine tasks in the initial state of the problem of Fig. 1 over the time. Blank positions represent 0.

	0	1	2	3	4	5	6	7	8	9	10	11	12	13	14
t_{01}	0.69	0.60	0.51	0.44	0.39	0.39	0.43	0.47							
t_{02}					0.55	0.47	0.44	0.47	0.55	0.61	0.63	0.67			
t_{03}								0.69	0.69	0.69	0.75	0.81	0.87	0.87	0.93
t_{11}	0.87	0.78	0.66	0.57	0.50	0.46	0.46	0.50							
t_{12}			1.0	1.0	0.94	0.76	0.62	0.54	0.54	0.54					
t_{13}						0.42	0.42	0.47	0.55	0.61	0.67	0.75	0.87		
t_{21}	1.0	1.0	0.96	0.83	0.65	0.50	0.42	0.39							
t_{22}				0.39	0.39	0.39	0.44	0.55	0.72	0.87	0.96				
t_{23}							0.60	0.56	0.56	0.61	0.82	0.88	0.88		

search depicted in Figure 1. As the task t_{02} has 8 possible start times or reservations, and assuming that there is no reason to believe that one reservation is more likely to be selected than another, each reservation is assigned an equal probability to be selected, in this case *1/8*. Given that the task t_{12} has duration of 3 time units, this task will demand to the resource R_1 at interval $6 \le t < 7$ if its start time is 4, 5 or 6. Hence, the individual demand of the task t_{12} for resource R_1 at this interval is estimated as $D_{02}(R_1,t) = \sigma_{02}(6)+\sigma_{02}(5)+\sigma_{02}(6) = 3/8$. On the other hand, the aggregate demand $D_{aggr}(R,\tau)$ for a resource is obtained by adding the individual demands of all tasks over the time. From the aggregate demand of a resource a contention peak is identified. This is an interval of the aggregate demand of duration equal to the average duration of the tasks requiring the same resource with the highest demand. We assume the value of the contention peak as the integral of the aggregate demand over this time interval. Then, the task with the largest contribution to the contention peak of the most contented resource is determined as the most critical and then it is selected first for reservation. This is the heuristic referred in [4] as *ORR* (*Operation Resource Reliance*).

At the same time, the value ordering heuristic is also computed from the profile demands for the resources. Given a task t_{ij} that demands the resource R_p, the heuristic consists of estimating the survivability of the reservations. The survivability of a reservation $\langle st_{ij}=T \rangle$ is the probability that the reservation will not conflict with the resource requirements of other tasks, that is, the probability that none of the other tasks require the resource during the interval $[T, T+du_{ij}-1]$. When the task demands are for only one resource, this probability is estimated as

$$\left(1-\frac{AVG\left(D^{aggr}\left(R_p,\tau\right)-D_{ij}\left(R_p,\tau\right)\right)}{AVG\left(n_p\left(\tau\right)-1\right)}\right)^{AVG\left(n_p(\tau)-1\right)\bullet du_{ij}\bullet\left(AVG(du)^{-1}\right)} \tag{1}$$

where du stands for the average duration of the tasks, $n_p(\tau)$ is the number of tasks that might demand the resource R_p at time τ and $AVG(f(\tau))$ represents the average value of $f(\tau)$ in the interval $[T, T+du_{ij}-1]$. When $AVG(n_p(\tau)-1)$ is 0, the value of expression (1) is 1 due to in this case no other task might require the resource R_p during the interval $[T, T+du_{ij}-1]$. Table 1 shows the survivability of all the reservations possible for all nine tasks of the problem of Figure 1 in the initial state. In principle, the value ordering heuristic consists of trying first those reservations with large survivabilities. Although, in [4] some more refinements are introduced, and the resulting heuristic is referred as *FSS* (*Filtered Survivable Schedules*). For a full description of these heuristics, we refer to the interested reader to [4].

4 Heuristic Chromosomes

As we have pointed out in the introduction, the main contribution of this work is to exploit the former heuristic information to obtain a good initial population as an alternative to the random generation. Our proposed strategy is composed by two main steps: First, we introduce the heuristic information corresponding to the initial state of the search space into a matrix representation, and then this matrix codification is translated to the permutation codification used by the GA.

Matrix Codification

We start with this chromosome representation because of its ability to represent the heuristic information provided by the former heuristics. In principle, a chromosome is represented by a *Resource Matrix* (*RM*). This is a $N \times M$ matrix where the ith row represents a partial schedule for the tasks requiring the resource R_i. From a *RM* matrix a schedule can be built by merging all of the partial schedules so that every precedence constraint is also satisfied. Unfortunately sometimes this is not possible because of the lack of compatibility among partial schedules and hence a number of them should be modified. In order to do that, we adopt a priority schema for resources so that partial schedules of low priority resources are modified first. To represent priorities we include into the chromosome representation a *Resource Priority* vector (*RP*) that codifies an ordering of resources, from high to low rescheduling priority. Figure 4 a) shows a *RM* representing partial schedules for the problem of Figure 1, and Figure 4 b) shows a *RP* vector of priorities among resources.

In order to build heuristic chromosomes we establish a probabilistic framework in which partial schedules are built from the survivability profiles of the tasks and the vector *RP* is also probabilistically obtained from the contention peaks of the resources.

To obtain partial schedules we define a probability distribution over the space of partial schedules of each resource R so that each partial schedule $\{t_{k0},\dots,t_{kM-1}\}$ is given the probability of start times held $\tau_{k0} < \tau_{k1} < \dots < \tau_{kM-1}$, τ_{ki} being a start time for task t_{ki} selected from a probability distribution given by the normalized survivability

$$\Pr_j(\{\tau_j\}) = surv_j(\tau_j) \Big/ \sum_{\tau} surv_j(\tau), \tag{2}$$

where $surv_i(\tau_i)$ stands for the survivability of task t_i at time τ_i; hence the probability of a partial schedule can be calculated as

$$\Pr(\{t_{k0},\dots,t_{kM-1}\}) = \sum_{\tau_{k0}<\tau_{k1}<\dots<\tau_{kM-1}} \left(\prod_{j=0}^{M-1} \Pr_j(\{\tau_j\}) \right). \tag{3}$$

The intuition behind this strategy is that this partial schedule gives to every task a chance to be scheduled at a time with a large survivability.
At the same time, to obtain the vector *RP* we define a probability distribution over the resource permutation space in which a permutation $\{R_{k0},...,R_{kN-1}\}$ is given a probability

$$\Pr(\{R_{k0},...,R_{kN-1}\}) = \prod_{i=0}^{N-2}\left(cp(R_{ki}) \Big/ \sum_{j=i}^{N-1} cp(R_{kj}) \right), \tag{4}$$

where $cp(R)$ is the value of the contention peak of resource R. Therefore, high contented resources are given a high chance of no rescheduling and hence their most promising schedules, as established in *RM*, are more likely maintained. Former expressions (3) and (4) may be scaled in order to remark the difference between low and high probabilities.

Permutation Codification
The GA we use in this work exploits the codification schema proposed in [1], that is the permutation with repetition. Accordingly to this codification, an individual represents a permutation of the whole set of operations of the problem in which the subset of tasks of the same job maintain their precedence. Consider for example the problem depicted in Figure 1 and the permutation of its tasks $(t_{01}\ t_{21}\ t_{11}\ t_{12}\ t_{22}\ t_{02}\ t_{03}\ t_{23}\ t_{13})$. From this permutation an individual is obtained by replacing the identifier of each task by the number of its job; therefore we obtain *(0 2 1 1 2 0 0 2 1)*. So, this representation should be understood to mean the following: the first 0 represents the first task of the job J_0, the first 2 is the first task of the job J_2, the second 2 is the second task of the job J_2, and so on. The main advantage of this codification is that every individual produced by the genetic operators is feasible, as we will see in the following paragraphs.
In order to generate chromosomes from the former heuristic information we need a strategy to translate a matrix *RM* and a vector *RP* into an individual. To do that, we have envisaged a strategy that merges the rows of the matrix *RM* to obtain a permutation of the tasks. As we have pointed out, two or more of the partial schedules expressed by the arrows of *RM* might result to be not compatible each one to the others. When this happens, a partial schedule should be modified to restore compatibility, i.e. to maintain the precedence among the operations of the same job. The selection of this partial schedule is done accordingly to the priorities expressed by the vector *RP*: schedules of low priority resources are modified first in order to maintain consistency. The rationale behind this strategy is to maintain the partial schedules obtained from the heuristic information as long as possible, and give a higher chance to critical resources when solving conflicts among partial schedules. Figure 4 shows a matrix *RM*, a vector *RP* and the subsequent chromosome. As we can

$$\begin{pmatrix} t_{11} & t_{01} & t_{22} \\ t_{21} & t_{13} & t_{02} \\ t_{03} & t_{12} & t_{23} \end{pmatrix} \qquad (R_0 \quad R_1 \quad R_2) \qquad \begin{matrix} (t_{11}\, t_{01}\, t_{21}\, t_{22}\, t_{12}\, t_{13}\, t_{02}\, t_{03}\, t_{23}) \\ (1\ 0\ 2\ 2\ 1\ 1\ 0\ 0\ 2) \end{matrix}$$

a) *RM* b) *RP* c) permutation with repetition chromosome

Figure 4. A heuristic chromosome obtained from a matrix *RM* of heuristic partial schedules and a resource priority vector *RP*.

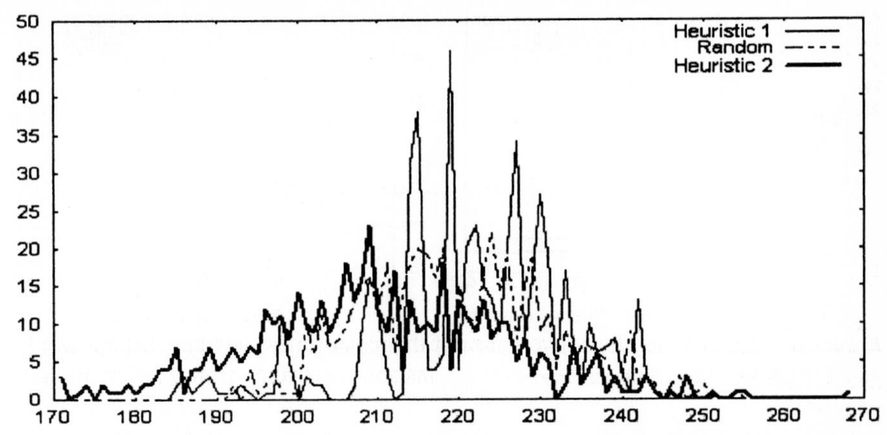

Figure 5. Makespan profiles of three populations of 1600 individuals generated by means of three different strategies: at random, heuristic 1, and heuristic 2.

see the partial schedule of resource R_2 expressed by the last row of *RM* is not maintained in the chromosome due to task t_{12} must be placed before task t_{13} in order to preserve precedence constraints.

5 Experimental Results

In this section we present experimental results about the quality of the heuristic individuals as well as the convergence of the GA when starting from different initial populations. In any case we consider the problem instance *enddr2-10-by-5-9* from the benchmark proposed in [4]. First, we compare a population generated by means of the proposed strategy (heuristic 2) against two populations generated at random and by means of the former one (heuristic 1) respectively. We generated 1600 individuals in any case and shown in Figure 5 the makespan profiles. As we can observe, the new strategy produces the most uniform distribution and, at the same time, the highest concentration of good individuals. Hence, this strategy looks to be the more appropriate because it translates into a higher diversity of the population and produces a bigger percentage of good individuals.

On the other hand, we shown in Figure 6 results about the convergence of the GA when solving the same problem instance *enddr2-10-by-5-9* starting from different initial populations. Here we consider a random initial population and two heuristic ones, generated by means of the heuristic 1 and heuristic 2 strategies respectively. When using heuristic 1, we generate only 30 out of the 100 individuals of the population; the reminder ones are generated at random in order to maintain an acceptable level of diversity in the population. When using heuristic 2, we generate one half of the population heuristically and the other half at random. In this case, we have scaled the former expressions (3) and (4) so that the minimum values are 0. As we can observe in Figure 6, both of the heuristic strategies outperforms the random generation, the heuristic 2 being better than the heuristic 1.

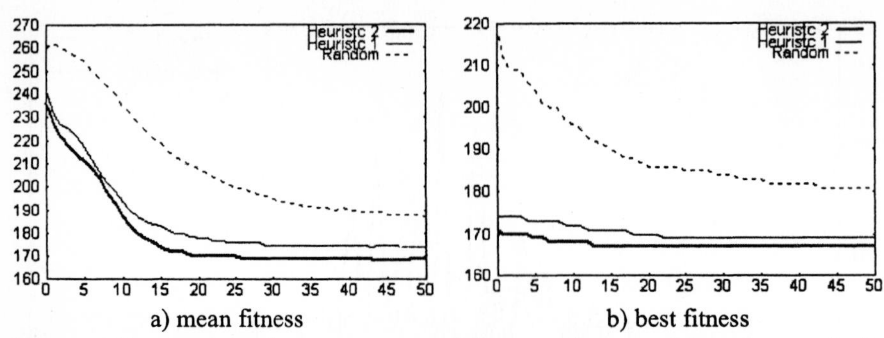

a) mean fitness b) best fitness

Figure 6. Results about the convergence of the mean fitness and best fitness of a GA when solving the `enddr2-10-by-5-9` instance starting from different initial populations.

6 Conclusions

In this paper we have presented a strategy to initialize a GA that is a generalization of the strategy we previously proposed in [5]. We shown how to exploit this strategy to introduce knowledge into a population of individuals by generating a matrix representation that is further translated into the permutation schema proposed in [1]. It often happens that during this process some of the heuristic information is lost, hence as a further work we plan to exploit the proposed matrix representation in the GA, what will require, for example, to envisage specific genetic operators for this codification.

References

1. Bierwirth, C., A Generalized Permutation Approach to Jobshop Scheduling with Genetic Algorithms. OR Spectrum, vol. 17 (1995) 87-92.
2. Grefenstette, J. J.: Incorporating problem specific knowledge in genetic algorithms. In: Genetic Algorithms and Simulated Annealing. Morgan Kaufmann (1987) 42-60.
3. Mattfeld, D. C.: Evolutionary Search and the Job Shop. Investigations on Genetic Algorithms for Production Scheduling. Springer-Verlag, November 1995.
4. Sadeh, N., Fox, M.S.: Variable and Value Ordering Heuristics for the Job Shop Scheduling Constraint Satisfaction Problem. Artificial Intelligence, Vol. 86 (1996) 1-41.
5. Varela, R., Gómez, A., Vela, C. R., Puente, J., and Alonso, C.: Heuristic Generation of the Initial Population in Solving Job Shop Problems by Evolutionary Strategies. In: Foundations and Tools for Neural Modeling, LNCS, Procs. of IWANN'99 (Vol I), Springer-Verlag, Alicante, Spain, (June 1999) 690-697.
6. Varela, R., Vela, C. R., Puente, J., Gómez, A. and Vidal, A. M.: Solving Job-Shop Scheduling Problems by Means of Genetic Algorithms. In: The Practical Handbook of Genetic Algorithms, Ch. 8. Ed. Lance Chambers, Chapman & Hall/CRC (2001) 275-293.

Evolving High-Posterior Self-Organizing Maps

Jorge Muruzábal
Statistics and Decision Sciences Group
University Rey Juan Carlos, 28936 Móstoles, Spain
j.muruzabal@escet.urjc.es

Abstract. Bayesian inference for neural networks has received a good deal of attention in recent years. Unlike standard methods, the bayesian approach provides the analyst with the richness (and complexity) of a probability distribution over the space of network weights (and possibly other quantities of interest). These *posterior* distributions prompt an optimization problem that may be suitable for evolutionary algorithms. This possibility is obviously of foremost interest when no alternative global functions are available for optimization. Some preliminary results related to one of such cases, namely, the self-organizing map, are presented in this paper. Specifically, a familiar "steady-state" diffusion genetic algorithm is described and tested.

1 Introduction

The bayesian approach to inference constitutes an interesting alternative to classical methods. One of the distinctive features of this approach is the availability of one or more posterior distributions describing the underlying uncertainty. While these posteriors can be summarized in various ways, a natural idea is to compute *maximum a posteriori* values. Thus, an optimization problem is explicitly brought about by the bayesian paradigm, yet few studies have been conducted to explore the feasibility of an evolutionary approach [6]. Additional research is clearly needed to clarify the prospect of the idea.

Various neural network schemes have been recently subject to bayesian modelling [3, 5, 9]. The bayesian approach not only provides a more solid theoretical framework, it also yields alternative fitting mechanisms. In this paper I focus on the self-organizing map (SOM) model [3, 5, 7]. SOMs provide a powerful tool to explore the structure of high-dimensional data sets. The SOM's utility relies heavily on the degree of achieved self-organization, a notion easy to understand but complex to capture formally [1]. When using the standard fitting algorithm, self-organization remains implicit and thus needs to be checked in each individual case. Not surprisingly, an effort has been made to design priors and other constructs that explicitly place a bias towards the SOM's self-organization property [3, 5]. The present paper is based on Utsugi's simpler formulation [3]; Bishop, Svensén and Williams' GTM idea [5] may be amenable to a similar treatment.

Bayesian models for neural networks are often burdened by severe posterior multimodality. Thus, when designing an evolutionary algorithm for posterior optimization, it is crucial to prevent the familiar premature convergence problem as

J. Mira and A. Prieto (Eds.): IWANN 2001, LNCS 2084, pp. 701-708, 2001.

much as possible. The diffusion genetic algorithm (DGA), see e.g. [11], is known to mitigate this problem by promoting some speciation abilities; while obviously not the only possible approach, it will be considered below. The present DGA introduces "posterior-minded" genetic operators specifically tailored to the SOM's problem and bayesian model. The resulting algorithm is used to explore several search problems arising from different choices for some key (hyper)parameters controlling the fitness landscape.

Previous work involving evolutionary algorithms within the SOM model has also been relatively scarce. Contributions tend to either concentrate on the evolution of learning laws themselves (rather than SOMs) [12], or else tackle the issue of optimal topology/dimensionality (relying on conventional training for fitness evaluation) [13]. While direct evolution of SOMs seems to be a novel idea, the present research is only preliminary in nature and primarily aims to provide a first exploration of the potential of evolutionary optimization in the bayesian neural network paradigm.

The paper is organized as follows. Section 2 briefly reviews basic aspects of the SOM model. Section 3 describes the bayesian approach and the emerging posterior of interest. Details of the DGA are provided in Section 4. Section 5 presents some experimental results and Section 6 closes with some discussion and prospects for future research.

2 Self-Organizing Maps

The SOM is a biologically-inspired neural network model that has proved useful for multivariate data analysis tasks such as clustering and outlier detection, [7, 8, 10]. The neural structure is a lattice of interconnected neurons (or simply units), each endowed with an associated pointer in data space, say $w \in \mathbb{R}^m$. Each SOM has a unique dimension, topology and neighbourhood system. In practice, 1-D or 2-D SOMs are nearly always considered for simplicity. In the 2-D case, topologies can be either rectangular, hexagonal or otherwise. In a 2-D SOM with rectangular topology, an interior unit may have 4 or 8 immediate neighbours. All SOMs considered below share the 4-neighbour *squared* topology with $r = k^2$ neurons in total.

SOM training is based on input data vectors $x^{(1)}, x^{(2)}, \ldots, x^{(n)} \in \mathbb{R}^m$. A trained SOM (fitted by the standard algorithm or otherwise) should satisfy two basic desiderata: (i) the density of the pointer cloud should "mimic" the underlying density of the data; and (ii) pointers should exhibit topological order or self-organization. Intuitively, what we seek in (ii) is a *smooth* configuration: pointers from units nearby *on the lattice* should also lie close in $\mathbb{R}^m$. As is well-known, there is no global function of data $x^{(l)}$ and pointers w_i whose maximation leads to a well-organized SOM [7].

Once trained, SOMs can not be inspected directly unless $m \leq 3$, but we can always project the set of fitted pointers onto 2-D space via Sammon's mapping (see Figs. 2, 3 below and [14]). Since pointers "inherit" the connectivity pattern, a "large" number of overcrossing connections in this 2-D image is symptomatic of poor self-organization. It is also customary to associate each input data vector with the pointer that lies closest to it. The *quantization error* (QE) of each $x^{(l)}$ is defined as the Euclidean distance to that pointer. QEs can be seen as *residuals* in that larger values tend to signal outlying data [10].

3 Bayesian Modelling

Bayesian models for neural networks often have the following structure: (i) Data $\mathbf{x}$ are assumed a sampling distribution $P(\mathbf{x}/\mathbf{w}, \beta)$, where $\mathbf{w}$ is the object of interest (for instance, our set of network pointers w_i) and $\beta > 0$ is a scale parameter. (ii) A prior density is assumed on $\mathbf{w}$, say $P(\mathbf{w}/\alpha)$, where $\alpha > 0$ is a *regularization* parameter. This prior usually reflects a body of assumptions about the ideal network configuration. (iii) A second-stage prior distribution (or hyperprior) is assumed over the nuisance parameters or hyperparameters, say $P(\alpha, \beta)$. The joint distribution of data and parameters is then $P(\mathbf{x}, \mathbf{w}, \alpha, \beta) = P(\mathbf{x}/\mathbf{w}, \beta)P(\mathbf{w}/\alpha)P(\alpha, \beta)$. Several posterior distributions, including the full $P(\mathbf{w}, \alpha, \beta/\mathbf{x})$ and the marginal $P(\mathbf{w}/\mathbf{x}) = \int P(\mathbf{w}, \alpha, \beta/\mathbf{x}) d\alpha d\beta$, can be derived from here.

In practice, these posteriors tend to be complex, multimodal functions and hence are usually hard to tackle analytically. The Markov chain Monte Carlo (MCMC) approach [2, 4] is one of the most popular resources to bypass this difficulty. The MCMC idea shares indeed some features with the evolutionary approach and therefore deserves further attention. In this paper, however, I will explore for simplicity the conditional posterior $P(\mathbf{w}/\mathbf{x}, \alpha_0, \beta_0) \propto P(\mathbf{w}, \alpha_0, \beta_0/\mathbf{x})$ for prespecified α_0 and β_0. This typically implies a trial and error process where several (α, β) combinations are tried out in turn.

Focusing now on our SOM case, it is natural to assume a Gaussian mixture model for the data $\mathbf{x}$ (containing the n vectors $x^{(l)}$), namely,

$$P(\mathbf{x}/\mathbf{w}, \beta) = \prod_{l=1}^{n} \sum_{i=1}^{r} \frac{1}{r} f(x^{(l)}/w_i, \beta),$$

where r is the total number of units, the w_i denote the SOM pointers and $f(\cdot/\xi, \beta)$ is the density of a $N_m(\xi, \beta I)$ variate [3, 5]. As regards $P(\mathbf{w}/\alpha)$, Utsugi [3] proposes a Gaussian *smoothing* density based on independent coordinates:

$$P(\mathbf{w}/\alpha) = \prod_{j=1}^{m} \left(\frac{\alpha}{2\pi}\right)^{\frac{R}{2}} \sqrt{\Delta} \exp\left\{-\frac{\alpha}{2} \left\|Dw^{(j)}\right\|^2\right\},$$

where R and Δ are respectively the rank and product of positive eigenvalues of $D^T D$, $w^{(j)} \in \mathbb{R}^r$ collects the j-th coordinates from all pointers w_i, and D is a $R \times r$ matrix ($R \leq r$) designed to measure smoothness along a single dimension. For example, in the 1-D (linear) SOM, D typically includes $R = r - 2$ rows like $(0\ 0\ 0\ ...\ 0\ 1\ {-2}\ 1\ 0\ ...\ 0\ 0)^T$; hence, the prior says

$$E(w_i^{(j)}) = E[\frac{1}{2}(w_{i-1}^{(j)} + w_{i+1}^{(j)})]$$

for all $2 \leq i \leq r-1$ and $1 \leq j \leq m$. For 2-D SOMs a similar prior can be constructed in terms of a similar D matrix with $R = (k-2)^2$ rows (again, one for each interior unit), each computing the difference between the value at the target unit and the average of the values at the four immediate neighbours on the lattice. This is the choice used below; the resulting log posterior

$$\Psi(\mathbf{w}) = \log P(\mathbf{w}/\mathbf{x}, \alpha, \beta) = \sum_{l=1}^{n} \log \sum_{i=1}^{r} \exp\left\{-\frac{\beta}{2}\left\|x^{(l)} - w_i\right\|^2\right\} - \frac{\alpha}{2}\sum_{j=1}^{m}\left\|Dw^{(j)}\right\|^2$$

is taken as the basic fitness function to be maximized (for predetermined α and β). Hyperparameters α and β play the role of weight factors affecting a single summand each. They clearly set up different search processes priming the fit over the smoothness component or viceversa.

4 A SOM-Oriented Genetic Algorithm

Optimization of the above log posterior is addressed by an adapted version of Davidor's DGA [11]. Population members (SOMs in our case) are located at the nodes of a $K \times K$ *toroidal* lattice; every individual has an 8-node neighbourhood. The algorithm works according to the so-called "steady-state" dynamics: only a few individuals are produced at each generation. Crossover and replacement apply to neighbouring individuals only. At each cycle, a random individual is selected from the torus (call it the winner) and its *local fitness differential* (based on Ψ above) determines the number of offspring to be produced. Mating partners are selected with rank-based selection probabilities. Each partner is mated in turn with the winner by applying specialized crossover and mutation operators (see below). All offspring are evaluated and compared to the current individuals in the neighbourhood (including the winner). These 9 individuals are cycled through from a random start and replacement occurs if and only if the offspring finds an individual with lower Ψ.

Each SOM $\mathbf{w}$ is represented directly as a vector of rm real numbers. Ordering is inmaterial, but the genetic operators work on the basis of the SOM's spatial arrangement and hence must "know" how pointers are distributed over the genotype. Three genetic operators are considered: "single-point" crossover (SPC) takes a pair of SOMs and produces a single SOM; fit-mutation (FMUT) and smooth-mutation (SMUT) slightly modify a given SOM. SPC works in three steps. The idea is to bring together those subsets of pointers from different SOMs that separately fit the data well on different portions of data space. To this end, the selected parent is first "aligned" with respect to the winner. After all, a squared SOM is rotation-invariant[1]; therefore, the second SOM should be "well-rotated" prior to exchanging information. Once the two maps are aligned, a further random rotation is applied to both[2]. Finally, a random path partitions

[1] Rotations are of course meant on the SOM *lattice*; thus, only $c\frac{\pi}{2}$ rotations, with $c = 1, 2, 3$, are considered.

[2] Since the actual crossover path that follows is always selected along the NS direction on the lattice, this prior rotation implements paths along all possible directions.

the SOM lattice into two pieces: pointers (from the first SOM) associated to one of the pieces are then "glued" to pointers (from the second SOM) associated to the other piece.

FMUT is designed to ensure that the evolutionary process has some means to fit the data appropriately. It computes the largest QE in the current SOM and shifts the associated pointer (and its four immediate neighbours) towards the responsible datum by a user-input amount (typically halfway in our experiments). Operator SMUT takes care of the complementary task of endowing the process with some smoothing abilities. SMUT selects a random interior pointer and "smooths it out" by replacing it with the average of its four neighbours.

While these operators may cooperate together in various ways, the following schedule was used below: during the first half of the run (comprising, say, T cycles) SPC acts first, then FMUT and SMUT are applied in order. During the second half FMUT is no longer applied, so SMUT works directly on SPC's output.

A final important issue refers to SOM initialization. A procedure called *random-hyperplane* initialization (similar to initialization based on principal components [14]) places all pointers evenly spaced on the hyperplane determined by m random data vectors $x^{(l)}$. The resulting SOMs are completely flat and hence unlikely to fit the data well; thus, a few shifts by FMUT are performed prior to insertion into the initial population.

5 Experimental Results

In this paper I restrict attention to two real cases, the *stack loss* ($m = 3, n = 21$) and *milk container* ($m = 8, n = 86$) data sets. These data present outliers of different nature and have been previously analyzed, see e.g. [10].

The primary goal of the present analysis is to test the usefulness of the prior $P(\mathbf{w}/\alpha)$ and the three ad-hoc genetic operators described above. We begin with the stack loss data. Fig. 1 presents a SOM trained traditionally and the best SOM found by our DGA under the "naive" choice $\alpha = 1$ and $\beta = 1$ ($k = 4$, $K = 7$, $T = 50$). The genetic SOM looks flatter and presents better fitness Ψ. It thus appears that the genetic strategy works well in this simple scenario.

We now switch to the milk container data. We first consider $\alpha = 10$ and $\beta = 1$ (emphasizing SOM smoothness), see Fig. 2 ($k = 6$, $K = 7$, $T = 500$). It is seen that the best genetic SOM leads to the correct detection of all outliers while providing adequate smoothness. The alternative choice $\alpha = 1$ and $\beta = 10$ (not shown here) confirms previous findings [3]: putting significant weight on the fit component may make the final SOMs prone to suffer from lack of organization. For this reason, I return in what follows to the previous choice $\alpha = 10$ and $\beta = 1$.

Consider next the case of a larger torus ($k = 5$, $K = 15$, $T = 500$); this should promote some useful variability. We find that this is indeed the case: the three best maps (Fig. 3) look indeed quite different. An analysis of the final fitness distribution reveals two well-separated patches of "emanating" high-

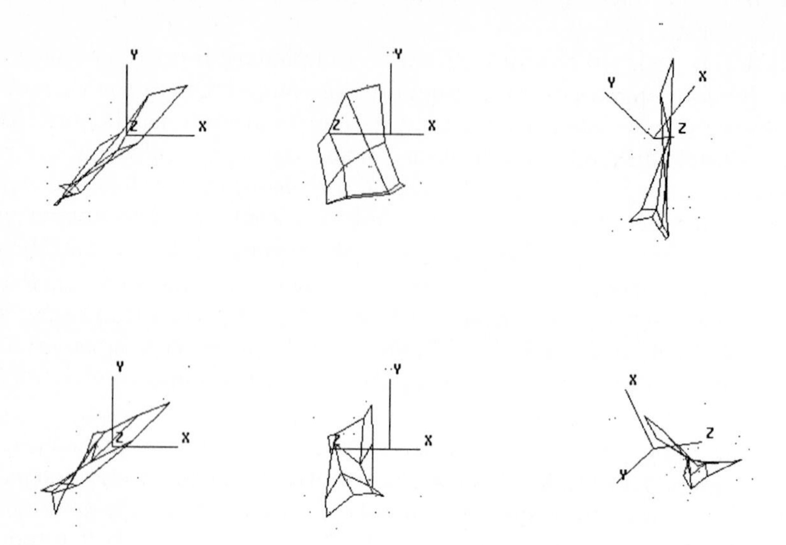

Fig. 1. Three 3-D views of a 2-D SOM trained traditionally (bottom row) and the best SOM found by the genetic strategy (top row) for the stack loss data. SOMs are shown with connections to aid visualization. The first two columns share the same point of view, the third shows different orientation to expose maximum curvature in each case. Data vectors are superimposed for reference.

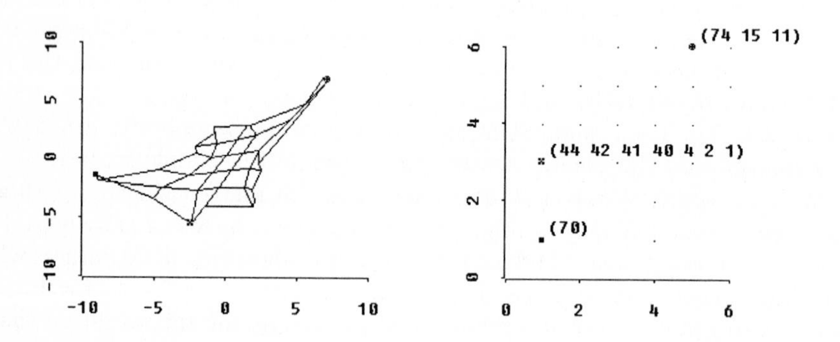

Fig. 2. An analysis of the milk container data under a strong bias for smoothness. Left: best projected SOM (with connections) and some pointers highlighted with different symbols. Right: the lattice of neurons showing outlying data associated with the highlighted pointers (empty units have been removed).

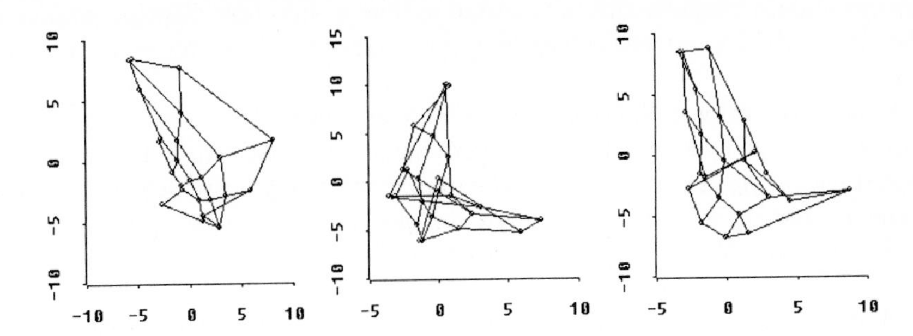

Fig. 3. Three projected SOMs for the milk container data found using a larger torus.

posterior SOMs. Hence, the DGA-based strategy is also successful in locating several (near) modes of the target distribution.

We finally try fitting larger SOMs ($k = 10$, $K = 10$, $T = 1000$). Recall that we would ordinarily want the SOM to *summarize* the data, so it is rare to use more neurons than training items – this example is introduced because it makes the contribution of individual pointers much smaller [6]. The best map found (not shown) is rather organized and leads again to the successful detection of all outliers. However, pointers do not cluster well and many units remain empty. It follows that the underlying prior does not exercise enough pressure on the algorithm in this case.

6 Discussion and Concluding Remarks

An evolutionary approach to neural network posterior optimization has been introduced and illustrated in two real data sets of small dimensionality. Maps of increasingly high posterior density have been obtained, and the analysis of structure and outliers based on these maps led to sensible conclusions. Results are consistent over independent runs of the DGA. Nice acceptance rates of about 90% (meaning only one out of ten proposals does not make it into the population) suggest that the proposed genetic operators are helpful.

The user has a number of choices concerning torus and map size, run length and model hyperparameters. As regards the latter, relatively larger values of α are preferred in principle (although genuine structure may be oversmoothed); relatively larger values of β may jeopardize self-organization. In the future, the familiar adaptation ideas from Evolution Strategies may be incorporated to modify on-line the values of α and β. This can also be done via MCMC simulation [4].

While it is hoped that the above ideas are useful for future developments in the SOM framework, a couple of caveats should be kept in mind. Firstly, the present genetic strategy is intrinsically parallel and rather flexible, yet it is also relatively slow. Posterior density evaluation is indeed expensive; future versions should consider operators that somehow simplify offspring evaluation. Likewise, FMUT's current design may be replaced with some less-demanding, random variant.

Finally, the prior in SOM space used here adequately takes care of the overall smoothness, but it does not seem to provide a strong enough bias to attain the desired *clustering* of pointers when the map is relatively large. Again, the design of the D matrix should be improved, or else we can switch to GTM's latent variable setting [5] and explore priors of a different nature.

Acknowledgement. The author is supported by grants HID98-0379-C02-01 and TIC98-0272-C02-01 from the spanish CICYT agency.

References

1. Bauer, H.-U., Herrmann, M., and Villmann, T. (1999). Neural Maps and Topographic Vector Quantization. *Neural Networks*, Vol. 12, 659–676.
2. Gilks, W.R., Richardson, S. and Spiegelhalter, D. J. (1996). *Markov Chain Monte Carlo in Practice.* Chapman and Hall.
3. Utsugi, A. (1997). Hyperparameter Selection for Self-Organizing Maps. *Neural Computation*, Vol. 9, No. 3, 623–635.
4. Utsugi, A. (2000). Bayesian Sampling and Ensemble Learning in Generative Topographic Mapping. *Neural Processing Letters*, Vol. 12, No. 3, 277–290.
5. Bishop, C. M., Svensén, M., and Williams, K. I. W. (1998). GTM: The Generative Topographic Mapping. *Neural Computation*, Vol. 10, No. 1, 215–235.
6. Franconi, L. and Jennison, C. (1997). Comparison of a Genetic Algorithm and Simulated Annealing in an Application to Statistical Image Reconstruction. *Statistics and Computing*, Vol. 7, 193–207.
7. Kohonen, T. (1997). *Self-Organizing Maps* (2nd Ed.). Springer Verlag.
8. Kraaijveld, M. A., Mao, J. and Jain, A. K. (1995). A Nonlinear Projection Method based on Kohonen's Topology Preserving Maps. *IEEE Transactions on Neural Networks*, Vol. 6, No. 3, 548–559.
9. Müller, P. and Rios, D. (1998). Issues in Bayesian Analysis of Neural Network Models. *Neural Computation*, Vol. 10, 571–592.
10. Muruzábal, J. and Muñoz, A. (1997). On the Visualization of Outliers via Self-Organizing Maps. *Journal of Computational and Graphical Statistics*, Vol. 6, No. 4, 355–382.
11. Davidor, Y. (1991). A Naturally Occurring Niche & Species Phenomenon: the Model and First Results. *Proceedings of the Fourth International Conference on Genetic Algorithms*, 257–263.
12. Harp, S. A. and Samad, T. (1991). Genetic Optimization of Self-Organizing Feature Maps. *Proceedings of the International Joint Conference on Neural Networks*, 341–346.
13. Polani, D. and Uthmann, T. (1993). Training Kohonen Feature Maps in Different Topologies: an Analysis using Genetic Algorithms. *Proceedings of the Fifth International Conference on Genetic Algorithms*, 326–333.
14. Kohonen, T., Hynninen, J., Kangas, L. and Laaksonen, J. (1995). SOM_PAK. The Self-Organizing Map Program Package. Technical Report, Helsinki University of Technology, Finland.

Using Statistical Techniques to Predict GA Performance

Rafael Nogueras and Carlos Cotta

Dpto. de Lenguajes y Ciencias de la Computación, Univ. de Málaga
Campus de Teatinos (3.2.49), 29071 - Málaga - Spain
ccottap@lcc.uma.es

Abstract. The design of models efficiently predicting the performance of a particular genetic algorithm on a given fitness landscape is a very important issue of practical interest. Virtual Genetic Algorithms (VGAs) constitute a statistical approach aimed at this objective. This work describes different improvements to the standard VGA model. These improvements are based on the use of a more representative dataset for the statistical analysis, the partitioning of this dataset into separate prediction models, and the utilization of a more sophisticated statistical model to grasp the distribution of fitnesses. The empirical evaluation of this enhanced model shows a more accurate fitness prediction. Furthermore, fast qualitative assessment of parameter changes is shown to be possible.

1 Introduction

Despite the algorithmic simplicity of the genetic-algorithm (GA) template, the computational cost of running a particular instance of this template can be overwhelming. This is specifically true in situations in which the fitness function involves hard calculations, e.g., the simulation of a physical system [7], complex mathematical expressions [3], etc. In these scenarios, it is very important to minimize the number of fitness calculations. This implies a careful instantiation of the basic template, i.e., choosing adequate representation/operators, and finely adjusting of the GA parameters (population size, operator probabilities, etc.) for this purpose.

This instantiation/parameterization of the GA is to some extent supported by several studies regarding the choice of adequate parameters (e.g., [1, 4, 10]). Nevertheless, there exist strong theoretical limitations to the generalizability of these studies [11]. Hence, a costly trial-and-error procedure is commonly used in many situations. This fact clearly underpins the need for models predicting either qualitatively or quantitatively the expected performance of a GA. In this sense, Grefenstette [6] has used a simple yet powerful statistical approach termed Virtual Genetic Algorithm (VGA). This approach is revisited in this work, proposing several modifications of the basic model that will redound in an improvement of its predictive capabilities.

The remainder of the paper is organized as follows. First, the basic VGA model is described and exemplified in Section 2. The limitations of this basic

J. Mira and A. Prieto (Eds.): IWANN 2001, LNCS 2084, pp. 709-716, 2001.

model give rise to several enhancements that are discussed in Section 3. Next, an extensive empirical evaluation of the different variants of VGA is presented in Section 4. This evaluation involves different target problems ranging from numerical optimization to combinatorial optimization. Finally, Section 5 summarizes our results, and outlines future research.

2 The Basic Virtual Genetic Algorithm

The VGA model [6] constitutes a very interesting approach to GA-performance prediction. This model resembles a standard GA, but with an important difference: individuals do not carry genetic information any more; on the contrary, they are used as place-holders for fitness information. This fitness information is transmitted from parents to offspring, using a statistical model to simulate the effects of genetic operators. More precisely, this statistical model is utilized to predict the fitness of a new individual in terms of the fitnesses of its parents. Hence, the reproduction (recombination plus mutation) and evaluation phases of a standard GA are joined in one single VGA phase, as show below:

1. Initialize population.
2. If termination condition satisfied, go to 3. Otherwise do
   (a) Select parents from population.
   (b) Apply virtual genetic operators to selected parents and perform virtual evaluations of them using the statistical model.
   (c) Insert descendants into population.
   (d) Go to 2.
3. End.

The statistical model proposed by Grefestette is based on using a linear regression approach for each genetic operator, whose parameters are obtained from previously generated datasets. These datasets (one per operator) are created by randomly constructing N full individuals ($2N$ in the case of recombination), applying the corresponding genetic operator to them, and evaluating both the parents and the descendant in each case. As a result, N pairs (f_p, f_o) are obtained, where f_p is the fitness of the parent (or the parents' mean fitness in case of recombination), and f_o is the descendant's fitness. Subsequently, a least squares fit is sought to a function $f_o = a + bf_p$.

Obviously, such a model could not grasp the stochastic nature of genetic operators (recall that two applications of an operator to the same parents will likely result in different descendants). For this reason, the linear model above is not used to generate the descendant's fitness but to calculate the center of a Gaussian distribution of descendant fitnesses. The variance of this distribution is also calculated via a linear regression on the dataset. The whole process is as follows:

1. Perform a least squares fit of the linear model $f_o = a + bf_p$ to the N pairs (f_p, f_o).

2. Sort the dataset according to f_p, and group the pairs in bins of size T.
3. Calculate the standard deviation σ_f in each bin.
4. Perform a least squares fit of the N/T pairs $(\bar{f}_p, \sigma_f)$ to the linear model $\sigma_f = \sigma_a + \sigma_b \bar{f}_p$, where $\bar{f}_p$ is the parent mean fitness in each bin.
5. Output the model for the operator ω: $\mathcal{M}(\omega) = \langle a, b, \sigma_a, \sigma_b \rangle$.

The so-obtained model is thus used to estimate the fitness of the descendant as a function of the parents' mean fitness as $f_o = N(a + bf_p, \sigma_a + \sigma_b f_p)$. This estimation is done in cascade for each genetic operator, e.g., assuming the typical situation of using recombination (ω_X) and mutation (ω_m), the parents mean fitness is used as an input to the $\mathcal{M}(\omega_X)$ model, and the output of this model is fed to $\mathcal{M}(\omega_m)$.

An experimental example of the basic VGA model is shown in Fig. 1 (left). This example correspond to an elitist generational GA/VGA (*popsize* = 100, $p_X = .9$, $p_m = 0.01$, *maxevals* = 10000), using ranking selection and SPX crossover, with application to a 100-bit OneMax problem. Both algorithms have been run 50 times, and the dataset for the VGA comprises 10^4 pairs (notice that this dataset and the corresponding statistical parameters are generated just once, being using in all subsequent runs).

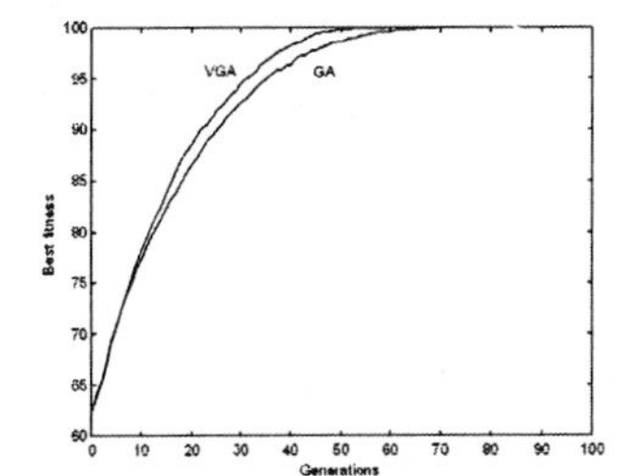

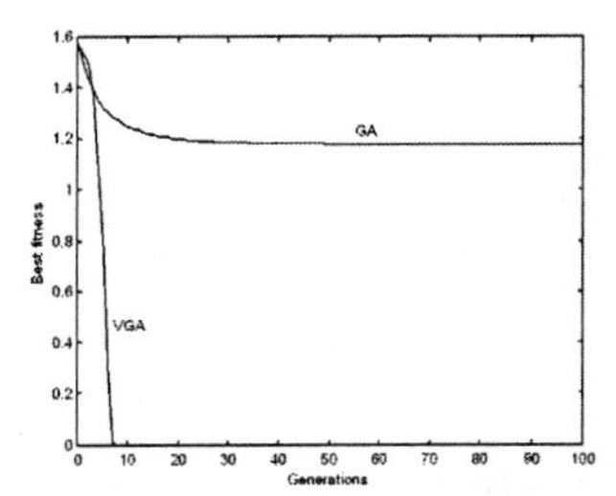

Fig. 1. Comparison of GA vs. VGA in the OneMax problem (left) and in the brachystochrone design problem (right).

As it can be seen the approximation of the VGA to the GA is fairly good. However, consider the situation shown in Fig. 1 (right). It corresponds to the application of the VGA to a different problem: the design of a brachystochrone [9]. In this case, the approximation is very unsatisfactory; the VGA fitness quickly drop to zero while the GA stabilizes around 1.2. This example illustrates the limitations of this basic model, and gives rise to different improvements that will be discussed in next section.

3 Enhanced Virtual Genetic Algorithms

As shown in the previous section, the approximation provided by the basic VGA (B-VGA) is far from satisfactory on some test problems. This is due to several reasons which will be explored below.

First of all, the acquisition of the dataset used to parameterize the statistical model is an important issue. In the basic VGA model, this dataset has been generated by using randomly constructed individuals. Hence, fitness pairs are concentrated in a little region of the fitness space. The B-VGA is grounded on the assumption that the fitness distribution in this region is very similar to the fitness distribution across the whole fitness space. This is approximately true in the OneMax problem (see Fig. 2-left), but very inaccurate in the case of the brachystochrone design problem (see Fig. 2-right). This fact partially explains the behavior of the B-VGA on this problem: the random-sampling dataset match the situation of the initial population, so it can model the behavior of the algorithm during the first generations; however, the B-VGA is unable to grasp the GA dynamics in further generations.

This effect can be mitigated by using a more representative dataset. Such a dataset can be obtained by means of a pilot run of a GA. The actual parameters of this pilot-GA are not very important, since its unique purpose is providing a trace of pairs (f_p, f_o) across a larger region of fitness space. The cost of this pilot run is roughly the same than generating the dataset though random sampling (selection and replacement are the unique overhead of the GA with respect to random sampling, and their computational cost is negligible when compared to reproduction and -mainly- fitness evaluation). A similar strategy was proposed in [5], in conjunction with a simulated dynamics model based on equations of motion for estimating the population mean fitness. However, the reported results are worse than using randomly constructed solutions. Moreover, the mutation rate is considered an internal parameter of the mutation operator, and hence the dataset must be generated for each value of this parameter. In contrast, we simply consider minimal mutations, and hence no regeneration of the dataset is required, being the mutation rate a parameter of the VGA.

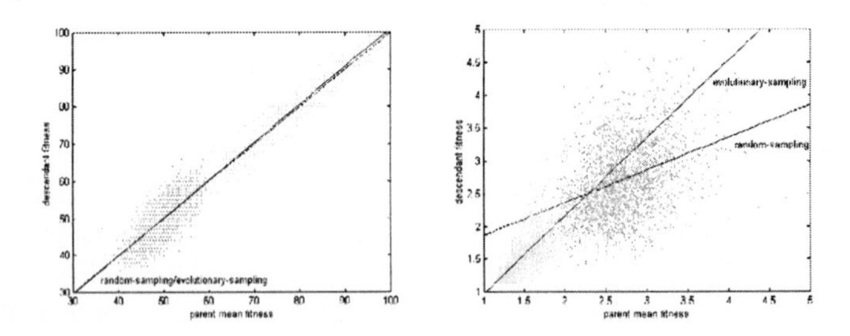

Fig. 2. (Left) Random-sampling (dark dots) vs. evolutionary-sampling (light dots) in the OneMax problem. (Right) Random-sampling vs. evolutionary-sampling in the brachystochrone design problem.

The second issue to be tackled is the asymmetry in the distribution of fitnesses for a particular operator. As an example, consider the case of a mutation operator. It is usually expected that a mutated individual be slightly better or slightly worse than the original individual (i.e., the strong causality principle

[8]). Nevertheless, a remarkable fitness difference between parent and descendant is eventually possible. Furthermore, the better the original solution, the more likely this difference will be in the non-improving direction. The brachystochrone design problem is a good example of this: a good track design can be fine-tuned by mutation, but it can also be totally disrupted. This is illustrated in Fig. 3 (left). As a consequence, the fitness variance of descendants can be high at some point, but this variance only takes place on one side of the distribution. However, the B-VGA model does not take this effect into account, resulting in the generation of virtual offspring much better than the parents, an event that does not match the sought distribution.

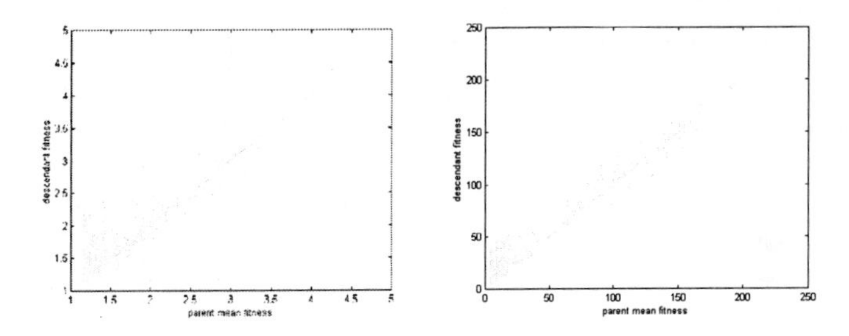

Fig. 3. Fitness distribution for bit-flip mutation in the Brachystochrone design problem (left) and in the Rastrigin function.

In order to deal with this asymmetry, the dataset is partitioned right after having calculated the parameters of the linear model $f_o = a + bf_p$. This partitioning produces two datasets U and L, such that $U = \{(f_p, f_o) \mid f_o > a + bf_p\}$, and $L = \{(f_p, f_o) \mid f_o \leq a + bf_p\}$. Each of these two datasets is separately processed so as to obtain the corresponding σ parameters. An additional precision is also introduced here: in the B-VGA model, standard deviations refer to the mean descendant fitness in each bin; however, a much more significant measure is the deviation with respect to the predicted descendant fitness –using the linear model– in each case. As a matter of fact, the linear model constitutes the center for the Gaussian distribution, and depending on the b parameter, there may exist a large difference between the two deviation measures.

Hence, two sigma models are obtained: $\mathcal{M}_U^\sigma(\omega) = \langle \sigma_a^U, \sigma_b^U \rangle$, and $\mathcal{M}_L^\sigma(\omega) = \langle \sigma_a^L, \sigma_b^L \rangle$. Subsequently, an improved VGA model can predict the descendant's fitness as $f_o = a + bf_p + \Delta$ where

$$\Delta = |N(0, \sigma_a^U + \sigma_b^U f_p)|, \quad \text{or} \quad \Delta = -|N(0, \sigma_a^L + \sigma_b^L f_p)|, \tag{1}$$

depending on a random test based on the relative weight $|U|/(|L| + |U|)$.

According to the considerations above, the pseudocode for the whole process is shown below. In order to make the VGA more robust, upper and lower limits of the fitness distribution are also calculated so as to truncate the Gaussian distribution, avoiding extreme values.

1. Sort the dataset S according to f_p, and perform a least squares fit to the linear model $f_o = a + bf_p$.
2. Partition the dataset into two subsets $S^U = \{(f_p, f_o) \in S \mid f_o > a + bf_p\}$ and $S^L = \{(f_p, f_o) \in S \mid f_o \leq a + bf_p\}$. Let $p = |S^U|/|S|$.
3. For each subset do
   (a) Group the pairs in bins of size T.
   (b) Perform a least squares fit to $\sigma_f^{U|L} = \sigma_a^{U|L} + \sigma_b^{U|L} \bar{f}_p$, where $\bar{f}_p$ is the mean of f_p in each bin, and $\sigma_f^{U|L}$ is the deviation of f_o with respect to $a + bf_p$.
   (c) For S^U calculate the maximum $f_p^{\max} = \max(f_p)$ in each bin (idem with the minimum $f_p^{\min} = \min(f_p)$ for S^L).
   (d) Perform a least squares fit to $f_p^{\max} = \lambda_a^U + \lambda_b^U \bar{f}_p$ (respectively to $f_p^{\min} = \lambda_a^L + \lambda_b^L \bar{f}_p$).
4. Output the model $\mathcal{M}(\omega) = \langle a, b, \sigma_a^U, \sigma_b^U, \sigma_a^L, \sigma_b^L, \lambda_a^U, \lambda_b^U, \lambda_a^L, \lambda_b^L, p \rangle$.

This VGA model will be termed *enhanced VGA*, and will be denoted as E-VGA. Next section is devoted to an empirical assessment of this new model.

4 Experimental Results

This section provides some examples of the application of the E-VGA model for several problems. Unless otherwise noted, experiments have been done with an elitist generational GA/VGA ($popsize = 100$, $p_X = .9$, $p_m = 0.01$, $maxevals = 10000$) using ranking selection. All results correspond to series of 50 runs.

First of all, Fig. 4 (left) shows the results for the brachystochrone design problem ($n = 8$, $length = 128$ bits, $x_0 = 4$, $y_0 = 2$) using SPX crossover. As it can be seen, the improvement of the E-VGA with respect to the B-VGA is conclusive. The same successful results are achieved on the Rastrigin function ($n = 8$, $length = 128$; Fig. 4-middle), and on a 15-machine 30-task flowshop scheduling problem taken from the OR-Library (Fig. 4-right). In this latter case, the UCX recombination operator [2] has been used.

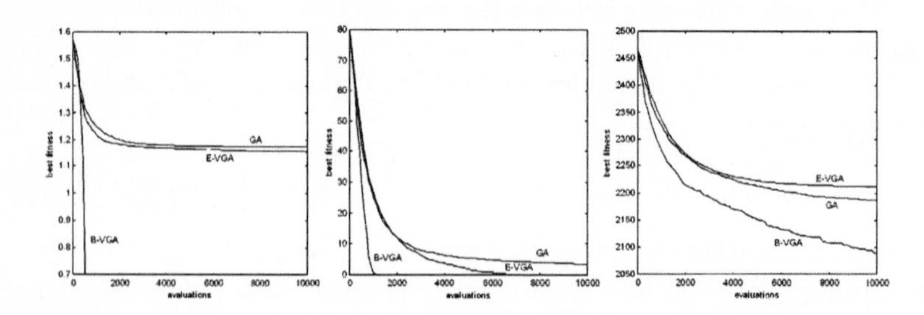

Fig. 4. Comparison of GA vs. E-VGA in the brachystochrone design problem (left), in the Rastrigin function (middle), and in a flowshop scheduling problem (right).

As important as obtaining an accurate performance prediction (and actually a relaxed view of this goal) is the capability of assessing GA performance from a qualitative perspective, i.e., estimate how a particular GA will perform with respect to another particular GA. To be precise, let A and B be two configurations; a head-to-head comparison between the VGA^A and the VGA^B should reflect the same relative properties than the GA^A vs. GA^B comparison. This is shown in Fig. 5 for GA/E-VGA in the Rastrigin function with three configurations: $A = \{popsize = 100, p_X = .9, p_m = 0.01\}$, $B = \{popsize = 50, p_X = .5, p_m = 0.1\}$, and $C = \{popsize = 80, p_X = .7, p_m = 0.05\}$.

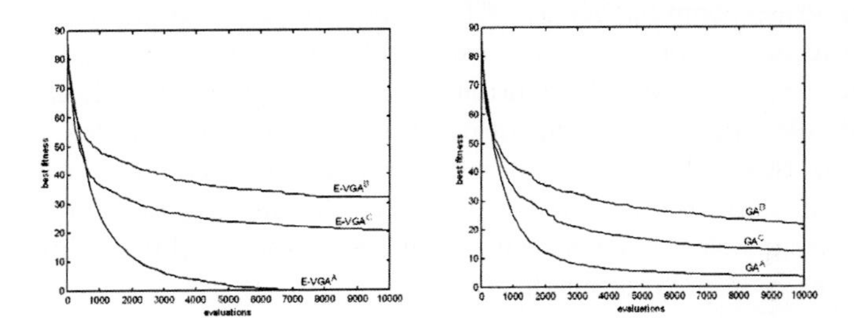

Fig. 5. Several configurations in the Rastrigin function with SPX crossover for the E-VGA (left) and for the GA (right).

As it can be seen, the relative behavior of the different E-VGAs satisfactorily matches the relative behavior of the real GA. Hence, it is possible to analyze several configurations, testing them on the E-VGA, and using the real GA only upon selection of the best one. Furthermore, the computational cost of the E-VGA is a small fraction of the cost of the real GA (up to a roughly 1/20-th in a flowshop scheduling problem; see Table 1), thus providing an important speed-up in the GA parameterization/tuning phase.

Table 1. Speed-up of the E-VGA vs. GA in several problems for increasing instance sizes. The experiments have been done on an AMD K6-II 400 MHz, 128 MB RAM, under Linux Mandrake 7.0.

Problem	Instance base-size (S)	Speed-Up					
		S×1	S×2	S×4	S×8	S×16	S×32
OneMax	100 bits	1.32	1.67	2.36	2.92	3.70	5.40
Rastrigin	8 variables	2.50	3.32	4.38	5.20	6.42	9.22
Brachystochrone	8 variables	3.17	4.08	5.36	6.29	7.51	10.40
Scheduling	30 tasks	4.36	6.83	10.08	13.23	16.57	19.23

As expected, this speed-up is low in simple problems such as OneMax, being higher in complex problems such as brachystochrone design or flowshop scheduling. Furthermore, the speed-up is positively correlated with the instance size, i.e., the larger the problem instance, the higher the speed-up.

5 Conclusions

Virtual genetic algorithms are a very interesting tool for predicting GA performance. This paper has shown the importance of the underlying statistical model used in the algorithm. It has been empirically corroborated that the basic VGA model is somewhat limited due to the strong assumptions about the fitness distribution on which it is grounded. A much more flexible model has been proposed in order to grasp some properties of the fitness distribution such as asymmetry, or statistical anisotropy. The obtained results are conclusive, showing a much more accurate performance estimation. Moreover, high computational speed-ups are shown, allowing fast assessment of different sets of parameters.

Future work will deal with other mechanisms for partitioning the dataset. The utilization of more sophisticated tools for predicting and/or simulating fitness distributions (e.g., artificial neural networks) is another interesting extension. Work is already in progress in this line.

References

1. Th. Bäck. Optimal mutation rates in genetic search. In S. Forrest, editor, *Proceedings of the Fifth International Conference on Genetic Algorithms*, pages 2–8, San Mateo, CA, 1993. Morgan Kaufmann.
2. C. Cotta and J.M. Troya. Genetic forma recombination in permutation flowshop problems. *Evolutionary Computation*, 6 (1):25–44, 1998.
3. A.E. Eiben and Th. Bäck. Empirical investigation of multiparent recombination operators in evolution strategies. *Evolutionary Computation*, 5(3):347–365, 1997.
4. D.E. Goldberg, K. Deb, and J.H. Clark. Genetic algorithms, noise, and the sizing of populations. *Complex Systems*, 6:333–362, 1992.
5. J.J. Grefenstette. Predictive models using fitness distributions of genetic operators. In L.D. Whitley and M.D. Vose, editors, *Foundations of Genetic Algorithms III*, pages 139–161, San Mateo, CA, 1995. Morgan Kaufmann.
6. J.J. Grefenstette. Virtual genetic algorithms: First results. Technical Report AIC-95-013, Navy Center for Applied Research in Artificial Intelligence, 1995.
7. Kh. Rasheed. Guided crossover: A new operator for genetic algorithm based optimization. In *Proceedings of the 1999 Congress on Evolutionary Computation*, pages 1535–1541, Washington D.C., 1999. IEEE NCC - EP Society - IEE.
8. I. Rechenberg. *Evolutionsstrategie*. Frommann-Holzboog Verlag, Stuttgart, 1994.
9. H.J. Sussmann. From the brachystochrone to the maximum principle. In *Proceedings of the 35th IEEE Conference on Decision and Control*, pages 1588–1594, New York NY, 1996. IEEE Publications.
10. C.-F Tsai, C.G.D. Bowerman, J.I. Tait, and C. Bradford. A fuzzy Taguchi controller to improve genetic algorithm parameter selection. In G.D. Smith, N.C. Steele, and R.F. Albrecht, editors, *Artificial Neural Nets and Genetic Algorithms 3*, pages 175–178, Wien New York, 1998. Springer-Verlag.
11. D.H. Wolpert and W.G. Macready. No free lunch theorems for optimization. *IEEE Transactions on Evolutionary Computation*, 1(1):67–82, 1997.

Multilevel Genetic Algorithm for the Complete Development of ANN

Julian Dorado, Antonino Santos, and Juan R. Rabuñal

Univ. da Coruña, Facultad Informática, Campus Elviña, 15071 A Coruña, Spain
{ julian, nino, juanra } @udc.es

Abstract. The utilization of Genetic Algorithms (GA) in the development of Artificial Neural Networks is a very active area of investigation. The works that are being carried out at present not only focus on the adjustment of the weights of the connections, but also they tend, more and more, to the development of systems which realize tasks of design and training, in parallel. To cover these necessities and, as an open platform for new developments, in this article it is shown a multilevel GA architecture which establishes a difference between the design and the training tasks. In this system, the design tasks are performed in a parallel way, by using different machines. Each design process has associated a training process as an evaluation function. Every design GA interchanges solutions in such a way that they help one each other towards the best solution working in a cooperative way during the simulation.

1 Introduction

It could be said that the NNs field has reached a considerable degree of maturity after half a century of development. Despite this maturity, almost every network model, ranging from the first ones (McCulloch and Pitts [1] and Rosenblatt [2]) to the latest ones, pose a number of problems which have not been solved satisfactorily up to date.

Perhaps the most relevant one is the selection or adjustment of the optimal network architecture [3]. The designer, when faced with a NN problem, needs to choose a network model in the first instance. Once the model has been selected, there are many process element architectures which could solve the problem. Most NN models do not provide methods for improving the network's structure during weight training, therefore, although the training process may converge, the resulting network might not be optimal for the problem to be solved. One possible solution are incremental or constructive models [4]. But they are especially oriented towards classification problems, and they present a high degree of dependence with regard to the initial architecture. This makes them prone to falling into local minima.

The lack of automatic processes for adjusting network architectures obliges the designer to refine the architecture through a trial and error process which may occasionally be tedious and time-consuming.

Another typical problem with developing networks is that of building the training set from the available data. The selection of the training sets has a considerable impact in the network's final performance, given that the training set has the knowledge provided to the network for its learning. If this set has any kind of flaw, the network's learning will not be perfect. Once again, in this case, there is no standard automatic method for carrying out this task. Only in certain cases is a statistical study carried out in order to guarantee the quality of our data [5] or to

J. Mira and A. Prieto (Eds.): IWANN 2001, LNCS 2084, pp. 717-724, 2001.

reduce their dimensionality [6]. Once again, the designer is faced with the realization of a series of tasks in order to achieve the network's convergence. These problems have not been too pressing until the present day, and NNs have been successfully developed in different application fields. The direct consequence is that, for each NN developed, the designer must carry out a series of manual tests which do not guarantee the achievement of an optimal design.

However, in recent years, the need to develop more powerful systems arises, systems which are capable of solving time problems. Recurrent NNs are the best option in the NN field [7]. This kind of network is more complex than traditional ones, so that the problems of network development are more acute. Their structure has a much higher number of connections, which complicates both the training process [8] and the architecture adjustment. Moreover, the previous approaches to the selection of the training set cannot be used with time-series, since they are based on the independence among observations. In the case of time- series, we have continuous variables in time, which cannot be segmented, with a high number of values [9]. The aim of selecting the training set is to find out which segments in the series are representative, so that they will be the only ones used for the learning process. The advantages entailed by this kind of approach would be that the training stage is shortened, and the ANN generalization would also improve.

2 Proposed System

The use of Evolutionary Computation techniques is an option for avoiding the manual treatment of the problems explained in the previous section. Evolutionary Computation has been used recently in the NN field [10]. Initially, it was used as an alternative training method for gradient algorithms [11], and gradually, its use has spread to NN design and adjustment tasks [12].

Our research group has been working for several years in the solution of these problems by means of GA proving to be very efficient in both fields [13], [14]. This paper explores the possibility of integrating the various systems developed into a single one [15], which will distinguish between network design and adjustment tasks. The design tasks will be distributed into a computer network using a co-operative approach in order to achieve the integral and automatic NN development.

2.1 Design and Training Modules

Two systems for network design have been developed until the present day: an architecture adjustment design system and a training sets selection system from time series. The first one adjusts network architecture characteristics such as the number of layers, the number of neurons, the pruning of connections [16] and the type of activation function for each neuron in the network [13]. The second builds the training set from the original time series. The segments of the series which obtain the best performance of the trained network are automatically selected. The two systems work independently at present. Thus, the system which designs the architecture trains the networks with the whole time series, while the one which builds the training set works with a fixed network architecture. The systems developed until now have used a common model structured into two levels: design level and training level.

The training level is constituted by a module which carries out the NN training with GAs. This module needs a network architecture and a training set as input. The resulting output is the error value of the trained network for that training set. This module carries out the adjustment of the weights of the feedforward or recurrent networks using static training sets or time series.

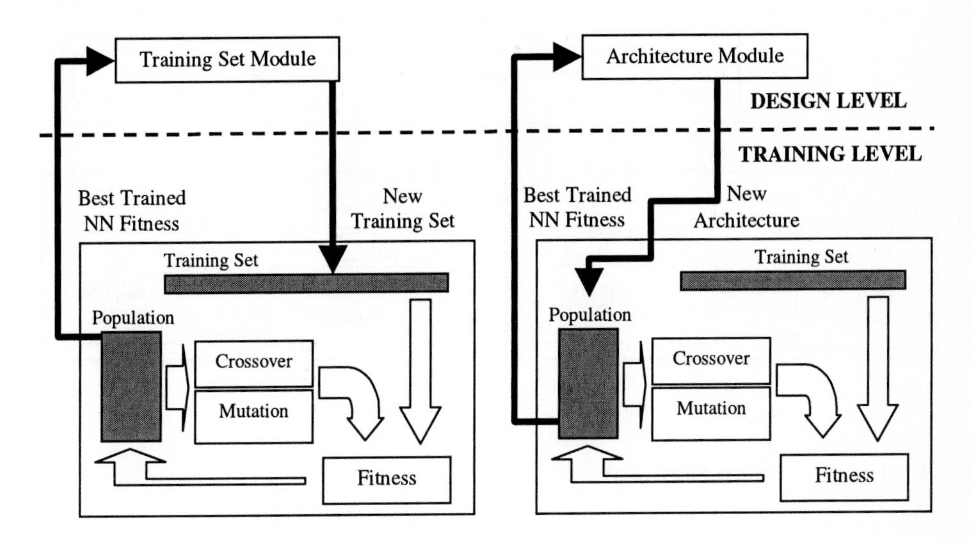

Fig. 1. Two independent systems for NN design

The design level modules also use GAs for adjusting NN design parameters (architecture adjustment or training set selection). Each individual of the GAs in the design level represents a design characteristic for a network (an architecture or a training set). In order to evaluate this individual, a simulation is carried out in the previous training module, where a NN will be trained by fixing the characteristic chosen by the design level GA. The error level of the trained network provides the design level GA individual's fitness.

2.2 Integration

The architecture design and training set construction modules consume many computational resources and require a considerable simulation time. This makes the integration of both design systems into a single application working on a single computer difficult, due to the needs of memory and processing power for GA simulation. This work explores the possibility of distributing processing among several terminals, due to the fact that the development of a NN may be divided into different tasks which require little communication among them. Several cheap computers (PCs) linked to a network by means of the TCP/IP protocol are used for this purpose. The final system will be a lot cheaper than a high-range server, and it will be easily scalable according to the various needs to be solved eventually.

In order to carry out the implementation, a server program must be used which organizes the work of the two design modules that used to work separately. Communications must also be implemented for the information exchange between the client and the servers. Each design system works as a client, it is executed in a network computer and it sends its results to the computer in which the server program which allows the data exchange between both systems is located (see Fig. 2). This structure can be generalized, allowing the integration of new design modules which increase the automation in NN development.

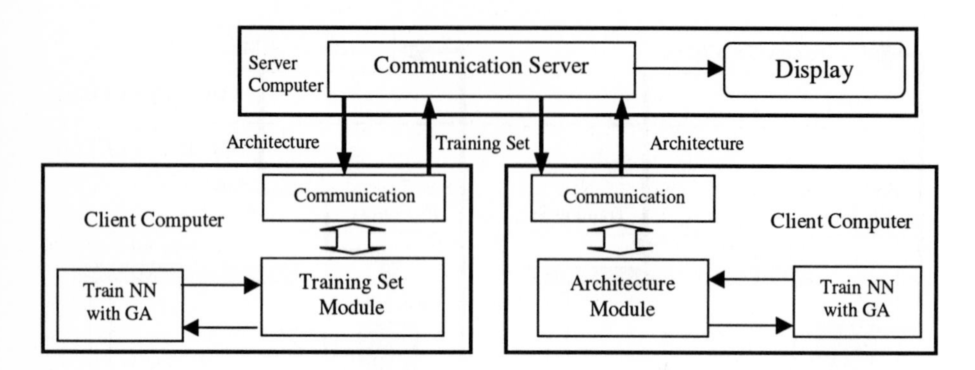

Fig. 2. Distributed proposed system

3 Co-operative Strategy

The goal of the integration of the two NN design modules is not only to reduce the network development time, but also to improve the performance of the networks achieved. This is the reason why the integration has been carried out via a co-operative approach. A periodical exchange of improvements between the two design modules is proposed via the server module, so that the improvements achieved by one of the modules are automatically used by the other and vice versa.

Thus, when the architecture design module finds an architecture which is better adapted, the server module is informed, receiving all the data belonging to this individual. The server module contacts the one which is in charge of designing the training set, thus the network architecture used for evaluating the training sets is modified, while this module makes the most of the advancements achieved by the other one. In case that the training sets module is the one which finds a better training set, this new training set will be moved through the server module to the architecture design module, which will use it to evaluate architectures from that moment on.

As a consequence of this co-operative strategy, the evolution of the adjustment of each design level GA is much better than it would be functioning individually, since each GA makes the most of the improvements found by the other one in order to improve its own adjustment level.

This entails a faster functioning of both design level GAs. The simulation time will be more reduced than expected by the parallelization process, creating a synergy in the general functioning. On the other hand, the networks generated by the training modules will cause lesser errors, since the two facets of network design are concurrently adjusted.

4 Results

Two real time-series have been used in order to test the system. These time-series are typical in the time-series analysis bibliography: sun spots and tobacco production in the USA. The sun spots series comprises 285 observations, and the tobacco production one comprises 113. Data are normalized to a [0,1] range, leaving a 10% higher and lower offset.

Several tests with different configurations have been carried out for design and training GAs. As a result of the performance observed in these tests, both types of GAs are independently configured. The design level GA parameters have the following values: 100 individuals in the population, 1% ratios of crossover and mutation, Montecarlo selection, one point crossover type, and substitution of the worst individuals. In the case of the training GAs, the crossover and mutation ratios are modified at 5%.

Each design level GA will perform 200 iterations per simulation. Although this is a small number of iterations for a GA, it is hoped that the results obtained will prove the potential of the implemented system. The training GA will perform 500 iterations per simulation. This GA's goal is not to achieve totally trained networks, but to search for a value which may serve as an indication of the aptness of a given architecture or training set. For this reason, the number of iterations will be small, so that a great number of networks may be evaluated in a short time. However, if the value is very small, the network might not start the convergence process, and, consequently, no real indicator of the aptness of the architecture will be achieved.

Tests have been carried out in order to adjust this value starting at 100 iterations, the result being that it was clearly insufficient, since no improvements occurred in the network's behavior at that number of cycles. Iteration values were then tested from 100 to 100 until 1000 iterations per network were reached. The convergence process can be clearly observed from 500 iterations. Obviously, this parameter must be adjusted in order to adapt it to the problem which is being solved.

4.1 Tests

In order to validate the functioning of the co-operative approach, the results obtained after using separately the architecture adjustment system and the selection of training test one are examined in comparison with the integrated system presented in this paper. The mean square error (MSE) and the mean error (ME) at short term prediction of time-series are shown for each test as a measure of the error obtained with the network development.

First the architecture adjustment system is used, obtaining the error levels shown in the table 1. Next, the selection system of training sets is used, starting from the best architecture obtained with the previous system. This justifies the improved performance achieved with this system. The integrated system obtains a better error level in the networks developed for both series. This improvement can be justified by saying that the functioning of the integrated system automates the process carried out in the first two steps by alternating the adjustment of the two design facets of the network an indefinite number of times (at least, as many times as data exchanges occur among the systems). The table 2 shows the simulation times which are necessary to obtain the errors shown in the previous table.

Table 1. Prediction errors

System \ Series	*Sun Spots*		*Tobacco Production*	
	MSE	ME	MSE	ME
Architecture adjustment system	0,0086	0,0723	0,0084	0,0681
Training set system	0,0077	0,0687	0,0071	0,0618
Integrated system	0,0074	0,0647	0,0065	0,0560

Table 2. Simulations times

	Systems	*Sun Spots*	*Tobacco Production*
1	Architecture adjustment system	2:03:10	43:55
2	Training set system	1:16:31	35:34
3	Architecture + Training set system	3:19:41	1:19:29
4	Parallel design system	2:03:10	43:55
5	Integrated system	45:30	30:35

The results of the simulation times produce some very interesting data about performance. If we have two separate design systems (lines 1 and 2), their use must be consecutive, therefore the development times for the networks correspond to the addition of the individual times (line 3). The most simple parallelization is carried out, so that each system simulates itself independently in a different processor. Thus independent results are obtained for each system, waiting for the one which takes the longest time (line 4), and then the result of the system which provides the best error level is selected. In the integrated system ,where both design systems work in a co-operative way, the simulation times decrease (more than 30%) when using more reduced training sets on architectures which are more and more optimized.

Finally, the last two tables show the specific parameters which have been adjusted in each NN design systems. In the case of the architecture design adjustment, the number of neurons is shown (recurrent networks with an unspecified number of layers are used), the number of connections which result from pruning, and the activation functions selected by each neuron (L: lineal, H: tangent hyperbolic, and S: sigmoid). In the case of the selection system of training sets, the percentage of the series which uses the best selection in the network training is used.

Table 3. Best results from independent system

Series \ Systems	*Architecture System*			*Training Sets System*
	Neurons	Connections	Activation Functions	% series
Sun spots	6	26	L L S L L H	14%
Tobacco prod.	7	14	L L L L L H L	42%

Table 4. Best architecture and training sets from integrated system

Series	*% series*	*Neurons*	*Connections*	*Activation Functions*
Sun spots	32%	6	15	L L H L H L
Tobacco production	53%	5	14	L S L H L

Two main points should be highlighted about the system's functioning. The first would be the error levels obtained by the selection system of training sets which correspond to a test phase. This is due to the fact that in this system, the NNs are trained only with part of the available data (less than 55% of the series). After training, a test is carried out with the whole series in order to obtain the error value provided by the network compared to real data. This is the error shown in the results section. As a consequence, apart from reducing the size of the training set, the error values achieved are much more relevant, since they naturally include a test phase.

The second point would relate to the development of the integrated system. This poses the problem of the data exchange between the design modules. The way in which this problem is tackled will affect performance dramatically. When an improvement occurs in one of the design modules, the architecture used in the training sets module or the training set used in the architecture module are updated. This update entails changes in the structure of the networks of their training modules, changes which take considerable time during the simulation. If these exchanges occur too often, the work of the modules may be blocked, thus degrading the performance of the integrated system. In order to control the negative effect of this problem, the number of updates in the system has been limited in time, so that each subsystem accumulates the improvements. Thus, if several improvements occur in a short time-span, only the last one is sent to be used by the other module. This problem is more serious at the start of the simulation, when the number of improvements per time-unit is greater. When the simulation finds itself at a further stage, the number of improvements occurring in each module is more limited.

5 Conclusions

The integration of the various approaches to the development of NNs has been achieved, so that they co-operate in the search for an optimal solution. The main point within the system is the co-operative feature, which allows the advancement in the search for the final solution through the exchange of individuals among the various GAs. Thus, the improvements achieved by one system are immediately used by the rest, in order to increase the adjustment of their own solutions, while the convergence process is accelerated.

Another important result of the system is the minimization of the parameters to be configured in order to develop a NN. In the integrated system, most of the configuration tasks refer to the GAs parameters, while the NN configuration is automatic with regard to the architecture and to the selection of the training set. Moreover, the error level has decreased, together with the recurrent NNs training times for the prediction of time-series. Thus, the model of distributed system presented in this paper opens up the possibility of developing (not only training) NNs automatically, in very short time-spans, and via cheap computer network.

6 Future Works

We are currently working on the development of two facets of the system. On the one hand, the implementation of distributed GAs for each of the GAs used in the system (both training and design ones), so that the available equipment can be extensively used in a wide computer network. In this case, the problems of communication of solutions among design level GAs increase, and generally, the simulation time increases, therefore the development of specific operators becomes necessary.

On the other hand, we are working on the distribution of the NN evaluation. The evaluation of the NNs consumes most of the simulation time in the system. It must be taken into account that the algorithm's initialization phase carries out the evaluation of each and every individual in the population generated at random. The way to distribute the evaluation of the individuals in different computers of the network is being researched, so that the terminal which carries out the GA simulation works as a server, using a certain number of other computers as clients in the NN evaluation.

It is hoped that the combination of these distributed simulation techniques will allow the drastic reduction of the networks development time. Thus, more complex problems will be tackled by using variations of the distribution model presented in this paper.

References

1. McCulloch, W.S. & Pitts, W.: "A Logical Calculus of Ideas Immanent in Nervous Activity". Bulletin of Mathematical Biophysics. Nº 5. Pp. 115-133. 1943.
2. Rosenblatt, F.: "The Perceptron: A Probabilistic Model for Information Storage and Oganization in the Brain". Psylogical Review. Nº 65. 1958.
3. Yee, P.: "Clasification Experiments involving Backpropagation and RBF Networks". Communications Research Laboratory. Report nº 249. McMaster university.1992.
4. Specht, D.F. (1990) "Probabilistic neural networks," Neural Networks, 3, 110-118.
5. Dorado, J., Santos, A. y Pazos, A.: "Methodology for the Construction of more Efficient Artificial Neural Networks by Means of Studying and Selecting the Training Set". International Conference on Neural Networks ICNN96. pp. 1285-1290. 1996.
6. I.T. Jolliffe, "Principal Component Analysis", Springer Verlag, 1986.
7. Williams, R.J. & Zipser, D.: "A Learning Algorithm for Continually Running Fully Recurrent Neural Networks". Neural Computation nº 1. Pp. 270-280. 1989.
8. Hee Yeal, & y Sung Yan B.: "An Improved Time Series Prediction by Applying the Layer-by-Layer Learning Method to FIR Neural Networks". Neural Networks. Vol. 10 Nº. 9. Pp. 1717-1729. 1997.
9. Vassilios Petridis & Athanasios Kehagias: "A Recurrent Network Implementation of Time Series Classification". Neural Computation. Nº. 8. Pp. 357-372. 1996.
10. Montana, D.J. & Davis, L.: "Training Feedforward NN Using Genetic Algorithms". Proc. of Eleventh International Joint Conference on Artificial Intelligence". Pp. 762-767. 1989.
11. Joya, G., Frias, J.J., Marín, M.M. & Sandoval, F. (1993): "New Learning Strategies from the Microscopy Level of an ANN". Electronics Letters. Vol. 29. Nº 20. Pp. 1775-1777.
12. Radcliffe, N.J.: "Genetic Set Recombination and its Application to NN Topology Optimization". Neural Computing and Applications. Vol. 1. Nº 1. Pp. 67-90. 1993.
13. A. Pazos, J. Dorado, A. Santos,J. R. Rabuñal y N. Pedreira. "AG paralelo multinivel para el desarrollo de RR.NN.AA.". CAEPIA-99. Murcia. Spain. 1999.
14. J. Dorado, A. Santos, A. Pazos, J. R. Rabuñal y N. Pedreira. "Automatic Selection of the Training Set with Genetic Algorithm for Training Artificial Neural Networks". Gecco 2000. Las Vegas, NV, EE.UU. 2000
15. McDonnell, J.R. & Waagen, D.: "Evolving Recurrent Perceptrons for Time Series Modeling". IEEE Transactions on Neural Networks. Vol. 5 Nº 1. Pp. 24-38. 1996.
16. Karnin, E.D. (1990): "A Simple Procedure for Pruning Back-Propagation Trained Neural Networks". IEEE Transactions on Neural Networks. Vol. 1 Nº 2. Pp. 239-242.

Graph Based GP Applied to Dynamical Systems Modeling

A.M. López[1], H. López[1], L. Sánchez[2]

[1] Electrical Engineering Department, Oviedo University, Spain
[2] Computer Science Department, Oviedo University, Spain
e-mail: antonio@isa.uniovi.es

Abstract. Model construction is usually guided by a trial-error process, where each iteration is divided into two steps: (i) collect or refine the set of equations that direct the system behaviour, normally in differential form, solving them using, most of the time, the $\mathcal{S}$ transform, and (ii) fit a set of properties (parameters) in the model obtained using observations taken from the real system.
There have been many attempts to automate this process. We will extend an approach based on a search of a model of the system in a block diagram representation, where the trial-error process is solved with Genetic Programming. Some modifications over this approach are made to allow a more general family of models and to enhance its efficiency.

1 Introduction

Genetic Programming has been applied to system modeling in different ways. One of the initial solutions was the proposed in [7], where a structured Genetic Algorithm, in a tree based representation, is used to solve identification problems. Quadratic functions of two input variables are used as the function set. In [1] a similar approach is described, but using a set of mathematics operators as the function set. Both works introduce dynamics into the solutions by means of old values of input variables.
A different approach is the described in [8], where an attempt is made to model a linear time invariant system by means of its impulsorial response.
Symbolic regression has also been applied to system modeling in [17]. In [15] a similar approach is followed, introducing dynamics by means of a special recursive node. In [2], GP is used to get a symbolic expression of an equation in differential form.
Up to our knowledge, the most complete application is described in [12–14], where the problem is addressed as a search of a model of a system using a block diagram representation, being the function set composed, among others, of continuous time blocks defined in the domain $\mathcal{S}$. We will extend this work to a more general class of systems and compare the efficiencies of both schemes.

J. Mira and A. Prieto (Eds.): IWANN 2001, LNCS 2084, pp. 725-732, 2001.

2 Tool Definition

The work described in [13] is based on a search of a continuous model of a system using a tree based block diagram representation (see figure 1). Some continuous time linear blocks of different orders and non-linear blocks are used as the function set. This set is complemented with arithmetic operators and a special node playing the role of a feedback node.
Continuous time blocks evaluation using sampled data are solved using their $\mathcal{Z}$ transform. Parameter tuning is made by means of well known parameter search methods. Our modifications to this method cover: (a) base class system and constructive block definition (b) individual representation (c) genetic operators definition (d) individual evaluation and (e) parameter tuning.

Base Class System & Constructive Block Definition Continuous dynamic non linear time invariant systems are the base class. Non linearity features are limited to dead zones and saturations, common features of most physical systems. Possible associations of different elements are restricted to be a linear (static or dynamic) block, with an optional preceding dead zone and an optional following saturation, all of them embodied in the constructive block (see figure 1). Dynamic blocks are reduced to first order and second order blocks and all gains are static.

Individual Representation Tree based individual representation, used in [13], is not generic; some systems can not be represented. We decided to use graphs instead. Graph based representation has been used previously in GP [4, 16], but the application field is completely different from ours. The representation we propose (see figure 1), mixes a link nodes list with ideas from [13] and [15].
An special feedback node is used, hanging from it the input branch and the feedback branch. It also contains a pointer to another node in the graph from which the feedback signal will be taken. This pointed node will play, together with its own function, the role of the bifurcation node. The terminal nodes of the feedback branch are "neglected" taking at each iteration as their actual value the last value returned by the bifurcation node. They could be seen as recessive nodes, being valid if they are moved to any point in another individual as a result of a genetic operator application.

Genetic Operators Definition Genetic operators are coherent with the new representation. Subtree and Internal [9] crossover are randomly combined. Both subtree and node mutation are applied. After crossover or mutation, a reparation is made if necessary over the individuals to get them valid.

Individual Evaluation Continuous systems evaluation using sampled data implies a previous conversion to discrete time. Different conversions can be used. $\mathcal{Z}$ transformation of continuous time blocks fails due to an attenuation factor induced by the sampling. So, approximate transformations (approximations of $\mathcal{Z} = e^{ST_m}$) need to be considered to find a good transformation for our purposes. Tustin transformation is selected to be such conversion method in basis of it accuracy and the conservation of stability, cascade and static gain properties [6].

Feedback evaluation presents the causality problem. To prevent infinite recursions, an implicit unit delay is assumed in the feedback branch.

Parameter Tuning Each model is dependent of a series of parameters which must be fitted using data sampled from the real system. In [13], deterministic optimization methods, whose result depends on the initial search point, are used. We will empirically show that this limitation is relevant and results can be improved with GA's.

3 Test Suite

A test suite (see figure 3) has been selected to explore the efficiency of the redefined scheme in comparison with the Marenbach scheme. A tray is made to cover a class of physical systems, from the simplest linear first order to a more complex non-linear one: System I and System II are linear systems of first and second order respectively. The aim is to evaluate the efficiency of the tool against a linear system, comparing the effect in the search of an evolutive parameter tuning with a classic parameter tuning, which is expected to perform better for such class of systems. In System III, System IV and System V, non-linear features are gradually introduced, in order to test if an evolutive parameter tuning performs better than a deterministic one as it is hoped.
Results will be evaluated both from its numerical efficiency and its morphological composition. Numerical comparison will be centered in the efficiency of Simplex and GA parameter tuning. Both methods will be applied to identify each parameter in the models contained in the test suite. Morphological comparison will be made by looking at the results of the application of Marenbachs's and Redefined schemes to the test suite. It will be centered in the similarity between the real model and the obtained one.

3.1 Parameter tuning

We will contrast the effectiveness of a deterministic evolutive method, Nelder and Mead's Simplex, and GAs for identification problems in which the structure is known. Next, we will combine the best of these methods with GP search to induce both parameters and structure.
The upper part of Figure 2 contains the mean square error obtained by means of both optimization methods applied to the test suite. To minimize the effect of the starting point, simplex was repeated 20 times, and the best solution selected. Simplex did not perform correctly in Test III, IV and V, where non linear features are involved. Results are very dispersed, while tuned values obtained by means of a GA are more concentrated. As an example, the lower part of figure 2 shows the results obtained by means of Simplex and GA tuning applied to System V, composed of a series association of a dead zone, a linear first order system and a saturation. Each row of both tables contain the values for the dead zone (start and stop value), the first order linear system (static gain and time constant) and the saturation (maximum and minimum).

3.2 Structure identification

Models obtained from the application of the Marenbach and redefined schema are shown in figure 3, columns 2 and 3 respectively. It can be seen how the redefined scheme guides to simpler solutions, closer to the real model, while the Marenbach scheme guides to overfitted solutions.
Results obtained by means of the Marenbach schema are quite complex compared with the real models. When applied to linear systems (System I and System II), non linear blocks are neglected, but the result is composed of different linear blocks combined by arithmetical operators. It could be attributed to the effort made by the tool at reducing the error as much as possible, while it is necessary to have in mind that the conversion made to evaluate each system introduces an error [11].
But our main interest is focused in non linear systems, where there is a lack of general methods to model such class of systems. Models obtained by means of the application of the Marenbach schema to non linear models (System III, System IV and System V) are quite far from the real model. Strange associations of linear and non-linear blocks are found in the solutions when they are compared with the real models.
Results obtained from the application of the redefined scheme to the test suite are very close to the real models. No strange associations of blocks are present in the solutions. The main difference between the obtained models and the real ones is the order of the dynamic linear components, which is higher than the order of those blocks in the real system. This can be attributed to the evaluation error, compensated at this scheme by this effect. Nevertheless, after a study of each linear component, it can be seen how can be easily reduced to a equivalent linear component of lower (first) order.
Almost, parameter tuning puts each parameter in a close environment of the real value, which is of high interest. Tuned values are not really the right ones, but the process could be completed by means of an identification to obtain the real model of the system. This identification would be quite simple once known the structure of the model and an starting point close to the minimum.

4 Conclusions

A new scheme is proposed to model physical systems based on Genetic Programming. This new scheme proposes a new graph based individual representation, together with a definition of genetic operators to fit the new representation. Also, parameter tuning is based on a Genetic Algorithm. Although the idea was proposed before and discarded due to the computational requirements, this paper shows how analytical methods fail, being an evolutive method the solution for the task. Obviously, an effort must be made in the future to reduce the computational effort making the tool more adequate for a profesional use. Last modification is the use of a constructive block to settle possible associations of linear and non-linear components, which is shown to avoid randomly combinations of components, guiding the search in the right direction.

The scheme has been applied to a test suite composed of systems of different orders with different non linearities involved. It has been shown how the proposed scheme performs better for such systems than the Marenbach scheme. Models obtained are structurally close to the real system and parameter values are in a narrow environment of the real values. The process could be complemented, as a final step with the aim of getting the right model, with a more precise identification being given the structure of the model and the tuned values for each parameter as a starting point.
Future work will be centered at reducing the computational effort of the evolutive parameter tuning and at exploring the efficiency of the tool against more complex systems.

References

1. V M Babovic, B Brozkova, M Mata, V Baca, and S Vanecek. System identification and modelling of vltava river system. In *Proceedings of the 2nd DHI Software User Conference*, 1991.
2. Saso Dzeroski, Ljupeo Todorovski, and Igor Petrovski. Dynamical system identification with machine learning. In Justinian P. Rosca, editor, *Proceedings of the Workshop on Genetic Programming: From Theory to Real-World Applications*, pages 50–63, Tahoe City, California, USA, 9 July 1995.
3. Gary J. Gray, David J. Murray-Smith, Yun Li, and Ken C. Sharman. Nonlinear model structure identification using genetic programming. In John R. Koza, editor, *Late Breaking Papers at the Genetic Programming 1996 Conference Stanford University July 28-31, 1996*, pages 32–37, Stanford University, CA, USA, 28–31 July 1996. Stanford Bookstore.
4. Frederic Gruau. Genetic synthesis of modular neural networks. In Stephanie Forrest, editor, *Proceedings of the 5th International Conference on Genetic Algorithms, ICGA-93*, pages 318–325, University of Illinois at Urbana-Champaign, 17-21 July 1993. Morgan Kaufmann.
5. Hugo Hiden, Mark Willis, Ben McKay, and Gary Montague. Non-linear and direction dependent dynamic modelling using genetic programming. In John R. Koza, Kalyanmoy Deb, Marco Dorigo, David B. Fogel, Max Garzon, Hitoshi Iba, and Rick L. Riolo, editors, *Genetic Programming 1997: Proceedings of the Second Annual Conference*, pages 168–173, Stanford University, CA, USA, 13-16 July 1997. Morgan Kaufmann.
6. Constantine H. Houpis and Gary B. Lamont. *Digital Control Systems.* McGraw Hill, 1992.
7. Hitoshi Iba, Takio Karita, Hugo de Garis, and Taisuke Sato. System identification using structured genetic algorithms. In Stephanie Forrest, editor, *Proceedings of the 5th International Conference on Genetic Algorithms, ICGA-93*, pages 279–286, University of Illinois at Urbana-Champaign, 17-21 July 1993. Morgan Kaufmann.
8. Martin A. Keane, John R. Koza, and James P. Rice. Finding an impulse response function using genetic programming. In *Proceedings of the 1993 American Control Conference*, volume III, pages 2345–2350, Evanston, IL, USA, 1993.
9. Kenneth E. Kinnear, Jr. Alternatives in automatic function definition: A comparison of performance. In Kenneth E. Kinnear, Jr., editor, *Advances in Genetic Programming*, chapter 6, pages 119–141. MIT Press, 1994.

10. John R. Koza. *Genetic Programming: On the Programming of Computers by Means of Natural Selection.* MIT Press, Cambridge, MA, USA, 1992.
11. A. M. López, H. López, and G. Ojea. Dinamyc system modelling using genetic programming with a genetic algorithm for parameter adjustment. In *Proceedings of the* 7th *IEEE International Conference on Emergent Technologies and Factory Automation*, 1999.
12. Peter Marenbach. Using prior knowledge and obtaining process insight in data based modelling of bioprocesses. *System Analysis Modelling Simulation*, 31:39–59, 1998.
13. Peter Marenbach, Kurt D. Betterhausen, and Stephan Freyer. Signal path oriented approach for generation of dynamic process models. In John R. Koza, David E. Goldberg, David B. Fogel, and Rick L. Riolo, editors, *Genetic Programming 1996: Proceedings of the First Annual Conference*, pages 327–332, Stanford University, CA, USA, 28–31 July 1996. MIT Press.
14. Hartmut Pohlheim and Peter Marenbach. Generation of structured process models using genetic algorithms. In T. C. Fogarty, editor, *Evolutionary Computing*, number 1143 in Lecture Notes in Computer Science, pages 102–109. Springer-Verlag, University of Sussex, UK, 1-2 April 1996.
15. Ken C. Sharman and Anna I. Esparcia-Alcazar. Genetic evolution of symbolic signal models. In *Proceedings of the Second International Conference on Natural Algorithms in Signal Processing, NASP'93*, Essex University, UK, 15-16 November 1993.
16. Astro Teller and Manuela Veloso. PADO: Learning tree structured algorithms for orchestration into an object recognition system. Technical Report CMU-CS-95-101, Department of Computer Science, Carnegie Mellon University, Pittsburgh, PA, USA, 1995.
17. A. H. Watson and I. C. Parmee. Identification of fluid systems using genetic programming. In *Proceedings of the Second Online Workshop on Evolutionary Computation (WEC2)*, number 2, pages 45–48, http://www.bioele.nuee.nagoya-u.ac.jp/wec2/, 4–22 March 1996. Nagoya University, Japan.

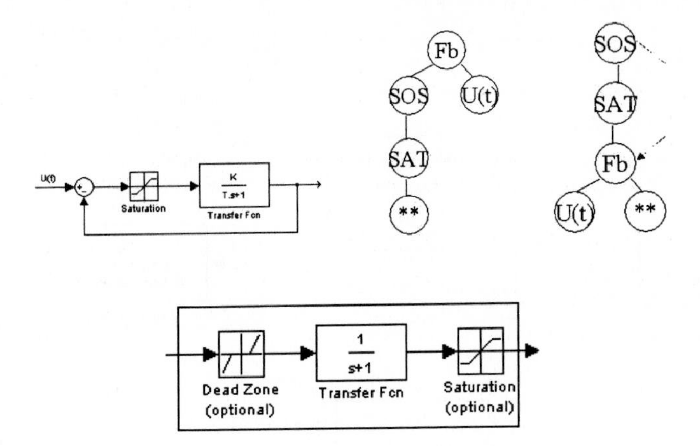

Fig. 1. System and Tree Representation. Upper part: System Model (left), Marenbach Scheme Tree Representation (center) and Redefined Scheme Graph Representation (right). Lower part: Constructive Block.

System	I	II	III	IV	V
Simplex	0.85	0.54	0.16	0.28	0.14
GA	0.85	0.6	0.17	0.29	0.17

	MSE	DZST	DZSP	LFOSG	LFOTC	STN	SMX
	1.943	0.25	7.59	6.19	5.27	3.12	4.43
	1.94	0.25	14.37	12	8.41	7.5	8.75
S	1.94	0.25	16.12	11.6	9.5	8.1	10.75
	1.94	0.25	21.12	13.63	10.76	4.4	13.6
I	1.94	0.25	22.71	17.9	14.24	10.62	16.24
	1.94	0.25	14.67	11.87	8.8	6.87	8.67
M	1.94	0.25	19.36	13.83	10.37	10.7	11.02
	0.2	-9.22	15.28	1.96	4.64	-2.7	2.84
P	1.94	0.25	6.4	5.28	3.89	2.56	4.13
	0.73	-0.69	2.08	2.5	1.86	-2.9	3.19
L	1.94	0.25	17.79	16.07	14.56	8.04	19.1
	0.14	-7.85	1.98	1.92	4.72	-1.94	1.86
E	0.19	-2	10.29	2.07	4.41	-2.93	3.11
	1.94	0.25	15.52	12.85	10.65	6.61	11.8
X	1.94	0.25	17.52	14.72	11.92	6.52	13.46
	1.94	0.25	12.7	9.35	6.6	5.22	8.17
	1.94	0.25	12.3	0.45	6.95	6.35	8
	1.94	0.25	10.56	6.91	5.37	2.01	7.2
	1.94	0.25	22.95	19.62	15.74	8.43	17.65
	1.94	0.25	12.56	8.77	8.01	6.65	7.98
	0.19	-2.74	3.06	2.26	5.07	-2.1	2.74
G	0.17	-2.42	2.42	2.26	5.39	-2.42	2.42
	0.17	-2.42	2.42	2.26	5.39	-2.42	2.42
A	0.2	-2.1	2.1	3.22	8.62	-4.03	2.1
	0.2	-3.71	4.03	2.9	4.75	-2.42	2.74

Fig. 2. Parameter tuning mean square error for the test suite (upper part), and details of the experimentation for System V (lower part). MSE = Mean Square Error. DZST = Dead Zone Start. SZSP = Dead Zone Stop. LFOSG = Linear First Order Static Gain. LFOTC = Linear First order Time Constant. STN = Saturation Minimum. STX = Saturation Maximum.

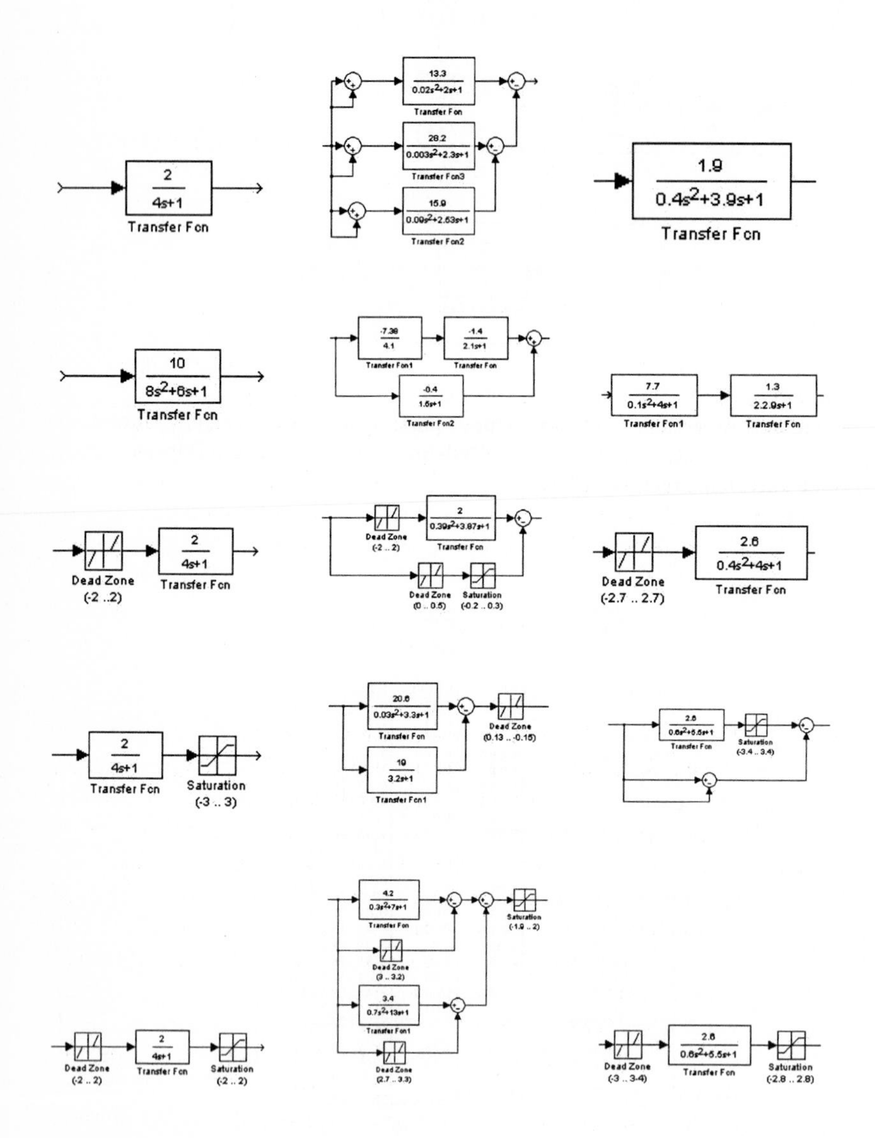

Fig. 3. Test Suite and Modeling Results. Column 1: Tests Suite, System I to V. Column 2: Results of Marenbach Scheme. Column 3: Results of Redefined Scheme.

Nonlinear System Dynamics in the Normalisation Process of a Self-Organising Neural Network for Combinatorial Optimisation

Terence Kwok and Kate A. Smith

School of Business Systems, Faculty of Information Technology, Monash University
Clayton, Victoria 3168, Australia
{terence.kwok, kate.smith}@infotech.monash.edu.au

Abstract. The weight normalisation process used for constraint satisfaction in a self-organising neural network (SONN) for combinatorial optimisation is investigated in this paper. The process relies on the mutual interaction of neuronal weights for computation, and we present a theoretical model to capture its long-term equilibrium dynamics. By solving the equilibrium states numerically, we reveal some nonlinear system phenomena hidden in the normalisation process: fixed point, symmetry-breaking bifurcation and cascades of period-doubling bifurcations to chaos. This leads to a new perspective of the weight normalisation within the SONN as a computational process based on nonlinear dynamics.

1 Introduction

The development of Hopfield-based neural networks has enabled the computation of solutions to NP-hard combinatorial optimisation problems (COP's) with potential for rapid speed by utilising parallel architectures for distributed processing. This powerful property of collective computation can also be realised in self-organising neural networks (SONN's), which is manifested as the self-organisation principle of Kohonen's self-organising feature map (SOFM) [1]. Earlier efforts of combining SOFM with the elastic net method [2] have seen a new neural approach to solve COP's. But the dependence of this technique on Euclidean geometry has restricted the applicability of this approach to geometric problems such as the travelling salesman problem (TSP) [3], [4]. In order to solve a broader class of "0-1" optimisation problems, Smith *et al* [5], [6] developed a more general SONN which operates within feasible permutation matrices rather than the Euclidean space of the elastic net. This SONN has subsequently incorporated a weight normalisation procedure introduced by Guerrero *et al* [7], resulting in a SONN with weight normalisation (SONN-WN). Such a modification is primarily for improving the efficiency of constraint satisfactions by promoting network convergence towards 0-1 solutions with reduced oscillations. A similar normalisation has been used for the Hopfield networks [8]. The SONN-WN also has in common with simulated annealing [9] the process of annealed convergence via the reduction of a temperature parameter. Experimental results of the SONN-WN

J. Mira and A. Prieto (Eds.): IWANN 2001, LNCS 2084, pp. 733-740, 2001.

show a high sensitivity of feasibility to annealing schedules, but a clear ability to improve the feasibility of the solutions [10]. A deeper understanding of the normalisation process is thus required.

In this paper we focus on the weight dynamics within a typical neuron of the SONN-WN. In Sect. 2, the SONN-WN architecture and algorithm are given in the context of solving the *N*-queen problem. An equilibrium model is presented in Sect. 3 to capture the nonlinear interactions among the weights due to updating and normalisation. Because of the strong nonlinearity, numerical computations are used to solve for the system states in Sect. 4. Discussions are concluded in Sect. 5.

2 The SONN-WN: Architecture and Algorithm

The SONN-WN is outlined here as an implementation to solve the *N*-queen problem. It should be noted that the SONN approach is general and has been used to solve a variety of COP's [5]-[7],[11].

The *N*-queen problem is an example of NP-hard constraint satisfaction problems. The aim is to place *N* queens onto an $N \times N$ chessboard without attacking each other. This is enforced by having 1) one queen in each row; 2) one queen in each column; 3) one queen on each diagonal (there are more than two diagonals), and 4) exactly *N* queens on the chessboard. Mathematically, the state of the chessboard can be represented by an $N \times N$ matrix $\mathbf{X}$ with elements of 1 wherever a queen is placed, and 0 otherwise. This forms a 0-1 optimisation problem minimising the following cost (*f*) which measures violation of the constraints described above,

$$f = \frac{1}{2}\sum_{i,j}\sum_{l \neq j} x_{ij}x_{il} + \frac{1}{2}\sum_{i,j}\sum_{k \neq i} x_{ij}x_{kj} + \frac{1}{2}\left(\sum_{i,j} x_{ij} - N\right)^2$$
$$+\frac{1}{2}\left(\sum_{i,j}\sum_{\substack{p \neq 0 \\ 1 \leq i-p \leq N \\ 1 \leq j-p \leq N}} x_{ij}x_{i-p,j-p} + \sum_{i,j}\sum_{\substack{p \neq 0 \\ 1 \leq i-p \leq N \\ 1 \leq j+p \leq N}} x_{ij}x_{i-p,j+p}\right). \quad (1)$$

To solve the *N*-queen problem of minimising constraint violation by (1), we use a SONN-WN based on that used in [7]. Fig. 1 shows its architecture comprising an input layer of *N* nodes, and an output layer of *N* nodes, representing the *N* columns and *N* rows of the chessboard respectively. The weight between an input node *j* and an output node *i* is given by W_{ij} and represents the continuous relaxation of the decision variable x_{ij} in (1). The training set consists of one input pattern for each column (each queen in this example). The input pattern corresponding to a column j^* is a vector of *N* components where the j^*-th component is 1, and the remaining components are 0. When the input pattern corresponding to a column j^* is presented, the net input to each node *i* of the output layer is the potential

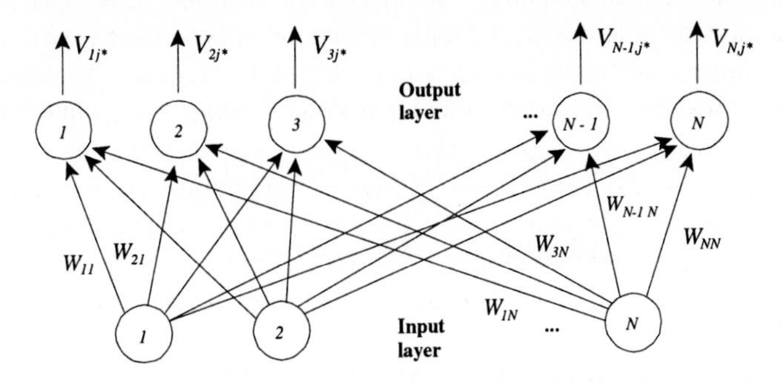

Fig. 1. Architecture of the SONN

$$V_{ij*} = \sum_{k,l=1}^{N} (1-\delta_{ik})(\delta_{i+j*,k+l} + \delta_{i-j*,k-l})W_{kl} + B\sum_{k=1}^{N}(1-\delta_{ik})W_{kj*} \qquad \textbf{(2)}$$

which is the diagonal and column contribution to the objective function f of placing a queen on column $j*$ of row i. $\delta_{ik} = 1$ if $i = k$ and 0 otherwise. $V_{ij}*$ represents the partial cost of assigning a queen to position (i, $j*$), and is related to the partial derivative of the objective function (1). B is a penalty parameter arbitrarily chosen to be 1 in this paper. The row constraints are enforced by the normalisation procedure described below during the weight updates.

The competition between the output nodes gives the winning node i_0 corresponding to the row with minimum potential, i.e $i_0 = arg\ min_i\ V_{ij}*$. We then define a neighbourhood of the winning node as those with the least potential (cost)

$$N(i_0, j^*) = \{i_0, i_1, i_2, ..., i_\eta\} \qquad \textbf{(3)}$$

where $V_{i_0,j*} \le V_{i_1,j*} \le V_{i_2,j*} \le ... \le V_{i_\eta,j*} \le ... \le V_{i_{N-1},j*}$ and $\eta \ge 1$ is the neighbourhood size. Thus the neighbourhood is defined according to relative cost, rather than spatially as in Kohonen's SOFM [1]. Once both the winning node and its neighbourhood have been determined, the weights are updated based on a Kohonen's weight adaptation rule. In order to explicitly enforce the constraint of one queen in each row, the weight normalisation for each output node as described in [7] is used:

$$W_{ij} \leftarrow \exp[-(1-W_{ij})/T] \Big/ \sum_{j=1}^{N} \exp[-(1-W_{ij})/T] \qquad \textbf{(4)}$$

where T is the temperature parameter, and is lowered as the learning process proceeds. It is crucial to choose an appropriate initial temperature $T(0)$ and a "cooling rate" such that high quality solutions are obtained efficiently. The normalisation operation guarantees that when convergence is completed, only one queen is assigned to each row. During the learning process, the neighbourhood size, the magnitude of the weight adaptations, and T are gradually decreased. The complete algorithm follows:

1. Randomly initialise the weights around 0.5.

2. Randomly choose a column j^* and present its corresponding input pattern.
3. Compute the potential V_{ij^*} for each output node i according to (2).
4. Determine the winning node i_0, and its neighbouring nodes according to (3).
5. Update weights, W_{ij^*}, connecting input node j^* with every output node i:

$$\Delta W_{ij^*}(t) = \alpha(i,t)(1 - W_{ij^*}) \quad \forall i \in N(i_0, j^*) \tag{5}$$

$$\Delta W_{ij^*}(t) = 0 \quad \forall i \notin N(i_0, j^*) \tag{6}$$

where

$$\alpha(i,t) = \beta(t)\exp\left(-\left|V_{i_0,j^*} - V_{i,j^*}\right| / \left|V_{i_0,j^*} - V_{i_{N-1},j^*}\right|\right). \tag{7}$$

The updated weights are:

$$W_{ij^*} \leftarrow W_{ij^*} + \Delta W_{ij^*}\,. \tag{8}$$

6. Normalise W_{ij}'s to enforce the one-queen-per-row constraint using (4).
7. Repeat from Step 2 until all columns have been selected as input patterns. This is one training epoch. Repeat from Step 2-6 for L_1 epochs, then anneal T to encourage 0-1 solution and decrease $\beta(t)$, according to the following:

$$T(t + L_1) = rT(t) \tag{9}$$

where $T(t)$ is the temperature at epoch t, and $0 < r \leq 1$ is the cooling rate. $\beta(t)$ also decays in a similar manner with r_β.

8. Repeat Step 7 L_2 times. Reduce η linearly. Halt when $\eta < 0$.

3 An Equilibrium Model of Weight Dynamics

In this section, we derive a theoretical model to approximate the weight dynamics described by Step 5 and 6 in the last section. The focus is on the weights of a typical winning neuron, and we seek the equilibrium states of those weights under repeated updating and normalisation. For simplicity, the following aspects of the SONN-WN are not included: competition among neurons to be the winner, effects of neighbourhood updating, real-time annealing by decreasing T and β. These simplifications enable us to focus on the interaction among weights without the influence of other complicated spatio-temporal dynamics of the SONN.

Fig. 2 shows how the weights of a given neuron are updated and normalised in the model. w_j represents the weight, with $j = 1\ldots N$. In Fig. 2, w_1 is updated at stage $t + ½$, and is "normalised" using an approximation to (4) at stage $t + 1$. To express mathematically the first stage using (5) and (7),

$$w_{j^*}(t + \tfrac{1}{2}) = w_{j^*}(t) + \beta\left(1 - w_{j^*}(t)\right) \tag{10}$$

with other w_j's being unchanged from t to $t + ½$. For the second stage,

stage	w_i					process
	$j = 1$	2	3	…	N	
t	◆	◆	◆	…	◆	———
$t + ½$	❖	◆	◆	…	◆	updating w_1
$t + 1$	●	◆	◆	…	◆	approx. normalisation

symbols: ◆ = initial weight; ❖ = updated weight; ● = "normalised"

Fig. 2. Schematic diagram of the equilibrium model showing weight updating for $j^* = 1$, followed by an approximated normalisation

$$w_j(t+1) = \exp\left[-\left(1 - w_j(t+\tfrac{1}{2})\right)/T\right] \Big/ \sum_{k=1}^{N} \exp\left[-\left(1 - w_k(t)\right)/T\right], \quad j = 1 \ldots N \tag{11}$$

which can be simplified to

$$w_j(t+1) = \exp\left(w_j(t+\tfrac{1}{2})/T\right) \Big/ \sum_{k=1}^{N} \exp\left(w_k(t)/T\right), \quad j = 1 \ldots N. \tag{12}$$

The difference between (11) and (12) is that the former is more suitable for computational purpose to prevent overflow problems, and the latter is a theoretically equivalent version to be used in this analysis for its algebraic simplicity. Note that both (11) and (12) are approximations to the exact normalisation procedure in (4), but the deviation tends to zero as $\beta \to 0$. The two-stage process described by (10)-(12) is repeated for $j^* = 1 \ldots N$ (an epoch), with j^* chosen in the order: 1, 2, 3, …, N, until all the weights at stage $t + 1$ are "normalised". It should be noticed that (10) is used whenever ◆ changes into ❖, and (12) is used whenever ❖ changes into ●. For asynchronous updating, $w_k(t+1)$ is used whenever available in the denominator of (12), while for synchronous updating, $w_k(t)$ is used for all k's.

Although this is a one-neuron model, the fact that the neuron belongs to a population of N neurons is also incorporated. Let P be the probability that a given neuron becomes the winner, and is hence updated. If the probability of being a winner is the same for every neuron, then $P = 1/N$. This greatly simplifies analysis and is justified for a one-neuron equilibrium. However, when the full SONN is used to solve COP's, each neuron has a different P depending on the cost potential structure. Here for the winning neuron, the average change to each w_j after each epoch is (using (12)):

$$Av[w_j(t+1) - w_j(t)] = \frac{\exp\left[\left(w_j(t) + Av[\Delta w_j(t)]\right)/T\right]}{\sum_{k=1}^{N} \exp\left(w_k(t)/T\right)} - w_j(t) \tag{13}$$

where $Av[\ldots]$ denotes the average of the argument, and

$$\Delta w_j(t) = w_j(t + \tfrac{1}{2}) - w_j(t) \,. \tag{14}$$

Using (10), (13) and (14), it follows that

$$Av[\dot{w}_j(t)] = \frac{\exp\{[w_j(t) + P\beta(1 - w_j(t))]/T\}}{\sum_{k=1}^{N} \exp(w_k(t)/T)} - w_j(t) \tag{15}$$

where $Av[\dot{w}_j(t)]$ is the average rate of change of w_j at any time t. For equilibrium states, we set $Av[\dot{w}_j(t)] = 0$ in (15). Let $\rho = P\beta$, it follows that

$$\hat{w}_j(t) = \frac{\exp\{[\hat{w}_j(t) + \rho(1 - \hat{w}_j(t))]/T\}}{\sum_{k=1}^{N} \exp(\hat{w}_k(t)/T)} \tag{16}$$

where $\hat{w}_j(t)$ denotes weights at equilibrium states, and is time dependent for unstable equilibria. The equilibrium states given by (16) are the main results of this section. It describes the behaviour of the equilibrium model of a one-neuron system capturing the updating and normalisation characteristics of the SONN-WN. Because of the model's nonlinearity, the equilibrium states are solved numerically as follows.

4 The Equilibrium States: Computational Explorations

We present here the equilibrium states obtained computationally by solving (16). In Fig. 3 we plot the weight values solved numerically by an iterative method given by the procedure depicted in Fig. 2, with asynchronous updating. After 30,000 iterations (or epochs), the last 50 values were plotted to show the convergence states, with T varying along the horizontal axis. Fig. 3a shows the case of $N = 1$ and $\beta = 0.95$. A bifurcation point can be observed at $T = 0.95$, where the system becomes unstable and switches into a 2-cycle. Theoretical analysis of (16) shows that bifurcation occurs at $T = \beta$ for this case. Fig. 3b and c correspond to the case of $N = 2$ with $\beta = 0.5$. In Fig. 3b, the two weights w_1 and w_2 have the same equilibrium value before a symmetry-breaking bifurcation at around $T = 0.47$. Note that for $0.23 < T < 0.47$, the equilibrium state is still a fixed point in (w_1, w_2) space. A close-up is shown in Fig. 3c, where the system clearly undergoes successive period-doubling bifurcations for each weight with T as the bifurcation parameter. The use of other values of N and β also shows similar behaviours (not shown). Such a cascade of period-doubling bifurcations leading to chaos is a universal phenomenon that occurs to nonlinear systems of diverse origins [12]. From these results, it is clear that with the weight normalisation process, the SONN-WN can be seen as a dynamical system exhibiting nonlinear phenomena such as bifurcation and chaos.

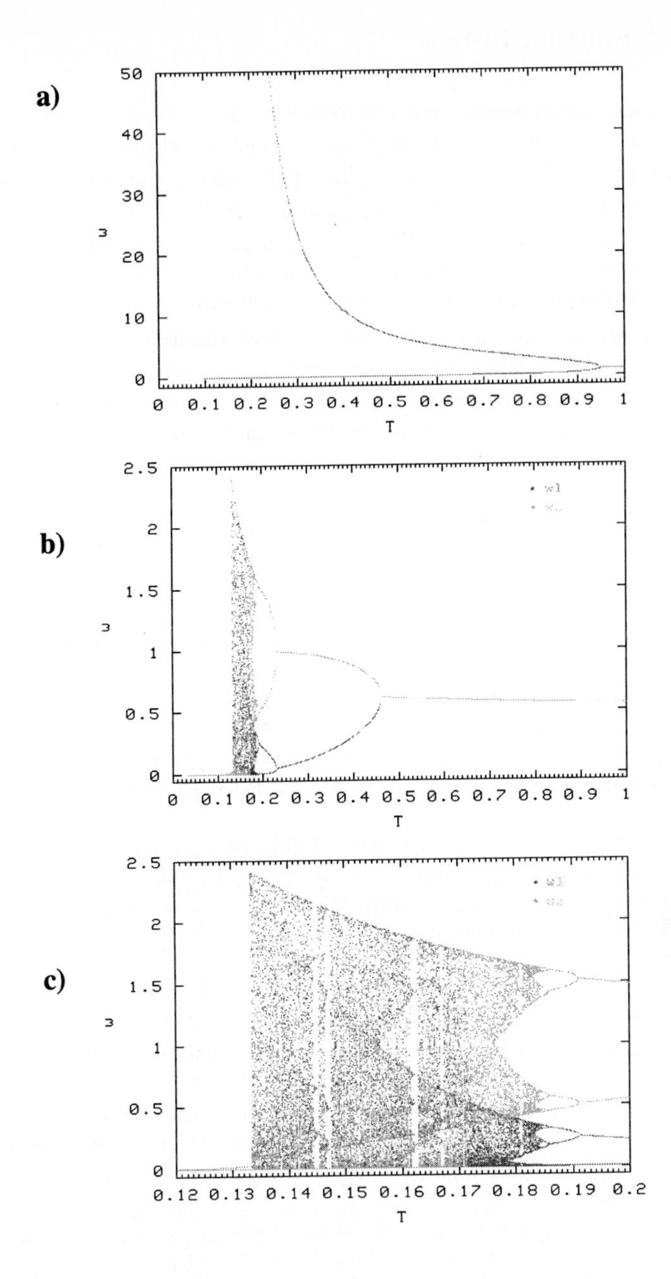

Fig. 3. Bifurcation diagrams of w_1 and w_2 against T. a) $N = 1$, $\beta = 0.95$; b) $N = 2$, $\beta = 0.5$; c) close-up of b)

5 Discussions and Conclusions

For the weight normalisation process in the SONN-WN, we have derived a theoretical model to capture its long-term equilibrium dynamics. The model focuses on the interactions among weights within a neuron, in which parameter T and β determines the nature of the equilibrium dynamics. We have computationally solved for the equilibrium states for $N = 1$ and 2. With T as the bifurcation parameter controlling the dynamical stability of the system, we have found a wide range of nonlinear phenomena exhibited by the equilibrium states: fixed-point, symmetry-breaking bifurcation, and cascades of period-doubling bifurcations to chaos. These findings give us a new dynamical perspective to the weight normalisation process, as well as the potential for a new breed of SONN's exploiting the rich behaviours of nonlinear dynamics. Future research should investigate how the new dynamics can be exploited for improved optimization.

References

1. Kohonen, T.: Self-organized Formation of Topologically Correct Feature Maps. Biol. Cybern., Vol. 43. (1982) 59-69
2. Durbin, R., Willshaw, D.: An Analogue Approach to the Travelling Salesman Problem Using an Elastic Net Method. Nature, Vol. 326. (1987) 689-691
3. Fort, J.C.: Solving a Combinatorial Problem via Self-organizing Process: An Application of the Kohonen Algorithm to the Travelling Salesman Problem. Biol. Cybern., Vol. 59. (1988) 33-40
4. Favata, F., Walker, R.: A Study of the Application of Kohonen-type Neural Networks to the Travelling Salesman Problem. Biol. Cybern., Vol. 64. (1991) 463-468
5. Smith, K.A., Palaniswami, M., Krishnamoorthy, M.: A Hybrid Neural Approach to Combinatorial Optimization. Computers Ops. Res., Vol. 23. (1996) 597-610
6. ________: Neural Techniques for Combinatorial Optimization with Applications. IEEE Trans. Neural Networks, Vol. 9, No. 6. (1998) 1301-1318
7. Guerrero, F., Smith, K.A., Lozano, S.: Self-organizing Neural Approach for Solving the Quadratic Assignment Problem. Proc. Int. Workshop on Soft Computing in Industry (1999) 13-18
8. Van den Bout, D.E., Miller, T.K.: Improving the Performance of the Hopfield-Tank Neural Network Through Normalization and Annealing. Biol. Cybern., Vol. 62. (1989) 129-139
9. Kirkpatrick, S., Gelatt, C.D., Vecchi, M.P.: Optimisation by Simulated Annealing. Science, Vol. 220. (1983) 671-680
10. Kwok, T., Smith, K.A.: Improving the Optimisation Performance of a Self-organizing Neural Network with Weight Normalisation. Proc. Int. ICSC Congress on Intelligent Systems & Applications (ISA'2000), Paper no. 1513-285. (2000)
11. Guerrero, F., Lozano, S., Smith, K.A., Kwok, T.: Manufacturing Cell Formation Using a New Self-organizing Neural Network. Proc. 26th Int. Conf. Computers & Industrial Engineering, Vol. 1. (1999) 668-672
12. Peitgen, H.-O., Jürgens, H., Saupe, D.: Chaos and Fractals: New Frontiers of Science. Springer-Verlag, Berlin Heidelberg New York (1992)

Continuous Function Optimisation via Gradient Descent on a Neural Network Approximation Function

Kate A. Smith [1] and Jatinder N. D. Gupta [2]

[1] School of Business Systems, Monash University, Clayton, Victoria 3168, Australia
kate.smith@infotech.monash.edu.au

[2] Department of Management, Ball State University, Muncie, Indiana 47306, USA
jgupta@gw.bsu.edu

Abstract. Existing neural network approaches to optimisation problems are quite limited in the types of optimisation problems that can be solved. Convergence theorems that utilise Liapunov functions limit the applicability of these techniques to minimising usually quadratic functions only. This paper proposes a new neural network approach that can be used to solve a broad variety of continuous optimisation problems since it makes no assumptions about the nature of the objective function. The approach comprises two stages: first a feedforward neural network is used to approximate the optimisation function based on a sample of evaluated data points; then a feedback neural network is used to perform gradient descent on this approximation function. The final solution is a local minima of the approximated function, which should coincide with true local minima if the learning has been accurate. The proposed method is evaluated on the De Jong test suite: a collection of continuous optimisation problems featuring various characteristics such as saddlepoints, discontinuities, and noise.

1 Introduction

The need for rapid and accurate optimisation techniques frequently arises in many areas of engineering, business and industry. Neural networks hold much potential due to their hardware implementability, and much research has been conducted over the last 15 years to apply neural networks to solve optimisation problems efficiently and effectively [1]. Two main neural approaches have emerged over this period: Hopfield-type networks which iterate to minimise an energy function representing the problem [2], and approaches based on the concepts of self-organisation and elastic nets [3]. Each of these approaches has previously suffered from limitations: the Hopfield approach has had problems ensuring feasibility of the final solution and frequently becomes trapped in poor quality local minima, while the self-organising approaches have suffered from a lack of generalisability to problems not embedded in the Euclidean plane. These limitations have now mostly been alleviated through modification of the original algorithms [1].

One limitation that these approaches have not been able to overcome, however, is their applicability only to certain types of optimisation problems. The Hopfield-type networks based on energy minimisation approaches require a quadratic formulation,

J. Mira and A. Prieto (Eds.): IWANN 2001, LNCS 2084, pp. 741-748, 2001.

and while many practical problems can be formulated in this manner [4], this is still quite a tight restriction. Elastic net methods also minimise an energy function and are likewise restricted. For optimisation problems where the objective function might be non-quadratic, non-smooth, discontinuous, non-differentiable, or even difficult to represent mathematically, these neural approaches cannot be applied.

This paper proposes a new neural approach for such difficult optimisation problems which makes no restrictions on the types of objective function required. The approach is to approximate these functions with a multilayer feedforward neural network (MFNN) which provides a smooth, continuous and differentiable representation of the objective function which is learnt with backpropagation through repeated presentation of sample points and their known objective values. This representation can then be optimised through a recurrent architecture incorporating any number of optimisation techniques which are suitable for continuous and differentiable functions. Thus the proposed approach involves two stages: a *feedforward learning stage* where the MFNN learns to represent the function based on sample points, followed by a *feedback optimisation stage* where this representation is optimised by iterating a randomly chosen starting point through a feedback mechanism.

The proposed approach is evaluated in this paper on the well-known De Jong test suite [5]. This suite is a collection of five continuous optimisation problems designed to exhibit a range of difficulties and challenges for optimisation algorithms. Some of the test functions are discontinuous, while other contain saddlepoints and many local minima. One of the functions also contains gaussian noise. Thus the functions enable researchers to see the strengths and limitations of their optimisation algorithms.

In Section 2 we present the proposed approach, Gradient Descent on a Neural Network Approximation Function (GDNNAF – pronounced "good enough"), as a neural-based heuristic method for minimising any continuous function. The De Jong test suite is described in Section 3, and then used in Section 4 to evaluate the merits of the GDNNAF approach. Conclusions are presented in Section 5, where the performance of the approach on the De Jong test suite functions is used to suggest promising research directions.

2 Gradient Descent on a Neural Network Approximation Function

In this section we present the details of the proposed two-stage GDNNAF approach to optimisation. The mathematical foundations of the approach as well as the algorithm are presented.

2.1 Feedforward Stage - Function Approximation

It is well known that MFNN's are able to approximate noise-free functions with a high degree of accuracy [6]. Suppose we are trying to approximate a function $h(\boldsymbol{x})$ where $\boldsymbol{x}$ is an n-dimensional vector. In this paper we place no restrictions on the function h

other than the requirement that we known the value of this function at N sample points: ie. there are N vectors $\boldsymbol{x}^1$, $\boldsymbol{x}^2$, ..., $\boldsymbol{x}^N$ for which $h(\boldsymbol{x}^1), h(\boldsymbol{x}^2), \ldots, h(\boldsymbol{x}^N)$ can be evaluated. These N pairs of inputs and known outputs can be fed into a MFNN consisting of n inputs, m hidden neurons and a single output neuron. The number of hidden neurons m needs to be determined and will depend upon the complexity of the function being approximated.

The weights connecting the input layer to the hidden layer are denoted by the matrix $\boldsymbol{W}$ and the weights connecting the hidden layer to the output neuron are denoted by the vector $\boldsymbol{V}$. These weights are initialised to random values. The following pseudocode can then be used to represent the function $h(\boldsymbol{x})$ using a standard supervised learning procedure:

1. Select a sample point $\boldsymbol{x}^i$ and present this point to the MFNN, calculating the value of the output neuron
2. Update the values of the weights $\boldsymbol{W}$ and $\boldsymbol{V}$ to correct the error between this output and the known value of $h(\boldsymbol{x}^i)$ (using the backpropagation learning rule or another technique)
3. Repeat from 1 with a new sample point until all N sample points have been presented
4. Calculate the total error E over all N sample points and repeat the entire procedure until E is below some pre-defined tolerance level.

Neural approximators have some distinct advantages including their ability to avoid the "curse of dimensionality" and their convergence properties, making them very attractive as approximators for high dimensional functions [7]. The resulting approximation is based on a differentiable, smooth, and continuous function, making optimisation of this function by conventional techniques possible. This approximation function is given by:

$$\hat{h}(\mathbf{x}) = f_o\left(\sum_{j=1}^{m} V_j^* f_h\left(\sum_{i=1}^{n} W_{ij}^* x_i\right)\right) \tag{1}$$

where $f_o(.)$ and $f_h(.)$ are the activation functions at the output and hidden neurons respectively, and $\boldsymbol{W}^*$ and $\boldsymbol{V}^*$ are the final weights learnt from the approximation procedure outlined above.

2.2 Feedback Stage - Function Minimisation

Once the original function (which may be non-differentiable, non-smooth, discontinuous or unspecified) has been approximated by the MFNN, resulting in the function (1) an optimisation technique can be applied to minimise the function (1). In this paper, we use a gradient descent technique, but there are natural extensions to global search techniques.

For the approximation function $\hat{h}(\mathbf{x})$ given by (1) we have:

$$\frac{\partial \hat{h}(\mathbf{x})}{\partial x_i} = \sum_{j=1}^{m} V_j^* W_{ij}^* f_o'(net_o) f_h'(net_j) \tag{2}$$

where

$$net_j = \sum_{i=1}^{n} W_{ij}^* x_i \qquad u_j = f_h(net_j) \tag{3}$$

$$net_o = \sum_{j=1}^{m} V_j^* u_j \qquad \hat{h}(\mathbf{x}) = f_o(net_o) \tag{4}$$

Then the gradient vector is given by

$$\nabla \hat{h}(\mathbf{x}) = (\frac{\partial \hat{h}(\mathbf{x})}{\partial x_1}, \frac{\partial \hat{h}(\mathbf{x})}{\partial x_2}, \cdots, \frac{\partial \hat{h}(\mathbf{x})}{\partial x_n}) \tag{5}$$

and can be used for gradient descent on (1), or incorporated into any optimisation technique using gradient information (including simulated annealing).

The mapping which achieves gradient descent can be defined as $\boldsymbol{x}(t+1)=\Gamma(\boldsymbol{x}(t))$ where α is the step size of the gradient descent and

$$\Gamma(x_i(t)) = x_i(t) - \alpha \sum_{j=1}^{m} V_j^* W_{ij}^* f_o'(net_o) f_h'(net_j) \tag{6}$$

Thus the minimisation of the approximated function $\hat{h}(\mathbf{x})$ can be seen as a two step process: a feedforward step to calculate the values of $\hat{h}(\mathbf{x}(t))$ and the hidden neuron outputs $u_j(t)$, followed by a feedback step which updates the input vector $\boldsymbol{x}(t+1)$ based on the mapping Γ and producing gradient descent on the approximated function.

3 De Jong's Test Suite

The DeJong test suite [5] is a classic set of continuous functions considered by many to be the minimum standard for performance comparisons of optimisation algorithms. DeJong selected a set of five functions that isolate the common difficulties found in many optimization problems.

1. *The Sphere Function:* $F_1(\mathbf{x}) = \sum_{i=1}^{3} x_i^2 \quad -5.12 \le x_i \le 5.12$

This three-dimensional function is a simple, non-linear, convex, unimodal, symmetric function with a single local minimum.

2. *Rosenbrock's Saddle:* $F_2(\mathbf{x}) = 100(x_1^2 - x_2)^2 + (1 - x_1)^2 \quad -2.048 \le x_i \le 2.048$

This two-dimensional function has a saddle point and causes difficulty due to a sharp, narrow ridge that runs along the top of the saddle in the shape of a parabola.

3. *The Step Function:* $F_3(\mathbf{x}) = \sum_{i=1}^{5} \mathrm{int}(x_i) \quad -5.12 \le x_i \le 5.12$

This five-dimensional function consists of many flat plateaus with uniform, steep ridges and poses difficulties for algorithms that require gradient information to determine a search direction.

4. *The Quartic Function:* $F_4(\mathbf{x}) = \sum_{i=1}^{30} i x_i^4 + Gauss(0,1) \quad -1.28 \le x_i \le 1.28$

This 30-dimensional function includes a random noise variable and tests an algorithm's ability to locate the global optimum for a simple unimodal function that is padded heavily with noise.

5. *Shekel's Foxholes:*

$$F_5(\mathbf{x}) = 0.002 + \sum_{j=1}^{25} \frac{1}{j + \sum_{i=1}^{2} (x_i - a_{ij})^6} \qquad -65.536 \le x_i \le 65.536$$

This two-dimensional function contains 25 foxholes of varying depth surrounded by a flat surface. Many algorithms will become stuck in the first foxhole they fall into.

4 Results

For the first three functions, 1000 randomly generate variables $\mathbf{x}^i$ were evaluated and used as the training data for the MFNN. For functions F_4 and F_5, 5000 samples were used due to the higher dimensionality (F_4) and complexity (F_5) of these functions. Twenty percent of the training data was reserved as a test set to ensure the MFNN learnt the general characteristics of the function, rather than merely memorising the data. A three layer MFNN trained with the backpropagation learning rule was used for all functions, with the learning rate = 0.1, momentum factor = 0.1, and various activation functions and hidden neurons as shown in Table 1. Here we report the best results obtained from numerous experiments, as measured by the coefficient of determination (R^2) on the test set. The logistic activation function is $1/(1+\exp(-net_o))$ while the Gaussian complement function is $1-\exp(-net_o^2)$.

Table 1. Results from MFNN learning of De Jong functions

Function	# hidden neurons	$f_h(.)$	$f_o(.)$	R^2 (test)
F_1	30	tanh	linear	0.9990
F_2	30	tanh	logistic	0.9965
F_3	31	tanh	linear	0.9917
F_4	70	Gauss comp	Gauss comp	0.8467
F_5	65	Gauss comp	Gauss comp	0.1109

The learning results show that the first three functions can be approximated well by a MFNN, the noise within the fourth function creates some difficulties (although the performance is still quite acceptable), but the Shekel's Foxholes problem is extremely difficult to approximate with a neural network. Since the natural tendency of the MFNN is to find a smooth interpolation of the training data, the approximation function fails to reflect the many local minima of this function, and the R^2 value is poor. Figure 1 plots the exact function F_5 based on the 5000 sample points and the MFNN approximation function, and clearly show the effect of the smoothing tendencies of the MFNN. Figure 2 plots function F_2 (the only other two-dimensional function that can be easily visualised), and show a much better approximation.

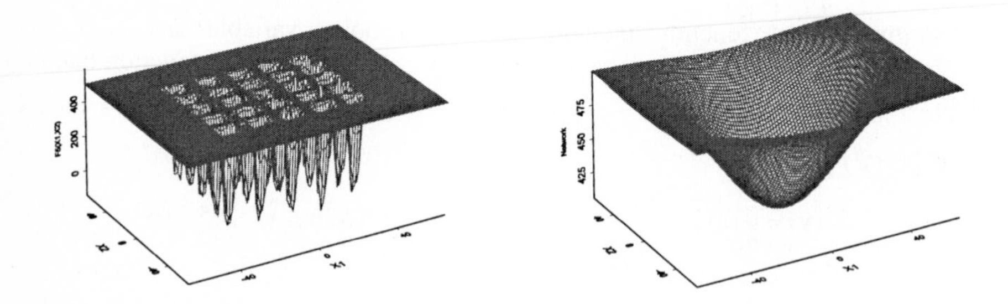

Fig. 1. De Jong function F_5 (left) and MFNN approximation to F_5 (right)

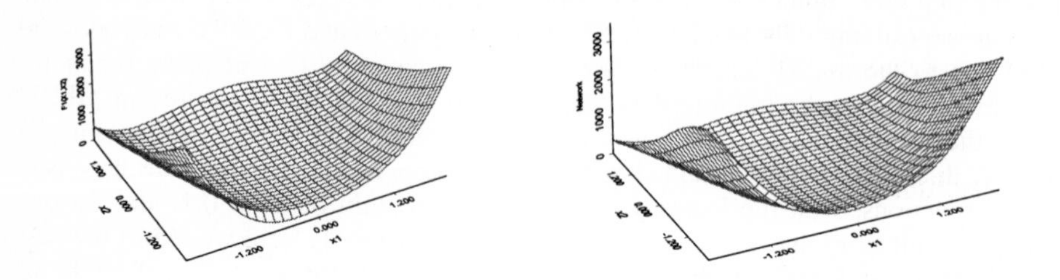

Fig. 2. De Jong function F_2 (left) and MFNN approximation to F_2 (right)

The results from the gradient descent of these approximated functions are shown in Table 2. It should be noted that the gradient descent method successfully converges to the local minima of the approximated function, and where a function has been well approximated by the MFNN in stage 1, this local minima is located in the vicinity of the true local minima of the function. A step size of α=0.001 was used for all functions. The initial starting point for iteration of the gradient descent was randomly chosen from within the feasible region. The gradient descent was also bounded so that if the trajectory left the feasible region in any direction, then the partial derivative of the function in that direction is set temporarily to zero.

Table 2. Results from gradient descent on neural network approximation functions

Function	Exact Minimum $\breve{\mathbf{x}}$	$\hat{h}(\breve{\mathbf{x}})$	Approximated Minimum $\hat{\mathbf{x}}$	$\hat{h}(\hat{\mathbf{x}})$
F_1	F_1(0,0,0)=0	0.2226	(0.001,-0.04,0.01)	-0.867
F_2	F_2(1,1)=0	0.2080	(-1.51,-1.53)	0.0007
F_3	F_3([-5.12,-5], …)=0	0.001	(-5.12, …)	0.001
F_4	F_4(0, … ,0)=0	0.346	all $\lvert x_i \rvert$<0.01	0.000
F_5	F_5(-32,-32)=1	1.000	(3.79,1,12)	0.814

Table 2 shows that in all cases, the gradient descent on the MFNN approximation produces a local minimum with a lower approximated objective function value $\hat{h}(\hat{\mathbf{x}})$ than the true local minimum value $\hat{h}(\breve{\mathbf{x}})$. This is because the approximation function does not acknowledge the true minimum of the function, but has its own minimal value. The exact minimum and the approximated minimum are quite close in all cases except F_2 (where the approximated function does not exactly follow the contour around the saddle point) and F_5 (where the gradient descent produces the minimal point of the smoothed function.

In order to improve the solution quality obtained we need to ensure that the MFNN more closely approximates the true function. Other neural network architectures or configurations could assist this approximation. Another alternative is to use these initial results to isolate the vicinity of the local minima, and repeat the procedure with a new sample of points chosen from within this narrow region. The MFNN can then be re-trained to learn the shape of the surface around this area. In this way the region of interest can be iteratively investigated.

5. Conclusions

In this paper we have presented a new approach to continuous function optimisation via a two-stage neural network. Stage one is a feedforward neural network used to find a smooth approximation to the function based only on a sample of points. Stage 2 uses this approximation function to represent the function, and gradient descent is employed to iterate a starting point through a feedback network and arrive at a local minimum of the approximation function.

We have used the De Jong test suite to evaluate the strengths and limitations of the approach. It can be seen from these results that when the MFNN is able to learn the data effectively, the approach has no problem finding local minima which lie in the vicinity of the true minima of the function. This is the case for the simple quadratic function, step function and noisy functions evaluated. The only limitation appears to be the difficulty in recognising large discontinuities in functions due to the tendency of the MFNN to produce smooth functions from the training data.

The advantages of the approach include its broad applicability to a wide range of functions, without the need for quadratic forms required by other neural network approaches. In addition, we do not even require an explicit analytical expression for the function in order to find approximate local minima, since the method is a data-driven one that only requires a sample of evaluated points.

In future research we will focus on replacing the gradient descent within stage 2 to consider global optimisation methods such as simulated annealing.

References

[1] K. A. Smith, "Neural Networks for Combinatorial Optimisation: a review of more than a decade of research", *INFORMS Journal on Computing*, vol. 11, no. 1, pp. 15-34, 1999.

[2] J. J. Hopfield and D. W. Tank, " 'Neural' Computation of Decisions in Optimization Problems', *Biological Cybernetics*, vol. 52, pp. 141-152, 1985.

[3] R. Durbin and D. Willshaw, "An analogue approach to the travelling salesman problem using an elastic net method", *Nature*, vol. 326, pp. 689-691, 1987.

[4] K. Smith, M. Palaniswami and M. Krishnamoorthy, "Neural Techniques for Combinatorial Optimisation with Applications", *IEEE Transactions on Neural Networks*, vol. 9, no. 6, pp. 1301-1318, 1998.

[5] K. A. De Jong, "An analysis of behavior of a class of genetic adaptive systems", Doctoral Dissertation, University of Michigan, *Dissertation Abstracts International*, vol. 36, no. 10, 5140B, 1975.

[6] K. Hornik, M. Stinchcombe and H. White, "Multilayer feedforward networks are universal approximators", *Neural Networks*, vol. 2, pp. 359-366, 1989.

[7] R. Barron, "Universal approximation bounds for superpositions of a sigmoidal function", *IEEE Transactions on Information Theory*, vol. 39, pp. 359-366, 1989.

An Evolutionary Algorithm for the Design of Hybrid Fiber Optic-Coaxial Cable Networks in Small Urban Areas

Pablo Cortés, Fernando Guerrero, David Canca, and José M. García

Dpto. Ingeniería de Organización, Universidad de Sevilla
pca@esi.us.es

Abstract. Telecommunication is one of the fastest growing business sectors. Future networks will need to integrate a wide variety of services demanding different qualities and capacities from the network. In this paper, network architecture based on hybrid fiber optic-coaxial cable (HFC) is proposed to develop cable integrated telematic services. An evolutionary algorithm is presented to solve the problem in suitable computation times when dealing with real times civil works problems. Finally we present the results over both problem library and real life scenarios.

1. Problem Description

Telecommunication is one of the fastest growing business sectors of modern Information Technologies. In the past it was enough to get telephone access, but today telecommunications include a vast variety of modern technologies and services. Future networks will need to integrate a wide variety of services demanding different qualities and capacities from the network. In this paper, network architecture based on hybrid fiber optic-coaxial cable (HFC) is proposed to develop cable integrated telematic services.

Nowadays architectures exclusively based on fiber optic, as FTTH (fiber to the home), even its more reduced version FTTC (fiber to the curb), are not profitable because of the amount of investment required. However the fiber optic allows reaching longer distances without regenerating the signal at the same time as transporting great volumes of data at high speed. In this context, fiber optic is specially indicated to constitute backbone links, as well as trunk and distribution links inside the metropolitan network where it substitutes with advantage the coaxial cable. Finally, the coaxial is indicated for the latest network stage correspondent to the subscriber access.

Telecommunication backbones connect with the urban node through the headend, where other analog and digital flows are received and inserted in the metropolitan network. After that a partially meshed fiber optic trunk network connects the head with the primary optical nodes over synchronous digital hierarchy (SDH), attending to the European transmission standard. The distribution network is constructed over fiber optic also, and connects the primary optical nodes with the optical network units (ONU) constituting the latest optic stage. Previously to the ONUs could be located

J. Mira and A. Prieto (Eds.): IWANN 2001, LNCS 2084, pp. 749-756, 2001.

secondary nodes acting as splitters with the aim of dividing the downstream signal from one fiber into several fibers. Finally, the feeder network corresponding to the subscriber access stage will be done by means of short length coaxial and lower quality than fiber.

Most of the bibliography based on telecommunication networks attends to large urban areas or inter-municipality communication backbones. Here, we deal with the problem of HFC network deployment in small urban areas. In this context we will have to balance profitability investment with survivability level, and for the smallest urban areas we will have to substitute survivability level to get profitability allowing the network deployment. For this situation small network deployment can be indicated such as star-star architectures, as figure 1 depicts.

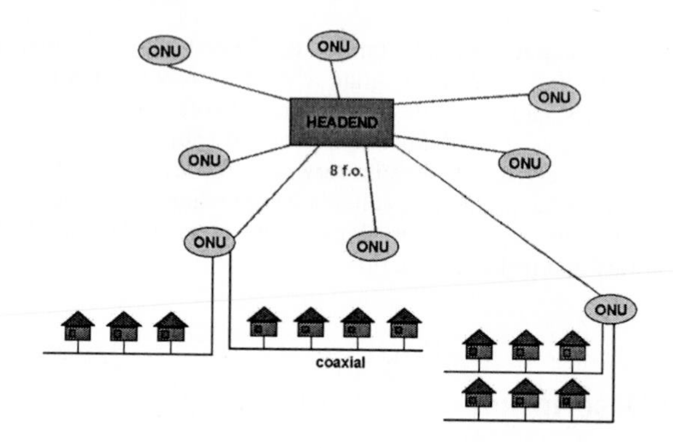

Fig. 1. Simple star architecture

The case can be well suited by the Steiner problem, as we will see in next section.

The rest of the paper deals with the problem formulation in section 2. Section 3 presents an evolutionary algorithm to solve the problem adaptively. Section 4 evaluates the utility of the approach into a problem library and shows the results for a real case corresponding to a municipality in the south of Spain. Finally section 5 summarizes the main results and conclusions.

2. Problem Formulation

The problem we have to deal with try to minimize the HFC network deployment cost in a small urban area. We suppose a set of nodes representing the headend and ONUs as known data. The problem of headend and ONUs location has been extensively dealt in the bibliography, examples are [1] and [2]. The potential network where the fibers will be placed is represented by a graph constructed from the urban street network

This situation is depicted by the next figure 2, where headend and ONUs correspond to terminal nodes and the rest of nodes (simple intersections in the urban street network) correspond to Steiner nodes attending to the Steiner problem nomenclature. The arcs represent feasible options where a conduit could be placed.

The fiber optic placement over the physical network gives place to the logical network displayed in figure 2, in which each of the own fiber optic cables, communicating the headend with the ONU, is laid inside the shared conduits.

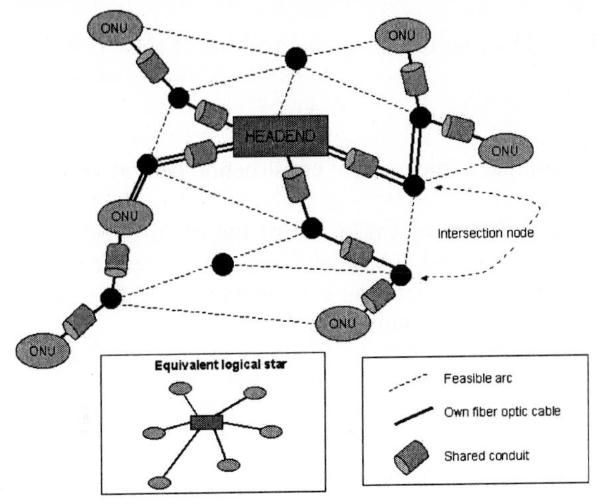

Fig. 2. Logical star network

As the conduit construction is the most relevant part of the telecommunication network setting up, costs can be directly approached by means of the arc lengths. As we have noted, this problem is stated as the Steiner problem in networks that can be summarized as [3]:

- Given: A non-directed graph $G = (N,A)$ with $|N|$ nodes and $|A|$ arcs with costs c_{ij}, for all $(i,j)\in A$, and a subset $I\subseteq N$ with $|I|$ nodes called terminals (in our case headend and ONUs). The rest of the nodes are called Steiner nodes.
- Find: A network $G_I\subseteq G$ joining all the terminal nodes in I at minimum cost. This network can include some of the Steiner nodes but has not to include all the Steiner nodes.

The network G_I will link the entire ONUs with the headend while the arcs will represent the facilities to place the conduits and lay the cables.

To formulate the problem we make use of the following notation:

Parameters:

- N set of nodes in the graph
- A set of arcs in the graph. We will separate the set A into the subsets A^+ for increasing arcs, i.e., if $i<j \Rightarrow (i,j)\subset A^+$, and A^- for decreasing arcs, i.e., if $i>j \Rightarrow (i,j)\subset A^-$.
- I set of terminal nodes in the network, i.e., the headend and ONUs. So $|I|$ represents the number of terminals
- J set of Steiner nodes in the network. So, $|J|$ represents the number of Steiner nodes.

Variables:

• y_{ij} binary variable taking value equal to one if there exits flow through the arc *(i,j)*, zero otherwise.
• x_{ij} flow variable, representing the amount of flow through the arc *(i,j)* in the direction $i \rightarrow j$.
• x_{ie} flow variable, representing the amount of flow between the node *i* and the headend, *e*.

Data:
• c_{ij} fixed cost due to the conduit construction between nodes *i* and *j*.

To model the problem we make use of the multi-injection formulation [4]. This formulation can be analyzed as a weak form of the Steiner problem as different from formulation in [5] that represents an equivalent but stronger form more useful when a branching method have to be implemented.

Model:

$$Z = MIN \sum_{(i,j)\in A^+} c_{ij} y_{ij}$$

$st:$

$$\sum_{j\in I(i)} x_{ij} - \sum_{j\in I(i)} x_{ji} = \begin{cases} 1 & \forall i \in I \\ 0 & \forall i \in J \\ -|I| & i = e \end{cases} \qquad (1)$$

$$|I| \cdot y_{ij} \geq x_{ij} + x_{ji} \qquad \forall (i,j) \in A^+ \qquad (2)$$

$$x_{ij} \geq 0 \qquad \forall (i,j) \in A$$

$$y_{ij} = \{0,1\} \qquad \forall (i,j) \in A^+$$

Constraint (1) is the flow balance equation. In the upper part, one unit of flow is sent from each of the ONUs, $i \in I$. In the medium part shows how the inflow equates the outflow at each node Steiner of the network. Finally, the total amount of flow homing into the headend *e* must be equal to the total amount of flow injected into the network, i.e., the number of ONUs. In the summations, the set *I(k)* indicates all the incident nodes into the node *i*. Constraint (2) is used to set to one the value of the binary variable y_{ij} when there exists positive flow through the arc *(i,j)*.

The model uses the x_{ij} variables to obtain a completely connected graph and the y_{ij} variables to evaluate the cost function. The dimension of the final model are $|N|+\frac{1}{2}|A|$ constraints, $\frac{1}{2}|A|$ integer variables and $|A|$ continuous variables.

3. Evolutionary Algorithm

One of the most popular approaches to deal with the Steiner problem is the *tree heuristic* known to have a worst-case analysis [6]. This approach involves an error not upper than 100%. This is one of the better known bounded heuristic approaches.

We present here an evolutionary algorithm that makes use of the tree heuristic to present a new algorithm that beats it, guaranteeing the worst case deviation and capable of adapting to the any kind of urban network. This algorithm constructs sub-networks by means of neighborhoods of size/radius α containing, at least, one terminal node. Applying the tree heuristic internally inside the sub-network, a feasible solution is obtained for the local Steiner problem. After that, the method is applied, again, among the different neighborhoods, shaping a Steiner super-tree. The neighborhood size term, characterized in the parameter α, turns to critical when an universal application is intended, and when managing with bounded computation times. Here the evolutionary algorithm fits the α parameter allowing the convergence to a good solution.

This new approach makes use of three clear ideas. First, when the number of terminals and the size of the problem are reduced, the local tree heuristic solution will be nearer to the local optimum solution (the heuristic works better with small problems). Second, when the separate sets have to be linked across one only arc, the solution so obtained will express better one of the principal Steiner ideas, a big flow over one arc cost the same that a small flow over the same arc. Finally, as the sum of minimums is always lower or equal than the minimum of the sum, and if the minimums are feasible the sum of minimums is feasible too, then it can be concluded that the worst-case deviation is maintained.

In this context we set as the population individuals any feasible •-value charactering a neighborhood. So the genetic encoding is defined by each •-value implying a different Steiner problem solution. The characteristics for the evolutionary algorithm are as follows.

- *Uniform crossover operators.* Calculated as the arithmetic average between each pair of •-parameters.
- *Mutation operator.* A mutation process will modify one of the •-parameters in the population by setting it to random value between the maximum arc length and the minimum arc length.
- *Random parents selection.* All the individuals from the size *N* population are randomly selected. This will enrich the population genetic variety.
- *Ranking based replacement.* We propose the use of a hypergeometric function to let more probability of replacement to the individuals with worse fitness and less probability of replacement to the individuals with better fitness. So, the individual in ranking position-*i*, have a replacement probability given by $q(1-q)^i$, being *q* the replacement probability of the worst individual.
- *Based on population entropy stop criterion.* We follow [7] as a good procedure to control the convergence in genetic algorithms. The maximum entropy limitation is given by: $S = \sum_{i=1}^{N} \alpha_i \cdot \ln(\alpha_i)$

4. Results

All the tests have been run on a 200 MHz PC Pentium MMX workstation. The OR-library available in [8] was used to test the heuristics computationally. Our algorithm was put in competition with the tree heuristic and was compared with the available optimum values. Table 1 summarizes the results as well as the error percentage with respect to the optimum solution. The referred parameters in the table (nodes, terminals and arcs) are relative to the results after preprocessing the Winter rules [3], so nodes and arcs are specified with respect to the reduced graph corresponding to each ID problem. Keys *tree* and *eval* have been used to refer the *tree heuristic* and the *evolutionary algorithm* respectively.

Table 1. Problem library computational results

ID	Nodes	% terms	arcs	Optimum	tree	% error	eval	α	% error
Steinb1	13	61,5	19	82	82	0,0	82	2-12	0,0
Steinb2	15	73,3	21	83	85	2,4	89	8	7,2
Steinb3	20	75,0	25	138	138	0,0	138	1-8	0,0
Steinb4	40	22,5	80	59	62	5,1	62	1-3	5,1
Steinb5	39	30,8	80	61	61	0,0	62	8-10	1,6
Steinb6	45	55,6	87	122	126	3,3	124	4-6	1,6
Steinb7	22	50,0	33	111	112	0,9	112	4	0,9
Steinb8	26	57,7	38	104	105	1,0	105	4	1,0
Steinb9	27	85,2	35	220	221	0,5	220	2-4	0,0
Steinb1	55	23,6	121	86	91	5,8	90	4	4,7
Steinb1	63	30,2	129	88	90	2,3	90	1-10	2,3
Steinb1	63	57,1	125	174	174	0,0	174	5-10	0,0
Steinb1	36	38,9	56	165	172	4,2	172	5-14	4,2
Steinb1	42	50,0	65	235	238	1,3	238	2	1,3
Steinb1	47	80,9	67	318	321	0,9	321	5-12	0,9
Steinb1	77	22,1	166	127	137	7,9	137	1-2:5-10	7,9
Steinb1	74	31,1	153	131	133	1,5	133	2	1,5
Steinb1	82	54,9	165	218	222	1,8	222	5:7-10	1,8
Steinc1	143	3,5	260	85	85	0,0	85	6-9	0,0
Steinc2	128	7,8	234	144	144	0,0	150	8	4,2
Steinc3	178	42,1	295	754	774	2,7	775	4	2,8
Steinc4	193	52,8	314	1079	1096	1,6	1086	7	0,6
Steinc5	223	80,7	341	1579	1582	0,2	1581	8	0,1
Steinc6	366	1,4	837	55	60	9,1	55	4	0,0
Steinc7	383	2,6	866	102	114	11,8	103	7:10	1,0
Steinc8	387	20,4	867	509	532	4,5	529	8	3,9
Steinc9	418	29,7	903	707	728	3,0	724	7	2,4
Steinc1	427	56,7	891	1093	1116	2,1	1114	8-10	1,9
Steinc1	499	1,0	2005	32	37	15,6	33	7-8	3,1
Steinc1	499	2,0	2065	46	48	4,3	46	2:6-8	0,0
Steinc1	498	16,7	2026	258	273	5,8	277	3	7,4
Steinc1	499	25,1	1968	323	342	5,9	339	8	5,0
Steinc1	500	50,0	1814	556	572	2,9	565	3	1,6
Steinc1	500	1,0	3517	11	13	18,2	12	2-4	9,1
Steinc1	500	2,0	3463	18	20	11,1	20	1-2	11,1
Steinc1	500	16,6	3495	113	126	11,5	126	1	11,5
Steinc1	500	25,0	3349	146	159	8,9	159	2-4	8,9
Steinc2	500	50,0	3099	267	268	0,4	269	1:4	0,7
Steind1	272	1,8	504	106	107	0,9	107	5-15	0,9
Steind2	283	3,5	519	220	235	6,8	228	1-3:8-	3,6
Steind3	350	42,3	585	1565	1612	3,0	1600	10-14	2,2
Steind4	359	57,7	590	1935	1970	1,8	1969	10-14	1,8
Steind5	470	80,2	708	3250	3269	0,6	3268	4	0,6
Steind6	759	0,7	1730	67	74	10,4	71	9-10	6,0
Steind7	749	1,3	1722	103	105	1,9	103	5	0,0
Steind8	802	20,7	1778	1072	1144	6,7	1140	5-6	6,3
Steind9	802	30,7	1769	1448	1534	5,9	1528	2	5,5
Steind1	836	58,0	1781	2110	2165	2,6	2163	1	2,5
Steind1	993	0,5	4442	29	31	6,9	29	6-10	0,0
Steind1	1000	1,0	4437	42	52	23,8	43	8-10	2,4
Steind1	998	16,7	4354	500	554	10,8	549	2	9,8
Steind1	998	25,1	4309	667	723	8,4	723	1	8,4
Steind1	996	50,0	3916	1116	1150	3,0	1152	4-5	3,2
Steind1	1000	0,5	8048	13	15	15,4	13	2-4	0,0
Steind1	1000	1,0	8061	23	30	30,4	25	2	8,7
Steind1	1000	16,7	7755	223	257	15,2	254	2-3	13,9
Steind1	1000	25,0	7552	310	348	12,3	347	2-4	11,9
Steind2	1000	50,0	6887	537	546	1,7	545	1-4	1,5

We have to note at the respect of the results the following considerations in relation with the problem of designing HFC telecommunication networks. Firstly, errors low than 5% have to be considered as successes when we are dealing with conduit installation costs, because of upper deviations in real time civil works will certainly take place. Secondly, for small urban areas graphs upper than 1,000 nodes will not be common, so we can consider the library quite representative in relation with our problem

The average error was estimated in 5.6 % for *tree* and 3.6 % for *eval*. *Eval* reached a total of 10 optimums (42 successes) for 6 optimums (34 successes) in *tree* case. The maximum error was calculated in 30.4% for *tree* and 13.9% for *eval*.

We have evaluated our algorithm in real life scenarios also. Here is represented a case study in which the model and algorithm are applied to the Montequinto area. Montequinto is a residential urban area sited near the metropolitan area of Seville, the largest city in the south of Spain, where advanced telecommunication services will probably be welcomed. Montequinto has a population around 25,000 inhabitants.

Figure 3 shows the HFC network solution for the Montequinto area. As primary design decision a simple star architecture was selected to distribute telecommunication services. The figure reveals the conduit physical network, and its equivalent logical star representation.

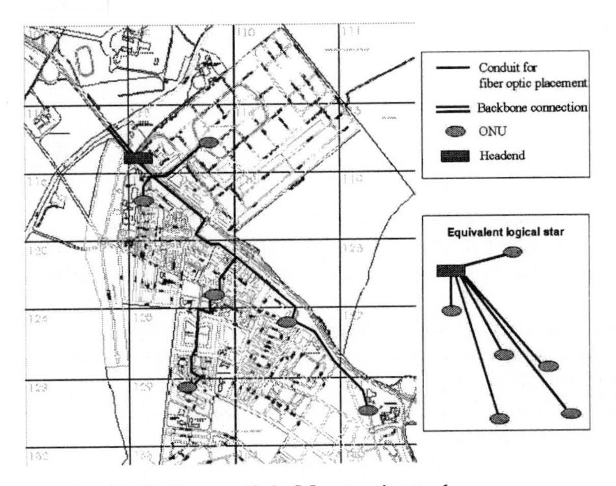

Fig. 3. HFC network in Montequinto urban area

The figure depicts the tree topology corresponding to the conduit network. This tree topology corresponds to a logical simple star when we represent the fiber network. In this small urban area context, a sub-headend, smaller than typical headends and dependent on the Seville headend, has been located near the motorway to provide easy connection between the trunk network and the inter-urban backbone. To distribute the telecommunication services six ONUs are displayed over the municipality. The connection among the sub-headend and the ONUs depicts the optical trunk-distribution network.

5. Conclusions

Most of the bibliography based on distributing telecommunication networks attends to large urban areas or inter-municipality communication backbones. Here, we deal with a small urban area where network deployment has to be not too large trying to maximize the profitability investment. In this context simple star architectures are indicated to broach the situation. Steiner problem is accurate to model the trunk and optical distribution network deployment. A multi-injection formulation is introduced to model the problem as a single commodity flow problem instead of a multicommodity flow problem, reducing the number of constraints and variables. An evolutionary algorithm has been presented showing a good behavior beating traditional generally accepted approaches. The evolutionary algorithm is used to set the neighborhood parameter allowing a universal and adaptive application. Finally, we test the model and algorithm within problem library and illustrate a real life application by means of a case study relative to a residential urban area near to Seville.

References

1. Belvaux, G., Boissin, N., Sutter, A. and Wolsey L.A.: Optimal placement of add/drop multiplexers: static and dynamic models. Eur J Opl Res 108 (1998) 26-35
2. Sutter, A., Vanderbeck, F. and Wolsey, L.A.: Optimal placement of add/drop multiplexers: heuristic and exact algorithms. Opns Res 46 (1998) 719-728
3. Winter, P.: Steiner problem in networks. Networks 17 (1987) 129-167
4. Cortes, P., Larrañeta, J., Onieva, L. and Garcia, J.M.: Multi-injection model to solve the Steiner problem. In: Sforza, A. (ed.): Simulation and Optimisation in Operations Management. Edizioni Scientifiche Italiane, Napoli (1999) 83-84
5. Wong, R.T.: A dual ascent approach for Steiner tree problems on a directed graph. Math Program 28 (1984) 271-287
6. Magnanti, T.L., Wolsey, L.A.: Optimal trees. In: Ball, M.O. et al (eds.): Network Models. North-Holland, Elsevier (1995) 503-615
7. Grefenstette, J.J.: Incorporating Problem-Specific Knowledge into Genetic Algorithms. In Davis, L. (ed.): Genetic Algorithms and their Applications. Morgan Kaufmann, Los Angeles (1987)
8. Beasley, J.E.: OR-Library. http://mscmga.ms.ic.ac.uk/info.html (2001).

Channel Assignment for Mobile Communications Using Stochastic Chaotic Simulated Annealing

Sa Li[1] and Lipo Wang[2]

[1] School of Electrical and Electronic Engineering, Nanyang Technological University, Nanyang Avenue, Singapore 639798
p149552752@ntu.edu.sg

[2] School of Electrical and Electronic Engineering, Nanyang Technological University, Nanyang Avenue, Singapore 639798
elpwang@ntu.edu.sg
http://www.ntu.edu.sg/home/elpwang

Abstract. The channel assignment problem (CAP), the task to assign the required number of channels to each radio cell in such a way that interference is precluded and the frequency spectrum is used efficiently, is known to be an NP-complete optimization problem. In this paper, we solve CAP using a stochastic chaotic neural network that we proposed recently. The performance of stochastic chaotic simulated annealing (SCSA) is compared with other algorithms in several benchmark CAPs. Simulation results showed that this approach is able to further improve on results obtained by other algorithms.

1 Introduction

In recent years, there is a steady increasing demand for cellular mobile telephone communication systems. But the usable frequency spectrum is limited. Thus optimal frequency channel assignment is becoming more and more important and it can greatly enhance the traffic capacity of a cellular system.

Gamst and Rave defined a general form of channel assignment problems in an arbitrary inhomogeneous cellular radio network [1]: minimize the span of channels subject to demand and interference-free constraints. The CAP is usually solved by graph coloring algorithms [3]. Kunz used the Hopfield neural network for solving the CAP [2]. The algorithm minimizes an energy or cost function representing interference constraints and channel demand. Since then, various techniques have been explored for the application of neural networks to CAP. Funabiki used a parallel algorithm which does not require a rigorous synchronization procedure [10]. Chan et al proposed an approach based on cascaded multilayered feedforward neural networks which showed good performance in dynamic CAP [11]. Kim et al proposed a modified Hopfield network without fixed frequencies [12]. Smith and Palaniswami reformulated the CAP as a generalized quadratic assignment problem [8]. They then found remarkably good

J. Mira and A. Prieto (Eds.): IWANN 2001, LNCS 2084, pp. 757-764, 2001.

solutions to CAPs using simulated annealing, a modified Hopfield neural network, and a self-organizing neural network. Potential interferences considered in their paper come from the cochannel constraint (CCC), the adjacent channel constraint (ACC), and the co-site constraint (CSC) [1].

Simulated annealing is a general method for obtaining approximate solutions of combinatorial optimization problems [6]. Since the optimization processing is undertaken in a stochastic manner, it is also called stochastic simulated annealing (SSA). Convergence to globally optimal solutions is guaranteed if the cooling schedule is sufficiently slow, i.e., no faster than logarithmically [7]. In recent years, a large amount of work has been done in chaotic simulated annealing (CSA) ([12]-[21]). CSA can search efficiently because of its reduced search spaces. Chen and Aihara proposed a transiently chaotic neural networks (TCNN) which adds a large negative self-coupling with slow damping in the Euler approximation of the continuous Hopfield neural network so that neurodynamics eventually converge from strange attractors to an equilibrium point [13]. The TCNN showed good performance in solving traveling salesman problem (TSP). But CSA is deterministic and is not guaranteed to settle down at a global minimum no matter how slowly the annealing takes place. In [15], we generalized CSA by adding a decreasing random noise into the neuron inputs in the TCNN to combine the best feasures of both SSA and CSA, thereby abtaining stochastic chaotic simulated annealing (SCSA). In this paper, we show that SCSA can lead to further improvements on solutions for CAP.

This paper is organized as followings. Section 2 reviews the static channel assignment problem and its mathemmatical formulation as given in [8]. In section 3, the SCSA is reviewed. In the section 4, we apply SCSA to several benchmarking CAPs. Finally in section 5, we conclude this paper.

2 Static Channel Assignment Problem

In this section, we review the formulation of CAP as given by Smith and Palaniswami [8]. Suppose the number of available channels in an N-cell static mobile communication system is M and the number of channels requiremed in cell i is D_i. The minimum distances in the frequency domain by which two channels must be separated in order to guarantee an acceptably low signal/interference ratio in each region are stored in the compatibility matrix $C = \{C_{ij}\}$, an $n \times n$ symmetric matrix, where n is the number of cells in the mobile networks and C_{ij} is the minimum frequency separation between cell i and one in cell j. In most of cases, the number of available channels is less than the lower bound, i.e., the minimum number of channels required for an interference-free assignment. A useful version of the CAP is defined as follows [8]:

minimize severity of interferences
subject to demand constraints

The output of neuron j, k is:

$$X_{j,k} = \begin{cases} 1, \text{ if cell j is assigned to channel } k & ; \\ 0, \text{ otherwise} . \end{cases}$$

A cost tensor $P_{j,i,m+1}$ is used to measure the degree of interference between cells j and i caused by such assignments that $X_{j,k} = X_{i,l} = 1$ [8], where $m = |k - l|$ is the distance in the channel domain between channels k and l. The cost tensor P can be calculated recursively as follows:

$$P_{j,i,m+1} = \max\,(0, P_{j,i,m} - 1), \qquad for \;\; m = 1, \cdots, M-1 \;, \tag{1}$$

$$P_{j,i,1} = C_{ji}, \qquad \forall j, i \neq j \;, \tag{2}$$

$$P_{j,j,1} = 0, \qquad \forall j \;. \tag{3}$$

Thus the CAP can be formulated to minimize the total interference of all assignments in the network:
minimize

$$F(X) \;=\; \sum_{j=1}^{N}\sum_{k=1}^{M} X_{j,k} \sum_{i=1}^{N}\sum_{l=1}^{M} P_{j,i,(|k-l|+1)} X_{i,l} \;, \tag{4}$$

subject to

$$\sum_{k=1}^{M} X_{j,k} \;= D_j, \qquad \forall j = 1, \cdots, N \;. \tag{5}$$

where $F(X)$ is the total interference.

3 Stochastic Chaotic Simulated Annealing

The stochastic chaotic simulated annealing (SCSA) is formulated as follows [15]:

$$x_{jk}(t) = \frac{1}{1 + e^{-y_{jk}(t)/\varepsilon}} \;, \tag{6}$$

$$y_{jk}(t+1) = k y_{jk}(t) + \alpha\Big(\sum_{i=1, i\neq j}^{N} \; \sum_{l=1, l\neq k}^{M} w_{jkil} + I_{ij}\Big) - z_{jk}(t)(x_j k(t) - I_0) + n(t) \;, \tag{7}$$

$$z_{jk}(t+1) = (1-\beta) z_{jk}(t) \qquad (i = 1, \cdots, n) \;, \tag{8}$$

$$A[n(t+1)] = (1-\beta) A[n(t)] \;, \tag{9}$$

where

x_{jk} : output of neuron j,k ;
y_{jk} : input of neuron j,k ;
$w_{jkil} = w_{iljk};\; w_{jkjk} = 0;\; \sum_{l=1, l\neq k}^{M} w_{jkil} + I_{ij} = -\partial E / \partial x_{jk}$:
connection weight from neuron j,k to neuron i,l ;
I_{jk} : input bias of neuron j, k ;

k : damping factor of nerve membrane ($0 \leq k \leq 1$);
α : positive scaling parameter for inputs ;
β : damping factor of the time dependent ($0 \leq \beta \leq 1$);
$z_{jk}(t)$: self-feedback connection weight or refractory strength ($z(t) \geq 0$) ;
I_0 : positive parameter;
ε : steepness parameter of the output function ($\varepsilon > 0$) ;
E : energy function;
$n(t)$: random noise injected into the neurons, in [-A, A] with a uniform distribution;
$A[n]$: the noise amplitude.

If n(t)=0 for all t, then the above noisy chaotic neural network reduces to the transiently chaotic neural network of Chen and Aihara [13].

The corresponding energy function E for CAP is given by (1) and (2):

$$E = \frac{W1}{2}\sum_{j=1}^{N}(\sum_{k=1}^{M} X_{jk} - D_j)^2 + \frac{W2}{2}\sum_{j=1}^{N}\sum_{k=1}^{M} X_{jk}\sum_{i=1}^{N}\sum_{l=1}^{M} P_{j,i,(|k-l|+1)}X_{il} \quad (10)$$

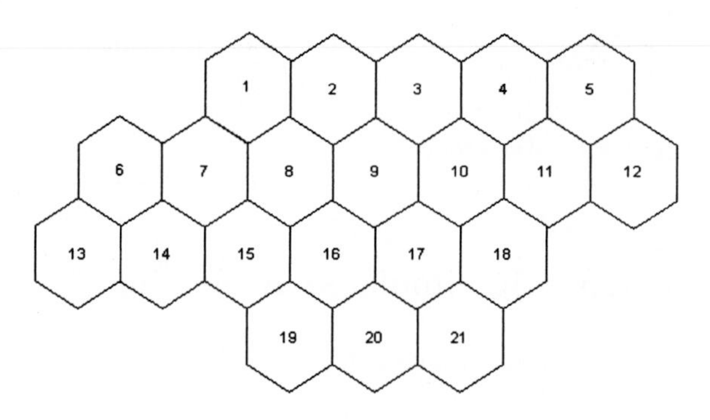

Fig. 1. Layout of a 21-cell hexagonal network

4 Application of SCSA to Benchmark CAPs

4.1 Descriptions of Various Problems

First we use the data set suggested by Sivarajan [3], denoted as EX1. The number of cells is $N = 4$, the number of channels available is $M = 11$, the demand of channels is given by $D^T = (1, 1, 1, 3)$. We also use a slightly larger extension of EX1, denoted as EX2 [8]:

$N = 5, M = 17, D^T = (2, 2, 2, 4, 3)$.

The Second example considered is the 21-cell cellular system (HEX1-HEX4) found in [4] (Fig.1). We used two sets of demands for the 21-cell as follows.

$D_1^T = (2, 6, 2, 2, 2, 4, 4, 13, 19, 7, 4, 4, 7, 4, 9, 14, 7, 2, 2, 4, 2)$;

$D_2^T = (1, 1, 1, 2, 3, 6, 7, 6, 10, 10, 11, 5, 7, 6, 4, 4, 7, 5, 5, 5, 6)$.

The details of HEX1-HEX4 are shown in Table 1 [8]. There are two different compatability matrices for HEX problems by considering the first two rings of cells around a particular cell as interferences [8]. The first one does not include ACC, so the relative off-diagonal terms C_{ij} are 1. The second one includes ACC, the off-diagonal terms of C_{ij} in second matrix are 1 or 2 respectively corresponding to CCC and ACC.

Table 1. Descriptions for hexagonal CAPs

Problem	N	M	D	co-channel	adjacent	C_{ii}
HEX1	21	37	D_1	yes	no	2
HEX2	21	91	D_1	yes	yes	3
HEX3	21	21	D_2	yes	no	2
HEX4	21	56	D_2	yes	yes	3

The final set of CAP is generated from the topographical data of an actual 24×21 km area around Helsinki, Finland [22]. Kunz calculated the traffic demand and interference relationships between the 25 regions around the base stations of this area. The compatability matrix is abtained from Kunz data as C_3 [8]. The demand vector is:

$D_3^T = (10, 11, 9, 5, 9, 4, 5, 7, 4, 8, 8, 9, 10, 7, 7, 6, 4, 5, 5, 7, 6, 4, 5, 7, 5)$.

This benchmarking CAP is divided into four classes by considering only the first 10 regions (KUNZ1), 15 regions (KUNZ2), 20 regions (KUNZ3), and the entire area (KUNZ4) (Table 2) [8].

4.2 The Simulation Results Of SCSA

We chose a set of parameters for each CAP as in Table 3.

Table 2. Descriptions for KUNZ problems

Problem	N	M	C	D
KUNZ1	10	30	$[C_3]_{10}$	$[D_3]_{10}$
KUNZ2	15	44	$[C_3]_{15}$	$[D_3]_{15}$
KUNZ3	20	60	$[C_3]_{20}$	$[D_3]_{20}$
KUNZ4	25	73	C_3	D_3

Our results are presented in Table 4. For comparison, Table 4 also includes results given in [8], i.e., the performances of GAMS/MINOS-5 (labeled GAMS), the traditional heuristics of steepest descent (SD), simulated annealing (SA), the original Hopfield network (HN) (with no hill-climbing), hill-climbing Hopfield

Table 3. The parameters for various CAPs

Problem	K	ε	I_0	α	β	z(0)	n(0)	W1	W2
EX1	0.9	1/250	0.65	0.0045	0.0005	0.1	0.5	1.0	0.02
EX2	0.9	1/250	0.65	0.0045	0.0005	0.1	0.5	1.0	0.02
HEX1	0.9	1/250	0.05	0.05	0.0005	0.08	0.08	1.0	0.25
HEX2	0.9	1/250	0.08	0.05	0.0005	0.08	0.08	1.0	0.25
HEX3	0.9	1/250	0.05	0.05	0.0005	0.08	0.08	1.0	0.2
HEX4	0.9	1/250	0.07	0.05	0.0005	0.08	0.08	1.0	0.3
KUNZ1	0.9	1/250	0.06	0.05	0.0004	0.08	0.08	1.0	0.45
KUNZ2	0.9	1/250	0.01	0.05	0.0005	0.08	0.08	1.0	0.45
KUNZ3	0.9	1/250	0.05	0.05	0.0005	0.08	0.08	1.0	0.45
KUNZ4	0.9	1/250	0.05	0.05	0.0005	0.08	0.08	1.0	0.45

network (HCHN), and the self-organizing neural network (SONN). Each of the heuristics is run from ten different random initial conditions. In Table 4, "Min" means the minimum total interference (eq. 4) found during these ten times, and "Av" is the average total interference [8]. We also computed the standard deviations (STDD) during these ten runs, as shown in Table 5.

Table 4. The results of SCSA and other techniques

	GAMS	SD		SA		HN		HCHN		SONN		SCSA	
problem	Min	Av.	Min	Av.	Min	Av.	Min	Av.	Min	Av.	Min	Av.	Min
EX1	2	0.6	0	0.0	0	0.2	0	0.0	0	0.4	0	0.0	0
EX2	3	1.1	0	0.1	0	1.8	0	0.8	0	2.4	0	0.0	0
HEX1	54	56.8	55	50.7	49	49.0	48	48.7	48	53.0	52	47.7	47
HEX2	27	28.9	25	20.4	19	21.2	19	19.8	19	28.5	24	18.5	18
HEX3	89	88.6	84	82.9	79	81.6	79	80.3	78	87.2	84	77.3	76
HEX4	31	28.2	26	21.0	17	21.6	20	18.9	17	29.1	22	17.2	16
KUNZ1	28	24.4	22	21.6	21	22.1	21	21.1	20	22.0	21	20.0	19
KUNZ2	39	38.1	36	33.2	32	32.8	32	31.5	30	33.4	33	30.3	30
KUNZ3	13	17.9	15	13.9	13	13.2	13	13.0	13	14.4	14	13.0	13
KUNZ4	7	5.5	3	1.8	1	0.4	0	0.1	0	2.2	1	0.0	0

The results in Table 4 show that the SCSA is able to further improve on results obtained by other approaches.

5 Conclusion

In this paper, we showed that stochastic chaotic simulated annealing (SCSA) [15] is very effective in solving combinatorial optimization problems, such as channel assignment problems in radio network planning. This approach has both noisy nature and chaotic

Table 5. The standard deviations for various CAPs

Problem	HEX1	HEX2	HEX3	HEX4	KUNZ1	KUNZ2	KUNZ3	KUNZ4
STDD	0.265	0.173	0.267	0.354	2.095	0.173	0.0	0.0

searching characteristics, so it can search in a smaller space and continue to search after the disappearance of chaos. Implementation of SCSA to solve other practical optimization problems will be studied in future work.

References

1. A. Gamst and W. Rave, "On frequency assignment in mobile automatic telephone systems," *Proc. GLOBECOM'82* , pp. 309-315, 1982.
2. D. Kunz, "Channel assignment for cellular radio using neural networks", *IEEE Trans. Veh. Technol.*, vol. 40, no. 1, pp. 188-193, 1991.
3. K. N. Sivarajan, R. J. McEliece and J.W. Ketchum, "Channel assignment in cellular radio", *Proc. 39th IEEE Veh. Technol. Soc. Conf.*, pp. 846-850, May, 1989.
4. A. Gamst, "Some lower bounds for a class of frequency assignment problems", *IEEE Trans. Veh. Tech.*, vol. VT-35, pp. 8-14, Feb., 1986.
5. W. K. Hale, "Frequency assignment: theory and application", *Pro. IEEE*, vol. 68, pp. 1497-1514, Dec., 1980.
6. M. Duque-Anton, D. Kunz, and B. Ruber, "Channel assignment for cellular radio using simulated annealing", *IEEE Trans. Veh. Technol.*, vol. 42, no. 1, February, 1993.
7. S. Geman and D. Geman, "Stochastic relaxation, Gibbs distributions, and the Bayesian restoration of images", *IEEE Trans. Pattern Analysis and Machine Intelligence*, vol. 6, pp. 721-741, 1984.
8. K. Smith and M. Palaniswami, "Static and dynamic channel assignment using neural network", *IEEE Journal on Selected Areas in Communications*, vol. 15, no. 2, Feb., 1997.
9. N. Funabiki and Y. Takefuji, "A neural network parallel algorithm for channel assignment problems in cellular radio networks", *IEEE Trans. Veh. Technol.*, vol. 41, no. 4, November, 1992.
10. P. Chan, M. Palaniswami and D. Everitt, "Neural network-based dynamic channel assignment for cellular mobile communication systems", *IEEE Trans. Veh. Technol.*, vol. 43, no. 2, May, 1994.
11. J. Kim, S. H. Park, P. W. Dowd and N. M. Nasrabadi, " Cellular radio channel assignment using a modified Hopfield network", *IEEE Trans. Veh. Technol.*, vol. 46, no. 4, November, 1997.
12. L. Chen and K. Aihara, "Chaotic simulated annealing by a neural network model with transient chaos", *Neural Networks*, vol. 8, no. 6, pp. 915-930, 1995.
13. L. Chen and K. Aihara, "Transient chaotic neural networks and chaotic simulated annealing", in M. Yamguti(ed.), *Towards the Harnessing of Chaos.* Amsterdam, Elsevier Science Publishers B.V. pp. 347-352, 1994.
14. L. Chen and K. Aihara, "Global searching ability of chaotic neural networks", *IEEE Trans. Circuits and Systems-I: Fundamental Theory and Applications*, vol. 46, no. 8, August, pp. 974-993, 1999.

15. L. Wang and F. Tian, "Noisy chaotic neural networks for solving combinatorial optimization problems", *Proc. International Joint Conference on Neural Networks.* (IJCNN 2000, Como, Italy, July 24-27, 2000)
16. F. Tian, L. Wang, and X. Fu, "Solving channel assignment problems for cellular radio networks using transiently chaotic neural networks", *Proc. International Conference on Automation, Robotics, and Computer Vision.* (ICARCV 2000, Singapore)
17. L. Wang, "Oscillatory and chaotic dynamics in neural networks under varying operating conditions", *IEEE Transactions on Neural Networks,* vol. 7, no. 6, pp. 1382-1388, 1996.
18. L. Wang and K. Smith, "On chaotic simulated annealing", *IEEE Transactions on Neural Networks,* vol. 9, no. 4, pp. 716-718, 1998.
19. L. Wang and K. Smith, "Chaos in the discretized analog Hopfield neural network and potential applications to optimization", *Proc. International Joint Conference on Neural Networks,* vol. 2, pp. 1679-1684, 1998.
20. T. Kwok, K. Smith and L. Wang, "Solving combinatorial optimization problems by chaotic neural networks", C. Dagli et al.(eds) *Intelligent Engineering Systems through Artificial Neural Networks* vol. 8, pp. 317-322, 1998.
21. T. Kwok, K. Smith and L. Wang, "Incorporating chaos into the Hopfield neural network for combinatorial optimization", *Proc. 1998 World Multiconference on Systemics, Cybernetics and Informatics*, N. Callaos, O. Omolayole, and L. Wang, (eds.) vol. 1, pp. 646-651, 1998.
22. T. Kohonen, "Self-organized formation of topologically correct feature maps", *Biol. Cybern.*, vol. 43, pp. 59-69, 1982.

Seeing is Believing: Depictive Neuromodelling of Visual Awareness

Igor Aleksander, Helen Morton, and Barry Dunmall

Intelligent and Interactive Systems Group, Department of Electrical and Electronic Engineering, Imperial College of Science, Technology and Medicine, London SW7, 2 AZ
{I.Aleksander, Helen.Morton, B.Dunmall}@ic.ac.uk

Abstract. The object of seeing is for the brain to create inner states that accurately model the world and recall it for purposeful use. In this descriptive paper we present virtual neuro-architectures called 'depictive' which have been developed to create hypotheses for the mechanisms necessary for such depiction and explain some elements of verbally induced visual working memory. Early work on applications to understanding visual deficits in Parkinsons' sufferers is included.

1 Introduction

This work discusses the development of hypotheses regarding the generation of visual awareness in the modular architectures found in the living brain. These architectures are modelled using digital neuromodels [1]. The accent is not on the neural modules but the behaviour emerging from the interaction of such modules. As an example we work with the perception and recall of both broad shape of a pattern and the detail which makes it up. This is best illustrated by fig. 1.

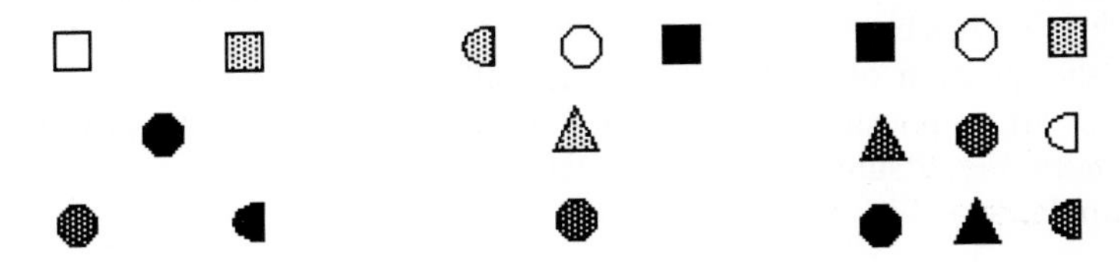

Fig. 1. From left to right these **objects** are called 'cross', 'tee' and 'square'. The detailed **shapes** from which the object is composed have both shape and colour.

These are typical figures that are used in visual working memory experiments with human beings [2]. The subject is allowed to observe one or more objects, and is then questioned about the detail in some position within the object (e.g. "What is the top left shape of the cross?") . In this paper we are not concerned with the prowess of individuals in executing the task, we enquire into the possible neurological mechanisms which allow the observer to have such an ability at all. We develop a hypothesis about this operation that involves eye movement and attention. This is

J. Mira and A. Prieto (Eds.): IWANN 2001, LNCS 2084, pp. 765–771, 2001.

based on what is known about the neuroscience of the system. An implementation of the proposed scheme is tested through neuromodelling and it is shown that the hypothesis is supported. The key questions that this procedure seeks to answer are, how does the system go from a retinotopic image to a coherent sensation of **both** object **and** the detailed makeup by its shapes? How is the process of recall organised at the two levels of object and shape? How does attention operate in visual working memory.

2 Neuroscientific Highlights of Vision Mechanisms

The features that can be extracted from the literature on the neuroscience of vision and that form the elements of this work feature in fig. 2 and are the following.

- The prime anatomical area that causes eye movement whether saccadic or smooth is the superior colliculus. ([3] is a good reference for pursuing this material).
- The superior colliculus receives direct input from the retina which is capable of causing movement to direct the fovea to areas of high contrast in the in the retinotopic image (including the perifoveal fields).
- The image pathways diverge into specialisms through primary areasV1 into V2, V3 etc..
- Importantly, in the parietal cortex, the prefrontal and the frontal cortex (broadly dubbed 'the extrastriate cortex' following the practice in [5]), there are projections from the motor activity which includes that caused by the superior colliculus.
- A most important piece of evidence is the discovery of cells that, on receiving positional signals originating from the muscular output, only fire for specific positions of the eyes. These are called 'gaze-locked' and their presence in areas such as V3 has been known for over ten years [4].
- On the question of memory, it is generally proposed that prefrontal structures which are responsible for perception are also involved in the process of visual memory. Fig. 2 summarises this structure and indicates what needs to be built in a simulation.

Pathway X is an important feature of the hypotheses set out in this paper as it serves several purposes. The salient among these is that higher visual awareness, postulated to take place in the extrastriate areas [5] (as the main 'purpose' of these areas) also influences the superior colliculus. For example, this is thought to be at work in context-dependent tasks such as face exploration or visual planning as may be involved in solving stacking problems "in one's head" [6]. The further postulated function of this pathway is discussed in part 4 of this paper.

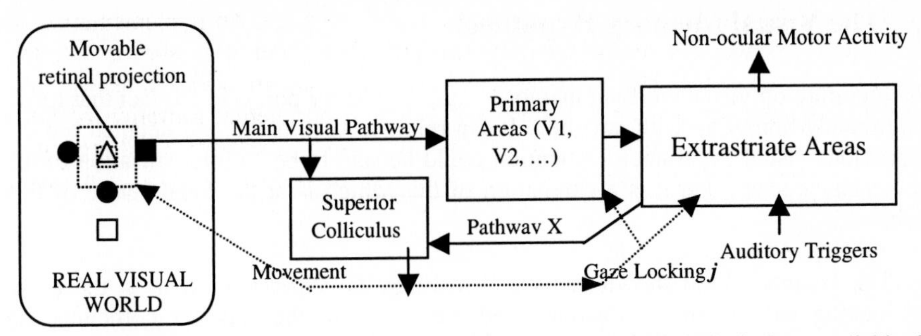

Fig. 2. A monocular diagram that summarises the neurophysiological data available for neuromodelling

3. The Depictive Hypothesis

The material of this section is an elaboration of a previously enunciated hypothesis about the asynchrony of visual awareness [7].

- Any sensory activity which has the potential of being reported (whether the agent can find the actual verbal expression or not) must be supported by rich neural activity which is at least on a one-one basis with world events that the organism is aware of. We call such neural support a *depiction*. (See Appendix A for a brief elaboration of this argument)
- Such depictions can only occur where there are gaze-locked cells as other areas cannot be compensated for eye and other (e.g., head,) movements. We call the vector of all gaze-locking information 'referent ***j***'.
- Neurons in areas receiving ***j*** need not be physically adjacent in order to achieve binding. For example a shape sensitive neuron and a colour sensitive neuron will encode information about the same element of the external visual world if they are indexed by the same values of ***j***. We note that this is a a resolution of the extreme postions on binding existing between the binding theory of Crick and Koch [5] which espouses the existence of tuned 40 Hz signals between binding neurons, and the hypothesis by Zeki and Bartels [8] that there is no need for binding and that the elements of a visual sensation arise asynchronously (as microconsciousnesses).
- We note that the area of depiction can be defined as the total area of visual awareness available to the observer at any point in time. We further note that, in reality, this involves not only eye and head movement but bodily movement as well.

In this paper we consider eye movement alone to establish how this contributes both to depiction and visual working memory. This opens the way to the study of other motor activity in a similar way.

4. The Visual Memory Hypothesis

The literature on visual working memory (e.g.[2]) clearly points to the fact that recall is intimately bound up with attention. It can relate to both broad and detailed reports of previous visual experience. Attention could be said to be an internalisation of the movement process, and it is a simulation of this which is at the focal centre of this paper.

- The feedback loop shown in fig.2 involving the superior colliculus, the gaze-locking path to the extrastriate, and the return to the superior colliculus via pathway X forms a learning state machine which allows for the inward exploration of a visual scene .
- Recall of both detail and shape of images such as those shown in fig.1 is achieved by the retention of the gaze locking signals as a subset of the state variables of this state machine.
- In the absence of visual input the state machine becomes autonomous and generates gaze-locking signals which now become described as inner attentional control patterns.
- Both foveal and perifoveal information is therefore retained. The state machine when triggered by auditory input (e.g. 'imagine the tee shape') is able to imagine the overall perifoveal shape, and highlight the detail of it through inner gaze locking to the extent that the detail can be decoded and output as a voice signal.

5. The Experimental System

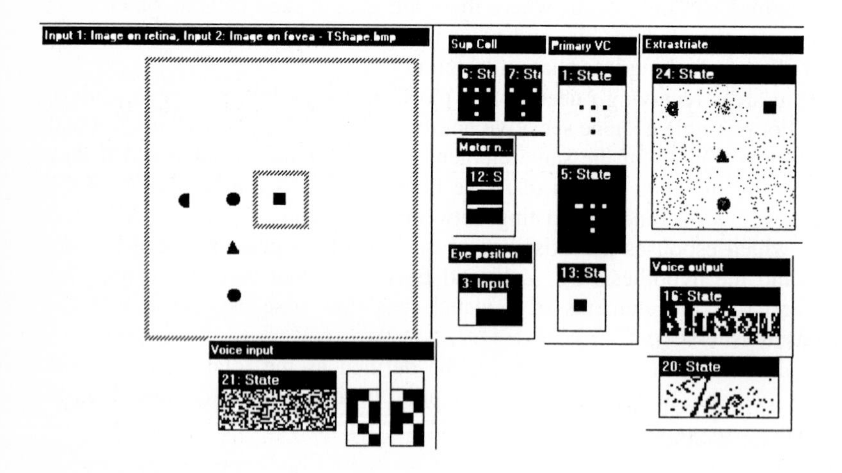

Fig. 3. The NRM virtual vision system during perception (only 14 of the 24 areas are shown for clarity). **Every dot is the output of a neuron.** Input 1 is the visual world containing the foveal and perifoveal areas, 6 and 7 are parts of the superior colliculus, while 1,5,and 13 are the primary areas. 24 is the crucial extrastriate 'awareness' area.

Without going into detail, a simulation comprising 24 neural areas and a total of approximately 360,000 neurons has been set up using the Neural Representation Modeller (www.sonnet.co.uk/nts). Only 14 of these areas are shown in figure 3. Here the superior colliculus, under the influence of perifoveal inputs is controlling the movement of the 'eye' in order to highlight and identify verbally both the overall shape (20) and the foveal detail (16).

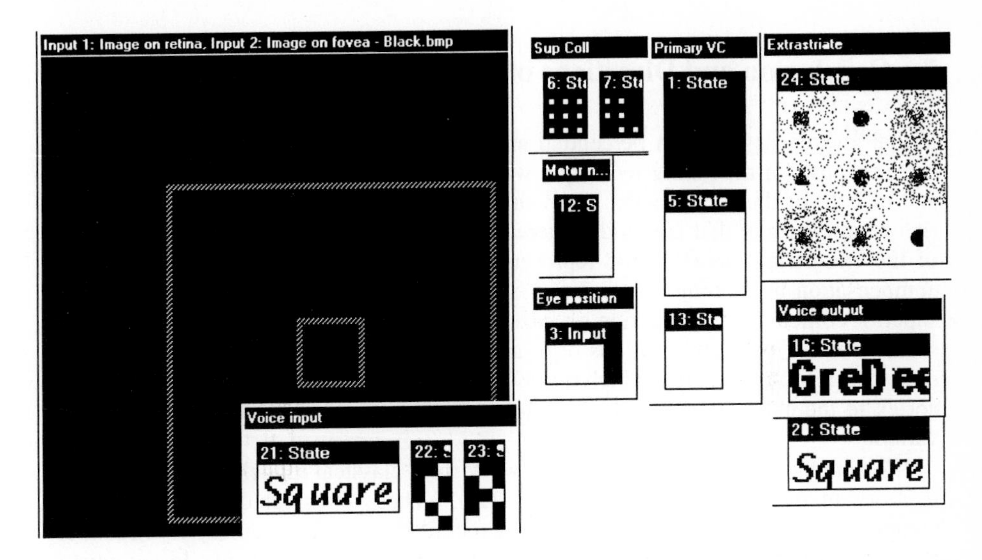

Fig. 4. This shows the NRM system in visual memory mode.

Fig. 4 shows the behaviour of the system where there is no visual input (blindfold or eyes closed) but it is asked to imagine the 'square'(at 21). There is no activity in the primary areas but the superior colliculus activity is sustained in the loop by pathway X. 24 shows that there is a vague awareness of the whole shape, but attention as mediated by the superior colliculus and gaze locking enables the detail (bottom right of 24) to be described in 16 (this appears as 'Green Dee' as the experiment is being done in colour.

6. Application: Visuo-Cognitive Deficits in Parkinson's Disease

It has been shown through measurements of eye movements that there is a distinct difference between unaffected subjects and PD affected ones in the strategic use of such movements in the solving of visual problems [6]. Unaffected subjects move their eyes to imaginary targets where objects have to be 'parked' temporarily in order to find a suitable sequence of moves that solves the problem. It is well known that PD is the result of dopamine deficit in the substantia nigra, which directly affects the basal ganglia which, in turn, have strong inhibitory synaptic connections to the superior colliculus. It is also known that hallucinations are not uncommon in PD sufferers. This suggests that the feedback loop discussed in this paper may be prone to falling into hallucinatory state-space minima and be less responsive to retinal signals as a result the reduced inhibition from the basal ganglia in the superior colliculus. We are initiating discussions with PD sufferers to obtain an understanding of their visualization difficulties in solving visual puzzles, and using the model discussed in this paper to attempt to predict (in area 24 of the model) what distortions PD-afflicted people might suffer.

7. Conclusion and Directions of This Work

Visual awareness has both perceptual and imagining components. In this paper we have proposed a modelling technique which addresses hypotheses about the sources and of both of these elements. The work is based on a depictive hypothesis of very rich neural activity that is world-centered, making perception (neural activity in part of the extrastriate areas) world-representative depending on eye movement and its compensation in creating the depiction. Central to this operation is the action of the superior colliculus and its action on gaze-locked cells that ensure awareness in extra-striate areas of the cortex.. It has been argued that visual working memory is due to sustained feedback activity which involves the superior colliculus and the extrastriate cortex to the extent that gaze-locking information becomes encoded in the feedback variables.

We note the distinct departure in this work from classical models of visual working memory in the cognitive sciences (e.g. [2]). These classical descriptions invoke computational terms such as 'central executive', 'buffer memory', 'circulatory registers', 'attention window' and so on. Of course such components do not exist in this computational form in the brain, they are merely descriptions of functions or algorithms that a computer would need to have visual working memory. The process of depictive neuromodelling creates models which are based on existing elements of neural processes such as the superior colliculus, and the other areas involved in the specific model discussed here. These have emergent stability and they contribute to explanations of real functions within the brain. This is particularly important when creating hypotheses that attempt to achieve a better understanding of departures from the normal, as was suggested above in the case of Parkinson's disease.

Much work remains to be done on visual working memory, in particular when more complex verbal descriptions of events are involved. This is the subject of current work.

Acknowledgements

We wish to thank the Wellcome Trust for granting us a Wellcome Showcase Award for the development of this work in the area of investigating mental deficits. The invaluable contribution of our collaborators in the Neurosciences division of the Imperial College School of Medicine, namely Chris Kennard and Tim Hodgson, is much appreciated.

Appendix A

Depiction and coherence: To support the hypothesis of depiction, we define an elemental event of visual awareness as the change in the visual world with the smallest geometrical dimensions that the observer can report. A tiny fly landing on a wall may be an example of such an event. Were the fly smaller it would not be seen. This change must cause an elemental change in the activity of at least one neuron otherwise, without which, it could not be seen. The same can be said for a second elemental event in the visual world an so on by induction leading to the conclusion that neural events stand in at least one-one relation to visual events in the external world. The implication of this is the support for many physiological findings that accurate depiction occurs in the extrastriate parts of the brain. Note that *depiction* does not imply 'pictures in the head' that could be discovered by a super-accurate brain scanner. The elemental events could be arbitrarily distributed in a geometrical sense. It is also known that the same event is broken down in V1 and partly depicted in deeper areas of the cortex, such as V3, V4 .. etc.. Therefore the 'at least one-one' description is stressed. The coherence of the experience is guaranteed by the hypothesized referent j .

References

1. Aleksander, I., Dunmall, B., Del Frate, V., Neurocomputational Models of Visualisation: A preliminary report, Proc. IWANN 99, Alicante – Spain (1999)
2. Logie, R. H., Visuo-spatial Processing in Working Memory. Quarterly Journal of Experimental Psychology, **38A** (1986) 229-247
3. Kandel, E. R., Schwartz, J. H. and Jessel, T. M., Principles of Neural Science, Fourth Edition, McGraw Hill, (2000).
4. Galletti, C. and Battaglini, P. P., Gaze-Dependent Visual Neurons in Area V3A of Monkey Prestriate Cortex. Journal of Neuroscience, **6**, (1989), 1112-1125
5. Crick, F., and Koch, C., Are we Aware of Neural Activity in Primary Visual Cortex? Nature, **375**, (1995)
6. Hodgson, T.L., Dittrich, W. H., Henderson, L and Kennard, C., Neuropsychologia, **37: 8**, (1999), 927-938
7. Aleksander, I. and Dunmall, B.: An extention to the hypothesis of the asynchrony of visual consciousness. Proc R Soc Lond B **267**, (2000) 197-200
8. Zeki, S., and Bartels, A., The asynchrony of consciousness. Proc R Soc Lond B **265**, (2000) 1583-1585.

DIAGEN-WebDB: A Connectionist Approach to Medical Knowledge Representation and Inference

José Mira, Rafael Martínez, José Ramón Álvarez, and Ana E. Delgado

Dpto. Inteligencia Artificial. Fac. Ciencias, UNED, Senda del Rey 9, 28040 Madrid, Spain
{jmira, rmtomas, jras, adelgado}@dia.uned.es

Abstract. In this paper we explore the relationships between symbolic and connectionist models of medical knowledge in the diagnosis task. First we reassume the motivations of the ground-work stages of connectionism. Then a relational network is obtained from the natural language description of the diagnosis task and subsequently this network is transformed into a connectionist one via the dual graph. Finally we comment on the symbiosis between symbolic and neural computation. The aim of the paper is to explore some of the similarities and differences between the two basic approach to artificial intelligence.

Keywords: Knowledge modelling, connectionism, diagnosis, re-usable components (entities, relations and methods), symbolic-connectionist bridge.

1 Introduction and Purpose

The purpose of Artificial Intelligence (AI) is to model and make computable a substantial quantity of non-analytical human knowledge concerning problem-solving tasks. The tasks considered by AI are "high-level", corresponding to what human experts generally call cognitive processes, and classified into three broad groups: (1) formal domains, (2) scientific-technical domains like medical diagnosis, and (3) basic and genuine human functions like seeing, hearing, walking or understanding natural language [1].

For the modelling and programming of the solution of these tasks there are two paradigms, which constitute two ways of analysing, modelling a process, and –essentially- two methods of synthesizing a solution:

(1) Symbolic paradigm (thick "grain").

(2) Connectionist paradigm (small "grain"), anticipating that at the level of physical processors all computing are connectionist.

Symbolic representations of entities, methods and tasks are born at the knowledge level and at the outer observer's domain, where the programmer, the program interpreter and the knowledge lies.

J. Mira and A. Prieto (Eds.): IWANN 2001, LNCS 2084, pp. 772-782, 2001.

AI began being neural computation when in 1943 Warren S. McCulloch and Walter Pitts introduced the first formal model of the brain [2]. That is to say, the first studies where related to the modelling of knowledge and synthesis of these models by means of networks of artificial "neurons". In 1956 the term AI was coined by John McCarthy and the 70's and early 80's where dominated by the symbolic paradigm that, since the introduction of the knowledge level by Newell [3], looks after recurrent abstractions in the modelling of tasks and methods. A taxonomy of these abstractions are "generic tasks", libraries of "problem-solving-methods (PSMs)" to decompose these tasks, and ontologies of entities and relations for modelling the domain knowledge.

The last two decades witnesses a very strong rebirth of connectionism as an alternative or complement to symbolic AI. In particular, from the proposals put forth by Rumelhart et al around 1985. Unfortunately, in the "fiorello" of connectionism most of the initial motivations had been forgotten and, very often, only the simplest computational models of the past (adders followed by sigmoid) are used again and again.

The purpose of this paper is to reassume some of the motivations of the groundwork stages of connectionism and consider three basic questions: (1) How can medical knowledge be represented in a connectionist network?. (2) How can this representation be used in inference?. (3) How can a symbolic-connectionist bridge be build?.

2 The Connectionist Paradigm

"Under the term of neural computation underlies the more general concept of connectionism which is a hierarchical and recurrent way to look at neurophysiology and computation through a huge three-dimensional net of nodes and arcs with the following characteristics [4]:

1. Modularity of "small grain" and massive parallelism.
2. A parallel graph or a finite state modular automaton as underlying formal model.
3. A great number of autonomous modules with mechanisms of local inference and dialogue.
4. A high degree of connectivity (convergent/divergent active data routing) with delays and receptive fields with complex spatio-temporal organizations.
5. A multiplicity of local and distal feedback loops.
6. A cooperative architecture organized in terms of functional groups (layers", "columns", "barrels", dynamic groups of Luria, Edelman's groups, dynamic binding, ...) with specialization of the modules within each group.
7. A topology preserving organization, so that what is near in the initial data field is also close in its posterior and subsequent architectural representations (coding without losing the reference of its origins).
8. A library of different styles of computation for each level.
9. a) Possibility of structural learning by means of modification, creation or suppression of local processes in arcs and/or nodes.

b) Possibility of functional learning changing the values of parameters in arcs or nodes.

10. Embodiment of some kind of self-organization, fault tolerance and emergency mechanisms".

An early example of connectionist representation of a diagnosis problem is the causal net CASNET, proposed by Weiss and Kulikowski in 1982 [5], and illustrated in figure 1. Here, each node-concept is represented by a generalized "neuron" (a rule) and the links which corresponded in the original work to relations of inclusion, membership or causality, are here always relations of influence, dependence or gradual causality.

Observations include symptoms, signs and tests. They constitute the input entities to the first layer. Physiopathological states are intermediate inference elements which describe the internal conditions which appear in the patient and which help to characterize the degree of normality or pathology in an organ. Finally the third layer of inferential elements correspond pathologies selectors. This selection process emerges from the net of causal connections (rules or relations) which include the observations and the physiopathological states. These diagnoses (y_1= no glaucoma, y_2 = closed angle glaucoma, y_3 = open angle glaucoma and y_4 = other types of glaucoma) consist of diseases characterized by certain classes of configurations of the inferential state of network.

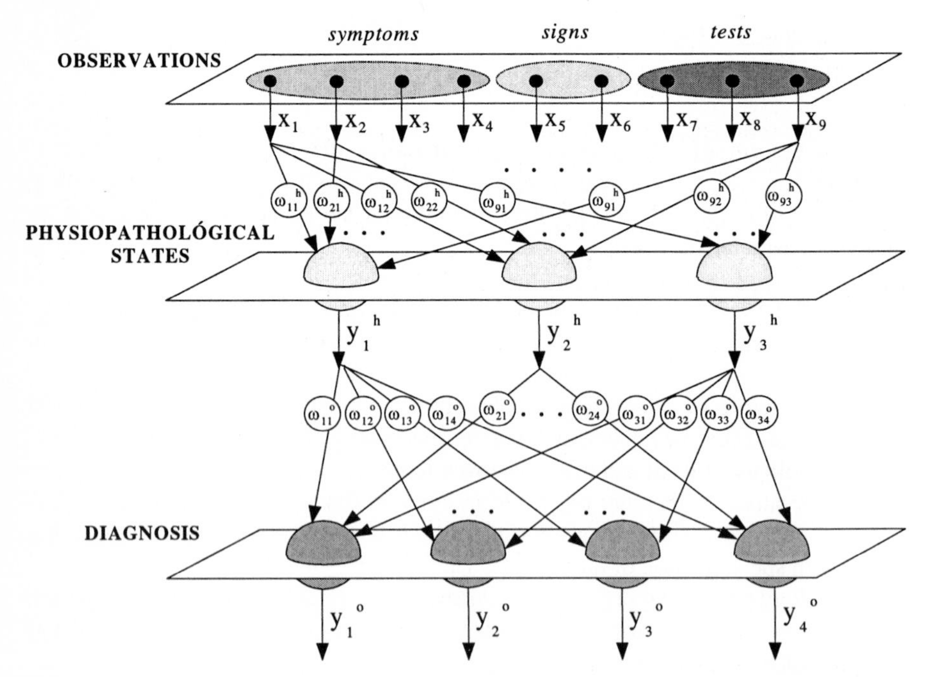

Fig. 1. Connectionist representation. As in the original model we use three functional groups ("layers") of inferential elements associated to the three levels in which authors of CASNET structured the inferential knowledge for the diagnosis of glaucoma: observations, physiopathological states and diagnosis.

3 Connectionist Entities and Relations in DIAGEN-WebDB

The way the medical expert's knowledge is analysed and modelled determines what we actually consider knowledge as represented by the connectionist model that we obtain and, finally, implement. In other words, if we look after a connectionist model of the domain knowledge, this perspective will drive the modelling process from the outset. In this paper concerning the knowledge modelling in the DIAGEN project, we build our network of inferential verbs based on the natural language description of the diagnosis tasks. Nouns and verbs used by the medical expert in the description of his/her reasoning are associated to entities and relations, reducing the granularity of the inferential elements and allowing for a greater degree of personalisation of the connectionist method used in DIAGEN [6]. Each physician can in principle model and edit the network of verbs and nouns that best fit his/her reasoning.

We consider three kinds of reusable components: *entities*, *roles* and *relations*. *Entities* are magnitudes that play origin and/or destination *roles* in *relations.* A *fact* describes the state of knowledge of an entity; it is an instantiation of the entity in the sense that the characteristics describing the magnitude are given values in the *fact.* *Connections* are instances of *relations* between two given *entities* (in the *effective roles* of origin and destination).

3.1 Entities and Facts

The *entities* play the same role as the physical magnitudes in an analytic model of ANN. They represent separate concepts that the human expert considers necessary and sufficient to describe his/her knowledge. They constitute the universe of discourse for the relations (generalized "neurons"), and are associated to names whose referent can be physical or mental. Examples in medicine are fever, sign, finding, diagnosis, etc. The entities are described by means of a set of attributes, such as name, description, units, and possible values as shown in Table 1.

Table 1. Frames for the description of entities and facts (instantiations of entities).

Entities
name: a short label used to name the entity
description: explanatory text (useful in help screens, for example)
unit: name of the measuring units
value_range: list of labels, or numeric value range, that the measurement component of the facts of the entity can take

Facts
value: measured value
entity: entity the fact belongs to
initial_instant: start of the interval in which the state is valid
duration: period of time in which the state is valid
qualifier: qualitative component of the state

An important characteristic of the model is the emphasis on the specification of the state of an entity, what we have denoted *fact*, by means of qualitative components about the knowledge of this entity which indicate, for example, the state of "belief", "provisionalness", "probability", etc. The time interval of validity of the state of an entity is also part of the fact. The set of all the facts gives the global knowledge state of the system during the period defined by the intersection of their respective validity time intervals [7].

In the study of the diagnosis tasks undertaken in DIAGEN we have found a set of entities, of different degrees of generality (organized according to a hierarchy), such as: *Observation*, *Individual-Finding*, *Finding*, *Syndrome*, *Qualitative-Observation*, *Physiopathological-Process*, *Diagnosis-Therapy*, *State-Evolution*, *Intermediate-Diagnosis*, *Surgery*, *Surgery-Result*, *Test*, *Observable*, *Symptom* and *Sign*, which are common to all the clinical specialties covered in DIAGEN

3.2 Relations, Roles and Connections

The relations, as well as the entities, are obtained from the task description in natural language according to schemes such as the following:

"*Origin* is related to *Destination* when *Origin* satisfies a condition and *Destination* another. The fact of being related results in an *Action* on *Destination*."

"If both *Condition on Origin* and *Condition on Destination* are satisfied then *Action on Destination* is taken".

The relations are thus oriented and binary, defined partly intentionally through conditions on origin and destination) and partially extensionally as the set of potentially-related pairs. These relations are instantiated as *connections* for certain domain entities. In a given domain, the relation roles are specialized in what we denote as *effective roles*, with more restricted features.

The names of the relations ("suggests", "confirms", "requires", ...) again reflect standard inference-process components, and though coined by each expert, they are sufficiently agreed-on to be accepted in verbal communication and to be reused. They are elements that are reasonably invariant within a given domain.

When describing the domain knowledge, the expert specifies what the system control should be, depending on certain components of the fact (suggested, stated, unknown, etc.). It is the expert who, when modelling them, determines how the states will be used in the relations.

Activation of a *Connection* takes place when there are *facts* that play the effective roles of origin and destination of that connection. Activation implies the execution of the action defined in the corresponding relation. The result of an action is the modification of the state of the fact that has played the effective role of destination.

Table 2 shows the attributes that characterize the relations, and their associated roles. The connection-description frame (instantiation of relations) contains fields that correspond to the effective roles; these are just a refinement of the roles that appear in the corresponding relation.

As a result of the study of analysis tasks in the clinical specialties considered in DIAGEN,a set of basic relations has been found to be sufficient for the modelling of the medical expert's chain of reasoning, for example *suggests*, *confirms*, *discards*, *abstracts-as*, *requires* and *explains*. The relational network obtained from natural language description of task and method is of the type shown in figure 2, where the grain size is diminished to a maximum in order to personalize the method for each expert.

Table 2. Frames for the description of relations (a) and roles (b)

(a) Relations

name: a short label used for referring to the relation.
description: a longer explanatory text.
origin_role: a reference to the origin role of the relation.
destination_role: a reference to the destination role of the relation.
action: modifications of the value and state of the destination that take place when the relation is activated. An action, in turn, is described by the following fields:
description: descriptive text, useful for explanatory purposes.
qualifier: state of the value the *destination* entity of the activated relation must end up in.
value: value of the measurement component of the *destination* entity after the activation of the relation.
amount: number of relations of the same type with the same destination, whose role conditions must be simultaneously satisfied for the action to be taken.

(b) Roles

name: a short label used for referring to the role
description: descriptive text, used for explanatory purposes
units: condition on the valid units for the entities in this role
value-range: range of acceptable values for this role
qualifier: condition on the state of the value of the entities in this role

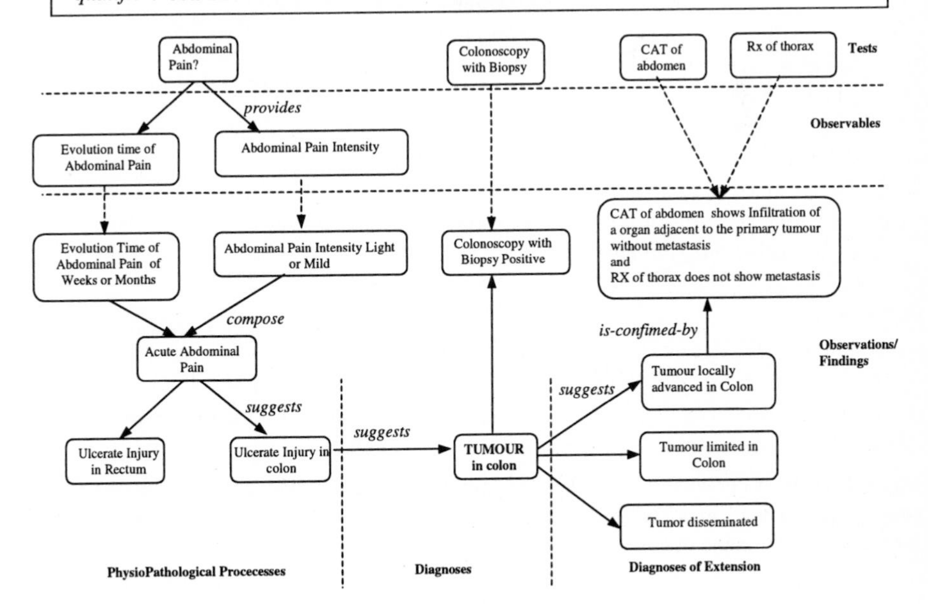

Fig. 2. Example of partial relational network in cancer of colon diagnostic domain.

4 From the Relational to the Connectionist Network

The relational network of figure 2 represents the medical knowledge in the same way as described by the physician. In this network the *nodes* represent concepts and the *arcs* represent relations. If we now want a connectionist representation of the same knowledge we have to build up the dual graph. That is to say, *nodes* must now represent verbs (inference rules) and *arcs* are associated to effective *roles*, the input and output "signals" of the inferential "neurons". In a connectionist network all the inferential elements compute in parallel on its input roles and it is from this network of concurrent local inferences from which the global functionality emerges. The specific subset of inferential "neurons" that are effectively active at each time interval is a direct consequence of the pattern of afferent information to the network and the subsequent evaluations of the activation conditions in each one of the nodes. That is to say, when the input roles are played by the set of pairs magnitude-value that are compatible with these roles. In fact, each role constitute the description in extenso of all the pairs magnitude-value that fulfil the compatibility test for this role. Also, these pairs are mutually exclusive so that in each time interval only one pair is able to activate the nodes, as far as the corresponding role is concerned. It is this flow of data that couples ("coordinate") the different local inferences and captures the coordinative interrelations between the different "small grain" component of the global inference.

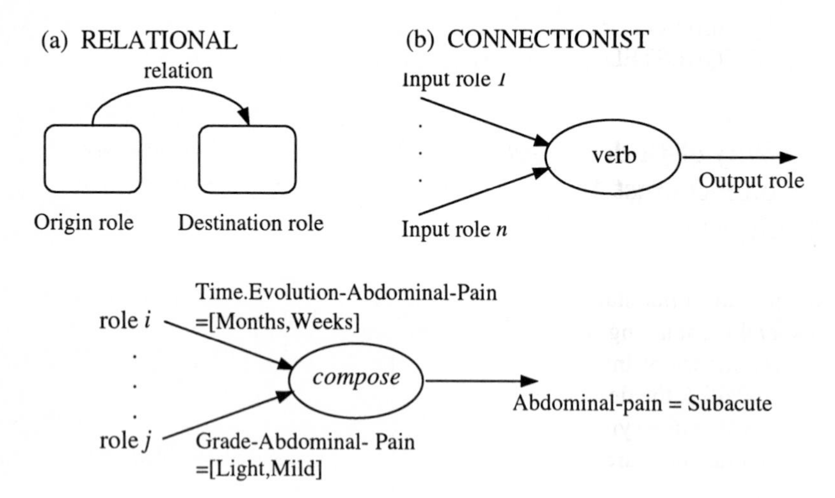

Fig. 3. Change of knowledge representation form (a) symbolic (relational) to (b) connectionist (dual graph), where "arcs" represent input and output roles and "nodes" corresponds to local inferences. (c) Example of connectionist node.

From the computational point of view, the nodes in this connectionist network are associated to conditional rules (IF-THEN) where the input lines are premises linked by the logic operator AND and each premise (each role) express a condition over a pair magnitude-value. As conclusion a new value is assigned to the entity referenced by the output role:

```
IF
    (Time-Evolution-Abdominal-Pain=Months)OR
    (Time-Evolution-Abdominal-Pain=Weeks)AND
    (Grade-Abdominal-Pain=Light)OR
    (Grade-Abdominal-Pain=Mild)
THEN
    Abdominal-Pain ← Subacute
END-IF
```

In a connectionist network there are a great diversity of local and distal feedback loops. In these cases, the output roles includes its own conditions but the pairs magnitude-value of one output are feedback as inputs of the same or other inferences. In these cases, although the physician don't mention it in a explicit manner, we have to reintroduce these feedback inputs. In figure 4 we show a part of the complete connectionist network result of the transformation of figure 2. The network is surrounded by an environment that includes the user of the system. This user contributes to the global inference by providing the tests and additional information requirements asked by different nodes. For example, given the output "Colonoscopy-Biopsy?=REQUESTED", provides the input "Colonoscopy-Biopsy?=REALISED".

5 A Way of Understanding the Bridge between Symbolic and Connectionist Methods of Knowledge Representation and Inference

Following an initial stage of competition between connectionist and symbolic paradigms, and recognizing the limitations in both, the current state of knowledge points to the advantages of integrating the two such that when the knowledge concerning a task is modelled, the decision is made as to which combination of task, data and prior knowledge favour a symbolic representation, which a connectionist and, finally, when hybrid architectures are best recommended to cope with an specific data-knowledge trade off.

In this paper we have presented an alternative characterized by the following decisions.

1. We use the formal model underlying connectionism but extend the local computation from analogical and logic to inferential. So the nodes are now conditionals on structured objects without sacrificing the essential characteristics of connectionism (modularity, "small grain", parallelism, …).

2. The connectionist network is "knowledge-based", in the sense that all the prior knowledge obtained from the natural language description of the task has been injected in the architecture of the net. The verbs that the human expert uses in the description of his/her reasoning are inferential "neurons" and the nouns are the entities that plays different input and output roles in one or several "neurons" of the net.
3. The passage from the natural language description of the intuitive relational network to the connectionist network is obtained by means of the dual graph. So the conversion process between symbolic and connectionist representations is reversible.

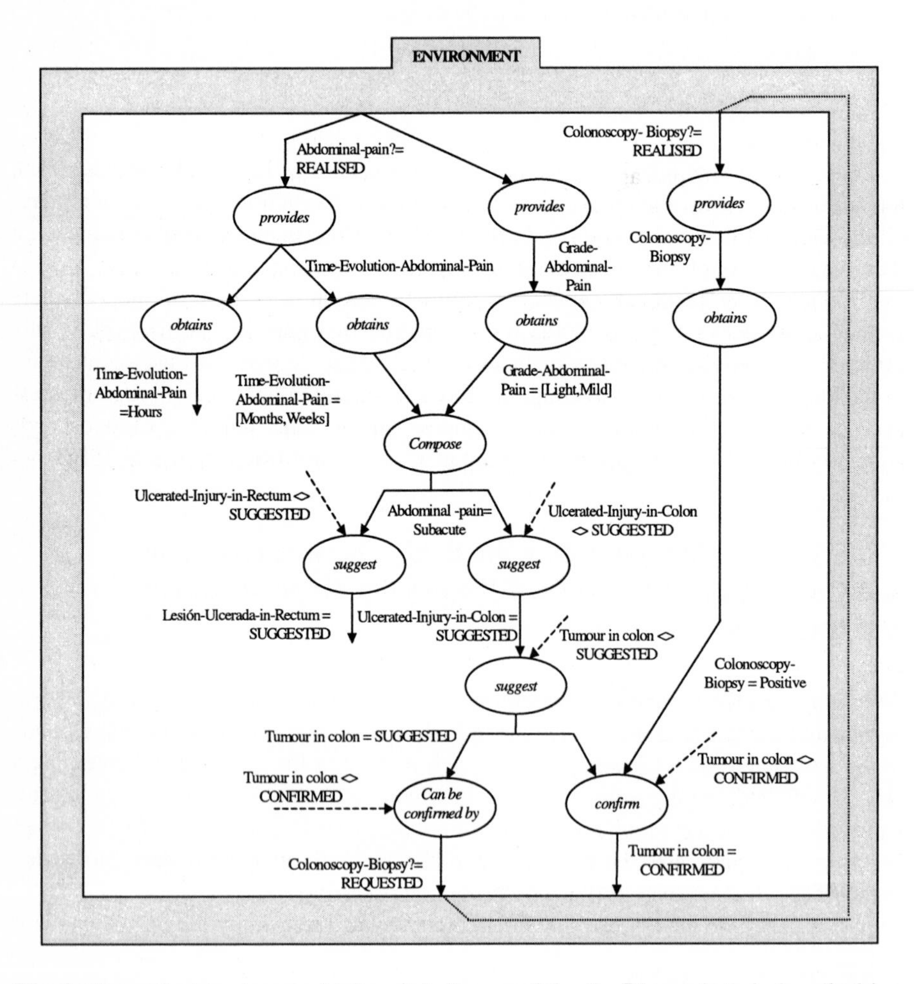

Fig. 4. Connectionist network of inferential "neurons" for the Diagnosis task described in a symbolic manner in figure 2. Dashed lines represent connections for which the origins are not explicit in the figure.

6 Conclusions

There are not essential differences between symbolic and connectionist representations at the level of physical processors. The differences between number and message appears at the knowledge level when we define granularity and assign meaning to the different entities and relations of the problem-solving-method used to decompose the tasks. In the connectionist approach we use all the knowledge we have available to specify the structure of the network, either during the design process or during the learning phase.

In this paper we have presented a symbolic model of medical knowledge and the corresponding transformation of this model into a connectionist network by means of: (1) decomposition of thick grain relations (those that appear in the natural language description of the task) into small "grain" inferences ("neurons") and (2) build-up of the dual graph, where input and output labelled lines are roles (pairs magnitude-value) in such a way that the difference between "passage of numbers" (connectionist) and "passage of messages" (symbolic) is blurred. All that is needed in connectionist computation is to lower the specification level.

A preliminary implementation of the network has been made using a relational data base support [7], because the components of the connectionist network can be easily associated to tables, and the inferential "neuron" operations are also easily associated to SQL commands of type "select" or "update". This implementation enables the management via Web using HTML interfaces.

The work continues to cope with learning understood as (1) modification, creation or suppression of local processes in arcs and/or nodes (structural learning) and (2) changing the values of "belief", "provisionalness" and "probability" in entities and facts (functional learning).

Acknowledgments

The authors would like to acknowledge the financial support of the Spanish CICYT under project TIC-97-0604.

References

1. Mira, J., Delgado, A.E.: Computación Neuronal. In: Aspectos Básicos de la Inteligencia Artificial Cap. 11. Sanz y Torres, S.L. Madrid, (1995) 485-575.
2. McCulloch, R (ed.): Collected Works of W.S. McCulloch. Vol. **I** to **IV**. Intersystems Pub. Cal. WA. (1989).
3. Newell, A.: The Knowledge Level. AI Magazine, summer: (1981) 1-20.
4. Mira, J., Delgado, A.E.: Neural Computation: From Neuroscience to Technology and Back Again.Proceedings VIth Brazilian Symposium on Neural Networks. IEEE Computer Society. USA (2000) 15-22.

5. Weiss, S.M., Kulikowski, C.A, Amarel, S., Safir, A.: "A Model-Based Method for Computer-Aided Medical Decision-Makig". Artificial Intelligence, **11** (1978) 145-172.
6. Mira, J.: Report of project DIAGEN. CICYT. TIC 97-0604 (1997).
7. Mira, J., Älvarez, J.R., Martínez, R.: Knowledge Edition and Reuse in DIAGEN: A relational Approach. IEE Proc.-Softw. Vol.147, N0. 5 (2000) 151-162

Conceptual Spaces as Voltage Maps

Janet Aisbett and Greg Gibbon

School of Information Technology, The University of Newcastle
Callaghan, Australia 2308.
{mgjea, mgggg}@cc.newcastle.edu.au

Abstract. Conceptual spaces have been proposed as a meso level representation, intermediate between symbolic and connectionist representations. We define a conceptual space to be a set of images or "voltage maps" on a compact sub plane, and equip it with pseudo-physiological notions of distance and betweenness. While our meso level representation is easily linked to higher and lower representations, we argue that its natural notion of geometry provides powerful additional tools for knowledge modeling and reasoning. As illustration, we offer an explanation of multi-dimensional experimental results which suggest distances follow different order Minkowski measures according to whether the dimensions are integral or separable.

1 Introduction

For over a decade Gärdenfors has advocated a level of representation lying between connectionist and symbolic representations (eg. [5]). Symbolic representation cannot tie meaning to the symbols, and cannot easily accommodate the notion of similarity that appears to be a fundamental component of reasoning. On the other hand, connectionist representations are often difficult to construct and to interpret, and assume structured input. The intermediate representation proposed by Gärdenfors is the *conceptual space*, built on dimensions that represent qualities of objects or concepts, such as height or happiness.

It is not clear how to formally represent a conceptual space, let alone how to organise knowledge within such a space. Gärdenfors implies that conceptual space is a set of spaces, each having orthogonal quality dimensions. He leaves the nature of dimensions as open as possible, preferring them to support elements of geometry through a *betweenness* relation rather than requiring them to have a metric. In earlier work [2] we argued that a metric is needed to embed similarity into the formal structure of conceptual spaces, and that the trivial metric is still useful on domains in which distance is not otherwise sensibly defined. We also argued for decoupling the definitions of betweenness and distance, and showed how a formal definition of conceptual space could describe logic programming, dynamical and neural systems.

In this paper, we introduce an important formulation of conceptual spaces which we call "voltage maps" and which are motivated by topographic representations in cortical layers. Rather than model the complex microstructure of cortical sheets required to understand the system dynamics, a gross approach is taken. Our representation nevertheless has natural links to symbolic and sub conceptual layers. It has a physiologically-inspired metric defined in terms of energy consumption. It also has an additive structure that allows for aggregation

J. Mira and A. Prieto (Eds.): IWANN 2001, LNCS 2084, pp. 783–790, 2001.

of like concepts and for composition of unlike concepts. It is granular, which fits with hierarchical concept descriptions and allows for learning. Finally, it has a natural way of showing when a dimension does not apply to an object, thus overcoming an inelegancy in Gärdenfors' development of conceptual spaces.

Psychological distance is an important tool for exploring cognitive processes (eg. [8, 10]). We illustrate the power of our representation by offering an explanation of why Euclidean distance appears to fit data from some multi-dimensional experiments involving so-called integral dimensions, whereas the city block metric fits data from experiments involving separable dimensions.

2 "Voltage map" representation of conceptual space

In analogy with cortical maps or sensor arrays, a low resolution approach to modelling cortical sheets suggest dimensions in conceptual space should be associated with a set of image pixels, each capable of acquiring an intensity (a voltage) over a continuous range of values above a resting level. In this section we formally define conceptual spaces as voltage maps, and distance as a voltage difference. Whilst we work in a continuous setting, the discrete version is obvious. We contrast our definition of conceptual distance with that usually defined in feature or vector spaces, then end the section with definitions of *properties* and *concepts*, which are crucial cognitive representational elements because they support generalisation.

So conceptual space is a set of voltage patterns, or images, on a compact plane in which locations are described as pixels, and values of a dimension are restrictions of images to the connected set of pixels which represents the dimension. The use of phase or frequency [4, 6] to bind physically separated representations of related events or objects suggests using multiple copies of this space, or, equivalently, using frequency or colour to distinguish different components of complex objects. Thus a simple object is a monochrome sub image, and a complex object is a coloured sub image. White noise represents parts of the plane which are irrelevant to the object description, and is the physiological equivalent of ∞ (point at infinity) adjoined to the set of realisable absolute voltage levels.

Distance is defined via the energy needed to transform voltage patterns. Specifically, we assume each value on a dimension is associated with the energy required to raise the voltage on its pixels. As energy is proportional to the square of voltage, the distance between two objects f and g is the equivalent voltage difference, that is, the square root of the energy needed to raise the lower voltage to the higher one[1], namely $\int |f^2(x) - g^2(x)|^{1/2} dx$. This reference to energy also suggests that objects do not aggregate according to the addition inherited from the reals; rather, adding two voltage functions (objects) f and g gives the object $(f^2 + g^2)^{1/2}$.

---

[1] The distance between objects not defined on the same subset of the image plane depends on the treatment of unmatched parts of the objects, and is discussed in [1, 2]. Here, we take the simple approach of assuming that objects not defined on the same pixels are incomparable, ie. $|f^2(x) - \infty| = \infty$ unless $f(x) = \infty$, when the difference is 0.

Betweenness is defined pixelwise, so that an object f is between g and h if and only if the voltage $f(x)$ is between $g(x)$ and $h(x)$ for each pixel x of the sub image which defines the object f.

The following definition formalises this and introduces domains as the union of integral dimensions.

Definition 1

- A conceptual space C is a family of piecewise differentiable functions $f: X \rightarrow (R \cup \infty)^n$, where X is a connected bounded subset of R^2 and n is the maximum number of components from which an object may be cognitively assembled. A function $f \in$ C is also described as a *state* of the conceptual space, or as an *object*.
- A *dimension* D_i is the restriction of the functions in C to a connected subset $X_i \subseteq X$. Call X_i the *region of definition* of the ith. dimension, and set Γ to be the index set for the dimensions.
- A dimension D_i is *separable* if and only if $X_i \cap X_j \neq \emptyset \Rightarrow i = j$ for all $j \in \Gamma$; otherwise D_i is *integral*.
- Each dimension D_i is in a *domain* which is the family of functions C restricted to $\cup_{k \in \Gamma(i)} X_k$ where $\Gamma(i) = \{j \in \Gamma: X_i \cap X_j \neq \emptyset\}$. So a separable dimension is a domain. The region of definition of the domain is $\cup_{k \in \Gamma(i)} X_k$.
- The *conceptual distance* $d(f, g)$ between objects f and g is $\int_X |f^2(x) - g^2(x)|^{1/2} dx$.
- An object $g \in$ C is between objects f and h if and only if (a) either $f(x) < g(x) < h(x)$ or $h(x) < g(x) < f(x)$ for every $x \in \{x: g(x) \neq \infty\}$ and (b) $g(x) = \infty \Rightarrow f(x) = \infty = g(x)$.
- C is equipped with an addition, $\oplus$: C $\times$C $\rightarrow$ C, defined by $f \oplus g = (f^2 + g^2)^{1/2}$

Many fields of study employ a feature or vector space whenever representation requires a distance. Distance on these multi-dimensional spaces is calculated with a city block or Euclidean treatment of component distances, or sometimes with a higher order Minkowski. In our setting, the distance between points f and g in a single domain is $\int_{X(i)} |f^2(x) - g^2(x)|^{1/2} dx$ where $X(i)$ is the region of definition of the domain. The city block distance between f and g in the context of a set of (disjoint) domains $\Gamma^* = \{r(1), \ldots r(j)\}$ is $\sum_i \int_{Xr(i)} |f^2(x) - g^2(x)|^{1/2} dx$. The difference between this and the conceptual distance is the term $\int_{X - \cup Xr(i)} |f^2(x) - g^2(x)|^{1/2} dx$. When Γ^* includes all cognitively recognised domains, this difference is the absolute local voltage difference integrated over the "unexplained" areas of X, that is, those that lie outside the region of definition of any dimension. The conceptual distance is always greater than or equal to distance calculated over recognised domains. Learning may comprise identifying dimensions in the "unexplained" areas, thus reducing the difference between the two forms of distance.

Properties and concepts are key notions in any knowledge representation scheme. In our formulation, properties are sets of images on subsets of the plane, as depicted in Figure 1 and explained below. Gärdenfors argued a special place for convex properties because, generally, if two values of a dimension are seen as belonging to a property, then so are all intermediate values. Concepts are images on subsets of the image plane which restrict to the set of properties which define them. Thus, although concepts include unions of properties (Gärdenfors' definition of a concept), we allow that they may be larger than this.

Concepts are not "bundles of features" [7] but may be liable to interpretation beyond the interpretation of their identified dimensions.

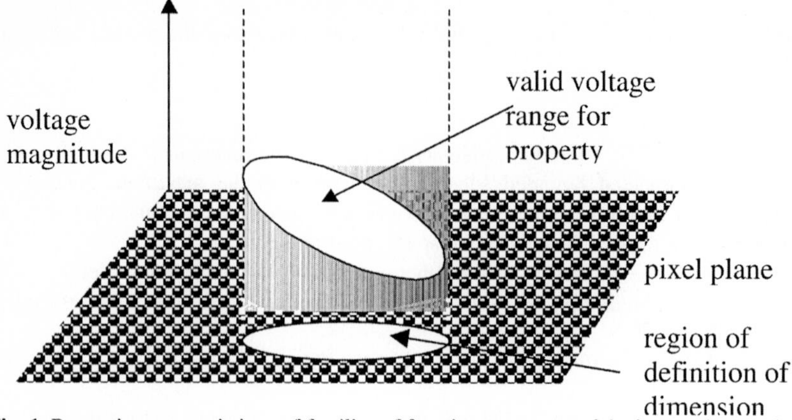

Fig. 1. Properties are restrictions of families of functions onto part of the image plane and part of the voltage range. See text.

We define convexity and connectedness using betweenness, slightly generalising the usual topological definitions.

Definition 2

- A (non singleton) subset A of a family of functions is *connected* if for any pair $f, h \in A$ there exists $n > 0$ and a sequence $g_0 = f, g_1, g_2, ..g_{n-1}, g_n = h$ contained in A such that for $0 \leq i \leq n$, every function g between g_i and g_{i+1} is also in A.
- A space A is *convex* if it is a singleton or if, for any pair $f, h \in A$, all functions g between f and h are in A.
- A *property* of a domain is a connected subset of that domain, and a natural property is a convex subset of a domain.
- A *concept* is a subset c of & such that, given any domain with region of definition X_i, either $c|_{X_i} \subseteq p_i$ for some property p_i or $c|_{X_i} = \infty$. We say the set $\{i: c|_{X_i} \neq \infty\}$ is the set of domains *relevant to the concept*.

Given any connected subset X^* of X, then any pair of functions f, g in & which do not intersect on X^* define a convex set, namely, the sets of functions whose restriction to X^* lie between the bounding functions f and g. Natural properties on a given domain are defined using a pair of bounding functions. As Figure 2 suggests, natural properties can sometimes be thought of as patterns between bounding functions, and sometimes as a range of average voltage magnitudes.

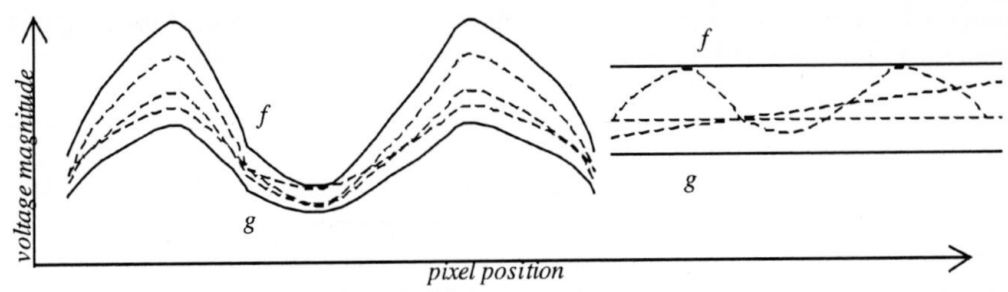

Fig. 2. Properties defined as families of functions on a subset of the plane which take values between two bounding functions f and g. The left hand family members exhibit similar patterns. The right hand family may take any pattern of values within the range $(f(x_0), g(x_0))$.

Concepts for which different domains are relevant are composed when added (eg *red ⊕ car)*, whereas the effect of addition of concepts involving integral domains (eg. *red ⊕ orange)* is to alter the property (see discussion of Figure 3 below).

3 Connection to lower and higher levels of representation

Subconceptual-conceptual. Definition 1 only limited the voltage maps to piecewise differentiable functions. Realisable functions might, however, be determined by structure in a subconceptual layer. Taking a localist perspective, the voltage map could be envisaged to result from neural columns responding to perceptual data and raising the energy in the regions of definition X_j of the dimensions. Such activity might be modelled as Gaussians whose 2D variances σ_j^2 determine the rate of fall-off with distance from activation centres y_j. The frequencies (colour) of the voltage might also be determined by the pre-conceptual level, if components of complex objects are distinguished at this level.

In this scenario, energy E_j received at the image plane as a result of stimulus of the jth. perceptual dimension by a component of a complex object indexed by angular frequency w_j would raise the voltage pattern from the resting voltage to the pattern $\sqrt{(E_j\ exp(-(x-y_j)^2/\sigma_j)/\sigma_j^2))}.\ exp(-iw_jt)$. Then $f \in$ & if and only if f is of the form

$$\Sigma_{j \in \Gamma} \sqrt{(E_j\ exp(-(x-y_j)^2/\sigma_j)/\sigma_j^2))}.\ exp(-iw_jt). \quad (1)$$

A related description of conceptual space generation is provided by Balkenius in [3]. His Eqn (1) models properties as collections of tuned detectors each with a normalised Gaussian tuning curve centred on a value taken by the property in that dimension. He justifies this approach as being consistent with neurological representation by place.

If the areas of activation for centres y_i and y_j have significant intersection, as in the right side of Figure 3, then the dimensions are integral. Thus separable properties might be those defined over areas in X which receive significant activity from just one centre, for some level of significance related to the ambient noise level. However, the Figure also suggests that it may be possible to learn to separate integral dimensions by learning to use smaller regions of definition to distinguish dimensions.

Symbolic-conceptual The symbolic level might connect to conceptual space through distinguished dimensions which would form a *symbol space,* a subspace of conceptual space. Interconnections between symbol and non-symbol dimensions would associate each symbol with patterns of voltage, and cause symbols to stimulate patterns, as well as patterns to stimulate symbols. The association of dimensions of conceptual space with the symbolic level could therefore be thought of as pattern recognition, for example through an ART network. Specifically, an instantiation depicted by the symbol α (eg. a ground predicate in first order logic or a symbol in a propositional calculus) would be associated with a set of voltage maps $\{f_a: a \in \Psi(\alpha)\}$. $\Psi(\alpha)$ might be a singleton, but in general would have more than one representative. Then $\alpha \Rightarrow \beta$ if and only if for any $a \in \Psi(\alpha)$ there exists $b \in \Psi(\beta)$ such that for all $x \in X$ either $f_a(x) = f_b(x)$ or $f_b(x) = \infty$.

The complexity of the relationship between a symbol and the raw" perceptual information delivered by neural columns will depend in part on the size of the sub image used in recognising the symbol. Thus a colour property could conceivably involve regions of the conceptual space concerned with colour, but also with texture or shape. Different symbols might access the same part of the image plane. On the other hand, there might be conceptual patterns which have not been lexically indexed but which are physiologically recognised, in that, say, their presence may stimulate an action.

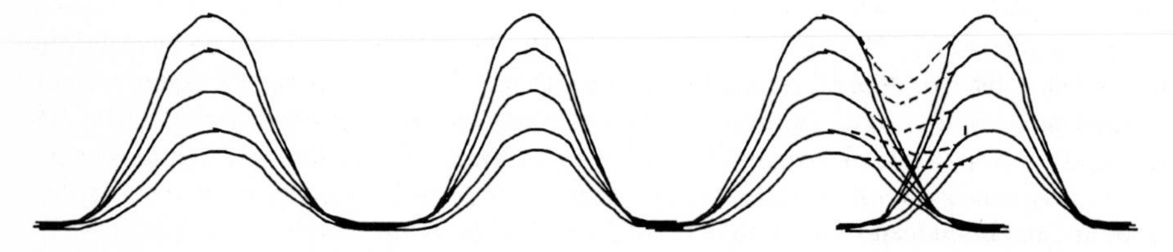

Fig. 3. Separable dimensions may be generated by one activation centre, whereas pixels in integral dimensions are activated by multiple centers. See text.

4 A possible explanation of experimental results on conceptual distances with separable and non-separable dimensions

An intriguing aspect of human and animal judgements of similarity and relative distance between concepts is the way that multi-dimensional judgements combine. Sometimes, dimensions are treated separably, in which case a city block metric $d = \Sigma d_i$ describes the way distance measures d_i on the individual dimensions are perceived in a multi-dimensional setting. In other cases, a Euclidean distance or even higher order Minkowski $d = \{\Sigma d_i^r\}^{1/r}$, $r \geq 2$, fits the data better (Nosofsky 1992). This psychological evidence can be explained simply by assuming that integral domains involve overlapping regions of definition on the image plane X.

Specifically, let $\{p_i: i \in \Gamma^*\}$ be a set of properties, where p_i is defined on domain X_i, and suppose f and g are two objects defined with these properties (ie f restricts to a function $f^i \in p_i$ on each X_i when $i \in \Gamma^*$ and $f(x) = \infty$ whenever $x \in X_i$ for $i \notin \Gamma^*$; and similarly for g). So $X_i \cap X_j = \varnothing$ whenever $i \neq j$. The distance between f and g, $d(f, g)$, is $\int_X |f^2(x) - g^2(x)|^{1/2}dx$. The distance between them in the context of the properties, $\sum_{i\in\Gamma^*} d(f^i, g^i)$, is $\sum_{i\in\Gamma^*} \int_{Xi} |f^2(x) - g^2(x)|^{1/2}dx$. Since $f = g = \infty$ on domains not in Γ^*, the difference between the conceptual distance and the city block metric defined on the properties is the distance between the functions on the "unexplained" areas of X, outside the region of definition of the domains. This distance is usually taken to be 0.

In contrast to the case for domains (including separable dimensions), there is "double counting" if a Minkowski metric is fitted to integral dimensions i and j for which $X_i \cap X_j \neq \varnothing$. This inflates distances computed using the city block metric over those computed using conceptual distance, and enables the latter to appear to be a Minkowski metric of order greater than one. This is true whatever the precise form of the conceptual distance. Specifically, if metric d_r is defined to be $d_r(f, g)=(\sum_{i\in\Gamma^*} d_i(f^i, g^i)^r)^{1/r}$ for $r \geq 1$, then $d_r(f, g)=(\sum_{i\in\Gamma^*} (\int_{Xi} |f^2(x) - g^2(x)|^{1/2}dx)^r)^{1/r})$. In the limit of $X_i \equiv X_0$ for all i,

$$d_r(f, g)=(k\,(\int_{X0} |f^2(x) - g^2(x)|^{1/2}dx)^r)^{1/r} = (k)^{1/r}\, d(f, g)$$

which approaches $d(f, g)$ for large r. That is, the conceptual distance can appear to be a high order Minkowski distance if the supports of properties overlap on X.

More generally, let A_i denote the distance between f and g in the context of the dimension i (ie., $A_i = \int_{Xi}|f^2(x) - g^2(x)|^{1/2}dx$) and let $A_{i,j}$ denote the "double counting" component $\int_{Xi\cap Xj} |f^2(x) - g^2(x)|^{1/2}dx$. Then the Euclidean metric $d_2(f, g)$ is by definition $(\sum_i A_i^{\,2})^{1/2}$ and the conceptual distance is $d(f, g) = \sum_i A_i - \sum_{i<j} A_{i,j}$. In the case of two dimensions, then clearly if $A_{1,2} = A_1 + A_2 - (A_1^{\,2} + A_2^{\,2})^{1/2}$ (that is, if $A_{1,2}$ is the difference between the city block and the Euclidean metrics) then $d(f, g) = d_2(f, g)$ ie. conceptual distance appears to be Euclidean. Although such assumptions are contrived, they are easily fulfilled, as illustrated in Figure 4.

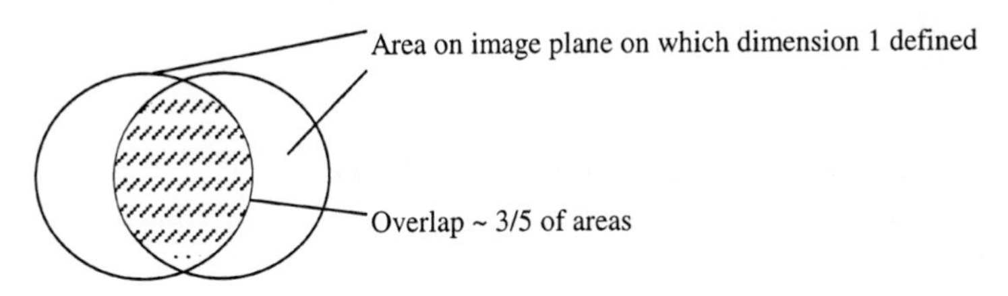

Fig. 4. Conceptual distance can appear Euclidean on dimensions with overlapping regions of definition.

5 Conclusion and Future Work

Gärdenfors identified the major advantages of conceptual spaces to be their ability to represent similarity; the grounding they give to symbols and expressions via constructions in the space; and the organisational advantage afforded by domains, which limits the combinatorial explosion of the frame problem [5]. He did not, however, formally define a space, or discuss how to represent complex objects or objects to which only some quality dimensions are relevant. The representation using voltage maps that we have developed

(a) has natural links to symbolic and sub conceptual representations, as well as to multi-dimensional feature spaces traditionally used to support distance notions;
(b) has a distance measure which restricts to the single dimensional measure on both separable and integral dimensions and yet can explain psychological experiments that suggest the former combine as a city block metric and the latter is Euclidean;
(c) has an additive structure which allows composition and aggregation;
(d) naturally represents that a dimension does not apply to an object;
(e) being image-based, has an intrinsic granularity which fits with hierarchical concept descriptions, and allows for learning.

Dynamics for general conceptual spaces were introduced in [2]. These can be implemented in the voltage map representation as association maps between pixels, as in a network such as that described by Omori et al [9]. We are exploring the enhancement provided by geometry in translation of an ontology from a medical sub domain.

References

1. Aisbett, J. and Gibbon, G (1994) A tunable distance measure for coloured solid models *Artificial Intelligence* 65, 143-164.
2. Aisbett, J. and Gibbon, G (2000) A general formulation of conceptual spaces, submitted to *Artificial Intelligence* .
3. Balkenius, C. (1998) Are there dimensions in the brain? In *Spinning ideas: electronic essays dedicated to Peter Gärdenfors on his fiftieth birthday.* http://www.lucs.lu.se/spinning/categories/cognitive/Balkenius/index.html
4. Freeman, W (1994) Qualitative Overview of Population Neurodynamics *Neural modeling and neural networks*, 1994, 185-215
5. Gärdenfors, P (2000) *Conceptual Spaces: The Geometry of Thought.* MIT Press.
6. Hummel, J. and Biederman, I. (1992) Dynamic binding in a neural network for shape recognition *Psychological Review 99*, 3, 480-517.
7. Margolis, E and Laurence S (1999) (ed) *Concepts: Core Readings* MIT Press
8. Nosofsky, R (1992) Similarity, scaling and cognitive process models *Annual Review of Psychology* 43, 25-53.
9. Omori, T, Mochizucki, A. et al (1999) Emergence of symbolic behavior from brain like memory with dynamic attention *Neural Networks* 12, 1157-1172.
10. Shepard, R. (1987) Toward a universal law of generalization for psychological science. *Science*, 237, 1317-132

Determining Hyper-planes to Generate Symbolic Rules

Guido Bologna

Computer Science Centre
University of Geneva, 24 Rue General Dufour, Geneva 1211, Switzerland
Guido.Bologna@cui.unige.ch

Abstract. In this work the purpose is to determine discriminant hyper-planes of a neural network in order to extract possible valuable knowledge by means of symbolic rules. We define a special neural network model denoted to as *Discretized Interpretable Multi Layer Perceptron* (DIMLP). As a result, rules are extracted in polynomial time with respect to the size of the problem and the size of the network. Further, the degree of matching between extracted rules and neural network responses is 100%. Our network model was tested on 7 classification problems of the public domain. It turned out that DIMLPs were significantly more accurate than C4.5 decision trees on average.

1 Introduction

Artificial neural networks are robust models used in classification problem domains. One major drawback is the fact that "knowledge" embedded therein is cryptically coded as a large number of weight and activation values. Recently, several authors proposed to explain neural network responses with symbolic rules given as: *"if tests on antecedents are true then conclusion"* [4], [5], [7], [8], [9], [12], [13], and [14]. A taxonomy describing techniques used to extract symbolic rules from neural networks is proposed by Andrews et al. [1]. Basically, rule extraction methods are grouped into three methodologies: *pedagogical*, *decompositional*, and *eclectic*. In the pedagogical methodology, symbolic rules are generated according to an empirical analysis of input-output patterns. This is also called the black-box approach. In decompositional techniques, symbolic rules are determined by inspecting the weights at the level of each hidden neuron and each output neuron. Finally, the eclectic methodology both combines elements of the pedagogical and decompositional methodologies. Golea showed that generating symbolic rules from neural networks is an NP-hard problem [6].

In this work we introduce the *Discretized Interpretable Multi Layer Perceptron* (DIMLP). Our purpose is to demonstrate that its own complex decision mechanism can be made explicit by means of symbolic rules. Therefore, turning our model into a potential tool for knowledge discovery. A special feature of the rule extraction technique applied to DIMLP networks is that rules are generated by determining discriminant hyper-plane frontiers. As a result, the

J. Mira and A. Prieto (Eds.): IWANN 2001, LNCS 2084, pp. 791-798, 2001.

degree of matching between network responses and rule classifications is 100%. Moreover, the computational complexity of the rule extraction algorithm scales in polynomial time with the size of the network and the size of the classification problem. Finally, continuous attributes do not need to be binary transformed as it is done in the majority of rule extraction techniques.

DIMLP networks were tested on 7 classification problems. Results showed that they were significantly more accurate than C4.5 decision trees. In the remaining sections, section 2 introduces the DIMLP model, section 3 presents the learning algorithm, section 4 describes the rule extraction technique, section 5 presents the results, followed by the conclusion.

2 The DIMLP Model

In the Discretized Interpretable Multi Layer Perceptron[1] there are an input layer, one or more hidden layers, and an output layer. As an example figure 1 illustrates a DIMLP network with two hidden layers. Note that neurons are not fully connected between the input layer and the first hidden layer. Moreover, the activation function of neurons in the first hidden layer is the staircase activation function instead of the sigmoid function.

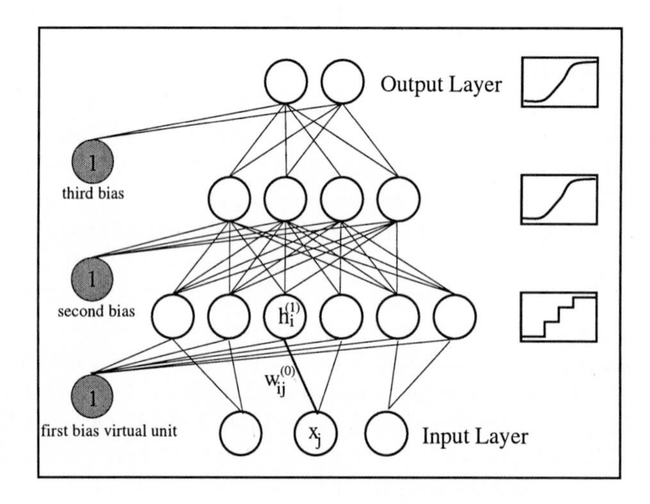

Fig. 1. A DIMLP network.

For clarity, let us first introduce several notations. The incoming signal into a neuron is given by the well-known weighted sum of inputs and weights. Vector $x = (x_0, x_1, \dots, x_n)$ specifies the input layer, and $h^{(l)}$ is the vector of activations of hidden layer l. For simplicity, the vector of activations of the output layer

[1] A C++ software can be retrieved from ftp://ftp.comp.nus.edu.sg/pub/rstaff/guido.

will be denoted to as $h^{(L+1)}$, with L the number of hidden layers. Note that the index of a bias virtual neuron is zero and its own activation is equal to one. Further, symbol $w_{ij}^{(l)}$ designates weight values between hidden layer $l-1$ and hidden layer l, where index i corresponds to the neuron of the lth layer. Symbol $w_{ij}^{(0)}$ corresponds to a weight between the input layer and the first hidden layer. The activation of a neuron after the first hidden layer is the sigmoid function given as

$$act(t) = \sigma(t) = \frac{1}{1+\exp(-t)}. \tag{1}$$

For neurons of the first hidden layer the activation is a staircase activation function that approximates a sigmoid. More precisely, let $K = (A, B) \subset \Re$ be an open interval and let $[a, b] \subset (A, B)$ be a sub-interval where the sigmoid function is quantized. We split $[a, b]$ into q equal segments of length $\delta = (b-a)/q$. Each point a_i of the division is defined to as $a_i = a + \delta i$, $0 \leq i \leq q$. The staircase activation function is defined on $[a, b]$ as

$$stair(t) = \sum_{i=0}^{q} step(t - a_i) \cdot [\sigma(a_i) - \sigma(a_{i-1})]; \tag{2}$$

where $\sigma(a_{-1})$ is equal to zero and $step(t)$ denotes a step function defined as

$$step(t) = \begin{cases} 1 & t \geq 0 \\ 0 & t < 0 \end{cases}. \tag{3}$$

Finally, for neurons of the first hidden layer the activation function is defined on K as

$$act(t) = \begin{cases} 0 & t < a \\ 1 & t > b \\ stair(t) & a \leq t \leq b \end{cases}, \tag{4}$$

with a and b equal to 5 in our implementation.

3 Learning

The training phase is carried out by varying the weights in order to minimize the *Mean Squared Error* (MSE) function given as

$$MSE = \frac{1}{2} \sum_{p,i} (h_i^{L+1} - t_i)^2; \tag{5}$$

where p denotes training examples, i is the index of output neurons and t_i are target values for supervised learning. Although (5) is not differentiable with staircase activation functions we use a back-propagation algorithm [?]. More precisely, during training the gradient is determined in each layer by the use of sigmoid functions, whereas the error related to the stopping criterion is calculated using staircase activation functions in the first hidden layer. More precisely, the training algorithm is given as:

1. Forward an example using sigmoid activation functions in the first hidden layer.
2. Compute the gradient of the error.
3. Modify weights using the gradient.
4. If all examples have been presented compute the mean squared error of all examples using staircase activation functions in the first hidden layer.
5. If stop criterion reached goto 1; else stop.

Generally, because the staircase activation function represents a quantization of the sigmoid function, the difference of the responses given by a network with staircase activation functions and a network with sigmoid activation functions tends to zero when the number of stairs is sufficiently large. As an heuristic from experience, between 30 and 100 stairs are sufficient to learn a large number of data sets.

4 Rule Extraction

In a DIMLP network the key idea behind rule extraction is the precise localization of discriminant frontiers. In a standard multi-layer perceptron discriminant frontiers are not linear [2]; further, their precise localization is not straightforward. The use of staircase activation functions turns discriminant frontiers into well-determined hyper-planes. A mathematical explanation is given below.

Corollary 1. *In a DIMLP network with one hidden layer or more the use of the staircase activation function builds a lattice of hyper-rectangles. In a hyper-rectangle the activation of all neurons of the output layer is constant.*

Proof. The activation of a neuron of the first hidden layer is

$$h_i^{(1)} = \sum_j act(w_{ij}^{(0)} x_j). \tag{6}$$

Therefore, a neuron of this layer has constant activation on each interval $[a_i, a_{i+1}]$. For all above layers, including the output layer, the activation is

$$h_k^{(l)} = \sum_i \sigma(w_{ki}^{(l-1)} h_i^{(l-1)}). \tag{7}$$

For $l = 2$, the activation of a neuron is constant on each hyper-rectangle formed by intervals $[a_i, a_{i+1}]$ in all dimensions of the input space. For $l \geq 3$, the same property holds because the image by a function of a domain with a constant value is again a domain with a constant value.

Corollary 2. *An axis-parallel hyper-plane between two hyper-rectangles is a possible discriminant frontier.*

Proof. From corollary 1, on each hyper-rectangle of the input space the outputs of a DIMLP network do not vary. Therefore, if two contiguous hyper-rectangles are associated with a different class the discriminant frontier is an axis-parallel hyper-plane between the two hyper-rectangles.

Definition 1. *A possible discriminant hyper plane is denoted to as a virtual hyper-plane.*

Corollary 3. *For a staircase activation function with q stairs of length δ, virtual hyper-planes lie in*

$$v_k^{(i)} := \frac{a + \delta k - w_{i0}^{(0)}}{w_{i1}^{(0)}}; \tag{8}$$

with $0 \leq k \leq q$.

Proof. We have

$$a + \delta k = w_{i0}^{(0)} + w_{i1}^{(0)} x_i. \tag{9}$$

Rearranging terms we obtain the corollary.

The purpose of the rule extraction algorithm is to cover with a minimal number of symbolic rules all hyper-rectangles containing training and testing examples. Because hyper-planes are precisely determined by weight values connecting an input neuron to a hidden neuron (cf. (8)), the degree of matching between network responses and extracted rules denoted also to as *fidelity* is 100%. The rule extraction algorithm checks whether a hyper-plane frontier is effective or not in a given region of the input space. As an example, figure 2 illustrates rule extraction with the use of a Karnaugh map. Note that Karnaugh maps cannot be used with more than 6 logical variables. Therefore, a more general algorithm should be defined. We decided to use decision trees as an important component of our rule extraction technique, because they are often faster than other algorithms. For a more detailed explanation of this algorithm see [2], and [3].

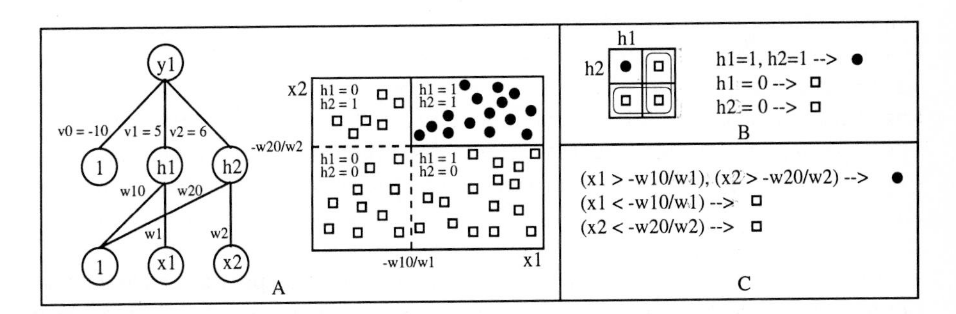

Fig. 2. An example of rule extraction from a network with step activation functions. Each rectangle is characterized by a distinct vector of binary activations of the hidden layer (00, 01, 10, and 11; respectively). A Karnaugh map is used to represent each hyper-rectangle. Finally, three rules are obtained after simplification.

Generally, the search for the minimal covering is an NP-hard problem. However, as our rule extraction algorithm uses several heuristics the overall computational complexity is polynomial with respect to the number of inputs, stairs and examples [2], and [3]. Therefore, even for reasonably large data sets and large DIMLP networks the rule extraction problem is tractable.

5 Experiments

Seven classification problems were selected from the University of California public domain. C4.5 decision trees with default learning parameters were also included in the experiments. Table 1 gives the characteristics of the data sets, as well as DIMLP neural architectures.

Table 1. Databases and neural network architectures.

Problem	Nb. Cases	Nb. Inputs	Nb. Classes	DIMLP Arch.
Cancer-w	699	9	2	9-9-20-2
Credit-g	1000	13	2	61-61-6-2
Diabetes	768	8	2	8-8-30-2
Heart-c	302	13	2	22-22-5-2
Labor	57	16	2	29-29-3-2
Sonar	208	60	2	60-60-3-2
Vehicle	846	18	4	18-18-10-4

5.1 Methodology

Generally, in order to discover new knowledge the optimal description of a data set is given by the minimal number of rules with the maximal predictive accuracy. For DIMLP networks and C4.5 decision trees the evaluation of our results was related to these two measures. Predictive accuracy and quality of rules in terms of number of extracted antecedents per rule set were evaluated by cross-validation. Results are given on the average calculated after ten repetitions of ten-fold cross-validation for all data sets. Further, the training phase of DIMLP networks was stopped according to the minimal error measured on an independent validation set. More precisely, for each cross-validation trial the proportions of the training set, validation set and testing set were 8/10, 1/10 and 1/10, respectively.

The number of neurons in the hidden layers was based on the heuristic that the number of weights must be less than twice the number of examples, and three neurons in the second hidden layer being a minimum. Weight values between the input layer and the first hidden layer were initialized with values such as for a given input neuron i the value of the virtual hyper-plane in the middle (cf. (8) with $k = 25$ and 50 stairs in the staircase activation function) was the average value of variable i in the training set. All other weights were set with the value $1/\sqrt{S}$, with S the number of incoming connections into a neuron. Finally, default learning parameters were used. Those parameters in our implementation of back-propagation are: $learning\ parameter = 0.1$; $momentum = 0.6$; $flat\ spot\ elimination = 0.01$.

5.2 Results

Average predictive accuracies and average number of antecedents per extracted rule set are given in table 2. We computed the standard error of the difference between the averages of the neural networks and those of decision trees. The t statistic for testing the null hypothesis that the two means are equal was then obtained and a two-tailed test was conducted. For predictive accuracy averages if the null hypothesis was rejected at the level of 1%, we emphasized in table 2 the method that gave the highest value, whereas for the number of extracted antecedents we emphasized the method that gave the lowest value.

Table 2. Predictive accuracies and number of antecedents per rule set.

Problem	DIMLP	C4.5Rules	DIMLP	C4.5Rules
Cancer-w	**96.7 ± 0.4**	95.7±0.5	**16.3±0.9**	21.4±1.7
Credit-g	**73.4±1.4**	71.5±0.9	163.5±7.8	**57.9±6.7**
Diabetes	**75.8±0.8**	74.1±0.9	62.4±4.0	**25.6±3.4**
Heart-c	**83.2±1.7**	78.5±2.1	39.1±2.4	**31.4±1.6**
Labor	**92.7±3.3**	78.5±4.2	6.9±0.5	**4.9±0.3**
Sonar	**80.6±1.9**	71.9±2.8	48.9±1.9	**26.7±1.8**
Vehicle	**76.2±1.3**	72.4±1.1	240.5±10.2	**127.7±10.8**

5.3 Discussion of Results

On average DIMLP networks have been significantly more accurate than C4.5 decision trees on all 7 problems. The reason for this dichotomy could reside in the fact that the inherent classification mechanism of neural networks is based on a multi-variate search technique, whereas those related to decision trees is based on a uni-variate search technique. Therefore, during the training phase a decision tree may miss rules involving multiple attributes which are weakly predictive separately but become strongly predictive in combination [11]. On the other hand, a neural network may fail to discern a strongly relevant attribute among several irrelevant ones. Finally, the number of extracted antecedents from C4.5 decision trees was significantly lower than those generated from DIMLPs on 6 data sets. This suggested that on those data sets more accurate classifiers were determined at the cost of more complex extracted rule sets.

6 Conclusion

DIMLP is a hybrid model having both a neural network structure and a symbolic module explaining its responses with 100% fidelity. This explanation capability by means of symbolic rules is a starting point to knowledge discovery. DIMLP networks can be used to learn any data set and there is no need to quantize the input variables as it is done in the majority of rule extraction techniques applied to neural networks. Here, on 7 real world classification problems DIMLP was significantly more accurate than C4.5 on average. Finally, the algorithmic complexity of the rule extraction algorithm is polynomial with respect to the number of examples, the dimensionality of the problem and the size of the network. Thus, it does not suffer from the typical exponential scaling presented by many decompositional rule extraction techniques.

References

1. Andrews, R., Diederich, J., Tickle, A.B.: Survey and Critique of Techniques for Extracting Rules from Trained Artificial Neural Networks. Knowledge-Based Systems, vol. 8, no. 6 (1995), 373–389.
2. Bologna, G.: Symbolic Rule Extraction from the DIMLP Neural Network. In Wermter, S., & Sun, R. (Eds.), Neural Hybrid Systems (2000), 240–254. Springer Verlag.
3. Bologna, G.: Rule Extraction from a Multi Layer Perceptron with Staircase Activation Functions. Proc. of the International Joint Conference on Neural Networks, **3** (2000), 419–424.
4. Craven, M., Shavlik J.: Extracting Tree-Structured Representations of Trained Networks. In Touretzky, Mozer, Hasselmo (Eds.), Proc. of Neural Information Systems, **8** (1996), 24–30.
5. Fu L.M.: A Neural-Network Model for Learning Domain Rules Based on Its Activation Function Characteristics. IEEE Transactions on Neural Networks, **9** (1998), (5) 787–795.
6. Golea, M.: On the Complexity of Rule Extraction from Neural Networks and Network Querying. Proc. of the Rule Extraction from Trained Artificial Neural Networks Workshop, Society for The Study of Artificial Intelligence and Simulation of Behaviour, Workshop Series (AISB'96). University of Sussex, Brighton, UK (1996), 51–59.
7. Ishikawa, M.: Rule Extraction by Successive Regularization. Neural Networks, **13** (2000), 1171–1183.
8. Krishnan, R., Sivakumar, G., Bhattacharya, P.: A Search Technique for Rule Extraction from Trained Neural Networks. Pattern Recognition Letters, **20** (1999), 273–280.
9. Maire, F.: A Partial Order for the M-of-N Rule Extraction Algorithm. IEEE Transactions on Neural Networks, **8** (1997), 1542–1544.
10. Quinlan, J.R.: C4.5: Programs for Machine Learning. Morgan Kaufman (1993).
11. Quinlan, J.R.: Comparing Connectionist and Symbolic Learning Methods. In Computational Learning Theory and Natural Learning, R. Rivest (eds.) (1994), 445–456.
12. Setiono, R.: Generating Concise and Accurate Classification Rules for Breast Cancer Diagnosis. Artificial Intelligence in Medicine, **18** (2000), 205–219.
13. Taha, I.A., Ghosh, J.: Symbolic Interpretation of Artificial Neural Networks. IEEE Transactions on Neural Networks, **11** (3) (1999), 448–463.
14. Vaughn, M.: Derivation of the Multi-Layer Perceptron Constraints for Direct Network Interpretation and Knowledge Discovery. Neural Networks, **12** (1999), 1259–1271.

Automatic Symbolic Modelling of Co-evolutionarily Learned Robot Skills

Agapito Ledezma, Antonio Berlanga and Ricardo Aler

Universidad Carlos III de Madrid Avda. de la Universidad, 30, 28911, Leganés (Madrid). Spain

Abstract Evolutionary based learning systems have proven to be very powerful techniques for solving a wide range of tasks, from prediction to optimization. However, in some cases the learned concepts are unreadable for humans. This prevents a deep semantic analysis of what has been really learned by those systems. We present in this paper an alternative to obtain symbolic models from subsymbolic learning. In the first stage, a subsymbolic learning system is applied to a given task. Then, a symbolic classifier is used for automatically generating the symbolic counterpart of the subsymbolic model.
We have tested this approach to obtain a symbolic model of a neural network. The neural network defines a simple controller of an autonomous robot. A competitive coevolutive method has been applied in order to learn the right weights of the neural network. The results show that the obtained symbolic model is very accurate in the task of modelling the subsymbolic system, adding to this its readability characteristic.

1 Introduction

The use of evolutionary computation (EC) techniques for software development suffers in some aspects from analogous problems to other software development methodologies or paradigms. In particular, we will focus in this paper in the declarative representation of the evolutionary generated descriptions; that is, how we (humans) interpret the output of the EC systems (their generated knowledge).

In the case of the application we present here, robot control, there are many types of knowledge that could be acquired by means of EC in order to build such systems. Examples are the internal model of robots, models of other robots, communication strategies, or reasoning heuristics. One way of automating this task consists on learning those models by either applying genetic algorithms [1], evolutionary strategies [2], classifier systems [3], or genetic programming [4]. Another view of this type of tasks is centered on the representation structure of the output: the systems can generate rules [5], neural networks [6], etc. When the output is represented in terms of subsymbolic structures (such as neural networks), it is very difficult to interpret the results in order to extract general conclusions on the correctness of the learned knowledge, its possible drawbacks, or the definition of improvements.

J. Mira and A. Prieto (Eds.): IWANN 2001, LNCS 2084, pp. 799-806, 2001.

The ability to transform a procedural description of the reasoning process of a given control skill into a declarative representation allows to more easily share knowledge, or reason about other robots behaviors. Specifically, one of our goals was the study of automatic ways of extracting knowledge (models) from non-symbolic representations, such as neural networks. This has been already studied by some authors by analysing the internal structure of the neural network [7]. We propose an alternative that consists on modeling their behavior by observing how they "solve problems": what output they generate from what input.

In this paper, Section 2 describes the task that we have used as the testbed. Section 3 presents our learning approach to symbolic modelling. Section 4 describes the way in which experiments were defined, and presents the obtained results. Finally, Section 5 discusses the obtained results.

2 Co-evolution of skills for robot control

Problems related with robotics have been one of the main fields of application of evolutive computation. A wide variety of robotic controllers, to solve specific tasks, have been investigated; robot planning [8], wall following task [4], collision avoidance [9], etc. The traditional evolutive computation techniques have several disadvantages. Coevolution has been proposed as a way to evolve a learner and a learning environment simultaneously such that open-ended progress arises naturally, via a competitive arms race, with minimal inductive bias [10,11]. The viability of an arms race relies on the sustained learnability [12,13] of environments. The capability to obtain the ideal learner, the better environments where the learning takes place, is the main advantage of the coevolutive method.

In this work, the task faced by the autonomous robot is to reach a goal in a complex bidimensional environment while avoiding obstacles found in its path. In the proposed model, the robot starts without information about the right associations between environmental signals and actions responding to those signals. The number of inputs (robot sensors), the range of the sensors, the number of outputs (number of robot motors) and its description is the only starting information. From the initial situation the robot is able to learn through experience the optimal associations between inputs and outputs.

The input sensors considered in this approach are the ambient and proximity sensors, s_i.

The Neural Network outputs are the wheel velocities v_1 and v_2. The velocity of each wheel is calculated by means of a linear combination of the sensor values, equation 1, using those weights (Equation 1):

$$v_j = f(\sum_{i=1}^{5} w_{ij} s_i) \qquad (1)$$

w_{ij} are the weights to be learned, s_i are sensor input values and f is a function for constraining the maximum velocity values of the wheels.

Weight values depend on problem features. To find them automatically, an evolutionary strategy (ES) with uniform coevolution (UC) is used [6] . In this approach each individual is composed of a 20 dimensional-real valued vector, representing each one of the above mentioned weights and their corresponding variances. The individual represents the robot behavior resulting from applying the weights to the equation 1. The evaluation of behaviors is used as the fitness function for the ES.

From all the general controllers obtained for navigation purposes using UC, a controller has been selected for automatic adquisition of its model. The main characteristic of this controller is that it only determines the speed of wheel v_2. The speed of wheel v_1 is fixed to the maximum velocity.

3 Automatic acquisition of models

The behaviour of a reactive robot can be understood in terms of its inputs (sensors readings) and outputs. Therefore, there is a clear analogy with a classification task in which each input parameter of the robot will be represented as an attribute that can have as many values as the corresponding input parameter. In terms of a classification task, this allows to define a class for each possible output. Therefore, the task of modelling (generating a declarative representation of a robot behavior) has been translated into a classification task.

For this problem, any classification technique c could be employed: instance based learning [14], learning decision trees [15], learning rules [16,17], or neural networks [18]. However, we want to obtain a declarative symbolic representation. This constrains the type of technique to be used to those that generate symbolic representations, such as decision trees, or rules.

In a previous paper we have presented results for agents whose outputs is discrete [19]. Given that the outputs of the robot control task are wheel velocities, which are continuous values, two different approaches can be used: either discretize the output and use a typical symbolic classifier (like C4.5 [17]), or use a symbolic algorithm that is able to deal with continous outputs (like regression trees [20,21]). Here, we have followed both approaches.

In the preliminary results presented here, the robot to be modelled is controlled by a neural network. The symbolic techniques to model this robot are C4.5 [17,22][1] and M5 [21]. C4.5 generates rules and M5 generates regression trees. The latter are also rules, whose then-part is a linear combination of the values of the input parameters.

The actual learning task is as follows:

- Inputs:
  - Set of attributes that model the input parameters (sensors) of robot r_1
  - For each attribute, the set of values that its corresponding input parameter can have (in this case, they are continuous variables)
  - Set of possible outputs in the case of discrete classes, and continuous range in the case of continuous classes

[1] We have used WEKA's C4.5 rules implementation [22] rather the original Quinlan's algorithm

  - Set of training instances T
  - A classification technique c
- Output: a declarative classifier that provides the same (or approximate) output as the robot r would provide given the same input instances

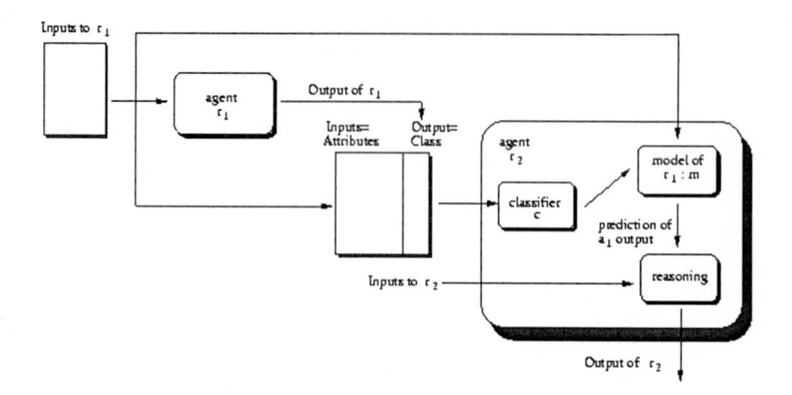

Figure 1. Architecture of the modelling of robots behavior.

4 Experimental Setup and Validation

The general framework is described in Figure 1 which shows the interrelation between the robot r_1, the modeler r_2 that tries to learn and reason about a model of r_1, the classification technique c used for modelling its behavior, and the obtained classifier m (model of r_1). This classifier m should model the behavior of robot r_1, in such a way that if one presents the same set of input patterns (sensory data) to both r_1 and m the error between the output provided by r_1 and m should be minimal.

To validate m (i.e. how closely r_2 knowledge models r_1 behaviour) we carried out ten-fold cross-validation. Testing data, which is different from the training data used in the previous section, was obtained in a similar way, by running r_1 and logging its inputs and outputs. In C4.5, the closeness of the performance of both r_1 and r_2 is measured as the number of examples in which the predictions of r_2 and r_1 differ (for the same sensory input). In the experiment that uses M5 to model r_1 behavior we use the correlation coefficient to measure the model error. The correlation coefficient is the measure of the correlation between the predic ted values and the real values of test instances. If correlation coefficient is 1, the predicted and real values are perfectly correlated. If the correlation coef ficient value is close to 0 there are no correlation. A -1 value means that they are inversely correlated. The next two subsections explain the validation carried out for the m obtained by C4.5 and M5, respectively.

4.1 Generating rules with C4.5

As C4.5 can only predict discrete outputs, wheel velocities have been discretized into five classes (see Table 1).

Table 1. Velocity Range

Class	Velocity	Number of instances
Slow	-1.000 to -0.500	54
Middle-slow	-0.499 to -0.003	26
Null	0	1
Middle-high	0.010 to -0.500	36
High	0.501 to 1.000	46
		165

The total number of testing instances is 165, which are distributed in five classes (see Table 1). Testing results are shown in Table 2. The first column shows the 10-fold crossvalidation modeling accuracy of pure C4.5, whose output is a decision tree. The second column shows the accuracy for C4.5 when it generates a set of rules. In short, the model m generated by C4.5-RULES is able to guess the output of the neural net 88 times out of 100, which is a quite good result.

Table 2. Results using C4.5.

	Hits/c4.5	Hits/c4.5 Rules
Cross-Validation	86.66%	88.48 %
Mean Absolute Error	0.0642	0.0534
Root Mean Squared Error	0.2221	0.2089

The rules learned can be seen in Table 3.

4.2 Generating regression trees with M5

The total number of instances is 976 with continuos classes (the class is the velocity of one wheel of r_1). We applied the M5 algorithm to generate a regression tree. The results of 10-fold crossvalidation are: Correlation Coeficient: 0.995, Mean Absolute error: 0.034 and Root Mean Square Error: 0.064. The regression tree predicts almost perfectly r_1 neural network.

The regression tree obtained is shown in Table 4. Each rule from a regression tree corresponds to a Linear Model (table 5) that estimate the class value (velocity of wheel v_2).

Table 3. Rules generating by C4.5 RULES.

SENSOR-4	$>$ -0.046587 AND
SENSOR-4	$>$ 0.362908: slow (30.0)
SENSOR-1	$>$ 0.186667 AND
SENSOR-2	$\leq$ 0.453333: slow (21.0/1.0)
SENSOR-4	$\leq$ -0.424854 AND
SENSOR-3	$\leq$ 0.16 AND
SENSOR-1	$\leq$ 0.053333 AND
SENSOR-2	$\leq$ 0.28: high (43.0)
SENSOR-4	$>$ -0.046587 AND
SENSOR-3	$\leq$ 0.026667: middle-slow (16.0)
SENSOR-4	$\leq$ 0.060036 AND
SENSOR-3	$\leq$ 0.506667 AND
SENSOR-2	$\leq$ 0.293333 AND
SENSOR-5	$>$ 0.058883: middle-high (28.0)
SENSOR-4	$\leq$ 0.060036 AND
SENSOR-3	$>$ 0.413333: middle-slow (8.0/1.0)
SENSOR-4	$\leq$ 0.060036 AND
SENSOR-1	$\leq$ 0.053333 AND
SENSOR-2	$\leq$ 0.293333: middle-high (6.0)
SENSOR-3	$\leq$ 0.026667 AND
SENSOR-1	$>$ 0.053333: middle-slow (4.0/1.0)
SENSOR-3	$>$ 0.026667: slow (4.0)
SENSOR-5	$\leq$ 0.786346: high (3.0)
: null (2.0/1.0)	

Table 4. Rules from the regression tree.

Sensor1	*Sensor2*	*Sensor3*	*Sensor4*	*Sensor5*	Model
$\leq$ 0.0333	-	$\leq$ 0.233	$\leq$ -0.841	$\leq$0.29	LM1
$\leq$ 0.0333	-	$\leq$ 0.233	$>$ -0.841 and $\leq$ -0.743	$\leq$0.29	LM2
$\leq$ 0.0333	-	$\leq$ 0.233	$>$ -0.743	$\leq$0.29	LM3
$\leq$ 0.0333	-	$\leq$ 0.233	$\leq$-0.59	$>$0.29 and $\leq$0.761	LM4
$\leq$ 0.0333	-	$\leq$ 0.233	$\leq$-0.59	$>$0.761 and $\leq$0.975	LM5
$\leq$ 0.0333	-	$\leq$ 0.233	$\leq$-0.59	$>$0.975	LM6
$\leq$ 0.0333	-	$\leq$ 0.233	$>$-0.59 and $\leq$ -0.453	-	LM7
$\leq$ 0.0333	-	$\leq$ 0.233	$>$-0.453	-	LM8
$\leq$ 0.0333	-	$>$ 0.233	$\leq$-0.161	-	LM9
$>$ 0.0333 and $\leq$0.22	-	-	$\leq$-0.161	-	LM10
$>$0.22 and $\leq$0.587	-	$\leq$0.213	$\leq$-0.161	-	LM11
$>$0.22 and $\leq$0.587	-	$>$0.213	$\leq$-0.161	-	LM12
$>$0.587	-	-	$\leq$-0.161	-	LM13
$\leq$0.193	-	-	$>$-0.161 and $\leq$0.134	-	LM14
$\leq$0.193	-	-	$>$0.134 and $\leq$0.711	-	LM15
$\leq$0.193	-	-	$>$0.711	-	LM16
$>$0.193	-	-	$>$-0.161	-	LM16

5 Conclusions

In this paper, we have presented an approach that allows to acquire a declarative representation of the behavior of a robot, by observing what output it produces from the inputs it receives. That is, instead of inspecting the robot internal model, it is considered as a black box and observed by another agent/robot. In particular, we have first used C4.5 to acquire a symbolic model (set of rules) of a neural-net based robot. Results show that C4.5 is quite good at modeling neural robots. The model obtained by C4.5 could be used by an opponent robot, either directly, or even better, reasoning about the model, taking advantage of its symbolic representation.

Then, we have used M5 to obtain a regression tree that approximates even better the target robot. However, although there is greater accuracy in this second case, the knowledge obtained is not so easily understandable.

It is important to remark that this method will only be applied successfully to reactive agents. If the agent to model is not reactive (i.e. its output depends on something else, like memory, besides the sensors), models would be quite inaccurate.

Table 5. Linear Models.

Model	Prediction	Independent factor	*Sensor1*	*Sensor2*	*Sensor3*	*Sensor4*	*Sensor5*
LM1:	class =	0.36	-0.0936	+0.0711	-0.895	-0.723	- 0.276
LM2:	class =	0.358	-0.0936	+0.129	-1.25	-0.668	-0.35
LM3:	class =	0.0201	-0.0936	+0.0288	-1.2	-1.15	-0.423
LM4:	class =	0.201	-0.0936	+0.0751	-1.29	-0.969	-0.531
LM5:	class =	0.886	-0.0936	+0.0365	-1.11	-0.427	-0.859
LM6:	class =	0.933	-0.0936	+0.0365	-1.11	-0.427	-0.943
LM7:	class =	0.0814	-0.0936	+0.0257	-0.312	-1	-0.134
LM8:	class =	-0.0041	-0.0936	+0.0202	-1.22	-1.2	-0.453
LM9:	class =	-0.00956	-0.0936	+0.0155	-1.14	-1.12	-0.166
LM10:	class =	-0.073	-2.6	+0.0441	-0.964	-1.03	-0.264
LM11:	class =	-0.149	-2.23	+0.0065	-0.407	-0.895	-0.0413
LM12:	class =	-0.427	-1.33	+0.0065	-0.424	-0.488	-0.0413
LM13:	class =	-0.719	-0.62	+0.0065	-0.226	-0.305	-0.0413
LM14:	class =	-0.0313	-2.87	+0.0897	-1.14	-1.13	-0.33
LM15:	class =	-0.0435	-2.25	+0.0879	-1.11	-1.11	-0.329
LM16:	class =	-0.408	-0.757	+0.0364	-0.38	-0.652	
LM17:	class =	-0.839	-0.126	+0.013	-0.129	-0.135	

References

1. Vicente Matellán, José Manuel Molina, and Camino Fernández, "Genetic learning of fuzzy reactive controllers," *Robotics and Autonomous Systems*, vol. 25, no. 1-2, pp. 33–41, October 1998.
2. Antonio Berlanga, Pedro Isasi, Araceli Sanchis, and José M. Molina, "Neural networks robot controller trained with evolutionary strategies," in *Proceedings of the Congress on Evolutionary Computation*, Peter J. Angeline, Zbyszek Michalewicz, Marc Schoenauer, Xin Yao, and Ali Zalzala, Eds., Mayflower Hotel, Washington D.C., USA, 6-9 July 1999, vol. 1, pp. 413–419, IEEE Press.
3. Araceli Sanchis, José M. Molina, Pedro Isasi, and Javier Segovia, "Rtcs: a reactive with tags classifier system," *Journal of Intelligent and Robotic Systems*, vol. 27, no. 4, pp. 379–405, 2000.
4. John R. Koza, *Genetic Programming: On the Programming of Computers by Means of Natural Selection*, MIT Press, Cambridge, MA, USA, 1992.
5. Vicente Matellán, José M. Molina, Javier Sanz, and Camino Fernández, "Learning fuzzy reactive behaviors in autonomous robots," in *Proceedings of the Fourth European Workshop on Learning Robots*, Alemania, 1995.

6. Antonio Berlanga, Araceli Sanchis, Pedro Isasi, and José M. Molina, "A general coevolution method to generalize autonomous robot navigation behavior," in *Proceedings of the Congress on Evolutionary Computation*, La Jolla, San Diego (CA) USA, July 2000, pp. 769–776, IEEE Press.
7. Jude W. Shavlik and Geoffrey G. Towell, *Machine Learning. A Multistrategy Approach.*, vol. IV, chapter Refining Symbolic Knowledge using Neural Networks, pp. 405–429, Morgan Kaufmann, 1994.
8. Simon G. Handley, "The automatic generations of plans for a mobile robot via genetic programming with automatically defined functions," in *Advances in Genetic Programming*, Kenneth E. Kinnear, Jr., Ed., chapter 18, pp. 391–407. MIT Press, 1994.
9. K. Solano, R. A. Chavarriaga, R. A. Nuñez, and C. A. Peña-Reyes, "Controladores adaptables basados en mecanismos de inferencia difusa," in *Memorias del Segundo Congreso Asociación Colombiana de Automática*, 1997.
10. Sevan G. Ficici and Jordan B. Pollack, "Challenges in coevolutionary learning: Arms-race dynamics, open-endedness, and mediocre stable states," in *Proceedings of the 6th International Conference on Artificial Life (ALIFE-98)*, Christoph Adami, Richard K. Belew, Hiroaki Kitano, and Charles Taylor, Eds., Cambridge, MA, USA, June 27–29 1998, pp. 238–247, MIT Press.
11. S. G. Ficici and J. B. Pollack, "Statistical reasoning strategies in the pursuit and evasion domain," in *Proceedings of the 5th European Conference on Advances in Artificial Life (ECAL-99)*, Dario Floreano, Jean-Daniel Nicoud, and Francesco Mondada, Eds., Berlin, Sept. 13–17 1999, vol. 1674 of *LNAI*, pp. 79–88, Springer.
12. Jordan B. Pollack, Alan D. Blair, and Mark Land, "Coevolution of a backgammon player," in *Proceedings of Artificial Life V*, C. G. Langton, Ed., Cambridge, MA, 1996, MIT Press.
13. Jordan B. Pollack and Alan D. Blair, "Co-evolution in the successful learning of backgammon strategy," *Machine Learning*, vol. 32, no. 1, pp. 225–240, 1998.
14. David W. Aha, Dennis Kibler, and Marc K. Albert, "Instance-based learning algorithms," *Machine Learning*, vol. 6, no. 1, pp. 37–66, jan 1991.
15. J. R. Quinlan, "Induction of decision trees," *Machine Learning*, vol. 1, no. 1, pp. 81–106, 1986.
16. Ryszard S. Michalski, "A theory and methodology of inductive learning," *Artificial Intelligence*, vol. 20, 1983.
17. J. Ross Quinlan, *C4.5: Programs for Machine Learning*, Morgan Kaufmann, San Mateo, CA, 1993.
18. D.E. Rummelhart, J.L. McClelland, and the PDP Research Group, *Parallel Distributed Processing Foundations*, The MIT Press, Cambridge, MA, 1986.
19. Ricardo Aler, Daniel Borrajo, Inés Galván, , and Agapito Ledezma, "Learning models of other agents," in *Proceedings of the Agents-00/ECML-00 Workshop on Learning Agents,*, Barcelona, Spain, June 2000, pp. 1–5.
20. L. Breiman, J.H. Friedman, K.A. Olshen, and C.J. Stone, *Classification and Regression Tress*, Wadsworth & Brooks, Monterey, CA (USA), 1984.
21. J. Ross Quinlan, "Combining instance-based and model-based learning," in *Proceedings of the Tenth International Conference on Machine Learning*, Amherst, MA, June 1993, pp. 236–243, Morgan Kaufmann.
22. E. Frank and I. Witten, "Generating accurate rule sets without global optimization," in *Proceedings of the Fifteenth International Conference on Machine Learning*. 1998, pp. 144–151, Morgan Kaufmann.

ANNs and the Neural Basis for General Intelligence

J.G. Wallace and K. Bluff

Swinburne University of Technology, P.O. Box 218, Hawthorn, Victoria, Australia 3122
jgwallace@swin.edu.au

Abstract. The existence of 'general intelligence' or '*g*' has long been the subject of controversy. Recent work suggests that direct investigation of the neural basis for *g* may break the deadlock and that a specific region of the lateral frontal cortex underpins performance in novel problem solving and other tasks with high *g* correlation. As a contribution to developing a model of *g* in terms of component frontal functions we present an early version of a theory of the evolution of a universal problem solver represented in specific neuronal circuitry in the frontal cortex. The theory draws on concepts derived from research on ANNs.

1 Introduction

The existence of 'general intelligence' was proposed by Spearman [1] on the basis of the pattern of almost universal positive correlation between measures of success in highly varied cognitive tests. Spearman attributed these results to the existence of a general or '*g*' factor underlying the range of performance. Factor analysis reveals that the tasks most highly correlated with *g* are largely tests of novel problem solving. The main alternative hypothesis, proposed by Thomson [2], asserts that performance on any task draws on a large set of component factors and the universal positive correlation does not arise from *g* but the components shared by any two tasks. Tasks with high apparent *g* correlations are those drawing most widely on the range of cognitive functions.

Due to the limitations of factor analysis it has not been possible to conclusively evaluate the relative merits of these two broad hypotheses by means of correlational data. Recent work suggests that direct investigation of the neural basis for *g* may break the deadlock. Duncan et al. [3] investigated the hypothesis that high *g* correlations should be characterized by specific recruitment of prefrontal cortex (PFC) and not by an increasingly diverse pattern of neural activation as required by the Thomson view. Their method was based on the psychometric finding that tasks with very different surface content can share the property of high *g* correlation. In their first experiment they gathered positron emission tomography (PET) data from subjects performing both spatial and verbal problem solving tasks whose *g* correlations were known to be high. The strongest relative activations on the spatial tasks occurred bilaterally in the lateral prefrontal cortex and in a discrete region of the

J. Mira and A. Prieto (Eds.): IWANN 2001, LNCS 2084, pp. 807-813, 2001.

medial frontal gyrus/anterior cingulate while the verbal tasks highlighted the lateral frontal cortex (PFC) of the left hemisphere. These data suggest that *g* reflects the function of a specific neural system including as one major part a specific region of the lateral frontal cortex. This outcome is supported by an analysis of other imaging findings which reveals that diverse forms of demand, including task novelty, response competition, working memory load and perceptual difficulty, produce broadly similar lateral frontal activations closely resembling the current experimental results. Duncan et al. conclude that "To show that *g* is associated with a relatively restricted neural system is not, of course, to show that it cannot be divided into finer functional components. For the future, indeed, a central problem will be development of more detailed models of *g* in terms of component frontal functions and their interactions".

Most linguists claim that the acquisition of language is done by specific neuronal circuitry within the brain and not by the general purpose problem solving ability of the brain. The findings of Duncan et al. suggest that general purpose problem solving ability may itself be significantly attributable to specific neuronal circuitry. In the case of language, innate specific neuronal circuitry provides a universal grammar comprising a mechanism to generate a search space for all candidate mental grammars and a learning procedure that specifies how to evaluate the sample sentences with a view to selecting a mental grammar. Nowak et al. [4] have recently formulated a mathematical theory covering the evolutionary selective pressures that act on the design of universal grammar (UG). We will present an early version of a theory of the evolution of a universal problem solver (UPS) represented in specific neuronal circuitry in the frontal cortex. Just as natural neural architectures and processes form a fertile source of ideas in the construction of artificial neural networks (ANNs) so research on ANNs may provide interesting concepts for consideration in neuroscience. Our account of the mechanisms underlying the evolution and operation of the UPS provides an example.

2 Evolution of an Architecture for *g*

One approach to the UPS is in terms of the varying degrees of flexibility in cortical operations which endow organisms with differing abilities to cope with the experience of novelty in their environmental interaction. The operation of instinct represents one extreme with hard wired connections between situations and behavioural responses. Humans are viewed as the other extreme with the greatest ability to employ learning to enable adaptation to novel environmental events. Evolutionary selection pressures have, however, set limits to this flexibility. Specialist cortical areas have evolved to constrain and facilitate responses to the complex stream of environmental input. The degree of specialisation and extent of constraint vary from the relatively low levels in the language areas underpinning the UG and its space of mental grammars to the high levels associated with automatisation of responses to previously problematic situations and the resulting transfer of cortical activity from prefrontal cortex to temporal/parietal areas.

A UPS to meet the evolutionary requirements must combine a continuing ability to cope with novel problems with mechanisms capable of dealing with complexity by rendering it tractable. Any UPS, like any UG, is derived from a space of UPSs varying in their effectiveness in providing these characteristics. We present a variant which deals with novel problems by an application of ANNs. Complexity is tackled by adopting a modular ANN approach which seeks to segment the stream of environmental input into tractable units. Critical to success is the ability to construct a range of modules which provide an optimal profile on the specificity-generality dimension in relation to environmental requirements.

3 A Universal Problem Solver

Our UPS comprises a range of ANNs interfacing a profile of sensori-perceptual inputs to a repertoire of actions.
SP: the set of sensori-perceptual inputs where each is regarded as precisely defined in terms of a specific neural source, $SP = \{SP_1 \text{ --- } SP_n\}$
A: the set of actions available to the system where each is regarded as precisely defined in specific motor neuron terms, $A = \{A_1 \text{ --- } A_m\}$
CN: the set of connection nets connecting SP to A, $CN = \{CN_1 \text{ --- } CN_j\}$
Each connection net is an ANN.

Previous work has emphasized application of the combination of ANNs in pattern recognition and in the solution of single problems. In an ensemble based approach a set of ANNs is trained in parallel on essentially the same task and their outputs are combined to yield a more reliable estimate. In a modular approach each ANN becomes a specialist in an aspect of the task and an overall outcome is derived by hierarchical combination. A hybrid system can reflect both approaches with, for example, an ensemble with each member composed of a set of modules. Since the UPS requires tackling multiple problems over time the CN architecture and processes represent a considerable extension and generalisation of typical combinations of ANNs.

Defining a CN architecture and processes involves evaluating a number of fundamental qualitative variables derived from the universe of UPSs. Consistent with the high *g* loading of competitive response tasks the relational architecture between CNs is based on a competitive principle.

Although locally hierarchical the overall structure of the UPS is heterarchical. This results from the effect on the connectivity of CN to SP and A of the selection pressures derived from the sequence of problems registering on the organism. Members of CN connecting SP to A are modified as a result of feedback on performance effectiveness. This results in the evolution of less flexible cortical areas more highly specialised or 'tilted' in terms of their range of effective application while the *g*/PFC functions continue to deal with the demands of novel experience.

A critical qualitative variable between members of the universe of UPSs is the learning mechanism underlying experience based modification. In our UPS learning comprises both intragenerational development and intergenerational evolution. To cover the requirements we include two broad types of nets. Combination of the members in both SP and A involves modular networks constructed of adaptive resonance theory (ART) NNs. Bartfai and White [5] combined ART networks in a systematic way into a modular architecture capable of learning class hierarchies from arbitrary sequences of input patterns while retaining the ability of single ART networks to engage in autonomous, fast and stable learning by self-organisation. The particular variant adopted in combining members of SP is HART-S, hierarchical ART with splitting. This involves an ART-based modular network developing cluster hierarchies by first creating large, general classes which are subsequently refined in lower layers by splitting up each super-class into subclasses by clustering features that are different from the super-class. At the beginning of the developmental and evolutionary processes the profile of active feature detectors is analysed in terms of large general classes. As Figure 1 indicates, the classes are directly connected to a range of uncombined actions in A. Individual actions initially have a randomly assigned probability of occurrence if stimulated. These SP •A connections provide the basis for the initial behaviour of the organism. In the early stages novel problem

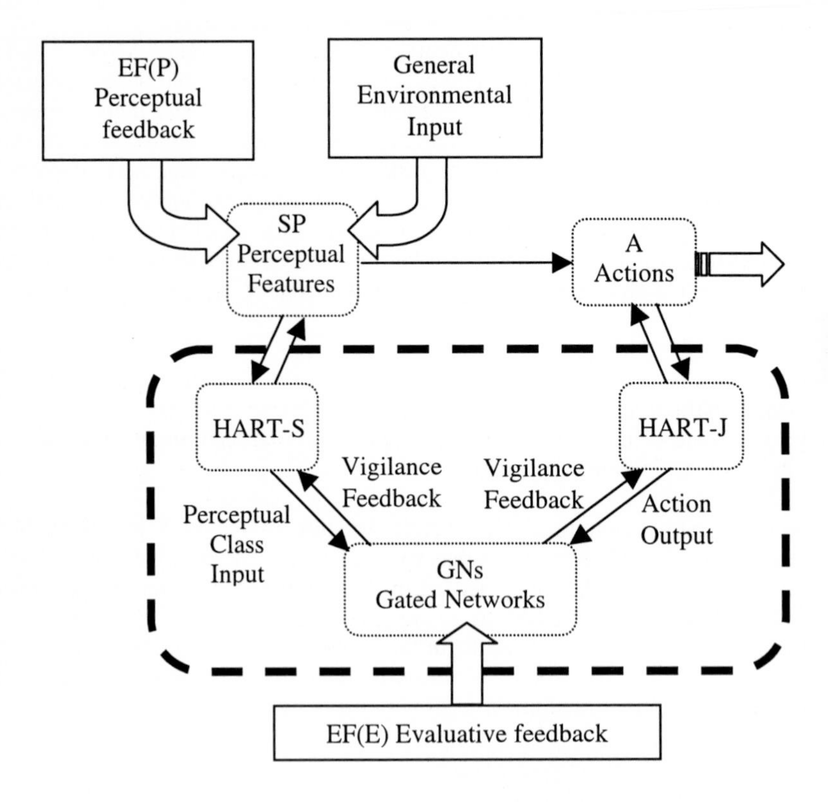

Fig. 1. The CN system. The heavy dashed line indicates the boundary of CN

solving is the dominant activity. CN enters a problem solving mode on the basis of evaluative environmental feedback, EF(E). EF(E) reflects disparities between expectations and situations produced by actions and emotional and motivational effects.

Combination of actions results from processing of the action stream by HART-J, hierarchical ART with joining. A HART-J network has several layers of ART modules. Each layer by learning to cluster the 'prototypes' developed by the previous layer builds successively larger classes in upper layers by joining the lower level, more specific ones. With experience of consistently co-occurring features and actions HART-S and HART-J construct categories reflecting the nature of environmental interaction.

It remains to learn the connnectivity between the two categorical systems which resolves novel problems and produces effective behaviour. The second broad type of net included in CN has been selected for this task. Consistent with the high *g* loading of competitive response tasks, it combines modular ANNs with gating networks. As Figure 2 indicates, gating networks determine the relative contribution of individual ANNs to performance on the basis of their differential contribution to success. Performance feedback, also, determines replacement of single ANNs by subsidiary gating networks. Gating nets and individual ANNs (GNs) receive input from HART-

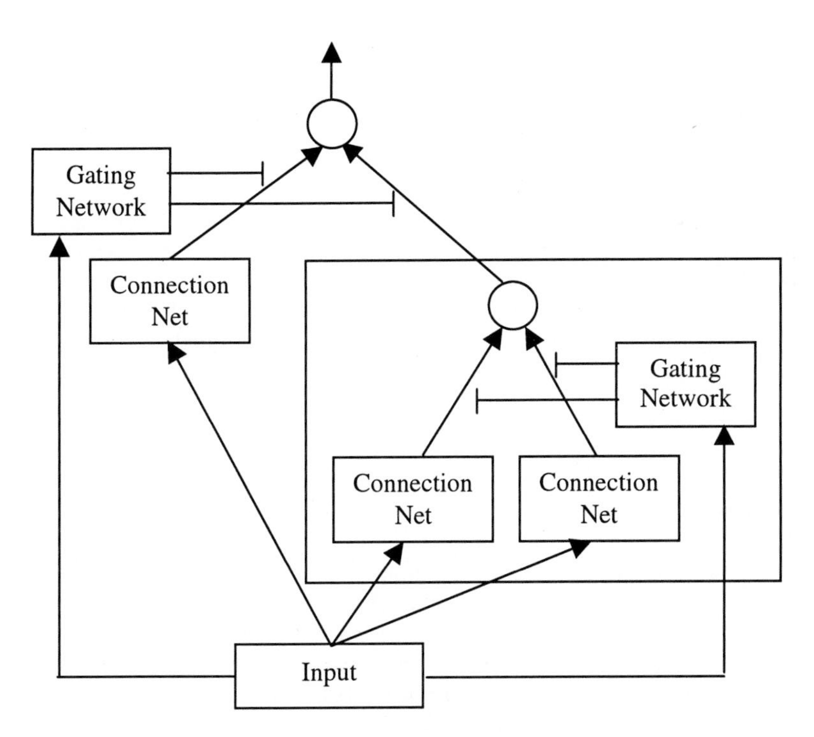

Fig. 2. The form of a simple hierarchical structure of gated nets in CN

S and HART-J in addition to feedback on the success achieved by actions on the environment. This enables segmentation of HART-S and HART-J input and definition of the appropriate connectivity between them indicated by feedback. This process results in the construction of more complex SP•A connection networks exhibiting differing degrees of flexibility. This ranges from a high degree of specialisation linked with narrow applicability to a structural and process 'tilt' providing relative compatibility with a broad content area but still exhibiting a considerable degree of versatility. In cortical terms this process underlies the modularity of areas subserving distinct functions, as new connectivity generated by problem solving in CN is transferred from PFC to other cortical areas.

Learning in individual ANNs involves backpropagation and a form of slow learning derived from a model of the information processing functions of astrocyte-neuron interaction [6].

The high degree of flexibility necessary to cope with novel problems characterises the PFC. In our model the orienting subsystem of ART modules is relied upon to detect novelty. The level of difference between new and previously coded input sufficiently significant to represent novelty is defined by a vigilance parameter (v) which is allowed to vary between modules. Novelty detection results in the deployment of fresh inexperienced network capacity within ART modules and through an escalation process within the HART-S and HART-J networks. It, also, may result in the introduction of fresh gating networks in GNs through interaction with HART-S and HART-J. Feedback from GNs controls the occurrence and magnitude of changes to v during novel problem solving.

Intergenerational evolution is critical in UPS. The key issue is determination of the flexibility or specificity – generality profile of the cortical architecture of the next generation in the light of the experience of previous generations. In our approach ANNs are hard coded in operation with representation of every weight and connection. In intergenerational transition, however, cloning of ANNs into the next generation without architectural modification is combined with a softer, probabilistic process in defining the profile of specialised, tilted and uncommitted net architectures to form the initial cortical structure. Further details of this process are provided elsewhere [7].

4 Conclusion

In seeking to prompt discussion of PFC component functions and interactions which may underlie *g* we have drawn on ANN research relevant to task novelty, response competition and perceptual difficulty, all areas associated with a high *g* loading. The space of UPSs is great and even our metalevel sketch of a UPS involves a daunting number of qualitative variables and parameters. Modelling is challenging but the power of advanced computation is reassuring.

As an encouraging postscript, experimental results supporting the existence in PFC of localised specific neural mechanisms consistent with the type of component functions included in our model continue to appear. Freedman et al. [8] have identified neurons in primate PFC sensitive to specific visual category membership while Hasegawa et al. [9] have detected PFC neurons correlated with performance in a probabilistic rather than deterministic manner and reflecting past or predicted much more than current performance.

References

1. Spearman, C.: The Abilities of Man. Macmillan, New York (1927)
2. Thomson, G.H.: The Factorial Structure of Human Ability. 5th edn. Univ. of London Press, London (1951)
3. Duncan, J., Seitz, R.J., Kolodny, J., Bor, D., Herzog, H., Ahmed, A., Newell, F.N., Emslie, H.: A Neural Basis for General Intelligence. Science, Vol. 289, American Assoc for the Advancement of Science, New York (2000) 457-460
4. Nowak, M.A., Komarova, N.L., Niyogi, P.: Evolution of Universal Grammar. Science, Vol. 291, American Assoc for the Advancement of Science, New York (2001) 114-118
5. Bartfai, G., White, R.: Adaptive Resonance Theory-based Modular Networks for Incremental Learning of Hierarchical Clusterings. Connection Science, Vol. 9, Issue 1, Carfax Publishing Company (1997) Abingdon, Oxfordshire 87-113
6. Wallace, J.G., Bluff, K.: Slow Learning and Fast Evolution: An Approach to Cytoarchitectonic Parcellation. Lecture Notes in Computer Science, Vol. 1607. Springer-Verlag, Berlin Heidelberg New York (1999) 34-42
7. Wallace, J.G., Bluff, K.: Neuro-architecture-motivated ANNs and Cortical Parcellation. IJCNN2000 Proceedings of the IEEE-INNS-ENNS International Joint Conference on Neural Networks, Vol. 5, IEEE Computer Society, Los Alamitos, California (2000) 647-651
8. Freedman, D.J., Riesenhuber, M., Poggio, T., Miller, E.K.: Categorical Representation of Visual Stimuli in the Primate Prefrontal Cortex. Science, Vol. 291, American Assoc for the Advancement of Science, New York (2001) 312-316
9. Hasegawa, R.P., Blitz, A.M., Geller, N.L., Goldberg, M.E.: Neurons in Monkey Prefrontal Cortex That Track Past or Predict Future Performance. Science, Vol. 290, American Assoc for the Advancement of Science, New York (2000) 1786-1789

Knowledge and Intelligence

Juan Carlos Herrero

e-mail: jcherrer@arrakis.es

Abstract. Knowledge and intelligence are phenomena we find in nature, usually attributed to men and women. From a scientific viewpoint, we do not know what is the nature of knowledge and intelligence. We know that it has to do with the brain, but we do not know yet how a brain has knowledge and has intelligence. There are also tests and exams to assess intelligence and knowledge, in terms of IQ and scores; this is accepted as the best way to carry out the evaluation, despite it is well known they are not all the truth: life itself is a great test on knowledge and intelligence to all of us. We properly may ask: What is knowledge? What is intelligence? Engineering also tries to build artificial devices that have knowledge and intelligence, but a suitably methodology is not yet available. This methodology should somehow set knowledge and intelligence as requirements, provide models of knowledge and intelligence, and then an assessment method in order to verify the requirements were accomplished. Precisely this is not the state of the art.

1 Introduction

If you do not mind, "Artificial Intelligence" ("AI") is used in this paper in its pure sense in order to name a kind of activities involving science and engineering that aim to make artificial devices that have intelligence and have knowledge, by means of vonNeuman computers or artificial neural networks.

What is intelligence?

What is knowledge?

We use these two words dayly, we all understand something about them, and we can look up dictionaries and encyclopaediae and find a lot of really interesting descriptions, explanations, and references to what really intelligent people and people who possessed a great amount of knowledge said about knowledge and intelligence. There are also tests and exams, in order to establish IQ and scores. You had better know they are not all the truth, because life itself is a great test for knowledge and intelligence.

AI has no definition for knowledge and intelligence, so AI has no definition for the stuff AI is handling. This has two main consequences. Firstly, although knowledge and intelligence are phenomena we find in nature, for instance attributed to man and women, AI has no model which describes and establishes knowledge and intelligence in this sense. Secondly, because in order for AI to develop as any other engineering, a methodology that starting from a definite set of specifications provided us with a detailed description of the process one has to follow in order to achieve the

J. Mira and A. Prieto (Eds.): IWANN 2001, LNCS 2084, pp. 814–821, 2001.

corresponding implementation is needed. In fact, those specifications are supposed to be the intended knowledge and intelligence the final implementation must display, and the methodology must teach how to model knowledge and intelligence. Thereafter, a suitable verification and validation process must assess the compliance with specifications (the intended knowledge and intelligence). This is the way to assure no disagreement between analysis and synthesis, and to assure that our work is done.

This is clearly not the state of the art. Without knowledge and intelligence models, a methodology, and verification and validation process, how can knowledge and intelligence be assessed?

Although most criticized, the only approach to an assessment belongs to the early years and was set up by Turing. It refers more to intelligence and leaves the decision to the sincere appreciation and cleverness of an observer. Place an observer in front of a human being and simultaneously in front of an artificial device whose intelligence we try to assess: if the observer cannot distinguish who is who by means of questions, it must be admitted that such artificial device is intelligent. If play is fair, this method is perfect. But observer's subjectivity may be misleading and in fact leads to the current state of the art: intelligence is attributed to men, women, computing systems, neural nets, mosquitoes, buildings, cards, electrical appliances... Maybe they are all intelligent (I wouldn't believe that, would you?), I am only saying that all this is said without a definition of intelligence, without a process of assessment on it which rendered the objective reasons to state it.

When it comes to knowledge, we find a similar situation. Any given vonNeumann computer circuit as well as any neural network circuit is designed so that output voltage depends on input voltage variations. But, what causes those input voltage variations? Let us suppose the circuit belongs to a computer that is programmed to solve some kind of diagnosis which is usually solved by a human expert. If we say, for instance, that some input is "temperature is 39 Celsius degrees", how is this implemented? We may have the usual case of someone typing "t=39" at a keyboard when some message is displayed on the computer screen. The keyboard turns the typing into voltage variations.

So, what is really the input and what is really the computation?

If we consider keyboard is the input device, computer input consist of voltage variations. If we want input to be temperature, we have to include at least the typist into a computational device ensemble, so that computation is carried out by at least the typist with the help of the computer or by the computer assisted by a human being (with a thermometer, for example). If we do not consider the typist as part of the computation, input (voltage variation) conveys no knowledge at all to the computer because the only being who knows what the typing means is he who typed. (Therefore, if we do not consider the typist as part of the computation, diagnosis cannot be carried out either.) We have to admit that nowadays most of computation cannot be carried out without human assistance. Thus, if for the simple reason that someone typed "t=39" at the keyboard, someone says this computer knows that "temperature is 39 Celsius degrees", we can affirm this is subjective and arbitrary.

The point here is that computer input has no causal relationship with environment temperature, being that causal relationship an objective reason which justifies our assessment of it. If temperature value were supplied to the computer by means of a sensor this would establish a causal link between the suitable events in the environment and the computer input, and the computation has causally to do with environment temperature. So we are not subjective if we say the changes the computer undergoes due to interaction between sensor and environment are causally related to temperature, there is not human assistance.

2 The Scientific Scenario

Let us make clear the scientific scenario where we introduce knowledge and intelligence. We need it, all the more because we intend to make artificial counterparts of knowledge and intelligence, as we just have said. We shall try to sketch the whole picture, but let us start from the very beginning.

The scientific scenario has three basic kind of players, namely: facts, observer, and description. Let us call them "1", "2", and "3", respectively. Figure 1 represents the players and how it all works: the observer observes the facts and writes up a description about them. But there are a few things to say about these players.

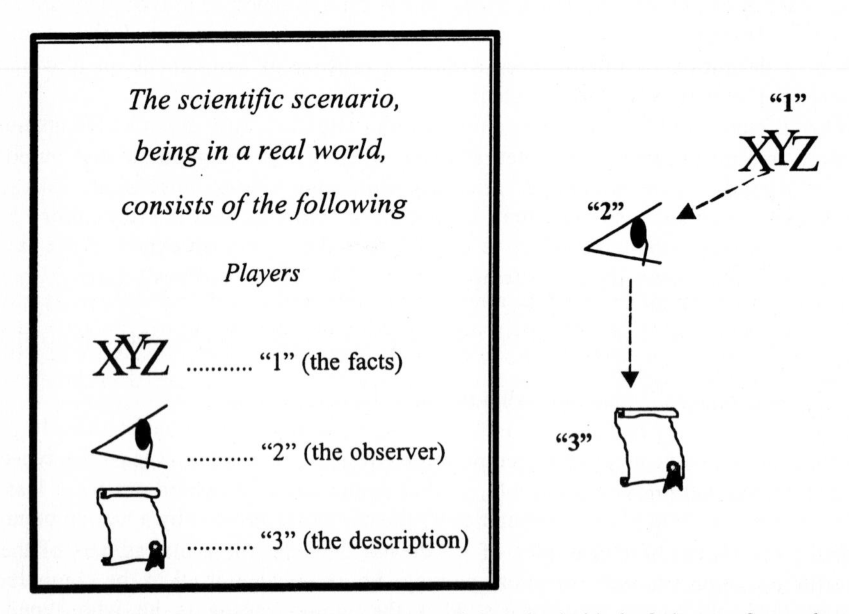

Fig. 1. Scientific scenario (on the left) consists of three basic kind of players: facts, observer and description. On the right: the observer observes the facts and writes up a description about them.

Implicitly, we admit the observer lives in the real world. This is not that easy, I mean the observer thought once: I think, therefore I am. He thought that maybe all that he preceived could be mere imaginations of him: all the observations about the world, the nature, and all his experiments. But in these perceptions there were other people who were like he thought he also was; they said they had what he thought similar experiences in their relationship with the same world which he thought he lived in. So he thought it could also be that instead of everything but him were a mere imagination, he were but just another part of what he called the real world. He was not arrogant at all, so he thought this was a better choice.

Thus, he thought that facts happened in the real world and therefore his perceptions were causally related to these facts. However, he clearly noted that he had perceptions that other people had not. So, he thought that there must be facts in the real world that were only perceived by him. Therefore, if he was part of the real world he had to find a place in the real world where all these facts only perceived by him happen. He called it "B", temporarily, because he had not found the place yet. (See figure 2.a).

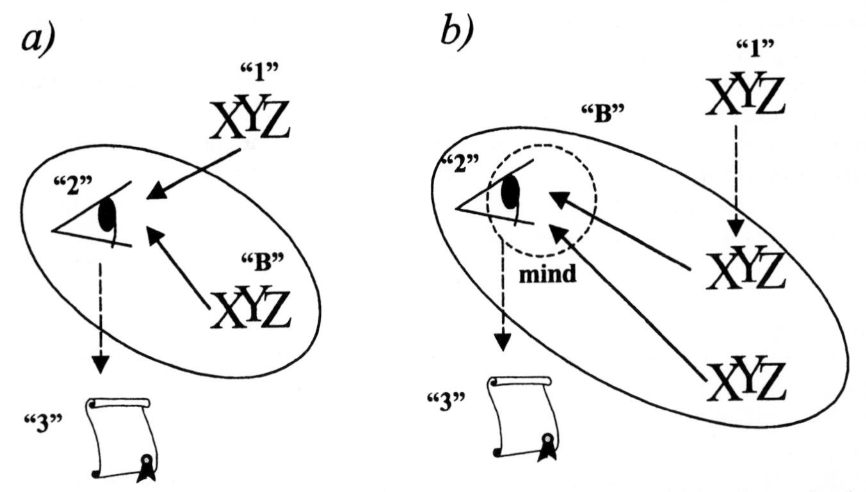

Fig. 2. *a)* **First approach**: Facts "1" are perceived by the observer "2" and can also be perceived by other observers. Perceptions are causal relationship with facts, this is represented by arrows. Oval encloses facts that are perceived only by the observer "2". *b)* **A more suitable approach**: Oval "B" encloses facts that are perceived only by the observer; some of these facts have causal relationships with facts that other observers can also perceive. In this way, only facts within the oval are perceived by the observer "2". The dashed circle represents the mind, the place where all perceptions take place, causally related to "B" facts. The observer only has access to this dashed circle.

Since he thought he is part of the real world, there must be a place where all his perceptions happen; he called it mind. At once, he realized that there should be minds like his, belonging to other people he perceived, according to what they said they were their experiences too (and he also supposed that all the people were like him).

It seemed to him a little bit strange that some of his perceptions were caused by "B" facts and some other not, so he thought that maybe all his perceptions were caused by "B" facts. Although "B" facts are perceived only by him, it could well be that part of "B" facts had also causal relationships with facts that any other people could observe; thus, following the causal chain, he could have perceptions of facts that other people could also perceive. But in this way, he found that "B" facts could well be the cause of all perceptions that take place in his mind. (See figure 2.b).

It could also be that some of those facts that only the observer can perceive had causal relationships with facts that any other observer could perceive, but this causal relationships occured in the past. The corresponding perceptions were called memories by the observer. It seemed to him that it was not unusual that their perceptions remained as memories. He then thought that he could also find memories of facts only perceived by him.

The observer is a scientist and engineer (maybe a philosopher too, but with a scientific viewpoint). For all he knows about himself, he defines himself as intelligent; this is not an act of arrogance at all, as we know, all the more because he believes that the other people are also intelligent. He decided to give name to all his perceptions, no matter the origin: he called them knowledge. He also noted that knowledge has not been there all the time, specially that he usually has new knowledge from time to time. So, he imagined that something causes the new knowledge to appear from time to time and he called it intelligence. At least, this is what he means when he talks about knowledge and intelligence. Besides, since he locates knowledge and intelligence in his mind and knowledge has causal relationships with "B" facts, he concluded that intelligence should also have causal relationships with "B" facts.

Descriptions express the knowledge the observer has about facts, but by no means descriptions are the knowledge the observer has about these facts.

Let us see some examples. One of the most famous is that of a wise and tenacious observer who after observing facts for a long time, sleeping from time to time, eventually wrote up a description called "Philosophiae Naturalis Principia Mathematica". The facts were related to movements of celestial bodies beside all that he had read and thought about it. So, "1" are movements of celestial bodies, the celestial bodies itself, books that he read about it, and so on. "B" are the effects of "1" upon the observer; these effects are also facts that cause the perceptions or knowledge in the observer's mind; some of them remain. More knowledge was generated in the observer's mind due to his intelligence. "3" expresses knowledge about "1"; "3" expresses knowledge belonging to the observer, i.e. "3" belongs to "2". By the way, "2" was Isaac Newton.

Another example. In this case, "1" are facts related to patients. "2" is the doctor. "B" are the effects of these facts upon the observer (e.g., effects of light upon his eyes and subsequent effects upon optical nerves, and so on, etc.) beside effects of past observations; in the doctor's mind, "B" causes all that he knows right now beside all that he learnt during his life. "3" expresses knowledge about "1". "3" could well be a diagnosis, let us suppose that this diagnosis is fully explained. So "3" expresses the observer's knowledge, i.e. "3" belongs to "2".

Now it is time to talk about AI's observer (figure 3). Let us call this case "C", and the previous one "A". "C" wants to observe "A", because he wants to build "4", an artificial device which has knowledge and intelligence. He has two possibilities: C's "1" (the facts) can be either A's "B" or A's "3". He does not know how C's "1" could be A's mind or perceptions, i.e. he does not know how to observ A's mind directly. However, he finds out a great thing along his experience as observer of similar cases: A's "B" is precisely A's brain, so he discovers that the brain is the place where the facts that cause perceptions take place; therefore, brain causes mind. A momentous discovery, and yet he cannot observe A's mind or perceptions.

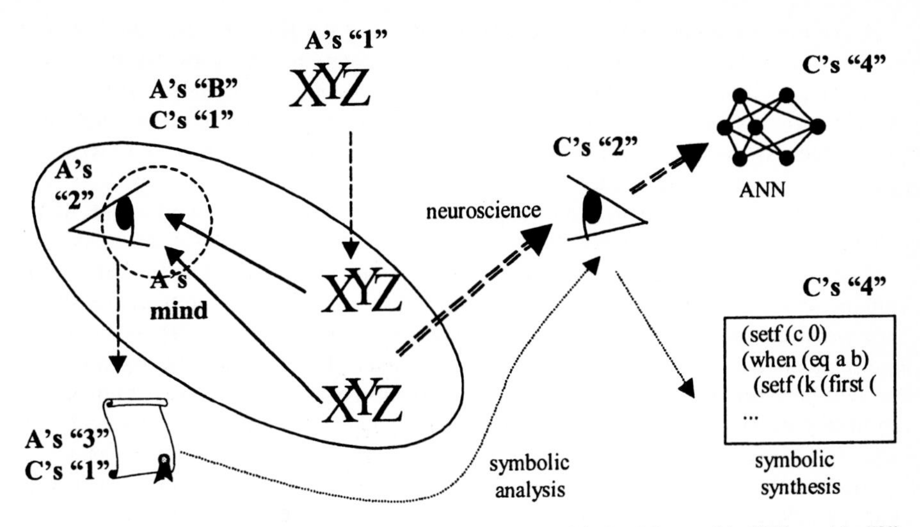

Fig. 3. The observer "C" now may look at two different kind of facts: A's "B" or A's "3". Then, the observer "C" introduces the 4th player: the so-called intelligent machine. This try changes forever the scientific scenario.

If "C" observes A's "B", he is doing neuroscience. What he builds from here is an artificial neural net or ANN (C's "4"). The problem here is that C does not know how A's "B" is related to A's mind, in other words, how A's brain is related to A's perceptions or knowledge. He knows that they are causally related, but he does not know how. He should investigate. But for the present, he is trying to study the behaviour of A's brain and the consequences that this has upon A's knowledge (in a civilized way that causes no harm at all).

If "C" observes A's "3", he obtains knowledge about knowledge, not directly but because of C's intelligence. Knowledge about knowledge? Two knowledges here: the former belongs to "C", the latter belongs to "A". Note that A's "3" expresses somehow the medical knowledge that A has. So an intelligent observer like "C" tries to figure out something about A's knowledge, and this will be knowledge to "C" afterwards. But now, "C" does not know nothing about structures and processes that could support A's knowledge and intelligence, or underlying models. So he has to imagine those processes or figure them out. He is working on it with the help of a vonNeumann computer and several programming languages and environments. He is doing "symbolic analysis", then "symbolic synthesis" (C's "4").

Both choices' artificial systems are very opposite. He found the number of inputs to an ANN has to be extremely high and parallel: for instance, if he tries to reproduce the visual path (he has to do it step by step, by chaining each step's output to the next step's input) he has to build more than 100,000,000 parallel inputs that have to be mapped at once into just about one million inputs to the next step. Besides, from this moment on the path becomes really entangled: some geometry seems to be preserved, but the elements are interconnected in a very entangled and dense way; and the worse part is that he really does not know if there is a last step.

He knows that perceptions are causally related to these facts, but he cannot figure out how to reproduce perceptions or their causal relationships with these facts by artificial means. He could not even find them in the neuroscientific counterpart: things like the greeness of grass, the hardness of stones, the coldness of snow, undoubtedly being greeness, hardness and coldness perceptions causally related to the effects of grass, stones and snow upon the brain. But where are greeness, hardness and coldness? How are they caused by facts in the brain? (And by the way: where is he who perceives? How it happens?)

Greeness, hardness and coldness are perceptions, so they are knowledge. He thought about more perceptions; he thought: if I explain this to someone, will he understand me? If I ask him, how will he know what is the answer? He will know if he did understand, or if he did not understand, or even if he do not know the answer... that will be a perception too for him. This perception of what we know is knowledge too, like greeness, hardness, and coldness. All science, he thought, is based on this kind of perceptions: measure and observation, either directly or through devices.

The good part is that he is observing brain's facts, because he knows they are causally related to knowledge. So he tries to figure out and find out knowledge models from here.

On the other hand, the number of inputs in vonNeumann computers is extremely lower. In the symbolic analysis, he is not observing facts at all, but he has knowledge about knowledge. It is completely more difficult here, because he has to imagine what are the facts. We also have to take into account the comments that we set out in the introduction of this paper, about the computation with human assistance.

3 Evolution

In an intuitive sense, we find intelligence in living creatures and intelligence is there as a result of evolution, brought about by interaction between those creatures' lineage and their environment. In an objective way, we find intelligence and knowledge related to the brain. Hence, in order to properly understand the phenomena of the brain and the brain itself we have to consider it in its environment and consider too that a working brain is always within a body, that is the way the brain interacts with the environment.

Each individual interacts with the environment through what we call knowledge and intelligence, that are causally related to facts in the brain. A lot of those brain's facts are causally related to facts in the environment, and only to these facts, now or in the past. But facts in the brain have also to do with the structure and functionality of the brain as a living tissue, a tissue that modifies itself, and this has to do not only with environment facts, but also with the genes of the individual. So, those individuals that went by, may transmit through their genes some of the structure and functionality of the brain, and therefore some of what made them succeed in the struggle for life, due to their knowledge and intelligence.

As we said above, life itself is a great test for knowledge and intelligence. This always was true. Well, since when? When intelligence began? How living creatures started to have knowledge? As Charles Darwin itself asked: "How it comes that certain colours, sounds, and forms should give pleasure to man and the lower animals, - that is, how the sense of beauty in its simplest form was first acquired,- we do not know any more than how certain odours and flavours were first rendered agreeable."

4 Acknowledgement

To my father, all the time.
I also would like to express my gratitude to José Mira and Ana E. Delgado for their support and suitable viewpoint on neural computation based on scientific discoveries about the brain along the past 20th century, from Cajal to J. Gonzalo, and beyond.

References

1. Darwin, C. On the Origin of Species by Means of Natural Selection, or the Preservation of Favoured Races in the Struggle for Life. (London: John Murray, Albemarle Street, 1859). See also 7th edition.
2. Gonzalo, I. Allometry in the Justo Gonzalo's Model of Sensorial Cortex. In Biological and Artificial Computation: From Neuroscience to Technology. Mira, Moreno-Díaz & Cabestany (Eds.) (Springer-Verlag, 1997) 169-177.
3. Herrero, J.C. and Mira, J. SCHEMA: A Knowledge Edition Interface for Obtaining Program Code from Structured Descriptions of PSM's. Two Cases Study. *Applied Intelligence* 10 (2/3), pp. 139-153.
4. Herrero, J.C. and Mira, J. Causality Levels in SCHEMA: A Knowledge Edition Interface. *IEE Proceedings-Software* (Accepted). UK
5. Russell, B. The Analysis of Matter. (T.J. Press Ltd, Padstow, Cornwall, 1927).
6. Russell, B. History of Western Philosophy. (George Allen & Unwin, 1946).
7. Russell, B. Human Knowledge: Its Scope and Limits. (George Allen & Unwin, 1948).
8. Russell, B. An Inquiry into Meaning and Truth. The William James lectures for 1940, delivered at Harvard University. (George Allen & Unwin, 1950).
9. Russell, B. My Philosophical Development. (George Allen & Unwin, 1959).
10. Turing, A.M. Computing Machinery and Intelligence. Mind 49 (1950) 433-460.

Conjecturing the Cognitive Plausibility of an ANN Theorem-Prover

Iara M. O. Vilela[1] and Priscila M.V.Lima[2]

[1] Programa de Engenharia de Sistemas e Computação- COPPE
Universidade Federal do Rio de Janeiro
Caixa postal 68511
21.945-970, Rio de Janeiro, RJ, Brazil
iara@cos.ufrj.br

[2] Instituto de Computação
Universidade Federal Fluminense
Rua Passo da Pátria, 156 - bl.E – sala 350
24 .210-240, Niterói, RJ, Brazil
priscila@cos.ufrj.br

Abstract. Many models in Artificial Neural Networks Systems seek inspiration from cognitive and biological mechanisms. Taking the opposite direction, this work conjectures on the cognitive plausibility of a propositional version of neural engine that finds proof by refutation using the Resolution Principle. We construct a parallel between the main characteristics of the computional system and several aspects of theories found in the psychology and neurocognitive literature. This way, the identification of an artificial neural explainer, already hypothesized by psychological and neurocognitive works, is of fundamental contribution to the development of the field of artificial cognition.

1. Introduction

Drawing inspiration from real life organisms is a natural way of designing computational mechanisms, specially in some fields of Artificial Intelligence such as Artificial Neural Networks. While some systems like ART [4], [20],..., are directly inspired by Nature, other systems do not. However, cognitive plausibility can be found even in systems that originally did not intend to incorporate this characteristic. As an example we can mention the analysis of [1] in [21].

ARQ-PROP-II [8][9] consists of a propositional version of a neural engine for finding proofs by refutation using the Resolution Principle. Such a neural architecture does not require special arrangements or different modules in order to do forward or backward reasoning, being, so to speak, driven by the goal posed to it. Besides that, the neural engine is capable of performing monotonic reasoning with both complete and incomplete knowledge in an integrated fashion. In order to do so, it was necessary to provide the system with the ability to create new sentences (clauses). ARQ-PROP-II is the first neural mechanism to our knowledge that does not require that the clauses

J. Mira and A. Prieto (Eds.): IWANN 2001, LNCS 2084, pp. 822-829, 2001.

of the knowledge base be either pre-encoded as constraints or learnt via examples, although the addition of these features to the system is not an impossibility.

One interesting question is the cognitive plausibility (or not) of this mechanism. The remainder of this work is organized in the following way: the next section briefly describes ARQ-PROP-II, stressing the model's features that are the subject of section 3´s analysis. Section 4 concludes the paper, pointing out ongoing and future work.

2. ARQ-PROP-II: Main Characteristics

ARQ-PROP-II consists of the propositional version of a neural engine that is capable of performing reasoning with both complete and incomplete knowledge in an integrated fashion. Before discussing this issue, it is necessary to describe the methodology used. The basis of this neural architecture is the description of the problem-solver in propositional logic, more specifically in Penalty Logic [17]. Such description must take into account that the inference rule is Resolution [18] and that there is a proof-area of limited size in the engine. The set of propositional constraints can then be seen as a logical formula, and solving a problem can be seen as finding a *model* for a logical sentence. In propositional logic that would consist of the assertion of truth-values (*true*, *false*)to the propositional symbols that appear in the formula in question in such a way that the formula as a whole becomes true. Satisfiability is converted to energy minimization [16]. If a logical formula is converted to an equivalent in clausal form [18], the result being a conjunction of disjunctions, it is possible to associate energy with the truth-value H of the negation of this conjunction of disjunctions. Nevertheless, energy calculated this way would only have two possible values: one, meaning solution not found (if the network has not yet reached global minimum), and zero when a model has been found. Intuitively, it would be better to have more "clues", or degrees of "non-satisfiability", on whether the network is close to a solution or not. So, instead of making $\boldsymbol{E} = \boldsymbol{H}$, we will consider $\boldsymbol{E} = \boldsymbol{H}^*$, the sum of truth-values of the negation of the disjuncts .

An issue to point out is that the resulting network of the above mapping may have hyper-arcs, i. e. connections involving more than two neurons. Concerning the direction of connections, there is no reason for them to be asymmetric. As the resulting energy equation has no information about states of units at previous time steps, there is no possibility of a neuron influencing itself, that is, there are no w_{ii} different from 0. Besides that, there is no obligation that all weights be different from 0.

Similarly to Hopfield networks [6], higher-order networks may also get stuck in local minima. However, it is possible to combine the behaviour of these networks with Simulated Annealing. Sejnowski [19] demonstrated that hyper-arcs Boltzmann Machines also converge to energy global minima. Before that, Geman and Geman [3] had developed a proof for the topic using *Markov Random Fields* (*MRFs*) as representation language. This convergence proof was based on the equivalence between *MRF*s and *Gibbs Random Fields* (*GRFs*).

In summary, ARQ-PROP-II consists of an artificial neural network that presents the following characteristics: **a)** it produces proofs by refutation with limited depth in propositional logic; **b)** the inference rule used is the Resolution Principle; **c)** its energy equation is the result of a modified conversion of satisfiability (finding a model to) of the propositional specification of the theorem-prover with the characteristics above to a numeric representation; **d)** the theorem-prover does not require that knowledge be pre-compiled into the network's weights; in other words, it is "empty of knowledge"; **e)** it uses a sort of "knowledge window" as the place where the knowledge base to be reasoned about is stored; **f)** reasoning is monotonic; **g)** it copes with incomplete knowledge, as long as reasoning is monotonic; **h)** it is goal-driven, as opposed to task-driven, allowing for several aspects of reasoning (deduction, abduction, prediction, planning, ...) to be integrated in a same mechanism; **i)** the fact that it is goal-driven implies in that no specific reasoning direction is obligatory; **j)** in other words, forward and backward reasoning take place depending on the task to be performed and are not defined into the mechanism itself.

3. Cognitive Plausibility of ARQ-PROP-II

This section analyses each one of the ARQ-PROP-II's characteristics selected in section 2 as of importance for our argumentation.

3.1. The Artificial Neural Explainer and Gazzaniga´s Interpreter

One basic characteristic of the artificial neural explainer ARQ-PROP-II is the fact that the knowledge base used by it does not have to be pre-compiled into the weights of the ANN. In other words, we can say that it is *empty of knowledge*, that is, the mechanism's structure is independent from the information that will be used to generate a proof (or explanation) for a fact (or query). The explainer generates a most consistent sequence of inference steps once a *window of knowledge*, that is, a subset of the total knowledge of the system, together with the sentence to be explained, are established.

Researchers in Neurobiology [7], Neuropsychology [10] and even Artificial Intelligence [5] agree that the brain is a modular processing structure. This way, several and relatively independent parallel streams would be responsible for the global functioning of the brain. Then, Gazzaniga [2] raises the question of how this modular structure can be felt as something unified. The results of experiments on split brain patients evidentiate performance differences between the right and left hemispheres of the human brain. From the analysis of these experiments, Gazzaniga hypothesizes the existence of an *interpreter*. According to him, this interpreter is situated in the left hemisphere, where the language is processed. Such a mechanism would be responsible for producing an explanation about the behaviour of the remaining areas of the brain, whereby these areas would be unified and given internal consistency and integrity.

It is worth noting that Gazzaniga identifies a bias in this proposed interpreter. In the case of a contradiction between the behaviour of the afore mentioned partial components of the brain, the solution chosen to accommodate the conflicting parts may not be in agreement with pieces of information already store by the individual. Such a behaviour is exemplified by how memory is differently recalled by each hemisphere: the left one is more prone to erroneously "recall" something in accordance with previous schemes than the right hemisphere. This indicates that the interpreter is led to construct coherent explanations rather than sticking to what has really been observed. Therefore, it is possible to infer that the explanation of the result of the processing of the other so-called modules of the brain is independently produced by the interpreter, which uses the knowledge given by modules in question in its window.

This description of Gazzaniga's interpreter is in accordance with the fact that ARQ-PROP-II is empty of knowledge and independent from the process that has acquired and selected the knowledge base that will fill its window.

3.2. Bidirectional Reasoning and Piaget´s Reversibility

An explanation generated from a given knowledge base may be used for several purposes. It may just be an explanation with complete knowledge, which would be classified as plain deduction, or constitute the sequence of steps of a plan. In case knowledge has to be completed with the hypothesis of a fact, then abduction would be said to have taken place. Conversely, if the piece of knowledge to be added to the so-called explanation were a conclusion, then the task performed could be classified as prediction. The fact that ARQ-PROP-II is goal-driven, as opposed to task-driven, means that the mechanism can be used for all the purpose mentioned before, without prior classification. The construction of the explanation depends only on its initial elements which can be situated in any point of the chain of reasoning. This section seeks to investigate evidences of the circumstances that the human reasoning presents similar characteristics.

According to Piaget [15] a cognitive structure is in *equilibrium* when fully developed. This equilibrium is characterized by the *reversibility* of the operations that constitute the structure in question. This means that for every operation there exist another one that allows the system to go back to its previous state. In other words, every operation has to be accompanied by its reverse as a component of the same operatory structure. It is acceptable to assume that the mechanism for producing explanations in propositional logic is part of the human cognitive structures. From the psychological point of view, and applying the principle proposed by Piaget, we argue that when such a mechanism is a cognitive structure in equilibrium, it is, then, reversible.

An explanation in logic can be seen as an ordered sequence of logical sentences connected by inference steps. These inferences steps depend on the inference rule(s) used. We propose that each inference step be seen as the application of an *piagetian operation*. In order to argue this point of view it is necessary be reversible. If this is the case, the reasoning sequence can be traversed in both directions. Therefore, human reasoning must embed the capability for bidirectional processing. Anticipating a fact has, indeed, the same psychological structure than reconstructing this same fact. It is possible to predict a consequence of a chain of events by using part of our knowledge about the subject in question. That is exactly what a financial consultant does. On the other hand, it is sometimes necessary to construct an explanation for a final observation. For example, a detective combines knowledge about a fact that has been observed and needs to be explained (a robbery or a murder) with the hypothesis about something that is not of his knowledge in order to complete the explanation for the crime. Yet another situation is that of an manager, who finds himself in the middle of a process, having to analyze past events while predicting future actions.

Piaget [12] [13] has shown a child is incapable of performing logical inference until he/she acquires the ability of systematically combining separate elements. Consider that an explanation that uses logic is composed by a combinatory of logical sentences linked by inferences steps. It is not reasonable to suppose that the mechanism that generates such an explanation presents a single processing direction. The flexibility for working in more than one direction is specially welcome if we suppose that the reasoning module is in principle empty of knowledge and independent from the other modules.

3.3. Resolution Steps as the Links in the Chain of Propositions

As mentioned in section 2, ARQ-PROP-II uses the Resolution Principle as sole inference rule. Each inference step connect a pair of literals that may be considered equal apart from their opposite signs. As there is no obligatory order between the signs of the literals being paired, Resolution may be considered a commutative operation. Moreover, Inverse Resolution [11] is also subject of studies. We argue that a resolution step connects two elements by what can be said a *reversibility point*, or *link*. These links connect the sentences that compose the explanation (or proof) in a sort of chain that can be traversed from one point to another in various directions.
Besides that it is possible to determine which sentence(s) is (are) missing so that the chain is complete. This(these) gap(s), if it(they) exists(exist), constitutes(constitute) what has to be conjectured (or hypothesized) in order to complete the explanation. The homogeneity of constructing an explanation using resolution steps allows for that to be done without incorporating to other reasoning resources into the system. Hence, the reversible logical operatory structure, generated by a combinatory of elements, that has been foreseen in Piaget's work, may now take the form of the result of computational mechanism.

3.4. Discussing Other Features

We have already pointed out that ARQ-PROP-II is empty of knowledge, in the sense that its knowledge base is independent from its behaviour. We have called the neural structure where this knowledge base is placed as *knowledge window*. One point that ARQ-PROP-II does not address is the location and the selection process of the contents this knowledge window. But, the literature in psychology can give us a hint on how this question is approached in humans.

The human being is constantly confronted with problems to be solved. Besides these problems of various types, each individual has to cope with his/her own internal conflicts. These circumstances of emotional nature tend to associate different affective values to pieces of information contained in the individual's knowledge about the world. These values can intensify or diminish the set of conflicts that he/she has to deal with. This affects the selection and the value that is attributed to the different elements that compose the knowledge window that is active in a given moment. Some elements are emphasized while others are ignored. There are also some elements that are included even though they might bear no direct relation with the problem being tackled. For example, an authoritarian or affective argument may be used to enforce a weak explanation.

Piaget [14]considers that the cognitive structures supply the intelligent production with (conscious and unconscious) elements, while affection supplies its energetic composition. Hence, motivation and emotion attributes more or less strength to the different elements of the cognitive structure that are involved in the process of reasoning.

Based on Leon Festinger's theory of *cognitive dissonance*, Gazzaniga [2] discusses possible psychological evidence for his proposal of an independent interpreter. According to Festinger's theory, discrepancy between various beliefs about a given situation lead to the search of a certain consistency by somehow changing information stored. This consistency may be obtained out more than one strategy: either by ignoring and even negating certain circumstances or by reinforcing others. This way, it is possible to accommodate "uncomfortable" incoherence, that is, to solve the situation named by Festinger as cognitive dissonance. According to him, the situation mentioned tends to be reduced otherwise the individual will present symptoms of psychological discomfort.

The factors that affect the amount of cognitive dissonance are related to the number and importance of dissonant beliefs. In order to solve this problem we can act in three different ways. The first one is to lower the importance of the dissonant elements. A second way is to heighten the number of consonant elements. Finally, a third strategy is to change the dissonant elements so they become consistent with the remaining elements. For example, imagine that someone pays a high price for an article that later presents some defect. This defect (the dissonant in question) may be considered of no importance (lowering the value of the dissonant), some extra advantages of the article may be found (augmenting the number of consonants), or the defect at first sight can be re-interpreted as an advantage (altering the dissonant). We introduce a further strategy which consists of including an a priori non-related, but reinforcing, sentence. In the example, this could be illustrated by the fact that the appearance of the article in a movie would give it status even though the movie contained no mention to its quality.

Devaluing the dissonant augmenting the number of consonant leads to the strengthening of the current belief, strategies that are in accordance with the monotonicity of ARQ-PROP-II. Only the modification of the dissonant would lead to a reconfiguration of knowledge. As this is a process of much higher cost than the two previous ones, there is a tendency in the human cognition to avoid it. This situation occurs only when the dissonant elements equally strong to compete between themselves and strong enough to compete with the other elements. In this case, the psychological discomfort would be so high as to enforce a change of beliefs. In computational terms, this process would require nonmonotonic tools.

In ARQ-PROP-II, we may consider that the initial value of certain nodes of the artificial neural networks are clamped because of the high "affective" value, given by need, desire, conviction, etc. This could be the psychological justification for the choice of the components of both the proof area (goal, facts that must take part in the explanation...), and of the contents of the knowledge window. Besides that, we can associate on another level the discomfort caused by cognitive dissonance in Gazzaniga's interpreter with the degree of non-satisfiability of the constraints that specify the theorem-prover defined in ARQ-PROP-II.

4. Conclusion

This work has examined the cognitive plausibility of the neural theorem-prover ARQ-PROP-II. Initially, some characteristics of the artificial module were presented so that we could compare them with natural cognitive elements. The first of the topics analyzed was the fact that the neural theorem-prover is empty of knowledge, being this way an interpreter instead of working with previously compiled knowledge. With respect to this, we would like to point out that interpreters were the first mechanisms used in conventional computing for the processing of logical inference. However, ours is different in the sense these early interpreters were neither goal-driven, nor based on a neural platform. As reasoning in ARQ-PROP-II is goal-driven, it conveys the capability for bidirectional processing, which constitute the second topic discussed. The fact that the inference rule is the Resolution Principle, and how it may allow for bidirecionality, consists of our third point of discussion. Other characteristics raised were monotonicity and the existence of a knowledge window.

In section 3, the plausibility of these properties of the artificial model were analyzed from the point of view of cognitive theories. The degree of non-satisfiability of the artificial model that has to be minimized has a parallel in Festinger's theory of tendency to diminish cognitive dissonance. The same theory can be applied to explain the tendency to prefer monotonic reasoning as the producer of a first solution. Besides, this can also serve as explanation and inspiration for the selection of the contents of the knowledge window. At this point, it is interesting to observe that we have used the same psychological theory to analyze both the behaviour of the theorem-prover itself, which is based on *model theory*, and the specification of what a good explanation should be, which is based on *proof theory*. Going back to natural cognition, this observation provides food for thought on the integration of model and proof theory in humans.

Piaget's theory of reversibility of operatory structures in equilibrium can be a match for the bidirecionality of ARQ-PROP-II, if resolution steps are considered as the reversible operations involved in the construction of the combinatory chain of sentences. The combination of these parallels makes ARQ-PROP-II compatible with Gazzaniga's concept of an independent interpreter module.

The conjectures herein made are to be further developed by experimentation involving ARQ-PROP-II and non logical neural reasoners, hoping to add more insight on knowledge selection and the boundary between the usefulness of the results of monotonic and the need of nonmonotonicity.

References

1. DeGregorio, M.: Integrating Inference and neural classification in a hybrid system for recognition tasks. Mathware & Soft Computing, Vol 3 (1996) 271-279
2. Gazzaniga, M.S. , LeDoux, J.E. :The Integrated Mind. Plenum Press, New York (1979)
3. Geman, S., Geman, D. : Stochastic relaxation, Gibbs distributions, and the Bayesian restoration of images. IEEE Trans. on Pattern Analysis and Machine Intelligence, PAMI-6 (1984) 721-741
4. Grossberg, S. :The Attentive Brain. American Scientist 83 (1995) 438-449

5. Grossberg, S. :The Complementary Brain: A Unifying View of Brain Specialization and Modularity. Technical Report CAS/CNS-TR-98-003 (2000)
6. Hopfield, J. J.: Neural networks and physical systems with emergent collective computational abilities. Proceedings of the National Academy of Sciences USA, 79 (1982) 2554-2558.
7. Kandel, E. R., Schwartz, J. H. Jessell, T.M. (eds.): Principles of Neural Sciences 4th edition. New York: McGraw-Hill. (2000)
8. Lima P. M. V.: Resolution-based Inference on Artificial neural Networks. Ph.D. Thesis, Department of Computing, Imperial College of Science, Technology and Medicine, London, UK (2000a).
9. Lima P. M. V.: A Neural Propositional Reasoner that is Goal-Driven and Works without Pre-Compiled Knowledge In: Proccedings of the VIth Brazilian Symposium on Neural Networks, Rio de Janeiro,RJ, Brazil, (2000b) 261-266
10. Luria, A. R. : Higher Cortical Functions in Man. Basic Books, Inc., Publishers, New York (1980)
11. Muggleton, S. : Inductive Logic Programming. New Generation Computing , Vol. 8 (1991) 295-318
12. Piaget, J. : De la Logique de l'Enfant à Logique de l'Adolescent Press Universitaires de France (1970)
13. Piaget, J. , Fraisse, P.(eds.): Traité de Psychologie Experimentale. Vol VII: l'Intelligence (1963)
14. Piaget, J.: Inconscient Affectif et Inconscient Cognitif. In: Problèmes de Psychologie Génetique. Éditions Denoel Paris (1972).
15. Piaget, J.: The Psychology of Intelligence. New York: Harcourt Brace & Co.. (1950)
16. Pinkas, G. : Energy minimization and the satisfiability of propositional calculus. Neural Computation Vol. 3, Morgan Kaufmann (1991) 282-291.
17. Pinkas, G.: Reasoning, Non-Monotonicity and Learning in Connectionist Networks that Capture Propositional Knowledge. Artificial Intelligence, Elsevier Science Publishers, 77, (1995) 203-247.
18. Robinson, J. A.: Logic: Form and Function. Edinburgh University Press (1979).
19. Sejnowski, T. J.: Higher-Order Boltzmann Machines. Proceedings of The American Institute of Physics Vol. 151, Snowbird, Utah (1986).
20. Vilela, I.M.O. : An Integrated Approach of Visual Computational Modelling. In: Proccedings of the Vith Brazilian Sympoosium on Neural Networks, Rio de Janeiro,RJ, Brazil, (2000) 293
21. Vilela, I.M.O.: Abordagem Integrada para Modelagem Computacional de Percepção Visual. MSc Thesis, Programa de Engeharia de Sistemas e Computação, COPPE – Universidade Federal do Rio de Janeiro, Brazil (1998)

Author Index

QM LIBRARY
(MILE END)

Lecture Notes in Computer Science

For information about Vols. 1–1990
please contact your bookseller or Springer-Verlag

Vol. 1991: F. Dignum, C. Sierra (Eds.), Agent Mediated Electronic Commerce. VIII, 241 pages. 2001. (Subseries LNAI).

Vol. 1992: K. Kim (Ed.), Public Key Cryptography. Proceedings, 2001. XI, 423 pages. 2001.

Vol. 1993: E. Zitzler, K. Deb, L. Thiele, C.A.Coello Coello, D. Corne (Eds.), Evolutionary Multi-Criterion Optimization. Proceedings, 2001. XIII, 712 pages. 2001.

Vol. 1994: J. Lind, Iterative Software Engineering for Multiagent Systems. XVII, 286 pages. 2001. (Subseries LNAI).

Vol. 1995: M. Sloman, J. Lobo, E.C. Lupu (Eds.), Policies for Distributed Systems and Networks. Proceedings, 2001. X, 263 pages. 2001.

Vol. 1997: D. Suciu, G. Vossen (Eds.), The World Wide Web and Databases. Proceedings, 2000. XII, 275 pages. 2001.

Vol. 1998: R. Klette, S. Peleg, G. Sommer (Eds.), Robot Vision. Proceedings, 2001. IX, 285 pages. 2001.

Vol. 1999: W. Emmerich, S. Tai (Eds.), Engineering Distributed Objects. Proceedings, 2000. VIII, 271 pages. 2001.

Vol. 2000: R. Wilhelm (Ed.), Informatics: 10 Years Back, 10 Years Ahead. IX, 369 pages. 2001.

Vol. 2001: G.A. Agha, F. De Cindio, G. Rozenberg (Eds.), Concurrent Object-Oriented Programming and Petri Nets. VIII, 539 pages. 2001.

Vol. 2002: H. Comon, C. Marché, R. Treinen (Eds.), Constraints in Computational Logics. Proceedings, 1999. XII, 309 pages. 2001.

Vol. 2003: F. Dignum, U. Cortés (Eds.), Agent Mediated Electronic Commerce III. XII, 193 pages. 2001. (Subseries LNAI).

Vol. 2004: A. Gelbukh (Ed.), Computational Linguistics and Intelligent Text Processing. Proceedings, 2001. XII, 528 pages. 2001.

Vol. 2006: R. Dunke, A. Abran (Eds.), New Approaches in Software Measurement. Proceedings, 2000. VIII, 245 pages. 2001.

Vol. 2007: J.F. Roddick, K. Hornsby (Eds.), Temporal, Spatial, and Spatio-Temporal Data Mining. Proceedings, 2000. VII, 165 pages. 2001. (Subseries LNAI).

Vol. 2009: H. Federrath (Ed.), Designing Privacy Enhancing Technologies. Proceedings, 2000. X, 231 pages. 2001.

Vol. 2010: A. Ferreira, H. Reichel (Eds.), STACS 2001. Proceedings, 2001. XV, 576 pages. 2001.

Vol. 2011: M. Mohnen, P. Koopman (Eds.), Implementation of Functional Languages. Proceedings, 2000. VIII, 267 pages. 2001.

Vol. 2012: D.R. Stinson, S. Tavares (Eds.), Selected Areas in Cryptography. Proceedings, 2000. IX, 339 pages. 2001.

Vol. 2013: S. Singh, N. Murshed, W. Kropatsch (Eds.), Advances in Pattern Recognition – ICAPR 2001. Proceedings, 2001. XIV, 476 pages. 2001.

Vol. 2014: M. Moortgat (Ed.), Logical Aspects of Computational Linguistics. Proceedings, 1998. X, 287 pages. 2001. (Subseries LNAI).

Vol. 2015: D. Won (Ed.), Information Security and Cryptology – ICISC 2000. Proceedings, 2000. X, 261 pages. 2001.

Vol. 2016: S. Murugesan, Y. Deshpande (Eds.), Web Engineering. IX, 357 pages. 2001.

Vol. 2018: M. Pollefeys, L. Van Gool, A. Zisserman, A. Fitzgibbon (Eds.), 3D Structure from Images – SMILE 2000. Proceedings, 2000. X, 243 pages. 2001.

Vol. 2019: P. Stone, T. Balch, G. Kraetzschmar (Eds.), RoboCup 2000: Robot Soccer World Cup IV. XVII, 658 pages. 2001. (Subseries LNAI).

Vol. 2020: D. Naccache (Ed.), Topics in Cryptology – CT-RSA 2001. Proceedings, 2001. XII, 473 pages. 2001

Vol. 2021: J. N. Oliveira, P. Zave (Eds.), FME 2001: Formal Methods for Increasing Software Productivity. Proceedings, 2001. XIII, 629 pages. 2001.

Vol. 2022: A. Romanovsky, C. Dony, J. Lindskov Knudsen, A. Tripathi (Eds.), Advances in Exception Handling Techniques. XII, 289 pages. 2001

Vol. 2024: H. Kuchen, K. Ueda (Eds.), Functional and Logic Programming. Proceedings, 2001. X, 391 pages. 2001.

Vol. 2025: M. Kaufmann, D. Wagner (Eds.), Drawing Graphs. XIV, 312 pages. 2001.

Vol. 2026: F. Müller (Ed.), High-Level Parallel Programming Models and Supportive Environments. Proceedings, 2001. IX, 137 pages. 2001.

Vol. 2027: R. Wilhelm (Ed.), Compiler Construction. Proceedings, 2001. XI, 371 pages. 2001.

Vol. 2028: D. Sands (Ed.), Programming Languages and Systems. Proceedings, 2001. XIII, 433 pages. 2001.

Vol. 2029: H. Hussmann (Ed.), Fundamental Approaches to Software Engineering. Proceedings, 2001. XIII, 349 pages. 2001.

Vol. 2030: F. Honsell, M. Miculan (Eds.), Foundations of Software Science and Computation Structures. Proceedings, 2001. XII, 413 pages. 2001.

Vol. 2031: T. Margaria, W. Yi (Eds.), Tools and Algorithms for the Construction and Analysis of Systems. Proceedings, 2001. XIV, 588 pages. 2001.

Vol. 2032: R. Klette, T. Huang, G. Gimel'farb (Eds.), Multi-Image Analysis. Proceedings, 2000. VIII, 289 pages. 2001.

Vol. 2033: J. Liu, Y. Ye (Eds.), E-Commerce Agents. VI, 347 pages. 2001. (Subseries LNAI).

Vol. 2034: M.D. Di Benedetto, A. Sangiovanni-Vincentelli (Eds.), Hybrid Systems: Computation and Control. Proceedings, 2001. XIV, 516 pages. 2001.

Vol. 2035: D. Cheung, G.J. Williams, Q. Li (Eds.), Advances in Knowledge Discovery and Data Mining – PAKDD 2001. Proceedings, 2001. XVIII, 596 pages. 2001. (Subseries LNAI).

Vol. 2037: E.J.W. Boers et al. (Eds.), Applications of Evolutionary Computing. Proceedings, 2001. XIII, 516 pages. 2001.

Vol. 2038: J. Miller, M. Tomassini, P.L. Lanzi, C. Ryan, A.G.B. Tettamanzi, W.B. Langdon (Eds.), Genetic Programming. Proceedings, 2001. XI, 384 pages. 2001.

Vol. 2039: M. Schumacher, Objective Coordination in Multi-Agent System Engineering. XIV, 149 pages. 2001. (Subseries LNAI).

Vol. 2040: W. Kou, Y. Yesha, C.J. Tan (Eds.), Electronic Commerce Technologies. Proceedings, 2001. X, 187 pages. 2001.

Vol. 2041: I. Attali, T. Jensen (Eds.), Java on Smart Cards: Programming and Security. Proceedings, 2000. X, 163 pages. 2001.

Vol. 2042: K.-K. Lau (Ed.), Logic Based Program Synthesis and Transformation. Proceedings, 2000. VIII, 183 pages. 2001.

Vol. 2043: D. Craeynest, A. Strohmeier (Eds.), Reliable Software Technologies – Ada-Europe 2001. Proceedings, 2001. XV, 405 pages. 2001.

Vol. 2044: S. Abramsky (Ed.), Typed Lambda Calculi and Applications. Proceedings, 2001. XI, 431 pages. 2001.

Vol. 2045: B. Pfitzmann (Ed.), Advances in Cryptology – EUROCRYPT 2001. Proceedings, 2001. XII, 545 pages. 2001.

Vol. 2047: R. Dumke, C. Rautenstrauch, A. Schmietendorf, A. Scholz (Eds.), Performance Engineering. XIV, 349 pages. 2001.

Vol. 2048: J. Pauli, Learning Based Robot Vision. IX, 288 pages. 2001.

Vol. 2051: A. Middeldorp (Ed.), Rewriting Techniques and Applications. Proceedings, 2001. XII, 363 pages. 2001.

Vol. 2052: V.I. Gorodetski, V.A. Skormin, L.J. Popyack (Eds.), Information Assurance in Computer Networks. Proceedings, 2001. XIII, 313 pages. 2001.

Vol. 2053: O. Danvy, A. Filinski (Eds.), Programs as Data Objects. Proceedings, 2001. VIII, 279 pages. 2001.

Vol. 2054: A. Condon, G. Rozenberg (Eds.), DNA Computing. Proceedings, 2000. X, 271 pages. 2001.

Vol. 2055: M. Margenstern, Y. Rogozhin (Eds.), Machines, Computations, and Universality. Proceedings, 2001. VIII, 321 pages. 2001.

Vol. 2056: E. Stroulia, S. Matwin (Eds.), Advances in Artificial Intelligence. Proceedings, 2001. XII, 366 pages. 2001. (Subseries LNAI).

Vol. 2057: M. Dwyer (Ed.), Model Checking Software. Proceedings, 2001. X, 313 pages. 2001.

Vol. 2059: C. Arcelli, L.P. Cordella, G. Sanniti di Baja (Eds.), Visual Form 2001. Proceedings, 2001. XIV, 799 pages. 2001.

Vol. 2062: A. Nareyek, Constraint-Based Agents. XIV, 178 pages. 2001. (Subseries LNAI).

Vol. 2064: J. Blanck, V. Brattka, P. Hertling (Eds.), Computability and Complexity in Analysis. Proceedings, 2000. VIII, 395 pages. 2001.

Vol. 2066: O. Gascuel, M.-F. Sagot (Eds.), Computational Biology. Proceedings, 2000. X, 165 pages. 2001.

Vol. 2068: K.R. Dittrich, A. Geppert, M.C. Norrie (Eds.), Advanced Information Systems Engineering. Proceedings, 2001. XII, 484 pages. 2001.

Vol. 2070: L. Monostori, J. Váncza, M. Ali (Eds.), Engineering of Intelligent Systems. Proceedings, 2001. XVIII, 951 pages. 2001. (Subseries LNAI).

Vol. 2072: J. Lindskov Knudsen (Ed.), ECOOP 2001 – Object-Oriented Programming. Proceedings, 2001. XIII, 429 pages. 2001.

Vol. 2073: V.N. Alexandrov, J.J. Dongarra, B.A. Juliano, R.S. Renner, C.J.K. Tan (Eds.), Computational Science – ICCS 2001. Part I. Proceedings, 2001. XXVIII, 1306 pages. 2001.

Vol. 2074: V.N. Alexandrov, J.J. Dongarra, B.A. Juliano, R.S. Renner, C.J.K. Tan (Eds.), Computational Science – ICCS 2001. Part II. Proceedings, 2001. XXVIII, 1076 pages. 2001.

Vol. 2077: V. Ambriola (Ed.), Software Process Technology. Proceedings, 2001. VIII, 247 pages. 2001.

Vol. 2081: K. Aardal, B. Gerards (Eds.), Integer Programming and Combinatorial Optimization. Proceedings, 2001. XI, 423 pages. 2001.

Vol. 2082: M.F. Insana, R.M. Leahy (Eds.), Information Processing in Medical Imaging. Proceedings, 2001. XVI, 537 pages. 2001.

Vol. 2083: R. Goré, A. Leitsch, T. Nipkow (Eds.), Automated Reasoning. Proceedings, 2001. XV, 708 pages. 2001. (Subseries LNAI).

Vol. 2084: J. Mira, A. Prieto (Eds.), Connectionist Models of Neurons, Learning Processes, and Artificial Intelligence. Proceedings, 2001. Part I. XXVII, 836 pages. 2001.

Vol. 2085: J. Mira, A. Prieto (Eds.), Bio-Inspired Applications of Connectionism. Proceedings, 2001. Part II. XXVII, 848 pages. 2001.

Vol. 2089: A. Amir, G.M. Landau (Eds.), Combinatorial Pattern Matching. Proceedings, 2001. VIII, 273 pages. 2001.

Vol. 2091: J. Bigun, F. Smeraldi (Eds.), Audio- and Video-Based Biometric Person Authentication. Proceedings, 2001. XIII, 374 pages. 2001.

Vol. 2092: L. Wolf, D. Hutchison, R. Steinmetz (Eds.), Quality of Service – IWQoS 2001. Proceedings, 2001. XII, 435 pages. 2001.

Vol. 2097: B. Read (Ed.), Advances in Databases. Proceedings, 2001. X, 219 pages. 2001.